KIRK-OTHMER

ENCYCLOPEDIA OF
CHEMICAL
TECHNOLOGY

FOURTH EDITION

VOLUME 6

CHLOROCARBONS AND CHLOROHYDROCARBONS—C_2
TO
COMBUSTION TECHNOLOGY

EXECUTIVE EDITOR
Jacqueline I. Kroschwitz

EDITOR
Mary Howe-Grant

KIRK-OTHMER

ENCYCLOPEDIA OF CHEMICAL TECHNOLOGY

FOURTH EDITION

VOLUME **6**

CHLOROCARBONS AND CHLOROHYDROCARBONS—C_2
TO
COMBUSTION TECHNOLOGY

A Wiley-Interscience Publication
JOHN WILEY & SONS
New York • Chichester • Brisbane • Toronto • Singapore

Copyright © 1993 by John Wiley & Sons, Inc.

Library of Congress Cataloging-in-Publication Data

Encyclopedia of chemical technology / executive editor, Jacqueline
  I. Kroschwitz; editor, Mary Howe-Grant.—4th ed.
    p. cm.
  At head of title: Kirk-Othmer.
  "A Wiley-Interscience publication."
  Includes index.
  Contents: v. 6. Chlorocarbons and chlorohydrocarbons to combustion technology.
  ISBN 0-471-52674-6
  1. Chemistry, Technical—Encyclopedias. I. Kirk, Raymond E.
(Raymond Eller), 1890–1957. II. Othmer, Donald F. (Donald
Frederick), 1904– . III. Kroschwitz, Jacqueline I., 1942– .
IV. Howe-Grant, Mary, 1943– . V. Title: Kirk-Othmer encyclopedia
of chemical technology.
  TP9.E685 1992 v. 6   91-16789
  660'.03—dc20

Printed in the United States of America

10 9 8 7 6 5 4 3 2 1

CONTENTS

EDITORIAL STAFF
FOR VOLUME 6

Executive Editor: **Jacqueline I. Kroschwitz**
Editor: **Mary Howe-Grant**
Assistant Editor: **Cathleen A. Treacy**
Editorial Supervisor: **Lindy J. Humphreys**
Copy Editor: **Linda M. Kim**

CONTRIBUTORS
TO VOLUME 6

Valery Addes, *Inland Steel Company, East Chicago, Illinois,* Carbonization (under Coal conversion processes)

Robert M. Baldwin, *Colorado School of Mines, Golden, Colorado,* Liquefaction (under Coal conversion processes)

Bruce Barden, *GenCorp Polymer Products, Columbus, Mississippi,* Coated fabrics

Gary Beebe, *Rohm and Haas Company, Bristol, Pennsylvania,* Colorants for plastics

Joseph F. Bieron, *Occidental Chemical Corporation, Grand Island, New York,* Benzyl chloride, benzal chloride, and benzotrichloride (under Chlorocarbons and chlorohydrocarbons)

Gary Blair, *Haarman & Reimer Corporation, Elkhart, Indiana,* Citric acid

Alan Bleier, *Oak Ridge National Laboratory, Oak Ridge, Tennessee,* Colloids

Robert F. Brady, Jr., *U.S. Naval Research Laboratory, Washington, D.C.,* Coatings, marine

James G. Bryant, *Standard Chlorine of Delaware, Inc., Delaware City,* Chlorinated benzenes (under Chlorocarbons and chlorohydrocarbons)

Shiao-Hung Chiang, *University of Pittsburgh, Pennsylvania,* Cleaning and desulfurization (under Coal conversion processes)

James T. Cobb, Jr., *University of Pittsburgh, Pennsylvania,* Cleaning and desulfurization (under Coal conversion processes)

K. J. Coeling, *Nordson Corporation, Amherst, Ohio,* Spray coating (under Coating processes)

Edward D. Cohen, *E. I. du Pont de Nemours & Co., Inc., Wilmington, Delaware,* Survey (under Coating processes)

John C. Crano, *PPG Industries, Inc., Monroeville, Pennsylvania,* Photochromic (under Chromogenic materials)

Jean-Roger Desmurs, *Rhone-Poulenc Recherches, Saint-Fons, France,* Chlorophenols

T. Dombrowski, *Englehard Corporation, Iselin, New Jersey,* Survey (under Clays)

H. G. Drickamer, *University of Illinois, Urbana,* Piezochromic (under Chromogenic materials)

Richard W. Drisko, *U.S. Naval Civil Engineering Laboratory, Port Hueneme, California,* Coatings, marine

Cecil Dybowski, *University of Delaware, Newark,* Chromatography

Robert G. Eilerman, *Givaudan-Roure Corporation, Clifton, New Jersey,* Cinnamic acid, cinnamaldehyde, and cinnamyl alcohol

Richard A. Eppler, *Eppler Associates, Cheshire, Connecticut,* Colorants for ceramics

Curtis E. Gidding, *DuCoa, Highland, Illinois,* Choline

Michael Grant, *Inland Steel Company, East Chicago, Illinois,* Carbonization (under Coal conversion processes)

Charles B. Greenberg, *PPG Industries, Inc., Pittsburgh, Pennsylvania,* Electrochromic; Thermochromic (both under Chromogenic materials)

Edgar B. Gutoff, *Consultant, Brookline, Massachusetts,* Survey (under Coating processes)

J. C. Hickman, *The Dow Chemical Company, Midland, Michigan,* Tetrachloroethylene (under Chlorocarbons and chlorohydrocarbons)

F. Galen Hodge, *Haynes International, Inc., Kokomo, Indiana,* Cobalt and cobalt alloys

Kelvin L. Houghton, *ICI Chemicals & Polymers Ltd., Runcorn, Cheshire, United Kingdom,* Chlorinated paraffins (under Chlorocarbons and chlorohydrocarbons)

Dennis Kaegi, *Inland Steel Company, East Chicago, Illinois,* Carbonization (under Coal conversion processes)

Mary A. Kaiser, *E. I. du Pont de Nemours & Co., Inc., Newark, Delaware,* Chromatography

Jon Kapecki, *Eastman Kodak Company, Rochester, New York,* Color photography

Chris Kneupper, *The Dow Chemical Company, Freeport, Texas,* Allyl chloride (under Chlorocarbons and chlorohydrocarbons)

A. B. Krewinghaus, *Shell Development Company, Houston, Texas,* Gasification (under Coal conversion processes)

Ramesh Krishnamurti, *Occidental Chemical Corporation, Grand Island, New York,* Ring-chlorinated toluenes (under Chlorocarbons and chlorohydrocarbons)

Henry C. Lin, *Occidental Chemical Corporation, Grand Island, New York,* Benzyl chloride, benzal chloride, and benzotrichloride; Ring-chlorinated toluenes (both under Chlorocarbons and chlorohydrocarbons)

Gary W. Loar, *McGean-Rohco, Inc., Cleveland, Ohio,* Chromium compounds

Uday Mahagaokar, *Shell Development Company, Houston, Texas,* Gasification (under Coal conversion processes)

Daniel Marmion, *Allied Signal, Buffalo, New York,* Colorants for foods, drugs, cosmetics, and medical devices

C. E. McDonald, *E. I. du Pont de Nemours & Co., Inc., Deepwater, New Jersey,* Chlorosulfuric acid

Stanley H. Mervis, *Polaroid Corporation, Cambridge, Massachusetts,* Color photography, instant

James A. Mertens, *Dow Chemical U.S.A., Midland, Michigan,* Dichloroethylene; Trichloroethylene (both under Chlorocarbons and chlorohydrocarbons)

Matt C. Miller, *Dow Chemical U.S.A., Freeport, Texas,* Ethyl chloride (under Chlorocarbons and chlorohydrocarbons)

Kurt Nassau, *Consultant, Lebanon, New Jersey,* Color

Billie J. Page, *McGean-Rohco, Inc., Cleveland, Ohio,* Chromium compounds

Sarma V. Pisupati, *Pennsylvania State University, University Park,* Combustion science and technology

Serge Ratton, *Rhone-Poulenc Specialties Chimiques, Saint-Fons, France,* Chlorophenols

Warren Rehman, *Kraft General Foods Corporation, White Plains, New York,* Coffee

H. Wayne Richardson, *CP Chemicals, Inc., Sumter, South Carolina,* Cobalt compounds

Douglas S. Richart, *Mortow International, Inc., Reading, Pennsylvania,* Powder technology (under Coating processes)

W. Frank Richey, *The Dow Chemical Company, Freeport, Texas,* Chlorohydrins

James Rodgers, *Eastman Kodak Company, Rochester, New York,* Color photography

Lester Saathoff, *The Dow Chemical Company, Freeport, Texas,* Allyl chloride (under Chlorocarbons and chlorohydrocarbons)

Stephen H. Safe, *Texas A&M University, College Station,* Toxic aromatics (under Chlorocarbons and chlorohydrocarbons)

Alan W. Scaroni, *Pennsylvania State University, University Park,* Combustion science and technology

Paul Sennett, *Engelhard Corporation, Iselin, New Jersey,* Uses (under Clays)

Reza Sharifi, *Pennsylvania State University, University Park,* Combustion science and technology

J. F. Smullen, *Hershey Foods Corporation, Hershey, Pennsylvania,* Chocolate and cocoa

Gayle Snedecor, *The Dow Chemical Company, Freeport, Texas,* Other chloroethanes (under Chlorocarbons and chlorohydrocarbons)

Philip Staal, *Haarmann & Reimer Corporation, Elkhart, Indiana,* Citric acid

Howard D. Stahl, *Kraft General Foods Corporation, White Plains, New York,* Coffee

Clare A. Stewart, Jr., *Consultant, Wilmington, Delaware,* Chloroprene (under Chlorocarbons and chlorohydrocarbons)

Hardarshan Valia, *Inland Steel Company, East Chicago, Illinois,* Carbonization (under Coal conversion processes)

Karl S. Vorres, *Argonne National Laboratory, Argonne, Illinois,* Coal

Vivian K. Walworth, *Polaroid Corporation, Concord, Massachusetts,* Color photography, instant

Gerald Wasserman, *Kraft General Foods Corporation, White Plains, New York,* Coffee

Jack H. Westbrook, *Sci-Tech Knowledge Systems, Ballston Spa, New York,* Chromium and chromium alloys

Peter Whitman, *Kraft General Foods Corporation, White Plains, New York,* Coffee

Zeno W. Wicks, Jr., *Consultant, Las Cruces, New Mexico,* Coatings

B. L. Zoumas, *Hershey Foods Corporation, Hershey, Pennsylvania,* Chocolate and cocoa

NOTE ON CHEMICAL ABSTRACTS SERVICE REGISTRY NUMBERS AND NOMENCLATURE

Chemical Abstracts Service (CAS) Registry Numbers are unique numerical identifiers assigned to substances recorded in the CAS Registry System. They appear in brackets in the *Chemical Abstracts* (CA) substance and formula indexes following the names of compounds. A single compound may have synonyms in the chemical literature. A simple compound like phenethylamine can be named β-phenylethylamine or, as in *Chemical Abstracts*, benzeneethanamine. The usefulness of the *Encyclopedia* depends on accessibility through the most common correct name of a substance. Because of this diversity in nomenclature careful attention has been given to the problem in order to assist the reader as much as possible, especially in locating the systematic CA index name by means of the Registry Number. For this purpose, the reader may refer to the CAS Registry Handbook—Number Section which lists in numerical order the Registry Number with the *Chemical Abstracts* index name and the molecular formula; eg, **458-88-8**, Piperidine, 2-propyl-, (*S*)-, $C_8H_{17}N$; in the *Encyclopedia* this compound would be found under its common name, coniine [*458-88-8*]. Alternatively, this information can be retrieved electronically from CAS Online. In many cases molecular formulas have also been provided in the *Encyclopedia* text to facilitate electronic searching. The Registry Number is a valuable link for the reader in retrieving additional published information on substances and also as a point of access for on-line data bases.

In all cases, the CAS Registry Numbers have been given for title compounds in articles and for all compounds in the index. All specific substances indexed in *Chemical Abstracts* since 1965 are included in the CAS Registry System as are a large number of substances derived from a variety of reference works. The CAS Registry System identifies a substance on the basis of an unambiguous computer-language description of its molecular structure including stereochemical detail. The Registry Number is a machine-checkable number (like a Social Security number) assigned in sequential order to each substance as it enters the registry system. The value of the number lies in the fact that it is a concise and unique means of substance identification, which is independent of, and therefore bridges, many systems of chemical nomenclature. For polymers, one Registry Number may be used for the entire family; eg, polyoxyethylene (20) sorbitan monolaurate has the same number as all of its polyoxyethylene homologues.

Cross-references are inserted in the index for many common names and for some systematic names. Trademark names appear in the index. Names that are incorrect, misleading, or ambiguous are avoided. Formulas are given very frequently in the text to help in identifying compounds. The spelling and form used, even for industrial names, follow American chemical usage, but not always the usage of *Chemical Abstracts* (eg, *coniine* is used instead of *(S)-2-propylpiperidine, aniline* instead of *benzenamine*, and *acrylic acid* instead of *2-propenoic acid*).

There are variations in representation of rings in different disciplines. The dye industry does not designate aromaticity or double bonds in rings. All double bonds and aromaticity are shown in the *Encyclopedia* as a matter of course. For example, tetralin has an aromatic ring and a saturated ring and its structure

appears in the *Encyclopedia* with its common name, Registry Number enclosed in brackets, and parenthetical CA index name, ie, tetralin [*119-64-2*] (1,2,3,4-tetrahydronaphthalene). With names and structural formulas, and especially with CAS Registry Numbers, the aim is to help the reader have a concise means of substance identification.

CONVERSION FACTORS, ABBREVIATIONS, AND UNIT SYMBOLS

SI Units (Adopted 1960)

The International System of Units (abbreviated SI), is being implemented throughout the world. This measurement system is a modernized version of the MKSA (meter, kilogram, second, ampere) system, and its details are published and controlled by an international treaty organization (The International Bureau of Weights and Measures) (1).

SI units are divided into three classes:

BASE UNITS

length	meter[†] (m)
mass	kilogram (kg)
time	second (s)
electric current	ampere (A)
thermodynamic temperature[‡]	kelvin (K)
amount of substance	mole (mol)
luminous intensity	candela (cd)

SUPPLEMENTARY UNITS

plane angle	radian (rad)
solid angle	steradian (sr)

[†]The spellings "metre" and "litre" are preferred by ASTM; however, "-er" is used in the *Encyclopedia*.

[‡]Wide use is made of Celsius temperature (t) defined by

$$t = T - T_0$$

where T is the thermodynamic temperature, expressed in kelvin, and $T_0 = 273.15$ K by definition. A temperature interval may be expressed in degrees Celsius as well as in kelvin.

DERIVED UNITS AND OTHER ACCEPTABLE UNITS

These units are formed by combining base units, supplementary units, and other derived units (2–4). Those derived units having special names and symbols are marked with an asterisk in the list below.

Quantity	Unit	Symbol	Acceptable equivalent
*absorbed dose	gray	Gy	J/kg
acceleration	meter per second squared	m/s^2	
*activity (of a radionuclide)	becquerel	Bq	1/s
area	square kilometer	km^2	
	square hectometer	hm^2	ha (hectare)
	square meter	m^2	
concentration (of amount of substance)	mole per cubic meter	mol/m^3	
current density	ampere per square meter	$A//m^2$	
density, mass density	kilogram per cubic meter	kg/m^3	g/L; mg/cm^3
dipole moment (quantity)	coulomb meter	C·m	
*dose equivalent	sievert	Sv	J/kg
*electric capacitance	farad	F	C/V
*electric charge, quantity of electricity	coulomb	C	A·s
electric charge density	coulomb per cubic meter	C/m^3	
*electric conductance	siemens	S	A/V
electric field strength	volt per meter	V/m	
electric flux density	coulomb per square meter	C/m^2	
*electric potential, potential difference, electromotive force	volt	V	W/A
*electric resistance	ohm	Ω	V/A
*energy, work, quantity of heat	megajoule	MJ	
	kilojoule	kJ	
	joule	J	N·m
	electronvolt[†]	eV[†]	
	kilowatt-hour[†]	kW·h[†]	
energy density	joule per cubic meter	J/m^3	
*force	kilonewton	kN	
	newton	N	$kg·m/s^2$

[†]This non-SI unit is recognized by the CIPM as having to be retained because of practical importance or use in specialized fields (1).

Quantity	Unit	Symbol	Acceptable equivalent
*frequency	megahertz	MHz	
	hertz	Hz	1/s
heat capacity, entropy	joule per kelvin	J/K	
heat capacity (specific), specific entropy	joule per kilogram kelvin	J/(kg·K)	
heat transfer coefficient	watt per square meter kelvin	W/(m²·K)	
*illuminance	lux	lx	lm/m²
*inductance	henry	H	Wb/A
linear density	kilogram per meter	kg/m	
luminance	candela per square meter	cd/m²	
*luminous flux	lumen	lm	cd·sr
magnetic field strength	ampere per meter	A/m	
*magnetic flux	weber	Wb	V·s
*magnetic flux density	tesla	T	Wb/m²
molar energy	joule per mole	J/mol	
molar entropy, molar heat capacity	joule per mole kelvin	J/(mol·K)	
moment of force, torque	newton meter	N·m	
momentum	kilogram meter per second	kg·m/s	
permeability	henry per meter	H/m	
permittivity	farad per meter	F/m	
*power, heat flow rate, radiant flux	kilowatt	kW	
	watt	W	J/s
power density, heat flux density, irradiance	watt per square meter	W/m²	
*pressure, stress	megapascal	MPa	
	kilopascal	kPa	
	pascal	Pa	N/m²
sound level	decibel	dB	
specific energy	joule per kilogram	J/kg	
specific volume	cubic meter per kilogram	m³/kg	
surface tension	newton per meter	N/m	
thermal conductivity	watt per meter kelvin	W/(m·K)	
velocity	meter per second	m/s	
	kilometer per hour	km/h	
viscosity, dynamic	pascal second	Pa·s	
	millipascal second	mPa·s	
viscosity, kinematic	square meter per second	m²/s	
	square millimeter per second	mm²/s	

Quantity	Unit	Symbol	Acceptable equivalent
volume	cubic meter	m^3	
	cubic decimeter	dm^3	L (liter) (5)
	cubic centimeter	cm^3	mL
wave number	1 per meter	m^{-1}	
	1 per centimeter	cm^{-1}	

In addition, there are 16 prefixes used to indicate order of magnitude, as follows:

Multiplication factor	Prefix	Symbol	Note
10^{18}	exa	E	
10^{15}	peta	P	
10^{12}	tera	T	
10^{9}	giga	G	
10^{6}	mega	M	
10^{3}	kilo	k	
10^{2}	hecto	h^a	[a]Although hecto, deka, deci, and centi
10	deka	da^a	are SI prefixes, their use should be
10^{-1}	deci	d^a	avoided except for SI unit-multiples
10^{-2}	centi	c^a	for area and volume and nontech-
10^{-3}	milli	m	nical use of centimeter, as for body
10^{-6}	micro	μ	and clothing measurement.
10^{-9}	nano	n	
10^{-12}	pico	p	
10^{-15}	femto	f	
10^{-18}	atto	a	

For a complete description of SI and its use the reader is referred to ASTM E 380 (4) and the article UNITS AND CONVERSION FACTORS which appears in Vol. 24.

A representative list of conversion factors from non-SI to SI units is presented herewith. Factors are given to four significant figures. Exact relationships are followed by a dagger. A more complete list is given in the latest editions of ASTM E 380 (4) and ANSI Z210.1 (6).

Conversion Factors to SI Units

To convert from	To	Multiply by
acre	square meter (m^2)	4.047×10^3
angstrom	meter (m)	1.0×10^{-10}[†]
are	square meter (m^2)	1.0×10^{2}[†]
astronomical unit	meter (m)	1.496×10^{11}

[†]Exact.

To convert from	To	Multiply by
atmosphere, standard	pascal (Pa)	1.013×10^5
bar	pascal (Pa)	$1.0 \times 10^{5\dagger}$
barn	square meter (m^2)	$1.0 \times 10^{-28\dagger}$
barrel (42 U.S. liquid gallons)	cubic meter (m^3)	0.1590
Bohr magneton (μ_B)	J/T	9.274×10^{-24}
Btu (International Table)	joule (J)	1.055×10^3
Btu (mean)	joule (J)	1.056×10^3
Btu (thermochemical)	joule (J)	1.054×10^3
bushel	cubic meter (m^3)	3.524×10^{-2}
calorie (International Table)	joule (J)	4.187
calorie (mean)	joule (J)	4.190
calorie (thermochemical)	joule (J)	$4.184^\dagger$
centipoise	pascal second (Pa·s)	$1.0 \times 10^{-3\dagger}$
centistokes	square millimeter per second (mm^2/s)	$1.0^\dagger$
cfm (cubic foot per minute)	cubic meter per second (m^3/s)	4.72×10^{-4}
cubic inch	cubic meter (m^3)	1.639×10^{-5}
cubic foot	cubic meter (m^3)	2.832×10^{-2}
cubic yard	cubic meter (m^3)	0.7646
curie	becquerel (Bq)	$3.70 \times 10^{10\dagger}$
debye	coulomb meter (C·m)	3.336×10^{-30}
degree (angle)	radian (rad)	1.745×10^{-2}
denier (international)	kilogram per meter (kg/m)	1.111×10^{-7}
	tex‡	0.1111
dram (apothecaries')	kilogram (kg)	3.888×10^{-3}
dram (avoirdupois)	kilogram (kg)	1.772×10^{-3}
dram (U.S. fluid)	cubic meter (m^3)	3.697×10^{-6}
dyne	newton (N)	$1.0 \times 10^{-5\dagger}$
dyne/cm	newton per meter (N/m)	$1.0 \times 10^{-3\dagger}$
electronvolt	joule (J)	1.602×10^{-19}
erg	joule (J)	$1.0 \times 10^{-7\dagger}$
fathom	meter (m)	1.829
fluid ounce (U.S.)	cubic meter (m^3)	2.957×10^{-5}
foot	meter (m)	$0.3048^\dagger$
footcandle	lux (lx)	10.76
furlong	meter (m)	2.012×10^{-2}
gal	meter per second squared (m/s^2)	$1.0 \times 10^{-2\dagger}$
gallon (U.S. dry)	cubic meter (m^3)	4.405×10^{-3}
gallon (U.S. liquid)	cubic meter (m^3)	3.785×10^{-3}
gallon per minute (gpm)	cubic meter per second (m^3/s)	6.309×10^{-5}
	cubic meter per hour (m^3/h)	0.2271

†Exact.
‡See footnote on p. xiv.

To convert from	To	Multiply by
gauss	tesla (T)	1.0×10^{-4}
gilbert	ampere (A)	0.7958
gill (U.S.)	cubic meter (m^3)	1.183×10^{-4}
grade	radian	1.571×10^{-2}
grain	kilogram (kg)	6.480×10^{-5}
gram force per denier	newton per tex (N/tex)	8.826×10^{-2}
hectare	square meter (m^2)	$1.0 \times 10^{4\dagger}$
horsepower (550 ft·lbf/s)	watt (W)	7.457×10^2
horespower (boiler)	watt (W)	9.810×10^3
horsepower (electric)	watt (W)	$7.46 \times 10^{2\dagger}$
hundredweight (long)	kilogram (kg)	50.80
hundredweight (short)	kilogram (kg)	45.36
inch	meter (m)	$2.54 \times 10^{-2\dagger}$
inch of mercury (32°F)	pascal (Pa)	3.386×10^3
inch of water (39.2°F)	pascal (Pa)	2.491×10^2
kilogram-force	newton (N)	9.807
kilowatt hour	megajoule (MJ)	$3.6^{\dagger}$
kip	newton(N)	4.448×10^3
knot (international)	meter per second (m/S)	0.5144
lambert	candela per square meter (cd/m^3)	3.183×10^3
league (British nautical)	meter (m)	5.559×10^3
league (statute)	meter (m)	4.828×10^3
light year	meter (m)	9.461×10^{15}
liter (for fluids only)	cubic meter (m^3)	$1.0 \times 10^{-3\dagger}$
maxwell	weber (Wb)	$1.0 \times 10^{-8\dagger}$
micron	meter (m)	$1.0 \times 10^{-6\dagger}$
mil	meter (m)	$2.54 \times 10^{-5\dagger}$
mile (statute)	meter (m)	1.609×10^3
mile (U.S. nautical)	meter (m)	$1.852 \times 10^{3\dagger}$
mile per hour	meter per second (m/s)	0.4470
millibar	pascal (Pa)	1.0×10^2
millimeter of mercury (0°C)	pascal (Pa)	$1.333 \times 10^{2\dagger}$
minute (angular)	radian	2.909×10^{-4}
myriagram	kilogram (kg)	10
myriameter	kilometer (km)	10
oersted	ampere per meter (A/m)	79.58
ounce (avoirdupois)	kilogram (kg)	2.835×10^{-2}
ounce (troy)	kilogram (kg)	3.110×10^{-2}
ounce (U.S. fluid)	cubic meter (m^3)	2.957×10^{-5}
ounce-force	newton (N)	0.2780
peck (U.S.)	cubic meter (m^3)	8.810×10^{-3}
pennyweight	kilogram (kg)	1.555×10^{-3}
pint (U.S. dry)	cubic meter (m^3)	5.506×10^{-4}

†Exact.

To convert from	To	Multiply by
pint (U.S. liquid)	cubic meter (m^3)	4.732×10^{-4}
poise (absolute viscosity)	pascal second (Pa·s)	$0.10^{\dagger}$
pound (avoirdupois)	kilogram (kg)	0.4536
pound (troy)	kilogram (kg)	0.3732
poundal	newton (N)	0.1383
pound-force	newton (N)	4.448
pound force per square inch (psi)	pascal (Pa)	6.895×10^3
quart (U.S. dry)	cubic meter (m^3)	1.101×10^{-3}
quart (U.S. liquid)	cubic meter (m^3)	9.464×10^{-4}
quintal	kilogram (kg)	$1.0 \times 10^{2\dagger}$
rad	gray (Gy)	$1.0 \times 10^{-2\dagger}$
rod	meter (m)	5.029
roentgen	coulomb per kilogram (C/kg)	2.58×10^{-4}
second (angle)	radian (rad)	$4.848 \times 10^{-6\dagger}$
section	square meter (m^2)	2.590×10^6
slug	kilogram (kg)	14.59
spherical candle power	lumen (lm)	12.57
square inch	square meter (m^2)	6.452×10^{-4}
square foot	square meter (m^2)	9.290×10^{-2}
square mile	square meter (m^2)	2.590×10^6
square yard	square meter (m^2)	0.8361
stere	cubic meter (m^3)	$1.0^{\dagger}$
stokes (kinematic viscosity)	square meter per second (m^2/s)	$1.0 \times 10^{-4\dagger}$
tex	kilogram per meter (kg/m)	$1.0 \times 10^{-6\dagger}$
ton (long, 2240 pounds)	kilogram (kg)	1.016×10^3
ton (metric) (tonne)	kilogram (kg)	$1.0 \times 10^{3\dagger}$
ton (short, 2000 pounds)	kilogram (kg)	9.072×10^2
torr	pascal (Pa)	1.333×10^2
unit pole	weber (Wb)	1.257×10^{-7}
yard	meter (m)	$0.9144^{\dagger}$

†Exact.

Abbreviations and Unit Symbols

Following is a list of common abbreviations and unit symbols used in the *Encyclopedia*. In general they agree with those listed in *American National Standard Abbreviations for Use on Drawings and in Text (ANSI Y1.1)* (6) and *American National Standard Letter Symbols for Units in Science and Technology (ANSI Y10)* (6). Also included is a list of acronyms for a number of private and government organizations as well as common industrial solvents, polymers, and other chemicals.

Rules for Writing Unit Symbols (4):

1. Unit symbols are printed in upright letters (roman) regardless of the type style used in the surrounding text.
2. Unit symbols are unaltered in the plural.
3. Unit symbols are not followed by a period except when used at the end of a sentence.
4. Letter unit symbols are generally printed lower-case (for example, cd for candela) unless the unit name has been derived from a proper name, in which case the first letter of the symbol is capitalized (W, Pa). Prefixes and unit symbols retain their prescribed form regardless of the surrounding typography.
5. In the complete expression for a quantity, a space should be left between the numerical value and the unit symbol. For example, write 2.37 lm, *not* 2.37lm, and 35 mm, *not* 35mm. When the quantity is used in an adjectival sense, a hyphen is often used, for example, 35-mm film. *Exception:* No space is left between the numerical value and the symbols for degree, minute, and second of plane angle, degree Celsius, and the percent sign.
6. No space is used between the prefix and unit symbol (for example, kg).
7. Symbols, not abbreviations, should be used for units. For example, use "A," not "amp," for ampere.
8. When multiplying unit symbols, use a raised dot:

$$\text{N·m}\quad\text{for}\quad\text{newton meter}$$

In the case of W·h, the dot may be omitted, thus:

$$\text{Wh}$$

An exception to this practice is made for computer printouts, automatic typewriter work, etc, where the raised dot is not possible, and a dot on the line may be used.
9. When dividing unit symbols, use one of the following forms:

$$\text{m/s}\quad or\quad\text{m·s}^{-1}\quad or\quad\frac{\text{m}}{\text{s}}$$

In no case should more than one slash be used in the same expression unless parentheses are inserted to avoid ambiguity. For example: write:

$$\text{J/(mol·K)}\quad or\quad\text{J·mol}^{-1}\cdot\text{K}^{-1}\quad or\quad\text{(J/mol)/K}$$

but *not*

$$\text{J/mol/K}$$

10. Do not mix symbols and unit names in the same expression. Write:

$$\text{joules per kilogram} \quad or \quad \text{J/kg} \quad or \quad \text{J·kg}^{-1}$$

but *not*

$$\text{joules/kilogram} \quad nor \quad \text{joules/kg} \quad nor \quad \text{joules·kg}^{-1}$$

ABBREVIATIONS AND UNITS

A	ampere	AOAC	Association of Official Analytical Chemists
A	anion (eg, HA)		
A	mass number	AOCS	Americal Oil Chemists' Society
a	atto (prefix for 10^{-18})		
AATCC	American Association of Textile Chemists and Colorists	APHA	American Public Health Association
		API	American Petroleum Institute
ABS	acrylonitrile–butadiene–styrene	aq	aqueous
abs	absolute	Ar	aryl
ac	alternating current, *n*.	*ar*-	aromatic
a-c	alternating current, *adj*.	*as*-	asymmetric(al)
ac-	alicyclic	ASHRAE	American Society of Heating, Refrigerating, and Air Conditioning Engineers
acac	acetylacetonate		
ACGIH	American Conference of Governmental Industrial Hygienists		
		ASM	American Society for Metals
ACS	American Chemical Society	ASME	American Society of Mechanical Engineers
AGA	American Gas Association	ASTM	American Society for Testing and Materials
Ah	ampere hour		
AIChE	American Institute of Chemical Engineers	at no.	atomic number
		at wt	atomic weight
AIME	American Institute of Mining, Metallurgical, and Petroleum Engineers	av(g)	average
		AWS	American Welding Society
		b	bonding orbital
AIP	American Institute of Physics	bbl	barrel
		bcc	body-centered cubic
AISI	American Iron and Steel Institute	BCT	body-centered tetragonal
		Bé	Baumé
alc	alcohol(ic)	BET	Brunauer-Emmett-Teller (adsorption equation)
Alk	alkyl		
alk	alkaline (not alkali)	bid	twice daily
amt	amount	Boc	*t*-butyloxycarbonyl
amu	atomic mass unit	BOD	biochemical (biological) oxygen demand
ANSI	American National Standards Institute		
		bp	boiling point
AO	atomic orbital	Bq	becquerel

C	coulomb	DIN	Deutsche Industrie Normen
°C	degree Celsius		
C-	denoting attachment to carbon	dl-; DL-	racemic
		DMA	dimethylacetamide
c	centi (prefix for 10^{-2})	DMF	dimethylformamide
c	critical	DMG	dimethyl glyoxime
ca	circa (approximately)	DMSO	dimethyl sulfoxide
cd	candela; current density; circular dichroism	DOD	Department of Defense
		DOE	Department of Energy
CFR	Code of Federal Regulations	DOT	Department of Transportation
cgs	centimeter-gram-second	DP	degree of polymerization
CI	Color Index	dp	dew point
cis-	isomer in which substituted groups are on same side of double bond between C atoms	DPH	diamond pyramid hardness
		dstl(d)	distill(ed)
		dta	differential thermal analysis
cl	carload		
cm	centimeter	(E)-	entgegen; opposed
cmil	circular mil	ϵ	dielectric constant (unitless number)
cmpd	compound		
CNS	central nervous system	e	electron
CoA	coenzyme A	ECU	electrochemical unit
COD	chemical oxygen demand	ed.	edited, edition, editor
coml	commercial(ly)	ED	effective dose
cp	chemically pure	EDTA	ethylenediaminetetraacetic acid
cph	close-packed hexagonal		
CPSC	Consumer Product Safety Commission	emf	electromotive force
		emu	electromagnetic unit
cryst	crystalline	en	ethylene diamine
cub	cubic	eng	engineering
D	Debye	EPA	Environmental Protection Agency
D-	denoting configurational relationship		
		epr	electron paramagnetic resonance
d	differential operator		
d	day; deci (prefix for 10^{-1})	eq.	equation
d-	dextro-, dextrorotatory	esca	electron spectroscopy for chemical analysis
da	deka (prefix for 10^{1})		
dB	decibel	esp	especially
dc	direct current, n.	esr	electron-spin resonance
d-c	direct current, adj.	est(d)	estimate(d)
dec	decompose	estn	estimation
detd	determined	esu	electrostatic unit
detn	determination	exp	experiment, experimental
Di	didymium, a mixture of all lanthanons	ext(d)	extract(ed)
		F	farad (capacitance)
dia	diameter	F	faraday (96,487 C)
dil	dilute	f	femto (prefix for 10^{-15})

FAO	Food and Agriculture Organization (United Nations)	hyd	hydrated, hydrous
		hyg	hygroscopic
		Hz	hertz
fcc	face-centered cubic	i (eg, Pr^i)	iso (eg, isopropyl)
FDA	Food and Drug Administration	i-	inactive (eg, i-methionine)
		IACS	International Annealed Copper Standard
FEA	Federal Energy Administration	ibp	initial boiling point
FHSA	Federal Hazardous Substances Act	IC	integrated circuit
		ICC	Interstate Commerce Commission
fob	free on board		
fp	freezing point	ICT	International Critical Table
FPC	Federal Power Commission	ID	inside diameter; infective dose
FRB	Federal Reserve Board		
frz	freezing	ip	intraperitoneal
G	giga (prefix for 10^9)	IPS	iron pipe size
G	gravitational constant = 6.67×10^{11} N·m^2/kg^2	ir	infrared
		IRLG	Interagency Regulatory Liaison Group
g	gram		
(g)	gas, only as in H$_2$O(g)	ISO	International Organization Standardization
g	gravitational acceleration		
gc	gas chromatography	ITS-90	International Temperature Scale (NIST)
gem-	geminal		
glc	gas–liquid chromatography	IU	International Unit
		IUPAC	International Union of Pure and Applied Chemistry
g-mol wt; gmw	gram-molecular weight		
GNP	gross national product	IV	iodine value
gpc	gel-permeation chromatography	iv	intravenous
		J	joule
GRAS	Generally Recognized as Safe	K	kelvin
		k	kilo (prefix for 10^3)
grd	ground	kg	kilogram
Gy	gray	L	denoting configurational relationship
H	henry		
h	hour; hecto (prefix for 10^2)	L	liter (for fluids only) (5)
ha	hectare	l-	$levo$-, levorotatory
HB	Brinell hardness number	(l)	liquid, only as in NH$_3$(l)
Hb	hemoglobin	LC$_{50}$	conc lethal to 50% of the animals tests
hcp	hexagonal close-packed		
hex	hexagonal	LCAO	linear combination of atomic orbitals
HK	Knoop hardness number		
hplc	high performance liquid chromatography	lc	liquid chromatography
		LCD	liquid crystal display
HRC	Rockwell hardness (C scale)	lcl	less than carload lots
		LD$_{50}$	dose lethal to 50% of the animals tested
HV	Vickers hardness number		

LED	light-emitting diode	N-	denoting attachment to nitrogen
liq	liquid		
lm	lumen	n (as n_D^{20})	index of refraction (for 20°C and sodium light)
ln	logarithm (natural)		
LNG	liquefied natural gas	^{n} (as Bu^n),	
log	logarithm (common)	n-	normal (straight-chain structure)
LPG	liquefied petroleum gas		
ltl	less than truckload lots	n	neutron
lx	lux	n	nano (prefix for 10^9)
M	mega (prefix for 10^6); metal (as in MA)	na	not available
		NAS	National Academy of Sciences
M	molar; actual mass		
$\overline{M}_w$	weight-average mol wt	NASA	National Aeronautics and Space Administration
$\overline{M}_n$	number-average mol wt		
m	meter; milli (prefix for 10^{-3})	nat	natural
		ndt	nondestructive testing
m	molal	neg	negative
m-	meta	NF	*National Formulary*
max	maximum	NIH	National Institutes of Health
MCA	Chemical Manufacturers' Association (was Manufacturing Chemists Association)	NIOSH	National Institute of Occupational Safety and Health
MEK	methyl ethyl ketone	NIST	National Institute of Standards and Technology (formerly National Bureau of Standards)
meq	milliequivalent		
mfd	manufactured		
mfg	manufacturing		
mfr	manufacturer		
MIBC	methyl isobutyl carbinol	nmr	nuclear magnetic resonance
MIBK	methyl isobutyl ketone		
MIC	minimum inhibiting concentration	NND	New and Nonofficial Drugs (AMA)
min	minute; minimum	no.	number
mL	milliliter	NOI-(BN)	not otherwise indexed (by name)
MLD	minimum lethal dose		
MO	molecular orbital	NOS	not otherwise specified
mo	month	nqr	nuclear quadruple resonance
mol	mole		
mol wt	molecular weight	NRC	Nuclear Regulatory Commission; National Research Council
mp	melting point		
MR	molar refraction		
ms	mass spectrometry	NRI	New Ring Index
MSDS	material safety data sheet	NSF	National Science Foundation
mxt	mixture		
μ	micro (prefix for 10^{-6})	NTA	nitrilotriacetic acid
N	newton (force)	NTP	normal temperature and pressure (25°C and 101.3 kPa or 1 atm)
N	normal (concentration); neutron number		

NTSB	National Transportation Safety Board	qv	quod vide (which see)
O-	denoting attachment to oxygen	R	univalent hydrocarbon radical
o-	ortho	(*R*)-	rectus (clockwise configuration)
OD	outside diameter	*r*	precision of data
OPEC	Organization of Petroleum Exporting Countries	rad	radian; radius
o-phen	*o*-phenanthridine	RCRA	Resource Conservation and Recovery Act
OSHA	Occupational Safety and Health Administration	rds	rate-determining step
owf	on weight of fiber	ref.	reference
Ω	ohm	rf	radio frequency, *n.*
P	peta (prefix for 10^{15})	r-f	radio frequency, *adj.*
p	pico (prefix for 10^{-12})	rh	relative humidity
p-	para	RI	Ring Index
p	proton	rms	root-mean square
p.	page	rpm	rotations per minute
Pa	pascal (pressure)	rps	revolutions per second
PEL	personal exposure limit based on an 8-h exposure	RT	room temperature
		RTECS	Registry of Toxic Effects of Chemical Substances
pd	potential difference	ˢ (eg, Buˢ); *sec-*	secondary (eg, secondary butyl)
pH	negative logarithm of the effective hydrogen ion concentration	S	siemens
		(*S*)-	sinister (counterclockwise configuration)
phr	parts per hundred of resin (rubber)	*S-*	denoting attachment to sulfur
p-i-n	positive-intrinsic-negative	*s-*	symmetric(al)
pmr	proton magnetic resonance	s	second
p-n	positive-negative	(s)	solid, only as in $H_2O(s)$
po	per os (oral)	SAE	Society of Automotive Engineers
POP	polyoxypropylene		
pos	positive	SAN	styrene-acrylonitrile
pp.	pages	sat(d)	saturate(d)
ppb	parts per billion (10^9)	satn	saturation
ppm	parts per million (10^6)	SBS	styrene–butadiene–styrene
ppmv	parts per million by volume	sc	subcutaneous
ppmwt	parts per million by weight	SCF	self-consistent field; standard cubic feet
PPO	poly(phenyl oxide)		
ppt(d)	precipitate(d)	Sch	Schultz number
pptn	precipitation	SFs	Saybolt Furol seconds
Pr (no.)	foreign prototype (number)	SI	Le Système International d'Unités (International System of Units)
pt	point; part		
PVC	poly(vinyl chloride)		
pwd	powder	sl sol	slightly soluble
py	pyridine	sol	soluble

soln	solution	*trans-*	isomer in which
soly	solubility		substituted groups are
sp	specific; species		on opposite sides of
sp gr	specific gravity		double bond between C
sr	steradian		atoms
std	standard	TSCA	Toxic Substances Control
STP	standard temperature and		Act
	pressure (0°C and 101.3	TWA	time-weighted average
	kPa)	Twad	Twaddell
sub	sublime(s)	UL	Underwriters' Laboratory
SUs	Saybolt Universal seconds	USDA	United States Department
syn	synthetic		of Agriculture
^{t} (eg, Bu^t),		USP	*United States*
t-, tert-	tertiary (eg, tertiary		*Pharmacopeia*
	butyl)	uv	ultraviolet
T	tera (prefix for 10^{12}); tesla	V	volt (emf)
	(magnetic flux density)	var	variable
t	metric ton (tonne)	*vic-*	vicinal
t	temperature	vol	volume (not volatile)
TAPPI	Technical Association of	vs	versus
	the Pulp and Paper	v sol	very soluble
	Industry	W	watt
TCC	Tagliabue closed up	Wb	weber
tex	tex (linear density)	Wh	watt hour
T_g	glass-transition	WHO	World Health
	temperature		Organization (United
tga	thermogravimetric		Nations)
	analysis	wk	week
THF	tetrahydrofuran	yr	year
tlc	thin layer chromatography	(*Z*)-	zusammen; together;
TLV	threshold limit value		atomic number

Non-SI (Unacceptable and Obsolete) Units		Use
Å	angstrom	nm
at	atmosphere, technical	Pa
atm	atmosphere, standard	Pa
b	barn	cm^2
bar†	bar	Pa
bbl	barrel	m^3
bhp	brake horsepower	W
Btu	British thermal unit	J
bu	bushel	m^3; L
cal	calorie	J
cfm	cubic foot per minute	m^3/s
Ci	curie	Bq
cSt	centistokes	mm^2/s
c/s	cycle per second	Hz
cu	cubic	exponential form

†Do not use bar (10^5Pa) or millibar (10^2Pa) because they are not SI units, and are accepted internationally only for a limited time in special fields because of existing usage.

Non-SI (Unacceptable and Obsolete) Units		Use
D	debye	$C \cdot m$
den	denier	tex
dr	dram	kg
dyn	dyne	N
dyn/cm	dyne per centimeter	mN/m
erg	erg	J
eu	entropy unit	J/K
°F	degree Fahrenheit	°C; K
fc	footcandle	lx
fl	footlambert	lx
fl oz	fluid ounce	m^3; L
ft	foot	m
ft·lbf	foot pound-force	J
gf den	gram-force per denier	N/tex
G	gauss	T
Gal	gal	m/s^2
gal	gallon	m^3; L
Gb	gilbert	A
gpm	gallon per minute	(m^3/s); (m^3/h)
gr	grain	kg
hp	horsepower	W
ihp	indicated horsepower	W
in.	inch	m
in. Hg	inch of mercury	Pa
in. H_2O	inch of water	Pa
in.-lbf	inch pound-force	J
kcal	kilo-calorie	J
kgf	kilogram-force	N
kilo	for kilogram	kg
L	lambert	lx
lb	pound	kg
lbf	pound-force	N
mho	mho	S
mi	mile	m
MM	million	M
mm Hg	millimeter of mercury	Pa
mμ	millimicron	nm
mph	miles per hour	km/h
μ	micron	μm
Oe	oersted	A/m
oz	ounce	kg
ozf	ounce-force	N
η	poise	Pa·s
P	poise	Pa·s
ph	phot	lx
psi	pounds-force per square inch	Pa
psia	pounds-force per square inch absolute	Pa
psig	pounds-force per square inch gage	Pa
qt	quart	m^3; L
°R	degree Rankine	K
rd	rad	Gy
sb	stilb	lx
SCF	standard cubic foot	m^3
sq	square	exponential form
thm	therm	J
yd	yard	m

BIBLIOGRAPHY

1. The International Bureau of Weights and Measures, BIPM (Parc de Saint-Cloud, France) is described in Appendix X2 of Ref. 4. This bureau operates under the exclusive supervision of the International Committee for Weights and Measures (CIPM).
2. *Metric Editorial Guide (ANMC-78-1)*, latest ed., American National Metric Council, 5410 Grosvenor Lane, Bethesda, Md. 20814, 1981.
3. *SI Units and Recommendations for the Use of Their Multiples and of Certain Other Units (ISO 1000-1981)*, American National Standards Institute, 1430 Broadway, New York, N.Y. 10018, 1981.
4. Based on *ASTM E 380-89a (Standard Practice for Use of the International System of Units (SI))*, American Society for Testing and Materials, 1916 Race Street, Philadelphia, Pa. 19103, 1989.
5. *Fed. Regist.*, Dec. 10, 1976 (41 FR 36414).
6. For ANSI address, see Ref. 3.

R. P. LUKENS
ASTM Committee E-43 on SI Practice

C

Continued

CHLOROCARBONS AND CHLOROHYDROCARBONS

ETHYL CHLORIDE

Ethyl chloride [*75-00-3*] (chloroethane), C_2H_5Cl, is a colorless, mobile liquid of bp 12.4°C, that has a nonirritating ethereal odor and a pleasant taste. It is flammable and burns with a green-edged flame, producing hydrogen chloride fumes, carbon dioxide, and water. Ethyl chloride has primarily been used in the manufacture of tetraethyllead [*78-00-2*] (TEL), an antiknock additive in engine fuel, but it also serves as an ethylating agent, solvent, refrigerant, and local and general anesthetic. It is less toxic than the chloromethanes.

Ethyl chloride became an important large-volume chemical in 1922 when manufacture of tetraethyllead began in the United States. It had previously been manufactured primarily for use as an anesthetic and refrigerant; before 1922 annual production had not exceeded several hundred metric tons in any of the producing countries. Use of ethyl chloride as a starting material for TEL makes it an automotive chemical, and increases in its production were linked with the growth of the automobile industry. Prior to World War II, annual output of ethyl chloride in the United States exceeded 23,000 t, but only 230–275 t were used for purposes other than the manufacture of TEL. During the war, ethyl chloride production increased approximately fivefold. United States output in postwar years more than doubled. From 1960 to 1979, production of ethyl chloride remained fairly constant with 90% of the ethyl chloride produced going into the manufacture of TEL. To reduce automotive emissions, legislative actions in the mid-1970s mandated the use of catalytic converters that could not tolerate the lead in TEL. The manufacture of automotive engines that could use TEL was gradually phased out until the end of 1978, after which no new car manufactured for sale in the United States was allowed to use TEL as a fuel. Because of this phasing out of leaded fuels, production of ethyl chloride has reduced steadily since 1979 and imports of ethyl chloride have been essentially zero since 1983. Since 40% of its production went to TEL in 1986, ethyl chloride demand is expected to continue to diminish.

Physical and Chemical Properties

The physical properties of ethyl chloride are listed in Table 1. At 0°C, 100 g ethyl chloride dissolve 0.07 g water and 100 g water dissolve 0.447 g ethyl chloride. The solubility of water in ethyl chloride increases sharply with temperature to 0.36 g/100 g at 50°C. Ethyl chloride dissolves many organic substances, such as fats, oils, resins, and waxes, and it is also a solvent for sulfur and phosphorus. It is miscible with methyl and ethyl alcohols, diethyl ether, ethyl acetate, methylene chloride, chloroform, carbon tetrachloride, and benzene. Butane, ethyl nitrite, and 2-methylbutane each have been reported to form a binary azeotrope with ethyl chloride, but the accuracy of this data is uncertain (1).

The C–Cl bond in ethyl choride (0.176 nm) is slightly shorter than the corresponding bond in methyl chloride (0.1786 nm). Ethyl chloride displays thermal stability similar to that of methyl chloride. It is practically unchanged on heating to 400°C where decomposition to ethylene and hydrogen chloride begins (2). This decomposition is nearly complete at 500–600°C on pumice or about 300°C with the chlorides of nickel, cobalt, iron, or lead, but not with chlorides of sodium, potassium, or silver. Several inorganic salts, eg, lithium chloride and calcium sulfate; metals, eg, platinum and iridium; and oxides, eg, aluminum oxide and silica, also catalyze the cracking of ethyl chloride (3,4).

Ethyl chloride can be dehydrochlorinated to ethylene using alcoholic potash. Condensation of alcohol with ethyl chloride in this reaction also produces some diethyl ether. Heating to 625°C and subsequent contact with calcium oxide and water at 400–450°C gives ethyl alcohol as the chief product of decomposition. Ethyl chloride yields butane, ethylene, water, and a solid of unknown composition when heated with metallic magnesium for about six hours in a sealed tube. Ethyl chloride forms regular crystals of a hydrate with water at 0°C (5). Dry ethyl chloride can be used in contact with most common metals in the absence of air up to 200°C. Its oxidation and hydrolysis are slow at ordinary temperatures. Ethyl chloride yields ethyl alcohol, acetaldehyde, and some ethylene in the presence of steam with various catalysts, eg, titanium dioxide and barium chloride.

Table 1. Physical Properties of Ethyl Chloride

Property	Value
melting point, °C	−138.3
boiling point at 101 kPa[a], °C	12.4
surface tension, air, mN/m (= dyn/cm)	
5°C	21.20
10°C	20.64
specific gravity, vapor at 101 kPa[a] (air = 1)	2.23
specific gravity, liquid	
0/4°C	0.92390
20/4°C	0.8970
refractive index, n^{20}_D	1.3676
specific heat, liquid from −48.8 to 45°[b], J/(kg·K)[c]	$1612 + 2.72\,t + 1.46 \times 10^{-2}\,t^2$
specific heat, vapor at 101 kPa[a], 40°C, J/mol[c]	1.017
critical temperature, °C	186.6
critical pressure, MPa[d]	5.27
thermal conductivity, W/(m·K)	
liquid	0.1467
vapor at bp	0.0095
coefficient of volume expansion, 0–15°, av	0.00156
dielectric constant	
liquid, 170°C	6.29
vapor, 23.5°C	1.01285
dipole moment, C·m[e]	6.672×10^{-30}
heat of combustion, kJ/mol[c]	1327
heat of formation, kJ/mol[c]	
liquid	132.4
vapor	107.7
latent heat of evaporation at bp, J/g[c]	383.4
latent heat of fusion, J/g[c]	69.09
heat of adsorption, on homogeneous carbon, kJ/mol[c]	10.47
flash point, OC, °C	−43
flash point. CC, °C	−50
ignition temperature, °C	519
explosive limits in air, vol %	3.16−15
explosive limits in oxygen, vol %	4.0−67.2
explosive limits with nitrous oxide (N_2O), vol %	2.1−32.8
viscosity, mPa·s (= cP)	
liquid	
5°C	0.292
20°C	0.260
vapor	
12.4°C	0.093×10^{-3}
35°C	0.0165×10^{-3}
vapor pressure, kPa[a]	
−30°C	15.2
−10°C	40.5
0°C	61.86
10°C	92.3
20°C	134.8
60°C	456.0
100°C	1165

[a]To convert kPa to mm Hg, multiply by 7.5.
[b]For example, specific heat at −30°C, 1542.5 J/kg; at 20°C, 1672 J/kg.
[c]To convert J to cal, divide by 4.184.
[d]To convert MPa to atm, divide by 0.101.
[e]To convert C·m to debye, divide by 3.336×10^{-30}.

When ethyl chloride is chlorinated under light, both ethylidene and ethylene chlorides are formed; the latter in smaller quantity (6). Chlorination in the presence of antimony pentachloride at 100°C produces ethylene chloride almost exclusively. Photochemical bromination of ethyl chloride yields a series of bromochloroethanes, eg, CH$_3$CHBrCl, CH$_3$CBr$_2$Cl, and CHBr$_2$CBr$_2$Cl (7). In contact with iron wire at 100°C, ethyl chloride can be brominated to ethyl bromide and ethylene bromide. When a mixture of ethyl chloride and ethylene bromide is maintained for a prolonged period at 25°C in the presence of aluminum chloride, a redistribution of halogen atoms occurs forming ethyl bromide, ethylene chlorobromide, and ethylene chloride (8). Hydriodic acid reacts at 130°C with ethyl chloride to give ethyl iodide. Vapor-phase fluorination with nitrogen-diluted fluorine below 60°C and in contact with copper gauze results in formation of carbon tetrafluoride, monochlorotrifluoromethane, dichlorodifluoroethylene, and several chlorofluoroethanes (9). The chlorine in ethyl chloride can be replaced by fluorine by reaction with hydrogen fluoride in the presence of antimony fluoride (see FLUORINE COMPOUNDS, ORGANIC).

Reaction of ethyl chloride with an alcoholic solution of ammonia yields ethylamine, diethylamine, triethylamine, and tetraethylammonium chloride (10,11) (see AMINES, LOWER ALIPHATIC).

In the presence of Friedel-Crafts catalysts, gaseous ethyl chloride reacts with benzene at about 25°C to give ethylbenzene, three diethylbenzenes, and other more complex compounds (12) (see XYLENES AND ETHYLBENZENE). Aromatic compounds can generally be ethylated by ethyl chloride in the presence of anhydrous aluminum chloride (see FRIEDEL-CRAFTS REACTIONS). Ethyl chloride combines directly with sulfur trioxide to give ethyl chlorosulfonate, C$_2$H$_5$OSO$_2$Cl, and 2-chloroethylsulfonic acid, CH$_2$ClCH$_2$SO$_2$OH (13).

Manufacture

Three industrial processes have been used for the production of ethyl chloride: hydrochlorination of ethylene, reaction of hydrochloric acid with ethanol, and chlorination of ethane. Hydrochlorination of ethylene is used to manufacture most of the ethyl chloride produced in the United States. Because of its prohibitive cost, the ethanol route to ethyl chloride has not been used commercially in the United States since about 1972. Thermal chlorination of ethane has the disadvantage of producing undesired by-products, and has not been used commercially since about 1975.

Hydrochlorination of Ethylene.   The exothermic vapor-phase reaction between ethylene [74-85-1] and hydrogen chloride [7647-01-0] can be carried out at 130–250°C under a variety of catalytic conditions. Yields are reported to be greater than 90% of theoretical (14).

$$CH_2\!=\!CH_2 + HCl \rightleftarrows C_2H_5Cl \qquad \Delta H = -56.1 \text{ kJ } (-13.4 \text{ kcal})$$

At 200–250°C equilibrium conversion falls off; nevertheless, the process is usually conducted at the higher temperature to achieve a practical rate of reaction. The higher temperature accelerates the reaction, but also causes the formation of polymerization products which ultimately destroy the catalyst. In the United States, ethyl chloride is produced mainly by reaction of ethylene and hydrogen chloride under 0.1–0.3 MPa (1–3 atm) pressure at normal temperature in a 2% solution of aluminum chloride in ethyl chloride (15–17). Other variations are reaction at 175–400°C in contact with a thorium salt, eg, thorium oxychloride on silica gel (18); use of 1,1,2-trichloroethane as a solvent and aluminum chloride catalyst for reaction at $-5°$ to 55°C at 0.1–0.9 MPa (1–9 atm) pressure (19); and reaction at high pressures in contact with a peroxygen catalyst (20). Use of ^{60}Co gamma radiation also produces ethyl chloride and n-butyl chloride from ethylene and hydrogen chloride (21).

The hydrogen chloride needed for hydrochlorination of ethylene may be a by-product of other chlorocarbon processes such as the cracking of 1,1,2,2-tetra-chloroethane to trichloroethylene [79-01-6] (22). In one important form of this tandem procedure (23,24), two gas streams are supplied to the process: one rich in ethane, the other rich in ethylene. Chlorination of the ethane-rich stream is carried out at high temperature. This stage is noncatalyzed. Hydrogen chloride from the ethane chlorination along with unreacted ethylene is passed to the reactor where the ethylene-rich stream is hydrochlorinated under pressure at a lower temperature in the presence of a granular catalyst. Ethylene dichloride [107-06-2] is a by-product of the process.

Significant quantities of ethyl chloride are also produced as a by-product of the catalytic hydrochlorination over a copper chloride catalyst, of ethylene and hydrogen chloride to produce 1,2-dichloroethane, which is used as feedstock in the manufacture of vinyl choride (see VINYL POLYMERS). This ethyl chloride can be recovered for sale or it can be concentrated and catalytically cracked back to ethylene and hydrogen chloride (25). As the market for ethyl chloride declines, recovery as an intermediate by-product of vinyl chloride manufacture may become a predominant method of manufacture of ethyl chloride.

Chlorination of Ethane. Ethane [74-84-0] may be chlorinated thermally, catalytically, photochemically, or electrolytically. Monochlorination is favored because ethyl chloride chlorinates at about one-fourth of the rate at which it is itself produced from ethane.

Thermal chlorination of ethane is generally carried out at 250–500°C. At ca 400°C, a free-radical chain reaction takes place:

$$Cl_2 \rightarrow 2\ Cl\cdot$$

$$Cl\cdot\ +\ C_2H_6 \rightarrow HCl\ +\ C_2H_5\cdot$$

$$C_2H_5\cdot\ +\ Cl_2 \rightarrow C_2H_5Cl\ +\ Cl\cdot\ (\text{chain carrier})$$

The chlorine and ethane are brought together in a fluid bed of finely divided, inert, solid heat-transfer medium, eg, sand, at 380–440°C; the linear velocity of the gas is sufficient to maintain the finely divided solid in suspension within the reactor (26).

The chlorination of ethane may be catalyzed by bringing the reacting gases into contact with metal chlorides (27), or crystalline carbon (graphite) (28).

Photochemical chlorination is not used industrially. Electrolytic chlorination, which involves passing a low voltage current through a catalytic mixture of $AlCl_3$–$NaAlCl_4$, has not been used on a large scale (29).

Reaction of Ethyl Alcohol and Hydrochloric Acid. For many years this reaction was the only established technical process for ethyl chloride, but it was abandoned because of the high cost of ethyl alcohol [64-17-5] when petrochemicals became available. Zinc and other metallic chlorides have been used as catalysts and ethyl chloride is recovered by distillation (30,31).

$$C_2H_5OH + HCl \rightarrow H_2O + C_2H_5Cl$$

From Diethyl Sulfate. Several processes (32–34) have been proposed for manufacture of ethyl chloride based on the following reaction:

$$(C_2H_5O)_2SO_2 + 2\ HCl \rightarrow 2\ C_2H_5Cl + H_2SO_4$$

Other Processes For Ethyl Chloride. 1,2-Dichloroethane and ethylene, after at least 3 min in the presence of anhydrous calcium sulfate at 250–350°C and pressures of 0.69–2.76 MPa (100–400 psi), yield a mixture of ethyl chloride and vinyl chloride (35).

$$CH_2ClCH_2Cl + CH_2{=}CH_2 \rightarrow C_2H_5Cl + CH_2{=}CHCl$$

Vinyl chloride [75-01-4] can be reduced to ethyl chloride at elevated temperatures by reaction with excess hydrogen in contact with a hydrogenation catalyst (36).

In the presence of a heavy-metal chloride and water at 80–240°C, and under sufficient pressure to maintain the water in the liquid phase, diethyl ether reacts with hydrochloric acid to give ethyl chloride (37).

Economic Aspects

Selected United States' production, sale, and price statistics for ethyl chloride from 1956–1987 are given in Table 2. The economic history of ethyl chloride is entirely dominated by the fact that its principal application is in the manufacture of tetraethyllead (TEL). Use of jet engines in aircraft, more widespread use of diesel engines in transportation, and the development of alternative methods for increasing the octane content of gasoline kept production relatively constant through the 1960s and the early 1970s. Essentially no growth was seen from 1975 through 1979. During the 1980s ethyl chloride production declined steadily (see Table 2). This rate of decline is expected to reduce in the near future as the amount of ethyl chloride used to manufacture products other than TEL becomes more significant. The only important demand for ethyl chloride, other than its use in TEL manufacture, arises from the ethylcellulose industry (see CELLULOSE ETHERS).

Table 2. Ethyl Chloride Production and Price Statistics[a]

Year	Production, 10^3	Sales quantity, 10^3 t	Unit value, $/kg
1956	293.4	66.3	0.154
1960	247.9	86.7	0.176
1965	311.7	124.5	0.154
1970	308.2	124.0	0.132
1975	261.5	127.3	0.202
1979	264.5	81.8	0.374
1981	147.4	66.0	
1983	128.0	44.2	0.396
1985	77.5	37.5	0.352
1986	74.3	48.5	0.330
1987	70.2		

[a]According the U.S. International Trade Commission.

Standards

Good technical-grade ethyl chloride should not contain more than the following quantities of the indicated impurities: water, 15 ppm; acid (as HCl), 120 ppm; residue on evaporation at 110°C, 50 ppm. Ethyl chloride does not require added stabilizers.

Health and Safety Factors

Ethyl chloride is handled and transported in pressure containers under conditions similar to those applied to methyl chloride. In the presence of moisture, ethyl chloride can be moderately corrosive. Carbon steel is used predominantly for storage vessels and prolonged contact with copper should be avoided.

Ethyl chloride is readily absorbed into the body through mucous membranes, lungs, and skin. Although rapidly excreted by the lungs, its high solubility in blood prolongs total elimination from the body (38). Ethyl chloride is apparently not metabolized to any significant degree (39). Recovery of consciousness after exposure to ethyl chloride often entails an unpleasant hangover period (38). Experiments with animals provide evidence of kidney irritation and promotion of fat accumulation in the kidneys, cardiac muscle, and liver. Concentrations of 15–30 vol % in air are quickly fatal to animals; a concentration of 2% causes some unsteadiness; exposure to 1% concentration has no observable effect. A short-term systemic inhalation study indicated that exposure to five times the ACGIH TLV and OSHA PEL values of 1000 ppm ethyl chloride (5000 ppm) were well tolerated despite an unusually long exposure period (40). An earlier inhalation study on rats and rabbits found no effect during or after exposure to 25.4 mg/L (ca 10,000 ppm) ethyl chloride, 7.5 hours a day, five days a week for 6.5 months. A report from Russian literature, however, indicates that adverse effects in the liver, lungs, and blood occur at much lower levels. Based on the limited available data, the Environmental Protection Agency stated that ethyl chloride was one of the least toxic of the chloroethanes (41). A more recent lifetime inhalation study in rats and mice by the National Toxicology Program showed a high incidence of uterine tumors in female mice exposed to very high (15,000 ppm) concentrations of ethyl chloride (42).

NFPA classifies ethyl chloride as follows (43):

Flammability = 4, ie, very flammable gas, very volatile, and materials that in the form of dusts or mists form explosive mixtures when dispersed in air

Health = 2, ie, hazardous to health, but may be entered freely with self-contained breathing apparatus

Reactivity = 0, ie, is normally stable when under fire-exposure conditions and is not reactive with water

Uses

Tetraethyllead can be manufactured by the reaction of ethyl chloride with lead-sodium alloy (see LEAD COMPOUNDS).

$$4\ PbNa\ +\ 4\ C_2H_5Cl \rightarrow Pb(C_2H_5)_4\ +\ 3\ Pb\ +\ 4\ NaCl$$

Ethylcellulose [9004-57-3], produced by the reaction of ethyl chloride with alkali cellulose, is used mainly in the plastics and lacquer industries (44) (see CELLULOSE ETHERS).

Ethyl chloride is used to some extent as an ethylating agent in the synthesis of dyestuffs and fine chemicals. Benzene can be ethylated by the reaction with ethyl chloride in the presence of a Friedel-Crafts catalyst (see ALKYLATION; FRIEDEL-CRAFTS REACTIONS). In one process (45), the hydrogen chloride liberated from the Friedel-Crafts reaction reacts with ethylene to produce more ethyl chloride, which is recycled to the main reactor. Ethylbenzene for production of styrene used in high tonnage in the manufacture of polystyrene (see STYRENE PLASTICS), is normally made by reaction of ethylene with benzene, or by reforming petroleum cycloparaffins (see BTX PROCESSES; XYLENES AND ETHYLBENZENE). Ethyl chloride is used as a solvent in the polymerization of olefins using Friedel-Crafts catalysts (46,47), and as a polymerization activator to produce polyquinoline from quinoline at high temperature (121–160°C) (48).

Ethyl chloride can also be used as a feedstock to produce 1,1,1-trichloroethane by thermal chlorination at temperatures of 375–475°C (49), or by a fluidized-bed reactor at similar temperatures (50).

Other minor uses of ethyl chloride include: blowing agents for thermoplastic foam (51) and styrene polymer foam (52), the manufacture of polymeric ketones used as lube oil detergents (53), the manufacture of acetaldehyde (qv) (54), as an aerosol propellant (55), as a refrigerant (R-160), in the preparation of acid dyes (56), and as a local or general anesthetic (57,58).

BIBLIOGRAPHY

"Ethyl Chloride" under "Chlorine Compounds, Organic" in *ECT* 1st ed., Vol. 3, pp. 751–760, by R. Herzog, I. M. Skinner, G. W. Thomson, and Hymin Shapiro, Ethyl Corp.; "Ethyl Chloride" under "Chlorocarbons and Chlorohydrocarbons" in *ECT* 2nd ed., Vol. 5, pp. 140–147, by D. W. F. Hardie, Imperial Chemical Industries, Ltd; in *ECT* 3rd ed., Vol. 5, pp. 714–721, by T. E. Morris and W. D. Tasto, Dow Chemical U.S.A.

1. L. H. Horsley and co-workers, "Azeotropic Data," in *Advances in Chemistry Series*, Vols. 1 and 2, Nos. 6 and 35, American Chemical Society, Washington, D.C., 1952 (Vol. 1) and 1962 (Vol. 2).
2. D. H. R. Barton and K. E. Howlett, *J. Chem. Soc.*, 165 (1949).
3. G. M. Schwab and H. Noller, *Z. Electrochem.* **58**, 762 (1954).
4. A. Heinzelmann, R. Letterer, and H. Noller, *J. Monatsh. Chem.* **102**, 1750 (1971).
5. P. Villard, *Ann. Phys.* **11**, 384 (1897).
6. J. D'Ans and J. Kautzsch, *J. Prakt. Chem.* **80**, 310 (1909).
7. J. Denzel, *Ann. Chem. Liebigs* **195**, 189 (1879).
8. G. Calingaert and co-workers, *J. Am. Chem. Soc.* **62**, 1546 (1940).
9. L. A. Bigelow, *Chem. Rev.* **40**, 51 (1947).
10. C. E. Groves, *J. Chem. Soc.* **13**, 331 (1860).
11. A. W. Hofmann, *Ber. Dtsch. Chem. Ces.* **3**, 109 (1870).
12. M. Blau and J. E. Willard, *J. Am. Chem. Soc.* **75**, 330 (1953).
13. Th. Von Purgold, *Z. Chem. Ind. U.S.S.R.* **8**, 669 (1868).
14. C. Trabalka and K. Alexandru, *Synth. Ethyl Chloride*, 68354 (1979).
15. Can. Pat. 448,020 (Apr. 20, 1948), E. V. Fasce (to Standard Oil Development Co.).
16. U.S. Pat. 3,345,421 (Nov. 24, 1961), M. D. Brown (to Halcon International, Inc.).
17. R. V. Chandhari and L. K. Doraiswamy, *Chem. Eng. Sci.* **29**, 349 (1974).
18. Can. Pat. 464,069 (Mar. 28, 1950), D. C. Bond and M. Savoy (to The Pure Oil Co.).
19. U.S. Pat. 2,140,927 (Sept. 29, 1936), J. E. Pierce (to The Dow Chemical Company).
20. Can. Pat. 464,488 (Apr. 18, 1950), W. E. Hanford and J. Harman (to Canadian Industries Ltd.).
21. A. Terakawa, J. Nakanishi, and T. Kiruyama, *Bull. Chem. Soc. Jpn.* **39**, 892 (1966).
22. Brit. Pat. 505,196 (Nov. 5, 1937), A. A. Levine (to E. I. du Pont de Nemours & Co., Inc.).
23. U.S. Pat. 2,246,082 (Aug. 22, 1939), W. E. Vaughn and F. F. Rust (to Shell Development Co.).
24. *Ind. Eng. Chem.* **47**, 984 (1955); *Pet. Refiner* **34**, 149 (1955).
25. U.S. Pat. 4,849,562 (July 18, 1989), C. Buhs, E. Dreher, and G. McConchie (to The Dow Chemical Company).
26. Brit. Pat. 667,185 (Mar. 9, 1949), P. A. Hawkins and R. T. Foster (to Imperial Chemical Industries, Ltd.}.
27. U.S. Pat. 2,140,547 (Aug. 26, 1936), J. H. Reilly (to The Dow Chemical Company).
28. Brit. Pat. 483,051 (Oct. 8, 1936), G. W. Johnson (to I. G. Farbenindustrie A.G.).
29. Belg. Pat. 654,985 (Apr. 28, 1965), (to Imperial Chemical Industries, Ltd.).
30. U.S. Pat. 2,516,638 (Mar. 3, 1947), J. L. McCrudy (to The Dow Chemical Company).
31. Fr. Pat. 858,724 (Dec. 2, 1940), (to Société anon. des Matières colorantes et Produits chimiques de Saint-Denis).
32. Brit. Pat. 566,147 (Dec. 28, 1942), E. G. Galitzenstein and C. Woolf (to Distillers Co. Ltd.).
33. A. P. Giraitis, *Erdol Kohle* **9**, 791 (1951).
34. U.S. Pat. 2,125,284 (Nov. 11, 1935), L. C. Chamberlain and J. L. Williams (to The Dow Chemical Company).
35. U.S. Pat. 2,681,372 (Jan. 16, 1951), P. W. Trotter (to Ethyl Corp.).

36. Brit. Pat. 470,817 (Feb. 18, 1936), G. W. Johnson (to I. G. Farbenindustrie A.G.).

37. U.S. Pat. 2,084,710 (Aug. 21, 1935), H. N. Spurlin (to Hercules Powder Co.).

38. J. I. Murray Lawson, *Br. J. Anaesth.* **37,** 667 (1965).

39. F. A. Patty, P. Irish, and D. Fassett, eds., *Industrial Hygiene and Toxicology,* 2nd ed., Vol. 2, Wiley-Interscience, Inc., New York, 1963, p. 1275.

40. T. Landry, K. Johnson, J. Phillips, and S. Weiss, *Fundam. Appl. Toxicol.* **13**(3), 516–522 (1989).

41. *Health Effects Assessment for Ethyl Chloride,* Report EPA/600/8-88/036, United States Environmental Protection Agency, Environmental Criteria Assessment Office, Cincinnati, Ohio, 1987.

42. *Toxicology and Carcinogenesis Studies of Chloroethane (ethyl chloride) in Rats and Mice,* Report NTP/346, The National Toxicology Program, 1989.

43. *Fire Protection Guide on Hazardous Materials,* 9th ed., National Fire Protection Agency, Washington, D.C., 1986.

44. S. L. Bass, A. J. Barry, and A. E. Young, in E. Ott, ed., *Cellulose and Cellulose Derivatives,* Interscience Publishers, Inc., New York, 1946, p. 758 ff.

45. Brit. Pat. 581,145 (July 30, 1942), (to Standard Oil Development Co.).

46. U.S. Pat. 2,387,784 (Dec. 28, 1940), R. M. Thomas and H. C. Reynolds (to Standard Oil Development Co.).

47. G. P. Below and co-workers, *Kinet. Katal.* **8,** 265 (1967).

48. R. F. Smirnov, B. I. Tikhomirov, and A. I. Yakubehik, *Vysokomol. Soedln Ser. B* **13,** 895 (1971).

49. U.S. Pat. 3,706,816 (Sept. 22, 1969), A. Campbell and R. A. Carruthers (to Imperial Chemical Industries, Ltd.).

50. U.S. Pat. 3,012,081 (Sept. 29, 1960), F. Conrad and A. J. Haefner (to Ethyl Corp.).

51. J. Korb and W. Harfmann, *Method and Apparatus for Preparing Thermoplastic Foam,* 9007407 A1, 1990.

52. U.S. Pat. 4,636,527 (Jan. 13, 1986), J. Kennedy and K. Suh (to The Dow Chemical Company).

53. U.S. Pat. 4,169,859 (Oct. 2, 1979), T. Clough (to Atlantic Richfield Co.).

54. U.S. Pat. 3,939,209 (Feb. 17, 1976), M. Sze and R. Wang (to The Lumus Co.).

55. Ger. Pat. 2,404,778 (Aug. 14, 1975), F. Huber (to Dynamit Nobel A.-G.)

56. Pol. Pat. 129377 (Dec. 31, 1986), C. Przybylski, J. Gmaj, L. Jaworski, and W. Kania.

57. J. D. Rochford and B. T. Broadbent, *Br. Med. J.* **7,** 664 (1943).

58. C. W. Lincoln, *Anesth. Anal.* **20,** 328 (1941).

MATT C. MILLER
Dow Chemical U.S.A.

OTHER CHLOROETHANES

1,1-DICHLOROETHANE

1,1-Dichloroethane [75-34-3], CH_3CHCl_2, ethylidene chloride, ethylidene dichloride, is a colorless liquid with an ethereal odor. It is miscible with most organic solvents and all chlorinated solvents. It is employed as a solvent, but its largest industrial use is as an intermediate in the production of 1,1,1-trichloroethane.

Physical and Chemical Properties

The properties of 1,1-dichloroethane are listed in Table 1. 1,1-Dichloroethane decomposes at 356–453°C by a homogeneous first-order dehydrochlorination, giving vinyl chloride and hydrogen chloride (1,2). Dehydrochlorination can also occur on activated alumina (3,4), magnesium sulfate, or potassium carbonate (5). Dehydrochlorination in the presence of anhydrous aluminum chloride (6) proceeds readily. The 48-h accelerated oxidation test with 1,1-dichloroethane at reflux temperatures gives a 0.025% yield of hydrogen chloride as compared to 0.4% HCl for trichloroethylene and 0.6% HCl for tetrachloroethylene. Reaction with an amine gives low yields of chloride ion and the dimer 2,3-dichlorobutane, $CH_3CHClCHClCH_3$. 2-Methyl-1,3-dioxaindan [14046-39-0] can be prepared by a reaction of catechol [120-80-9] with 1,1-dichloroethane (7).

Manufacture

1,1-Dichloroethane is produced commercially from hydrogen chloride and vinyl chloride at 20–55°C in the presence of an aluminum, ferric, or zinc chloride catalyst (8,9). Selectivity is nearly stoichiometric to 1,1-dichloroethane. Small amounts of 1,1,3-trichlorobutane may be produced. Unreacted vinyl chloride and HCl exit the top of the reactor, and can be recycled or sent to vent recovery systems. The reactor product contains the Lewis acid catalyst and must be separated before distillation. Spent catalyst may be removed from the reaction mixture by contacting with a hydrocarbon or paraffin oil, which precipitates the metal chloride catalyst into the oil (10). Other inert liquids such as siloxanes and perfluorohydrocarbons have also been used (11). 1,1-Dichloroethane is also one of the intermediate products of high temperature thermal chlorination of ethane or ethyl chloride. In ethane chlorination, the reaction proceeds through ethyl chloride as an intermediate (12). 1,1-Dichloroethane itself is usually an intermediate in the production of vinyl chloride and of 1,1,1-trichloroethane by thermal chlorination or photochlorination (13).

Environmental Concerns

The energy requirements for desorbing 1,1-dichloroethane from activated carbon in a stripping–adsorption process for water purification have been calculated at

Table 1. Properties of 1,1-Dichloroethane

Property	Value
melting point, °C	−96.7
boiling point, °C	57.3
density at 20°C, g/L	1.1747
n_D^{22}	1.4166
viscosity at 20°C, mPa·s (=cP)	0.377
surface tension at 20°C, mN/m (=dyn/cm)	23.34
specific heat at 20°C, J/(g·K)^a	
liquid	1.087
gas	0.824
latent heat of vaporization at 20°C, J/g^a	280.3
critical temperature, °C	261.5
critical pressure, MPa^b	5.06
flash point (closed cup), °C	−12.0
explosive limits in air at 25°C, vol %	5.4–11.4
autoignition temperature, °C	458
heat of combustion, kJ/g^a	12.57
dielectric constant, liquid at 20°C	10.9
vapor pressure, kPa^c	
10°C	15.37
20°C	24.28
30°C	36.96
solubility at 20°C, g	
dichloroethane in 100 g H$_2$O	0.55
H$_2$O in 100 g dichloroethane	0.097
binary azeotropes, bp, °C	
with 1.9% H$_2$O	53.3^d
with 11.5% ethanol	54.6

^{a}To convert J to cal, multiply by 0.239.
^{b}To convert MPa to atm, multiply by 9.87.
^{c}To convert kPa to mm Hg, multiply by 7.5.
^{d}At 97 kPa^c.

112 kJ/kg (14). Chlorinated hydrocarbons such as 1,1-dichloroethane may easily be removed from water by air or steam stripping.

Toxicity. 1,1-Dichloroethane, like all volatile chlorinated solvents, has an anesthetic effect and depresses the central nervous system at high vapor concentrations. The 1991 American Conference of Governmental Industrial Hygienists (ACGIH) recommends a time-weighted average (TWA) solvent vapor concentration of 200 ppm and a permissible short term exposure level (STEL) of 250 ppm for worker exposure. The oral LD$_{50}$ of 1,1-dichloroethane in rats is 14.1 g/kg, classifying it as essentially nontoxic by oral ingestion (15)

1,2-DICHLOROETHANE

1,2-Dichloroethane [107-06-2], ethylene chloride, ethylene dichloride, CH$_2$Cl-CH$_2$Cl, is a colorless, volatile liquid with a pleasant odor, stable at ordinary temperatures. It is miscible with other chlorinated solvents and soluble in common organic solvents as well as having high solvency for fats, greases, and waxes. It is most commonly used in the production of vinyl chloride monomer.

Physical and Chemical Properties

The physical properties of 1,2-dichloroethane are listed in Table 2.

Pyrolysis. Pyrolysis of 1,2-dichloroethane in the temperature range of 340–515°C gives vinyl chloride, hydrogen chloride, and traces of acetylene (1,18) and 2-chlorobutadiene. Reaction rate is accelerated by chlorine (19), bromine, bromotrichloromethane, carbon tetrachloride (20), and other free-radical generators. Catalytic dehydrochlorination of 1,2-dichloroethane on activated alumina (3), metal carbonate, and sulfate salts (5) has been reported, and lasers have been used to initiate the cracking reaction, although not at a low enough temperature to show economic benefits.

Hydrolysis. Heating 1,2-dichloroethane with excess water at 60°C in a nitrogen atmosphere produces some hydrogen chloride. The rate of evolution is dependent on the temperature and volume of the aqueous phase. Hydrolysis at 160–175°C and 1.5 MPa (15 atm) in the presence of an acid catalyst gives ethylene glycol, which is also obtained in the presence of aqueous alkali at 140–250°C and up to 4.0 MPa (40 atm) pressure (21).

Oxidation. Atmospheric oxidation of 1,2-dichloroethane at room or reflux temperatures generates some hydrogen chloride and results in solvent discoloration. A 48-h accelerated oxidation test at reflux temperatures gives only 0.006% hydrogen chloride (22). Addition of 0.1–0.2 wt. % of an amine, eg, diisopropylamine, protects the 1,2-dichloroethane against oxidative breakdown. Photooxidation in the presence of chlorine produces monochloroacetic acid and 1,1,2-trichloroethane (23).

Corrosion. Corrosion of aluminum, iron, and zinc by boiling 1,2-dichloroethane has been studied (22). Dry and refluxing 1,2-dichloroethane completely consumed a 2024 aluminum coupon in a 7-d study, whereas iron and zinc were barely attacked. Aluminum was attacked less than iron or zinc by refluxing with 1,2-dichloroethane containing 7% water. Corrosion rates in μm/yr (mils penetration per year or mpy) in dry solvent are 0.254 (0.01) for iron and 3.05 (0.12) for zinc. In the wet solvent, the corrosion rate for iron increases to 145 μm/yr (5.7 mpy) and for zinc to 1.2 mm/yr (47 mpy). Corrosion rate for aluminum in the wet solvent is 2.36 mm/yr (92 mpy) as compared to complete dissolution in the dry solvent.

Nucleophilic Substitution. The kinetics of the bimolecular nucleophilic substitution of the chlorine atoms in 1,2-dichloroethane with NaOH, $NaOC_6H_5$, $(CH_3)_3N$, pyridine, and CH_3COONa in aqueous solutions at 100–120°C has been studied (24). The reaction of sodium cyanide with 1,2-dichloroethane in methanol at 50°C to give 3-chloropropionitrile proceeds very slowly. Dimethyl sulfoxide as a solvent for the reaction greatly enhances nucleophilic substitution of the chlorine atom. Further reaction of sodium cyanide at room temperature gives acrylonitrile (qv), $CH_2{=}CHCN$ (25). 1,2-Dichloroethane reacts with toluene in the presence of Friedel-Crafts catalysts such as $AlBr_3$, $AlCl_3$, $GaCl_3$, and $ZrCl_3$ (26).

Table 2. Properties of 1,2-Dichloroethane[a]

Property	Value
melting point, °C	− 35.3
boiling point, °C	83.7
density at 20°C, g/L	1.2529
n_D^{20}	1.4451
viscosity at 20°C, mPa·s (= cP)	0.84
surface tension at 20°C, mN/m (= dyn/cm)	31.28
specific heat at 20°C, J/(g·K)[a]	
liquid	1.288
gas	1.066
latent heat of vaporization at 20°C, J/g[b]	323.42
latent heat of fusion, J/g[b]	88.36
critical temperature, °C	290
critical pressure, MPa[c]	5.36
critical density, g/L	0.44
flash point °C	
closed cup	17
open cup	21
explosive limits in air at 25°C, vol %	6.2–15.6
autoignition temperature in air, °C	413
thermal conductivity, liquid at 20°C, W/(m·K)[d]	0.143
heat of combustion, kJ/g[b]	12.57
heat of formation, kJ/mol)[b]	
liquid	157.3
vapor	122.6
dielectric constant	
liquid, 20°C	10.45
vapor, 120°C	1.0048
dipole moment, C·m[e]	5.24×10^{-30}
coefficient of cubical expansion, mL/g, 0–30°C	0.00116
vapor pressure, kPa[f]	
10°C	5.3
20°C	8.5
30°C	13.3
solubility at 20°C, g	
1,2-dichloroethane in 100 g H_2O	0.869
H_2O in 100 g 1,2-dichloroethane	0.160
azeotropes[g], bp, °C	
with 10.5% H_2O	72
with 5% H_2O and 17% ethanol	66.7

[a]See Ref. 16 for additional property data.
[b]To convert J to cal, divide by 4.184.
[c]To convert MPa to atm, multiply by 9.87.
[d]To convert W/(m·K) to Btu·ft)/(h·ft²·°F), divide by 1.73.
[e]To convert C·m to debyes, multiply by 3×10^{29}.
[f]To convert kPa to mm Hg, multiply by 7.5.
[g]See Ref. 17 for additional binary azeotropes.

Ammonolysis of 1,2-dichloroethane with 50% aqueous ammonia at 100°C is a primary commercial process for producing ethyleneamines (27).

Manufacture

1,2-Dichloroethane is produced by the vapor- (28) or liquid-phase chlorination of ethylene. Most liquid-phase processes use small amounts of ferric chloride as the catalyst. Other catalysts claimed in the patent literature include aluminum chloride, antimony pentachloride, and cupric chloride and an ammonium, alkali, or alkaline-earth tetrachloroferrate (29). The chlorination is carried out at 40–50°C with 5% air or other free-radical inhibitors (30) added to prevent substitution chlorination of the product. Selectivities under these conditions are nearly stoichiometric to the desired product. The exothermic heat of reaction vaporizes the 1,2-dichloroethane product, which is purified by distillation.

$$H_2C{=}CH_2 + Cl_2 \xrightarrow[\text{50–150°C}]{\text{FeCl}_3} ClH_2C{-}CH_2Cl \tag{1}$$

Oxychlorination of ethylene has become the second important process for 1,2-dichloroethane. The process is usually incorporated into an integrated vinyl chloride plant in which hydrogen chloride, recovered from the dehydrochlorination or cracking of 1,2-dichloroethane to vinyl chloride, is recycled to an oxychlorination unit. The hydrogen chloride by-product is used as the chlorine source in the chlorination of ethylene in the presence of oxygen and copper chloride catalyst:

$$2\,H_2C{=}CH_2 + 4\,HCl + O_2 \xrightarrow[\text{270°C}]{\text{CuCl}_2} 2\,ClH_2C{-}CH_2Cl + H_2O \tag{2}$$

Reactor designs have included fixed and fluidized beds. A fluidized-bed oxychlorination reactor developed by B. F. Goodrich is claimed to provide very good temperature control (31). A large number of patents deal with the catalyst technology (32–41), which usually includes $CuCl_2$ and minor amounts of by-product inhibitors, such as potassium, sodium, lithium, or magnesium. To reduce oxidation to carbon dioxide and carbon monoxide, plant designs may include two or three reactors in series, with the HCl and oxygen feeds split to the secondary reactor(s) to decrease the C_2H_4:oxygen ratio in each reactor. By-products of this reaction include carbon dioxide, carbon monoxide, ethyl chloride, 1,1,2-trichloroethane, and trichloroacetaldehyde (chloral). The reactor products are usually condensed, unreacted HCl and water are separated from the organics, and the 1,2-dichloroethane is purified by distillation. Unreacted ethylene and oxygen can be recycled to the reactor, or sent to a vent recovery system; however, a purge is needed in the recycle system to prevent inert gas buildup in the recycle stream. Air or pure oxygen can be used; oxygen-based systems lose less ethylene during the inert purge.

Economic Aspects and Uses

A significant portion (88%) of U.S. 1,2-dichloroethane production is converted to vinyl chloride monomer (see VINYL POLYMERS) (42). Since it has very few solvent or emissive uses, 1,2-dichloroethane has not faced the regulatory pressures of other chlorinated hydrocarbons. As can be seen in Table 3, dramatic growth has been seen for conversion of 1,2-dichloroethane to vinyl chloride. Other uses for 1,2-dichloroethane have been growing more slowly or declining. 1,2-Dichloroethane is also a starting material for chlorinated solvents such as 1,1,1-trichloroethane, vinylidene chloride, trichloroethylene, and perchloroethylene. Other uses include as a reactant to prepare ethyleneamines. Table 4 shows that production capacity is fairly evenly split between the United States and Western Europe, although a significant amount of 1,2-dichloroethane is also produced in Japan. Most producers of poly(vinyl chloride) resins have back integrated and produce 1,2-dichloroethane for captive use.

Table 3. U.S. Uses of 1,2-Dichloroethane, 10^3 t[a]

Use	1968	1987
vinyl chloride	1	6,300
other chlorohydrocarbons	305	360
ethylenediamine	123	90
TEL[b] antiknock mixtures	101	9
Total production	_2.41 $\times$ 10^3_	_7.15 $\times$ 10^3_

[a]Ref. 42. Courtesy of SRI International.
[b]TEL = tetraethyllead.

Table 4. World 1,2-Dichloroethane Production Capacity,[a] t

Country or region	1988
North America	9,445
Western Europe	9,830
Japan	3,068
other	8,351
Total	_30,691_

[a]Ref. 42. Courtesy of SRI International.

Environmental Concerns

Removal of metal chlorides from the bottoms of the liquid-phase ethylene chlorination process has been studied (43). A detailed summary of production methods, emissions, emission controls, costs, and impacts of the control measures has been made (44). Residues from this process can also be recovered by evaporation, decomposition at high temperatures, and distillation (45). A review of the by-prod-

ucts produced in the different manufacturing processes has also been performed
(46). Several processes have been developed to limit ethylene losses in the inerts
purge from an oxychlorination reactor (47,48).

Toxicity.   1,2-Dichloroethane at high vapor concentrations (above 200 ppm)
can cause central nervous system depression and gastrointestinal upset characterized by mental confusion, dizziness, nausea, and vomiting. Liver, kidney,
and adrenal injuries may occur at the higher vapor levels. The recommended
1991 AGCIH vapor exposure TWA standard for 1,2-dichloroethane was 10 ppm,
with a STEL guideline of 40 ppm. The odor threshold for 1,2-dichloroethane is
50–100 ppm and thus odor does not serve as a good warning against possible
overexposure.

1,2-Dichloroethane is one of the more toxic chlorinated solvents by inhalation (49). The highest nontoxic vapor concentrations in chronic exposure studies
with various animals range from 100 to 200 ppm (50,51). 1,2-Dichloroethane exhibits a low single-dose oral toxicity in rats; LD_{50} is 680 mg/kg (49). Repeated skin
contact should be avoided since the solvent can cause defatting of the skin, severe
irritation, and moderate edema. Eye contact may have slight to severe effects.

1,1,1-TRICHLOROETHANE

1,1,1-Trichloroethane [71-55-6], methyl chloroform, CH_3CCl_3, is a colorless, nonflammable liquid with a characteristic ethereal odor. It is miscible with other
chlorinated solvents and soluble in common organic solvents. The compound was
first prepared by Regnault about 1840.

1,1,1-Trichloroethane is among the least toxic of the chlorinated solvents
used in industry today. The commercial metal-cleaning grades contain added inhibitors that make usage acceptable for all common metals including aluminum.
It has excellent solvency for various greases, oils, tars, and waxes and a wide
range of organic materials (see SOLVENTS, INDUSTRIAL). Emissive uses of 1,1,1-
trichloroethane will decline, whereas its use as a chemical intermediate, especially for the production of fluorocarbons, should increase.

Physical and Chemical Properties

The physical properties of 1,1,1-trichloroethane are given in Table 5.

Pyrolysis.   The pyrolysis of 1,1,1-trichloroethane at 325–425°C proceeds by
a simultaneous unimolecular and radical-chain mechanism to yield 1,1-dichloro-
ethylene and hydrogen chloride (52). 1,1,1-Trichloroethane vapors mixed with air
and passed over hot metal surfaces form relatively little phosgene at temperatures
up to 370°C (53).

Hydrolysis.   1,1,1-Trichloroethane heated with water at 75–160°C under
pressure and in the presence of sulfuric acid or a metal chloride catalyst decomposes to acetyl chloride, acetic acid, or acetic anhydride (54). However, hydrolysis
under normal use conditions proceeds slowly. The hydrolysis is 100–1000 times
faster with trichloroethane dissolved in the water phase than vice versa. Reflux-

Table 5. Properties of 1,1,1-Trichloroethane

Property	Value
melting point, °C	−33.0
boiling point, °C	74.0
density at 20°C	1.3249
n_D^{20}	1.4377
viscosity at 20°C, mPa·s (=cP)	0.858
surface tension at 20°C, mN/m (=dyn/cm)	25.54
specific heat at 20°C, J/(g·K)^a	
liquid	1.004
gas	0.782
latent heat of vaporization at 20°C, J/g^a	248.11
critical temperature, °C	311.5
critical pressure, MPa^b	4.48
flash point (closed cup), °C	none
explosive limits in air at 25°C, vol %	8.0–10.5
autoignition temperature, °C	537
heat of combustion, kJ/g^a	6.69
dielectric constant, liquid at 20°C	7.5
vapor pressure^c, kPa^d	
20°C	13.3
40°C	31.7
solubility at 20°C, g	
trichloroethane in 100 g H$_2$O	0.095
H$_2$O in 100 g trichloroethane	0.034
binary azeotropes, bp, °C	
with 4.3% H$_2$O	65.0
with 23% methanol	55.5
with 17.4% ethanol	64.4
with 17% isopropyl alcohol	68.2
with 17.2% t-butyl alcohol	70.2

^{a}To convert J to cal, divide by 4.184.
^{b}To convert MPa to atm, multiply by 9.87.
^{c}Antoine constants for 1,1,1-trichloroethane: A = 7.76632; B = 1204.66; C = 226.671, where $\log_{10}$ pressure (kPa) = $A - B/T + C$; T = temperature in °C.
^{d}To convert kPa to mm Hg, multiply by 7.5.

ing 1,1,1-trichloroethane with ferric and gallium chloride hydrate salts gives hydrogen chloride, acetic acid, 1,1-dichloroethylene, and the dehydrated salts (55).

Oxidation.   1,1,1-Trichloroethane is stable to oxidation when compared to olefinic chlorinated solvents like trichloroethylene and tetrachloroethylene. Use of a 48-h accelerated oxidation test gave no hydrogen chloride, whereas trichloroethylene gave 0.4 wt % HCl and tetrachloroethylene gave 0.6 wt % HCl (22).

Corrosion.   The corrosion rates of 1,1,1-trichloroethane with metals in dry and wet environments have been reported (22). Refluxing uninhibited 1,1,1-trichloroethane reacts vigorously with aluminum to give aluminum chloride, 2,2,3,3-tetrachlorobutane, 1,1-dichloroethylene, and hydrogen chloride. Adequate metal inhibitors, however, prevent this reactivity and allow the solvent to be used in aluminum metal-cleaning applications. Dry, uninhibited 1,1,1-trichloroethane is not very corrosive with iron or zinc; corrosion rate with iron is <2.54 μm/yr (<0.1 mpy) and with zinc <25.4 μm/yr (<1.0 mpy). Addition of 7% water increases corrosion rates to 254 μm/yr (10.0 mpy) for iron and >254 μm/yr (>10.0 mpy) for zinc. The presence of both water and ethanol increases iron or tin attack at reflux. Highest metal loss with tin-plated iron coupons occurred at a solvent composition of 30% 1,1,1-trichloroethane, 55% ethanol, and 15% water (56). Addition of water and/or ethanol increases the solubility of the metal chloride products obtained from the reacting metal surface.

Inhibition. Organic inhibitors for proprietary grades of 1,1,1-trichloroethane are (1) acid acceptors, namely epoxide compounds used to neutralize small amounts of hydrogen chloride normally formed during solvent use; and (2) metal stabilizers, namely organic compounds that deactivate metal surfaces and remove or complex trace amounts of metal chloride salts that might form. Several hundred United States and foreign patents have been issued on 1,1,1-trichloroethane stabilization, usually with the aid of a Lewis base such as an amino acid or carbonyl compound (57) (see CORROSION AND CORROSION INHIBITORS).

Dehydrochlorination. 1,1,1-Trichloroethane over activated alumina (3) or anhydrous aluminum chloride at 0°C (6) gives rapid hydrogen chloride evolution and 1,2-dichloroethylene, which may form polymer (6). Aluminum fluoride is the most active metal fluoride catalyzing dehydrochlorination (58). Several chlorinated solvents, including 1,1,1-trichloroethane, can be dehydrochlorinated on hopcalite catalyst (MnO_2–CuO–Cu) at 305–315°C (59). Other catalytic materials include $MgSO_4$ and K_2CO_3 (5) and molecular sieves containing the cations H^+, Mg^{2+}, Li^+, Na^+, or K^+ (60).

Miscellaneous. 1,1,1-Trichloroethane reacts with olefins, $CH_2{=}CRR'$, in the presence of $P(O)[(NCH_3)]_3$ and $FeCl_2$ at 130°C, to give compounds of the type $CH_3CCl_2CH_2CRR'Cl$ (61). Fluorination of 1,1,1-trichloroethane with anhydrous hydrogen fluoride at 144°C gives both 1,1-dichloro-1-fluoroethane [17171-00-6], HCFC-141b, and 1-chloro-1,1-difluoroethane [75-68-3], HCFC-142b (62). The use of 1,1,1-trichloroethane as a feedstock to chlorofluorocarbon compounds that are not fully halogenated (hydrochlorofluorocarbons, HCFCs) will probably grow, as these compounds have low ozone depletion potential (ODP) and will replace to some extent fully halogenated chlorofluorocarbons in many applications. Reactivity with amines (22) and mercaptan groups is low.

Manufacture

In the most important process, vinyl chloride obtained from 1,2-dichloroethane is hydrochlorinated to 1,1-dichloroethane which is then thermally or photochemically chlorinated:

$$CH_2{=}CHCl + HCl \rightarrow CH_3CHCl_2 \xrightarrow{Cl_2 \text{ vapor phase}} CH_3CCl_3 + HCl \qquad (3)$$

The ultraviolet lamps used in the photochlorination process serve to dissociate the chlorine into free radicals and start the radical-chain reaction. Other radical sources, such as 2,2'- azobisisobutyronitrile, have been used (63,64). Primary by-products of the photochlorination process include 1,1,2-trichloroethane (15–20%), tetrachloroethanes, and pentachloroethane. Selectivity to 1,1,1-trichloroethane is higher in vapor-phase chlorination. Various additives, most containing iodine or an aromatic ring in the molecule, have been used to increase the selectivity of the reaction to 1,1,1-trichloroethane (65–67). A review of efforts to increase selectivity has been published (68). The by-product HCl can be recycled to the 1,1-dichloro-ethane process. In the thermal chlorination process, primary by-products include vinyl chloride, vinylidene chloride, tetrachloroethanes, and pentachloroethane. If the only desired product is 1,1,1-trichloroethane, the vinyl and vinylidene chloride from the reaction are hydrochlorinated with $FeCl_3$ catalyst to 1,1-dichloroethane and 1,1,1-trichloroethane, respectively. Carbon formation can be significant in the thermal chlorination process. Alternatively, chloroethane or ethane may be fed to this process, producing *in situ* the 1,1-dichloroethane intermediate (12,69).

In a second process, hydrogen chloride is added to 1,1-dichloroethylene in the presence of a $FeCl_3$ catalyst:

$$CH_2=CCl_2 + HCl \xrightarrow[30°C]{FeCl_3} CH_3CCl_3 \qquad (4)$$

Unreacted 1,1-dichloroethylene exits the reactor as vapor and can be condensed and recycled to the reactor. Product 1,1,1-trichloroethane exits the reactor as a liquid, along with the Lewis acid catalyst, and can be removed from the catalyst by flash distillation. Selectivity is high; however, some dehydrochlorination of the product can occur in the distillation step.

A process of minor importance utilizes a continuous noncatalytic chlorination of ethane which produces 1,1,1-trichloroethane and a number of other products, depending on the reaction conditions.

Stabilized 1,1,1-dichloroethane is stable to normal shipping conditions, and has been shipped in metal and lined drums, tank trucks, rail cars, barges, and ships. Care must be taken that the container is free from previously contained material because of the high solvency of 1,1,1-trichloroethane. Shipping containers should also be dry.

Economic Aspects

The principal U.S. producers of 1,1,1-trichloroethane include The Dow Chemical Company, PPG Industries Inc., and Vulcan Materials Co. Several European and Japanese companies also produce large amounts annually. Over 70% of the production is based on the vinyl chloride–1,1-dichloroethane process, 20% on the 1,1-dichloroethylene process, and about 10% on the direct chlorination of ethane.

The estimated U.S. demand (Bureau of the Census) for 1,1,1-trichloroethane in 1990 was 280,000 t. Total world demand for 1,1,1-trichloroethane in 1987 was 578,000 t (70). All nonessential emissive uses of 1,1,1-trichloroethane will be phased out by the year 2000. U.S. production capacity was estimated to be about 470,000 metric tons and production capacity outside the United States was estimated to be approximately 454,000 t in 1989 (71).

Health and Safety Factors

1,1,1-Trichloroethane is among the least toxic of the industrial chlorinated solvents (72). The acute oral LD_{50} dosages for 1,1,1-trichloroethane for guinea pigs, mice, rabbits, and rats range from 5 to 12 g/kg (73). Injection of radioactive $CH_3^{14}CCl_3$ in doses of 700 mg/kg into rats demonstrated that 99% of the tracer solvent was eliminated by respiration. Vapor inhalation causes depression of the central nervous system (dizziness, light-headedness). The LC_{50} for rats is 18,000 ppm/3 h and 14,000 ppm/7 h. The 1991 TWA for 1,1,1-trichloroethane suggested by the ACGIH was 350 ppm with a recommended STEL of 450 ppm.

Entry into a tank that has contained any chlorinated or any easily evaporated solvent requires special procedures to ensure worker safety. The heavier vapors tend to concentrate in unventilated spaces. The proper tank entry procedure requires positive ventilation, testing for residue solvent vapor and oxygen levels, and the use of respiratory equipment and rescue harness. Monitoring the tank from outside is also important. The use of an appropriate gas mask is permissible in vapor concentrations of less than 2% and when there is no deficiency of atmospheric oxygen, but not for exposures exceeding one-half hour. Skin exposure to 1,1,1-trichloroethane can cause irritation, pain, blisters, and even burning. Eye exposure may produce irritation, but should not cause serious injury.

Environmental Concerns

1,1,1-Trichloroethane has been targeted for production phaseout by the year 2000 for emissive uses by the 1987 Montreal Protocol on Substances that Deplete the Ozone Layer. This treaty addresses compounds that may contribute to depletion of the earth's stratospheric ozone layer. The 1990 U.S. Clean Air Act accelerated this production phaseout, and other countries have also accelerated phaseout schedules. 1,1,1-Trichloroethane has been given an ozone depletion potential (ODP) of 0.15. CFC-12, for comparison, has an ODP of 1.0. Even though 1,1,1-trichloroethane has a relatively low ODP, it was targeted for inclusion in the Montreal protocol because of its large production for primarily emissive uses (Table 6). Primary replacements for 1,1,1-trichloroethane in metal- and circuit board-cleaning systems include aqueous mixtures, such as surfactants, and hydrocarbons, with their concomitant increases in flammability risks. In addition, increasing emphasis is being placed on recovery of spent solvents via carbon adsorption (74,75), vacuum distillation (76–78), or extraction (79). Steam or air stripping is a proven method to remove chlorinated hydrocarbons from water solutions. Reducing fugitive emissions to soils and groundwaters (80) and the fate of these solvents in the environment (72–83) have also been studied.

Uses

Inhibited grades of 1,1,1-trichloroethane are used in hundreds of different industrial cleaning applications. 1,1,1-Trichloroethane is preferred over trichloroethy-

Table 6. U.S. 1,1,1-Trichloroethane Demand, 10^3 t[a]

	1986	1987	1988	1989	1990
metal cleaning	141	136	150	154	141
adhesives	23	25	26	25	25
coatings	20	20	20	20	20
aerosols	18	23	23	27	27
electronics	18	20	23	26	26
HCFCs	23	23	27	33	34
other	22	26	13	9	9
Total	*265*	*273*	*282*	*294*	*282*

[a]Ref. 71. Courtesy of Chem. Systems, Inc.

lene or tetrachloroethylene because of its lower toxicity. Additional advantages of 1,1,1-trichloroethane include optimum solvency, good evaporation rate, and no fire or flash point as determined by standard test methods. Common uses include cleaning of electrical equipment, motors, electronic components and instruments, missile hardware, paint masks, photographic film, printed circuit boards, and various metal and certain plastic components during manufacture (see METAL SURFACE TREATMENTS).

1,1,1-Trichloroethane and other chlorinated solvents are used for vapor degreasing (84–90). Other uses include cold metal cleaning, printed circuit board cleaning, and as a solvent for inks, coatings, adhesives, and aerosols. 1,1,1-Trichloroethane is an excellent solvent for development of photoresist polymers used in printed circuit board manufacture (see INTEGRATED CIRCUITS; PHOTOCONDUCTIVE POLYMERS).

1,1,2-TRICHLOROETHANE

1,1,2-Trichloroethane [79-00-5], vinyl trichloride, $CH_2ClCHCl_2$, is a colorless, nonflammable liquid with a pleasant odor, miscible with chlorinated solvents, and (as is 1,1,1-trichloroethane) soluble in the other common organic solvents.

Physical and Chemical Properties

The physical properties of 1,1,2-trichloroethane are listed in Table 7.

Dehydrochlorination 1,1,2-Trichloroethane is easily dehydrochlorinated by a number of catalytic reagents to give 1,1-dichloroethylene and some 1,2-dichloroethylene. Refluxing with aqueous and methanolic solutions of NaOH, $Ca(OH)_2$, and $Mg(OH)_2$ and water gives 1,1-dichloroethylene (91–93). The rate of reaction is faster with the 1,1,2-trichloroethane than with 1,1,1-trichloroethane. Base-catalyzed dehydrochlorination gives primarily 1,1-dichloroethylene, whereas thermal or acid catalyzed dehydrochlorination gives approximately equal mixtures of 1,1-dichloroethane, *cis*-1,2,-dichloroethylene and *trans*-1,2-dichloroethylene. Careful attention should be given when strongly basic solutions are used

Table 7. Properties of 1,1,2-Trichloroethane

Property	Value
melting point, °C	−36.5
boiling point, °C	113.8
density at 25°C	1.430
viscosity at 25°C, mPa·s (=cP)	1.69
surface tension at 25°C, mN/m (=dyn/cm)	33
specific heat at 25°C, J/(g·K)^a	
liquid	1.086
gas	0.668
latent heat of vaporization at 25°C, J/g^a	301.4
critical temperature, °C	333
critical pressure, MPa^b	5.141
flash point (closed cup), °C	none
explosive limits in air at 25°C, vol %	8.3–12.1
autoignition temperature, °C	none
vapor pressure^c, kPa^d	
25°C	2.98
50°C	10.13
solubility at 20°C, g	
trichloroethane in 100 g H$_2$O	0.095

^{a}To convert J to cal, divide by 4.184.
^{b}To convert MPa to atm, multiply by 9.87.
^{c}Antoine constants for 1,1,1-trichloroethane: $A = 7.76632$; $B = 1204.66$; $C = 226.671$, where $\log_{10}$ pressure (kPa) = $A - B/T + C$; T = temperature in °C.
^{d}To convert kPa to mm Hg, multiply by 7.5.

because the product 1,1-dichloroethylene can dehydrochlorinate as well to chloroacetylene, which is highly flammable and can form explosive metal acetylides. Anhydrous aluminum chloride also gives rapid dehydrochlorination (6,91). In this reaction the 1,1,1-isomer reacts faster. 1,1,2-Trichloroethane mixed with nitrogen and 0.1% chlorine can be dehydrochlorinated at 450°C in the presence of NaCl particles to give a 1:1 mixture of 1,1-dichloroethylene and *trans*-1,2-dichloroethylene (94). Other catalytic materials include MgSO$_4$ and K$_2$CO$_3$ (5), molecular sieves containing metal cations (60), alkali metal hydroxides on silica gel (95), and zinc dust (96). In thermal dehydrochlorination up to 500°C *cis*- and *trans*-1,2-dichloroethylene, 1,1-dichloroethylene, and hydrogen chloride were obtained (1).

    Miscellaneous Reactions. Chlorinolysis of mixtures containing 1,1,2-trichloroethane at 550°C was found to give primarily perchloroethylene and hexachloroethane (97).

Manufacture

1,1,2-Trichloroethane is produced in the United States directly or indirectly from ethylene, for example, by chlorination of 1,2-dichloroethane, a product from ethylene (98,99).

$$CH_2{=}CH_2 + Cl_2 \rightarrow ClCH_2CH_2Cl + Cl_2 \xrightarrow{\Delta} CH_2ClCHCl_2 + HCl \qquad (4)$$

Oxychlorination of ethylene with hydrogen chloride and oxygen at 280–370°C on a fluidized $CuCl_2$–KCl (on attapulgite) catalyst bed yields 1,2-dichloroethane and 1,1,2-trichloroethane, along with some higher chlorinated ethanes (33). A typical process consists of a reactor, HCl stripping tower, 1,2-dichloroethane recycle tower, product purification tower, and HCl purification tower.

1,1,2-Trichloroethane is also a coproduct in the thermal and photochemical chlorination of 1,1-dichloroethane to produce 1,1,1-trichloroethane. Vapor chlorination favors the 1,1,1-isomer, whereas reaction in the liquid phase may give much higher ratios of 1,1,2-trichloroethane. Y-type zeolites have been used in vapor-phase chlorination of 1,1-dichloroethane to produce 1,1,2-trichloroethane in high selectivity (100).

Other routes to 1,1,2-trichloroethane are chlorination of acetylene in the presence of HCl (101) and chlorination of vinyl chloride at room temperatures with $FeCl_3$ (102–104), hydrochlorination of cis- and trans-1,2-dichloroethylene with $FeCl_3$ catalyst (105), vapor-phase oxychlorination of 1,2-dichloroethane using a Cu–K catalyst on zeolite (106), and chlorination of mixed hydrocarbons using a catalyst (107).

Health, Safety, and Environmental Factors

1,1,2-Trichloroethane is much more toxic than 1,1,1-trichloroethane in acute exposure studies (108). The 1991 ACGIH recommended TWA value for 1,1,2-trichloroethane is 10 ppm.

An acute lethal dose (LC_{50}) for vapor exposure to 1,1,2-trichloroethane in the rat is 2000 ppm for a 4-h exposure. The same lethal effect occurs at 18,000 ppm vapor during 3 h exposure to 1,1,1-trichloroethane. The oral LD_{50} for 1,1,2-trichloroethane in rats is 0.1–0.2 g/kg, classifying it as moderately toxic (109). Liver and kidney damage occurs at even lower dosages. Skin adsorption is a possible route of overexposure.

1,1,2-trichloroethane may be removed from water by several methods such as evaporation or air or steam stripping. Elastomeric thin-film pervaporation membranes may also be used (110).

Uses

The principal use of 1,1,2-trichloroethane is as a feedstock intermediate in the production of 1,1-dichloroethylene. 1,1,2-Trichloroethane is also used where its high solvency for chlorinated rubbers, etc, is needed, as a solvent for pharmaceutical preparation, and in the manufacture of electronic components.

1,1,1,2-TETRACHLOROETHANE

1,1,1,2,-Tetrachloroethane [630-20-6], CCl_3CH_2Cl, is used primarily as a feedstock for the production of solvents such as trichloroethylene and tetrachloroethylene.

Physical and Chemical Properties

The physical properties of 1,1,1,2-tetrachloroethane are listed in Table 8.

Pyrolysis   Thermal decomposition of 1,1,1,2-tetrachloroethane produces tetrachloroethylene (by disproportionation), hydrogen chloride, and trichloroethylene via dehydrochlorination (111). The yield of the latter is increased in the presence of ferric chloride (112). Other catalytic materials include $FeCl_3-KCl$ mixture (113), $AlCl_3$ (6), the complex of $AlCl_3$ with nitrobenzene (114), activated alumina (3), $Ca(OH)_2$ (115,116), and NaCl (94).

Oxidation.   Oxidation of 1,1,1,2-tetrachloroethane in the presence of ionizing radiation gives dichloroacetyl chloride, $Cl_2CHCOCl$ (117). The gas-phase photochlorination of 1,1,1,2-tetrachloroethane in the absence and presence of oxygen has been studied (115).

Manufacture

1,1,1,2-Tetrachloroethane is often an incidental by-product in the manufacture of chlorinated ethanes. It can be prepared by heating the 1,1,2,2-isomer with anhydrous aluminum chloride or chlorination of 1,1-dichloroethylene at 40°C (118). Hydrochlorination of trichloroethylene using a $FeCl_3$ catalyst may also be used.

Table 8. Properties of 1,1,1,2-Tetrachloroethane

Property	Value
melting point, °C	− 68.7
boiling point, °C	130.5
density at 20°C	1.5465
n_D^{20}	1.4822
vapor pressure, kPa[a]	
20°C	1.246
100°C	40.99
dipole moment, C·m[b]	4.0×10^{-30}
heat capacity, J/(g·K)[c]	
20°C	0.887
100°C	0.995
flash point, °C	none
autoignition temperature, °C	none
flammability limits	none
viscosity at 20°C, mPa·s (= cP)	1.509
surface tension at 20°C, mN/m (= dyn/cm)	32.9
latent heat of vaporization at 20°C, J/g[c]	243.50
solubility at 20°C, g	
H_2O in 100 g 1,1,1,2-tetrachloroethane	0.056

[a]To convert kPa to mm Hg, multiply by 7.5.
[b]To convert C·m to debyes, multiply by 3×10^{29}.
[c]To convert J to cal, divide by 4.184.

Toxicity

Rats exposed to 1000 ppm of vapors for 4–7 h/d for eight days showed ataxia, decreased body weight and growth rate, and minimal central fatty metamorphosis of the liver (119). Trichloroethanol and trichloroacetic acid were urinary metabolites. Tetrachloroethylene did not show any of these effects.

1,1,2,2-TETRACHLOROETHANE

1,1,2,2-Tetrachloroethane [79-34-5], acetylene tetrachloride, $CHCl_2CHCl_2$, is a heavy, nonflammable liquid with a sweetish odor. It is miscible with the chlorinated solvents and shows high solvency for a number of natural organic materials. It is also a solvent for sulfur and a number of inorganic compounds, eg, sodium sulfite.

Physical and Chemical Properties

The physical properties of 1,1,2,2-tetrachloroethane are listed in Table 9.

Pyrolysis.   1,1,2,2-Tetrachloroethane, like the 1,1,1,2-isomer, is thermally degraded with or without a catalytic agent to give trichloroethylene, tetrachloroethylene, and hydrogen chloride (111–113). An effective catalyst may include ferric chloride (112), $FeCl_3$–KCl (113), activated alumina (3), aluminum chloride (6), aluminum chloride in nitrobenzene (113), calcium hydroxide (91,116), molecular sieves containing metal cations (60), or a Fe–Cr–K oxide catalyst (121). Thermal cracking of both 1,1,1,2- and 1,1,2,2-tetrachloroethane gives a 95% conversion to trichloroethylene and tetrachloroethylene (122).

Dehydrochlorination and Chlorination.   The simultaneous chlorination and dehydrochlorination of 1,1,2,2-tetrachloroethane proceeds via formation of labile intermediate, Cl_3CCHCl_2 (123). Chlorination of tetrachloroethane to hexachloroethane is accelerated by 315–354 nm light (124). Heating a mixture of tetrachloroethane vapors and chlorine over active charcoal at 400°C gives carbon tetrachloride and hydrogen chloride (125).

Miscellaneous.   Air oxidation of 1,1,2,2-tetrachloroethane under ionizing radiation gives dichloroacetyl chloride (117). Contact of 1,1,2,2-tetrachloroethane with strong alkali gives trichloroethylene. An excess of alkali causes the trichloroethylene to form dichloroacetylene which may cause an explosion at high concentrations.

Manufacture

1,1,2,2-Tetrachloroethane is produced by direct chlorination or oxychlorination utilizing ethylene as a feedstock. In most cases, 1,1,2,2-tetrachloroethane is not isolated, but immediately thermally cracked at 454°C to give the desired trichloroethylene and tetrachloroethylene products (122). A two-stage chlorination of

Table 9. Properties of 1,1,2,2-Tetrachoroethane[a]

Property	Value
melting point, °C	−42.5
boiling point, °C	146.3
density at 20°C, g/L	1.593
n_D^{20}	1.4942
viscosity at 20°C, mPa·s (=cP)	1.77
surface tension at 20°C, mN/m (=dyn/cm)	34.72
specific heat, J/(g·°C)[b]	
liquid at 20°C	1.13
gas at 146.3°C	0.920
latent heat of vaporization at 20°C, J/g[b]	230.5
critical temperature, °C	388
critical pressure, MPa[c]	3.99
thermal conductivity, W/(m·K)[d]	0.134
heat of combustion, vapor, kJ/g[b]	5.786
dielectric constant, liquid at 20°C	8.00
dipole moment, C·m[e]	6.17×10^{-30}
coefficient of cubical expansion, mL/g	0.00103
vapor pressure, kPa[f]	
0°C	0.176
20°C	0.647
100°C	25.2
solubility at 25°C, g	
1,1,2,2-tetrachloroethane in 100 g H_2O	0.32
H_2O in 100 g 1,1,2,2-tetrachloroethane	0.11
sulfur in 100 g 1,1,2,2-tetrachloroethane at 120°C	100
binary azeotropes[g], bp, °C	
with 31% H_2O	93.2

[a]For additional physical property data over large temperature ranges, see Ref. 120.
[b]To convert J to cal, divide by 4.184.
[c]To convert MPa to atm, multiply by 9.87.
[d]To convert W/(m·K) to (Btu·ft)/(h·ft^2·°F), divide by 1.73.
[e]To convert C·m to debyes, multiply by 3.0×10^{29}.
[f]To convert kPa to mm Hg, multiply by 7.5.
[g]See Ref. 17 for additional binary azeotropes.

1,2-dichloroethane to give 1,1,2,2-tetrachloroethane has been patented (126). High purity 1,1,2,2-tetrachloroethane is made by chlorinating acetylene.

Toxicity and Environmental Concerns

1,1,2,2-Tetrachloroethane has a TWA of 1 ppm as recommended by the ACGIH (1991). Skin adsorption may also pose an exposure hazard. 1,1,2,2-Tetrachloroethane is one of the most toxic chlorinated hydrocarbons (127,128). The liver is most affected.

The reported lethal oral dose for dogs is 0.3 mL/kg body weight. Rats survive a 4-h vapor exposure at 500 ppm but not 4 h at 1000 ppm (120). Cats and rabbits exposed to 100–160 ppm 1,1,2,2-tetrachloroethane vapors for eight to nine hours daily for four weeks did not show any organ damage (130). Injuries to workers have been reported at much lower vapor concentrations (127). Studies of 1,1,2,2-tetrachloroethane-^{14}C metabolism in the mouse showed that half of the dose was expired as carbon dioxide, and 30% as oxidized products in the urine within a 3-d period; 4% of the solvent was released through the lungs chemically unchanged and 15% remained in the body after three days (131). Silica gel has been used to remove acid and water from 1,1,2,2-tetrachloroethane (132). Steam and air stripping readily remove chlorinated hydrocarbons such as 1,1,2,2-tetrachloroethane from water solutions.

Uses

The only significant use of 1,1,2,2-tetrachloroethane is as a feedstock in the manufacture of trichloroethylene, tetrachloroethylene, and 1,2-dichloroethylene. Although it is an excellent solvent, its use should be discouraged in view of its high toxicity.

PENTACHLOROETHANE

Pentachloroethane [76-01-7], $CHCl_2CCl_3$, is a colorless, heavy, nonflammable liquid with a chloroformlike odor; it is miscible with common organic solvents.

Physical and Chemical Properties

Physical properties of pentachloroethane are listed in Table 10. The kinetics and mechanism of the pyrolysis of pentachloroethane in the temperature ranges of 407–430°C and 547–592°C have been studied (133–135). Tetrachloroethylene and hydrogen chloride are the two primary pyrolysis products, showing that dehydrochlorination is the primary reaction.

Various catalytic materials promote dehydrochlorination including $AlCl_3$ (6,91), $AlCl_3$–nitrobenzene complex (114), activated alumina (3), and $FeCl_3$ (112). Chlorination in the presence of anhydrous aluminum chloride gives hexachloroethane. Dry pentachloroethane does not corrode iron at temperatures up to 100°C. It is slowly hydrolyzed by water at normal temperatures and oxidized in the presence of light to give trichloroacetyl chloride.

Manufacture

Pentachloroethane can be made by chlorinating 1,1,2,2-tetrachloroethane under ultraviolet light (136), or trichloroethylene at 70°C in the presence of ferric chloride, sulfur, or ultraviolet light (137). Oxychlorination of ethylene gives pentachloroethane as well as lower chlorinated hydrocarbons (33). Chlorination of trichloroethylene can also give pentachloroethane in good yield.

Table 10. Properties of Pentachloroethane

Property	Value
melting point, °C	−29
boiling point, °C	161.95
density at 20°C, g/L	1.678
n_D^{20}	1.5035
viscosity at 20°C, mPa·s (=cP)	2.45
surface tension at 20°C, mN/m (=dyn/cm)	33.77
specific heat, liquid at 20°C, J/(g·°C)[a]	0.900
latent heat of vaporization at bp, J/g[a]	182.4
thermal conductivity, liquid at 20°C, W/(m·K)[b]	0.130
heat of combustion, kJ/g[a]	4.25
vapor pressure at 20°C, kPa[c]	0.444
solubility at 20°C, g	
pentachloroethane in 100 g H_2O	0.05
H_2O in 100 g 1,1,2,2-pentachloroethane	0.03
binary azeotropes,[d] bp, °C	
with 43.4% H_2O	95.197

[a]To convert J to cal, divide by 4.184.
[b]To convert W/(m·K) to (Btu·ft)/(h·ft^2·°F), divide by 1.73.
[c]To convert kPa to mm Hg, multiply by 7.5.
[d]See Ref. 17 for additional binary azeotropes.

Toxicity

The toxicity of pentachloroethane is similar to that of the tetrachloroethanes
(138). The strong narcotic effect of pentachloroethane is even greater than that
of chloroform. Significant pathological changes in the liver, lungs, and kidneys of
cats were observed at vapor concentrations of 121 ppm given 8–9 h daily for 23
days (130). Metabolism in the mouse is reported to give trichloroethanol and trichloroacetic acid in the urine. Expired air from the mouse contained trichloroethylene and tetrachloroethylene, indicating dechlorination as well as dehydrochlorination (139).

Uses

Pentachloroethane is a good solvent for cellulose acetate, certain cellulose ethers,
and for natural gums and resins, but its high toxicity has discouraged these uses.
Pentachloroethane is still used as an intermediate in some tetrachloroethylene
processes.

HEXACHLOROETHANE

Hexachloroethane [67-72-1], perchloroethane, CCl_3CCl_3, is a white crystalline
solid with a camphorlike odor. Hexachloroethane is nonflammable and has a num-

ber of minor industrial uses which are limited because of its toxic nature. Crystalline hexachloroethane is a minor product in many industrial chlorination processes of saturated and unsaturated C_2 hydrocarbons.

Physical and Chemical Properties

Physical properties of hexachloroethane are listed in Table 11. Hexachloroethane is thermally cracked in the gaseous phase at 400–500°C to give tetrachloroethylene, carbon tetrachloride, and chlorine (140). The thermal decomposition may occur by means of radical-chain mechanism involving $-C_2Cl_5$, $-Cl$, or CCl_3 radicals. The decomposition is inhibited by traces of nitric oxide. Powdered zinc reacts violently with hexachloroethane in alcoholic solutions to give the metal chloride and tetrachloroethylene; aluminum gives a less violent reaction (141). Hexachloroethane is unreactive with aqueous alkali and acid at moderate temperatures. However, when heated with solid caustic above 200°C or with alcoholic alkalis at 100°C, decomposition to oxalic acid takes place.

Trichloromethyl Free Radical.   Degradation of carbon tetrachloride by photochemical, x-ray, or ultrasonic energy produces the trichloromethyl free radical which on dimerization gives hexachloroethane. Chloroform under strong x-ray irradiation also gives the trichloromethyl radical intermediate and hexachloroethane as final product.

Table 11. Properties of Hexachloroethane

Property	Value
sublimation point, °C	185.0
boiling point, °C	186.0
density at 20°C	2.094
crystal structure	
rhombic	≤46°C
triclinic	46–71°C
cubic	>71°C
specific heat, liquid at 25°C, J/(g·°C)[a]	0.728
latent heat of vaporization at bp, J/g[a]	194
heat of combustion, kJ/g[a]	3.073
vapor pressure, kPa[b]	
20°C	0.028
100°C	5.07
150°C	34.8
solubility at 22.3°C, g	
hexachloroethane in 100 g H_2O	0.005
binary azeotropes,[c] bp, °C	
with 30% phenol	173.7

[a]To convert J to cal, divide by 4.184.
[b]To convert kPa to mm Hg, multiply by 7.5.
[c]See Ref. 17 for additional binary azeotropes.

Manufacture

Hexachloroethane is formed in minor amounts in many industrial chlorination processes designed to produce lower chlorinated hydrocarbons, usually via a sequential chlorination step. Chlorination of tetrachloroethylene, in the presence of ferric chloride, at 100–140°C is one convenient method of preparing hexachloroethane (142). Oxychlorination of tetrachloroethylene, using a copper chloride catalyst (143) has also been used. Photochemical chlorination of tetrachloroethylene under pressure and below 60°C has been studied (144) and patented as a method of producing hexachloroethane (145), as has recovery of hexachloroethane from a mixture of other perchlorinated hydrocarbon derivatives via crystallization in carbon tetrachloride. Chlorination of hexachlorobutadiene has also been used to produce hexachloroethane (146).

Toxicity

Hexachloroethane is considered to be one of the more toxic chlorinated hydrocarbons. The 1991 ACGIH recommended time-weighted average (TWA) for hexachloroethane was 1 ppm or 10 mg/m^3 of air. Skin adsorption is a route of possible exposure hazard. The primary effect of hexachloroethane is depression of the central nervous system (147). Pentachloroethane and tetrachloroethylene are primary metabolites of hexachloroethane in sheep (148).

Uses

Hexachloroethane, like carbon tetrachloride and 1,1,1-trichloroethane, can be used to formulate extreme pressure lubricants (149,150). For example, lubricating oils containing 0.02–3.0 wt % (as halogen) of hexachloroethane reduce the abrasion of exhaust valve seats in internal combustion engines (151) (see LUBRICATION AND LUBRICANTS).

Hexachloroethane has been suggested as a degasifier in the manufacture of aluminum and magnesium metals. Hexachloroethane has been used as a chain-transfer agent in the radiochemical emulsion preparation of propylene tetrafluoroethylene copolymer (152). It has also been used as a chlorinating agent in the production of methyl chloride from methane (153).

Other uses of hexachloroethane are as moth repellent, plasticizer for cellulose esters, anthelmintic in veterinary medicine, rubber accelerator, and as a component in fungicidal and insecticidal formulations. Hexachloroethane reacts with silumin (an aluminum/silicon alloy) at 483 K to generate an intense white smoke, which is useful in certain pyrotechnics (154).

BIBLIOGRAPHY

"Chlorocarbons and Chlorohydrocarbons, Other Chloroethanes" under "Chlorine Compounds, Organic" in *ECT* 1st ed., Vol. 3, "Ethylidene Chloride, Ethylene Chloride," pp.

760–764, "1,1,1-Trichloroethane," pp. 764–765, "1,1,2-Trichloroethane," pp. 765–767 by J. Conway, Carbide and Carbon Chemicals Corp., "1,1,2,2-Tetrachloroethane," pp. 767–771, "Pentachloroethane," pp. 771–773, by J. Searles and H. A. McPhail, E. I. du Pont de Nemours & Co., Inc., "Hexachloroethane," pp. 773–774 by J. Werner, General Aniline & Film Corp., General Aniline Works Division; "Chlorocarbons and Chlorohydrocarbons, Other Chloroethanes," in *ECT* 2nd ed., Vol. 5, pp. 149–170, by D. W. F. Hardie, Imperial Chemical Industries, Ltd.; in *ECT* 3rd ed., Vol. 5, pp. 722–742, by W. I. Archer, Dow Chemical U.S.A.

1. D. H. R. Barton, *J. Chem. Soc.*, 148 (1949).
2. H. Hartmann, H. Heydtmann, and G. Rinck, *Z. Physik Chem. (Frankfurt)* **28,** 71 (1961).
3. S. Kiyonori and A. Shuzo, *Nippan Kagaku Kaishi* (10), 1045 (1974).
4. S. Kiyonori, *Bull. Chem. Soc. Jpn.* **47**(10), 2406 (1974).
5. P. Andreu and co-workers, *Am. Quim.* **65**(11), 931 (1969).
6. N. K. Taikova, A. E. Kulikova, and E. N. Zil'berman, *Zh. Org. Khim.* **4**(11), 1880 (1968).
7. T. Isobe, H. Seino, and S. Watanabe, *Yukagaku* **34**(5), 349–351, (1985).
8. U.S. Pat. 2,007,144 (June 2, 1934), H. S. Nutting, P. S. Petrie, and M. E. Huscher (to The Dow Chemical Company).
9. USSR Pat. 470,512 (May 15, 1975), Yu. A. Treger, I. Mokroisova, and S. M. Velichko.
10. U.S. Pat. 4,412,086 (Oct. 25, 1983), W. Q. Beard Jr., and R. L. Wilson (to Vulcan Materials Co. USA).
11. Fr. Pat. 2,490,214 (Mar. 19, 1982), A. P. Mantulo, I. N. Novikov, and I. N. Feldman (to USSR).
12. O. A. Zaidman, E. V. Sonin, and Y. A. Treger, *Khim. Prom-st. (Moscow)* (3), 7–9 (1990).
13. R. Muradka, *Asahi Garasu Kenkyu Hokoku* **16,** 123 (1966).
14. F. J. Weissenhorn, *Chem.-Ztg.* **107**(7–8), 218–221 (1983).
15. P. G. Stecher, ed., *The Merck Index*, 8th. ed., Merck & Co. Inc., Rahway, N.J., 1968.
16. R. W. Gallant, *Hydrocarbon Process.* **45**(7), 111 (1966).
17. L. E. Horsley, *Azeotropic Data, Advances in Chemistry Series*, No. 6, American Chemical Society, Washington, D.C., 1952; *Azeotropic Data-II*, No. 35, 1962.
18. K. A. Holbrook, R. W. Walker and W. R. Watson, *J. Chem. Soc.* **B**, 577 (1971).
19. P. G. Ashmore, J. W. Gardner, and A. J. Owen, *J. Chem. Soc., Faraday Trans. 1* **78**(3), 657–676 (1982).
20. S. Inokawa and co-workers, *Kogyo Kagoku Zasshi* **67**(10), 1540 (1964).
21. U.S. Pat. 2,148,304 (Feb. 21, 1939), J. D. Ruys and H. R. McCombie (to Shell Development Co.).
22. W. L. Archer and E. L. Simpson, *I and EC Prod., Rand D* **16**(2), 158 (June 1977).
23. Y. Chen and X. Jin, *Zheijiang Gongxueyuan Xuebao* (1), 47–53 (1991).
24. K. Okamoto and co-workers, *Bull. Chem. Soc. Jpn.* **40**(8), 1917 (1967).
25. G. E. Ham and J. Stevens, *J. Org. Chem.* **27**, 4638 (1962); U.S. Pats. 3,206,499 (Sept. 14, 1965) and 3,206,499 (Sept. 14, 1965), G. E. Ham (to The Dow Chemical Company).
26. S. Kunichika, S. Oka, and T. Sugiyama, *Bull. Inst. Chem. Res. Kyoto Univ.* **48**(6), 276 (1970).
27. Z. Leszczynski, J. Strzelecki, and D. Zelazko, *Przemyst. Chem.* **44**(6), 330 (1965).
28. E. Lundberg, *Kem. Tidskr.* **96**(10), 34–36, 38 (1984).
29. Ger. Pat. 3,245,366 (June 14, 1984), J. Hundeck, H. Scholz, and H. Hennen (to Hoechst A.-G. Fed. Rep. Ger.).
30. Jpn. Pat. 57,109,727 (July 8, 1982) (to Ryo-Nichi Co., Ltd. Japan).
31. *Hydrocarbon Process. Petrol. Refiner* **44**, 289 (1965).
32. Fr. Pat. 1,577,105 (Aug. 1, 1965), H. Riegel (to Lummus Co.).

33. Fr. Pat. 1,555,518 (Jan. 31, 1969), A. Antonini, P. Joffre, and F. Laine (to Products Chimiques Pechiney Saint-Gobain).

34. Jpn. Pat. 7,133,010 (Sept. 27, 1971), K. Miyauchi, Y. Sato, and S. Okamoto (to Mitsui Toatsu Chemicals Co.).

35. Ger. Pat. 2,106,016 (Sept. 16, 1971), C. H. Cather (to PPG Industries Inc.).

36. Bel. Pat. 900,647 (Mar. 21, 1985) (to BASF A.-G. Fed. Rep. Ger.).

37. E. Cavaterra, *Hydrocarbon Process., Int. Ed.* **67**(12), 63–67 (1988).

38. U.S. Pat. 4,446,249 (May 1, 1984), J. S. Eden (to Goodrich, B. F., Co. USA).

39. Ger. Pat. 3,607,449 (Sept. 10, 1987), H. D. Eichhorn, W. D. Mross, and H. Schachner (to BASF A.-G. Fed. Rep. Ger.).

40. Ger. Pat. 3,522,473 (Jan. 2, 1987), H. D. Eichhorn, W. D. Mross, and H. Schachner (to BASF A.-G. Fed. Rep. Ger.).

41. Jpn. Pat. 57,136,928 (Aug. 24, 1982) (to Kanegafuchi Chemical Industry Co., Ltd. Japan).

42. Z. Sedaghat-Pour, "Ethylene Dichloride," *Chemical Economics Handbook*, 651.5000, Stanford Research Institute, Menlo Park, Calif. 1989.

43. U.S. Pat. 4,614,643 (Sept. 30, 1986), E. P. Doane (to Stauffer Chemical Co. USA).

44. J. A. Key, C. W. Stuewe, and R. L. Standifer, *Technical report*, U.S. Environmental Protection Agency, Office of Air Quality Planning and Standards, Washington, D.C., EPA-450/3-80-028c, 363 pp. 1980.

45. G. Scharein, *Hydrocarbon Process., Int. Ed.* **60**(9), 193–194 (1981).

46. J. Schulze and M. Weiser, *Chem. Ind. (Duesseldorf)* **36**(8), 468–474 (1984).

47. Rom. Pat. 89,942 (Aug. 30, 1986), N. Brindas, A. Emanoil, and N. Chiroiu (to Combinatul Chimic, Rimnicu-Vilcea Rom).

48. Belg. Pat. 890,813 (Apr. 1982), B. Gorny and H. Mathais (to Produits Chimiques Ugine Kuhlmann Fr).

49. D. D. Irish, in F. A. Patty, ed., *Industrial Hygiene and Toxicology*, 3rd Revised Ed., John Wiley & Sons, Inc., New York, 1963, pp. 3491–3497.

50. D. D. McCollister and co-workers, *Arch. Ind. Health* (13), 1 (1956).

51. H. C. Spencer and co-workers, *A.M.A. Arch. Ind. Hyg. Occupational Med.* **4**, 482 (1951).

52. D. H. R. Barton and P. F. Onyon, *J. Am. Chem. Soc.* **72**, 988 (1950).

53. W. B. Crummett and V. A. Stanger, *Ind. Eng. Chem.* **48**, 434 (1956).

54. S. C. Stowe, C. F. Raley, W. L. Howard, and J. D. Burger, unpublished data, The Dow Chemical Company, Midland, Mich., 1992.

55. M. E. Hill, *J. Org. Chem.* **25**, 1115 (1960).

56. W. L. Archer, *Aerosol Age* **12**(8), 16 (1967).

57. U.S. Pat. 3,444,248 (May 13, 1969), W. L. Archer (to The Dow Chemical Company); U.S. Pat. 3,452,108 (June 24, 1969), W. L. Archer (to The Dow Chemical Company); U.S. Pat. 3,452,109 (June 24, 1969), W. L. Archer (to The Dow Chemical Company); U.S. Pat. 3,454,659 (July 8, 1969), W. L. Archer (to The Dow Chemical Company); U.S. Pat. 3,468,966 (Sept. 23, 1960), W. L. Archer (to The Dow Chemical Company); U.S. Pat. 3,472,903 (Oct. 14, 1960), W. L. Archer (to The Dow Chemical Company); U.S. Pat. 3,546,305 (Dec. 8, 1970), W. L. Archer (to The Dow Chemical Company); U.S. Pat. 3,681,469 (Aug. 1, 1972), W. L. Archer (to The Dow Chemical Company).

58. S. Okazaki and M. Komata, *Nippon Kagaku Kaishi* (3), 459 (1973).

59. J. K. Musick and F. W. Williams, *Am. Soc. Mech. Eng. Pap.* **75** ENAs, 17 (1975).

60. I. Mochida and Y. Yoneda, *J. Org. Chem.* **33**(5), 2161 (1968).

61. T. Sato, M. Seno, and T. Asahara, *Seisan-Kenkyu* **24**(6), 230 (1972).

62. J. H. Brown and W. B. Whalley, *J. Soc. Chem. Ind.* **67**, 332 (1948).

63. USSR Pat. 726,824 A1 (Jan. 7, 1982), Y. A. Treger, E. R. Berlin, and I. Y. Mokrousova (to USSR).

64. Jpn. Pat. 59 044,290 (Oct. 29, 1984) (to Tokuyama Soda Co., Ltd. Japan).
65. Ger. Pat. 3,011,689 (Dec. 4, 1980), I. Goto, S. Okado, and S. Yaba (to Asahi Glass Co., Ltd. Japan).
66. V. A. Aver'yanov and Y. A. Treger, *Khim. Prom-st. (Moscow)* (5), 266–269 (1988).
67. Jpn. Pat. 55 079,329 (June 14, 1980) (Asahi Glass Co., Ltd. Japan).
68. V. A. Aver'yanov and Y. A. Treger, *Khim. Prom-st. (Moscow)* (4), 248–250 (1987).
69. P. B. Abramov, V. V. Astashkin, and I. Y. Mokrousova, *Khim. Prom-st., Ser.: Khlornaya Prom-st.* (3), 36–38 (1981).
70. E. Linak, H. J. Lutz, and E. Nakamura, "C2 Chlorinated Solvents," *Chemical Economics Handbook 623.3000 E.*, Stanford Research Institute, Menlo Park, Calif., Dec. 1988.
71. *Process Evaluation Research Planning*, Petrochemical Reports, Chem. Systems, Inc., Tarrytown, N. Y., 1991.
72. D. M. Aviado and co-workers, *Methyl Chloroform and Trichloroethylene in the Environment*, CRC Press Inc., Cleveland, Ohio, 1976, pp. 5–44.
73. T. R. Torkelson and co-workers, *Am. Ind. Hyg. Assoc. J.* **19,** 353 (1958).
74. R. E. Kenson *Environ. Prog.* **4**(3), 161–164 (1985).
75. Jpn. Pat. 57,165,016 (Oct. 9, 1982) (to Mitsubishi Heavy Industries, Ltd. Japan).
76. I. Koike, *Nenryo oyobi Nensho* **48**(8), 561–569 (1981).
77. G. W. Bohnert and D. A. Carey, Technical report, KCP-613-4479, Order No. DE91007175; *Energy Res. Abstr. 1991*, **16**(4), Abstr. No. 10294 (1991).
78. T. C. Keener, *ASTM Spec. Tech. Publ.* **1043**, 104–112 (1989).
79. Austrian Pat. 373,858 (Feb. 27, 1984), R. Marr, M. Siebenhofer, and W. Rueckl (to VOEST-ALPINE A.-G Austria).
80. J. R. Berlow and E. Eby, Technical report, EPA-530/SW-90/012E, Order No. PB90-166430; *Gov. Rep. Announce. Index (U.S.) 1990* **90**(10), Abstr. No. 024,374 (1989).
81. T. M. Vogel and P. L. McCarty, *Environ. Sci. Technol.* **21**(12), 1199–1204 (1987).
82. Y. Makide and F. S. Rowland, *Proc. Natl. Acad. Sci. U.S.A.* **78**(10), 5933–5937 (1981).
83. G. M. Klecka, S. J. Gonsior, and D. A. Markham, *Environ. Toxicol. Chem.* **9**(12), 1437–1451 (1990).
84. W. G. Rollo and A. O'Grady, *Can. Paint Finish*, 15 (Oct. 1973).
85. W. L. Archer, *Cleaning Stainless Steel, ASTM STP538*, American Society for Testing and Materials, Philadelphia, Pa., 1973, pp. 54–64.
86. W. L. Archer, *Met. Prog.*, 133 (Oct. 1974).
87. L. E. Musgrave, Technical report No. 2231, under contract AT (29-1)-1106, U.S. Atomic Energy Commission, Washington, D.C., Aug. 8, 1974.
88. P. Goerlich, *Ind.-Lackier-Betr.* **43**(11), 383 (1975).
89. J. C. Blanchet, *Surfaces* **14**(94), 51 (1975).
90. L. Skory, J. Fulkerson, and D. Ritzema, *Prod. Finish.* **38**(5) (1974).
91. A. Suzuki, H. Iwata, and J. Nakamura, *Kogyo Kagaku Zasski* **69**(10), 1903 (1966).
92. J. Svoboda, I. Ondrus, and J. Mazanec, *Petrochemia* **20**(3–4), 120–124 (1980).
93. U.S. Pat. 5,107,040 (Apr. 21, 1992), T. G. Snedecor (to The Dow Chemical Company)
94. Ger. Pat. 1,928,199 (Dec. 4, 1969), S. Berkowitz (to FMC Corp.).
95. A. P. Khardin, A. V. Spitsyn, and P. Y. Gokhberg, *Khim. Prom-st. (Moscow)* (4), 208–209 (1982).
96. T. Alfrey, H. C. Haas, and C. W. Lewis, *J. Am. Chem. Soc.* **74**, 2097 (1952).
97. F. Gajewski, J. Ogonowski, and J. Rakoczy, *Przem. Chem.* **62**(2), 95–97 (1983).
98. U.S. Pat. 3,919,337 (Nov. 11, 1975), S. C. Gordon and A. N. Theodore (to Diamond Shamrock Corp.).
99. J. S. Miller and J. Perkowski, *Przem. Chem.* **69**(5), 207–209 (1990).
100. U.S. Pat. 4,605,801 (Aug. 12, 1986), R. L. Juhl, M. S. Johnson, and T. E. Morris (to Dow Chemical Co. USA).

101. USSR Pat. 791,721 (Dec. 30, 1980), Y. A. Pazderskii, G. D. Ivanyk, and Y. A. Treger (to USSR).

102. USSR Pat. 910,573 (Mar. 7, 1982), O. A. Zaidman, E. V. Sonin, and I. G. Plakhova (to USSR).

103. E. Huang, Z. Xu, and S. Wang, *Huaxue Fanying Gongcheng Yu Gongyi* 4(4), 60–66 (1988).

104. Rom. Pat. 92,593 (Oct. 30, 1987), N. Chiroiu, H. P. Ichim, and G. Lumezeanu (to Combinatul Chimic, Rimnicu Vilcea Rom).

105. USSR Pat. 742,421 (June 25, 1980), A. P. Khardin, A. V. Spitsyn, and O. I. Tuzhikov (to USSR).

106. L. M. Kartashov, I. M. Alekseeva, and Y. A. Treger, *Khim. Prom-st., Ser.: Khlornaya Prom-st.* (5), 16–18 (1980).

107. USSR Pat. 882,989 (Nov. 23, 1981), B. P. Kuchkov, V. S. Sitanov, and A. V. Spitsyn (to USSR).

108. Ref. 49, pp. 3510–3513.

109. J. E. Wahlberg, *Ann. Occup. Hyg.* **19,** 115 (1976).

110. J. Kaschemekat, J. G. Wijmans, and R. W. Baker, in R. H. Bakish, ed., *Proc. Int. Conf. Pervaporation Processes Chem. Ind.*, 3rd ed., Bakish Materials Corp., Englewood, N.J., 1988, pp. 405–412.

111. D. H. R. Bartonand and K. E. Howlett, *J. Chem. Soc.*, 2033 (1951).

112. Fr. Pat. 2,057,606 (June 25, 1971) (to Toa Gosei Chemical Industry Co. Ltd.); U.S. Pat. 3,732,322 (to Toa Gosei Chemical Industry Co., Ltd.).

113. Jpn. Pat. 75 34,003 (Nov. 5, 1975), T. Uchino, K. Sato, and M. Takeuchi (to Asahi Glass Co.).

114. U.S. Pat. 3,304,336 (Feb. 14, 1967), W. A. Callahan (to Detrex Chemical Ind.).

115. D. Gillotay and J. Olbregts, *Int. J. Chem. Kinet* 8(1), 11 (1976).

116. Z. Roh, *Chem Prum.* **25**(6), 294 (1975).

117. USSR Pat. 195,445 (Mar. 25, 1976), V.A. Poluektov and co-workers.

118. Ger. Pat. 530,649 (July 4, 1929) (to I.G. Farbenindustri A.G.).

119. W. N. Piper and G. L. Sparschu, unpublished data, The Dow Chemical Company, Midland, Mich., 1969.

120. R. W. Gallant, *Hydrocarbon Process.* **46**(12), 119 (1967).

121. R. B. Valitov and co-workers, *Neftekhimiya* **15**(6), 917 (1975).

122. S. Tsuda, *Chem. Eng.* **74** (May 4, 1970).

123. N. N. Lebedev, V. F. Shvets, and V. A. Averyanov, *Kinet. Katal* 12(3), 560 (1971).

124. J. A. Pearce, *Can J. Res.* 24F, 369 (1946).

125. Fr. Pat. 836,979 (Jan. 31, 1939) (to I.G. Farbenindustrie A.G.).

126. Bel. Pat. 602,840 (Apr. 20, 1961) (to PPG Industries Inc.).

127. Ref. 49, pp. 3513–3516.

128. J. B. Sherman, *J. Trop. Med. Hyg.* **56,** 129 (1953).

129. H. F. Smyth, Jr., *Am. Ind. Hyg. Assoc. Quart.* **17,** 129 (1956).

130. H. B. Lehmann and F. Flury, *Toxicology and Hygiene of Industrial Solvents*, Trans. B.Y.F. King and H. F. Smyth, Williams and Wilkins, Baltimore, Md., 1943.

131. S. Yllner, *Acta Pharmacol. Toxicol.* **29,** 499 (1971).

132. Jpn. Pat. 59,112,928 (June 29, 1984) (to Toa Gosei Chemical Industry Co., Ltd. Japan).

133. T. J. Houser and R. B. Berstein, *J. Am. Chem. Soc.* **80,** 4439 (1958).

134. T. J. Houser and T. Cuzcano, *Int. J. Chem. Kinet.* **7**(3), 331 (1975).

135. V. A. Averyanov and G. F. Lebedeva, *Kinet. Katal.* **16**(4), 1073 (1975).

136. Ger. Pat. 248,982 (May 7, 1991) (to Salzbergiverk Neustassfurt and Teilnehmer).

137. Ger. Pat. 843,843 (Apr. 14, 1942), R. Decker and H. Holz (to Wacker Chemie, GmbH).

138. Ref. 49, pp. 3519.

139. S. Yllner, *Acta Pharmacol. Toxicol.* **29**, 481 (1971).
140. J. Puyo and co-workers, *Bull. Soc. Lorraine, Sci.* **2**(2), 75 (1962).
141. A. Lamouroux and J. Meyer, *Mem. Poudres* **39**, 435 (1957).
142. Rom. Pat. 78,711 (June 30, 1982), P. Dumitru, V. Udrescu, and E. Georgescu (to Central de Cercetari, Rimnicu-Vilcea Rom).
143. U.S. Pat. 4,990,696 (Feb. 5, 1991), J. E. Stauffer (to USA).
144. J. Perkowski, *Przem. Chem.* **63**(6), 306–309 (1984).
145. U.S. Pat. 2,440,731 (May 4, 1948), W.H. Vining and O.W. Cass (to E.I. du Pont de Nemours & Co., Inc.).
146. Ger. Pat. 3,606,746 (Sept. 3, 1987), G. Sticken (to Huels A.-G Fed. Rep. Ger.).
147. Ref. 49, pp. 3521–3525.
148. J. S. L. Fowler, *Br. J. Pharmac.* **35**, 530 (1969).
149. Brit. Pat. 841,788 (July 20, 1960), J. S. Elliott and E. D. Edwards (to C.C. Wakefield and Co. Ltd.).
150. W. Davey, *J. Inst. Petrol.* **31**, 73 (1945).
151. Jpn. Pat. 17,803 (Feb. 16, 1974), K. Sugiura and T. Miyagawa (to Nippon Oil Co.).
152. Jpn. Pat. 22,083 (Mar. 8, 1975), N. Suzuki, J. Okamoto, and O. Matsuda (to Japan Atomic Energy Res. Institute).
153. U.S. Pat. 4,990,696 (Feb. 5, 1991), J. E. Stauffer.
154. F. R. Hartley, S. G. Murray, and M. R. Williams, *Propellants Explos. Pyrotech* **9**(3), 108–114 (1984).

GAYLE SNEDECOR
The Dow Chemical Company

DICHLOROETHYLENE

1,1-Dichloroethylene [75-35-4] is more commonly known as vinylidene chloride and is covered in an article in the *Encyclopedia* by that title.

1,2-Dichloroethylene [540-59-0] (1,2-dichloroethene) is also known as acetylene dichloride, dioform, α,β-dichloroethylene, and *sym*-dichloroethylene. It exists as a mixture of two geometric isomers: *trans*-1,2-dichloroethylene [156-60-5] (**1**) and *cis*-1,2-dichloroethylene [156-59-2] (**2**).

The isomeric mixture is a colorless, mobile liquid with a sweet, slightly irritating odor resembling that of chloroform. It decomposes slowly on exposure to light, air, and moisture. The mixture is soluble in most hydrocarbons and only slightly soluble in water. The cis–trans proportions in a crude mixture depend on the production conditions. The isomers have distinct physical and chemical properties and can be separated by fractional distillation.

Physical and Chemical Properties

1,2-Dichloroethylene consists of a mixture of the cis and trans isomers, as manufactured. The physical properties of both isomeric forms are listed in Table 1. Binary and ternary azeotrope data for the cis and trans isomers are given in Table 2.

Manufacturing and Processing

1,2-Dichloroethylene can be produced by direct chlorination of acetylene at 40°C. It is often produced as a by-product in the chlorination of chlorinated compounds

Table 1. Physical Properties of the Isomeric Forms of 1,2-Dichloroethylene

Property	Trans	Cis
mol wt	96.95	96.95
mp, °C	−49.44	−81.47
bp, °C	47.7	60.2
density, g/mL	1.2631	1.2917
15°C	1.44903	1.45189
20°C	1.44620	1.44900
viscosity, mPa s($=$cP)		
−50°C	1.005	1.156
−25°C	0.682	0.791
0°C	0.498	0.577
10°C	0.447	0.516
20°C	0.404	0.467
surface tension at 20°C, mN/m ($=$dyn/cm)	25	28
latent heat of vaporization[a], kJ/kg[b]	297.9	311.7
heat capacity at 20°C, kJ/(kg·K)[b]	1.158	1.176
vapor pressure, kPa[c]		
−20°C	5.3	2.7
−10°C	8.5	5.1
0°C	15.1	8.7
10°C	24.7	14.7
20°C	35.3	24.0
30°C	54.7	33.3
40°C	76.7	46.7
47.7°C	101	66.7
60.25°C		
soly of the isomer in water at 25°C, g/100 g	0.63	0.35
soly of water in the isomer at 25°C, g/100 g	0.55	0.55
steam distillation point at 101 kPa,[c] °C	45.3	53.8
flash point, °C	4	6
explosion limit in air, vol %[d]		5.6–12.8

[a]At the boiling point.
[b]To convert J to cal, divide by 4.184.
[c]To convert kPa to mm Hg, multiply by 7.5.
[d]A cis–trans mixture (1).

Table 2. Azeotropes of *Trans* and *Cis*-1,2-Dichloroethylene Isomers

Second component	Bp, °C	Binary azeotropes			
		Trans isomer in mixture, wt %	Bp of azeotrope, °C	Cis isomer in mixture, wt %	Bp of azeotrope, °C
methanol	64.5			87	51.5
ethanol	78.2	94.0	46.5	90.2	57.7
water	100.0	98.1	45.3	96.65	55.3

Ternary azeotropes[a]				
Ethanol in mixture, wt %	Water in mixture, wt %	Trans isomer in mixture, wt %	Cis isomer in mixture, wt %	Bp of azeotrope, °C
1.4	1.1	94.5		44.4
6.65	2.85		90.5	53.8

[a]1,2-Dichloroethylene, ethanol, and water.

(2) and recycled as an intermediate for the synthesis of more useful chlorinated ethylenes (3). 1,2-Dichloroethylene can be formed by continuous oxychlorination of ethylene by use of a cupric chloride–potassium chloride catalyst, as the first step in the manufacture of vinyl chloride [75-01-4] (4).

The trans isomer is more reactive than the cis isomer in 1,2-addition reactions (5). The cis and trans isomers also undergo benzyne, C_6H_4, cycloaddition (6). The isomers dimerize to tetrachlorobutene in the presence of organic peroxides. Photolysis of each isomer produces a different excited state (7,8). Oxidation of 1,2-dichloroethylene in the presence of a free-radical initiator or concentrated sulfuric acid produces the corresponding epoxide [60336-63-2], which then rearranges to form chloroacetyl chloride [79-04-9] (9).

The unstabilized grade of 1,2-dichloroethylene hydrolyzes slowly in the presence of water, producing HCl. Although unaffected by weak alkalies, boiling with aqueous NaOH may give rise to an explosive mixture because of monochloroacetylene [593-63-5] formation.

Storage and Handling

1,2-Dichloroethylene is usually shipped in 208-L (55 gal) and 112-L (30 gal) steel drums. Because of the corrosive products of decomposition, inhibitors are required for storage. The stabilized grades of the isomers can be used or stored in contact with most common construction materials, such as steel or black iron. Contact with copper or its alloys and with hot alkaline solutions should be avoided to preclude possible formation of explosive monochloroacetylene. The isomers do have explosive limits in air (Table 1). However, the liquid, even hot, burns with a very cool flame which self-extinguishes unless the temperature is well above the flash point. A red label is required for shipping 1,2-dichloroethylene.

Health and Safety

1,2-Dichloroethylene is toxic by inhalation and ingestion and can be absorbed by the skin. It has a TLV of 200 ppm (10). The odor does not provide adequate warning of dangerously high vapor concentrations. Thorough ventilation is essential whenever the solvent is used for both worker exposure and flammability concerns. Symptoms of exposure include narcosis, dizziness, and drowsiness. Currently no data are available on the chronic effects of exposure to low vapor concentrations over extended periods of time.

1,2-Dichloroethylene was selected in April 1990 for National Toxicological Program (NTP) carcinogenesis studies; there is no data available as of summer 1992.

1,2-Dichloroethylene appeared frequently in the 1980s literature largely because of its presence at ground water cleanup sites. The continued presence of 1,2-dichloroethylene may be a result of the biotransformation of tetrachloroethylene and trichloroethylene, which are much more common industrial solvents and are likely present because of past disposal practices (11,12).

Uses

1,2-Dichloroethylene can be used as a low temperature extraction solvent for organic materials such as dyes, perfumes, lacquers, and thermoplastics (13–15). It is also used as a chemical intermediate in the synthesis of other chlorinated solvents and compounds (2).

Recently several patents have been issued (16–18) describing the use of 1,2-dichloroethylene for use in blends of chlorofluorocarbons for solvent vapor cleaning. This art is primarily driven by the need to replace part of the chlorofluorocarbons because of the restriction on their production under the Montreal Protocol of 1987. Test data from the manufacturer show that the cleaning ability of these blends exceeds that of the pure chlorofluorocarbons or their azeotropic blends (19).

BIBLIOGRAPHY

"Chlorocarbons and Chlorohydrocarbons-Dichloroethylenes" under "Chlorine Compounds, Organic", in *ECT* 1st ed., Vol. 3, pp. 786–787, by J. Werner, General Aniline & Film Corp., Aniline Works Division; "Chlorocarbons and Chlorohydrocarbons-Dichloroethylenes" in *ECT* 2nd ed., Vol. 5, pp. 178–183, by D. W. F. Hardie, Imperial Chemical Industries, Ltd; "1,2-Dichloroethylene" under "Chlorocarbons, -Hydrocarbons" in *ECT* 3rd ed., Vol. 5, pp. 742–745, by V. L. Stevens, The Dow Chemical Company.

1. *NFPA Bulletin*, 325 M, National Fire Protection Association, 1984.
2. M. D. Rosenzweig, *Chem. Eng.* **105**, (Oct. 18, 1971).
3. Jpn. Pat. 7,330,249 (Sept. 18, 1973), H. Takenobu and co-workers (to Central Glass Co., Ltd.).
4. Czarny and co-workers, *Przem. Chem.* **65**(12), 659–661 (1986).
5. G. Berens and co-workers, *J. Am. Chem. Soc.* **97**, 7076 (1975).

6. M. Jones, *Tetrahedron Lett.* (53), 5593 (1968).

7. R. Ausubel, *J. Photochem.* **4**, 2418 (1975).

8. R. Ausubel, *Int. J. Chem. Kinet.* **7**, 739 (1975).

9. U.S. Pat. 3,654,358 (Apr. 4, 1977), J. Gaines (to The Dow Chemical Company).

10. *1990–1991 Threshold Limit Values for Chemical Substances and Physical Agents and Biological Indices*, American Conference of Governmental Industrial Hygienists, Cincinnati, Ohio, 1991, p. 18.

11. J. T. Wilson and B. H. Wilson, *Appl. Environ. Microbiol.*, 242–243 (Jan. 1985).

12. T. M. Vogel and P. L. McCarty, *Appl. Environ. Microbiol.*, 1080–1083 (May 1985).

13. C. Marsden, *Solvents Guide*, 2nd ed., Interscience Publishers, New York, 1963, p. 181.

14. L. Scheflan, *The Handbook of Solvents*, D. Van Nostrand Co., New York, 1953, p. 266.

15. G. Hawley, *The Condensed Chemical Dictionary*, Van Nostrand Reinhold Co., Inc., New York, 1977, p. 279.

16. Jpn. Pat. 02135290 (May 24, 1990), A. Asano and co-workers (to Asahi Glass Co., Ltd.).

17. U.S. Pat. 4961870 (Oct. 9, 1990), J. G. Burt and J. P. Burns (to Allied Signal).

18. U.S. Pat. 4808331 (Feb. 28, 1989), K. D. Cook and co-workers (to E. I. du Pont de Nemours & Co., Inc.).

19. *Freon SMT Cleaning Agent*, product brochure, E. I. du Pont de Nemours & Co., Inc., June 1988.

JAMES A. MERTENS
Dow Chemical USA

TRICHLOROETHYLENE

Trichloroethylene [*79-01-6*], trichloroethene, CHCl=CCl$_2$, commonly called "tri," is a colorless, sweet smelling (chloroformlike odor), volatile liquid and a powerful solvent for a large number of natural and synthetic substances. It is nonflammable under conditions of recommended use. In the absence of stabilizers, it is slowly decomposed (autoxidized) by air. The oxidation products are acidic and corrosive. Stabilizers are added to all commercial grades. Trichloroethylene is moderately toxic and has narcotic properties.

Trichloroethylene was first prepared by Fischer in 1864. In the early 1900s, processes were developed in Austria for the manufacture of tetrachloroethane and trichloroethylene from acetylene. Trichloroethylene manufacture began in Germany in 1920 and in the United States in 1925. Early uses of trichloroethylene were as an extraction solvent for natural fats and oils, such as palm, coconut, and soybean oils. It was later used for decaffeination of coffee, but this use has essentially been replaced by steam processes today. The demand for trichloroethylene was stimulated by the development of the vapor-degreasing process during the 1920s and by the growth of the dry-cleaning industry during the 1930s. By the mid-1950s, perchloroethylene had replaced trichloroethylene in dry-cleaning, and metal cleaning became the principal use for trichloroethylene.

The demand for trichloroethylene grew steadily until 1970. Since that time trichloroethylene has been a less desirable solvent because of restrictions on emissions under air pollution legislation and the passage of the Occupational Safety and Health Act. Whereas previously the principal use of trichloroethylene was for vapor degreasing, currently 1,1,1-trichloroethane is the most used solvent for vapor degreasing. The restrictions on production of 1,1,1-trichloroethane [71-55-6] from the 1990 Amendments to the Montreal Protocol on substances that deplete the stratospheric ozone and the U.S. Clean Air Act 1990 Amendments will lead to a phase out of 1,1,1-trichloroethane by the year 2005, which in turn will likely result in a slight resurgence of trichloroethylene in vapor-degreasing applications. The total production, however, will probably stay relatively low because regulations will require equipment designed to assure minimum emissions.

Physical and Chemical Properties

The physical properties of trichloroethylene are listed in Table 1. Trichloroethylene is immiscible with water but miscible with many organic liquids; it is a versatile solvent. It does not have a flash or fire point. However, it does exhibit a flammable range when high concentrations of vapor are mixed with air and exposed to high energy ignition sources.

The most important reactions of trichloroethylene are atmospheric oxidation and degradation by aluminum chloride. Atmospheric oxidation is catalyzed by free radicals and accelerated with heat and with light, especially ultraviolet. The addition of oxygen leads to intermediates (**1**) and (**2**).

| (1) | (2) |

Compound (**1**) decomposes to form dichloroacetyl chloride, which in the presence of water decomposes to dichloroacetic acid and hydrochloric acid (HCl) with consequent increases in the corrosive action of the solvent on metal surfaces. Compound (**2**) decomposes to yield phosgene, carbon monoxide, and hydrogen chloride with an increase in the corrosive action on metal surfaces.

In the presence of aluminum, oxidative degradation or dimerization supply HCl for the formation of aluminum chloride, which catalyzes further dimerization to hexachlorobutene. The latter is decomposed by heat to give more HCl. The result is a self-sustaining pathway to solvent decomposition. Sufficient quantities of aluminum can cause violent decomposition, which can lead to runaway reactions (1,2). Commercial grades of trichloroethylene are stabilized to prevent these reactions in normal storage and use conditions.

Amine-stabilized products, once the predominant grade, are sold today only in limited amounts. Most vapor-degreasing grades contain neutral inhibitor mixtures (3–5) including a free-radical scavenger, such as an amine or pyrrole, to prevent the initial oxidation reaction. Epoxides, such as butylene oxide and epichlorohydrin, are added to scavenge any free HCl and AlCl₃. Concern over the toxicity of these epoxides has eliminated the use of epichlorohydrin in the United States during the 1980s and may restrict butylene oxide in the future.

Table 1. Properties of Trichloroethylene

Property	Value
molecular weight	131.39
melting point, °C	−86.5
boiling point, °C	87.3
specific gravity, liquid	
20/4°C	1.464
100/4°C	1.322
vapor density at bp, kg/m^3	4.61
n_D	
liquid, 20°C	1.4782
vapor, 0°C	1.001784
viscosity, mPa·s (= cP) liquid	
20°C	0.57
60°C	0.42
vapor at 100°C	0.01246
surface tension at 20°C, mN/m(= dyn/cm)	29.3
heat capacity, J/(kg·K)[a]	
liquid at 20°C	938
vapor at 100°C	693
critical temperature, °C	300.2
critical pressure, MPa[b]	4.986
thermal conductivity, W/(m·K)	
liquid at 20°C	0.115
vapor at bp	0.00851
coefficient of cubical expansion, liquid at	0.00119
0−40°C	
dielectric constant, liquid at 16°C	3.42
dipole moment, C·m[c]	3.0×10^{-30}
heat of combustion, kJ/g[a]	−6.56
heat of formation, kJ/mol[a]	
liquid	−42.3
vapor	−7.78
latent heat of evaporation at bp, kJ/kg[a]	238
explosive limits, vol % in air	
25°C	8.0-saturation
100°C	8.0−44.8
vapor pressure[d], kPa[e]	
Antoine constants	
A	5.75373
B	1076.67
C	199.991
solubility, g	
H_2O in 100 g trichloroethylene	
0°C	0.010
20°C	0.0225
60°C	0.080
trichloroethylene in 100 g H_2O	
20°C	0.107
60°C	0.124

[a]To convert J to cal, divide by 4.184.
[b]To convert MPa to atm, divide by 0.101.
[c]To convert C·m to debye, divide by 3.336×10^{-30}.
[d]$\log_{10} P = A - B/(T + C)$, T in °C.
[e]To convert kPa to mm Hg, multiply by 7.5.

Trichloroethylene is not readily hydrolyzed by water. Under pressure at 150°C, it gives glycolic acid, $CH_2OHCOOH$, with alkaline hydroxides. Strong alkalies dehydrochlorinate trichloroethylene with production of spontaneously explosive and flammable dichloroacetylene. Reaction with sulfuric acid (90%) yields monochloroacetic acid, $CH_2ClCOOH$. Hot nitric acid reacts with trichloroethylene violently, producing complete oxidative decomposition. Under carefully controlled conditions, nitric acid gives trichloronitromethane (chloropicrin) and dinitrochloromethane (6). Dichloroacetylene, C_2Cl_2, can also be formed from trichloroethylene in the presence of epoxides and ionic halides (7).

In the presence of catalysts, trichloroethylene is readily chlorinated to pentachloro- and hexachloroethane. Bromination yields 1,2-dibromo-1,1,2-trichloroethane [13749-38-7]. The analogous iodine derivative has not been reported. Fluorination with hydrogen fluoride in the presence of antimony trifluoride produces 2-chloro-1,1,1-trifluoroethane [75-88-7] (8). Elemental fluorine gives a mixture of chlorofluoro derivatives of ethane, ethylene, and butane.

Liquid trichloroethylene has been polymerized by irradiation with [60]Co γ-rays or 20-keV x-rays (9). Trichloroethylene has a chain-transfer constant of <1 when copolymerized with vinyl chloride (10) and is used extensively to control the molecular weight of poly(vinyl chloride) polymer.

A variety of trichloroethylene copolymers have been reported, none with apparent commercial significance. The alternating copolymer with vinyl acetate has been patented as an adhesive (11) and as a flame retardant (12, 13). Copolymerization with 1,3-butadiene and its homologues has been reported (14–16). Other comonomers include acrylonitrile (17), isobutyl vinyl ether (18), maleic anhydride (19), and styrene (20).

Terpolymers have been made with vinyl chloride–vinylidene chloride (21) and vinyl acetate–vinyl alcohol (22).

Manufacture

As late as 1968, 85% of the production capacity in the United States was based on acetylene, but rising acetylene [74-86-2] costs reduced this figure to 8% by 1976 (23), and now most trichloroethylene is made from ethylene [74-85-1] or 1,2-dichloroethane [107-06-2].

From Acetylene. The acetylene-based process consists of two steps. First acetylene is chlorinated to 1,1,2,2-tetrachloroethane [79-34-5]. The reaction is exothermic (402 kJ/mol = 96 kcal/mol) but is maintained at 80–90°C by the vaporization of solvent and product. Catalysts include ferric chloride and sometimes phosphorus chloride and antimony chloride (24).

The product is then dehydrohalogenated to trichloroethylene at 96–100°C in aqueous bases such as $Ca(OH)_2$ (25) or by thermal cracking, usually over a catalyst (24) such as barium chloride on activated carbon or silica or aluminum gels at 300–500°C. The yield of trichloroethylene (23) is about 94% based on acetylene. A significant disadvantage of the alkaline process is the loss of chlorine as calcium chloride. In thermal cracking the chlorine can be recovered as hydrochloric acid, an important feedstock in many chemical processes. Since it poisons the catalysts during thermal cracking, all ferric chloride must be removed from the tetrachloroethane feed (24). Tetrachloroethane can also be cracked to trichloroethylene without catalysts at 330–770°C, but considerable amounts of tarry by-products are formed.

Chlorination of Ethylene. Dichloroethane, produced by chlorination of ethylene, can be further chlorinated to trichloroethylene and tetrachloroethylene. The exothermic reaction is carried out at 280–450°C. Temperature is controlled by a fluidized bed, a molten salt bath, or the addition of an inert material such as perchloroethylene. The residence time in the reactor varies from 2 to 30 seconds, depending on conditions (24). Catalysts include potassium chloride and aluminum chloride (26), Fuller's earth (27), graphite (28), activated carbon (29), and activated charcoal (27).

Maximum conversion to trichloroethylene (75% of dichloroethane feed) is achieved at a chlorine to dichloroethane ratio of 1.7:1. Tetrachloroethylene conversion reaches a maximum (86% conversion of dichloroethane) at a feed ratio of 3.0:1 (24).

Oxychlorination of Ethylene or Dichloroethane. Ethylene or dichloroethane can be chlorinated to a mixture of tetrachoroethylene and trichloroethylene in the presence of oxygen and catalysts. The reaction is carried out in a fluidized-bed reactor at 425°C and 138–207 kPa (20–30 psi). The most common catalysts are mixtures of potassium and cupric chlorides. Conversion to chlorocarbons ranges from 85–90%, with 10–15% lost as carbon monoxide and carbon dioxide (24). Temperature control is critical. Below 425°C, tetrachloroethane becomes the dominant product, 57.3 wt % of crude product at 330°C (30). Above 480°C, excessive burning and decomposition reactions occur. Product ratios can be controlled but less readily than in the chlorination process. Reaction vessels must be constructed of corrosion-resistant alloys.

Other Routes. A unique process that produces vinyl chloride, trichloroethylene, dichloroethane, and trichloroethane simultaneously has been developed by Produits Chemiques Pechiney-Saint-Gobain in France (31). Dichloroethylene is chlorinated directly at low temperature to tetrachloroethane, which is then thermally cracked to give trichloroethylene and hydrochloric acid. The dichloroethylene feed is coproduced with vinyl chloride in a hot chlorination reactor, using chlorine and ethylene as feedstocks.

A Japanese process developed by Taogosei Chemical Co. chlorinates ethylene directly in the absence of oxygen at 811 kPa (8 atm) and 100–130°C (32). The products are tetrachlorethanes and pentachloroethane [76-01-7], which are then thermally cracked at 912 kPa (9 atm) and 429–451°C to produce a mixture of trichloroethylene, perchloroethylene [127-18-4], and hydrochloric acid.

Shipping and Storage

Shipment of trichloroethylene is usually by truck or rail car and also in 208-liter (55-gallon) steel drums. Mild steel tanks, if appropriately equipped with vents and vent driers to prevent the accumulation of water, are adequate for storage. Precautions, such as diking, should be taken to provide for adequate spill containment at the storage tank. Seamless black iron pipes are suitable for transfer lines, gasketing should be of Teflon, Viton, or other solvent impermeable material.

Centrifugal or positive-displacement pumps made from cast iron, steel, or stainless steel are suitable for use. Aluminum should never be used as a construction material for any halogenated hydrocarbon. Glass containers, amber or green, are suitable for small quantities, such as in a laboratory, but care should be taken for spill containment in the event of breakage.

Containers should bear warning labels against breathing vapors, ingesting the liquid, splashing solvent in eyes or on skin and clothing, and using it near an open flame, or where vapors will come in contact with hot metal surfaces (>176°C). Precautions in handling any waste products in conformance with federal, state, and local regulations should be included.

Although the flammability hazard is very low, ignition sources should not be present when trichloroethylene is used in highly confined or unventilated areas. Tanks in which flammable concentrations could develop should be grounded to prevent build-up of static electric charges. Under no circumstances should welding or cutting with a torch take place on any storage container or process equipment containing trichloroethylene.

Economic Aspects

In 1990, worldwide capacities and production figures in metric tons were Western Europe capacity 266,000, production 180,000; Japan capacity 90,000, production 64,000; and United States capacity 145,000, production 88,600 (23).

United States production and price statistics are presented in Table 2. The demand for trichloroethylene in the United States has been shrinking sharply since 1970 because of pressures from environmental and safety legislation. Similar pressures have weakened the demand in Europe and Japan.

In 1966, the Los Angeles Air Pollution Control Board designated trichloroethylene as a photochemically reactive solvent that decomposes in the lower atmosphere, contributing to air pollution. In 1970 all states were required to submit pollution control plans to EPA to meet national air quality standards. These plans, known as State Implementation Plans (SIPS), controlled trichloroethylene as a

Table 2. United States Trichloroethylene Production and Prices[a]

Year	Production, 10^3 t	Price, ¢/kg
1960	160.4	28.05
1965	197.5	22.55
1970	277.6	23.1
1975	133.0	39.05
1980	121.1	59.95
1985	79.5	84.7
1986	77.3	84.7
1987	88.6	84.7

[a]Ref. 23.

volatile organic compound (VOC). They were designed to have each state achieve the National Ambient Air Quality Standard (NAAQS) for ozone. The regulations were established to control the emission of precursors for ozone, of which trichloroethylene is one.

For worker exposure to trichloroethylene vapor, OSHA set a maximum eight-hour time-weighted average (TWA) concentration of 100 ppm. This severely restricted certain applications, and many organizations converted to other chlorinated solvents. As a result, U.S. production of trichloroethylene declined about 70% from a peak in 1970 (Table 2). In 1989, OSHA lowered the permissible exposure limit (PEL) from 100 ppm eight-hour TWA to 50 ppm eight-hour TWA (33). This added further pressure for some users to consider changing to alternative solvents.

In addition to environmental and safety factors, some of the early decline in manufacture was hastened by a series of plant shutdowns between 1971 and 1973 resulting primarily from the high costs of the acetylene-based process. No new production capacity is planned in the United States for the foreseeable future.

Shortages, together with rapidly escalating fuel and feedstock prices, have led to a dramatic increase in the price of trichloroethylene, which more than doubled between 1972 and 1976 and doubled again between 1975 and 1985. The price stayed flat during the late 1980s. During the 1990s, the price will likely depend on energy demands and the availability of trichloroethylene.

Specifications and Standards

Commercial grades of trichloroethylene, formulated to meet use requirements, differ in the amount and type of added inhibitor. The grades sold in the United States include a neutrally inhibited vapor-degreasing grade and a technical grade for use in formulations. U.S. Federal Specification O-T-634b lists specifications for a regular and a vapor-degreasing grade.

Apart from added stabilizers, commercial grades of trichloroethylene should not contain more than the following amounts of impurities: water 100 ppm; acidity, ie, HCl, 5 ppm; insoluble residue, 10 ppm. Free chlorine should not be detectable. Test methods have been established by ASTM to determine the following characteristics of trichloroethylene: acid acceptance, acidity or alkalinity, color, corrosivity on metals, nonvolatile-matter content, pH of water extractions, relative evaporation rate, specific gravity, water content, water-soluble halide ion content, and halogen content (34).

The passage of the Resource Conservation and Recovery Act in 1978 and its implementation in 1980 generated an increase in the recycling of trichloroethylene, which, in turn, defined the need for specifications for recycled solvent. The ASTM is currently working on a set of consensus specifications for recycled solvent.

Health and Safety Factors (Toxicity)

Trichloroethylene is acutely toxic, primarily because of its anesthetic effect on the central nervous system. Exposure to high vapor concentrations is likely to cause headache, vertigo, tremors, nausea and vomiting, fatigue, intoxication, unconsciousness, and even death. Because it is widely used, its physiological effects have been extensively studied.

Exposure occurs almost exclusively by vapor inhalation, which is followed by rapid absorption into the bloodstream. At concentrations of 150–186 ppm, 51–70% of the trichloroethylene inhaled is absorbed. Metabolic breakdown occurs by oxidation to chloral hydrate [302-17-0], followed by reduction to trichloroethanol [115-20-8], part of which is further oxidized to trichloroacetic acid [76-03-9] (35–37). Absorbed trichloroethylene that is not metabolized is eventually eliminated through the lungs (38). The OSHA permissible exposure limit (PEL) eight-hour TWA concentration has been set at 50 ppm for eight-hour exposure (33).

It is estimated that concentrations of 3000 ppm cause unconsciousness in less than 10 minutes (39). Anesthetic effects have been reported at concentrations of 400 ppm after 20-min exposure. Decrease in psychomotor performance at a trichloroethylene concentration of 110 ppm has been reported in one study (33), whereas other studies find no neurotoxic effects at concentrations of 200 ppm (40–43).

Victims of overexposure to trichloroethylene should be removed to fresh air, and medical attention should be obtained immediately. A self-contained positive pressure breathing device should be used wherever high vapor concentrations are expected, eg, when cleaning up spills or when accidental releases occur.

The distinctive odor of trichloroethylene may not necessarily provide adequate warning of exposure, because it quickly desensitizes olfactory responses. Fatalities have occurred when unprotected workers have entered unventilated areas with high vapor concentrations of trichloroethylene or other chlorinated solvents. For a complete description of proper entry to vessels containing any chlorinated solvent, see ASTM D4276-84, Standard Practice for Confined Area Entry (34).

Ingestion of large amounts of trichloroethylene may cause liver damage, kidney malfunction, cardiac arrhythmia, and coma (38); vomiting should not be induced, but medical attention should be obtained immediately.

Protective gloves and aprons should be used to prevent skin contact, which may cause dermatitis (44–46). Eyes should be washed immediately after contact or splashing with trichloroethylene.

The National Cancer Institute reported in 1975 that massive oral doses of trichloroethylene caused liver tumors in mice but not in rats (47). Trichloroethylene was tested again in the 1980s by the National Toxicology Program (NTP) with similar results (48,49). The EPA has classified trichloroethylene as B2, a probable human carcinogen. The International Agency for Research on Cancer (IARC) classifies it as group 3, ie, unclassifiable as to human carcinogenicity. Teratogenicity studies conducted with trichloroethylene by The Dow Chemical Company showed no significant effects on fetal development (50). During the 1980s several epidemiology studies were conducted on worker populations exposed to trichloroethylene (51–53). Each of these studies failed to show a positive link between human exposure in the work place and cancer.

During the 1980s a significant amount of work was done on developing methods for treatment of contaminated groundwater and also on setting standards for trichloroethylene under the Safe Drinking Water Act. The EPA has set a maximum contaminant level goal (MCLG) at 0 based on the animal carcinogenic effects (54). The maximum contaminant level (MCL) is currently set at five micrograms per liter.

Uses

Approximately 85% of the trichloroethylene produced in the United States is consumed in the vapor degreasing of fabricated metal parts (see METAL SURFACE TREATMENTS); the remaining 15% is divided equally between exports and miscellaneous applications (23). The Western European consumption was 95% in vapor degreasing and 5% for other uses. Japanese consumption is similar to the United States at 83% and 17% (23). A variety of miscellaneous applications include use of trichloroethylene as a component in adhesive and paint-stripping formulations, a low temperature heat-transfer medium, a nonflammable solvent carrier in industrial paint systems, and a solvent base for metal phosphatizing systems. Trichloroethylene is used in the textile industry as a carrier solvent for spotting fluids and as a solvent in waterless preparation dying and finishing operations. Many of these uses have gradually switched over to alternatives, such as 1,1,1-trichloroethane, because of the environmental regulations enforced on trichloroethylene.

Trichloroethylene was approved for use for many years as an extraction solvent for foods. In late 1977, the Food and Drug Administration (FDA) banned its use as a food additive, directly or indirectly, prohibiting the use in hop extraction, decaffeination of coffee, isolation of spice oleoresins, and other applications. The FDA also banned the use of trichloroethylene in cosmetic and drug products (23).

Trichloroethylene is widely used as a chain-transfer agent in the production of poly(vinyl chloride). An estimated 5500 metric tons are consumed annually in this application.

Trichloroethylene is being evaluated by the industry as a precursor in the production of hydrochlorofluorocarbons (HCFC), the replacement products for the chlorofluorocarbons implicated in the depletion of the stratospheric ozone. At this time it is too early to project any estimates or probabilities for potential volume changes as a result of this opportunity (23).

BIBLIOGRAPHY

"Chlorine Compounds, Organic (Chlorocarbons and Chlorohydrocarbons-Trichloroethylene)" in *ECT* 1st ed., Vol. 3, pp. 788–794, by J. Searles and H. A. McPhail, E. I. du Pont de Nemours & Co., Inc.; "Chlorocarbons and Chlorohydrocarbons (Trichloroethylene)" in *ECT* 2nd ed., Vol. 5, pp. 183–195, by D. W. F. Hardie, Imperial Chemical Industries Ltd; "Chlorocarbons Hydrocarbons (Trichloroethylene)" in *ECT* 3rd ed., Vol. 5, pp. 745–753, by W. C. McNeil, Jr., Dow Chemical U.S.A.

1. L. Metz and A. Roedig, *Chem. Ing. Technick* **21,** 191 (1949).
2. W. L. Archer and E. L. Simpson, *Chem. Prof. Polychloroethanes Polychloroalkenes I&EC Prod. Res. Dev.* **167,** 158–162 (June 1977).
3. U.S. Pat. 2,795,623 (June 11, 1957), F. W. Starks (to E. I. du Pont de Nemours & Co., Inc.).

4. U.S. Pat. 2,818,446 (Dec. 31, 1957), F. W. Starks (to E. I. du Pont de Nemours & Co., Inc.).

5. Brit. Pat. 794,700 (May 7, 1958), H. B. Copelin (to E. I. du Pont de Nemours & Co., Inc.).

6. R. B. Burrows and L. Hunter, *J. Chem. Soc.*, 1357 (1932).

7. D. B. Robinson and G. E. Green, *Chem. Ind.*, 214 (Mar. 4, 1972).

8. A. J. Rudge, *The Manufacture and Use of Flourine and its Compounds*, Oxford University Press (for Imperial Chemical Industries Ltd.), Cambridge, Mass., 1962, p. 71.

9. H. L. Cornish, Jr., *U.S. At. Energy Comm.* TID-21388, 1964.

10. J. Pichler and J. Rybicky, *Chem. Prum.* **16,** 559 (1966).

11. Jpn. Pat. 72 45,415 (Nov. 16, 1972), Kimimura, Takayoshi, and S. Wataru (to Hoechst Gosel Co. Ltd.).

12. U.S. Pat. 3,846,508 (Nov. 5, 1974), D. H. Heinert (to The Dow Chemical Company).

13. U.S. Pat. 3,907,872 (Sept. 23, 1975), D. H. Heinert (to The Dow Chemical Company).

14. Ger. Pat. 719,194 (Mar. 26, 1942), H. Kopff and C. Rautenschauch (to I.G. Farbenindustrie, AG).

15. Z. Jedlinski and E. Grzywa, *Polimery* **11,** 560 (1966).

16. Pol. Pat. 53,152 (Feb. 28, 1967), E. Grzywa and Z. Jedlinski (to Zaklady Chemiczne "Oswiecim").

17. S. U. Mullik and M. A. Quddus, *Pak. J. Sci. Ind. Res.* **12**(3), 181 (1970).

18. T. A. DuPlessis and A. C. Thomas, *J. Polym. Sci. Polym. Chem. Ed.* **11,** 2681 (1973).

19. R. A. Siddiqui and M. A. Quddus, *Pak. J. Sci. Ind. Res.* **14**(3), 197 (1971).

20. H. Asai, *Nippon Kagaku Zasshi* **85,** 252 (1964).

21. E. Krotki and J. Mitus, *Polimery* **9,** 155 (1964).

22. Jpn. Pat. 71 01,719 (Jan. 16, 1971), Shimokawa and Wataru (to Hekisto Gosei Co. Ltd.).

23. E. Linak with H. J. Lutz and E. Nakamura, "C$_2$ Chlorinated Solvents," in *Chemical Economics Handbook,* Stanford Research Institute, Menlo Park, Calif., Dec. 1990, pp. 632.30000a–632.3001Z.

24. L. M. Elkin, *Process Economics Program*, Chlorinated Solvents, Report No. 48, Stanford Research Institute, Menlo Park, Calif., Feb. 1969.

25. Ger. Pat. 901,774 (Nov. 3, 1940), (to Wacker Chemie, GmbH).

26. U.S. Pat. 2,140,548 (Dec. 30, 1938), J. H. Reilly (to The Dow Chemical Company).

27. Brit. Pat. 673,565 (June 11, 1952), (to Diamond Alkali).

28. U.S. Pat. 2,725,412 (Nov. 29, 1955), F. Conrad (to Ethyl Chemical Co.).

29. Neth. Appl. 6,607,204 (Nov. 28, 1966), F. Sanhaber (to Donau Chemic).

30. Fr. Pat. 1,435,542 (Mar. 7, 1966), A. C. Schulz (to Hooker Chemical).

31. M. D. Rosenzweig, *Chem. Eng.* **78**(24), 105 (Oct. 18, 1971).

32. S. Tsuda, *Chem. Eng.* **77**(10), 74 (May 4, 1970).

33. *Fed. Reg.* **54**(12), 2332, 2955 (Jan. 19, 1989).

34. *1990 Annual Book of ASTM Standards*, Section 15, Philadelphia, Pa., 1990.

35. B. Soucek and D. Vlachove, *Br. J. Ind. Med.* **17,** 60 (1960).

36. V. Bartonicek, *Br. J. Ind. Med.* **19,** 134 (1962).

37. M. Ogata, Y. Takatsuka, and K. Tomokuni, *Br. J. Ind. Med.* **28,** 386 (1971).

38. D. M. Avaido and co-workers, *Methyl Chloroform and Trichloroethylene in the Environment*, CRC Press, Cleveland, Ohio, 1976.

39. E. O. Longley and R. Jones, *Arch. Environ. Health* **7,** 249 (1963).

40. R. D. Steward and co-workers, *Arch. Environ. Health* **20,** 64 (1970).

41. G. J. Stopps and W. McLaughlin, *Am. Ind. Hyg. Assoc. J.* **29,** 43 (1967).

42. R. J. Vernon and R. K. Ferguson, *Arch. Environ. Health* **18,** 894 (1964).

43. R. K. Ferguson and R. J. Vernon, *Arch. Environ. Health* **29,** 462 (1970).

44. K. Kadlec, *Cesk. Dermatol.* **38,** 395 (1963).

45. S. M. Peck, *J. Am. Med. Assoc.* **125,** 190 (1944).
46. J. M. Schirren, *Berufs-Dermatosen* **19,** 240 (1971).
47. *Carcinogenesis Bioassay of Trichloroethylene*, NCI-CG-TR-2, U.S. Dept. of HEW, Washington, D.C., Feb. 1976, p. 197.
48. *Carcinogenesis Bioassay of Trichloroethylene in F344 Rats and B6C3F1 Mice*, NTD 81–84 NIH Publication No. 82-1799, National Toxicology Program (NTP), Research Triangle Park, N.C., 1982.
49. *Toxicological and Carcinogenesis of Trichloroethylene in Four Strains of Rats* (ACI, August, Marchall, Osborne-Mendel), NTP TR 273, NIH Publication No. 88-2529, National Toxicological Program (NTP), Research Triangle Park, N.C., 1988.
50. B. A. Schwetz, B. K. Leong, and P. J. Gehring, *Toxicol. Appl. Pharmacol.* **32,** 84 (1975).
51. S. Shindel and S. Ulrich, *Report of Epidemiologic Study: Warner Electric Brake & Clutch Co., South Beloit, Ill., Jan. 1957 to July 1983*, Ergotopology Investigative Medicine for Industry, Milwaukee, Wis., Aug. 1984.
52. F. D. Schaumburg, "Banning Trichloroethylene: Responsible Action or Overkill?," *Environ. Sci. Technol.* **24**(1), (1990).
53. L. P. Brown, D. G. Farrar, and C. G. DeRooij, *Health Risk Assessment of Environmental Exposure to Trichloroethylene, Regulatory, Toxicol. Pharmacol.* **11,** 24–41 (1990).
54. *Fed. Reg.* **50** FR 46880 Part III, 46880 (Nov. 13, 1985).

JAMES A. MERTENS
Dow Chemical U.S.A.

TETRACHLOROETHYLENE

Tetrachloroethylene [*127-18-4*], perchloroethylene, $CCl_2{=}CCl_2$, is commonly referred to as "perc" and sold under a variety of trade names. It is the most stable of the chlorinated ethylenes and ethanes, having no flash point and requiring only minor amounts of stabilizers. These two properties combined with its excellent solvent properties account for its dominant use in the dry-cleaning industry as well as its application in metal cleaning and vapor degreasing.

Tetrachloroethylene was first prepared in 1821 by Faraday by thermal decomposition of hexachloroethane. Tetrachloroethylene is typically produced as a coproduct with either trichloroethylene or carbon tetrachloride from hydrocarbons, partially chlorinated hydrocarbons, and chlorine. Although production of tetrachloroethylene and trichloroethylene from acetylene was once the dominant process, it is now obsolete because of the high cost of acetylene. Demand for tetrachloroethylene peaked in the 1980s. The decline in demand can be attributed to use of tighter equipment and solvent recovery in the dry-cleaning and metal cleaning industries and the phaseout of CFC 113 (trichlorotrifluoroethane) under the Montreal Protocol.

Physical and Chemical Properties

The physical properties of tetrachloroethylene are listed in Table 1. It dissolves a number of inorganic materials including sulfur, iodine, mercuric chloride, and

Table 1. Properties of Tetrachloroethylene

Property	Value
molecular weight	165.83
melting point, °C	−22.7
boiling point at 101 kPa[a], °C	121.2
specific gravity, liquid, at °C	
10/4	1.63120
20/4	1.62260
30/4	1.60640
120/4	1.44865
vapor density at bp at 101 kPa, kg/m^3	5.8
viscosity, mPa·s ($=$ cP)	
liquid, °C	
15	0.932
25	0.839
50	0.657
75	0.534
vapor at 60°C	9900
surface tension, mN/m ($=$ dyn/cm)	
15°C	32.86
30°C	31.27
thermal capacity, kJ/(kg·K)[b]	
liquid at 20°C	0.858
vapor at 100°C	0.611
thermal conductivity, mW/(m·K)	
liquid	126.6
vapor at bp	8.73
heat of combustion	
constant pressure with formation of aq HCl, kJ/mol[b]	679.9
constant volume at 18.7°C, kJ/mol[b]	831.8
latent heat of vaporization at 121.2°C, kJ/mol[b]	34.7
critical temperature, °C	347.1
critical pressure, MPa[c]	9.74
latent heat of fusion, kJ/mol[b]	10.57
heat of formation, kJ/mol[b]	
vapor	−25
liquid	12.5
n_D at 20°C	1.50547
dielectric constant at 1 kHz, 20°C	2.20
electrical conductivity at 20°C, 10^{15} $(\Omega \cdot m)^{-1}$	55.8
coefficient of cubical expansion at 15−90°C, av	0.001079
vapor pressure, kPa[c], at °C	
−20.6	0.1333
13.8	1.333
40.0	5.466
60.0	13.87
80.0	30.13
100.0	58.46
121.2	101.3
solubility at 25°C, mg	
C_2Cl_4 in 100 g H_2O	15
H_2O in 120 g C_2Cl_4	8

[a]To convert kPa to mm Hg, multiply by 7.5.
[b]To convert kJ to kcal, divide by 4.184.
[c]To convert MPa to atm, divide by 0.101.

appreciable amounts of aluminum chloride. Tetrachloroethylene dissolves numerous organic acids, including benzoic, phenylacetic, phenylpropionic, and salicylic acid, as well as a variety of other organic substances such as fats, oils, rubber, tars, and resins. It does not dissolve sugar, proteins, glycerol, or casein. It is miscible with chlorinated organic solvents and most other common solvents. Tetrachloroethylene forms approximately sixty binary azeotropic mixtures (1).

Stabilized tetrachloroethylene, as provided commercially, can be used in the presence of air, water, and light, in contact with common materials of construction, at temperatures up to about 140°C. It resists hydrolysis at temperatures up to 150°C (2). However, the unstabilized compound, in the presence of water for prolonged periods, slowly hydrolyzes to yield trichloroacetic acid [76-03-9] and hydrochloric acid. In the absence of catalysts, air, or moisture, tetrachloroethylene is stable to about 500°C. Although it does not have a flash point or form flammable mixtures in air or oxygen, thermal decomposition results in the formation of hydrogen chloride and phosgene [75-44-5] (3).

Under ultraviolet radiation in the presence of air or oxygen, tetrachloroethylene undergoes autoxidation to trichloroacetyl chloride [76-02-8]. This reaction, which accounts for the slow decomposition of tetrachloroethylene under prolonged storage in the presence of light and air or oxygen, is inhibited in commercial products by the addition of amines or phenols as stabilizers. Peroxy compounds (1) and (2) are intermediates of this autoxidation. Compound (1) rearranges to form trichloroacetyl chloride and oxygen, whereas compound (2) breaks down to form two molecules of phosgene.

$$CCl_2=CCl_2 + O_2 \longrightarrow \begin{bmatrix} Cl \\ Cl- \\ Cl- \\ Cl \end{bmatrix} O^+ - O^- \end{bmatrix} + \begin{bmatrix} Cl \\ Cl-O \\ Cl-O \\ Cl \end{bmatrix}$$

$$(1) \qquad\qquad (2)$$

Reaction with hydrogen at 220°C in the presence of reduced nickel catalyst results in total decomposition to hydrogen chloride and carbon. An explosive reaction occurs with butyllithium in petroleum ether solution (4). Tetrachloroethylene also reacts explosively with metallic potassium at its melting point, however it does not react with sodium (5).

Photochlorination of tetrachloroethylene, observed by Faraday, yields hexachloroethane [67-72-1]. Reaction with aluminum bromide at 100°C forms a mixture of bromotrichloroethane and dibromodichloroethane [75-81-0] (6). Reaction with bromine results in an equilibrium mixture of tetrabromoethylene [79-28-7] and tetrachloroethylene. Tetrachloroethylene reacts with a mixture of hydrogen fluoride and chlorine at 225–400°C in the presence of zirconium fluoride catalyst to yield 1,2,2-trichloro-1,1,2-trifluoroethane [76-13-1] (CFC 113) (7).

Tetrachloroethylene reacts with formaldehyde and concentrated sulfuric acid at 80°C to form 2,2-dichloropropanoic acid [75-99-0] (8). Copolymers with styrene, vinyl acetate, methyl acrylate, and acrylonitrile are formed in the presence of dibenzoyl peroxide (9,10).

Tetrachloroethylene is heated at 110–120°C with *o*-benzenedithiol, in the presence of sodium ethoxide, to form 2,2′-bis-1,3-benzdithiolene (11).

The addition of stabilizers to tetrachloroethylene inhibits corrosion of aluminum, iron, and zinc which otherwise occurs in the presence of water (12).

Manufacture

Many processes have been used to produce tetrachloroethylene. One of the first was chlorination of acetylene (C$_2$H$_2$) to form tetrachloroethane, followed by de-hydrochlorination to trichloroethylene. If tetrachloroethylene was desired, the trichloroethylene was further chlorinated to pentachloroethane and dehydro-chlorinated. This process is no longer used in the United States; Hooker Chemical closed down the last plant in 1978.

In Japan, Toagosei is reported to produce trichloroethylene and tetrachloro-ethylene by chlorination of ethylene followed by dehydrochlorination. In this proc-ess the intermediate tetrachloroethane is either dehydrochlorinated to tri-chloroethylene or further chlorinated to pentachloroethane [76-01-7] followed by dehydrochlorination to tetrachloroethylene. Partially chlorinated by-products are recycled and by-product HCl is available for other processes.

The following processes are commonly used today.

Chlorination of Ethylene Dichloride. Tetrachloroethylene and trichloro-ethylene can be produced by the noncatalytic chlorination of ethylene dichloride [107-06-2] (EDC) or other two-carbon (C2) chlorinated hydrocarbons. This process is advantageous when there is a feedstock source of mixed C2 chlorinated hydro-carbons from other processes and an outlet for the by-product HCl stream. Product ratios of tri- and tetrachloroethylene are controlled by adjusting the Cl$_2$:EDC ratio to the reactor. Partially chlorinated by-products are recycled to the chlorinator. The primary reactions are

$$CH_2ClCH_2Cl + 3\ Cl_2 \longrightarrow Cl_2C{=}CCl_2 + 4\ HCl$$
$$CH_2ClCH_2Cl + 2\ Cl_2 \longrightarrow CHCl{=}CCl_2 + 3\ HCl$$

Chlorination of C1–C3 Hydrocarbons or Partially Chlorinated Derivatives. Tetrachloroethylene and carbon tetrachloride are produced with or without a ca-talyst at high temperatures (550–700°C) from light hydrocarbon feedstocks or their partially chlorinated derivatives. This is one of the most versatile processes, allowing for a wide range of mixed chlorinated hydrocarbon wastes from other processes to be used as feedstocks. However, the large quantities of HCl produced requires integration with other HCl consuming processes. As with the previous process, product distribution is controlled by controlling feedstock ratios, and par-tially chlorinated by-products are recycled to the chlorinator. As examples, reac-tion of EDC or propane are shown in the following.

$$3 \; CH_2ClCH_2Cl + 11 \; Cl_2 \longrightarrow 2 \; Cl_2C{=}CCl_2 + 2 \; CCl_4 + 12 \; HCl$$

$$CH_3CH_2CH_3 + 8 \; Cl_2 \longrightarrow Cl_2C{=}CCl_2 + CCl_4 + 8 \; HCl$$

Oxychlorination of C2 Chlorinated Hydrocarbons. Tetrachloroethylene and trichloroethylene can be produced by reaction of EDC with chlorine or HCl and oxygen in the presence of a catalyst. When hydrochloric acid is used, additional oxygen is required. Product distribution is varied by controlling reactant ratios. This process is advantageous in that no by-product HCl is produced, and it can be integrated with other processes as a net HCl consumer. The reactions may be represented as follows:

$$CH_2ClCH_2Cl + Cl_2 + O_2 \longrightarrow Cl_2C{=}CCl_2 + 2 \; H_2O$$

$$CH_2ClCH_2Cl + \tfrac{1}{2} \; Cl_2 + \tfrac{3}{4} \; O_2 \longrightarrow CHCl{=}CCl_2 + \tfrac{3}{2} \; H_2O$$

Shipping and Storage

Tetrachloroethylene is shipped by barge, tank car, tank truck, and 55-gallon (208-L) steel drums. It may be stored in mild steel tanks that are dry, free of rust, and equipped with a chemical (such as calcium chloride) vent dryer and controlled evaporation vent. Appropriate secondary containment including dikes and sealed surfaces should be provided in accordance with federal and local standards to prevent potential groundwater contamination in the event of a leak. Piping and centrifugal or positive displacement pumps should be constructed of ductile iron or carbon steel with gasket materials made of impregnated cellulose fiber, cork base materials, or Viton resin.

Economic Aspects

World capacity and demand for tetrachloroethylene were approximately 1100 and 845 thousand metric tons in 1974, respectively. Although demand increased into the mid-1980s, since then demand for tetrachloroethylene has decreased significantly as a result of the phaseout of chlorofluorocarbons, the use of more efficient dry-cleaning equipment, and increased reclamation of waste solvent. World capacity and demand as of 1988 are provided in Table 2. Several United States'

Table 2. Tetrachloroethylene World Production Capacity and Demand[a] by Region, 10^3 t

Area	Capacity	Demand
United States	327	252
Western Europe	599	257
Japan	96	101
Total	*1012*	*610*

[a]In 1987 (13).

manufacturers have shut down facilities in the last fifteen years. Current manufacturers and their capacities are listed in Table 3. United States' production and sales history is shown in Table 4.

Specifications

Commercial grades of tetrachloroethylene include a vapor degreasing grade; a dry-cleaning grade; an industrial grade for use in formulations; a high purity, low

Table 3. United States Tetrachloroethylene Producers and Their Capacities[a], 10^3 t

Producer	1988 Capacity[b]
Dow Chemical Co.	
Pittsburg, Calif.	22.7[c]
Plaquemine, La.	40.8
Occidental Chemical Corp.	
Deer Park, Tex.	81.6
PPG Industries, Inc.	
Lake Charles, La.	90.7
Vulcan Materials Co.	
Geismar, La.	68.0
Wichita, Kans.	22.7

[a]Ref. 13.
[b]Capacities are flexible, depending on feedstocks and operating conditions.
[c]In 1990, Dow announced its intention to shut down this facility.

Table 4. United States Tetrachloroethylene Production and Sales[a], 10^3 t

Year	Production	Sales	Price[b] $/ton
1974	333.1	321.5	243
1976	303.4	259.7	391
1978	333.4	249.1	276
1980	347.1	268.1	507
1982	265.3	229.0	441
1984	260.0	196.7	683
1986	187.8	219.9	683
1988	225.8	256.3	683
1989	218.3	197.1	683
1990[c]	132.3		683

[a]Courtesy of the U.S. International Trade Commission.
[b]List price: tanks, industrial-grade, consumers, delivered.
[c]1990 production through third quarter report.

residue grade; and a grade specifically formulated for use as a transformer fluid. The various grades differ in the amount and type of added stabilizers. U.S. Federal Specification OT-236A covers tetrachloroethylene.

ASTM has established standard test methods to determine acid acceptance, acidity, alkalinity, color, corrosivity to metals, nonvolatile matter content, pH of water extractions, relative evaporation rate, boiling point range, specific gravity, water content, water-soluble halide ions, and halogens (14). Typical commercial grades should not contain more than 50 ppm water, 0.0005 wt % acidity (as HCl), or 0.001 wt % insoluble residue.

Uses

Approximately 50% of the demand for tetrachloroethylene is in the dry-cleaning industry where about 80% of all dry cleaners use it as their primary cleaning agent. Use as a feedstock for chlorofluorocarbon production accounts for 30% of current demand. Metal cleaning and miscellaneous applications represent 12 and 8% of demand, respectively. The miscellaneous applications include such varied uses as transformer insulating fluid, chemical maskant formulations, and as a process solvent for desulfurizing coal.

Toxicity

Overexposure to tetrachloroethylene by inhalation affects the central nervous system and the liver. Dizziness, headache, confusion, nausea, and eye and mucous tissue irritation occur during prolonged exposure to vapor concentrations of 200 ppm (15). These effects are intensified and include incoordination and drunkenness at concentrations in excess of 600 ppm. At concentrations in excess of 1000 ppm the anesthetic and respiratory depression effects can cause unconsciousness and death. A single, brief exposure to concentrations above 6000 ppm can be immediately dangerous to life. Reversible changes to the liver have been reported following prolonged exposures to concentrations in excess of 200 ppm (16–22). Alcohol consumed before or after exposure may increase adverse effects.

The OSHA permissible exposure limit (PEL) for tetrachloroethylene is 25 ppm (8-h TWA) (23). The American Conference of Governmental Industrial Hygienists (ACGIH) threshold limit value (TLV) is 50 ppm. In addition they recommend a 15 minute, short-term exposure limit (STEL) of 200 ppm (24). The odor threshold for tetrachloroethylene ranges from about 5 to 70 ppm (15). Therefore odor alone does not provide adequate warning of potential overexposure in the workplace. Air sampling of the work environment should be performed in order to determine the need for protective equipment. Fatalities have occurred when workers have entered unventilated tanks or equipment containing high vapor concentrations of tetrachloroethylene without utilizing a self-contained breathing apparatus (25). Victims of overexposure should be removed from the area, given artificial respiration or oxygen if necessary, and a physician should be consulted (26).

Repeated exposure of skin to liquid tetrachloroethylene may defat the skin causing dermatitis. When frequent or prolonged contact is likely, gloves of Viton, nitrile rubber, or neoprene should be used, discarding them when they begin to deteriorate. Tetrachloroethylene can cause significant discomfort if splashed in the eyes. Although no serious injury results, it can cause transient, reversible corneal injury. If contact with skin or eyes occurs, follow standard first-aid practices.

Ingestion of small amounts of tetrachloroethylene is not likely to cause permanent injury; however, ingestion of large amounts may result in serious injury or even death. All containers should be properly labeled. If solvent is swallowed, consult a physician immediately. Do not induce vomiting. If solvent is aspirated it is rapidly absorbed through the lungs and may cause systemic effects and chemical pneumonia.

Exposure to tetrachloroethylene as a result of vapor inhalation is followed by absorption into the bloodstream. It is partly excreted unchanged by the lungs (17,18). Approximately 20% of the absorbed material is subsequently metabolized and eliminated through the kidneys (27–29). Metabolic breakdown occurs by oxidation to trichloroacetic acid and oxalic acid.

Three significant studies have been conducted on the potential carcinogenic effects of tetrachloroethylene in laboratory animals (30–32). Two of these studies showed increases in observed liver and/or kidney tumors at high dosage levels. The third study showed no significant differences between exposed and control groups of animals at inhalation exposure levels up to 600 ppm. Tetrachloroethylene is classified in Group 2B, a "possible human carcinogen" by the International Agency for Research on Cancer (IARC). The National Toxicology Program (NTP) lists tetrachloroethylene as "reasonably anticipated to cause cancer in humans." Pharmacokinetic studies suggest the effects observed in laboratory animals are not directly applicable to humans (33,34). During the early 1990s, the Environmental Protection Agency (EPA), under its *Guidelines for Carcinogenic Risk Assessment*, had not made a final decision on the classification for this chemical (35).

No teratogenic effects were observed in mice and rats exposed to vapor concentrations of 300 ppm. Exposure levels having no effect on the mother are not anticipated to affect the fetus (36).

Environmental Regulations

Tetrachloroethylene is subject to inventory and release reporting under Title III of the Superfund Amendments and Reauthorization Act of 1986 (SARA). Tetrachloroethylene waste is considered hazardous waste under the Resource Conservation and Recovery Act of 1984 (RCRA). The preferred methods of disposal are through licensed reclaimers or permitted incinerators. The EPA revised the reportable quantity (RQ) for tetrachloroethylene to 100 lbs in 1989. Although tetrachloroethylene does not contribute to smog formation, and the EPA recommended exemption from Volatile Organic Compounds (VOC) regulations in 1983 (37), it continues to be controlled as a VOC. Under the Clean Air Act Amendment of 1990, tetrachloroethylene is considered a hazardous air pollutant. Under this act, the EPA will develop standards to control tetrachloroethylene emissions in dry-cleaning and metal cleaning applications. Under the Safe Drinking Water Act, EPA has established a maximum contaminant level (MCL) of 0.005 mg/L and a goal of 0 mg/L for tetrachloroethylene (38). Packed tower aeration and granular activated carbon are considered the best available technologies for removal of tetrachloroethylene from drinking water.

BIBLIOGRAPHY

"Tetrachloroethylene" under "Chlorine Compounds, Organic" in *ECT* 1st ed., Vol. 3, pp. 794–798, by J. Searles and H. A. McPhail, E. I. du Pont de Nemours & Co., Inc.; "Tetrachloroethylene" under "Chlorocarbons and Chlorohydrocarbons" in *ECT* 2nd ed., Vol. 5, pp. 195–203, by D. W. F. Hardie, Imperial Chemical Industries, Ltd.; in *ECT* 3rd ed., Vol. 5, pp. 754–762, by S. L. Keil, The Dow Chemical Company.

1. L. H. Horsley and co-workers, *Adv. Chem. Ser.* **6,** 32 (1952).
2. W. L. Howard and T. L. Moore, unpublished data, The Dow Chemical Company, 1966.
3. R. P. Marquardt, unpublished data, The Dow Chemical Company, 1964.
4. W. R. H. Hurtley and S. Smiles, *J. Chem. Soc.*, 2269 (1926).
5. L. D. Rampino, *Chem. Eng. News* **36,** 62 (1958).
6. A. Besson, *Compt. Rend.* **118,** 1347 (1894).
7. U.S. Pat. 2,850,543 (Sept. 2, 1958), C. Woolf (to Allied Chemical Corp.).
8. M. J. Prins, *Rec. Trav. Chim.* **51,** 473 (1932).
9. K. W. Doak, *J. Am. Chem. Soc.* **70,** 1525 (1948).
10. F. R. Mayo, F. M. Lewis, and C. Walling, *J. Am. Chem. Soc.* **70,** 1529 (1948).
11. C. S. Marvel, F. D. Hager, and D. D. Coffman, *J. Am. Chem. Soc.* **49,** 2328 (1927).
12. W. L. Archer and E. L. Simpson, *I&EC Prod. R&D* **16,** 158 (June 1977).
13. *Chemical Economics Handbook*, SRI International, Menlo Park, Calif., 1988.
14. *1990 Annual Book of ASTM Standards*, Section 15, Vol. 15.05, ASTM, Easton, Md., 1990.
15. V. K. Rowe and co-workers, *Arch. Ind. Hyg. Occup. Med.* **5,** 556 (1952).
16. R. D. Stewart, *Arch. Environ. Health* **2,** 516 (1961).
17. R. D. Stewart and co-workers, *Arch. Environ. Health* **20,** 224 (1970).
18. R. D. Stewart and co-workers, report number NIOSH-MCOW-ENVM-PCE-74-6, The Medical College of Wisconsin, Milwaukee, Wis., 1974, 172 pp.
19. C. P. Carpenter, *J. Ind. Hyg. Toxic.* **19,** 323 (1937).
20. P. D. Lamson, *Am. J. Hyg.* **9,** 430 (1929).
21. R. Patel, *J. Am. Med. Assoc.* **223,** 1510 (1973).
22. T. C. Tuttle, *Final Report for Contract HSM99-73-35*, Westinghouse Behavioral Services Center, Columbia, Md., p. 124, 1976.
23. *Fed. Reg.* **54**(12), 2332 (Jan. 19, 1989).
24. *1990–1991 Threshold Limit Values for Chemical Substances and Physical Agents*, American Conference of Governmental Industrial Hygienists, 1990.
25. *Standard Practice for Confined Area Entry, D4276-84*, American Society for Testing and Materials, ASTM, Philadelphia, Pa., 1984.
26. *Specialty Chlorinated Solvents Product Stewardship Manual*, 1991 ed, The Dow Chemical Company, Midland, Mich., form 100-6170-90HYC.
27. S. Yllner, *Nature (London)* **191,** 82 (1961).
28. M. Ikeda and co-workers, *Br. J. Ind. Med.* **29,** 328 (1972).
29. M. Ogata and co-workers, *Br. J. Ind. Med.* **28,** 386 (1971).
30. National Cancer Institute, NCI-CG-TR-13, 1977.
31. National Toxicology Program (NTP), TR-311, Research Triangle Park, N.C., 1985.

32. L. W. Rampy, J. F. Quast, B. K. J. Leong, and P. J. Gehring, *Proceedings of the First International Congress on Toxicology*, Academic Press, New York, 1978.
33. D. G. Pegg, J. A. Zempel, W. H. Braun, and P. G. Watanabe, *Toxicol. Appl. Pharmacol.* **51**, 465–474 (1979).
34. A. M. Schumann, J. F. Quast, and P. G. Watanabe, *Toxicol. Appl. Pharmacol.* **55**, 207–219 (1980).
35. *Final Report EPA/600/8-83/0005F*, Environmental Protection Agency, Washington, D.C., 1985.
36. B. A. Schwetz, B. K. J. Leong, and P. G. Gehring, *Toxicol. Appl. Pharmacol.* **32**, 84–96 (1975).
37. *Fed. Reg.* **48**(206), 49,097 (Oct. 24, 1983).
38. *Fed. Reg.* **56**(20), 3536 (Jan. 30, 1991).

J. C. HICKMAN
The Dow Chemical Company

ALLYL CHLORIDE

Efficient and economical synthesis of allyl chloride or 3-chloropropene [*107-05-1*] was made possible by the discovery in the late 1930s of a direct high temperature (300–500°C) chlorination reaction by the Shell Development Co. (1–4). This synthesis allows good yields and use of common inexpensive raw materials such as propylene and chlorine. Although World War II delayed commercial implementation of this chemistry, particularly in Europe, Shell Chemical Co. was able to begin commercial production of allyl chloride in 1945 at their refinery site near Houston, which is now Deer Park, Texas (5). In 1955, The Dow Chemical Company began commercial manufacture of allyl chloride in Freeport, Texas. Initially, for both companies, the allyl chloride product was largely converted to allyl alcohol (qv) and then to glycerol (qv); however, the emergence of epoxy resins (qv) in the 1950s caused a shift to production of epichlorohydrin from allyl chloride. Both epoxy resins and glycerol can be easily produced from epichlorohydrin (6).

The direct high temperature chlorination of propylene continues to be the primary route for the commercial production of allyl chloride. The reaction results in allyl chloride selectivities of 75–80% from propylene and about 75% from chlorine. Additionally, a significant by-product of this reaction, 1,3-dichloropropene, finds commercial use as an effective nematocide when used in soil fumigation. Overall efficiency of propylene and chlorine use thus is significantly increased. Remaining by-products include 1,2-dichloropropane, 2-chloropropene, and 2-chloropropane.

A second method for synthesis of allyl chloride is thermal dehydrochlorination, ie, cracking, of 1,2-dichloropropane, but this method is generally less satisfactory because of low allyl chloride selectivity (50–60%) and operating temperatures of 500–600°C (4,7–10). The by-products of cracking are 1-chloropropene and 2-chloropropene, which have no significant commercial use.

The oxychlorination of propylene to allyl chloride, using hydrogen chloride and oxygen, has also been demonstrated. However, with inferior yields, less than satisfactory catalyst life, and a complex processing scheme, (11–20) this route to allyl chloride is not utilized commercially.

Physical Properties

Allyl chloride is a colorless liquid with a disagreeable, pungent odor. Although miscible in typical compounds such as alcohol, chloroform, ether, acetone, benzene, carbon tetrachloride, heptane, toluene, and acetone, allyl chloride is only slightly soluble in water (21–23). Other physical properties are given in Table 1.

Chemical Properties

Allyl chloride exhibits reactivity as an olefin and as an organic halide. Its activity as a chloride is enhanced by the presence of the double bond, but its activity as an olefin is somewhat less than that of propylene. Allyl chloride participates in

Table 1. Physical Properties of Allyl Chloride

Property	Value
molecular weight	76.53
freezing point, °C	−134.5
boiling point at 101.3 kPa[a], °C	45.1
specific gravity at 20/4°C	0.938
liquid density at 25°C, kg/m^3	931
flash point[b] (tag closed cup), °C	−29
flammable limits (by volume in air), %	3.3–11.1
critical temperature, °C	240.7
critical pressure, kPa[a]	4710
heat of combustion, kJ/g	24.8
heat of vaporization at 20°C, J/g[c]	357
specific heat, liquid, at 20°C, J/(g·°C)[c]	1.32
solubility at 20°C, in water, %	0.33
solubility at 20°C, water in allyl chloride, %	0.08
azeotrope with water (22%), °C	43
refractive index at 15°C	1.4153
electrical resistivity, Ω/cm^3	107
vapor density (air = 1)	2.64[d]
autoignition point, °C	392
liquid viscosity, mPa·s (= cP)	
0°C	0.4070
25°C	0.3136
50°C	0.2519
expansion coefficient[e] (0–30°C)	1.41K^{-1}
vapor pressure[f] (p in kPa[a], T in K)	$\log p = 19.403 - 2098.0/T$ $-4.2114 \times \log T$

[a]To convert kPa to atm, divide by 101.3.
[b]Ref. 24.
[c]To convert J to cal, divide by 4.184.
[d]Ref. 23.
[e]Ref. 22.
[f]Ref. 25.

most types of reactions characteristic of either functional group; reactions can be directed by control of conditions, selection of reagents, and provision of suitable catalysts. Allyl chloride does not polymerize well by free-radical techniques (see ALLYL MONOMERS AND POLYMERS).

Addition to the Double Bond. Chlorine, bromine, and iodine react with allyl chloride at temperatures below the inception of the substitution reaction to produce the 1,2,3-trihalides. High temperature halogenation by a free-radical mechanism leads to unsaturated dihalides CH_2=CHCHClX. Hypochlorous and hypobromous acids add to form glycerol dihalohydrins, principally the 2,3-dihalo isomer. Dehydrohalogenation with alkali to epichlorohydrin [106-89-8] is of great industrial importance.

$$CH_2\!=\!CHCH_2Cl \xrightarrow{HOCl} ClCH_2\overset{\overset{\displaystyle OH}{\displaystyle |}}{C}HCH_2Cl + HOCH_2\overset{\overset{\displaystyle Cl}{\displaystyle |}}{C}HCH_2Cl \xrightarrow{base} \overset{\displaystyle O}{\overset{\displaystyle \diagup\!\!\diagdown}{CH_2\!-\!CH}}CH_2Cl$$

Hydrogen halides normally add to form 1,2-dihalides, though an abnormal addition of hydrogen bromide is known, leading to 3-bromo-1-chloropropane [109-70-6]; the reaction is believed to proceed by a free-radical mechanism. Water can be added by treatment with sulfuric acid at ambient or lower temperatures, followed by dilution with water. The product is 1-chloro-2-propanol [127-00-4].

Replacement of Chlorine Allyl choride can be hydrolyzed to allyl alcohol [107-18-6] under either alkaline or acidic conditions. Other simple replacements of Cl are by I, CN, SCN, NCS, SH, and others.

Formation of Allyl Esters. Allyl esters are formed by reaction of allyl chloride with sodium salts of appropriate acids under conditions of controlled pH. Esters of the lower alkanoic, alkenoic, alkanedioic, cycloalkanoic, benzenecarboxylic, alkylbenzene carboxylic, and aromatic dicarboxylic acids may be prepared in this manner (25). More information can be found about the reactivity of allyl compounds (see ALLYL ALCOHOL AND MONOALLYL DERIVATIVES).

Formation of Amines. Mono-, di-, and triallyl amines are prepared by reaction with ammonia. The ratio of reagents determines product distribution; with sufficient time and excess of allyl chloride, tetraallylammonium chloride [13107-10-3] and triallylamine [102-70-5] predominate. Mixed amines are prepared in similar fashion by using a substituted amine in place of ammonia; they may also be prepared with allylamine [107-11-9] and a suitable organic chloride.

Alkylation. Several alkylation reactions are known; either the olefin or chloro- group may be involved. The reactions of allyl chloride with benzene are typical of reactions involving the double bond. In the presence of ferric or zinc chloride, the products are 2-chloropropylbenzene [10304-81-1] and 1,2-diphenyl-propane [5814-85-7]:

$$C_6H_6 + ClCH_2CH\!=\!CH_2 \longrightarrow [C_6H_5CH_2CH\!=\!CH_2] + HCl \longrightarrow C_6H_5CH_2CHClCH_3 \xrightarrow{C_6H_6}$$

$$C_6H_5CH_2\overset{\overset{\displaystyle C_6H_5}{\displaystyle |}}{C}HCH_3 + HCl$$

Several allylation reactions are known, frequently using an organometallic derivative of the compound being allylated, or a strongly electropositive metal in conjunction with the reactants. Grignard reactions are in this group.

Allyl chloride reacts with sodamide in liquid ammonia to produce benzene; when sodamide is in excess, hexadiene dimer is the principal product, with some trimer and tetramer (C_{24}, six double bonds). Allylation at carbon atoms alpha to polar groups is used in the preparation of α-allyl-substituted ketones and nitriles. Preparation of β-diketone derivatives, methionic acid derivatives, and malonic ester, cyanoacetic ester, and β-keto-ester derivatives, etc, involving substitution on an alpha carbon between two polar carbonyl groups, is particularly facile.

$$CH_2{=}CHCH_2Cl + C_2H_5OOC{=}\overline{C}H{=}COOC_2H_5 \rightarrow CH_2{=}CHCH_2CH(COOC_2H_5)_2$$

Manufacture and Processing

Substitutive chlorination of propylene is the commercial route to allyl chloride. For this reaction $\Delta H°_{298} = -113$ kJ/mol (-27 kcal/mol).

$$CH_2{=}CH{-}CH_3 + Cl_2 \rightarrow CH_2{=}CH{-}CH_2Cl + HCl$$

Reaction Mechanism. High temperature vapor-phase chlorination of propylene [115-07-1] is a free-radical mechanism in which substitution of an allylic hydrogen is favored over addition of chlorine to the double bond. Abstraction of allylic hydrogen is especially favored since the allyl radical intermediate is stabilized by resonance between two symmetrical structures, both of which lead to allyl chloride.

$$CH_2{=}CH{-}CH_3 + Cl\cdot \longrightarrow [CH_2{=}CH{-}\overset{\cdot}{C}H_2 \longleftrightarrow \overset{\cdot}{C}H_2{-}CH{=}CH_2] + HCl$$
$$CH_2{=}CH{-}\overset{\cdot}{C}H_2 + Cl_2 \longrightarrow CH_2{=}CH{-}CH_2Cl + Cl\cdot$$

Abstraction of other hydrogens occurs to a very small degree and leads to small amounts of 2-chloropropene [557-98-2] and 1-chloropropene [590-21-6]. Significant competing reactions include the addition reaction forming 1,2-dichloropropane [78-87-5] (eq. 1, $\Delta H°_{298} = -186$ kJ/mol (-44.5 kcal/mol)), which is important below 300°C but is merely a by-product reaction above this temperature, and the secondary chlorination reaction, which produces 90% 1,3-dichloropropene [542-75-6] (eq. 3, $\Delta H°_{298} = -125$ kJ/mol (-29.9 kcal/mol)) and 10% 3,3-dichloropropene [563-57-5] (eq. 2, $\Delta H°_{298} = -117$ kJ/mol (-27.9 kcal/mol)).

$$CH_2{=}CH{-}CH_3 + Cl\cdot \longrightarrow \overset{\displaystyle Cl}{\overset{|}{CH_2{-}\overset{\cdot}{C}H}}{-}CH_3 \xrightarrow{Cl_2} \overset{\displaystyle Cl \quad Cl}{\overset{|\quad\;|}{CH_2{-}CH}}{-}CH_3 + Cl\cdot \tag{1}$$

$$CH_2{=}CH{-}CH_2Cl + Cl\cdot \longrightarrow [CH_2{=}CH{-}\overset{\cdot}{C}HCl \longleftrightarrow \overset{\cdot}{C}H_2{-}CH{=}CHCl] + HCl$$

$$CH_2{=}CH{-}\overset{\cdot}{C}HCl + Cl_2 \longrightarrow CH_2{=}CH{-}CHCl_2 + Cl\cdot \tag{2}$$

$$\overset{\cdot}{C}H_2{-}CH{=}CHCl + Cl_2 \longrightarrow ClCH_2{-}CH{=}CHCl + Cl\cdot \tag{3}$$

As shown, in the case of chlorination of allyl chloride, the resonance states of the chloroallyl radical intermediates are not symmetrical and their propagation reactions lead to the two different dichloropropene isomers in an approximate 10:90 ratio (26). In addition, similar reactions result in further substitution and addition with products such as trichloropropanes, trichloropropenes, tetrachloropropanes, etc in diminishing amounts. Propylene dimerization products such as 1,5-hexadiene, benzene, 1-chloropropane, 2-chloropropane, high boiling tars, and coke are also produced in small amounts.

Reaction Conditions.   Typical industrial practice of this reaction involves mixing vapor-phase propylene and vapor-phase chlorine in a static mixer, followed immediately by passing the admixed reactants into a reactor vessel that operates at 69–240 kPa (10–35 psig) and permits virtual complete chlorine conversion, which requires 1–4 s residence time. The overall reactions are all highly exothermic and as the reaction proceeds, usually adiabatically, the temperature rises. Optimally, the reaction temperature should not exceed 510°C since, above this temperature, pyrolysis of the chlorinated hydrocarbons results in decreased yield and excessive coke formation (27).

Two variables of primary importance, which are interdependent, are reaction temperature and chlorine:propylene ratio. Propylene is typically used in excess to act as a diluent and heat sink, thus minimizing by-products (eqs. 2 and 3). Since higher temperatures favor the desired reaction, standard practice generally involves preheat of the reactor feeds to at least 200°C prior to combination. The heat of reaction is then responsible for further increases in the reaction temperature toward 510°C. The chlorine:propylene ratio is adjusted so that, for given preheat temperatures, the desired ultimate reaction temperature is maintained. For example, at a chlorine:propylene molar ratio of 0.315, feed temperatures of 200°C (propylene) and 50°C (chlorine) produce an ultimate reaction temperature of approximately 500°C (10). Increases in preheat temperature toward the ultimate reactor temperature, eg, in attempts to decrease yield of equation 1, must be compensated for in reduced chlorine:propylene ratio, which reduces the fraction of propylene converted and, thus allyl chloride quantity produced. A suitable economic optimum combination of preheat temperature and chlorine:propylene ratio can be readily determined for individual cases.

Pressure and residence time have relatively little effect on reaction selectivity, at least within the ranges normally encountered. Poor mixing and excessive residence time result in increased carbonization of the reactor.

Reactor Design.   Industrial practice typically involves adiabatic reactor operation since the high reaction velocity effectively prohibits cooling of such magnitude to allow isothermal operation. Different reactor designs can affect the reaction selectivity because of the temperature effect on yield (28). Over the years, many reactor designs have been proposed (29–35), some of which are quite complex. Multiple reactor designs have also been suggested (34,36).

The feed streams should be reasonably pure to limit yield losses and protect the purity of the final products. Typically, polymer-grade propylene with 99.5% purity is employed; propane impurity can react to undesirable 1-chloropropane (bp 46.6°C), which is very difficult to separate from allyl chloride (bp 45°C). Both propylene and chlorine should be dry to prevent corrosion in downstream equipment where mixtures with HCl occur.

Product Recovery. The allyl chloride product is recovered through the use of several fractional distillation steps. Typically, the reactor effluent is cooled and conducted into an initial fractionator to separate the HCl and propylene from the chloropropenes, dichloropropanes, dichloropropenes, and heavier compounds. The unconverted propylene is recycled after removal of HCl, which can be accomplished by adsorption in water or fractional distillation (33,37,38) depending on its intended use. The crude allyl chloride mixture from the initial fractionator is then subjected to a lights and heavies distillation; the lighter (than allyl chloride) compounds such as 2-chloropropene, 1-chloropropene, and 2-chloropropane being the overhead product of the first column. Allyl chloride is then separated in the second purification column as an overhead product. Product purities can exceed 99.0% and commercial-grade allyl chloride is typically sold in the United States in purities about 99.5%.

Materials of Construction. Generally, carbon steel is satisfactory as a material of construction when handling propylene, chlorine, HCl, and chlorinated hydrocarbons at low temperatures (below 100°C) in the absence of water. Nickel-based alloys are chiefly used in the reaction area where resistance to chlorine and HCl at elevated temperatures is required (39). Elastomer-lined equipment, usually PTFE or Kynar, is typically used when water and HCl or chlorine are present together, such as adsorption of HCl in water, since corrosion of most metals is excessive. Stainless steels are to be avoided in locations exposed to inorganic chlorides, as stainless steels can be subject to chloride stress-corrosion cracking. Contact with aluminum should be avoided under all circumstances because of potential undesirable reactivity problems.

Storage and Shipment

Storage. Purified and dry allyl chloride can be safely stored in carbon steel vessels. Use of lined vessels is recommended if slight discoloration or trace presence of metals is undesirable for its intended use. In any event, the presence of air should be avoided for safety (flammability) reasons through the use of an inert gas pad. Tank vents should be treated, eg, by incineration, prior to venting to the atmosphere. Some commercial producers intentionally add about 0.1% propylene oxide as a stabilizer to prevent discoloration; however, this is usually unnecessary if product purity is sufficiently high.

Shipment. The use of vapor balancing or closed-loop systems is recommended when transferring to or from transportation containers in order to minimize vent flow and use of fresh pad gas. Typically, tank trucks or railroad cars are used to transport allyl chloride. Allyl chloride is also sold in drums; however, because of the vapor exposure potential and the empty drum disposal problem, it is highly recommended that large volume consumers employ bulk shipment whenever possible. Marine shipments of allyl chloride are not common and International Maritime Organization (IMO) container tanks are recommended if necessary. Loading and unloading connections (liquid transfer and vapor return) should be made with dry-disconnect fittings to minimize liquid and vapor spillage upon disconnection. Rail cars should be pressure-type specification DOT 105A100W cars made of carbon steel lined with a high baked phenolic coating. Tank trucks are typically single-compartment, stainless steel specification MC307 tank trailers. Both rail cars and tank trucks should be padded with an inert gas such as nitrogen.

Economic Aspects

Producers. In the years since 1945, production capacities and the number of producing companies have substantially increased; however the high temperature chlorination reaction has remained the exclusive technique for commercial production of allyl chloride. Production facilities thought to be in existence in 1990 are listed in the following, in order of estimated production capacities (40–48).

Company	Location
The Dow Chemical Company	Freeport, Tex.
	Stade, Germany
Shell Chemical Co.	Pernis, Holland
	Norco, La.
Solvay & Cie	Tavaux, France
	Rheinberg, Germany
Kashima Chemical	Kashima, Japan
Chinese National Technical Import	Qilu, China
Daiso	Mizushima, Japan
Organika-Zachem	Bydgoszcz, Poland
Sumitomo Chemical	Niihama, Japan
Alclor Quimica	Maceio, Brazil
MTT Co.	Tokuyama, Japan
Spolek Pro Chemickou	Usti nad Labem, Czechoslovakia

Allyl chloride capacity is also thought to exist in several plants within the former Soviet Union and Romania; however actual production is believed to be relatively small.

Production figures are not published by these producers, so precise production amounts are not available; however, it is roughly estimated that global production in 1989–1990 was 500,000–600,000 t/yr. Approximately 90% of this allyl chloride production is used captively to synthesize epichlorohydrin. The remainder is sold on the merchant market with bulk list U.S. prices in 1989–1990 of $1.63/kg. Some of the producers listed above and several additional companies have announced their intentions to expand or build allyl chloride capacity.

Specifications, Standards, and Quality Control

Sales Specifications. The generic sales specifications for commercial-grade allyl chloride typical in 1990 were

Property	Typical sales specification
purity	99.0 wt % min
color	100 APHA max
acid as HCl	100 ppm max
water	200 ppm max
specific gravity, 20/20°C	0.936–0.940

Statistical quality control is used to first measure and then continuously improve product quality. For example, The Dow Chemical Company's average 1989 performance compared to the typical sales specification were purity, = 99.65 wt %; color, APHA = 4; acid (as HCl) = 7.3 ppm; and water = 26 ppm. Averages of properties were based on rail car and tank truck shipment samples during 1989.

Analytical and Test Methods

The Dow Chemical Company has published a summary of analytical techniques for allyl chloride (49) and a brief summary of each method is given here.

Purity. Gas chromatographic analysis is performed utilizing a wide-bore capillary column (DB-1, 60 m × 0.32 mm ID × 1.0 μm film) and a flame ionization detector in an instrument such as a Hewlett-Packard 5890 gas chromatograph. A calibration standard is used to determine response factors for all significant impurities, and external standard calculation techniques are used to estimate the impurity concentrations. Allyl chloride purity is determined by difference.

Color. Color is determined through the use of a Nessler-type visual color comparator as described in ASTM Designation D1209.

Acid (as HCl). The sample is mixed with an equal volume of 90% isopropyl alcohol and titrated with alcoholic potassium hydroxide to the phenolphthalein end point.

Water. Water is determined by Karl Fischer titration using ASTM Designation D1744.

Specific Gravity. The specific gravity is determined by weighing a known volume of sample in a Lipkin bicapillary pycnometer as described in ASTM Designation D941.

Health and Safety Factors

Health Hazards. Allyl chloride is a toxic, highly flammable compound that is severely irritating to the skin and mucous membranes. Allyl chloride is considered to be moderately to highly toxic (LD_{50} = 275–700 mg/kg body weight) via oral exposure. Amounts incidental to industrial handling are unlikely to cause injury. Large amounts, however, can cause injury, even death (24,50).

Allyl chloride is toxic through liquid contact with the skin and can cause severe irritation, resulting in deep-seated pain and delayed burns. Effects may be more severe in instances where liquid is confined to the skin. The LD_{50} for toxicity through skin absorption in rabbits is 400–2200 mg/kg body weight (24,50). Contact with the eyes can cause permanent eye injury, even blindness.

The vapors of allyl chloride are very irritating to the eyes, nose, and throat. Lung injury may be delayed in onset. Liver and kidney injury can result from exposure to vapors; kidney injury is expected to be most severe in acute exposures. High concentrations of vapor can be lethal. Following chronic exposures to the vapors, liver injury would be expected to occur first (23).

Exposure Limits. The American Conference of Governmental Industrial Hygienists (ACGIH) has recommended a threshold limit value (TLV) of 1 ppm allyl chloride in air based on a time-weighted average (TWA) of an eight-hour work day, with a short-term exposure limit (STEL) of 2 ppm. OSHA has established its permissible exposure limit (PEL) at this same level (24,50). The National Institute for Occupational Safety and Health (NIOSH) recommends that exposure to allyl chloride be controlled to a concentration no greater than 1 ppm of air by volume, which is the TWA for up to a 10-h workday in a 40-h work week, or a ceiling concentration of no more than 3 ppm for any 15-min period (51).

Allyl chloride has a disagreeable, pungent odor. The odor threshold has been estimated at approximately 3–6 ppm (51). Olfactory detection of odor is thus not an adequate warning of overexposure.

Personal Protective Equipment. Personal protective and emergency safety equipment should not be relied on as the primary protection from allyl chloride. Prevention of exposure should be considered the preferred precautionary measure. Where the exposure guideline may be greatly exceeded, an approved positive-pressure air supplied or self-contained breathing apparatus with full facepiece should be used (51).

Emergency Response to Fires, Spills, and Leaks. Vapors of allyl chloride are heavier than air and may travel a considerable distance to sources of ignition. Combustion products of allyl chloride may be more toxic than the allyl chloride itself. Emergency response personnel should wear full protective clothing and self-contained breathing apparatus. Alcohol foam, carbon dioxide, and dry chemicals are effective extinguishing agents for allyl chloride fires. Water may be used to keep fire-exposed containers cool, and water spray may be used to flush burning spills away from exposure to ignition sources. Allyl chloride floats on water, making water alone potentially inadequate for fire fighting.

Spills should be confined and prevented from entering water sources. Smother with foam and take up residue with an absorbent and put into drums for disposal. The suggested method of disposal is incineration at an approved waste handling facility in a system equipped with a combustion gas scrubber system (23).

Uses and Derivatives

Allyl chloride is typically used to make intermediates for downstream derivatives such as resins and polymers. Allyl chloride is very important in the production of epichlorohydrin [106-89-8], which is used as a basic building block for epoxy resins (qv). Synthetic glycerol [56-81-5] is also a very important derivative of allyl chloride with epichlorohydrin, an intermediate in this process (see CHLOROHYDRINS; GLYCEROL). Allyl chloride is a starting material for allyl ethers of phenols, bisphenol A, novolak phenolic resins (qv), and the like. Allylic esters, which can be cross-linked, are also reaction products of allyl chloride. Other compounds made from allyl chloride are quaternary amines used as chelating agents and quaternary ammonium salts, which are used in water clarification and sewage sludge flocculation (23). Sodium allyl sulfonate [2495-39-8] can be made from allyl chloride and is sometimes used as a metal brightener in electroplating baths. Poly(allyl chloride)s have been used as plasticizers and flexibilizers in other polymers such as poly(vinyl chloride). Ziegler catalysts have been modified with allyl chloride to be further used in the production of low molecular-weight olefin polymers (23).

BIBLIOGRAPHY

"Allyl Chloride" under "Chlorine Compounds, Organic" in *ECT* 1st ed., Vol. 3, pp. 800–806, by H. G. Vesper, Shell Development Co.; "Allyl Chloride" under "Chlorocarbons and Chlorohydrocarbons" in *ECT* 2nd ed., Vol. 5, pp. 205–214, by B. H. Pilorz, Shell Chemical Co.; "Allyl Chloride" under "Chlorocarbons, -Hydrocarbons" in *ECT* 3rd ed., Vol. 5, pp. 763–773, by A. DeBenedictis, Shell Chemical Co.

1. E. C. Williams, *Ind. Eng. Chem.* **16**, 630–632 (Dec. 10, 1938).
2. U.S. Pat. 2,130,084 (Sept. 13, 1938), H. P. A. Groll, G. Hearne, J. Burgin, and D. S. LaFrance (to Shell Development Co.).
3. H. P. A. Groll and G. Hearne, *Ind. Eng. Chem.* **31**(12), 1530–1537 (Dec. 1939).
4. E. C. Williams, *Trans. AIChE* **37**, 157–207 (1941); *Chem. Met. Eng.* **47**, 834–838 (Dec. 1940).
5. A. W. Fairbairn, H. A. Cheney, and A. J. Cherniavsky, *Chem. Eng. Progr.* **43**(6), 280–290 (June 1947).
6. *Chem. Eng. News*, 48–53 (July 31, 1967).
7. U.S. Pat. 2,207,193 (July 9, 1940), H. P. A. Groll (to Shell Development Co.).
8. D. H. R. Barton and A. J. Head, *Trans. Faraday Soc.* **46**, 114–124 (1950).
9. K. A. Holbrook and J. S. Palmer, *Trans. Faraday Soc.* **67**(1), 80–87 (1971).
10. U.S. Pat. 4,319,062 (Mar. 9, 1982), T. S. Boozalis, J. B. Ivy, and G. G. Willis (to The Dow Chemical Company).
11. U.S. Pat. 2,966,525 (Dec. 27, 1960), D. E. Steen (to Monsanto Chemical Co.).
12. Brit. Pat. 1,016,094 (Jan. 5, 1966), (to Toyo Soda).
13. Brit. Pats. 1,157,584 (July 9, 1969); 1,174,509 (Dec. 17, 1969); 1,175,952 (Jan. 1, 1970); U.S. Pat. 3,489,816 (Jan. 13, 1970), L. Hornig, L. Hirsh, G. Mau, and T. Quadflieg (to Farbwerke Hoechst AG).
14. Brit. Pat. 1,251,535 (Oct. 27, 1971), (to Showa Denko K.K.).
15. Brit. Pat. 1,252,578 (Nov. 10, 1971), (to Deutsche Texaco).
16. U.S. Pat. 3,855,321 (Dec. 17, 1974), H. C. Bach and H. E. Hinderer (to Monsanto Co.).
17. K. Fujimoto, H. Takashima, and T. Kunugi, *J. Catal.* **43**, 234–243 (1976).
18. *Res. Discl.* **175**, 31 (1978).
19. *Ibid.*, pp. 44–45.
20. U.S. Pat. 4,244,892 (Jan. 13, 1981), N. M. O. Guseinov and co-workers.
21. N. I. Sax and R. J. Lewis, Sr., rev., *Hawley's Condensed Chemical Dictionary*, 11th ed., Van Nostrand Reinhold Co., New York, 1987.
22. *Ullmanns' Encyclopedia of Industrial Chemistry*, 5th compl. rev. ed., Vol. A1, VCH Publishers, Deerfield Beach, Fla., 1985.
23. *Allyl Chloride*, Technical Bulletin 296-676-86, The Dow Chemical Company, Midland, Mich., 1986.

24. *Allyl Chloride*, Material Safety Data Sheet, The Dow Chemical Company, Midland, Mich., June 14, 1990.
25. U.S. Pat. 2,939,879 (June 7, 1960), A. De Benedictis (to Shell Oil Co.).
26. G. W. Hearne, T. W. Evans, H. L. Yale, and M. C. Hoff, *J. Am. Chem. Soc.* **75** 1392–1394 (1953).
27. L. M. Porter and F. F. Rust, *J. Am. Chem. Soc.* **78** 5571–5573 (1956).
28. J. M. Smith, *Chemical Engineering Kinetics*, 3rd ed., McGraw-Hill Book Co., Inc., New York, 1981, pp. 229–246.
29. U.S. Pat. 2,643,272 (June 23, 1953), A. E. Lacomble, G. W. Hearne, and D. S. LaFrance (to Shell Development Co.).
30. U.S. Pat. 2,763,699 (Sept. 18, 1956), C. P. van Dijk, F. J. F. van der Plas (to Shell Development Co.).
31. Brit. Pat. 901,680 (July 25, 1962), (to Columbia-Southern Chemical Corp.).
32. U.S. Pat. 3,054,831 (Sept. 18, 1962), R. H. Samples and L. E. Hilbert (to Union Carbide Corp.).
33. U.S. Pat. 3,356,749 (Dec. 5, 1967); 3,472,902 (Oct. 14, 1969), C. P. van Dijk (to Pullman Inc.).
34. Jpn. Pat. (Kokoku) 48 26732 (Aug. 15, 1973), H. Yamamoto, T. Nakahata, and Y. Nakamura (to Asahi Denka Kogyo).
35. Jpn. Pat. (unexamined publication, Kokai) 61 40232 (Feb. 26, 1986), A. Kataoka, H. Miki, Y. Waizumi, T. Oishi, and Y. Hiraiki (to Sumitomo Chemical Co.).
36. Brit. Pat. 761,831 (Nov. 21, 1956), (to Shell Chemical Co.).
37. Brit. Pat. Appl. 2,140,014A (Nov. 21, 1984), W. Madej and co-workers (to Blachownia Institute of Heavy Organic Synthesis).
38. Brit. Pat. Appl. 2,142,626A (Jan. 23, 1985), J. Wasilewski and co-workers (to Blachownia Institute of Heavy Organic Synthesis, Organika-Zachem).
39. Jpn. Pat. (unexamined publication, Kokai) 60 252434 (Dec. 13, 1985), H. Miki, Y. Izumi, and T. Oishi (to Sumitomo Chemical Co.).
40. *Chem. Purchas.* 67–73 Apr. 1980.
41. *Jpn. Chem. Week*, 3 (July 12, 1984).
42. *Jpn. Chem. Week*, 2 (Dec. 20, 1984).
43. *Jpn. Econ. J.*, 17 (May 14, 1985).
44. *Chem. Week*, 26–27 (Feb. 19, 1986).
45. *Przem. Chem.* **66**(2), 71–73 (1987).
46. *Jpn. Chem. Week*, (Apr. 16, 1987).
47. *Jpn. Chem. Week* **30**(1508) (Mar. 9, 1989).
48. *Eur. Chem. News, Intl. Proj. Rev. Part 1,* 28 (Mar. 1990).
49. *Allyl Chloride, Dry* Analytical Method 06768B, The Dow Chemical Company, Midland, Mich. Apr. 29, 1982.
50. *Allyl Chloride*, Material Safety Data Sheet, Shell Oil Co., Houston, Tex., Jan. 10, 1991.
51. *Occupational Exposure to Allyl Chloride*, HEW Publication No. 76-204, GPO, (NIOSH) Criteria for a Recommended Standard, 1976.

CHRIS KNEUPPER
LESTER SAATHOFF
The Dow Chemical Company

CHLOROPRENE

Chloroprene (2-chloro-1,3-butadiene), [126-99-8] was first obtained as a by-product from the synthesis of divinylacetylene (1). When a rubbery polymer was found to form spontaneously, investigations were begun that promptly defined the two methods of synthesis that have since been the basis of commercial production (2), and the first successful synthetic elastomer, Neoprene, or DuPrene as it was first called, was introduced in 1932. Production of chloroprene today is completely dependent on the production of the polymer. The only other use accounting for significant volume is the synthesis of 2,3-dichloro-1,3-butadiene, which is used as a monomer in selected copolymerizations with chloroprene.

The original commercial production was from acetylene through monovinylacetylene [689-97-4], C_4H_4.

$$2 \; HC\equiv CH \xrightarrow{\text{(CuCl)}} HC\equiv C-CH=CH_2 \xrightarrow[\text{(CuCl)}]{\text{HCl}} H_2C=CCl-CH=CH_2$$

Since the 1960s, because of an increasing price for acetylene and decreasing price for butadiene, 1,3-butadiene [106-99-0], C_4H_6, has displaced acetylene as the feedstock.

Physical Properties

Selected physical properties of chloroprene are listed in Table 1. When pure, the monomer is a colorless, mobile liquid with slight odor, but the presence of small traces of dimer usually give a much stronger, distinctive odor similar to terpenes and inhibited monomer may be colored from the stabilizers used. Ir and Raman spectroscopy of chloroprene (4) have been used to estimate vibrational characteristics and rotational isomerization.

Chemical Properties

The chemical properties of chloroprene are generally similar to those of butadiene. Chloroprene has very low reactivity with nucleophilic reagents; the chlorine atom is very difficult to replace. Toward electrophilic reagents, eg, chlorine or maleic anhydride, it is somewhat less reactive than butadiene (5,6) though it does oxidize with air. It is markedly more reactive in free-radical additions and polymerization. Q and e values of 7.3 and -0.02 have been given compared to 2.4 and -1.05 for butadiene (7,8), indicating higher reactivity in copolymerizations without strong alternating tendency.

Dimerization. Presumably because it is more active as a dienophile, chloroprene reacts with itself to form dimers, isomeric chlorovinylchlorocyclohexenes, at a considerably faster rate than that at which butadiene is converted to vinylcyclohexene. At the same time, 1,2-divinyl,1,2-dichlorocyclobutanes are also formed, amounting initially to about 60% of the dimer mixture. After a period of

Table 1. Physical Properties of Chloroprene[a]

Property	Value
molecular formula	C_4H_5Cl
mol wt	88.54
melting point, °C	-130 ± 2
boiling point at 101 kPa[b], °C	59.4
critical temperature, °C	261.7
vapor pressure (T in K, p in kPa[b])	$\log_{10} p = 6.652 - 1545/T$
viscosity at 25°C, mPa·s($=$ cP)	0.394
density at 20°C, g/mL	0.9585
average coefficient of volumetric expansion (20–61°C), K^{-1}	0.001235
refractive index, n_D^{20}	1.4583
flash point (ASTM, open cup), °C	-20
latent heat of vaporization, kJ/g[c]	
0°C	0.3328
60°C	0.3027
specific heat, kJ/(kg·K)[c]	
liquid at 20°C	1.314
gas at 100°C	1.0383
thermal conductivity (where t is °C) mW/(m·K),	$2.410 \times 10^{-5} + 0.160 \times 10^{-5} t$
dielectric constant at 27°C	4.9

[a]Ref. 3.
[b]To convert kPa to mm Hg, multiply by 7.5.
[c]To convert J to cal, divide by 4.184.

time, depending on the temperature, the composition of the crude dimers shifts, reflecting the slow decomposition of the *cis*-substituted cyclobutane isomer to form the *trans*-substituted isomer, a vinyldichlorocyclohexene, and 1,6-dichlorocyclooctadiene, and probably to regenerate some monomer. At higher temperature (>100°C) the trans isomer undergoes a similar decomposition. Except for these decomposition processes, the ratio of products is relatively independent of temperature. The proportion of vinylcyclohexenes is significantly increased at high pressure, however. The effect has been attributed to a more compact four-center transition state for the formal Diels-Alder products vs a less compact two-center one for the cyclobutanes (9,10). The rate of dimerization is about 1%/h at the normal boiling point and about 0.35%/d at room temperature, so that low temperature storage is required to maintain purity of the monomer for polymerization and some dimer will be formed during polymerization by any of the normal processes.

Polymerization. Chloroprene is normally polymerized with free-radical catalysts in aqueous emulsion, limiting the conversion of monomer to avoid formation of cross-linked insoluble polymer. At a typical temperature of 40°C, the polymer is largely head-to-tail in orientation and trans in configuration, but modest amounts of head-to-head, cis, 1,2, and 3,4 addition units can also be detected. A much more regular and highly crystalline polymer can be made at low temperature (11). Chloroprene can also be polymerized with cationic polymerization catalysts, giving a polymer with significantly more 1,2 and 3,4 structures, frequently with partial or extensive elimination of HCl and cross-linking during the process. Anionic catalysts are generally ineffective. A largely cis polymer of chloroprene has been prepared by an indirect route (12).

Isomers and Analogues.   Structural isomers and analogues are significant because they occur as impurities of the manufacturing process and are also common comonomers.

4-Chloro-1,2-butadiene [25790-55-0] is mainly of historical interest (2). It is formed from vinylacetylene and HCl in the absence of an isomerization catalyst. In the usual process for chloroprene using cuprous chloride, a portion of this isomer may be formed initially and then isomerize, but most of the chloroprene is apparently formed directly by the addition.

1-Chloro-1,3-butadiene [627-22-5] is present as an impurity by any of the synthetic methods customarily used. It is formed as a by-product in the hydrochlorination of vinylacetylene or in the dehydrochlorination of 3,4-dichloro-1-butene. It can also be introduced as an impurity from the chlorination of butadiene or formed by dehydrochlorination of 1,4-dichlorobutenes. Cis and trans isomers are both present in varying proportion. Both copolymerize with chloroprene but at a lower rate, so that monomer recovered from partial polymerization is less pure than the original mixture.

2,3-Dichloro-1,3-butadiene [1653-19-6] is a favored comonomer to decrease the regularity and crystallization of chloroprene polymers. It is one of the few monomers that will copolymerize with chloroprene at a satisfactory rate without severe inhibition. It is prepared from by-products or related intermediates. It is also prepared in several steps from chloroprene beginning with hydrochlorination. Subsequent chlorination to 2,3,4-trichloro-1-butene, followed by dehydrochlorination leads to the desired monomer in good yield if polymerization is prevented.

$$H_2C{=}CH{-}CCl{=}CH_2 + HCl \xrightarrow{(CuCl)} ClCH_2{-}CH{=}CCl{-}CH_3$$

$$ClCH_2{-}CH{=}CCl{-}CH_3 + Cl_2 \longrightarrow ClCH_2{-}CHCl{-}CCl{=}CH_2 + HCl$$

$$ClCH_2{-}CHCl{-}CCl{=}CH_2 + NaOH \longrightarrow H_2C{=}CCl{-}CCl{=}CH_2 + NaCl + H_2O$$

Similarly, preparation of dichlorobutenes from butadiene and Cl_2 is accompanied by formation of tetrachlorobutanes, which can be dehydrochlorinated to form the desired product.

$$ClCH_2{-}CH{=}CH{-}CH_2Cl + Cl_2 \longrightarrow ClCH_2{-}CHCl{-}CHCl{-}CH_2Cl$$

$$ClCH_2{-}CHCl{-}CHCl{-}CH_2Cl + 2\ NaOH \longrightarrow CH_2{=}CCl{-}CCl{=}CH_2 + 2\ NaCl + 2\ H_2O$$

It is preferable to prepare the 1,2,3,4-tetrachlorobutane [3405-32-1] from *trans*-1,4-dichloro-2-butene [110-57-6], which gives mainly the meso tetrachloride [28507-96-2], which in turn gives a better yield and higher isomeric purity of the resulting monomer (13).

Manufacture

The vinylacetylene route to chloroprene has been described elsewhere (14). It is no longer practical because of costs except where inexpensive by-product acetylene and existing equipment are available (see ACETYLENE-DERIVED CHEMICALS). In the production of chloroprene from butadiene, there are three essential steps, chlorination, isomerization, and caustic dehydrochlorination of the 3,3-dichloro-1-butene, as shown by the following equations:

Chlorination

$$CH_2{=}CH{-}CH{=}CH_2 + Cl_2 \rightarrow ClCH_2{-}CH{=}CH{-}CH_2Cl + CH_2{=}CH{-}CHCl{-}CH_2Cl$$

Isomerization

$$ClCH_2{-}CH{=}CH{-}CH_2Cl \xrightarrow{\text{CuCl}} CH_2{=}CH{-}CHCl{-}CH_2Cl$$

Dehydrochlorination

$$CH_2{=}CH{-}CHCl{-}CH_2Cl + NaOH \rightarrow CH_2{=}CH{-}CCl{=}CH_2 + NaCl + H_2O$$

Chlorination of Butadiene.   Butadiene and chlorine combine under almost any conditions; reaction occurs by a variety of mechanisms. In gas-phase chlorination, the process is a free-radical chain reaction and leads to a near-equilibrium mixture of 1,4-dichloro-*cis* and *trans*-2-butene isomers. The reaction is highly exothermic and relatively unselective. Good yields have been claimed (15–19) as long as a reasonable excess of butadiene is used, reactants are well mixed before appreciable reaction takes place, and temperature is maintained high enough to avoid condensation of liquid products but not so high that extensive dimerization of butadiene and carbonization occurs. Generally, one mole of chlorine is mixed with 4 to 15 moles of preheated butadiene or butadiene and diluent. The gases are allowed to react in a tubular (plug flow) reactor with or without surface cooling or in a partially back-mixed reactor followed by a plug-flow zone. Most of the reaction occurs at 200–350°C at 100–200 kPa (1–2 atm). Yields of 85 to 92% based on butadiene are expected. By-products include HCl, 1-chloro-1,3-butadiene [627-22-5], trichlorobutenes and tetrachlorobutanes, butadiene dimer, and higher boiling products, several of which involve two moles of butadiene to one of chlorine. The products are obtained by cooling the gas stream and distilling the condensate, recycling uncondensed excess butadiene for further reaction.

Liquid-phase chlorination of butadiene in hydroxylic or other polar solvents can be quite complicated in kinetics and lead to extensive formation of by-products that involve the solvent. In nonpolar solvents the reaction can be either free radical or polar in nature (20). The free-radical process results in excessive losses to tetrachlorobutanes if near-stoichiometric ratios of reactants are used or polymer if excess of butadiene is used. The "ionic" reaction, if a small amount of air is used to inhibit free radicals, can be quite slow in a highly purified system but is accelerated by small traces of practically any polar impurity. Pyridine, dipolar aprotic solvents, and oil-soluble ammonium chlorides have been used to improve the re-action (21). As a commercial process, the use of a solvent requires that the prod-ucts must be separated from solvent as well as from each other and the excess butadiene which is used, but high yields of the desired products can be obtained without formation of polymer at higher butadiene to chlorine ratio.

Refining and Isomerization. Whatever chlorination process is used, the crude product is separated by distillation. In successive steps, residual butadiene is stripped for recycle, impurities boiling between butadiene ($-5°C$) and 3,4-di-chloro-1-butene [760-23-6] (123°C) are separated and discarded, the 3,4 isomer is produced, and 1,4 isomers (140–150°C) are separated from higher boiling by-products. Distillation is typically carried out continuously at reduced pressure in corrosion-resistant columns. Ferrous materials are avoided because of catalytic effects of dissolved metal as well as unacceptable corrosion rates. Nickel is sat-isfactory as long as the process streams are kept extremely dry.

Commercial production of dichlorobutenes was originally established to pro-duce hexamethylene diamine by cyanating the 1,4-dichlorides and hydrogenating the product. Unless the 1,4 dichlorides are desired for this or other purposes, these must be isomerized to obtain a satisfactory yield of the desired isomer. When the isomers are heated with cuprous chloride and any of a variety of solubilizing agents (22), rapid exchange occurs and a vapor richer in the more volatile 3,4-dichloride can be removed. The crude condensed isomerization product can be mixed with the original chlorination product for refining or distilled separately, recycling residual 1,4-dichlorides to the isomerization process. The isomerization process is very nearly quantitative except for a small loss to 1-chloro-1,3-buta-diene. The equilibrium concentrations are approximately 21% 3,4; 7% cis-1,4; and 72% trans-1,4 isomers in the liquid phase, 52%, 6%, and 42% in the vapor phase.

Dehydrochlorination. In the dehydrochlorination process as first described (23), 3,4-dichloro-1-butene is added at a slow, continuous rate to a well-stirred reactor along with a sodium hydroxide solution at about 10 wt % concentration, typically at about 85°C, distilling off a chloroprene stream containing water and a small amount of unreacted dichlorobutene. At lower sodium hydroxide concen-tration, an undesirably large volume of waste salt solution is generated. At higher concentration, reaction is actually slower, presumably because of the limited sol-ubility of organics in the reaction medium. The maximum rate of reaction and effective conversion are related to the steam distillation temperature of the water-insoluble organic mixture as well as the conversion of NaOH to NaCl. Variations of this basic process that have been described include the use of intense agitation or surfactants, addition of various catalytic species, and decantation or stripping of products in a separate step so that reaction can be carried out more selectively at lower temperature (24–26). Reaction has also been described using butanol or other solvents to provide a homogeneous, low polarity medium for reaction (27). In each case, careful attention to inhibition of polymerization is required (28), and materials resistant to wet dichlorobutene corrosion must be used (29).

Except for the solvent process above, the crude product obtained is a mixture of chloroprene, residual dichlorobutene, dimers, and minor by-products. Depend-ing on the variant employed, this stream can be distilled either before or after decantation of water to separate chloroprene from the higher boiling impurities. When the concentration of 1-chloro-1,3-butadiene is in excess of that allowed for polymerization, more efficient distillation is required since the isomers differ by only about seven degrees in boiling point. The latter step may be combined with repurifying monomer recovered from polymerization. Reduced pressure is used for final purification of the monomer. All streams except final polymerization-grade monomer are inhibited to prevent polymerization.

Waste Disposal.   The waste brine from the dehydrochlorination process, after removal of separate-phase organics, is a relatively pure solution of sodium chloride and hydroxide in water. It can be electrochemically regenerated, recovering half of the chlorine initially used, or neutralized and purified further and discharged to naturally occurring seawater or underground brines. The organic wastes from the various purification steps are high in chlorine content and in some cases quite toxic. These are generally incinerated, scrubbing the flue gases to recover HCl.

Storage, Handling, and Shipment

Uninhibited chloroprene suitable for polymerization must be stored at low temperature ($<10°C$) under nitrogen if quality is to be maintained. Otherwise, dimers or oxidation products are formed and polymerization activity is unpredictable. Insoluble, autocatalytic "popcorn" polymer can also be formed at ambient or higher temperature without adequate inhibition. For longer term storage, inhibition is required. Phenothiazine [92-84-2], tert-butylcatechol [2743-78-1], picric acid [88-89-1], and the ammonium salt of N-nitroso-N-phenylhydroxylamine [135-20-6] have been recommended.

Because chloroprene is a flammable, polymerizable liquid with significant toxicity, it must be handled with care even in the laboratory. In commercial quantities, precaution must be taken against temperature rise from dimerization and polymerization and possible accumulation of explosive vapor concentrations. Storage vessels for inhibited monomer require adequate cooling capacity and vessel pressure relief facilities, with care that the latter are free of polymer deposits. When transportation of monomer is required, it is loaded cold ($<-10°C$) into sealed, insulated vessels with careful monitoring of loading and arrival temperature and duration of transit.

Economic Aspects

Chloroprene production can be approximately equated to the amount of polymer produced. Table 2 lists estimated volumes of dry polychloroprene consumed annually for the world excluding China, Russia, and Eastern Bloc countries. This should be increased about 10% to include polymer sold as latex, yield losses, and other uses. Somewhat lower figures given elsewhere (30) are apparently for free-world consumption, not including exports to Eastern Bloc countries. About half of this volume is consumed in the United States. Production in the United States is by Du Pont and Mobay. Production elsewhere is by Distugil (France), Bayer AG, and Knapsack-Griesheim AG (Germany), Denki Kagaku, Showa Denko, and Toyo Soda (Japan), as well as Du Pont (Northern Ireland). Production capacity

Table 2. World Production ^{a} of Dry
Polychloroprene

Year	Annual consumption, 10^3t
1940	2.6
1950	51
1960	135
1970	254
1980	314
1989	321

^{a}Excluding Russia, China, and Eastern Bloc
countries.

also exists in the former Soviet republics and China, but utilization is unknown
at this time.

Compared to natural rubber or general-purpose SBR synthetic rubber, polychloroprene [9010-98-4] has typically been 25 to 75% higher in cost, slightly more
when compounded formulations are calculated on a volume basis. Because of this
and unsatisfactory properties at low temperature except in specialized formulations, it has been used mainly where material cost is not controlling or where its
superior properties with regard to thermal stability, aging, flexing, and oil resistance, and resistance to abrasion are needed (see ELASTOMERS, SYNTHETIC–
POLYCHLOROPRENE). Principal areas of use are in molded goods, insulation of
electrical wiring, belts, and hoses. In recent years, much of the growth in these
markets, particularly wire coatings, has been filled with thermoplastic compositions of other polymers if the dimensional characteristics of a thermoset are not
needed. Polychloroprene foams have also found uses where an intermediate level
of fire protection is required. It is consumed if a sufficiently vigorous fire occurs,
but in contrast to general-purpose compositions, it chars and extinguishes instead
of melting and initiating a fire with some of the common sources of ignition.

Specifications and Quality Control

Polymerization-grade chloroprene is typically at least 99.5% pure, excluding inert
solvents that may be present. It must be substantially free of peroxides, polymer,
and inhibitors. A low, controlled concentration of inhibitor is sometimes specified.
It must also be free of impurities that are acidic or that will generate additional
acidity during emulsion polymerization. Typical impurities are 1-chlorobutadiene
and traces of chlorobutenes (from dehydrochlorination of dichlorobutanes produced from butenes in butadiene), 3,4-dichlorobutene, and dimers of both chloroprene and butadiene. Gas chromatography is used for analysis of volatile impurities. Dissolved polymer can be detected by turbidity after precipitation with
alcohol or determined gravimetrically. Inhibitors and dimers can interfere with
quantitative determination of polymer either by precipitation or evaporation if
significant amounts are present.

Safety and Health Factors

Uncontrolled polymerization and fire are the most important acute hazards in handling chloroprene. Flammable limits in air are 1.9 to 10%. It is detectable by odor at about 1 ppm in air, or lower if appreciable dimer impurities are present. A dose of several grams can produce anesthesia, but a single, brief exposure is not otherwise expected to result in lasting health effects. Long-term exposure must be more strictly controlled. Although chloroprene is mutagenic to bacteria, animal testing indicates that it does not have carcinogenic, embryotoxic, or reproductive effects. It is, however, physiologically active. Inhalation, ingestion, or absorption through the skin can result in reduced blood pressure, loss of appetite, headache, indigestion, or abnormal urine characteristics. Dermatitis, conjunctivitis, and hepatic or renal damage can also occur (31). The maximum exposure limit is 10 ppm in air. 1,3-Butadiene and 1,4-dichlorobutenes used in manufacture must be treated as suspect human carcinogens. The 1,4-dichlorobutenes are also acutely toxic and irritating, particularly to mucous membranes.

BIBLIOGRAPHY

"Chloroprene" under "Chlorocarbons and Chlorohydrocarbons" in *ECT* 2nd ed., Vol. 5, pp. 215–231, by P. S Bauchwitz; in *ECT* 3rd ed., Vol. 5, pp. 773–785, by P. R. Johnson, E. I. du Pont de Nemours & Co., Inc.

1. U.S. Pat. 1,950,431 (Mar. 13, 1934), W. H. Carothers and A. M. Collins (to E. I. du Pont de Nemours & Co., Inc.). Divinylacetylene is an extremely unstable substance that should not be handled except with adequate barricade protection.
2. W. H. Carothers, I. Williams, A. M. Collins, and E. J. Kirby, *J. Am. Chem. Soc.* **53,** 4203 (1931).
3. P. S. Bauchwitz, J. B. Finlay, and C. A. Stewart, Jr., in E. C. Leonard, ed., *Vinyl and Diene Monomers, Part II,* John Wiley & Sons, Inc., New York, 1971, pp. 1149–1183.
4. D A. C. Compton, W. O. George, J. E. Goodfield, and W. F. Maddams, *Spectrochim. Acta, Part A,* **37A**(3), 147–161 (1981).
5. A. A. Petrov and V. O. Babayan, *Zh. Obshch. Khim.* **34**(8), 2633 (1964).
6. D. D. Craig, J. J. Shipman, and R. B. Fowler, *J. Am. Chem. Soc.* **83,** 2885 (1961).
7. G. E. Ham, ed., *Copolymerization (High Polymers)* Vol. XVIII, Wiley-Interscience, New York, 1964.
8. R. Z. Greenley, *J. Macromol. Sci. Chem.* **A14**(4), 427–443 (1980). Greenley gives values of 10.5 and 1.2 for chloroprene, 1.7 and −0.5 for butadiene.
9. N. C. Billingham, P. A. Leeming, R. S. Lehrle, and J. C. Robb, *Nature* **213,** 494 (1967).
10. C. A. Stewart, Jr., *J. Am. Chem. Soc.* **93,** 4815 (1971); *Ibid.,* **94,** 635 (1972).
11. J. T. Maynard and W. E. Mochel, *J. Polym. Sci.* **13,** 235, 251 (1954).
12. C. A. Aufdermarsh, Jr. and R. Pariser, *J. Polym. Sci. A* **2,** 4727 (1964).
13. U.S. Pat. 3,901,950 (Aug. 29, 1975), J. H. Richards and C. A. Stewart, Jr. (to E. I. du Pont de Nemours & Co., Inc.).
14. Ref. 3, pp. 1151–1153.
15. P. M. Colling, "The Vapor Phase Chlorination of 1,3-Butadiene," Ph.D. dissertation, University of Texas, Austin, Tex., June 1963.
16. Brit. Pat. 798,027, 798,028 (July 16, 1958), F. J. Bellringer and H. P. Crocker (to Distillers Co., Ltd.).
17. Brit. Pat. 661,806 (Nov. 28, 1951), (to E. I. du Pont de Nemours & Co., Inc.).

18. Brit. Pat. 914,920 (Jan. 9, 1963), P. G. Caudle and H. P. Crocker (to Distillers Co., Ltd.).
19. A. J. Besozzi, W. H. Taylor, and C. W. Capp, *Preprint, Division of Petroleum Chemistry,* American Chemical Society, New York, Aug. 27, 1972.
20. M. L. Poutsma, *Science,* **157,** 997–1005 (1967).
21. Ger. Pat. 2,349,984 (Apr. 11, 1974), J. H. Richards and C. A. Stewart, Jr. (to E. I. du Pont de Nemours & Co., Inc.).
22. U.S. Pat. 3,819,730 (June 6, 1975), B. T. Nakata and E. D. Wilhoit (to E. I. du Pont de Nemours & Co., Inc.).
23. U.S. Pat 2,430,016 (Nov. 4, 1947), G. W. Hearne and D. S. LaFrance (to Shell Development Co.).
24. Ger. Pat. 1,212,513 (Mar. 17, 1966), R. Lauterbach and H. Schwarz (to Farbenfabriken Bayer AG).
25. U.S. Pat. 3,981,937 (Sept. 21, 1976), J. B. Campbell and R. E. Tarney (to E. I. du Pont de Nemours & Co., Inc.).
26. U.S. Pat. 3,755,476 (Aug. 28, 1973), J. W. Crary and R. E. Tarney (to E. I. du Pont de Nemours & Co., Inc.).
27. U.S. Pat. 4,104,316 (Aug. 1, 1978), G. Scharfe and R. Wenzel (to Bayer AG).
28. U.S. Pat. 2,948,761 (Aug. 9, 1960), P. A. Jenkins (to Distillers Co., Ltd.).
29. Brit. Pat. 985,289 (Mar. 3, 1965), C. W. Clapp, F. C. Newman, and F. J. Bellringer (to Distillers Co., Ltd.).
30. *Chem. Week* (May 8, 1991).
31. *Material Safety Data Sheet,* E. I. du Pont de Nemours & Co., Inc. (Oct. 16, 1989).

CLARE A. STEWART, JR.
Consultant

CHLORINATED PARAFFINS

Chlorinated paraffins with the general molecular formula $C_x H_{(2x-y+2)} Cl_y$ have been manufactured on a commercial basis for over 50 years. The early products werē based on paraffin wax feedstocks and were used as fire retardants and plasticizers in surface coatings and textile treatments and as extreme pressure–antiwear additives in lubricants. The development of chlorinated paraffins into new and emerging technologies was constrained principally because of the limitations of grades based on paraffin wax and the lack of suitable alternative feedstocks to meet the demands of the new potential markets.

In the early 1960s the petroleum industry employing molecular sieve technology made available a low cost and plentiful supply of normal paraffin fractions of very high purity. This enabled chlorinated paraffin manufacturers to exploit new applications with a range of products specifically designed to meet the technical and commercial requirements.

The principal feedstocks used today are the normal paraffin fractions C10–C13, C12–C14, C14–C17, and C18–C20 together with paraffin wax fractions of C24–C30, precise compositions may vary depending on petroleum oil source. Chlorination extent generally varies from 30 to 70% by weight. The choice of paraffinic feedstock and chlorine content is dependent on the application.

The availability of alpha olefins has enabled some manufacturers to offer a range of chlorinated alpha olefins alongside their existing range of chlorinated

paraffins. Chlorinated alpha olefins are virtually indistinguishable from chlorinated paraffins but do offer the manufacturer a single-carbon number paraffinic
feedstock and even greater flexibility in the product range.

Chemical and Physical Properties

By virtue of the nature of the paraffinic feedstocks readily available, commercial
chlorinated paraffins are mixtures rather than single substances. The degree of
chlorination is a matter of judgment by the manufacturers on the basis of their
perception of market requirements; as a result, chlorine contents may vary from
one manufacturer to another. However, customers purchasing requirements often
demand equivalent products from different suppliers and hence similar products
are widely available.

The physical and chemical properties of chlorinated paraffins are determined by the carbon chain length of the paraffin and the chlorine content. This
is most readily seen with respect to viscosity (Fig. 1) and volatility (Fig. 2); increasing carbon chain length and increasing chlorine content lead to an increase
in viscosity but a reduction in volatility.

Chlorinated paraffins vary in their physical form from free-flowing mobile
liquids to highly viscous glassy materials. Chlorination of paraffin wax (C24–C30)
to 70% chlorine and above yields the only solid grades. Physical properties of some
commercially available chlorinated paraffins are listed in Table 1.

A key property associated with chlorinated paraffins, particularly the high
chlorine grades, is nonflammability, which has led to their use as fire-retardant
additives and plasticizers in a wide range of polymeric materials. The fire-
retardant properties are considerably enhanced by the inclusion of antimony
trioxide.

Chlorinated paraffins are relatively inert and exhibit excellent resistance to
chemical attack and are hydrolytically stable. They are soluble in chlorinated

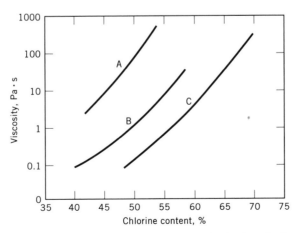

Fig. 1. Viscosity of chlorinated paraffins at 25°C. Paraffin feedstock: A, wax; B, C14–C17;
C, C10–C13. To convert Pa·s to P, multiply by 10.

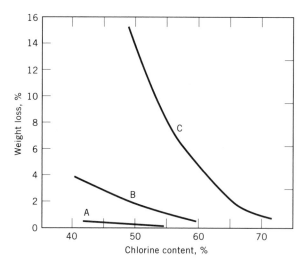

Fig. 2. Volatility of chlorinated paraffins at 180°C after four hours. Paraffin feedstock: A, wax; B, C14–C17; C, C10–C13.

Table 1. Physical Properties of Selected Commercial Chlorinated Paraffins

Paraffin carbon chain length	Nominal chlorine contents, %w/w	Color hazen (APHA)	Viscosity,[a] mPa·s (= cP)	Density,[a] g/mL	Thermal stability,[b] %w/w HCl	Volatility,[c] %w/w	Refractive index
C10–C13	50	100	80	1.19	0.15	16.0	1.493
	56	100	800	1.30	0.15	7.0	1.508
	60	135	3500	1.36	0.15	4.4	1.516
	63	125	11,000	1.41	0.15	3.2	1.522
	65	150	30,000	1.44	0.20	2.5	1.525
	70	200	800[d]	1.50	0.20	0.5	1.537
C14–C17	40	80	70	1.10	0.2	4.2	1.488
	45	80	200	1.16	0.2	2.8	1.498
	52	100	1600	1.25	0.2	1.4	1.508
	58	150	40,000	1.36	0.2	0.7	1.522
C18–C20	47	150	1700	1.21	0.2	0.8	1.506
	50	250	18,000	1.27	0.2	0.7	1.512
Wax	42	250	2500	1.16	0.2	0.4	1.506
>20	48	300	28,000	1.26	0.2	0.3	1.516
	70	100[e]	f	1.63	0.2		

[a]At 25°C unless otherwise noted.
[b]Measured in a standard test for four hours at 175°C.
[c]Measured in a standard test for four hours at 180°C.
[d]At 150°C
[e]10 g in 100 mL toluene solvent.
[f]Solid, softening point = 95–100°C.

solvents, aromatic hydrocarbons, esters, ketones, and ethers but only moderately soluble in aliphatic hydrocarbons and virtually insoluble in water and lower alcohols.

Although considered to possess good thermal stability, chlorinated paraffins, if held at high temperatures for prolonged periods, first darken in color and then release detectable quantities of hydrochloric acid. Manufacturers often quote a thermal stability index which is a measure of the quantity of hydrochloric acid released expressed as a percentage by weight after heating the product for four hours at 175°C. Degradation of chlorinated paraffins can also be accelerated at elevated temperatures in the presence of iron, zinc, and dehydrochlorination catalysts. The thermal stability of chlorinated paraffins can be improved by the inclusion of epoxidized compounds which typically are epoxy esters, antioxidants of the hindered phenol-type, and metal soaps.

Manufacture

Chlorinated paraffins are manufactured by passing pure chlorine gas into a liquid paraffin at a temperature between 80 and 100°C depending on the chain length of the paraffin feedstock. At these temperatures chlorination occurs exothermically and cooling is necessary to maintain the temperature at around 100°C. Catalysts are not usually necessary to initiate chlorination, but some manufacturers may assist the process with ultraviolet light. Failure to control the reactive exotherm during chlorination may lead to a colored and unstable product. The reaction is terminated by stopping the flow of chlorine when the desired degree of chlorination has been achieved. This is estimated by density, viscosity, or refractive index measurements. The reactor is then purged with air or nitrogen to remove excess chlorine and hydrochloric acid gas. Small quantities of a storage stabilizer, typically epoxidized vegetable oil, may be mixed in at this stage or later in a blending vessel.

In general terms, for each ton of chlorinated paraffin produced, approximately one-half ton of hydrochloric acid is generated. Thus materials of construction must be resistant to acid attack. Reactor vessels were traditionally lined with lead or ceramics but glass-lined mild steel is now preferred. Ancillary equipment such as stirrers, pumps, valves, and pipelines should be of corrosion/acid-resistant material. Good housekeeping is vital as minute traces of metal chlorides entering the process can cause dehydrochlorination leading to discoloration of the chlorinated paraffin. A typical system employed in commercial production is shown in Figure 3.

In order to operate an economically viable chlorinated paraffin business, it is essential to have a profitable outlet for the surplus hydrochloric acid, either through direct sales into the market, or preferably via an oxychlorination unit in an integrated vinyl chloride/chlorinated solvent unit, while still maintaining the option of direct sales.

Applications

Chlorinated paraffins are versatile materials and are used in widely differing applications. As cost-effective plasticizers, they are employed in plastics particu-

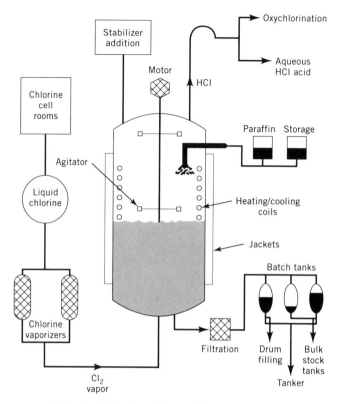

Fig. 3. Chlorinated paraffin manufacture.

larly PVC, rubbers, surface coatings, adhesives, and sealants. Where required they impart the additional features of fire retardance, and chemical and water resistance. In conjunction with antimony trioxide, they constitute one of the most cost-effective fire-retardant systems for polymeric materials, textiles, surface coatings, and paper products. Chlorinated paraffins are also employed as components in fat liquors used in the leather industry, as extreme pressure additives in metal-working lubricants, and as solvents in carbonless copying paper.

Plasticized PVC. Chlorinated paraffins are employed as secondary plasticizers with fire-retardant properties in PVC and can be used as partial replacements for primary plasticizers (qv) such as phthalates (1) and phosphate esters (2).

By selection of those chlorinated paraffins specifically developed for the PVC industry to match the properties of primary plasticizers, reductions in costs can be achieved without significant change in properties. However, certain aspects can be improved by the inclusion of chlorinated paraffin such as flame resistance, chemical and water resistance, low temperature performance, and the viscosity aging stability in plastisols.

Metal-Working Lubricants. A range of chlorinated paraffins are used as components of straight and emulsifiable metal-working lubricants as well as gear oils for industrial and automotive applications. In heavy-duty industrial gears, hypoid gears, metal cutting, and allied operations where high pressures and rubbing action are encountered, hydrodynamic lubrication cannot be maintained. In order to maintain lubrication under such conditions, extreme pressure (EP) or antiwear additives must be added to the lubricant. Such additives contain one or more of the elements of chlorine, sulfur, or phosphorus. Chlorinated paraffins are cost-effective extreme pressure additives and are either used alone or in combination with additives containing sulfur and/or phosphorus according to the application. High chlorine content lubricants are used for severe metal-forming operations such as deep drawing and stamping. This area is the principal outlet for chlorinated paraffins in the United States, accounting for approximately 50% of total sales.

The selection of chlorinated paraffin and the level of additives to a lubricating oil depends on the type of application and the severity of the operation. An approximate guide for the formulation of straight-cutting oils for metal-working is as follows:

Fluid type	Additives
heavy-duty broaching	60% chlorine + 10% fatty acid
very heavy-duty cutting	30% chlorine + 1% sulfur + 10% fatty acid
medium to heavy-duty	5–10% chlorine + 1–0.5% sulfur + 1–5% fatty oil
light to medium-duty	2–5% chlorine + 0.2% sulfur + 1–5% fatty oil

Metal-forming operations such as deep drawing, stamping, wire drawing, etc, are extremely severe and require large amounts of chlorinated paraffins; often the high chlorine containing grades are preferred. In some applications a mid-range chlorine content grade may be used neat. After deep drawing and stamping of mild steel, components are frequently left unprotected in storage. This can result in corrosion problems. For these applications chlorinated paraffins containing corrosion inhibitors and special stabilizers are available.

Paints. Chlorinated paraffins are used as plasticizers for paints based on many types of resins, particularly chlorinated rubber and vinyl copolymers. Chlorinated rubber-based paints are employed in aggressive marine and industrial environments and vinyl copolymer principally for the protection of exterior masonry. The excellent chemical resistance of chlorinated paraffins and their ability to withstand prolonged contact with water makes them ideally suited as plasticizers for the most demanding applications.

Adhesives and Sealants. Various grades of chlorinated paraffins are used as nonvolatile inert fire-retardant plasticizers and modifying resins in adhesives and sealants (3). They find wide application in polysulfide, polyurethane, acrylic, and butyl sealants for use in building and construction. The low volatility high chlorine types are also employed in sealants for double- and triple-glazed windows.

Flame-Retardant Applications. The flame resistance of polyolefins, unsaturated polyester, rubber, and many other synthetic materials can be improved by the inclusion of chlorinated paraffins. The solid 70% chlorine product is the preferred choice in most polymeric systems, but the liquid grades are widely used in rubbers, polyurethane, and textile treatments.

Chlorinated paraffins and modified types are used as solvents in carbonless copying paper production based on the encapsulation of a solution of reactive dyes. Chlorinated paraffins fullfill the technical requirements for a solvent including excellent solvency for the dyes; they do not react with the dyes nor encapsulation material, are immiscible with water, and have low volatility and low odor.

Fat Liquors for Leather.   The addition of a chlorinated paraffin to a sulfated or sulfonated oil offers alternatives to natural oils as fat liquors for leather.

Storage and Transportation

Liquid chlorinated paraffins are shipped in drums usually lacquer-lined mild steel or polyethylene and in road or rail barrels. Where appropriate larger quantities can be shipped by sea either in deck tanks of conventional cargo ships or in chemical parcel tankers for larger consignments.

The high viscosity of a number of grades generally precludes consideration of bulk supplies unless special transport, heating, pumping, and storage can be made.

Road and rail barrels are usually constructed of stainless steel or lacquer-lined mild steel and may require some provision for heating during cold weather. This is best achieved by submerged coils circulating hot water at about 40°C or low pressure steam, but care must be taken to ensure the surface temperature of the coils does not rise excessively otherwise discoloration of the product may occur.

The main points to be considered when designing a bulk storage installation are (1) the viscosity of all grades of chlorinated paraffin varies sharply with a change in temperature; (2) chlorinated paraffins should not be exposed to temperatures in excess of 40°C for prolonged periods of time; (3) chlorinated paraffin stability can be affected by contact with zinc and iron, therefore tanks of mild steel should generally be lined and galvanized steel avoided. Stainless steel, lacquer, or glass-lined mild steel tanks are recommended. The preferred linings are of the heat-cured phenol–formaldehyde type; and (4) chlorinated paraffins swell certain types of rubber and therefore rubber joints should be avoided. Polytetrafluorethylene joints are recommended.

As for storage tanks, stainless steel and lacquer-lined mild steel are suitable materials of construction for pipe lines. For pumps, valves, etc, various alloys are suitable, including phosphor bronze, gun metal, Monel, stainless steel, and certain nickel steel alloys. Alloys with high proportions of zinc and tin together with copper and aluminum are not recommended.

Economic Aspects

The global market for chlorinated paraffins excluding the former Soviet Union and the People's Republic of China is around 300,000 t.

The largest single market is the United States at approximately 40,000 t. Europe as a whole is approximately two and a half times greater than the United States mainly because of the extensive use of chlorinated paraffins as secondary plasticizers in plasticized PVC, which is virtually absent in the United States.

Although chlorinated paraffins are manufactured throughout the world, the principal centers of production are Europe, North America, and Japan. ICI is the leading supplier having its main production centered in Europe, but it has a number of plants around the world including North America.

During the 1980s some rationalization of capacity occurred in the United States reducing the number of producers to four with the purchase of Neveille's chlorinated paraffin business by Dover (ICC). In Europe, Dynamit Nobel ceased as a supplier. However, elsewhere new companies have commenced manufacture to take advantage of the opportunities in new and emerging economies, particularly in Asia. Some suppliers of chlorinated paraffins are listed in Table 2.

Table 2. Chlorinated Paraffin Suppliers

Company	Location
Europe	
ICI	UK, Germany, France, Spain
Hoechst	Germany
Hüls	Germany
BASF Schwarzheide	Germany
Caffaro	Italy
(ATO) Rio Rodano	Spain
Rhône Poulenc	France
M&T[a]	France
North America	
Occidental Chemical	United States
Dover Chemical (ICC)	United States
Keil Chemical Div Ferro Corp.	United States
Witco Corp.	United States
ICI	Canada
Plasticlor SA	Mexico
Others	
Ajinomoto	Japan
Tosoh	Japan
Ashai Denka	Japan
Hardy	Taiwan
Koruma	Turkey
NCP	South Africa
AECI	South Africa
ICI	Australia
Hoechst	Brazil

[a]Solid grade only.

In the United States approximately 50% of the 40,000 t of chlorinated paraffins consumed domestically are used in metal-working lubricants. Approximately 20% are consumed as plastic additives, mainly fire retardants, and similarly 12% in rubber. The remainder as plasticizers in paint (9%) and caulks, adhesives, and sealants at 6%.

Health and Safety Factors

A substantial body of information on the toxicological and environmental effects of chlorinated paraffins has been compiled over the past 20 years, and research is still continuing in both areas.

Toxicity. The acute toxicity of chlorinated paraffins has been tested in a range of animals and was found to be very low (4). A comprehensive study (5) demonstrated that the toxicity of chlorinated paraffins was related to carbon chain length and to a lesser degree chlorine content. The shorter chain-length chlorinated paraffins were more toxic than the longer chain chlorinated paraffins.

Subchronic studies in mice, rats, and rabbits determined the liver as the primary organ for attack by chlorinated paraffins. Mutagenicity, and reproductive and teratology studies revealed no abnormal effects. However, the National Toxicology Program (NTP) in the United States concluded that there was sufficient evidence of carcinogenicity from lifetime studies in rats and mice with one chlorinated paraffin having chain length of 12 carbon atoms and chlorine content 58% by weight. It is listed in the *Fifth Annual Report on Carcinogens*. Parallel studies on a long-chain chlorinated paraffin of average chain length (23 carbon atoms and 42% chlorine) showed no statistical increase in tumors. The NTP studies were reviewed by the International Agency for Research on Cancer (IARC) who concluded that the short-chain chlorinated paraffin was a possible human carcinogen (Cat II B). More recently, an extensive series of experiments (6) have been conducted to study further the biochemistry of the carcinogenic effects of the short-chain chlorinated paraffin C12 and 58% chlorine in rats and mice. The results support an earlier hypothesis (4) that the mechanism responsible for the occurrence of tumors in the liver of rats and mice is of a nongenotoxic nature and is associated with liver growth. This work shows that the effect is accompanied by peroxisome proliferation, and these effects are unlikely to occur in humans.

Because of the nature of some applications in which chlorinated paraffins are used, skin contact is inevitable and therefore an important potential route into the body. Skin absorption studies (7) have shown that chlorinated paraffins are very poorly absorbed through the skin and should not cause significant systemic concentrations.

Environmental. In general, chlorinated paraffins biodegrade; the rate is determined by chlorine content and carbon chain length. Microorganisms previously acclimatized to specific chlorinated paraffins show a greater ability to degrade the compounds than nonacclimatized organisms. Mammals and fish have been shown to metabolize chlorinated paraffins (8).

The acute toxicity of chlorinated paraffins to mammals, birds, and fish is very low (8), but over longer periods of exposure certain chlorinated paraffins have proved to be toxic to some aquatic species. However, the very low water solubility of chlorinated paraffins has made studies on aquatic species complicated. Laboratory experiments in which the chlorinated paraffins had been artificially solubilized showed only the short-chain grades to be toxic at low concentration; other longer-chain grades showed no adverse effects on the majority of aquatic species tested. The degree of solubilization achieved in the laboratory is unlikely ever to be experienced in the environment and is of doubtful environmental relevance (9).

In the United States further information and advice is readily available from the Chlorinated Paraffin Manufacturers Association (CPIA) based in Washington D.C.

BIBLIOGRAPHY

"Chlorinated Paraffins" under "Chlorine Compounds, Organic" in *ECT* 1st ed., Vol. 3, pp. 781–786, by H. M. Roberts, Imperial Chemical Industries Ltd.; "Chlorinated Paraffins" under "Chlorocarbons and Chlorohydrocarbons" in *ECT* 2nd ed., Vol. 5, pp. 231–240, by D. W. F. Hardie, Imperial Chemical Industries Ltd.; in *ECT* 3rd ed., Vol. 5, pp. 786–791, by B. A. Schenker, Diamond Shamrock Corp.

1. H. J. Caesar "Chlorinated Paraffins as Secondary Plasticizers in PVC," *Chem. Ind.* (Aug. 1978).
2. H. J. Caesar and P. J. Davis "Flame Retardant Vinyl Compounds," *33rd Annual Technical Conference,* Atlanta, Ga., May 6, 1975.
3. K. L. Houghton and M. E. Moss "Chlorinated Paraffins as Plasticizers in Polymer Sealant Systems," *ASC Supplier Short Course,* Nashville, Tenn., May 14–17, 1990.
4. R. D. N. Birtley and co-workers, *Toxicol. Appl. Pharmacol.,* **54,** 514 (1980).
5. D. M. Serrone and co-workers, *Fd. Chem. Toxic.* **25**(7), 553–562 (1987).
6. C. R. Elcombe and co-workers, *Mutagenesis* **5**(5), 515–518 (1990).
7. R. C. Scott, *Arch. Toxicol.,* (63), 425–426 (1989).
8. J. R. Madely and R. D. N. Birtley, *Environ. Sci. Technol.* **14,** 1215 (1980).
9. I. Campbell and G. McConnell, *Environ. Sci. Technol.* **14,** 1209 (1980).

KELVIN L. HOUGHTON
ICI Chemicals and Polymers Ltd.

CHLORINATED BENZENES

The chlorination of benzene can theoretically produce 12 different chlorobenzenes. With the exception of 1,3-dichlorobenzene, 1,3,5-trichlorobenzene, and 1,2,3,5-tetrachlorobenzene, all of the compounds are produced readily by chlorinating benzene in the presence of a Friedel-Crafts catalyst (see FRIEDEL-CRAFTS REACTIONS). The usual catalyst is ferric chloride either as such or generated *in situ* by exposing a large surface of iron to the liquid being chlorinated. With the exception of hexachlorobenzene, each compound can be further chlorinated;

therefore, the finished product is always a mixture of chlorobenzenes. Refined products are obtained by distillation and crystallization.

Chlorobenzenes were first synthesized around the middle of the nineteenth century; the first direct chlorination of benzene was reported in 1905 (1). Commercial production was begun in 1909 by the former United Alkali Co. in England (2). In 1915, the Hooker Electrochemical Co. at Niagara Falls, New York, brought on stream its first chlorobenzenes plant in the United States with a capacity of about 8200 metric tons per year.

The Dow Chemical Company started production of chlorobenzenes in 1915 (3). Chlorobenzene was the first and remained the dominant commercial product for over 50 years with large quantities being used during World War I to produce the military explosive picric acid [88-89-1].

The Dow Chemical Company in the mid-1920s developed two processes which consumed large quantities of chlorobenzene. In one process, chlorobenzene was hydrolyzed with ammonium hyroxide in the presence of a copper catalyst to produce aniline [62-53-3]. This process was used for more than 30 years. The other process hydrolyzed chlorobenzene with sodium hydroxide under high temperature and pressure conditions (4,5) to product phenol [108-95-2]. The I.G. Farbenwerke in Germany independently developed an equivalent process and plants were built in several European countries after World War II. The ICI plant in England operated until its closing in 1965.

In the 1930s, the Raschig Co. in Germany developed a different chlorobenzene–phenol process in which steam with a calcium phosphate catalyst was used to hydrolyze chlorobenzene to produce phenol (qv) and HCl (6). The recovered HCl reacts with air and benzene over a copper catalyst (Deacon Catalyst) to produce chlorobenzene and water (7,8). In the United States, a similar process was developed by the Bakelite Division of Union Carbide Corp., which operated for many years. The Durez Co. licensed the Raschig process and built a plant in the United States which was later taken over by the Hooker Chemical Corp. who made significant process improvements.

Although Dow's phenol process utilized hydrolysis of the chlorobenzene, a reaction studied extensively (9,10), phenol production from cumene (qv) became the dominant process, and the chlorobenzene hydrolysis processes were discontinued.

With the discontinuation of some herbicides, eg, 2,4,5-trichlorophenol [39399-44-5], based on the higher chlorinated benzenes, and DDT, based on monochlorobenzene, both for ecological reasons, the production of chlorinated benzenes has been reduced to just three with large-volume applications of (mono)chlorobenzene, o-dichlorobenzene, and p-dichlorobenzene. Monochlorobenzene remains a large-volume product, considerably larger than the other chlorobenzenes, in spite of the reduction demanded by the discontinuation of DDT.

Physical and Chemical Properties

The important physical properties of chlorobenzenes appear in Table 1. Only limited information is available for some chlorobenzenes:

Table 1. Physical Properties of Chlorobenzenes

	Chloro-benzene [108-90-7]	1,2-Di-chloro-benzene [95-50-1]	1,3-Di-chloro-benzene [541-73-1]	1,4-Di-chloro-benzene [106-46-7]	1,2,4-Tri-chloro-benzene [120-82-1]	1,2,3,4-Tetra-chloro-benzene [634-66-2]	1,2,4,5-Tetra-chloro-benzene [95-94-2]	Hexachloro-benzene [1118-74-1]
CAS Registry Number								
mol wt	112.56	147.005	147.005	147.005	181.45	215.90	215.9	284.80
mp, °C	−45.34	−16.97	−24.76	53.04	17.15	46.0	139.5	228.7
bp at 101.3 kPa[a], °C	131.7	180.4	173.0	174.1	213.8	254.9	248.0	319.3
critical temperature, °C	359.2	417.2	415.3	407.5	453.3	450	489.8	551
critical pressure, kPa[a]	4519	4031	4864	4109	3718	3380	3380	2847
critical density, kg/L	0.3655	0.411	0.458	0.411	0.447	0.40	0.475	0.518
liquid density, kg/L	1.10118	1.3022	1.2828	1.2475	1.44829	1.70	1.833(s)	1.596
viscosity, mPa·s (=cP)	0.756	1.3018	1.0254			3.37		
heat capacity for liquid, J/g[b]	1.339	1.159		1.188	1.008	1.259	1.142	
heat of fusion, J/g[b]	90.33	86.11	85.98	123.8	85.78	64.52	112.2	89.62
heat of vaporization, J/g[b]	331.1	311.0	296.8	297.4	280.0	268.9	221.8	190.8
flash point[c], °C	28	71		67	99		none	
standard heat of formation of liquid[d], J/g[b]	−95.90	−125.23	−145.73	−284.6(s)	−263.1			−460(s)
thermal conductivity of liquid, W/(m·K)	0.127	0.121		0.105	0.108			
refractive index of liquid, n_D^{25}	1.5219	1.5492	1.54337	1.52849 (55°C)	1.56933			
dielectric constant of liquid	5.621	9.93	5.04	2.41	2.24			
surface tension, mN/m (=dyn/cm)	32.65	36.61	36.20	31.4	38.54	21.6		

[a] To convert kPa to mm Hg, multiply by 7.5.
[b] To convert J to cal, divide by 4.184.
[c] ASTM method D56-70, closed cup.
[d] Ref. 11.

89

Chlorobenzene	CAS Registry Number	Mol wt	Melting point, °C	Normal boiling point, °C
1,2,3-trichloro	[87-61-6]	181.45	53.5	218.5
1,3,5-trichloro	[108-70-3]	181.45	63.5	208.5
1,2,3,5-tetrachloro	[634-90-2]	215.9	51	246
pentachloro	[608-93-5]	250.35	85	276

Vapor pressure as a function of the temperature is correlated by the Antoine equation:

$$\log_{10} P(\text{kPa}) = A - B/(T + C) - 0.875097 \qquad (1)$$

$$(\log_{10} P(\text{mm Hg}) = A - B/(T + C)$$

where T is the temperature in °C, and A,B,C are the Antoine constants (Table 2).

Table 2. Antoine Constants[a] for Chlorobenzenes

Chlorobenzene	A	B	C
monochlorobenzene	7.046324	1482.156	224.115
1,2-dichlorobenzene	7.143024	1703.916	219.352
1,3-dichlorobenzene	7.072644	1629.811	215.821
1,4-dichlorobenzene	7.002424	1578.149	208.84
1,2,4-trichlorobenzene	7.136684	1790.267	206.283
1,2,3,4-tetrachlorobenzene	7.159274	1930.023	196.213
1,2,4,5-tetrachlorobenzene	7.284164	2003.495	207.038
hexachlorobenzene	6.66747	1654.17	117.536

[a]See equation 1; 1 kPa = 7.5 mm Hg; log kPa = log mm Hg − 0.875097.

Nitration of chlorobenzenes, mostly monochlorobenzene in the United States, with nitric acid has wide industrial applications.

$$\text{Cl}_n\text{C}_6\text{H}_{6-n} + \text{HNO}_3 \rightarrow \text{Cl}_n\text{C}_6\text{H}_{5-n}\text{NO}_2 + \text{H}_2\text{O}$$

$$n = 1 \text{ to } 5$$

Nitrated monochlorobenzene is used as a building block to produce many other products. There is also some commercial nitration of o-dichlorobenzene in the United States and Western Europe.

Manufacture

The production of any chlorinated benzene is a multiple-product operation. Plants for chlorobenzene must produce HCl and some other chlorinated benzenes. Only limited control can be exercised over the product ratios. Chlorinated benzenes can be produced by the vapor-phase chlorination of benzene using air and HCl as chlorinating agents. This was the first stage of the Raschig phenol process (7,8). The energy costs are so high, this process could never have been considered in the past to commercially produce chlorobenzenes as main products. Chlorine and benzene react in the vapor phase at 400 to 500°C to give a different distribution of products (12), but such a process is much more costly than conventional liquid-phase operations.

All of the chlorobenzenes are now produced by chlorination of benzene in the liquid phase. Ferric chloride is the most common catalyst. Although precautions are taken to keep water out of the system, it is possible that the $FeCl_3 \cdot H_2O$ complex catalyst is present in most operations owing to traces of moisture in benzene entering the reactor. This $FeCl_3 \cdot H_2O$ complex is probably the most effective catalyst (13).

The liquid-phase chlorination of benzene is an ideal example of a set of sequential reactions with varying rates from the single-chlorinated molecule to the completely chlorinated molecule containing six chlorines. Classical papers have modeled the chlorination of benzene through the dichlorobenzenes (14,15). A reactor system may be simulated with the relative rate equations and flow equation. The batch reactor gives the minimum ratio of $(C_6H_{5-n}Cl_{n+1}:(C_6H_{6-n})Cl_n$. This can be approximated by either a plug flow reactor or a multistage stirred reactor. A single-stage, stirred reactor will produce the highest $(C_6H_{5-n})Cl_{n+1}:(C_6H_{6-n})Cl_n$ ratio. If chlorobenzene (mono) is the desired product, control over the dichlorobenzenes to chlorobenzene ratio is effected primarily by controlling the extent of chlorination. The low di:mono ratio is obtained at the expense of energy used in recycling the unreacted benzene.

In the liquid-phase chlorination, 1,3-dichlorobenzene is found only in a small quantity, and 1,3,5-trichlorobenzene and 1,2,3,5-tetrachlorobenzene are undetectable. The ratios of 1,4- to 1,2-dichlorobenzene with various catalysts are shown in Table 3. Iodine plus antimony trichloride is effective in selectively chlorinating 1,2,4-trichlorobenzene to 1,2,4,5-tetrachlorobenzene (22), however, 1,2,4,5-tetrachlorobenzene is of limited commercial significance.

The chlorination reaction is exothermic. The heat liberated is about 1.83 kJ/g Cl_2 (437 cal/g Cl_2). Heat is removed in some cases by circulating the reaction liquid through a suitable cooler (see HEAT EXCHANGE TECHNOLOGY). In other cases, chlorination occurs at the boiling point. The heat of the reaction is removed from the reactor by the vaporizing liquid. The latter procedure has the disadvantage

Table 3. Ratio of 1,4- to 1,2-Dichlorobenzene with Various Catalysts

Catalyst	Ratio	Reference
$FeCl_3$	1.4	
several	1.2–1.7	16
$FeCl_3$, $AlCl_3$, or $SbCl_3$ and organic sulfur compounds	2.2–3.3	17
FeS and organic sulfur compounds	2.0–2.65	18
$FeCl_3$ or $SbCl_3$ and sulfur	2–4	19
$SbCl_3$ and sulfur	3–5	20
$SnCl_4$ and/or $TiCl_4$ and $AlCl_3$	2.3	21

of operating at a higher temperature but has the advantage of allowing a low inventory reactor system, which saves equipment costs, reduces operating hazards, and makes heat recovery possible.

Benzene chlorination reactors are subject to design and operating hazards. Stagnant areas must be avoided in reactor design as they allow chlorination to the tetra- and pentachlorobenzenes. These compounds have low solubility in the liquid and can cause plugging. Another hazard is the equivalent of spontaneous combustion. The temperature can rise locally to a point where the reaction $C_6H_6 + 3\ Cl_2 \rightarrow 6\ C + 6\ HCl$ can occur, primarily in the vapor phase. The exothermic reaction proceeds out of control and releases large amounts of HCl gas. This phenomenon can also occur when the chlorine concentration builds up in the reactor if the normal chlorination catalyst is inactivated by a cause such as an operational error that allows a sudden input of water.

Because HCl is constantly present in most parts of the equipment, corrosion is always a potential problem. Chlorine and benzene, or any recycled material, must be free of water to trace amounts to prevent corrosion and deactivation of the catalyst. The reactor product contains HCl and iron. In some plants, the product is neutralized with aqueous NaOH before distillation. In others, it is handled in a suitably-designed distillation train, which includes a final residue from which $FeCl_3$ can be removed with the high boiling tars.

Chlorobenzene mixtures behave in distillation as ideal solutions. In a continuous distillation train, heat may be conserved by using the condensers from some units as the reboilers for others thereby, saving process energy.

The dichlorobenzene isomers have very similar vapor pressures making separation by distillation difficult. Crystallization is generally used in combination with distillation to obtain the pure 1,2 and 1,4-dichlorobenzene isomers. The small quantity of 1,3-dichlorobenzene isomer produced is not generally isolated as a pure product. Environmental concerns have led to the use of improved crystallization systems that contain the products with minimal losses to the environment.

HCl is a constant by-product in the manufacture of chlorobenzenes. It is usually recovered by passing the gas stream through a scrubber tower over which a reactor mixture containing chlorination catalyst is circulated. This removes any unreacted chlorine that may have passed through the reactors. The HCl is then passed through one or more scrubbing towers in which high boiling chlorobenzenes are used as the solvent to remove the organic content. The absorbent in the final tower is refrigerated to the lowest possible temperature.

The HCl gas is absorbed in water to produce 30–40% HCl solution. If the HCl must meet a very low organic content specification, a charcoal bed is used ahead of the HCl absorber, or the aqueous HCl solution product is treated with charcoal. Alternatively, the reactor gas can be compressed and passed to a distillation column with anhydrous 100% liquid HCl as the distillate; the organic materials are the bottoms and are recirculated to the process. Any noncondensible gas present in the HCl feed stream is vented from the distillation system and scrubbed with water.

Any plant at times produces unwanted isomers. This requires an incinerator, capable of burning chlorinated hydrocarbons to HCl, H_2O, and CO_2 equipped with an efficient absorber for HCl (see INCINERATORS). An alternative to burning is dechlorination using hydrogen over a suitable catalyst. The ultimate product could be benzene.

$$C_6H_{6-n}Cl_n + n\,H_2 \rightarrow C_6H_6 + n\,HCl$$

Dechlorination can be done in the vapor phase with palladium, platinum, copper, or nickel catalysts (23–26) or in the liquid phase with palladium catalysts (27). The vapor-phase dechlorination of 1,2,4-trichlorobenzene is reported to give good yields of 1,3-dichlorobenzene (24,26).

Another alternative to burning is rearrangement of the undesired isomers. This technique is practiced extensively in the petroleum industry, for example, in the production of xylene isomers using an HF–BF$_3$ catalyst-extractor system (28) (see BTX PROCESSING). Polychlorinated benzenes are considerably more resistant to rearrangement than are the isomeric hydrocarbon mixtures. Some patents have been issued to cover rearrangements using an aluminum chloride catalyst (29–31). A HF–SbCl$_5$ catalyst system is also reported to be effective in converting dichlorobenzenes to 1,3-dichlorobenzene (32). To date, there have been no reported commercial operations using these technologies.

Storage, Shipment, and Handling

Chlorobenzenes are stored in manufacturing plants in liquid form in steel containers. Mono-, 1,2-di- and 1,2,4-trichlorobenzenes are liquids at room temperature and are shipped in bulk in aluminum tank trucks and steel or stainless steel tank cars. In situations where chlorobenzenes are contained in aluminum at elevated temperatures, the product must be clean, ie, nonacidic and the moisture, low. The use of aluminum with a mixture of chlorobenzenes and strong oxidizers should be avoided. Mixtures of orthodichlorobenzene and chlorinated olefins react with aluminum, leading to catastrophic failure of aluminum tanks containing such a mixture. 1,4-Dichlorobenzene is shipped either in molten form in insulated steel tank cars with heater coils, or as flake or granular solid in suitably sealed containers, such as paper bags, fiber packs, or drums. Phenolic linings in all vessels offer protection over a wide range of conditions for all chlorobenzenes as well as the vessels themselves. For drums, the phenolic coating should be modified with epoxy for maximum impact resistance. 1,4-Dichlorobenzene has different labeling classifications depending on its intended use. Regulatory requirements change so the latest regulations should be checked and observed.

Chlorobenzenes are generally considered nonflammable materials with the exception of monochlorobenzene, which has a flash point of 34.5°C and is a flammable solvent based on DOT standards.

Chlorobenzenes are stable compounds and decompose slowly only under excess heating at high temperatures to release some HCl gas and traces of phosgene. It is possible, under certain limited conditions of incomplete combustion or pyrolysis, to form polychlorinated dibenzo-p-dioxins (PCDDs) and dibenzofurans (PCDFs) from chlorobenzenes (CHLOROCARBONS AND CHLOROHYDROCARBONS, TOXIC AROMATICS).

Health and Safety Aspects

In general, all of the chlorobenzenes are less toxic than benzene. Liquid chlorobenzenes produce mild to moderate irritation upon skin contact. Continued contact may cause roughness or a mild burn. Solids cause only mild irritation. Absorption through the skin is slow. Consequently, with short-time exposure over a limited area, no significant quantities enter the body.

Contact with eye tissue at normal temperatures causes pain, mild to moderate irritation, and possibly some transient corneal injury. Prompt washing with large quantities of water is helpful in minimizing the adverse effects of eye exposure.

The data from some single-dosage oral toxicity tests, expressed as LD_{50}, are reported in Table 4. The values reported on the order of 1 g/kg or greater indicate a low acute oral toxicity. In animals, continued ingestion of chlorobenzenes over a long time can cause kidney and liver damage.

The threshold limit value (TLV), the vapor concentration in ppm by volume, to which humans may be exposed for an eight-hour working day for many years without adverse effects, is also reported in Table 4. The saturated vapor concentration of the chlorobenzenes at 20°C listed in Table 4 are well above the TLV values; therefore well-designed ventilation is required for working areas. A few kidney and liver damage cases reported may have been caused by repeated exposure to some chlorobenzenes. Fires involving chlorobenzenes liberate HCl and possibly phosgene. Under certain limited conditions of incomplete combustion or pyrolysis, it has been reported that chlorobenzenes form 2,3,7,8-tetrachlorodibenzo-p-dioxin. When chlorobenzenes are involved in a fire, the proper protective equipment must be used for personnel involved in fighting the fire.

Table 4. Toxicity of Chlorinated Benzenes

Chlorobenzene	Fish[a] toxicity, mg/L[b]	LD_{50}, g/kg	TLV (inhal), ppm[c]	Saturated concentration, ppm by vol at 20°C
monochloro	<3[d,e]	2.9 (rat)	75	11,900
	16	2.8 (rabbit)		
1,2-dichloro	3	0.8–2.0 (guinea pig)	50	1,125
1,4-dichloro	0.7[e,f]	3.8 (rat)	75	1,570
	5	2.8 (guinea pig)		
1,2,4-trichloro	2	1 (rat)	[g]	260
1,2,4,5-tetrachloro	<1	<1 (rat)	[g]	

[a]Fathead unless otherwise noted.
[b]No observed adverse effect at this concentration in H_2O; 72 h static test (33) unless otherwise noted.
[c]Volume per volume of air.
[d]Rainbow trout.
[e]96 h dynamic test (33).
[f]Bluegill.
[g]No TLV suggested.

Toxicity to fish is included in the data listed in Table 4. Marine life, particularly fish, may suffer damage from spills in lakes and streams. The chlorobenzenes, because they are denser than water, tend to sink to the bottom and may persist in the area for a long time. However, some data indicate that dissolved 1,2,4-trichlorobenzene can be biodegraded by microorganisms from wastewater treatment plants and also has a tendency to slowly dissipate from water by volatilization (34).

Most recently, the main health emphasis has been on carcinogenic potential. *p*-Dichlorobenzene was listed by the National Toxicology Program (NTP) as a product that could possible cause cancer. The results were based on mice and rats having large quantities of the chemical forced into their stomachs. The U.S. EPA subsequently reviewed the data and observed that there is no evidence of para-induced cancer. It is believed that the chemical complexes with a protein found only in the male rat, thereby not likely to pose a risk to humans. The male rat protein alpha 2u globulin involvement is linked to the hyaline droplet nephropathy, which is seen in most of the chlorobenzene studies. The Science Advisory Board of the U.S. EPA reaffirmed EPA's position that the data on *p*-dichlorobenzene demonstrate the phenomenon cannot be used for evaluating human carcinogenic risk from chlorinated benzenes (35). Domestically, the U.S. Consumer Product Safety Commission (CPSC), and internationally, the International Programme of Chemical Safety (IPCS) under the World Health Organization (WHO), in considering the mechanistic data also concluded that *p*-dichlorobenzene does not produce a human carcinogenic risk.

The listing by the NTP triggers labeling requirements in an organization such as the Occupational Safety and Health Administration (OSHA). However, the EPA classification may trigger different requirements such as no labeling, as in the case of *p*-dichlorobenzene. States develop their own labeling requirements, therefore the possible use of *p*-dichlorobenzene and state requirements for proper labeling must be considered.

Economic Aspects

Total production of chlorobenzenes in the three principal producing regions of the world amounted to aproximately 400 thousand metric tons in 1988: the United States, 46%, Western Europe, 34%, and Japan, the remainder. Monochlorobenzene accounted for over 50% of the total production of chlorinated benzenes. The largest use of monochlorobenzene worldwide is for the production of nitrochlorobenzene: 41% for the United States' demand, 70% for the Western European demand, and 89% for the Japanese demand in 1988. Nitrochlorobenzenes are used to make dye and pigment intermediates, rubber processing chemicals, pesticides, pharmaceuticals, and other organic intermediates. Solvent use of monochlorobenzene in the United States is much greater than in Western Europe and Japan because of its use in herbicide formulations and other agriculture products.

o-Dichlorobenzene is consumed for 3,4-dichloroaniline, the base material for several herbicides, in the United States and Western Europe and is emulsified in Japan for garbage treatment. The greatest market worldwide for *p*-dichlorobenzene is for deodorant blocks and moth control. A growing use for p-dichlorobenzene is the manufacture of poly(phenylene sulfide) (PPS) resins.

With the exception of use in the manufacture of polymers, markets for chlorobenzenes are mature, and demand is expected to show little if any growth in the next few years.

The chlorobenzene operations in the United States were developed primarily for the manufacture of phenol, aniline, and DDT. However, with the process changes in the production of phenol and aniline, the phase-out of DDT production, and changes in the herbicide and solvent markets, the U.S. production of chlorinated benzenes has shrunk by more than 50% since the total production peaked in 1969. U.S. production of monochlorobenzene peaked in the 1960s and decreased to a low of 101 million kg in 1986 with an 11% and 9% increase, respectively, in 1988 and 1989.

Commercial chlorination of benzene today is carried out as a three-product process (monochlorobenzene and o- and p-dichlorobenzenes). The most economical operation is achieved with a typical product split of about 85% monochlorobenzene and a minimum of 15% dichlorobenzenes. Typically, about two parts of p-dichlorobenzene are formed for each part of o-isomer. It is not economical to eliminate the coproduction of the dichlorobenzenes. To maximize monochlorobenzene production (90% monochlorobenzene and 10% dichlorobenzene), benzene is lightly chlorinated; the density of the reaction mixture is monitored to minimize polychlorobenzene production and the unreacted benzene is recycled.

Producing the chlorobenzenes higher than mono- can pose significant process problems because production must match the market or the unwanted material must be destroyed. Use must be found for the HCl by-product and Cl_2 must be available at a reasonable price.

In 1988, the United States consumption of monochlorobenzene was 120 million kilograms; 42% for the production of nitrochlorobenzenes, 28% for solvent uses, and the remaining 30% for other applications such as diphenyl ether, ortho- and para-phenylphenols, sulfone polymers, and diphenyldichlorosilane, an intermediate for specialty silicones.

The principal use of o-dichlorobenzene is to manufacture 3,4-dichloroaniline, which is a raw material for several herbicides and for the production of 3,4,4'-trichlorocarbanilide (TCC), a bacteriostat used in deodorant soaps. Some is exported, but the amount is expected to decline as Brazil brings on increased capacity. A modest decline in U.S. consumption between 1989 and 1994 is expected. About 11,400 t were consumed in 1988.

The largest single market and a growing outlet for p-dichlorobenzene in the United States is the production of poly(phenylene sulfide) (PPS) resin. Of 42 million kilograms of p-dichlorobenzene consumed in the United States in 1988, 21% was for PPS. The second largest consumption in the United States of p-dichlorobenzene (16%) is the room deodorant market which is static and likely to remain unchanged. Moth control (11%) is also expected to remain static. However, when the room deodorant and moth control markets in the United States are added together they become the largest consumption, similar to the world market. Exports accounted for about 30% and about 21% remained in inventory.

Prices for the chlorobenzenes fluctuate widely. Some prices fluctuate with the price of benzene. Typical prices have been given (36).

Western Europe has a capacity of around 228 thousand metric tons as of January 1989; 76% of that capacity is located in Germany. Most of the capacity is captively consumed in products similar to the United States and also production of other products no longer produced in the United States.

The production of chlorobenzenes in Eastern Europe is concentrated in the former Soviet Union, Poland, and Czechoslovakia. The estimated capacity is 200–250 thousand metric tons; the former Soviet Union has most (230 thousand tons) of this capacity. There is trade between Eastern and Western Europe on monochlorobenzene and the dichlorobenzenes, but the net trade balance is probably even at about 20 thousand metric tons. Eastern Europe exported 20 thousand metric tons of monochlorobenzene principally to Germany, France, and the United States.

Japan, as of January 1, 1989, had a total capacity of 28 thousand metric tons of monochlorobenzene and 49 thousand tons of dichlorobenzenes. The Japanese prices have remained fairly constant since 1985. The Japanese consumption of p-dichlorobenzene is 81% for moth control, 11% for PPS resins, and 8% for dyestuffs. There has been very little export from Japan of chlorobenzenes and imports have been mainly p-dichlorobenzene from the United States, Germany, France, and the United Kingdom.

Brazil has two small producers of chlorobenzenes. One producer has a capacity of 4.8 thousand metric tons. The other producer's facility has a rated capacity of 28 thousand metric tons, which produces mono and ortho for local consumption, and the para may be used in Brazil and possibly exported. A third plant with a 400 metric ton capacity is believed to be on standby.

Canada has no known basic producers of chlorobenzene. There is one company that isolates small quantities of ortho and para from purchased mixed dichlorobenzenes. Some of the isolated product is exported. The primary portion of Canada's chlorobenzenes comes from the United States.

Specifications, Analyses, and Quality Control

Trade specifications for the chlorobenzenes are subject to modification by agreement with the customer of each producer's standards. All of the chlorobenzenes show readily separated and identifiable peaks by glc. This method is used exclusively for plant, quality control, and for sales specifications.

Typical analyses in wt % are chlorobenzene: benzene <0.05, dichlorobenzenes <0.1; and 1,4-dichlorobenzene: chlorobenzene and trichlorobenzenes <0.1, 1,2- and 1,3-dichlorobenzene: each <0.5.

1,2-Dichlorobenzene is sold as two grades: technical: chlorobenzene <0.05, trichlorobenzenes <1.0, 1,2-dichlorobenzene 80, and other isomers <19.0; and purified, produced by redistilling the technical product in a very efficient still: chlorobenzene <0.05, 1,2,4 trichlorobenzene <0.2; and 1,2-dichlorobenzene 98.0.

Uses

Monochlorobenzene. The largest use of monochlorobenzene in the United States is in the production of nitrochlorobenzenes, both ortho and para, which are separated and used as intermediates for rubber chemicals, antioxidants (qv), dye and pigment intermediates, agriculture products, and pharmaceuticals (Table 5). Since the mid-1980s, there have been substantial exports of both o-nitrochlorobenzene, estimated at 7.7 million kg to Europe and p-nitrochlorobenzene, estimated at 9.5 million kg to the Far East. Solvent use of monochlorobenzene accounted for about 28% of the U.S. consumption. This application involves solvents

Table 5. Derivatives of Nitrochlorobenzenes

Derivative	CAS Registry Number	Reactant(s)	Intermediate for
First-step derivatives of p-nitrochlorobenzene			
p-nitrophenol (PNP)	[100-02-7]	caustic hydrolysis	parathion acetyl-p-aminophenol, dyes
p-nitroaniline (PNA)	[100-01-6]	ammonia	p-phenylenediamine gasoline antioxidants, dyes, rubber chemicals
p-nitrophenetole (PNPt)	[100-29-8]	sodium ethylate	p-phenetidine used as ethoxyquin intermediate
4-nitrodiphenylamine (4 NDPA)	[836-30-6]	aniline	N-phenyl-p-phenylenediamine rubber chemicals
p-chloroaniline (PCA)	[106-47-8]	hydrogen	agriculture chemicals, carbanilide bacteriostats
4,4'-dinitrodiphenyl ether (DNDPO)	[101-63-3]	sodium phenate	oxybisaniline used as polymer intermediate
First-step derivatives of o-nitrochlorobenzene			
o-nitroaniline (ONA)	[88-74-4]	ammonia	fungicide (benomyl), stabilizers, benzotriazole
o-nitrophenol (ONP)	[88-75-5]	caustic hydrolysis	carbofuran and agriculture chemicals
o-nitroanisole (ONAS)	[91-23-6]	sodium methylate	3,3'-dimethoxybenzidine (a pigment intermediate)
3,3'-dichlorobenzidine	[91-94-1]	self	yellow pigments
o-chloroaniline (OCA)	[95-51-2]	hydrogen	agriculture chemicals

for herbicide production and the solvent for diphenylmethane diisocyanate manufacture and other chemical intermediates.

Other applications for monochlorobenzene include production of diphenylether, *ortho*- and *para*-phenylphenol, 4,4'-dichlorodiphenylsulfone, which is a primary raw material for the manufacture of polysulfones, diphenyldichlorosilane, which is an intermediate for specialty silicones, Grignard reagents, and in dinitrochlorobenzene and catalyst manufacture.

o-Dichlorobenzene. The principal use of *o*-dichlorobenzene in the United States is the manufacture of 3,4-dichloroaniline [95-76-1], a raw material used in the production of herbicides. A small amount of 3,4-dichloroaniline is used to produce 3,4,4'-trichlorocarbanilide [101-20-2] (TCC) used as a bacteriostat in deodorant soaps.

p-Dichlorobenzene. *p*-Dichlorobenzene's largest and growing outlet is in the manufacture of poly(phenylene sulfide) resin (PPS). Other applications include room deodorant blocks and moth control, a market which is static and likely to remain unchanged but combined is currently a larger outlet than PPS. Small amounts of *p*-dichlorobenzene are used in the production of 1,2,4-trichlorobenzene, dyes, and insecticide intermediates. Exports have been a principal factor in U.S. production with about 25% exported in 1988.

m-Dichlorobenzene. Isolation of pure *m*-dichlorobenzene [541-73-1], produced at ∼ 1% in the mixed dichlorobenzenes, is not economical. It is produced by rather exotic chemistry and has established only very specialized uses, believed to be only a few hundred kg per year, because of its high cost and the lack of commercial availability. However, there is potential for *m*-dichlorobenzene in some new experimental agricultural chemicals. There are a number of patents that cover its production (24,26,29–32), but only limited commercial production has been reported to date.

Other Chlorobenzenes. The market for the higher chlorobenzenes (higher than di) is small in comparison to the combined mono- and dichlorobenzenes. Trichlorobenzenes are used in some pesticides, as a dye carrier, in dielectric fluids, as an organic intermediate and a chemical manufacturing solvent, in lubricants, and as a heat-transfer medium. These are small and decreasing markets.

BIBLIOGRAPHY

"Chlorinated Benzenes" are treated in *ECT* 1st ed. under "Chlorine Compounds, Organic," Vol. 3, "Monochlorobenzene," pp. 812–817, by L. A. Kolker, Kolker Chemical Works, Inc., and N. Poffenberger, The Dow Chemical Company; "o-Dichlorobenzene," pp. 817–818, by N. Poffenberger, The Dow Chemical Company; "p-Dichlorobenzene," pp. 819–822, by Axel Heilborn, Niagara Alkali Co.; "Chlorinated Benzenes" under "Chlorocarbons and Chlorohydrocarbons" in *ECT* 2nd ed., Vol. 5, pp. 253–267, by D. W. F. Hardie, Imperial Chemical Industries Ltd.; in *ECT* 3rd ed., Vol. 5, pp. 797–808, by Chi-I Kao and N. Poffenberger, Dow Chemical U.S.A.

1. J. B. Cohen and P. Hartley, *J. Chem. Soc.*, 87, 1360 (1905).
2. D. W. F. Hardie, "A History of the Chemical Industry in Widnes," Imperial Chemical Industries Ltd., 1950, p. 155.
3. M. Campbell and H. Hatton, *Herbert H. Dow: Pioneer in Creative Chemistry,* Appleton-Century-Crosts, Inc., New York, 1951, p. 1114.

4. U.S. Pat. 1,607,618 (Nov. 23, 1926), W. J. Hale and E. C. Britton (to The Dow Chemical Company).
5. W. J. Hale and E. C. Britton, *Ind. Eng. Chem.* **20,** 114 (1928).
6. R. M. Crawford, *Chem. Eng. News* **25**(1), 235 (1947).
7. Gen. Pat. 539,176 (Nov. 12, 1931), W. Prohl (to F. Raschig GmbH).
8. Gen. Pat. 575,765 (Apr. 13, 1933), W. Prahl and W. Mathes (to F. Raschig GmbH).
9. L. Luttrighaus and D. Ambrose, *Chem. Ber.* **89,** 463 (1956).
10. J. D. Roberts and A. T. Bottini, *J. Am. Chem. Soc.* **79,** 1458 (1957).
11. D R. Stull, E. F. Westrum, and G. C. Sinke, *The Chemical Thermodynamics of Organic Compounds,* John Wiley & Sons, Inc., New York, 1969.
12. Brit. Pat. 388,818 (Mar. 6, 1933), T. S. Wheeler (to ICI).
13. H. van den Berg and R. M. Westerink, *Ind. Eng. Chem. Fund.* **15**(3), 164 (1976).
14. M. F. Bourion, *Ann. Chim. Paris* **14**(9), 215 (1920).
15. R. B. MacMullin, *Chem. Eng. Prog.* **44**(3), 183 (1948).
16. H. F. Wiegandt and P. R. Lantos, *Industrial Engineering Chemistry* **43,** 2167 (1951).
17. U.S. Pat. 3,226,447 (Dec. 28, 1965), G. H. Bing and R. A. Krieger (to Union Carbide Australia Ltd.).
18. Neth. Pat. 7413614 (Oct. 16, 1974), S. Robota, R. Paolieri, and J. G. McHugh (to Hooker Chemicals).
19. U.S. Pat. 1,946,040 (Feb. 6, 1934), W. C. Stoesser and F. B. Smith (to The Dow Chemical Company).
20. U.S. Pat. 2,976,330 (Mar. 21, 1961), J. Guerin (to Société Anonyme).
21. U.S. Pat. 3,636,171 (Jan. 18, 1972), K. L. Krumel and J. R. Dewald (to The Dow Chemical Company).
22. U.S. Pat. 3,557,227 (Jan. 19, 1971), M. M. Fooladi (to Sanford Chemical Co.).
23. U.S. Pat. 2,826,617 (Mar. 11, 1958), H. E. Redman and P. E. Weimer (to Ethyl Corp.).
24. U.S. Pat. 2,943,114 (June 28, 1960), H. E. Redman and P. E. Weimer (to Ethyl Corp.).
25. U.S. Pat. 2,886,605 (May 12, 1959), H. H. McClure, J. S. Melbert, and L. D. Hoblit (to The Dow Chemical Company).
26. U.S. Pat. 2,866,828 (Dec. 30, 1958), J. A. Crowder and E. E. Gilbert (to Allied Chemical Corp.).
27. U.S. Pat. 2,949,491 (Aug. 16, 1960), J. J. Rucker (to Hooker Chemical Corp.).
28. S. Ariki and A. Ohira, *Chem. Econ. Eng. Rev.,* **5**(7), 39 (1973).
29. U.S. Pat. 2,666,085 (Jan. 12, 1954), J. T. Fitzpatrick (to Union Carbide Corp.).
30. U.S. Pat. 2,819,321 (Jan. 7, 1958), B. O. Pray (to Columbia Southern Chemical Corp.).
31. U.S. Pat. 2,920,109 (Jan. 5, 1960), J. W. Angelkorte (to Union Carbide Corp.).
32. Yu G. Erykolov and co-workers, *Zh. Org. Khim* **9,** 348 (1973).
33. *Standard Methods for Examination of Water and Wastewater,* 14th ed., American Public Health Association, Washington, D.C., 1975, p. 800.
34. P. Simmons, D. Branson, and R. Bailey, *"Biodegradability of 1,2,4-Trichlorobenzene",* paper presented at the *1976 Association of Textile Chemicals and Colorist International Technical Conference,* Chicago, Ill., 1976.
35. J. A. Barter and R. S. Nair, *Review of the Scientific Evidence on the Human Carcinogenic Potential of Para-Dichlorobenzene,* Chlorobenzene Producers Association, Washington, D.C., 1990.
36. W K. Johnson with A. Leder and Y. Sakuma, "CEH Product Review", *Chlorobenzenes Chemical Economics Handbook,* SRI International, Menlo Park, Calif., Oct. 1989.

James G. Bryant
Standard Chlorine of Delaware, Inc.

RING-CHLORINATED TOLUENES

The ring-chlorinated derivatives of toluene form a group of stable, industrially important compounds. Many chlorotoluene isomers can be prepared by direct chlorination. Other chlorotoluenes are prepared by indirect routes involving the replacement of amino, hydroxyl, chlorosulfonyl, and nitro groups by chlorine and the use of substituents, such as nitro, amino, and sulfonic acid, to orient substitution followed by their removal from the ring.

The first systematic study of the reaction of chlorine with toluene was carried out in 1866 by Beilstein and Geitner. During the next 40 years, many studies were performed to isolate and identify the various chlorination products (1). During the early 1930s, Hooker Electrochemical Co. (Hooker Chemicals & Plastics Corp.) and the Heyden Chemical Corp. (Tenneco) began the manufacture of chlorotoluenes. Hooker Electrochemical Co. was later acquired by Occidental Petroleum Corp. and became the Occidental Chemical Corp. In the mid-1970s, Heyden exited chlorotoluenes production; Occidental thus is the sole U.S. producer of chlorotoluenes.

Mono- and dichlorotoluenes are used chiefly as chemical intermediates in the manufacture of pesticides, dyestuffs, pharmaceuticals, and peroxides, and as solvents. Total annual production was limited prior to 1960 but has expanded greatly since that time. Chlorinated toluenes are produced in the United States, Germany, Japan, and Italy. Since the number of manufacturers is small and much of the production is utilized captively, statistics covering production quantities are not available. Worldwide annual production of o- and p-chlorotoluene is estimated at several tens of thousands of metric tons. Yearly productions of polychlorotoluenes are in the range of 100–1000 tons.

MONOCHLOROTOLUENES

Physical Properties

o-Chlorotoluene [95-49-8] (1-chloro-2-methylbenzene, OCT) is a mobile, colorless liquid with a penetrating odor similar to chlorobenzene. It is miscible in all proportions with many organic liquids such as aliphatic and aromatic hydrocarbons, chlorinated solvents, lower alcohols, ketones, glacial acetic acid, and di-n-butylamine; it is insoluble in water, ethylene and diethylene glycols, and triethanolamine.

p-Chlorotoluene [106-43-4] (1-chloro-4-methylbenzene, PCT) and m-chlorotoluene [108-41-8] (1-chloro-3-methylbenzene, MCT) are mobile, colorless liquids with solvent properties similar to those of the ortho isomer.

Ortho and p-chlorotoluene form binary azeotropes with various organic compounds including alcohols, acids, and esters (2). Oxygen indexes, the minimum percentage of oxygen in an oxygen–nitrogen atmosphere required to sustain combustion after ignition, for the chlorotoluene isomers are ortho 19.2, meta 19.7, and para 19.1 (3). Ortho and p-chlorotoluene form stable ionic complexes with antimony pentachloride (4). They also form complexes with a number of organometallic derivatives, such as those of chromium (5), cobalt (6), iron (7), etc, many of which have synthetic utility. Physical properties of the monochlorotoluene isomers, mol wt 126.59, appear in Table 1 (8–13).

Chemical Properties

The monochlorotoluenes are stable to the action of steam, alkalies, amines, and hydrochloric and phosphoric acids at moderate temperatures and pressures. Three classes of reactions, those involving the aromatic ring, the methyl group, and the chlorine substituent, are known for monochlorotoluenes.

Reactions of the Aromatic Ring. Ring chlorination of o-chlorotoluene yields a mixture of all four possible dichlorotoluenes, the 2,3-, 2,4-, 2,5-, and 2,6-isomers as shown in equation 1 (14).

Table 1. Physical Properties of the Monochlorotoluenes, C_7H_7Cl

Property	Ortho	Meta	Para
mp, °C	−35.6	−47.8	7.5
bp, °C	159.2	161.7	162.4
flash point, °C	47	47	49
density[a], kg/m^3			
20°C	1082.5	1072.2	1069.7
25°C	1077.6		1065.1[24.4]
30°C	1072.7		
refractive index[a], n_D^t			
20°C	1.52680	1.5214[19]	1.5211
25°C	1.52221		1.5193[24.4]
surface tension[a], mN/m (= dyn/cm)	334.4[20]		322.4[25]
	323.3[30]		292.2[30]
dielectric constant at 20°C	4.73	5.55	6.20
viscosity (dynamic), mPa·s (= cP)			0.09
dipole moment, C·m[b]	4.80×10^{-30}	5.97×10^{-30}	
heat of vaporization, kJ/mol[c]	43.01	42.18	42.475
vapor density (air = 1)			4.37
vapor pressure, °C at kPa[d]			
0.13	5.4	4.8	5.5
1.3	43.2	43.2	43.8
5.3	72.0	73.0	73.5
13.3	94.7	96.3	96.6
53.3	137.1	139.7	139.8

[a]Superscript indicates temperature.
[b]To convert C·m to debye, divide by 3.336×10^{-30}.
[c]To convert kJ to kcal, divide by 4.184.
[d]To convert kPa to mm Hg, multiply by 7.5.

$$(1)$$

The principal isomer, 2,5-dichlorotoluene, constitutes up to 60% of the product mixture (15,16). Similarly, nitration of o-chlorotoluene produces a mixture of the four corresponding nitrochlorotoluene isomers. Nitration of p-chlorotoluene gives a mixture of 66% 4-chloro-2-nitrotoluene [89-59-8] and 34% of 4-chloro-3-nitrotoluene [89-60-1], $C_7H_6ClNO_2$, (17). Chlorosulfonation of o-chlorotoluene produces 2-chloro-5-chlorosulfonyltoluene [6291-02-7], (4-chloro-3-methylbenzenesulfonyl chloride), $C_7H_6Cl_2O_2S$, as the principal product (18). Sulfonation of p-chlorotoluene with 20% oleum gives the 2-sulfonic acid derivative in 68% yield (19). Trifluoromethylation of monochlorotoluenes has been achieved by reaction with carbon tetrachloride and hydrogen fluoride (20). Chloromethylation of o-chlorotoluene gives 2-chloro-4-chloromethyltoluene [2719-40-6] as the sole product. With p-chlorotoluene, a mixture of 4-chloro-2-chloromethyltoluene [34060-72-5] and 4-chloro-3-chloromethyltoluene [34896-68-9], $C_8H_8Cl_2$, is formed in a 63:37 ratio, respectively (21).

Reactions of the Methyl Group. Monochlorotoluenes are widely used to synthesize compounds derived from reactions of the methyl group. Chlorination under free-radical conditions leads successively to the chlorinated benzyl, benzal, and benzotrichloride derivatives (see CHLOROCARBONS AND CHLOROHYDROCARBONS—BENZYL CHLORIDE, BENZAL CHLORIDE, AND BENZOTRICHLORIDE). Oxidation to form chlorinated benzaldehydes and benzoic acids can be performed under both liquid- and vapor-phase conditions (22,23). Catalytic ammoxidation under vapor-phase conditions with oxygen and ammonia produces chlorobenzonitriles (24). Reaction of p-chlorotoluene with cyanogen chloride at 650–700°C gives p-chlorophenylacetonitrile [140-53-4], C_8H_6ClN, as shown in equation 2 (25). Side-chain bromination of p-chlorotoluene by bromine catalyzed by lanthanum triacetate is a facile process (26).

$$(2)$$

Halogen Reactions. Hydrolysis of chlorotoluenes to cresols has been effected by aqueous sodium hydroxide. Both displacement and benzyne formation are involved (27,28). o-Chlorotoluene reacts with sodium in liquid ammonia to afford a mixture of 67% of o-toluidine [95-53-4] and 33% of m-toluidine [108-44-1], C_7H_9ClN, as shown in equation 3 (29).

$$
\underset{\text{Cl}}{\overset{\text{CH}_3}{\bigcirc}} \xrightarrow{\text{Na, NH}_3(l)} \underset{\substack{\\ 67\%}}{\overset{\text{CH}_3}{\bigcirc}}\text{NH}_2 \quad + \quad \underset{\substack{\\ 33\%}}{\overset{\text{CH}_3}{\bigcirc}}\text{NH}_2 \tag{3}
$$

With hydrogen sulfide at 500–600°C, monochlorotoluenes form the corresponding thiophenol derivatives (30). In the presence of palladium catalysts and carbon monoxide, monochlorotoluenes undergo carbonylation at 150–300°C and 0.1–20 MPa (1–200 atm) to give carboxylic acids (31). Oxidative coupling of p-chlorotoluene to form 4,4'-dimethylbiphenyl can be achieved in the presence of an organonickel catalyst, generated *in situ,* and zinc in dipolar aprotic solvents such as dimethylacetamide (32,33). An example is shown in equation 4.

$$
\underset{\text{Cl}}{\overset{\text{CH}_3}{\bigcirc}} \xrightarrow[\text{CH}_3,\text{ CON(CH}_3)_2]{\text{NiCl}_2,\text{ (C}_6\text{H}_5)_3\text{P, Zn}} \text{CH}_3\text{—}\bigcirc\text{—}\bigcirc\text{—CH}_3 \tag{4}
$$

Dehalogenation of monochlorotoluenes can be readily effected with hydrogen and noble metal catalysts (34). Conversion of p-chlorotoluene to p-cyanotoluene is accomplished by reaction with tetraethylammonium cyanide and zero-valent Group (VIII) metal complexes, such as those of nickel or palladium (35). The reaction proceeds by initial oxidative addition of the aryl halide to the zerovalent metal complex, followed by attack of cyanide ion on the metal and reductive elimination of the aryl cyanide. p-Methylstyrene is prepared from p-chlorotoluene by a vinylation reaction using ethylene as the reagent and a catalyst derived from zinc, a triarylphosphine, and a nickel salt (36).

Preparation

Monochlorotoluenes have been prepared by chlorinating toluene with a wide variety of chlorinating agents, catalysts, and reaction conditions. The ratio of ortho and para isomers formed can vary over a wide range. Particular attention has been given to studies aimed at increasing the para isomer content owing to its greater commercial significance. The meta isomer must be prepared by indirect means since only a small amount, <1%, is formed by direct chlorination.

Chlorinations with Elemental Chlorine. Reaction of toluene with chlorine in the presence of certain Lewis acid catalysts including the chlorides of aluminum, tin, titanium, and zirconium give monochlorotoluene mixtures that contain more than 70% of the ortho isomer (37,38). A number of catalyst systems have been developed to enhance the formation of p-chlorotoluene in toluene chlorination. Monochlorotoluenes containing 45–55% of the p-isomer are obtained through the use of certain specific metal sulfides or cocatalyst systems consisting of specific metal salts and sulfur, inorganic sulfides, or divalent sulfur compounds with or without other functional groups (39–45). A growing number of heterogeneous processes that employ zeolite-type catalysts for chlorination of aromatics have been discovered (see CATALYSIS). The majority of these catalysts are synthetic zeolites, specifically the L-type zeolites (46). It is sometimes possible to achieve highly regioselective chlorination of arenes, such as toluene, by the use of certain specific types of zeolites. A catalyst system comprised of TSZ-506, which is a synthetic zeolite, and monochloroacetic acid affords p-chlorotoluene with a selectivity of 75% relative to the ortho isomer at an operating temperature of 70°C (47). One common problem during zeolite-catalyzed chlorination is the structural breakdown of the zeolite lattice because of reaction with hydrogen chloride liberated in the reaction. However, modifications in the synthetic procedure have enabled the preparation of newer types of zeolites that are more resistant to structural deterioration (48).

Noncatalytic ring chlorination of toluene in a variety of solvents has been reported. Isomer distributions vary from approximately 60% ortho in hydroxylic solvents, eg, acetic acid, to 60% para in solvents, eg, nitromethane, acetonitrile, and ethylene dichloride (49,50). Reaction rates are relatively slow and these systems are particularly appropriate for kinetic studies.

Chlorination with Other Reagents. Chlorotoluenes can also be obtained in good yields by the reaction of toluene with stoichiometric proportions of certain Lewis acid chlorides such as iron(III) chloride, as the chlorinating agent (51). Generally, the product mixture contains p-chlorotoluene as the principal component. Several modifications have been proposed to improve product yields (52,53).

Toluene chlorination has also been effected with hydrogen chloride as the chlorinating agent. The reaction is catalyzed by nitric acid under aqueous conditions to give a good conversion and yield on monochlorotoluenes (54). Oxychlorination of toluene with oxygen and hydrogen chloride in the vapor phase over supported copper and palladium catalysts yields chlorotoluene mixtures containing up to 60% of p-chlorotoluene along with varying amounts of side-chain chlorinated products (55,56).

Other methods for preparing p-chlorotoluene include α-elimination from an organotellurium(IV) halide (57), palladium-catalyzed decarbonylation of 4-methylbenzoyl chloride (58), and desulfonylation of p-toluenesulfonyl chloride catalyzed by chlorine (59) or chlorotris(triphenylphosphine)rhodium (60).

Pure monochlorotoluene isomers are prepared by diazotization of the corresponding toluidine isomers followed by reaction with copper(I) chloride (Sandmeyer reaction). This is the preferred method of obtaining m-chlorotoluene.

The rate of chlorination of toluene relative to that of benzene is about 345 (61). Usually, chlorination is carried out at temperatures below 70°C with the reaction proceeding at a profitable rate even at 0°C. The reaction is exothermic with ca 139 kJ (33 kcal) of heat produced per mole of monochlorotoluene formed. Chlorine efficiency is high, and toluene conversion to monochlorotoluene can be carried to about 90% with the formation of only a few percent of dichlorotoluenes. In most catalyst systems, decreasing temperatures favor formation of increasing amounts of p-chlorotoluene. Concentrations of required catalysts are low, generally on the order of several tenths of a percent or less.

Only trace amounts of side-chain chlorinated products are formed with suitably active catalysts. It is usually desirable to remove reactive chlorides prior to fractionation in order to minimize the risk of equipment corrosion. The separation of o- and p-chlorotoluenes by fractionation requires a high efficiency, isomer-separation column. The small amount of m-chlorotoluene formed in the chlorination cannot be separated by fractionation and remains in the p-isomer fraction. The toluene feed should be essentially free of paraffinic impurities that may produce high boiling residues that foul heat-transfer surfaces. Trace water contamination has no effect on product composition. Steel can be used as construction material for catalyst systems containing iron. However, glass-lined equipment is usually preferred and must be used with other catalyst systems.

Both batch and continuous processes are suitable for commercial chlorination. The progress of the chlorination is conveniently followed by specific gravity measurements.

Handling and Shipment

Monochlorotoluenes are shipped in bulk in steel tank cars and tank trucks. Drum shipments are made using lined or unlined steel drums. Aluminum tanks can be used to store only acid-free material. Under DOT regulations, for transport of over 415 L (110 gal) of monochlorotoluenes, freight classification is combustible liquid NOS, and for truck transport, chemical NOI. The storage vessels are vented to a safe atmosphere and should be protected with suitable diking. Protection against static charge is essential when transferring material. Suitable ventilation should be provided and sources of ignition avoided as the vapor forms flammable mixtures with air.

Identification and Analysis

A number of analytical methods have been developed for the determination of chlorotoluene mixtures by gas chromatography. These are used for determinations in environments such as air near industry (62) and soil (63). Liquid crystal stationary columns are more effective in separating m- and p-chlorotoluene than conventional columns (64). Prepacked columns are commercially available. Zeolites have been examined extensively as a means to separate chlorotoluene mixtures (see MOLECULAR SIEVES). For example, a Y-type zeolite containing sodium and copper has been used to separate m-chlorotoluene from its isomers by selective absorption (65). The presence of benzylic impurities in chlorotoluenes is determined by standard methods for hydrolyzable chlorine. Proton (66) and carbon-13 chemical shifts, characteristic ir absorption bands, and principal mass spectral peaks are available along with sources of reference spectra (67).

Health and Safety Factors

Inhalation of high concentrations of monochlorotoluenes will cause symptoms of central nervous system depression. Inhalation studies produced an LC_{50} (rat, 4 h) of 7119 ppm for o-chlorotoluene (68). o- and p-Chlorotoluene are both considered moderately toxic by ingestion (Table 2). A study of the relationship between the electronic structure and toxicity parameters for a series of mono-, di-, and tri-chlorotoluenes has been reviewed (72). A thin-layer chromatographic method has been developed to assess the degree of occupational exposure of workers to chlorotoluenes by determining p-chlorohippuric acid [13450-77-6], $C_9H_8ClNO_3$, (N-(4-chlorobenzoyl)glycine) in urine samples (73).

Table 2. Toxicity Parameters for Monochlorotoluenes

Parameter	Ortho	Para
LD_{50} (rat), mg/kg	2350^a	2100^b
TLV^c, ppm	50	
PEL^d, ppm	50	

[a]Ref. 69.
[b]Ref. 70.
[c]ACGIH, 259 mg/m^3, 8 h TWA (71).
[d] OSHA, 250 mg/m^3, 8 h TWA.

A study to isolate and examine the genetic characteristics of bacteria that metabolize chlorotoluenes, such as OCT, PCT, and 2,6-dichlorotoluene, has been reported (74). Two products were isolated from a study of the metabolism of PCT by *Pseudomonas putida*: (+)-*cis*-4-chloro-2,3-dihydroxy-1-methylcyclohex-4,6-diene and 4-chloro-2,3-dihydroxy-1-methylbenzene (75). Enzymatic dehydrogenation of the former compound to the latter was also demonstrated.

HIGHER CHLOROTOLUENES

Dichlorotoluenes

There are six possible dichlorotoluene isomers, $C_7H_6Cl_2$, (mol wt 161.03) all of which are known. Only the 2,4-, 2,5-, and the 3,4-isomers are available from direct chlorination of monochlorotoluenes. Physical properties of the dichloro- and other higher chlorotoluenes are given in Table 3.

2,4-Dichlorotoluene (2,4-dichloro-1-methylbenzene) constitutes 80–85% of the dichlorotoluene fraction obtained in the chlorination of PCT with antimony trichloride (76) or zirconium tetrachloride (77) catalysts. It is separated from 3,4-dichlorotoluene (1,2-dichloro-4-methylbenzene), the principal contaminant, by distillation. Chlorination of OCT with sulfuryl chloride gives mainly 2,4-dichlorotoluene and small amounts of the 2,3 isomer (78). 2,5-Dichlorotoluene (1,3-dichloro-2-methylbenzene) is formed in up to 60% yield in the sulfide-cocatalyzed chlorination of OCT. Purification by recrystallization gives 99% pure product (15,16).

Chlorination of OCT with chlorine at 90°C in the presence of L-type zeolites as catalyst reportedly gives a 56% yield of 2,5-dichlorotoluene (79). Pure 2,5-dichlorotoluene is also available from the Sandmeyer reaction on 2-amino-5-chlo-

Table 3. Physical Properties of the Higher Chlorotoluenes

Toluene	CAS Registry Number	Mp, °C	Bp, °C	n_D^t	Density at 20°C, kg/m^3
2,3-dichloro	[32768-54-0]	5	208.3	1.5511[20]	
2,4-dichloro	[95-73-8]	−13.5	201.1	1.5480[22]	1249.8
2,5-dichloro	[19398-61-9]	5	201.8	1.5449[20]	1253.5
2,6-dichloro	[118-69-4]		200.6	1.5507[20]	1268.6
3,4-dichloro	[95-75-0]	−15.3	208.9	1.5471[20]	1256.4
3,5-dichloro	[25186-47-4]	26	201.2	1.5438[20]	
2,3,4-trichloro	[7359-72-0]	43–44	244		
2,3,5-trichloro	[56961-86-5]	45–46	229–231		
2,3,6-trichloro	[2077-46-5]	45–46	118[a]		
2,4,5-trichloro	[6639-30-1]	82.4	229–230[b]		
2,4,6-trichloro	[23749-65-7]	38			
3,4,5-trichloro	[21472-86-6]	45–45.5	246–247[c]		
2,3,4,5-tetrachloro	[1006-32-2]	98.1			
2,3,4,6-tetrachloro	[875-40-1]	92	266–276		
2,3,5,6-tetrachloro	[1006-31-1]	93–94			
pentachloro	[877-11-2]	224.5–225.5	301		

[a]At 2.4 kPa (18 mm Hg).
[b]At 95.4 kPa (716 mm Hg).
[c]At 102.4 kPa (768 mm Hg).

rotoluene. 3,4-Dichlorotoluene (1,2-dichloro-4-methylbenzene) is formed in up to 40% yield in the chlorination of PCT catalyzed by metal sulfides or metal halide–sulfur compound cocatalyst systems (80).

2,3-Dichlorotoluene (1,2-dichloro-3-methylbenzene) is present in about 10% concentration in reaction mixtures resulting from chlorination of OCT. It is best prepared by the Sandmeyer reaction on 3-amino-2-chlorotoluene.

2,6-Dichlorotoluene (1,3-dichloro-2-methylbenzene) is prepared from the Sandmeyer reaction on 2-amino-6-chlorotoluene. Other methods include ring chlorination of *p*-toluenesulfonyl chloride followed by desulfonylation (81), and chlorination and dealkylation of 4-*tert*-butyltoluene (82) or 3,5-di-*tert*-butyltoluene (83,84).

Trichlorotoluenes

The chlorination of toluene and *o*- and *p*-chlorotoluenes produces a mixture of trichlorotoluenes, (C$_7$H$_5$Cl$_3$, (mol wt 195.48): the 2,3,6-isomer (1,2,4-trichloro-3-methylbenzene) and 2,4,5-trichlorotoluene (1,2,4-trichloro-5-methylbenzene) containing small amounts of 2,3,4-trichlorotoluene (1,2,3-trichloro-4-methylbenzene) and 2,4,6-trichlorotoluene (1,3,5-trichloro-2-methylbenzene). When toluene is chlorinated in the presence of iron(III) chloride catalyst, a mixture containing nearly equal amounts of 2,4,5- and 2,3,6-trichlorotoluenes is produced (38,85). Chlorination of OCT yields a mixture containing >60% of 2,3,6-trichlorotoluene

(86). Reaction of p-toluenesulfonic acid with chlorine and antimony trichloride in chloroform and then sulfuric acid at reflux affords 2,3,6-trichlorotoluene in 89% yield (87). Metal sulfide-catalyzed chlorination of PCT gives trichlorotoluene fractions containing more than 75% of the 2,4,5-isomer (eq. 5) (88). The other chlorotoluenes are available from the Sandmeyer reaction on the corresponding amines. A gas chromatographic study has been conducted to determine the isomer selectivity of stationary phases of different polarity with respect to various chlorotoluenes including the trichlorotoluene isomers (89).

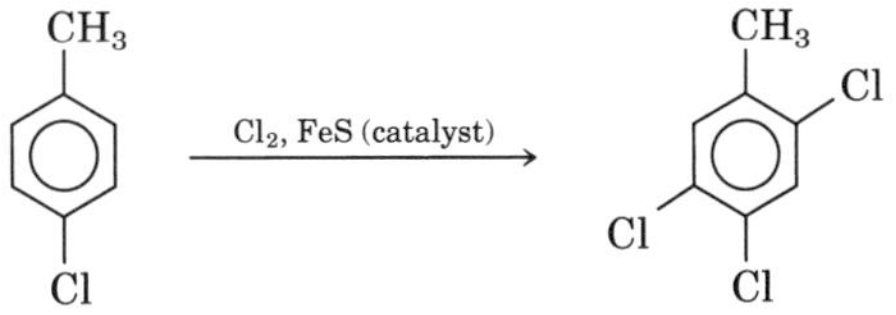

(5)

Gas-phase ammoxidation of trichlorotoluenes in the presence of catalyst affords the corresponding benzonitrile derivatives (90). In a 28-day feeding study, 2,3,6-trichlorotoluene showed only mild toxicological changes when administered to rats (91).

Tetra- and Pentachlorotoluenes

2,3,4,6-Tetrachlorotoluene, $C_7H_4Cl_4$ (mol wt 229.93) (1,2,3,5-tetrachloro-4-methylbenzene), is prepared from the Sandmeyer reaction on 3-amino-2,4,6-trichlorotoluene. 2,3,4,5-Tetrachlorotoluene (1,2,3,4-tetrachloro-5-methylbenzene) is the principal isomer in the further chlorination of 2,4,5-trichlorotoluene. Exhaustive chlorination of p-toluenesulfonyl chloride, followed by hydrolysis to remove the sulfonic acid group yields 2,3,5,6-tetrachlorotoluene (1,2,4,5-tetrachloro-3-methylbenzene) in good yield (92). Pentachlorotoluene (pentachloromethylbenzene), $C_7H_3Cl_5$ (mol wt 264.37), is formed in 90% yield by the ferric chloride-catalyzed chlorination of toluene in carbon tetrachloride or hexachlorobutadiene solution (93). Oxidation of pentachlorotoluene with excess sulfur trioxide, followed by hydrolysis of the intermediate pentachlorobenzyl disulfooxonium hydroxide inner salt produces pentachlorobenzyl alcohol in 91% yield (94). Gas chromatographic separation selectivities of stationary phases of different polarities toward tetrachlorotoluene isomers and pentachlorotoluene have been examined (89).

Uses

Chlorotoluenes are used as intermediates in the pesticide, pharmaceutical, peroxide, dye, and other industries. Many side chain-chlorinated derivatives are converted to end products. p-Chlorotoluene is used primarily in the manufacture of p-chlorobenzotrifluoride [98-56-6], a key intermediate in dinitroaniline and diphenyl ether herbicides (95). Other applications include manufacture of p-chlorobenzyl chloride, p-chlorobenzaldehyde, p-chlorobenzoyl chloride, p-chlorobenzoic acid, and 2,4- and 3,4-dichlorotoluenes. *p*-Chlorotoluene is an intermediate for a novel class of polyketone polymers (96).

Chlorotoluene isomer mixtures, especially those containing a relatively high amount of *o*-chlorotoluene, are widely used as solvents in industry for such purposes as metal-cleaning formulations, railroad industrial cleaners, diesel fuel additives, carbon removal procedures, paint thinners, and agricultural chemicals. Halso 99 and Halso 125 are examples of such solvents.

2,4-Dichlorotoluene is an intermediate for manufacture of herbicides. It is also used to obtain 2,4-dichlorobenzyl chloride and 2,4-dichlorobenzoyl chloride. 2,6-Dichlorotoluene is applied as a herbicide and dyestuff intermediate. 2,3,6-Trichlorotoluene is used as a herbicide intermediate. The other polychlorotoluenes have limited industrial application.

BIBLIOGRAPHY

"Ring-Chlorinated Toluenes" under "Chlorocarbons, Chlorohydrocarbons" in *ECT* 3rd ed., Vol. 5, pp. 819–827, by S. Gelfand, Hooker Chemical & Plastics Corp.

1. J. B. Cohen and H. D. Dakin, *J. Chem. Soc.* **79**, 1111 (1901).
2. L. H. Horsley and co-workers, *Azeotropic Data III, Advances in Chemistry Series*, No. 116, American Chemical Society, Washington, D.C., 1973, pp. 197–198.
3. G. L. Nelson and J. L. Webb, *J. Fire Flammability* **4**, 325 (1973).
4. R. G. Makitra, Ya. M. Tsikanchuk, and D. K. Tolopko, *J. Gen. Chem. U.S.S.R.* **45**, 1883 (1975).
5. R. S. Bly, K.-K. Tse, and R. K. Bly, *J. Organomet. Chem.* **117**, 35 (1976).
6. V. Galamb, G. Palyi, F. Ungvary, L. Marko, R. Boese, and G. Schmid, *J. Am. Chem. Soc.* **108**, 3344 (1986).
7. A. S. Abd-El-Aziz, C. C. Lee, A. Piorko, and R. G. Sutherland, *Synth. Commun.* **18**, 291 (1988).
8. J. Timmerman, *Physico-Chemical Constants of Pure Organic Compounds*, Elsevier Science Publishing Co., Inc., New York, 1950, pp. 297–298.
9. V. Sedivec and J. Flek, *Handbook of Analysis of Organic Solvents*, John Wiley & Sons, Inc., New York, 1976, pp. 164–168, 398.
10. K. Raznjevic, *Handbook of Thermodynamic Tables and Charts*, McGraw-Hill Book Co., New York, 1976, tables 27-1 and 30-2.
11. R. R. Dreisbach, *Physical Properties of Chemical Compounds I, Advances in Chemistry Series*, No. 15, American Chemical Society, Washington, D.C., 1955, p. 139.
12. A. L. McClellan, *Tables of Experimental Dipole Moments*, W. H. Freeman and Company, San Francisco, Calif., 1963, p. 243.
13. R. M. Stephenson and S. Malanowski, *Handbook of the Thermodynamics of Organic Compounds*, Elsevier Science Publishing Co., Inc., New York, 1987, p. 227.
14. Eur. Pat. Appl. EP 46,555 (Mar. 3, 1982), G. M. Petruck and R. Wambach (to Bayer A-G).
15. Ger. Offen, 2,523,104 (Nov. 25, 1976), H. Rathjen (to Bayer A-G).
16. U.S. Pat. 4,031,146 (June 21, 1977), E. P. DiBella (to Tenneco Chemicals Inc.).
17. Jpn. Kokai 75 151,828 (Dec. 6, 1975), M. Matsui, T. Kitsukawa, K. Sato, and T. Ogawa (to Mitsubishi Chem. Ind. Co., Ltd.).
18. Ger. Offen. 2,721,429 (Nov. 16, 1978), H. U. Balnk (to Bayer A-G).
19. Y. Muramoto and H. Asakura, *Nippon Kagaku Kaishi* **6**, 1070 (1975).
20. Ger. Offen. 2,837,499 (Mar. 20, 1980), A. Marhold and E. Klauke (to Bayer A-G).
21. E. Kuimova and B. M. Mikhailov, *J. Org. Chem. USSR* **7**, 1485 (1971).

22. Swiss. Pat. CH 645,335 (Sept. 28, 1984), J. Beyrich and W. Regenass (to Ciba-Geigy A-G).
23. B. Chopra and V. Ramakrishnan, *Indian Chem. J. Annu.* **38**, (1972).
24. Jpn. Kokai 81 18,951 (Feb. 23, 1981), K. Sempuku (to Yuki Gosei Kogyo Co., Ltd.).
25. R. A. Grimm and J. E. Menting, *Ind. Eng. Chem. Prod. Res. Div.* **14**, 158 (1975).
26. M. Ouertani, P. Girard, and H. H. Kagan, *Bull. Soc. Chim. Fr.* **9–10**, 327 (1982).
27. A. L. Bottini and J. D. Roberts, *J. Am. Chem. Soc.* **79**, 1458 (1957).
28. M. Zoratti and J. F. Bunnett, *J. Org. Chem.* **45**, 1769 (1980).
29. R. Levine and E. R. Biehl, *J. Org. Chem.* **40**, 1835 (1975).
30. M. G. Voronkov and co-workers, *J. Org. Chem. USSR* **11**, 1118 (1975).
31. Eur. Pat. Appl. EP 283,194 (Sept. 21, 1988), K. Suto, K. Nakasa, M. Kudo, and M. Yamamoto (to Nihon Nohyaku Co., Ltd.).
32. U.S. Pat. 4,263,466 (Apr. 21, 1981), I. Colon, L. M. Maresca, and G. T. Kwiatkowski (to Union Carbide Corp.).
33. R. Vanderessa, J. J. Brunet, and P. Caubere, *J. Organomet. Chem.* **264**, 263 (1984).
34. M. Kraus and V. Bazant, in J. W. Hightower, ed., *Proceedings of the Fifth International Conference on Catalysis*, Palm Beach, Fla., North-Holland Publishing Co., Amsterdam, The Netherlands, 1972.
35. U.S. Pat. 4,499,025 (Feb. 12, 1985), J. B. Davison, R. J. Jasinski, and P. J. Peerce-Landers (to Occidental Chemical Corp.).
36. U.S. Pat. 4,334,081 (June 8, 1982), I. Colon (to Union Carbide Corp.).
37. I.G. Farben Industries, *Reports of the Intermediate Products Commission*, PB-17658, National Technical Information Service, Springfield, Va., 1935–1936, frames 2247–2256.
38. U.S. Pat. 3,000,975 (Sept. 19, 1961), E. P. DiBella (to Heyden Newport Chemical Corp.).
39. Neth. Pat. 6,511,484 (Mar. 3, 1966), (to Hooker Chemicals & Plastics Corp.).
40. U.S. Pats. 4,031,142, 4,031,147 (June 21, 1977), J. C. Graham (to Hooker Chemicals & Plastics Corp.).
41. U.S. Pat. 4,024,198 (May 17, 1977), H. E. Buckholtz and A. C. Bose (to Hooker Chemicals & Plastics Corp.).
42. U.S. Pats. 4,069,263, 4,069,264 (Jan. 17, 1978), H. C. Lin (to Hooker Chemicals & Plastics Corp.).
43. Jpn. Kokai JP 60136576 (July 20, 1985), J. Kiji, H. Konishi, and M. Shimizu (to Ihara Chemical Industry Co., Ltd.).
44. Ger. Offen. DE 3,432,095 (Mar. 6, 1986), H. Wolfram (to Hoechst (A-G).
45. U.S. Pat. 4,851,596 (July 25, 1989), M. Franz-Josef, F. Helmut, R. Kai,and W. Karlfried (to Bayer A-G).
46. D. W. Breck, *Zeolite Molecular Sieves-Structure, Chemistry, and Use*, John Wiley & Sons, Inc., New York, 1974, p. 257.
47. Eur. Pat. Appl. EP 154,236 (Sept. 11, 1985), T. Suzuki and L. Komatsu (to Ihara Chemical Industry Co., Ltd.).
48. U.S. Pat. 4,794,201 (Dec. 27, 1988), Y. Higuchi and Suzuki (to Ihara Chemical Industry Co., Ltd.).
49. L. M. Stock and A. Himoe, *Tetrahedron Lett.* (13), 9 (1960).
50. L. M. Stock and A. Himoe, *J. Am. Chem. Soc.* **83**, 4605 (1961).
51. P. Kovacic, in G. A. Olah, ed., *Friedel-Crafts & Related Reactions*, Vol. IV, Interscience Publishers, Inc., a division of John Wiley & Sons, Inc., New York, 1965, Chapt. XLVIII, pp. 111–127.
52. Jpn. Kokai 74 76,828 (July 24, 1974), (to International Minerals & Chemical Corp.).
53. Ger. Offen. 2,230,369 (Jan. 18, 1973), K. Sawazaki, H. Fujii, and M. Dehura (to Nikkei Kako Co., Ltd. and Sugai Chem. Ind. Ltd.).

54. C. M. Selwitz and V. A. Notaro, *Prepr. Div. Pet. Chem. ACS* **17**(4), E37–46 (1972).
55. Jpn. Kokai 73 81,822 (Nov. 1, 1973), R. Fuse, T. Inoue, and T. Kato (to Ajinomoto Co., Inc.).
56. A. B. Salomonov, P. P. Gertsen, and A. N. Ketov, *Zh. Prikl. Khim.* **43**, 1612 (1970).
57. S. Uemura and S. Fukuzawa, *J. Organomet. Chem.* **268**, 223 (1984).
58. J. W. Verbicky, Jr., B. A. Dellacoletta, and L. Williams, *Tetrahedron Lett.* **23**, *371 (1982).*
59. B. Miller, *J. Org. Chem.* **38**, 1243 (1973); U.S. Pat. 3,844,917 (Oct. 29, 1974).
60. J. Blum, *Tetrahedron Lett.* (26), 3041 (1966).
61. P. B. D. DeLaMare and P. W. Robertson, *J. Chem. Soc.*, 279 (1943).
62. T. Bernath, *Gas Waerme Int.* **31**, 338 (1982).
63. D. R. Thielen, P. S. Foreman, A. Davis, and R. Wyeth, *Environ. Sci. Technol.* **21**, 145 (1987).
64. H. Kelker and E. Von Schivizhoffen, in J. C. Giddings and R. A. Kelker, eds., *Advances in Chromatography*, Vol. 6, Marcel Dekker, Inc., New York, 1968, pp. 247–297.
65. Jpn. Kokai JP 59,176,223 (Oct. 5, 1984) (to Toray Industries, Inc.).
66. J. G. Lindberg, G. Y. Sugiyama, and R. L. Mellgren, *J. Magn. Reson.* **17**, 112 (1975).
67. J. G. Grasselli and W. M. Richey, eds., *Atlas of Spectral Data and Physical Constants for Organic Compounds*, 2nd ed., Vol. IV, CRC Press Inc., Cleveland, Ohio, 1975, pp. 652–653.
68. Hazleton Laboratories, Project No. 157-147/148; May 10, 1972.
69. Younger Laboratories, Inc., Project No. Y-76-31; Feb. 27, 1976.
70. Springborn Institute, Project No. 3090. Dec. 31, 1980.
71. *Documentation of the Threshold Limit Values for Substances in Workroom Air with Supplements for those Substances Added or Changed Since 1971*, American Conference of Government Industrial Hygienists, 3rd ed., 1971, second printing, 1974, pp. 302–303.
72. I. P. Ulanova, P. N. Dyachkov, and A. I. Khalepo, *Pharmacochem. Libr.* **8** (QSAR Toxicol. Xenobiochem.), 83 (1985).
73. J. Gartzke, D. Burck, P. Schmidt, and G. G. Avilova, *Z. Klin. Med.* **40**, 1701 (1985).
74. P. A. Vandenbergh and R. H. Olsen, *Appl. Environ. Microbiol.* **42**, 737 (1981).
75. D. T. Gibson and co-workers, *Biochemistry* **7**, 3795 (1968).
76. U.S. Pat. 4,006,195 (Feb. 1, 1977), S. Gelfand (to Hooker Chemicals & Plastics Corp.).
77. U.S. Pat. 3,366,698 (Jan. 30, 1968), E. P. DiBella (to Tenneco Chemicals Inc.).
78. T. Tkaczynski, Z. Winiarksi, and W. Markowski, *Przem. Chem.* **58**, 669 (1979).
79. Jpn. Kokai JP 59 206,322 (Nov. 22, 1984), (to Ihara Chemical Industry Co., Ltd.).
80. U.S. Pat. 4,031,145 (June 21, 1977), E. P. DiBella (to Tenneco Chemicals, Inc.).
81. U.S. Pat. 4,721,822 (Jan. 26, 1988), A. Leone-Bay, P. E. Timony, and L. Glaser (to Stauffer Chemical Co.).
82. Brit. Pat. 1,110,030 (Apr. 18, 1968), C. F. Kohll, H. D. Scharf, and R. Van Helden (to Shell Int'l Res. Maat. N. V.).
83. Neth. Appl. 6,907,390 (Nov. 17, 1970), D. A. Was (to Shell Int'l. Res. Maat. N. V.).
84. Jpn. Kokai JP 61 36,234 (Feb. 20, 1986), T. Irie and S. Doi (to Nitto Chemical Industry Co., Ltd.).
85. U.S. Pat. 3,219,688 (Nov. 23, 1965), E. D. Weil and co-workers (to Hooker Chemicals & Plastics Corp.).
86. H. C. Brimelow, L. Jones, and T. P. Metcalfe, *J. Chem. Soc.*, 1208 (1951).
87. F. F. Shcherebina, D. N. Tmenov, T. V. Lysukho, and N. P. Belous, *Zh. Prikl. Khim.* **53**, 2737 (1980).
88. U.S. Pat. 3,692,850 (Sept. 19, 1972), E. P. DiBella (to Tenneco Chemicals, Inc.).
89. V. S. Kozlova and A. N. Korol, *Zh. Anal. Khim.* **34**, 2406 (1979).
90. Jpn. Kokai JP 60 67,454 (Apr. 17, 1985), (to Nippon Kayaku Co., Ltd.).

91. I. Chu, S. Y. Shen, D. C. Villeneuve, V. E. Secours, and V. E. Valli, *J. Environ. Sci. Health, Part B* **B19,** 183 (1984).
92. R. Nishiyama and co-workers, *Yuki Gosei Kagaku Kyokai Shi* **23,** 515, 521 (1965).
93. Jpn. Kokai 70 28,367 (Sept. 16, 1970), M. Ishida (to Kureha Chemical Ind. Co., Ltd.).
94. V. Mark and co-workers, *J. Am. Chem. Soc.* **93,** 3538 (1971).
95. F. M. Ashton and A. S. Crafts, *Mode of Action of Herbicides*, John Wiley & Sons, Inc., New York, 1973, pp. 10–24, 438–448.
96. U.S. Pat. 3,914,298 (Oct. 21,1975), K. J. Dahl (to Raychem Corp.).

HENRY C. LIN
RAMESH KRISHNAMURTI
Occidental Chemical Corporation

BENZYL CHLORIDE, BENZAL CHLORIDE, AND BENZOTRICHLORIDE

The chlorination of toluene in the absence of catalysts that promote nuclear substitution occurs preferentially in the side chain. The reaction is promoted by free-radical initiators such as ultraviolet light or peroxides. Chlorination takes place in a stepwise manner and can be controlled to give good yields of the intermediate chlorination products. Small amounts of sequestering agents are sometimes used to remove trace amounts of heavy-metal ions that cause ring chlorination.

Experimental data taken from the chlorination of toluene in a continuous stirred tank flow reactor at 111°C and irradiated with light of 500 nm wavelength yield a product distribution shown in Table 1 (1).

Nearly all of the benzyl chloride [100-44-7], benzal chloride [98-87-3], and benzotrichloride [98-07-7] manufactured is converted to other chemical intermediates or products by reactions involving the chlorine substituents of the side chain. Each of the compounds has a single primary use that consumes a large portion of the compound produced. Benzyl chloride is utilized in the manufacture of benzyl butyl phthalate, a vinyl resin plasticizer; benzal chloride is hydrolyzed

Table 1. Distributions of Reactor Products[a] from Batch Chlorination of Toluene

$\dfrac{\text{Mol Cl}_2}{\text{Mol reactant in pdt}}$	Toluene	Benzyl chloride	Benzal chloride	Benzotrichloride
0.30	0.717	0.271	0.012	
0.51	0.507	0.480	0.013	
0.82	0.250	0.685	0.065	
0.98	0.138	0.744	0.118	
1.19	0.040	0.729	0.231	
1.32	0.030	0.672	0.325	
1.53		0.482	0.503	0.015
1.95		0.105	0.842	0.053
2.18		0.016	0.774	0.210

[a]Mole fractions.

to benzaldehyde; benzotrichloride is converted to benzoyl chloride. Benzyl chloride is also hydrolyzed to benzyl alcohol, which is used in the photographic industry, in perfumes (as esters), and in peptide synthesis by conversion to benzyl chloroformate [501-53-1] (see BENZYL ALCOHOL AND β-PHENETHYL ALCOHOL; CARBONIC AND CARBONOCHLORIDIC ESTERS).

Several related compounds, primarily ring-chlorinated derivatives, are also commercially significant. p-Chlorobenzotrichloride is converted to p-chlorobenzotrifluoride, an important intermediate in the manufacture of dinitroaniline herbicides.

Physical Properties

Benzyl chloride [(chloromethyl)benzene, α-chlorotoluene], $C_6H_5CH_2Cl$, is a colorless liquid with a very pungent odor. Its vapors are irritating to the eyes and mucous membranes, and it is classified as a powerful lacrimator. The physical properties of pure benzyl chloride are given in Table 2 (2–7). Benzyl chloride is insoluble in cold water, but decomposes slowly in hot water to benzyl alcohol. It is miscible in all proportions at room temperature with most organic solvents. The flash point of benzyl chloride is 67°C (closed cup); 74°C (open cup); autoignition temperature is 585°C; lower flammability limit: 1.1% by volume in air. Its volume coefficient of expansion is 9.72×10^{-4}.

Benzal chloride [(dichloromethyl)benzene, α,α-dichlorotoluene, benzylidene chloride], $C_6H_5CHCl_2$, is a colorless liquid with a pungent, aromatic odor. Benzal chloride is insoluble in water at room temperature but is miscible with most organic solvents.

Benzotrichloride [(trichloromethyl)benzene, α,α,α-trichlorotoluene, phenylchloroform], $C_6H_5CCl_3$, is a colorless, oily liquid with a pungent odor. It is soluble in most organic solvents, but it reacts with water and alcohol. For benzotrichloride the flash point is 127°C (Cleveland open cup) and the autoignition temperature is 211°C (8).

Binary azeotropic systems are reported for all three derivatives (9). The solubilities of benzyl chloride, benzal chloride, and benzotrichloride in water have been calculated by a method devised for compounds with significant hydrolysis rates (10).

Chemical Properties

The reactions of benzyl chloride, benzal chloride, and benzotrichloride may be divided into two classes: (1) reactions of the side chain containing the halogen; and (2) reactions of the aromatic ring.

Reactions of the Side Chain. Benzyl chloride is hydrolyzed slowly by boiling water and more rapidly at elevated temperature and pressure in the presence of alkalies (11). Reaction with aqueous sodium cyanide, preferably in the presence of a quaternary ammonium chloride, produces phenylacetonitrile [140-29-4] in high yield (12). The presence of a lower molecular-weight alcohol gives faster rates and higher yields. In the presence of suitable catalysts benzyl chloride reacts with

Table 2. Physical Properties of Benzyl Chloride, Benzal Chloride, and Benzotrichloride

Property	Benzyl chloride	Benzal chloride	Benzotrichloride
mol wt	126.58	161.03	195.48
freezing point, °C	−39.2	−16.4	−4.75
boiling point, °C	179.4	205.2	220.6
density, kg/m^3	1113.5$^4_4$	1256$^{14}_{14}$	1374$^{20}_4$
	1104$^{15}_{15}$		
	1100$^{15}_{20}$		
refractive index, n^t_D	1.54124^{15}		
	1.5392^{20}	1.5502^{20}	1.55789^{20}
surface tension, mN/m(= dyn/cm)	19.50$^{179.5}$	20.20$^{203.5}$	38.03^{20}
	0.03765^{20}		
dipole moment[a], C·m	6.24×10^{-30}	6.9×10^{-30}	7.24×10^{-30}
diffusion of vapor in air, D_o, cm^2/s	0.066		
vapor density(air = 1)	4.34		6.77
heat of combustion, kJ/mol[b]	3708[c]	3852[d]	3684[d]
specific heat at 25°C, J/kg·K)[b]	1444	1377	1206
heat of vaporization, kJ/mol[b]	50.1[e]	50.4[f]	52[g]
vapor pressure, °C at kPa[h]			
0.13	22.0	35.4	45.8
0.67	47.8	64.0	73.7
1.33	60.8	78.7	87.6
5.33	90.7	112.1	119.8
8.00	100.5	123.6	130.0
13.3	114.2	138.3	144.3
26.7	134.0	160.7	165.6
53.3	155.8	187.0	189.2

[a]In dilute benzene solution. To convert C·m to debye, divide by 3.336×10^{-30}.
[b]To convert J to cal, divide by 4.184.
[c]At constant volume.
[d]At constant pressure.
[e]At 25°C.
[f]At 72°C.
[g]At 80°C.
[h]To convert kPa to mm Hg, multiply by 7.50.

carbon monoxide to produce phenylacetic acid [103-82-2] (13–15). With different catalyst systems in the presence of calcium hydroxide, double carbonylation to phenylpyruvic acid [156-06-9] occurs (16). Benzyl esters are formed by heating benzyl chloride with the sodium salts of acids; benzyl ethers by reaction with sodium alkoxides. The ease of ether formation is improved by the use of phase-transfer catalysts (17) (see CATALYSIS, PHASE-TRANSFER).

The benzylation of a wide variety of aliphatic, aromatic, and heterocyclic amines has been reported. Benzyl chloride is converted into mono-, di-, and tri-benzylamines by reaction with ammonia. Benzylaniline [103-32-2] results from the reaction of benzyl chloride with aniline. Reaction with tertiary amines yields quaternary ammonium salts; with trialkylphosphines, quaternary phosphonium salts; and with sulfides, sulfonium salts are formed.

Benzyl chloride readily forms a Grignard compound by reaction with magnesium in ether with the concomitant formation of substantial coupling product, 1,2-diphenylethane [103-29-7]. Benzyl chloride is oxidized first to benzaldehyde [100-52-7] and then to benzoic acid. Nitric acid oxidizes directly to benzoic acid [65-85-0]. Reaction with ethylene oxide produces the benzyl chlorohydrin ether, $C_6H_5CH_2OCH_2CH_2Cl$ (18). Benzylphosphonic acid [10542-07-1] is formed from the reaction of benzyl chloride and triethyl phosphite followed by hydrolysis (19).

Benzyl chloride reacts with alkali hydrogen sulfides, sulfides, and polysulfides to yield benzenethiol, dibenzyl sulfide, and dibenzyl polysulfide, respectively. With sodium cyanate it forms benzyl isocyanate (20).

Benzyl chloride reacts with benzene in the presence of a Lewis acid catalyst to give diphenylmethane [101-81-5]. It undergoes self-condensation to form polymeric oils and solids (21). With phenol, benzyl chloride produces a mixture of o- and p-benzylphenol.

Benzal chloride is hydrolyzed to benzaldehyde under both acid and alkaline conditions. Typical conditions include reaction with steam in the presence of ferric chloride or a zinc phosphate catalyst (22) and reaction at 100°C with water containing an organic amine (23). Cinnamic acid in low yield is formed by heating benzal chloride and potassium acetate with an amine as catalyst (24).

Benzotrichloride is hydrolyzed to benzoic acid by hot water, concentrated sulfuric acid, or dilute aqueous alkali. Benzoyl chloride [98-88-4] is produced by the reaction of benzotrichloride with an equimolar amount of water or an equivalent of benzoic acid. The reaction is catalyzed by Lewis acids such as ferric chloride and zinc chloride (25). Reaction of benzotrichloride with other organic acids or with anhydrides yields mixtures of benzoyl chloride and the acid chloride derived from the acid or anhydride (26). Benzotrifluoride [98-08-8] is formed by the reaction of benzotrichloride with anhydrous hydrogen fluoride under both liquid- and vapor-phase reaction conditions.

Aromatic Ring Reactions. In the presence of an iodine catalyst chlorination of benzyl chloride yields a mixture consisting mostly of the ortho and para compounds. With strong Lewis acid catalysts such as ferric chloride, chlorination is accompanied by self-condensation. Nitration of benzyl chloride with nitric acid in acetic anhydride gives an isomeric mixture containing about 33% ortho, 15% meta, and 52% para isomers (27); with benzal chloride, a mixture containing 23% ortho, 34% meta, and 43% para nitrobenzal chlorides is obtained.

Chlorosulfonation of benzotrichloride with chlorosulfonic acid (28) or with sulfur trioxide (29) gives m-chlorosulfonyl benzoyl chloride [4052-92-0] in high yield. Nitration with nitronium fluoroborate in sulfolane gives 68% m-nitrobenzotrichloride [709-58-0] along with 13% of the ortho and 19% of the para isomers (30).

Nitrobenzotrichloride is also obtained in high yield with no significant hydrolysis when nitration with a mixture of nitric and sulfuric acids is carried out below 30°C (31). 2,4-Dihydroxybenzophenone [131-56-6] is formed in 90% yield by the uncatalyzed reaction of benzotrichloride with resorcinol in hydroxylic solvents (32) or in benzene containing methanol or ethanol (33). Benzophenone derivatives are formed from a variety of aromatic compounds by reaction with benzotrichloride in aqueous or alcoholic hydrofluoric acid (34).

Benzotrichloride with zinc chloride as catalyst reacts with ethylene glycol to form 2-chloroethyl benzoate [7335-25-3] (35). Perchlorotoluene is formed by chlorination with a solution of sulfur monochloride and aluminum chloride in sulfuryl chloride (36).

Manufacture

Benzyl chloride is manufactured by the thermal or photochemical chlorination of toluene at 65–100°C (37). At lower temperatures the amount of ring-chlorinated by-products is increased. The chlorination is usually carried to no more than about 50% toluene conversion in order to minimize the amount of benzal chloride formed. Overall yield based on toluene is more than 90%. Various materials, including phosphorus pentachloride, have been reported to catalyze the side-chain chlorination. These compounds and others such as amides also reduce ring chlorination by complexing metallic impurities (38).

Under typical liquid-phase chlorination conditions the maximum conversion to benzyl chloride of about 70% is reached after reaction of about 1.1 moles of chlorine per mole of toluene (39). Higher yields of benzyl chloride have been claimed: 80% for low temperature chlorination (40); 80–85% for light-catalyzed chlorination in the vapor phase (41) and 93.6% for continuous chlorination above 125°C in a column packed with glass rings (42).

In commercial practice, chlorination may be carried out either in batches or continuously. Glass-lined or nickel reactors may be used. Because certain metallic impurities such as iron catalyze ring chlorination and self-condensation, their presence must be avoided. The crude product is purged of dissolved hydrogen chloride, neutralized with alkali, and distilled. Chlorine efficiency is high; muriatic acid made by absorbing the by-product hydrogen chloride in water is usually free of significant amounts of dissolved chlorine.

An 80% yield of benzyl chloride is obtained with sulfuryl chloride as chlorinating agent. Yields of >70% of benzyl chloride are obtained by the zinc chloride-catalyzed chloromethylation of benzene but formation of bis-chloromethyl ether presents a health hazard for this reaction pathway.

Benzyl chloride undergoes self-condensation relatively easily at high temperatures or in the presence of trace metallic impurities. The risk of decomposition during distillation is reduced by the use of various additives including lactams (43) and amines (44,45). Lime, sodium carbonate, and triethylamine are used as stabilizers during storage and shipment. Other soluble organic compounds that are reported to function as stabilizers in low concentration include DMF (46), arylamines (47), and triphenylphosphine (48).

Benzal chloride can be manufactured in 70% yield by chlorination with 2.0–2.2 moles of chlorine per mole of toluene. The benzal chloride is purified by distillation. Benzal chloride is also formed by the reaction of dichlorocarbene (:CCl$_2$) with benzene (49).

Further chlorination at a temperature of 100–140°C with ultraviolet light yields benzotrichloride. The chlorination is normally carried to a benzotrichloride content of greater than 95% with a low benzal chloride content. After purging with inert gas to remove hydrogen chloride, the crude product is utilized directly or purified by distillation. Under batch conditions chlorine efficiency during the latter stages of the chlorination is low. Product quality and chlorine efficiency can be improved by carrying out the chlorination continuously in a multistage system (50). Additives such as phosphorus trichloride are used to complex metallic impurities. Contaminants or reaction conditions that cause darkening and thereby reduce light penetration must be avoided if the chlorination is to be efficient (51). The radiation-initiated chlorination of toluene has also been investigated (52–54).

An understanding of competing reactions in the manufacturing process is important if by-products are to be minimized. Three competing reactions are possible under conditions of the reaction.

Free-radical substitution of the side chain of toluene

Addition to the aromatic ring

Electrophilic substitution on the aromatic ring

An extensive kinetic study of the photochlorination of toluene in a continuous annular reactor has investigated the parameters that effect the product distribution from these reactions (39). Chlorination on the aromatic ring can occur by either addition followed by elimination of HCl or electrophilic aromatic substitution. Both reactions occur at low (40°C) temperature and are promoted by high concentration of chlorine. Electrophilic substitution is catalyzed by traces of metals like iron and aluminum. Formation of ring-chlorinated compounds is markedly increased by lowering the temperature to 40°C and chlorinating in the dark. These products contribute to a high boiling fraction that reduces the yield of side-chain chlorination products.

Free-radical chlorine substitution of the methyl group hydrogens is promoted by elevated temperature (80–130°C), a radical producing light source, and free-radical catalysts like peroxides. Oxygen inhibits the reaction. The ratio of benzyl to benzal to benzotrichloride depends on the ratio of chlorine to toluene in the reaction. From analyses of the product distribution for the free-radical chlorination at 100°C and irradiation with blue light, the relative rates are $k_1/k_2 = 5.9$ and $k_2/k_3 = 5.2$. Blue light (energy maximum at about 425 nm) gives a higher rate of chlorination than ultraviolet (about 370 nm) because it more effectively penetrates a solution containing free chlorine.

Handling and Shipment

As is the case during manufacture, contact with those metallic impurities that catalyze Friedel-Crafts condensation reactions must be avoided. The self-condensation reaction is exothermic and the reaction can accelerate producing a rapid buildup of hydrogen chloride pressure in closed systems.

Benzyl chloride is available in both anhydrous and stabilized forms. Both forms can be shipped in glass carboys, nickel and lined-steel drums, and nickel tank trucks and tank cars. Stabilized benzyl chloride can be shipped in unlined and lacquer-lined drums, and tank trucks or cars of construction other than nickel. Glass-lined tanks are the first choice for bulk storage of anhydrous benzyl chloride; lead-lined, nickel, or ceramic tanks can also be used.

Benzyl chloride is classified by DOT as chemicals NO1BN, poisonous, corrosive and a hazardous substance (100 lbs-45.45 kg). Benzal chloride is classified as poisonous and a hazardous substance (5000 lbs-2270 kg). Benzotrichloride is classified under DOT regulation as a corrosive liquid NOS and a hazardous substance (10 lbs-4.5 kg). The Freight Classification Chemical NOI applies. It is shipped in lacquer-lined steel drums and nickel-lined tank trailers. Benzal chloride is handled in a similar fashion.

Economic Aspects

Plant capacities for the production of benzyl chloride in the western world totaled 144,200 t/yr in 1989. Monsanto, with plants in Belgium (23,000 t/yr) and Bridgeport, New Jersey (40,000 t/yr) is the largest producer. Bayer in West Germany (20,000 t/yr) and Tessenderlo Chemie in Belgium (18,000 t/yr) are also principal producers.

In the United States, in addition to Monsanto, Akzo, which took over part of Stauffer, is the only other producer of benzyl chloride (9,000 t/yr). Velsicol was a producer but shut down in 1986 because of a declining market forecast. Total western world production in 1988 was approximately 92,700 t, with U.S. production at 26,500 t or 54% of capacity.

Benzotrichloride is produced from total side-chain chlorination of toluene or of residual products from benzyl chloride production. In Western Europe, Bayer has the largest capacity (14,000 t/yr), and there are only two significant producers in the United States: Occidental Chemical in Niagara Falls, New York (20,000 t/yr), and Velsicol Chemical (11,000 t/yr). Total capacity in the western world is 68,000 t/yr and production of benzotrichloride in 1988 was estimated at 31,500 t.

Benzyl chloride and butyl alcohol react with phthalic anhydride in one step to yield benzyl butyl phthalate [85-68-7], a plasticizer made by Monsanto and known by its trade name Santicizer 160.

Benzotrichloride is a chemical intermediate used to produce two significant products. Partial hydrolysis or reaction with benzoic acid yields benzoyl chloride, whereas chlorination and subsequent reaction with hydrogen fluoride yields p-chlorobenzotrifluoride [98-56-6].

Identification and Analysis

The side-chain chlorine contents of benzyl chloride, benzal chloride, and benzotrichlorides are determined by hydrolysis with methanolic sodium hydroxide followed by titration with silver nitrate. Total chlorine determination, including ring chlorine, is made by standard combustion methods (55). Several procedures for the gas chromatographic analysis of chlorotoluene mixtures have been described (56,57). Proton and ^{13}C nuclear magnetic resonance shifts, characteristic infrared absorption bands, and principal mass spectral peaks have been summarized including sources of reference spectra (58). Procedures for measuring trace benzyl chloride in air (59) and in water (60) have been described.

A gas chromatographic determination of benzotrichloride and related compounds in the work environment, after adsorption on a polymeric adsorbant and desorption with CCl_4 has been reported (61). Trace amounts of benzyl chloride, benzal chloride, and benzotrichloride in environmental samples can be analyzed by Method 8120 of *EPA Manual SW-846* with modifications (62).

Health and Safety Factors

Benzyl chloride is a severely irritating liquid and causes damage to the eyes, skin, and respiratory tract including pulmonary edema. Other possible effects of overexposure to benzyl chloride are CNS depression, liver, and heart damage. Table 3 lists some exposure limits.

Table 3. Toxicology of Side-Chain Chlorinated Toluenes

	Benzyl chloride	Benzal chloride	Benzotrichloride
LD_{50} (rat), mg/kg	1000[a,b]		6000[b,c]
LD_{50} (mice), mg/kg		467[d,e]	
LC_{50} (mice, inhalation 2 h), ppm	80[f]		
LC_{50} (rat, inhalation), ppm	150[f]	82[g]	30[f]

[a]Administered subcutaneously in oil.
[b]Slightly toxic.
[c]Ref. 63.
[d]Moderately toxic.
[e]Ref. 64.
[f]Ref. 65.
[g]Ref. 66.

Benzyl chloride induced a positive mutagenic response in the Ames Assay in strain TA 100 with and without rat liver S-9 metabolic activation. Benzyl chloride also induced *in vitro* cellular transformation in Syrian hamster embryo cultures and DNA alkylation in several organs of the male mouse following iv administration. In summary, IARC states there is limited evidence that benzyl chloride is carcinogenic in experimental animals; epidemiological data was inadequate to evaluate carcinogenicity to humans (67).

Other toxicological effects that may be associated with exposure to benzyl chloride based on animal studies are skin sensitization and developmental embryo and/or fetal toxicity. A 1980 OSHA regulation has established a national occupational exposure limit for benzyl chloride of 5 mg/m^3 (1 ppm). Concentrations of 160 mg/m^3 (32 ppm) in air cause severe irritation of the eyes and respiratory tract (68).

Vapors of both benzal chloride and benzotrichloride are strongly irritating and lacrimatory. Reported toxicities appear in Table 3. Also, for benzotrichloride, the lowest published lethal dose (frog) is 2150 mg/kg (69) and the toxic dose level (inhalation rats) is 125 ppm/4 h (69).

For all three compounds, biological data relevant to the evaluation of carcinogenic risk to humans are summarized in the World Health Organization International Agency for Research on Cancer monograph (70).

Uses

Nearly all uses and applications of benzyl chloride are related to reactions of the active halide substituent. More than two-thirds of benzyl chloride produced is used in the manufacture of benzyl butyl phthalate, a plasticizer used extensively in vinyl flooring and other flexible poly(vinyl chloride) uses such as food packaging. Other significant uses are the manufacture of benzyl alcohol [100-51-6] and of benzyl chloride-derived quaternary ammonium compounds, each of which consumes more than 10% of the benzyl chloride produced. Smaller volume uses include the manufacture of benzyl cyanide [140-29-4], benzyl esters such as benzyl acetate [140-11-4], butyrate, cinnamate, and salicylate, benzylamine [100-46-9], and benzyldimethylamine [103-83-8], and p-benzylphenol [101-53-1]. In the dye industry benzyl chloride is used as an intermediate in the manufacture of triphenylmethane dyes (qv). First generation derivatives of benzyl chloride are processed further to pharmaceutical, perfume, and flavor products.

Nearly all of the benzal chloride produced is consumed in the manufacture of benzaldehyde. Benzaldehyde (qv) is used in the manufacture of perfume and flavor chemicals, dyes, and pharmaceuticals. The principal part of benzotrichloride production is used in the manufacture of benzoyl chloride (see BENZOIC ACID). Lesser amounts are consumed in the manufacture of benzotrifluoride, as a dyestuff intermediate, and in producing hydroxybenzophenone ultraviolet light stabilizers (see UV STABILIZERS). Benzotrifluoride is an important intermediate in the manufacture of herbicides, pharmaceuticals, antimicrobial agents, and the lampreycide, 4-nitro-3-(trifluoromethyl)phenol [88-30-2].

Benzyl-derived quaternary ammonium compounds are used widely as cationic surface-active agents and as germicides, fungicides, and sanitizers. Benzyl alcohol is used in a wide spectrum of applications including pharmaceuticals and perfumes, as a solvent, and as a textile dye assistant.

Derivatives

Ring-Substituted Derivatives.   The ring-chlorinated derivatives of benzyl chloride, benzal chloride, and benzotrichloride are produced by the direct side-chain chlorination of the corresponding chlorinated toluenes or by one of several indirect routes if the required chlorotoluene is not readily available. Physical constants of the main ring-chlorinated derivatives of benzyl chloride, benzal chloride, and benzotrichloride are given in Table 4.

The 2- and 4-monochloro and 2,4- and 3,4-dichlorobenzyl chloride, benzal chloride, and benzotrichlorides are manufactured by side-chain chlorination of the appropriate chlorotoluene. *p*-Chlorobenzotrichloride (1-chloro-4-trichloromethylbenzene) can be prepared by peroxide-catalyzed chlorination of *p*-toluenesulfonyl chloride or di-*p*-toluylsulfone (71). 2,4-Dichlorobenzotrichloride (1,3-dichloro-4-trichloromethylbenzene) is obtained by the chlorination of 2-chloro-4-chlorosulfonyltoluene (72).

3,4-Dichlorobenzyl chloride (1,2-dichloro-4-chloromethylbenzene) containing some 2,3-dichlorobenzyl chloride is produced by the chloromethylation of *o*-dichlorobenzene in oleum solution (73). Chlorination of 2-chloro-6-nitrotoluene at 160–185°C gives a mixture of 2,6-disubstituted benzal chloride and 2,6-dichlorobenzyl chloride (74).

The ring-chlorinated benzyl chlorides are used in the preparation of quaternary ammonium salts and as intermediates for pharmaceuticals and pesticides. *p*-Chlorobenzyl chloride is an intermediate in the manufacture of the rice herbicide, Saturn ((*S*-4-chlorobenzyl)-*N*,*N*-diethylthiolcarbarmate [28249-77-6]) (75). The *o*- and *p*-chlorobenzal chlorides (1-chloro-2-and 4-dichloromethylbenzenes) are starting materials for the manufacture of *o*- and *p*-chlorobenzaldehydes.

The *o*- and *p*-monochloro- and 2,4- and 3,4-dichlorobenzotrichlorides are intermediates in the manufacture of the corresponding chlorinated benzoic acids and benzoyl chlorides. Fluorination of the chlorinated benzotrichlorides produces the chlorinated benzotrifluorides, intermediates in the manufacture of dinitroaniline and diphenyl ether herbicides (76).

2,6-Dichlorobenzal chloride is used in the manufacture of 2,6-dichlorobenzaldehyde and 2,6-dichlorobenzonitrile (77). With the exception of certain products used in the manufacture of herbicides, the volume of individual compounds produced is small, amounting to no more than several hundred tons annually for any individual compound.

Side-Chain Chlorinated Xylene Derivatives.   Only a few of the nine side-chain chlorinated derivatives of each of the xylenes are available from direct chlorination. All three of the monochlorinated compounds, α-chloro-*o*-xylene (1-(chloromethyl)-2-methylbenzene [552-45-4]), α-chloro-*m*-xylene (1-(chloromethyl)-3-methylbenzene [620-19-9]), α-chloro-*p*-xylene (1-(chloromethyl)-4-methylbenzene [104-82-5]) are obtained in high yield from partial chlorination of the xylenes. 1,3-Bis(chloromethyl)benzene [626-16-4] can be isolated in moderate yield from chlorination mixtures (78,79).

Table 4. Physical Constants of the Main Ring-Chlorinated Derivatives of Benzyl Chloride, Benzal Chloride, and Benzotrichloride

Benzene derivative	Common name	CAS Registry Number	Mp,°C	Bp,°C	n_D^{20}	Density, kg/m³
1-chloro-2-(chloromethyl)	o-chlorobenzyl chloride	[611-19-8]	-17	217	1.5330	1270
1-chloro-3-(chloromethyl)	m-chlorobenzyl chloride	[620-20-2]		215-216[a]		1269.5
1-chloro-4-(chloromethyl)	p-chlorobenzyl chloride	[104-83-6]	31	222	1.5554	
1-chloro-2-(dichloromethyl)	o-chlorobenzal chloride	[88-66-4]		228.5	1.5670[b]	1399
1-chloro-3-(dichloromethyl)	m-chlorobenzal chloride	[15145-69-4]		235-237		
1-chloro-4-(dichloromethyl)	p-chlorobenzal chloride	[13940-94-8]		236[c]		
2,4-dichloro-1-(chloromethyl)	2,4-dichlorobenzyl chloride	[94-99-5]	-2.6	248	1.5761	1407
1,3-dichloro-2-(chloromethyl)	2,6-dichlorobenzyl chloride	[2014-83-7]	39-40	117-119[d]		
1,2-dichloro-4-(chloromethyl)	3,4-dichlorobenzyl chloride	[102-47-6]	37-37.5	241		1412
1-chloro-2-(trichloromethyl)	o-chlorobenzotrichloride	[2136-89-2]	29.4	264.3	1.5836	1519
1-chloro-3-(trichloromethyl)	m-chlorobenzotrichloride	[2136-81-4]		255	1.4461	1495
1-chloro-4-(trichloromethyl)	p-chlorobenzotrichloride	[5216-25-1]		245	1.4463	1495
1,3-dichloro-2-(dichloromethyl)	2,6-dichlorobenzal chloride	[81-19-6]		250		
1,2-dichloro-4-(dichloromethyl)	3,4-dichlorobenzal chloride	[56961-84-3]		257		1518
2,4-dichloro-1-(dichloromethyl)	2,4-dichlorobenzal chloride	[134-25-8]	47-48	155-159[e]		
1,2-dichloro-4-(trichloromethyl)	3,4-dichlorobenzotrichloride	[13014-24-9]	25.8	283.1	1.5886	1591

[a]At 100.4 kPa (753 mm Hg).
[b]At 16°C.
[c]At 100.7 kPa (755 mm Hg).
[d]At 1.87 kPa (14 mm Hg).
[e]At 2.67 kPa (20 mm Hg).

123

The fully side-chain chlorinated products, 1,3-bis(trichloromethyl)benzene [*881-99-1*] and 1,4-bis(trichloromethyl)benzene [*68-36-0*], are manufactured by exhaustive chlorination of meta and para xylenes. For the meta compounds, ring chlorination cannot be completely eliminated in the early stages of the reaction. The xylene hexachlorides are intermediates in the manufacture of the xylene hexafluorides and of iso- and terephthaloyl chloride [*100-20-9*] (see PHTHALIC ACIDS).

1-(Dichloromethyl)-2-(trichloromethyl)benzene [*2741-57-3*], the end product of exhaustive side-chain chlorination of *o*-xylene (80) is an intermediate in the manufacture of phthalaldehydic acid [*119-67-5*].

BIBLIOGRAPHY

"Benzyl Chloride, Benzal Chloride, and Benzotrichloride" under "Chlorine Compounds, Organic" in *ECT* 1st ed., Vol. 3, pp. 822–826 by R. L. Clark and C. P. Neidig, Heyden Chemical Corp.; "Benzyl Chloride, Benzal Chloride, and Benzotrichloride" under "Chlorocarbons and Chlorohydrocarbons" in *ECT* 2nd ed., Vol. 5, pp. 281–289, by H. Sidi, Heyden Newport Chemical Corp.; "Benzyl Chloride, Benzal Chloride, and Benzotrichloride" under "Chlorocarbons, -Hydrocarbons (Benzyl)" in *ECT* 3rd ed., Vol. 5, pp. 828–837, by S. Gelfand, Hooker Chemical Corp.

1. J. S. Ratcliffe, *Br. Chem. Eng.* **11,** 1535 (1966).
2. *Handbook of Chemistry and Physics*, 58th ed., CRC Press Inc., Cleveland, Ohio, 1977–1978, pp. C-522, 523, 527, 528, 738, D-198.
3. *International Critical Tables*, Vol. 5, McGraw-Hill Book Co., New York, 1929, pp. 62, 111, 169.
4. R. R. Dreisbach, in *Advances in Chemistry Series*, American Chemical Society, Washington, D.C., 1955, pp. 141–143.
5. A. L. McClellan, *Tables of Experimental Dipole Moments*, W. H. Freeman and Co., San Francisco, Calif., 1963, pp. 232, 237, 238, 243.
6. J. Timmermans and Mme. Hennant-Roland, *J. Chim. Phys.* **32,** 501 (1935).
7. D. R. Stull, *Ind. Engr. Chem.* **39,** 525 (1947).
8. Occidental Chemical Corp. MSDS; M7608, Feb. 19, 1991.
9. L. H. Horsley and co-workers, *Azeotropic Data III*, no. 116 in *Advances in Chemistry Series*, American Chemical Society, Washington, D.C., 1973.
10. K. Ohnishi and K. Tanabe, *Bull. Chem. Soc. Jpn.* **44,** 2647 (1971).
11. U.S. Pat. 3,557,222 (Jan. 19, 1971), H. W. Withers and J. L. Rose (to Velsicol Chemical Corp.).
12. Brit. Pat. 1,336,883 (Nov. 14, 1973), H. Coates, R. L. Barker, R. Guest, and A. Kent (to Albright & Wilson, Ltd.).
13. J. K. Stille and P. K. Wong, *J. Org. Chem.* **40,** 532 (1975).
14. Ger. Offen. 2,259,072 (June 20, 1974), M. E. Chahawi and H. Richtzenhain (to Dynamit Nobel AG).
15. Ger. Offen. 2,035,902 (Feb. 4, 1971), M. Foa, L. Cassar, and G. P. Chiusoli (to Montecatini Edison SPA).
16. U.S. Pat. 4,689,431 (Aug. 25, 1987), M. Tanaka and K. Oktsuka (to Nissan Chemical Industries, Ltd.).
17. H. H. Freedman and R. A. DuBois, *Tetrahedron Lett.* **38,** 3251 (1975).
18. Jpn. Kokai 75 62,942 (May 29, 1975), S. Komori.
19. Brit. Pat. 1,366,600 (Sept. 11, 1974), F. J. Harris and H. L. Brown (to Scottish Agric. Ind. Ltd.).

20. Ger. Offen. 2,449,607 (Apr. 30, 1975), Y. Inamoto and co-workers (to Kao Soap Co., Ltd.).
21. H. C. Haas, D. I. Livingston, and M. Saunders, *J. Polym. Sci.* **15**, 503 (1955).
22. U.S. Pat. 3,542,885 (Aug. 18, 1970), A. J. Deinet (to Tenneco Chemicals Inc.).
23. Jpn. Pat. 69 12,132 (June 2, 1969), H. Funamoto (to Kureha Chem. Ind. Co. Ltd.).
24. Jpn. Kokai 73 81,830 (Nov. 30, 1973), K. Shinoda and K. Kobayashi (to Kureha Chem. Ind. Co. Ltd.).
25. Jpn. Kokai 54 019929 (Feb. 15, 1979), (to Nikkei Kako, KK).
26. Jpn. Kokai 61 155350 (July 15, 1986), (to Ihara-Nikkei Kagaku).
27. F. DeSarlo and co-workers, *J. Chem. Soc.*, B719 (1971).
28. U.S. Pat. 3,290,370 (Dec. 12, 1966), E. D. Weil and R. J. Lisanke (to Hooker Chemical Corp.).
29. U.S. Pat. 3,322,822 (May 30, 1967), S. Gelfand (to Hooker Chemical Corp.).
30. G. Grynkiewicz and J. H. Ridd, *J. Chem. Soc.*, B716 (1971).
31. U.S. Pat. 3,182,091 (May 4, 1965), O. Scherer, H. Hahn, and N. Munch (to Farb. Hoeschst Akt.).
32. U.S. Pat. 3,769,349 (Oct. 30, 1973), M. Yukutomi, Y. Tanaka, S. Genda, and M. Kitauri (to Kyodo Chemical Co. Ltd.).
33. Ger. Offen. 2,208,197 (Aug. 30, 1973), B. Lachmann and H. J. Rosenkrantz (to Bayer AG).
34. Ger. Offen. 2,451,037 (Apr. 29, 1976), K. Eiglmeier (to Hoechst AG).
35. U.S. Pat. 3,050,549 (Aug. 21, 1962), S. Gelfand (to Hooker Chemical Corp.).
36. M. Ballester, C. Molinet, and J. Castaner, *J. Am. Chem. Soc.* **82**, 4254 (1960).
37. *Faith, Keyes, and Clark's Industrial Chemicals* 4th ed., John Wiley & Sons, Inc., New York, 1975, pp. 145–148.
38. U.S. Pat. 2,695,873 (Nov. 30, 1954), A. J. Loverde (to Hooker Electrochemical Co.).
39. H. G. Haring and H. W. Knol, *Chem. Process. Eng.* **45**, 540, 619, 690 (1964); 46, 38 (1965).
40. G. Benoy and L. DeMayer, *Compt. Rend. 27th Congr. Intern. Chim. Ind.*, Brussels, Belgium, 1954; *Industrie Chim. Belg.* **20**, Spec. No. 160-2 (1955).
41. G. V. Asolkar and P. C. Guha, *J. Indian Chem. Soc.* **23**, 47 (1946).
42. A. Scipioni, *Ann. Chim. (Rome)* **41**, 491 (1951).
43. U.S. Pat. 3,715,283 (Feb. 6, 1973), W. Bockmann (to Bayer Akt.).
44. Czeck. Pat. 159,100 (June 15, 1975), J. Best and M. Soolek.
45. Brit. Pat. 1,410,474 (Oct. 15, 1975), C. H. G. Hands (to Albright and Wilson Ltd.).
46. Jpn. Kokai 73 05,726 (Jan. 24, 1972), N. Kato and Y. Sato (to Mitsui Toatsu Chemicals Inc.).
47. Jpn. Kokai 73 05,725 (Jan. 24, 1972), N. Kato and Y. Sato (to Mitsui Toatsu Chemicals Inc.).
48. U.S. Pat. 3,535,391 (Oct. 20, 1970), G. D. Kyker (to Velsicol Chemical Co.).
49. Brit. Pat. 1,390,394 (Apr. 9, 1975), A. D. Forbes, R. C. Pitkethly, and J. Wood (to Brit. Petrol. Co. Ltd.).
50. Ger. Offen. 2,152,068 (Apr. 26, 1973), W. Bockmann and R. Hornung; D.T. 2,227,337 (Aug. 28, 1975), (to Bayer AG).
51. Jpn. Kokai 76 08, 223 (Jan. 23, 1976), M. Fuseda and K. Ezaki (to Hodogaya Chemical Co. Ltd.).
52. J. Y. Yang, C. C. Thomas, Jr., and H. T. Cullinan, *Ind. Eng. Chem. Process Res. Develop.* **9**, 214 (1970).
53. H. T. Cullinan, Jr. and co-workers, in Ref. 47, p. 222.
54. B. F. Ives, H. T. Cullinan, Jr., and J. Y. Yang, *Nucl. Technol.* **18**, 29 (1973).
55. W. Kirsten, *Anal. Chem.* **25**, 74 (1953).
56. D. A. Solomons and J. S. Ratcliffe, *J. Chromatog.* **76**, 101 (1973).

57. R. Ramakrishnan and N. Subramanian, *J. Chromatog. 114,* 247 (1975).
58. J. G. Grasselli and W. M. Richey, eds., *Atlas of Spectral Data and Physical Constants for Organic Compounds,* 2nd ed., Vol. IV, CRC Press Inc., Cleveland, Ohio, 1975.
59. B. B. Baker, Jr., *J. Am. Ind. Hyg. Assoc.* **35,** 735 (1974).
60. G. A. Junk and co-workers, *J. Chromatog. 99,* 745 (1974).
61. H. Matsushita and S. Kanno, *Ind. Health* **17,** 199–206, 1979.
62. V. Lopez-Avila, N. S. Dodhiwala, J. Milones, and W. F. Beckert, *J. Assoc. Off. Anal. Chem.* **72,** 593–602 (1989).
63. N. I. Sax, *Dangerous Properties of Industrial Materials,* 4th ed., Van Nostrand Reinhold Co., New York, 1975.
64. V. V. Stankevich and V. I. Osetrov, *Gigiena i Fisiol. Tr. Proizv. Toksikol., Klinika Prof. Zabolevanii,* 96 (1963).
65. *IARC Monogr. Eval. Carcinog. Risk Chem. Man 11,* 217–223 (1976); *Toxbib.* **77,** 50224 (1977).
66. T. V. Mikhailova, *Gig. Tr. Prof. Zabol* **8,** 14 (1964).
67. *IARC Monogr. Eval. Carcinog. Risk Chem. Man* **29,** 59 (1982).
68. W. F. von Oettingen, *The Halogenated Aliphatic, Olefinic, Cyclic, Aromatic and Aliphatic-Aromatic Hydrocarbons including the Halogenated Insecticides, their Toxicity and Potential Dangers,* DHEW (PHS) Publication No. 414, Washington D.C., U.S. Government Printing Office, 1955, pp. 300–302.
69. H. E. Christensen, ed., *Registry of Toxic Effects of Chemical Substances,* U.S. Dept. of Health, Education, and Welfare, Rockville, Md., 1976.
70. *IARC Monograph* **29,** 49–80 (1982).
71. Jpn. Kokai 75 25,534 (Mar. 18, 1975), K. Kobayashi, N. Ishimo, and T. Nobeoka (to Fuso Chemical Co. Ltd.).
72. U.S. Pat. 3,230,268 (Jan. 18, 1966), K. Kobayashi and N. Ishino (to Fuso Chemical Co. Ltd.).
73. Brit. Pat. 951,302 (Mar. 4, 1964), (to Monsanto Canada Ltd.).
74. Ger. Pat. 1,237,552 (Mar. 30, 1967), J. T. Hackmann, J. Yates, T. J. Wilcox, P. T. Haken, and D. A. Wood (to Shell Research Ltd.).
75. U.S. Pat. 3,914,270 (Oct. 21, 1975), K. Makoto, H. Kamata, and K. Masuro (to Kumiai Chem. Ind. Co. Ltd.).
76. F. M. Ashton and A. S. Crafts, *Mode of Action of Herbicides,* John Wiley & Sons, New York, 1973, pp. 10–24, 438–448.
77. U.S. Pat. 3,458,560 (July 29, 1969), R. A. Carboni (to E. I. du Pont de Nemours & Co., Inc.).
78. U.S. Pat. 2,994,653 (Apr. 27, 1959), G. A. Miller (to Diamond Alkali Co.).
79. E. Clippinger, *ACS Petrol. Div. Prep.* **15**(1), B 37 (1970).
80. Ger. Offen. 2,535,969 (Feb. 17, 1977), P. Riegger, H. Richtzenhain, and G. Zoche (to Dyanmit Nobel AG).

HENRY C. LIN
JOSEPH F. BIERON
Occidental Chemical Corporation

TOXIC AROMATICS

Chlorinated biphenyls, chlorinated naphthalenes, benzene hexachloride, [608-73-1], and chlorinated derivatives of cyclopentadiene are no longer in commercial use because of their toxicity. However, they still impact on the chemical industry because of residual environmental problems. This article discusses the toxicity and environmental impact of these materials.

Polychlorinated Biphenyls

Polychlorinated biphenyls (PCBs) typify halogenated aromatic hydrocarbons (HAHs), industrial compounds or by-products that have been widely identified in the environment and chemical waste dumpsites (1–8). Other HAHs include the polychlorinated dibenzo-*p*-dioxins (PCDDs), dibenzofurans (PCDFs), diphenyl ethers (PCDEs), naphthalenes (PCNs), and benzenes (PCBzs). PCBs were used in industry as heat-transfer fluids, organic diluents, lubricant inks, plasticizers, fire retardants, paint additives, sealing liquids, immersion oils, adhesives, de-dusting agents, waxes, and as dielectric fluids for capacitors and transformers. After the initial detection of PCBs in the environment in the late 1960s, several studies confirmed their widespread occurrence throughout the global ecosystem. These studies led to the initial ban on all open uses of PCBs in the early 1970s (8–17) and a later ban on their closed uses as dielectric fluids in transformers and capacitors. New transformers and capacitors, as well as PCB-containing electrical equipment, are now filled with alternative fluids.

Chemistry and Environmental Impact. PCBs are synthesized by the chlorination of biphenyl and the resulting products are designated according to their percent (by weight) chlorine content (2). For example, Aroclors 1221, 1242, and 1260 contain 21, 42, and 60 wt % chlorine. The commercial Aroclors were produced by the Monsanto Chemical Corp. and similar PCB mixtures were manufactured worldwide by other chemical companies. Over 600 million kg of commercial PCBs were produced in the United States and the estimated worldwide production is approximately double this quantity (Table 1). Properties of the commercial PCBs varied from highly fluid liquids (Aroclor 1221) to viscous liquids or solids. All of these preparations contained a complex mixture of isomers and congeners and as

Table 1. Estimated Production and Disposition of PCBs[a]

	United States, 10^6 kg	Worldwide, 10^6 kg
production/use	610	1200
mobile environmental reservoir	82	400
static reservoirs		
in service	340	
dumps	130	
Total static	*470*	*800*

[a]Ref. 8.

the degree of chlorination increased there was a corresponding increase in the relative concentrations of the more highly chlorinated congeners. There are 209 possible PCBs and the properties of these commercial mixtures and the individual PCBs have been extensively investigated. More recent studies indicate that the commercial PCBs contained 132 different compounds (18).

Environmental problems associated with PCBs are the result of a number of factors. Several open uses of PCBs have resulted in their direct introduction into the environment, eg, organic diluents; careless PCB disposal practices have resulted in significant releases into aquatic and marine ecosystems; higher chlorinated PCBs are very stable in their persistence in different environmental matrices; and by a variety of processes (Fig. 1) PCBs are transported throughout the global ecosystem and preferentially bioconcentrate in higher trophic levels of the food chain.

PCBs have been identified in ambient air samples from diverse locations (15–17,18). In one study of the atmospheric levels of several halogenated aromatic hydrocarbons around Kobe, Japan (19), the average concentration of PCBs was 2800 pg/m^3, whereas the PCDD and PCDF levels were 8.6 and 8.8 pg/m^3, respectively. The PCB levels were probably the result of emissions from sites where commercial PCBs were spilled or dumped, whereas the other compounds were from combustion-derived sources. PCBs are also routinely identified in aquatic and marine sediments at highly variable levels dependent on the proximity to a point source pollution problem.

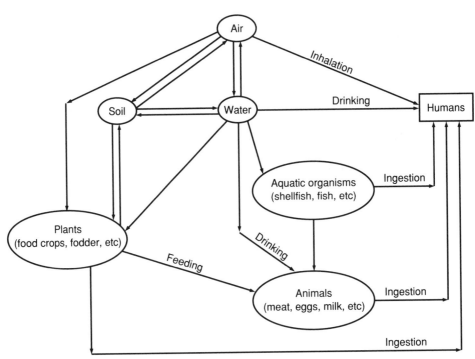

Fig. 1. Transport routes for PCBs and related halogenated aromatic hydrocarbons in the environment.

The identification of PCB residues in fish, wildlife, and human tissues has been reported since the 1970s (9–13,20–26). The results of these analytical studies led to the ultimate ban on further use and production of these compounds. The precise composition of PCB extracts from biota samples is highly variable and depends, in part, on the specific analyte and the commercial PCB preparations associated with a contaminated area (14). PCBs found in a composite human milk sample from Michigan (26) were highly complex, and the congener composition and their relative concentrations did not resemble any of the commercial PCB preparations. This fact raises obvious problems with regard to the hazard assessment of PCB mixtures (27).

Commercial PCBs: Toxic and Biochemical Effects. PCBs and related halogenated aromatic hydrocarbons elicit a diverse spectrum of toxic and biochemical responses in laboratory animals dependent on a number of factors including age, sex, species, and strain of the test animal and the dosing regimen (single or multiple) (27–32). In Bobwhite and Japanese quail, the LC_{50} dose for several different commercial PCB preparations ranged from 600 to 30,000 ppm in the diet; the LC_{50} values for mink that were fed Aroclors 1242 and 1254 were 8.6 and 6.7 ppm in the diet, respectively (8,28,33). The toxic responses elicited by most PCB preparations are also observed for other classes of HAHs (27–32) and include a progressive weight loss not simply related to decreased food consumption and accompanied by weakness, debilitation, and ultimately death, ie, a wasting syndrome; lymphoid involution, thymic and splenic atrophy with associated humoral and/or cell-mediated immunosuppression and/or associated bone marrow and hematologic dyscrasia; a skin disorder called chloracne accompanied by acneform eruptions, alopecia, edema, hyperkeratosis, and blepharitis resulting from hypertrophy of the Meibomian glands; hyperplasia of the epithelial lining of the extrahepatic bile duct, the gall bladder, and urinary tract; hepatomegaly and liver damage accompanied by necrosis, hemorrhage, and intrahepatic bile duct hyperplasia; hepatotoxicity also manifested by the development of porphyria and altered metabolism of porphyrins; teratogenesis, developmental and reproductive toxicity observed in several animal species; carcinogenesis as caused by PCBs in laboratory animals and primarily associated with their effects as promoters; and endocrine and reproductive dysfunction, ie, altered plasma levels of steroid and thyroid hormones with menstrual irregularities, reduced conception rate, early abortion, excessive menstrual and postconceptional hemorrhage, and anovulation in females, and testicular atrophy and decreased spermatogenesis in males.

The biochemical responses elicited by PCBs are also numerous and include the induction of CYP1A1 and CYP1A2 gene expression and the associated monooxygenase enzyme activities, ie, aryl hydrocarbon hydroxylase (AHH) and ethoxyresorufin O-deethylase (EROD), and several other cytochrome P-450 dependent monooxygenases; the induction of steroid metabolizing enzymes, DT diaphorase, UDP glucuronosyl transferase, epoxide hydrolase, glutathione (S)-transferase, and δ-aminolevulinic acid synthetase; increased Ah receptor binding activity; decreased uroporphinogen decarboxylase activity; and decreased vitamin A levels (27).

Structure–Function Relationships. Since PCBs and related HAHs are found in the environment as complex mixtures of isomers and congeners, any meaningful risk and hazard assessment of these mixtures must consider the qualitative and quantitative structure–function relationships. Several studies have investigated the structure–activity relationships for PCBs that exhibit 2,3,7,8-tetrachlorodibenzo-p-dioxin [1746-01-6] (1) (TCDD)-like activity (27,28,34–43).

(1)

Figure 2 illustrates the two primary classes of PCBs that exhibit this type of activity, namely the coplanar PCBs and their monoortho coplanar analogues. The coplanar PCBs, 3,4,4′,5-tetraCB, 3,3′,4,4′-tetraCB, 3,3′,4,4′,5-pentaCB, and 3,3′,4,4′,5,5′-hexaCB, which are substituted in both para, at least two meta, and no ortho positions, are the most toxic members of the class of halogenated aromatics. The relative toxic and biochemical potencies of the coplanar PCB congeners exhibit considerable variations that are dependent on the specific response and the test species.

The data show that 3,3′,4,4′,5-pentaCB is the most toxic coplanar PCB congener and the 2,3,7,8-TCDD/3,3′,4,4′,5-pentaCB potency ratios are 66/1 (body weight loss, rat); 8.1/1 (thymic atrophy, rat); 10/1 (fetal thymic lymphoid development, mouse); 125/1 (AHH induction, rat); 3.3/1 (AHH induction, hepatoma H-4-II E cells, rat); and 100/1 (embryo hepatocytes, chick). Both the 3,3′,4,4′-tetra- and 3,3′,4,4′,5,5′-hexaCB congeners are considerably less toxic than 3,3′,4,4′,5-pentaCB and their relative potencies are highly variable. Results from in vivo studies in the rat have shown that 3,3′,4,4′-tetraCB is > 30 times less toxic than 3,3′,4,4′,5,5′-hexaCB, whereas in most of the in vitro assays these compounds exhibit similar potencies or the reverse order of potency. Using a potency scheme relative to TCDD, toxic equivalence factors (TEFs) of 0.1, 0.05, and 0.01 for 3,3′,4,4′,5-pentaCB, 3,3′,4,4′,5,5′-hexaCB, and 3,3′,4,4′-tetraCB, respectively, have been assigned (27) (TEF for TCDD = 1.0). Similarly, the relative potencies for the monoortho coplanar PCBs (Fig. 2) were also dependent on the test animal/cell and the response; however, for risk assessment purposes, a TEF value of 0.001 was provisionally assigned to this group of PCB congeners (27). Other structural classes of PCBs also exhibit TCDD-like activities (44); however, the potential contribution of these congeners to the TCDD-like activity of commercial mixtures and PCBs in environmental samples is minimal (27).

The TEF values for PCBs and related halogenated aromatics can be utilized for the hazard and risk assessment of these compounds in environmental mixtures. This subgroup of congeners constitutes only a small fraction of the total number of possible PCBs. Therefore, the proposed TEFs for PCBs do not account for the potential toxicity of the non-TCDD-like congeners or their interactive effects. There are several reports showing that some members of this structural class of PCBs elicit biochemical and toxic responses (30). For example, several PCB congeners resemble both phenobarbital (PB) and dexamethasone as inducers of hepatic microsomal cytochrome P-450 isozymes, ie, cytochromes b/e and cytochrome p, respectively (30). Both 2,2′,4,4′,5,5′-hexachlorobiphenyl [35065-27-1] and hexabromobiphenyl have been characterized as PB-type inducers of hepatic drug-metabolizing enzymes (30). Moreover, like PB, 2,2′,4,4′,5,5′-hexabromobi-

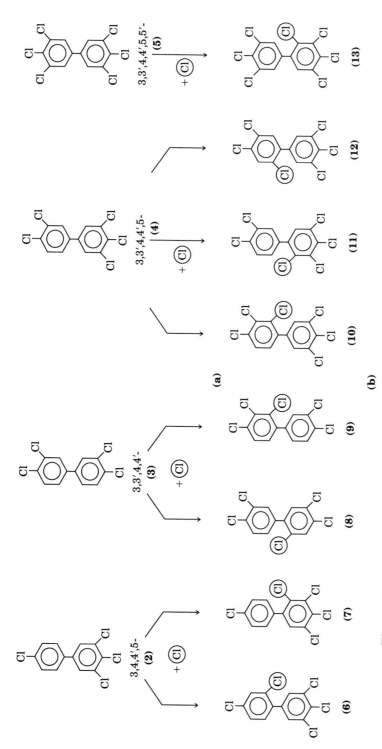

Fig. 2. Toxic PCB congeners: coplanar (**a**) and monoortho coplanar compounds (**b**). See Table 2.

131

Table 2: Coplanar Polychlorinated Biphenyls

PCB isomer	Structure number	Molecular formula	CAS Registry Number
Tetrachlorobiphenyls		$C_{12}H_6Cl_4$	[26914-33-0]
3,4,4',5	(2)		[70362-50-4]
3,3',4,4'	(3)		[32598-13-3]
Pentachlorobiphenyls		$C_{12}H_5Cl_5$	[25429-29-2]
3,3',4,4',5	(4)		[57465-28-8]
2,3',4,4',5'	(6)		[65370-44-3]
2,3,4,4',5	(7)		[74472-37-0]
2,3,4,4',5	(8)		[31508-00-6]
2,3,3',4,4'	(9)		[32598-14-4]
Hexachlorobiphenyls		$C_{12}H_4Cl_6$	[26601-64-9]
3,3'4,4',5,5'	(5)		[32774-16-6]
2,3,3',4,4',5'	(10)		[69782-90-7]
2,3,3',4,4',5	(11)		[38380-08-4]
2,3',4,4',5,5'	(12)		[52663-72-6]
Heptachlorobiphenyl		$C_{12}H_3Cl_7$	[28655-71-2]
2,3,3',4,4',5,5'	(13)		[39635-31-9]

phenyl [59080-40-9] promoted diethylnitrosamine-initiated enzyme altered foci in Sprague-Dawley rats using a two-stage hepatocarcinogenesis protocol (45). These data suggest that the corresponding PCB congeners may also exhibit comparable tumor-promoting activities, and it is clear that future studies should focus on the development of hazard and risk assessment approaches for those congeners not covered in the current TEF schemes (27).

Human Health Effects.   Any assessment of adverse human health effects from PCBs should consider the route(s) of and duration of exposure; the composition of the commercial PCB products, ie, degree of chlorination; and the levels of potentially toxic PCDF contaminants. As a result of these variables, it would not be surprising to observe significant differences in the effects of PCBs on different groups of occupationally-exposed workers.

The accidental leakage of a PCB-containing heat-transfer fluid into rice oil resulted in serious poisoning incidents in Japan (Yusho poisoning, 1966–1968) and Taiwan (Yucheng poisoning, 1978–1979) (46–50). Many of the poison victims received a relatively high oral dose of PCBs over a limited time period (from weeks to months) and these groups were used as a benchmark for the effects of PCBs on humans. However, it is clear from the results of several studies that the principal etiologic agents in the Yusho/Yucheng accidents were not the PCBs but the unusually high levels of PCDFs found as contaminants in the PCB-containing fluid (51–56).

Several studies have reported relatively high levels of PCBs in the serum or adipose tissues of occupationally-exposed individuals, eg, > 3000 ppb in the serum

(57,58). Not surprisingly, after these exposures were terminated, the PCB serum concentrations tended to decrease (59–61).

Chloracne and related skin problems have been observed in several groups of workers and it was suggested that the air concentrations of commercial PCBs > 0.2 mg/m^3 were associated with this effect (62). It was also reported that after occupational exposure to PCBs was terminated there was a gradual decrease in the severity and number of dermatological problems in the exposed workers, and this paralleled a decrease in their serum levels of PCBs (61).

The effects of occupational exposure to PCBs on the concentrations of several serum clinical, chemical, and hematological parameters have been reported (58). Mildly elevated SGOT and γ-glutamyl transpeptidase (GGTP) suggest some liver damage and induction of hepatic monooxygenase enzymes; these results are similar to those observed in animal studies. In one study, it was reported that as PCB serum levels decreased over time the GGTP serum levels also decreased to normal values. A relatively high incidence of pulmonary dysfunction in capacitor-manufacturing workers has been reported (62) with symptoms including coughing, 13.8%; wheezing, 3.4%; tightness in the chest, 10.1%; and upper respiratory or eye irritation, 48.2%. The pulmonary toxicity of PCBs in laboratory animals has not been widely reported (30).

Retrospective mortality studies (63) in 2567 workers (> 3 months employment) from two capacitor manufacturing plants indicated that the mortality of the workers in both plants was lower than the control group, and there were no significant increases in either liver or rectal cancer. An update of the mortality study (64) in which seven additional years had elapsed, and therefore there were more deaths in the exposed group, did not alter the initial findings. Otherwise, workplace studies report various effects of PCBs on the incidence of cancer at different organ sites; however, it is apparent that there are no consistent increases in any one cancer in all the epidemiological studies. It is apparent from most reports that workplace exposure to relatively high levels of PCBs results in limited and moderate toxicity in humans. These toxic symptoms appear to be reversible after exposure to PCBs is terminated and this is accompanied by a decline in serum levels of PCBs.

Environmental exposures to PCBs are significantly lower than those reported in the workplace and are therefore unlikely to cause adverse human health effects in adults. However, it is apparent from the results of several recent studies on children that there was a correlation between *in utero* exposure to PCBs, eg, cord blood levels, and developmental deficits (65–68) including reduced birth weight, neonatal behavior anomalies, and poorer recognition memories. At four years of age, there was still a correlation between prenatal PCB exposure levels and short-term memory function (verbal and quantitative). In these studies the children were all exposed to relatively low environmental levels of PCBs. Although these effects may be related to other contaminants, it is clear that this is an area of concern regarding the potential adverse human health impacts of PCBs.

Polychlorinated Naphthalenes

Polychlorinated naphthalenes (PCNs) are halogenated aromatic hydrocarbons that are no longer produced. They can be synthesized by the chlorination of naphthalene. The commercial products were graded and sold according to their chlorine content (wt %), and used as waxes and impregnants (for protective coatings), water repellents, and wood preservatives (3,6,7).

Commercial PCNs were produced by several companies, eg, Koppers Chemical Co., Halochem, Prodelec, Bayer, and ICI, and marketed under a number of trade names including Halowaxes, Nibren waxes, Seekay waxes, and Clonacire waxes. However, the exact yearly or total production figures are obscure. In 1972, the estimated market for PCNs was less than 2300 t. This figure subsequently decreased as a result of the periodic reported toxicities that have accompanied the production and use of PCNs and the decreased overall utilization of HAHs because of their unacceptable environmental properties.

Like the PCBs, most of the commercial PCNs were complex mixtures of isomers and congeners, although two products, namely monoPCN and Halowax 1051/N-Wax 80, contained primarily 1-chloronaphthalene [90-13-1], $C_{10}H_7Cl$, and octachloronaphthalene [2234-13-1], $C_{10}Cl_8$, respectively.

The environmental impact of PCNs has not been extensively investigated and PCNs are not routinely measured in analytical studies of extracts from environmental samples. However, PCNs have been identified in birds of prey in Britain (69) and The Netherlands (70), in a drainage ditch in Florida, and in sediments from San Francisco Bay (71).

Animal and Human Toxicology.   The mammalian toxicology of PCNs has not been studied in detail; however, it is believed that these compounds elicit mixture- and structure-dependent biochemical and toxic responses resembling those reported for PCBs and other toxic HAHs (32). For example, the effects of the commercial Halowax PCNs as microsomal enzyme inducers were dependent on their degree of chlorination. Halowax 1000, the PCN with the lowest degree of chlorination (26 wt %), did not induce AHH activity (72). In contrast, Halowax 1099 (52 wt % Cl) enhanced microsomal AHH activity and cytochrome P-450 content. At dose levels of 600 μmol/kg, Halowaxes 1099, 1013, 1014, and 1051 significantly induced AHH activity (72). Subsequent studies on a limited number of PCN congeners showed that octachloronaphthalene, 1,2,3,4,5,6,7-hepta-, 1,2,3,4,5,6,8-hepta-, and 1,2,3,4,5,6-hexachloronaphthalene induced rat hepatic microsomal AHH activity, whereas the 1-chloro, 2-chloro-, 1,8-dichloro-, 1,5-dichloro-, 2,7-dichloro-, and 1,2,3,4-tetrachloronaphthalene congeners were inactive. The most active compounds were substituted in three of four of the lateral 2, 3, 6, or 7 positions and the SARs for PCNs were comparable to those reported for other HAHs (73,74).

There have been several reported accidental exposures to commercial PCNs. One of the earliest incidents, the poisoning of cattle, was first reported in 1941 in New York State, and became known as X-Disease because of its unknown etiology. Eventually it was traced to the use of PCNs as high pressure lubricants in feed pelleting machines which resulted in contamination of the feed and ingestion of the PCNs by the animals (75,76). The symptoms exhibited by the cattle included a thickening of the skin referred to as hyperkeratosis, excess lacrimation and salivation, anorexia, depression, and a decrease in plasma vitamin A. A similar outbreak in cattle was reported in Germany in 1947 (75–79). An isolated case of PCN poisoning involving chickens, referred to as Chick Edema Disease (77), occurred in 1957 and was the result of contamination of the feed by a mixture of the penta- and hexachlorinated isomers and congeners.

Human incidents have been reported in workers involved in the production or uses of PCNs. In the United States as well as in Germany and Australia, the severity of the PCN-induced toxicosis was higher after exposure to the higher chlorinated PCN mixtures. In humans the inhalation of hot vapors was the most important route of exposure and resulted in symptoms including rashes or chloracne, jaundice, weight loss, yellow atrophy of the liver, and in extreme cases, death (75,77–79).

Lindane and Hexachlorocyclopentadiene

Both lindane [58-89-9] (14) and hexachlorocyclopentadiene [77-47-4] (15) are halogenated hydrocarbons; unlike the PCBs and PCNs, they do not contain an aromatic ring.

(14)   (15)

Lindane is one of eight different hexachlorocyclohexane (HCH), $C_6H_6Cl_6$, isomers and its *Chemical Abstract* name is $1\alpha,2\alpha,3\beta,4\alpha,5\alpha,6\beta$-hexachlorocyclohexane [58-89-9] (γ-HCH or γ-BHC, benzene hexachloride) (80). Commercial products containing lindane are marketed as either a mixture of isomers or as the pure γ-BHC isomer. Not unexpectedly, lindane is a highly stable lipophilic compound and it has been used extensively worldwide as an insecticide. In contrast, hexachloropentadiene, C_5Cl_6, is an extremely reactive industrial intermediate used as a chemical intermediate in the synthesis of a broad range of cyclodiene-derived pesticides, which include endosulfan, endrin, heptachlor, and several different organohalogen flame retardants (81).

Chemistry and Environmental Impact. Lindane is produced by the photocatalyzed addition of chlorine to benzene to give a mixture of isomers. The active γ-HCH isomer can be preferentially extracted and purified. Composition of the technical-grade product is α (65–70%), β (7–10%), γ (14–15%), δ (7%), and ϵ (1–2%). Lindane has been produced worldwide for its use as an insecticide and for other minor uses in veterinary, agricultural, and medical products.

The relatively high stability and lipophilicity of lindane and its global use pattern has resulted in significant environmental contamination by this hydrocarbon. For example, lindane has been identified in low ppb levels in sediments in the Elbe River, Elbe Estuary, and Corpus Christi Bay (82,83), and in birds from the Falkland Islands (84) and the Shenandoah Valley, Virginia (85) (20 ppb median level, 86% detection frequency). In the National Pesticide Monitoring program (U.S.), lindane concentrations in fish were 30 ppm (wet weight basis) and the frequency of detection was 16% (86). In contrast, relatively high levels of α-hexachlorocyclohexane [319-84-6] were observed in this study, accompanied by higher frequencies of detection. The mean adipose tissue concentrations of lindane in humans (wet weight basis) has been reported as 19.5 ppb (Germany) and 6.7 ppb (The Netherlands) (87). In surveys of adipose tissue in Canadians, β-hexachlorocyclohexane [319-85-7] is the dominant HCH contaminant with mean levels of 31 ppb with a detection frequency of 100% (88). In another Canadian study, comparable results were obtained and the β-HCH was detected in all samples, whereas the detection frequencies of lindane varied from 0–19% (89). In contrast, studies in the early 1990s showed that β-HCH levels in adipose tissue from individuals in several European countries, such as Italy, Poland, and Spain, were considerably higher (2.26, 0.221, and 2.99 ppm, respectively) than observed in the Canadian studies (90–92). The results suggest that overall lindane contamination may be higher in Europe than in North America. The residues of lindane are probably derived from various foods that contain this compound (93). The environmental impact of lindane on exposed populations has not been determined.

The highly reactive hexachlorocyclopentadiene is rapidly degraded in the environment and is not routinely detected as an environmental pollutant (81).

Animal and Human Toxicity. The acute toxicity of lindane depends on the age, sex, and animal species, and on the route of administration. The oral LD_{50} in mice, rats, and guinea pigs is 86, 125–230, and 100–127 mg/kg, respectively. In contrast, most of the other isomers were considerably more toxic (94,95). Some of the other toxic responses caused by lindane in laboratory animals include hepato- and nephotoxicity, reproductive and embryotoxicity, mutagenicity in some short-term *in vitro* bioassays, and carcinogenicity (80). The mechanism of the lindane-induced response is not known. Only minimal data are available on the mammalian toxicities of hexachlorocyclopentadiene.

The effects of occupational exposure to lindane have been investigated extensively (96–100). These studies indicated that occupational exposure to lindane resulted in increased body burdens of this chemical; however, toxic effects associated with these exposures were minimal and no central nervous system disorders were observed. This is in contrast to the polyneuropathies that are often observed after exposure to other haloorganic solvents.

BIBLIOGRAPHY

"Chlorocarbons and Chlorohydrocarbons, Chlorinated Biphenyl and Related Compounds," are treated under "Chlorinated Diphenyls" under "Chlorine Compounds, Organic" in *ECT* 1st ed., Vol. 3, pp. 826–832, by C. F. Booth, Monsanto Chemical Co.; in *ECT* 3rd ed., Vol. 5, pp. 844–848, by R. E. Hatton, Monsanto Co.; "Chlorinated Derivatives of Cyclopentadiene," under "Chlorocarbons and Chlorohydrocarbons," in *ECT* 2nd ed., Vol. 5, pp. 240–252, by R. R. Whetson, Shell Development Co.; in *ECT* 3rd ed., Vol. 5, pp. 791–797, by J. E. Stevens, Hooker Chemical & Plastics Corp.; "Chlorinated Naphthalenes" under "Chlorine Compounds, Organic" in *ECT* 1st ed., Vol. 3, pp. 832–837, by J. Werner, General Aniline & Film Corp., General Aniline Works Division; "Chlorinated Naphthalenes" under "Chlorocarbons and Chlorohydrocarbons" in *ECT* 2nd ed., Vol. 5, pp. 297–303, by D. W. F. Hardie, Imperial Chemical Industries Ltd.; in *ECT* 3rd ed., Vol. 5, pp. 838–843, by H. Dressler, Koppers Co., Inc.; "Benzene Hexachloride" under "Chlorine Compounds, Organic" in *ECT* 1st ed., Vol. 3, pp. 808–812, by J. J. Jacobs, Consulting Chemical Engineer; "Benzene Hexachloride" under "Chlorocarbons and Chlorohydrocarbons" in *ECT* 2nd ed., Vol. 5, pp. 267–281, by D. W. F. Hardie, Imperial Chemical Industries Ltd.; in *ECT* 3rd ed., Vol. 5, pp. 808–818, by J. G. Colson, Hooker Chemicals and Plastics Corp.

1. C. Rappe, H. R. Buser, and H.-P. Bosshardt, *Ann. N.Y. Acad. Sci.* **320,** 1 (1979).
2. O. Hutzinger, S. Safe, and V. Zitko, *The Chemistry of PCBs*, CRC Press, Boca Raton, Fla., 1974.
3. U. A. Th. Brinkman and A. De Kok, in R. D. Kimbrough, ed., *Halogenated Biphenyls, Terphenyls, Naphthalenes, Dibenzodioxins and Related Products*, Elsevier/North-Holland, Amsterdam, The Netherlands, 1980, p. 1.
4. I. Pomerantz and co-workers, *Environ. Health Perspect.* **24,** 133 (1978).
5. C. Rappe and H. R. Buser, in Ref. 3, p. 41.
6. K. Ballschmiter, C. Rappe, and H. R. Buser, in R. D. Kimbrough and A. A. Jensen, eds., *Halogenated Biphenyls, Terphenyls, Naphthalenes, Dibenzodioxins and Related Products*, 2nd ed., Elsevier/North-Holland, Amsterdam, The Netherlands, 1989, p. 47.
7. P. De Voogt and U. A. Th. Brinkman, in Ref. 6, p. 1.
8. L. Hansen, in S. Safe and O. Hutzinger, eds., *Polychlorinated Biphenyls (PCBs): Mammalian and Environmental Toxicology*, Vol. 1, Springer-Verlag Publishing Co., Heidelberg, Germany, 1987, p. 15.
9. R. W. Risebrough, P. Rieche, S. G. Herman, D. B. Peakall, and M. N. Kirven, *Nature* **220,** 1098 (1968).
10. L. Fishbein, *J. Chromatogr.* **68,** 345 (1972).
11. K. Ballschmiter, H. Buchert, and S. Bihler, *Z. Fresenius Anal. Chem.* **306,** 323 (1981).
12. K. Ballschmiter and co-workers, *Z. Fresenius Anal. Chem.* **309,** 1 (1981).
13. M. Wasserman, D. Wasserman, S. Cucos, and H. J. Miller, *Ann. N.Y. Acad. Sci.* **320,** 69 (1979).
14. S. Safe, L. Safe, and M. Mullin, in Ref. 8, p. 133.
15. G. R. Harvey and W. G. Steinhauer, *Atmos. Environ.* **8,** 777 (1974).
16. S. Tanabe, H. Hidaka, and R. Tatsukawa, *Chemosphere* **12,** 277 (1983).
17. E. Atlas and C. S. Giam, *Science* **211,** 163 (1981).
18. D. E. Schulz, G. Petrick, and J. C. Duinker, *Environ. Sci. Technol.* **23,** 852 (1989).
19. T. Nakano, M. Tsuji, and T. Okiuino, *Chemosphere* **16,** 1781 (1987).
20. K. Wickstrom, H. Pyysalo, and M. Perttila, *Chemosphere* **10,** 999 (1981).
21. P. Olsen, H. Settle, and R. Swift, *Aust. Wildl. Res.* **7,** 139 (1980).
22. J. R. Wharfe and W. L. F. Van Den Broek, *Mar. Pollut. Bull.* **9,** 76 (1978).
23. J. Mowrer and co-workers, *Bull. Environ. Contam. Toxicol.* **18,** 588 (1977).
24. M. G. Castelli, G. P. Martelli, C. Spagone, L. Capellini, and R. Fanelli, *Chemosphere* **12,** 291 (1983).
25. S. Tanabe, N. Kanna, A. Subramanian, S. Watanabe, and R. Tatsukawa, *Environ. Pollut.* **47,** 147 (1987).
26. S. Safe, L. Safe, and M. Mullin, *J. Agric. Food Chem.* **33,** 24 (1985).
27. S. Safe, *CRC Crit. Rev. Toxicol.* **21,** 51 (1990).
28. A. Parkinson and S. Safe, in Ref. 8, p. 49.
29. A. Poland and J. C. Knutson, *Ann. Rev. Pharmacol. Toxicol.* **22,** 517 (1982).

30. S. Safe, *CRC Crit. Rev. Toxicol.* **13,** 319 (1984).
31. A. Poland, W. F. Greenlee, and A. S. Kende, *Ann. N.Y. Acad. Sci.* **320,** 214 (1979).
32. J. A Goldstein and S. Safe, in Ref. 6, p. 239.
33. R. K. Ringer, R. J. Aulerich, and M. R. Bleavins, in M. A. Q. Khan, ed., *Halogenated Hydrocarbons: Health and Ecological Effects*, Pergamon Press, Inc., Elmsford, N.Y., 1981, p. 329.
34. A. Parkinson and co-workers, *J. Biol. Chem.* **258,** 5967 (1983).
35. J. A. Goldstein and co-workers, *Toxicol. Appl. Pharmacol.* **36,** 81 (1976).
36. J. A. Goldstein, *Ann. N.Y. Acad. Sci.* **320,** 164 (1979).
37. H. Yoshimura, S. Yoshihara, N. Ozawa, and M. Miki, *Ann. N.Y. Acad. Sci.* **320,** 179 (1979).
38. A. Poland and E. Glover, *Mol. Pharmacol.* **13,** 924 (1977).
39. J. A. Goldstein, P. Hickman, H. Bergman, J. D. McKinney, and M. P. Walker, *Chem. Biol. Interact.* **17,** 69 (1977).
40. T. Sawyer and S. Safe, *Toxicol. Lett.* **18,** 87 (1982).
41. A. Parkinson, R. Cockerline, and S. Safe, *Biochem. Pharmacol.* **29,** 259 (1980).
42. A. Parkinson, L. Robertson, L. Safe, and S. Safe, *Chem. Biol. Interact.* **30,** 271 (1981).
43. B. Leece, M. A. Denomme, R. Towner, S. M. A. Li, and S. Safe, *J. Toxicol. Environ. Health* **16,** 379 (1985).
44. D. Davis and S. Safe, *Toxicol.* **63,** 97 (1990).
45. R. K. Jensen, S. D. Sleight, J. I. Goodman, S. D. Aust, and J. E. Trosko, *Carcinogenesis* **3,** 1183 (1982).
46. K. Higuchi, *PCB Poisoning and Pollution*, Kodansha, Ltd. and Academic Press, Tokyo, London, New York, 1976, Chapt. 1, p. 5.
47. M. Kuratsune, T. Yoshimura, J. Matsuzaka, and A. Yamaguchi, *Environ. Health Perspect.* **1,** 119 (1972).
48. M. Kuratsune, in Ref. 3, p. 28.
49. S.-T. Hsu and co-workers, *Environ. Health Perspect.* **59,** 5 (1985).
50. M. Kuratsune and R. E. Shapiro, *PCB Poisoning in Japan and Taiwan*, Alan R. Liss, New York, 1984.
51. M. Morita, J. Nakagawa, K. Akiyama, S. Mimura, and N. Isono, *Bull. Environ. Cont. Toxicol.* **18,** 67 (1977).
52. H. Miyata, A. Nakamura, and T. Kashimoto, *J. Food Hyg. Soc. Japan* **17,** 227 (1976).
53. H. Miyata, T. Kashimoto, and N. Kunita, *J. Food Hyg. Soc. Japan* **19,** 260 (1977).
54. P. H. Chen, K. T. Chang, and Y. D. Lu, *Bull. Environ. Contam. Toxicol.* **26,** 489 (1981).
55. T. Kashimoto and co-workers, *Arch. Environ. Health* **36,** 321 (1981).
56. N. Kunita and co-workers, *Am. J. Ind. Med.* **5,** 45 (1984).
57. A. B. Smith and co-workers, *Br. J. Ind. Med.* **39,** 361 (1982).
58. S. Safe, in Ref. 8, p. 1.
59. M. Takamatsu and co-workers, *Environ. Health Perspect.* **59,** 91 (1985).
60. R. W. Lawton, M. R. Ross, J. Feingold, and J. F. Brown, *Environ. Health Perspect.* **60,** 165 (1985).
61. I. Hara, *Environ. Health Perspect.* **59,** 85 (1985).
62. R. Warshaw, A. Fischbein, J. Thornton, A. Miller, and I. J. Selikoff, *Ann. N.Y. Acad. Sci.* **320,** 277 (1979).
63. D. P. Brown and M. Jones, *Arch. Environ. Health* **36,** 120 (1981).
64. D. P. Brown, *Arch. Environ. Health* **43,** 333 (1987).
65. J. L. Jackson, S. W. Jacobson, and H. E. B. Humphrey, *J. Pediatr.* **116,** 38 (1990).

66. W. J. Rogan and co-workers, *J. Pediatr.* **109,** 335 (1986).
67. B. C. Gladen and co-workers, *J. Pediatr.* **113,** 991 (1988).
68. S. W. Jacobson, G. G. Fein, J. L. Jacobson, and P. M. Schwartz, *Child Dev.* **56,** 853 (1985).
69. M. Cooke, D. J. Roberts, and M. E. Tillett, *Sci. Total Environ.* **15,** 237 (1980).
70. J. H. Koeman, H. C. W. Van Velzen-Blad, R. De Bries, and J. G. Vos, *J. Reprod. Fert. Suppl.* **19,** 353 (1973).
71. L. M. Law and D. F. Goerlitz, *Pesticide Monit. J.* **8,** 33 (1974).
72. R. Cockerline, M. Shilling, and S. Safe, *Gen. Pharmacol.* **12,** 83 (1981).
73. M. A. Campbell, S. Bandiera, L. Robertson, A. Parkinson, and S. Safe, *Toxicology* **22,** 123 (1981).
74. *Ibid.,* **26,** 193 (1983).
75. W. Hansel and K. McEntee, *J. Dairy Sci.* **38,** 875 (1955).
76. F. P. Flinn and N. E. Jarvic, *Proc. Soc. Exptl. Biol. Med.* **35,** 118 (1936).
77. P. Olafson and K. McEntee, *Cornell Vet.* **41,** 107 (1951).
78. M. Kuratsune, *Environ. Health Perspect.* **1,** 129 (1972).
79. G. L. Sparschu, F. L. Dunn, and V. K. Rowe, *Food Cosmet. Toxicol.* **9,** 405 (1971).
80. *Evaluation of the Carcinogenic Risk of Chemicals to Humans,* IARC Monographs, 1979, p. 195.
81. Y. H. Atallah, D. M. Whitacre, and R. G. Butz, in M. A. Q. Khan and R. H. Stanton, eds., *Toxicology of Halogenated Hydrocarbons,* Pergamon Press, Inc., Elmsford, N.Y., 1981, p. 344.
82. G. Eder, R. Sturm, and W. Ernst, *Chemosphere* **16,** 2487 (1987).
83. L. E. Ray, H. E. Murray, and C. S. Giam, *Chemosphere* **12,** 1039 (1983).
84. K. Ballschmiter, Ch. Scholz, H. Buchert, and M. Zell, *Fres. Zeit. Anal. Chem.* **309,** 1 (1981).
85. A. K. Blumton and co-workers, *Bull. Environ. Contam. Toxicol.* **45,** 697 (1990).
86. C. J. Schmitt, J. L. Zajicek, and M. A. Ribick, *Arch. Environ. Contam. Toxicol.* **14,** 225 (1985).
87. H. Geyer, I. Scheunert, and F. Korte, *Reg. Toxicol. Pharmacol.* **6,** 313 (1986).
88. J. Mes, L. Marchand, and D. J. Davies, *Bull. Environ. Contam. Toxicol.* **45,** 681 (1990).
89. D. T. Williams, G. L. LeBel, and E. Junkins, *J. Assoc. Offic. Anal. Chem.* **71,** 410 (1988).
90. S. Focardi, C. Fossi, C. Leonzio, and R. Romei, *Bull. Environ. Contam. Toxicol.* **36,** 644 (1986).
91. G. A. Szymczynski, S. M. Waliszewski, M. Tuzewski, and P. Pyda, *J. Environ. Sci. Health* **21,** 5 (1986).
92. M. Camps and co-workers, *Bull. Environ. Contam. Toxicol.* **42,** 195 (1989).
93. V. Leoni and S. U. D'Arca, *Sci. Total Environ.* **5,** 253 (1976).
94. G. Czegledi-Janko and P. Avar, *Br. J. Ind. Med.* **27,** 283 (1970).
95. C. M. Ginsburg and W. Lowry, *Pharmacol. Ther. Pediatr. Dermatol.* **1,** 74 (1983).
96. K. Baumann, K. Behling, H. L. Brassow, and K. Stapel, *Int. Arch. Occup. Environ. Health* **48,** 165 (1981).
97. A. Zesch, K. Nitzsche, and M. Lange, *Arch. Dermatol. Res.* **273,** 43 (1982).
98. J. Angerer, R. Maap, and R. Heinrich, *Int. Arch. Occup. Environ. Health* **52,** 59 (1983).
99. S. K. Nigam and co-workers, *Int. Arch. Occup. Environ. Health* **57,** 315 (1986).
100. L. Drummond, E. M. Gillanders, and H. K. Wilson, *Br. J. Ind. Med.* **45,** 493 (1988).

STEPHEN H. SAFE
Texas A & M University

CHLOROCARBONS AND CHLOROHYDROCARBONS, VINYL CHLORIDE. See Vinyl polymers.

CHLOROCARBONS AND CHLOROHYDROCARBONS, VINYLIDENE CHLORIDE. See Vinylidene chloride and poly(vinylidene chloride).

CHLOROFLUOROCARBONS. See Fluorine compounds, organic, fluorinated aliphatic compounds; Refrigeration and refrigerants.

CHLOROHYDRINS

A chlorohydrin has been defined (1) as a compound containing both chloro and hydroxyl radicals, and chlorohydrins have been described as compounds having the chloro and the hydroxyl groups on adjacent carbon atoms (2). Common usage of the term applies to aliphatic compounds and does not include aromatic compounds. Chlorohydrins are most easily prepared by the reaction of an alkene with chlorine and water, though other methods of preparation are possible. The principal use of chlorohydrins has been as intermediates in the production of various oxirane compounds through dehydrochlorination.

Properties

Ethylene chlorohydrin [107-07-3], $HOCH_2CH_2Cl$, is the simplest chlorohydrin. It may also be called 2-chloroethanol, 2-chloroethyl alcohol, or glycol chlorohydrin. Ethylene chlorohydrin is a liquid at 15°C and 101.3 kPa (1 atm) (Table 1). This polar compound is miscible with water [7732-18-5] and ethanol [64-17-5] and is slightly soluble in ethyl ether [60-29-7] (5).

Table 2 gives physical property data for propylene chlorohydrins. 2-Chloro-1-propanol [78-89-7], $HOCH_2CHClCH_3$, is also named 2-propylene chlorohydrin, 2-chloropropyl alcohol, or 2-chloro-1-hydroxypropane. 1-Chloro-2-propanol [127-00-4], $ClCH_2CHOHCH_3$, also known as sec-propylene chlorohydrin, 1-chloroiso-propyl alcohol, and 1-chloro-2-hydroxypropane, is a colorless liquid, miscible in water, ethanol, and ethyl ether.

3-Chloro-1,2-propanediol [96-24-2], $HOCH_2CHOHCH_2Cl$, a liquid with n_D^{20} = 1.4831 (6), boils at 213°C and 101.3 kPa (1 atm) with decomposition. It can be distilled at 114–120°C at 1.87 kPa (14 mm Hg). Synonyms for this compound include 3-chloro-1,2-dihydroxypropane, glycerol monochlorohydrin, α-chlorohydrin, and 3-chloropropylene glycol. It is miscible in water, ethanol, ethyl ether, and acetone [67-64-1] (8) and is soluble in hot benzene [71-43-2].

3-Chloro-1,2-propanediol has a mol wt of 110.48 and a specific gravity at 20°C of 1.3218. Its flash point is 135°C (9). Its heat of formation at 298 K is −525.8 kJ/mol (−125.7 kcal/mol) and the heat of combustion at constant volume is 15.2 kJ/g (3.63 kcal/g) (8).

Physical property data for dichloropropanols, $C_3H_6Cl_2O$, appear in Table 3. 1,2-Dichloro-3-propanol [616-23-9] $ClCH_2CHClCH_2OH$, is also known as 1,2-dichlorohydrin, β-dichlorohydrin, or 1,2-dichloro-3-hydroxypropane. It is miscible in ethanol, ether, acetone, and benzene and is slightly soluble in H_2O (5).

1,3-Dichloro-2-propanol [96-23-1], $ClCH_2CHOHCH_2Cl$, has a vapor pressure at 28°C of 0.13 kPa (0.98 mm Hg) (10). Other names for it include 1,3-

Table 1. Physical Properties of Ethylene Chlorohydrin[a]

Property	Value
molecular formula	C_2H_5ClO
molecular weight	80.51
boiling point at 101.3 kPa[b], °C	128.7
melting point, °C	−67.5
density at 20°C, g/cm^3	1.2015–1.2025
vapor pressure at 20°C, kPa[b]	0.65
specific thermal capacity at 20°C, kJ/(kg·K)[c]	1.965
viscosity at 20°C, mPa·s (=cP)	3.43
refractive index, n_D^{20}	1.4418–1.442
flash point, °C	55
autoignition temperature, °C	425
explosive limits in air, vol %	5–16
specific heat of vaporization, kJ/kg[c]	552
heat of formation[d] at 298 K, kJ/mol[c]	−294.3
heat of combustion[d], kJ/kg[c]	15,080

[a]Ref. 3 unless otherwise noted.
[b]To convert kPa to mm Hg, multiply by 7.5.
[c]To convert kJ to kcal, divide by 4.184.
[d]Ref. 4.

Table 2. Physical Properties of Propylene Chlorohydrins, C_3H_7ClO

Property	2-Chloro-1-propanol	1-Chloro-2-propanol	Reference
mol wt	94.54	94.54	
boiling point, °C	133–134	126–127	
specific gravity at 20°C	1.103	1.115	6
refractive index at 20°C	1.4362	1.4362	6
vapor density (air=1)	3.3	3.3	
flash point, °C	44[a]	52	7

[a]ASTM D3278.

Table 3. Physical Properties of Dichloropropanols

Property	1,2-Dichloro-3-propanol	1,3-Dichloro-2-propanol
mol wt	128.99	128.99
boiling point, °C	183–185	174.3[a]
mp, °C		−4
specific gravity	1.3607[b]	1.3506[c]
heat of combustion at constant volume, kJ/g	13.3[d]	
refractive index	1.4819[b]	1.4802[c]
flash point, °C		85
vapor density (air = 1)		4.4[e]

[a]Ref. 6.
[b]At 20°C (5).
[c]At 17°C.
[d]To convert kJ to kcal, divide by 4.184 (8).
[e]Ref. 7.

dichlorohydrin, glycerol dichlorohydrin, and 1,3-dichloro-2-hydroxypropane. It is very soluble in water and ethanol, miscible in ethyl ether; and soluble in acetone (5).

Composition and bp data for selected binary azeotropes of chlorohydrins are given in Table 4 (11).

Chemistry

SYNTHESIS OF CHLOROHYDRINS

Hypochlorination. Both ethylene chlorohydrin and propylene chlorohydrin were prepared by Wurtz (12) as a result of the reactions of HCl [7647-01-0] with the corresponding glycol under pressure. Shortly afterward, Carius (13) synthesized ethylene chlorohydrin by reaction of hypochlorous acid with ethylene [74-85-1]. The first detailed investigation of the formation of ethylene chlorohydrin by the reaction of ethylene with hypochlorous acid [7790-92-3] was performed by Gomberg (14).

Hypochlorous acid is most readily made by the reaction of chlorine [7782-50-5] with water: $Cl_2 + H_2O \rightleftarrows HCl + HOCl$. However, since the equilibrium constant for this reaction is only 4.2×10^{-4} (15), the amount of hypochlorous acid, as compared to that of chlorine, is quite small. Thus it was expected that the addition of chlorine to the double bond to produce 1,2-dichloroethane [107-06-2] would be the principal reaction. However, when the reaction is done under well-stirred conditions so that the ethylene can only react with the chlorine in solution and not with gaseous chlorine, the main product is the ethylene chlorohydrin. Very little 1,2-dichloroethane is observed until a concentration of chlorohydrin in the reaction solution reaches 6–8%. These results suggested that the reaction of the ethylene with the hypochlorous acid is significantly faster than the reaction of ethylene with dissolved chlorine. Subsequent studies have shown that in an

Table 4. Binary Azeotropes of Chlorohydrins

A component	B component	Bp, °C[a]	Wt % A
2-Chloroethanol		128.6	100
	water	98[b]	42
	2-methoxyethanol [109-86-4]	130	69
	2-ethoxyethanol [110-80-5]	136	15
	3-methyl-1-butanol [123-51-3]	128	75
	chlorobenzene [108-90-7]	120	42
	cyclohexene [110-83-8]	81	11
	4-methyl-3-penten-2-one [141-79-7]	130	33
	cyclohexane [110-82-7]	78.5	10
	2-hexanone [591-78-6]	129	75
	butyl acetate [123-86-4]	125.6	31
	o-chlorotoluene [95-49-8]	128	75
	toluene [108-88-3]	107	24.4
1-Chloro-2-propanol		127	100
	water	95.4	54
	chlorobenzene	122	55
	butyl acetate	125.5	25
	toluene	109	15
	heptane [142-82-5]	96.5	17
	o-xylene [95-47-6]	125.5	85
	perchloroethylene [127-18-4]	113	28
2-Chloro-1-propanol		133.7	100
	water	96	49
	chlorobenzene	126	36
	o-xylene	130.5	70
	isobutyl ether [628-55-7]	120	25
	perchloroethylene	115	13
	butyl ether [142-96-1]	130.5	70
1,3-Dichloro-2-propanol		175.8	100
	water	99	23.2
	o-dichlorobenzene [95-50-1]	170.5	60
	p-dichlorobenzene [106-46-7]	168	45
	o-chlorotoluene	158	15
	ethoxybenzene [103-72-1]	169	37
	1,3,5-trimethylbenzene [108-67-8]	161.5	32
	d-limonene [5989-27-5]	166	67
2,3-Dichloro-1-propanol		182.5	100
	o-dichlorobenzene	174	40
	p-dichlorobenzene	171	30
	p-methylanisole [104-93-5]	175.5	32
	mesitylene [108-67-8]	163	18
	d-limonene	169	40
	ethyl butyl ether [628-81-9]	180	53
1,3-Dichloro-2-methyl-2-propanol [597-32-0]		174	100
	water	98.3	35.2
1-Chloro-2-methyl-2-propanol [558-42-9]		126.7	100
	water	93–94	66

[a]At 101.3 kPa (1 atm) unless otherwise noted.
[b]At 99.7 kPa (0.98 atm).

143

aqueous system chlorohydrins are not formed by the addition of hypochlorous acid but rather by the reaction of the olefinic compound with chlorine and water, successively (2).

$$Cl_2 + \underset{/}{\overset{\backslash}{C}} = \underset{\backslash}{\overset{/}{C}} \longrightarrow \underset{|}{\overset{Cl}{\underset{|}{-C}}}\overset{+}{-}\underset{\backslash}{\overset{/}{C}}- + Cl^- \xrightarrow{H_2O} \underset{\underset{+}{OH_2}}{\overset{Cl}{\underset{|}{-C}}-\underset{|}{\overset{|}{C}}-} \xrightarrow{-H^+} \underset{OH}{\overset{Cl}{\underset{|}{-C}}-\underset{|}{\overset{|}{C}}-}$$

Chlorohydrins from Epoxides. Traditionally epoxides have been manufactured by the dehydrochlorination of chlorohydrins. However, the reverse reaction may be used as a source of chlorohydrins, especially in the case of ethylene chlorohydrin from ethylene oxide [75-21-8], which is now produced by the direct oxidation of the olefin. A study of the reaction of hydrogen chloride with propylene oxide [75-56-9] showed that an anhydrous system at low temperatures (<0°C) gives the highest yield of chlorohydrin with best isomeric selectivity (16).

Various techniques are mentioned in the literature for the reaction of ethylene oxide and HCl. These include gas-phase reaction (17); reaction of gaseous HCl with ethylene oxide (18); and in solution, especially with ethylene chlorohydrin as solvent (19–21).

Glycerol dichlorohydrin (1,3-dichloro-2-propanol) may be synthesized by the reaction of HCl with epichlorohydrin (chloromethyloxirane [106-89-8]). A patent describes a continuous process using the dichlorohydrin as the reaction solvent to suppress by-product formation (22).

Chlorohydrins via Enzyme Technology. During the 1980s a series of patents assigned to Cetus Corp. (23–25) appeared and describe the synthesis of chlorohydrins by the use of a halogenating enzyme, an oxidizing agent, and a halide ion source. A preferred embodiment involves chloroperoxidase derived from the microorganism *Caldariomyces fumago,* hydrogen peroxide [7722-84-1] as the oxidizing agent, and sodium chloride [7647-14-5] as the halide ion source. The hydrogen peroxide and sodium chloride are mixed in a buffered aqueous solution, the enzyme is added, and the olefin is added to the system either as a gas or a liquid.

Chlorohydrins from Chromyl Chloride. Several olefins have been treated with chromyl chloride [14977-61-8], CrO_2Cl_2, in carbon tetrachloride [56-23-5]. Chlorohydrins were formed in 35–50% yields from propylene [115-07-1], 1 butene [106-98-9], 1-pentene [109-67-1], and 1-hexene [592-41-6]. In each case the hydroxyl group of the chlorohydrin was located almost exclusively in the primary position, which is in contrast to the main product from the reaction of olefins with aqueous chlorine (26).

REACTIONS OF CHLOROHYDRINS

Dehydrochlorination to Epoxides. The most useful chemical reaction of chlorohydrins is dehydrochlorination to form epoxides (oxiranes). This reaction was first described by Wurtz in 1859 (12) in which ethylene chlorohydrin and propylene chlorohydrin were treated with aqueous potassium hydroxide [1310-58-3] to form ethylene oxide and propylene oxide, respectively. For many years

both of these epoxides were produced industrially by the dehydrochlorination reaction. In the past 40 years, the ethylene oxide process based on chlorohydrin has
been replaced by the direct oxidation of ethylene over silver catalysts. However,
such epoxides as propylene oxide (qv) and epichlorohydrin are still manufactured
by processes that involve chlorohydrin intermediates.

The conversion of chlorohydrins into epoxides by the action of base is an
adaptation of the Williamson synthesis of ethers. In the presence of hydroxide
ion, a small proportion of the alcohol exists as alkoxide, which displaces the chloride ion from the adjacent carbon atom to produce a cyclic ether (2).

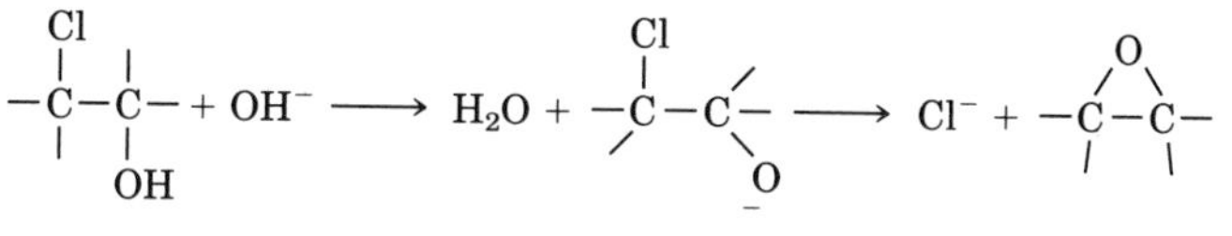

$$\underset{\underset{OH}{|}}{\overset{\overset{Cl}{|}}{-C-C-}} + OH^- \longrightarrow H_2O + \underset{\underset{O}{\diagdown}}{\overset{\overset{Cl}{|}}{-C-C}}\diagup \longrightarrow Cl^- + \overset{O}{\overset{\diagup\diagdown}{-C-C-}}$$

The dehydrochlorination of chlorohydrins to epoxides exhibits second-order
kinetics in that the rate is dependent on both the concentration of the chlorohydrin
and the hydroxide ion (27). This suggests a rapid formation of the alkoxide ion in
the presence of base followed by the slower, rate-determining elimination of the
chloride ion with accompanying ring closure (28). Intramolecular displacements
leading to epoxides are generally thousands of times faster than the intermolecular attack of alkoxides on alkyl chlorides under comparable conditions (29). Although ring strain causes the epoxide formation to be less favored energetically
than is the case with a noncyclic ether, far less restriction of motion is necessary
in closing a small ring than in bringing together two free molecules into a single
activated complex. Thus the ease of formation of epoxides from chlorohydrins is
an entropy effect.

Epoxide formation from chlorohydrins is marked by an increase in rate with
alkyl substitution (28) as shown in Figure 1. This phenomenon has been explained
on the basis that steric crowding in the chlorohydrin is somewhat relieved as the
epoxide is formed, so that the greatest relief of strain results from ring closure of
the most crowded chlorohydrin (28).

Formation of Mustard Gas from Ethylene Chlorohydrin. Ethylene chlorohydrin is readily converted to bis(2-chloroethyl) sulfide [505-60-2], the so-called
mustard gas, $C_4H_8Cl_2S$, used widely in World War I (14). The preparation involves
the addition of a 70–80% aqueous solution of ethylene chlorohydrin to solid sodium sulfide hydrate [1313-84-4]. There is little temperature change because al-

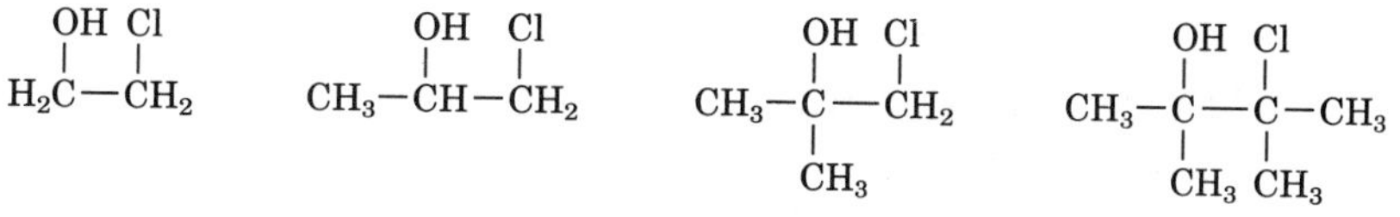

$\underset{H_2C-CH_2}{\overset{\overset{OH}{\mid}\;\;\overset{Cl}{\mid}}{}}$	$\underset{CH_3-CH-CH_2}{\overset{\overset{OH}{\mid}\;\;\overset{Cl}{\mid}}{}}$	$\underset{\underset{CH_3}{\mid}}{\underset{CH_3-C-CH_2}{\overset{\overset{OH}{\mid}\;\;\overset{Cl}{\mid}}{}}}$	$\underset{\underset{CH_3\;CH_3}{\mid\;\;\;\;\mid}}{\underset{CH_3-C-C-CH_3}{\overset{\overset{OH}{\mid}\;\;\overset{Cl}{\mid}}{}}}$

Relative rate
of epoxide   *1*   *21*   *250*   *1370*
formation

Fig. 1. Effect of alkyl substitution on chlorohydrin epoxidation.

though the reaction of the chlorohydrin with the sulfide is exothermic, the process of dissolving the sodium sulfide is endothermic. When the reaction is over, the excess sodium sulfide is neutralized with 90% sulfuric acid [7664-93-9]. Concentrated hydrochloric acid is added to the neutralized solution, and the precipitated sodium salts are filtered and washed with acid. When the clear, yellow hydrochloric acid solution of the thiodiglycol is heated to 60–75°C, dichloroethyl sulfide, $ClCH_2CH_2SCH_2CH_2Cl$, separates as a heavy, yellow oil. Yields of up to 98% based on starting chlorohydrin are reported (14). In view of the ease of manufacture of this dangerous war gas from ethylene chlorohydrin, the U.S. State Department imposed export controls on ethylene chlorohydrin to Iran, Iraq, and Syria in 1986 (30).

Hydrolysis to Glycols.   Ethylene chlorohydrin and propylene chlorohydrin may be hydrolyzed in the presence of such bases as alkali metal bicarbonates, sodium hydroxide, and sodium carbonate (31–33). In water at 97°C, 1-chloro-2-propanol forms acid, acetone, and propylene glycol [57-55-6] simultaneously; the kinetics of production are first order in each case, and the specific rate constants are nearly equal. The relative rates of solvolysis of 2-chloroethanol, 1-chloro-2-propanol, and 1-chloro-2-methyl-2-propanol in water at 97°C are 1.0:0.81:5.5 (34). Glycerol monochlorohydrin (3-chloro-1,2-propanediol) is readily hydrolyzed to glycerol when treated with 3% excess sodium bicarbonate for 30 minutes at 150°C. In a continuous system, monochlorohydrin was fed simultaneously with a 10% NaOH, 1% Na_2CO_3 solution to a stirred autoclave to produce glycerol in about 90% yield (35).

Formation of Cyclic Carbonates.   In the absence of water, chlorohydrins such as 2-chloroethanol and 1-chloro-2-propanol react with an alkali carbonate or bicarbonate to produce cyclic carbonates such as ethylene carbonate [96-49-1] and propylene carbonate [108-32-7] in yields of up to 80% (36). An improved method involves the reaction of a chlorohydrin and CO_2 in the presence of an amine, which gives cyclic carbonate selectivities of 90–95% (37). Cyclic carbonates are produced in high yield under very mild conditions by treating chlorohydrins with tetramethylammonium hydrogen carbonate in acetonitrile [75-05-8] under a CO_2 atmosphere (38).

Esterification.   Chlorohydrins can react with salts of carboxylic acids to form esters. For example, 2-hydroxyethyl benzoate [134-11-2] was prepared in 92% yield by heating sodium benzoate [532-32-1] with an excess of ethylene chlorohydrin in the presence of a small amount of diethylamine [109-89-7] at 140°C for four hours (38).

Etherification.   A mixture of ethylene chlorohydrin in 30% aqueous NaOH may be added to phenol at 100–110°C to give 2-phenoxyethanol [122-99-6] in 98% yield (39). A cationic starch ether is made by reaction of a chlorohydrin–quaternary ammonium compound such as 3-chloro-2-hydroxypropyl trimethylammonium chloride [101396-91-2] with a starch slurry at pH 11–12 (40).

Oxidation.   Monochloroacetic acid [79-11-8] may be synthesized by the reaction of ethylene chlorohydrin with nitric acid [7697-37-2]. Yields of greater than 90% are reported (41). Beta-chlorolactic acid (3-chloro-2-hydroxypropanoic acid) [1713-85-5] is produced by the reaction of nitric acid with glycerol monochlorohydrin (42). Periodic acid [10450-60-9] and glycerol monochlorohydrin gives chloroacetaldehyde [107-20-0] in 50% yield (43).

Quaternization. Choline chloride [67-48-1] was prepared in nearly quantitative yield by the reaction of trimethylamine [121-44-8] with ethylene chlorohydrin at 90–105°C and 981–1471 kPa (10–15 kg/cm^2) pressure (44). Precursors to quaternary ammonium amphoteric surfactants have been made by reaction of ethylene chlorohydrin with tertiary amines containing a long chain fatty acid group (45).

Manufacture and Processing

For many years ethylene chlorohydrin was manufactured on a large industrial scale as a precursor to ethylene oxide, but this process has been almost completely displaced by the direct oxidation of ethylene to ethylene oxide over silver catalysts. However, since other commercially important epoxides such as propylene oxide and epichlorohydrin cannot be made by direct oxidation of the parent olefin, chlorohydrin intermediates are still important in the manufacture of these products.

PROPYLENE CHLOROHYDRIN

Chlorohydrination in Chlorine and Water. The hypochlorination of propylene gives two isomers: 90% 1-chloro-2-propanol and 10% 2-chloro-1-propanol.

$$CH_3-CH=CH_2 + Cl_2 + H_2O \longrightarrow \underset{\underset{OH}{|}\,\underset{Cl}{|}}{CH_3-CH-CH_2} + \underset{\underset{Cl}{|}\,\underset{OH}{|}}{CH_3-CH-CH_2} + HCl$$

The principal by-products are the result of direct chlorine addition to give 1,2-dichloropropane [78-87-5] and ether formation. The ether product is dichloropropyl ether [108-60-1] or 2,2'-oxybis(1-chloropropane).

$$CH_3-CH=CH_2 + \underset{\underset{OH}{|}\,\underset{Cl}{|}}{CH_3-CH-CH_2} + Cl_2 \longrightarrow \underset{\underset{Cl}{|}\,\underset{CH_3}{|}}{CH_2-CH-O}\underset{\underset{CH_3}{|}\,\underset{Cl}{|}}{-CH-CH_2} + HCl$$

Commercial chlorohydrin reactors are usually towers provided with a chlorine distributor plate at the bottom, an olefin distributor plate about half way up, a recirculation pipe to allow the chlorohydrin solution to be recycled from the top to the bottom of the tower, a water feed into the recirculation pipe, an overflow pipe for the product solution, and an effluent gas takeoff (46). The propylene and chlorine feeds are controlled so that no free gaseous chlorine remains at the point where the propylene enters the tower. The gas lift effect of the feeds provides the energy for the recirculation of the reaction solution from the top of the tower.

Chlorohydrination occurs as the propylene and dissolved chlorine pass up through the tower. It is important that no significant amount of free chlorine remain in the effluent gas as this could cause explosive reactions. After scrubbing through a solution of NaOH and a sufficient amount is bled off to prevent accumulation of inert gases, the effluent gas is mixed with fresh propylene and fed back to the reactor. Fresh water is fed into the recirculation leg of the tower at a rate sufficient to maintain the chlorohydrin concentration in the circulating liquid at 4–4.5% by weight. An equivalent amount of chlorohydrin solution overflows for further processing. The reactions are exothermic, which maintains a temperature in the reaction column of 30–40°C without external heating or cooling. A diagram of a typical chlorohydrin reactor for the manufacture of propylene oxide is shown in Figure 2 (47).

In addition to maintaining the circulation of the reaction liquid, the inert gas is important in temperature control, and it prevents the formation of a separate nonaqueous phase by carrying off substantial quantities of dichloropropane in the effluent.

Since the formation of the chlorohydrin is accompanied by the production of an equimolar quantity of hydrogen chloride, the reaction solution is strongly acidic and corrosive. The first chlorohydrin reaction towers were built of stoneware or of mild steel and lined with rubber and ceramic tiles. More recently corrosion-resistant reinforced plastics have been used with good results, but operating pressures must be maintained at or near atmospheric.

Yields of propylene chlorohydrin range from 87–90% with dichloropropane yields of 6–9%. The dichloropropane is not only a yield loss but also represents a disposal problem as few uses are known for this material. Since almost all the propylene chlorohydrin is dehydrochlorinated to propylene oxide with lime or so-

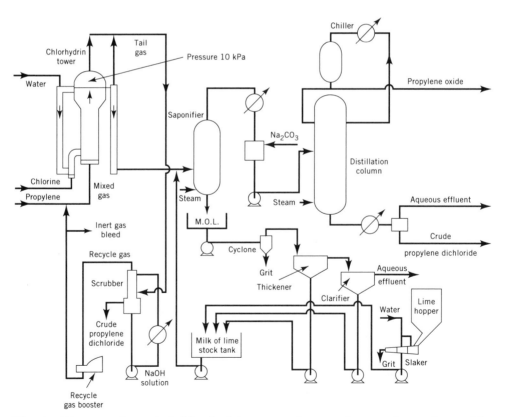

Fig. 2. Diagram of a typical chlorohydrin reactor for manufacture of propylene oxide. M.O.L. = milk of lime. To convert kPa to mm Hg, multiply by 7.5.

dium hydroxide, none of the chlorine appears in the final product. Instead, it ends up as dilute calcium or sodium chloride solutions, which usually contain small amounts of propylene glycol and other organic compounds that can present significant disposal problems.

Chlorohydrination with Nonaqueous Hypochlorous Acid. Because the presence of chloride ions has been shown to promote the formation of the dichloro by-product, it is desirable to perform the chlorohydrination in the absence of chloride ion. For this reason, methods have been reported to produce hypochlorous acid solutions free of chloride ions. A patented method (48) involves the extraction of hypochlorous acid with solvents such as methyl ethyl ketone [78-93-3], acetonitrile, and ethyl acetate [141-78-6]. In one example hypochlorous acid was extracted from an aqueous brine with methyl ethyl ketone in a 98.9% yield based on the chlorine used. However, when propylene reacted with a 1 M solution of hypochlorous acid in either methyl ethyl ketone or ethyl acetate, chlorohydrin yields of only 60–70% were obtained (10).

Chlorohydrination with *tert*-Alkyl Hypohalites. Olefins react with ethyl hypochlorite [624-85-1] to form the corresponding chlorohydrin (49). In 1938 both Shell Development Co. (50) and Arthur D. Little, Inc. (51) patented the preparation of chlorohydrins by the reactions of olefins with tertiary alkyl hypochlorites. Examples with ethylene and propylene in the Shell patent reported chlorohydrin yields of greater than 95% with *tert*-butyl hypochlorite [507-40-4].

Almost 40 years later the Lummus Co. patented an integrated process involving the addition of chlorine along with the sodium chloride and sodium hydroxide from the cathode side of an electrolytic cell to a tertiary alcohol such as tertiary butanol to produce the tertiary alkyl hypochlorite. The hypochlorite phase separates, and the aqueous brine solution is returned to the electrolytic cells. The *tert*-alkyl hypochlorite reacts with an olefin in the presence of water to produce a chlorohydrin and the tertiary alcohol, which is returned to the chlorinator. With propylene, a selectivity to the chlorohydrin of better than 96% is reported (52). A series of other patents covering this technology appeared during the 1980s (53–56).

MANUFACTURE OF GLYCEROL MONOCHLOROHYDRINS

From Allyl Alcohol. The reaction of allyl alcohol [107-18-6] with chlorine and water gives a mixture of glycerol monochlorohydrins consisting of 73% 3-chloropropane-1,2-diol and 27% of 2-chloropropane-1,3-diol (57). In a recycle reaction system in which allyl alcohol is fed as a 4.5–5.5 wt % solution, chlorine is added at a rate of 7–9 moles per hour. The reaction time is about five seconds, the reaction temperature 50–60°C and the recycle ratio is 10–20:1. Under these conditions monochlorohydrins have been obtained in 88% yield with 9% dichlorohydrins (58) (see ALLYL ALCOHOL AND DERIVATIVES).

From Glycerol. A procedure for synthesizing alpha-monochlorohydrin (3-chloro-1,2-propanediol) in 85–88% yields by the reaction of glycerol [56-81-5] with aqueous hydrochloric acid in the presence of a catalytic amount of acetic acid has been developed (59). An anhydrous procedure that involves the reaction of glycerol and HCl gas in the presence of acetic acid has also been described (60).

MANUFACTURE OF GLYCEROL DICHLOROHYDRINS

From Allyl Chloride. The hypochlorination of allyl chloride [107-05-1] gives a mixture of the glycerol dichlorohydrins, 2,3-dichloropropanol and 1,3-dichloropropanol in about 7:3 ratio. Because of the poor solubility of allyl chloride in water, it is essential to minimize the formation of an organic phase in which direct chlorination of the allyl chloride results in the unwanted by-product 1,2,3-trichloropropane [96-18-4].

Many techniques have been developed to accomplish this, for example, the use of a cooled recirculating system in which the chlorine is dissolved in one part and the allyl chloride is dissolved and suspended in another (61). The streams are brought together in the main reaction zone and thence to a separator to remove water-insoluble products. Another method involves maintaining any organic phase present in the reaction zone in a highly dispersed condition (62). A continuous reactor consists of a recycle system in which make-up water and allyl chloride in a volume ratio of 10–50:1 are added upstream from a centrifugal pump so that the allyl chloride is completely dispersed in the aqueous phase as particles less than 100 mμ in diameter by the time of chlorine addition is reached. Dichlorohydrin yields of greater than 92% based on the allyl chloride feed are reported.

Emulsifiers may also be used to disperse the allyl chloride (63). Finely emulsified water and allyl chloride are fed into a tubular reactor along with gaseous chlorine. This is a single-pass reactor with a residence time of at least 15 seconds and a linear velocity of the reaction mixture of at least 0.5 m/s. The emulsion is prepared by feeding water and allyl chloride in a weight ratio of at least 25:1 along with nonionic or anionic surfactants (0.2–0.5 wt % with respect to allyl chloride) through a static mixer. Maximum dichlorohydrin yields of greater than 94% are reported, compared to an 84.5% yield in the same reactor without the inclusion of an emulsifier.

The reaction of allyl chloride and chlorine in water produces trichloropropane as a by-product even in the aqueous phase, along with tetrachloropropyl ether. For maximum dichlorohydrin yield it is necessary to run the reaction at low concentrations of chloride ion and of chlorohydrin, that is, with high water dilution. However, high dilution results in an aqueous effluent that contains minor amounts of these by-products that require significant treatment to reduce them to levels acceptable in outfalls to rivers, lakes, and other public waterways.

One patent (64) describes an extraction method to remove both trichloropropane and tetrachloropropyl ether from the dichlorohydrin solution by the use of carbon tetrachloride as a solvent. In this way the by-products are removed from the aqueous phase into an organic phase from which they can be separated by distillation and disposed of in a safe and proper manner.

From Allyl Alcohol. An alternative route to dichlorohydrins from allyl chloride begins with the hydrolysis to allyl alcohol. Significant yields of 2,3-dichloropropanol can be obtained from the reaction of chlorine with allyl alcohol if the reaction is performed in the presence of concentrated hydrochloric acid (65). Several patents for the manufacture of 2,3-dichloropropanol by the chlorination of allyl alcohol at low temperature, $-30°$ to $+20°$C, in 25–40% HCl solution have appeared (66–68). Product yields as high as 98% are claimed.

Economic Aspects

The most important chemical reaction of chlorohydrins is dehydrochlorination to produce epoxides. In the case of propylene oxide, The Dow Chemical Company is the only manufacturer in the United States that still uses the chlorohydrin technology. In 1990 the U.S. propylene oxide production capacity was listed as 1.43×10^6 t/yr, shared almost equally by Dow and Arco Chemical Co., which uses a process based on hydroperoxide intermediates (69,70). More recently, Dow Europe SA, announced a decision to expand its propylene oxide capacity by 160,000 metric tons per year at the Stade, Germany site. This represents about a 40% increase over the current capacity (71).

Epichlorohydrin (chloromethyloxirane), which has a production capacity in the United States of 291,000 t/yr, is manufactured by the chlorohydrination of allyl chloride and subsequent dehydrochlorination of the glycerol dichlorohydrin isomers (69). Dow and Shell Chemical are the two producers of epichlorohydrin in the United States.

The merchant market for chlorohydrins is small, primarily for specialty applications. Ethylene chlorohydrin is sold in the United States by BASF Corp., Parsippany, N.J., available in 230 kg net lined steel drums. Glycerol monochlorohydrin (3-chloro-1,2-propanediol) is available from Dixie Chemical Co., Houston, Tex., in lined steel drums (227.3 kg net); from Raschig Corp., Richmond, Va.; and from Henley Chemicals, Inc., Montvale, N.J., in steel drums (240 kg net). Glycerol dichlorohydrin (1,3-dichloro-2-propanol) is not currently being produced for the U.S. merchant market but has been available in the past at a selling price of $5–6/kg.

Health and Safety Factors

In general, chlorohydrins are relatively toxic irritants. They are harmful if swallowed, inhaled, or absorbed through the skin. They cause irritation to the eyes, skin, mucous membrane, and upper respiratory tract.

For handling chlorohydrins, chemical safety goggles, chemical-resistant gloves, OSHA/MSHA approved respirators, and other protective clothing are required. In case of contact, one should immediately flush eyes or skin with copious amounts of water for at least 15 minutes and remove contaminated clothing and shoes. If inhaled, the person should be moved to fresh air (72).

Chlorohydrins are combustible and should be stored away from heat and open flame in a cool, dry place. These materials are generally incompatible with strong oxidizing agents and strong bases. Under fire conditions toxic fumes of hydrogen chloride, phosgene, and carbon monoxide may be generated.

Toxicity of 2-Chloroethanol. Ethylene chlorohydrin is an irritant and is toxic to the liver, kidneys, and central nervous system. In addition, it is rapidly absorbed through the skin (73). The vapor is not sufficiently irritating to the eyes and respiratory mucous membranes to prevent serious systemic poisoning. Contact of the liquid in the eyes of rabbits causes moderately severe injury, but in humans corneal burns have been known to heal within 48 hours. Several human fatalities have resulted from inhalation, dermal contact, or ingestion. One fatality was caused by exposure to an estimated 300 ppm in air for 2.25 hours. In another fatal case, autopsy revealed pulmonary edema and damage to the liver, kidneys, and brain (73).

Toxic amounts can be absorbed through the skin without causing dermal irritation. The dermal LD_{50} for rabbits is 68 mg/kg. Two-year dermal studies showed no evidence of carcinogenicity in rats given 50 to 100 mg/kg/d or mice given 15 mg per animal per day (73) (Table 5). A ceiling limit for exposure to

Table 5. Toxicity Data for Selected Chlorohydrins[a,b]

Compound and animal test	LD_{50}, mg/kg	LC_{50}	LCL_0, ppm/4 h[c]	LDL_0
2-Chloroethanol				
oral (rat)	71			
inhalation (rat)		290 mg/m^3		
skin (human)			305[d]	
skin (rabbit)	67			
2-Chloro-1-propanol				
oral (rat)	218			
inhalation (rat)			500	
oral (dog)				200 mg/kg
skin (rabbit)	529			
1,3-Dichloro-2-propanol				
oral (rat)	110			
inhalation (rat)				125 ppm/4 h
skin[e] (rabbit)	800			
2,3-Dichloropropanol				
oral (rat)	90			
inhalation (rat)			500	
skin[f] (rabbit)	200			
3-Chloro-1,2-propanediol				
oral (rat)	26[g]			
skin (rat)	1057–1849[h]			
inhalation (rat)		125 ppm/4 h[i]		

[a]Ref. 74.
[b]Toxicity parameters are defined as follows: LD_{50} = lethal dose 50% kill; LCL_0 = lowest published lethal concentration; LC_{50} = lethal concentration 50% kill; LDL_0 = lowest published lethal dose.
[c]Unless otherwise stated.
[d]305 ppm/2 h.
[e]Also causes moderately severe injury to rabbit's eyes; (75).
[f]Similar to acetone in rabbit eyes, reversible.
[g]Ref. 72.
[h]Ref. 76.
[i]Ref. 77.

ethylene chlorohydrin of 1 ppm is recommended by the American Conference of Governmental Industrial Hygienists. Protection against skin absorption is strongly recommended.

Ethylene chlorohydrin is classified as a Class B poison in the Code of Federal Regulations. The NFPA Hazard Classification gives the material a health hazard rating of 3, a flammability rating of 2, and a reactivity rating of 0. Fire-extinguishing agents include water, alcohol or polymer foam, dry chemical powder or carbon dioxide (4).

Uses of Chlorohydrins

From a volume standpoint almost all of the chlorohydrins produced are immediately converted into epoxides such as propylene oxide and epichlorohydrin. The small quantity of various chlorohydrins sold in the merchant market are used in specialty applications.

Ethylene chlorohydrin may be used in the manufacture of dye intermediates, pharmaceuticals, plant-protection agents, pesticides, and plasticizers (3).

Glycerol monochlorohydrin has been found to be an effective toxicant-chemosterilant specifically for the Norway rat (78). This compound was approved for commercial use in the United States by the Environmental Protection Agency in 1982 and is sold under the trade name Epibloc by Pestcon Systems Inc. (79). Glycerol monochlorohydrin is also used to make guaiacol glycerol ether, an expectorant used in cough remedies, by its reaction with sodium guaiacolate (6). Another application is as an intermediate in the production of x-ray contrast media.

BIBLIOGRAPHY

"Epichlorohydrin" in *ECT* 1st ed., Vol. 3, pp. 865–869, P. H. Williams, Shell Development Co.; "Glycerol Chlorohydrins" in *ECT* 1st ed., Vol. 3, pp. 857–865, P. H. Williams, Shell Development Co.; "Propylene, Trimethylene, and Tetramethylene Chlorohydrins" in *ECT* 1st ed., Vol. 3, pp. 856–857, M. G. Gergel and M. Revelise, Columbia Organic Chemicals Co.; "Chlorohydrins" in *ECT* 2nd ed., Vol. 5, pp. 304–324, G. D. Lichtenwalter and G. H. Riesser, Shell Chemical Co.; in *ECT* 3rd ed., Vol. 5, pp. 848–864, by G. H. Riesser, Shell Development Co.

1. R. Grant and C. Grant, *Grant & Hackh's Chemical Dictionary,* 5th ed., McGraw-Hill Book Co., New York, 1987, p. 132.
2. R. T. Morrison and R. N. Boyd, *Organic Chemistry,* 5th ed., Allyn and Bacon, Inc., Boston, Mass. 1987, p. 323.
3. *Ethylene Chlorohydrin,* Technical Bulletin, BASF Corp., Parsippany, N.J., 1989.
4. *Chemical Hazard Response Information System (CHRIS),* Vol. II, U.S. Coast Guard, Washington, D.C., 1984.
5. J. G. Grasselli, ed., *Atlas of Spectral Data and Physical Constants for Organic Compounds,* CRC Press, Cleveland, Ohio, 1973.
6. M. Windholz, ed., *The Merck Index,* Merck & Co., Inc., Rahway, N.J., 1976.
7. *Catalog Handbook of Fine Chemicals,* Aldrich Chemical Co., Inc., Milwaukee, Wis., 1990–1991.

8. F. Richter, ed., *Beilsteins Handbuch der Organischen Chemie,* Vol. III, Springer-Verlag, Berlin, Germany, 1958, p. 2151.

9. *3-Chloropropane-1,2-diol,* Technical Data Sheet, Raschig Corp., 1988.

10. I. Kh. Bikbulatov and co-workers, *Zh. Prik. Khim.* **58,** 2499 (1985).

11. L. H. Horsley, *Azeotropic Data,* Advances in Chemistry Series, No. 6, American Chemical Society, Washington, D.C., 1952.

12. A. Wurtz, *Ann.* **110,** 125 (1859).

13. L. Carius, *Ann.* **124,** 265 (1862).

14. M. Gomberg, *J. Am. Chem. Soc.* **41,** 1414 (1919).

15. F. A. Cotton and G. Wilkinson, *Advanced Inorganic Chemistry,* Interscience Publishers, New York, 1962.

16. C. A. Stewart and C. A. VanderWerf, *J. Am. Chem. Soc.* **76,** 1259 (1954).

17. Eur. Pat. Appl. 10,013 (Apr. 16, 1980), W. Jequier and co-workers, (to Produits Chimiques Ugine Kuhlmann).

18. V. Parausanu and co-workers, *Rev. Chim. (Bucharest)* **21,** 743 (1970).

19. Jpn. Pat. 69 18,841 (Aug. 16, 1969), K. Miyauchi, S. Senoya, and M. Yoshino (to Japan Oils and Fats Co., Ltd.).

20. Jpn. Pat. 83 146,521 (Sept. 1, 1983), (to Nisso Petrochemical Industries Co., Ltd.).

21. Rom. Pat. 60,731 (Apr. 15, 1976), C. Roncea and co-workers (to Institutul de Cercetari pentru Chimizarea Petrolului).

22. Ger. Offen. 2,719,463 (Nov. 2, 1978), H. Erpenbach, K. Gehrmann and H. Joest (to Hoechst A-G).

23. U.S. Pat. 4,247,641 (Jan. 27, 1981), S. L. Neidleman, W. F. Amon, Jr., and J. Geigart (to Cetus Corp.).

24. U.S. Pat. 4,426,449 (Jan. 17, 1984), J. Geigart and S. L. Neidleman (to Cetus Corp.).

25. U.S. Pat. 4,587,217 (May 6, 1986), J. Geigart and S. L. Neidleman (to Cetus Corp.).

26. S. J. Cristol and K. R. Eilar, *J. Am. Chem. Soc.* **72,** 4353 (1950).

27. S. Winstein and H. J. Lucas, *J. Am. Chem. Soc.* **61,** 1576 (1939).

28. E. S. Gould, *Mechanism and Structure in Organic Chemistry,* Holt, Rinehart, and Winston, New York, 1959, p. 567.

29. J. E. Stevens, C. L. McCabe, and J. C. Warner, *J. Am. Chem. Soc.* **70,** 2449 (1948).

30. *Fed. Reg.* **51**(108), 20467 (1986).

31. U.S. Pat. 1,442,386 (Jan. 11, 1923), G. O. Curme, Jr. and C. O. Young (to Carbide and Carbon Chemicals Corp.).

32. U.S. Pat. 1,695,250 (Dec. 11, 1928), G. O. Curme, Jr. (to Carbide and Carbon Chemicals Corp.).

33. P. Sherwood, *Petroleum Refiner* **28,** 120 (1949).

34. I. K. Gregor, N. V. Riggs, and V. R. Stimson, *J. Chem. Soc.,* 76 (1956).

35. E. C. Williams, *AIChE Trans.* **37,** 157 (1941).

36. U.S. Pat. 1,907,891 (May 9, 1933), G. Steimmig and M. Wittwer (to I. G. Farben).

37. U.S. Pat. 3,923,842 (Dec. 2, 1975), Y. Wu (to Phillips Petroleum Co.).

38. C. Venturello and R. D'Aliosio, *Synthesis,* 33 (1985).

39. N. K. Bliznyvk and co-workers, *Zh. Obsch. Khim.* **36,** 480 (1966).

40. U.S. Pat. 2,876,217 (Mar. 3, 1959), E. F. Paschall (to Corn Products Co.).

41. U.S. Pat. 2,455,405 (Dec. 7, 1948), L. A. Burrows and M. F. Fuller (to E. I. du Pont de Nemours & Co., Inc.).

42. C. F. Koelsch, *J. Am. Chem. Soc.* **52,** 1105 (1930).

43. L. F. Hatch and H. E. Alexander, *J. Am. Chem. Soc.* **67,** 688 (1945).

44. Ger. Pat. 105,207 (Apr. 12, 1974), E. Fibitz and B. Nussbuecker.

45. N. Parris, J. K. Weil, and W. M. Linfield, *J. Am. Oil Chem. Soc.* **53,** 97 (1976).

46. A. J. Gait, in E. G. Hancock, ed., *Propylene and Its Industrial Derivatives,* John Wiley & Sons, Inc., New York, 1973, pp. 274–279.

47. A. C. Fyvie, *Chem. Ind.,* 384–388 (Mar. 7, 1964).
48. U.S. Pat. 3,578,400 (May 11, 1971), J. A. Wojtowicz, M. Lapkin, and M. S. Puar (to Olin Corp.).
49. S. Goldschmidt, R. Endres, and R. Dirsch, *Ber.* **58,** 572 (1925).
50. U.S. Pat. 2,106,353 (Jan. 25, 1938), S. L. Langedijk (to Shell Development Co.).
51. U.S. Pat. 2,107,789 (Feb. 8, 1938), C. G. Harford (to Arthur D. Little, Inc.).
52. U.S. Pat. 4,008,133 (Feb. 15, 1977), A. P. Gelbein and J. T. Kwon (to The Lummus Co.).
53. U.S. Pat. 4,126,526 (Nov. 21, 1978), J. T. Kwon and A. P. Gelbein (to The Lummus Co.).
54. U.S. Pat. 4,342,703 (Aug. 3, 1982), J. T. Kwon (to The Lummus Co.).
55. U.S. Pat. 4,443,620 (Apr. 17, 1984), A. P. Gelbein and A. S. Nislick (to The Lummus Co.).
56. U.S. Pat. 4,496,777 (Jan. 29, 1985), G. D. Suciu, J. T. Kwon, and A. M. Shaban (to The Lummus Co.).
57. P. B. D. DeLaMare and J. G. Pritchard, *J. Chem. Soc.,* 3990 (1954).
58. J. Myszkowski and A. Z. Zidinski, *Przemysl Chem.* **44,** 249 (1965).
59. T. H. Rider and A. J. Hill, *J. Am. Chem. Soc.* **52,** 1521 (1930).
60. H. Gilman, ed., *Organic Syntheses,* Collective Vol. I, John Wiley & Sons, Inc., New York, 1941, p. 294.
61. C. S. Miner and N. N. Dalton, eds., *Glycerol,* Reinhold Publishing Corp., New York, 1953, p. 83.
62. U.S. Pat. 2,714,121 (May 25, 1951), J. Anderson, G. F. Johnson, and W. C. Smith (to Shell Development Co.).
63. Brit. Pat. 2,029,821 (Mar. 26, 1980), C. Civo, M. Petri, M. Lazzari, and A. Bigozzi (to Euteco SPA).
64. U.S. Pat. 4,900,849 (Sept. 9, 1988) D. I. Saletan (to Shell Oil Co.).
65. H. R. Ing, *J. Chem. Soc.* 1393 (1948).
66. Fr. Demande 2,565,229 (Dec. 6, 1985), N. Nagato, H. Mori, K. Maki, and R. Ishioka (to Showa Denko K.K.).
67. Jpn. Pat. 87 19,544 (Jan. 28, 1987), S. Takakuwa, T. Nakada, and K. Nagao (to Osaka Soda Co.).
68. Jpn. Pat. 88 290,835 (Nov. 28, 1988), N. Nagato, H. Mori, and R. Ishioka (to Showa Denko K.K.).
69. H. H. Szmant, *Organic Building Blocks of the Chemical Industry,* John Wiley & Sons, Inc., New York, 1989, p. 281.
70. *Chem. Mark. Rep.,* (Jan 8, 1990).
71. *Chem. Week,* (Sept. 4, 1989).
72. R. E. Lenga, ed., *The Sigma-Aldrich Library of Chemical Safety Data,* Sigma-Aldrich Corp., 1985.
73. N. H. Proctor, J. P. Hughes, and M. L. Fischman, *Chemical Hazards of the Workplace,* J. B. Lippincott Co., Philadelphia, Pa. 1988, p. 240.
74. N. I. Sax and R. J. Lewis, Sr., *Dangerous Properties of Industrial Materials,* Van Nostrand Reinhold, New York, 1989.
75. W. M. Grant, *Toxicology of the Eye,* Charles C Thomas, Springfield, Ill., 1974, p. 266.
76. *3-Chloropropane-1,2-diol,* Material Safety Data Sheet, Raschig Corp., 1989.
77. *3-Chloro-1,2-propanediol,* Material Safety Data Sheet, Dixie Chemical Co., 1986.
78. R. V. Andrews and R. W. Belknap, *J. Hyg.* **91,** 359 (1983).
79. *J. Commerce.,* 22B (Aug. 6, 1982).

W. FRANK RICHEY
The Dow Chemical Company

CHLOROPHENOLS

The chlorophenols make up an important class of industrial chemical compounds. They are used as either intermediates in the synthesis of agrochemicals, dyestuffs, and pharmaceuticals or directly in formulations.

Physical Properties

The main characteristics and physical properties of the chlorophenols are brought together in Table 1. With the exception of o-chlorophenol, they are all solids at room temperature. The refractive indexes of the monochlorophenols, C_6H_5ClO, are as follows: ortho, 1.5524; meta, 1.5565; para, 1.5579. The pK_a values of chlorophenols depend on the number and the position of the substituents.

Preparation

Chlorination of Phenols. Industrially, the phenols are chlorinated without solvent. Chlorine reacts rapidly with phenol and with the chlorophenols, which makes it difficult to determine the relative reaction rates because of the super-chlorination that sometimes results from an unsatisfactory chlorine dispersion. Studies have yielded the relative reaction rates indicated in Figure 1.

Monochlorophenols. Chlorination of phenol [*108-95-2*] between 50 and 120°C gives a para/ortho ratio of 1.65. To improve the selectivity in the para position, it is possible to use dialkyl sulfides, diaryl sulfides (12), or alkyl and aryl sulfides combined. Sulfides are active only at low temperatures ($\leq 50°C$), because at high temperatures the active species decomposes into sulfur and chlorine.

The use of phosphine has also been described (13). The para/ortho ratios obtained with sulfur catalysis vary between 1.8 and 2.1, according to the nature of the disulfide, but the best para isomer selectivity is obtained by using sulfuryl chloride as the chlorinating agent. Sulfuryl chloride has also been used in the presence of sulfides and Lewis acids (14–16). Ortho chlorination is generally favored by using a nonpolar solvent (17). o-Chlorophenol can be obtained with a selectivity in the vicinity of 90% by chlorinating the phenol in a halogenated solvent with traces of amines present (18).

Other chlorinating agents, such as pentachlorocyclohexadienone, have been subjected to laboratory study to make it possible to select each of the isomers (19).

Table 1. Physical Properties of the Chlorophenols

Compound	CAS Registry Number	Mp, °C	Bp, °C	pK_a Water[a]	pK_a Methanol/water[b]	Density, g/mL
2-chlorophenol[c]	[95-57-8]	8.7	175–176	8.5–8.52	9.13	$1.2634^{20°}$
3-chlorophenol	[108-43-0]	32.8	215–217	8.97–9[d]	9.53	1.27
4-chlorophenol[e]	[106-48-9]	40–41	219	9.37–9.44[f]	9.70	1.265
2,3-dichlorophenol	[576-24-9]	58	206	7.4–7.71	8.52	$1.388^{50°}$
2,4-dichlorophenol	[120-83-2]	42.8	210	7.9[g]–7.9	8.51	
2,5-dichlorophenol	[583-78-8]	58	212–213	7.5–7.51	7.69	
2,6-dichlorophenol	[87-65-0]	67	219–220	6.8[h]–6.80	7.15	
3,4-dichlorophenol	[95-77-2]	65	253	8.6[f]–8.62	8.87	
3,5-dichlorophenol	[591-35-5]	68	233	8.2[i]–8.25	8.54	
2,3,4-trichlorophenol	[15950-66-0]	83.5		6.97[j]–6.97	7.34	
2,3,5-trichlorophenol	[933-78-8]	62	255	6.43	6.92	
2,3,6-trichlorophenol	[933-75-5]	101	272	5.8[j]–5.80	6.10	
2,4,5-trichlorophenol	[95-95-4]	68	245–246	6.72–7.3[k]	7.20	
2,4,6-trichlorophenol	[88-06-2]	68	244.5	5.99–6.2[h]	6.51	1.49
3,4,5-trichlorophenol	[609-19-8]	101	275	7.55–7.8[f]	7.57	
2,3,4,5-tetrachlorophenol	[4901-51-3]	115–117		5.64[j]–5.64	5.92	
2,3,4,6-tetrachlorophenol	[58-90-2]	69–70		5.22[f]–5.22	5.53	
2,3,5,6-tetrachlorophenol	[935-95-5]	115		5.02[j]–5.03	5.76	
pentachlorophenol	[87-86-5]	190	309–310	4.74–4.8	4.93	1.98

[a] At 25°C from Ref. 1 unless otherwise noted.
[b] 60% $CH_3OH/40\%$ H_2O at 20°C (2).
[c] pK_a in pyridine = 12.1 (3).
[d] Ref. 4.
[e] pK_a in DMSO = 16.1 (5)
[f] Ref. 6.
[g] Ref. 7.
[h] Ref. 8.
[i] Ref. 9.
[j] Ref. 10.
[k] Ref. 11.

157

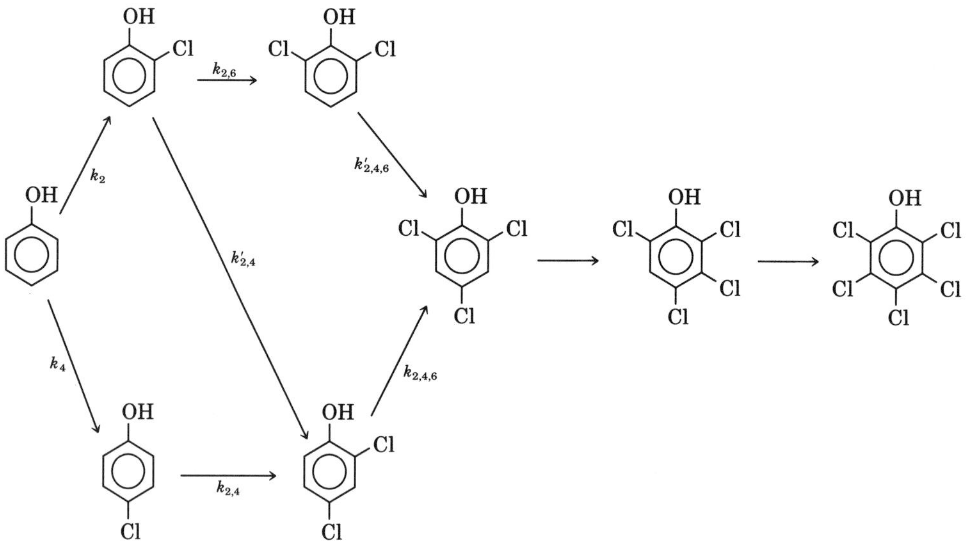

Fig.1. Kinetics of the chlorination of phenol. Relative rate constants are $k_2 = 1090$; $k_{2,6} = 16$; $k_4 = 1910$; $k_{2,4,6} = 0.9$; $k_{2,4} = 124$; $k'_{2,4,6} = 0.7$; and $k'_{2,4} = 61$.

The use of 2,3,4,5,6,6-hexachlorocyclohexa-2,4-dien-1-one [21306-21-8] makes chlorination possible in the ortho position.

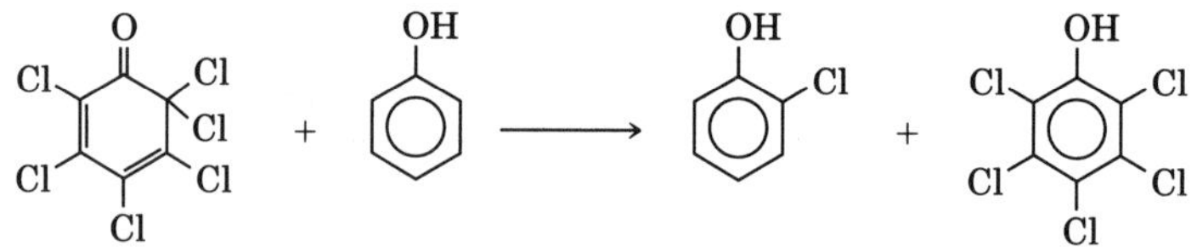

The 2,3,4,4,5,6-hexachlorocyclohexa-2,5-dien-1-one [599-52-0] gives the para isomer.

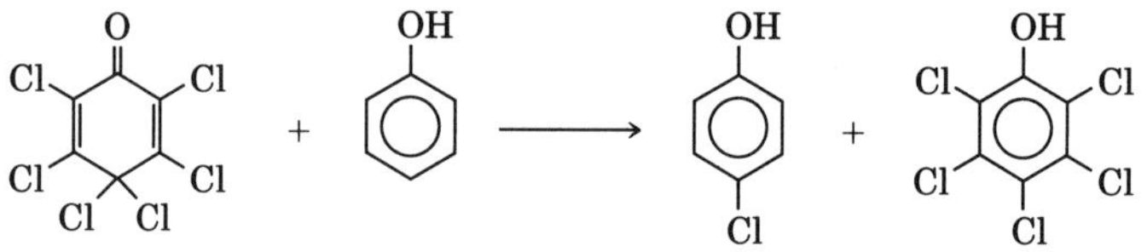

Dichlorophenols. Among all the dichlorophenols, $C_6H_4Cl_2O$, it is 2,4-dichlorophenol that is produced in greatest quantity. 2,4-Dichlorophenol is used in manufacturing 2,4-dichlorophenoxyacetic acid [94-75-7] (2,4-D) and 2-(2,4-dichlorophenoxy)propionic acid [720-36-5] (2,4-DP). Industrially, 2,4-dichlorophenol can be obtained by chlorinating phenol, *p*-chlorophenol, *o*-chlorophenol, or a mixture of these compounds in cast-iron reactors. The chlorinating agent may be chlorine or sulfuryl chloride in combination with a Lewis acid. For example:

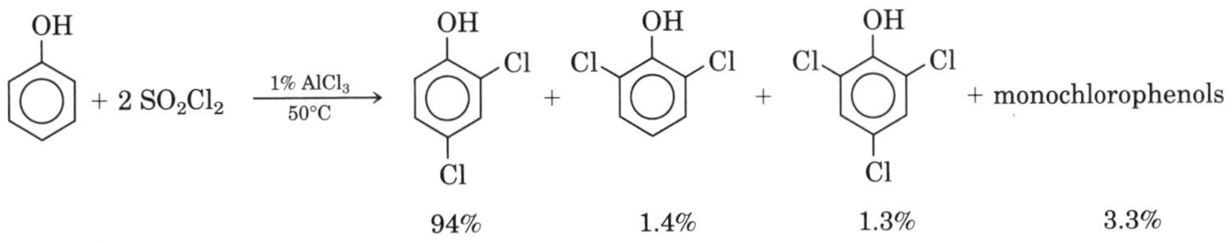

| | 94% | 1.4% | 1.3% | 3.3% |

Chlorination with SO_2Cl_2, which is favorable to the para isomer at the mono-chlorination stage, gives an excellent yield of 2,4-dichlorophenol. Starting with o-chlorophenol, it is possible to attain a selectivity for 2,4-dichlorophenol of 98%, if chlorination is carried out in liquid SO_2 at low temperature (20). 2,6-Dichloro-phenol is also used as an intermediate. It is obtained by chlorinating o-chloro-phenol in the presence of a catalytic quantity of an amine, with or without a solvent medium (21,22), giving a yield of 90%.

2,4,6-Trichlorophenol. Although 2,4,6-trichlorophenol, $C_6H_3Cl_3O$, can be prepared directly from phenol, the two real precursors are 2,4-dichlorophenol and 2,6-dichlorophenol. The chlorination of these two chlorophenols presents several problems. First, in the chlorination of 2,4-dichlorophenol there is the formation of 2,4,5-trichlorophenol, which is exceedingly troublesome even if present only in trace quantities. The formation of 2,4,5,6,6-pentachloro-2-cyclohexen-1-one is a problem for the 2,6 isomer. Still another problem consists in stopping the process precisely at the tri stage and preventing the formation of tetrachlorophenol or gem-chlorinated cyclohexadienones.

In the chlorination of 2,4-dichlorophenol it has been found that traces of amine (23), onium salts (24), or triphenylphosphine oxide (25) are excellent cat-alysts to further chlorination by chlorine in the ortho position with respect to the hydroxyl function. During chlorination (80°C, without solvent) these catalysts cause traces of 2,4,5-trichlorophenol ($\sim$500–1000 ppm) to be transformed into tetrachlorophenol. Thus these techniques leave no 2,4,5-trichlorophenol in the final product, yielding a 2,4,6-trichlorophenol of outstanding quality. The possi-bility of chlorination using SO_2Cl_2 in the presence of Lewis catalysts has been discussed (26), but no mention is made of 2,4,5-trichlorophenol formation or content.

Chlorination of 2,6-dichlorophenol by chlorine at 70°C gives a yield of only 85%. Fifteen percent of the mixture is made up of 2,4,5,6,6-pentachloro-2-cyclo-hexen-1-one, the formation of which can be explained by the following mechanism:

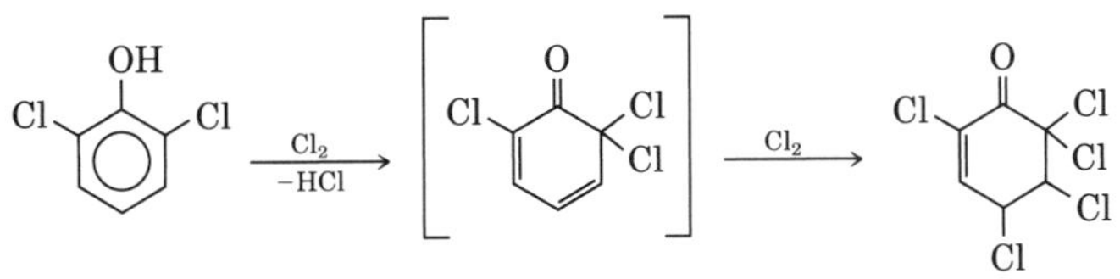

To avoid the formation of this kind of by-product, a direct attack by chlorine in the para position must be encouraged. This can be achieved by using catalysis based on a strong acid (27), or on a sulfur (28), an amine (29), or an onium salt (30). The yields can go as high as 98%.

The catalytic systems described thus far have the advantage of preventing large quantities of gem-chlorinated cyclohexadienones from forming. This type of by-product can, however, always be eliminated with reducing agents (25,31,32) or acids (33).

Tetrachlorophenol. Pure 2,3,4,6-tetrachlorophenol, $C_6H_2Cl_4O$, is not sold commercially because of the difficulty encountered in adjusting the chlorination at the tetra stage. It is found in combination with pentachlorophenol. However, there are patents which describe the use of zirconium tetrachloride (34) or of aluminum trichloride-amine complex (35) as the catalyst to get a yield of 85% for this chlorophenol.

Pentachlorophenol. Because of the high melting temperature of pentachlorophenol, C_6HCl_5O, its preparation makes it necessary to raise the temperature progressively throughout chlorination. The presence of Lewis acid catalysts is essential. The most commonly used of these are $AlCl_3$ and $FeCl_3$.

Much research has been done to improve pentachlorophenol selectivity, either by using a solvent or by seeking more effective catalysts (36–40). A recent study has demonstrated the catalytic role of gem-chlorinated cyclohexadienones in the process of by-product formation, in particular for the formation of chlorinated phenoxyphenols. Insofar as the gem-chlorinated cyclohexadienones constitute an essential intermediate step to the heavier chlorophenols, their presence during chlorination also explains the quantities of chlorinated phenoxyphenols present in pentachlorophenol.

Hydrodechlorination. The polychlorophenols can be broken back down into lighter chlorophenols by catalytic hydrogenation with Pd (41,42), CO (43) in liquid or in gaseous phase (44). This technique is particularly valuable for giving access to phenols chlorinated in the meta position (3-chlorophenol, 3,5-dichlorophenol), because certain conditions yield a regioselective hydrodechlorination (45–47).

Sandmeyer Reaction. This general reaction allows the phenol function to be introduced. The technique complements chlorination insofar as it makes it possible to produce chlorophenols chlorinated in the meta position from the corresponding meta-chlorinated anilines.

Polyhalogenobenzene Hydrolysis. The chlorobenzenes can be transformed into chlorophenols by hydrolysis in a liquid-phase basic medium. The two most commonly used techniques are treatment in aqueous alkali medium at a temperature between 200 and 350°C (48), or a milder hydrolysis (200–250°C) treatment with dilute sodium hydroxide in the presence of copper. The hydrolysis may be carried out in the vapor phase (250–400°C) on solid catalysts based on rare-earth phosphates (49) or copper-bearing silica.

Sulfonation–Desulfonation of Chlorobenzenes. Sulfonation of chlorobenzenes can also be used to produce chlorophenols. Sulfonation is carried out at 60–80°C using oleum at 15–20%. The subsequent desulfonation usually calls on aqueous alkali solutions at 15–20% at temperatures between 170 and 230°C.

Analysis and Specifications

The light chlorophenols normally have a purity greater than 98.5%, but they often reach over 99%, or even 99.5% with direct phenol chlorination. The APHA color test is always below 100. For 2,4,6-trichlorophenol, the 2,4,5-trichlorophenol content constitutes an essential quality index and should be under 20 mg/kg. None of the light chlorophenols contain any polychlorodibenzoparadioxins or polychlorodibenzofurans.

Polychlorophenoxyphenols are the principal impurity in mixtures of tetrachlorophenols and pentachlorophenols. Traces of polychlorodibenzoparadioxins and polychlorodibenzofurans can also be present if the chlorination is not conducted correctly. 2,3,7,8-Tetrachlorodibenzoparadioxin [1746-01-6], which is highly toxic, has never been detected in any products derived by chlorination.

Chlorophenol Analysis. The chlorophenols can be analyzed by acidimetric titration of the hydroxyl function (50). This overall method yields only an approximate evaluation for mixtures. To analyze chlorophenol mixtures, gas chromatography has been the reference method used, as it made it possible to separate and quantify the various chlorophenols (51), but this technique can be a source of errors: the gem-chlorinated cyclohexadienones that may be present along with the chlorophenols are broken back down into lighter chlorophenols under the analysis conditions usually employed.

Therefore, hplc methods seem more effective. By using a combined uv and electrochemical detection technique (52), the gem-chlorinated cyclohexadienones, the chlorophenols, and the phenoxyphenols present in the chlorination mixtures can be determined with great accuracy.

All the chlorophenols can be separated using C_{18}-grafted silica columns. In NH_2-grafted columns, the elution depends on the pK_a. An electrochemical detector in oxidation mode, more sensitive than uv detectors, is generally used to detect very low quantities, especially in analyzing 2,4,5-trichlorophenol.

Polychlorodibenzodioxins, polychlorodibenzofurans, and polychlorophenoxy-phenols formed during thermal or chemical breakdown of chlorophenols can be analyzed by hplc (with uv detection in concentrations as low as ~1 mg/kg) (53). To increase sensitivity and to lower detection thresholds, samples are placed in an alkaline medium, extracted with hexane, and separated in a liquid chromatography column to bunch the products in homogeneous groups. Final detection is effected after gpc separation by electron capture or by mass spectrometry. The detection limit for 2,3,7,8-tetrachlorodibenzodioxin is less than 1 μg/kg.

Health and Safety

Effects in Animals. The LD in rats for all light chlorophenols, irrespective of the administration route, lies between 130 and 4000 mg/kg body weight. The toxicity of these compounds in order of increasing strength is: tetrachlorophenols > monochlorophenols > dichlorophenols > trichlorophenols when the chlorophenol is administered either orally or by subcutaneous injection.

The principal symptoms of chlorophenols at lethal doses are general effects on the central nervous system. Chlorophenols are of medium irritation to the skin and eyes. The effect increases with the number of chlorine atoms in the phenol nucleus. No sensitizing effect has been observed in chlorophenols. Long-term studies have demonstrated effects on the liver and kidneys which accumulate high concentrations of chlorophenols. A carcinogenicity study of 2,4-dichlorophenol run for two years in rats and mice proved negative. The toxicology of chlorophenols is made more complicated by the presence of microcontaminants such as polychlorophenoxyphenols and polychlorodibenzofurans in the technical products.

Effects in Humans. In chlorophenol production, irritation symptoms of the nose, eyes, respiratory tract, and skin resulting in chloroacne have been observed. The results of epidemiology studies on the long-term effects of chlorophenols are quite contradictory and have not allowed the experts to reach any firm conclusions (54).

Economic Aspects

Overall, the chlorophenol market is in decline. Table 2 gives worldwide production figures for 1989, excluding China, India, and Russia. Part of Western Europe's production is exported to Russia for reasons of quality. The main producers are brought together in Table 3 according to the nature of their chlorophenol production.

Rhône-Poulenc, with a capacity of around 20,000 t/yr, is the world's leading producer of light chlorophenols. Excluding the unknown factors for which no statistics are available (China, Russia), the market for pentachlorophenol can be estimated at ~25,000 t/yr. The principal producers of pentachlorophenol are given in Table 4.

Table 2. Worldwide Chlorophenol Production in 1989[a], t/yr

Region	Mono (para + ortho)	Di 2,4	Tri	Penta
Western Europe	8,000	14,000	2,500	4,000
United States, Canada, Brazil		19,000		15,000
Asia	2000–3000	11,000[b]	1,000	6,000[c]
Total	*10,000–11,000*	*44,000*	*3,500*	*25,000*

[a]Excluding China, India, and Russia.
[b]Asia and other miscellaneous countries.
[c]Southeast Asia, South America, and Africa.

Table 3. Producers of Light Chlorophenols

Chlorophenol	Producers	Country
o-chlorophenol	Rhône-Poulenc	France
p-chlorophenol	Rhône-Poulenc	France
	Coalite	UK
	Inui	Japan
2,4-dichlorophenol	Dow[a]	United States, Brazil
	Rhône-Poulenc	France
	Coalite	UK
2,4,6-trichlorophenol	Coalite	UK
	Rhône-Poulenc	France
	Inui	Japan

[a]Dow produces 2,4-dichlorophenol chiefly for a captive market in the United States and Brazil.

Table 4. Pentachlorophenol Producers

Producer	Production, t/yr	Product
Vulcan (USA)	12,000	pentachlorophenol
Idacon (USA)	8,000	pentachlorophenol
Rhône-Poulenc	4,000	pentachlorophenol, Na pentachlorophenate
Chapman	1,500	Na pentachlorophenate

Applications

The main applications of mono-, di-, or trichlorophenols are in agrochemicals and for pentachlorophenol in wood protection.

2-Chlorophenol is used chiefly in the manufacture of an insecticide [*41198-08-7*] (55,56)

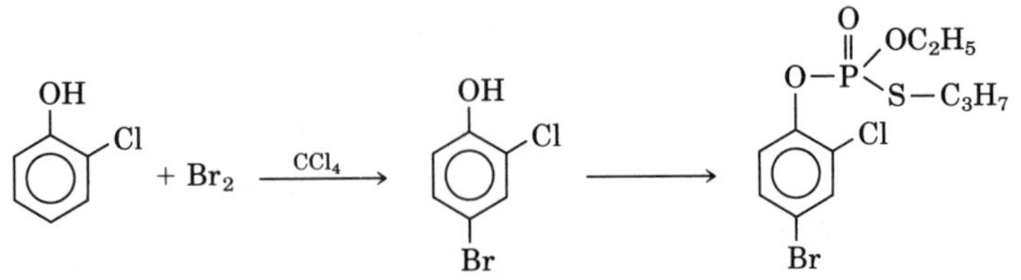

One of the most important applications of 4-chlorophenol is in the synthesis of derivatives of quinizarin [81-64-1], anthraquinone dyes (see DYES, ANTHRA-QUINONE).

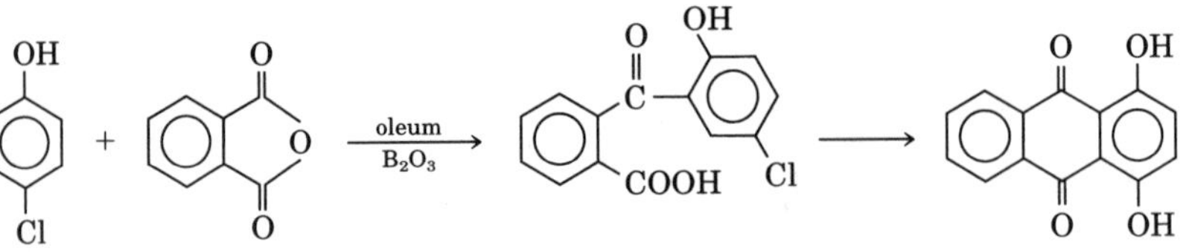

4-Chlorophenol also enters into the synthesis of a biocide (57), 2-benzyl-4-chlorophenol [120-32-1].

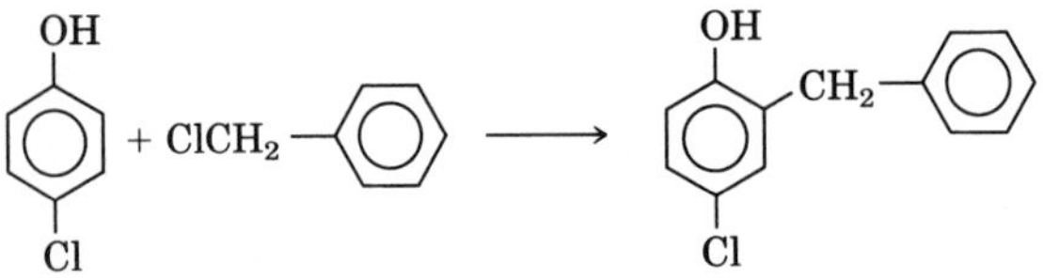

4-Chlorophenol is one of the raw materials used for the synthesis of two fungicides, dichlorophen [97-23-4] and triadimefon [43121-43-3] (58,59).

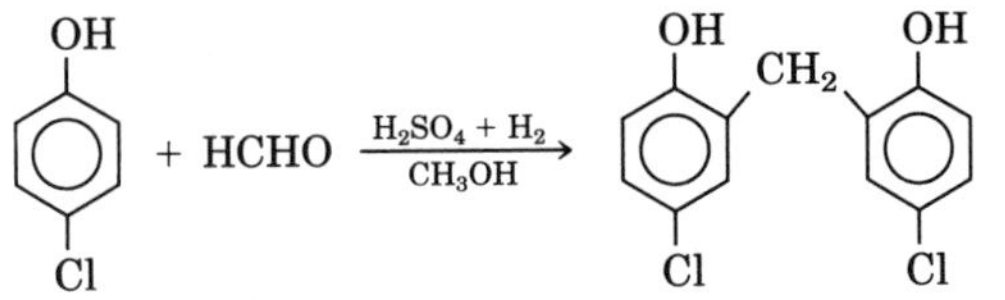

dichlorophen

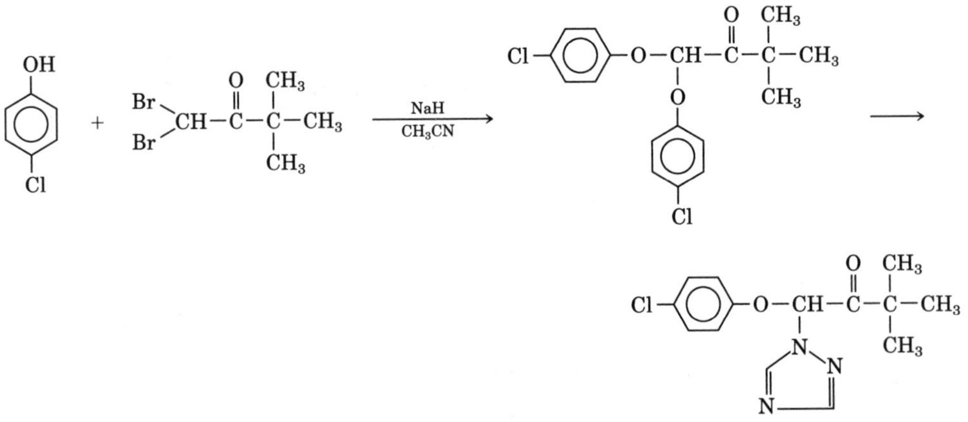

triadimefon

Another application of 4-chlorophenol is in the synthesis of a drug, ethyl α, α-dimethyl-4-chlorophenoxy acetate [637-07-0] (60), used as a cholesterol-reducing agent. This synthesis involves reaction with acetone and chloroform, followed by ethanol esterification.

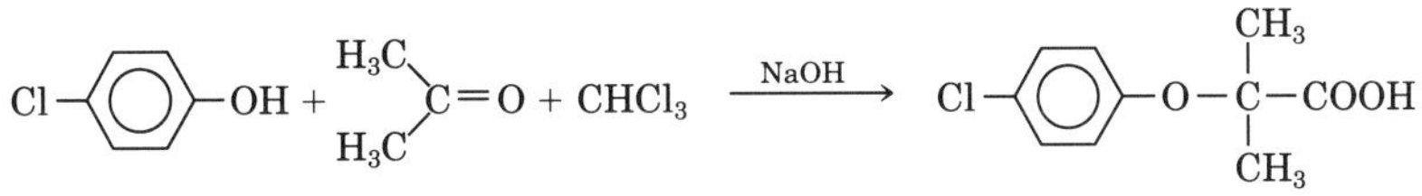

In addition to the use of 2,4-dichlorophenol in the synthesis of 2,4-D herbicides (acid 2,4-D, acid 2,4-DP, acid 2,4-DB), it is also found in the selective postemergence herbicide, diclofop-methyl [51338-27-3] (61) and as a selective preemergence herbicide, oxadiazon [19666-30-9] (62). A postemergence herbicide is applied between the emergence of a seedling and the maturity of a crop plant.

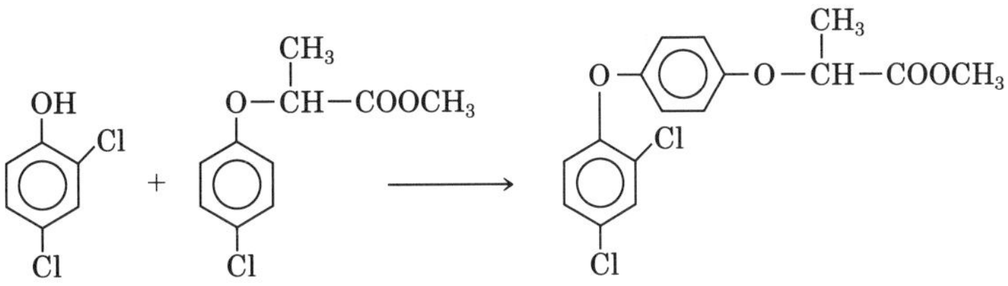

diclofop-methyl

A preemergence herbicide is used before emergence of seedlings above ground.

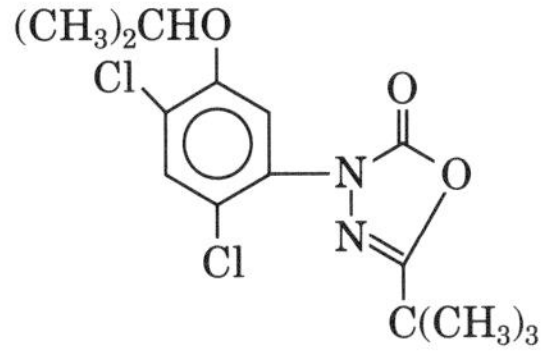

oxadiazon

2,4,6-Trichlorophenol has had two main applications. It is still widely used in the synthesis of a fungicide, prochloraz [67747-09-5] (63).

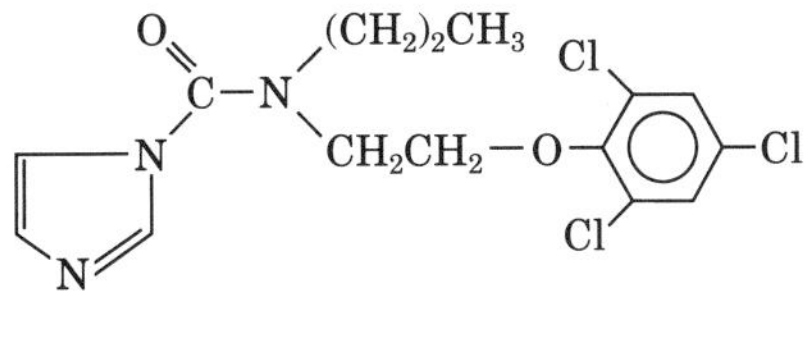

It is also used to manufacture chloranile, [118-75-2], a coloring agent, but a new process to synthesize this product has greatly reduced this market. This new process, with hydroquinone as raw material (64–67), has the advantage of giving a product of much higher quality than can be obtained with 2,4,6-trichlorophenol.

Pentachlorophenol is used in three types of products for treating and preserving wood. One is the provisional protection of freshly cut wood from blue rot by soaking it in an aqueous solution containing 3–5% sodium pentachlorophenate. The protection lasts 6–11 months. Sodium pentachlorophenate [131-52-2] is formed by the reaction of pentachlorophenol with NaOH. The second use is the long-term protection of wood from fungi by soaking it or permeating it in an autoclave with solvents containing pentachlorophenol at 3–5% concentration. Protection lasts 25–40 years. This is the only use of pentachlorophenol in the United States. Last is the treatment of heavy-duty textiles used in manufacturing cables, rigging, and tarpaulins. The treatment is based on the lauric ester of pentachlorophenol. This market is not large and no longer exists in the United States.

BIBLIOGRAPHY

"Chlorophenols" under "Phenol and Phenols," *ECT* 1st ed., Vol. 10, pp. 317–320, by C. Golumbic, Bureau of Mines, U.S. Department of the Interior; "Chlorophenols" in *ECT* 2nd ed., Vol. 5 pp. 325–339, by J. D. Doedens, The Dow Chemical Company; in *ECT* 3rd ed., Vol. 5, pp. 864–872, by E. R. Freiter, Dow Chemical U.S.A.

1. J. Drahonovsky and Z. Vacek, *Collect. Czech. Chem. Commun.* **36**, 3431 (1971).
2. S. Li, M. Paleologou, and W. C. Purdy *J. Chromatogr. Sci.* **29**, 66 (1991).
3. M. Nigretto and M. Josefowicz, *Electrochim. Acta* **18**, 148 (1973).
4. M. Bergon, N. Ben Hamida, and J. Calmon, *J. Agric. Food Chem.* **33**, 577 (1985).
5. J. Courtot-Coupez and M. Le Demezei, *Bull. Soc. Chim. Fr.*, 1033 (1969).
6. P. D. Bolton, J. Ellis, and F. M. Hall, *J. Chem. Soc. (B)*, 1252 (1970).
7. O. M. Dmitrieva, K. A. V'yunov, A. I. Ginak, and E. G. Sochilin, *J. Org. Chem. USSR* **17**(1), 57 (1981).
8. A. Fisher, G. J. Leary, R. D. Topsom, and J. Vaughan, *J. Chem. Soc. (B)*, 686 (1967).
9. P. D. Bolton, F. M. Hall, and J. Kudrynski, *Aust. J. Chem.* **21**, 1541 (1968).
10. K. Ugland, E. Lundanes, and T. Griebrokk, *Chromatography* **213**, 83 (1981).
11. V. A. Dadali, B. V. Panchenko, and L. M. Litvinenko *J. Org. Chem. USSR* **16**(8), 1725 (1980).
12. W. D. Watson, *Tetrahedron Lett.* **30**, 2591 (1976).
13. Jpn. Pat. 61/207351 (Mar. 12, 1985), H. Mise, T. Tsuji, T. Kameyama, and T. Tamura, (to Kenmyo Yakuhin Kogyo K. K.).
14. W. D. Watson, *J. Org. Chem.* **50**, 2145 (1985).
15. Belg. Pat. 827912 (Aug. 25, 1971), W. D. Watson (to The Dow Chemical Company).
16. Neth. Pat. 7504410 (July 14, 1975), W. D. Watson (to The Dow Chemical Company).
17. G. Dutruc-Rosset, Ph.D. dissertation, *n° 79-09*, Lyon, France, 1979.
18. Ger. Pat. 3318791 (May 26, 1982), I. Szekely (to CIBA-GEIGY).
19. A. Guy, M. Lemaire, and J. P. Guette *Tetrahedron* **38**(15), 2339 (1982).
20. U.S. Pat. 2,759,981 (June 24, 1950), B. O. Pray and D. N. Sukow (to Columbia Southern Chemical Corp.).
21. Eur. Pat. 196260 (Mar. 12, 1985), J. C. LeBlanc and S. Ratton (to Rhône-Poulenc).
22. Eur. Pat. 283416 (Mar. 5, 1987), J. C. LeBlanc, S. Ratton, B. Besson, and J. R. Desmurs (to Rhône-Poulenc Chimie).
23. Eur. Pat. 216714 (Sept. 19, 1985), J. C. LeBlanc and S. Ratton (to Rhône-Poulenc Chimie).
24. Eur. Pat. 299891 (July 17, 1987), J. R. Desmurs, B. Besson, and I. Jouve (to Rhône-Poulenc Chimie).

25. Eur. Pat. 243038 (Oct. 28, 1987), G. M. Cole, T. H. Jackson, R. W. Taylor, and P. J. Foreman (to Schering Agrochemicals Ltd.).

26. Jpn. Pat. 62/175429 (Jan. 27, 1987), K. Takase (to Nippon Kayaku Co.).

27. Fr. Pat. 2601001 (July 2, 1986), S. Ratton and J. R. Desmurs (to Rhône-Poulenc Chimie).

28. Brit. Pat. 2177396 (June 27, 1985), S. Ratton (to Rhône-Poulenc Specialties Chimiques).

29. Eur. Pat. 283411 (Mar. 5, 1987), B. Besson, J. R. Desmurs, and I. Jouve (to Rhône-Poulenc Chimie).

30. Eur. Pat. 299890 (July 17, 1987), J. R. Desmurs, B. Besson, I. Jouve (to Rhône-Poulenc Chimie).

31. Brit. Pat. 2135310 (Feb. 23, 1983), K. Gladwin (to Coalite).

32. Eur. Pat. 262061 (Sept. 25, 1986), J. R. Desmurs and S. Ratton (to Rhône-Poulenc Chimie).

33. Eur. Pat. 262063 (Sept. 25, 1986), J. R. Desmurs and S. Ratton (to Rhône-Poulenc Chimie).

34. Fr. Pat. 2636942 (Sept. 23, 1988), J. R. Desmurs and I. Jouve (to Rhône-Poulenc Chimie).

35. Fr. Pat. 2634759 (July 29, 1988), J. R. Desmurs and I. Jouve (to Rhône-Poulenc Chimie).

36. Brit. Pat. 2226314 (Dec. 23, 1988), J. R. Desmurs and I. Jouve (to Rhône-Poulenc Chimie).

37. U.S. Pat. 4,294,996 (June 9, 1980), W. H. Wetzel, H. L. Pan, R. J. Goodwin, and J. E. Wilkinson (to Reichhold Chemical).

38. U.S. Pat. 4,160,114 (July 20, 1977), F. J. Shelton, W. H. Wetzel, J. E. Wilkinson, and R. J. Goodwin (to Reichhold Chemicals).

39. Brit. Pat. 1213090 (Nov. 5, 1968), (to Monsanto Corp.).

40. Fr. Pat. 2649696 (July 11, 1989), J. R. Desmurs and I. Jouve (to Rhône-Poulenc Chimie).

41. Ger. Pat. 2443152 (Sept. 10, 1974), K. Wedemeyer, W. Kiel, and W. Evertz (to Bayer).

42. Ger. Pat. 2344925 (Sept. 6, 1979), W. Kiel, K. Wedemeyer, and W. Evertz (to Bayer).

43. Ger. Pat. 2344926 (Sept. 6, 1973), K. Wedemeyer, E. Koppelmann, and W. Evertz (to Bayer).

44. Fr. Pat. 2161861 (Dec. 17, 1971), G. Rivier (to Progil).

45. Eur. Pat. 55196 (Dec. 24, 1980), G. Cordier (to Rhône-Poulenc).

46. Eur. Pat. 55197 (Dec. 24, 1980), G. Cordier (to Rhône-Poulenc).

47. Eur. Pat. 55198 (Dec. 24, 1980), G. Cordier (to Rhône-Poulenc).

48. SU Pat. 1392068 (Mar. 31, 1986), M. B. Skibinskaya and co-workers.

49. Jpn. Pat. 63/267740 (Apr. 27, 1987), H. Sato and Y. Kawashima (to Idemitsu Kosan).

50. W. Selig, *Mikrochim. Acta* **2**(3,4), 259 (1971).

51. K. Abrahamsson and T. Minxie, *J. Chromatogr.* **279,** 199 (1983).

52. H. Kempf, A. Chamard, J. R. Desmurs, J. Dananche, and G. Bauer, *Poster Iières Journées d'Electrochimie*, Dijon, France, 1987.

53. L. Castle, *J. Chem. Soc., Chem. Commun.* **16,** 704 (1978).

54. *International Programme on Chemical Safety*, Environmental Criteria 93, 1989.

55. F. Buholzer, *Proc. Br. Insectic. Fungic. Conf.* **2,** 659 (1975).

56. Ger. Pat. 3309459 (Mar. 19, 1982), V. Dittrich (to CIBA-GEIGY).

57. U.S. Pat. 1,967,825 (Jan. 7, 1932), E. Klarmann and L. William (to Lehn & Fink).

58. U.S. Pat. 3,912,752 (Jan. 11, 1972), W. Meiser, K. H. Buchel, W. Kramer, and F. Grewe (to Bayer).

59. U.S. Pat. 2,334,408 (July 26, 1941), W. S. Gump and M. Luthy (to Burton T. Bush).

60. Brit. Pat. 860303 (June 20, 1958), W. Glynne, M. Jones, J. M. Thorp, and W. S. Waring (to Imperial Chemical Industry).
61. Ger. Pat. 2646124 (Oct. 13, 1967), H. Boesenger, W. Becker, and K. Matterstoc (to Hoechst).
62. Fr. Pat. 1394774 (Dec. 13, 1963), (to Rhône-Poulenc).
63. Ger. Pat. 2429523 (June 21, 1973), (to Boots).
64. Ger. Pat. 3707148 (Feb. 6, 1987), O. Arndt and T. Papenfuhs (to Hoechst).
65. Ger. Pat. 3703567 (Feb. 6, 1987), O. Arndt and T. Papenfuhs (to Hoechst).
66. Eur. Pat. 326455 (Jan. 27, 1988), J. R. Desmurs and I. Jouve (to Rhône-Poulenc Chimie).
67. Eur. Pat. 326456 (Jan. 27, 1988), J. R. Desmurs and I. Jouve (to Rhône-Poulenc Chimie).

JEAN-ROGER DESMURS
Rhône-Poulenc Recherches

SERGE RATTON
Rhône-Poulenc Specialites Chimiques

CHLOROPHYLL.   See DYES, NATURAL.

CHLOROPRENE.   See CHLOROCARBONS AND CHLOROHYDROCARBONS.

CHLOROSULFURIC ACID

Although chlorosulfuric acid [7790-94-5], $ClSO_3H$, is the *Chemical Abstracts* name, chlorosulfonic acid is the commercial designation by which this compound is more widely known. Other synonyms include sulfuric chlorohydrin, sulfuric acid chlorohydrin, monochlorosulfuric acid, chlorohydrated sulfuric acid, monochlorosulfonic acid, and chlorohydrosulfurous acid.

Chlorosulfuric acid is a clear to straw-colored liquid with a pungent odor. It is a highly reactive compound that reacts violently with water to produce heat and dense white fumes of hydrochloric acid and sulfuric acid. Chlorosulfuric acid reacts with most organic materials, in some cases with charring. It is used principally in organic synthesis as a sulfating, sulfonating, or chlorosulfonating agent. The main application for chlorosulfuric acid is as an intermediate in the production of synthetic detergents, drugs, and dyestuffs (see DETERGENCY; DYES AND DYE INTERMEDIATES; and PHARMACEUTICALS). This acid is preferred in many applications because it yields the desired isomers. It has also been used as a smoke-forming agent in warfare (see CHEMICALS IN WAR).

Chlorosulfuric acid preparation and properties were described in 1854 (1), but the structure was debated for many years until it was shown in 1941 by magnetic susceptibility measurements that the chlorine was bonded directly to the sulfur atom (2). The chlorosulfuric acid structure ($Cl-SO_2-OH$) is analogous to the structure of sulfuric acid, the chlorine replacing one of the hydroxyl groups (see SULFURIC ACID AND SULFUR TRIOXIDE). This structure has been substantiated using Raman spectra (3,4).

Physical Properties

Chlorosulfuric acid is actually an equilibrium mixture of chlorosulfuric acid and minor amounts of hydrogen chloride, sulfur trioxide, and some related compounds. Heating chlorosulfuric acid results in formation of sulfuryl chloride [7791-25-5], Cl_2O_2S, sulfuric acid [7664-93-9], pyrosulfuryl dichloride [7791-27-7], $Cl_2O_5S_2$, and pyrosulfuric acid [7783-05-3], $H_2O_7S_2$. There is also evidence of the formation of higher polyacids such as $H(SO_3)_4Cl$ (see SULFUR COMPOUNDS; SULFURIC AND SULFUROUS ESTERS). Heating beyond the boiling point results in decomposition into sulfur dioxide [7446-09-5], chlorine, and water. Distillation tends to degrade the acid rather than purify it; therefore, the physical constants reported reflect the influence of varying amounts of these impurities. The physical property values given in Table 1 are considered to be the most reliable available. In addition, values have been reported for heats of formation (5), vapor pressure data (6), infrared spectra (7), and thermal constants of mixtures with sulfur trioxide [7446-11-9] (8).

Chlorosulfuric acid is miscible with sulfur trioxide, sulfuric acid, and pyrosulfuryl chloride in all proportions. Mixtures with sulfur trioxide are used as smoke-forming agents. The properties of such mixtures have been described (3,15,16). Mixtures of chlorosulfuric acid and pyrosulfuryl chloride form an azeotrope when distilled (17).

Chlorosulfuric acid is soluble in sym-tetrachloroethane, $C_2H_2Cl_4$, chloroform, and dichloromethane, ie, halocarbons that also contain hydrogen, but is only slightly soluble in carbon disulfide and carbon tetrachloride, a halocarbon not containing hydrogen. It is soluble in liquid sulfur dioxide, nitrobenzene, acetic acid and acetic anhydride, and trifluoroacetic acid and its anhydride, as well as sulfuryl chloride. Chlorosulfuric acid reacts with alcohols, ketones (qv), diethyl ether, and dimethyl sulfoxide, although some literature references report the use of the latter two as solvents. Caution should be used when working with any solvent because a reaction may occur when the temperature is increased or in the presence of catalysts. The solubility of hydrogen chloride in chlorosulfuric acid is indicated to be 0.51 wt % at 20°C and 101.3 kPa (760 mm Hg) (18), but this decreases rapidly with increasing temperature.

Chemical Properties

Chlorosulfuric acid is a strong acid containing a relatively weak sulfur–chlorine bond. Many salts and esters of chlorosulfuric acid are known, most of them are relatively unstable or hydrolyze readily in moist air.

Table 1. Physical Properties of Chlorosulfuric Acid

Property	Value
mol wt	116.531
mp, °C	-81 to -80
bp, °C	151–152
vapor pressure, in Pa[a], T in K	$\log P = 11.496 - 2752/T$
vapor density	
at 216°C, kg/m³[b]	2.4
specific gravity at 15.6°C	1.752
density, kg/m³	
from 0–100°C[c]	$1784.7 - 1.616T + 1.21T^2 \times 10^{-3} - 4.1T^3 \times 10^{-6}$
at -10°C	1800
at -70°C[d]	1900
viscosity, mPa·s ($=$ cP)	
at -31.6°C	10.0
at -17.8°C	6.4
at 15.6°C	3.0
at 49°C	1.7
specific heat, J/(kg·K)[e]	1.18×10^3
heat of formation, $\Delta H_f, 298$, J/mol[e]	-597.1×10^3
heat of vaporization, J/g[e]	452–460
heat of solution in water, J/mol[e]	168.6×10^3
index of refraction, n_D, at 14°C	1.437
dielectric constant at 15°C[f]	60 ± 10
electrical conductivity,	
(ohm·cm)$^{-1}$ at 25°C[g]	0.2–0.3×10^{-3}

[a]To convert Pa to mm Hg, multiply by 0.0075.
[b]The calculated value is 4.04 kg/m³.
[c]Ref. 9.
[d]Ref. 10.
[e]To convert J to cal, divide by 4.184.
[f]Ref. 11.
[g]Ref. 12. For the value in sulfuric acid or in liquid HCl, see References 13 and 14, respectively.

Strong dehydrating agents such as phosphorous pentoxide or sulfur trioxide convert chlorosulfuric acid to its anhydride, pyrosulfuryl chloride [7791-27-7], $S_2O_5Cl_2$. Analogous trisulfuryl compounds have been identified in mixtures with sulfur trioxide (3,19). When boiled in the presence of mercury salts or other catalysts, chlorosulfuric acid decomposes quantitatively to sulfuryl chloride and sulfuric acid. The reverse reaction has been claimed as a preparative method (20), but it appears to proceed only under special conditions. Noncatalytic decomposition at temperatures at and above the boiling point also generates sulfuryl chloride, chlorine, sulfur dioxide, and other compounds.

In organic reactions, chlorosulfuric acid is a powerful sulfating and sulfonating agent, a fairly strong dehydrating agent, and a specialized chloridating agent. In most of its applications it is used to form sulfates, sulfonates, sulfonyl chlorides, and occasionally other chlorine derivatives with organic compounds such as hydrocarbons (qv), alcohols, phenols, and amines (qv). Reactions of chlorosulfuric acid are the result of attachment of a $-SO_3H$ or a $-SO_2Cl$ group to give a sulfonate, sulfate, or a sulfonyl chloride. The general reactions are

$$RH + ClSO_3H \longrightarrow RSO_3H + HCl$$

$$ROH + ClSO_3H \longrightarrow ROSO_3H + HCl$$

$$RH + 2 ClSO_3H \longrightarrow RSO_2Cl + H_2SO_4 + HCl$$

The acid is rather slow to react with aliphatic hydrocarbons unless a double bond or other reactive group is present. This permits straight-chain fatty alcohols such as lauryl alcohol [112-53-8], $C_{12}H_{26}O$, to be converted to the corresponding sulfate without the degradation or discoloration experienced with the more vigorous reagent sulfur trioxide. This is important in shampoo base manufacture (see HAIR PREPARATIONS).

In the presence of excess acid, a sulfonyl chloride group ($-SO_2Cl$) can be attached to an aromatic group, ie, chlorosulfonation can occur,

$$C_6H_6 + 2 ClSO_3H \longrightarrow C_6H_5SO_2Cl + HCl + H_2SO_4$$

or attachment can be to an alkoxy group in a process called chlorosulfation.

$$(CH_3)_2CHOH + 2 ClSO_3H \longrightarrow (CH_3)_2CHOSO_2Cl + HCl + H_2SO_4$$

In the presence of a large excess of acid, sulfones such as diphenyl sulfone [127-63-9], $(C_6H_5)_2SO_2$, can be formed (see SULFOLANES AND SULFONES). Sulfamation forms a $-C-N-S-$ bond as in sodium cyclohexylsulfamate [139-05-9], $C_6H_{11}NHSO_3Na$, (see SULFAMIC ACID AND SULFAMATES). Reviews of chlorosulfuric acid reactions are available (21,22).

The fluorine analogue of chlorosulfuric acid, fluorosulfuric acid [7789-21-1], FSO_3H, is considerably more stable than chlorosulfuric acid because of the stronger fluorine–sulfur bond (see FLUORINE COMPOUNDS, INORGANIC-SULFUR, FLUOROSULFURIC ACID). Bromosulfuric acid [25275-22-3], $BrSO_3H$, decomposes in air at $-30°C$, and the iodine equivalent has not been synthesized (23).

Manufacture

Modern plants manufacture chlorosulfuric acid by direct union of equimolar quantities of sulfur trioxide and dry hydrogen chloride gas. The reaction takes place spontaneously with evolution of a large quantity of heat. Heat removal is necessary to maintain the temperature at 50–80°C and thus minimize unwanted side reactions. The sulfur trioxide may be in the form of 100% liquid or gas, as obtained from boiling oleum, ie, fuming sulfuric acid, or may be present as a dilute gaseous mixture as obtained directly from a contact sulfuric acid plant (24). The hydrogen chloride gas can be in the form of 100% gas or in a diluted form.

Processes for the manufacture of high quality acid are described in a number of patents (24–34). The most common features of these processes are continuous-flow operation, two or more vessels in series for gas–liquid contacting, heat exchangers for controlling temperatures, and the use of excess chlorosulfuric acid as a solvent during at least part of the reaction. If the product is to be used only as a smoke-forming agent and not as a chemical intermediate, or wherever else quality is not critical, then an adiabatic or high temperature operation can be used. The design and operation of this type of plant is described in several reports (35,36). In some older, discontinued processes, hydrogen chloride reacts directly with oleum, or an alkali or alkaline-earth chloride reacts with oleum, followed by distillation.

Chlorosulfuric acid attacks brass, bronze, lead, and most other nonferrous metals. From a corrosion standpoint, carbon steel and cast iron are acceptable below 35°C provided color and iron content is not a concern. Stainless steels (300-series) and certain aluminum alloys are acceptable materials of construction, as is Hastelloy. Glass, glass-lined steel, or Teflon-lined piping and equipment are the preferred materials at elevated temperatures and/or high velocities or where trace iron contamination is a problem, such as in the synthetic detergent industry.

Economic Aspects

There are 20 manufacturers of chlorosulfuric acid in Europe, Asia, and Australia, plus manufacturers in Brazil and Mexico (37). The two United States manufacturers are E. I. du Pont de Nemours & Co., Inc. having a capacity in excess of 30,000 t/yr (38), and Gabriel Chemical Co. having a capacity of 13,600 t/yr (39). The United States and Canadian consumption is about 27,000 t/yr. Pricing from 1988 through the early 1990s has held constant at $386/t, down 6% from previous years (38).

Detergent and other surfactants (qv) manufacturing is the leading consumer of chlorosulfuric acid at approximately 40%; pharmaceuticals (qv) is next at 20%.

Specifications and Standards

No formal industrywide specifications for chlorosulfuric acid exist. Each producer or user establishes individual specifications as needed. However, typical commercial chlorosulfuric acid meets the specifications given in Table 2. The U.S. military specification MILC 379A applies to a mixture of chlorosulfuric acid and sulfur trioxide.

The acid may be shipped in tank cars, tank trucks, iso-tainers or drums via common carrier (40). The shipping hazard class is Corrosive Material, the DOT labels and placards required are Corrosive and Poison. The UN number is 1754. When iron content and color are not of concern, the acid may be stored and shipped in steel equipment. The iron content is generally 25–50 ppm and the color is pale yellow to amber.

Analytical and Test Methods

Total acidity and total chlorides can be determined by conventional techniques after hydrolyzing a sample. Satisfactory procedures for determining hydrogen

Table 2. Specifications for Chlorosulfuric Acid

Property	Value	Typical analysis
appearance	clear, mobile liquid, colorless to slightly straw colored	
color, APHA[a]	100	<10
turbidity, APHA[a]	16	<5
assay, ClSO$_3$H, wt %[b]	98.5	99.4
total Cl as HCl, wt %[b]	30.8	31.2
iron as Fe, ppm[a]	5	1[c]
free SO$_3$, wt %[a]	0.7	0.4
sulfuric acid, wt %[a]	1.5	0.4
aluminum as Al		<1[c]

[a]Value given is maximum value.
[b]Value given is minimum value.
[c]Value is given in ppm.

chloride and free-sulfur trioxide are described in the literature (18,41). Small amounts of both hydrogen chloride and sulfur trioxide can be found in the same sample because of the equilibrium nature of the liquid. Procedures for the direct determination of pyrosulfuryl chloride have also been described (42,43), but are not generally required for routine analysis. Small concentrations of sulfuric acid can be determined by electrical conductivity.

Spot tests for determining chlorosulfuric acid are based on the use of powdered tellurium, which gives a cherry-red color, and powdered selenium, which gives a moss-green color in the presence of the acid.

Health and Safety Factors

Safety. Chlorosulfuric acid is a strong acid and the principal hazard is severe chemical burns when the acid comes into contact with body tissue. The vapor is also hazardous and extremely irritating to the skin, eyes, nose, and respiratory tract. Exposure limits for chlorosulfuric acid have not been established by OSHA or ACGIH. However, chlorosulfuric acid fumes react readily with moisture in the air to form hydrochloric and sulfuric acid mists, which do have established limits. The OSHA 8-h TWA limits and ACGIH TLV–TWA limits are sulfuric acid = 1 mg/m^3; hydrochloric acid = 5 ppm or 7 mg/m^3 (ceiling limit).

Personal protective equipment should be used whenever contact with the acid could be encountered. Chemical safety goggles, hard hat with brim, safety shoes, and acid-resistant clothing, ie, wool or polyester, should be the minimum requirement when working near chlorosulfuric acid. Where the potential for exposure is higher as in loading, sampling, valve operation, etc, the additional use of an acid-proof jacket, pants, and gauntlet gloves should be a required minimum. For emergencies or where there is a high potential for exposure, the protective equipment should include a complete acid suit with hood, gloves, and boots. In some instances, respiratory protective equipment may be needed.

Fire Hazard. Although chlorosulfuric acid itself is not flammable, it may cause ignition by contact with combustible materials because of the heat of reaction. Open fires, open lights, and matches should not be used in or around tanks or containers where hydrogen gas may be collected because of the action of chlorosulfuric acid on metals. Water, carbon dioxide, and dry-chemical fire extinguishers should be kept readily available.

Storage and Handling. The acid should never be allowed to stand in a line completely sealed between two closed valves or check valves. Excessive pressure caused by thermal expansion of the liquid can cause leaks or pipe ruptures. All lubricants and packing materials in contact with chlorosulfuric acid must be chemically resistant to the acid. Flanged connections are recommended over screwed fittings and flange guards should be used.

Spills and Waste Disposal. Chlorosulfuric acid spills generally have a liquid and vapor hazard associated with them. The liquid is an extremely strong acid; fumes react with moisture in the air to form dense clouds of hydrochloric acid and sulfuric acid mists. When the fumes pose the greater danger, the use of a fog nozzle to dilute the acid to nonfuming strengths or the use of a foam to blanket the acid is recommended.

Depending on the magnitude of the spill, control can be achieved by absorption (qv) into absorbents such as expanded clay, diatomaceous earth, or sand (see CLAYS; DIATOMITE). The materials can be removed from the area for controlled dilution with water and/or neutralization with alkali. The use of limestone, ashes, or sand–soda ash mixtures have the advantage of both absorption and partial neutralization. SPILL-X-A (Ansul Fire Protection Co., Marinette, Wisconsin), a magnesium oxide-based absorbent and neutralizer, reduces vapor emissions and neutralizes and solidifies spills of fuming acids such as sulfur trioxide, 65% oleum, and chlorosulfuric acid (44).

Spills can also be diluted with large volumes of water. Care should be taken, however, because chlorosulfuric acid reacts violently with water liberating heat, hydrochloric acid, and sulfuric acid mists and steam. The water should be applied from a safe distance upwind of the spill using a fog nozzle. Remaining traces of acid should be neutralized with soda ash, caustic soda, or lime before disposal.

Uses

Surfactant manufacturing from chlorosulfuric acid includes sulfates of olefins or unsaturated oils, sulfates of polyoxypropylene glycol, sulfonates of long-chain alcohols, particularly lauryl sulfonate, and alkylated diphenyl ether. Pharmaceuticals such as sulfa drugs (see ANTIBACTERIAL AGENTS, SULFONAMIDES), synthetic sweeteners (qv), anticoagulants (see BLOOD COAGULANTS AND ANTICOAGULANTS), phenolphthalein, substituted sulfuric acids and salts, diuretics, and active chlorine agents for disinfection, are made from chlorosulfuric acid. Dyes and pigments (qv) from chlorosulfuric acid include acid dyes, vat dyes, monoazo dyes (see AZO DYES), phthalocyanine dyes, and surface treatment of polyethylene or polyester fibers and films.

Chlorosulfuric acid is used as a sulfonating agent in the preparation of resin-based ion exchange (qv) materials. Sulfonation of glycol phthalate or benzyl chloride–naphthalene resins with chlorosulfuric acid produces water-soluble resins (qv). Thermosetting resins are produced by reaction of indoles and formaldehyde (qv) using chlorosulfuric acid as a catalyst.

Chlorosulfuric acid exhibits catalytic properties in the following reactions: esterification of aliphatic acids in both liquid and vapor phase, alkylation of olefinic hydrocarbons, preparation of alkyl halides from olefinic halides and isoparaffins with tertiary hydrogen, and preparation of unsaturated ketones from olefins and anhydrides of fatty acids. Miscellaneous uses include preparation of pesticides, plasticizers (qv), tanning agents, textile and paper specialties, fluorocarbons, rubber and plastic release agents (qv), as a vulcanization aid for isoolefin copolymers, condensing agents, source of anhydrous hydrogen chloride, and as a separating agent for mixtures of sulfur dioxide and chlorine.

BIBLIOGRAPHY

"Chlorosulfuric Acids" under "Chlorosulfonic Acid" in *ECT* 1st ed., Vol. 3, pp. 885–889 by D. P. Shedd, Monsanto Chemical Co.; in *ECT* 2nd ed., Vol. 5, pp. 357–363, by J. R. Donovan, Monsanto Chemical Co.; "Chlorosulfuric Acid" in *ECT* 3rd ed., Vol. 5, pp. 873–880, by H. O. Burrus, E. I. du Pont de Nemours & Co., Inc.

1. A. W. Williamson, *Proc. Royal Soc. (London)* **7**, 11 (1854).
2. S. S. Dharmatti, *Proc. Indian Acad. Sci. Sect. A* **13**, 359 (1941).
3. H. Gerding, *J. Chem. Phys.* **46**, 118 (1948).
4. R. J. Gillespie and E. A. Robinson, *Can. J. Chem.* **40**, 644 (1962).
5. G. W. Richards and A. A. Woolf, *J. Chem. Soc.* **A**(7), 1118 (1967).
6. L. P. Ryadneva and A. S. Lenskii, *Zh. Prikl. Khim.* **36**, 2413 (1963).
7. R. Savoie and P. A. Giguere, *Can. J. Chem.* **42**, 277 (1964).
8. A. S. Lenskii and co-workers, *Zh. Neorgan. Khim.* **9**, 1147 (1964).
9. T. E. Thorpe, *J. Chem. Soc.* **37**, 327 (1880).
10. M. Schmidt and G. Talsky, *Chem. Ber.* **92**, 1539 (1959).
11. R. J. Gillespie and F. M. White, *Trans. Faraday Soc.* **54**, 1846 (1958).
12. P. Walden, *Z. Anorg. Allgem. Chem.* **29**, 371 (1902).
13. J. Barr and co-workers, *Can. J. Chem.* **39**, 1266 (1961).
14. M. E. Peach and T. C. Waddington, *J. Chem. Soc.*, 2680 (1962).
15. E. W. Balson and N. K. Adam, *Trans. Faraday. Soc.* **44**, 412 (1948).
16. R. J. McCallum and E. L. Tollefson, *Can. J. Res. Sect. F.* **26**, 241 (1948).
17. C. R. Sanger and E. R. Riegel, *Proc. Am. Acad. Arts Sci.* **47**, 673 (1912); *Z. Anorg. Allgem. Chem.* **76**, 79 (1912).
18. E. Korinth, *Agnew. Chem.* **72**, 108 (1960).
19. R. J. Gillespie and E. A. Robinson, *Can. J. Chem.* **39**, 2179 (1961); **40**, 675 (1962).
20. U.S. Pat. 1,554,870 (Sept. 22, 1925), R. H. McKee and C. M. Salls.
21. K. E. Jackson, *Chem. Revs.* **25**, 81 (1939).
22. E. E. Gilbert, *Sulfonation and Related Reactions,* Interscience Publishers, a division of John Wiley & Sons, Inc., New York, 1965.
23. M. Schmidt and G. Talsky, *Z. Anorg. Allgem. Chem.* **303**, 210 (1960).
24. U.S. Pat. 2,311,619 (Feb. 16, 1943), N. A. Laury (to American Cyanamid Co.).
25. U.S. Pat. 1,013,181 (Jan. 2, 1912), A. Klages and H. Vollberg.
26. U.S. Pat. 1,422,335 (July 11, 1922), T. L. Briggs (to General Chemical Co.).
27. U.S. Pat. 2,377,642 (June 5, 1945), R. B. Mooney and G. E. Wentworth (to Imperial Chemical Industries Ltd.); Brit. Pat. 561,841 (1945), (to Imperial Chemical Industries Ltd.).

28. Ger. Pat. 543,758 (May 25, 1929), K. Dachlauer (to I. G. Farbenindustrie AG).
29. Ger. Pat. 914,733 (July 8, 1954), H. Beyer.
30. Jpn. Pat. (May 1, 1957), M. Kawamoto and E. Ejiri (to Mitsubishi Chemical Industries Co.).
31. USSR Pat. 113,664 (Aug. 20, 1958), A. S. Lenskii and co-workers.
32. Neth. Appl. 6,154,410 (May 9, 1966), (to BASF A.G.).
33. Jpn. Pat. 70 24,648 (Aug. 17, 1970), (to Mitsubishi Chemical Industries Co.).
34. Jpn. Pat. 76 10,840 (Apr. 7, 1976), (to Mitsubishi Chemical Industries Co.).
35. R. E. Richardson and co-workers, *U.S. Dept. Comm. Office Tech. Serv. PB Rept. 218,* 1945.
36. W. A. M. Edwards and co-workers, *U.S. Dept. Comm. Office Tech. Serv. PB Rept. 34005, BIOS Final Report 243, Item 22,* May 1946; *U.S. Dept. Comm. Office Tech. Serv. PB Rept. L34005-S, FIAT Tech. Bull. T12,* Mar. 1947.
37. *Directory of World Chemical Producers,* 1989/1990 ed., Chemical Information Services, Ltd., Oceanside, N.Y.
38. *Chem. Mark. Rep.,* 27 (May 21, 1990).
39. *Chem. Mark. Rep.,* 25 (Aug. 27, 1990).
40. *Chlorosulfonic Acid, Properties, Uses, Storage & Handling,* E52057, E. I. du Pont de Nemours & Co., Inc., Wilmington, Del., Nov. 1982.
41. W. Seaman and co-workers, *Anal. Chem.* **22,** 549 (1950).
42. J. H. Payne, Jr., Ph.D. dissertation, Purdue University, Lafayette, Ind., June 1947.
43. G. V. Zavorov, *Zavodsk, Lab.* **27,** 1208 (1961).
44. J. A. Engman, "Clean Up and Control of Large Fuming Acid Spills," *Proceedings of the Hazardous Materials Central Management Conference,* Tower Conference Management Co., Glen Ellyn, Ill., Mar. 1991.

General Reference

J. W. Mellor, *A Comprehensive Treatise on Inorganic and Theoretical Chemistry,* Vol. 10, Longmans, Green & Co., London, 1920, pp. 684–692.

C. E. McDONALD
E. I. du Pont de Nemours & Co., Inc.

CHOCOLATE AND COCOA

The name *Theobroma cacao,* food of the gods, indicating both the legendary origin and the nourishing qualities of chocolate, was bestowed upon the cacao tree by Linnaeus in 1720. All cocoa and chocolate products are derived from the cocoa bean, the seed of the fruit of this tree. Davila Garibi, a contemporary Mexican scholar, has traced the derivation of the word from basic root words of the Mayan language to its adoption as chocolate in Spanish (1).

The terms cocoa and cacao often are used interchangeably in the literature. Both terms describe various products from harvest through processing. In this article, the term cocoa will be used to describe products in general and the term cacao will be reserved for botanical contexts. Cocoa traders and brokers frequently use the term raw cocoa to distinguish unroasted cocoa beans from finished products; this term is used to report statistics for cocoa bean production and consumption.

Standards for Cocoa and Chocolate

In the United States, chocolate and cocoa are standardized by the U.S. Food and Drug Administration under the Federal Food, Drug, and Cosmetic Act. The current definitions and standards resulted from prolonged discussions between the U.S. chocolate industry and the Food and Drug Administration (FDA). The definitions and standards originally published in the *Federal Register* of December 6, 1944, have been revised only slightly.

The Food and Agricultural Organization (FAO) and the World Health Organization (WHO) jointly sponsor the Codex Alimentarius Commission, which conducts a program for developing worldwide food standards. The Codex Committee for Cocoa Products and Chocolate has developed standards for chocolate (Codex Standard 87-1981), and cocoa powders and dry cocoa–sugar mixtures (Codex Standard 105-1981). As a member of the Codex Alimentarius Commission, the United States is obligated to consider all Codex standards for acceptance.

The FDA announced in the *Federal Register* of January 25, 1989 a proposal to amend the U.S. chocolate and cocoa standards of identity. The proposed amendments respond principally to a citizen petition submitted by the Chocolate Manufacturers Association (CMA) and, to the extent practicable, will achieve consistency with the Codex standards. The proposed amendments would allow for the use of nutritive carbohydrate sweeteners, neutralizing agents, and emulsifiers; reduce slightly the minimum milkfat content and eliminate the nonfat milk solids-to-milkfat ratios in certain cocoa products including milk chocolate; update the language and format of the standards; and provide for optional ingredient labeling requirements. FDA has also received a proposal to establish a new standard of identity for white chocolate. Comments regarding the proposal amendments are under review by FDA, and a final ruling is expected to be issued in the near future.

White Chocolate. There is at present no standard of identity in the United States for white chocolate. Virtually all current uses of the term white chocolate do not meet the standards for chocolate, which prescribes the presence of ground cacao nibs. This restrictive requirement has acted as a practical deterrent to companies developing and marketing white chocolate-type products in the United States. When such products have been introduced and marketed in the United States, companies have had to label them with fanciful names to avoid the standardized labeling issues. In other countries where a standard of identity for white chocolate exists and where a minimum amount of cocoa butter is required by law, consumers have available to them a variety of easily recognizable products.

White chocolate has been defined by the European Economic Community (EEC) Directive 75/155/EEC as free of coloring matter and consisting of cocoa butter (not less than 20%); sucrose (not more than 55%); milk or solids obtained by partially or totally dehydrated whole milk, skimmed milk, or cream (not less than 14%); and butter or butter fat (not less than 3.5%).

Cocoa Beans

The cocoa bean is the basic raw ingredient in the manufacture of all cocoa products. The beans are converted to chocolate liquor, the primary ingredient from which all chocolate and cocoa products are made. Figure 1 depicts the conversion of cocoa beans to chocolate liquor, and in turn to the chief chocolate and cocoa products manufactured in the United States, ie, cocoa powder, cocoa butter, and sweet and milk chocolate.

Significant amounts of cocoa beans are produced in about 30 different localities. These areas are confined to latitudes 20° north or south of the equator. Although cocoa trees thrive in this very hot climate, young trees require the shade of larger trees such as banana, coconut, and palm for protection.

New cocoa hybrids and selections have been developed in Malaysia and other countries that produce significantly higher yields in select soil and climate conditions. In addition, high density plantings have demonstrated higher and earlier yield in Malaysia and the Philippines. Low or no shade cocoa has also proven to increase yields. However, both high density and reduced shade cocoa production requires additional inputs of management and nutrition. Additional inputs to control pests and diseases also may be required.

A cocoa tree produces its first crop in three to four years and a full crop after six to seven years. A full grown tree can reach a height of 12 to 15 m but is normally trimmed to 5–6 m to permit easy harvest. Because of differences in climate the crop is not confined to one short season but may extend for several months. Indeed some areas have cocoa pods almost all year long with one or two minor peaks. Many areas have peak harvests; for example, in West Africa there is one large main crop (80% or more of total crop) from September to March and a small or medium-sized crop in May. Brazil on the other hand has two crops a year that are almost equal in size.

Fermentation (Curing).   Prior to shipment from producing countries, most cocoa beans undergo a process known as curing, fermenting, or sweating. These terms are used rather loosely to describe a procedure in which seeds are removed from the pods, fermented, and dried. Unfermented beans, particularly from Haiti and the Dominican Republic, are used in the United States.

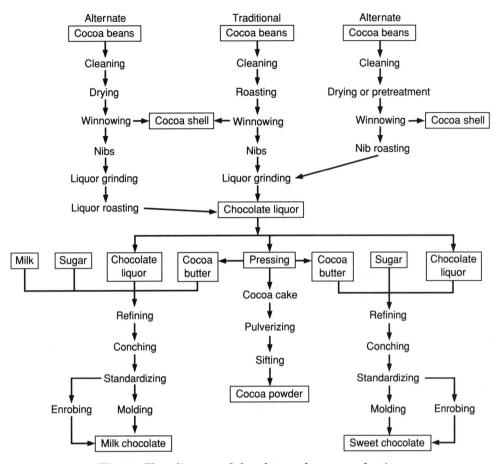

Fig. 1. Flow diagram of chocolate and cocoa production.

The age-old process of preparing cocoa beans for market involves specific steps that allegedly promote the activities of certain enzymes. Various methods of fermentation are used to the same end.

Fermentation plays a principal role in flavor development of beans by mechanisms that are not well understood (2). Because freshly harvested cocoa beans are covered with a white pulp, rich in sugars, fermentation begins almost immediately upon exposure to air. The sugars are converted to alcohol, and finally to acetic acid, which drains off, freeing the cotyledon from the pulpy mass. The acetic acid and heat formed during fermentation penetrate the skin or shell, killing the germ and initiating chemical changes within the bean that play a significant role in the development of flavor and color. During this initial stage of fermentation, the beans acquire the ability to absorb moisture, which is necessary for many of the chemical reactions that follow.

Commercial Grades.   Most cocoa beans imported into the United States are one of about a dozen commercial varieties that can be generally classified as Criollo or Forastero. Criollo beans have a light color, a mild, nutty flavor, and an odor somewhat like sour wine. Forastero beans have a strong, somewhat bitter flavor and various degrees of astringency. The Forastero varieties are more abundant and provide the basis for most chocolate and cocoa formulations. Table 1 shows the main varieties of cocoa beans imported into the United States. The varieties are usually named for the country or port of origin.

Table 1. Main Varieties of Cocoa Beans Imported into the United States

Africa	South America	Pacific	West Indies
Ivory Coast	Bahia (Brazil)	Malaysia	Sanchez (Dominican Republic)
Accra (Ghana)	Arriba (Ecuador)	New Guinea	Grenada
Lagos	Venezuelan	Indonesia	Trinidad
Nigeria		Samoa	
Fernando Po			
Sierra Leone			

Bean Specifications. Cocoa beans vary widely in quality, necessitating a system of inspection and grading to ensure uniformity. Producing countries have always inspected beans for proper curing and drying as well as for insect and mold damage. Recently, a procedure for grading beans has been established at an international level. This ordinance, reached primarily through the efforts of FAO, has been adopted by Codex as the model ordinance for inspection and grading of beans. It classifies beans into two principal categories according to the fraction of moldy, slaty, flat, germinated, and insect-damaged beans (3).

Cocoa beans are sometimes evaluated in the laboratory to distinguish and characterize flavors. Beans are roasted at a standardized temperature for a specific period of time, shelled, usually by hand, and ground or heated slightly to obtain chocolate liquor. The liquor's taste is evaluated by a panel of experts who characterize and record the particular flavor profile. The Chocolate Manufacturers' Association of the United States recently formed a committee to standardize this laboratory evaluation (3).

Blending. Most chocolate and cocoa products consist of blends of beans chosen for flavor and color characteristics. Cocoa beans may be blended before or after roasting, or nibs may be blended before grinding. In some cases finished liquors are blended. Common, or basic beans, are usually African or Brazilian and constitute the bulk of most blends. More expensive flavor beans from Venezuela, Trinidad, Ecuador, etc are added to impart specific characteristics. The blend is determined by the end use or type of product desired.

Production. Worldwide cocoa bean production has increased significantly over the past 10 years from approximately 1.6 million t in the 1979–1980 crop year to over 2.4 million t in 1990. The production share by country has also changed dramatically in the last 10 years. The big gainers are Malaysia, Indonesia, and the Ivory Coast. The large gains in Malaysia and Indonesia have helped to diversify production and partially shield the market from adverse weather-induced supply shocks. The biggest losers in production share have been Brazil and Ghana. Table 2 lists production statistics for these countries.

Consumption. Worldwide cocoa bean consumption has increased significantly over the past 10 years from approximately 1.5 million t in the 1979–1980

crop year to almost 2.3 million t today. This growth was uneven between East and West. North America and Western Europe increased grind by approximately 200% over this time period whereas Russia and Eastern Europe dropped 11%. Table 3 gives the annual tonnage of cocoa bean grind by the United States and other leading countries.

Marketing. Most of the cocoa beans and products imported into the United States are done so by New York and London trade houses. The New York Sugar, Coffee, and Cocoa Exchange provides a mechanism by which both chocolate manufacturers and trade houses can hedge their cocoa bean transactions. Additional information on the functions of the New York Cocoa Exchange is available (3).

Table 2. Production of Raw Cocoa Beans, 10^3 t

Region	1979–1980	1989–1991	1990–1991[a]
Ivory Coast	379	710	740
Brazil	294	370	369
Ghana	285	295	250
Malaysia	35	255	265
Nigeria	172	170	170
Indonesia	7	110	122
Other	459	510	511
Total	*1631*	*2420*	*2427*

[a] Estimated.

Table 3. Grind of Raw Cocoa Bean, 10^3 t

Region	1979–1980	1984–1985	1990–1991[a]
United States	137	206	280
United Kingdom	62	95	125
Germany	158	205	280
The Netherlands	129	165	256
Eastern Europe and Russia	201	262	179
other consuming countries	259	282	354
producing countries	537	648	799
Total	*1483*	*1863*	*2273*

[a]Estimated.

Chocolate Liquor

Chocolate liquor is the solid or semisolid food prepared by finely grinding the kernel or nib of the cocoa bean. It is also commonly called chocolate, unsweetened chocolate, baking chocolate, or cooking chocolate. In Europe chocolate liquor is often called chocolate mass or cocoa mass.

Cleaning. Cocoa beans are imported in the United States in 70-kg bags. The beans can be processed almost immediately or stored for later use. They are usually fumigated prior to storage.

The first step in the processing of cocoa beans is cleaning. Stones, metals, twigs, twine, and other foreign matter are usually removed by passing beans in a large thin layer over a vibrating screen cleaner. Large objects are retained as the beans fall through a lower screen. The second screen removes sand and dirt that have adhered to the beans. Strategically placed magnets are commonly used to remove small pieces of metal.

Roasting. The chocolate flavor familiar to the consumer is primarily developed during roasting, which promotes reactions among the latent flavor precursors in the bean. Good flavor depends on the variety of bean and the curing process used. The bacterial or enzymatic changes that occur during fermentation presumably set the stage for the production of good flavor precursors.

Although flavor precursors in the unroasted cocoa bean have no significant chocolate flavor themselves, they react to form highly flavored compounds. These flavor precursors include various chemical compounds such as proteins, amino acids, reducing sugars, tannins, organic acids, and many unidentified compounds.

The natural moisture of the cocoa bean combined with the heat of roasting cause many chemical reactions other than flavor changes. Some of these reactions remove unpleasant volatile acids and astringent compounds, partially break down sugars, modify tannins and other nonvolatile compounds with a reduction in bitterness, and convert proteins to amino acids that react with sugars to form flavor compounds, particularly pyrazines (4). To date, over 300 different compounds, many of them formed during roasting, have been identified in the chocolate flavor (5)

Roasting is essentially a cooking process developed by craftsmen who were guided by their senses of smell and taste and their knowledge of how beans of differing degrees of roast behaved in the subsequent processes. The ease and efficiency with which the processes of winnowing, grinding, pressing, and conching can be performed is affected by the degree of roast.

Roasting conditions can be adjusted to produce different flavors. Low, medium, full, and high roasts can be developed by varying time, temperature, and humidity in the roaster. Low roasts produce mild flavors and light color; high roasts produce strong flavors and dark color (6).

Roasters have evolved from the coke-fired rotary drum type to continuous feed roasters. It is traditional to roast cocoa beans with the shell still on. However, other methods of roasting include nib roasting, in which the shell is first removed by a rapid or moist heating step, and liquor roasting (Fig. 1). The newer nib and liquor roasters are designed to subject the cocoa to more uniform heat conditions in addition to minimizing the loss of cocoa butter to the shell. Nib and liquor roasters do not need the high temperatures necessary for whole bean roasting and, therefore, can be considered more energy efficient. Roasting times vary from about 30 to 60 minutes. Actual temperature of the bean in the roaster is difficult to measure but probably ranges from as low as 70°C to as high as 180°C.

Winnowing. Winnowing, often called cracking and fanning, is one of the most important operations in cocoa processing. It is a simple process that involves separating the nib, or kernel, from the inedible shell. Failure to remove shell results in lower quality cocoa and chocolate products, more wear on nib grinding machines, and lower efficiency in all subsequent operations.

Because complete separation of shell and nib is virtually impossible, various countries have established maximum allowable limits of shell by weight; U.S. manufacturers average from 0.05 to 1%.

The analysis of cocoa shell (7) is given in Table 4. In the United States, shells are often used as mulch or fertilizer for ornamental and edible plants, as animal feed, and as fuel for boilers.

Grinding. The final step in chocolate liquor production is the grinding of the kernel or nib of the cocoa bean. The nib is a cellular mass containing about

Table 4. Analyses of Cocoa Shell from Roasted Cocoa Beans

Component	Shell, %
water	3.8
fat	3.4
ash	
total	8.1
water-soluble	3.5
water-insoluble	4.6
silica, etc	1.1
alkalinity (as K_2O)	2.6
chlorides	0.07
iron (as Fe_2O_3)	0.03
phosphoric acid (as P_2O_5)	0.8
copper	0.004
nitrogen	
total nitrogen	2.8
protein nitrogen	2.1
ammonia nitrogen	0.04
amide nitrogen	0.1
theobromine	1.3
caffeine	0.1
carbohydrates	
glucose	0.1
sucrose	0.0
starch (taka-diastase method)	2.8
pectins	8.0
fiber	18.6
cellulose	13.7
pentosans	7.1
mucilage and gums	9.0
tannins	
tannic acid (Lowenthal's method)	1.3
cocoa-purple and cocoa-brown	2.0
acids	
acetic (free)	0.1
citric	0.7
oxalic	0.32
extracts	
cold water	20.0
alcohol, 85%	10.0

50 to 56% cocoa fat (cocoa butter). Grinding liberates the fat locked within the cell wall while producing temperatures as high as 110°C.

Nibs are usually ground while they are still warm from roasting. The original horizontal three-tier stone mills and vertical disk mills have been replaced by modern horizontal disk mills, which have much higher outputs and are capable of grinding nibs to much greater fineness. Two modern machines in particular account for a large percentage of liquor grinding. One uses a pin mill mounted over a roller refiner. The pin mill grinds the nibs to a coarse but fluid liquor. The liquor is delivered to a roll refiner that reduces the particle size to a very fine limit. The second type is a vertical or horizontal ball mill. Coarsely ground nib is fed to the base of a vertical cylinder, which contains small balls in separate compartments. A central spindle causes the balls to rotate at very high speeds, grinding the liquor between them and against the internal wall of the cylinder (8).

Cocoa Powder

Cocoa powder (cocoa) is prepared by pulverizing the remaining material after part of the fat (cocoa butter) is removed from chocolate liquor. The U.S. chocolate standards define three types of cocoas based on their fat content. These are breakfast, or high fat cocoa, containing not less than 22% fat; cocoa, or medium fat cocoa, containing less than 22% fat but more than 10%; and low fat cocoa, containing less than 10% fat.

Cocoa powder production today is an important part of the cocoa and chocolate industry because of increased consumption of chocolate-flavored products. Cocoa powder is the basic flavoring ingredient in most chocolate-flavored cookies, biscuits, syrups, cakes, and ice cream. It is also used extensively in the production of confectionery coatings for candy bars.

Cocoa Powder Manufacture.   When chocolate liquor is exposed to pressures of 34–41 MPa (5000–6000 psig) in a hydraulic press, and part of the fat (cocoa butter) is removed, cocoa cake (compressed powder) is produced. The original pot presses used in cocoa production had a series of pots mounted vertically one above the other. These have been supplanted by horizontal presses that have four to twenty-four pots mounted in a horizontal frame. The newer presses are capable of complete automation, and by careful selection of pressure, temperature, and time of pressing, cocoa cake of a specified fat content can be produced.

Cocoa powder is produced by grinding cocoa cake. Cocoa cake warm from the press breaks easily into large chunks but is difficult to grind into a fine powder. Cold, dry air removes the heat generated during most grinding operations. Because the finished cocoa powder still contains fat, great care must be taken to prevent the absorption of undesirable odors and flavors.

Commercial cocoa powders are produced for various specific uses and many cocoas are alkali treated, or Dutched, to produce distinctive colors and flavors. The alkali process can involve the treatment of nibs, chocolate liquor, or cocoa with a wide variety of alkalizing agents (9).

Cocoa powders not treated with alkali are known as cocoa or natural cocoa. Natural cocoa has a pH of about 4.8 to 5.8 depending on the type of cocoa beans used. Alkali processed cocoa ranges in pH from about 6 to as high as 8.5.

Cocoa Butter

Cocoa butter is the common name given to the fat obtained by subjecting chocolate liquor to hydraulic pressure. It is the main carrier and suspending medium for cocoa particles in chocolate liquor and for sugar and other ingredients in sweet and milk chocolate.

The FDA has not legally defined cocoa butter, and no standard exists for this product under the U.S. Chocolate Standards. For the purpose of enforcement, the FDA defines cocoa butter as the edible fat obtained from cocoa beans either before or after roasting. Cocoa butter as defined in the *U.S. Pharmacopeia* is the fat obtained from the roasted seed of *Theobroma cacao Linne.*

The Codex Committee on Cocoa and Chocolate Products defines cocoa butter as the fat produced from one or more of the following: cocoa beans, cocoa nibs, cocoa mass (chocolate liquor), cocoa cake, expeller cake, or cocoa dust (fines) by a mechanical process and/or with the aid of permissible solvents (10). It further states that cocoa butter shall not contain shell fat or germ fat in excess of the proportion in which they occur in the whole bean.

Codex has also defined the various types of cocoa butter in commercial trade (10). Press cocoa butter is defined as fat obtained by pressure from cocoa nib or chocolate liquor. In the United States, this is often referred to as prime pure cocoa butter. Expeller cocoa butter is defined as the fat prepared by the expeller process. In this process, cocoa butter is obtained directly from whole beans by pressing in a cage press. Expeller butter usually has a stronger flavor and darker color than prime cocoa butter and is filtered with carbon or otherwise treated prior to use. Solvent extracted cocoa butter is cocoa butter obtained from beans, nibs, liquor, cake, or fines by solvent extraction (qv), usually with hexane. Refined cocoa butter is any of the above cocoa butters that has been treated to remove impurities or undesirable odors and flavors.

Composition and Properties.   Cocoa butter is a unique fat with specific melting characteristics. It is a solid at room temperature (20°C), starts to soften around 30°C, and melts completely just below body temperature. Its distinct melting characteristic makes cocoa butter the preferred fat for chocolate products.

Cocoa butter is composed mainly of glycerides of stearic, palmitic, and oleic fatty acids (see FATS AND FATTY OILS). The triglyceride structure of cocoa butter has been determined (11,12) and is as follows:

tri-saturated, 3%;

mono-unsaturated (oleo-distearin), 22%;

oleo-palmitostearin, 57%;

oleo-dipalmitin, 4%;

di-unsaturated (stearo-diolein), 6%;

palmito-diolein, 7%;

tri-unsaturated, tri-olein, 1%.

Although there are actually six crystalline forms of cocoa butter, four basic forms are generally recognized as alpha, beta, beta prime, and gamma. The γ (gamma) form, the least stable, has a melting point of 17°C. It changes rapidly to the α (alpha) form which melts at 21–24°C. At normal room temperature the β' (beta prime) form changes to the β (beta) form, melting at 27–29°C, and finally, the β form is reached. It is the most stable form with a melting point of 34–35°C (13).

Since cocoa butter is a natural fat, derived from different varieties of cocoa beans, no single set of specifications or chemical characteristics can apply. Codex has attempted to define the physical and chemical parameters of the various types of cocoa butter (14) (Table 5).

Substitutes and Equivalents.   In the past 25 years, many fats have been developed to replace part or all of the added cocoa butter in chocolate-flavored products. These fats fall into two basic categories commonly known as cocoa butter substitutes and cocoa butter equivalents. Neither can be used in the United States in standardized chocolate products, but they are used in small amounts, usually up to 5% of the total weight of the product, in some European countries.

Cocoa butter substitutes of all types enjoy widespread use in the United States chiefly as ingredients in chocolate-flavored products. Cocoa butter equivalents are not widely used because of their higher price and limited supply.

Cocoa butter substitutes do not chemically resemble cocoa butter and are compatible with cocoa butter only within specified limits. Cocoa butter equivalents are chemically similar to cocoa butter and can replace cocoa butter in any proportion without deleterious physical effects (15,16).

Cocoa butter substitutes and equivalents differ greatly with respect to their method of manufacture, source of fats, and functionality; they are produced by several physical and chemical processes (17,18). Cocoa butter substitutes are produced from lauric acid fats such as coconut, palm, and palm kernel oils by fractionation and hydrogenation; from domestic fats such as soy, corn, and cotton seed oils by selective hydrogenation; or from palm kernel stearines by fractionation.

Table 5. Properties and Composition of Cocoa Butter[a]

Characteristic	Press cocoa butter	Expeller cocoa butter	Refined cocoa butter
refractive index n_D^{40}, °C	1.456–1.458	1.453–1.459	1.453–1.462
melting behavior			
slip point, °C	30–40	30–34	30–34
clear melting point, °C	31–35	31–35	31–35
free fatty acids			
mol % oleic acid	0.5–1.75	0.5–1.75	0–1.75
saponification value			
mg KOH/g fat	192–196	192–196	192–196
iodine value, Wijs	33.8–39.5	35.6–44.6	35.7–41.0
unsaponifiable matter			
petroleum ether % m/m	not more than 0.35%	not more than 0.40%	not more than 0.50%

[a]Contaminants not to exceed 0.5 mg/kg of arsenic, 0.4 mg/kg of copper, 0.5 mg/kg of lead, and 2.0 mg/kg of iron; Ref. 15.

Cocoa butter equivalents can be produced from palm kernel oil and other specialty fats such as shea and illipe by fractional crystallization; from glycerol and selected fatty acids by direct chemical synthesis; or from edible beef tallow by acetone crystallization.

In the early 1990s, the most frequently used cocoa butter equivalent in the United States was derived from palm kernel oil but a synthesized product was expected to be available in the near future.

Sweet and Milk Chocolate

Most chocolate consumed in the United States is consumed in the form of milk chocolate and sweet chocolate. Sweet chocolate is chocolate liquor to which sugar and cocoa butter have been added. Milk chocolate contains these same ingredients and milk or milk solids (Fig. 2).

U.S. definitions and standards for chocolate are quite specific (19). Sweet chocolate must contain at least 15% chocolate liquor by weight and must be sweetened with sucrose or mixtures of sucrose, dextrose, and corn syrup solids in specific ratios. Semisweet chocolate and bittersweet chocolate, though often referred to as sweet chocolate, must contain a minimum of 35% chocolate liquor. The three products, sweet chocolate, semisweet chocolate, and bittersweet chocolate, are often simply called chocolate or dark chocolate to distinguish them from milk chocolate. Table 6 gives some typical formulations for sweet chocolates (5).

Sweet chocolate can contain milk or milk solids (up to 12% max), nuts, coffee, honey, malt, salt, vanillin, and other spices and flavors as well as a number of specified emulsifiers. Many different kinds of chocolate can be produced by careful selection of bean blends, controlled roasting temperatures, and varying amounts of ingredients and flavors (20).

The most popular chocolate in the United States is milk chocolate. The U.S. Chocolate Standards state that milk chocolate shall contain no less than 3.66 wt % of milk fat and not less than 12 wt % of milk solids. In addition, the ratio of nonfat milk solids to milk fat must not exceed 2.43:1 and the chocolate liquor content must not be less than 10% by weight. Some typical formulations of milk chocolate and some compositional values are shown in Table 7 (5).

Production. The main difference in the production of sweet and milk chocolate is that in the production of milk chocolate, water must be removed from the milk. Many milk chocolate producers in the United States use spray-dried milk powder. Others condense fresh whole milk with sugar, and either dry it, producing milk crumb, or blend it with chocolate liquor and then dry it, producing milk chocolate crumb. These crumbs are mixed with additional chocolate liquor, sugar, and cocoa butter later in the process (21). Milk chocolates made from crumb typically have a more caramelized milk flavor than those made from spray-dried milk powder.

Mixing. The first step in chocolate processing is the weighing and mixing of ingredients. This is usually a fully automated process carried out in a batch or continuous processing system. In batch processing, all the ingredients for one batch are automatically weighed into a mixer and mixed for a specific period of time. The mixture is conveyed to storage hoppers directly above the refiners. In

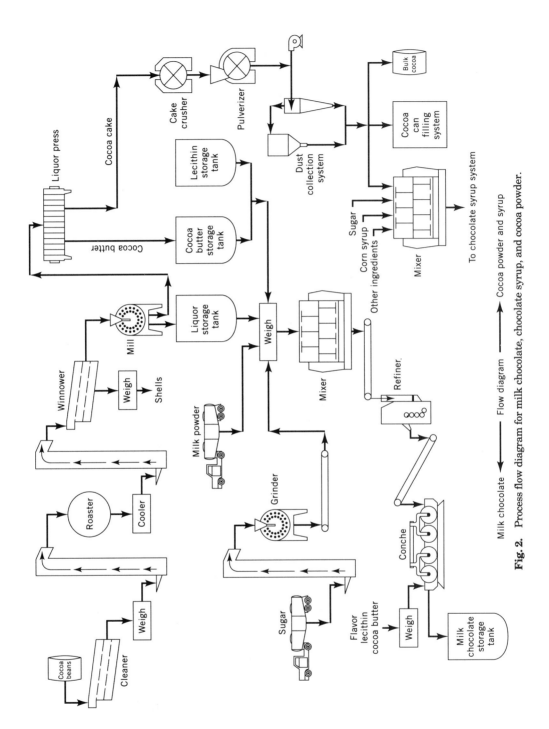

Fig. 2. Process flow diagram for milk chocolate, chocolate syrup, and cocoa powder.

Milk chocolate ──── Flow diagram ────▶ Cocoa powder and syrup

188

Table 6. Typical Formulations for Sweet (Dark) Chocolates

Ingredient	Formulation, %		
	1	2	3
chocolate liquor	15.0	35.0	70.0
sugar	60.0	50.4	29.9
added cocoa butter	23.8	14.2	
lecithin	0.3	0.3	
vanillin	0.9	0.1	0.1
Total fat	*32.0*	*33.0*	*37.1*

Table 7. Typical Formulations for Milk Chocolate

Ingredient	Formulation, %		
	1	2	3
chocolate liquor	11.0	12.0	12.0
dry whole milk	13.0	15.0	20.0
sugar	54.6	51.0	45.0
added cocoa butter	21.0	21.6	22.6
lecithin	0.3	0.3	0.3
vanillin	0.1	0.1	0.1

the continuous method, ingredients are metered into a continuous kneader, which produces a constant supply to the refiners (22). The continuous process requires very accurate metering and rigid quality control procedures for all raw materials.

Refining.   The next stage in chocolate processing is a fine grinding in which a coarse paste from the mixer is passed between steel rollers and converted to a drier powdery mass. Refining breaks up crystalline sugar, fibrous cocoa matter, and milk solids.

Tremendous advances have been made in the design and efficiency of roll refiners. The methods currently used for casting the rolls have resulted in machines capable of very high output and consistent performance. The efficiency of the newer refiners has also been improved by hydraulic control of the pressure between the rolls and thermostatic control of cooling water to the rolls.

Modern 5-roll refiners with 2-m wide rollers can process 2200 kg of paste per hour. Output is dependent on particle size.

Particle size is extremely important to the overall quality of sweet and milk chocolate. Hence the refining process, which controls particle size, is critical. Fine chocolates usually have no particles larger than 25 or 30 μm. This is normally accomplished by passing the paste through refiners more than once. However, smooth chocolates can be produced with only a single pass through the refiners if the ingredients are ground prior to mixing. Particle size has a significant effect on both texture and flavor of the finished chocolate.

Conching.   After refining, chocolate is subjected to conching, a step critical to the flavor and texture development of high quality chocolates. Conching is a mixing–kneading process allowing moisture and volatile components to escape while smoothing the chocolate paste. It is one of the less satisfactorily explained parts of the chocolate making process and can embrace a wide range of phenomena, ranging from the relatively simple process of reliquifaction of a newly refined chocolate paste to complex and often controversial processes of flavor development, gloss development, agglomerate reduction, viscosity reduction, and modification of the melting quality.

The name conche derives from the seashell shape of the first really effective conching machine, which consisted of a tank with curved ends and a granite bed on which the chocolate paste from the refiners was slowly pushed back and forth by a granite roller. This longitudinal conche, the development of which is commonly attributed to Rodolph Lindt of Switzerland in 1879, is still used and many experts consider it best for developing subtle flavors.

Several other kinds of conches also are used today. The popular rotary conche can handle chocolate paste in a dry stage direct from the refiners (23). The recently developed continuous conche actually liquifies and conches in several stages and can produce up to 3600 kg of chocolate per hour in a floor area of only 34 m^2.

Conching temperatures range from 55–85°C for sweet chocolate and from 45–55°C for milk chocolate. Higher temperatures are used for milk chocolate if caramel or butterscotch flavors are desired (24).

Conching time varies from a few hours to many days and many chocolates receive no conching. Nonconched chocolate is usually reserved for inexpensive candies, cookies, and ice cream. In most operations, high quality chocolate receives extensive conching for as long as 120 hours.

Flavors, emulsifiers, or cocoa butter are often added during conching. The flavoring materials most commonly added in the United States are vanillin, a vanillalike artificial flavor, and natural vanilla (25) (see FLAVORS AND SPICES). Cocoa butter is added to adjust viscosity for subsequent processing.

Several chemical changes occur during conching including a rise in pH and a decline in moisture as volatile acids (acetic) and water are driven off. These chemical changes have a mellowing effect on the chocolate (26).

Standardizing. In standardizing or finishing, emulsifiers and cocoa butter are added to the chocolate to adjust viscosity to final specifications.

Lecithin (qv), a natural phospholipid possessing both hydrophilic and hydrophobic properties, is the most common emulsifier in the chocolate industry (5). The hydrophilic groups of the lecithin molecules attach themselves to the water, sugar, and cocoa solids present in chocolate. The hydrophobic groups attach themselves to the cocoa butter and other fats such as milk fat. This reduces both the surface tension, between cocoa butter and the other materials present, and the viscosity. Less cocoa butter is then needed to adjust the final viscosity of the chocolate.

The amount of lecithin required falls within a narrow range of about 0.2–0.6% (27). It can have a substantial effect on the amount of cocoa butter used, reducing the final fat content of chocolate by as much as 5%. Because cocoa butter is usually the most costly ingredient in the formulation of chocolate, the savings to a large manufacturer can be substantial.

Lecithin is usually introduced in the standardizing stage but can be added earlier in the process. Some lecithin is often added during mixing or conching. The addition at this point has the added advantage of reducing the energy necessary to pump the product to subsequent operations since the product viscosity is reduced.

Chocolate does not behave as a true liquid owing to the presence of cocoa particles and the viscosity control of chocolate is quite complicated. This non-Newtonian behavior has been described (28). When the square root of the rate of shear is plotted against the square root of shear stress for chocolate, a straight line is produced. With this Casson relationship method (29) two values are obtained, Casson viscosity and Casson yield value, which describe the flow of chocolate. The chocolate industry was slow in adopting the Casson relationship but this method now prevails over the simpler MacMichael viscometer. Instruments such as the Carri-Med Rheometer and the Brookfield and Haake Viscometers are now replacing the MacMichael.

At this stage of manufacture, chocolate may be stored for future use in bulk liquid form if usage is expected to be within one to two weeks, or at 43–50°C in a hot water jacketed agitated tank or in solid block form where it can be stored for as long as 6 to 12 months. Blocks typically weigh between 3 and 30 kg. Storage conditions for block chocolate should be cool and dry, ie, 7 to 18°C and 40 to 45% relative humidity. If chocolate has been stored in block form, it can be remelted to temperatures up to 50°C and then processed in the same manner as freshly made liquid chocolate.

At this stage the chocolate is ready for forming into its final shape after it is tempered. The two most common forms are molding or enrobing.

Tempering. The state, or physical structure, of the fat base in which sugar, cocoa, and milk solids are suspended is critical to the overall quality and stability of chocolate. Production of a stable fat base is complicated because the cocoa butter in solidified chocolate exists in several polymorphic forms. Tempering is the process of inducing satisfactory crystal nucleation of the liquid fat in chocolate.

Nucleation tempering of the still molten fat is necessary because the cocoa butter, if left to itself, can solidify in a number of different physical forms, ie, into an unstable form if cooled rapidly, or into an equally unacceptable super stable form if cooled too slowly, as commonly happens when a chocolate turns gray or white after being left in the sun. The coarse white fat crystals that can form in the slowly cooled center of a very thick piece of chocolate are similarly in a super stable form known in the industry as fat bloom.

Control of the polymorphic forms in cocoa butter is further complicated by the presence of other fats such as milk fat. The fat in a chocolate can be likened to the mortar between the bricks in a mason's wall. The solid particles in a well-conched chocolate bed down better than the solids in a coarsely refined and poorly mixed one (30).

A stable crystalline form for chocolate depends primarily on the method used to cool the fat present in the liquid chocolate. To avoid the grainy texture and poor color and appearance of improperly cooled chocolate, the chocolate must be tempered or cooled down so as to form cocoa butter seed crystals (31). This is usually accomplished by cooling the warm (44–50°C) liquid chocolate in a water jacketed tank, which has a slowly rotating scraper or mixer. As the chocolate cools, the fat begins to solidify and form seed crystals. Cooling is continued to around 26–29°C, during which time the chocolate becomes more viscous. If not further processed quickly, the chocolate will become too thick to process.

In another method of tempering, solid chocolate shavings are added as seed crystals to liquid chocolate at 32–33°C. This is a particularly good technique for a small confectionery manufacturer, who does not produce his own chocolate. However, the shavings are sometimes difficult to disperse and may cause lumps in the finished product (20). Most companies use continuous thin-film heat exchangers for the tempering process.

Molding. The liquid tempered chocolate is deposited into a metal or plastic (polycarbonate) mold in the shape of the final product. There are three basic types of molding: solid (or block), shell, and hollow.

Solid chocolate, eg, Hershey's Milk Chocolate Bar, is the most common molding. The chocolate, either milk or dark, is deposited into a mold and the mold passes through a cooling tunnel with a residence time in the tunnel of approximately 25 minutes. When the molds emerge from the tunnel, the chocolate is solidified. In addition to solidifying, the chocolate also contracts if it is correctly tempered, thus facilitating the removal of the bar from the mold. The demolded bar is then wrapped and packaged for shipment to the consumer.

Shell molding is a process by which a liquid or soft center is incorporated inside a chocolate shell. Modern equipment codeposits both the center filling and the chocolate shell in one step. The old method, and still the most common, is to form a hollow shell of chocolate in the mold followed by a short cooling tunnel. A filling is then deposited into the shell and sometimes followed by further cooling. A layer of chocolate then is deposited on top of the filling; this layer welds itself to the originally formed shell thus completely encapsulating the filling. The molds are cooled for a third and final time after which the pieces can be removed and wrapped.

Hollow molding as the name implies is a molded product with a hollow center such as Easter eggs, bunny rabbits, and Santa Claus. The molds used in hollow molding are divided in two halves and connected by a hinge. Chocolate is deposited into one half of the mold. The mold is then closed and rotated so that the liquid chocolate completely coats the inside surface of the mold. After cooling the molds are opened and the piece removed.

Enrobing. A preformed center such as nougat, fondant, fudge, cookies, etc, is placed on a conveyor belt and passed through a curtain of liquid tempered chocolate. The weight and thickness of the coating adhering to the center is controlled by an air curtain and vibration mechanism located immediately after the chocolate curtain. The now chocolate-coated centers pass into a cooling tunnel with a dwell time between 5 and 10 minutes. Upon emergence from the cooling tunnel the chocolate-coated pieces are ready for wrapping and packing.

Theobromine and Caffeine

Chocolate and cocoa products, like coffee, tea, and cola beverages, contain alkaloids (qv) (1). The predominant alkaloid in cocoa and chocolate products is theo-

Table 8. Variations in Theobromine and Caffeine Content of Various Chocolate Liquors

Country of origin	Theobromine, %	Caffeine, %	Total alkaloid, %	Theobromine-to-caffeine ratio
New Guinea	0.818	0.329	1.15	2.49:1
New Guinea	0.926	0.330	1.26	2.81:1
Malaysia	1.050	0.252	1.30	4.17:1
Malaysia	1.010	0.228	1.24	4.43:1
Bahia	1.210	0.183	1.39	6.61:1
Main Lagos	1.730	0.159	1.89	10.90:1
Light Lagos	1.230	0.137	1.37	8.98:1
Sanchez	1.570	0.177	1.75	8.87:1
Sanchez (small)	1.250	0.261	1.51	4.79:1
Fernando Po	1.470	0.064	1.53	23.00:1
Tabascan	1.410	0.113	1.52	12.50:1
Trinidad	1.240	0.233	1.47	5.32:1
average	1.240	0.206	1.45	7.91:1
maximum	1.730	0.330	1.89	23.00:1
minimum	0.818	0.064	1.15	2.49:1

bromine [83-67-0], though caffeine [58-08-2] is also present in smaller amounts. Concentrations of both alkaloids vary depending on the origin of the beans. Published values for the theobromine and caffeine content of chocolate vary widely because of natural differences in cocoa beans and differences in analytical methodology. This latter problem has been alleviated by the recent introduction of high pressure liquid chromatography (hplc) which has greatly improved the accuracy of analyses. Hplc values for theobromine and caffeine in a number of chocolate liquor samples have been published (32) (Table 8). Of the 12 varieties tested, the ratio of theobromine to caffeine varied widely from 2.5:1 for New Guinea liquor to 23.2:1 for that obtained from Fernando Po. Total alkaloid content, however, remained fairly constant, ranging from 1.15 to 1.89%.

The theobromine and caffeine contents of several finished chocolate products as determined by hplc at Hershey's laboratories are presented in Table 9.

Table 9. Theobromine and Caffeine Content of Finished Chocolate Products

Product	Theobromine, %	Caffeine, %
baking chocolate	1.386	0.164
chocolate flavored syrup	0.242	0.019
cocoa, 15% fat	2.598	0.247
dark sweet chocolate	0.474	0.076
milk chocolate	0.197	0.022

Nutritional Properties of Chocolate Products

Chocolate and cocoa products supply proteins, fats, carbohydrates, vitamins, and minerals. The Chocolate Manufacturers' Association of the United States (McLean, Virginia) completed a nutritional analysis from 1973 to 1976 of a wide variety of chocolate and cocoa products representative of those generally consumed in the United States. Complete nutritional data for the various products analyzed are given in Tables 10 to 14, for analyses conducted by Philip Keeney's laboratory at Pennsylvania State University, and in Table 15, for analyses done at South Dakota University. Where possible, data on more than one sample of a given variety or type of product are presented.

Table 10. Amino Acid Content of Cocoa and Chocolate Products, mg/g

	Whole beans[a]	Chocolate liquor[b] Natural	Dutch	Cocoa[c] Natural	Dutch	Sweet chocolate[d]	Milk chocolate 12% MS[e]	20% MS[f]
tryptophan	1.2	1.3				0.6		
threonine	3.5	3.9	3.6	7.7	8.0	1.5	1.8	2.8
isoleucine	3.3	3.8	4.0	7.0	7.4	1.4	2.2	3.5
leucine	5.3	6.0	6.3	11.5	11.3	2.3	3.8	6.1
lysine	4.8	5.1	5.1	8.7	8.3	1.9	2.4	3.9
methionine	0.7	1.1	0.9	2.0	1.7	0.4	0.9	1.4
cystine	1.4	1.1	1.0	2.1	2.1	0.4	0.3	0.4
phenylalanine	4.1	4.9	5.3	9.9	9.7	1.7	2.2	3.6
tyrosine	2.6	3.5	3.6	7.8	8.0	1.2	1.9	3.0
valine	5.1	5.8	6.3	11.1	10.9	2.1	2.7	4.3
arginine	5.0	5.3	5.1	11.3	11.3	1.9	1.2	1.9
histidine	1.6	1.7	1.7	3.4	3.0	0.6	0.7	1.0
alanine	3.8	4.3	4.1	8.7	8.4	1.5	1.4	2.3
aspartic acid	9.2	10.0	9.8	19.1	18.3	3.9	3.5	5.5
glutamic acid	12.8	14.1	14.1	28.0	26.2	5.7	8.5	13.7
glycine	4.0	4.4	4.5	8.3	8.5	1.6	1.0	1.6
proline	3.4	3.7	3.9	7.6	7.5	1.4	3.4	5.5
serine	3.7	4.1	4.0	6.8	8.2	1.5	1.9	3.0
Total AA recovered[g]	*75.5*	*84.1*	*83.3*	*162.0*	*158.8*	*31.6*	*39.8*	*63.4*

[a]Whole beans = 48% fat, 5% moisture, 10% shell.
[b]Chocolate liquor = 55% fat.
[c]Cocoa = 13% fat.
[d]Sweet chocolate = 35% chocolate liquor, 35% total fat.
[e]12% MS milk chocolate = 12% whole milk solids, 10% liquor, 32% total fat.
[f]20% MS milk chocolate = 20% whole milk solids, 13% liquor, 33% total fat.
[g]Total AA recovered = sum of individual amino acids (qv).

Table 11. Composition of Cocoa Beans and their Products, Whole Weight Basis in %

	Total solids	Total protein[a]	Cocoa protein[b,c]	Fat	Ash	Total carbohydrates[d]
whole cocoa beans						
Ghana	92.9	10.1	10.1	47.8	2.7	30.3
	94.0	9.8	9.8	51.6	2.6	28.0
	94.5	10.2	10.2	46.4	2.9	33.0
	94.7	10.3	10.3	46.3	3.1	33.0
Bahia	94.0	10.0	10.0	49.3	2.7	30.0
	94.1	10.2	10.2	48.6	2.7	30.6
	95.1	10.2	10.2	48.2	2.7	32.0
	94.9	10.2	10.2	48.4	2.7	31.6
chocolate liquor						
natural	98.4	9.4	9.4	56.2	2.4	28.5
	98.5	9.5	9.5	54.1	2.6	30.5
	98.9	10.1	10.1	57.0	2.4	27.4
	98.5	10.2	10.2	55.1	2.6	28.6
Dutch	98.6	9.2	9.2	55.4	3.8	28.5
	99.2	9.4	9.4	56.0	3.8	28.1
cocoa						
natural	96.3	18.4	18.4	12.8	4.6	56.9
	96.2	18.4	18.4	16.4	4.8	52.9
	97.4	19.8	19.8	12.7	4.5	56.5
Dutch	97.1	17.5	17.5	12.0	8.3	55.9
	97.4	18.3	18.3	14.3	7.4	53.7
sweet chocolate						
	99.6	3.4	3.4	35.1	1.0	59.4
	99.3	3.8	3.8	36.5	1.0	57.3
	99.5	3.6	3.6	35.0	1.0	59.2
milk chocolate						
12% whole milk solids	99.2	4.2	1.0(3.2)	34.7	0.9	59.2
	99.5	4.3	1.1(3.1)	30.2	1.0	63.8
	99.6	4.5	1.1(3.4)	32.3	0.9	61.6
	99.5	4.0	1.4(2.6)	29.6	1.0	64.6
20% whole milk solids	98.8	6.6	1.3(5.2)	34.4	1.5	56.1
	99.5	6.5	1.2(5.2)	33.1	1.4	58.3
	99.4	6.8	1.4(5.4)	30.5	1.5	60.4

[a]Total protein = milk protein + cocoa protein.
[b]Cocoa protein = (total nitrogen − milk nitrogen) × 4.7.
[c]Milk protein = milk nitrogen × 6.38 appears in parentheses.
[d]Total carbohydrate by difference using cocoa nitrogen × 5.63.

Table 12. Vitamin Content of Various Samples of Cocoa Beans and Chocolate Products,[a] Whole Weight Basis, mg/100 g

	B_1	B_2	Pantothenic acid	Niacin	B_6
whole cocoa beans					
Ghana	0.21	0.16	0.24	0.19	0.22
	0.17	0.18	0.35	1.07	0.21
	0.19	0.18	0.57	0.91	0.18
	0.16	0.15	0.32	0.52	0.01
Bahia	0.14	0.18	0.34	0.46	0.61
	0.17	0.18	0.35	1.13	0.16
	0.13	0.27	0.61	1.00	0.16
	0.16	0.16	0.38	0.81	0.09
chocolate liquor	0.08	0.17	0.20	0.88	0.09
	0.11	0.16	0.27	1.02	0.20
	0.08	0.15	0.17	1.01	0.16
	0.05	0.11	0.15	0.29	0.02
cocoa	0.05	0.19	0.33	1.34	0.17
	0.13	0.23	0.35	1.53	0.17
	0.15	0.22	0.32	1.37	0.24
milk chocolate	0.07	0.10	0.37	0.14	0.02
	0.11	0.24	0.37	0.38	0.02
	0.07	0.16	0.45	0.21	0.07
	0.10	0.25	0.61	0.24	0.08
	0.15	0.33	0.32	1.11	0.20

[a]Vitamin A and C, negligible amounts present.

Table 13. Tocopherols of Chocolate of Cocoa Beans and Chocolate Products, mg/100 g

	Total tocopherol	Alpha tocopherol
Bahia-Ghana beans	10.3	1.0
liquor, natural	10.9	1.1
liquor, Dutch	10.0	0.8
cocoa butter, natural	19.2	1.2
cocoa butter, Dutch	18.7	1.1
cocoa, natural	2.3	0.2
cocoa, Dutch	2.2	0.2
dark chocolate	6.0	0.7
milk chocolate, 12% milk	5.6	0.7
milk chocolate, 20% milk	6.3	0.7

Table 14. Fatty Acid Composition of Raw Cocoa Beans and Cocoa Butter[a]

	Fatty acid[b], mol %					
	14:0	16:0	18:0	18:1	18:2	20:0
cocoa beans						
Ghana	0.16	28.31	34.30	34.68	2.55	
	0.53	30.20	31.88	33.55	3.84	
	0.19	31.72	32.57	32.82	2.70	
	0.23	31.50	32.39	33.06	2.82	
Bahia	0.15	29.29	31.70	35.24	3.62	
	0.12	26.68	32.06	37.90	3.24	
	0.25	33.99	28.80	33.62	3.34	
	0.19	30.91	30.37	35.22	3.31	
natural cocoa butter	0.15	27.08	32.64	35.61	3.63	0.89
	0.19	27.68	32.64	35.03	3.63	0.83
	0.14	28.42	32.55	34.71	3.23	0.95
	0.14	27.29	32.41	35.36	3.70	1.10
Dutch cocoa butter	0.16	27.23	32.69	35.54	3.31	1.07
	0.15	26.63	34.24	34.68	3.52	0.78
	0.15	26.47	33.53	35.45	3.40	1.00

[a]Calculated from peak areas of the gas chromatograms.
[b]Fatty acid is designated by chain length followed by sites of unsaturation.

Table 15. Mineral Element Content[a] of Cocoa and Chocolate Products, mg/100 g

Product	Ca	Fe	Mg	P[b]	K	Na	Zn	Cu	Mn
raw Accra nibs	59.56	2.50	232.16	385.33	626.70	11.98	3.543	1.930	1.600
raw Bahia nibs	52.73	2.45	229.11	383.33	622.55	13.55	3.423	1.940	2.060
natural cocoa	115.93	11.34	488.51	7716.66	1448.56	20.12	6.306	3.620	3.770
Dutch cocoa	111.41	15.52	475.98	7276.00	2508.58	81.14	6.370	3.610	3.750
chocolate liquor	59.39	5.61	265.23	3996.66	679.61	18.89	3.530	2.050	1.850
12% milk chocolate	106.41	1.23	45.56	159.00	156.64	80.09	0.773	1.020	0.282
20% milk chocolate	174.00	1.40	52.26	207.96	346.33	115.40	1.240	0.126	0.139
dark chocolate	26.33	2.34	93.70	142.90	302.53	18.63	1.500	0.432	0.345

[a]Mean values from duplicate analyses of each of three samples by atomic absorption spectrophotometry.
[b]Total phosphorus-ash below 550°C (AOAC procedure).

Economic Aspects

Chocolate consumption on a global basis was approximately $20 billion in 1990. In the United States, Hershey, Mars, and Nestlé control about 70% of the market. For Europe, Nestlé (including Rowntree), Mars, Jacob Suchard (Philip Morris), Cadbury, and Ferrero control over 70% of the chocolate trade. In Japan, Lotte,

Meiji, Fujiya, Morinaga, and Ezaki Glico sell 88% of the chocolate. Per capita chocolate consumption for some leading countries include England, 19 pounds; Germany, 16 pounds; United States, 12 pounds; and Japan, 3 pounds.

The leading chocolate companies continue to pursue a global confectionery business strategy with an increase in the early 1990s of confectionery business activity in the Eastern Bloc countries, Russia, China, and South America. Generally as per capita income increases, chocolate consumption increases, and sugar consumption decreases. Consumer demographics, the declining child population, and the increase in consumer awareness of health issues play important roles in the economics of chocolate consumption. Chocolate confectionery business trends during the early 1990s include product down-sizing leading to snack size finger foods, increased emphasis on specialty chocolates with concentration on dessert chocolates, and chocolate brand equity spread into beverages, baked goods, frozen novelties, and even sugar confections.

BIBLIOGRAPHY

"Chocolate and Cocoa" in *ECT* 1st ed., Vol. 3, pp. 889–918, by W. Tresper Clarke, Rockwood & Co.; in *ECT* 2nd ed., Vol. 5, pp. 363–402, by B. D. Powell and T. L. Harris, Cadbury Brothers Ltd.; in *ECT* 3rd ed., Vol. 6, pp. 1–19, by B. L. Zoumas, E. J. Finnegan, Hershey Foods Corp.

1. W. T. Clarke, *The Literature of Cacao,* American Chemical Society, Washington, D.C., 1954.
2. *Report of the Cocoa Conference, Cocoa, Chocolate and Confectionery* Alliance, London, 1957.
3. C. E. Taneri, *Manuf. Confect.* **52**(6), 45 (1972).
4. G. A. Reineccius, P. G. Keeney, and W. Weissberger, *J. Agric. Food Chem.* **20**(2), 202 (1972).
5. L. R. Cook, *Chocolate Production and Use,* Magazines for Industry, Inc., New York, 1972.
6. H. R. Riedl, *Confect. Prod.* **40**(5), 193 (1974).
7. A. W. Knapp and A. Churchman, *J. Soc. Chem. Ind. (London)* **56**, 29 (1937).
8. A. Szegvaridi, *Manuf. Confect.* **50,** 34 (1970).
9. H. J. Schemkel, *Manuf. Confect.* **53**(8), 26 (1973).
10. *Report of Codex Committee on Cocoa Products and Chocolate, Codex Alimentarious Commission,* 10th Session, Geneva, Switzerland, 1974.
11. T. P. Hilditch and W. J. Stainsby, *J. Soc. Chem. Ind.* **55,** 95T (1936).
12. M. L. Meara, *J. Chem. Soc.,* 2154 (1949).
13. S. J. Vaeck, *Manuf. Confect.* **40**(6), 35 (1960).
14. Codex Standards for Cocoa Products and Chocolate, Cocoa Butter Standard 86-1981, *Codex Alimentarius,* Vol. VII, 1st ed., Joint FAO/WHO Food Standards Program, 1981.
15. J. Robert Ryberg, *Cereal Sci. Today* **15**(1), 16 (1970).
16. K. Wolf, *Manuf. Confect.* **57**(4), 53 (1977).
17. P. Kalustian, *Candy Snack Industry* **141**(3), 1976.
18. B. O. M. Tonnesmann, *Manuf. Confect.* **57**(5), 38 (1977).
19. Code of Federal Regulations, No. 21, Part 14, *Cacao Products,* Apr. 1, 1974.
20. B. W. Minifie, *Chocolate, Cocoa, and Confectionery: Science and Technology,* AVI, Westport, Conn., 1970.
21. B. Christiansen, *Manuf. Confect.* **56**(5), 69 (1976).

22. H. R. Riedl, *Confect. Prod.* **42**(41), 165 (1976).
23. E. M. Chatt, in Z. J. Kertesz, ed., *Economic Crops,* Vol. 3, Interscience Publishers, Inc., New York, 1953, p. 185.
24. L. R. Cook, *Manuf. Confect.* **56**(5), 75 (1975).
25. H. C. J. Wijnougst, *The Enormous Development in Cocoa and Chocolate Marketing Since 1955,* Mannheim, Germany, 1957, p. 161.
26. J. Kleinert, *Manuf. Confect.* **44**(4), 37 (1964).
27. R. Heiss, *Twenty Years of Confectionery and Chocolate Progress,* AVI, Westport, Conn., 1970, p. 89.
28. E. H. Steiner, *Inter. Choc. Rev.* **13,** 290 (1958).
29. N. Casson, *Brit. Soc. Rheo. Bull. ns* 52, (Sept. 1957).
30. M. G. Reade, UK, personal communication, 1990.
31. W. N. Duck, Ref. 3, p. 22.
32. W. R. Kreiser and R. A. Martin, *J. Assoc. Off. Analy. Chem.* **61**(6), (1978).

B. L. ZOUMAS
J. F. SMULLEN
Hershey Foods Corporation

CHOLESTEROL. See CARDIOVASCULAR AGENTS; FAT SUBSTITUTES.

CHOLINE

Choline base [*123-41-1*], [(CH$_3$)$_3$NCH$_2$CH$_2$OH]$^+$OH$^-$, trimethyl(2-hydroxyethyl)-ammonium hydroxide, derives its name from bile (Greek *cholē*) from which it was first obtained. This so-called free-choline is a colorless, hygroscopic liquid with an odor of trimethylamine. The quarternary ammonium compound (**1**) choline [*62-49-7*] or a precursor is needed in the diet as a constituent of certain phospholipids universally present in protoplasm.

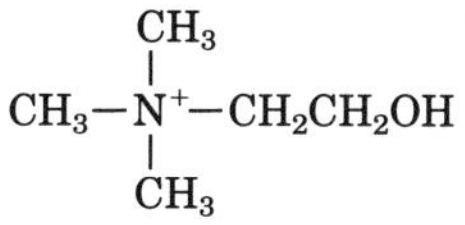

$$CH_3-\underset{\underset{CH_3}{|}}{\overset{\overset{CH_3}{|}}{N^+}}-CH_2CH_2OH$$

(1)

This makes choline an important nutritional substance. It is also of great physiological interest because one of its esters, acetylcholine [*51-84-3*], appears to be responsible for the mediation of parasympathetic nerve impulses and has been postulated to be essential to the transmission of all nerve impulses. Acetylcholine and other more stable compounds that simulate its action are pharmacologically important because of their powerful effect on the heart and on smooth muscle. Choline is used clinically in liver disorders and as a constituent in animal feeds.

Choline was isolated from ox bile in 1849 by Strecker. During 1900 to 1920, observations led to interest in the vasodepressor properties of the esters of choline, and in the 1920s it was shown that acetylcholine was presumably the "vagus-substance." The nutritional importance of choline was recognized in the 1930s, when it was found that choline would prevent fatty infiltration of the liver in rats. Subsequent observations showed that choline deficiency could produce cirrhosis (1) or hemorrhagic kidneys (2) in experimental animals under various conditions.

Physical and Chemical Properties

Choline is a strong base ($pK_B = 5.06$ for $0.0065–0.0403$ M solutions) (3). It crystallizes with difficulty and is usually known as a colorless deliquescent syrupy liquid, which absorbs carbon dioxide from the atmosphere. Choline is very soluble in water and in absolute alcohol but insoluble in ether (4). It is stable in dilute solutions but in concentrated solutions tends to decompose at $100°C$, giving ethylene glycol, poly(ethylene glycol), and trimethylamine (5).

Biological Functions

In nutrition, the most important function of choline appears to be the formation of lecithin (phosphatidylcholine) (**2**) and other choline-containing phospholipids.

$$
\begin{array}{l}
\quad\quad\quad\overset{\displaystyle O}{\overset{\|}{}} \\
H_2C-OC(CH_2)_mCH_3 \\
\;|\quad\quad\overset{\displaystyle O}{\overset{\|}{}} \\
HC-OC(CH_2)_nCH_3 \\
\;|\quad\quad\quad\overset{\displaystyle O}{\overset{\|}{}} \\
H_2C-O-P-OCH_2CH_2\overset{+}{N}(CH_3)_3 \\
\quad\quad\quad\;|\\
\quad\quad\quad OH
\end{array}
$$

phosphorylcholine portion; $m, n = 10–16$

(**2**)

Lecithin (qv) may be regarded as a triglyceride in which one of the fatty acid residues has been replaced by a phosphoric acid derivative of choline (phosphorylcholine) via cytidine diphosphate choline. The replacement changes the physical properties of the fat so that lecithin is readily dispersible with water. This property is important in the transport of fats in the blood. In choline deficiency, fats tend to accumulate in the liver presumably because they are not transformed into lecithin and hence are not carried away from the liver by the circulating blood. Owing to this effect, choline is said to have a lipotropic (fat-moving) action. Other known lipotropic substances, such as methionine, $CH_3SCH_2CH_2CH(NH_2)COOH$, and betaine, $(CH_3)_3N^+CH_2COO^-$, furnish labile methyl groups that unite with

2-aminoethanol to form choline in the body. Choline itself can also yield labile methyl groups for the methylation of other organic compounds (2,6–17).

Fatty infiltration of the liver has been observed to precede cirrhosis in experimental animals receiving diets low in choline and other substances that can furnish labile methyl groups, and can thus serve as precursors of choline.

Figure 1 shows some of the biological reactions involving labile methyl groups. The groups can originate from serine, formaldehyde, or formate by enzymatic reactions involving tetrahydrofolic acid, FAH_4, so that the compound N^5,N^{10}-methylenetetrahydrofolic acid, $5,10\text{-}CH_2FAH_4$, is formed. This undergoes hydrogenation to form 5-methyltetrahydrofolic acid, $5\text{-}CH_3FAH_4$, from which the methyl group is transferred to a vitamin B_{12} compound, shown in the diagram as CH_3B_{12}. This compound methylates homocysteine to produce methionine, which may become activated by adenosine triphosphate (ATP) with the formation of S-adenosylmethionine. The methyl group attached to sulfur in this compound can be transferred to various receptor molecules. One of these is 2-aminoethanol, $HOCH_2CH_2NH_2$, which is thereby converted to choline. Betaine is formed by dehydrogenation of choline, and can furnish a methyl group to homocysteine in an alternative pathway, catalyzed by methylpherase, for methionine biosynthesis.

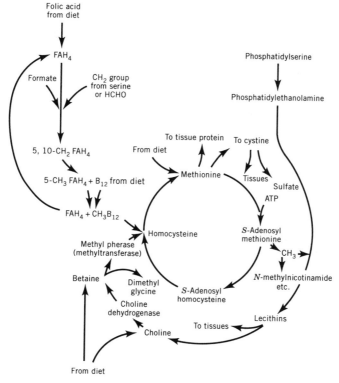

Fig. 1. The choline and methionine cycles showing the origin and disposition of labile methyl groups. FAH_4 = tetrahydrofolic acid; CH_3B_{12} = methylated vitamin B_{12}; and ATP = adenosine triphosphate.

Occurrence

Choline occurs widely in nature and, prepared synthetically, it is available as an article of commerce. Soybean lecithin and egg-yolk lecithin have been used as natural sources of choline for supplementing the diet. Other important natural-food sources include liver and certain legumes (18–22).

Preparation

Choline is not usually encountered as the free base but as a salt, most commonly, the chloride, $[(CH_3)_3N(CH_2CH_2OH)]^+Cl^-$. As a quaternary ammonium hydroxide, choline reacts with hydrochloric acid to form the chloride and water, whereas primary, secondary, and tertiary amines combine with hydrochloric acid to form hydrochlorides.

An earlier procedure for the production of choline and its salts from natural sources, such as the hydrolysis of lecithin (23), has no present-day application. Choline is made from the reaction of trimethylamine with ethylene oxide [75-21-8] or ethylene chlorohydrin [107-07-3].

$$N(CH_3)_3 + ClCH_2CH_2OH \longrightarrow (CH_3)_3N^+CH_2CH_2OH\ Cl^-$$

The chlorohydrin process (24) has been used for the preparation of acetyl-β-alkylcholine chloride (25). The preparation of salts may be carried out more economically by the neutralization of choline produced by the chlorohydrin synthesis. A modification produces choline carbonate as an intermediate that is converted to the desired salt (26). The most practical production procedure is that in which 300 parts of a 20% solution of trimethylamine is neutralized with 100 parts of concentrated hydrochloric acid, and the solution is treated for 3 h with 50 parts of ethylene oxide under pressure at 60°C (27).

Choline Salts

Choline Chloride. This compound [67-48-1] is a crystalline deliquescent salt, usually with a slight odor of trimethylamine (6). It is very soluble in water, freely soluble in alcohol, slightly soluble in acetone and chloroform, and practically insoluble in ether, benzene, and ligroin. Its aqueous solutions are neutral to litmus and are stable (4). The specific gravity of these solutions is a straight-line function between pure water and the value of 1.10 for the 80% solution, which represents the approximate limit of solubility. Choline chloride absorbs moisture from the atmosphere at relative humidities greater than 20% at 25.5°C.

Choline Dihydrogen Citrate. This compound [77-91-8] is a white, crystalline, granular substance possessing an acid taste, mp 105–107.5°C, and is freely soluble in water, very slightly soluble in alcohol, and practically insoluble in benzene, chloroform, and ether. The pH of a 25% solution is about 4.25.

Choline dihydrogen citrate $(CH_3)_3N(CH_2CH_2OH)C_6H_7O_7$, is prepared by methods similar to those for preparing choline chloride. It has the same pharmacological action as the chloride, but contains a lower proportion of choline. It is not as deliquescent as the chloride, and absorbs moisture from the atmosphere only at relative humidities greater than 56% at 25.5°C. It is more palatable than the chloride.

Tricholine Citrate Concentrate. This compound [546-63-4] is a clear, faintly yellow to light-green syrupy aqueous liquid containing 65.0 ± 2.0% tricholine citrate. It usually has a slight amine odor. It should have a pH of 9.0–10.0 and should contain not more than 0.2% trimethylamine, 0.5% ethylene glycol, 10 ppm of formaldehyde, and 0.1% residue on ignition. Its limit for heavy metals is 20 ppm and it should contain more than 0.2% chlorides or sulfates.

Choline Bitartrate. This substance [87-67-2] is a white crystalline material possessing an acid taste. It melts at 149–153°C. Analysis by cobaltous chloride shows more than 99% as the bitartrate. Free ethylene glycol is less than 0.25%, with free alkali at 0.0%.

Others. Other choline salts available as commercial products include choline bicarbonate [78-73-9], choline salicylate [2016-36-6], and the bronchodilator choline theophyllinate [4499-40-5].

Analysis

In biological materials, various nonspecific precipitants have been used in the gravimetric determination of choline, including potassium triiodide, platinum chloride, gold chloride, and phosphotungstic acid (28). Choline may also be determined spectrophotometrically and by microbiological, enzymatic, and physiological assay methods.

Choline reineckate is used in the spectrophotometric determination of choline. Ammonium reineckate [13573-16-5] forms a water-insoluble complex with choline. The complex is soluble in acetone and a widely used method for determination of choline is by light absorption of acetone solutions at 520 μm (8,29–31). The sensitivity of the assay is as little as 100 ppm choline in plant and animal tissue (32).

The use of mutant 34486 of *Neurospora crassa* for the microbiological assay of choline has been described (8). A physiological method has also been used in which the choline is extracted after hydrolysis from a sample of biological material and acetylated. The acetylcholine is then assayed by a kymographic procedure, in which its effect in causing contraction of a piece of isolated rabbit intestine is measured (33).

Enzymatic methods have been described (34) as well as gc, lc, radiochemical, and fluorometric procedures for choline analysis (32).

Uses

Choline has a low toxicity, eg, LD_{50} (rat, oral) = 3–6 g/kg, (35,36). It is used clinically and as a dietary supplement for poultry.

As a therapeutic agent, choline is administered orally in the form of syrups or elixers containing the chloride, citrates or bitartrate, or in the form of compressed tablets or capsules of the dihydrogen citrate. Choline is also given in small doses as a nutritional supplement in combination with a variety of other materials. In dry pharmaceutical-dosage forms, the dihydrogen citrate is usually preferred because of its lower tendency to absorb atmospheric moisture. Both salts have been used parenterally.

In the feeding of animals, choline is often added to chicken and turkey feeds as a dietary supplement (37–39). Its use in cattle feed has also been studied (40,41).

Choline in the form of choline base(hydroxide) is a strong organic base with a pH of approximately 14. This product can have industrial applications where it is important to replace inorganic bases with organic materials. Choline base is currently used in the formulation of photoresist stripping products for use in the printed wire board industry. Dilute aqueous solutions (5%) of choline base that have very low concentrations of metallic ions have been utilized for applications in the semiconductor industry.

Economic Aspects

The world market for choline chloride used in animal feeds is estimated at 113,000 t on a 100% basis. The market for good grade choline chloride is a small market by comparison and is utilized mainly in the supplementation of infant formulas. Other choline salts are utilized solely in the human vitamin supplementation markets and are also small compared to animal feed usage.

There are nine primary producers of choline chloride within the world. These are listed in Table 1. There are also small producers located in Taiwan and the People's Republic of China.

Market prices for choline products in 1992 were as follows: 70% choline chloride(FeedGrade), $0.77/kg; choline dihydrogen citrate, $7.20/kg; and choline hydrogen tartrate, $9.50/kg.

Derivatives

Important derivatives of choline are acetylcholine, acetyl-β-methylcholine, and carbamylcholine. Many other choline derivatives have been synthesized and studied, but have not been found satisfactory for clinical use.

Table 1. Choline Chloride Producers

Company	Country
DuCoa	United States
Bioproducts, Inc.	United States
Chinook	Canada
I.C.I.	United Kingdom
U.C.B.	Belgium
Akzo	The Netherlands
BASF	Germany
Mitsubishi	Japan

Acetylcholine [*51-84-3*] occurs as the bromide [*66-23-9*] (Pragmoline) and the chloride [*60-31-1*] (Acecoline). The chloride is a hygroscopic, crystalline powder. It is very soluble in cold water and alcohol but is practically insoluble in diethyl ether. It is decomposed by hot water and alkalies.

Acetylcholine bromide can be prepared by direct reaction of trimethylamine and β-bromoethyl acetate in benzene (42).

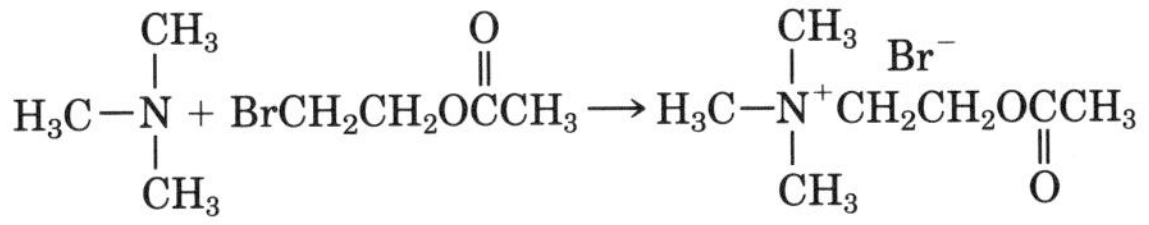

$$
\begin{array}{c}
\text{CH}_3 \\
| \\
\text{H}_3\text{C}-\text{N} + \text{BrCH}_2\text{CH}_2\text{OCCH}_3 \longrightarrow \text{H}_3\text{C}-\text{N}^+\text{CH}_2\text{CH}_2\text{OCCH}_3 \\
| \\
\text{CH}_3
\end{array}
$$

acetylcholine bromide

Acetylcholine is the product of the reaction between choline and acetyl coenzyme A in the presence of choline acetylase (41).

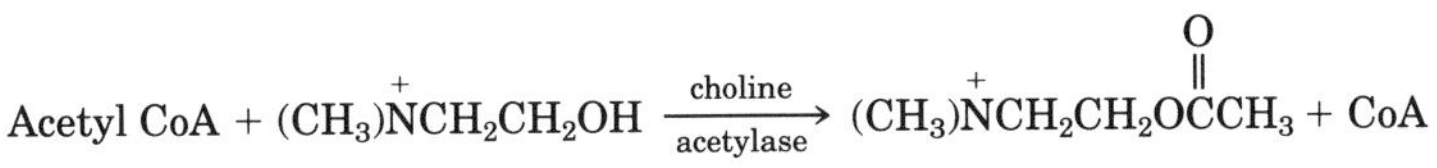

$$
\text{Acetyl CoA} + (\text{CH}_3)\overset{+}{\text{N}}\text{CH}_2\text{CH}_2\text{OH} \xrightarrow[\text{acetylase}]{\text{choline}} (\text{CH}_3)\overset{+}{\text{N}}\text{CH}_2\text{CH}_2\text{OCCH}_3 + \text{CoA}
$$

acetylcholine

Acetylcholine is a neurotransmitter at the neuromuscular junction in autonomic ganglia and at postganglionic parasympathetic nerve endings (see NEUROREGULATORS). In the CNS, the motor-neuron collaterals to the Renshaw cells are cholinergic (43). In the rat brain, acetylcholine occurs in high concentrations in the interpeduncular and caudate nuclei (44). The LD_{50} (subcutaneous) of the chloride in rats is 250 mg/kg.

Acetyl-β-methylcholine chloride [*62-51-1*], commonly called methacholine chloride, is a parasympathomimetic bronchoconstrictor with clinical efficacy in bronchial asthma (45,46).

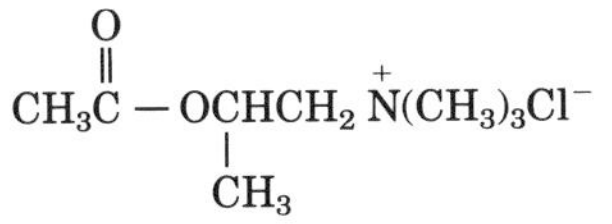

$$
\underset{\underset{\text{CH}_3}{|}}{\text{CH}_3\text{C}} - \text{OCHCH}_2\,\overset{+}{\text{N}}(\text{CH}_3)_3\text{Cl}^-
$$

Carbamylcholine chloride [*51-83-2*] is also called carbachol (47). Its principal use is as a parasympathomimetic in veterinary practice for large animals.

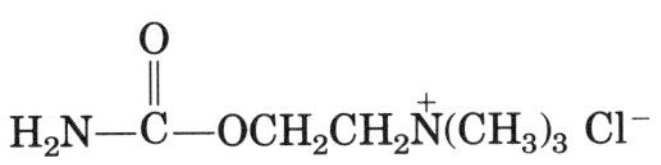

$$
\text{H}_2\text{N}-\overset{\text{O}}{\overset{||}{\text{C}}}-\text{OCH}_2\text{CH}_2\overset{+}{\text{N}}(\text{CH}_3)_3 \; \text{Cl}^-
$$

BIBLIOGRAPHY

"Choline" in *ECT* 1st ed., Vol. 3, pp. 919–927, by T. H. Jukes, American Cyanamid Co., Lederle Laboratories Div. and G. H. Schneller, American Cyanamid Co., Calco Chemical

Div.; in *ECT* 2nd ed., Vol. 5, pp. 403–413, by T. H. Jukes, University of California, Berkeley; *ECT* 3rd ed., Vol. 6, pp. 19–28, by T. H. Jukes, University of California, Berkeley.

1. C. L. Connor and I. L. Chaikoff, *Proc. Soc. Exptl. Biol. Med.* **39,** 356 (1938).
2. W. H. Griffith and D. J. Mulford, *J. Am. Chem. Soc.* **63,** 929 (1941).
3. C. W. Prince and W. C. M. Lewis, *Trans. Faraday Soc.* **29,** 775 (1933).
4. S. Budavari, ed., *The Merck Index,* 11th ed., Merck & Co., Inc., Rahway, N.J., 1989, p. 342.
5. E. Kahane and J. Lévy, *Biochimie de la Choline et de ses Dérivés,* Hermann, Paris, 1938.
6. *The United States Pharmacopeia,* 22nd rev. ed. (USP XXII), The United States Pharmacopeial Convention, Inc., Rockville, Md., 1990, p. 1736.
7. *Choline: Functions and Requirements,* DuCon, Highland, Ill., 1992.
8. W. H. Sebrell and R. S. Harris, eds., *The Vitamins,* 2nd ed., Vol. 3, Academic Press, Inc., New York, 1971, pp. 2–154.
9. H. Blumberg and E. V. McCollum, *Science* **93,** 598 (1941).
10. G. O. Broun and R. O. Muether, *J. Am. Med. Assoc.* **118,** 1403 (1942).
11. V. du Vigneaud, S. Simmonds, J. P. Chandler, and M. Cohn, *J. Biol. Chem.* **165,** 639 (1946).
12. S. S. Kewar, J. H. Mangum, K. G. Scringeour, J. D. Brodies, and F. M. Huennekens, *Arch. Biochem. Biophys.* **116,** 305 (1966).
13. C. Entenman, I. L. Chaikoff, and M. Montgomery. *J. Biol. Chem.* **155,** 573 (1944).
14. J. V. Lowry, L. L. Ashburn, and W. H. Sebrell, *Quart. J. Studies Alc.* **6,** 271 (1945).
15. A. E. Schaefer and J. L. Knowles, *Proc. Soc. Exper. Biol. Med.* **77,** 655 (1951).
16. A. W. Moyer and V. du Vigneaud, *J. Biol. Chem.* **143,** 373 (1942).
17. D. Stetten, Jr., *J. Biol. Chem.* **140,** 143 (1941).
18. A. Z. Hodson, *J. Nutr.* **29,** 137 (1945).
19. J. M. McIntire, B. S. Schweigert, and C. A. Elvehjem, *J. Nutr.* **28,** 219 (1944).
20. R. W. Engle, *J. Nutr.* **24,** 441 (1943).
21. S. M. McIntire, B. S. Schweigert, and C. A. Elvehjan, *J. Nutr.* **28,** 219 (1944).
22. J. J. Wurtman, *Nutrition and the Brain,* Vol. 5, 1979.
23. Ger. Pat. 193,449 (Dec. 22, 1906), J. D. Reidel (Aktiengesellschaft Berlin).
24. U.S. Pat. 2,774,759 (1956), Blackett and Soliday (to American Cyanamid Co.).
25. U.S. Pat. 2,198,629 (Apr. 30, 1940), R. T. Major and H. T. Bonnett (to Merck & Co., Inc.).
26. Fr. Pat. 736,107 (Apr. 29, 1932), F. Korner.
27. U.S. Pat. 2,137,314 (Nov. 22, 1938), H. Ulrich and E. Ploetz (to I. G. Farbenindustrie A.G.).
28. I. Sakakibaia and T. Yoshinaga, *J. Biochem. (Japan)* **23,** 211 (1936).
29. D. Glick, *J. Biol. Chem.* **156,** 643 (1945).
30. I. Hanin, *Handbook of Chemical Assay Methods,* Raven Press Publishers, New York, 1974.
31. S. E. Valdes Martinez, *Analyst (London)* **108**(1290), 1114–1119 (1983).
32. A. S. Atwal, N. A. M. Eskin, and M. Vaisey-Genser, *Cereal Chem.* **57,** 367–369 (1980).
33. J. P. Fletcher, C. H. Best, and O. M. Solandt, *Biochem. J.* **29,** 2278 (1935).
34. M. Takayama, S. Itoh, T. Nagasaki, and I. Tanimizu, *Clin. Chim. Acta* **79,** 93–98 (1977).
35. H. C. Hodge, *Proc. Soc. Exptl. Biol. Med.* **58,** 212 (1945).
36. M. W. Neumann and H. C. Hodge, *Proc. Soc. Exptl. Biol. Med.* **58,** 87 (1945).
37. V. K. Tsiagbe, C. W. Kang, and M. L. Sunde, *Poult. Sci.* **61**(10), 2060–2064 (1982).
38. M. L. Sunde, *Proc. Cornell Nutr. Conf. Feed Manuf.,* 67–73 (1982).
39. Jpn. Kokai 108 787 (1981), Yakult.
40. R. A. Erdman, R. D. Shaver, and J. H. Vandersall, *J. Dairy Sci.* **67**(2), 410–415 (1984).

41. T. S. Rumsey and R. R. Oltjen, *J. Anim. Sci.* **41,** 416 (1975).
42. E. Fourneau and J. Page, *Bull. Soc. Chim. (France)* **15,** 544 (1914).
43. J. R. Cooper, F. E. Bloom, and R. H. Roth, *The Biochemical Basis of Neuropharmacology,* 3rd ed., Oxford University Press, New York, 1978.
44. M. A. Lipton, A. DiMascio, and K. F. Killam, eds., *Psychopharmacology—A Generation of Progress,* Raven Press, New York, 1978.
45. Ref. 4, p. 14.
46. Ref. 4, p. 935.
47. Ref. 6, p. 838.

CURTIS E. GIDDING
DuCoa

CHROMATOGRAPHY

Chromatography is a technique for separating and quantifying the constituents of a mixture. Separation techniques are essential for the characterization of the mixtures that result from most chemical processes. Chromatographic analysis is used in many areas of science and engineering: in environmental studies, in the analysis of art objects, in industrial quality control (qv), in analysis of biological materials, and in forensics (see BIOPOLYMERS, ANALYTICAL TECHNIQUES; FINE ART EXAMINATION AND CONSERVATION; FORENSIC CHEMISTRY). Most chemical laboratories employ one or more chromatographs for routine analysis (1).

The first scientific reports demonstrating chromatographic phenomena appeared in the 1890s. However, the era of analytical chromatography began in 1903 when a paper was published describing the separation and identification of the components of a mixture of structurally similar yellow and green chloroplast pigments in leaf extracts. A solution of these extracts in carbon disulfide was passed through a column packed with chalk (2). The application of a pure solvent to the development of a chromatogram was significant, as was the explanation of adsorption (qv) as the mechanism by which separation occurred and the realization that this technique was potentially valuable as a means for identifying compounds other than by color. In 1906 the term chromatography was coined for these processes, coming from the combination of two Greek roots "chroma," meaning color, and "graphe," meaning writing (3).

Chromatographic separations rely on fundamental differences in the affinity of the components of a mixture for the phases of a chromatographic system. Thus chromatographic parameters contain information on the fundamental quantities describing these interactions and these parameters may be used to deduce stability constants, vapor pressures, and other thermodynamic data appropriate to the processes occurring in the chromatograph.

The importance of chromatographic processes to science can be gauged in many ways. For example, the 1952 Nobel Prize in chemistry was awarded for the development of liquid–liquid partitioning, or liquid–liquid chromatography (4). A key element in the development of this technique was the realization that both liquid phases need not move simultaneously to effect separation. In addition to liquid–liquid chromatography, this led to the development of gas–liquid chromatographic techniques. From a practical standpoint the importance of chromatography is evidenced by the number of chromatographic instruments that exist in laboratories (1).

Chromatography is not restricted to the analytical laboratory. Preparative chromatography is often the preferred tool for process separations. In addition, on-line chromatographic devices are used as detectors in many chemical processes such as reaction kinetics. Preparative chromatography is performed primarily to purify materials for subsequent use, eg, for additional analyses or for creating high purity materials required by some processes. Whereas all forms of chromatography are used for preparative purposes, liquid chromatography has been of highest value, especially in biological and pharmaceutical applications. Usually the column capacity is stretched to the limit to allow the greatest quantity of sample to be added to the mobile phase. The detector for preparative chromatography should be nondestructive and allow adequate discrimination between the product and its impurities.

Principles

The principle of chromatographic separation is quite straightforward. A mixture is allowed to come into contact with two phases, one referred to as the stationary phase and the other as the mobile phase. The stationary phase is contained in a column or sheet through which the mobile phase moves in a controlled manner relative to the stationary phase, carrying with it any material that may prefer to mix with it. For a comparison with other methods, see Table 1 which lists various separation techniques, the phenomena upon which they are based, and the corresponding methods of separation. In addition, there are separation techniques based on molecular geometry, which exploit differences within a single phase, for example, molecular sieving, field flow fractionation, gel-filtration chromatography, size exclusion chromatography, gas diffusion, inclusion complexes, ultra filtration, and dialysis (qv) (see MOLECULAR SIEVES; INCLUSION COMPOUNDS).

Because of differences in affinity of the mixture's constituents for the mobile phase, as compared to the stationary phase, these constituents tend to be swept along with the mobile phase at different rates. This selective interaction is known as partitioning. In adsorption chromatography the constituents in the dissolved sample compete with the mobile phase for the active sites on the stationary phase. To remove constituents adsorbed on the stationary phase, the mobile-phase chromatographic strength is increased by modifying the mobile phase to have a greater affinity for the stationary phase than the adsorbed sample constituents. To determine the effects of adsorption or partitioning on retention of substances on the column, a detector measures either the time required to travel a given distance or the distance traveled in a fixed time. The detector may be as simple

Table 1. Separation Techniques

		Methods	
Phenomenon	Heat[a]	Physicochemical[a] interaction	Nonuniformities of concentration[b]
adsorption, surface activity		gas−solid chromatography, liquid−solid chromatography, thin-layer chromatography, supercritical fluid chromatography	foam fractionation
electromigration			electrophoresis electrodecantation
exchange partition coefficient	zone refining zone melting	ion exchange gas−liquid chromatography, extraction, liquid−liquid chromatography, paper chromatography	
volatility	distillation		

[a]These methods create a second phase having a different concentration.
[b]These methods exploit differences within a single phase.

as the human nose or the human eye or as complex as a microsensor. A plot of detector response versus time of travel for a fixed distance is called a chromatogram. In preparative chromatography a device may be attached to the end of a column to collect the separated components of a mixture.

The nature of the stationary and mobile phases in a particular chromatographic experiment determines the efficacy of component separation in a particular mixture. A wide variety of combinations of stationary and mobile phases is used as is shown in Figure 1. The stationary phase may be a solid or a liquid supported on a solid. The mobile phase may be a gas, a liquid, or a material such as a supercritical fluid (see SUPERCRITICAL FLUIDS). A particular chromatographic technique is specified by naming the mobile phase, followed by the stationary. Thus, gas−liquid chromatography (glc) is a system that uses a gaseous mobile phase in contact with a film of liquid stationary phase.

In the analytical chromatographic process, mixtures are separated either as individual components or as classes of similar materials. The mixture to be separated is first placed in solution, then transferred to the mobile phase to move through the chromatographic system. In some cases, irreversible interaction with the column leaves material permanently attached to the stationary phase. This process has two effects: because the material is permanently attached to the stationary phase, it is never detected as leaving the column and the analysis of the mixture is incomplete; additionally, the adsorption of material on the stationary phase alters the ability of that phase to be used in future experiments. Thus it is extremely important to determine the ultimate fate of known materials when used in a chromatographic system and to develop a feeling for the kinds of materials in an unknown mixture before use of a chromatograph.

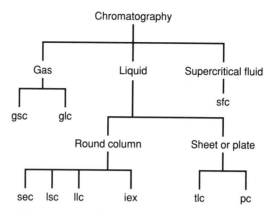

Fig. 1. Classification of chromatographic systems where gsc is gas–solid chromatography; glc, gas–liquid chromatography; sec, size-exclusion chromatography; lsc, liquid–solid chromatography; llc, liquid–liquid chromatography; iec, ion-exchange chromatography; tlc, thin-layer chromatography; pc, paper chromatography; and sfc, supercritical-fluid chromatography.

Development of the Chromatogram. The term development describes the process of performing a chromatographic separation. There are several ways in which separation may be made to occur, eg, frontal, displacement, and elution chromatography. Frontal chromatography uses a large quantity of sample and is usually unsuited to analytical procedures. In displacement and elution chromatography, much smaller amounts of material are used.

Passing impure material through a packed bed is frequently used for purifying large quantities of fluids such as gases and solvents. In such techniques the primary concern is removal of impurities through retention on the column packing or stationary phase. To carry out this procedure, a stationary phase that selectively retains the impurities, usually via adsorption, is sought. A significant factor is the amount of impurity the stationary phase can contain before it is saturated. The higher the capacity, the better the material for purification. The two principal applications of this technique are the purification of gas or solvent and selective sorption of trace materials from a fluid.

The three principal types of chromatography are frontal, displacement, and elution. Elution is by far the most common. Frontal chromatography is a technique in which the sample is introduced onto a column continuously. In essence the sample that is collected at the end of the column is the mobile phase, free of materials that adsorb/absorb on the stationary phase. Once the bed, ie, the stationary phase, is saturated and can no longer remove the impurities, the material coming off the column contains these materials. Using an appropriate detector, the condition at which this transition occurs can be determined, thereby determining the capacity of the column. This technique is called frontal chromatography or frontal analysis. Figure 2**a** shows an example of an integral chromatogram from a frontal analysis.

In displacement chromatography a small sample is displaced by the much more strongly held mobile phase, so that the sample is gradually pushed through the column as the mobile phase advances. As this happens, the components are

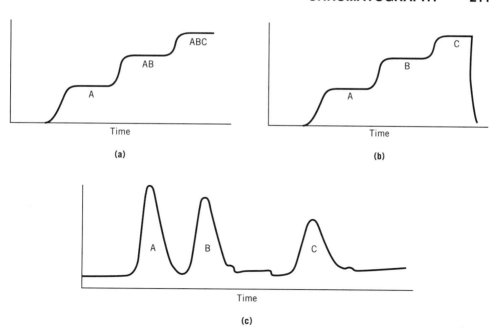

Fig. 2. Chromatograms of a mixture containing three components A, B, and C, where A is less sorbed than B, and B is less sorbed than C; (**a**) frontal analysis; (**b**) displacement analysis; and (**c**) differential elution chromatogram.

dispersed into bands that can then either be excised to obtain the pure material or displaced from the column. Such techniques are useful for the generation of quantities of pure material, particularly in application to problems in environmental analyses. Figure 2**b** shows an integral chromatogram obtained by displacement analysis.

Both frontal and displacement chromatographies suffer a significant disadvantage in that once a column has been used, part of the sample remains on the column. The column must be regenerated before reuse. In elution chromatography all of the sample material is usually removed from the column during the chromatographic process, allowing reuse of the column without regeneration. Most analytical applications of chromatography employ elution methods where a small sample is put onto the column, at the column head as a plug or a band. The sample is applied, sometimes by injection, while a stream of eluent, the mobile phase, is moving through the column. Because of the difference in affinities of the sample's components for the stationary phase, constituents travel through the column at different rates and elute at different times. Figure 2**c** shows a typical differential elution chromatogram.

Gas Chromatography

The most frequently used chromatographic technique is gas chromatography (gc) for which instrumentation was first offered commercially in 1955 by Burrell Corp.,

Perkin-Elmer, and Podbielniak. Five additional companies offered instrumentation in 1956. Gas chromatographs were the most frequently mentioned analytical instrumentation planned for purchase in surveys in 1990, and growth in sales is projected to remain around 6% through 1995 (1,5).

Gas chromatography, depending on the stationary phase, can be either gas-liquid chromatography (glc) or gas–solid chromatography (gsc). The former is the most commonly used. Separation in a gas–liquid chromatograph arises from differential partitioning of the sample's components between the stationary liquid phase adsorbed on a porous solid, and the gas phase. Separation in a gas–solid chromatograph is the result of preferential adsorption on the solid or exclusion of materials by size.

A second way of classifying gas chromatographic separations is by use. If the desired result is an analysis of the sample, the technique is analytical gas chromatography. If the desired result is the production of purer materials through fractionation in the chromatographic column, the technique is known as preparative gas chromatography. If the analytical chromatography is performed as part of the control of a manufacturing process, the technique is known as process gas chromatography.

Columns.   The chromatographic column is often described as the heart of the chromatographic system, because it is the single part of the system that must be present to effect a separation. Columns come in a wide variety of sizes and shapes. They are frequently tubes made of various materials coated on or bonded to stationary phase. In general, the larger the diameter of the column, the poorer the separation of the components. Large (>2 mm internal diameter) columns are often used in preparative chromatography, whereas medium (diameter between 1 and 2 mm) columns are used for analytical chromatography. Typically such columns are coated with a liquid phase, which is either chemically bonded or coated on porous particles that are presumed to be inert, ie, the solid supports do not affect the separation process. A schematic of a packed column is shown in Figure 3.

Capillary Chromatography.   Capillary, or wall-coated open tubular (WCOT), columns are fine tubes having internal diameters in the range of 0.20–0.75 mm. Use of these columns affects high resolution gas chromatography, ie, remarkably efficient separations of components. Columns having diameters in the range of 0.20–0.35 mm generally give the best separations. Capillary columns require the use of a sample splitter, a device to allow a representative aliquot of

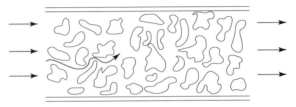

Fig. 3.   A packed chromatographic column showing the thin column walls and the irregularly shaped solid support coated with a liquid phase. The arrows indicate the movement of the mobile phase.

sample onto the column, because the sample capacity of a capillary column is not very great. Capillary columns are small and thus have a small (ca 100 ng/component) capacity. Sample size is therefore a critical factor in achieving the high resolutions of which such columns are capable. Columns of 0.32 mm internal diameter are popular because such columns retain the high resolution feature, but accept a larger (ca 500 ng/component) capacity. A 0.53-mm diameter column can usually be installed on a classical gas chromatograph without any modification of the system and accepts samples having up to ca 2000 ng/component.

Columns having internal diameters in the range of 0.53–0.75 mm are known as wide bore open tubular columns. The 0.75-mm diameter column is particularly useful for such coupled tandem techniques as gas chromatography–mass spectrometry (gc–ms), gas chromatography–Fourier transform infrared spectroscopy (gc–ftir), and gas chromatography–nuclear magnetic resonance spectroscopy (gc–nmr), in which a greater quantity of material is needed for detection (see ANALYTICAL METHODS, HYPHENATED INSTRUMENTS). These columns are also used with detectors of low sensitivity such as the thermal conductivity detectors. Capacities up to 15,000 ng can be achieved using 0.75-mm diameter columns.

The Liquid Phase. The stationary phase in an open tubular column is generally coated or chemically bonded to the wall of the capillary column in the same way the phase is attached to the support of a packed column. These are called nonbonded and bonded phases, respectively. In capillary columns there is no support material or column packing.

The greater the thickness of the liquid-phase (film) coating, the more coating per unit length and the greater the retarding potential of the column for components attracted to the stationary phase. Thick films (0.5–5.0 μm) are usually used to separate low boiling materials. For most other applications, columns having coatings of 0.2–0.4 μm are employed.

Chemically bonded phases are usually more resilient than nonbonded phases, tending not to wash out as large amounts of solvent pass through the column, and having much better thermal stability than do the nonbonded phases. Frequently a chemically bonded phase can be identified to effect a given separation at the same efficiency as a nonbonded one, thus the bonded phases are generally preferred

In some cases, increased stability of the stationary phase can be achieved by acting on it chemically. The phase can be cross-linked to give more mechanical, chemical, and thermal stability. Free-radical initiators such as peroxides, ozone, and azo-compounds are used to initiate cross-linking. Gamma-irradiation is also sometimes used. Cross-linked phases are very stable and can withstand solvent washing to clean the interior of the chromatographic column. Not all phases are capable of being cross-linked, however, and care should be taken to determine if the phase is bonded or cross-linked, or nonbonded, before attempting to clean the column.

Several hundred types of liquid phases are commercially available. These have been used individually or in combination with other liquid phases, inorganic salts, acids, or bases. The selection of stationary phases for a particular application is beyond the scope of this article, however, it is one of the most important chromatographic tasks. Stationary phase selection is discussed at length in books, journal articles, and catalogs from vendors. See general references for examples.

The Support Material.   The support is the inert frame onto which the liquid phase is applied. The most common support materials for packed-column gas chromatography are the diatomaceous earths. These are the remains of diatoms, single-cell algae. The porous siliceous material has pores approximately 1 μm in diameter (see DIATOMITE). These materials are typically treated with sodium carbonate to approximately 900°C. Following this treatment, they are sieved to obtain material of reasonably uniform dimensions. Supports are classified by the particle size or the screen mesh through which the particles pass. Most chromatographic packings are 80/100 mesh (177–149 μm) where the first number indicates the grid through which the material passes and the second number is the grid which stops it.

Once separated, the supports are washed using acid (HCl) and silanized, ie, treated with dimethyldichlorosilane [75-78-5], (DMCS), to reduce the polarity. Silanizing replaces adjacent SiOH groups with nonpolar CH_3 caps. In addition to diatomaceous earths, supports of carbon (qv), halocarbons, eg, Teflon, and glass beads are in use by various chromatographers.

$$
\begin{array}{ccc}
\underset{\text{surface}}{
\begin{array}{c}
\text{OH} \quad\ \text{OH} \\
| \qquad\ | \\
-\text{Si}-\text{O}-\text{Si}- \\
| \qquad\ |
\end{array}}
+ (CH_3)_2SiCl_2 \longrightarrow
& \underset{\text{surface}}{
\begin{array}{c}
CH_3\ \ CH_3 \\
\diagdown\ \ \diagup \\
\text{Si} \\
\diagup\ \ \diagdown \\
\text{O} \qquad \text{O} \\
| \qquad\ | \\
-\text{Si}-\text{O}-\text{Si}- \\
| \qquad\ |
\end{array}}
& +\ 2\,\text{HCl}
\end{array}
$$

The Stationary Phase in Gas–Solid Chromatography.   In gas–solid chromatography the packing itself is responsible for separation, because it interacts directly with the sample through its surface or its pore structure. Gas–solid chromatography has some advantages: (1) there is no liquid phase that can bleed, ie, evaporate or degrade, at higher temperatures that may interfere with detection of the sample, particularly at lower levels of sample; (2) the column usually has greater thermal stability than gas–liquid columns; and (3) there are certain unique properties to this system. The disadvantages of gas–solid chromatography mainly have to do with (1) the strong interactions between the sample and the surface of many packings, especially for polar samples; (2) the inhomogeneous surface of the solid; (3) the small sample capacity relative to liquid-phase systems; and (4) limitations on kinds of samples that may be analyzed.

Separation by Molecular Size.   The most common type of solid-phase packing in gas–solid chromatography is a molecular sieve. Molecular sieves (qv) are zeolites or carbon sieves that have a regular pore structure and are used almost exclusively for the separation of small molecules such as permanent gases, eg, oxygen, nitrogen, carbon monoxide, argon, and nitric oxide, or low carbon-number hydrocarbons (qv).

Porous organic polymers have been used for separating low molecular-weight mixtures containing halogenated or sulfur-containing compounds, water, alcohols, glycols (qv), free fatty acids (qv), ketones (qv), esters (see ESTERS, ORGANIC), and aldehydes (qv). Such porous polymers usually have a maximum operating temperature lower than many common liquid phases.

Silica, alumina, and other metal oxides and salts have been used as the stationary phase in gas–solid chromatographic systems. The applicability of these materials is limited by the difficulty of producing a consistent, resilient, reproducible material.

Column Tubing. The chromatographic column is contained in a tubing, the composition of which may have a dramatic effect on the separation process, because the sample components may also interact with the walls of the tube. Some of the materials used for columns are

Material	Uses/Comments
stainless steel	adequate for packed columns for hydrocarbons, permanent gases, and nonpolar materials
glass-lined stainless steel	combines innate inertness of glass with the durability of stainless steel, however, cannot easily determine if lining is broken
glass	inert, but inflexible; for packed columns, must be custom fitted to instrument
copper	good heat-transfer properties, but not as inert as stainless steel or glass
nickel	more inert than copper
Teflon	good for reactive systems such as sulfur gas analyses; columns do not pack well
aluminum	more active than stainless steel; used for high temperature applications
fused silica	most inert; commonly used for capillary columns; fused silica is usually protected by a silicon nitride and polyimide coating to prevent breakage

For sensitive compounds such as certain pharmaceuticals (qv), steroids, and pesticides, the standard practice is to use columns packed in glass tubes. The surface of glass is more nearly inert than are the surfaces of metal tubes. Glass columns also have the advantage of being transparent, giving a visible means of examination for column degradation or contamination. Packing efficiency is also easy to monitor in glass columns. Column voids may be visually detected. Glass, however, is more fragile than metal. Packing and chromatograph assembly of glass containing columns must be carried out with the utmost care to ensure that the column is not damaged in either of these operations.

Tubes made of metal such as stainless steel, nickel, copper, or aluminum are much more resistant than glass to damage in handling. Stainless steel is often preferred because it is less active than other metals. If corrosive samples are used, however, it is sometimes necessary to contain the column in a tubing made of a material such as Teflon. Columns made of Teflon generally are difficult to pack. Additionally, connections to the tubing that are free of leaks, which may degrade the performance of the chromatographic system, are difficult to make.

For capillary columns fused silica is the material of choice for the column container. It has virtually no impurities (<1 ppm metal oxides) and tends to be quite inert. In addition, fused silica is relatively easily processed and manufacture of columns from this material is reproducible. In trace analysis, inertness of tubing is an important consideration to prevent all of the tiny amounts of sample from becoming lost through interaction with the wall during an analysis.

Fused-silica columns are externally coated using a protective polyimide layer to improve strength and durability, and to provide a measure of protection against reaction of the silica with water in the environment. The fused-silica column is an inherently straight wire of material, ie, its resting state is straight, not coiled. To use the material in a chromatographic oven, it must be wound onto a frame that secures it in the coiled configuration, after which it is inserted into the oven. The process of creating a fused-silica column is complex and requires sophisticated, expensive equipment, a high temperature (2000°C) furnace, and a laser-based system for determining the trueness of diameter of the ultimate product. On the other hand, capillary columns made of glass are easily made on an inexpensive drawing machine in the chromatography laboratory.

In addition to fused-silica and glass capillary columns, there are several designs for glass capillary columns, although few are widely used. One system is the so-called porous layer open tubular (PLOT) column. These are made from glass that has been pulled to a capillary with an internal layer of solid packing, often similar to the material used in gas–solid chromatography. Subsequently this material is removed by etching to produce a column with many pores at the surface of the capillary. Support-coated open tubular (SCOT) columns are glass capillary columns having a coating on the column wall. Because this phase, frequently a liquid one, is different from the wall material, additional mechanisms for separation are possible. Micropacked columns are glass capillaries packed using small-mesh particles similar to those in liquid-chromatographic columns.

The Mobile Phase. The purpose of the mobile phase, also called the carrier gas, is to transport the sample through the chromatographic column. The selection of carrier gas is often dictated by the type of detector attached to the chromatographic system. To achieve the best performance, gases of relatively higher molecular weight such as nitrogen, carbon dioxide, or argon are used as a carrier. If these gases do not permit sufficiently high gas velocities, then a lower molecular-weight gas such as helium or hydrogen should be used. The purity of the carrier gas is an important consideration because gas passes through the column and impurities could interfere with chromatographic separation. Gases used are generally of 99.995% purity or better. Two particularly troubling contaminants are water and air, which can affect the stability of the liquid phase in a packed column. The best compromise for column performance and safety is helium for capillary columns and either helium or nitrogen for packed columns.

Detectors. The function of the gc detector is to sense the presence of a constituent of the sample at the outlet of the column. Selectivity is the property that allows the detector to discriminate between constituents. Thus a detector selective to a particular compound type responds especially well to compounds of that type, but not to other chemical species. The response is the signal strength generated by a given quantity of material. Sensitivity is a measure of the ability of the detector to register the presence of the component of interest. It is usually given as the quantity of material that can be detected having a response at twice the noise level of the detector.

By far the most used detector is the thermal conductivity detector (TCD). Detectors like the TCD are called bulk-property detectors, in that the response is to a property of the overall material flowing through the detector, in this case the thermal conductivity of the stream, which includes the carrier gas (mobile phase) and any material that may be traveling with it. The principle behind a TCD is that a hot body loses heat at a rate that depends on the composition of the material. Most materials have lower thermal conductivities than helium, the typical carrier gas, and most organic materials have similar thermal conductivities, thus the TCD is often used for quantitative analysis of separation of organics. Of course, the thermal conductivities of organic compounds are not exactly the same, so very accurate results require an evaluation of the response factor for each material, essentially an evaluation of the thermal conductivity of each material.

Another type of detector, one most extensively used in capillary chromatography, is the flame-ionization detector (FID). The principle behind its operation is the detection of a current from ions formed when organic materials are burned in a small hydrogen–oxygen flame at the end of the column. Typically a voltage is applied across this region and the small current carried by the ions is detected using a sensitive electrometer. One of the primary advantages of the FID is that it is, in general, more sensitive than the TCD. In addition, it does not respond to materials such as water, carbon dioxide, carbon monoxide, and most simple sulfur-containing gases. This is an advantage in the analysis of certain samples, for example traces of organics in water. The principal disadvantage of the FID is that it destroys the separated material in the process of detection.

A third detector type is the electron-capture detector (ECD) which is very selective for the detection of highly electronegative compounds in the effluent such as chlorinated hydrocarbons, many pesticides, and polychlorinated biphenyls (see CHLOROHYDROCARBONS). Its principle of action is the interaction of such compounds with electrons emitted from a radioactive source. A detector sensitive to these beta emissions that is positioned across the stream from a source senses a drop in emissions when the stream contains compounds that capture the emitted electrons. Thus the response to the passage of materials is a loss of signal. The sources generally used in these detectors are nickel-63 or tritium. One disadvantage is the radioactive source, which requires special handling. Also, the linear response range is not very great and the detector is subject to high background noise, unless care is taken to eliminate column, carrier-gas, or sample contaminants.

The flame-photometric detector (FPD) is selective for organic compounds containing phosphorus and sulfur, detecting chemiluminescent species formed in a flame from these materials. The chemiluminescence is detected through a filter by a photomultiplier. The photometric response is linear in concentration for phosphorus, but it is second order in concentration for sulfur. The minimum detectable level for phosphorus is about 10^{-12} g/s; for sulfur it is about 5×10^{-11} g/s.

The alkali flame-ionization detector (AFID), sometimes called a thermionic (TID) or nitrogen–phosphorus detector (NPD), has as its basis the fact that a phosphorus- or nitrogen-containing organic material, when placed in contact with an alkali salt above a flame, forms ions in excess of thermal ionic formation, which can then be detected as a current. Such a detector at the end of a column then reports on the elution of these compounds. The mechanism of the process is not clearly understood, but the enhanced current makes this type of detector popular for trace analysis of materials such as phosphorus-containing pesticides.

The mass spectrometer (ms) is a common adjunct to a chromatographic system (see MASS SPECTROMETRY). The combination of a gas chromatograph for component separation and a mass spectrometer (gc/ms) for detection and identification of the separated components is a powerful tool, particularly when the data are collected using an on-line data-handling system. Qualitative information inherent in the separation can be coupled with the identification of structure and relatively straightforward quantification of a mixture's components.

Infrared (ir) spectrometers are gaining popularity as detectors for gas chromatographic systems, particularly because the Fourier transform infrared (ftir) spectrometer allows spectra of the eluting stream to be gathered quickly. Gc/ir data are valuable alone and as an adjunct to gc/ms experiments. Gc/ir is a definitive tool for identification of isomers (see INFRARED AND RAMAN SPECTROSCOPY).

Plasma atomic emission spectrometry is also employed as a detection method for gc (see PLASMA TECHNOLOGY). By monitoring selected emission lines a kind of selective detection based on elemental composition can be achieved (see SPECTROSCOPY).

Theory.   Most theoretical models of gas chromatographic processes are based on analogy to processes such as distillation (qv) or countercurrent extraction experiments (6). The separation process is viewed as a type of successive partitioning of the components of a mixture between the stationary and mobile phases similar to the partitioning that occurs in distillation columns. In those experiments an important parameter is the number of theoretical plates of which the column may be considered to be composed; the greater the number of theoretical plates, the greater the efficiency of the column for achieving separations of similar components. In gas chromatography, the equivalent measure of efficacy is the height equivalent to theoretical plates (HETP), which measures the ultimate ability of the column to separate like components. This quantity depends on many instrumental parameters such as wall or particle diameter, type of carrier gas, flow rate, liquid-phase thickness, etc. The theoretical expression relating these various parameters is called the van Deemter equation, which relates the change in efficiency to the flow.

$$ \text{HETP} = A + B/\mu + C\mu $$

where A is the eddy diffusion or multipath effect term, B is the molecular diffusion term, μ is the linear gas velocity or flow rate, and C is the resistance to mass transfer term.

The expanded version of the van Deemter equation is used to help understand the relationships between the packing parameters and the gas flow.

$$ \text{HETP} = 2\lambda d_p + \frac{2\gamma D_{(g)}}{\mu} + \frac{8k' d_f^2}{\pi^2 (l + k')^2 D_{(\text{liq})}} \mu $$

where λ is a packing constant, d_p is the average particle diameter of the solid support, γ is the tortuosity factor used to account for the tortuosity of the gas channels in the column, $D_{(g)}$ is the diffusivity of the solute in the gas phase, k' is the capacity factor, d_f is the effective thickness of the liquid film coated on the support, and $D_{(liq)}$ is the diffusivity of the solute in the liquid phase. A plot of HETP versus μ gives a hyperbola with a minimum HETP. This minimum is the optimal flow rate where the column operates most efficiently.

Inlet Systems. The inlet or injector is the means by which the sample is introduced onto the chromatographic column. The process of sample introduction requires one to create a representative aliquot of the sample at the beginning of the column without degradation or without discrimination among the components of the sample. Most inlets operate on the principle that a sample can be vaporized quickly, assuming it is not already a gaseous material, after being squirted out of a microliter syringe into a small, heated volume, usually at about 50°C hotter than the maximum temperature of the column during the experiment. This vaporized material is quickly swept as a narrow band onto the column by a flow of carrier gas. Once on the column, the components interact with the stationary phase and begin to travel along with the carrier gas at differing rates, depending on the strengths of interaction with the stationary phase. Sample introduction is critical to the proper operation of a gas chromatograph. In order that reliable data be obtained, a representative sample must reach the column in a narrow band so that all components begin the process at the same time. Selective interaction of the components with the stationary phase then allows different components to reach the detection region at different times. Whereas direct injection from a syringe is the most widely used technique for sample introduction in gc, other means such as injection valves, pyrolyzers, headspace samplers, thermal desorbers, and purge-and-trap samplers are found in various applications.

Inlets for syringe sampling are divided into two main categories: one for packed-column and the other for capillary-column devices. For packed columns, all material injected is carried by the mobile phase onto the column. The inlet is usually an open tube, but sometimes, albeit rarely, the inlet itself may be packed, eg, to assure that the first centimeters of the column do not become contaminated with degradation products or nonvolatile materials that may affect the efficacy of the column.

Capillary columns require increased care in injection because of the much smaller capacity of the column. There are four different inlet designs: direct, split/splitless, programmed-temperature vaporization, and cool on-column injectors. Direct inlets are generally used with capillary columns of larger diameter and work much as direct injectors for packed columns. Split/splitless injectors operate in two modes. In the split mode, most of the sample introduced into the inlet goes out a vent that has less resistance to flow than the column. Thus, in the split mode, a smaller amount of sample actually enters the column than was introduced into the inlet. Splitters usually remove 90–99% of the volume of the material injected. A primary problem with this injection mode is that the material finding its way onto the column is not always representative of the sample. Discrimination resulting from differential boiling points of the components is such that certain components of the mixture are more likely to be vented than introduced onto the column. Such a problem is not related to the reproducibility of the injection itself.

In the splitless mode, the vent is turned off and everything injected goes onto the column. After a short period, the vent is opened and any residual solvent is vented. The splitless mode is found particularly in trace analytical schemes (see TRACE AND RESIDUE ANALYSIS). Splitless sample injection is an art, and it requires practice to ensure reproducible introduction of sample onto the column. This type of injection is usually used for qualitative analysis.

Programmed-temperature vaporizers are flexible sample-introduction devices offering a variety of modes of operation such as split/splitless, cool-sample introduction, and solvent elimination. Usually the sample is introduced onto a cool injection port liner so that no sample discrimination occurs as in hot injections. After injection, the temperature is increased to vaporize the sample.

Cool on-column injection is used for trace analysis. All of the sample is introduced without vaporization by inserting the needle of the syringe at a place where the column has been previously stripped of liquid phase. The injection temperature must be at or below the boiling point of the solvent carrying the sample. Injection must be rapid and no more than a very few, usually no more than two, microliters may be injected. Cool on-column injection is the most accurate and reproducible injection technique for capillary chromatography, but it is the most difficult to automate.

Temperature Considerations. The inlet, detector, and the oven compartment where the column is kept, are usually controlled at different temperatures, because each part serves a different function that is best performed in a specified temperature range. In practice, the maximum oven temperature expected to be reached in the course of an analysis that is high enough to achieve the desired result in minimum time is chosen. This temperature should also be low enough to minimize the probability of column liquid-phase degradation. Generally, retention time is halved for every 30°C the temperature is increased. The injection port's temperature is usually slightly higher than the maximum oven temperature, but low enough to minimize thermal degradation or thermal rearrangement of sample components. Ideally, the thermal energy in the injection port will cause instantaneous vaporization without causing a loss of separation efficiency by spreading the sample over a large volume. The detector temperature is usually 10–30°C higher than the injector, but low enough to avoid thermal degradation of the column's liquid phase in that part of the column near the detector.

For materials with a wide boiling range, temperature programming is often used. The initial temperature, if possible, should be near the boiling point of the most volatile component; the final temperature should be near or, if possible, slightly higher than the least volatile component. The heating rate from low temperature to high temperature is usually empirically determined to obtain the most efficient separation in the shortest possible analysis time. In the consideration of heating rate, cool-down time must be counted as part of the analysis time when doing multiple-sample analyses.

Liquid Chromatography

Liquid chromatography (lc) refers to any chromatographic process in which the mobile phase is a liquid. Traditional column chromatography, thin-layer chromatography (tlc), paper chromatography (pc), and high performance liquid chromatography (hplc) are all members of this class of processes. Modern liquid chromatographic techniques originated in the late 1960s and early 1970s. Developments in hplc were driven by improvements in instrumentation, column packings, and theoretical understanding of the various separation processes involved. For example, use of pressurized mobile phases in place of gravity-driven ones in chromatography greatly shortened the time necessary for a separation. Other terms used for hplc are high speed or high pressure chromatography.

Liquid chromatography is complementary to gas chromatography because samples that cannot be easily handled in the gas phase, such as nonvolatile compounds or thermally unstable ones, eg, many natural products, pharmaceuticals, and biomacromolecules, are separable by partitioning between a liquid mobile phase and a stationary phase, often at ambient temperature. Developments in the technology of lc have led to many separations, done by gc in the past, to be carried out by liquid chromatography.

An advantage of liquid chromatography is that the composition of the mobile phase, and perhaps of the stationary phase, can be varied during the experiment to provide greater efficacy of the separation. There are many more combinations of mobile and stationary phases to effect a separation in lc than one would have in a similar gas chromatographic experiment, where the gaseous mobile phase often serves as little more than a convenient carrier for the components of the sample.

In classical column chromatography the usual system consisted of a polar adsorbent, or stationary phase, and a nonpolar solvent, mobile phase, such as a hydrocarbon. In practice, the situation is often reversed, in which case the technique is known as reversed-phase lc.

Paper chromatography originated in the 1940s and tlc in the 1950s. In these techniques a chamber is usually used to isolate the column, which is a piece of filter paper in pc and a glass plate coated with an adsorbent such as silica gel in tlc, from the laboratory environment. The chromatogram is developed by allowing a mobile phase to creep through the column, carrying with it materials soluble to various extents. After this process has proceeded sufficiently, the column is removed from the solvent tank and the mobile phase evaporated. The separated components are visualized elsewhere, for example under an ultraviolet lamp in which various fluorescent bands indicate how fluorescent materials are separated by the movement of the solvent. Paper chromatography is sometimes described as a type of liquid–liquid chromatography, because the paper inherently contains bound water that acts as a stationary liquid phase. In tlc the usual mechanism for separation is partitioning resulting from adsorption on the stationary phase.

Paper and thin-layer chromatography may be further classified as either one- or two-dimensional, by direction, eg, ascending, descending, ascending/descending, and by capacity as either analytical or preparative. In one-dimensional pc or tlc, a small spot of sample is applied, usually from a micropipet, at a point near the edge from which solvent is to enter the paper or plate. In traditional ascending chromatography, the plate or paper is dipped in the mobile phase in the bottom of the chamber and the mobile phase is allowed to rise through the column by capillary action. When the mobile phase nears the top, the plate or paper is removed from the chamber and allowed to dry. Sometimes at this point a visualizing agent is sprayed or dipped to allow detection of the components. The distance from the origin to the migrated spot is used to calculate a retardation factor (r_f), the ratio of the distance a component travels to the distance the solvent front has traveled. Retardation factors are always less than or equal to one and are used to characterize the partitioning of a component for a particular solvent and stationary phase. In two-dimensional chromatography, after the plate or paper is removed from one solvent and allowed to dry, it may be placed in another tank with a different solvent entering the paper or plate at 90° to the direction of travel of the first solvent. The result is a further resolution of components that may have had similar partitioning in the first solvent. One-dimensional tlc plates have a very high capacity, eg, up to 20 samples and standards applied, versus one-at-a-time single spots in two-dimensional tlc, or in column chromatography (hplc or gc). A critical advantage of tlc is that a fresh plate is used for each analysis. Another important advantage is that many visualizing techniques are available that are not available using lc detectors. In addition, the spots or, more frequently, the band in prep tlc, can be physically cut from the plate.

Columns. As for gc, the column is the heart of the chromatographic system. Columns for modern lc can be packed with a variety of materials: inert particles bonded to a liquid phase (liquid–liquid chromatography); a porous gel as for size-exclusion chromatography (sec) or gel-permeation chromatography (gpc); an ion-exchange resin as for ion-exchange chromatography (iec); or an affinity adsorbent as for affinity chromatography (ac). Most column tubing is about 4.5 mm in internal diameter having a 0.16 cm wall; the columns are generally 10 to 15 cm in length. Tubing materials for lc are

Material	Uses/Comments
316 stainless steel	general-utility material; good for high pressure systems
poly(ether ether ketone) (PEEK)	inert to almost all organic solvents except methylene chloride, thetrahydrofuran, dimethylsulfoxide, and concentrated nitric and sulfuric acids; holds to 34 MPa (5000 psi); good for metal-free biological systems
tefzel	common for metal-free applications; inert
titanium	withstands pressures to 34 MPa (5000 psi); corrosion-resistant; expensive
fused silica	used for capillary lc
glass	limited pressure range
glass-lined stainless steel	difficult to know when glass is broken

The interior surface of the stainless steel tubing is usually polished to allow the particles to pack efficiently. An important consideration is uniformity of the particles, which are generally about 5 μm in diameter for typical liquid chromatographic experiments.

Packings. Most packings for lc are made of chemically modified silica gel having functional groups covalently attached to the surface of the particles.

Reversed-phase packings have covalently bonded octadecyl groups (a C-18 phase) or octyl groups (a C-8 phase) at the surface to provide a nonpolar environment. Sometimes the further reaction of these materials with other reagents to attach trimethylsilyl groups (endcapping) is attempted. This treatment is generally supposed to cover the regions of the surface that are not covered by the first treatment, eliminating interactions that may degrade the efficiency of the column. In addition to the C-18 and C-8 column packings, other species used for chemically binding to the support particles include phenyl, nitro, and amino groups. For size-exclusion chromatography, porous polymers such as polystyrene–divinylbenzene are sometimes used, as are the usual treated and untreated silicas. Silica columns are limited to pH below about 8.

The Mobile Phase. The great power of lc to separate the components of a mixture lies in the differential solubility of the components in the mobile liquid phase and the stationary phase. In isocratic lc, the composition of the mobile phase remains constant throughout the course of the experiment. However, the effective separating power of lc can often be enhanced by changing the composition of the mobile phase during the course of the experiment. This process, known as gradient lc, is analogous to programming the column temperature in gc. In gradient lc, the composition is deliberately altered over the course of the experiment to achieve more rapid separation. Frequently the switch is from a weak solvent for a given material to a strong one. The change can be made in a single step or by slowly varying the composition of the mobile phase with time during the separation process. Most processes involve two solvents or solvent mixtures, although there are some cases in which three solvents are used. Obviously, the more solvents used, the more complex the program of mixing. An important consideration in gradient lc is the selection of a detector. The detector must be compatible with all the solvents used in the separation process.

Solvent-delivery systems ensure uniform transfer of the mobile phase to the column. These devices must give reproducibly uniform flow without pulsations in order to ensure reproducible retention times and peak areas for analyses. Because of the small diameter of the particles used in modern hplc, there is a high resistance to flow through the column, and high pressure pumps are required (see PUMPS). To obtain the best signal from a detector, it is important that the detector be insensitive to pump strokes at all flows. The pump materials must not only be able to withstand such pressures, but must not be affected by the solvents in the system. For gradient elution lc, the pump should have a small mixing (hold-up) volume. This minimizes memory effects from solvent changes. Pump parameters for liquid chromatography are

Property	Comment
noise	pulsation from piston movement
drift	flow change as a function of time
accuracy	ability to deliver a set flow
precision (short-term)	constancy of volume output over a short time period, eg, 15 min
resettability	ability to match exactly the flow parameters for run to run

There are two approaches to carrying out a gradient elution: high pressure mixing and low pressure mixing. In the high pressure mode two or more pumps are programmed to pump mobile phase components into a mixing chamber before the mixed mobile phase enters the head of the liquid chromatographic column. The output of each is independently controlled, allowing almost any kind of gradient to be formed at the outlet of the mixing chamber. In the low pressure approach, the gradient is formed by mixing two or more solvents at atmospheric pressure, then pumping the mixture onto the column using a single high pressure pump. A primary advantage of the low pressure approach is lower cost because only a single pump is used instead of the two or more required in the high pressure mode.

Reciprocating pumps are those most commonly used in high performance lc. The single-piston type usually has inlet and outlet check valves with some mechanism such as variable stroke frequency to minimize the effect of pump pulsations. Dual-piston pumps operate with the pistons 180° out of phase to minimize pulsations. For this system to work optimally, the piston units must be identical.

Sample introduction onto liquid chromatographic columns is usually accomplished with a sampling valve. A sample loop of volume between 5 and 50 µL attached to the sampling valve is filled or partly filled with sample solution. At the time of introduction of the sample, the sample valve is either manually or pneumatically actuated so flow of solvent through the sample loop moves sample onto the column. Because the sample loops themselves are sometimes made of small-diameter tubing, the sample is often prefiltered to remove particles that may clog these loops. Prefilling also assures that the inlet frits and columns are not clogged by the sample.

Detectors.   Lc detectors must be compatible with the solvent system (mobile phase) and are optimized for sensitivity, stability, and speed of response. They are designed to retain the quality of the separation. No versatile, universal detector is in use for lc, as, eg, the flame-ionization or thermal-conductivity detectors are for gc. Instead, the most common detector found in lc is the ultraviolet (uv) detector, a selective detector that measures the absorption of radiation at a specified wavelength. These devices may be set at a fixed wavelength or the wavelength may be variable and only sensitive to materials that absorb radiation in the range of the detector. Selective detectors are also known as being solute-property detectors. Each measures some property of only the sample component. Uv detectors are relatively insensitive to temperature or flow changes, but the response can be sensitive to solvent composition, which can effect sample absorption characteristics, as in gradient–elution chromatography.

The fluorescence detector, perhaps the most sensitive of the commonly used detectors in lc, is limited in its utility to the detection of materials that fluoresce or have derivatives that fluoresce. These detectors find particular use in analysis of environmental and food samples, where measurements of trace quantities are required.

Electrochemical detectors sense electroreducible and electrooxidizable compounds at low concentrations. For these detectors to work efficiently, the mobile phase (solvent) must be conductive and not subject to electrochemical decomposition.

Another classification of detector is the bulk-property detector, one that measures a change in some overall property of the system of mobile phase plus sample. The most commonly used bulk-property detector is the refractive-index (RI) detector. The RI detector, the closest thing to a universal detector in lc, monitors the difference between the refractive index of the effluent from the column and pure solvent. These detectors are not very good for detection of materials at low concentrations. Moreover, they are sensitive to fluctuations in temperature.

Conductivity detectors, commonly employed in ion chromatography, can be used to determine ionic materials at levels of parts per million (ppm) or parts per billion (ppb) in aqueous mobile phases. The infrared (ir) detector is one that may be used in either nonselective or selective detection. Its most common use has been as a detector in size-exclusion chromatography, although it is not limited to sec. The detector is limited to use in systems in which the mobile phase is transparent to the ir wavelength being monitored. It is possible to obtain complete spectra, much as in some gc–ir experiments, if the flow is not very high or can be stopped momentarily.

Affinity Chromatography. This technique involves the use of a bioselective stationary phase placed in contact with the material to be purified, the ligate. Because of its rather selective interaction, sometimes called a lock-and-key mechanism, this method is more selective than other lc systems based on differential solubility. Affinity chromatography is sometimes called bioselective adsorption.

Chiral Chromatography. Chiral chromatography is used for the analysis of enantiomers, most useful for separations of pharmaceuticals and biochemical compounds (see BIOPOLYMERS, ANALYTICAL TECHNIQUES). There are several types of chiral stationary phases: those that use attractive interactions, metal ligands, inclusion complexes, and protein complexes. The separation of optical isomers has important ramifications, especially in biochemistry and pharmaceutical chemistry, where one form of a compound may be bioactive and the other inactive, inhibitory, or toxic.

Ion-Exchange Chromatography. In iec, the column contains a stationary phase having ionic groups such as a sulfonate or carboxylate. The charge of these groups is compensated by counterions such as sodium or potassium. The mobile phase is usually an ionic solution, eg, sodium chloride, having pH and salt concentrations that act as the separation variables, having ions similar to the counterions. Ionic samples are introduced into the mobile phase, and retardation in movement results from ion exchange with the stationary phase of the form:

$$M^+ + SO_3Na \rightleftarrows Na^+ + SO_3M$$

where M^+ is the ion in the sample to be analyzed and the sulfonate groups are assumed to be a part of the stationary phase. The more the ion interacts with the exchanger, the more strongly it is retained. For cation-exchange chromatography, positively charged ions are separated, as shown in the equation. In anion-exchange chromatography, negatively charged ions in the sample interact with and bind to cationic stationary phases (see ION EXCHANGE).

Ion chromatography (ic), a novel form of ion-exchange chromatography, is a technique in which a weak ion-exchange column is used for separation. Detection is usually done conductimetrically. After passing through the weak ion-exchange column, the eluent passes through a subsequent column called a stripper column, in which the stream, usually made acidic or basic in the ion-exchange column, is neutralized. The stripper column may be a hollow fiber suppressor as well as a column, or the suppressor may be absent, with high conductivity being suppressed electronically. This stream then gives no conductimetric response in this condition; however, when added ions are present, such as happens when sample is passing through the stripper column, the conductivity of the solution changes and a signal is detected. Ic is a powerful technique for examining low concentrations of anions and cations. It has the advantage over selective ion-electrode analysis in that it simultaneously gives information on many ions in a single experiment (see ELECTROANALYTICAL TECHNIQUES).

Ion-pair chromatography (ipc), a variant of iec, is also sometimes called paired-ion chromatography (pic), soap chromatography, extraction chromatography, or chromatography with a liquid ion exchanger. In this technique the mobile phase consists of a solution of an aqueous buffer and an organic cosolvent containing an ion of charge opposite to the charge on the sample ion. The sample ion and the solvated ion form an ionic pair that is soluble in the stationary phase. Thus retention is determined by the ability to form the ion pair as well as the solubility of the complex in the stationary phase.

In size-exclusion chromatography (sec) or gel-permeation chromatography (gpc), the material with which the column is packed has pores in a certain range of sizes. Molecules or solvent-molecule complexes too large to pass through these pores pass rapidly through the column, whereas molecules or complexes of suffiently small size are retained and are the last to exit the column. Molecules of intermediate size are partially retained and elute from the chromatographic column at intermediate times. SEC is extremely useful as a tool for characterization of polymer materials because the retention mechanism is reproducible enough to give good comparative data. SEC can also give valuable information about the distribution of sizes of molecules in a sample.

Supercritical-Fluid Chromatography

Supercritical-fluid chromatography (sfc), developed in the late 1960s, was not used extensively until the early 1980s. This technique is the link between gc and lc, because its mobile phase, a supercritical fluid, has physicochemical properties intermediate between a gas and a liquid (see SUPERCRITICAL FLUIDS). The physicochemical properties of the mobile phase are strong factors determining the selectivity, sensitivity toward a component, and efficiency of separation in the chromatographic process. Supercritical fluids, for example, can be viewed as dense gases that cannot become liquid. The density of a supercritical material increases continuously with pressure at constant temperature and its solvating power increases with pressure, because the solubility of materials in a solvent usually increases with density. Hence this can be used as a powerful means of changing retention. Carbon dioxide is the mobile phase most often used in sfc.

Sfc can be performed with either capillary or lc-like packed columns. Carbon dioxide is compatible with chromatographic hardware, is readily available, and is noncorrosive. The most important detector for sfc is the flame-ionization detector because the mobile phase does not give a significant background signal. Most early applications of sfc were in the separation of petroleum (qv) products, however, as of the 1950s many of those separations are carried out using hplc. More recent applications of sfc include separations in fields as diverse as natural products, drugs, foods, pesticides, herbicides (qv), surfactants (qv), and polymers. These are a direct result of the advantages that sfc has over other forms of chromatography because of low operating temperature, selective detection, and sensitivity to molecular weight.

BIBLIOGRAPHY

"Chromotography, Affinity" in *ECT* 3rd ed., Vol. 6, pp. 35–54, by A. H. Nishikawa, Hoffmann-LaRoche Inc.

1. G. Wilkinson, *Today's Chemist* **29** (Dec. 1990).
2. M. S. Tswett, *Proc. Warsaw Soc. Nat. Sci.* (1903).
3. M. S. Tswett, *Ber. Deut. Bot. Ges.* **24,** 384 (1906).
4. A. J. P. Martin and R. L. M. Synge, *Biochem. J. (London)* **35,** 1358 (1941).
5. *Res. Dev.*, 19 (Jan. 1991).
6. R. L. Grob, *Modern Practice of Gas Chromatography*, Wiley-Interscience, John Wiley & Sons, Inc., New York, 1985.

General References

L. R. Snyder, J. L. Glajsch, and J. J. Kirkland, *Practical HPLC Method Development*, Wiley-Interscience, John Wiley & Sons, Inc., New York, 1988.

L. R. Snyder and J. J. Kirkland, *Introduction to Modern Liquid Chromatography*, Wiley-Interscience, John Wiley & Sons, Inc., New York, 1989.

M. S. Klee, *GC Inlets—An Introduction*, Hewlett Packard, Avondale, Pa., 1990.

J. M. Miller, *Chromatography: Concepts and Contrasts*, Wiley-Interscience, John Wiley & Sons, Inc., New York, 1988.

W. Jennings, *Gas Chromatography with Glass Capillary Columns*, Academic Press, Inc., New York, 1980.

M. Lee, F. Yang, and K. D. Bartle, *Open Tubular Column Gas Chromatography*, Wiley-Interscience, John Wiley & Sons, Inc., New York, 1984.

R. L. Grob, *Chromatographic Analysis of the Environment*, Marcel Dekker, New York, 1983.

R. P. W. Scott, *Small Bore Liquid Chromatography Columns—Their Properties and Uses*, Wiley-Interscience, John Wiley & Sons, Inc., New York, 1984.

N. A. Parris, *Instrumental Liquid Chromatography*, Elsevier, Amsterdam, The Netherlands, 1984.

R. L. Grob and M. A. Kaiser, *Environmental Problem Solving Using Gas and Liquid Chromatography*, Elsevier, Amsterdam, The Netherlands, 1982.

B. Fried and J. Sherma, *Thin-Layer Chromatography, Techniques and Applications*, Marcel Dekker, New York, 1986.

C. Horvath and J. Nikelly, *Analytical Biotechnology: Capillary Electrophoresis and Chromatography*, ACS Books, Washington, D.C., 1990.

N. Grinberg, *Modern Thin-Layer Chromatography*, Marcel Dekker, New York, 1990.

S. Ahuja, *Chiral Separations by Liquid Chromatography*, ACS Books, Washington, D.C., 1991.

H. F. Walton and R. D. Rocklin, *Ion Exchange in Analytical Chemistry*, CRC Press, Boca Raton, Fla., 1990.

B. J. Hunt and S. R. Holding, *Size Exclusion Chromatography*, Blackie, Glasgow, Scotland, 1989.

H. J. Cortes, *Multidimensional Chromatography: Techniques and Applications*, Marcel Dekker, New York, 1990.

P. R. Haddad and P. E. Jackson, *Ion Chromatography: Principles and Applications*, Elsevier, Amsterdam, The Netherlands, 1990.

MARY A. KAISER
E. I. du Pont de Nemours & Co., Inc.
CECIL DYBOWSKI
University of Delaware

CHROMATOGRAPHY, AFFINITY.   See CHROMATOGRAPHY.

CHROME DYES.   See AZO DYES; DYES, APPLICATION AND EVALUATION.

CHROMIUM AND CHROMIUM ALLOYS

Chromium [*7440-47-3*], Cr, also loosely called chrome, is the twenty-first element in relative abundance with respect to the earth's crust, ranking with V, Zn, Ni, Cu, and W, yet is the seventh most abundant element overall because Cr is concentrated in the earth's core and mantle (1,2). It has atomic number 24 and belongs to Group 6 (VIB) of the Periodic Table and is positioned between vanadium and manganese. Other Group 6 members are molybdenum and tungsten. On a tonnage basis, chromium ranks fourth among the metals and thirteenth of all mineral commodities in commercial production.

Chromium was first isolated and identified as a metal in 1789 by Vauquelin who was working with a rare mineral, Siberian red lead or crocoite [*14654-05-08*], $PbCrO_4$ (3). The name chromium comes from the Greek word *chroma* meaning color and resulted from the wide variety of brilliant colors displayed by compounds of the new metal. An early application of chromium compounds, particularly chrome yellow, $PbCrO_4$, was as pigments (qv). Basic chromium sulfate was used for tanning hides. The reaction of chromium and collagen raises the hydrothermal stability of the leather (qv) and renders it resistant to bacterial attack. The most important application of chromium, ie, use as an alloying element, gradually developed during the nineteenth century, and in 1821 led ultimately to chromium steels (see STEEL) (4,5). The first significant structural application of chromium steels was in the famous Eads Bridge across the Mississippi (1867–1874). Further technological developments included improved oxidation resistance and hardenability, and the superior corrosion resistance of a ferritic 12.8% chromium–iron and of austenitic alloys of 18Cr–8Ni. The oxide, chromite, was employed as a furnace refractory as early as 1879 (see CHROMIUM COMPOUNDS).

Occurrence and Mining

The only commercial ore, chromite [1308-31-2], which is also called chromite ore, chrome ore, and chrome, has the ideal composition $FeO \cdot Cr_2O_3$, ie, 68 wt % Cr_2O_3, 32 wt % FeO, or ca 46 wt % chromium. Actually the Cr:Fe ratio varies considerably and the ores are better represented as $(Fe,Mg)O \cdot (Cr,Fe,Al)_2O_3$. Table 1 gives the classification of chromite ores.

Chromite deposits occur in olivine- and pyroxene-type rocks and derivatives. Geologically these appear in stratiform deposits several feet thick covering a very wide area and are usually mined by underground methods in such countries as South Africa, Zimbabwe, India, and Finland. Podiform deposits, ie, isolated lenticular, tabular, or pod-shaped bodies ranging in size from a kilogram to several million tons are mined by both surface and underground methods in Russia and Albania, depending on size and occurrence, but these account for only about 10% of the world's chromium ore resources. Lateritic and placer deposits are not commercially significant. Most chromite ores are rich enough for hand sorting. However, fines or lower grade ores can be effectively concentrated by gravity separation methods yielding products as high as 50% Cr_2O_3 with the Cr:Fe ratio of the original ore usually unchanged (see MINERAL RECOVERY AND PROCESSING). Decreasing world supplies (7) of high grade lumpy ore and increasing availability of high grade fines and concentrates have led to an increased use of three agglomeration methods: (1) briquetting with a binder; (2) production of an oxide pellet by kiln firing; and (3) production of a prereduced pellet by furnace treatment.

Table 1. Classification of Chromite Ores[a]

Class	Geologic deposit type	Cr_2O_3 composition, wt %	Cr:Fe ratios	Principal use
high chromium	podiform, stratiform	46–55	>2:1	metallurgy
high iron	stratiform	40–46	1.5–2:1	metallurgy, chemistry
high aluminum[b]	podiform	33–38	2.0–2.5:1	refractory

[a]Ref. 6.
[b]Ore also contains from 22–34 wt % Al_2O_3.

Properties

Chemical Properties. The valence states of chromium are +2, +3, and +6, the latter two being the most common. The +2 and +3 states are basic, whereas the $+6$ is acidic, forming ions of the type CrO_4^{2-} (chromates) and $(Cr_2O_7)^{2-}$ (dichromates). The blue–white metal is refractory and very hard.

Oxidation tests in oxygen at atmospheric pressure on a chromium specimen containing 0.04% carbon showed the formation of an oxide film 0.15 μm thick in 2 h at 700°C and 2.4 μm thick in 1 h at 900°C (8). Chromium-containing alloys are resistant to sulfidation at high sulfur partial pressures, but these alloys may sulfidize rapidly at low sulfur partial pressures (9,10). Sulfidation of metals at high temperatures is of increasing importance in industrial applications using sulfur-bearing fuels or atmospheres. Furthermore sulfidation rates can be several orders of magnitude greater than oxidation rates.

Chromium is highly acid-resistant and is only attacked by hydrochloric, hydrofluoric, and sulfuric acids. It is also resistant to other common corroding agents including acetone, alcohols, ammonia, carbon dioxide, carbon disulfide, foodstuffs, petroleum products, phenols, sodium hydroxide, and sulfur dioxide.

Physical and Mechanical Properties. The physical properties of chromium are listed in Table 2 (8,11–14).

Perhaps more so than any other common metal, the mechanical properties of chromium (8,14–17) depend on purity, history, grain size, strain rate, and surface condition. Most reported mechanical properties for chromium are those of an ill-defined dilute alloy of unique history and metallurgical condition. More meaningful data are those reported for swaged iodide chromium as shown in Table 3.

The ductile-to-brittle transition temperature (DBTT) is dependent on purity, history, grain size, etc. Furthermore, the potential utility of the metal is impaired by the fact that the ductility below this transition is essentially nil. To achieve measurable ductility, impurities should be below O, 2000 ppm; N, 100 ppm; C, 100 ppm; H, 20 ppm; Si, 1500 ppm; S, 150 ppm.

Production

An overview of the production processes (8,13,18–20) leading to chromium metal and the various chromium compounds is shown in Figure 1. Very little chromite is processed all the way to ductile chromium; most can be used in an intermediate form. For example, the chromite ore itself, mixed with small amounts of lime or magnesia, is made into refractory brick; ferrochromium is used directly in steel making; chrome alum is used as a mordant and in tanning leather; pigments, metal finishing agents, wood preservatives, etc, are made from sodium dichromate [10588-01-9], $Na_2Cr_2O_7$, and chromic acid [11115-74-5], in addition to being the main source of chromium for electroplating (qv), is also used for metal finishing, wood (qv) preservatives, organic syntheses, and the manufacture of catalysts (see CATALYSIS; METAL SURFACE TREATMENT).

Ferrochromium. Ferrochromium, also called ferrochrome and typically classified as low carbon, high carbon, or charge-grade (charge chrome), is usually made by reduction of chromite with coke in a three-phase electric submerged arc furnace. This process leads inevitably to a high carbon ferrochromium [11114-46-8], the use of which was historically restricted to high carbon steels. However, it is also used in argon–oxygen decarburization (AOD) and similar alloy and stainless steels processing. Care is taken to keep sulfur content low as it embrittles

Table 2. Physical Properties of Chromium

Property			Value	
at no.			24	
at wt			51.996	
isotopes				
mass	50	52	53	54
relative abundance, %	4.31	83.76	9.55	2.38
crystal structure			bcc	
lattice parameter, a_o, nm			0.2888–0.2884	
density at 20°C, g/mL			7.19	
mp, °C			1875	
bp, °C			2680	
vapor pressure, at 1610 °C, Pa[a]			130	
heat of fusion, kJ/mol[b]			14.6	
latent heat of vaporization at bp, kJ/mol[b]			305	
specific heat at 25°C, J/(mol·K)[b]			23.9	
linear coefficient of thermal expansion at 20°C, K^{-1}			6.2×10^{-6}	
thermal conductivity at 20°C, W/(m·K)			91	
electrical resistivity at 20°C, $\mu\Omega\cdot m$			0.129	
superconducting transition temperature, K			.08	
antiferromagnetic (Néel) transition temperature, K			311	
specific magnetic susceptibility at 20°C			3.6×10^{-6}	
total emissivity at 100°C[c]			0.08	
reflectivity, R, %				
at 30 nm			67	
at 50 nm			70	
at 100 nm			63	
at 400 nm			88	
refractive index, α, for $\lambda = 257-608$ nm			1.64–3.28	
standard electrode potential, V				
$Cr \rightarrow Cr^{3+} + 3\,e^{-}$			-0.74	
$Cr^{4+} \rightarrow Cr^{6+} + 2\,e^{-}$			$+0.95$	
ionization potential, V				
1st			6.74	
2nd			16.6	
half-life of ^{51}Cr isotope, days			27.8	
thermal neutron scattering cross section, m^2			6.1×10^{-28}	
elastic modulus, GPa[d]			250	
compressibility[e] at 10–60 TPa			70×10^{-3}	

[a]To convert Pa to mm Hg, multiply by 0.0075.
[b]To convert J to cal, divide by 4.184.
[c]Nonoxidizing atmosphere.
[d]To convert GPa to psi, multiply by 145,000.
[e]99% Cr; to convert TPa to megabars, multiply by 10.

Table 3. Mechanical Properties of Room Temperature Swaged Iodide Chromium[a]

Condition	Yield strength, 0.2% offset, MPa[b]	Ultimate strength, MPa[b]	Elongation, %[c]	Reduction of area, %
wrought	362	413	44	78
recrystallized		282	0	0

[a]Pure chromium made by the iodide process (99.996% Cr).
[b]To convert MPa to psi, multiply by 145.
[c]Percent elongation in a 6 mm gauge length.

both Cr metal and the Fe–Ni and Ni-base alloys to which Cr is added. Plasma smelting, a newer production process, is being developed in Sweden and South Africa (21,22). By permitting the use of chromite fines as feed without the need for an agglomeration step, it appears that ferrochromium can be produced economically from what were formerly noneconomic chromite deposits.

Low carbon ferrochromium cannot be made by carbonaceous reduction unless accompanied by top blowing with oxygen. Aluminum, or especially silicon, is frequently used as the reducing agent. When silicon is employed, high silicon ferrochromium that is practically carbon-free is first produced in a submerged arc furnace, and then treated in an open arc-type furnace using a synthetic slag containing chromium(III) oxide [1308-38-9], Cr_2O_3. A ferrochromium of very low (0.01 wt %) carbon content is produced in a solid-state process by heating high carbon ferrochromium and oxidized ferrochromium in a high vacuum. The carbon is removed as carbon monoxide (Simplex process). In the other smelting processes molten ferrochromium is tapped from the furnace, cast into chills, broken into lumps, and graded. The compositions of several grades of ferrochromium are given in Table 4.

Chromium Metal by Pyrometallurgical Reduction. The principal pyrometallurgical process for commercial chromium metal is the reduction of Cr_2O_3 by aluminum.

$$Cr_2O_3 + 2\,Al \longrightarrow 2\,Cr + Al_2O_3$$

Chromium oxide is mixed with aluminum powder, placed in a refractory-lined vessel, and ignited with barium peroxide and magnesium powder. The reaction is exothermic and self-sustaining. Chromium metal of 97–99% purity is obtained, the chief impurities being aluminum, iron, and silicon (Table 4).

Commercial chromium metal may also be produced from the oxide by reduction with silicon in an electric-arc furnace.

$$2\,Cr_2O_3 + 3\,Si \longrightarrow 4\,Cr + 3\,SiO_2$$

The product is similar to that obtained by the aluminothermic process; however, the aluminum content is lower and silicon may run as high as 0.8%.

Chromium oxide may also be reduced with carbon at low pressure.

$$Cr_2O_3 + 3\,C \longrightarrow 2\,Cr + 3\,CO$$

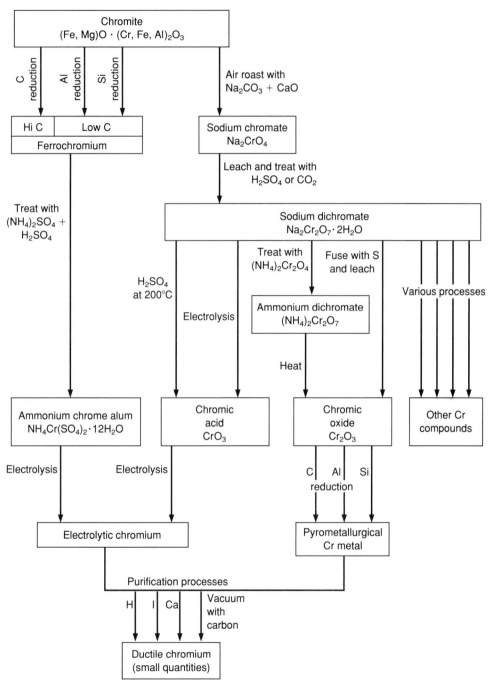

Fig. 1. Simplified flow chart for the production of metallic chromium and chromium compounds from chromite.

Table 4. Composition of Chromium Ferroalloys and Chromium Metal, wt %[a]

Material[b]	Grade	Chromium[c]	Carbon[d]	Silicon[d]	Sulfur[d]	Phosphorus[d]	Nitrogen[d]
ferrochromium							
high carbon	A	52–58	6.0–8.0	6.0	0.040	0.030	0.050
	B	55–64	4.0–6.0	8.0–14.0	0.040	0.030	0.050
	C	62–72	4.0–9.5	3.0	0.060	0.030	0.050
low carbon	A	60–67	0.025	1.0–8.0	0.025	0.030	0.12
	B	67–75	0.025	1.0	0.025	0.030	0.12
	C	67–75	0.050	1.0	0.025	0.030	0.12
	D	67–75	0.75	1.0	0.025	0.030	0.12
vacuum	E	67–72	0.020	2.0	0.030	0.030	0.050
low carbon	F	67–72	0.010	2.0	0.030	0.030	0.050
	G	63–68	0.050	2.0	0.030	0.030	5.0–6.0
nitrogen bearing		62–70	0.10	1.0	0.025	0.030	1.0–5.0
ferrochromium– silicon	A	34–38	0.060	38–42	0.030	0.030	0.050
	B	38–42	0.050	41–45	0.030	0.030	0.050
chromium metal	A[e]	99.0	0.050	0.15	0.030	0.010	0.050
	B[f]	99.4	0.050	0.10	0.010	0.010	0.020

[a]Ref. 23.
[b]Chemical requirements in addition to those listed here are specified by ASTM.
[c]Minimum, except where range of values indicating minimum and maximum appears.
[d]Maximum, except where range of values indicating minimum and maximum appears.
[e]Presumably electrolytic.
[f]Presumably aluminothermic.

Briquets of mixed, finely divided oxide and carbon are heated to 1275–1400°C in a refractory container. The minimum pressure is about 40 Pa (0.3 mm Hg) for reduction at 1400°C. Lower pressures or higher temperatures cause excessive volatilization of chromium. The result is a high purity, low interstitial product.

Electrowinning of Chromium. *Chrome Alum Electrolysis.* In the chrome alum process (Fig. 2) typified by the 2000 tons per year Elkem Metals Co. plant at Marietta, Ohio, high carbon ferrochromium is leached with a hot solution of reduced anolyte plus chrome alum mother liquor and makeup sulfuric acid. The slurry is then cooled to 80°C by the addition of cold mother liquor from the ferrous ammonium sulfate circuit, and the undissolved solids, mostly silica, are separated by filtration. The chromium in the filtrate is then converted to the nonalum form by conditioning treatment for several hours at elevated temperature.

Ammonium chrome alum [10022-47-6], $NH_4Cr(SO_4)\cdot12H_2O$, exists in either a violet or green modification, the properties of which are quite different with regard to conductivity, solubility, and ionization. Above 50°C, the green complex is more stable; at room temperature it changes slowly to the violet form with a change in pH. In solutions of ammonium chromium alum, the chromium(III) ion can exist in a variety of forms, depending on time, temperature, and past conditions. At higher temperatures, a variety of green nonalum ions such as $[Cr(H_2O)_5(SO_4)]^+$, $[Cr(H_2O)_5(OH)]^{2+}$, and $[(SO_4)(H_2O)_4Cr-O-Cr(H_2O)_5]^{2+}$ form, whereas the violet hexaaquachromium(III) ion [14873-01-9], $[Cr(H_2O)_6]^{3+}$, pre-

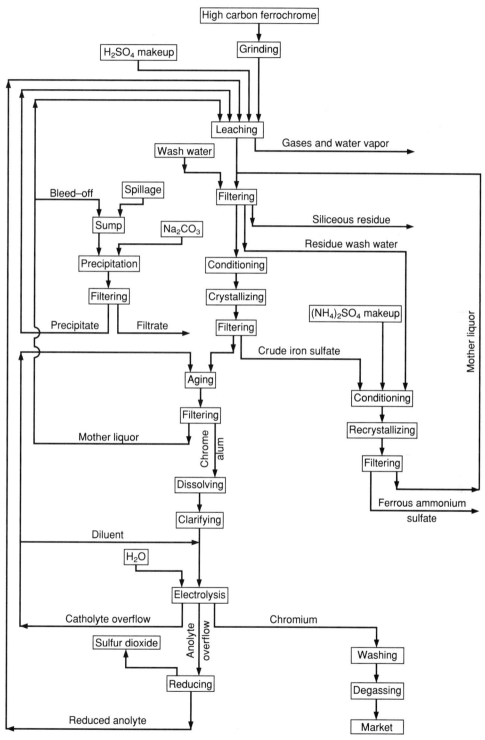

Fig. 2. Flow sheet for production of electrolytic chromium by the chrome alum process at the Marietta Plant, Elkem Metals Co., Marietta, Ohio.

dominates in cool, dilute solutions of moderate acidity. It is only the violet form that permits crystallization of the desired ammonium chromium alum.

Once the green, nonalum modifications of the chromium ion have formed, reversion to the hexaaquo form on cooling is sufficiently slow that on chilling to 5°C a crude ferrous ammonium sulfate can be crystallized, removing nearly all the iron from the system. This crude iron salt is treated with makeup ammonium sulfate, heated again to retain the chromium impurities in the green noncrystal-lizable form, and then cooled to separate the bulk of the iron as a technical ferrous ammonium sulfate, which is sold for fertilizer and other purposes. The mother liquor from this crystallization is returned to the filtration step.

The mother liquor from the crude ferrous sulfate crystallization contains nearly all the chromium. It is clarified and aged with agitation at 30°C for a considerable period to reverse the reactions of the conditioning step. Hydrolysis reactions are being reversed; therefore, the pH increases. Also, sulfate ions are released from complexes and the chromium is converted largely to the hexaaquo ion. Ammonium chrome alum then precipitates as a fine crystal slurry. It is fil-tered and washed and the filtrate sent to the leach circuit; the chrome alum is dissolved in hot water, and the solution is used as cell feed.

The principal electrolytic cell reactions are shown in Figure 3. A diaphragm-type cell prevents the sulfuric acid and chromic acid formed at the anode from mixing with the catholyte and oxidizing the divalent chromium. Electrolyte is continuously fed to the cells to maintain the proper chromium concentration. The catholyte pH is controlled by adjusting the flow through the diaphragms into the anolyte compartments. Control of the pH between narrow limits governs the suc-cessful electrodeposition of chromium as well as the preservation of divalent chro-mium at the cathode.

The analyses of the solutions in the electrolytic circuit and cell operating data are given in Tables 5 and 6, respectively. The current efficiency of 45% shown in Table 6 includes low efficiencies that always prevail during the startup of a reconditioned cell. The 2.1–2.4 pH range used in the plant also results in some-what lower current efficiency but provides a safe operating latitude.

At the end of the 72-h cycle, the cathodes are removed from the cells, washed in hot water, and the brittle deposit, 3–6 mm thick, is stripped by a series of air hammers. The metal is then crushed by rolls to 50-mm size and again washed in hot water. The metal contains about 0.034% hydrogen and, after drying, is de-hydrogenated by heating to at least 400°C in stainless steel cans. Composition limits for electrolytic chromium are shown in Table 4.

Chromic Acid Electrolysis.   Alternatively, as shown in Figure 1, chromium metal may be produced electrolytically or pyrometallurgically from chromic acid, CrO_3, obtained from sodium dichromate by any of several processes. Small amounts of an ionic catalyst, specifically sulfate, chloride, or fluoride, are essential to the electrolytic production of chromium. Fluoride and complex fluoride cata-lyzed baths have become especially important in recent years. The cell conditions for the chromic acid process are given in Table 7.

The low current efficiency of this process results from the evolution of hy-drogen at the cathode. This occurs because the hydrogen deposition overvoltage on chromium is significantly more positive than that at which chromous ion depo-sition would be expected to commence. Hydrogen evolution at the cathode surface

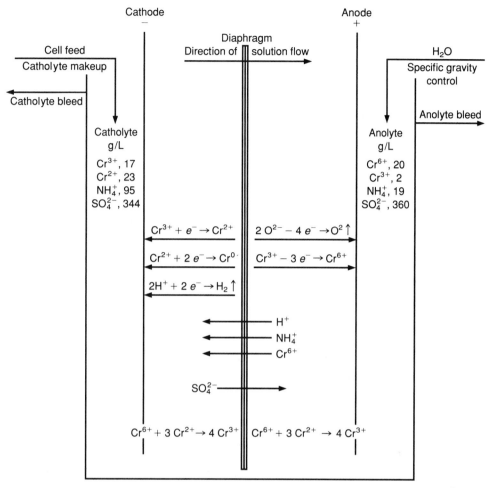

Fig. 3. Principal electrolytic cell reactions and electrolyte concentrations in chromium production by the chrome alum process.

Table 5. Solution Concentrations in Electrolysis of Chrome Alum, g/L

Solution	Total Cr	Cr^{6+}	Cr^{3+}	Cr^{2+}	Fe	NH_3	H_2SO_4
cell feed	130	0	130	0	0.2	43	3
circulating mixture	65	0	63	2	0.1	68	1
catholyte	24	0	11.5	12.5	0.035	84	
anolyte	15	13	2	0	0.023	24	280

Table 6. Operating Parameters for Electrowinning of Chromium from Chrome Alum

Parameter	Value
cathode current density, A/m^2	753
cell potential, V	4.2
current efficiency, %	45
electrical consumption, MJ/kg^a	67
pH of catholyte	2.1–2.4
catholyte temperature, °C	53 ± 1
deposition time, h	72
cathode material	Type 316 stainless steel
anode material, wt %	1–99, Ag–Pb

^{a}To convert J to cal, divide by 4.184.

Table 7. Cell Conditions for Chromic Acid Process for Electrowinning of Chromium^a

Parameter	Value
bath composition, CrO$_3$, g/L^b	300
temperature, °C	84–87
current density, A/m^2	9500
current efficiency, %	6–7
plating time, h	80–90
production rate, g/wk	1000

^{a}From Ref. 15.
^{b}Also contains 4 g/L SO$_4^{2-}$.

also increases the pH of the catholyte beyond 4, which may result in the precipitation of $Cr(OH)_3$ and $Cr(OH)_2$, causing a partial passivation of the cathode and a reduction in current efficiency. The latter is also inherently low, as six electrons are required to reduce hexavalent ions to chromium metal.

Plating variables for this process may be summarized as: higher (87°C) operating temperatures enable the oxygen content of the metal to be reduced to 0.01%; the CrO_3:SO_4 ratio should be below 100 to obtain low oxygen metal; current efficiencies > 8% are associated with high oxygen contents; and better current efficiencies are obtained at low current densities.

The metal obtained by this process contains less iron and oxygen than that from the chrome alum electrolyte. The gas content is 0.02 wt % O, 0.0025 wt % N, and 0.009 wt % H. If desired, the hydrogen content can be lowered still further by a dehydrogenation treatment.

Purification. The metal obtained from both electrolytic processes contains considerable oxygen, which is believed to cause brittleness at room temperature. For most purposes the metal as plated is satisfactory. However, if ductile metal is desired, the oxygen can be removed by hydrogen reduction, the iodide process, calcium refining, or melting in a vacuum in the presence of a small amount of carbon.

In the hydrogen reduction process, the electrolytic metal is heated to about 1500°C in a closed-circuit stream of dry, pure hydrogen. Although this process reduces the oxygen content considerably and the nitrogen content somewhat less so, it has little effect on the other impurities. A typical product from this process might contain 0.005 wt % oxygen and 0.001 wt % nitrogen.

In the iodide purification process (Van Arkel process), the impure chromium and iodine are sealed in an evacuated bulb containing an electrically heated wire. The bulb is heated to the temperature of formation of chromium(III) iodide [13569-75-0], CrI_3. This chromic iodide reacts with more of the impure metal to form chromium(II) iodide [13478-28-9], CrI_2, which diffuses to the hot wire where it decomposes and deposits chromium. The freed iodine forms additional chromous iodide, and the process continues (cyclic process). Iodination at 900°C and decomposition at 1000–1300°C may also be carried out separately (straight-flow process) but with lower efficiency. Analyses of metal treated by iodide purification gives 0.0008 wt % oxygen, 0.003 wt % nitrogen, 0.00008 wt % hydrogen, and 0.001 wt % carbon; the other impurities were very low. This metal is produced by Materials Research Corp. under license from the Chromalloy Corp. under the trade name Iochrome.

In the calcium refining process, the chromium reacts with calcium vapor at about 1000°C in a titanium-lined bomb, which is first evacuated and then heated to the proper temperature. A pressure of about 2.7 Pa (20 μm Hg) is maintained during heating until the calcium vapor reaches the cold end of the bomb and condenses. This allows the calcium vapor to pass up through the chromium metal where it reacts with the oxygen. Metal obtained by this process contains 0.027 wt % oxygen, 0.0018 wt % nitrogen, 0.008 wt % carbon, 0.012 wt % sulfur, and 0.015 wt % iron.

Purification using carbon is accomplished by addition to the melt or to the solid charge before vacuum melting. Pressure rises as a result of the reaction of the carbon with dissolved oxygen. Completion of the deoxidation process is marked by a rapid pressure drop indicating when the evolution of CO is complete.

The vacuum melting process can upgrade chromium at a modest cost; the other purification processes are very expensive. Thus iodide chromium is about 100 times as expensive as the electrolytic chromium and, therefore, is used only for laboratory purposes or special biomedical applications.

Consolidation and Fabrication. Chromium metal may be consolidated by powder metallurgy techniques or by arc melting in an inert atmosphere (8,13,24,25) (see METALLURGY; METALLURGY, POWDER).

For powder metal consolidation the metal is first ball-milled using chromium-plated balls, then pressed at 300–500 MPa (3000–5000 atm) with or without a binder such as a wax. The binder is then removed by heating to about 300°C. The sintering operation that follows is carried out at high temperatures, 1450–1500°C, in a slow stream of purified hydrogen, helium, or argon. Sometimes sintering and purification can be combined, and in this case a large volume of purified hydrogen is needed to ensure adequate oxygen removal.

In a chemical vapor deposition (CVD) variant of conventional powder metallurgy processing, fine chromium powder is obtained by hydrogen reduction of CrI_2 and simultaneously combined with fine thorium(IV) oxide [1314-20-1], ThO_2, particles. This product is isostatically hot pressed to 70 MPa (700 atm) and 1100°C for 2 h. Compacts are steel clad and hot rolled to sheets (24).

Vacuum melting of chromium must take place in highly refractory crucibles such as pure zirconia, beryllia, or alumina lined with thoria. Under proper conditions, the nitrogen content of chromium-base alloys can be lowered to 0.01 wt % or less, and carbon can be used as a deoxidizer to obtain alloys having 0.01 wt % carbon and oxygen. Volatilization of some chromium is a drawback in this process. Arc melting into a water-cooled copper mold in an inert atmosphere has been successfully used for chromium and has the advantage that no refractory material comes in contact with the molten metal. The chromium must be given an oxygen purification before melting if low oxygen material is required. It is claimed that addition of a small amount of yttrium or other rare-earth element during melting to act as a scavenger improves the workability and mechanical properties.

The initial cast structure of arc melted ingots must be carefully broken down by hot working in order to permit subsequent warm working. Forging, swaging, and extrusion are all possible. Ease of working increases from forging to extrusion, particularly in the case of hydrostatic extrusion. Hot working usually takes place in a steel or stainless sheath to protect the metal from contamination and this sheath is later removed by acid dissolution on completion of the working. The working process seems to make the material more ductile, over and above benefits resulting from break-up of the as-cast structure (25). Electropolishing is often used to further improve the apparent ductility of sheet and wire samples. Elimination of the surface layer in this way removes any air-contaminated layer of chromium and also minimizes the effects of any notches.

Electroplating, Chromizing, and Other Chromium-Surfacing Processes

Electroplating. Chromium is electroplated onto various substrates in order to realize a more decorative and corrosion- or wear-resistant surface (24–32). About 80% of the chromium employed in metal treatment is used for chromium plating; over 50% is for decorative chromium plating (see METAL SURFACE TREATMENTS). Hard chromium plating differs from decorative plating mostly in terms of thickness. Hard chromium plate may be 10 to several 100 μm thick, whereas the chromium layers in a decorative plate may be as thin as 0.25 μm, which corresponds to about two grams Cr per square meter of surface.

Hard plating is noted for its excellent hardness, wear resistance, and low coefficient of friction. Decorative plating retains its brilliance because air exposure immediately forms a thin, invisible protective oxide film. The chromium is not applied directly to the surface of the base metal but rather over a nickel (see NICKEL AND NICKEL ALLOYS) plate, which in turn is laid over a copper (qv) plate. Because the chromium plate is not free of cracks, pores, and similar imperfections, the intermediate nickel layer must provide the basic protection. Indeed, optimum performance is obtained when a controlled but high density (40–80 microcrack intersections per linear millimeter) of microcracks is achieved in the chromium leading to reduced local galvanic current density at the imperfections and increased cathode polarization. A duplex nickel layer containing small amounts of sulfur is generally used. In addition to applications in the automotive and plumbing fields, chromium plate is being used as a substitute for tin plate on steel for canning purposes. Since 1926, commercial chromium plating has used a chromic acid bath, Cr^{6+}, to which various catalyzing anions have been added. However, much attention is being given to the development of trivalent plating baths (see ELECTROPLATING).

Chromizing. The other principal method of obtaining a chromium-rich surface on steel is by chromizing (29). The material to be treated is embedded in a mixture of ferrochromium powder, a chromium halide, alumina, and sometimes NH_4Cl. The chromium is diffused in by a furnace treatment at about 1100°C to produce an effective stainless steel surface where the mean composition is about 18 wt % Cr and the thickness is controlled by the time of treatment. Chromizing is an economical process for improving the corrosion resistance of steel parts where cut edges and appearance are not important considerations, eg, automotive exhaust systems, heat exchangers, and silos.

Other Surface Processes. Whereas sputtering, ion implantation, chemical vapor deposition (CVD), metal spraying, cladding, and weld overlayment have been used for chromium, only the last two have commercial significance for protecting steels (29,33,34). Stainless clad steel, single- or double-faced, has been prepared since the mid-1900s by hot rolling a duplex or triplex metal sandwich. It is an attractive means of conserving expensive stainless steel. Weld overlayment is used where clad is either not available or unreliable or where wear resistance is an important consideration. Flame and arc welding (qv) are both practiced. The most popular alloys deposited are the Co–Cr-based Stellites and similar compositions, but chromium carbides and oxides can also be deposited (see METAL SURFACE TREATMENTS).

For protection of Fe-,Co-, or Ni-based superalloys (qv) (33–35), the same coating processes as used for steels are applied to obtain Cr-rich coatings. Chromium by itself is protective up to above 800°C where vaporization of the oxide begins to be significant. To extend protection above 800°C Si or Al must be added to the coating. Chromium contents of 20 wt % or more are used to prevent depletion of Cr in the substrate alloy. The types of coatings used are classified as: Cr–Al, Ni–Cr–B–Si, Ce–Cr–Al, Ni–Cr–Si, and Fe–Cr–Al, Co–Cr–Al, Ni–Cr–Al, each with or without the adherence-promoting addition of a fraction of a percent yttrium. The latter are referred to as FeCrAlY, CoCrAlY, and Ni-CrAlY coatings.

Columbium (niobium)-based alloys, used as rocket components having service temperatures up to 1370°C, are protected by silicide coatings containing 20% Cr and 20% Fe (36).

Economic Aspects

During much of the nineteenth century, the United States was the principal world producer of chromite ore (37). However in the latter twentieth century the United States has become completely dependent on imports from South Africa and Turkey (chromite); South Africa, Zimbabwe, Turkey, and Yugoslavia (ferrochromium); and the Philippines (chromite for refractory brick).

Since the 1970s, imports to the United States and other highly developed countries have shifted from chromite to ferrochromium as countries having low cost energy and natural resources have installed the furnace capacity to convert the ore to ferrochromium. Supply routes from overseas sources are long and thus chromium-consuming industries maintain large stocks. A study by the National Materials Advisory Board of national contingency plans for chromium utilization during crisis has been published (39), and a U.S. congressionally-mandated stockpile conversion program whereby a large part of the chromite ore in the National Defense Stockpile is being converted to ferrochromium in a seven-year program running from 1987 to 1993 has been implemented (20).

The distribution of chromium consumption in the United States by physical form is shown in Table 8. Growth has been modest since the 1970s, and there have been few changes in the consumption pattern, which for the 1980s was 79.5% in stainless and heat-resisting steel, 8.2% in full-alloy steel, 1.7% in carbon steel, 1.6% in high strength low alloy steel, 1.1% in tool steel, 1.7% in cast iron, 2.9% in superalloys, 0.2% in structural and hard-facing welding materials, 0.8% in cutting materials and magnetic, aluminum, copper, nickel, and other alloys, and 2.19% in miscellaneous and unspecified uses (40). A 3–4% annual growth in chromium consumption leading to a total U.S. primary chromium demand of 1,000,000 metric tons in the year 2000 has been projected (20). Chromium is absolutely essential to the production of stainless steel, which accounts for the largest use of this metal. Moreover, a chromium-free stainless steel is unlikely. Thus chromium importance can be judged in view of the importance of stainless steel to the U.S. industrial economy.

The cost of chromium ore is determined by operating, ie, mining and beneficiation, and transportation costs, whereas the price of the ore is affected by chromium and carbon contents and particle size. Lumpy or coarse grades usu-

Table 8. Chromium Consumption in the United States[a,b]

Material	Contained chromium, 10^3 t	
	1980	1989
chromium ore		
chemical	68.2	137.3[c]
metallurgical	128.5	
refractory	35.6	10.7
Total	*232.3*	*148.1*
chromium ferroalloy and metal		
high carbon ferrochromium	183.4	168.7
low carbon ferrochromium	30.0	15.6
ferrochromium–silicon	7.8	5.4
other[d]	7.8	4.7
Total	*229.0*	*194.4*

[a]Excluding scrap.
[b]Ref. 40.
[c]Includes metallurgical.
[d]Chromium ferroalloys and metal not specifically listed.

ally command a premium price. The average 1989 values on the basis of contained chromium, were \$347/t for chromium ore, \$2058/t for ferrochromium, and \$8294/t for chromium metal. The added value, particularly for the step from the ore to ferrochromium, is quite large. There is considerable incentive for countries to market ferrochromium rather than ore. However, smelting of ferrochromium is a relatively energy intensive process requiring about 2800 kW/t of alloy. In the United States, stringent air pollution controls have also added to the smelting cost (see EXHAUST CONTROL, INDUSTRIAL). Imports of chromite ore dropped from a little over two million tons in 1979 to just under 237,000 tons in 1989. This decline was compensated for by increases in ferrochromium imports. The world price of all chromium materials has increased by 10 to 15% per year since 1963. In general demand for stainless steel has driven the price of ferrochromium, which in turn affects the price of the ore, even when ore costs are relatively constant.

CHROMIUM ALLOYS

Metallurgically Important Chromium Compounds

There are a number of metallic compounds of chromium that are used either as the compound itself or as metallurgical constituents in Cr-bearing alloys. Trichromium dicarbide [12012-35-0], Cr_3C_2, is important as a wear-resistant gauge material; chromium boride(1:1) [12006-79-0], CrB, for oil well drilling; chromium dioxide [12018-01-8], CrO_2, for magnetic tape; and $Cr_xMn_{2x}Sb$ as a magnetic material having unique characteristics. The intermetallic compounds trichromium aluminum [12042-09-0], Cr_3Al, trichromium silicon [12018-36-9], Cr_3Si, and dichromium titanium [12018-27-8], Cr_2Ti, are encountered in developmental oxidation-resistant coatings. The carbides (qv) chromium carbide(23:6), [12105-81-6], $Cr_{23}C_6$, and chromium carbide(7:3) [12075-40-0], Cr_7C_3; chromium iron(1:1) [12052-89-0], CrFe (σ phase), and chromium iron molybdenum(12:36:10) [12053-58-6], $Cr_{12}Fe_{36}Mo_{10}$ (χ phase), are found as constituents in many alloy steels; Cr_2Al_{13} and CoCr are found in aluminum and cobalt-based alloys, respectively. The chromium-rich interstitial compounds, Cr_2H, chromium nitrogen(2:1) [12053-27-9], Cr_2N, and $Cr_{23}C_6$, play an important role in the effect of trace impurities on the properties of unalloyed chromium. The intermetallic and the interstitial compounds of chromium are stabilized by electronic and/or spatial factors and are not to be regarded as simple ionic or covalent compounds.

Chromium-Based Alloys

Alloying does not solve the problem of chromium's lack of resistance to gaseous embrittlement. Alloying with yttrium improves the resistance to embrittlement by high temperature exposure to oxygen but not to nitrogen-bearing atmospheres, and a barrier coating approach must be used (8,15–17). Furthermore, although solid solution additions can improve high temperature strength by three- to four-fold over unalloyed chromium, these additions have also resulted in increases in the ductile-to-brittle transition temperature (DBTT). A better combination of properties has been achieved through precipitation or dispersion hardening. The second phases may be oxides, carbides, or borides.

Nitrogen embrittlement appears to be the result of dislocation pinning by the presence of small Cr_2N particles on certain crystallographic planes rather than by elemental nitrogen retained in solid solution. On the other hand, the improvement of ductility experienced using oxide, carbide, or boride particles is apparently achieved through the generation and multiplication of free dislocations by one of several mechanisms. Chromium–thorium oxide dispersions prepared by CVD have been found not only to lower the DBTT relative to unalloyed chromium but largely to preserve this improvement even after a one hour anneal at 1200°C (24). Silicides (16), rare earths (Y and Y + La) (16), and a Cu–30Pd alloy (41) seem to improve protection against nitrogen embrittlement (100–200 h at 1150°C).

The leading developmental chromium-based alloys are shown in Table 9. Creep rupture results indicate a temperature advantage of 110–140°C over the strongest superalloys with 100 h stress-rupture strengths as high as 140 MPa (20,300 psi) at 1150°C. Further improvements in low temperature toughness and high temperature nitridation resistance are still being sought.

Table 9. Developmental Chromium-Based Alloys

Country	Common name	Composition, wt %
United States	C-207	Cr–7.5W–0.8Zr–0.2Ti–0.1C–0.15Y
	CI-41	Cr–7.1Mo–2Ta–0.09C–0.1(Y + La)
	IM-15	Cr–1.7Ta–0.1B–0.1Y
	Chrome-30	Cr–6MgO–0.5Ti
	Chrome-90	Cr–3MgO–2.5V–0.5Si
	Chrome-90S	Cr–3MgO–2.5V–1Si–0.5Ti–2Ta–0.5C
	TD-CVD-Cr[a]	Cr–3ThO$_2$
Australia	Alloy E	Cr–2Ta–0.5Si–0.1Ti
	Alloy H	Cr–2Ta–0.5Si–0.5RE[b]
	Alloy J	Cr–2Ta–0.5Si
Russia	VKh-1I	Cr–(0.3–1.0Y)–0.02C
	VKh-2	Cr–(0.1–0.35)V–0.02C
	VKh-2I	Cr–(0.3–1.0)Y–(0.1–0.35)V–(0.1–0.2Ti)–0.02C
	VKh-4	Cr–32Ni–1.5W–0.3V–0.2Ti–0.08C

[a]CAS Registry Number is [*12650-51-8*].
[b]RE = rare earth.

Stainless Steel

The stainless quality is conferred on steels that contain enough chromium to form a protective surface film. About 12 wt % chromium is required for protection in mild atmospheres or in steam. At 18–20 wt % chromium, sufficient protection is achieved for satisfactory performance in a wide variety of more destructive environments, including those occurring in the chemical, petrochemical, and the power-generating industries. Stainless grades having 25 wt % chromium or more and containing other alloying elements such as molybdenum provide even higher corrosion resistance. In certain stainless steels, the chromium depresses the martensite transformation below room temperature. By thus stabilizing austenite the chromium permits achievement of desired mechanical properties without loss of corrosion resistance (42–48) (see STEEL).

Stainless steels are classified in terms of their microstructures as austenitic, martensitic, ferritic, duplex (austenite + ferrite), and precipitation hardening (PH). The microstructure type is determined by base composition and heat treatment and, in turn, dictates the properties, especially strength, toughness, and corrosion resistance. The compositions of the leading wrought grades of stainless steels are shown in Tables 10–16 and the classes are illustrated diagramatically in Figure 4. The popular grades of cast stainless steels have compositions slightly modified from those of the wrought counterparts. There are also closely related cast alloys designed to maximize heat resistance rather than corrosion resistance. Both are shown in Table 17. Because most stainless steels are based on the Fe–Cr–Ni system, the relationships are conveniently interpreted in reference to the metastable-phase diagram for such alloys having a carbon content of 0.1 wt %, rapidly cooled from 1000°C as shown in Figure 5. A vertical section through the diagram at 8% Ni is shown in Figure 6 (5).

Austenitic Stainless Steels.   The austenitic stainless steels have outstanding corrosion and oxidation resistance coupled with good formability. They cannot, however, be hardened by heat treatment which, in certain types, may impair corrosion resistance by allowing precipitation of chromium carbide near the grain boundaries (sensitization) and hence result in local impoverishment in chromium. Variants of the basic 18 Cr–8 Ni composition contain Mo for improved resistance to sulfuric acid, or Ti or Nb to prevent the undesirable precipitation of chromium carbide at grain boundaries. Sulfur or selenium additions improve machinability. The nitrogen strengthened Cr–Mn proprietary alloys, leading grades of which are shown in Table 11, have higher strength at both room and elevated temperatures than standard 300 grades. These materials also have improved corrosion resistance and low temperature toughness.

Ferritic Stainless Steels.   Ferritic stainless steels having low carbon and high chromium content can be heated to the melting point without transformation to the austenitic structure as shown in Figure 7. This fact ensures freedom from quench hardening, thereby facilitating fabrication, but eliminates the possibility of grain refinement by heat treatment. Such steels are widely used for architectural work, automotive trim, and equipment in the chemical and food industries. Types 430 and 446 are especially resistant to oxidation and hence are favored for furnace parts, heat exchangers, and other high temperature equipment. Table 14 lists the so-called super-ferritics, which offer superior toughness, weldability, and stress corrosion cracking resistance at strength levels comparable to those of conventional ferritics.

Martensitic Stainless Steels.   The martensitic stainless steels have somewhat higher carbon contents than the ferritic grades for the equivalent chromium level and are therefore subject to the austenite–martensite transformation on heating and quenching. These steels can be hardened significantly. The higher carbon martensitic types, eg, 420 and 440, are typical cutlery compositions,

Table 10. Austenitic Grades of Stainless Steels

UNS[a] no.	AISI[b] type	CAS Registry Number	C[c]	Mn[c]	Si[a]	Cr	Ni	Other[d]
						Composition, wt %		
S20100	201	[12725-21-2]	0.15	7.50	1.00	16.0–18.0	3.5–5.5	0.25 N[c]
S20200	202	[37285-92-0]	0.15	10.00	1.00	17.0–19.0	4.0–6.0	0.25 N[c]
S20500	205	[51258-26-5]	0.12–0.25	14.0–15.5	1.00	16.5–18.0	1.0–1.75	0.06 P[c], 0.25 N[c]
S30100	301	[12725-26-7]	0.15	2.00	1.00	16.0–18.0	6.0–8.0	
S30200	302	[12671-80-6]	0.15	2.00	1.00	17–19	8–10	
S30215	302B	[37241-57-9]	0.45	2.00	3.00	17–19	8–10	
S30300	303	[12725-27-8]	0.15	2.00	1.00	17–19	8–10	0.15 min S
S30323	303Se	[37268-91-0]	0.15	2.00	1.00	17–19	8–10	0.15 min Se, 0.020 P[c], 0.060 S[c]
S30400	304	[11109-50-5]	0.08	2.00	1.00	18–20	8–12	
S30403	304L	[12611-86-8]	0.03	2.00	1.00	18–20	8–12	
S30453	304LN	[39418-84-3]	0.03	2.00	1.00	18–20	8–12	0.10–0.16 N
530430	304Cu	[54938-23-7]	0.08	2.00	1.00	17–19	8–10	3.0–4.0 Cu
S30451	304N	[39403-18-4]	0.08	2.00	1.00	18–20	8.0–10.5	0.10–0.16 N
S30500	305	[12620-36-9]	0.12	2.00	1.00	17–19	10–13	
S30800	308	[12671-81-7]	0.08	2.00	1.00	19–21	10–12	
S30900	309	[12725-28-9]	0.20	2.00	1.00	22–24	12–15	
S30908	309S	[37241-61-5]	0.08	2.00	1.00	22–24	12–15	
S31000	310	[12725-29-0]	0.25	2.00	1.50	24–26	19–22	
S31008	310S	[37301-67-0]	0.08	2.00	1.50	24–26	19–22	
S31400	314	[37268-89-6]	0.25	2.00	3.00	23–26	19–22	
S31600	316	[11107-04-3]	0.08	2.00	1.00	16–18	10–14	2.0–3.0 Mo
S31603	316L	[11134-23-9]	0.03	2.00	1.00	16–18	10–14	2.0–3.0 Mo
S31653	316LN	[3848-85-4]	0.03	2.00	1.00	16–18	10–14	2.0–3.0 Mo, 0.10–0.16 N
S31651	316N	[39403-19-5]	0.08	2.00	1.00	16–18	10–14	2.0–3.0 Mo, 0.10–0.16 N
S31700	317	[127288-33-5]	0.08	2.00	1.00	18–20	11–15	3.0–4.0 Mo
S31703	317L	[39302-82-4]	0.03	2.00	1.00	18–20	11–15	3.0–4.0 Mo
S32100	321	[12611-78-8]	0.08	2.00	1.00	17–19	9–12	5×C min Ti

Table 10. (*Continued*)

UNS[a] no.	AISI[b] type	CAS Registry Number	Composition, wt %					
			C[c]	Mn[c]	Si[a]	Cr	Ni	Other[d]
N08330	330	[37245-99-1]	0.08	2.00	0.75–1.50	17–20	34–37	
S34700	347	[12725-20-1]	0.08	2.00	1.00	17–19	9–13	10×C min (Nb + Ta)
S34800	348	[12725-31-4]	0.08	2.00	1.00	17–19	9–13	10×C min (Nb + Ta); 0.10 Ta[c]
S38400	384	[57219-22-4]	0.08	2.00	1.00	15–17	17–19	

[a]Unified Numbering System; Ref. 49.
[b]American Iron and Steel Institute.
[c]Maximum value unless otherwise noted.
[d]Except as noted, all have a maximum of 0.045 wt % P and 0.030 wt % S.

Table 11. Mn–N-Rich Austenitic Stainless Steels

UNS[a] no.	CAS Registry Number	Common name (former)	Composition, wt %					
			C	Mn	Cr	Ni	N	Other
S24000	[39303-37-2]	Nitronic 33 (18-3 Mn)	0.05	12.0	18.0	3.2	0.32	
S21904	[66776-05-4]	Nitronic 40 (21-6-9 LC)	0.03	9.0	21.0	7.0	0.30	
S20910	[12724-48-0]	Nitronic 50 (22-13-5)	0.04	5.0	21.2	12.5	0.30	2.2 Mo, 0.2 Cb, 0.2 V
S21400	[12605-28-6]	USS Tenelon (XM-31)	0.10	15.0	18.0		0.40	
S21460		Cryogenic Tenelon (XM-14)	0.10	15.1	17.5	5.5	0.42	
S21600	[39403-17-3]	AL 216 (XM-17)	0.06	8.3	20.0	6.0	0.37	2.5 Mo

[a]Unified Numbering System.

Table 12. Martensitic or Hardenable Grades of Stainless Steel

UNS[a] no.	AISI type	CAS Registry Number	C[b]	Mn[b]	Si[b]	Cr	Ni	Other[b]
						Composition, wt %		
S40300	403	[39345-19-2]	0.15	1.00	0.50	11.5–13.0		
S41000	410	[12611-79-9]	0.15	1.00	1.00	11.5–13.5		
S41400	414	[37241-54-6]	0.15	1.00	1.00	11.5–13.5	1.25–2.5	
S41600	416	[37373-59-4]	0.15	1.25	1.00	12.0–14.0		0.15 S[c], 0.060 P
S41623	416Se	[37268-92-1]	0.15	1.25	1.00	12.0–14.0		0.06 S, 0.060 P 0.15 Se[c]
S42000	420	[37241-55-7]	0.15[c]	1.00	1.00	12.0–14.0		
S42020	420F	[20961-79-7]	0.30–0.40	1.25	1.00	12.0–14.0	0.50	0.60 Mo, 0.15 S[c], 0.060 P
S42200	422	[51835-85-9]	0.20–0.25	1.00	0.75	11.5–13.5	0.50–1.0	0.75–1.25 Mo, 0.75–1.25 W 0.040 P, 0.030 S, 0.15–0.30 V
S43100	431	[12793-24-7]	0.20	1.00	1.00	15.0–17.0	1.25–2.50	0.040 P, 0.030 S
S44002	440A	[56507-68-7]	0.60–0.75	1.00	1.00	16.0–18.0		0.75 Mo
S44003	440B	[64159-66-6]	0.75–0.95	1.00	1.00	16.0–18.0		0.75 Mo
S44004	440C	[12725-30-3]	0.95–1.25	1.00	1.00	16.0–18.0		0.75 Mo

[a]Unified Numbering System.
[b]Maximum value unless otherwise noted or range is given.
[c]Minimum value.

248

Table 13. Ferritic or Nonhardenable Grades of Stainless Steel

UNS[a] no.	AISI[b] type	CAS Registry Number	C[c]	Mn[c]	Si[c]	Cr	Mo	Ni	Other[d]
S40500		[37202-69-0]	0.08	1.00	1.00	11.5–14.5			0.10–0.30 Al
S40900	409	[39418-83-2]	0.08	1.00	1.00	10.5–11.75		0.5[c]	(6 × C,0.75 max)Ti, 0.045 P[c], 0.045 S[c]
S42900	429	[60998-97-2]	0.12	1.00	1.00	14.0–16.0			
S43000	430	[11109-52-7]	0.12	1.00	1.00	16.0–18.0			
S43020	430F	[57923-44-1]	0.12	1.25	1.00	16.0–18.0	0.60[e]		0.15 min S,0.060 P[c]
S43023	430FSe		0.12	1.25	1.00	16.0–18.0			0.06 S, 0.06 P, 0.15 min Se
S43400	434	[12725-32-5]	0.12	1.00	1.00	16.0–18.0	1.0		
S43600	436	[12741-28-5]	0.12	1.00	1.00	16.0–18.0	0.75–1.25		(Nb + Ti)=5 × C min–0.70 max
S43035	439	[51836-03-4]	0.07	1.00	1.00	17.0–19.0			0.15 Al, 12 × C min–1.10 Ti
S44200	442	[39442-86-9]	0.20	1.00	1.00	18.0–23.0		0.50	
S44600	446	[12629-05-9]	0.20	1.50	1.00	23.0–27.0			0.25 N[c]

Composition, wt %

[a]Unified Numbering System; Ref. 50.
[b]American Iron and Steel Institute.
[c]Maximum value.
[d]Except as noted, all have a maximum of 0.040 wt % P and 0.030 wt % S.
[e]Optional.

Table 14. Super-Ferritic Grades of Stainless Steel

UNS[a] no.	Common name	CAS Registry Number	Composition, wt %						
			C[b]	Mn[b]	Si[b]	Cr	Mo	Ni	Other[c]
S44400	18-2-Ti	[54824-47-4]	0.025	1.00	1.00	17.5–19.5	1.75–2.50	1.0[b]	0.025 Nb[b], C + N <0.030 desirable, Ti + Nb = 0.20 + 0.4 (C + N) min = 0.80 max
S44626	26-1-Ti	[57971-21-8]	0.060	0.75	0.75	25.0–27.0	0.75–1.50	0.50[b]	0.040 Nb[b], 0.2 to 1.0 Ti typical [7 × (C + N) min], C + N = 0.050 typical, Cu = 0.20[b], S = 0.020[b]
	26-1-Cb	[73695-09-7]	0.005	1.00	1.00	25.5–27.5	0.6–1.4		Nb = 13–29 × N %
	28-2-4-Cb		0.015	1.00	1.00	26.0–30.0	1.75–2.25	3.6–4.4	0.035 Nb[b], C + N ≤0.040 with Nb ≥0.2 + 12 (C + N)%
S44700	29-4	[60005-36-9]	0.010	0.30	0.20	28.0–30.0	3.5–4.2		0.020 Nb[b], C + N ≤0.025, 0.15 Cu[b,d]
S44800	29-4-2	[65107-55-3]	0.010	0.30	0.20	28.0–30.0	3.5–4.2		0.020 Nb[b], C + N ≤0.025, 0.15 Cu[b,d]

[a]Unified Numbering System; Ref. 49.
[b]Maximum value.
[c]Except as noted, all have a maximum of 0.040 wt % P, 0.030 wt % S.
[d]Materials all have a maximum of 0.025 wt % P and 0.020 wt % S.

Table 15. Duplex (Ferrite + Austenite) Grades of Stainless Steel

UNS[a] no.	AISI[b] type or common name	CAS Registry Number	Composition, wt %					
			C[c]	Mn[c]	Si[c]	Cr	Ni	Other[d]
S31200	44LN	[73559-69-0]	0.030	2.00	1.00	24.0–26.0	5.50–6.50	1.20–2.00 Mo, 0.14–0.20N, 0.045 P[c]
S31260	DP-3	[61584-44-9]	0.030	1.00	0.75	24.0–26.0	5.50–7.50	2.50–3.50 Mo, 0.20–0.80 Cu, 0.10–0.30 N, 0.10–0.50 W
S31500	315 or 3RE60	[37270-63-6]	0.030	1.20–2.00	1.4–20	18.0–19.0	4.25–5.25	2.5–3.0 Mo, 0.030 P[c]
S31803	2205	[71631-40-8]	0.030	2.00	1.00	21.0–23.0	4.50–6.50	2.5–3.5 Mo, 0.08–0.20 N, 0.020 S, 0.030 P
S32404	Uranus 50	[56508-08-8]	0.040	2.00	1.00	20.5–22.5	5.5–8.5	2.0–3.0 Mo, 1.0–2.0 Cu, 0.20 N, 0.030 P, 0.010 S
S32550	Ferrallium 255	[74010-05-2]	0.040	1.50	1.00	24.0–27.0	4.50–4.60	2.0–4.0 Mo, 1.5–2.5 Cu, 0.10–0.25 N
S32900	329	[39369-78-3]	0.20	2.00	0.75	25.0–30.0	3.0–6.0	1.0–2.0 Mo

[a]Unified Numbering System.
[b]American Iron and Steel Institute.
[c]Maximum value unless otherwise noted.
[d]Except as noted, all have a maximum of 0.040 wt % P and 0.030 wt % S.

Table 16. Precipitation Hardenable Stainless Steels

UNS[a] no.	Common name	CAS Registry Number	Composition, wt %							
			C[b]	Mn[b]	Si[b]	Cr	Ni	Mo	Al	Other[c]
S13800	PH 13-8 Mo	[39344-65-5]	0.05	0.10	0.10	12.25–13.25	7.5–8.5	2.0–2.5	0.90–1.35	0.010 N, 0.01 P[b], 0.008 S[b]
S15500	15-5 PH	[39403-20-8]	0.07	1.0	1.0	14.5–15.5	3.5–5.5			2.4–4.5 Cu, Nb + Ta = 0.15–0.45
S17400	17-4 PH	[12611-80-2]	0.07	1.0	1.0	15.5–17.5	3.0–5.0			3.0–5.0 Cu, Nb + Ta = 0.15–0.45
S17700	17-7 PH	[12742-98-2]	0.09	1.0	1.0	16.0–18.0	6.5–7.75		0.75–1.5	0.040 S[b]
S15700	PH 15-7 Mo	[12742-99-3]	0.09	1.0	1.0	14.0–16.0	6.5–7.75	2.0–3.0	0.75–1.5	

[a]Unified Numbering System.
[b]Maximum value unless otherwise noted.
[c]Except as noted, all have a maximum of 0.040 wt % P and 0.030 wt % S.

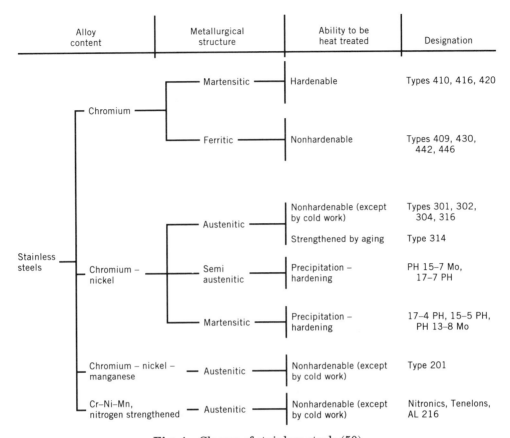

Fig. 4. Classes of stainless steels (50).

whereas the lower carbon grades are used for special tools, dies, and machine parts and equipment subject to combined abrasion and mild corrosion.

Leading grades of the so-called precipitation hardening (PH) steels are shown in Table 16. Hardened by precipitating a nickel-rich intermetallic compound in a martensite matrix, these steels are shipped either in the martensitic or austenitic condition. The austenite is transformed to martensite before the precipitation-aging heat treatment. Compared to standard stainless steels, especially the martensitic grades, the PH stainless steels offer high yield strength, some even above 1400 MPa (203,000 psi), good ductility, and corrosion resistance.

Although a minimum content of ~12 wt % Cr is required to impart the stainless characteristic to steels, much effort has been applied to develop new grades of stainless steel having significantly reduced chromium contents without unacceptable degradation of corrosion resistance and other properties. There has been some modest success in this endeavor (34,53–56).

Other Alloy Steels

In low alloy steels chromium contributes more to hardenability, tempering resistance, and toughness than to solid–solution hardening or oxidation resistance

Table 17. Cast Stainless Steels, Corrosion and Heat-Resistant Grades

UNS[a] no.	Common name	CAS Registry Number	Equivalent wrought-grade	Composition, wt %						
				C[b]	Mn[b]	Si[b]	Cr	Mo	Ni	Other[c]
				Corrosion resistant						
J91150	CA-15	[127570-06-3]	410	0.15	1.00	1.50	11.5–14.0	0.50[b]	1.0	
J92600	CF-8	[12750-50-4]	304	0.08	1.50	2.00	18.0–21.0		8.0–11.0	
J92900	CF-8M	[125561-71-9]	316	0.08	1.50	2.00	18.0–21.0	2.0–3.0	9.0–12.0	
J94202	CK-20	[132627-46-4]	310	0.20	2.00	2.00	23.0–27.0		19.0–22.0	
J95150	CN-7M	[12620-51-8]		0.07	1.50	1.50	19.0–22.0	2.0–3.0	17.5–30.5	3.0–4.0 Cu
J93404	Atlas 958	[110711-52-9]					25		7	0.25 N[b]
J92615	CC-50	[90351-80-7]	446	0.30	1.00	1.50	26.0–30.0	4.5	4.0	
				Heat resistant						
J92605	HC	[77062-65-8]		0.50	1.00	2.00	26.0–30.0		4[b]	
J93503	HH	[66677-80-3]		0.20–0.50	2.00	2.00	24.0–28.0		11.0–14.0	0.20 N[b]
J94224	HK	[53116-10-2]		0.20–0.60	2.00	2.00	24.0–28.0		18.0–22.0	
J94605	HT			0.35–0.75	2.00	2.50	13.0–17.0		33.0–37.0	

[a]Unified Numbering System.
[b]Maximum value unless otherwise noted.
[c]Except as noted, all have a maximum of 0.040 wt % P and 0.040 wt % S.

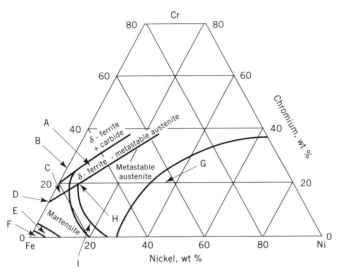

Fig. 5. Metastable Fe–Ni–Cr "ternary"-phase diagram where C content is 0.1 wt % and for alloys cooled rapidly from 1000°C showing the locations of austenitic, duplex, ferritic, and martensitic stainless steels with respect to the metastable-phase boundaries. For carbon contents higher than 0.1 wt %, martensite lines occur at lower alloy contents (43). A is duplex stainless steel, eg, Type 329, 327; B, ferritic stainless steels, eg, Type 446; C, δ ferrite + martensite; D, martensitic stainless steels, eg, Type 410; E, ferrite + martensite; F, ferrite + pearlite; G, high nickel alloys, eg, alloy 800; H, austenitic stainless steels, eg, Type 304; and I, martensite + metastable austenite.

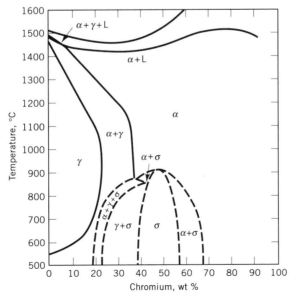

Fig. 6. Section of the Fe–Cr–Ni diagram at 8% nickel (5) where α, γ, and σ represent phases and L = liquid alloy.

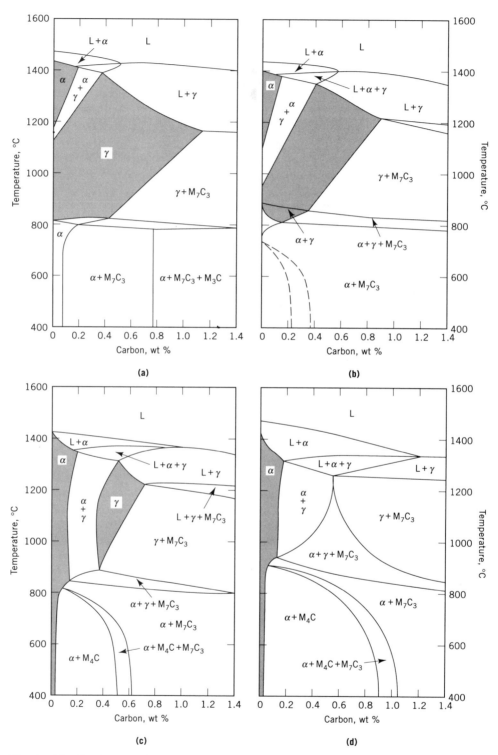

Fig. 7. Fe–Cr–C vertical sections (51,52) at chromium contents of (**a**) 8 wt %; (**b**) 12 wt %; (**c**) 15 wt %; and (**d**) 20 wt % where α and γ represent phases and L = liquid alloy.

(13,44). The marked effect of small additions of chromium on hardenability is shown in Figure 8. Whereas other alloying elements may show a similar or greater effect, chromium is one of the cheapest. The effect of chromium on resistance to tempering is shown in Figure 9. In the high chromium tool steel compositions chromium carbides contribute high abrasion resistance and improve the high hot hardness.

Wrought alloy steels, alloyed cast irons and steels, and tool steels account for 22–25% of the annual U.S. consumption of chromium. The wrought alloy steels are the largest category and are classified as shown in Table 18. In terms of chromium consumption, the most important classes are, in order, the CrMo, NiCrMo, and Cr steels. These may be further subdivided into carburizing and through-hardening grades. Such steels are extensively used in machinery, construction, and other structural work, and in machine parts such as bearings, gears, rolls, springs, and shafting. The AISI 500 series steels shown in Table 19, having higher chromium content, find application in the oil industry where resistance to corrosion and oxidation is especially important. Another class of Cr steels, closely related to the stainless and heat-resistant categories, are the valve steels that contain up to 23 wt % Cr and often have significant amounts of silicon.

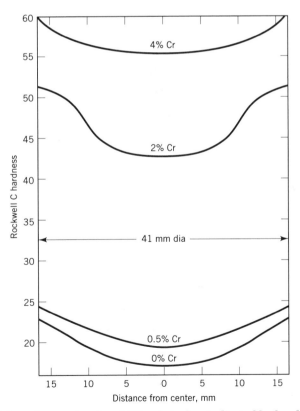

Fig. 8. Effect of chromium on hardenability of steel as indicated by hardness distribution across 41 mm rounds of oil-quenched 0.35% C steel (44).

Table 18. Basic Numbering System for Chromium-Bearing Low Alloy Steels

Alloy	AISI/SAE designation[a]	UNS[b] no.
nickel–chromium steels		
Ni 1.25; Cr 0.65	31XX	G31XXZ
Ni 3.50; Cr 1.57	33XX	G33XXZ
chromium–molybdenum steels		
Cr 0.50 and 0.95; Mo 0.25, 0.20, and 0.12	41XX	G41XXZ
nickel–chromium–molybdenum steels		
Ni 1.82; Cr 0.50 and 0.80; Mo 0.25	43XX	G43XXZ
Ni 1.05; Cr 0.45; Mo 0.20	47XX	G47XXZ
Ni 0.55; Cr 0.50 and 0.65; Mo 0.20	86XX	G86XXZ
Ni 0.55; Cr 0.50; Mo 0.25	87XX	G87XXZ
Ni 3.25; Cr 1.20; Mo 0.12	93XX	G93XXZ
Ni 1.00; Cr 0.80; Mo 0.25	98XX	G98XXZ
chromium steels		
Cr 0.27, 0.40, and 0.50	50XX	G50XXZ
Cr 0.80, 0.87, 0.92, 0.95, 1.00, and 1.05	51XX	G51XXZ
Cr 0.50	501XX	G509ZZ
Cr 1.02	511XX	G519ZZ
Cr 1.45	521XX	G529ZZ
chromium–vanadium steels		
Cr 0.80 and 0.95; V 0.10 and 0.15 (min)	61XX	G61XXZ
boron-treated chromium steels	XXBXX[c]	various

[a]SAE is the Society of Automotive Engineers. AISI is the American Iron and Steel Institute; the XX after the designation is left open for carbon content, thus 3110 equals a 0.10% carbon, nickel–chromium steel, and in this case XX = 0.10.
[b]UNS is the Unified Numbering System. The XX after the grade designation follows the AISI/SAE system and hence indicates carbon content; the Z is either an arbitrary number or 0. For a fuller description and tabulations of this new numbering system, see reference 49.
[c]The XX before the B is for the type numbers and after the B for carbon content.

Table 19. Composition of Heat-Resisting Chromium Steels

UNS[a] no.	AISI[b] type	CAS Registry Number	Composition, wt %						
			C	Mn[c]	Si[c]	P[c]	S[c]	Cr	Mo
S50100	501	[39367-01-6]	>0.10	1.00	1.00	0.04	0.03	4.0–6.0	0.40–0.65
S50200	502	[50946-74-2]	0.10[c]	1.00	1.00	0.04	0.03	4.0–6.0	0.40–0.65
S50300	503	[60746-70-5]	0.15[c]	1.00	1.00	0.04	0.04	6.0–8.0	0.45–0.65
S50400	504	[12604-41-0]	0.15[c]	1.00	1.00	0.04	0.04	8.0–10.0	0.90–1.10

[a]Unified Numbering System.
[b]American Iron and Steel Institute.
[c]Maximum value unless otherwise noted or range is given.

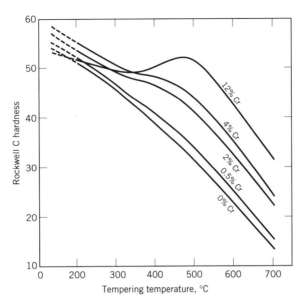

Fig. 9. Effect of chromium on tempering resistance of quenched 0.35% C steel (44).

A very different chromium-bearing alloy family is the Cr–Co–Fe magnetic alloys discovered in 1971. These alloys, containing about 30 wt % Cr, have magnetic properties comparable to the brittle Alnicos yet are cold formable. The magnetic properties are superior to the cobalt magnet steels, CuNiFe's, and Vicalloys. The most extensive application is as a cold-formed, cup-shaped magnet for telephone receivers (57).

Nonferrous Alloys

Nonferrous alloys account for only about 2 wt % of the total chromium used in the United States. Nonetheless, some of these applications are unique and constitute a vital role for chromium. For example, in high temperature materials, chromium in amounts of 15–30 wt % confers corrosion and oxidation resistance on the nickel-base and cobalt-base superalloys used in jet engines; the familiar electrical resistance heating elements are made of Ni–Cr alloy; and a variety of Fe–Ni and Ni-based alloys used in a diverse array of applications, especially for nuclear reactors, depend on chromium for oxidation and corrosion resistance. Evaporated, amorphous, thin-film resistors based on Ni–Cr with Al additions have the advantageous property of a near-zero temperature coefficient of resistance (58).

Chromium is a significant alloying addition in several other families of nonferrous alloys. In copper, chromium additions of less than 1 wt %, ie, wrought alloys, UNS C18200, C18400, and C18500, and cast alloys, UNS C81400 and C81500, and in cupro–nickels, additions from 0.5 to 2.8 wt % Cr, UNS C71900 and C72200, confer microstructural stability and workability while improving strength and maintaining relatively high conductivity. Similarly in aluminum alloys, chromium additions of less than 0.5 wt % contribute microstructural stability and some strengthening via $Al_{13}Cr_2$ precipitates in the 700 series of casting alloys and via $Al_{18}Mg_3Cr_2$ precipitates in the 5000 series of wrought alloys. Chromium contents of 2 to 11% are found in titanium sheet and forging alloys where the chromium acts as a beta stabilizer and strengthener.

Recovery and Reuse

As of this writing only about 20% of the chromium consumed in the United States is recycled and this comes largely from stainless steel scrap. Whereas much of the chromium formerly lost from plating operations is recovered, this does not amount to a significant tonnage. Improved recovery of the substantial chromium losses incurred in the past from refractory and foundry applications of chromite grain is being investigated (see RECYCLING).

Toxicity and Environmental Aspects

Chromium is generally recognized as being essential to human health, however, hexavalent chromium compounds are also known to be toxic, significantly more so than trivalent ones (59–65). Thus chromium releases into the environment are regulated by the Environmental Protection Agency (61,62), and workplace exposure is regulated by OSHA (64). The exposure limit set by OSHA for chromium metal is 1 mg/m^3 in an 8-h time-weighted average, and for chromic acid and chromates it is 0.1 mg/m^3, ceiling exposure limit. Much research is aimed at removing chromium from industrial waste solutions from the plating, etching, grinding, and tanning processes, not just to conserve chromium, but to protect the environment. The environmental and physiological aspects of chromium have been reviewed (65). The occurrence of chromium in the atmosphere, oceans, groundwater, and soils is treated as are the effects of chromium on algae, bacteria, fish, and humans.

BIBLIOGRAPHY

"Chromium and Chromium Alloys" in *ECT* 1st ed., Vol. 3, pp. 935–940, by J. J. Vetter, Diamond Alkali Co.; in *ECT* 2nd ed., Vol. 5, pp. 451–472, by F. E. Bacon, Union Carbide Corp.; in *ECT* 3rd ed., Vol. 6, pp. 54–82, by J. H. Westbrook, General Electric Co.

1. E. Matzat, and K. H. Shiraki, "Chromium," in K. H. Wedepohl and co-eds., *Handbook of Geochemistry*, Springer-Verlag, Heidelberg, Germany, 1974.
2. L. G. Liu, *Geochem. J.* **16**, 287–310 (1982).
3. J. O. Nriagu, *Adv. Environ. Sci. Tech.* **20**, 1–19 (1988).

4. A. B. Kinzel and W. Crafts, *The Alloys of Iron and Chromium*, Vol. 1., McGraw-Hill Book Co., New York, 1937.

5. A. B. Kinzel and R. Franks, *The Alloys of Iron and Chromium*, Vol. 2, McGraw-Hill Book Co., New York, 1940.

6. J. H. DeYoung, M. P. Lee, and B. R. Lipin, *International Strategic Minerals Inventory Summary Report-Chromium*, USGS Circular 930-B, 1984, 41 pp.

7. P. R. Grabfield, *Iron Steel Metall.*, 16–26 (Oct. 1975); 27–23 (Nov. 1975).

8. C. L. Rollinson, *The Chemistry of Chromium, Molybdenum, and Tungsten*, Vol. 21, Pergamon Press, Oxford, UK, 1975.

9. T. Narita, and T. Ishikawa, *Mat. Sci. Eng.* **87**, 51–61 (1987).

10. J. S. Dunning and S. C. Rhoads, *Effects of Al Additions on Sulfidation Resistance of Some Fe–Cr–Ni Alloys*, Bureau of Mines, 1989, 19 pp.

11. T. N. Irvine, ed., *Chromium: Its Physicochemical Behaviour and Petrologic Significance*, Carnegie Institution of Washington Conference, Geophysical Laboratory, Pergamon Press, Inc., Elmsford, N.Y., 1977.

12. *Gmelin Handbuch der Anorgansischen Chemie*, System-Nr. 52, Chrom (Chromium) Part A, 1, 1962, 438 pp.; Part A, 2, 1963, 330 pp. Verlag Chemie.

13. M. J. Udy, *Chromium, Metallurgy of Chromium and Its Alloys*, Vol. 2, Reinhold Publishing Corp., New York, 1956.

14. F. H. Perfect, *ASM Handbook*, 10th ed., Vol. 2, 1990, pp. 1107–1109.

15. *Ductile Chromium*, American Society for Metals, Cleveland, Ohio, 1957.

16. W. D. Klopp, in C. Sims and W. Hagel, eds., *The Superalloys*, John Wiley & Sons, Inc., New York, 1972, pp. 175–196.

17. C. White, in *Research and Development of High Temperature Materials for Industry*, Elsevier Applied Science Publishers, Ltd., Barking, UK, 1989, pp. 51–69.

18. M. J. Udy, *Chromium, Chemistry of Chromium and Its Compounds*, Vol. 1, Reinhold Publishing Corp., New York, 1956.

19. J. O. Nriagu, in *Chromium in the Natural and Human Environments*, J. O. Nriagu and E. Nieboer, eds. *Adv. Envir. Sci. Tech.* **20** 1–19 (1988).

20. J. F. Papp, *Minerals Yearbook*, U.S. Bureau of Mines, Oct. 1990, pp. 26.

21. B. Kjellberg, and S. E. Stenkvist, *DC Technology for Smelting Furnaces*, Paper in Proceedings of the 5th International Ferroalloys Congress, New Orleans, La. Apr. 23–26, 1989, pp. 88–94.

22. H. J. Kammeyer, K. U. Maske, and G. Pugh, *Open-Bath Production of Ferrochromium in a DC Plasma Furnace*, Paper in Proceedings of the 5th International Ferroalloys Congress, New Orleans, La., Apr. 23–26, 1989, Ferroalloys Association, 1989, pp. 95–102.

23. *Annual Book of Standards*, ASTM, Washington, D.C., 1990.

24. N. D. Veigel and co-workers. *Development of a Chromium–Thoria Alloy*, NASA Report. CR-72901, Mar. 10, 1971.

25. D. P. Shashkov, in J. L. Walter, J. H. Westbrook, and D. A. Woodford, eds., *Grain Boundaries in Engineering Materials*, Claitor's Publishing Division, Baton Rouge, La., 1975, pp. 657–668.

26. G. Dubpernell, *Electrodeposition of Chromium from Chromic Acid Solutions*, Pergamon Press, Oxford, 1977, 148 pp.

27. G. Dubpernell and F. A. Lowenheim, *Modern Electroplating*, 3rd ed., John Wiley & Sons, Inc., New York, 1974, pp. 87–151.

28. B. L. McKinney and C. L. Faust, *J. Electrochem. Soc.* **124**, 379C (1977).

29. R. M. Burns and W. W. Bradley, *Protective Coatings for Metals*, 3rd ed., American Chemical Society, Washington, D.C., 1975, pp. 243–255.

30. R. Winer, *Electrolytic Chromium Plating*, Finishing Publications Ltd., 1978.

31. J. P. Hoare, *Plating Surf. Finish.* **76**(9) 46–52, (1989).

32. G. Dubpernell, *Plating Surf. Finish.* **71**(6) 84–91, (1984).

33. G. H. Faber, in D. Coutsouradis and co-eds., *High Temperature Alloys in Gas Turbines*, Elsevier Applied Science Publishers, Ltd., Barking, UK, 1978, pp. 225–230.

34. C. J. McHargue, *J. Metals*, 30–36 (July 1983).

35. I. G. Wright, *Oxidation of Iron-, Nickel-, and Cobalt-Base Alloys*, MCIC Report 72-07, Battelle-Columbus Laboratories, Battelle Memorial Institute, 1972.

36. S. Gerardi, *ASM Metals Handbook*, 10th ed., Vol. 2, 1990, p. 565.

37. Roskill Information Services, *The Economics of Chromium*, 4th ed., Roskill Information Services, London, UK, 1982, 556 pp.

38. *Mineral Commodity Summaries 1990*, U.S. Bureau of Mines, Washington, D.C., 199 pp.

39. E. R. Parker, and co-workers, *Contingency Plans for Chromium Utilization*, NMAB Report 335, 1977 347 pp.

40. J. F. Papp, private communication, 1991.

41. A. Ankara, *Met. Technol.* **4**, 279 (1977).

42. M. A. Streicher, "Stainless Steels: Past, Present and Future," in *Stainless Steel '77*, Climax Molybdenum Corp., 1978.

43. M. O. Speidel, in R. W. Staehle and M. O. Speidel, eds., *Stress Corrosion of Austenitic Stainless Steels*, ARPA report, 1979.

44. H. W. Paxton, and E. D. Bain, *Alloying Elements in Steel*, 2nd ed., ASM, Cleveland, Ohio, 1966, 201 pp.

45. *Source Book on Stainless Steels*, American Society for Metals, Cleveland, Ohio, 1976, 408 pp.

46. D. Peckner and I. M. Bernstein, *Handbook of Stainless Steels*, McGraw-Hill Book Co., New York, 1977, 1200 pp.; *Miner. Yearb.* 1980 and 1984 ed., Chromium Mineral Industry Surveys, 1990.

47. F. B. Pickering, ed., *The Metallurgical Evolution of Stainless Steels*, American Society for Metals, Metals Park, Ohio, 1979.

48. R. A. Lula, ed., *Source Book on the Ferritic Stainless Steels*, American Society for Metals, Metals Park, Ohio, 1982.

49. *ARMCO Stainless Steels*, Armco Steel Corp., Middletown, Ohio, 1968.

50. *Unified Numbering System for Metals and Alloys*, SAE HE 1086 or ASTM DS-56A, 5th ed., 1989; Also available in electronic form on floppy disk.

51. W. Tofaute, A. Sponheuer, and H. Bennek, *Arch. Eisenhuttenwes.* **8**, 499 (1934).

52. W. Tofaute, C. Kuttner, and A. Buttinghaus, *Arch. Eisenhuttenwes.* **9**, 606 (1936).

53. C. J. Keith and V. K. Sharma, *Development of Chromium-Free Grades of Constructional Alloy Steels*, NTIS PB 83-243873, 1983, 60 pp.

54. J. J. Heger, *J. Test. Eval.* **14**(3), 160–162 (1986).

55. J. M. Oh, *J. Electrochem. Soc.* **135**, 749–755 (1988).

56. M. B. Cortie, *Mater. Soc.* **13**, pp. 13–31 (Nov. 1, 1989).

57. G. Y. Chin and J. H. Wernick, in M. Bever, ed., *Encyclopedia of Materials Science and Engineering*, Pergamon Press, Oxford, UK, 1986, pp. 2651–2653.

58. E. Schippel, *Thin Solid Films*, **146**(2), 133–138 (1987).

59. S. Langård, *Environmental and Biological Aspects of Chromium*, Elsevier, Amsterdam, The Netherlands, 1983.

60. D. Burrows, *Chromium: Metabolism and Toxicity*, CRC Press, Boca Raton, Fla., 1983, 184 pp.

61. *Health Effects Assessment for Trivalent Chromium*, Rep. EPA/540/1-86-035, U.S. Environment Protection Agency, Sept. 1984, 32 pp.

62. *Health Effects Assessment for Hexavalent Chromium*, Rep. EPA/540/1-86-019, U.S. Environment Protection Agency, Sept. 1984, 49 pp.

63. A. Yassi, and E. Nieboer, in J. O. Nriagu and E. Nieboer, eds., *Chromium in the Natural and Human Environments*, John Wiley & Sons, Inc., New York, 1988, Chapt. 17.

64. Occupational Safety and Health Administration (Dept. of Labor), *Air Contaminants* **54**(12), 2332–2983 (1989).

65. J. O. Nriagu and E. Nieboer, eds., Chromium in the Natural and Human Environments, in *Adv. Environ. Sci. Tech.* **20** (1988).

JACK H. WESTBROOK
Sci-Tech Knowledge Systems

CHROMIUM COMPOUNDS

The first chromium compound was discovered in the Ural mountains of Russia, during the latter half of the eighteenth century. Crocoite [*14654-05-8*], a natural lead chromate, found immediate and popular use as a pigment because of its beautiful, permanent orange-red color. However, this mineral was very rare, and just before the end of the same century, chromite was identified as a chrome bearing mineral and became the primary source of chromium [*7440-47-3*] and its compounds (1) (see CHROMIUM AND CHROMIUM ALLOYS).

Around 1800, the attack of chromite [*53293-42-8*] ore by lime and alkali carbonate oxidation was developed as an economic process for the production of chromate compounds, which were primarily used for the manufacture of pigments (qv). Other commercially developed uses were: the development of mordant dyeing using chromates in 1820, chrome tanning in 1828 (2), and chromium plating in 1926 (3) (see DYES AND DYE INTERMEDIATES; ELECTROPLATING; LEATHER). In 1824, the first chromyl compounds were synthesized followed by the discovery of chromous compounds 20 years later. Organochromium compounds were produced in 1919, and chromium carbonyl was made in 1927 (1,2).

Russia and the Republic of South Africa account for more than half the world's chromite ore production. Almost all of the world's known reserves of chromium are located in the southeastern region of the continent of Africa. South Africa has 84% and Zimbabwe 11% of these reserves. The United States is completely dependent on imports for all of its chromium (4). The chromite's constitution varies with the source of the ore, and this variance can be important to processing. Typical ores are from 20 to 26 wt % Cr, from 10 to 25 wt % Fe, from 5 to 15 wt % Mg, from 2 to 10 wt % Al, and between 0.5 and 5 wt % Si. Other elements that may be present are Mn, Ca, Ti, Ni, and V. All of these elements are normally reported as oxides; iron is present as both Fe(II) and Fe(III) (5,6).

Properties

Chromium compounds number in the thousands and display a wide variety of colors and forms. Examples of these compounds and the corresponding physical properties are given in Table 1. More detailed and complete information on sol-

Table 1. Physical Properties of Chromium Compounds[a]

Compound	CAS Registry Number	Formula	Appearance
chromium(0) hexacarbonyl	[13007-092-6]	$Cr(CO)_6$	colorless crystals
dibenzene chromium(0)	[1271-54-1]	$(C_6H_6)_2Cr$	brown crystals
bis(biphenyl) chromium(I) iodide	[12099-17-1]	$(C_{12}H_{10})_2CrI$	orange plates
chromium(II) acetate dihydrate	[628-52-4]	$Cr_2(C_2H_3O_2)_4\cdot2H_2O$	red crystals
chromium(II) chloride	[10049-05-5]	$CrCl_2$	white crystals
ammonium chromium(II) sulfate hexahydrate	[25638-51-1]	$(NH_4)_2Cr(SO_4)_2\cdot6H_2O$	blue crystals
chromium(III) chloride	[10025-73-7]	$CrCl_3$	bright purple plates
chromium(III) acetylacetonate	[13681-82-8]	$Cr(C_5H_7O_2)_3$	red-violet crystals
potassium chromium(III) sulfate dodecahydrate	[7788-99-0]	$KCr(SO_4)_2\cdot12H_2O$	deep purple crystals
chromium(III) chloride hexahydrate	[10060-12-5]	$[Cr(H_2O)_4Cl_2]Cl\cdot2H_2O$	bright green crystals
		$[Cr(H_2O)_6]Cl_3$	violet crystals
chromium(III) oxide	[1308-38-9]	Cr_2O_3	green powder or crystals
chromium(IV) oxide	[12018-01-8]	CrO_2	dark brown or black powder
chromium(IV) fluoride	[10049-11-3]	CrF_4	very dark greenish black powder
barium chromate(V)	[12345-14-1]	$Ba_3(CrO_4)_2$	black-green crystals
chromium(VI) oxide	[1333-82-0]	CrO_3	dark red crystals
chromium(VI) dioxide dichloride	[14977-61-8]	CrO_2Cl_2	cherry-red liquid
ammonium dichromate(VI)	[7789-09-5]	$(NH_4)_2Cr_2O_7$	red-orange crystals
potassium dichromate(VI)	[7778-50-9]	$K_2Cr_2O_7$	orange-red crystals
sodium dichromate(VI) dihydrate	[7789-12-0]	$Na_2Cr_2O_7\cdot2H_2O$	orange-red crystals
potassium chromate(VI)	[7789-00-6]	K_2CrO_4	yellow crystals
sodium chromate(VI)	[7775-11-3]	Na_2CrO_4	yellow crystals
potassium chlorochromate(VI)	[16037-50-6]	$KCrO_3Cl$	orange needles
silver chromate(VI)	[7784-01-2]	Ag_2CrO_4	maroon crystals
barium chromate(VI)	[10295-40-3]	$BaCrO_4$	pale yellow crystals
strontium chromate(VI)	[7789-06-2]	$SrCrO_4$	yellow crystals
lead chromate(VI)	[7758-97-6]	$PbCrO_4$	yellow crystals orange crystals red crystals

[a]Refs. 7–12. [b]Measurement taken at temperature in °C noted in subscript. [c]Explodes. [d]In vacuum. [e]Incongruent. [f]Loses all water at temperature indicated. [g]Calculated value.

Crystal system	Density[b] g/cm^3	Mp, °C	Bp, °C	Solubility
orthorhombic	1.77_{18}	148.5	210[c]	sl sol CCl_4; insol H_2O, $(C_2H_5)_2O$, C_2H_5OH, C_6H_6
cubic	1.519	284–285	sub 150[d]	insol H_2O; sol C_6H_6
	1.617_{16}	178	dec	sol C_2H_5OH, C_5H_5N
monoclinic	1.79			sl sol H_2O, C_2H_5OH; sol acids
tetragonal	2.88	815	1300	sol H_2O to blue soln, absorbs O_2
monoclinic				sol H_2O to blue soln, absorbs O_2
hexagonal	2.76_{15}	877	sub 947	insol H_2O; sol H_2O + Cr(II), Zn or Mg
monoclinic	1.34	216	340	insol H_2O; sol C_6H_6
cubic	1.826_{25}	89[e]	400[f]	sol H_2O
triclinic or monoclinic	1.835_{25}	95		sol H_2O, green soln turning green-violet
rhombohedral		90		sol H_2O, violet soln turning green-violet
rhombohedral	5.22_{25}	2330	3000	insol H_2O; sol hot 70% $HClO_4$ dec
tetragonal	4.98^g	dec	dec 300 to Cr_2O_3	sol acids with dec to Cr(III) and Cr(VI)
amorphous	2.89	ca 277	ca 400	sol H_2O, dec; insol organic solvents
same as $Ca_3(PO_4)_2$				sl dec H_2O; sol acids with dec to Cr(III) and Cr(VI)
orthorhombic	2.7_{25}	197	dec	v sol H_2O; sol CH_3COOH, $(CH_3CO)_2O$
	1.9145_{25}	−96.5	115.8	insol H_2O, hydrolyzes; sol CS_2, CCl_4
monoclinic	2.155_{25}	dec 180		sol H_2O
triclinic	2.676_{25}	398	dec 500	sol H_2O
monoclinic	2.348_{25}	356; 84.6[f]	dec 400	v sol H_2O
orthorhombic	2.732_{18}	975		sol H_2O
orthorhombic	2.723_{25}	792		sol H_2O
monoclinic	2.497_{39}	dec		sol H_2O, hydrolyzes
monoclinic	5.625_{25}			v sl sol H_2O; sol dilute acids
orthorhombic	4.498_{25}	dec		v sl sol H_2O; sol strong acids
monoclinic	3.895_{15}	dec		sl sol H_2O; sol dilute acids
orthorhombic				
monoclinic	6.12_{15}	844		insol H_2O; sol strong acids
tetragonal				

ubilities, including some solution freezing and boiling points, can be found in References 7–10, and 13. Data on the thermodynamic values for chromium compounds are found in References 7, 8, 10, and 13.

Chromium is able to use all of its $3d$ and $4s$ electrons to form chemical bonds. It can also display formal oxidation states ranging from Cr($-$II) to Cr(VI). The most common and thus most important oxidation states are Cr(II), Cr(III), and Cr(VI). Although most commercial applications have centered around Cr(VI) compounds, environmental concerns and regulations in the early 1990s suggest that Cr(III) may become increasingly important, especially where the use of Cr(VI) demands reduction and incorporation as Cr(III) in the product.

Preparation and chemistry of chromium compunds can be found in several standard reference books and advanced texts (7,11,12,14). Standard reduction potentials for select chromium species are given in Table 2 whereas Table 3 is a summary of hydrolysis, complex formation, or other equilibrium constants for oxidation states II, III, and VI.

Low Oxidation State Chromium Compounds. Cr(0) compounds are π-bonded complexes that require electron-rich donor species such as CO and C_6H_6 to stabilize the low oxidation state. A direct synthesis of $Cr(CO)_6$, from the metal and CO, is not possible. Normally, the preparation requires an anhydrous Cr(III) salt, a reducing agent, an arene compound, carbon monoxide that may or may not be under high pressure, and an inert atmosphere (see CARBONYLS).

$$CrCl_3 + 6\ CO \xrightarrow[\text{4°C, 101 kPa, H}_2\text{O}]{C_6H_5MgBr,\ (C_2H_5)_2O} Cr(CO)_6 + 3\ HCl \tag{1}$$

In equation 1, the Grignard reagent, C_6H_5MgBr, plays a dual role as reducing agent and the source of the arene compound (see GRIGNARD REACTION). The $Cr(CO)_6$ is recovered from an apparent phenyl chromium intermediate by the

Table 2. Standard Reduction Potentials for Chromium Species[a]

Half-cell reaction	$E°$, V
$Cr^{3+} + 3\ e^- \rightarrow Cr$	-0.74
$Cr(OH)^{2+} + H^+ + 3\ e^- \rightarrow Cr + H_2O$	-0.58[b]
$Cr^{2+} + 2\ e^- \rightarrow Cr$	-0.91
$Cr^{3+} + e^- \rightarrow Cr^{2+}$	-0.41
$Cr_2O_7^{2-} + 14\ H^+ + 6\ e^- \rightarrow 2\ Cr^{3+} + 7\ H_2O$	1.33
$Cr_2O_7^{2-} + 10\ H^+ + 6\ e^- \rightarrow 2\ Cr(OH)_2^+ + 3\ H_2O$	1.10[b]
$CrO_4^{2-} + 4\ H_2O + 3\ e^- \rightarrow Cr(OH)_3 + 5\ OH^-$	-0.13
$CrO_4^{2-} + e^- \rightarrow CrO_4^{3-}$	0.1[c]
$Cr^{6+} + e^- \rightarrow Cr^{5+}$	0.6[c,d]
$Cr^{5+} + e^- \rightarrow Cr^{4+}$	1.3[c,d]
$Cr^{4+} + e^- \rightarrow Cr^{3+}$	2.0[c,d]

[a]Ref. 12.
[b]Calculated from free energy data, Ref. 13.
[c]Ref. 13.
[d]In acid solutions.

Table 3. Hydrolysis, Equilibrium, and Complex Formation Constants

Reaction	$\log K$	Ref.
$Cr^{2+} + H_2O \rightleftharpoons Cr(OH)^+ + H^+$	5.3	15
$Cr^{3+} + H_2O \rightleftharpoons Cr(OH)^{2+} + H^+$	-4.2	16
$Cr^{3+} + 2\,H_2O \rightleftharpoons Cr(OH)_2^+ + 2\,H^+$	-10.4	16
$Cr^{3+} + 3\,H_2O \rightleftharpoons Cr(OH)_3 + 3\,H^+$	-18.7	16
$Cr^{3+} + 4\,H_2O \rightleftharpoons Cr(OH)_4^- + 4\,H^+$	-27.8	16
$2\,Cr^{3+} + 2\,H_2O \rightleftharpoons [Cr_2(OH)_2]^{4+} + 2\,H^+$	-5.3	16
$3\,Cr^{3+} + 4\,H_2O \rightleftharpoons [Cr_3(OH)_4]^{5+} + 4\,H^+$	-8.7	17
$4\,Cr^{3+} + 6\,H_2O \rightleftharpoons [Cr_4(OH)_6]^{6+} + 6\,H^+$	-13.9	17
$Cr^{3+} + 3\,C_2O_4^{2-} \rightleftharpoons [Cr(C_2O_4)_3]^{3-}$	15.4	16
$Cr^{3+} + H_2EDTA^{2-} \rightleftharpoons [CrEDTA]^- + 2\,H^+$	23.4^a	16
$Cr^{3+} + SO_4^{2-} \rightleftharpoons CrSO_4^+$	1.8	18
$H_2CrO_4 \rightleftharpoons HCrO_4^- + H^+$	0.61	12
$HCrO_4^- \rightleftharpoons CrO_4^{2-} + H^+$	-5.9	12
$2\,HCrO_4^- \rightleftharpoons Cr_2O_7^{2-} + H_2O$	2.2	12
$H_2CrO_4 + Cl^- \rightleftharpoons CrO_3Cl^- + H_2O$	1.0	13
$HCrO_4^- + HSO_4^- \rightleftharpoons CrSO_7^{2-} + H_2O$	0.60	13

^{a}H$_2$EDTA^{2-} is [(HOOCCH$_2$)$_2$ NCH$_2$CH$_2$N(CH$_2$COO)$_2$]$^{2-}$, dihydrogen ethylenediamine tetraacetate.

addition of water (19,20). Other routes to chromium hexacarbonyl are possible, and an excellent summary of chromium carbonyl and derivatives can be found in reference 2. The only access to the less stable $Cr(-II)$ and $Cr(-I)$ oxidation states is by reduction of $Cr(CO)_6$.

The preparation of disodium pentacarbonylchromide [51233-19-3], $Na_2[Cr(CO)_5]$, is performed in solvents such as liquid ammonia, diglyme, or tetrahydrofuran. The $Cr(0)$ in the $Cr(CO)_6$ solution is reduced to $Cr(-II)$ by the addition of Na, sodium amalgam, Li, Ca, or Ba. If $NaBH_4$ is used as the reducing agent, then the $Cr(-I)$ compound disodium decacarbonyldichromide [15616-67-8], $Na_2[Cr_2(CO)_{10}]$, is produced (21,22). The coordination number for chromium in the carbonyls and most of the derivatives is six, with octahedral geometry around the metal. However, the geometry of Cr in some organochromium(0) compounds, eg, $(C_6H_6)_2Cr$, is very different. In dibenzene chromium(0), the Cr atom is sandwiched between the two centers of high electron density provided by the benzene molecules. The π-orbitals of C_6H_6 donate electrons as the π^* orbitals simultaneously accept electrons. This back donation of electron density lowers the formal oxidation state of the metal (see ORGANOMETALLICS, METAL π-COMPLEXES).

The normal preparation of organochromium(0) compounds is indirect. First the organochromium(I) compound is formed

$$3\,CrCl_3 + AlCl_3 + 6\,C_6H_6 + 2\,Al \xrightarrow[C_6H_6]{AlCl_3} 3\,[(C_6H_6)_2Cr]^+ + 3\,AlCl_4^- \qquad (2)$$

then the salt is reduced using dithionite in the presence of base (23,24).

$$2\,[(C_6H_6)_2Cr]^+ + S_2O_4^{2-} + 4\,OH^- \longrightarrow 2\,(C_6H_6)_2Cr + 2\,SO_3^{2-} + 2\,H_2O \qquad (3)$$

The reductant of equation 3 can also be hypophosphite. Mixed organo–carbonyl compounds of Cr(0) and other oxidation states are also possible. These mixed compounds make the preparation of highly unstable chromium hydrides, eg, tricarbonyl(η^5-2,4-cyclopentadien-1-yl)hydrochromium [*36495-37-1*], $C_5H_5Cr(CO)_3H$, possible (25). Equation 2 represents a typical preparation for organochromium(I) compounds. The orange–yellow dibenzene chromium(I) cation forms sparingly soluble salts with large anions, eg, $B(C_6H_5)_4^-$.

Chromium(II) Compounds.   The Cr(II) salts of nonoxidizing mineral acids are prepared by the dissolution of pure electrolytic chromium metal in a deoxygenated solution of the acid. It is also possible to prepare the simple hydrated salts by reduction of oxygen-free, aqueous Cr(III) solutions using Zn or Zn amalgam, or electrolytically (2,7,12). These methods yield a solution of the blue $Cr(H_2O)_6^{2+}$ cation. The isolated salts are hydrates that are isomorphous with Fe^{2+} and Mg^{2+} compounds. Examples are chromous sulfate heptahydrate [*7789-05-1*], $CrSO_4 \cdot 7H_2O$, chromous chloride hexahydrate [*83082-80-8*], $CrCl_2 \cdot 6H_2O$, and $(NH_4)_2Cr(SO_4)_2 \cdot 6H_2O$.

The standard reduction potential of Cr^{2+} (Table 2) shows that this ion is a strong reducing agent, and Cr(II) compounds have been used as reagents in analytical chemistry procedures (26). The reduction potential also explains why Cr(II) compounds are unstable in aqueous solutions. In the presence of air, the oxidation to Cr(III) occurs by reaction with oxygen. However, Cr(II) also reacts with water in deoxygenated solutions, depending on acidity and the anion present, to produce H_2 and Cr(III) (27,28).

The anhydrous halides, chromium(II) fluoride [*10049-10-2*], CrF_2, chromium(II) bromide [*10049-25-9*], $CrBr_2$, chromium(II) chloride [*10049-05-5*], $CrCl_2$, and chromium(II) iodide [*13478-28-9*], CrI_2, are prepared by reaction of the hydrohalide and pure Cr metal at high temperatures, or anhydrous chromium(II) acetate [*15020-15-2*], $Cr_2(CH_3COO)_4$, at lower temperatures, or by hydrogen reduction of the Cr(III) halide at about 500–800°C (2,12). These halides generally display a coordination number of six, have a distorted octahedral geometry, are moisture sensitive, and are easily oxidized when exposed to humid air.

When organic acids, RCOOH, are added to aqueous Cr(II) solutions, compounds having the general formula $Cr_2(RCOO)_4L_2$ where L = H_2O are formed. The dimeric red molecules contain a quadruple Cr–Cr bond and are diamagnetic (29). They are stable in dry air but rapidly oxidize under humid conditions. Each Cr atom has a coordination number of six and an octahedral geometry. The $RCOO^{2-}$ anion serves to bridge the interpenetrating octahedra. Compounds containing quadruple Cr–Cr bonds and octahedral geometries are also obtained from XYZ ligands such as acetanilide [*103-84-4*], $CH_3CONHC_6H_5$, where the Y, in this case the carbonyl carbon, connects electron-rich centers X, the carbonyl oxygen, and Z, the nitrogen. The Cr–Cr bond is part of a five-membered ring system that has one Cr bonded to X and the other to Z (30,31).

Chromium(II) also forms sulfides and oxides. Chromium(II) oxide [*12018-00-7*], CrO, has two forms: a black pyrophoric powder produced from the action of nitric acid on chromium amalgam, and a hexagonal brown-red crystal made from reduction of Cr_2O_3 by hydrogen in molten sodium fluoride (32). Chromium(II) sulfide [*12018-06-3*], CrS, can be prepared upon heating equimolar quantities of pure Cr metal and pure S in a small, evacuated, sealed quartz tube at 1000°C for at least 24 hours. The reaction is not quantitative (33). The sulfide has a coordination number of six and displays a distorted octahedral geometry (34).

The Cr^{2+} ion is extensively hydrolyzed in aqueous solutions (Table 3) and is not easily complexed in this medium. However, many complexes, such as those of cyanide, bipyridine, phenanthroline, acetylacetone, and propylenediamine, have been prepared. The first three of these ligands form octahedral complexes, acetylacetone produces a square complex, and propylenediamine forms a trigonal bipyramidal complex (35).

Chromium(III) Compounds. Chromium(III) is the most stable and most important oxidation state of the element. The $E°$ values (Table 2) show that both the oxidation of Cr(II) to Cr(III) and the reduction of Cr(VI) to Cr(III) are favored in acidic aqueous solutions. The preparation of trivalent chromium compounds from either state presents few difficulties and does not require special conditions. In basic solutions, the oxidation of Cr(II) to Cr(III) is still favored. However, the oxidation of Cr(III) to Cr(VI) by oxidants such as peroxides and hypohalites occurs with ease. The preparation of Cr(III) from Cr(VI) in basic solutions requires the use of powerful reducing agents such as hydrazine, hydrosulfite, and borohydrides, but Fe(II), thiosulfate, and sugars can be employed in acid solution. Cr(III) compounds having identical counterions but very different chemical and physical properties can be produced by controlling the conditions of synthesis.

The anhydrous halides, chromium(III) fluoride [7788-97-8], CrF_3, chromium(III) chloride [10025-73-7], $CrCl_3$, chromium(III) bromide [10031-25-1], $CrBr_3$, and chromium(III) iodide [13569-75-0], CrI_3, can be made by the reaction of Cr metal and the corresponding halogen at elevated temperatures (12,36). Other methods of synthesis for the halides are also possible (36–38). All of the halides have a layer structure and contain Cr(III) in an octahedral geometry. They are only slightly soluble in water but dissolve slowly when Cr(II) or a reducing agent such as Zn or Mg is added.

An unusual crystal arrangement is exhibited by the isomorphous compounds $CrCl_3$ and CrI_3. The close-packed cubic array of Cl or I atoms has two-thirds of the octahedral holes between every other pair of chlorine or iodine planes filled with chromium atoms. Alternate layers of the halogen compounds are held together by van der Waals' forces (39,40).

The chemistry of Cr(III) in aqueous solution is coordination chemistry (see COORDINATION COMPOUNDS). It is dominated by the formation of kinetically inert, octahedral complexes. The bonding can be described by d^2sp^3 hybridization, and literally thousands of complexes have been prepared. The kinetic inertness results from the $3d^3$ electronic configuration of the Cr^{3+} ion (41). This type of orbital charge distribution makes ligand displacement and substitution reactions very slow and allows separation, persistence, and/or isolation of Cr(III) species under thermodynamically unstable conditions.

The simple hexaaquachromium(III) ion, $Cr(H_2O)_6^{3+}$, obtained from chromium(III) nitrate [26679-46-9], $Cr(NO_3)_3 \cdot 9H_2O$, crystals, or chromium(III) perchlorate [25013-81-4], $Cr(ClO_4)_3 \cdot 9H_2O$, is stable at room temperature. This violet ion displays strong dichroism, ie, solutions are blue by reflected light and reddish blue by transmitted light. If these solutions are heated, the color changes to green indicating hydrolysis (Table 3) and the formation of basic trivalent cations. A general equation for hydrolysis of the trivalent ion Cr^{3+} is

$$x\,Cr^{3+} + y\,H_2O \longrightarrow [Cr_x(OH)_y]^{(3x-y)+} + y\,H^+ \tag{4}$$

The term basic used with Cr^{3+} defines the hydroxyl ion's displacement of H_2O in the primary coordination sphere of Cr(III). This displacement effectively lowers the positive charge on the cation. Aqueous solutions of other Cr(III) salts that contain the hexaaqua ion also show some tendency by the anion of the salt to displace the coordinated water molecule, even without heating.

Figure 1 illustrates the complexity of the Cr(III) ion in aqueous solutions. The relative strength of anion displacement of H_2O for a select group of species follows the order perchlorate $\leq$ nitrate $<$ chloride $<$ sulfate $<$ formate $<$ acetate $<$ glycolate $<$ tartrate $<$ citrate $<$ oxalate (42). It is also possible for any anion of this series to displace the anion before it, ie, citrate can displace a coordinated tartrate or sulfate anion. These displacement reactions are kinetically slow, however, and several intermediate and combination species are possible before equilibrium is obtained.

The carboxylic acids or anions in the displacement series prevent the formation of basic complexes whenever present in large excess. This is not true of the acids of inorganic anions. Chromium(III) acetate [1066-30-4], $Cr(CH_3COO)_3$, is not isomorphous with the Cr(II) salt and shows no tendency to form Cr–Cr bonds. Rather, the structure depends on the ratio of acetate to Cr (43). The hydroxy carboxylates, eg, tartrate, may show bonding through both the alcohol and the carboxylic groups, yielding a cage-type structure (44).

The hydrolysis of Cr^{3+} (eq. 4) and the addition of less than equivalent amounts of hydroxide ion to aqueous solutions of Cr^{3+}, followed by aging, yields basic cationic polymers containing multiple chromium centers (45). The existence and formation of these polymers is of interest (17,46–49). Chromium polymers that consist of OH bridged octahedra linked via edges and/or faces, can be isolated by ion-exchange chromatography. The fraction of bridged polynuclear complexes present is proportional to the amount of hydroxide added (17,45).

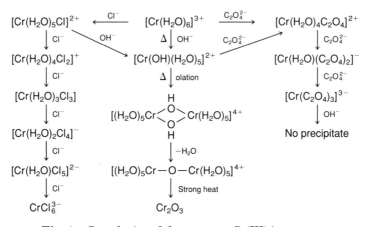

Fig. 1.   Complexity of the aqueous Cr(III) ion system.

When sufficient hydroxide is added to an aqueous solution of the trivalent chromium ion, the precipitation of a hydrous chromium(III) oxide, $Cr_2O_3 \cdot nH_2O$, of indefinite composition occurs. This compound is commonly misnamed as chromic or chromium(III) hydroxide [1308-14-1], $Cr(OH)_3$. A true hydroxide, chromium(III) hydroxide trihydrate [41646-40-6], $Cr(OH)_3 \cdot 3H_2O$, does exist and is prepared by the slow addition of alkali hydroxide to a cold aqueous solution of hexaaquachromium(III) ion (40). The fresh precipitate is amphoteric and dissolves in acid or in excess of hydroxide to form the metastable $Cr(OH)_4^-$. This ion decomposes upon heating to give the hydrous chromium(III) oxide. However, if the precipitate is allowed to age, it resists dissolution in excess hydroxide.

The trivalent chromium ion coordinates with almost all chelating agents (qv) and strong Lewis bases. Mixed ligand complexes of Cr(III) can yield stereoisomers. When the coordination number is 6, and J and Q are monodentate ligands, the formulas $[CrJ_4Q_2]^z$, $[CrJ_2Q_4]^z$, and $[CrJ_3Q_3]^z$, and the octahedral geometry of Cr^{3+}, allow for both cis and trans isomers. If only J is a bidentate ligand, then the formula $[CrJ_2Q_2]^z$ allows cis and trans isomers. However, if both J and Q are bidentate, then optical isomers are possible from the formulas $[CrJ_2Q]^z$, $[CrJQ_2]^z$, $[CrJ_3]^z$, and $[CrQ_3]^z$. The possible formation of binuclear complexes, eg, $[J_2Cr(OH)_2CrJ_2]^z$, allows for tartrate-type isomerism, ie, d, l, and $meso$. If the bidentate ligand is not symmetrical with respect to the coordination centers then structural isomers based on this asymmetry are also possible (50). The charge of the complex, z, depends on the charge of the ligands and can be positive, negative, or zero.

Chromium(III) oxide, Cr_2O_3, may be prepared by heating the hydrous chromium(III) oxide to completely remove water, as the final product of the calcination of chromium(VI) oxide, CrO_3, or by calcining chromium(III) salts that contain anions of volatile acids, eg, acetates. The Cr_2O_3 structure is isomorphous with α-alumina and α-Fe_2O_3. The best way to prepare pure chromium(III) oxide is by the decomposition of $(NH_4)_2Cr_2O_7$.

$$(NH_4)_2Cr_2O_7 \xrightarrow[200°C]{} Cr_2O_3 + N_2 + 4\,H_2O \qquad (5)$$

When Cr_2O_3 is introduced as an impurity into the α-Al_2O_3 lattice, as occurs in the semiprecious mineral ruby, the color is red rather than the normal green. This color anomaly is the result of ligand field splitting of the Cr(III) ion (51,52). Chromium(III) also colors other minerals (53).

Compounds that have the empirical formulas $MCrO_2$ and DCr_2O_4 where M is a monovalent and D a divalent cation, are known as chromites. These are actually mixed oxides and probably are better written as $M_2O \cdot Cr_2O_3$ and $DO \cdot Cr_2O_3$, respectively. The oxides of D are largely spinels, ie, the oxygen atoms define a close-packed cubic array having the octahedral holes occupied by the Cr(III) cation and the tetrahedral holes occupied by D (54). Chromite ore is an important member of this class of oxides.

Chromium(IV) and Chromium(V) Compounds. The formal oxidation states Cr(IV) and Cr(V) show some similarities. Both states are apparently intermediates in the reduction of Cr(VI) to Cr(III). Neither state exhibits a compound that has been isolated from aqueous media, and Cr(V) has only a transient existence in water (55). The majority of the stable compounds of both oxidation states contain either a halide, an oxide, or a mixture of these two. As of this writing, knowledge of the chemistry is limited.

Chromium(IV) fluoride [10049-11-3], CrF_4, and chromium(V) fluoride [14884-42-5], CrF_5, can be prepared by fluorinating Cr, CrF_3, or $CrCl_3$. The Cr(IV) compound is quite stable, but the fluoride of Cr(V) decomposes at 117°C and is easily hydrolyzed. The fluoride of Cr(IV) forms complexes of the type M_2CrF_6 and $DCrF_6$. These complexes are easily hydrolyzed. The K^+ salt, potassium hexafluorochromate(IV) [19652-00-7], K_2CrF_6, decomposes to potassium hexafluorochromate(III) [13822-82-7], K_3CrF_6, and CrF_5 when heated to 300°C. Although Cr(V) displays no other halides, chromium(IV) chloride [15597-88-3], $CrCl_4$, chromium(IV) bromide [51159-56-9], $CrBr_4$, and chromium(IV) iodide [23518-77-6], CrI_4, have been identified in the vapor phase of high temperature, high respective halogen vapor pressure systems.

The pure, crystalline chromium(V) oxide trifluoride can be prepared by the reaction of xenon(II) difluoride and chromium(VI) dioxide difluoride [7788-96-7], also known as chromyl fluoride, CrO_2F_2 (56):

$$XeF_2 + 2\ CrO_2F_2 \longrightarrow 2\ CrOF_3 + Xe + O_2 \qquad (6)$$

Other methods of preparation, eg, the reaction of ClF_3 or BrF_3 and CrO_3, yield the oxyfluoride contaminated with reactants and side reaction products. The crystal structure of $CrOF_3$ has been found to be an infinite three-dimensional array of corner shared $CrOF_5$ octahedra (57). The species $[CrOX_4]^-$, X = F, Cl, and Br, contains the oxochromium(V) cation [23411-25-8], CrO^{3+}, and the Cr exhibits a square pyramidal geometry. Compounds containing this cation are among the most stable Cr(V) compounds.

Chromium(IV) oxide [12018-01-8], CrO_2, is obtained from the hydrothermal decomposition of mixed oxides of Cr(III) and Cr(VI). A mixed oxide of the empirical formula Cr_xO_y, where the ratio of $2y$ to x is greater than 4 and less than 6, is heated at high pressure and in the presence of water to between 250 and 500°C (58). The resulting CrO_2 has an undistorted rutile structure, is ferromagnetic, and has metallic conductance. The chromium(V) oxide [12218-36-9], Cr_2O_5, is prepared by the thermal decomposition of CrO_3. It is always deficient in oxygen, giving an O to Cr mole ratio of about 2.4.

Both Cr(IV) and Cr(V) form mixed metal oxides. The blue-black and unstable tetrasodium chromate(IV) [50811-44-4], Na_4CrO_4, is formed when sodium chromite [12314-42-0], $NaCrO_2$, is heated in the presence of Na_2O to 1000°C. Compounds that have the formula D_2CrO_4 are prepared by heating the divalent metal's chromate(VI), the corresponding hydrate, and chromium(III) oxide to 1000°C. The emerald green, air-stable compounds of Sr and Ba contain the tetrahedral CrO_4^{4-} ions. Chromium(IV) mixed metal oxide species having the formulas $DCrO_3$, D_3CrO_5, and D_4CrO_6 are also known. The mixed metal oxides of chromium(V) are dark green, hygroscopic compounds of formula M_3CrO_4 or $D_3(CrO_4)_2$. The divalent cation's compound is prepared by heating a mixture of the divalent carbonate and the chromate in O_2 free nitrogen to temperatures at or above 1000°C. Both the monovalent and the divalent compounds contain the tetrahedral CrO_4^{3-} ion. Calcium chromate(V) [12205-18-4], $Ca_3(CrO_4)_2$, is isomorphous with

$Ca_3(PO_4)_2$, and the other metal chromate(V) compounds show some structural similarities to the phosphates (59,60). There is also a series of compounds that have the formula (RE)CrO_4, where RE = La, Pr, Nd, Y, and Sm through Lu (61).

Peroxy compounds of Cr(IV) and Cr(V) are known. The chromium(IV) diperoxide adduct with ammonia [7168-85-3], $Cr(O_2)_2 \cdot 3NH_3$, crystallizes as light brown needles that are unstable and may explode, if an ammoniacal solution of ammonium perchromate is heated to 50°C, and then cooled to 0°C (62). The crystals contain Cr(IV) having a coordination number of 7 and a pentagonal bipyramidal geometry (63). Potassium tetraperoxochromate(V) [12331-76-9], $K_3Cr(O_2)_4$, is obtained as stable red-brown crystals when H_2O_2 is added to a basic solution of K_2CrO_4 maintained at 0°C (64). The geometry of Cr(V) in these crystals is dodecahedral, and its coordination number is 8 (63).

Chromium(VI) Compounds. Virtually all Cr(VI) compounds contain a Cr–O unit. The chromium(VI) fluoride [13843-28-2], CrF_6, is the only binary Cr^{6+} halide known and the sole exception. This fluoride, prepared by fluorinating Cr at high temperature and pressure, easily disproportionates to CrF_5 and F_2 at normal pressures, even at −100°C. The fluorination of chromium(VI) oxide or the reaction of KrF_2 and CrO_2F_2 in liquid HF produces chromium(VI) oxide tetrafluoride [23276-90-6], $CrOF_4$ (65). Only fluorine displays an oxyhalide having this formula.

The other Cr(VI) halides have the formula CrO_2X_2, where X = F, Cl, or Br. The mixed oxyhalides CrO_2ClY, where Y = F or Br, have been prepared but are not well characterized (66). The formula CrO_2X_2 also describes nonhalide compounds, where X = ClO_4^-, NO_3^-, SO_3F^-, N_3^-, CH_3COO^-, etc (67). Compounds containing the theoretical cation CrO_2^{2+} are commonly named chromyl. All of the chromyl compounds are easily hydrolyzed to H_2CrO_4 and HX.

The primary Cr–O bonded species is chromium(VI) oxide, CrO_3, which is better known as chromic acid [1115-74-5], the commercial and common name. This compound also has the aliases chromic trioxide and chromic acid anhydride and shows some similarity to SO_3. The crystals consist of infinite chains of vertex-shared CrO_4 tetrahedra and are obtained as an orange-red precipitate from the addition of sulfuric acid to the potassium or sodium dichromate(VI). Completely dry CrO_3 is very dark red to red purple, but the compound is deliquescent and even traces of water give the normal ruby red color. Chromium(VI) oxide is a very powerful oxidizer and contact with oxidizable organic compounds may cause fires or explosions.

Chromium(VI) oxide dissolves in water to yield the theoretical H_2CrO_4, which is only superficially similar to H_2SO_4. The two acids are about the same size, and they both have a central atom that displays a formal oxidation state of VI and a tetrahedral geometry. However, H_2CrO_4 is a very weak acid compared to sulfuric acid; H_2CrO_4 is easily reduced, but sulfuric acid is very stable. Unlike H_2SO_4, the chromium(VI) acid cannot be isolated as a pure compound, and a $HCrO_4^-$ salt analogue of $NaHSO_4$ has not been prepared. The $HCrO_4^-$ ion shows a distinct tendency to dimerize to $Cr_2O_7^{2-}$ at low total Cr(VI) concentrations, but the corresponding anion $S_2O_7^{2-}$ has not been identified in dilute aqueous sulfuric acid solutions.

The hydrolysis equilibria for H_2CrO_4 given in Table 3 are only valid in HNO_3 or $HClO_4$ solutions. Other acids yield complexes such as those shown for chloride and bisulfate ions. The exact composition of chromate(VI) anion(s) present in aqueous solution is a function of both pH and hexavalent chromium concentration (68). However, at pH values above 8, virtually all the Cr(VI) is present as the CrO_4^{2-} anion. When the pH is between 2 and 6, an equilibrium mixture of $HCrO_4^-$ and $Cr_2O_7^{2-}$ is present; when the pH is below 1, the principal species is H_2CrO_4 (68,69). At very high Cr(VI) concentrations the polychromates $Cr_3O_{10}^{2-}$ and $Cr_4O_{13}^{2-}$ may be present, but this has not been confirmed. The salts of these ions, called trichromates and tetrachromates respectively, do exist (70).

When a warm solution of $K_2Cr_2O_7$ and HCl is allowed to cool, orange needles of $KCrO_3Cl$ precipitate. The fluoride, bromide, and iodide analogues can be prepared in a similar manner. The expected oxidation of the halides by Cr(VI) is kinetically hindered allowing for the formation of CrO_3Cl^-, CrO_3Br^-, and CrO_3I^- ions. All of these compounds display a distorted octahedral geometry, hydrolyze easily, and decompose if heated (71).

The chromate(VI) salts containing the tetrahedral CrO_4^{2-} ion are a very important class of Cr(VI) compounds. Only the alkali metal, ammonium ion, and magnesium chromates show considerable water-solubility. Some cations, eg, Ag^+, Ba^{2+}, and Pb^{2+}, are so insoluble that they precipitate from acidic Cr(VI) solutions, demonstrating the labile equilibria of H_2CrO_4. Salts of colorless cations generally have a pure yellow color, but there are some useful exceptions: silver chromate(VI) [7784-01-2], Ag_2CrO_4, is a maroon color and lead chromate(VI) [15804-54-3], $PbCrO_4$, displays colors that indicate its trimorphism. The stable form is monoclinic and has an orange-yellow color. An unstable tetragonal orange-red form is isomorphous with, and stabilized by, $PbMoO_4$. A second unstable yellow form is orthorhombic, isomorphous with and stabilized by $PbSO_4$. The diversity shown by the lead salt is the key to its versatility as a pigment.

The dichromate(VI) salts may be obtained by the addition of acid to the chromate(VI) salts. However, they are better prepared by adding one-half the acid equivalent of a metal hydrate, oxide, or carbonate to an aqueous solution of CrO_3, then removing the water and/or CO_2. Most dichromates(VI) are water-soluble, and the salts contain water(s) of hydration. However, the normal salts of K, Cs, and Rb are anhydrous. Dichromate(VI) compounds of the colorless cations are generally orange-red. The geometry of $Cr_2O_7^{2-}$ is described as two tetrahedral CrO_4 linked by the shared odd oxygen (72).

Chromate(VI) esters and salts of organic bases are known. The esters are generally very unstable, especially those of primary alcohols. The rapid formation of chromate(VI) esters is thought to be the first step in the Cr(VI) oxidation of alcohols and aldehydes (qv) 73–75 (see ALCOHOLS, HIGHER ALIPHATIC; ALCOHOLS, POLYHYDRIC). The adduct $CrO_3 \cdot 2L$ describes the formula of virtually all the organic base salts. Examples of organic bases for the adduct L are pyridine, picolines, lutidines, and quinoline. All organic chromates(VI) are photosensitive and decompose when exposed to light.

When hydrogen peroxide is added to an acid solution of Cr(VI), a deep blue color, indicating the formation of chromium(VI) oxide diperoxide [35262-77-2], $CrO(O_2)_2$, is observed. This compound is metastable and rapidly decomposes to Cr(III) and oxygen at room temperature. The reaction sequence is unique and can be used to qualitatively confirm the presence of Cr(VI). The $CrO(O_2)_2$ species can be extracted from the aqueous solution with ether and is stable in this solvent. If pyridine is added to the ether extract, then the oxodiperoxy-

(pyridine)chromium(VI) [*33361-75-0*], $C_5H_5N \cdot CrO(O_2)_2$, adduct is prepared. When the acid Cr(VI) solution is at 0°C or below, the green cationic species $Cr_2(O_2)^{4+}$ and $Cr_3(O_2)_2^{5+}$ are obtained. If H_2O_2 is added to a neutral or slightly acid solution of potassium, ammonium, or thallium dichromate, the blue-violet species $[CrO(O_2)_2OH]^-$ is formed. The salts of this anion are violently explosive (63).

Manufacture

The primary industrial compounds of chromium made directly from chromite ore are sodium chromate, sodium dichromate, and chromic acid. Secondary chromium compounds produced in quantity include potassium dichromate, potassium chromate, and ammonium dichromate. The secondary trivalent compounds manufactured in quantity are chrome acetate, chrome nitrate, basic chrome chloride, basic chrome sulfate, and chrome oxide.

Sodium Chromate, Dichromate, and Chromic Acid. The basic chemistry used to process chromite ore has not changed since the early nineteenth century. However, modern technologies have added many refinements to the manufacturing techniques (76,77), and plants have been adapted to meet health, safety, and environmental regulations. A generalized block flow diagram for the modern chromite ore processing plant is given in Figure 2. In the United States, chemical-grade ore from the Transvaal Region of The Republic of South Africa is employed. Historical procedures and equipment are discussed in Reference 78.

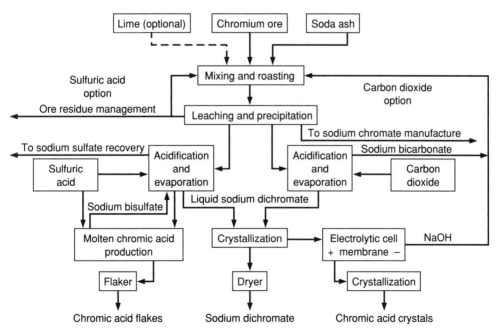

Fig. 2. Flow diagram for the production of sodium chromate, sodium dichromate, and chromic acid flake and crystals.

The chemical-grade ore, containing about 30% chromium, is dried, crushed, and ground in ball mills until at least 90% of its particles are less than 75 μm. It is then mixed with an excess of soda ash and, optionally, with lime and leached residue from a previous roasting operation. In American and European practice, a variety of kiln mixes have been used. Some older mixes contain up to 57 parts of lime per 100 parts of ore. However, in the 1990s manufacturers use no more than 10 parts of lime per 100 parts of the ore, and some use no lime at all (77). The roasting may be performed in one, two, or three stages, and there may be as much as three parts of leached residue per part of ore. These adaptations are responses to the variations in kiln roast and the capabilities of the furnaces used.

After thorough mixing, the mixture is roasted in a mechanical furnace, usually a rotary kiln. An oxidizing atmosphere is essential, and the basic reaction of a theoretical chromite is

$$4\,FeCr_2O_4 + 8\,Na_2CO_3 + 7\,O_2 \longrightarrow 2\,Fe_2O_3 + 8\,Na_2CrO_4 + 8\,CO_2 \qquad (7)$$

The temperature in the hottest part of the kiln is closely controlled using automatic equipment and a radiation pyrometer and generally is kept at about 1100–1150°C (see TEMPERATURE MEASUREMENT). Time of passage is about four hours, varying with the kiln mix being used. The rate of oxidation increases with temperature. However, the maximum temperature is limited by the tendency of the calcine to become sticky and form rings or balls in the kiln, by factors such as loss of Na_2O by volatilization, and by increased rate of attack on the refractory lining.

A gas-fired furnace with a revolving annular hearth also has been used to roast chrome ore (78). The mix is charged continuously at the outer edge of the hearth. A water-cooled helical screw moves it toward the inner edge where it is discharged. Mixes containing a much higher (28% Na_2CO_3) soda ash content can be handled in these furnaces. Also, the lower proportion of lime limits the formation of the suspected carcinogenic compound $Ca_3(CrO_4)_2$ (79).

Modern manufacturing processes quench the roast by continuous discharge into the leach water held in tanks equipped with agitators. At this point the pH of the leach solution is adjusted to between 8 and 9 to precipitate aluminum and silicon. The modern leaching operations are very rapid because no or little lime is used. After separation of the ore residue and precipitated impurities using rotary vacuum filters, the crude liquid sodium chromate may need to be treated to remove vanadium, if present, in a separate operation. The ore residue and precipitants are either recycled or treated to reduce hexavalent chromium to Cr(III) before disposal.

All stacks and vents attached to the process equipment must be protected to prevent environmental releases of hexavalent chromium. Electrostatic precipitators and baghouses are desirable on kiln and residue dryer stacks. Leaching operations should be hooded and stacks equipped with scrubbers (see AIR POLLUTION CONTROL METHODS). Recovered chromate values are returned to the leaching-water cycle.

Technical developments in the roasting and leaching area include refinements in pelletizing the mix fed to the kilns (80–82) and in the pre-oxidation of the ore prior to roasting (83). Both of these variants intend to increase the kiln capacity, the first through increasing the permissible fraction of soda ash in the mix, the second through increasing the effective rate of oxidation.

The neutralized, alumina-free sodium chromate solution may be marketed as a solution of 40° Bé (specific gravity = 1.38), evaporated to dryness, or crystallized to give a technical grade of sodium chromate or sodium chromate tetrahydrate [10034-82-9], $Na_2CrO_4 \cdot 4H_2O$. If the fuel for the kilns contains sulfur, the product contains sodium sulfate as an impurity. This compound is isomorphous with sodium chromate and hence difficult to separate. High purity sodium chromate must be made from purified sodium dichromate.

Sodium chromate can be converted to the dichromate by a continuous process treating with sulfuric acid, carbon dioxide, or a combination of these two (Fig. 2). Evaporation of the sodium dichromate liquor causes the precipitation of sodium sulfate and/or sodium bicarbonate, and these compounds are removed before the final sodium dichromate crystallization. The recovered sodium sulfate may be used for other purposes, and the sodium bicarbonate can replace some of the soda ash used for the roasting operation (76). The dichromate mother liquor may be returned to the evaporators, used to adjust the pH of the leach, or marketed, usually as 69% sodium dichromate solution.

Chromic acid may be produced by the reaction of sulfuric acid and sodium dichromate

$$Na_2Cr_2O_7 + 2\,H_2SO_4 \longrightarrow 2\,CrO_3 + 2\,NaHSO_4 + H_2O \tag{8}$$

This is the sulfuric acid option of Figure 2.

Traditionally, sodium dichromate dihydrate is mixed with 66° Bé (specific gravity = 1.84) sulfuric acid in a heavy-walled cast-iron or steel reactor. The mixture is heated externally, and the reactor is provided with a sweep agitator. Water is driven off and the hydrous bisulfate melts at about 160°C. As the temperature is slowly increased, the molten bisulfate provides an excellent heat-transfer medium for melting the chromic acid at 197°C without appreciable decomposition. As soon as the chromic acid melts, the agitator is stopped and the mixture separates into a heavy layer of molten chromic acid and a light layer of molten bisulfate. The chromic acid is tapped and flaked on water cooled rolls to produce the customary commercial form. The bisulfate contains dissolved CrO_3 and soluble and insoluble chromic sulfates. Environmental considerations dictate purification and return of the bisulfate to the treating operation.

Instead of the dihydrate and sulfuric acid, 20% oleum [8014-95-7] and anhydrous sodium dichromate may be used. In this case, the reaction requires little if any external heat, and liquid chromic acid is spontaneously produced. This procedure is the basis for a continuous process (84).

Molten chromic acid decomposes at its melting point at a significant rate. The lower oxides formed impart darkness and turbidity to the water solution. Accordingly, both temperature and time are important in obtaining a quality product.

Another process depends on the addition of a large excess of sulfuric acid to a concentrated solution or slurry of sodium dichromate. Under the proper conditions, a high purity chromic acid, may be precipitated and separated (77,85).

A newer technology for the manufacture of chromic acid uses ion-exchange (qv) membranes, similar to those used in the production of chlorine and caustic soda from brine (76) (see ALKALI AND CHLORINE PRODUCTS; CHEMICALS FROM BRINE; MEMBRANE TECHNOLOGY). Sodium dichromate crystals obtained from the carbon dioxide option of Figure 2 are redissolved and sent to the anolyte compartment of the electrolytic cell. Water is loaded into the catholyte compartment, and the ion-exchange membrane separates the catholyte from the anolyte (see ELECTROCHEMICAL PROCESSING).

When a potential is applied across the cell, the sodium and other cations are transported across the membrane to the catholyte compartment. Sodium hydroxide is formed in the catholyte compartment, because of the rise in pH caused by the reduction of water. Any polyvalent cations are precipitated and removed. The purified NaOH may be combined with the sodium bicarbonate from the sodium dichromate process to produce soda ash for the roasting operation. In the anolyte compartment, the pH falls because of the oxidation of water. The increase in acidity results in the formation of chromic acid. When an appropriate concentration of the acid is obtained, the liquid from the anolyte is sent to the crystallizer, the crystals are removed, and the mother liquor is recycled to the anolyte compartment of the cell. The electrolysis is not allowed to completely convert sodium dichromate to chromic acid (76). Patents have been granted for more electrolytic membrane processes for chromic acid and dichromates manufacture (86).

Other Chromates and Dichromates.   The wet operations employed in the modern manufacture of the chromates and dichromates are completely enclosed and all stacks and vents equipped with scrubbers and entrainment traps to prevent contamination of the plant and its environment. The continuous process equipment that is used greatly facilitates this task. The trapped material is recycled.

Potassium and ammonium dichromates are generally made from sodium dichromate by a crystallization process involving equivalent amounts of potassium chloride or ammonium sulfate. In each case the solubility relationships are favorable so that the desired dichromate can be separated on cooling, whereas the sodium chloride or sulfate crystallizes out on boiling. For certain uses, ammonium dichromate, which is low in alkali salts, is required. This special salt may be prepared by the addition of ammonia to an aqueous solution of chromic acid. Ammonium dichromate must be dried with care, because decomposition starts at 185°C and becomes violent and self-sustaining at slightly higher temperatures.

Potassium chromate is prepared by the reaction of potassium dichromate and potassium hydroxide. Sulfates are the most difficult impurity to remove, because potassium sulfate and potassium chromate are isomorphic.

Water-Soluble Trivalent Chromium Compounds.   Most water-soluble Cr(III) compounds are produced from the reduction of sodium dichromate or chromic acid solutions. This route is less expensive than dissolving pure chromium metal, it uses high quality raw materials that are readily available, and there is more processing flexibility. Finished products from this manufacturing method are marketed as crystals, powders, and liquid concentrates.

The general method of production for aqueous trivalent compounds involves dissolving a Cr(VI) source in an acid solution of the desired anion, eg, nitric acid, in a reactor constructed of acid-resistant materials. Next, the reducing agent is added at a controlled rate until the Cr(VI) has been reduced to Cr(III). For some reducing agents it is necessary to complete the reduction at boiling or under reflux conditions. A simplified, general flow diagram for this process is given in Figure 3.

The product use determines the Cr(VI) source and limits the choice of reducing agents. High purity trivalent chromium compounds are produced from chromic acid and a variety of reducing agents that yield either the anion needed or a minimum of side reaction products. When a clean product is not required and the presence of sodium does not affect the intended application, solutions of sodium dichromate are reduced using sugars, starches, and/or other materials. Sodium-free products employ chromic acid and the same reducing agents. The reduction of Cr(VI) with sugar can be written

$$4\ Cr_2O_7^{2-}\ +\ 32\ H^+\ +\ C_6H_{12}O_6\ \longrightarrow\ 8\ Cr^{3+}\ +\ 6\ CO_2\ +\ 22\ H_2O \tag{9}$$

Although equation 9 is written as a total oxidation of sugar, this outcome is never realized. There are many intermediate oxidation products possible. Also, the actual form of chromium produced is not as simple as that shown because of hydrolysis, polymerization, and anion penetration. Other reducing agents are chosen to enhance the performance of the product.

The final consideration for the manufacture of Cr(III) compounds is the mole ratio of acid to Cr. This ratio determines the basicity value of the product. Basicity can also be stated as the amount of positive charge on chromium(III) neutralized by hydroxide. For example, Cr^{3+} is 0% basic, $Cr(OH)^{2+}$ is 33.3% basic, and $Cr_2(OH)_3^{3+}$ is 50% basic. The basicity value can vary continuously from 0% to 100%. It is unlikely that these formulas represent actual cationic species, but are

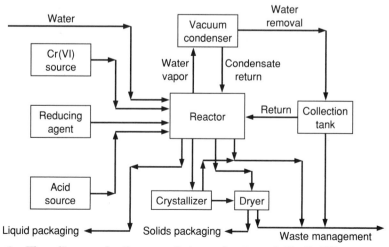

Fig. 3. Flow diagram for the manufacture of water-soluble Cr(III) compounds.

rather simplistic images of the average charge. These formulas can be used, however, to determine the mole ratio of acid needed for manufacture of the compound. For a monovalent anion, the mole ratio of acid to Cr for 0% basicity is 3, for 33.3% basicity it is 2, and for 50% basicity it is 1.5.

Basic chrome sulfate [12336-95-7], $Cr(OH)SO_4$, is manufactured as a proprietary product under various trade names for use in leather tanning. It is generally made by reduction of sodium dichromate in the presence of sulfuric acid, and contains sodium sulfate, small amounts of organic acids if carbohydrate reducing agents are used, plus various additives. When sulfur dioxide is employed as the reducing agent, a 33.3% basic chromic sulfate is automatically obtained

$$Na_2Cr_2O_7 + 3\ SO_2 + H_2O \longrightarrow 2\ Cr(OH)SO_4 + Na_2SO_4 \tag{10}$$

Pure sulfur dioxide is bubbled through the sodium dichromate solution in an acid-resistant tank, or sulfur burner gas is passed through a ceramic-packed tower countercurrent to descending dichromate solution. After reduction is complete, steam is bubbled through the solution to decompose any dithionate that may have formed, and to remove excess sulfur dioxide. Also, after reduction any desired additives, such as aluminum sulfate, are incorporated, and the solution is aged. It is then spray dried. Careful temperature control during drying is necessary to obtain a highly water-soluble, solid product.

The compounds are sold on a specification of chromic oxide content, 20.5–25% Cr_2O_3, and basicity, 30–58%. Solutions are also available.

Economic Aspects

In 1989, the total chrome ore consumption in the United States, including chemical, metallurgical (the principal use), and refractory grades, was 560,711 metric tons, having an average concentration of 42.6% Cr_2O_3. This quantity was the highest amount consumed for the years 1985 to 1989. The low point for this span occurred in 1986, when consumption was only 387,584 metric tons of a mean 40.2% Cr_2O_3. The Republic of South Africa was the largest U.S. supplier in 1989, contributing 70% of the total imports, and Turkey was second having about 19.7%. The world production of chromite in 1989 was 11,901,300 metric tons, down from the 1988 level of 12,166,910 metric tons (87).

The prices of some important chromium chemicals are given in Table 4, and production and shipment data for sodium chromate and dichromate are given in Table 5. Data for the production and shipment of chromic acid have not been available since 1972. However, traditionally CrO_3 has held at about 30–35% of sodium dichromate production. The estimated capacity for domestic production of sodium dichromate is 150,000 to 200,000 t/yr.

Table 4. 1991 Prices of Commercial Chromium Compounds[a]

Commercial name	CAS Registry Number	Molecular formula	$/kg
ammonium bichromate	[7789-09-5]	$(NH_4)_2Cr_2O_7$	4.40
chromic acid flake	[1333-82-0]	CrO_3	2.75
chrome fluoride	[7788-97-8]	CrF_3	5.83
chrome nitrate crystal	[26679-46-9]	$Cr(NO_3)_3 \cdot 9H_2O$	3.19
chrome oxide	[1308-38-9]	Cr_2O_3	4.29
potassium dichromate	[7778-50-9]	$K_2Cr_2O_7$	2.42
potassium chromate	[7789-00-6]	K_2CrO_4	1.25
sodium dichromate	[7789-12-0]	$Na_2Cr_2O_7 \cdot 2H_2O$	1.32
sodium chromate	[7775-11-3]	Na_2CrO_4	1.54

[a]Ref. 88.

Table 5. Production and Shipments of Sodium Dichromate and Hydrous Chromate, t[a]

Year	Total production	Shipments
1978	160,020	72,737
1980	140,142	91,926
1982	111,601	64,000[b]
1984	130,372	58,529
1988	104,000[b]	49,051

[a]Ref. 89.
[b]Estimated.

Specifications and Shipment

Chromates and dichromates are sold in both technical and reagent grades (90,91). Chlorides and sulfates are the principal impurities. Both manufacturers' and U.S. General Services Administration (GSA) specifications exist for the technical grades (92,93) and there are also producer specifications available for some trivalent chromium compounds (94). Specifications are shown in Tables 6 and 7.

Sodium dichromate, sodium chromate, and mixtures thereof are shipped as concentrated solutions in tank cars and trucks. The chloride and sulfate contents are usually somewhat higher than in the crystalline product. Sodium dichromate is customarily shipped at a concentration of 69% $Na_2Cr_2O_7 \cdot 2H_2O$, which is close to the eutectic composition freezing at $-48.2°C$.

Chromic acid is transported in steel drums and by rail in tank cars. Multiwall paper bags, fiber drums, as well as steel drums can be employed to ship the solid chromate salts, dichromate salts, and trivalent compounds. Trivalent chromium liquid concentrates are also available in polyethylene drums. The U.S. Department of Transportation (DOT) requires all packages having a capacity ≤416.4L (110 gallons) to be marked with the proper shipping name and identifi-

Table 6. U.S. Government Specifications for Chromium(VI) Compounds[a]

Specification	$Na_2Cr_2O_7 \cdot 2H_2O$	CrO_3[b]	Na_2CrO_4	$K_2Cr_2O_7$[c]	$(NH_4)_2Cr_2O_7$[d]
number	O-S-595B	O-C-303D	O-S-588C	O-P-559	O-A-498B
assay,[e] %	99.0[f]	99.5	98.5	99.0	99.7
Cl^- wt %[g]	0.1	0.1	0.1	0.1	0.005
SO_4^{2-}, wt %[g]	0.2	0.2	1.0	0.1	0.06
H_2O insol., wt %[g]	0.2	0.1		0.1	0.02
LOD[h] at 120°C, %[g]	12.5		0.5	0.2	

[a]Ref. 92.
[b]No more than 30% may pass a 600 μm (30 mesh) U.S. Sieve screen.
[c]All must pass 2000 μm (10 mesh) U.S. Sieve screen, and no more than 25% may pass a 149 μm (100 mesh) U.S. Sieve screen.
[d]Contains 13.5% NH_3 minimally. The pH of a 20% w/v soln is 3.2.
[e]Minimum value given.
[f]Actually % $Na_2Cr_2O_7$ after drying at 120°C.
[g]Maximum value given.
[h]LOD = loss on drying.

Table 7. Specifications for Trivalent Chromium Compounds[a]

Specification	Chrome alum	Basic chrome chloride	Chrome acetate
formula	$KCr(SO_4) \cdot 12H_2O$	$Cr_5(OH)_6Cl_9 \cdot xH_2O$[b]	$Cr(C_2H_3O_2)_3$
form	violet crystals	green powder	green liquid
Cr, wt %	10.2–10.6	29.0–33.0	11.2–11.8
basicity, %		33.0–43.0	−5.0–5.0
Cl^-, wt %		33.0–39.0	<0.05
SO_4^{2-}, wt %			<0.05
Fe, wt %	<0.01		
Cu, wt %	<0.001		
Pb, wt %	<0.005		
H_2O insolubles, wt %	<0.1	<0.25	

[a]Ref. 94.
[b]Where $8.8 \le x \le 12$.

cation number of the chemical contained. The Occupational Safety and Health Administration (OSHA) requires all compounds containing chromium to be labeled as hazardous and all Cr(VI) compounds are required to contain an additional cancer hazard warning.

Analytical and Test Methods

The classical wet-chemical qualitative identification of chromium is accomplished by the intense red-violet color that develops when aqueous Cr(VI) reacts with (S)-diphenylcarbazide under acidic conditions (95). This test is sensitive to 0.003

ppm Cr, and the reagent is also useful for quantitative analysis of trace quantities of Cr (96). Instrumental qualitative identification is possible using inductively coupled argon plasma–atomic emission spectroscopy (icap/aes) having a sensitivity of <10 ppb using the 205.552 nm line; using optical atomic emission spectroscopy (oaes) having an arc sensitivity of 1 ppm for the persistent emission line at 425.43 nm; and using neutron activation analysis (naa) having a sensitivity of < 0.5 microgram per sample.

The methods for quantitative analysis of chromium are dependent on the concentration and the nature of the chromium containing material. There are two types of samples: water-soluble and water-insoluble. The insoluble samples, eg, ores, refractories (qv), some organochromium compounds, and some pigments, need to be converted to water-soluble forms before analyzing. This can be accomplished by fusion using sodium peroxide or dissolution in an oxidizing acid mixture. Information on preparing chromium containing samples for analysis is available (93,97–100).

Wet-Chemical Determinations. Both water-soluble and prepared insoluble samples must be treated to ensure that all the chromium is present as Cr(VI). For water-soluble Cr(III) compounds, the oxidation is easily accomplished using dilute sodium hydroxide, dilute hydrogen peroxide, and heat. Any excess peroxide can be destroyed by adding a Ni^{2+} catalyst and boiling the alkaline solution for a short time (101). Appropriate aliquot portions of the samples are acidified and chromium is found by titration either using a standard ferrous solution or a standard thiosulfate solution after addition of potassium iodide to generate an iodine equivalent. The ferrous endpoint is found either potentiometrically or by visual indicators, such as ferroin, a complex of iron(II) and o-phenanthroline, and the thiosulfate endpoint is ascertained using starch as an indicator.

To determine moderate amounts of Cr(III) and Cr(VI) in samples that have both oxidation states present, Cr(VI) is analyzed by direct titration in one sample, and the total chromium is found in a second sample after oxidation of the Cr(III). The Cr(III) concentration is determined as the difference. Trace quantities of Cr(VI) in Cr(III) compounds can be detected and analyzed by (S)-diphenylcarbazide. Trace quantities of Cr(III) in Cr(VI) may be detected and analyzed either photometrically (102) or by ion chromatography using various modes of detection (103).

Instrumental Quantitative Analysis. Methods such as x-ray spectroscopy, oaes, and naa do not necessarily require pretreatment of samples to soluble forms. Only reliable and verified standards are needed. Other instrumental methods that can be used to determine a wide range of chromium concentrations are atomic absorption spectoscopy (aas), flame photometry, icap-aes, and direct current plasma–atomic emission spectroscopy (dcp-aes). These methods cannot distinguish the oxidation states of chromium, and speciation at trace levels usually requires a previous wet-chemical separation. However, the instrumental methods are preferred over (S)-diphenylcarbazide for trace chromium concentrations, because of the difficulty of oxidizing very small quantities of Cr(III).

Impurities in industrial chromium compounds include chloride, sulfate, insoluble matter, and trace metals. The chloride ion-selective electrode can be used to determine chloride values; sulfate is found by barium precipitation, after the reduction of Cr(VI) to Cr(III) for hexavalent chromium compounds; and a variety of methods are available to determine the specified trace metals, eg, aas and icap-aes. The standard methods of organizations and agencies such as the American Society for Testing and Materials (ASTM), the American Wood Preservers' Association, the General Services Administration (GSA), and the American Leather Chemists' Association (ALCA) contain procedures for the analysis of commercially available chromium formulations, pigments, and compounds. A wider variety of tests are required for reagent chemicals (90,91).

Health and Safety Factors

Acute and Chronic Toxicity. Although chromium displays nine oxidation states, the low oxidation state compounds, -II to I, all require special conditions for existence and have very short lifetimes in a normal environment. This is also true for most organochromium compounds, ie, compounds containing Cr–C bonds. Chromium compounds that exhibit stability under the usual ambient conditions are limited to oxidation states II, III, IV, V, and VI. Only Cr(III) and Cr(VI) compounds are produced in large quantities and are accessible to most of the population. Therefore, the toxicology of chromium compounds has been historically limited to these two states, and virtually all of the available information is about compounds of Cr(III) and/or Cr(VI) (59,104). However, there is some indication that Cr(V) may play a role in chromium toxicity (59,105–107). Reference 104 provides an overview and summary of the environmental, biological, and medical effects of chromium and chromium compounds as of the late 1980s.

The primary routes of entry for animal exposure to chromium compounds are inhalation, ingestion, and, for hexavalent compounds, skin penetration. This last route is more important in industrial exposures. Most hexavalent chromium compounds are readily absorbed, are more soluble than trivalent chromium in the pH range 5 to 7, and react with cell membranes. Although hexavalent compounds are more toxic than those of Cr(III), an overexposure to compounds of either oxidation state may lead to inflammation and irritation of the eyes, skin, and the mucous membranes associated with the respiratory and gastrointestinal tracts. Skin ulcers and perforations of nasal septa have been observed in some industrial workers after prolonged exposure to certain hexavalent chromium compounds (108–110), ie, to chromic acid mist or sodium and potassium dichromate.

Acute systemic poisoning by chromium compounds is rare (108), and only hexavalent compounds have been implicated. It has been suggested that the principal routes of exposure allow for a detoxification by reduction of Cr(VI) to Cr(III) by the body's sulfur-containing proteins, eg, glutathione, or the oxidizable compounds contained in the gastric juices and saliva (105,111,112). The target organ for acute systemic toxicity is the kidney. Usually, poisoning by Cr(VI) results in acute tubular necrosis of the kidney, the reported cause of death (108). The lethality of sodium chromate(VI), and the dichromates of sodium, potassium, and ammonium by inhalation (LC_{50} = ca 120 mg/m^3), by ingestion (LD_{50} = ca 54 mg/kg), and by dermal exposures (LD_{50} = ca 1.3 g/kg) have been determined for Fischer rats and New Zealand rabbits (113). The ingestion results were shown to be dependent on the chromate concentration at the time of exposure, indicating that higher concentrations lower the LD_{50} (113).

Prolonged contact with certain chromium compounds may produce allergic reactions and dermatitis in some individuals (114). The initial response is usually caused by exposure to Cr(VI) compounds, but once the allergy is established, it is extended to the trivalent compounds (111,115). There is also limited evidence of possible chromium associated occupational asthma, but there is insufficient data to estimate a dose for assumed chromium-induced asthma. Reference 116 provides a summary and discussion of chromium hypersensitivity.

Reproductive Toxicity. No data are available that implicate either hexavalent or trivalent chromium compounds as reproductive toxins, unless exposure is by way of injection. The observed teratogenic effects of sodium dichromate(VI), chromic acid, and chromium(III) chloride, administered by injection, as measured by dose-response relationships are close to the amount that would be lethal to the embryo, a common trait of many compounds (111). Reported teratogenic studies on hamsters (117,118), the mouse (119–121), and rabbits (122) have shown increased incidence of cleft palate, no effect, and testicular degeneration, respectively. Although the exposures for these experiments were provided by injections, in the final study (122) oral, inhalation, and dermal routes were also tried, and no testicular degeneration was found by these paths.

Carcinogenicity, Mutagenicity, and Genotoxicity. There is evidence that hexavalent chromium may be a carcinogen, but there is some doubt about which Cr(VI) compounds are implicated (111,123,124). The National Institute for Occupational Safety and Health (NIOSH) has classified the chromate and dichromate salts of lithium, potassium, ammonium, rubidium, cesium, and hydrogen, plus chromic acid, as noncarcinogenic Cr(VI) compounds (125). NIOSH considers all of the other Cr(VI) compounds carcinogenic (125). Manufacturing processing practices have reduced worker exposure to hexavalent chromium. No carcinogenic potency has been demonstrated for Cr(III) compounds (111,123,126,127).

The key to hexavalent chromium's mutagenicity and possible carcinogenicity is the ability of this oxidation state to penetrate the cell membrane. The Cr(VI) species promotes DNA strand breaks and initiates DNA–DNA and DNA-protein cross-links both in cell cultures and *in vivo* (105,112,128–130). The mechanism of this genotoxic interaction may be the interceullar reduction of Cr(VI) in close proximity to the nuclear membrane. When *in vitro* reductions of hexavalent chromium are performed by glutathione, the formation of Cr(V) and glutathione thiyl radicals are observed, and these are believed to be responsible for the formation of the DNA cross-links (112).

The relationship of chromium's oxidation state and the mechanism(s) that may lead to carcinogenicity are still ill defined, especially with regard to the body's defenses and hexavalent chromium's detoxification (130). Although the lack of genotoxic effects of Cr(III) has been demonstrated for normal exposure routes, there are reports of DNA interactions under certain specific conditions (112). Hexavalent chromium compounds have been intensely studied and are generally found to be mutagens as well as chromosome aberrants. Although the number of studies are small compared to Cr(VI), trivalent compounds routinely yield negative results as a mutagen or genotoxin. As of the late 1980s almost 90% of the published studies on Cr(VI) indicated positive mutagenic effects, but only 25% of the studies suggested that Cr(III) compounds were mutagens (131). The same review reported 96% positive genotoxic effects for Cr(VI) and 26% for Cr(III) compounds. The results obtained from studies of Cr(III) compounds may be inconclusive because some of the compounds used may have been contaminated with Cr(VI) (132–134).

Nutrition. Chromium, in the trivalent oxidation state, is recognized as an essential trace element for human nutrition, and the recommended daily intake is 50 to 200 micrograms (135). The transport of glucose via insulin's reaction with the cell membrane, a necessary mechanism of glucose metabolism, appears to be mediated by chromium (136,137). Increased coronary disease risk, glucose intolerance, elevated serum cholesterol and elevated insulin levels have been linked to chromium deficiency (138–141). Evidence is available that suggests dietary supplements of chromium(III) may improve glucose tolerance, and there is some indication that a correlation may exist between diabetes and chromium deficient diets (142,143).

Chromium Exposure Levels and U.S. Government Regulations. The level of exposure to chromium compounds for employees in industry and for the general population via waste disposal and industrial emissions is the subject of much regulation, research, and controversy. Some U.S. Government regulations, such as the Comprehensive Environmental Response, Compensation, and Liability Act (CERCLA), also known as the Superfund Act, make no distinction as to the oxidation state of chromium (144). However, there is valence distinction in other regulations.

The two categories of regulatory concern are the workplace and the environment. The latter concern is directed toward industrial emissions in the air, water, and land. Local and state as well as federal regulations have been promulgated for these areas. In addition, nonregulatory organizations, such as the American Conference of Governmental Industrial Hygienists (ACGIH), the International Agency for Research on Cancer (IARC), and the Agency for Toxic Substances and Diseases Registry (ATSDR), have published standards for exposure to chromium and chromium compounds.

Workplace. The Occupational Safety and Health Administration (OHSA) has established workplace permissible exposure limits (PEL) for chromium metal and three forms of chromium compounds. OSHA's PEL for chromic acid and chromates is 0.1 mg/m^3 CrO_3 as both a ceiling, ie, no exposure above this concentration is allowed, and an 8-h time-weighted average (TWA). Chromium metal and insoluble chromium salts have an 8-h TWA PEL of 1.0 mg/m^3 Cr, and the same standard is 0.5 mg/m^3 Cr for soluble Cr(III) and Cr(II) compounds (144).

The NIOSH recommended exposure limit for carcinogenic hexavalent chromium is 1 µg/m^3 Cr(VI) as a 10-h TWA, and for noncarcinogenic Cr(VI) the 10-h TWA is 25 µg/m^3 Cr(VI), including a 15-min maximum exposure of 50 µg/m^3 Cr(VI). According to NIOSH, the noncarcinogenic Cr(VI) compounds are chromic acid and the chromates and dichromates of sodium, potassium, lithium, rubidium, cesium, and ammonia. NIOSH considers any hexavalent chromium compound that does not appear on the preceding list carcinogenic (145).

Recommendations by the ACGIH are classified as threshold limit values (TLV) based on 8-h TWA. Chromium metal and alloys, Cr(II) compounds and Cr(III) compounds, including chromite ore, have a TLV of 0.5 mg/m^3 Cr in air. Water-soluble Cr(VI) compounds have a TLV of 0.05 mg/m^3 Cr. Certain water-insoluble Cr(VI) compounds, ie, the chromates of zinc, barium, calcium, lead, strontium, sintered chromic acid, and processing chromite ores, also have a TLV of 0.05 mg/m^3 as well as a human carcinogen designation (145).

Environment. There are no federal standards proclaimed for ambient air concentrations of chromium, but the EPA has published a notice of intent to list Cr(VI) or total Cr as a toxic air contaminant (146). However, several states have issued standards and these vary widely (147). In the United States, ambient air concentrations of Cr ranged from 5.2 ng/m^3 (24-h background level) to 156.8 ng/m^3 (urban annual average) for the time period 1977–1980 (137,147). These values refer to total chromium. Some hexavalent chromium ambient air concentrations have been measured, and the limited results appear to suggest an average range for Cr(VI) of 0.5 to 5 ng/m^3 for U.S. urban areas (147,148). Although natural sources may be responsible for part of the chromium value, all of the Cr(VI) and some of the Cr(III) measured is probably from anthropogenic sources. The most likely origins of Cr(VI) values are industrial emissions, and the most likely forms are aqueous aerosol fog, mists, or droplets and aerosol powders (124,137,149,150).

The EPA has set the National Interim Primary Drinking Water Standard at 50 µg/L total chromium and the current Maximum Contamination Level (MCL) is 120 µg/L. This agency has also issued a Cr(VI) ambient water quality standard of 50 µg/L and has proposed a Maximum Contamination Level Goal (MCLG) of 0.1 µg/L (151). Industrial discharges of total Cr(VI) are regulated by National Pollutant Discharge Elimination System (NPDES) permits, specific for the area that receives the waste or discharge.

Chromium containing solids from manufacturing and wastewater treatment sludges are classified as hazardous wastes and must be handled as such (152). These wastewater treatment sludges are F006 from electroplating wastewaters; K002, K003, K005, K006, and K008 from pigment producers' wastewaters; K086 wastewaters generated as a result of cleaning process equipment used to make chromium containing inks (qv) from pigments, driers, soaps (qv) and stabilizers; U032 wastewaters from the production of calcium chromate. These solids are characterized as D007 wastes because they exceed the Resource Conservation and Recovery Act (RCRA) threshold of 5.0 mg/L Cr as determined by the extraction procedure (EP) toxic characteristic leaching procedure (TCLP) tests and may contain both Cr(VI) and Cr(III) (153).

The EPA has established exposure levels for both Cr(III) and Cr(VI) for the general population (124,127). For exposures of short duration that constitute an insignificant fraction of the lifespan the acceptable intake subchronic (AIS) by ingestion is 979 mg/d for trivalent chromium and 1.75 mg/d for hexavalent chromium. There was insufficient data to calculate an AIS by inhalation for Cr(III), and the EPA believes this type of standard is inappropriate for hexavalent chromium (124). For lifetime exposures, an acceptable intake chronic (AIC) of 103 mg/d Cr(III) and 0.35 mg/d Cr(VI) is established for ingestion. The inhalation AIC is estimated to be 0.357 mg/d Cr(III). The EPA has calculated an inhalation cancer potency for Cr(VI) of 41 [mg/(kg·d)]$^{-1}$ risk for a lifetime exposure to 1 µg/m^3 hexavalent chromium (124,127,130,137).

Waste Management

Despite modern engineering designs, production and consumption waste by-products containing chromium are generated. These wastes have been traditionally managed by burial of dewatered sludges that are mainly the result of the reduction of Cr(VI) to Cr(III) and the latter's precipitation as a hydrated oxide from the treatment of wastewaters. Scrap iron (154), ferrous sulfate, sodium bisulfite, sulfur dioxide, sodium hydrosulfite, and sulfide wastes (155) have all been employed as reducing agents for waste streams containing hexavalent chromium. Following hexavalent chromium's reduction, lime or other alkali is added to raise the pH, causing hydrated chromium(III) oxide to precipitate. The slurry is allowed to segregate and the clear, purified supernatant is decanted and either recycled or discharged. Often this type of procedure allows several metals, such as Cu and Ni, to be removed with the chromium, because the hydrous oxide is a good collector.

At one time, it was thought that the low toxicity, low solubility Cr(III) compounds would remain in this stable oxidation state, and these wastes could be safely employed as landfill. However, studies suggest that under certain conditions, trivalent chromium may be oxidized to Cr(VI) by manganese dioxide (156,157) and/or hypochlorite and chlorine (158). These oxidations are partially dependent on the solubility of the trivalent chromium. Thus an excess of lime combined with fly ash, clay minerals, eg, kaolin, ferric compounds, or other proprietary compounds, is added to lower the mobility or solubility of Cr(III) (111,159–162). Also, if the sludge is high in oxidizable organic compounds, eg, tannery wastes, or burial is in an area containing a high percentage of naturally occuring oxidizable organic materials, eg, fluvic acids, interconvertibility of chromium is highly unlikely (111,163,164). In fact, because leather tanning employs Cr(III) exclusively, tannery wastes have been exempt from the RCRA hazardous waste classification even though 86% of the delisted wastes, K053–K059, yield > 5.0 mg/L in the EP/TCLP tests (153,165).

Where appropriate, the direct precipitation of hexavalent chromium with barium, and recovery of the Cr(VI) value can be employed (166). Another recycling (qv) option is ion exchange (qv), a technique that works for chromates and Cr^{3+} (161). Finally, recovery of the chromium as the metal or alloy is possible by a process similar to the manufacture of ferrochromium alloy and other metals (161).

Uses

Chromium compounds are essential to many industries. The percentage distribution of consumption for chromium compounds is wood preservation, 38; metal finishing, 15; leather tanning, 10; pigments, 8; chemical manufacturing, 8; oil drilling muds, 4; textiles, 3; magnetic tapes, 2; and other uses such as for catalysts, photography, etc, 11 (77,167).

Metal Finishing and Corrosion Control. The exceptional corrosion protection provided by electroplated chromium and the protective film created by applying chromium surface conversion techniques to many active metals, has made chromium compounds valuable to the metal finishing industry. Cr(VI) compounds have dominated the formulas employed for electroplating (qv) and surface conversion, but the use of Cr(III) compounds is growing in both areas because of the health and safety problems associated with hexavalent chromium and the low toxicity of trivalent chromium (see CORROSION AND CORROSION INHIBITORS; METAL SURFACE TREATMENTS; METALLIC COATINGS).

Electroplating of Chromium. Until the middle to late 1970s, all of the commercially electroplated chromium was produced from plating baths prepared from chromic acid. Although these baths still accounted for the majority of chromium electroplated products as of 1992 (168–171), decorative trivalent chromium baths are successfully operated in many installations (172).

Compositions and operating parameters for both Cr(VI) and Cr(III) baths are given in Table 8. Two types of trivalent baths result from different anode arrangement (173,174). The No. 1 bath uses a graphite anode that is in the bath during plating, and relies on proprietary additives combined with current density control to limit the anodic oxidation of trivalent to hexavalent chromium. Bath No. 2 employs an anode that is isolated from the plating bath by a hydrogen ion-selective membrane that allows only H^+ to pass, and therefore any anodic oxidation of Cr(III) is prevented (174). Small amounts of hexavalent chromium reduces the efficiency of trivalent baths.

Because the thickness of the plate deposited from trivalent baths is limited, these have only been employed for decorative applications. However, the bluish white deposit obtained from chromic acid baths can be closely matched by trivalent chromium baths (173).

Unlike most metals, chromium can be plated from solutions in which it is present as an anion in a high oxidation state. The deposition of chromium from chromic acid solutions also requires the presence of a catalyst anion, usually sulfate, although fluoride, fluosilicate, and mixtures of these two with sulfate have been extensively used. The amount of catalyst must be carefully regulated. Neither pure chromic acid or solutions containing excess catalysts produce a satisfactory plate. Even using carefully controlled temperature, current density, and

Table 8. Chemical Composition and Operating Parameters for Chromium Electroplating Baths

Parameter	Cr(III)[a] No. 1	Cr(III)[a] No. 2	Cr(VI)[a] decorative	Cr(VI)[b] functional
Cr, g/L	20–23	5–10	100–200	100–200
CrO_3, g/L			190–380	190–380
SO_4^{2-}, g/L	c	c	1.9–3.8	1.9–3.8
H_3BO_3, g/L	60–65	60–65		
temperature, °C	20–50	45–55	30–50	50–60
pH	2.3–2.9	3.5–3.9	<1	<1
anode type	internal	external	internal	internal
anode material	graphite	93%Pb/7%Sn	93%Pb/7%Sn	93%Pb/7%Sn
current density, A/dm²	4–15	4–15	17.5–30	3.6–36
deposit thickness, μm	0.05–0.5	0.05–0.5	0.05–0.5	2.5–500

[a]Ref. 172–174.
[b]Ref. 175.
[c]No specifications, but may be present.

bath composition, chromium plating is one of the most difficult electroplating operations. Throwing power and current efficiency are notably poor, making good racking procedures and good electrical practices essential.

In 1979, a viable theory to explain the mechanism of chromium electroplating from chromic acid baths was developed (176). An initial layer of polychromates, mainly $HCr_3O_{10}^-$, is formed contiguous to the outer boundary of the cathode's Helmholtz double layer. Electrons move across the Helmholtz layer by quantum mechanical tunneling to the end groups of the polychromate oriented in the direction of the double layer. Cr(VI) is reduced to Cr(III) in one-electron steps and a colloidal film of chromic dichromate is produced. Chromous dichromate is formed in the film by the same tunneling mechanism, and the Cr(II) forms a complex with sulfate. Bright chromium deposits are obtained from this complex.

Decorative chromium plating, 0.2–0.5 μm deposit thickness, is widely used for automobile body parts, appliances, plumbing fixtures, and many other products. It is customarily applied over a nonferrous base in the plating of steel plates. To obtain the necessary corrosion resistance, the nature of the undercoat and the porosity and stresses of the chromium are all carefully controlled. Thus microcracked, microporous, crack-free, or conventional chromium may be plated over duplex and triplex nickel undercoats.

Functional or hard chromium plating (169,175) is a successful way of protecting a variety of industrial devices from wear and friction. The most important examples are cylinder liners and piston rings for internal combustion engines. Functional chromium deposits must be applied to hard substrates, such as steel, and are applied in a wide variety of thicknesses ranging from 2.5 to 500 μm.

Black and colored plates can also be obtained from chromic acid baths. The plates are mostly oxides (177). Black chromium plating bath compositions are proprietary, but most do not contain sulfate. The deposit has been considered for use in solar panels because of its high absorptivity and low emissivity (175).

Chromium Surface Conversion. Converting the surface of an active metal by incorporating a barrier film of complex chromium compounds protects the metal from corrosion, provides an excellent base for subsequent painting, provides a chemical polish, and/or colors the metal. This conversion is normally accomplished by immersion, but spraying, swabbing, brushing, and electrolytic methods are also employed (178) (see METAL SURFACE TREATMENTS). The metals that benefit from chromium surface conversion are aluminum, cadmium, copper, magnesium, silver, and zinc. Zinc is the largest consumer of chromium conversion baths, and more formulations are developed for zinc than for any other metal.

The compositions of the conversion baths are proprietary and vary greatly. They may contain either hexavalent or trivalent chromium (179,180), but baths containing both Cr(III) and Cr(VI) are rare. The mechanism of film formation for hexavalent baths has been studied (181,182), and it appears that the strength of the acid and its identity, as well as time and temperature, influences the film's thickness and its final properties, eg, color. The newly prepared film is a very soft, easily damaged gel, but when allowed to age, the film slowly hardens, assumes a hydrophobic character and becomes resistant to abrasion. The film's structure can be described as a cross-linked Cr(III) polymer, that uses anion species to link chromium centers. These anions may be hydroxide, chromate, fluoride, and/or others, depending on the composition of the bath (183).

Clear-bright and blue-bright chromium conversion colors are thin films (qv), and may be obtained from both Cr(III) and Cr(VI) conversion baths. The perceived colors are actually the result of interference phenomena. Iridescent yellows, browns, bronzes, olive drabs, and blacks are only obtained from hexavalent conversion baths, and the colors are listed in the order of increasing film thickness. Generally, the thicker the film, the better the corrosion protection (see FILM DEPOSITION TECHNIQUES).

Oxide films on aluminum are produced by anodizing in a chromic acid solution. These films are heavier than those produced by chemical conversion and thinner and more impervious than those produced by the more common sulfuric acid anodizing. They impart exceptional corrosion resistance and paint adherence to aluminum and were widely used on military aircraft assemblies during World War II. The films may be dyed. A typical anodizing bath contains 50 to 100 g/L CrO_3 and is operated at 35–40°C. The newer processes use about 20 volts dc and adjust the time to obtain the desired film thickness (184).

Dichromates and chromic acid are used as sealers or after-dips to improve the corrosion resistance of various coatings on metals. For example, phosphate coatings on galvanized iron or steel as well as sulfuric acid anodic coatings on aluminum can be sealed by hexavalent chromium baths.

Chromium compounds are used in etching and bright-dipping of copper and its alloys. A typical composition for the removal of scale after heat-treating contains 30 g/L $Na_2Cr_2O_7 \cdot 2H_2O$ and 240 mL/L concentrated H_2SO_4. It is used at 50–60°C.

Chromates are used to inhibit metal corrosion in recirculating water systems. When methanol was extensively used as an antifreeze, chromates could be successfully used as a corrosion inhibitor for cooling systems in locomotive diesels and automobiles (185).

Steel immersed in dilute chromate solutions does not rust. The exact mechanism of the inhibition is not known, although it is agreed that polarization of the local anodes that serve as corrosion foci is important. In the inhibition of iron and steel corrosion a film of τ-Fe_2O_3, in which some Cr is present, appears to form. The concentration of chromate required to inhibit corrosion may range from 50 to 20,000 ppm, depending on conditions, and a pH of 8–9 is usually optimum. The inclusion of chromium compounds in formulations permits the use of such corrosive salts as zinc chloride and copper sulfate in steel cylinders.

Pigments. Chromium pigments can be divided into chromate color pigments based on lead chromate, chromium oxide greens, and corrosion inhibiting pigments based on difficultly soluble chromate. An excellent discussion of these pigments is given in Reference 186. An older reference is also useful (187) (see PIGMENTS, INORGANIC). Data for the domestic production of chromium pigments are given in Table 9. Prices for individual pigments are given in Table 10.

Chromate Pigments Based on Lead. Pigments based on lead can be further subdivided into primrose, lemon, and medium yellows, and chrome orange, molybdate orange [12709-98-7], and normal lead silicochromates. Although earlier emphasis was on pure lead compounds, modern pigments contain additives to improve working properties, hue, light fastness, and crystal size and shape and to maintain metastable structures (188).

Table 9. Chromium Pigment Production, 1977–1988, t[a]

Pigment	1977	1982	1988
chrome yellow and orange	31,940	18,500	21,528
molybdate orange	12,514	6,015	5,500[b]
chrome oxide green	7,980	3,917	5,000[b]
zinc yellow	2,300[b]		1,400[b]
other chrome colors	4,146	2,300[b]	2,700[b]

[a]Ref. 89.
[b]Values are estimates.

Table 10. Prices of Select Chromium Pigments[a], June 1991

Pigment[b]	CAS Registry Number	$/kg
chrome green(cp)	[7758-97-6]	3.70
chrome oxide green	[1308-38-9]	4.19
Guignet's green	[12001-99-9]	12.13
chrome yellow(cp)	[1344-37-2]	2.98
chrome orange	[1344-38-3]	2.05
lead silicochromate	[11113-70-5]	1.10
zinc yellow	[37300-23-5]	2.76

[a]Ref. 88.
[b]The designation (cp) indicates chemical purity.

The chemical composition and ASTM specifications (189) of these pigments is given in Table 11. Details for commercial procedures are not disclosed. The pigments are characterized as follows: Medium yellows are orange–yellows that are essentially pure monoclinic lead chromate. Light lemon or primrose yellows containing up to 40% lead sulfate have some or all of the lead chromate in the metastable orthorhombic form, which is stabilized by lead sulfate and other additives. The higher the orthorhombic content, the greener the shade. Chrome oranges are basic lead chromate [18454-12-1], $PbCrO_4 \cdot PbO$. Molybdate oranges are tetragonal solid solutions of lead sulfate, lead chromate, and lead molybdate. An aging step is required in precipitation to permit development of the orange tetragonal form. Lead silicochromate, essentially medium chrome yellow precipitated on silica, has been developed for use in traffic paints where the silica gives better abrasion resistance. Chrome green, not to be confused with chromic oxide green, is a mixture of a light chrome yellow, ie, lemon or primrose, and a blue, usually iron blue. The pigment may be produced by grinding, mixing in suspension, or precipitating the yellow on the blue. The last method is the preferred. The first is hazardous because the pigment, containing both oxidizing (chromate) and reducing (ferrocyanide) components, may undergo spontaneous combustion. Phthalocyanine blues have replaced iron blues to some extent and organic greens and chromic oxide have displaced chrome green, which has poor acid and alkali resistance.

Table 11. Chemical Composition and ASTM Specifications for Chromate Color Pigments[a]

		Composition, wt %		
			Actual	
Analyte[b]	Spec[c]	Theory	Min	Max
Primrose chrome yellow, D211-67 Type I				
$PbCrO_4$	50[d]	77.3	52.0	82.7
$PbSO_4$		22.7	4.2	25.9
TFM	8.0[e]			
Lemon chrome yellow, D211-67 Type II				
$PbCrO_4$	65[d]	72.7	52.4	68.8
$PbSO_4$		27.3	17.4	39.0
TFM	10.0[e]			
Medium chrome yellow, D211-67 Type III				
$PbCrO_4$	87[d]	100	82.4	98.2
TFM	10.0[e]			
Light chrome orange, D211-67 Type IV				
$PbCrO_4$	55[d]	59.2		
PbO		40.8		
TFM	10.0[e]			
Dark chrome orange, D211-67 Type V				
$PbCrO_4$	55[d]	59.2		
PbO		40.8		
TFM	3.0[e]			
Chrome yellow for green, D211-67 Type VI				
$PbCrO_4$	75[d]			
TFM	8.0[e]			
Pure chrome green, D212-80				
$PbCrO_4$	70[d]			
Molybdate orange, D2218-67				
$PbCrO_4$	70[d]	82.3		
$PbMoO_4$	8[d]	14.9		
$PbSO_4$		2.8		
TFM	12[e]			

[a]Ref. 189.
[b]TFM = total foreign materials or total of all substances that are not insoluble lead compounds.
[c]Spec = specification. [d]Value is minimum. [e]Value is maximum.

Chromium Oxide Greens.   The chromium oxide green pigments comprise both the pure anhydrous oxide, Cr_2O_3, and hydrated oxide, or Guignet's green (190). The following manufacturing processes appear to be in use.

An alkali dichromate is reduced in self-sustaining dry reaction by a reducing agent such as sulfur, carbon, starch, wood flour, or ammonium chloride. For pigment use, the reducing agent is generally sulfur. When a low sulfur grade is needed in the manufacture of aluminothermic chromium, a carbonaceous reducing agent is employed:

$$Na_2Cr_2O_7 + S \rightarrow Na_2SO_4 + Cr_2O_3 \tag{11}$$

$$Na_2Cr_2O_7 + 2\,C \rightarrow Na_2CO_3 + Cr_2O_3 + CO \tag{12}$$

The mixture is ignited with an excess of reducing agent in a reverberatory furnace or small kiln, transferred to leaching tanks, filtered, washed, dried, and pulverized. The product is $99 + \%$ Cr_2O_3, and the metallurgical grades contain less than 0.005% of sulfur.

Chromate–dichromate solutions are reduced by sulfur in a boiling alkaline suspension (191).

$$2\,Na_2CrO_4 + Na_2Cr_2O_7 + 6\,S + 2x\,H_2O \rightarrow 2\,Cr_2O_3 \cdot xH_2O + 3\,Na_2S_2O_3 \tag{13}$$

Excess NaOH is used to start the reaction and not over 35% of the chromium is added as dichromate. At the end of the reaction, the thiosulfate is removed by filtration and recovered. The hydrous oxide slurry is then acidified to pH 3–4 and washed free of sodium salts. On calcination at 1200–1300°C, a fluffy pigment oxide is obtained, which may be densified and strengthened by grinding. The shade can be varied by changes in the chromate:dichromate ratio, and by additives.

A dichromate or chromate solution is reduced under pressure to produce a hydrous oxide, which is filtered, washed, and calcined at 1000°C. The calcined oxide is washed to remove sodium chromate, dried, and ground. Sulfur, glucose, sulfite, and reducing gases may be used as reducing agent, and temperatures may reach 210°C and pressures 4–5 MPa (600–700 psi).

A number of manufacturers around the world are using the decomposition of ammonium dichromate to produce chrome oxide (eq. 5) (78). Generally, an excess of finely ground ammonium sulfate is mixed with sodium dichromate, and the dry mixture is heated to form chrome oxide and sodium sulfate, evolving nitrogen and steam.

$$(NH_4)_2SO_4 + Na_2Cr_2O_7 \rightarrow Cr_2O_3 + Na_2SO_4 + N_2 + 4\,H_2O \tag{14}$$

This is a favorable process because the side reaction products, nitrogen and water, are not pollutants and the sodium sulfate can be recovered and sold. Also, all of the wash water used to remove the sodium sulfate from the chrome oxide can be recycled.

Chromic oxide green is the most stable green pigment known. It is used where chemical and heat resistance are required and is a valuable ceramic color (see COLORANTS FOR CERAMICS). It is used in coloring cement (qv) and granulated rock for asphalt (qv) roofing. An interesting application is in camouflage paints, as the infrared reflectance of chromic oxide resembles green foliage. A minor use is in the coloring of synthetic gem stones (see GEMS, SYNTHETIC). Ruby, emerald, and the dichroic alexandrite all owe their color to chromic oxide (53).

Guignet's green, or hydrated chromic oxide green, is not a true hydrate, but a hydrous oxide, $Cr_2O_3 \cdot xH_2O$, in which x is about 2. It is obtained from the production of hydrous oxide at elevated temperature, and sometimes pressure, in a borax or boric acid melt. Although Guignet's green is permanent, it does not withstand use in ceramics. It has poor tinting strength but is a very clean, transparent, bluish green. It is used in cosmetics (qv) and metallic automotive finishes (see COLORANTS FOR FOOD, DRUGS, COSMETICS, AND MEDICAL DEVICES).

Corrosion Inhibiting Pigments.   Pigments inhibiting corrosion derive effectiveness from the low solubility of chromate. The principal pigment of this group is zinc chromate or zinc yellow. Others include zinc tetroxychromate, basic lead silicochromate, strontium chromate, and barium potassium chromate (192). The chemical composition and ASTM specifications of some of these pigments are shown in Table 12.

Zinc yellow became an important corrosion-inhibiting pigment for aircraft during World War II. However, the war production rate of 11,000 t/yr has not since been reached. Now, zinc yellow is widely used for corrosion inhibition on auto bodies, light metals, and steel, and in combination with red lead and ferric oxide for structural steel painting.

Zinc yellow is not a normal zinc chromate, having the empirical formula $K_2O \cdot 4ZnO \cdot 4CrO_3 \cdot 3H_2O$ [*12433-50-0*]. It belongs to the group of salts having the general formula $M(I)_2O \cdot 4M(II)O \cdot 4CrO_3 \cdot 3H_2O$ (193). The sodium zinc salt has occasionally been used as a pigment. The sodium copper salt has been tested as an antifouling marine pigment and is an ingredient of dips for auto bodies (see COATINGS, MARINE).

Zinc yellow is made by a variety of processes, all based on the reaction of zinc compounds, chromates, and potassium salts in aqueous solution. If products free of chloride and especially sulfate are desired, they are excluded from the system. In one process, for example, zinc oxide is swollen with potassium hydroxide and the chromates are added as a solution of potassium tetrachromate [*12422-53-6*] (194).

$$4\,ZnO \,+\, K_2Cr_4O_{13} \,+\, 3\,H_2O \rightarrow K_2O \cdot 4ZnO \cdot 4CrO_3 \cdot 3H_2O \qquad (15)$$

The final pH is 6.0–6.6. Care must be taken in washing to avoid hydrolysis and loss of chromate.

Zinc tetroxychromate [*13530-65-9*], approximately $4ZnO \cdot ZnCrO_4 \cdot xH_2O$, has a somewhat lower chromate solubility than zinc yellow and has been used in wash primers.

Strontium chromate [*12677-00-8*], $SrCrO_4$, is used increasingly despite its high cost. It works well on light metals, and is compatible with some latex emulsions where zinc compounds cause coagulation (see LATEX TECHNOLOGY). It is also an ingredient of some proprietary formulations for chrome plating.

Basic lead silicochromate [*11113-70-5*] (National Lead Co. designation Pigment M-50) is a composite in which basic lead chromate, ie, chrome orange, is precipitated onto a lead silicate–silica base. It does not have an appreciable chromate solubility and depends on lead oxide for its effectiveness.

Leather Tanning and Textiles.   Although chromium(VI) compounds are the most important commercially, the bulk of the applications in the textile and tan-

Table 12. Chemical Compositions and Analytical Specifications for Chromate Corrosion Inhibiting Pigments

Analyte[a]	Composition, wt %		
	Spec[b]	Theory	Typical
Basic lead silicochromate D 1648 – 81 Type 1			
CrO_3	5.1–5.7		5.4
PbO	46.0–49.0		47.0
SiO_2	45.5–48.5		47.0
SSD, μm	<8.5		
Basic lead silicochromate D 1648 – 81 Type 2			
CrO_3	6.3–7.2		
PbO	42.5–46.0		
SiO_2	47.5–50.5		
SSD, μm	<2.0		
Strontium chromate D 1649 – 82			
CrO_3	41[c]		
SrO	41[c]		
SO_3	0.2[d]		
Zinc yellow (zinc chromate) D 478 – 49 Type I			
CrO_3	41[c]	45.8	45.0
ZnO	35–40	37.2	36.0
K_2O	13[d]	10.8	10.0
SO_3	0.2[d]		0.05
Cl	0.1[d]		
Zinc yellow (zinc chromate) D 478 – 49 Type II			
CrO_3	41[c]	45.8	45.0
ZnO	35–40	37.2	36.0
K_2O	13[d]	10.8	10.0
SO_3	3.0[d]		1.0
Cl	0.8[d]		
[e]	1.0[d]		
Zinc tetroxychromate			
CrO_3			17.0
ZnO			71.0
H_2O			10.0

[a]SSD = selective surface diameter determined by ASTM D1366.
[b]Ref. 189.
[c]Value is minimum.
[d]Value is maximum.
[e]When SO_3 and Cl are below maximum, the expression $[(\%SO_3/3) + (\%Cl/0.8)]$ must be used to determine conformance to specifications.

ning industries depend on the ability of Cr(III) to form stable complexes with proteins, cellulosic materials, dyestuffs, and various synthetic polymers. The chemistry is complex and not well understood in many cases, but a common denominator is the coordinating ability of chromium(III) (see LEATHER; TEXTILES).

The chrome tanning is one step in a complicated series of leather operations leading from the raw hide to the finished products. Chrome tanning is the most important tannage for all hides except heavy cattle hides, which are usually vegetable tanned. In heavy shoe uppers and soles, a chrome tanned leather is frequently given a vegetable retan to produce chrome retan leather.

Sodium dichromate and various chromic salts are employed in the textile industry (195,196). The former is used as an oxidant and as a source of chromium, for example, to dye wool and synthetics with mordant acid dyes, oxidize vat dyes and indigosol dyes on wool, aftertreat direct dyes and sulfur dyes on cotton to improve washfastness, and oxidize dyed wool. Premetallized dyes are also employed. These are hydroxyazo or azomethine dyes in which chromium or other metals are combined in the dye (see AZINE DYES; AZO DYES).

Acid Black 63 [32517-36-5] (CI 12195) is a typical premetallized dye. The commercial product contains some of the 1:2 chelate shown.

Another use of chromium compounds is in the production of water- and oil-resistant coatings on textiles, plastic, and fiber glass. Trade names are Quilon, Volan, and Scotchgard (197,198) (see WATERPROOFING AND WATER/OIL REPELLANCY).

Wood Preservation. The use of chromium compounds in wood preservation is largely because of the excellent results achieved by chromated copper arsenate (CCA), available in three modifications under a variety of trade names. The treated wood (qv) is free from bleeding, has an attractive olive-green color, and is paintable. CCA is widely used, especially in treating utility poles, building lumber, and wood foundations. About 62% of all the chromic acid produced in the United States is consumed by the wood preservation industry (77,167) (see BUILDING MATERIALS, SURVEY).

Chromium compounds are also used in fire-retardant formulations where their function is to prevent leaching of the fire retardant from the wood and corrosion of the equipment employed.

Chromium-containing wood preservatives and their chemical compositions are listed in Table 13 (199). Chromium compounds have a triple function in wood preservation (200). Most importantly, after impregnation of the wood the Cr(VI) compounds used in the formulations react with the wood extractives and the other

Table 13. Chemical Composition and Specifications for Wood Preservatives[a]

Type	Component	Composition, wt %		
		Optimum	Min	Max
	Acid copper chromate (ACC)			
	CrO_3	68.2	63.3	
	CuO	31.8	28.0	
	Chromated copper arsenate (CCA)			
type A	CrO_3	65.5	59.4	69.3
	CuO	18.1	16.0	20.9
	As_2O_5	16.4	14.7	19.7
type B	CrO_3	35.3	33.0	38.0
	CuO	19.6	18.0	22.0
	As_2O_5	45.1	42.0	48.0
type C	CrO_3	47.5	44.5	50.5
	CuO	18.5	17.0	21.0
	As_2O_5	34.0	30.0	38.0
	Chromated zinc chloride (CZC)			
	CrO_3	20	19	
	ZnO	80	76	
	Fluor chrome arsenate phenol (FCAP)			
	CrO_3	37	33	41
	As_2O_5	25	22	28
	F	22	20	24
	DNP^b	16	14	18

[a]Ref. 199.
[b]DNP = dinitrophenol.

preservative salts to produce relatively insoluble complexes from which preservative leaches only very slowly. This mechanism has been studied in the laboratory (201–206) and the field (207). Finally, although most of the chromium is reduced to chromium(III), there is probably some slight contribution of the chromium(VI) to the preservative value (208).

Drilling Muds in the Petroleum and Natural Gas Industry. Since 1941, chromium chemicals have been used in the drilling of wells to combat fatigue corrosion cracking of drill strings, with about one metric ton of sodium chromate being used annually for an average West Texas well. Other early uses were in gas-condensate wells in Louisiana and East Texas.

However, the petroleum (qv) industry has turned to proprietary drilling-mud formulations, specially designed to suit the aqueous environment and rock strata in which the well is located (see PETROLEUM, DRILLING FLUIDS AND OTHER OIL RECOVERY CHEMICALS). In addition to heavy minerals, such as barite, and both soluble and difficultly soluble chromates for corrosion control, many of these formulations contain chromium lignosulfonates. The latter Cr(III) compounds are prepared like a tanning formula from sodium dichromate, using lignosulfonate waste from sulfite pulp (qv) mills as the reducing agent. This use amounts to about 4% of the total chromium compound consumption (209,210).

Acrylamide–polymer/Cr(III)carboxylate gel technology has been developed and field tested in Wyoming's Big Horn Basin (211,212). These gels economically enhance oil recovery from wells that suffer fracture conformance problems. The Cr(III) gel technology was successful in both sandstone and carbonate formations, and was insensitive to H_2S, high saline, and hard waters (212).

Miscellaneous Uses. A large number of chromium compounds have been sold in small quantities for a variety of uses, some of which are described in Table 14 (1,236–238).

Catalysts. A more important minor use of chromium compounds is in the manufacture of catalysts (Table 14). Chromium catalysts are used in a great variety of reactions, including hydrogenations, oxidations, and polymerizations (229–231). Most of the details are proprietary and many patents are available.

Chromia–alumina catalysts are prepared by impregnating τ-alumina shapes with a solution of chromic acid, ammonium dichromate, or chromic nitrate, followed by gentle calcination. Zinc and copper chromites are prepared by coprecipitation and ignition, or by thermal decomposition of zinc or copper chromates, or organic amine complexes thereof. Many catalysts have spinel-like structures (239–242).

Photosensitive Reactions. The reduction of chromium(VI) by organic compounds is highly photosensitive, and this property is used in photosensitive dichromate-colloid systems.

A dichromate-colloid system is applied to a metal pringing plate (243). This soluble material is exposed to an image, and, where light strikes, the photochemical reaction reduces the dichromate. The chromium(III) produced forms an insoluble complex with the colloid in a reaction similar to that of dye mordanting or leather tanning. The unreacted colloid is washed off exposing bare metal that can be etched. Some of the colloids used are shellac, glue, albumin, casein, gum arabic, and gelatin. The newer technology employs more consistent and readily controlled synthetic materials, such as poly(vinyl alcohol).

Batteries. The shelf life of dry batteries (qv) is increased from 50 to 80% by the use of a few grams of zinc chromate or dichromate near the zinc anode. This polarizes the anode on open circuit but does not interfere with current delivery.

Since World War II, the U.S. space program and the military have used small amounts of insoluble chromates, largely barium and calcium chromates, as activators and depolarizers in fused-salt batteries (214,244). The National Aeronautics and Space Administration (NASA) has also used chromium(III) chloride as an electrolyte for redox energy storage cells (245).

Magnetic Tapes. Chromium dioxide, CrO_2, is used as a ferromagnetic material in high fidelity magnetic tapes (qv). Chromium dioxide has several technical advantages over the magnetic iron oxides generally used (58,246).

Reagent-Grade Chemicals. Potassium dichromate is an important analytical standard, and other chromium chemicals, in reagent grades, find considerable laboratory use (90,91). This use, though small, is most important in wet analyses.

Alloys. A substantial amount of chromic oxide is used in the manufacture of chromium metal and aluminum–chromium master alloys.

Table 14. Chromium Compounds Properties and Uses

Name	CAS Registry Number	Molecular formula	Properties	Uses
		Cr(VI) compounds		
ammonium chromate	[7788-98-9]	$(NH_4)_2CrO_4$	yellow crystals, ρ 1.91, sol H_2O	textile printing, photography, dye mordant for wool, analytical reagent
barium chromate	[10294-40-3]	$BaCrO_4$	[a]	pyrotechnics, high temp. batteries, pigment for glasses and ceramics[b,c,d]
barium dichromate	[10031-16-0]	$BaCr_2O_7 \cdot 2H_2O$	bright red-yellow needles, $-2H_2O$ at 120°C, dec in H_2O	ceramics
barium potassium chromate	[27133-66-0]	$K_2Ba(CrO_4)_2$	yellow crystals, ρ 3.65	corrosion inhibiting pigment[e]
cadmium chromate	[14312-00-6]	$CdCrO_4$	yellow crystals, insol H_2O	catalysts, pigments
cadmium dichromate	[69239-51-6]	$CdCr_2O_7 \cdot H_2O$	orange crystals, sol H_2O	metal finishing
calcium chromate	[13765-19-0]	$CaCrO_4$	yellow crystals, sl sol H_2O	metal primers, high temp. batteries, corrosion inhibitor[b,c]
calcium dichromate	[14307-33-6]	$CaCr_2O_7 \cdot 4.5H_2O$	orange crystals, ρ 2.136, sol H_2O	metal finishing, catalyst, corrosion inhibitor[f]
cesium chromate	[13454-78-9]	Cs_2CrO_4	yellow crystals, ρ 4.237, sol H_2O	electronics
chromic chromate	[11056-30-7]	variable	brown, amorphous, and hydrated	catalysts, mordants

300

chromyl chloride	[14977-61-8]	CrO_2Cl_2		Etard reaction, oxidation of organics, catalyst-polymerization of olefins[g]
cobalt chromate	[13455-25-9]	$CoCrO_4$	[a]	ceramics[h]
copper chromate, basic copper dichromate	[12433-14-6] [13675-47-3]	$4CuO \cdot CrO_3 \cdot xH_2O$ $CuCr_2O_7 \cdot 2H_2O$	gray-black crystals, insol H_2O, sol acids brown, amorphous black crystals, ρ 2.283, very sol H_2O, sol acids and NH_4OH	fungicides, catalysts catalysts, wood preservatives
copper sodium chromate	[68399-60-0]	$Na_2O \cdot 4CuO \cdot 4CrO_2 \cdot 3H_2O$	maroon crystals, ρ 3.57, sl sol H_2O	antifouling pigment[i]
lithium chromate	[7789-01-7]	$Li_2CrO_4 \cdot 2H_2O$	yellow crystals, ρ 2.149, sol H_2O, transition to anhydrous at 74.6°C	corrosion inhibitor esp. in air-conditioner and nuclear reactors[j]
lithium dichromate	[10022-48-7]	$Li_2Cr_2O_7 \cdot 2H_2O$	orange-red crystals, ρ 2.34, very sol H_2O	corrosion inhibitor[k]
magnesium chromate	[16569-85-0]	$MgCrO_4 \cdot 5H_2O$	yellow crystals, ρ 1.954, sol H_2O, turns to 7 H_2O at 17.2°C	corrosion inhibitor in gas turbines, refractories
magnesium dichromate	[34448-20-9]	$MgCr_2O_7 \cdot 6H_2O$	orange-red crystals, ρ 2.002, sol H_2O, tr to 5H_2O at 48.5°C	catalyst, refractories[l]
mercuric chromate	[13444-75-2]	$HgCrO_4$	red crystals, slightly sol	antifouling formulations
mercurous chromate morpholine chromate	[13465-34-4] [36969-05-8]	Hg_2CrO_4 $(C_4H_{10}NO)_2CrO_4$	red crystals, very sl sol yellow oily material	antifouling formulations vapor-phase corrosion inhibitor in catalysts
nickel chromate	[14721-18-7]	$NiCrO_4$	maroon to black crystals, very sl sol	catalyst
pyridine–chromic acid adduct	[26412-88-4]	$CrO_3 \cdot 2C_5H_5N$	dark red crystals, explodes on warming	research oxidant

Table 14. *(Continued)*

Name	CAS Registry Number	Molecular formula	Properties	Uses
		Cr(VI) compounds (Continued)		
pyridine dichromate	[20039-37-6]	$(C_5H_5NH)_2Cr_2O_7$	orange crystals	photosensitizer in photoengraving
silver chromate	[7784-01-2]	Ag_2CrO_4	[a]	catalyst
strontium chromate	[7789-06-2]	$SrCrO_4$	[a]	corrosion-inhibiting pigment, plating additive
tetramminecopper(II) chromate	[13870-96-7]	$Cu(NH_3)_4CrO_4$	dark green needles	catalyst, gas absorbant
zinc sodium chromate	[68399-59-7]	$Na_2O \cdot 4ZnO \cdot 4CrO_3 \cdot 3H_2O$	yellow crystals, sl sol ρ 3.24	corrosion-inhibiting pigment
		Cr(III) compounds		
ammonium tetrathiocyanato diamminechromate(III)	[13573-16-5]	$NH_4(NH_3)_2Cr(SCN)_4$	red crystals	known as Reinecke's salt, analytical reagent for amines and alkaloids
basic chrome acetate	[39430-51-8]	$Cr_3(OH)_2(C_2H_3O_2)_7 \cdot xH_2O$	blue-green powder, sol H_2O	oil drilling muds, textile dye mordant, catalyst for organic oxidations[m,n]
basic chrome chloride	[50925-66-1]	$Cr_5(OH)_6Cl_9 \cdot xH_2O$	available as a green powder sol H_2O, hygroscopic	textile dye mordant, release adhesives, polymerization cross-linking agent[n]
basic chrome formate	[73246-98-7]	$Cr(OH)(OOCH)_2 \cdot 4H_2O$	green needles, example of rare crystalline basic Cr(III) salt, sol H_2O	skein printing of cotton tanning
chromic acetate	[1066-30-4]	$Cr(C_2H_3O_2)_3$	usually sold as a solution	printing and dyeing textiles[o]

302

Name	CAS Number	Formula	Properties	Uses
chromic acetylacetonate	[13681-82-8]	Cr(C$_5$H$_7$O$_2$)$_3$	[a]	preparation Cr complexes, catalysts, antiknock compounds
chromic ammonium sulfate	[10022-47-6]	NH$_4$Cr(SO$_4$)$_2$·12H$_2$O	violet crystals, ρ 1.72, mp 94°C, sol water	Cr electrowinning salt
chromic chloride	[10025-73-7]	CrCl$_3$	[a]	chromizing, Cr metal organochromium compounds[p]
chromic fluoborate	[27519-39-7]	Cr(BF$_4$)$_3$	available as a solution	Cr plating, in catalysts
chromic fluoride	[7788-97-8]	CrF$_3$	green crystals, ρ 3.78, insol H$_2$O	chromizing
chromic hydroxy dichloride	[14982-80-0]	Cr(OH)Cl$_2$	sold as a water solution, can be made as isopropanol solution	manufacturing Quilon, Volan, and Scotchgard[q]
chromic napthenate	[61788-69-0]	no definite formula	sold as soln in petroleum solvents	textile preservative
chromic nitrate	[26679-46-9]	Cr(NO$_3$)$_3$·9H$_2$O	violet crystals, ρ 1.80, mp 66.3°C, sol in H$_2$O	catalysts, textiles, manufacturing CrO$_2$[r]
chromic phosphate	[27096-04-4]	CrPO$_4$	green powder, also available as a solution in H$_3$PO$_4$	pigments, phosphate coatings, wash primers
chromic potassium oxalate	[15275-09-9]	K$_3$[Cr(C$_2$O$_4$)$_3$]·3H$_2$O	violet crystals	dye mordant
chromic potassium sulfate	[7788-99-0]	KCr$_2$(SO$_4$)$_3$·12H$_2$O	[a] also available as a green powder with < 12 H$_2$O[m]	hardening photographic emulsions, dietary supplement
chromic sulfate	[15005-90-0]	Cr$_2$(SO$_4$)$_3$·xH$_2$O	green amorphous powder	insolubilizing gelatin

303

Table 14. (Continued)

Name	CAS Registry Number	Molecular formula	Properties	Uses
		Cr(III) compounds (Continued)		
cobalt chromite	[12016-69-2]	$CoCr_2O_4$	turquoise blue crystals, spinel	ceramics, catalysts
copper chromite	[12018-10-9]	$CuCr_2O_4$	black crystals, distorted spinel	catalysts esp. automobile exhaust[r]
magnesium chromite	[12053-26-8]	$MgCr_2O_4$	brown crystals, spinel, ρ 4.415	refractory
zinc chromite	[12018-19-8]	$ZnCr_2O_4$	green crystals, spinel, ρ 5.30	catalyst[s]
		Other oxidation states		
chromium(0) hexacarbonyl	[13007-92-6]	$Cr(CO)_6$	[a]	synthesis of organo— chromium and hydride compounds, preparation of CrO^t
dicumene chromium(0)	[12001-89-7]	$[(CH_3)_2CHC_6H_5]_2Cr$	estd bp 300°C, explodes at 210°C	preparation of Cr carbides by vapor deposition[u]
chromium(II) chloride	[10049-05-5]	$CrCl_2$	[a]	chromizing, preparation of Cr metal[v]
chromium(IV) oxide	[12018-01-8]	CrO_2	[a]	magnetic tapes[w]
calcium chromate(V)	[12205-18-4]	$Ca_3(CrO_4)_2$	green crystals, similar to $Ba_3(CrO_4)_2$[a]	corrosion inhibiting pigment, suspect carcinogen[x]

[a]See Table 1. [b]Ref. 213. [c]Ref. 214. [d]Ref. 215. [e]Ref. 216. [f]Ref. 217; Ref. 218. [g]Ref. 219; Ref. 220. [h]Ref. 221. [i]Ref. 222. [j]Ref. 223. [k]Ref. 224. [l]Ref. 225. [m]Ref. 94. [n]Ref. 211. [o]Ref. 226. [p]Ref. 227; Ref. 228. [q]Ref. 197; Ref. 198. [r]Ref. 58. [s]Ref. 229; Ref. 230; Ref. 231. [t]Ref. 24; Ref. 233; Ref. 233. [u]Ref. 234. [v]Ref. 235. [w]Ref. 5; Ref. 236. [x]Ref. 59.

BIBLIOGRAPHY

"Chromium Compounds" in *ECT* 1st ed., Vol. 3, pp. 941–995, by J. J. Vetter, Diamond Alkalai Co., and C. Mueller, General Aniline & Film Corp.; in *ECT* 2nd ed., Vol. 5, pp. 473–516, by W. H. Hartford and R. L. Copson, Allied Chemical Corp.; in *ECT* 3rd ed., Vol. 6, pp. 82–120, by W. H. Hartford, Belmont Abbey College.

1. M. J. Udy, ed., *Chromium*, Vol. 1, Reinhold Publishing Co., New York, 1956, pp. 1–6.
2. C. L. Rollinson, in J. C. Bailer, Jr., H. J. Emeléus, R. Nyholm, and A. F. Trotman-Dickenson, eds., *Comprehensive Inorganic Chemistry*, Vol. 3, Pergamon Press, Oxford, UK, 1973, pp. 624–625.
3. U.S. Pat. 1,591,188 (Apr. 20, 1926), C. G. Fink (to United Chromium).
4. *Strategic Minerals — Extent of U.S. Reliance on South Africa*, U.S. General Accounting Office (GAO), Report to Congressional Requesters, GAO/NSIAD-88-201, Gaithersburg, Md., June 1988.
5. W. H. Hartford, in I. M. Kolthoff and P. J. Elving, eds., *Analytical Chemistry of the Elements*, Vol. 8, Part II, Wiley-Interscience, New York, 1963, p. 278.
6. Ref. 2, p. 625.
7. Ref. 1, pp. 113–250.
8. J. A. Dean, ed., *Lange's Handbook of Chemistry*, 13th ed., McGraw-Hill Book Co., New York, 1985.
9. W. F. Linke, ed., *Solubilities*, 4th ed., Vol. 1, D. Van Nostrand Co., Princeton, N.J., 1958.
10. R. C. Weast, ed., *CRC Handbook of Chemistry and Physics*, 65th ed., CRC Press Inc., Boca Raton, Fla., 1985.
11. F. Hein and S. Herzog, in G. Brauer, ed., *Handbook of Preparative Inorganic Chemistry*, Vol. 2, 2nd ed., Academic Press, Inc., New York, 1965, pp. 1334–1399.
12. F. A. Cotton and G. Wilkinson, *Advanced Inorganic Chemistry*, 5th ed., John Wiley & Sons, Inc., New York, 1988, pp. 679–697.
13. I. Dellin, F. M. Hall, and L. G. Hepler, *Chem. Rev.* **76,** 283,292 (1976).
14. "Chrom" in *Gmelins Handbuch der Anorganischen Chemie*, 8th ed., System no. 52, 1963–1965.
15. I. Nagypal and co-workers, *J. Chem. Soc. Dalton Trans.*, 1335 (1983).
16. J. Kragten, *Atlas of Metal Ligand Equilibria in Aqueous Solution*, Ellis Horwood Ltd., Coll House, UK, 1978, pp. 214–222.
17. H. Stünzi, L. Spiccia, F. P. Rotzinger, and W. Marty, *Inorg. Chem.* **28,** 66 (1989).
18. N. Tanaka, K. Ogino-Ebata, and G. Sato, *Bull. Chem. Soc. Japan* **31,** 366 (1966).
19. B. B. Owen, J. English, Jr., H. G. Cassidy, and C. V. Dundon, in L. F. Audrieth, ed., *Inorganic Syntheses*, Vol. 3, McGraw-Hill Book Co., New York, 1950, pp. 156–160.
20. Ref. 11, p. 1741.
21. R. B. King, *Advances in Organometallic Chemistry*, Vol. 2, Academic Press, Inc., New York, 1964, p. 182.
22. F. Calderazzo, R. Ercoli, and G. Natta, *Organic Syntheses via Metal Carbonyls*, Vol. 1, Wiley-Interscience, New York, 1968, p. 147.
23. Ref. 11, p. 1395.
24. Ref. 2, p. 648.
25. Ref. 2, p. 655.
26. I. M. Kolthoff and R. Belcher, *Volumetric Analysis*, Vol. 3, Interscience Publishers, Inc., New York, 1957, pp. 630–631.
27. Ref. 12, p. 684.
28. R. W. Kolaczkowski and R. A. Plane, *Inorg. Chem.* **3,** 322 (1964).
29. Ref. 12, pp. 685–686.
30. A. Bino, F. A. Cotton, and W. Kaim, *J. Am. Chem. Soc.* **101,** 2506 (1979).

31. J. J. H. Edema, S. Gambarotta, F. van Bolhuis, and A. L. Spek, *J. Am. Chem. Soc.* **111,** 2142 (1989).
32. Ref. 2, p. 660.
33. Ref. 11, p. 1346.
34. Ref. 12, p. 680.
35. Ref. 12, pp. 684–685.
36. B. J. Sturm, *Inorg. Chem.* **1,** 665 (1962).
37. G. W. Watt, P. S. Gentile, and E. P. Helvenston, *J. Am. Chem. Soc.* **77,** 2752 (1955).
38. G. B. Heisig, B. Fowkes, and R. Hedin, *Inorganic Syntheses*, Vol. 2, McGraw-Hill Book Co., New York, 1946, p. 193.
39. Ref. 2, p. 664.
40. Ref. 12, p. 682.
41. F. Baslo and R. G. Pearson, *Mechanisms of Inorganic Reactions*, 2nd ed., John Wiley & Sons, Inc., New York, 1967, pp. 141–145.
42. Ref. 2, p. 679.
43. J. E. Tackett, *Appl. Spectros.* **43,** 490 (1989).
44. G. L. Robbins and R. E. Tapscott, *Inorg. Chem.* **15,** 154 (1976).
45. J. A. Laswick and R. A. Plane, *J. Am. Chem. Soc.* **81,** 3564 (1959).
46. L. Spiccia and W. Marty, *Inorg. Chem.* **25,** 266 (1986).
47. D. Rai, B. M. Sass, and D. A. Moore, *Inorg. Chem.* **26,** 345 (1987).
48. L. Spiccia, H. Stoeckli-Evans, W. Marty, and R. Giovanoli, *Inorg. Chem.* **26,** 474 (1987).
49. L. Spiccia, *Inorg. Chem.* **27,** 432 (1988).
50. Ref. 2, pp. 670–672.
51. Ref. 12, pp. 689–690.
52. B. M. Lorffler and R. G. Burns, *Am. Sci.* **64,** 636 (1977).
53. W. H. Hartford, *Rocks Miner.* **52,** 169 (1977).
54. Ref. 2, p. 666.
55. D. M. L. Goodgame and A. M. Joy, *Inorg. Chim. Acta* **135,** 115 (1987).
56. Ref. 12, p. 692.
57. G. L. Gard, *Inorg. Chem.* **25,** 426 (1986).
58. U.S. Pat. 3,117,093 (Jan. 7, 1964), P. Arthur, Jr. and J. N. Ingram (to E. I. du Pont de Nemours & Co., Inc.).
59. W. H. Hartford, in D. M. Serrone, ed., *Proceedings Chromium Symposium 1986: An Update*, Industrial Health Foundation, Inc., Pittsburgh, Pa., 1986, p. 9.
60. R. Scholder and H. Suchy, *Z. Anorg. Allgem. Chem.* **308,** 295 (1961).
61. H. Schwartz, *Z. Anorg. Allgem. Chem.* **323,** 275 (1963).
62. Ref. 11, pp. 1392–1393.
63. Ref. 12, p. 696.
64. Ref. 11, pp. 1391–1392.
65. K. O. Christe and co-workers, *Inorg. Chem.* **25,** 2163 (1986).
66. Ref. 12, p. 694.
67. K. B. Wiberg, in K. B. Wiberg, ed., *Oxidation in Organic Chemistry, Part A*, Academic Press, Inc., New York, 1965, pp. 69–184.
68. T. Shen-yang and L. Ke-an, *Talanta* **33,** 775 (1986).
69. Ref. 12, p. 693.
70. Ref. 1, p. 137.
71. Ref. 2, p. 695.
72. P. Löfgren and K. Waltersson, *Acta Chem. Scand.* **25,** 35 (1971).
73. K. B. Wiberg and W. H. Richardson, *J. Am. Chem. Soc.* **84,** 2800 (1962).
74. K. B. Wiberg and P. A. Lepse, *J. Am. Chem. Soc.* **86,** 2612 (1964).
75. F. H. Westheimer, *Chem. Rev.* **45,** 419 (1949).

76. R. J. Barnhart, *AESF Second Chromium Colloquium*, American Electroplaters and Surface Finishers Society, Orlando, Fla., 1990, Session III, A. p. 1.

77. R. J. Barnhart, private communication, Mar. 16, 1992.

78. F. Ullman, *Enzyklopädie der Technischen Chemie*, Vol. 3, Urban and Schwarzenberg, Berlin, Germany, 1929, pp. 400–433.; F. McBerty and Wilcoxon, *FIAT Rev. Ger. Sci. PB22627, Final Report No. 796*, 1946; Ref. 1, pp. 262–282.

79. Ref. 59, p. 21.

80. U.S. Pat. 3,095,266 (June 25, 1963), W. B. Lauder and W. H. Hartford (to Allied Chemical Corp.).

81. U.S. Pat. 3,853,059 (Dec. 3, 1974), C. P. Bruen, W. W. Low, and E. W. Smalley (to Allied Chemical Corp.).

82. S. African Pat. 74-03,604 (Apr. 28, 1975), C. P. Bruen and co-workers (to Allied Chemical Corp.).

83. U.S. Pat. 3,816,095 (June 11, 1974), C. P. Bruen, W. W. Low, and E. W. Smalley (to Allied Chemical Corp.).

84. U.S. Pat. 1,873,589 (Aug. 23, 1932), P. R. Hines (to Harshaw Chemical Co.).

85. U.S. Pat. 3,065,055 (Nov. 20, 1962), T. S. Perrin and R. E. Banner (to Diamond Alkali Co.); U.S. Pat. 3,607,026 (Sept. 21, 1971), T. S. Perrin, R. E. Banner, and J. O. Brandstaetter (to Diamond Alkali Co.).

86. U.S. Pat. 5,094,729 (Mar. 10, 1992), H. Klotz, H. D. Pinter, R. Weber, H. Block, and N. Lönhoff (to Bayer Aktiengesellschaft); U.S. Pat. 5,096,547 (Mar. 17, 1992) H. Klotz, R. Weber, and W. Ohlendorf (to Bayer Aktiengesellschaft).

87. J. F. Papp, *Chromium 1989 Minerals Yearbook*, U.S. Department of the Interior, Bureau of Mines, Washington, D.C., 1989, pp. 4, 7, 18.

88. *Chem. Mark. Rep.*, June 17, 1991.

89. *Inorganic Chemicals*, Report MA28A(81)-1, p. 10, 1981; Report MA28A(86)-1, p. 8, 1986; Report MA28A(88)-1, p. 8, 1988; U.S. Bureau of the Census, Washington, D.C.

90. *Reagent Chemicals*, 6th ed., American Chemical Society, Washington, D.C., 1981, pp. 185–190, 430–431, 434–435.

91. J. Rosin, *Reagent Chemicals and Standards*, 5th ed., D. Van Nostrand Reinhold Co., New York, 1967.

92. *Federal Specifications* O-C-303D, Aug. 13, 1985; O-A-498B, United States General Services Administration (GSA), July 11, 1985; O-S-588C, June 20, 1985; O-S-595B, May 15, 1967; O-P-559, May 26, 1952.

93. W. H. Hartford, in F. D. Snell and L. C. Ettre, eds., *Encyclopedia of Industrial Chemical Analysis*, Vol. 9, John Wiley & Sons, Inc., New York, 1970, pp. 680–709.

94. *Chrome*, McGean Division of McGean-Rohco, Inc., Cleveland, Ohio, 1991.

95. F. Feigl, *Spot Tests*, 4th English ed., Elsevier Publishing Co., Amsterdam, The Netherlands, 1954, pp. 159–162.

96. F. D. Snell, *Photometric and Fluorometric Methods of Analysis of Metals*, John Wiley & Sons, Inc., New York, 1978, pp. 714–729.

97. E. D. Olsen and C. C. Foreback, in F. D. Snell and L. C. Ettre, in Ref. 93, Vol. 9, pp. 632–680.

98. W. H. Hartford, in Ref. 93, Vol. 9, pp. 176–213.

99. W. H. Hartford, in I. M. Kolthoff and P. J. Elving, eds., *Treatise on Analytical Chemistry*, Part II, Vol. 8, John Wiley & Sons, Inc., New York, 1963, pp. 273–369.

100. R. Bock, *A Handbook of Decomposition Methods in Analytical Chemistry*, John Wiley & Sons, Inc., New York, 1979.

101. Ref. 26, pp. 334–335.

102. Ref. 96, pp. 729–732.

103. V. D. Lewis, S. H. Nam, and I. T. Urasa, *J. Chromatogr. Sci.* **27,** 489 (1989).

104. E. Nieboer and A. A. Jusys, in J. O. Nriagu and E. Nieboer, eds., *Chromium in the Natural and Human Environments*, John Wiley & Sons, Inc., New York, 1988, p. 21.
105. F. L. Petrilli and S. DeFlora, in Ref. 59, p. 112.
106. W. E. Reinhart and S. C. Gad, *Am. Ind. Hyg. Assoc. J.* **47,** 696 (1986).
107. D. Steinhoff, S. C. Gad, G. K. Hatfield, and U. Mohr, *Exp. Pathol.* **30,** 129 (1986).
108. J. A. Hathaway, in Ref. 59, p. 87.
109. R. D. Harbison and W. E. Rinehart, eds., *Conclusions of the Expert Review Panel on Chromium Contaminated Soil in Hudson County, New Jersey*, Industrial Health Foundation, Pittsburgh, Pa., 1990, p. 30.
110. T. F. Mancuso, *Ind. Med. Surg.* **20,** 358 (1951).
111. Chrome Coalition, *Issues Document*, Feb. 1988, p. 14.
112. S. DeFlora and K. E. Wetterhan, *Life Chemistry Reports* **7,** 169 (1989).
113. S. C. Gad and co-workers, Ref. 59, p. 43.
114. L. Polak, in D. Burrows, ed., *Chromium Metabolism and Toxicity*, CRC Press, Inc., Boca Raton, Fla., 1983, pp. 51–136.
115. S. Fregert and H. Rorsman, *Arch. Dermatol.* **90,** 4 (1964).
116. A. T. Haynes and E. Nieboer, in Ref. 104, pp. 497–532.
117. T. F. Gale, *Environ. Res.* **16,** 101 (1978).
118. T. F. Gale and J. D. Bunch, *Teratology* **19,** 81 (1979).
119. B. R. G. Danielsson, E. Hassoun, and L. Dencker, *Arch. Toxicol.* **51,** 233 (1982).
120. N. Matsumoto, S. Iijima, and H. Katsunuma, *J. Toxicol. Sci.* **2,** 1 (1976).
121. S. Iijima, N. Matsumoto, C. C. Lu, and H. Katsunuma, *Teratology* **20,** 152 (1979).
122. J. Behari, S. V. Chandra, and S. K. Tandon, *Acta. Biol. Med. Germ.* **37,** 463 (1978).
123. *IARC Monographs on the Evaluation of Carcinogenic Risks to Humans*, supplement 7, World Health Organization, International Agency for Research on Cancer (IARC), 1987, pp. 165–168.
124. *Health Effects Assessment for Hexavalent Chromium*, EPA/540/1-86-019, United States Environmental Protection Agency (EPA), Sept. 1984; *Toxicological Profile for Chromium*, Agency for Toxic Substances and Disease Registry (ASTDR), ASTDR/TP-88/10, 1989.
125. NIOSH Publication No. 76-129, U.S. Department of Health, Education and Welfare, Cincinnati, Ohio, 1975.
126. *Fifth Annual Report on Carcinogens Summary 1989*, NTP 89-239, U.S. Department of Health and Human Services (DHHS), National Toxicology Program (NTP) of the National Institute of Environmental Health Sciences, Research Triangle Park, N.C., 1989, pp. 24–30.
127. *Health Effects Assessment for Trivalent Chromium*, EPA/540/1-86-035, United States Environmental Protection Agency (EPA), Sept. 1984.
128. F. L. Petrilli and S. DeFlora, in Ref. 59, p. 100.
129. E. Nieboer and S. L. Shaw, in Ref. 104, pp. 399–442.
130. A. Yassi and E. Nieboer, in Ref. 104, pp. 443–496.
131. Ref. 129, p. 411.
132. V. Bianchi and co-workers, *Mutat. Res.* **117,** 279 (1983).
133. Z. Elias and co-workers, *Carcinogenesis* **4,** 605 (1983).
134. P. Venier and co-workers, *Carcinogenesis* **6,** 1327 (1985).
135. W. Mertz, *Science* **212,** 1332 (1981).
136. W. Mertz, *Physiol. Ref.* **49,** 163 (1969).
137. *Health Assessment Document for Chromium*, EPA-600/8-83-014F, United States Environmental Protection Agency (EPA), 1984.
138. W. Mertz, *Contemp. Nut.* **7**(3), 2 (1982).
139. K. N. Jeejeebhoy and co-workers, *Am. J. Clin. Nutr.* **30,** 531 (1977).
140. H. Freund, S. Atamian and J. E. Fischer, *J. Am. Medical Assoc.* **241,** 496 (1979).

141. R. A. Anderson, *Sci. Total Environ.* **17,** 13 (1981).
142. R. S. Anderson, in Ref. 59, p. 238.
143. J. A. Fisher, *The Chromium Program*, Harper and Row Publishers, New York, 1990.
144. *U.S. Code of Federal Regulations* (CFR) is cited as Title CFR Part number: 40 CFR 302.4; 40 CFR 300.5; 40 CFR 401.15; 29 CFR 1910.1000.
145. NIOSH Publication No. 85-114, U.S. Department of Health and Human Services.
146. 40 CFR 61.
147. Ref. 111, p. 20.
148. California Air Resources Board and Department of Health Services, *Report to the Scientific Review Panel on Chromium*, Sacramento, Calif., 1985.
149. J. M. Pacyna and J. O. Nriagu, in Ref. 104, pp. 105–123.
150. C. Signeur, in Ref. 59, p. 415.
151. *Ambient Water Quality Criteria for Chromium*, EPA 440/5-80-35, United States Environmental Protection Agency (EPA), 1980; *Fed. Reg.* (FR), cited as Volume FR Number: 56 FR 3526; 56 FR 30266.
152. 52 CFR 8140.
153. 40 CFR 261.4; Ref. 6.
154. U.S. Pat. 3,027,321 (Mar. 27, 1962), R. P. Selm and B. T. Hulse (to Wilson & Co.).
155. U.S. Pat. 3,294,680 (Dec. 29, 1966), L. E. Lancy (to Lancy Laboratories).
156. R. J. Bartlett, in Ref. 59, p. 310.
157. R. J. Bartlett and B. R. James, *J. Environ. Quality* **8,** 31 (1979).
158. 55 FR 61; 51 FR 26420; 50 FR 46966.
159. R. A. Griffin, A. K. Au, and R. R. Frost, *J. Environ. Sci., Health* **A12**(8), 431 (1977).
160. D. Rai and L. E. Eary, in Ref. 59, p. 331.
161. *Treatment Technology Background Document*, United States Environmental Protection Agency (EPA), 1989.
162. *Proposed Best Demonstrated Available Technology (BDAT) Background Document for Chromium Wastes D007 and U032*, EPA/530-SW-90-011U, United States Environmental Protection Agency (EPA), 1989.
163. R. A. Saar and J. H. Weber, *Environ. Sci. Technol.* **16**(9), 510A (1982).
164. *Field Investigation and Evaluation of Land Treating Tannery Sludges*, EPA/600/52-86/033, United States Environmental Protection Agency (EPA), 1986.
165. 45 FR 72027; 45 FR 72035; 55 FR 11862.
166. U.S. Pat. 3,552,917 (Jan. 5, 1971), C. O. Weiss (to M&T Chemicals, Inc.); U.S. Pat. 3,728,273 (Apr. 17, 1973), C. P. Bruen and C. A. Wamser (to Allied Chemical Corp.).
167. *Chem. Mark. Rep.* Oct.
168. W. H. Hartford in A. J. Bard, ed., *Applied Electrochemistry of the Elements*, Marcel Dekker, New York, 1977.
169. G. Dubpernell in F. A. Lowenheim, ed., *Modern Electroplating*, 3rd ed., Wiley-Interscience, New York, 1974, pp. 87–151.
170. J. M. Hosdowich, in M. J. Udy, ed., *Chromium*, Vol. II, Reinhold Publishing Corp., New York, 1958, pp. 65–91.
171. F. A. Lowenheim and M. R. Moran, *Faith, Keyes, and Clark's Industrial Chemicals*, 4th ed., Wiley-Interscience, New York, 1975, pp. 716–721.
172. V. Opaskar and D. Crawford, *Metal Finishing* **89**(1), 49 (1989).
173. D. L. Snyder, in P. H. Langdon, ed., *Metal Finishing 59th Guidebook and Directory*, Metals and Plastics Publications, Inc., Hackensack, N.J., 1991, pp. 179–187.
174. D. Smart, T. E. Such, and S. J. Wake, *Bull. Inst. Met. Finish.* **61,** 105 (1983).
175. K. R. Newby, in Ref. 173, pp. 188–196.
176. J. P. Hoare, *J. Electrochem. Soc.: Electrochem. Sci. and Technol.* **126,** 190 (1979).
177. P. Caokan, G. Barnafoldi, and I. Royik, *Bull. Doc. Cent. Inform. Chrome Dur.*, Paris, France, June 1971, pp. 11–A50.

178. F. W. Eppensteiner and M. R. Jenkins, in Ref. 173, pp. 418–431.
179. U.S. Pat. 4,171,231 (Oct. 16, 1979), C. V. Bishop, T. J. Foley, and J. M. Frank (to R. O. Hull & Co., Inc.).
180. U.S. Pat. 4,705,576 (Nov. 10, 1987), K. Klos, K. Lindemann, and W. Birnstiel (to Elektro-Brite GmbH).
181. G. Jarrett, *Met. Finish.* **65**(3), 90 (1967).
182. L. F. G. Williams, *Surf. Technol.* **4**, 355 (1976).
183. C. V. Bishop, D. M. Burdt, and K. R. Romer, *Galvanotechnik* **71**, 1199 (1980).
184. D. C. Montgomery and C. A. Grubbs, in Ref. 173, pp. 409–416.
185. Ref. 1, pp. 406–422.
186. T. C. Patton, *Pigment Handbook*, Vol. 1, Wiley-Interscience, New York, 1973.
187. C. H. Love, *Important Inorganic Pigments*, Hobart, Washington, D.C., 1947.
188. Ref. 187, pp. 357–389.
189. *1984 Annual Book of ASTM Standards*, Section 6, Vol. 6.02, American Society for Testing and Materials (ASTM), Philadelphia, Pa., 1984.
190. Ref. 187, pp. 351–357.
191. U.S. Pat. 2,246,907 (July 30, 1940), O. F. Tarr and L. G. Tubbs (to Mutual Chemical Company of America).
192. Ref. 187, pp. 843–861.
193. W. H. Hartford, *J. Am. Chem. Soc.* **72**, 1286 (1950).
194. U.S. Pat. 2,415,394 (Feb. 4, 1947), O. F. Tarr and M. Darrin (to Mutual Chemical Company of America).
195. Ref. 1, pp. 283–301.
196. H. A. Lubs, *The Chemistry of Synthetic Dyes and Pigments*, Robert E. Krieger Publishing Co., Huntington, N.Y., 1972, pp. 153, 160, 161, 247, 258, 284, 426.
197. U.S. Pat. 2,662,835 (Dec. 16, 1953), T. S. Reid (to Minnesota Mining & Manufacturing Co.).
198. U.S. Pat. 2,683,156 (July 6, 1954), R. F. Iler (to E. I. du Pont de Nemours & Co., Inc.).
199. *American Wood-Preserver's Association, Standards 1984*, Section 2, Standard No. P5-83, American Wood-Preserver's Association (AWPA), Stevensville, Md., 1984.
200. W. H. Hartford, in D. D. Nicholas, ed., *Wood Deterioration and its Prevention by Preservative Treatments*, Vol. 2, Syracuse University Press, Syracuse, N.Y., 1973, pp. 1–120.
201. S. E. Dahlgren and W. H. Hartford, *Holzforschung*, **26**(2), 62 (1972).
202. *Ibid.* **26**(3), 105 (1972).
203. *Ibid.* **26**(4), 142 (1972).
204. *Ibid.* **28**(2), 58 (1974).
205. *Ibid.* **29**(3), 84 (1975).
206. *Ibid.* **29**(4), 130 (1975).
207. R. D. Arsenault, *Proc. Am. Wood Preserver's Assoc.* **71**, 126 (1975).
208. W. H. Hartford, *Proc. Am. Wood Preserver's Assoc.* **72**, 172 (1976).
209. W. F. Rogers, *Composition and Properties of Oil Well Drilling Fluids*, 3rd ed., Gulf Publishing Co., Houston, Tex. 1963, pp. 420–422.
210. W. G. Skelly and D. E. Dieball, *J. Soc. Petrol. Engineers* **10**(2), 140 (1970).
211. U.S. Pat. 4,683,949 (Aug. 4, 1987), R. D. Sydansk and P. A. Argabright (to Marathon Oil Co.).
212. R. D. Sydansk and P. E. Moore, *Enhanced Oil Recovery Symposium*, Casper, Wyo., May 3–4, 1988.
213. R. H. van Domelyn and R. D. Wehrle, *Proceedings of the 9th Intersociety Energy Conservation Engineering Conference*, American Society of Mechanical Engineers, New York, 1975, pp. 665–670.
214. F. Tepper, in Ref. 213, pp. 671–677.

215. R. H. Comyn, M. L. Couch, and R. E. McIntyre, *Report TR-635*, Diamond Ordnance Fuze Laboratories, 1958.
216. M. L. Kastens and M. J. Prigotsky, *Ind. Eng. Chem.* **41**, 2376 (1949).
217. W. H. Hartford, K. A. Lane, and W. A. Meyer Jr., *J. Am. Chem. Soc.* **72**, 3353 (1950).
218. S. Budavari, ed., *The Merck Index*, Merck and Company, Rahway, N.J., 1989, p. 253.
219. W. H. Hartford and M. Darrin, *Chem. Rev.* **58**, 1 (1958).
220. Ref. 218, p. 349.
221. U.S. Pat. 3,824,160 (July 30, 1974), W. H. Hartford (to Allied Chemical Corp.).
222. W. H. Hartford, K. A. Lane, and W. A. Meyer, Jr., *J. Am. Chem. Soc.* **72**, 1286 (1950).
223. U.S. Pat. 2,764,553 (Sept. 25, 1956), W. H. Hartford (to Allied Chemical Corp.).
224. W. H. Hartford and K. A. Lane, *J. Am. Chem. Soc.* **70**, 647 (1948).
225. R. L. Costa and W. H. Hartford, *J. Am. Chem. Soc.* **80**, 1809 (1848).
226. Ref. 1, pp. 219–221.
227. U.S. Pat. 3,305,303 (Feb. 21, 1967), W. H. Hartford and E. B. Hoyt (to Allied Chemical Corp.).
228. U.S. Pat. 3,309,172 (Mar. 14, 1967), W. H. Hartford and E. B. Hoyt (to Allied Chemical Corp.).
229. U.S. Pat. 3,532,457 (Oct. 6, 1970), K. H. Koepernik (to KaliChemie A.G.).
230. O. F. Joklik, *Chem. Eng.*, **80**(23), 49 (1973).
231. U.S. Pat. 3,007,905 (Nov. 7, 1961), G. C. Bailey (to Phillips Petroleum).
232. Ger. Pat. 1,007,305 (May 2, 1957), E. O. Fischer and W. Hafner (to Badische Anilin- und Soda-Fabrik AG).
233. A. F. Wells, *Structural Inorganic Chemistry*, 3rd ed., Oxford University Press, London, UK, 1975, p. 76.
234. W. H. Metzger, Jr., *Plating* **49**, 1176 (1962).
235. U.S. Pat. 3,414,428 (Dec. 3, 1968) and 3,497,316 (Feb. 24, 1970), W. R. Kelly and W. B. Lauder (to Allied Chemical Corp.).
236. *Chemical Week 1991 Buyer's Guide*, Vol. 147, No. 18, Chemical Week Associates, New York, 1990.
237. *Chem Sources-USA 1988*, 28th ed., Directories Publishing Co., Ormond Beach, Fla., 1988.
238. *1991 OPD Chemical Buyers Directory*, 78th ed., Schnell Publishing Co., New York, 1990.
239. Ref. 11, pp. 1672–1674.
240. F. Hanic, I. Horváth, G. Plesch, and L. Gáliková, *J. Solid State Chem.* **59**, 190 (1985).
241. F. Hanic, I. Horváth, and G. Plesch, *Thermochimica Acta*, **145**, 19 (1989).
242. Ref. 233, pp. 489–498.
243. Ref. 1, pp. 385–405.
244. R. H. Domelyn and R. D. Wehrle, in Ref. 213, pp. 665–670.
245. D. A. Johnson and M. A. Reid, *J. Electrochem. Soc.: Electrochem. Sci. Technol.* **132**(5), 1058 (1985).
246. U.S. Pat. 2,885,365 (May 5, 1959), A. L. Oppegard (to E. I. du Pont de Nemours & Co., Inc.).

BILLIE J. PAGE
GARY W. LOAR
McGean-Rohco, Inc.

CHROMOGENIC MATERIALS

ELECTROCHROMIC

The term electrochromism was apparently coined to describe absorption line shifts induced in dyes by strong electric fields (1). This definition of electrochromism does not, however, fit within the modern sense of the word. Electrochromism is a reversible and visible change in transmittance and/or reflectance that is associated with an electrochemically induced oxidation–reduction reaction. This optical change is effected by a small electric current at low d-c potential. The potential is usually on the order of 1 V, and the electrochromic material sometimes exhibits good open-circuit memory. Unlike the well-known electrolytic coloration in alkali halide crystals, the electrochromic optical density change is often appreciable at ordinary temperatures.

Coloration occurs both cathodically and anodically, as well as in both organic and inorganic materials. Compounds of all types may be classified within one or the other of two general groups based on the nature of charge balancing. In one group, an electrolyte separates a cathode–anode pair, one or both of which may be chromogenically active. Typically, the chromogenic material is a thin film on the cathode or anode. As charge neutrality must be preserved, and the electrochromic cathode or anode is a solid, insertion/extraction of ions, often H^+ or alkali, accompanies reduction–oxidation within the electrode surface layer. Insertion/extraction in the cathode or anode is the distinguishing feature of this group. The second group is best described by referring to the viologens, a family of halides of quaternary bases derived from the 4,4'-bipyridinium structure. Viologens are recognized as the first important organic electrochromic materials (2,3). Some of these color deeply within solution by simple reduction; others are distinguished by their deep coloration when electrodeposited from solution onto a cathode. These colorations typify the noninsertion group, though incidental insertion may accompany electrodeposition.

Members of the ion-insertion/extraction group, as inorganic or organic thin films, especially the former, have attracted the widest interest most recently. Tungsten trioxide was the earliest exploited inorganic compound (4), even before the mechanism of its electrochromic response was understood (5). It is still the best known of the important ion-insertion/extraction group.

Ongoing research with a wide variety of materials from both groups has been reviewed frequently (5–13). Much of the earliest published work followed from research on displays, but opportunities for switchable mirrors and windows have been highlighted as well. With one noteworthy exception, however, there has been no remarkable commercial success. In part, this is because the technology involves many complex scientific and engineering principles. Also, the competing liquid crystal technology has evolved successfully in some display applications. The one commercial exception is an electrochromic automotive rearview mirror which has been gaining popularity since 1988. The mirror contains an encapsulated solution of viologen, which undergoes optical switching without electrodeposition (14).

Oxidation–Reduction in the Noninsertion/Extraction Group

The best known examples in this group are organic dyes, and the vehicle in which the oxidation–reduction takes place is, in general, a liquid electrolyte. For displays, however, it is preferred that color not be developed within the liquid itself, but rather by electrodeposition. Otherwise, there is drifting of the coloration and poor memory, which are especially troublesome for displaying information with high resolution. Earlier examples, which depended on oxidation–reduction in solution or on a pH indicating effect at an electrode, were abandoned (6). The drifting, however, has proven to be acceptable for a mirror (14).

Organic Compounds. Viologens typically require a very low charge density of 2 mC/cm^2 to develop sufficient contrast for display applications. They are the only compounds of the group that have been studied extensively (11). The best known viologen is 1,1'-diheptyl-4,4'-bipyridinium dibromide [6159-05-3]. A cell is caused to switch from clear to bluish-purple when the divalent cation is univalently reduced in aqueous KBr solution and electrodeposited on the cathode as the bromide (2,3). The electrochromic response is visible with an applied potential more negative than -0.66 V vs SCE. The peak of visible absorption is at 545 nm. A widely accepted structure for the divalent diheptylviologen ion is shown below, where R = C$_7$H$_{15}$.

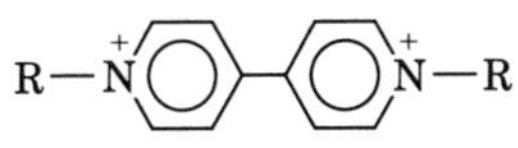

$$R-\overset{+}{N}\bigcirc\!\!-\!\!\bigcirc\overset{+}{N}-R$$

The added electron is delocalized on the monovalent radical ion to which it is reduced (3). There is no general agreement on the molecular representation of the reduced structure. Various other viologen compounds have been mentioned (9,12). Even a polymeric electrochromic device (15) has been made, though the penalty for polymerization is a loss in device speed. Methylviologen dichloride [1910-42-5] was dissolved in hydrated poly(2-acrylamido-2-methylpropanesulfonic acid) [27119-07-9], producing a tacky polymer electrolyte (16,17). Poly(2-acrylamido-2-methylpropanesulfonic acid) was also used to immobilize methylene blue [61-73-4] (15), which oxidizes from the colorless and neutral molecule by a one-electron transfer. Ion transport was not described for either polymeric example.

Other oxidation–reduction organic compounds have been used, in a solution or paste, which demonstrate various electrochromic colors. These include o-tolidine (4,4'-diamino-3,3'-dimethylbiphenyl) [119-93-7] (18) and 2-tert-butylanthraquinone [84-47-9] (19). However, even with the flexibility for molecular synthesis with organic materials, the experience with viologens typifies two important difficulties. Cycling instability associated with the onset of side reactions remains a consistent problem for displays, for which a lifetime $\geq 10^7$ cycles is generally required. Viologens suffer especially from a so-called recrystallization of the electrodeposit, which impairs erasure of the darkened state (20). Efforts to alter the molecular structure of the radical ion halide salt have been without far-reaching success for display devices. Besides this, susceptibility to degradative oxidation and photo-oxidation requires sealing out oxygen and minimizing exposure to uv frequencies of light. Clearly, this susceptibility to oxidation is an especially serious technical hurdle for switchable mirrors and windows exposed to outdoor conditions.

Inorganic Electrodeposition. From a comprehensive analysis (21,22) with a variety of electrodepositable metals, the reversible cathodic electroplating of silver [7440-22-4] has been determined to be best. A preferred aqueous solution for light shutters and displays contains 3.0–3.5 M AgI [7783-96-2] and 7 M NaI [7681-82-5]. The complementary I_3^- anodic oxidation product contributes to the change in optical density. For a targeted change in optical density of at least 3.0 for all visible wavelengths, the minimum charge density = 80 mC/cm^2. This is a comparatively high charge density, though for some applications the demand might not be so great. For the highest speeds and contrast intended at the time of the analysis, several plating cells were required, back to back. The technology has this problem as well as complexities with current distributions and a strong tendency toward nonuniformity. It has not matured for practical displays any more than the reversible, drift-prone, dye-electrolyte solutions it was intended to replace. In both cases, also, dependence on liquid-state electrolytes presents practical problems in cell assembly and sealing. The solid state is preferred.

Insertion/Extraction Compounds

The seminal work on these materials began at American Cyanamid Co. in the 1960s (4,23), though these workers did not author the ion-insertion/extraction model that has become widely accepted (5). Numerous patents were granted to American Cyanamid Co. as a result of its display-oriented work. Much of what others have written in the open literature either confirms or adds to what these teach. Important papers (16) about cathodic WO$_3$-based insertion devices and others (17,24,25) summarize this activity. The so-called amorphous, or poorly crystallized, tungsten oxide thin film, which developed as the most important material, is, like the viologens, of great interest because of its reversible clear-to-deep-blue coloration in transmission. Coloration efficiency is high.

An important way to assess the many insertion/extraction films known is to compare spectral coloration efficiencies, CE(λ), for the visible region.

$$CE(\lambda) = \Delta OD(\lambda)/q \qquad (1)$$

where $\Delta OD(\lambda)$ represents the change in single-pass, transmitted optical density at the wavelength of interest λ, because of a transfer of charge q, as C/cm^2. Adherence to Lambert's law must either be assumed or tested to avoid pitfalls with thin films; cathodic coloration is well described by $CE(\lambda)$ at moderate values of q for many inorganic amorphous films (26). The coloration efficiency is determined by spectroelectrochemistry, using a cell which employs a cathode–anode pair in a liquid electrolyte. The cell is operated such that only the electrode of interest is in the light path of a spectrophotometer. For example, a 1 N LiClO$_4$–propylene carbonate electrolyte solution was used with a Li anode counter electrode strip to measure the optical density change in cathodic tungsten oxide and other films during galvanostatic switching (27). Li$^+$ was alternately inserted into and extracted from the films. The films were deposited on conductive glass by reactive RF sputtering. Others have used similar techniques to obtain electrochromic cycling data, and propylene carbonate has been commonly used as well. Often films have been deposited by other common methods in vacuum and sometimes by anodization of the metal.

Cathodically Colored Inorganic Films. The generalized cathodic, monovalent ion-insertion reaction for inorganic thin films is

$$M_nO_m \cdot yH_2O + x\, e^- + x\, J^+ \rightarrow J_xM_nO_m \cdot yH_2O \tag{2}$$

where M is a multivalent cation of the electrochromic oxide with valence $2m/n$; both m and n are taken as integers here, though that may not always be so. J$^+$ is the ion being inserted, and usually $0 < x < 1$. The y moles of H$_2$O indicate that these materials are generally variably hydrated, depending on preparation technique. Hydration and porosity are required for rapid coloration (5,28–30). On the other hand, for amorphous tungsten oxide thin films, porosity and water content are known to be associated with accelerated dissolution in acid aqueous electrolyte (5,24,29,31). It is noteworthy that single-crystal WO$_3$ is essentially insoluble, while otherwise interesting MoO$_3$ films are soluble.

Dissolution of amorphous tungsten oxide films in sealed capsules has been reported to be 2.0–2.5 nm/d at 50°C in 10:1 glycerol/H$_2$SO$_4$ (24). Dissolution is much more facile in water–H$_2$SO$_4$, an even earlier electrolyte choice. With more recently and successfully developed sulfonic acid-functionalized polymer electrolytes, cell stability depends on minimizing the water content of the polymer (17). As with the film itself, however, some minimum water content is necessary in the electrolyte for achieving rapid electrochromic response. A balance must be struck for the water content of both film and electrolyte. One other critical issue with proton-based electrolytes, which applies when they are used with any ion-insertion/extraction film, is the likelihood for the water-containing electrolyte to contribute to electrochemical H$_2$ or O$_2$ gas evolution. Because of these problems collectively, most of the recently published work with amorphous tungsten oxide has been done with alkali-ion insertion from nonaqueous electrolytes. Generally, however, these electrolytes have relatively low ion conductivity.

Table 1 shows ion-insertion data for some of the better known vacuum-deposited thin films of this class. These grow as amorphous (α) or poorly crystallized films at indicated low substrate temperatures. The coloration efficiency data illustrate one reason why amorphous tungsten oxide remains of high interest, despite the constraints already mentioned. All of the alternatives, including polycrystalline c-WO$_3$, have lesser coloration efficiencies in the all important visible region. In the near infrared, however, c-WO$_3$ exhibits a relatively high reflectance when darkened (34,35). The data in Table 1 have been used to propose an all solid-state window based on Li$^+$ insertion/extraction in cathodically coloring WO$_3$ (27). A low coloration, insertion/extraction oxide film such as α-Nb$_2$O$_5$ is deposited on conductive glass and serves as the counter electrode. A transparent

Table 1. Some Cathodically Colored, Inorganic Insertion[a]/Extraction Films[b]

Film	CAS Registry Number	Growth parameter[c]	CE(λ),[d] cm^2/C	λ, nm
α-WO$_3$	[1314-35-8]	ambient	$30 < CE < 125$	400–700
α-WO$_3$[e]		240–660 nm	$30 < CE < 50$	633
α-WO$_3$[f]		150°C	55	633
		1500 nm		
c-WO$_3$	[1314-35-8]	310°C	$5 < CE < 50$	400–700
c-MoO$_3$	[1313-27-5]	351°C	$30 < CE < 90$	500–700
α-Nb$_2$O$_5$	[1313-96-8]	40°C	<12	400–700
α-TiO$_2$	[13463-67-7]	40°C	<5	400–700
α-Ta$_2$O$_5$	[1314-61-0]	40°C	<5	400–700

[a]Insertion ion is Li$^+$ unless otherwise noted.
[b]Ref. 27, unless otherwise noted.
[c]Besides growth temperature, the column includes film thickness when known.
[d]Given as a value at a λ or bracketed over the indicated λ range.
[e]Also H$^+$ as insertion ion; Ref. 32.
[f]Ref. 33.

thin film such as LiAlF$_4$ overlies it and is the nonliquid electrolyte. Deposited on top of the electrolyte are a Li$^+$-precharged WO$_3$ film first, and then a conductive and transparent film. During cycling, the α-Nb$_2$O$_5$ does not contribute much coloration on the WO$_3$-bleach cycle, as desired, when it assumes a cathodic role. As ideal as this seems in principle, however, it has not yet been reduced to a practical art.

By way of contrast, using proton conductivity, at least two other workable solid-state designs have actually been demonstrated at scale. The function of neither one is dictated by the coloration efficiency of a counter electrode. These use sulfonic acid-functionalized polymer electrolytes and depend on unique counter electrodes. One has a high surface area carbon paper counter electrode for reversible proton storage in a display configuration (16). The other has a very fine, reversibly oxidized copper grid that permits vision through a large-area transparency (36,37).

Anodically Colored Inorganic Films. The important electrochromic films of this class have been discussed in the open literature since 1978 and include Prussian blue (PB) (38) and the highly hydrated (h) oxides of iridium (39) and nickel (40). Data for these are shown in Table 2. Of lesser significance are the hydrous oxides of rhodium [12680-36-3] and cobalt [11104-61-3] (7,45,46). Like the cathodically coloring insertion films, the anodically coloring films depend, for useful darkening and bleaching rates, on having open porosity and hydration. Various colors have been reported qualitatively, though full coloration is limited sometimes by the onset of O$_2$ evolution (7). As Table 2 shows, Prussian blue is especially interesting because of its relatively high coloration efficiency. Except for Prussian blue, a common method of film preparation is potential cycling (pc) on either a bulk metal or a conductive film. Reactive sputtering (rs) and galvanostatic electrodeposition (ged) have also been used for these electrochromic films.

Table 2. Some Anodically Colored, Inorganic Insertion/Extraction Films

Film	CAS Registry Number	Growth method	Insertion ion	CE(λ), cm^2/C	λ, nm	References
h-IrO$_x$	[12645-46-4]	pc^a	H$^+$/OH$^-$	15	633	8
h-IrO$_x$		rs	H$^+$/OH$^-$	18,12	633,600	7,41
h-NiO$_x$	[11099-02-8]	ged	H$^+$/OH$^-$	~50	~440 peak	42,43
Prussian blue	[12240-15-1]^b	ged	K$^+$	68	633	44

^{a}~180 nm.
^{b}CI Pigment Blue 27.

In the case of Prussian blue, the film has been grown by electroless reduction, by the sacrificial anode method or by galvanostatic electrodeposition (47–51).

Despite the considerable progress made in the few years in which anodic insertion/extraction films have been known, neither film compositions, film properties, nor electrochemical reactions are sufficiently well characterized. There have been disagreements, as indicated for h-IrO$_x$ and h-NiO$_x$ in Table 2, as to whether H$^+$ is being extracted or OH$^-$ inserted during coloration. The general problem is best illustrated by the important example of Prussian blue. Early work (47–50) resulted in two different sets of equations for electrochromic reduction:

$$KFe^{3+}[Fe^{2+}(CN)_6] + e^- + K^+ \longrightarrow K_2Fe^{2+}[Fe^{2+}(CN)_6] \qquad (3)$$

$$Fe_4^{3+}[Fe^{2+}(CN)_6]_3 + 4e^- + 4\,K^+ \longrightarrow K_4Fe^{2+}[Fe_4^{2+}(CN)_6]_3 \qquad (4)$$

The compounds KFe^{3+}[Fe^{2+}(CN)$_6$] [25869-98-1] and Fe$_4^{3+}$[Fe^{2+}(CN)$_6$]$_3$ [14038-43-8] are both called Prussian blue. The first is known as the water-soluble form, though actually it only peptizes easily, and the second as the insoluble form. The reduced compounds K$_2$Fe^{2+}[Fe^{2+}(CN)$_6$] and K$_4$Fe$_4^{2+}$[Fe^{2+}(CN)$_6$]$_3$ are known as Everitt's salt [15362-86-4] and Prussian white [81681-39-2], respectively. A similar lack of specificity occurs when Prussian blue is oxidized to Berlin green [14433-93-3]. This has led to propositions that film composition depends on the K$^+$ concentration of the growth solution (50) and on cycling (52,53). A film that is in the insoluble form initially, prior to cycling, is said thereafter to develop an intermediate K$^+$ content between the two forms of Prussian blue. The radius of the hydrated insertion ion is believed to be the key factor that dictates the reversibility of ion injection/extraction for the zeolite structure of Prussian blue. Reversibility has been shown to be best for K$^+$, Rb$^+$, Cs$^+$ and NH$_4^+$; hydrated radii were determined to be in the range 0.118–0.125 nm (49).

Doped/Undoped Organic Films. This class of electrochromic materials is probably the youngest and least thoroughly explored from the practical viewpoint. There has been more interest generally in the very high, metal-like conductivity of the oxidized state of some of its members than in the insertion/extraction electrochromism accompanying oxidation–reduction. Of interest have been applications for lightweight and moldable batteries and also for antistatic and electromagnetic shielding. On the other hand, there is not enough reported in the open literature to permit good comparisons of coloration efficiencies. Also, though the films themselves are solid state, almost all electrochromic work has been done with liquid electrolytes. This suggests that research and development are still at a fundamental stage. Nevertheless, two excellent reviews (13,54) do emphasize optical properties and color switching. Some of this is summarized in Table 3, though the dopant column should be considered with care. Analytical work using the quartz–crystal microbalance (59,62,63) suggests that charge balancing may sometimes involve both anions and cations in complicated ways. The first four materials in Table 3 are best known for the e^- conductivity that is associated with a conjugated π-electron structure. These are usually deposited by electropolymerization. Tetrathiafulvalene [31366-25-3] (TTF) and 2,4,7-trinitro-9-fluorenylidene malononitrile [1172-02-7] (TNF-MN) demonstrate some variety in the growth method for this electrochromic class. TNF-MN is also different because coloration only occurs in the reduced state. Generally, the whole class is characterized by a wide variety of transmitted colors in both the oxidized and nonoxidized forms. Some of the colors reported are shown in Table 3, though these colors are known only qualitatively and without specification of viewing conditions. They are only indicated here to show the breadth possible. Even wider color variety is possible with structural substitutions (13). Some degree of predictability may be possible because solid-state colors seem to be similar to those reported for solutions (54,72).

The same color variety is not typical with inorganic insertion/extraction materials; blue is a common transmitted color. However, rare-earth diphthalocyanine complexes have been discussed, and these exhibit a wide variety of colors as a function of potential (73–75). Lutetium diphthalocyanine [12369-74-3] has been studied the most. It is an ion-insertion/extraction material that does not fit into any one of the groups herein but has been classed with the organics in reviews. Films of this complex, and also erbium diphthalocyanine [11060-87-0], have been prepared successfully by vacuum sublimation and even embodied in solid-state cells (76,77).

Table 3. Some Organic Insertion/Extraction Films

Film	CAS Registry Number	Growth method[a]	Possible dopants	Oxidizing color shift[b]	References
polyaniline	[25233-30-1]	ep	H^+, Br^-	lt yel–grn–bl	55–59
polypyrrole	[30604-81-0]	ep	ClO_4^-, BF_4^-, Li^+	yel/grn–gr/br	13,54,60–64
polythiophene	[25233-34-5]	ep	ClO_4^-, BF_4^-	red–bl	54,64–67
poly(isothia-naphthene)	[91201-85-3]	ep	ClO_4^-, BF_4^-	bl/blk–lt grn/yel	54,68,69
TTF-functionalized polymer		spin-cast	ClO_4^-, BF_4^-	orange–brown	70,71
TNF-MN	[1172-02-7]	vacuum	K^+	grn–transparent	72

[a]ep = electropolymerization.
[b]lt = light; yel = yellow; grn = green; bl = blue; gr = gray; br = brown; blk = black.

There is tangible support for ion insertion/extraction in some materials, such as polypyrrole, polyaniline, and Prussian blue, from analyses with the mass-sensitive quartz-crystal microbalance (59,62,63,78). This relatively new technique is developing as an important one for electrochromic materials generally. It augments the standard electrochemical analyses, especially cyclic voltammetry, that have been effectively used up to now. Its further use, for charge-balancing ion insertion/extraction, should give growth to the analytical technique, and also help speed along developments in electrochromism, which is still a young science.

BIBLIOGRAPHY

"Chromogenic Materials, Electrochromic and Thermochromic" in *ECT* 3rd ed., Vol. 6, pp. 129–142, by J. H. Day, Ohio University.

1. J. R. Platt, *J. Chem. Phys.* **34,** 862 (1961).
2. C. J. Schoot, J. J. Ponjeé, H. T. van Dam, R. A. van Doorn, and P. T. Bolwijn, *Appl. Phys. Lett.* **23,** 64 (1973).
3. H. T. van Dam and J. J. Ponjeé, *J. Electrochem. Soc.* **121,** 1555 (1974).
4. S. K. Deb, *Appl. Opt. Suppl.* **3,** 192 (1969); U.S. Pat. 3,521,941 (July 28, 1970), S. K. Deb and R. F. Shaw (to American Cyanamid Co.).
5. B. W. Faughnan and R. S Crandall, in J. I. Pankove, ed., *Display Devices, Topics in Applied Physics,* Vol. 40, Springer-Verlag, Berlin, Germany, 1980, p. 181.
6. I. F. Chang, in A. R. Kmetz and F. K. von Willisen, eds., *Nonemissive Electrooptic Displays,* Plenum Press, New York, 1976, p. 155.
7. W. C. Dautremont-Smith, *Displays* **3,** 3, 67 (1982).
8. G. Beni and J. L. Shay, *Adv. Image Pickup Display,* **5,** 83 (1982).
9. C. M. Lampert, *Sol. Energy Mater.* **11,** 1 (1984); U.S. DOE Contract W-7405-ENG-48, Springfield, Va., Oct. 1980.
10. S. A. Agnihotry, K. K. Saini, and S. Chandra, *Indian J. Pure Appl. Phys.* **24,** 19 (1986).
11. T. Oi, *Ann. Rev. Mater. Sci.* **16,** 185 (1986).
12. A. Donnadieu, *Mater. Sci. Eng.* **B3,** 185 (1989).
13. M. Gazard, in T. A. Skotheim, ed., *Handbook of Conducting Polymers,* Vol. I, Marcel Dekker, Inc., New York, 1986, p. 673.
14. U.S. Pat. 4,902,108 (Feb. 20, 1990), H. J. Byker (to Gentex Corp.).
15. J. M. Calvert, T. J. Manuccia, and R. J. Nowak, *J. Electrochem. Soc.* **133,** 951 (1986).
16. R. D. Giglia and G. Haacke, *S.I.D.Dig.* **12,** 76 (1981); *Proc. S.I.D.* **23,** 41 (1982).
17. J.-P. Randin, *J. Electrochem. Soc.* **129,** 1215 (1982).
18. I. F. Chang, B. L. Gilbert, and T. I. Sun, *J. Electrochem. Soc.* **122,** 955 (1975).
19. L. G. Van Uitert, G. J. Zydzik, S. Singh, and I. Camlidel, *Appl. Phys. Lett.* **36,** 109 (1980).
20. J. A. Barltrop and A. C. Jackson, *J. Chem. Soc. Perkin Trans. II,* 367 (1984).
21. S. Zaromb, *J. Electrochem. Soc.* **109,** 903, 912 (1962).
22. J. Mantell and S. Zaromb, *J. Electrochem. Soc.* **109,** 992 (1962).
23. S. K. Deb, *Philos. Mag.* **27,** 801 (1973).
24. J.-P. Randin, *J. Electron. Mater.* **7,** 47 (1978).
25. J.-P. Randin, *J. Electrochem. Soc.* **129,** 2349 (1982).
26. S. F. Cogan, N. M. Nguyen, S. J. Perrotti, and R. D. Rauh, *J. Appl. Phys.* **66,** 1333 (1989).
27. S. F. Cogan, E. J. Anderson, T. D. Plante, and R. D. Rauh, in C. M. Lampert, ed., *Proceedings of the SPIE,* Vol. 562, SPIE, Bellingham, Wash., 1985, p. 23.
28. R. Hurditch, *Electron. Lett.* **11,** 142 (1975).

29. B. Reichman and A. J. Bard, *J. Electrochem. Soc.* **126,** 583 (1979).
30. P. Schlotter and L. Pickelmann, *J. Electron. Mater.* **11,** 207 (1982).
31. T. C. Arnoldussen, *J. Electrochem. Soc.* **128,** 117 (1981).
32. O. Bohnke, C. Bohnke, G. Robert, and B. Casquille, *Solid State Ionics* **6,** 121 (1982).
33. P. Schlotter, *Sol. Energy Mater.* **16,** 39 (1987).
34. O. F. Schirmer, V. Wittwer, G. Baur, and G. Brandt, *J. Electrochem. Soc.* **124,** 749 (1977).
35. R. B. Goldner and R. D. Rauh, *Sol. Energy Mater.* **11,** 177 (1984).
36. C. B. Greenberg, in S. A. Marolo, ed., *Proceedings of the 15th Conference on Aerospace Transparent Materials and Enclosures,* II, WRDC-TR-89-4044, Wright-Patterson AFB, Dayton, Ohio, 1989, p. 1124; U.S. Pat. 4,768,865 (Sept. 6, 1988), C. B. Greenberg and D. E. Singleton (to PPG Industries, Inc.).
37. K.-C. Ho, D. E. Singleton, and C. B. Greenberg, *J. Electrochem. Soc.* **137,** 3858 (1990); in M. K. Carpenter and D. A. Corrigan, eds., *Proceedings of the Symposium on Electrochromic Materials,* The Electrochemical Society, Inc., Pennington, N.J., 1990, p. 349.
38. V. D. Neff, *J. Electrochem. Soc.* **125,** 886 (1978).
39. S. Gottesfeld, J. D. E. McIntyre, G. Beni, and J. L. Shay, *Appl. Phys. Lett.* **33,** 208 (1978); D. N. Buckley and L. D. Burke, *J. Chem. Soc. Faraday Trans. I* **71,** 1447 (1975).
40. L. D. Burke and D. P. Whelan, *J. Electroanal. Chem.* **109,** 385 (1980).
41. S. F. Cogan, T. D. Plante, R. S. McFadden, and R. D. Rauh, in C. M. Lampert and S. Holly, eds., *Proceedings of the SPIE,* Vol. 692, SPIE, Bellingham, Wash., 1986, p. 32.
42. M. K. Carpenter, R. S. Conell, and D. A. Corrigan, *Sol. Energy Mater.* **16,** 333 (1987).
43. S. Morisaki, K. Kawakami, and N. Baba, *Jpn. J. Appl. Phys.* **27,** 314 (1988).
44. H. Tada, Y. Bito, K. Fujino, and H. Kawahara, *Sol. Energy Mater.* **16,** 509 (1987).
45. L. D. Burke and E. J. M. O'Sullivan, *J. Electroanal. Chem.* **93,** 11 (1978).
46. L. D. Burke and O. J. Murphy, *J. Electroanal. Chem.* **109,** 373 (1980).
47. D. Ellis, M. Eckhoff, and V. D. Neff, *J. Phys. Chem.* **85,** 1225 (1981).
48. K. Itaya, H. Akahoshi, and S. Toshima, *J. Electrochem. Soc.* **129,** 1498 (1982).
49. K. Itaya, T. Ataka, and S. Toshima, *J. Am. Chem. Soc.* **104,** 4767 (1982).
50. K. Itaya, I. Uchida, and V. D. Neff, *Acc. Chem. Res.* **19,** 162 (1986).
51. Y. Yano, N. Kinugasa, H. Yoshida, K. Fujino, and H. Kawahara in Ref. 37, p. 125.
52. R. J. Mortimer and D. R. Rosseinsky, *J. Chem. Soc. Dalton Trans.,* 2059 (1984).
53. C. A. Lundgren and R. W. Murray, *Inorg. Chem.* **27,** 933 (1988).
54. A. O. Patil, A. J. Heeger, and F. Wudl, *Chem. Rev.* **88,** 183 (1988).
55. T. Kobayashi, H. Yoneyama, and H. Tamura, *J. Electroanal. Chem.* **177,** 281,293 (1984); *Ibid.,* **161,** 419 (1984).
56. A. G. MacDiarmid and coworkers, *Mol. Cryst. Liq. Cryst.* **121,** 173 (1985).
57. P. M. McManus, S. C. Yang, and R. J. Cushman, *J. Chem. Soc., Chem. Commun.,* 1556 (1985).
58. R. J. Cushman, P. M. McManus, and S. C. Yang, *J. Electroanal. Chem.* **291,** 335 (1986).
59. D. Orata and D. A. Buttry, *J. Am. Chem. Soc.* **109,** 3574 (1987).
60. A. F. Diaz, J. J. Castillo, J. A. Logan, and W. Y. Lee, *J. Electroanal. Chem.* **129,** 115 (1981).
61. A. F. Diaz and K. K. Kanazawa, in J. S. Miller, ed., *Extended Linear Chain Compounds,* Vol. 3, Plenum Press, New York, 1983, p. 417.
62. J. H. Kaufman, K. K. Kanazawa, and G. B. Street, *Phys. Rev. Lett,* **53,** 2461 (1984).
63. K. Naoi, M. M. Lien, and W. H. Smyrl, *J. Electroanal. Chem.* **272,** 273 (1989).
64. K. Kaneto, K. Yoshino, and Y. Inuishi, *Jpn. J. Appl. Phys.* **22,** L412 (1983).
65. T.-C. Chung, J. H. Kaufman, A. J. Heeger, and F. Wudl, *Phys. Rev. B* **30,** 702 (1984).

66. F. Garnier, G. Tourillon, M. Gazard, and J. C. Dubois, *J. Electroanal. Chem.* **148,** 299 (1983).

67. K. Kaneto, H. Agawa, and K. Yoshino, *J. Appl. Phys.* **61,** 1197 (1987).

68. M. Kobayashi, N. Colaneri, M. Boysel, F. Wudl, and A. J. Heeger, *J. Chem. Phys.* **82,** 5717 (1985).

69. M. Colaneri, M. Kobayashi, A. J. Heeger, and F. Wudl, *Synth. Met.* **14,** 45 (1986).

70. F. B. Kaufman, A. H. Schroeder, E. M. Engler, S. R. Kramer, and J. Q. Chambers, *J. Am. Chem. Soc.* **102,** 483 (1980).

71. F. B. Kaufman, A. H. Schroeder, E. M. Engler, and V. V. Patel, *Appl. Phys. Lett.* **36,** 422 (1980).

72. A. Yasuda and J. Seto, *J. Electroanal. Chem.* **247,** 193 (1988); in Ref. 37, p. 192.

73. P. N. Moskalev and I. S. Kirin, *Opt. Spectros.* **29,** 220 (1970); *Russ. J. Inorg. Chem.* **16,** 57 (1971); *Russ. J. Phys. Chem.* **46,** 1019 (1972).

74. G. A. Corker, B. Grant and N. J. Clecak, *J. Electrochem. Soc.* **126,** 1339 (1979).

75. M. M. Nicholson and F. A. Pizzarello, *J. Electrochem. Soc.* **126,** 1490 (1979).

76. N. Egashira and H. Kokado, *Jpn. J. Appl. Phys.* **25,** L462 (1986).

77. M. Starke, I. Androsch, and C. Hamann, *Phys. Stat. Sol. (A)* **120,** K95 (1990).

78. B. J. Feldman and O. R. Melroy, *J. Electroanal. Chem.* **234,** 213 (1987).

General Reference

N. Baba, M. Yamana, and H. Yamamoto, eds., *Electrochromic Display,* Sangyo Tosho Co. Ltd., Tokyo, 1991.

CHARLES B. GREENBERG
PPG Industries, Inc.

PHOTOCHROMIC

Photochromism, in its broadest sense, can be defined as a reversible change in the absorption system of a material induced by electromagnetic radiation. The broadness of this definition limits its utility. Therefore, for the purpose of this article the definition of photochromism is restricted to a reversible change in the color or darkening of a material caused by absorption of ultraviolet or visible light. The change in color, or darkening of the material, implies a change in the absorption spectrum in the visible range of light (400–700 nm). The change must be reversible, although there is no implication about the kinetics or speed of the reversion to the original state.

Schematically, the photochromic reaction can be stated by the simple equation 1:

$$A \underset{\substack{\Delta \text{ or} \\ h\nu'}}{\overset{h\nu}{\rightleftharpoons}} B \tag{1}$$

Substance A has an absorption spectrum in one or more regions of the ultraviolet or visible spectral range. Irradiation of A at a wavelength corresponding to one of the absorption bands results in formation of substance B, which has a visible absorption spectrum different from A. Most commonly, substance A is uncolored or only slightly colored, whereas substance B is colored or appears darker than A.

The reverse reaction, B returning to A, can be driven either by thermal or photochemical energy, or both. When the reversion is photochemically driven, the process is called optical bleaching. Optical bleaching is a general characteristic and is a factor in almost all photochromic systems, even those normally thought of as being thermally reversible.

Another important concept in the discussion of photochromic systems is fatigue. Fatigue is defined as a loss in photochromic activity as a result of the presence of side reactions that deplete the concentration of A and/or B, or lead to the formation of products that inhibit the photochemical formation of B. The inhibition can result from quenching of the excited state of A or screening of active light. Fatigue, therefore, is caused by the absence of total reversibility within the photochromic reaction (eq. 2).

$$A \underset{\substack{\Delta \text{ or} \\ h\nu'}}{\overset{h\nu}{\rightleftarrows}} B \longrightarrow C \tag{2}$$

Photochromic systems can be separated into two broad categories, organic and inorganic. The two types are vastly different in their observable characteristics and mechanisms, but there are several examples of both which fit the definition of photochromism given. The purpose of the discussion is to define, with the help of the examples, the principle characteristics of each photochromic system.

Inorganic Photochromic Systems

Silver Halide-Containing Glasses.
The most important examples of inorganic systems are those containing silver halide crystallites dispersed throughout a glass matrix. The first description of photochromic silver halide-containing glasses appeared in 1964 (1,2). In general, these systems are characterized by broad absorption of visible light by the colored species and excellent resistance to fatigue.

The principle behind the generation of a photochromic glass with silver halide is the controlled formation of silver halide particles or crystallites suspended throughout the glass matrix (3–5). The formation of crystallites of the correct size and concentration is the key to a useful photochromic system. The general procedure involves the initial melting of a glass-forming mixture which is then cooled to a solid glass shape. Rapid cooling to room temperature results in a nonphotochromic glass. Holding the solid at a temperature in the range of 500–600°C for several minutes to hours causes the nucleation and growth of silver halide crystallites, the active photochromic species. Again, the size of the crystallites is important. With a size of less than 10 nm, significant darkening upon exposure to sunlight is not achieved. Above 20 nm, the scattering of visible light becomes a problem, leading to haziness. Also, with the larger particles the rate of thermal fading slows to an unacceptable rate.

Copper (I) (cuprous) ion serves as a catalyst for both the photochemical darkening and thermal fading reactions (1). Therefore, a small amount of cuprous ion is normally added to the glass batch.

The darkening reaction involves the formation of silver metal within the silver halide particles containing traces of cuprous halide. With the formation of metallic silver, cuprous ions are oxidized to cupric ions (1,4). The thermal or photochemical (optical bleaching) reversion to the colorless or bleached state corresponds to the reoxidation of silver to silver ion and the reduction of cupric ion to reform cuprous ion.

One of the most important characteristics of the inorganic glass matrix for a photochromic system is the temperature dependence of the solubility of silver halides (4). It is required that the solubility of silver halides be high at the temperature used to melt the glass mixture, and relatively low at the intermediate temperature at which the silver halide crystallites are formed. The composition of a typical photochromic glass system, that marketed as Photogray Extra by Corning Inc., Corning, New York, is approximately as given in Table 1 (6).

The color of the darkened state is controlled by the size and shape of the minute silver specks formed during photochemical reduction, but the relationship is not well understood. Since the shapes of the silver particles vary considerably throughout the matrix, a broad absorption over the visible range results (4,5). The color can be modified from gray to brown by changing the heat treatment for silver halide nucleation and growth and thus changing the size/shape distribution of the crystallites (7). The color can also be shifted from gray to brown by the addition of trace amounts of palladium or gold (3–4 ppm) to the glass batch (6).

An alternative to the uniform distribution of silver halide throughout the glass is the diffusion of silver ion into the surface of the glass. This has been accomplished by immersion of the glass article into a silver-containing fused salt, for example, silver nitrate plus sodium nitrate, molar ratio = 17:83 (8). Heat treatment to allow crystallite formation is still essential and copper oxide is added to the glass batch to catalyze the photochromic reactions. In general, photo-

Table 1. Composition of a Silver Halide
Photochromic Glass[a]

Component	Wt %
SiO_2	55.8
B_2O_3	18.0
Al_2O_3	6.48
Li_2O	1.88
Na_2O	4.04
K_2O	5.76
ZrO_2	4.89
TiO_2	2.17
CuO	0.011
Ag	0.24
Cl	0.20
Br	0.13

[a]Ref. 6.

chromic glasses formed in this manner are not as active (do not get as dark) as the systems containing thoroughly dispersed silver, or have a slower thermal fade rate.

Thin films of photochromic glass containing silver halide have been produced by simultaneous vacuum deposition of silicon monoxide, lead silicate, aluminum chloride, copper (I) chloride, and silver halides (9). Again, heat treatment (120°C for several hours) after vacuum deposition results in the formation of photochromic silver halide crystallites. Photochemical darkening and thermal fade rates are much slower than those of the standard dispersed systems.

Other Inorganic Systems. An effective silver-free photochromic system can be obtained by the dispersion of crystallites containing cadmium halide and copper halide throughout an inorganic glass matrix (10,11). Heat treatment of the solid glass is again required to allow the formation of the metal halide crystallites. The mechanism of the darkening reaction is apparently the formation of colloidal copper metal particles by disproportionation of cuprous ion (11). The color of the darkened state, ranging from yellow-brown to green, is controlled by the type of heat treatment used for crystallite formation.

A typical range of glass compositions for this type of photochromic system is given in Table 2 (11).

Another inorganic photochromic glass system was prepared by the addition of europium (II) or cerium (III) to a soda–silica glass with an approximate composition of $Na_2O-2.5SiO_2$ (12). The concentration of the rare-earth ion was low (100 ppm). With europium (II)-doped glass, the photochemical darkening resulted in an amethyst color that faded rapidly thermally. These glasses were subject to fairly rapid fatigue, losing all photochromic behavior after a 20-h exposure to uv radiation centered at 332.5 nm. This was probably the result of the oxidation of Eu(II) to Eu(III). Interestingly, the photochromic behavior could be recovered by exposure of the exhausted glass to high energy uv light at 213.7 nm.

Photochromic silver–copper halide films were produced by vacuum evaporation and deposition of a mixture of the components onto a silicate glass substrate (13). The molar ratio of the components was approximately 9:1 (Ag:Cu) and film thicknesses were in the range of 0.45–2.05 μm. Coloration rate upon uv

Table 2. Composition of Copper/Cadmium Halide Photochromic Glass[a]

Component	Wt %
SiO_2	57.2–58.6
Al_2O_3	10.0–10.1
B_2O_3	21.5–22.7
Na_2O	8.1–8.2
F	1.05–1.14
Cl	0.43–0.45
CdO	0.43–0.45
SnO	0.20
CuO	0.12–0.15

[a]Ref. 11.

exposure was high but thermal fade rates were very slow when compared with standard silver halide glass photochromic systems.

Simultaneous deposition of cadmium chloride and copper chloride by vacuum evaporation onto fused silica or optical glass resulted in photochromic thin films (14). The thickness ranged from 0.25 to 1.3 μm.

Thin films of photochromic silver complex oxides were prepared by anodic oxidation of silver metal films (15). Complex oxides, such as Ag_2VO_4, Ag_4SiO_4, and Ag_2PO_4, darkened by exposure to visible light, but required heating to 150–250°C for thermal bleaching.

Organic Photochromic Systems

The organic photochromic systems that have been studied are numerous and it is helpful to classify them into a few categories by way of the general mechanism of the photochromic reaction in each category.

Photochromism Based on Geometric Isomerism. The simplest examples of a photochromic reaction involving reversible cis–trans isomerization is the photoisomerization of azobenzene [*103-33-3*], $C_{12}H_{10}N_2$ (16).

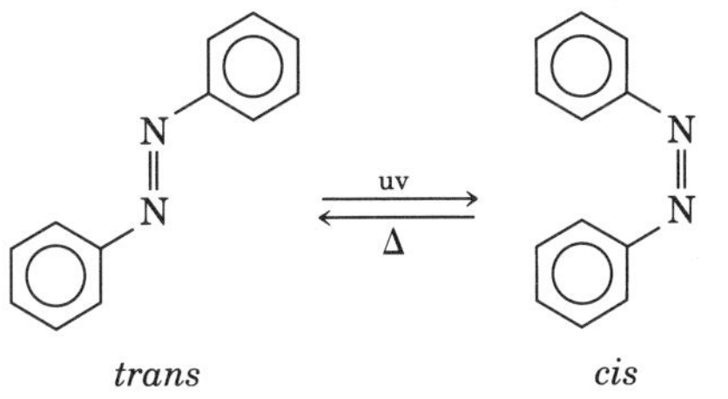

trans   *cis*

This facile reaction involves a modest change in the absorption of visible light, largely because of the visible absorption band of *cis*-azobenzene [*1080-16-6*] having a larger extinction coefficient than *trans*-azobenzene [*17082-12-1*]. Several studies have examined the physical property changes that occur upon photolysis of polymeric systems in which the azobenzene structure is part of the polymer backbone (17).

The cis–trans isomerization of stilbenes is technically another photochromic reaction (18). Although the absorption bands of the stilbene isomers, $C_{14}H_{12}$, occur at nearly identical wavelengths, the extinction coefficient of the lowest energy band of *cis*-stilbene [*645-49-8*] is generally less than that of *trans*-stilbene [*103-30-0*].

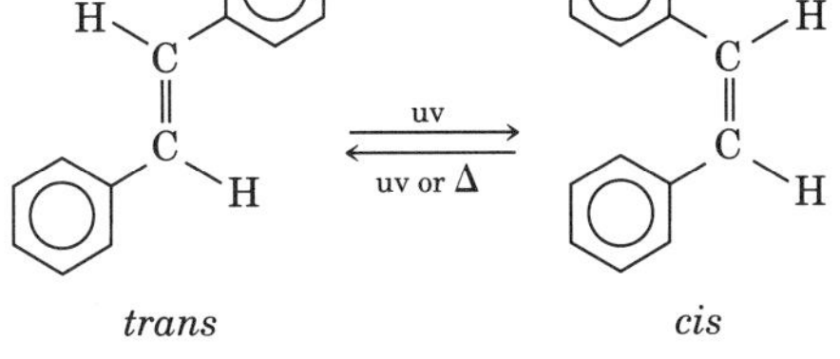

trans   *cis*

Photochromism Based on Cycloaddition Reactions. The photochemically reversible formation of endoperoxides is an example of this photochromic system (19,20). With some polycyclic aromatic compounds, the reaction with singlet oxygen to form the endoperoxide and the photochemical elimination of singlet oxygen can be accomplished essentially quantitatively. The reaction is also accompanied by a rather dramatic color change because of the disruption of the polycyclic chromophore during endoperoxide formation. A good example is the reversible formation of the 4b, 12b endoperoxide [74292-77-6] of dibenzo(a,j)perylene-8,16-dione [5737-94-0] (20). In this case, the endoperoxide $C_{28}H_{14}O_4$, is colorless and the parent compound, $C_{28}H_{14}O_2$, is red.

colorless	red

Photochromism Based on Tautomerism. Several substituted anils of salicylaldehydes are photochromic but only in the crystalline state. The photochromic mechanism involves a proton transfer and geometric isomerization (21). An example of a photochromic anil is N-salicylidene-2-chloroaniline [3172-42-7], $C_{13}H_{10}ClNO$.

Photochromism Based on Dissociation Processes. Both heterolytic and homolytic dissociation processes can result in the generation of a photochromic system. An example of an heterolytic process is the reversible formation of triphenylmethyl cation, $C_{19}H_{15}$, by photolysis of triphenylmethyl chloride [76-83-5] in acetonitrile (22).

The classical example of a photochromic process involving an homolytic dissociation is the formation of a red-purple free radical by photolysis of bis(2,4,5-triphenylimidazole) [63245-02-3], $C_{42}H_{30}N_4$ (23).

Photochromism Based on Triplet Formation. Upon absorption of light, many polycyclic aromatic hydrocarbons and their heterocyclic analogues undergo transitions to their triplet state which has an absorption spectrum different from that of the ground state (24). In rigid glasses and some plastics, the triplet state, which may absorb in the visible, has a lifetime of up to 20 seconds.

An example of such a polycyclic compound is 1,2;5,6-dibenzacridine [226-36-8], $C_{21}H_{13}N$, which, in a rigid matrix, absorbs uv radiation to form a triplet state absorbing strongly in the visible with a maximum at approximately 550 nm.

Photochromism Based on Redox Reactions. Although the exact mechanism of the reversible electron transfer is often not defined, several viologen salts (pyridinium ions) exhibit a photochromic response to uv radiation in the crystalline state or in a polar polymeric matrix, for example, poly(N-vinyl-2-pyrrolidinone) [9003-39-8] (25).

blue

In the example shown, the reduced form [49765-27-7] is blue with a visible absorption maximum at 610 nm. The rate of the reoxidation of the reduced form (cation radical, $C_{24}H_{22}N_2^+$) is usually, but not always, strongly dependent on the presence of oxygen.

The use of an electron-accepting counter ion leads to a photochromic system that is highly reversible under an inert atmosphere. An anion that has been used successfully is tetra-bis[3,5-di(trifluoromethyl)phenyl]borate anion [79230-20-9], $C_{32}H_{12}BF_{24}^-$ (26).

$$\left[\underset{CF_3}{\overset{CF_3}{\bigcirc}} B^- \right]_4$$

Photochromism Based on Electrocyclic Reactions. The most common general class of photochromic systems involves reversible electrocyclic reactions. Within this general class, the most well-studied compounds are the indolino spiropyrans and indolino spiroxazines.

Nitro-substituted indolino spirobenzopyrans or indolino spironaphthopyrans are photochromic when dissolved in organic solvents or polymer matrices (27). Absorption of uv radiation results in the colorless spiro compound [1498-88-0], $C_{19}H_{18}N_2O_3$, being transformed into the colored, ring-opened species. This colored species is often called a photomerocyanine because of its structural similarity to the merocyanine dyes (see CYANINE DYES). Removal of the ultraviolet light source results in thermal reversion to the spiro compound.

colorless ⇌ (uv / Δ) colored species

The nitro spiropyrans are susceptible to fatigue which has limited their application. Indolino spiroxazines exhibit photochromism by way of a mechanism that is very similar to that of the spiropyrans.

(1)

The spiroxazines, however, are much more resistant to fatigue. As measured by the quantum yield for photodegradation, indolino spironaphthoxazines, such as (1) [27333-47-7], $C_{22}H_{20}N_2O$, are two to three orders-of-magnitude more stable than indolino spirobenzopyrans (28).

Another class of photochromic compounds, which operate through an electrocyclic mechanism, is the fulgides; although, with this class, the colored species is formed through a ring formation rather than ring-opening (29,30). The reversion to the colorless species [59000-86-1], $C_{15}H_{16}O_4$, does not occur thermally at ambient temperatures but can efficiently be driven photochemically with visible light.

colorless   colored

Properly substituted, for example not having a labile hydrogen on the site of cyclization, fulgides have both high quantum yields for ring closure and opening and good fatigue resistance.

The photochemical ring closure of certain stilbenes, eg, the highly methyl substituted compound (**2**) [*108028-39-3*], $C_{22}H_{28}$, and their heterocyclic analogues is the basis for another class of photochromic compounds (31–33).

(**2**)

By changing the substituents on the ethylenic linkage and exchanging phenyl rings for heteroaromatic rings, photochromic systems that are thermally reversible are transformed into systems that are thermally irreversible but photochemically reversible. The transition between the benzothiophene-derivative isomers, $C_{22}H_{14}N_2S_2$, (**3**) [*129199-46-8*] and (**4**) [*129199-81-1*] offers an example.

(**3**)   (**4**)

The dihydroindolizines are photochromic compounds that undergo a photochemical ring-opening reaction to form a colored zwitterionic species (34). In general, these classes of compounds are very efficient photochromic systems, yielding a variety of species of different colors depending on the substitution pattern. The colored zwitterion reverts to the uncolored or slightly colored starting material, in this example [*82250-16-6*], $C_{24}H_{19}NO_4S$, either thermally or photochemically.

zwitterion

Applications

Although the proposed applications for photochromic systems are numerous, few have received broad use. By far, the most successful commercial application is the use of photochromic silver halide-containing glasses in prescription eyewear. The convenience of having lenses that darken automatically upon exposure to sunlight has proven appealing to spectacle wearers (35). With the increasing penetration of plastic lenses into the ophthalmic market, the desire for plastic photochromic ophthalmic lenses has also increased, and considerable effort has been spent on the discovery of photochromic systems for plastic eyewear.

In order to achieve an organic photochromic system for the eyewear market, two primary problems had to be solved. The first problem of fatigue resistance was alleviated with the discovery of indolino spiroxazines which are inherently more fatigue-resistant than other systems. The photostability of the spiroxazines can be improved even further by protecting them from oxygen (36,37) or by the addition of chemical stabilizers, specifically nickel complexes (38) or hindered amine light stabilizers (39). The second problem is caused by the relatively narrow absorption bands of activated organic photochromic compounds in the visible range of light. This problem can be handled by mixing compatible compounds which results in a much broadened visible light absorption (40).

Photochromic lenses for eyewear serve as variable density optical filters. Other applications for photochromic light filters have been proposed including glazing applications for solar attenuation, variable transmission camera lenses, and shields for protection against the light flash from a nuclear explosion.

Besides the use of photochromic systems in light filters, their color development has also received considerable attention. For example, the introduction of photochromic components into product labels, tickets, credit cards, etc adds a mechanism for verification of authenticity (41,42). The active components are invisible until activated with an ultraviolet light source, after which they are easily detected.

The color development of photochromic compounds can also be utilized as a diagnostic tool. The temperature dependence of the fading of 6-nitroindolinospiropyran served as the basis for a nondestructive inspection technique for honeycomb aerospace structures (43). One surface of the structure to be examined was covered with a paint containing the photochromic compound and activated to a violet color with ultraviolet light. The other side of the structure was then heated. The transfer of heat through the honeycomb structure caused bleaching of the temperature-dependent photochromic compound. Defects in the honeycomb where heat transfer was inhibited could be detected as darker areas.

Photochromic compounds that can be thermally faded have also been used in engineering studies to visualize flows in dynamic fluid systems (44,45).

Most photochromic compounds undergo large structural changes while being transformed from the uncolored to the colored form. This property has been used to examine the pore size of polymers by utilizing the relationship of pore size and the kinetics of the photochromic response (46).

BIBLIOGRAPHY

"Chromogenic Materials, Photochromic" in *ECT* 3d ed., Vol. 6, pp. 121–128, by R. J. Aranjo, Corning Glass Works.

1. W. H. Armistead and S. D. Stookey, *Science* **144,** 150 (1964).
2. U.S. Pat. 3,208,860 (Sept. 28, 1965), W. H. Armistead and S. D. Stookey (to Corning).
3. J. P. Smith, *J. Photogr. Sci.* **18,** 41 (1970).
4. R. J. Araujo, *Contemp. Phys.* **21,** 77 (1980).
5. H. J. Hoffmann, in C. M. Lampert and C. G. Granqvist, eds., *Large-Area Chromogenics: Materials and Devices for Transmittance Control,* Vol. IS 4, SPIE Institutes for Advanced Optical Technologies, Bellington, Wash., 1990, p. 86.
6. U.S. Pat. 4,251,278 (Feb. 17, 1981), G. B. Hares (to Corning).
7. U.S. Pat. 4,043,781 (Aug. 23, 1977), C. V. DeMunn, D. J. Kerko, R. A. Westwig, and D. B. Wrisley, Jr. (to Corning).
8. H. M. Garfinkel, *Appl. Opt.* **7,** 789 (1968).
9. M. Mizuhashi and S. Furuuchi, *Thin Solid Films* **30,** 259 (1975).
10. U.S. Pat. 3,325,299 (June 13, 1967), R. J. Araujo (to Corning).
11. D. M. Trotter, Jr., J. W. H. Schreurs, and P. A. Tick, *J. Appl. Phys.* **53,** 4657 (1982).
12. A. J. Cohen, *Science* **137,** 981 (1962).
13. A. F. Perveyev and A. V. Mikhaylov, *Sov. J. Opt. Technol.* **39,** 117 (1972).
14. H. Marquez, J. Ma Rincon, and L. E. Celeya, *Appl. Opt.* **29,** 3699 (1990).
15. T. H. Hirono, T. Yamada, and T. Nishi, *J. Appl. Phys.* **59,** 948 (1986).
16. H. Rau, in H. Durr and H. Bouas-Laurent, eds., *Photochromics, Molecules and Systems,* Elsevier, Amsterdam, The Netherlands, 1990, p. 165.
17. G. S. Kumar and D. C. Neckers, *Chem. Rev.* **89,** 1915 (1989).
18. J. Saltiel and Y.-P. Sun, in Ref. 16, p. 64.
19. H. D. Brauer and R. Schmidt, *Photochem. Photobiol.* **37,** 587 (1983).
20. Ger. Offen. 2,910,668 (Sept. 25, 1980), H. D. Brauer, R. Schmidt, and W. Drews.
21. M. D. Cohen and G. M. Schmidt, *J. Phys. Chem.* **66,** 2442 (1962).
22. L. E. Manring and K. S. Peters, *J. Phys. Chem.* **88,** 3516 (1984).
23. T. Hayashi and K. Maeda, *Bull. Chem. Soc. Jpn.* **33,** 565 (1960).
24. J. L. Kropp and M. W. Windsor, *U.S. Air Force Technical Report,* AFML-TR-68-220, Washington, D.C., Aug. 1968.
25. H. Kamogawa and T. Suzuki, *Bull. Chem. Soc. Jpn.* **60,** 794 (1987).
26. T. Nagamura, K. Sakai, and T. Ogawa, *J. Chem. Soc., Chem. Commun.,* 1035 (1988).
27. R. Guglielmetti, in Ref. 16, p. 314.
28. N. Y. C. Chu, *Proceedings of the 10th IUPAC Symposium on Photochemistry,* Interlaken, Switzerland, 1984.
29. H. G. Heller, *IEE Proc.* **130**(5), 209 (1983).
30. J. Whittal, in Ref. 16, p. 467.
31. S. Nakamura and M. Irie, *J. Org. Chem.* **53,** 6136 (1988).
32. Y. Nakayama, K. Hayashi, and M. Irie, *J. Org.Chem.* **55,** 2592 (1990).
33. K. Uchida, Y. Nakayama, and M. Irie, *Bull. Chem. Soc. Jpn.* **63,** 1311 (1990).

34. H. Durr, *Angew. Chem. Int. Ed.* **28,** 413 (1989).
35. *OMA* (Optical Manufacturers Association) *National Consumer Eyewear Study VI,* Falls Church, Va., 1990.
36. U.S. Pat. 4,166,043 (Aug. 28, 1979), D. R. Uhlmann, E. Snitzer, R. J. Hovey, N. Y. C. Chu, and J. T. Fournier (to American Optical).
37. U.S. Pat. 4,367,170 (Jan. 4, 1983), D. R. Uhlmann, E. Snitzer, R. J. Hovey, and N. Y. C. Chu (to American Optical).
38. U.S. Pat. 4,440,672 (Apr. 3, 1984), N. Y. C. Chu (to American Optical).
39. U.S. Pat. 4,720,356 (Jan. 19, 1988), N. Y. C. Chu (to American Optical).
40. U.S. Pat. 4, 968,454 (Nov. 6, 1990), J. C. Crano, P. L. Kwiatkowski, and R. J. Hurditch (to PPG Industries).
41. Eur. Pat. Appl. 328,320 A1 (Aug. 16, 1989), P. Wright (to Courtaulds).
42. PCT Int. Appl. WO 90/06539 A1 (June 14, 1990), S. Wallace (to Traqson Ltd.).
43. S. Allinikov, *U.S. Air Force Technical Report*, AFML-TR-70-246, Washington, D.C., Dec. 1990.
44. R. E. Falco and C. C. Chu, *Proc. SPIE (Int. Conf. Photomech. Spec. Met.)* **814**(2), 706 (1988).
45. Brit. Pat. Appl. 2,209,751 A (May 24, 1989), C. Trundle (to Plessey Co.)
46. W.-C. Yu, C. S. P. Sung, and R. E. Robertson, *Macromolecules* **21,** 355 (1988).

JOHN C. CRANO
PPG Industries, Inc.

PIEZOCHROMIC

In its most general sense piezochromism is the change in color of a solid under compression. There are three aspects of the phenomenon. The first is, in a sense, trivial, but it is very general. The color of a solid results from the absorption of light in selected regions of the visible spectrum by excitation of an electron from the ground electronic state to a higher level. If the two electronic energy levels are perturbed differently by pressure, compression results in a color change. This is the basic definition of pressure tuning spectroscopy. Examples include, among others, increased splitting of the d orbitals of transition-metal ions in complexes with pressure, the shift to higher energy of a color center (a vacancy containing an electron) in an alkali halide or glass environment, and a change in the relative energy of bonding and antibonding orbitals as pressure increases. This last phenomenon depends on the relative importance of intra- and intermolecular interactions. The compression of a bond increases the difference in energy between bonding and antibonding orbitals. On the other hand, the attractive van der Waals interactions between molecules are generally stronger around an excited molecule. Where the latter interaction dominates, one observes a shift to lower energy (red shift) of these excitations. Further examples of this include the energy associated with the transfer of an electron in an electron donor–acceptor complex and the difference in energy between the top of the valence band and bottom of the conduction band, ie, the absorption edge, in insulators and semiconductors.

These excitations are widely used to characterize electronic states and excitations, to test theories about electronic phenomena, and to delineate the nature of local sites in glasses, disordered solids, intercalates, etc. However, this aspect of changing color with pressure is so general as to be hardly satisfactory for defining piezochromism.

Phase Transitions. A second aspect involves a discontinuous change of color when a crystalline solid undergoes a first-order phase transition from one crystal structure to another. The most obvious example is the change of the absorption edge. For example, CdS changes from yellow to deep red at 2.7 GPa (27 kbar) when the crystal structure changes from wurtzite to sodium chloride (face centered cubic). CdSe, ZnS, ZnSe, and ZnTe undergo similar transitions with distinct color changes at pressures from 5–15 GPa (50–150 kbar). First-order phase transitions involving alterations in crystal structure only can change the electronic excitation energy associated with almost any kind of electronic process provided the two electronic states interact differently with the changing environment. This phenomenon is also of scientific interest. However, for most molecular crystals at modest pressures, the coupling between the electronic states of the molecule and the lattice modes is rather small so these perturbations are usually not large, with few exceptions.

Changes in Molecular Geometry

The phenomenon of most interest is a change in color of a solid as a result of a change in the molecular geometry of the molecules that make up the solid. The color change takes place because the change in geometry alters the relative energy of different electronic orbitals, and therefore the electronic absorption spectrum. Frequently it rearranges the order of these orbitals or provides new combinations of atomic orbitals because of symmetry changes. The rearrangements may be discontinuous at a given pressure, may occur over a modest range of pressures, or may occur gradually over the whole range of available pressure as for chemical equilibria in solution (1). A few examples, together with the principles or generalizations that arise from them, are discussed.

Piezochromism has been observed in a wide variety of materials. Three classes which illustrate well some of the generalizations that have been developed are organic molecules in crystals and polymer films, metal cluster compounds, and organometallic complexes of Cu(II).

The prototype of piezochromic organic molecules are the salicylidene anils, eg, *N*-salicylidene-2-chloroaniline [*3172-42-7*] (2–4). At ambient pressure in the crystalline state they are either photochromic or thermochromic, but never both, depending on the side groups on the aromatic rings (see CHROMOGENIC MATERIALS, PHOTOCHROMIC). When dissolved in a polymer film they are photochromic. The ground state has an OH group opposite a nitrogen on the adjacent ring, thus it is called the "enol" form. When heated, the thermochromic compounds exhibit an absorption in the visible spectrum which corresponds to a transfer of the H from O to N without other change of molecular geometry. This is the "*cis*-keto" form. The photochromic molecules, upon irradiation at low temperature, develop an absorption assigned to the "*trans*-keto" form.

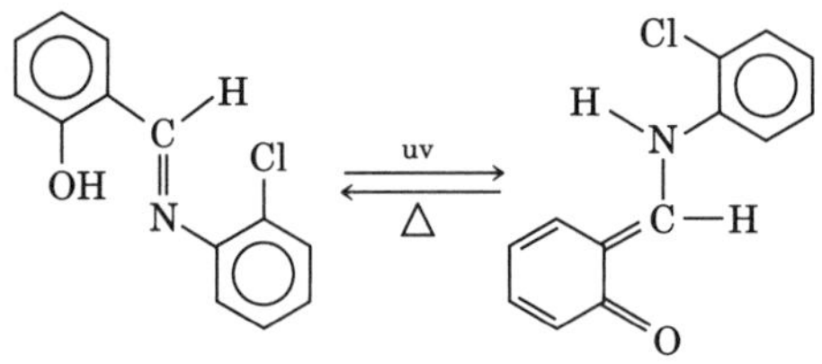

With increasing pressure at 25°C the *cis*-keto form is stabilized vis á vis both the enol and *trans*-keto isomers so that both thermochromic and photochromic materials exhibit the same type of piezochromism, but in different degrees. The conversion increases continuously with pressure, much like the changing of chemical equilibrium in liquid solution with pressure.

Two principles are illustrated here. In this case, increasing pressure and temperature favor the same process. With increasing temperature the possibility of crossing a barrier of a given height increases. With increasing pressure the potential wells associated with the two states are perturbed differently. This perturbation can either augment or oppose the effect of increasing temperature, ie, temperature and pressure are not in general conjugate variables as is frequently assumed. In the second place one can extract a volume decrease of ~2 cm^3/mol from the change in the enol–keto equilibrium with pressure. This is much larger than any change in bond length or molecular geometry induced by hydrogen transfer. This illustrates the point that changes in intermolecular interaction and packing are more likely to determine the amount of piezochromic or other reaction introduced by pressure than are differences in molecular geometry. Other piezochromic reactions have been observed in spiropyrans and bianthrones. The changes in molecular configuration are different, but no further principles are derived.

Compounds with metal–metal bonds stabilized by appropriate ligands constitute a second class of materials where a number of cases of piezochromic behavior have been observed (1,5). Compounds involving a Re–Re bond stabilized either by eight halides or by bridging (bidentate) ligands like the pivalate ion are two well-established cases. Octahalodirhenates, $Re_2X_8^{2-}$ (X = Cl,Br,I), in crystals with a number of counterions, exhibit an absorption peak which corresponds to an excitation from a bonding to antibonding orbital with angular momentum two around the bond (δ orbital). When pressure is applied to the iodide a new peak grows in at a lower energy. It is associated with the rotation of the iodides on opposing Re ions from an eclipsed (directly opposite) to a staggered (not necessarily 45°) position. The reduced repulsion between opposing I^- ions more than compensates for the weakened metal–metal bond. The amount of conversion is continuous with increasing pressure.

In the case of the bridged complexes, the process involves changing from a bidentate to a monodentate configuration. For these systems the mode of transformation is variable. In close-packed crystals the rearrangement is a first-order process, ie, it occurs discontinuously at a fixed pressure. For slightly less close-packed crystals the transformation occurs over some range of pressure, eg, 2–3 GPa (20–30 kbar). In the language of physics the process corresponds to a higher order phase transition. When the molecules are dissolved in polymer films the reaction is continuous (stochastic) over the entire range of available pressure. These studies illustrate the importance of the environment on the extent of cooperativity involved in pressure-induced molecular geometry changes and thus on piezochromism.

Complexes involving larger metal clusters, eg, Au_9 or Au_2Rh_4, also undergo piezochromic rearrangements with rather dramatic changes in the absorption spectrum, and well-defined changes in molecular structure (6,7).

Complexes of Cu(II) occur in a wide variety of distorted geometries. The d^9 configuration is stabilized by distortions from a high to a slightly lower symmetry, ie, the Jahn-Teller effect (1,5,8,9). Cu(II) complexed to organic molecules like ethylenediamine derivatives lie in a square planar configuration of nitrogens from the organic molecules. If the axial ligands are far off or not aligned the Cu(II) retains this four-coordinate geometry in the solid state. When these axial ligands are sufficiently close the system becomes six coordinate with definite changes in the ordering of the electronic orbitals and thus in the visible spectrum. In the crystalline state the rearrangement from four to six coordinate is apparently a first-order transition, whereas when dissolved in a polymeric matrix the transformation occurs over a range of 3–5 GPa (30–50 kbar). This is then another example of the importance of the environment on the pressure dependence of the rearrangement.

Cu(II) complexed to four Cl^- ions can adopt arrangements from tetragonal to square planar. The latter arrangement gives maximum Cu–Cl bonding; the former minimizes the Cl–Cl repulsion. In practice, complexes occur near both extremes and with almost all intermediate arrangements. The geometry of a given complex is determined by the counterion. Strongly hydrogen bonding counterions draw off electron density from the Cl^- and favor square planar symmetry. With pressure one finds that both nearly square planar and nearly tetrahedral complexes distort toward an intermediate symmetry with clear-cut changes in the electronic absorption spectrum. These transformations occur over relatively short ranges of pressure ~1.5–2.5 GPa (15–25 kbar). The principle illustrated here is that at high compression the economy of best geometric packing overcomes the weaker van der Waals and hydrogen bonding forces.

Piezochromic effects have been observed in a variety of other Cu(II) complexes. In some cases it can be shown that the structures of a series of related complexes follow a reaction pathway with the structure of one complex at, for example, 8 GPa (80 kbar) corresponding to that of a related complex at, for example, 2 GPa (20 kbar). The changes in color of the complex, of course, follow the same sequence.

In general there appear to be two modes by which pressure induced isomerizations and consequently, piezochromism, can occur. There can be a transformation from one distinct conformation to another as a result of the relative stabilization of the potential well associated with conformation B with respect to that of A. This process may be either an equilibrium process, if the energy barrier is small, or involve a first or higher order phase transition depending on the extent of cooperativity demanded by the type of transformation and the environment. Such transformations are characterized by a change in the electronic character of the ground state. If there is no distinct change in the electronic character of the ground state, a series of compounds may follow each other along a reaction pathway as they undergo a similar piezochromic transformation over different pressure ranges.

In the foregoing discussion polymers have been used as a medium for small molecules in comparison with the crystalline state. It has also been observed that there are changes in polymeric geometry and various rotational motions introduced by pressure (10–14). These are at times reflected in the absorption spectrum (usually in the ultraviolet) or in the emission spectrum and are a form of piezochromism.

The examples of piezochromism discussed so far involve rather well-established changes of molecular geometry. There are examples of pressure-induced changes in the electronic ground state with resultant changes in the electronic absorption spectrum where the changes in molecular geometry are not well established. One example involves intramolecular or intracomplex charge transfer (15–20). At ambient pressure this process takes place by optical excitation. At sufficiently high pressure the charge-transferred state may be stabilized sufficiently to become the ground state of the system with a different electronic absorption spectrum but where the geometry changes are not well defined. In most transition-metal complexes the metallic ion exists in the state of maximum multiplicity "high spin" according to Hund's rule. With compression the splitting of the d states may become sufficient to establish a spin paired "low spin" ground state with resultant changes in the electronic absorption spectrum (21–23). Again, the changes in molecular geometry are not well established.

Pressure can also induce a change in the spin state of a transition-metal ion in a molecule or crystal with resultant change in the spectrum. The usual change observed is from high to low spin, but the inverse transition has been observed in some cases.

The systematic study of piezochromism is a relatively new field. It is clear that, even within the restricted definition used here, many more systems will be found which exhibit piezochromic behavior. It is quite possible to find a variety of potential applications of this phenomenon. Many of them center around the estimation of the pressure or stress in some kind of restricted or localized geometry, eg, under a localized impact or shock in a crystal or polymer film, in such a film under tension or compression, or at the interface between bearings. More generally it conveys some basic information about inter- and intramolecular interactions that is useful in understanding processes at atmospheric pressure as well as under compression.

BIBLIOGRAPHY

1. H. G. Drickamer, in R. Pucci and J. Picatto, eds., *Molecular Systems under High Pressure,* North Holland Press, New York, 1991, p. 91 and references therein.
2. D. L. Fanselow and H. G. Drickamer, *J. Chem. Phys.* **61,** 4567 (1974).
3. E. N. Hochert and H. G. Drickamer, *J. Chem. Phys.* **67,** 6168 (1977).
4. Z. A. Dreger and H. G. Drickamer, *Chem. Phys. Lett.* **179,** 199 (1990).
5. H. G. Drickamer and K. L. Bray, *Intern. Rev. of Phys. Chem.* **8,** 41 (1989) and references therein.
6. J. L. Coffer, J. R. Shapley, and H. G. Drickamer, *Inorg. Chem.* **29,** 3900 (1990).

7. K. L. Bray, H. G. Drickamer, D. M. P. Mingos, M. J. Watson, and J. R. Shapley, *Inorg. Chem.* **30,** 864 (1991).

8. H. G. Drickamer and K. L. Bray, *Accts. Chem. Res.* **23,** 55 (1991) and references therein.

9. B. Scott and R. D. Willett, *J. Am. Chem. Soc.* **113,** 5253 (1991).

10. K. Song, R. D. Miller, G. M. Wallraff, and J. F. Rabolt, *Macromolecules* **24,** 4084 (1991).

11. J. F. Rabolt and co-workers, *Polym. Prep.* **31,** 262 (1991).

12. R. A. Nallicheri and M. E. Rubner, *Macromolecules* **24,** 517 (1991).

13. K. Song, H. Kuzmany, G. M. Wallraff, R. D. Miller, and J. F. Rabolt, *Macromolecules* **23,** 3870 (1990).

14. G. Cryssomallis and H. G. Drickamer, *J. Chem. Phys.* **71,** 4817 (1979).

15. W. H. Bentley and H. G. Drickamer, *J. Chem. Phys.* **42,** 1573 (1965).

16. V. C. Bastron and H. G. Drickamer, *J. Solid State Chem.* **3,** 550 (1971).

17. R. B. Ali, P. Banerjee, J. Burgess, and A. E. Smith, *Trans. Met. Chem.* **13,** 106 (1988).

18. R. B. Ali, J. Burgess, and Guardano, *Trans. Met. Chem.* **13,** 126 (1988).

19. J. Burgess, *Spectrochim Acta* **45A,** 159 (1989).

20. M. Matsui, K. Shibata, and H. Muramatsa, *Bull. Chem. Soc. Jpn.* **63,** 1845 (1990).

21. Y. Kitamara, T. Ito, and M. Kato, *Inorg. Chem.* **23,** 3826 (1984).

22. H. T. Macholt, R. Van Eldik, H. Kelm, and H. Elias, *Inorg. Chim. Acta.* **104,** 115 (1985).

23. K. L. Bray and H. G. Drickamer, *J. Phys. Chem.* **94,** 7037 (1990).

H. G. Drickamer
University of Illinois

THERMOCHROMIC

Thermochromism is the reversible change in the spectral properties of a substance that accompanies heating and cooling. Strictly speaking, the meaning of the word specifies a visible color change; however, thermochromism has come to also include some cases for which the spectral transition is either better observed outside of the visible region or not observed in the visible at all. Primarily, thermochromism occurs in solid or liquid phase, but it also describes a thermally dependent equilibrium between brown nitrogen dioxide [10102-44-0], NO_2, and colorless dinitrogen tetroxide [10544-72-6], N_2O_4, a rare example in the gas phase (1). Although in the last edition of the *Encyclopedia* even irreversible spectral changes were included, and many interesting materials fall in this category, the absence of reversibility is taken here as a criterion for exclusion from the group.

There are many materials, especially organic and metal-organic materials, which exhibit true thermochromism, with a variety of sometimes debatable structural transition mechanisms; it is difficult to summarize the whole with any continuity. For this reason, an effort is made to delineate the scope of the field by listing several thermochromic transitions (Table 1). Selected thermochromic material examples are accompanied in each instance by the corresponding transition stimulus for that case. Characteristically sharp transition temperatures, T_t, are indicated where appropriate. At the other extreme are examples of comparatively gradual transitions, associated for example with an equilibrium or a changing bandwidth. The sharpness of the transition is one aspect by which the several mechanisms could be classified. On the other hand, it is useful also to group materials into metal-complex, inorganic, and organic classes. In this way, the variety

Table 1. Some Typical Thermochromic Compounds and Their Transitions

Thermochromic material	CAS Registry Number	Thermochromic transition[a]	T_t^b,°C	References
Co^{2+} solutions and glasses		equilibrium shift, two coordinations		1–3
$[(C_2H_5)_2NH_2]_2CuCl_4$	[52003-09-5], [52003-08-4]	square planar to tetrahedral	50	4–6
$[(CH_3)_2NH_2]_3CuCl_5$	[52003-06-2]	variation in bandwidth		6
$Cu_4I_4(Py)_4$	[62121-41-9]	fluorescence variations		7
$Al_{2-x}Cr_xO_3$ (ruby)	[12174-49-1]	lattice expansion/ contraction		1,8,9
VO_2	[12036-21-4]	monoclinic/tetragonal	68	10–12
Cu_2HgI_4	[13876-85-2]	order/disorder	68	13–15
di-β-naphthospiropyran	[178-10-9]	close/open spiro ring		16
poly(xylylviologen dibromide)	[38815-69-9]	hydration/ dehydration	100	17
ETCD polydiacetylene[c]	[63809-82-5]	side group rearrangement	~115	18,19

[a]When applicable, expressed as a change upon heating; various colors have been reported, often qualitatively.
[b]Transition temperatures for sharp transitions.
[c]Urethane-substituted polymer of $(=C-C\equiv C-C=)_n$ where R = $(CH_2)_4OCONHCH_2CH_3$.

$$\begin{array}{cc} | & | \\ R & R \end{array}$$

of thermochromic changes in each of the three material classes can easily be realized.

Spectral Transitions in Metal Complexes

Crystal field theory, which is simpler to handle than the more comprehensive molecular orbital treatment, has been used to describe d–d orbital excitations of transition-metal ions and the effect on these excitations of ligand coordination geometry and field strength. Absorption bands in the visible region arise in energy states made nondegenerate by the crystal field. Color changes, such as induced by heating or cooling, are therefore a direct indication of change in the surrounding environment of the metal ion. Crystal field theory, even though it does not include charge-transfer processes, has proven to be qualitatively adequate for the $3d$ orbital transition metals because the $3d$ states are not well shielded from ligand field effects. Crystal field splittings are on the order of 10,000 cm^{-1}. The lanthanide $4f$ orbitals are better shielded and have crystal field splittings only on the order of 100 cm^{-1}. It is not surprising, therefore, that many good examples of thermochromism are to be found in $3d$ transition-metal complexes and other $3d$ metal compounds.

A simple and well-known example, from Table 1, is the case of anhydrous cobalt chloride [7646-79-9], $CoCl_2$, in alcohol solutions (1,2). At room temperature, Co^{2+} is predominately tetrahedrally coordinated, as dichlorobis(ethanol) cobalt (II) [15168-62-4], $Co(C_2H_5OH)_2Cl_2$, in the case of ethyl alcohol solution, and is colored blue. The tetrahedral absorption band peaks at about 660 nm. With cooling, this band shrinks in intensity. The solution begins to acquire a pink color indicative of the growing dominance of a weak octahedral coordination band, perhaps attributable to chloropentakis(ethanol) cobalt (II) [32354-52-2], $[Co(C_2H_5OH)_5Cl]^+$ (1). The temperature-dependent equilibrium in an ethyl alcohol–water solution is a classroom demonstration of a "thermometer" for observations below and above room temperature. By analogy, the proportioning between the blue tetrahedral and pink octahedral coordinations of Co^{2+} is also evident in molten inorganic glasses (3). This is clearly not associated with oxidation to Co^{3+} because the latter is unstable at the high temperatures of glass melting. Cobalt-containing glasses that are pink at room temperature become blue with heating. This color is likely to be retained with rapid quenching. However, if the glass is cooled slowly, equilibrium is more nearly approached in favor of octahedral coordination, although the unbluing may be slow and difficult to detect by the eye in any case. This sort of kinetic problem is a general one that sometimes makes difficult the distinction between processes that are reversible or irreversible.

Other Co^{2+} halide solutions have been reviewed (1), as have equilibria involving complexes and chelates of Cu^{2+} and Ni^{2+} in solutions. Good thermochromism is well known, for example, in aqueous solutions of copper chloride [7447-39-4], $CuCl_2$, and it has also been demonstrated in solutions of $Ni(ClO_4)_2$ (20). Good thermochromism is not restricted to transition-metal ions in solution. The example of $[(C_2H_5)_2NH_2]_2CuCl_4$ in Table 1 is solid-state and it has been reported to undergo a particularly pronounced, discontinuous, first-order, reversible, green-to-yellow color transition with heating (4–6). This is associated with a change in coordination geometry for the $CuCl_4{}^{2-}$ anion. Similar examples abound and many other compounds have been described in depth in an extensive overview about Cu^{2+} and Ni^{2+} salts that show discontinuous thermochromism (5). Various compounds containing N,N-diethylethylenediamine (21–25) or the isopropylammonium ion (6,26–28) are included.

The solid-state, transition-metal example in Table 1 of $[(CH_3)_2NH_2]_3CuCl_5$ illustrates another form of thermochromism: the color shifts gradually and continuously because of changes in bandwidth with either heating or cooling (6). It is not unique, as this behavior has been mentioned for the class of compounds $(RNH_3)_2CuCl_4$, where R = alkyl group (6), and also for compounds of the form $M(N,N$-diethylethylenediamine$)_2(X)_2$, where M = Cu^{2+} or Ni^{2+} and X is an anion that does not disorder easily (5). With easily disordered anions such as $ClO_4{}^-$ and $BF_4{}^-$, thermochromism occurs discontinuously at a reasonably well-defined T_t.

Halide complexes of Cu^+ with nitrogen base ligands are known to exhibit another form of reversible spectral change known as fluorescence thermochromism. The example of $Cu_4I_4(Py)_4$ from Table 1 is typical and shows red shifting in the visible emission spectrum while the sample is both cooled and irradiated with a 364 nm ultraviolet source (7).

Transitions in Inorganic Compounds

There are not many oxides and sulfides that may be classified as truly thermochromic; again, however, compounds of transition metals dominate. Ruby exhibits a well-known, reversible, ligand-field thermochromism at different temperatures depending on the concentration of Cr^{3+} in the Al_2O_3 lattice (1,8,9). This is a manifestation of change in the ligand field strength as dependent on lattice expansion/contraction. A whole family of oxides is known to undergo reversible, nonmetal–metal, thermoresistive transitions with heating (29–31). The shifts in band structure have frequently been debated. Sometimes these transitions are associated with symmetry changes. VO_2 is one of the best known of these compounds because its transition, which has as much as a fivefold switch in resistivity, occurs close to room temperature with a large, free-carrier, infrared change (10–12). There is also a smaller effect in the visible, as shown for easily grown thin films (12,32). Doping has been found to shift the resistivity transition temperature up and down (33–35), but this does not seem to have any great effect on the spectral range of switching for thin films made in various ways (36). This has discouraged interest in VO_2 films for large-area transparencies. Exploratory work on VO_2 films for optical storage and laser switching applications has also been discussed (37–39).

Another oxide that exhibits a nonmetal–metal transition is V_2O_3 [1314-34-7]. It undergoes a symmetry change at about $-123°C$ (monoclinic/rhombohedral) with as much as a 10^7-fold change in resistivity (40,41). Fe_3O_4 [1317-61-9] undergoes a small-order symmetry change at about $-154°C$ (42,43). It is orthorhombic below this temperature but a cubic spinel above it. Also, some of the Magnéli phases of vanadium and titanium, of the form M_nO_{2n-1} (M is the metal ion and $n = 3, 4, 5 \ldots$), have been observed to undergo relatively sharp thermal changes, especially as single crystals (29,31). However, except for VO_2, no remarkable spectral switching has been mentioned with any other member in this category of materials.

In Table 2, four sulfides that also undergo reversible nonmetal–metal transitions are shown. In three cases spectral changes are known, though only in one case for the visible region. With the $Sm_{1-x}Ln_x^{3+}S$ series of compounds a dramatic color change occurs at T_t with cooling. It is associated with expansion in the lattice

Table 2. Some Sulfide Compounds with Nonmetal–Metal Transitions

Compound	CAS Registry Number	Thermochromic transition[a]	Approximate T_t, °C	Spectral shift	References
Ag_2S	[21548-73-2]	monoclinic/cubic	178	ir	44–46
NiS(hex)	[16812-54-7]	antiferromagnetic/ paramagnetic	−9	ir	29,47–49
FeS(tet)	[1317-37-9]	tetragonal/ hexagonal	157		29,50,51
$Sm_{1-x}Ln_x^{3+}S^b$		Sm^{2+}/Sm^{3+}	c	visible	52,53

[a]Expressed as a change upon heating.
[b]Ln = Ce, Pr, Nd, Gd, Tb, Dy, Ho, Er, Tm, or Y.
[c]Ln_x^{3+}-dependent.

without change in the cubic structure. Qualitatively, the sulfide is black below the transition and metallic yellow above it. This remarkable example derives its behavior from samarium sulfide [29678-92-0], SmS, which undergoes a like black-to-yellow color transition when it is rubbed or taken to a relatively low applied pressure of 650 MPa (6400 atm). A large infrared spectral change accompanies the visible change. The high pressure form of SmS has been made stable at room temperature and atmospheric pressure by doping with Ln^{3+}. Doping promotes a shift from Sm^{2+} to Sm^{3+} by $4f$ electron delocalization, so that the sulfide can be switched back and forth below room temperature.

Thermochromic compounds such as Ag_2HgI_4 [12344-40-0] and $Cu_2 HgI_4$ have long been known (13). These compounds color reversibly exhibiting the discontinuous red shift of a charge-transfer band edge during heating (14). As for VO_2, the characteristic hysteresis of reflectance in the visible suggests application for infrared image recording (15).

Organic and Polymeric Compounds

Simple organic molecules tend to be colorless with electronic transitions in the uv, whereas visible absorption, or color, is usually associated with electronic excitations in extended and conjugated structures. Color is influenced considerably by the extent of conjugation, as well as by the molecular environment imparted by substituents. So, thermochromism arises from critical, thermally induced changes in the existing structure. Thousands of thermochromic organic examples are known (8). Three of these are in Table 1. The di-β-naphthospiropyran example represents a well-known and extensively reviewed family of compounds that develop color at the onset of a thermally induced ring opening (16). Heterolytic bond cleavage in the molecule results in polar or ionic resonance structures, with conjugation, and it is these attributes that have been associated with the appearance of color. The second organic example in Table 1, poly(xylylviologen dibromide), is conjugated between pyridinium rings. It is characterized by charge-transfer interactions with counterions. The charge-transfer energy levels in the solid state are sensitive to the molecular environment so that thermochromism occurs when the polymer is subjected to hydration/dehydration sequences (17). Similarly, for urethane-substituted ETCD polydiacetylene (Table 1), thermally induced transitions in the conformation of the unsaturated backbone have been associated with restructuring of side-group substituents (18,19). The occurrence is manifested as a change in color.

Certain poly(di-n-alkylsilanes) and germanes, when in the solid state in particular, also exhibit large spectral changes that have been associated with side-chain influence on the backbone, but these (the polymer being saturated) occur at uv rather than visible wavelengths (54–56). Poly(di-n-hexylsilane) [94904-85-5] is an example. Initially, crystallization of the n-hexyl substituent groups locks the backbone into a configuration that is characterized by a red shifted absorption band at 374 nm. With heating through about 41°C, there is a reversible relaxation to the higher energy 317 nm band associated with disordering.

The subject of thermochromism in organic and polymeric compounds has been reviewed in some depth previously (8,16,18), and these expansive overviews should be used by readers with deeper and more particular interest in the subject. Many more examples can be found in the reviews that further illustrate the pattern of association between thermochromism and molecular restructuring of one kind or another. The specific assignment of structures is still open to debate in many cases, and there are still not many actual commercial uses for these or any of the other thermally reversible materials discussed herein. Temperature indicators have been mentioned, though perhaps as much or more for irreversible materials.

BIBLIOGRAPHY

"Chromogenic Materials, Electrochromic and Thermochromic" in *ECT* 3rd ed., Vol. 6, pp. 129–142, by J. H. Day, Ohio University.

1. K. Sone and Y. Fukuda, *Inorganic Thermochromism*, Vol. 10, Springer-Verlag, New York, 1987, pp. 2,13.
2. W. C. Nieuwpoort, G. A. Wesselink, and E. H. A. M. Van der Wee, *Rec. Trav. Chim. Pays-Bas* **85**, 397 (1966).
3. W. A. Weyl, *Coloured Glasses*, The Society of Glass Technology, Sheffield, UK, 1951, p. 179.
4. D. R. Bloomquist, M. R. Pressprich, and R. D. Willett, *J. Am. Chem. Soc.* **110**, 7391 (1988).
5. D. R. Bloomquist and R. D. Willett, *Coord. Chem. Rev.* **47**, 125 (1982).
6. R. D. Willett, J. A. Haugen, J. Lebsack, and J. Morrey, *Inorg. Chem.* **13**, 2510 (1974).
7. H. D. Hardt and A. Pierre, *Inorg. Chim. Acta* **25**, L59 (1977).
8. K. Nassau, *The Physics and Chemistry of Color*, John Wiley & Sons, Inc., New York, 1983, pp. 77,109.
9. D. S. McClure, *J. Chem. Phys.* **36**, 2757 (1962).
10. F. J. Morin, *Phys. Rev. Lett.* **3**, 34 (1959).
11. A. S. Barker, Jr., H. W. Verleur, and H. J. Guggenheim, *Phys. Rev. Lett.* **17**, 1286 (1966).
12. H. W. Verleur, A. S. Barker, Jr., and C. N. Berglund, *Phys. Rev.* **172**, 172 (1968).
13. J. H. Day, *Chem. Rev.* **68**, 649 (1968).
14. H.-R.C. Jaw, M. A. Mooney, T. Novinson, W. C. Kaska, and J. I. Zink, *Inorg. Chem.* **26**, 1387 (1987).
15. J. S. Chivian, R. N. Claytor, D. D. Eden, and R. B. Hemphill, *Appl. Opt.* **11**, 2649 (1972).
16. J. H. Day, *Chem. Rev.* **63**, 65 (1963).
17. J. S. Moore and S. I. Stupp, *Macromolecules* **19**, 1815 (1986).
18. D. N. Batchelder, *Contemp. Phys.* **29**, 3 (1988).
19. M. F. Rubner, D. J. Sandman, and C. Velazquez, *Macromolecules* **20**, 1296 (1987).
20. T. R. Griffiths and R. K. Scarrow, *Trans. Farad. Soc.* **65**, 1727 (1969).
21. W. E. Hatfield, T. S. Piper, and U. Klabunde, *Inorg. Chem.* **2**, 629 (1963).
22. H. Yokoi, M. Sai, and T. Isobe, *Bull. Chem. Soc. Jpn.* **42**, 2232 (1969).
23. A. B. P. Lever, E. Mantovani, and J. C. Donini, *Inorg. Chem.* **10**, 2424 (1971).
24. L. Fabbrizzi, M. Micheloni, and P. Paoletti, *Inorg. Chem.* **13**, 3019 (1974).
25. J. R. Ferraro, L. J. Basile, L. R. Garcia-Ineguez, P. Paoletti, and L. Fabbrizzi, *Inorg. Chem.* **15**, 2342 (1976).
26. S. A. Roberts, D. R. Bloomquist, R. D. Willett, and H. W. Dodgen, *J. Am. Chem. Soc.* **103**, 2603 (1981).
27. D. R. Bloomquist, R. D. Willett, and H. W. Dodgen, *J. Am. Chem. Soc.* **103**, 2610 (1981).

28. D. R. Bloomquist and R. D. Willett, *J. Am. Chem. Soc.* **103,** 2615 (1981).
29. D. Adler, *Rev. Mod. Phys.* **40,** 714 (1968).
30. D. Adler, in J. I. Budnick and M. P. Kawatra, eds., *Conference on Dynamical Aspects of Critical Phenomena*, Gordon and Breach, London, 1972, p. 392.
31. J. M. Honig and L. L. Van Zandt, in R. A. Huggins, ed., *Annual Review of Material Science*, Vol. 5, Annual Reviews, Inc., Palo Alto, Calif., 1975, p. 225.
32. C. B. Greenberg, in S. A. Marolo, ed., *Proceedings of the 15th Conference on Aerospace Transparent Materials and Enclosures, II*, WRDC-TR-89-4044, Wright-Patterson AFB, Dayton, Ohio, 1989, p. 1124.
33. M. Nygren and M. Israelsson, *Mater. Res. Bull.* **4,** 881 (1969).
34. T. Horlin, T. Niklewski, and M. Nygren, *Mater. Res. Bull.* **7,** 1515 (1972).
35. J. M. Reyes, G. F. Lynch, M. Sayer, S. L. McBride, and T. S. Hutchinson, *J. Can. Ceram. Soc.* **41,** 69 (1972).
36. C. B. Greenberg, *Thin Solid Films* **110,** 73 (1983).
37. W. R. Roach, *Appl. Phys. Lett.* **19,** 453 (1971).
38. A. W. Smith, *Appl. Phys. Lett.* **23,** 437 (1973).
39. I. Balberg and S. Trokman, *J. Appl. Phys.* **46,** 2111 (1975).
40. M. Foëx, *Compt. Rend.* **223,** 1126 (1946).
41. J. Feinleib and W. Paul, *Phys. Rev.* **155,** 841 (1967).
42. E. J. W. Verwey, *Nature* **144,** 327 (1939).
43. P. A. Miles, W. B. Westphal, and A. von Hippel, *Rev. Mod. Phys.* **29,** 279 (1957).
44. M. H. Hebb, *J. Chem. Phys.* **20,** 185 (1952).
45. P. Brüesch and J. Wullschleger, *Solid State Commun.* **13,** 9 (1973).
46. T.-Y. Hsu, H. Buhay, and N. P. Murarka, in G. A. Tanton, ed., *Proceedings of the SPIE*, Vol. 259, SPIE, Bellingham, Wash., 1980, p. 38.
47. J. T. Sparks and T. Komoto, *J. Appl. Phys.* **34,** 1191 (1963); *Phys. Letters* **25A,** 398 (1967); *Rev. Mod. Phys.* **40,** 752 (1968).
48. A. S. Barker, Jr. and J. P. Remeika, *Phys. Rev. B* **10,** 987 (1974).
49. T. Ohtani, *J. Phys. Soc. Jpn.* **37,** 701 (1974).
50. M. Murakami, *J. Phys. Soc. Jpn.* **16,** 187 (1961).
51. E. F. Bertaut, P. Burlet, and J. Chappert, *Solid State Commun.* **3,** 335 (1965).
52. A. Jayaraman, E. Bucher, P. D. Dernier, and L. D. Longinotti, *Phys. Rev. Lett.* **31,** 700 (1973).
53. A. Jayaraman, P. D. Dernier, and L. D. Longinotti, *Phys. Rev. B* **11,** 2783 (1975); *High Temp.-High Press.* **7,** 1 (1975).
54. R. D. Miller, D. Hofer, J. Rabolt, and G. N. Fickes, *J. Am. Chem. Soc.* **107,** 2172 (1985).
55. J. F. Rabolt, D. Hofer, R. D. Miller, and G. N. Fickes, *Macromolecules* **19,** 611 (1986).
56. R. D. Miller and R. Sooriyakumaran, *J. Polym. Sci., Part A: Polym. Chem.* **25,** 111 (1987).

General References

D. S. McClure, *Electronic Spectra of Molecules and Ions in Crystals*, Academic Press, Inc., New York, 1959.
J. H. Day and R. D. Willett, in C. M. Lampert and C. G. Granqvist, eds., *Large-Area Chromogenics: Materials and Devices for Transmittance Control*, IS 4, SPIE Optical Engineering Press, Bellingham, Wash., 1990, containing some known errors in referencing.

CHARLES B. GREENBERG
PPG Industries, Inc.

CINNAMIC ACID, CINNAMALDEHYDE, AND CINNAMYL ALCOHOL

The earliest references to cinnamic acid, cinnamaldehyde, and cinnamyl alcohol are associated with their isolation and identification as odor-producing constituents in a variety of botanical extracts. It is now generally accepted that the aromatic amino acid L-phenylalanine [63-91-2], a primary end product of the Shikimic Acid Pathway, is the precursor for the biosynthesis of these phenylpropanoids in higher plants (1,2).

The widespread use of cinnamic derivatives has led to the pursuit of reliable methods for their direct synthesis. Commercial processes have focused on condensation reactions between benzaldehyde and a number of active methylene compounds for assembly of the requisite carbon skeleton. The presence of a disubstituted carbon–carbon double bond in the sidechain of these chemicals also gives rise to the existence of two distinct stereoisomers, the cis or (Z)- and trans or (E)-isomers:

(Z)-isomer   (E)-isomer

where X = COOH, CHO, or CH_2OH

A considerable range of products, including flavors, fragrances, agrochemicals, pharmaceuticals, and polymers, has been developed using these chemicals as either synthetic intermediates or ingredients (3).

Cinnamic Acid

3-Phenyl-2-propenoic acid [621-82-9], commonly referred to as cinnamic acid, is a white crystalline solid having a low intensity sweet, honeylike aroma. It has been identified as a principal constituent in the botanical exudates from Styrax (*Liquidamber orientalis*), Benzoin (*Styrax benzoin*), Peru Balsam (*Myroxylon pereirae*), and Tolu Balsam (*Myroxylon balsamum*) (4,5). In these, as well as numerous other natural products, it exists both as the free acid and in the form of one or more of its esters, as for example, methyl cinnamate, benzyl cinnamate [103-41-3], and cinnamyl cinnamate.

Physical and Chemical Properties. Cinnamic acid is generally encountered as the thermodynamically favored (E)-isomer. (E)-Cinnamic acid [140-10-3] is an off-white solid having the properties outlined in Table 1 (6,7).

For (Z)-cinnamic acid [102-94-3], three distinct polymorphic forms have been characterized. The most stable form, referred to as allocinnamic acid, has a melting point of 68°C, and the two metastable forms, isocinnamic acids, have melting

Table 1. Properties of (*E*)-Cinnamic Acid

Property	Value
molecular formula	$C_9H_8O_2$
mol wt	148.2
melting point, °C	133
boiling point, °C at 101.3 kPa^a	300
specific gravity at 25°C	1.245
solubility at 20°C	
g/L H_2O	0.5
g/L C_2H_5OH	189
dissociation constant, K_a, at 25°C	3.5×10^{-5}

^{a}To convert kPa to mm Hg, multiply by 7.5.

points of 58°C and 42°C, respectively. (*E*)-Cinnamic acid can be converted to the (*Z*)-isomer photochemically through irradiation of a solution with ultraviolet light.

Cinnamic acid undergoes reactions that are typical of an aromatic carboxylic acid. Using standard methodology, simple esters are easily prepared and salts are formed upon neutralization with the appropriate base. Hydrogenation of cinnamic acid under mild conditions leads to 3-phenylpropanoic acid [501-52-0] whereas under forcing conditions, such as under high pressure in presence of a nickel catalyst, complete saturation to 3-cyclohexylpropanoic acid [701-97-3] is readily accomplished (8).

Decomposition to styrene and carbon dioxide has been observed upon heating the acid to temperatures in excess of 150°C. The decarboxylation process can be accelerated with the addition of a bicyclic amine base (9).

Selective oxidation of either the aromatic ring or the side chain can also be accomplished. For example, epoxidation of the double bond of cinnamic acid is effected in excellent yield by treatment with potassium hydrogen persulfate (10).

Manufacture. The most widely employed method for the commercial synthesis of (*E*)-cinnamic acid utilizes benzaldehyde, acetic anhydride, and anhydrous sodium or potassium acetate in a condensation reaction commonly referred to as the Perkin reaction (11).

$$C_6H_5CHO + (CH_3CO)_2O \xrightarrow{CH_3COONa} C_6H_5CH{=}CHCOOH + CH_3COOH$$

In a typical process, a mixture of acetic anhydride, anhydrous sodium acetate, and benzaldehyde in a ratio of 1.8:1:1 is charged into a reactor equipped with a column suitable for fractional distillation. The reaction mixture is heated to 180–190°C and acetic acid is continuously removed over an 8–10-h period. This process yields over 80% cinnamic acid based on consumed benzaldehyde. Other catalysts such as pyridine, potassium acetate, potassium carbonate, and sodium borate have been examined in an attempt to maximize the yield and reduce byproduct formation. None of the alternative systems have given dramatically improved performance.

Treatment of benzal chloride [98-87-3] with anhydrous sodium acetate at 180–200°C provides another economically attractive route to cinnamic acid.

$$C_6H_5CHCl_2 + 2\ CH_3COONa \longrightarrow C_6H_5CH{=}CHCOOH + 2\ NaCl + CH_3COOH$$

The chloride is readily available as a by-product of benzyl chloride [100-44-7] production (see CHLOROCARBON AND CHLOROHYDROCARBONS-BENZYL CHLORIDE, BENZAL CHLORIDE, AND BENZOTRICHLORIDE). The yield is comparable to the Perkin-based process, but the difficulty associated with removal of trace halogenated impurities makes the resultant cinnamic acid less desirable for many applications.

Another potentially valuable method for the preparation of cinnamic acid involves treatment of benzaldehyde with ketene (12). The initially formed oligomer of β-hydroxy-β-phenylpropionic acid is thermally decomposed at 100–250°C in the presence of an acid or base catalyst.

$$C_6H_5CHO + H_2C{=}C{=}O \longrightarrow \left(O{-}\overset{C_6H_5}{\underset{}{\overset{|}{C}}}HCH_2\overset{O}{\overset{\|}{C}} \right)_n \longrightarrow C_6H_5CH{=}CHCOOH$$

Esters of cinnamic acid are used more extensively than the acid itself, and can be converted to the acid by standard hydrolysis protocols. The Claisen condensation between benzaldehyde and the appropriate acetate ester provides a direct, high yield route to the simple esters.

$$C_6H_5CHO + CH_3COOR \xrightarrow{\text{NaOR}} C_6H_5CH{=}CHCOOR$$

The catalyst of choice for this reaction is the corresponding sodium alkoxide.

Several newer methods take advantage of the highly selective nature of organopalladium reagents. A palladium acetate-triarylphosphine catalytic system has been employed to induce the coupling of bromobenzene with the desired acrylate ester (13).

$$C_6H_5Br + CH_2{=}CHCOOR + R'_3N \xrightarrow[\text{phosphine}]{\text{Pd (OOCCH}_3)_2} C_6H_5CH{=}CHCOOR + R'_3NH^+Br^-$$

Cinnamate ester yields of 70–95% have been realized, but the substrates are expensive when compared with those employed in the standard Claisen approach.

The oxidative carbonylation of styrene with carbon monoxide, oxygen, and an aliphatic alcohol in the presence of a palladium salt, a copper salt, and sodium propionate also provides the requisite cinnamate.

$$C_6H_5CH{=}CH_2 + CO + \tfrac{1}{2}O_2 + ROH \longrightarrow C_6H_5CH{=}CHCOOR + H_2O$$

Conditions that give selectivity as high as 95% have been defined (14).

Economic Aspects.   There are no published production figures for cinnamic acid. Most of the manufactured acid is consumed internally to generate a series of cinnamate esters for flavor and fragrance applications. With this in mind, it was possible to estimate a 1990 usage in the range of 175 metric tons. The cinnamic acid that does find its way into the marketplace has been sold for $12–14/kg in drum quantities.

Health and Safety. The Flavor and Extract Manufacturers' Association (FEMA) and the Research Institute for Fragrance Materials (RIFM) have developed procedures which employ expert panels to evaluate the safety in use of new and existing flavor and fragrance ingredients. The FEMA expert panel has given cinnamic acid GRAS (generally recognized as safe) status and FEMA No. 2288 has been assigned to this material (15). As a consequence, the FDA has approved it for food use. The acid is likewise devoid of any significant dermal irritation or sensitization and has been approved for fragrance use (16).

Uses. Although cinnamic acid is not considered an important odorant, it serves as a precursor for derivatives such as the esters (17) which have pleasant long-lasting aromas. Methyl cinnamate [103-26-4] enjoys the greatest usage and is found in flavor and fragrance compositions created for products which include soaps and cosmetics as well as beverages, baked goods, and convenience foods. Reported applications for cinnamic acid and its derivatives also include: use as a light penetration inhibitor in sunscreen formulations (18); for the preparation of herbicidal compositions (19); as a substrate in the formation of photopolymers (20–22); as a raw material in the synthesis of heterocyclic color complexes (23); and in the electroplating process for zinc (24).

One of the most interesting uses for cinnamic acid in recent years has been as a raw material in the preparation of L-phenylalanine [63-91-2], the key intermediate for the synthetic dipeptide sweetener aspartame (25). Genex has described a biosynthetic route to L-phenylalanine which involves treatment of immobilized cells of *R. rubra* containing the enzyme phenylalanine ammonia lyase (PAL) with ammonium cinnamate [25459-05-6] (26).

Cinnamaldehyde

3-Phenyl-2-propenal [104-55-2], also referred to as cinnamaldehyde, is a pale yellow liquid with a warm, sweet, spicy odor and pungent taste reminiscent of cinnamon. It is found naturally in the essential oils of Chinese cinnamon (*Cinnamomum cassia*, Blume) (75–90%) and Ceylon cinnamon (*Cinnamomum zeylanicum*, Nees) (60–75%) as the primary component in the steam distilled oils (27). It also occurs in many other essential oils at lower levels.

Physical and Chemical Properties. The (*E*)- and (*Z*)-isomers of cinnamaldehyde are both known. (*E*)-Cinnamaldehyde [14371-10-9] is generally produced commercially and its properties are given in Table 2. Cinnamaldehyde undergoes reactions that are typical of an α,β-unsaturated aromatic aldehyde. Slow oxidation to cinnamic acid is observed upon exposure to air. This process can be accelerated in the presence of transition-metal catalysts such as cobalt acetate (28). Under more vigorous conditions with either nitric or chromic acid, cleavage at the double bond occurs to afford benzoic acid. Epoxidation of cinnamaldehyde via a conjugate addition mechanism is observed upon treatment with a salt of *t*-butyl hydroperoxide (29).

Hydrogenation of cinnamaldehyde has been studied extensively since selectivity has often been an issue. Under mild conditions the carbonyl group is reduced

Table 2. Properties of (*E*)-Cinnamaldehyde

Property	Value
molecular formula	C_9H_8O
mol wt	132.2
boiling point, °C	
at 101 kPa^a	252
at 1.3 kPa^a	120
specific gravity at 20°C	1.049
refractive index at 20°C	1.6195
solubility	
in 50% C_2H_5OH	1:25
in ethyl ether	infinite

^{a}To convert kPa to mm Hg, multiply by 7.5.

giving cinnamyl alcohol, whereas at elevated temperatures complete reduction to 3-phenylpropanol [*122-97-4*] results. It is possible to saturate the double bond without concomitant reduction of the carbonyl group through selective hydrogenation with a ferrous chloride-activated palladium catalyst (30), thereby producing 3-phenylpropanal [*104-53-0*].

The formation of acetals with methanol, ethanol, or ethylene glycol in the presence of an acid catalyst such as hydrogen chloride or benzenesulfonic acid is straightforward. Sodium bisulfite and hydroxylamine form adducts with cinnamaldehyde that are used in typical quantitative analysis protocols.

Upon treatment with aluminum ethoxide, the aldehyde is converted to cinnamyl cinnamate [*122-69-0*] (Tishchenko reaction), a valuable perfumery ingredient.

Manufacture. Cinnamaldehyde is routinely produced by the base-catalyzed aldol addition of benzaldehyde [*100-52-7*] with acetaldehyde [*75-07-0*], a procedure which was first established in the nineteenth century (31). Formation of the (*E*)-isomer is favored by the transition-state geometry associated with the elimination of water from the intermediate. The commercial process is carried out in the presence of a dilute sodium hydroxide solution (ca 0.5–2.0%) with at least two equivalents of benzaldehyde and slow addition of the acetaldehyde over the reaction period (32).

$$C_6H_5CHO + CH_3CHO \xrightarrow{\text{NaOH}} C_6H_5CH{=}CHCHO + H_2O$$

In this manner, self-condensation of acetaldehyde is minimized and yields in the range of 77–85% are obtained. However, even with these precautions a detectable amount of 5-phenyl-2,4-pentadienal [*13466-40-5*] is invariably formed.

Another approach is based on the rearrangement of an acetylenic carbinol formed between benzaldehyde and acetylene.

$$C_6H_5CH(OH)C{\equiv}CH \longrightarrow C_6H_5CH{=}CHCHO$$

Isomerization catalyzed by silyl vanadates (33) gives cinnamaldehyde in high yield.

Economic Aspects. Since the 1970s cinnamaldehyde has been produced in significant quantities by Fritzsche Dodge & Olcott (FDO), Haarmann & Reimer (H&R), and Dutch State Mines (DSM). However, by the end of 1989 DSM was the only remaining producer for this material. Production statistics are listed in Table 3.

Table 3. Cinnamaldehyde Production

Year	Volume^a, 10^3t	Price,^{b} $/kg
1970	685	2.30–2.60
1980	850	3.30–3.65
1990	975	3.50–4.00

^{a}Estimates based on U.S. International Trade Commission figures and unpublished data.
^{b}Prices based on drum quantities.

Health and Safety. FEMA has examined cinnamaldehyde and established its GRAS status (No. 2286). The material has been used in some fragrance compositions, but RIFM (34) has noted its potential for sensitization and limited the use in perfumes for skin contact at 1% in the formula. Eugenol and limonene have been used in conjunction with cinnamaldehyde as quenchers to neutralize the irritation reaction that some individuals have toward this aldehyde.

Uses. Greater than 95% of the consumption of cinnamaldehyde occurs in flavor applications where a spicy, cinnamon character is required. It is used in a wide range of products including bakery goods, confection, and beverages as well as in toothpastes, mouthwashes, and chewing gum. It is also used effectively in air fresheners where odor neutralization can be accomplished by reaction with sulfur and nitrogen malodorants.

In electroplating processes, cinnamaldehyde is utilized as a brightener (35). Other applications include its efficacy as an animal repellent (36), its use in compositions to attract insects (37), and demonstration of a positive antifungal activity (38).

Cinnamaldehyde has been efficiently isolated in high purity by fractional distillation from cassia and cinnamon bark essential oils. This material has been utilized in several manufacturing protocols (39–41) for the preparation of natural benzaldehyde through a retro-aldol process. Since the late 1970s the demand for natural flavors has increased dramatically. This demand has led to a corresponding requirement for a more extensive line of readily available natural aroma chemicals for flavor creation.

Cinnamyl Alcohol

3-Phenyl-2-propen-1-ol [*104-54-1*], commonly referred to as cinnamyl alcohol, is a colorless crystalline solid with a sweet balsamic odor that is reminiscent of hyacinth. Its occurrence in nature is widespread as, for example, in Hyacinth absolute (*Hyacinthus orientalis*) (42), the leaf and bark oils of cinnamon (*Cinnamomum cassia, Cinnamomum zeylancium*, etc), and Guava fruit (*Psidium guajava L.*) (43). In many cases it is also encountered as the ester or in a bound form as the glucoside.

Physical and Chemical Properties. Although both the (*E*)- and (*Z*) [*4510-34-3*] isomers of cinnamyl alcohol are known in nature, (*E*)-cinnamyl alcohol [*4407-36-7*] is the only isomer with commercial importance. Its properties are summarized in Table 4.

Table 4. Properties of (*E*)-Cinnamyl Alcohol

Property	Value
molecular formula	$C_9H_{10}O$
mol wt	134.2
boiling point, °C at 1.33 kPa	117–118
melting point, °C	33
density, g/mL at 20°C	1.044
refractive index at 20°C	1.5819
solubility in 60% C_2H_5OH	1:2

When heated in the presence of a carboxylic acid, cinnamyl alcohol is converted to the corresponding ester. Oxidation to cinnamaldehyde is readily accomplished under Oppenauer conditions with furfural as a hydrogen acceptor in the presence of aluminum isopropoxide (44). Cinnamic acid is produced directly with strong oxidants such as chromic acid and nickel peroxide. The use of *t*-butyl hydroperoxide with vanadium pentoxide catalysis offers a selective method for epoxidation of the olefinic double bond of cinnamyl alcohol (45).

Halogens add to the double bond of the alcohol to afford the corresponding dihalo derivatives, eg, $C_6H_5CHXCHXCH_2OH$, where X = Cl or Br. The allylic chloride C_9H_9Cl [*2687-12-9*] can be obtained by treatment of the alcohol with hydrochloric acid, thionyl chloride, or carbon tetrachloride–triphenylphosphine as the halogen donor.

Manufacture. A limited amount of natural cinnamyl alcohol is produced by the alkaline hydrolysis of the cinnamyl cinnamate present in Styrax Oil. Thus treatment of the essential oil with alcoholic potassium hydroxide liberates cinnamyl alcohol of reasonable purity which is then subjected to distillation. This product is sometimes preferred in fine fragrance perfumery because it contains trace impurities that have a rounding effect in finished formulations.

One of the first practical methods for the manufacture of cinnamyl alcohol involved reduction of cinnamic aldehyde diacetate with iron filings in acetic acid. This approach suffered from low yields and liberation of a significant amount of the starting aldehyde.

The commercial production of cinnamyl alcohol is accomplished exclusively by the reduction of cinnamaldehyde.

$$C_6H_5CH{=}CHCHO \rightarrow C_6H_5CH{=}CHCH_2OH$$

The preferred method for many years has been the Meerwein-Ponndorf-Verley reaction (46). In a typical process, cinnamaldehyde is dissolved in two volumes of 2-propanol containing aluminum isopropoxide (5–8 mol %) and acetone formed is removed continuously at reflux. Purification affords the alcohol in 85–90% yield. The reduction is mild and highly chemoselective, attacking only the carbonyl group in an α,β-unsaturated aldehyde. A significant disadvantage of the Meerwein reduction is the waste treatment problem associated with disposal of large quantities of aluminum salts. Another process involves liquid-phase hydrogenation in the presence of a platinum catalyst. The reaction is carried out in a two-phase, eg, water–toluene, solvent system at 20–40°C and 3–6 MPa (435–870 psi) hydrogen pressure (47). The cinnamyl alcohol (ca 75–80% yield) is accompanied by 5–8% 3-phenylpropanol which must be removed by careful distillation.

Economic Aspects. The market prices for cinnamyl alcohol quoted in Table 5 have been adjusted to reflect an average price for the relative quantities of the different grades sold. As of this writing, DSM is the only significant supplier for this material.

Table 5. Cinnamyl Alcohol Production

Year	Volume^a, 10^3t	Price^b, $/kg
1970	95	3.50–3.85
1980	175	7.95–8.18
1990	260	7.70–8.55

^{a}Estimate based on U.S. International Trade Commission figures and unpublished data.
^{b}Based on drum quantities.

Health and Safety. Cinnamyl alcohol has been evaluated by FEMA and given GRAS status (FEMA No. 2294). Two of its esters, cinnamyl cinnamate (FEMA No. 2298) and cinnamyl acetate (FEMA No. 2293), are also used extensively in flavor and fragrance compositions. Cinnamyl alcohol has also been tested by RIFM (48) and found to be safe for use. There have been reported cases of irritation and several manufacturers market a desensitized alcohol for use in fragrance applications.

Uses. Cinnamyl alcohol and its esters, especially cinnamyl acetate, are widely employed in perfumery because of their excellent sensory and fixative properties. They are frequently used in blossom compositions such as lilac, jasmine, lily of the valley, hyacinth, and gardenia to impart balsamic and oriental notes to the fragrance. In addition, they are utilized as modifiers in berry, nut, and spice flavor systems. The value of cinnamyl alcohol has also been mentioned in a variety of applications which include the production of photosensitive polymers (49), the creation of inks for multicolor printing (50), the formulation of animal repellent compositions (51), and the development of effective insect attractants (52).

BIBLIOGRAPHY

"Cinnamic Acid, Cinnamaldehyde, and Cinnamyl Alcohol" in *ECT* 1st ed., Vol. 4, pp. 1–8, by W. F. Ringk, Benzol Products Co.; in *ECT* 2nd ed., Vol. 5, pp. 517–523, by W. F. Ringk, Benzol Products Co.; in *ECT* 3rd ed., Vol. 6, pp. 142–149, by W. F. Ringk.

1. P. Schreier, *Chromatographic Studies of Biogenesis of Plant Volatiles*, A. Hüthig Verlag, Heidelberg, 1984, pp. 53, 84–88.
2. M. Luckner, *Secondary Metabolism in Microorganisms, Plants and Animals*, Springer-Verlag, Berlin, 1984.
3. *Beilstein's Handbuch der Organische Chemie*, Vol. 6, 4th ed., Springer-Verlag, Berlin, p. 570; 1st Suppl., Vol. 6, p. 281; 2nd Suppl., Vol. 6, p. 525; 3rd Suppl., p. 2401; 4th Suppl., Vol. 6, p. 3799; Vol. 7, p. 348; 1st Suppl., Vol. 7, p. 187; 2nd Suppl., Vol. 7, p. 273; 3rd Suppl., Vol. 7, p. 1364; 4th Suppl., Vol. 7, p. 948; Vol. 9, p. 572; 1st Suppl., Vol. 9, p. 2670; 4th Suppl., Vol. 9, p. 2001.
4. M. R. I. Saleh, A-A. M. Habib, and N. El-Shaer, *J. Assoc. Off. Anal. Chem.* **63,** 1195 (1980).
5. E. Guenther, *The Essential Oils*, Vol. 5, D. Van Nostrand Co., New York, 1952, pp. 212, 220, and 243.
6. J. A. Dean, ed., *Lange's Handbook of Chemistry*, 13th ed., McGraw-Hill Book Co., New York, 1985, p. 7–240.
7. T. E. Furia and N. Bellanca, *Fenaroli's Handbook of Flavor Ingredients*, 2nd ed., Vol. 2, CRC Press, Cleveland, Ohio, 1975, p. 92.
8. V. Ipatiev, *Chem. Ber.* **42,** 2097 (1909).
9. U.S. Pat. 4,262,157 (Apr. 14, 1981), Y. Hari, Y. Nagano, and H. Taniguchi (to Abbott Laboratories).
10. R. Curci, M. Fiorentino, L. Troisi, J. Edwards, and R. Pater, *J. Org. Chem.* **45,** 4758 (1980).
11. J. R. Johnson, in R. Adams, ed., *Organic Reactions*, Vol. 1, John Wiley & Sons, Inc., New York, 1942, pp. 210–265.
12. Ger. Offen. 3,743,616 (Aug. 4, 1988), G. Ihl, G. Roscher, and N. Mayer (to Hoechst AG).
13. U.S. Pat. 3,783,140 (Jan. 1, 1974), R. Heck (to Hercules Inc.).
14. U.S. Pat. 4,620,027 (Oct. 28, 1986), C.-Y. Hsu (to Sun Refining and Marketing Co.).
15. *Flavor and Fragrance Materials—1991*, Allured Publishing Corp., Wheaton, Ill., 1991.
16. D. L. Opdyke, *Food Cosmet. Toxicol.* **16,** 687 (1978).
17. K. Bauer and D. Garbe, *Common Fragrance and Flavor Materials*, VCH Publishers, Weinheim, 1985, p. 80.
18. Jpn. Pat. 63 277,615 (Nov. 15, 1988), S. Oreal.
19. U.S. Pat. 3,183,075 (May 11, 1965), B. L. Walworth (to American Cyanamid Co.).
20. U.S. Pat. 3,307,941 (Mar. 7, 1967), R. W. Gundlach (to Xerox Corp.).
21. U.S. Pat. 3,387,976 (June 11, 1968), J. L. Sarkin (to Harris-Intertype Corp.).
22. U.S. Pat. 2,670,286 (Feb. 23, 1954), L. M. Minsk, W. P. VanDeusen, and E. M. Robertson (to Eastman Kodak Co.).
23. Jpn. Pat. 62 175,752 (Aug. 1, 1987), T. Ishikawa and N. Sakai (to Fuji Photo Film Co., Ltd.).
24. Belg. Pat. 872,662 (Mar. 30, 1979), D. Arcilesi (to M & T Chemicals, Inc.).
25. A. Klausner, *Bio/Technology* **3,** 301 (1985).
26. U.S. Pat. 4,504,582 (Mar. 12, 1985), W. E. Swann (to Genex Corp.).
27. R. O. B. Wijesekera, *CRC Crit. Rev. Food Sci. Nutr.* **10**(9), 1–30 (1978).
28. Eur. Pat. Appl. 170,520 (Feb. 5, 1986), H. Harada (to Sumitomo Chemical Industries Co., Ltd.).
29. G. B. Payne, *J. Org. Chem.* **25,** 275 (1960).

30. U.S. Pat. 3,372,199 (Mar. 5, 1968), P. N. Rylander and N. Himelstein (to Engelhard Industries, Inc.).
31. P. Z. Bedoukian, *Perfumery & Flavoring Synthetics*, 3rd ed., Allured Publications, Wheaton, Ill., 1986, pp. 98–105.
32. U.S. Pat. 2,529,186 (Nov. 7, 1950), H. H. Richmond (to United States Rubber Co.).
33. Ger. Offen. 2,353,145 (May 16, 1974), N. C. Hindley and D. A. Andrews (to Hoffmann-LaRoche).
34. D. L. Opdyke, *Food Cosmet. Toxicol.* **17,** 253 (1979).
35. Ger. Offen. 2,852,433 (June 21, 1979), D. A. Arcilesi (to M & T Chemicals, Inc.).
36. U.S. Pat. 4,097,607 (June 27, 1978), K. A. Larson.
37. Jpn. Pat. 75 42,053 (Apr. 16, 1975), J. Nakano (to Yamabum Yuka K.K.).
38. N. Kurita, M. Miyaji, R. Kurane, and Y. Takahara, *Agric. Biol. Chem.* **45,** 945 (1981).
39. U.S. Pat. 4,673,766 (June 16, 1987), K. T. Buck, A. J. Boeing, and J. E. Dolfini (to Mallinckrodt, Inc.).
40. U.S. Pat. 4,810,824 (Mar. 7, 1989), J. E. Dolfini and J. Glinka (to Mallinckrodt, Inc.).
41. U.S. Pat. 4,617,419 (Oct. 14, 1986), C. Wiener and A. O. Pittet (to International Flavors and Fragrances).
42. S. Arctander, *Perfume and Flavor Materials of Natural Origin*, published by author, Elizabeth, N.J., 1960, p. 302.
43. O. Nishimura, K. Yamaguchi, S. Mihara, and T. Shibamoto, *J. Agric. Food Chem.* **37,** 139 (1989).
44. Ger. Offen. 2,556,161 (Dec. 16, 1976), W. J. Ehmann and W. E. Johnson, Jr. (to SCM Corp.).
45. D. Huang and L. Huang, *Tetrahedron* **46,** 3135 (1990).
46. A. L. Wilds, in R. Adams, ed., *Organic Reactions*, Vol. 2, John Wiley & Sons, Inc., New York, 1944, p. 178.
47. U.S. Pat. 4,247,718 (Jan. 27, 1981), J. M. A. Dantzenberg, J. M. C. A. Mulders, and P. A. M. J. Stijfs (to Stamicarbon, B.V.).
48. D. L. Opdyke, *Food Cosmet. Toxicol.* **12,** 855 (1974).
49. Fr. Demande 2,009,112 (Jan. 30, 1970), L. Katz and co-workers (to GAF Corp.).
50. Hung. Pat. 32,145 (Jan. 28, 1984), G. Riachak.
51. Jpn. Pat. 81 65,803 (June 3, 1981), to Mikasa Chemical Co.).
52. U.S. Pat. 4,880,624 (Nov. 14, 1989), R. L. Metcalf and R. L. Lampman (to the University of Illinois).

Robert G. Eilerman
Givaudan-Roure Corporation

CINNAMON.   See Flavors and Spices.

CINNAMYL ALCOHOL.   See Cinnamic Acid, Cinnamaldehyde, and
Cinnamyl Alcohol.

CITRAL.   See Flavors and Spices; Perfumes; Terpenoids.

CITRIC ACID

Citric acid [77-92-9] (2-hydroxy-1,2,3-propanetricarboxylic acid), is a natural component and common metabolite of plants and animals. It is the most versatile and widely used organic acid in foods, beverages, and pharmaceuticals.

$$\begin{array}{c} CH_2-COOH \\ | \\ HO-C-COOH \\ | \\ CH_2-COOH \end{array}$$

Because of its functionality and environmental acceptability, citric acid and its salts (primariy sodium and potassium) are used in many industrial applications for chelation, buffering, pH adjustment, and derivatization. These uses include laundry detergents, shampoos, cosmetics, enhanced oil recovery, and chemical cleaning.

Citric acid specifications are defined in a number of compendia including *Food Chemicals Codex* (FCC), *United States Pharmacopeia* (USP), *British Pharmacopeia* (BP), *European Pharmacopeia* (EP), and *Japanese Pharmacopeia* (JP).

Historically, about AD 1200, the alchemist Vincentius Bellovacensis recognized that lemon and lime juices contained an acid substance. In 1784 Scheele first isolated crystalline citric acid from lemon juice. In 1834 Liebig recognized citric acid as a hydroxy tribasic acid, and in 1893 Wehmer indicated that certain fungi produce citric acid when grown on sugar solutions. The microbial fermentation of a carbohydrate substrate is virtually the exclusive commercial procedure to produce citric acid.

Physical Properties

Citric acid, anhydrous, crystallizes from hot aqueous solutions as colorless translucent crystals or white crystalline powder. Its crystal form is monoclinic holohedra. Citric acid is deliquescent in moist air. Some physical properties are given in Table 1 (1–3). The solubility of citric acid in water and some organic solvents is given in Table 2. The pH and specific gravity of aqueous solutions of citric acid are shown in Table 3.

Aqueous solutions of citric acid make excellent buffer systems when partially neutralized because citric acid is a weak acid and has three carboxyl groups, hence three pK_a's. At 20°C $pK_1 = 3.14$, $pK_2 = 4.77$, and $pK_3 = 6.39$ (2). The buffer range for citrate solutions is pH 2.5 to 6.5. Buffer systems can be made using a solution of citric acid and sodium citrate or by neutralizing a solution of citric acid with a base such as sodium hydroxide. In Table 4 stock solutions of 0.1 M (0.33 N) citric acid are combined with 0.1 M (0.33 N) sodium citrate to make a typical buffer solution.

Citric acid monohydrate [5949-29-1] has a molecular weight of 210.14 and crystallizes from cold aqueous solutions. When gently heated, the crystals lose their water of hydration at 70–75°C and melt in the range of 135–152°C. Rapid heating causes dehydration at 100°C to form crystals that melt sharply at 153°C. Citric acid monohydrate is available in limited commercial quantities since most applications now call for the anhydrous form.

Table 1. Physical Properties of Citric Acid, Anhydrous

Property	Value
molecular formula	$C_6H_8O_7$
mol wt	192.13
gram equivalent weight	64.04
melting point, °C	153
thermal decomposition temp., °C	175
density, g/mL	1.665
heat of combustion,[a] MJ/mol[b]	1.96
heat of solution, J/g[b]	117

[a]At 25°C.
[b]To convert J to cal, divide by 4.184.

Table 2. Solubility[a] of Citric Acid, Anhydrous

Temperature, °C	g/100 g satd soln
In water	
10	54.0
20	59.2
30	64.3
40	68.6
50	70.9
60	73.5
70	76.2
80	78.8
90	81.4
100	84.0
In organic solvents at 25°C	
amyl acetate	4.2
diethyl ether[b]	1.0
ethyl alcohol[b]	38.3

[a]Ref. 4.
[b]Absolute.

Table 3. pH and Specific Gravity of
Aqueous Citric Acid Solutions

Concentration, % w/w	pH	Specific gravity at 25°C
0.1	2.8	
0.5	2.4	
1.0	2.2	
5.0	1.9	
10.0	1.7	1.035
20.0		1.084
30.0	1.2	1.131
40.0		1.182
50.0	0.8	1.243
60.0		1.294

Table 4. Citric Acid Buffer Solutions

0.1 M Citric acid, mL	0.1 M Sodium citrate, mL	Buffer solution pH
46.5	3.5	3.0
33.0	17.0	4.0
20.5	29.5	5.0
9.5	41.5	6.0

Chemical Properties

Citric acid undergoes most of the reactions typical of organic hydroxy polycarboxylates.

Decomposition.   When heated above 175°C, citric acid decomposes to form aconitic acid [499-12-7], citraconic acid [498-23-7], itaconic acid [97-65-4], acetonedicarboxylic acid [542-05-2], carbon dioxide, and water, as shown in Figure 1.

Esterification.   Citric acid is easily esterified with many alcohols under azeotropic conditions in the presence of a catalyst such as sulfuric acid, p-toluenesulfonic acid, or sulfonic acid-type ion-exchange resin. Alcohols boiling above 150°C esterify citric acid without a catalyst (5–8).

$$
\begin{array}{c}
\text{CH}_2\text{—COOH} \\
| \\
\text{HO—C—COOH} \\
| \\
\text{CH}_2\text{—COOH}
\end{array}
\;+\; 3\,\text{ROH} \;\xrightarrow[\text{catalyst}]{\text{H}^+}\;
\begin{array}{c}
\text{CH}_2\text{—COOR} \\
| \\
\text{HO—C—COOR} \\
| \\
\text{CH}_2\text{—COOR}
\end{array}
\;+\; 3\,\text{H}_2\text{O}
$$

Alcohols typically used in citric acid esterification are methyl, ethyl, butyl, and allyl alcohols.

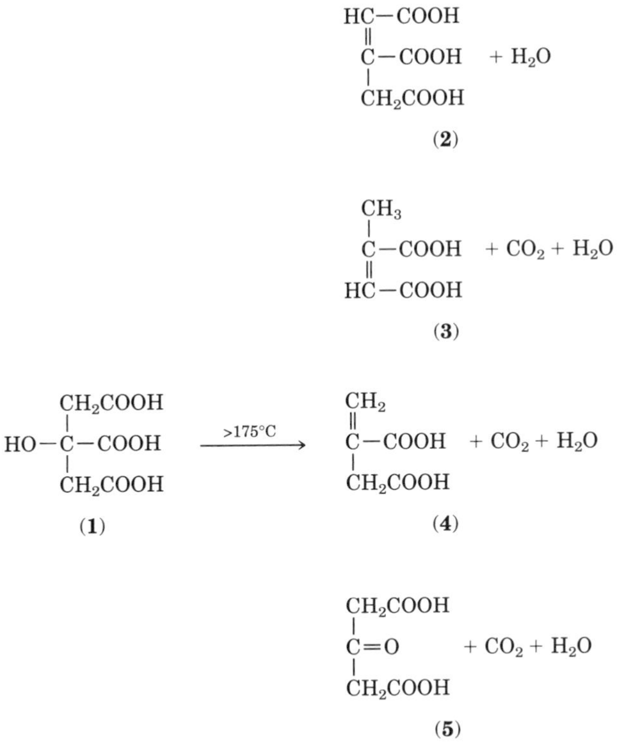

$$\begin{array}{l} \text{HC-COOH} \\ \quad \| \\ \text{C-COOH} \quad + H_2O \\ \quad | \\ \text{CH}_2\text{COOH} \end{array}$$

(2)

$$\begin{array}{l} \text{CH}_3 \\ \quad | \\ \text{C-COOH} \quad + CO_2 + H_2O \\ \quad \| \\ \text{HC-COOH} \end{array}$$

(3)

$$\begin{array}{l} \text{CH}_2\text{COOH} \\ \quad | \\ \text{HO-C-COOH} \xrightarrow{\;>175°C\;} \\ \quad | \\ \text{CH}_2\text{COOH} \end{array} \qquad \begin{array}{l} \text{CH}_2 \\ \quad \| \\ \text{C-COOH} \quad + CO_2 + H_2O \\ \quad | \\ \text{CH}_2\text{COOH} \end{array}$$

(1)   (4)

$$\begin{array}{l} \text{CH}_2\text{COOH} \\ \quad | \\ \text{C=O} \quad + CO_2 + H_2O \\ \quad | \\ \text{CH}_2\text{COOH} \end{array}$$

(5)

Fig. 1. Thermal decomposition of citric acid (**1**) to aconitic acid (**2**), citraconic acid (**3**), itaconic acid (**4**), and oxidation to acetonedicarboxylic acid (**5**).

Oxidation. Citric acid is easily oxidized by a variety of oxidizing agents such as peroxides, hypochlorite, persulfate, permanganate, periodate, hypobromite, chromate, manganese dioxide, and nitric acid. The products of oxidation are usually acetonedicarboxylic acid (**5**), oxalic acid (**6**), carbon dioxide, and water, depending on the conditions used (5).

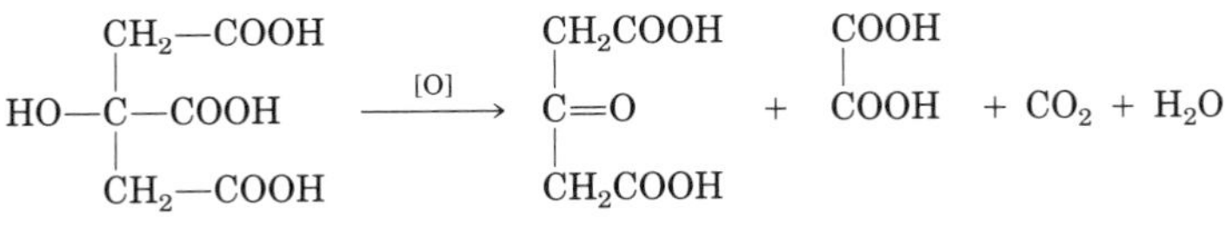

$$\begin{array}{l} \text{CH}_2\text{-COOH} \\ \quad | \\ \text{HO-C-COOH} \xrightarrow{\;[O]\;} \\ \quad | \\ \text{CH}_2\text{-COOH} \end{array} \quad \begin{array}{l} \text{CH}_2\text{COOH} \\ \quad | \\ \text{C=O} \\ \quad | \\ \text{CH}_2\text{COOH} \end{array} \quad + \quad \begin{array}{l} \text{COOH} \\ \quad | \\ \text{COOH} \end{array} \quad + CO_2 + H_2O$$

                                        (5)   (6)

Reduction. The hydrogenation of citric acid yields 1,2,3-propanetricarboxylic acid [99-14-9] (5).

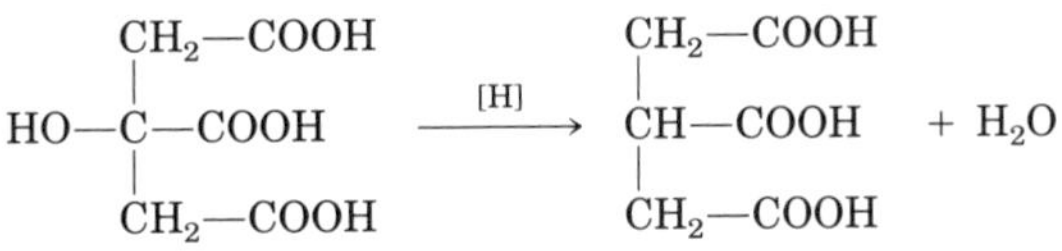

$$\begin{array}{l} \text{CH}_2\text{-COOH} \\ \quad | \\ \text{HO-C-COOH} \xrightarrow{\;[H]\;} \\ \quad | \\ \text{CH}_2\text{-COOH} \end{array} \quad \begin{array}{l} \text{CH}_2\text{-COOH} \\ \quad | \\ \text{CH-COOH} \quad + H_2O \\ \quad | \\ \text{CH}_2\text{-COOH} \end{array}$$

Hydrogenation of trisodium citrate over a Ni catalyst at 8.6 MPa (85 atm) and a temperature of 220–230°C results in hydrogenolysis fragments.

Salt Formation. Citric acid forms mono-, di-, and tribasic salts with many cations such as alkalies, ammonia, and amines. Salts may be prepared by direct neutralization of a solution of citric acid in water using the appropriate base, or by double decomposition using a citrate salt and a soluble metal salt.

Trisodium citrate is more widely used than any of the other salts of citric acid. It is generally made by neutralization of a water solution of citric acid using sodium hydroxide. The neutralization reaction is highly exothermic giving off 1109 J/g of citric acid. To conserve energy, the heat evolved can be used in the sodium citrate concentration and crystallization steps.

$$\begin{array}{c} CH_2-COOH \\ | \\ HO-C-COOH \\ | \\ CH_2-COOH \end{array} \; + \; 3\,NaOH \; \longrightarrow \; \begin{array}{c} CH_2-COONa \\ | \\ HO-C-COONa \cdot 2\,H_2O \; + \; H_2O \\ | \\ CH_2-COONa \end{array}$$

Other sources of sodium ion that are used to make sodium citrate are sodium carbonate and sodium bicarbonate. These reactions evolve large volumes of carbon dioxide gas, resulting in much foaming but less exotherm.

The mono- and disodium citrate salts are made by limiting the amount of sodium available by using only one mole of base for each mole of citric acid for the monosodium citrate and two moles for the disodium citrate. The result is primarily the mono or disalt with small amounts of the other forms and citric acid being present. Other salts that have been offered commercially are shown in Table 5.

Table 5. Salts of Citric Acid

Salt	CAS Registry Number	Molecular formula
ammonium citrate	[3012-65-5]	$(NH_4)_2HC_6H_6O_7$
calcium citrate	[813-94-5]	$Ca_3(C_6H_5O_7)_2$
calcium citrate tetrahydrate	[5785-44-4]	$Ca_3(C_6H_5O_7)_2 \cdot 4H_2O$
cobalt citrate	[866-81-9]	$Co_3(C_6H_5O_7)_2$
copper citrate	[866-82-0]	$C_6H_8O_7 \cdot 2Cu$
ferric ammonium citrate	[1185-57-5]	$C_6H_8O_7 \cdot xFe \cdot xNH_3$
ferric citrate	[2338-05-8]	$C_6H_8O_7 \cdot xFe$
lead citrate	[512-26-5]	$Pb_3(C_6H_5O_7)_2$
lithium citrate	[919-16-4]	$Li_3C_6H_5O_7$
magnesium citrate	[3344-18-1]	$Mg_3(C_6H_5O_7)_2$
manganese citrate	[5968-88-7]	$MnC_6H_5O_7$
nickel citrate	[6018-92-4]	$Ni_3(C_6H_5O_7)_2$
potassium citrate	[866-84-2]	$K_3C_6H_5O_7$
potassium citrate hydrate	[6100-05-6]	$K_3C_6H_5O_7 \cdot H_2O$
sodium citrate dihydrate	[6132-04-3]	$Na_3C_6H_5O_7 \cdot 2H_2O$
zinc citrate	[546-46-3]	$Zn_3(C_6H_5O_7)_2$

Chelate Formation. Citric acid complexes with many multivalent metal ions to form chelates (9,10). This important chemical property makes citric acid and citrates useful in controlling metal contamination that can affect the color, stability, or appearance of a product or the efficiency of a process.

Citric acid, with its one hydroxyl and three carboxyl groups, is a multidentate ligand. Two or more of these sites are utilized to form a ring structure. The normal molar ratio of metal-to-ligand is 1:1. With some metal ions, under certain conditions, more than one ring can be formed allowing a higher metal-to-ligand ratio (see CHELATING AGENTS).

When a metal ion is chelated by a ligand such as citric acid, it is no longer free to undergo many of its chemical reactions. A metal ion that is normally colored may, in the presence of citrate, have little or no color. Under pH conditions that may precipitate a metal hydroxide, the citrate complex may be soluble. Organic molecules that are catalytically decomposed in the presence of metal ions can be made stable by chelating the metal ions with citric acid.

Chelation is an equilibrium reaction. There are always some free-metal ions present as well as chelated metal ions. In a system where a metal salt is being reduced, such as in metal plating, the rate of the reaction forming the metal can be controlled by using the metal citrate chelate.

$$\text{metal ions} + \text{citrate ion} \rightleftharpoons \text{chelated metal ions}$$

The log function of the ratio of chelated metal ions to free-metal ions is expressed as the stability constant or formation constant as shown in Table 6. The higher the stability constant the greater the percentage of metal ions that are chelated (11).

Table 6. Stability Constants for Metal Citrates

Metal	Valence	$\log K$
Fe	+3	12.5
Al	+3	7.00
Pb	+2	6.50
Ni	+2	5.11
Co	+2	4.80
Zn	+2	4.71
Ca	+2	4.68
Cu	+2	4.35
Cd	+2	3.98
Mn	+2	3.67
Mg	+2	3.29
Fe	+2	3.08
Ba	+2	2.98

Stability constants are measured at their optimum pH. Conditional stability constants are measured at a specific pH. In general, stability constants for metal citrates are very low below pH 2–3, high at pH 3–10, and low above pH 10–12.

Corrosion. Aqueous solutions of citric acid are mildly corrosive toward carbon steels. At elevated temperatures, 304 stainless steel is corroded by citric acid, but 316 stainless steel is resistant to corrosion. Many aluminum, copper, and nickel alloys are mildly corroded by citric acid. In general, glass and plastics such as fiber glass reinforced polyester, polyethylene, polypropylene, poly(vinyl chloride), and cross-linked poly(vinyl chloride) are not corroded by citric acid.

Occurrence

Citric acid occurs widely in the plant and animal kingdoms (12). It is found most abundantly in the fruits of the citrus species, but is also present as the free acid or as a salt in the fruit, seeds, or juices of a wide variety of flowers and plants. The citrate ion occurs in all animal tissues and fluids (12). The total circulating citric acid in the serum of humans is approximately 1 mg/kg body weight. Normal daily excretion in human urine is 0.2–1.0 g. This natural occurrence of citric acid is described in Table 7.

Table 7. Natural Occurrence of Citric Acid

Fruits and vegetables		Animal tissues and fluids	
Plant	Citric acid, wt %	Location	Citric acid, ppm
lemons	4.0–8.0	human whole blood	15
grapefruit	1.2–2.1	human blood plasma	25
tangerines	0.9–1.2	red blood cells	10
oranges	0.6–1.0	human milk	500–1250
currants		urine	100–750
black	1.5–3.0	semen	2000–4000
red	0.7–1.3	thyroid gland	750–900
raspberries	1.0–1.3	kidney	20
strawberries	0.6–0.8	bone	7500
apples	0.008	saliva	4–24
potatoes	0.3–0.5	sweat	1–2
tomatoes	0.25	tears	5–7
asparagus	0.08–0.2		
turnips	0.05–1.1		
peas	0.05		
corn kernels	0.02		
lettuce	0.016		
eggplant	0.01		

Physiological Role of Citric Acid. Citric acid occurs in the terminal oxidative metabolic system of virtually all organisms. This oxidative metabolic system (Fig. 2), variously called the Krebs cycle (for its discoverer, H. A. Krebs), the

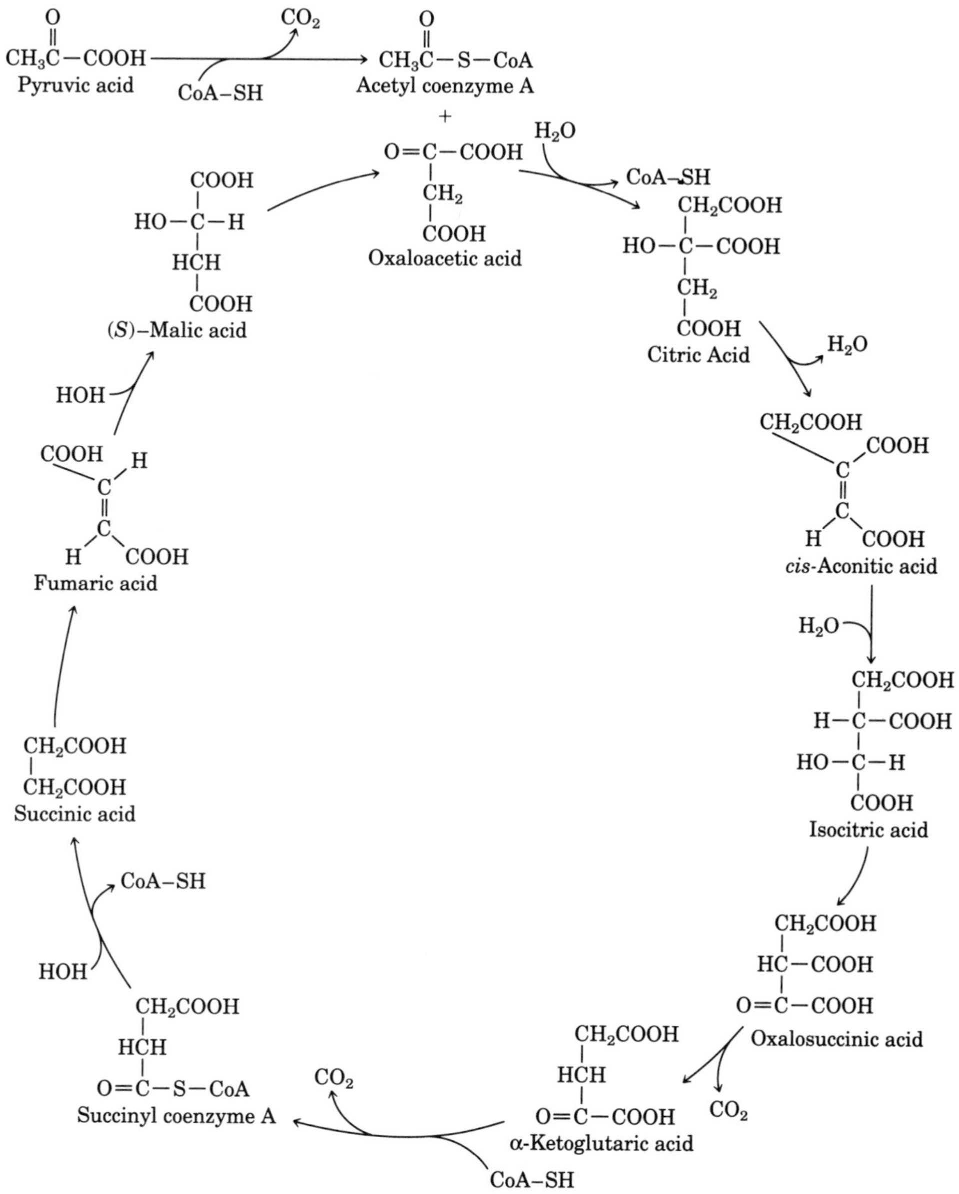

Fig. 2. Krebs (citric acid) cycle. Coenzyme A is represented CoA–SH. The cycle begins with the combination of acetyl coenzyme A and oxaloacetic acid to form citric acid.

tricarboxylic acid cycle, or the citric acid cycle, is a metabolic cycle involving the conversion of carbohydrates, fats, or proteins to carbon dioxide and water. This cycle releases energy necessary for an organism's growth, movement, luminescence, chemosynthesis, and reproduction. The cycle also provides the carbon-containing materials from which cells synthesize amino acids and fats. Many yeasts, molds, and bacteria conduct the citric acid cycle, and can be selected for their ability to maximize citric acid production in the process. This is the basis for the efficient commercial fermentation processes used today to produce citric acid.

Manufacturing and Processing

Historically, citric acid was isolated by crystallization from lemon juice and later was recognized as a microbial metabolite. This work led to the development of commercial fermentation technology (13). The basic raw materials for making citric acid include corn starch, molasses (sugar cane, beet sugar), and normal paraffin hydrocarbons.

Fermentation. The microbial production of citric acid on a commercial scale was begun in 1923 utilizing certain strains of *Aspergillus niger* to produce citric acid on the surface of a sucrose and salt solution. This tray fermentation technique is still used today, although it is being replaced by a submerged process known as deep tank fermentation (14–22).

In the deep tank submerged process, *Aspergillus niger* mold spores are grown under controlled aseptic conditions on a test-tube slant and transferred to a seed tank or inoculum which is added to a fermentor along with pasteurized syrup. The pH is adjusted and nutrients added. Sterile air is sparged into the fermentor while the sugar is converted to citric acid. The complete fermentation cycle can take as long as 15 days.

Recovery. Citric acid fermentation broth is generally separated from the biomass using filtration or centrifugation. The citric acid is usually purified using either a lime-sulfuric acid method or a liquid extraction process (23).

Lime-Sulfuric. Recovery of citric acid by calcium salt precipitation is shown in Figure 3. Although the chemistry is straightforward, the engineering principles, separation techniques, and unit operations employed result in a complex commercial process. The fermentation broth, which has been separated from the insoluble biomass, is treated with a calcium hydroxide (lime) slurry to precipitate calcium citrate. After sufficient reaction time, the calcium citrate slurry is filtered and the filter cake washed free of soluble impurities. The clean calcium citrate cake is reslurried and acidified with sulfuric acid, converting the calcium citrate to soluble citric acid and insoluble calcium sulfate. Both the calcium citrate and calcium sulfate reactions are generally performed in agitated reaction vessels made of 316 stainless steel and filtered on commercially available filtration equipment.

The citric acid solution is deionized at this stage to remove trace amounts of residual calcium, iron, other cationic impurities, and to improve crystallization. In some processes, trace-impurity removal and decolorization are accomplished with the aid of adsorptive carbon.

The aqueous citric acid solution is concentrated in a series of crystallization steps to achieve the physical separation of citric acid from remaining impurities. Standard evaporation, crystallization, and filtration equipment can be employed in this operation. The choice of crystallizer temperature dictates the formation of anhydrous or monohydrate citric acid. Above 37°C, the transition point,

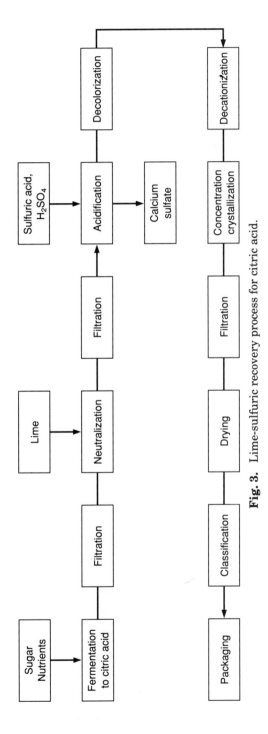

Fig. 3. Lime-sulfuric recovery process for citric acid.

anhydrous citric forms whereas below 37°C, the monohydrate will crystallize. The finished citric acid is dried and classified. Because anhydrous citric acid is hygroscopic, to protect against caking care must be taken to avoid handling, packaging, and storage of the crystals in areas of high temperature and high humidity.

   Liquid Extraction.   The recovery process, shown in Figure 4, was developed in the 1970s and involves the extraction of citric acid from fermentation broth using a mixture of trilaurylamine, n-octanol, and C_{10} or C_{11} hydrocarbon, followed by re-extraction of the citric acid from the solvent phase into water (24). Efficient citric acid extraction is achieved through a series of countercurrent steps that ensure intimate contact of the aqueous and nonaqueous phases. When transfer of the citric acid to the solvent phase is complete, the citric acid is re-extracted into water, again using a multistage countercurrent system. The two steps differ mainly in the temperature at which they are performed.

   The final processing steps are a wash of the aqueous citric acid solution by the hydrocarbon solvent, followed by passage of the acid solution through granular activated carbon columns. Effluent from the carbon columns is processed through a conventional sequence of evaporation/crystallization/drying and packaging steps to complete the manufacturing process.

   Citric acid is also commercially available as a 50% w/w solution made either by dissolving crystalline citric acid in water, or a combination or crystalline citric acid, and one of the citric acid process streams. There are several grades of citric acid solutions available, each made according to quality which is measured by color and trace impurities. The citric acid content of each grade can be identical, 50% w/w, which is near the solubility limit.

   By-Products.   The biomass from the fungal fermentation process is called mycellium and can be used as a supplement for animal feed since it contains digestable nutrients (25,26). The lime-sulfuric purification and recovery process results in large quantities of calcium sulfate cake, which is usually disposed of into a landfill but can find limited use in making plaster, cement, wallboard, or as an agricultural soil conditioner. The liquid extraction purification and recovery process has the advantage of little solid by-products.

   Energy.   In recent years the concern for energy conservation has resulted in many innovative process improvements to make the manufacture of citric acid more efficient. For example, heat produced by the exotherm of the neutralization of citric acid with lime is used in another part of the process where heat is required, such as the evaporation/crystallization step.

   Chemical Synthesis.   The chemical synthesis of citric acid was reported in 1880 (27). Since then, many different synthetic routes have been investigated, reported, and patented (28–36). However, none of these have proven to be commercially feasible.

Shipment and Storage

Crystalline citric acid, anhydrous, can be stored in dry form without difficulty, although conditions of high humidity and elevated temperatures should be avoided to prevent caking. Storage should be in tight containers to prevent ex-

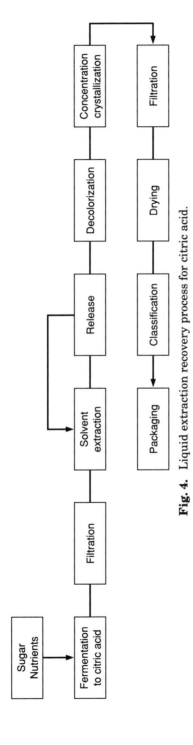

Fig. 4. Liquid extraction recovery process for citric acid.

posure to moist air. Several granulations are commercially available with the larger particle sizes having less tendency toward caking.

Liquid citric acid is commercially available in a variety of concentrations with 50% w/w being most common. Grades are available that vary in appearance, purity, and color. Packaging is usually in drums, tank trucks, or rail cars. Liquid citric acid should be kept above 0°C to prevent crystallization.

Solutions of citric acid are corrosive to normal concrete, aluminum, carbon steel, copper, copper alloys, and should not be used with nylon, polycarbonates, polyamides, polyimides, or acrylics.

Recommended materials of construction for pipes, tanks, and pumps handling citric acid solutions are 316 stainless steel, fiber glass-reinforced-polyester, polyethylene, polypropylene, and poly(vinyl chloride). At elevated temperatures, 304 stainless steel is not recommended (Table 8).

Although not as corrosive as the acid, the sodium and potassium salts of citric acid should be handled in the same type of equipment as the acid to avoid corrosion problems.

Table 8. Citric Acid[a] Corrosion Rates[b]

Material	Temperature, °C	Corrosion rate, mm/yr
316 stainless steel	25	0.03
	50	0.03
304 stainless steel	25	0.03
	50	0.23
carbon steel	25	4.6
	50	32.8

[a]50 wt % solution.
[b]Ref. 37.

Economic Aspects

Citric acid is manufactured in over 20 countries with 1990 worldwide production estimated at approximately 550,000 t, distributed as shown in Figure 5. Most of this production is used for foods and beverages; however, industrial applications, eg, detergents, metal cleaning, of citric acid are becoming more important on a worldwide basis.

It was estimated that 1990 U.S. citric acid and citrate salt consumption was 152,000 t. Citric acid represents approximately 90% of this volume. This citric acid/citrate use and its historical distribution in various markets is described in Table 9. From Table 9 it can be seen that although citric acid usage in the United

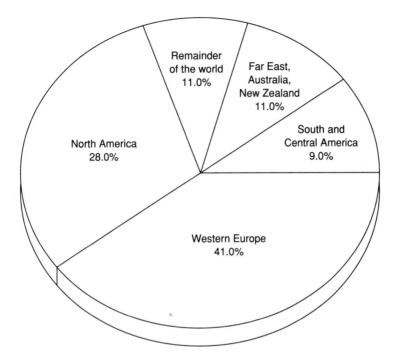

Fig. 5 Worldwide citric acid production in 1990 (37).

States has shown steady growth at an average annual rate of 4.4% from 1986–1990, the end use patterns have been quite stable.

Table 9. U.S. Citric Acid/Citrate Distribution by End Use, %[a]

Product	1990	1988	1986
beverages	45	43	44
foods	21	20	22
pharmaceuticals and cosmetics	8	8	8
household detergents and cleaning products	19	23	20
misc. nonfood	7	6	6
Total annual consumption	*152,000*	*148,000*	*128,000*

[a]Ref. 37.

The list price for citric acid at the end of 1992 was approximately $1.74/kg (38). The price for sodium citrate was $1.75/kg and for potassium citrate was $2.33/kg

Specifications, Standards, and Quality Control

Since citric acid is produced and sold throughout the world, it must meet the criteria of a variety of food and drug compendia (39–43).

Analytical and Test Methods.   Aqueous titration with 1 N NaOH remains the official method for assaying citric acid (39,40). Although not citrate-specific, the procedure is satisfactory in the absence of interfering substances. Low concentrations of citric acid can be determined by a spectrophotometic method based on the Furth and Herrmann reaction with pyridine and acetic anhydride (PAA) (44). This PAA method is citrate-specific at 420 nm and is sensitive for citrate ions at concentrations down to 5 ppm. The PAA method can be used to quantify citrate in foods, beverages, and industrial products such as detergents.

An enzymatic method (45), which is specific for the citrate moiety, can be used as a combined assay and identification test for citric acid and its common salts down to 20 ppm.

A high performance liquid chromotography (hplc) method to determine citric acid and other organic acids has been developed (46). The method is an isocratic system using sulfuric acid to elute organic acids onto a specific hplc column. The method is sensitive for citric acid down to ppm levels and is capable of quantifying citric acid in clear aqueous systems.

Health, Safety, and Environmental Considerations

Citric acid, as well as its common sodium and potassium salt forms, are Generally Recognized As Safe (GRAS) by the U.S. Food and Drug Administration as Multiple Purpose Food Substances (47). Citric acid is also approved by the Joint FAO/WHO Expert Committee on Food Additives for use in foods without limitation (48). The use of citric acid and certain of its salts and esters has been evaluated by a Special Committee on GRAS Substances (SCOGS) of the Federation of American Societies for Experimental Biology under contract with the FDA (49). The evaluation was based largely on two scientific literature reviews prepared for the FDA that summarize the world's applicable scientific literature from 1920 through 1973 (12).

Tests have shown that citric acid is not corrosive to skin but is a skin and ocular irritant (50). For these reasons it is recommended that individuals use appropriate personal protection to cover the hands, skin, eyes, nose, and mouth when in direct contact with citric acid solutions or powders.

Citric acid is biodegraded readily by many organisms under aerobic and anaerobic wastewater treatment conditions and in the natural environment (51).

Applications

Citric acid is utilized in a large variety of food and industrial applications because of its unique combination of properties. It is used as an acid to adjust pH, a buffer to control or maintain pH, a chelator to form stable complexes with multivalent metal ions, and a dispersing agent to stabilize emulsions and other multiphase systems (see DISPERSANTS). In addition, it has a pleasant, clean, tart taste making it useful in food and beverage products.

Food Uses.   *Beverages.*   Citric acid, sodium citrate, and potassium citrate are used extensively in carbonated and noncarbonated beverages (52). Juice-added beverages, low calorie beverages, and thirst quenchers, also known as isotonic drinks, use citric acid alone and in combination with citrate salts for flavoring and buffering properties and to increase the effectiveness of antimicrobial preservatives. The high solubility of citric acid is important in beverage syrups. The amount of acid used depends on the flavor desired in the product as well as taste evaluation and customer preference. The acid concentration of most fruit-flavored carbonated beverages (qv) falls in the range of 0.10–0.25% w/w.

Citric acid and its salts are used in dry beverage mixes, convenience teas, and cocktail mixes for pH control and flavor, and are used in wine coolers at 0.10–0.55%, combining well with fruity and light flavors.

Jams, Jellies, and Preserves. Citric acid is used in jams and jellies to provide tartness and to adjust the product pH for optimum gelation (53).

Candy. Citric acid is added in candy for tartness (54,55). To suppress the inversion of sucrose, it should be added after the cook, at levels from 0.5 to 2.0%. The pH of pectin gel candies is adjusted with citric acid for maximum gel strength.

Salads. A combination of citric acid and ascorbic acid is used as an alternative to sulfites in prevention of enzymatic browning in fresh prepared vegetables (56).

Frozen Food. The chelating and acidic properties of citric acid enable it to optimize the stability of frozen food products by enhancing the action of antioxidants and inactivating naturally present enzymes which could cause undesirable browning and loss of firmness (57,58).

Citric acid also inhibits color and flavor deterioration in frozen fruit. Here again the function is to inhibit enzymatic and trace metal-catalyzed oxidation.

Canned Fruits and Vegetables. The use of citric acid to bring the pH below 4.6 can reduce heat treatment requirements in canned fruits and vegetables. In addition, citric acid chelates trace metals to prevent enzymatic oxidation and color degradation, and enhances the flavor, especially of canned fruits.

Fats and Oils. The oxidation of fats and oils in food products can be prevented by the addition of citric acid to chelate the trace metals that catalyze the oxidation. Citric acid is also used in the bleaching clays and the degumming process during oil refining to remove chlorophyll and phospholipids (59–63).

Confections and Desserts. Citric acid and sodium citrate are utilized in the confection industry to optimize gel-setting characteristics, provide tartness, and enhance flavor.

Pasteurized Process Cheese. Sodium citrate is used in pasteurized process and sliced cheese as an emulsifying salt to stabilize the water and oil emulsion and improve process cheese body and texture (64).

Dairy Products. Sodium citrate is an important stabilizer used in whipping cream and vegetable-based dairy substitutes. Addition of sodium citrates to ice cream, ice milk, and frozen desserts before pasteurization and homogenization reduces the viscosity of the mix, making it easier to whip.

Seafood. Citric acid is used in combination with other preservatives/antioxidants to lower the pH to retard microbial growth, which can lead to spoilage, formation of off-flavors, and colors on fish and other seafood products.

Meat Products. Citric acid is used in cured meat products to increase the effectiveness of the antioxidant preservatives, as a processing aid, and a texture modifier. It is often encapsulated and released at a specific temperature from a controlled release matrix.

Medical Uses. Citric acid and citrate salts are used to buffer a wide range of pharmaceuticals at their optimum pH for stability and effectiveness (65–74). Effervescent formulations use citric acid and bicarbonate to provide rapid dissolution of active ingredients and improve palatability. Citrates are used to chelate trace metal ions, preventing degradation of ingredients. Citrates are used to prevent the coagulation of both human and animal blood in plasma and blood fractionation. Calcium and ferric ammonium citrates are used in mineral supplements.

Industrial Uses. *Laundry Detergents.* Sodium citrate is used in both liquid and powder laundry detergents as a builder (75–93). In many detergent applications, builder systems containing citrates are used as environmentally acceptable replacements for phosphates. Citrates chelate water hardness ions, disperse soil, and are used as processing aids. High water solubility and performance at both low and high pH and low and high temperatures are the keys to citrate use in detergents. In powder detergents citrates are used as auxiliary co-builders usually with zeolites and carbonates (see DETERGENCY).

Hard Surface Cleaners. Citric acid and sodium citrate are used in hard surface cleaners as an acid and chelator for dissolving hard water deposits and as a builder to increase the efficacy of the surfactants.

Reverse Osmosis Membrane Cleaning. Citric acid solutions are used to remove iron, calcium, and other cations that foul cellulose acetate and other membranes in reverse osmosis and electrodialysis systems. Citric acid solutions can solubilize and remove these cations without damaging the membranes (94–96).

Agricultural Use. Citric acid and its ammonium salts are used to form soluble chelates of iron, copper, magnesium, manganese, and zinc micronutrients in liquid fertilizers (97–103). Citric acid and citrate salts are used in animal feeds to form soluble, easily digestible chelates of essential metal nutrients, enhance feed flavor to increase food uptake, control gastric pH and improve feed efficiency.

Metal Cleaning. Citric acid, partially neutralized to ~pH 3.5 with ammonia or triethanolamine, is used to clean metal oxides from the water side of steam boilers and nuclear reactors with a two-step single fill operation (104–122). The resulting surface is clean and passivated. This process has a low corrosion rate and is used for both pre-operational mill scale removal and operational cleaning to restore heat-transfer efficiency.

High pressure sprays of heated neutralized citric acid solutions replace sandblasting techniques to clean stainless steel equipment and areas not easily accessible such as ship bilges.

Petroleum. Citric acid is added to hydrochloric acid solutions in acidizing limestone formations. Citric acid prevents the formation of ferric hydroxide gel in the spent acid solution by chelating the ferric ions present. Formation of the gel would plug the pores, preventing the flow of oil to the producer well (123–127).

A clear solution of aluminum citrate neutralized to pH 7 is used for *in situ* gelling of polymers in polymer flooding and well stimulation in enhanced oil recovery techniques (128–132). The citrate chelate maintains aluminum ion solubility and controls the rate of release of the aluminum cross-linker.

Citric acid is used to chelate vanadium catalyst in a process for removing hydrogen sulfide from natural and refinery gas and forming elemental sulfur, a valuable product (133).

Flue Gas Desulfurization. Citric acid can be used to buffer systems that can scrub sulfur dioxide from flue gas produced by large coal and gas-fired boilers generating steam for electrical power (134–143). The optimum pH for sulfur dioxide absorption is pH 4.5, which is where citrate has buffer capacity. Sulfur dioxide is the primary contributor to acid rain, which can cause environmental damage.

Mineral and Pigment Slurries. Citric acid can be used as a dispersing agent in slurries of ores, rocks, clays, and pigments during refining and transport. Citric acid controls swelling of clays and reduces pumping viscosity by contributing thixotropic properties to the dispersions (144–152).

Electrodeposition of Metals. Citric acid and its salts are used as sequestrants to control deposition rates in both electroplating and electroless plating of metals (153–171). The addition of citric acid to an electroless nickel plating bath results in a smooth, hard, nonporous metal finish.

Concrete, Mortar, and Plaster. Citric acid and citrate salts are used as admixtures in concrete, mortar, and plaster formulations to retard setting times and reduce the amount of water required to make a workable mixture (172–180). The citrate ion slows the hydration of Portland cement and acts as a dispersant, reducing the viscosity of the system (181). At levels below 0.1%, citrates accelerate the setting rate while at 0.2–0.4% the set rate is retarded. High early strength and improved frost resistance have been reported when adding citrate to concrete, mortar, and plaster.

Textiles. Citric acid acts as a buffer in the manufacture of glyoxal resins which are used to give textiles a high quality durable-press finish (see AMINO RESINS). It has been reported to increase the soil-release property of cotton with wrinkle-resistant finishes and is used as a buffer, a chelating agent, and a nonvolatile acid to adjust pH in disperse dying operations (182–193).

Plastics. Citric acid and bicarbonate are used as an effervescent blowing agent to foam polystyrene for insulated food and beverage containers replacing blowing agents such as chlorinated fluorocarbons (194–206).

Citric acid is used as a chelating agent in catalyst systems for making resins, and citrate esters are used as plasticizers (qv) in PVC film, especially in food packaging (207).

Paper. Citric acid is added to the pulp slurry prior to bleaching to sequester metal ions and prevent discoloration (208–211). Citrates are used in cigarette paper to control the burning rate to match that for tobacco.

Tobacco. Citric acid is a natural constituent of the tobacco leaf and during tobacco processing additional citric acid is added to enhance the flavor and to effect more complete combustion of tobaccos (212).

Cosmetics and Toiletries. Citric acid and bicarbonate are used in effervescent type denture cleansers to provide agitation by reacting to form carbon dioxide gas. Citric acid is added to cosmetic formulations to adjust the pH, act as a buffer, and chelate metal ions preventing formulation discoloration and decomposition (213–218).

Refractories and Molds.   Citric acid is used as a binder for refractory cements, imparting volume stability and strength in ceramic materials for electrical condensers, foundry and glassmaking molds, and sand molds for metal castings (219–223).

Derivatives

Salts.   The trisodium citrate salt is made by dissolving citric acid in water at a concentration of 50% w/w or higher. A 50% solution of sodium hydroxide is carefully added to pH 8.0–8.5. The reaction is exothermic and cooling is necessary to prevent boiling. The hot solution can be treated with activated carbon to remove impurities before evaporating and crystallizing. The concentrated slurry is filtered to separate the sodium citrate dihydrate [6132-04-3], which is washed with water, dried in a hot air dryer, classified, and packaged in bags, drums, or large sacks. The tripotassium salt of citric acid is made in a similar manner using potassium hydroxide. The product crystallizes as the monohydrate [6100-05-6].

Ammonium salts of citric acid are made by adding either aqueous or anhydrous ammonia to citric acid dissolved in water. They are usually used in the liquid form rather than isolated as a dry product. Citric acid salts are listed in Table 5. Solubility data is as follows (1).

Citrate salt	Water solubility, wt %
diammonium citrate	50
calcium citrate tetrahydrate	0.10
ferric ammonium citrate	very soluble
potassium citrate monohydrate	60
sodium citrate dihydrate	42

Esters.   The significant esters of citric acid are trimethyl citrate, triethyl citrate, tributyl citrate, and acetylated triethyl- and tributyl citrate. Many other esters are available but have not been used on a commercial scale. Citric acid esters are made under azeotropic conditions with a solvent, a catalyst, and the appropriate alcohol.

Catalysts used are usually acids such as sulfuric acid, *p*-toluenesulfonic acid, sulfonic acid ion-exchange resins, and others. The water from the reaction of the citric acid and the alcohol is continuously removed as the azeotrope until no more water is formed. At this point, the reaction is usually complete and the solvent and any excess alcohol is distilled off under mild vacuum. The catalyst is neutralized using carbonate or sodium hydroxide, leaving a crude product. If a pure product is desired, the ester can be distilled under high vacuum.

The properties of citric acid esters are described in Table 10.

Citric acid esters are used as plasticizers in plastics such as poly(vinyl chloride), poly(vinylidene chloride), poly(vinyl acetate), poly(vinyl butyral), polypropylene, chlorinated rubber, ethylcellulose, and cellulose nitrate. Most citrate esters are nontoxic and are acceptable by the FDA for use in food-contact packaging and for flavor in certain foods. As a plasticizer, citrate esters provide good heat

Table 10. Properties of Citric Acid Esters

Name	CAS Registry Number	Molecular weight	Density,[a] g/mL	Bp[b], °C
triethyl citrate	[77-93-0]	276.29	1.136	126–127
tri-n-butyl citrate	[77-94-1]	360.43	1.042	169–170
tricyclohexyl citrate	[4132-10-9]	438.57	1.7	57[c]
acetyl triethyl citrate	[77-89-4]	318.31	1.135	131–132
acetyl tri-n-butyl citrate	[77-90-7]	402.46	1.046	172–174
acetyl tri-2-ethylhexyl citrate	[144-15-0]	570.81	0.983	225

[a]At 25°C.
[b]At 133 Pa = 1 mm Hg.
[c]Melting point.

and light stability and excellent flexibility at low temperatures. Triethyl citrate, tri-n-butyl citrate, isopropyl citrate, and stearyl citrate are considered GRAS for use as food ingredients (224–228).

BIBLIOGRAPHY

"Citric Acid" in *ECT* 1st ed., Vol. 4, pp. 8–23, by G. B. Stone, Chas. Pfizer & Co., Inc.; in *ECT* 2nd ed., Vol. 5, pp. 524–540, by L. B. Lockwood and W. E. Irwin, Miles Chemical Co.; in *ECT* 3rd ed., Vol. 6, pp. 150–179, by E. F. Bouchard and E. G. Merritt, Pfizer Inc.

1. *The Merck Index*, 11th ed., Merck & Co., Rahway, N.J. 1989.
2. R. C. Weast, *CRC Handbook of Chemistry and Physics*, 69th ed., CRC Press, Boca Raton, Fla., 1988, 1989, p. 163.
3. *Perry's Chemical Engineering Handbook*, 6th ed., McGraw-Hill Book Co., Inc., New York, 1984.
4. A. Seidell, *Solubilities of Inorganic and Organic Compounds*, 3rd ed., Vol. 2, D. Van Nostrand Co., Inc., New York, 1941, pp. 427–429.
5. G. T. Blair and M. F. Zienty, *Citric Acid: Properties and Reactions*, Miles Laboratories, Inc., 1979.
6. U.S. Pat. 3,997,596 (Dec. 14, 1976), F. Smeets (to Citrex SA, Belgium).
7. *Citrest-Citric Acid Fatty Esters*, technical information, Cyclo Chemicals Corp., Miami, Fla.
8. C. J. Knuth and A. Bavley, *Plast. Technol.* **3,** 555 (1957).
9. P. W. Staal, *Chelation*, Technical Bulletin A-1014, Haarmann & Reimer Corp., Springfield, N.J., 1989.
10. M. K. Musho, *Citric Acid Chelation Chemistry*, Technical Bulletin A-1013, Haarmann & Reimer Corp., Springfield, N.J., 1989.
11. J. A. Dean, *Lange's Handbook of Chemistry*, 12th ed., McGraw-Hill Book Co., Inc., New York, 1979.
12. *Scientific Literature Review on GRAS Food Ingredients-Citrates*, PB-223 850, National Technical Information Service, Springfield, Va., Apr. 1973; *Scientific Literature Review on GRAS Food Ingredients-Citric Acid,* PB-241 967, National Technical Information Service, Springfield, Va., Oct. 1974.
13. G. T. Austin, *Shreve's Chemical Process Industries*, 5th ed., McGraw-Hill Book Co., Inc., New York, 1984.

14. L. M. Miall, in A. H. Rose, ed., *Primary Products of Metabolism*, Academic Press, Inc., New York, 1978.
15. J. N. Currie, *J. Biol. Chem.* **31,** 15 (1917).
16. H. Amelung, *Chem. Ztg.* **54,** 118 (1930)
17. L. H. C. Perquin, *Bijdrage Tot De Kennis Der Oxydative Dissimilatic Van Aspergillus niger van Tiegham*, Meinema, Delft, The Netherlands, 1938.
18. S. M. Martin and W. R. Waters, *Ind. Eng. Chem.* **44,** 2229 (1952).
19. D. S. Clark, *Can. J. Microbiol.* **8,** 133 (1962).
20. Brit. Pat. 653,808 (Mar. 23, 1951), R. L. Snell and L. B. Schweiger (to Miles Laboratories Inc.).
21. U.S. Pat. 3,285,831 (Nov. 15, 1966), E. J. Swarthout (to Miles Laboratories Inc.).
22. Brit. Pat. 1,145,520 (Mar. 19, 1969), M. A. Batti (to Miles Laboratories Inc.).
23. *Ullmanns Encyklopädie der Technischen Chemie*, 4th ed., Vol. 9, Urban & Schwarzenberg, Munich, Berlin, Germany, 1975, pp. 624–636.
24. *Code of Federal Regulations*, Title 21,§173.280, U.S. Government Printing Office, Washington, D.C., 1990.
25. *The Association of American Feed Control Officials*, Official Publication, 1977, Section 36, pp. 87–88.
26. H. C. DeRoo, *Conn. Agri. Exp. St. Bull.* 750 (1975).
27. E. Grimoux and P. Adam, *C. R. Acad. Sci. Paris* **90,** 1252 (1880); *Bull. Soc. Chim. Fr.* **36,** 18 (1881).
28. H. V. Pechmann and M. Dunschmann, *Ann. Chem.* **261,** 162 (1891); A. Haller and A. Held, *Ann. Chim. Phys.* **23,** 175 (1891).
29. W. T. Lawrence, *J. Chem. Soc.* **71,** 457 (1897); E. Ferrario, *Gazz. Chim. Ital.* **38,** 99 (1908).
30. E. Baur, *Chem. Ber.* **46,** 852 (1913).
31. H. Franzen and F. Schmitt, *Chem. Ber.* **58,** 222 (1925).
32. H. O. L. Fischer and G. Dangschat, *Helv. Chim. Acta* **17,** 1196 (1934); E. Baer, J. M. Grosheintz, and H. O. L. Fischer, *J. Am. Chem. Soc.* **61,** 2607 (1939).
33. F. Knopp and C. Martius, in Hoppe-Seyler, ed., *Z. Physiol. Chem.* **242,** 204 (1936); C. Martius, *Ibid.*, 279, 96 (1943).
34. A. M. Gakhokidze and A. P. Guntsadze, *J. Gen. Chem. USSR* **17,** 1642 (1947).
35. P. E. Wilcox, C. Heidelberger, and V. R. Potter, *J. Am. Chem. Soc.* **62,** 5019 (1950).
36. M. Taniyama, *Toho-Reiyon Kenkyu Hokoku* **1,** 40 (1954).
37. Haarmann & Reimer Corp. data 1990; Citric Acid, *Chemical Economics Handbook*, Stanford Research Institute, Menlo Park, Calif., May 1989.
38. *Chem. Mark. Rep.*, **242**(12), 32, 36, 37 (Sept. 21, 1992).
39. *Food Chemicals Codex*, 3rd ed., 3rd Suppl., National Academy Press, Washington, D.C., 1992.
40. *United States Pharmacopeia XXII*, United States Pharmacopeial Convention, Inc., Rockville, Md., 1990.
41. *British Pharmacopoeia*, British Pharmacopoeia Commission, London, 1988.
42. *The Pharmacopoeia of Japan*, The Society of Japanese Pharmacopoeia, Tokyo, Japan, 1987.
43. *European Pharmacopoeia*, 2nd ed., 1986.
44. C. G. Hartford, *Anal. Chem.* **34,** 426 (1962).
45. J. A. Taraborelli and R. P. Upton, *J. Am. Oil Chem. Soc.* **52,** 248 (1975).
46. G. D. Guerrand and co-workers, "Organic Acid Analysis," *J. Clin. Microbiol.* **16,** 355 (1982).
47. *Code of Federal Regulations*, Title 21,§182.1033,182.6033, U.S. Government Printing Office, Washington, D.C., 1990.

48. *FAO Nutrition Meetings Report*, Series No. 40 A,B,C, Food and Agriculture Organization of the United Nations World Health Organization, New York, 1967, p. 134.

49. *Tentative Evaluation of the Health Aspects of Citric Acid, Sodium Citrate, Potassium Citrate, Calcium Citrate, Triethyl Citrate, Isopropyl Citrate, and Stearyl Citrate as Food Ingredients*, PB280 954, National Technical Information Service, Springfield, Va., 1977.

50. Internal Reports: *Skin Corrosion Potential of Liquid Citric Acid 50%*, Jan. 19, 1979, *Dermal Irritation of Citric Acid in the Rabbit*, Dec. 20, 1990, *Ocular Irritation of Citric Acid in the Rabbit*, May 14, 1991, Miles Inc., Elkhart, Ind.

51. *Ecological Effects of Non-Phosphate Detergent Builders-Final Report on Organic Builders Other than NTA*, International Joint Commission, Windsor, Ontario, Canada, July 21, 1980.

52. *The Story of Soft Drinks*, National Soft Drink Association, Washington, D.C., 1982.

53. *Preservers Handbook*, Sunkist Growers, Inc., Ontario, Calif., 1964.

54. C. D. Barnett, *The Science & Art of Candy Manufacturing*, Harcourt Brace Jovanovich Publications, Duluth, Minn., 1978, p. 58.

55. A. F. Porter, *Spice Up Your Candy With Citric Acid*, Candy Industry, Aug. 1985.

56. N. A. Eskin, H. M. Henderson, and R. J. Townsend, *Biochemistry of Foods*, Academic Press, New York, 1981; *Code of Federal Regulations*, Title 21, Part 182, Subpart D, U.S. Government Printing Office, Washington, D.C., 1990.

57. K. H. Moledina and co-workers, *J. Food Sci.* **42,** 759 (1977).

58. C. E. Wells, D. C. Martin, and D. A. Tichenor, *J. Am. Dietetic Assoc.* **61,** 665 (1972).

59. L. L. Diosady, P. Sleggs, and T. Kaji, *JAOCS*, **59,**(7), 313–316 (1982).

60. A. Smiles, Y. Kakuda, and B. MacDonald, *JAOCS* **65,**(7), 1151–1155 (1988).

61. K. S. Law and K. G. Berger, *Citric Acid in the Processing of Oils and Fats*, No. 11, Porim Technology, Palm Oil Research Institute of Maylasia, July 1984.

62. D. D. Brooks and co-workers, "The Synergistic Effect of Neutral Bleaching Clay and Citric Acid: Chlorophyll Removal," paper presented at the *79th Annual AOCS Meeting*, Phoenix, Ariz. 1988.

63. S. K. Brophy and co-workers, "Chlorophyll Removal from Canola Oil: A New Concept," paper presented at the *80th Annual AOCS Meeting*, Cincinnati, Ohio, 1989.

64. *Code of Federal Regulations*, Title 21, Part 133, U.S. Government Printing Office, Washington, D.C., 1977.

65. U.S. Pat. 2,999,293 (Sept. 12, 1961), J. White and R. Kolb (to Warner Lambert Pharmaceutical Co.).

66. *Remingtons' Pharmaceutical Sciences*, 15th ed., Mack Publishing Co., Inc., Easton, Pa., 1975, p. 1574.

67. *The National Formulary*, 14th ed., American Pharmaceutical Association, Washington, D.C., 1975, pp. 389–390.

68. U.S. Pat. 3,956,156 (May 11, 1976), A. N. Osband, F. W. Gray, and J. C. Jervert (to Colgate-Palmolive Co.).

69. *Handbook of Non-Prescription Drugs*, 5th ed., American Pharmaceutical Association, Washington, D.C., 1977, pp. 3–17.

70. D. Entriken and C. Becker, *J. Am. Pharm. Assoc.* **43,** 693 (1954).

71. S. Bhattacharya and co-workers, *J. Indian Chem. Soc.* **31,** 231 (1954).

72. *USP XIX*, The United States Pharmacopeial Convention, Inc., Rockville, Md., 1975, pp. 33–35.

73. *Code of Federal Regulations*, Title 21,§182.1195, 182.5195, 182.6195 (calcium citrate); 182.1625, 182.6625 (potassium citrate), 182.5449 (manganese citrate), U.S. Government Printing Office, Washington, D.C., 1990.

74. *USP XXII, 1990*, USP Convention, Inc., Rockville, Md., 1989, p. 103.

75. *Environmental Impact of Citrates*, Information Sheet No. 2030, Pfizer Chemicals Division, New York, 1974.

76. D. L. Muck and H. L. Gewanter, "The Detergent Building Properties of Trisodium Citrate," paper presented at the *American Oil Chemist Society Meeting*, Apr. 1972.

77. U.S. Pat. 4,028,262 (June 7, 1977), B.-D. Cheng (to Colgate-Palmolive Co.).

78. U.S. Pat. 4,013,577 (Mar. 22, 1977) (to Colgate-Palmolive Co.).

79. U.S. Pat. 4,009,114 (Feb. 22, 1977), J. A. Yurke (to Colgate-Palmolive Co.).

80. Brit. Pat. 1,477,775 (June 26, 1977) (to Hoechst AG).

81. Brit. Pat. 1,427,071 (Mar. 3, 1976), M. Filcek and co-workers (to Benckiser GmbH).

82. U.S. Pat. 3,985,669 (Oct. 12, 1976), H. K. Krummel and T. W. Gault (to The Procter & Gamble Co.).

83. Ref. 68, p. 742.

84. U.S. Pat. 4,021,377 (May 3, 1977), P. J. Borchert and J. L. Neff (to Miles Laboratories, Inc.).

85. Belg. Pat. 848,533 (May 20, 1977) (to Henkel & Cie, GmbH).

86. U.S. Pat. 4,024,078 (May 7, 1977), A. Gilbert and J. W. Schuette (to The Procter & Gamble Co.).

87. *Citrosol-50 T,W and E*, Information Sheet No. 626, Pfizer Chemicals Division, New York, 1974.

88. U.S. Pat. 3,968,048 (July 6, 1976), J. A. Bolan (to The Drackett Co.).

89. U.S. Pat. 3,920,564 (Nov. 18, 1975), J. J. Greecsek (to Colgate-Palmolive Co.).

90. U.S. Pat. 4,379,080 (Apr. 5, 1983), A. P. Murphy (to The Procter & Gamble Co.).

91. U.S. Pat. 4,490,271 (Dec. 25, 1984), G. L. Spadini, A. L. Larabee, and D. K. K. Liu (to The Procter & Gamble Co.).

92. U.S. Pat. 4,605,509 (Aug. 12, 1986), J. M. Corkill, B. L. Madison, M. E. Burns (to The Procter & Gamble Co.).

93. U.S. Pat. 4,965,013 (Oct. 23, 1990), K. L. Pratt (to Miles Inc.).

94. Jpn. Pat. 76 18,280 (Feb. 13, 1976), T. Mizumoto and co-workers (to Ebara-Infilco Co., Ltd.).

95. Jpn. Pat. 75 153,778 (Dec. 11, 1975), T. Mizumoto and co-workers (to Ebara-Infilco Co., Ltd.).

96. K. J. McNulty and co-workers, *Laboratory and Field Evaluation of NS-100 Reverse Osmosis Membrane*, EPA-600/2-80-059, Industrial Environmental Research Laboratory, EPA, Cincinnati, Ohio, Apr. 1980.

97. U.S. Pat. 2,813,014 (Nov. 12, 1957), J. R. Allison and C. A. Hewitt (to Leffingwell Chem. Co.).

98. Brit. Pat. 827,521 (Feb. 3, 1960), I. S. Perold (to Union of South Africa, Dir. of Tech. Service, Dept. of Agriculture).

99. U.S. Pat. 3,869,272 (Mar. 4, 1975), R. J. Windgasen (to Standard Oil Co.).

100. A. Marchesini, P. Sequi, and G. A. Lanzani, *Agrochimica* **10**(2), 183 (1966).

101. F. L. Daniel, P. O. Ramaswani, and T. P. Mahadevan, *Madras Agr. J.* **55**(1), 31 (1968).

102. M. S. Omran and co-workers, *Egypt. J. Soil, Sci.* **27**(1), 31–42 (1987).

103. J. J. Mortvedt, *Solutions*, 64–79 (May–June 1979).

104. W. J. Blume, *Mater. Perform.* **16**(3), 15 (1977).

105. U.S. Pat. 3,806,366 (Apr. 23, 1974), D. B. Cofer and co-workers (to Southwire Co.).

106. U.S. Pat. 3,664,870 (May 23, 1972), A. W Oberhofer and co-workers (to Nalco Chemical Co.).

107. U.S. Pat. 3,072,502 (Jan. 8, 1963), S. Alfano (to Pfizer Inc.); 3,248,269 (Apr. 26, 1966), W. E. Bell (to Pfizer Inc.).

108. L. D. Martin and W. P. Banks, "Electrochemical Investigation of Passivating Systems," *Proceedings of the 35th International Water Conference*, Pittsburgh, Pa., Oct. 29–31, 1974.

109. G. W. Bradley and co-workers, "Investigation of Ammonium Citrate Cleaning Solvents," *Proceedings of the 36th International Water Conference*, Pittsburgh, Pa., Nov. 4–6, 1975.
110. U.S. Pat. 3,003,898 (Oct. 10, 1961), C. F. Reich (to The Dow Chemical Company).
111. U.S. Pat. 3,496,017 (Feb. 17, 1970), R. D. Weed (to U.S. Atomic Energy Commission).
112. U.S. Pat. 3,013,909 (Dec. 19, 1961), G. P. Pancer and J. L. Zegger (to U.S. Atomic Energy Commission).
113. McCollum and Logan, *Electrolytic Corrosion of Iron in Soils*, Technical Paper No. 25, Bureau of Standards, Washington, D.C., 1913, p. 7.
114. N. Hall and Hogaboom, "Metal Finishing," *21st Annual Guidebook Directory*, Westwood, N.J., 1953.
115. U.S. Pat. 2,558,167 (June 26, 1951), A. J. Beghin, P. F. Hamberg, and H. E. Smith (to Insl-X Corp.).
116. E. C. Wackenhuth, L. W. Lamb, and J. P. Engle, *Power Eng.*, 68 (Nov. 1973).
117. Jpn. Kokai 75 47,457 (Apr. 26, 1975), Y. Kudo and co-workers (to Mitsubishi Heavy Industries, Ltd.).
118. S. Arrington and G. Bradley, "Service Water System Cleaning with Ammoniated Citric Acid," paper presented at *Corrosion 87-NACE*, No. 387, San Francisco, Calif., 1987.
119. U.S. Pat. 4,190,463 (Feb. 26, 1980), R. I. Kaplan (to Nalco Chemical Co.).
120. U.S. Pat. 4,540,443 (Sept. 10, 1985), A. G. Barber (to Union Carbide Corp.).
121. C. A. Poulos, *Mater. Perform.* **23**(8), 19–21 (1984).
122. J. R. Gatewood and co-workers, *Mater. Perform.* **18**(7), 9–14 (1979).
123. R. T. Johansen, J. P. Powell, and H. N. Dunning, *U.S. Bur. Mines Inform. Circ.*, 7797 (1957).
124. *Pfizer Products for Petroleum Production*, Technical Bulletin No. 97, Pfizer Inc., New York, 1961.
125. U.S. Pat. 3,335,793 (Aug. 15, 1967), J. W. Biles and J. A. King (to Cities Service Oil Co.).
126. U.S. Pat. 3,402,137 (Sept. 17, 1968), P. W. Fischer and J. P. Gallus (to Union Oil Co. of California).
127. U.S. Pat. 3,732,927 (May 15, 1973), E. A. Richardson (to Shell Oil Co.).
128. U.S. Pat. 4,447,364 (May 8, 1984), P. W. Staal (to Miles Laboratories, Inc.).
129. C. W. Crowe, *J. Petrol. Technol.* 691–695, (Apr. 1985).
130. U.S. Pat. 4,151,098 (Apr. 24, 1979), W. R. Dill, J. A. Knox (to Halliburton Co.).
131. U.S. Pat. 3,952,806 (Apr. 27, 1976), J. C. Trantham (to Phillips Petroleum Co.).
132. J. C. Mack and M. L. Duvall, "Performance and Economics of Minnelusa Polymer Floods," paper presented at the *1984 Rocky Mountain Regional Meeting of Society of Petroleum Engineers,* Casper, Wyo., May 21–23, 1984.
133. U.S. Pat. 4,432,962 (Feb. 21, 1984), H. W. Gowdy and D. M. Fenton (to Union Oil of California).
134. D. R. George, L. Crocker, and J. B. Rosenbaum, *Min. Eng.* **22**(1), 75 (1970).
135. J. B. Rosenbaum, D. R. George, and L. Crocker, "The Citrate Process for Removing SO$_2$ and Recovering Sulfur from Waste Gases," paper presented at the *AIME Environmental Quality Conference,* Washington, D.C., June 7–9, 1971.
136. J. B. Rosenbaum and co-workers, *Sulfur Dioxide Emission Control by Hydrogen Sulfide Reaction in Aqueous Solution—The Citrate System*, PB221914/5, National Technical Information Service, Springfield, Va., 1973.
137. L. Korosy and co-workers, *Adv. Chem. Ser.* **139,** 192 (1975).
138. U.S. Pat. 3,757,488 (Sept. 11, 1973), R. R. Austin and A. L. Vincent (to International Telephone and Telegraph Corp.).
139. U.S. Pat. 3,933,994 (Jan. 20, 1976), G. L. Rounds (to Kaiser Steel Corp.).

140. Ger. Pat. 2,432,749 (Jan. 30, 1975), W. J. Balfanz, R. M. DePirro, and L. P. Van Brocklin (to Stauffer Chemical Co.).

141. T. Wasag, J. Galka, and M. Fraczak, *Ochr. Powietrza* **9**(3), 72 (1975).

142. P. M. Bever and G. E. Klinzing, *Environ. Prog.* **4**(1)1–6 (Feb. 1985).

143. B. K. Dutta and co-workers, *Ind. Eng. Chem. Res.* **26**(7), 1291–1296 (1987).

144. U.S. Pat. 4,042,666 (Aug. 16, 1977), H. L. Rice and R. A. Wilkins (to Petrochemicals Co., Inc.).

145. U.S. Pat. 3,663,284 (May 16, 1972), D. J. Stanicoff and H. J. Witt (to Marine Colloids, Inc.).

146. U.S. Pat. 2,952,580 (Sept. 13, 1960), J. Frasch.

147. U.S. Pat. 3,029,153 (Apr. 10, 1962) K. L. Hackley (to Champion Papers, Inc.).

148. U.S. Pat. 3,245,816 (Apr. 12, 1966), H. C. Schwaibe (to Mead Corp.).

149. V. Laskova, *Pap. Cellul.* **30**(7–8), 173 (1975).

150. U.S. Pat. 2,336,728 (Dec. 14, 1943), H. W. Hall (to the Dicalite Co.).

151. U.S. Pat. 4,144,083 (Mar. 13, 1979) W. F. Abercrombie, Jr. (to J. M. Huber Corp.).

152. U.S. Pat. 4,309,222 (Jan. 5, 1982) H. L. Hoyt, 4th (to Pfizer, Inc.).

153. C. W. Smith and C. B. Munton, *Met. Finish.* **39,** 415 (1941).

154. A. W. Hothersall, "The Adhesion of Electrodeposited Nickel to Brass," paper presented to *Electroplaters' and Depositors' Technical Society,* Northampton Polytechnic Institute, London, UK, May 18, 1932.

155. Fr. Pat. 813,548 (June 3, 1937) (to The Mound Co., Ltd.).

156. U.S. Pat. 2,474,092 (June 21, 1949), A. W. Liger (to Battelle Development Corp.).

157. C. W. Fleetwood and L. F. Yntema, *Ind. Eng. Chem.* **27,** 340 (1935).

158. J. Kashima and F. Fuhushima, *J. Electrochem. Assoc. Jpn.* **15,** 33 (1947).

159. W. E. Clark and M. L. Hold, *J. Electrochem. Soc.* **94,** 244 (1948).

160. N. N. S. Siddhanta, *J. Indian Chem. Soc. Ind News Ed.* **14,** 6 (1951).

161. C. N. Shen and H. P. Chung, *Chin. Sci.* **2,** 329 (1951).

162. K. G. Sodeberg and H. L. Pinkerton, *Plating* **37,** 254 (1930).

163. W. E. Brodt and L. R. Taylor, *Trans. Electrochem. Soc.* **73** (1938).

164. L. F. Yntema, *J. Am. Chem. Soc.* **54,** 3775 (1932).

165. U.S. Pat. 2,599,178 (June 3, 1952), M. L. Holt and H. J. Seim (to Wisconsin Aluminum Research Foundation).

166. Can. Pat. 443,256 (July 29, 1947), E. M. Wise and R. F. Vines (to The International Nickel Company of Canada, Ltd.).

167. L. E. Netherton and M. L. Holt, *J. Electrochem. Soc.* **95,** 324 (1949).

168. *Ibid.* 98, 106 (1951).

169. Ital. Pat. 444,078 (Jan. 12, 1949), A. Sacco and M. Gandusi.

170. K. S. Rajam, *Met. Finish.,* 41–45 (Oct. 1990).

171. U.S. Pat. 4,371,573 (Feb. 1, 1983), H. Januschkowetz and H. Laub (to Siemens AG).

172. *J. Am. Concr. Inst.* **60**(11), (Nov. 1963).

173. U.S. Pat. 2,174,051 (Sept. 26, 1939), K. Winkler.

174. U.S. Pat. 3,656,985 (Apr. 18, 1972), B. Bonnel and C. Hovasse (to Progil, France).

175. U.S. Pat. 2,542,364 (Feb. 20, 1951), F. A. Schenker and A. Ammann (to Kaspar Winkler Cie).

176. F. Tamas, NASNRC Publ. **1389,** 392 (1966).

177. F. Tamas and G. Liptay, *Proc. Conf. Silicate Ind.* 8,299 (1965).

178. E. C. Combe and D. C. Smith, *J. Appl. Chem.* (*London*) **16**(3), 73 (1966).

179. *Pfizer Organic Chelating Agents,* Technical Bulletin No. 32, Pfizer Inc., Chemicals Division, New York, 1972.

180. U.S. Pat. 4,004,066 (Jan. 18, 1977), A. J. DeArdo (to Aluminum Company of America).

181. N. B. Singh and co-workers, *Cem. Concr. Res.* **16,** 911–920 (1986).

182. U.S. Pat. 3,754,860 (Aug. 28, 1973), J. G. Frick, Jr., and co-workers (to U.S. Secretary of Agriculture).
183. U.S. Pat. 3,212,928 (Oct. 19, 1965), H. R. Hushebeck (to Joseph Bancroft and Sons Co.).
184. D. D. Gagliardi and F. B. Shippee, *Am. Dyestuff Rep.* **52,** 300 (1963).
185. L. Benisek, *J. Soc. Dyers Colour.,* 277 (Aug. 1971).
186. "Mordanting of Wool: A Dyeing Technique to Increase the Flame Resistance of Wool Shag Carpets," *Wool Facts,* Vol. 1, No. 1, Wool Bureau, Inc., New York, privately presented, Wool Bureau Technical Center, Woodbury, N.Y., 1971.
187. "Mordanting Wool with Zirconium," *Wool Facts,* Vol. 1, No. 5, Wool Bureau, Inc., New York. 1971.
188. R. R. Haynes, J. H. Mathews, and G. A. Heath, *Text. Chem. Color.* **1**(3),16/74 (1969).
189. U.S. Pat. 2,720,441 (Oct. 11, 1955), J. G. Wallace (to E. I. du Pont de Nemours & Co., Inc.).
190. U.S. Pat. 2,898,179 (Aug. 4, 1959), C. J. Rogers (to E. I. du Pont de Nemours & Co., Inc.).
191. *Replacement of Phosphates with Citric Acid in Nylon Carpet Dyeing,* Information Sheet No. 2025, Pfizer Chemicals Division, New York, 1973.
192. B. J. Harper and co-workers, *Text. Chem. Color.* **3**(5), 65/127–70/132 (May 1971).
193. B. A. Kottes Andrews and R. M. Reinhardt, "How Mixed Catalysts Differ," paper presented at the *Fourth Annual Natural Fibers Textile Conference,* New Orleans, La., Sept. 14–16, 1981.
194. Neth. Appl. 6,605,358 (Oct. 24, 1966) (to Koppers Co.).
195. U.S. Pat. 2,950,263 (Aug. 23, 1960), W. Abbotson, R. Hurd, and H. Jackson (to Imperial Chemical Ind., Ltd.).
196. U.S. Pat. 3,185,588 (May 25, 1965), J. Y. Resnick (to Int Res. & Dev. Co., New York).
197. U.S. Pat. 4,016,110 (Apr. 5, 1977), W. E. Cohrs and co-workers (to The Dow Chemical Company).
198. V. G. Kharakhash, *Plast. Massy* **6,** 40 (1971).
199. U.S. Pat. 3,523,988 (Aug. 11, 1970), Z. M. Roehr, R. Berger, and P. A. Plasse (to Roehr Metals & Plastics Co.).
200. U.S. Pat. 3,482,006 (Dec. 2, 1969), F. A. Carlson (to Mobil Oil Corp.).
201. U.S. Pat. 3,069,367 (Dec. 18, 1962), R. D. Beaulieu, P. N. Speros, and D. A. Popielski (to Monsanto Chemical Co.).
202. Ger. Pat. 11,144,911 (Mar. 7, 1963), F. Stastny, B. Ikert, and E. F. v. Behr (to Badische Anilin-und Soda-Fabrik).
203. U.S. Pat. 3,660,534 (May 25, 1972), F. E. Carrock and K. W. Ackerman (to Dart Indus.).
204. B. C. Mitra and S. R. Palit, *J. Indian Chem. Soc.* **50**(2), 141 (1973).
205. Jpn. Pat. 72 28,766 (July 29, 1972), K. Okuno, K. Itagaki, and H. Takashi (to Mitsubishi Chemical Industries Co.).
206. J. Jarusek and J. Mleziva, *Chem. Prumyst* **16**(11), 671 (1966).
207. E. H. Hull and K. K. Mathur, *Mod. Plast.,* 66–70 (May 1984).
208. U.S. Pat. 3,674,619 (July 4, 1972), H. I. Scher and I. S. Ungar (to Esso Research Engineering Co.).
209. Ger. Pat. 1,269,874 (June 6, 1968), R. H. McKillip and R. Henderson (to Olin Mathieson Chemical Corp.).
210. Brit. Pat. 1,079,762 (June 4, 1969), I. R. Horne and R. N. Lewis (to Bakelite Ltd.).
211. *Ind. Eng. Chem.* **53**(1), 28A (1961).
212. C. O. Jensen, *Chemical Changes During the Curing of Cigar Leaf Tobacco,* The Pennsylvania State College, *American Chemical Society,* State College, Pa., Sept. 1951.
213. L. MacDonald, *Am. Perfum.* **76**(7), 22 (1961).

214. M. Ash and I. Ash, *Formulary of Cosmetic Preparations,* Chemical Publishing Co., Inc., New York, 1977.
215. L. Smith and M. Weinstein, *Household Pers. Prod. Ind.* **14**(10), 54 (1977).
216. U.S. Pat. 3,718,236 (Feb. 27, 1973), E. M. Reyner and M. E. Reyner.
217. Ref. 87, p. 260.
218. E. W. Flick, *Cosmetic and Toiletry Formulations,* Noyes Publications, Park Ridge, N.J., 1984.
219. U.S. Pat. 3,333,972 (Aug. 1, 1967), J. T. Elmer and B. G. Atlman (to Kaiser Aluminum & Chemical Corp.).
220. Aust. Pat. 259,442 (Jan. 10, 1968), T. Chvatal.
221. Neth. Appl. 6,409,880 (Mar. 1, 1965), (to Sprague Electric Co.).
222. Fr. Appl. 2,172,160 (Nov. 2, 1973), F. Sembera.
223. Jpn. Pat. 75 03,020 (May 15, 1973), S. Sugiyama, Y. Heima, and M. Sugi (to Toshiba Mach. Co. Ltd.).
224. *Code of Federal Regulations,* Title 21,§175.105, 175.300, 175.320, 175,380, 175.390, 176.170, 177.1210, 178.3910, 181.27(acetyl triethyl citrate and acetyl tri-*n*-butyl citrate);175.105, 175.300, 175.320, 175.380, 175.390, 176.170, 177.1210,181.27 (triethyl citrate),175.105 (tri-*n*-butyl citrate), U.S. Government Printing Office, Washington, D.C., 1990.
225. *Code of Federal Regulations,* Title 21,§182.1911(triethyl citrate), U.S. Government Printing Office, Washington, D.C., 1990.
226. R. L. Hall and B. L. Oser, *Food Technol.* **19**(2), 151 (1965).
227. *Code of Federal Regulations,* Title 21,§182.6386(isopropyl citrate), 182.6511 (mono-isopropyl citrate), U.S. Government Printing Office, Washington, D.C., 1990.
228. *Code of Federal Regulations,* Title 21,§182.6851, U.S. Government Printing Office, Washington, D.C., 1990.

GARY BLAIR
PHILIP STAAL
Haarmann & Reimer Corporation

CITRUS JUICE. See FRUIT JUICES.

CITRONELLA OIL. See OILS, ESSENTIAL.

CITRONELLAL. See PERFUMES; TERPENOIDS.

CLATHRATION. See INCLUSION COMPOUNDS.

CLAYS

SURVEY

The terms clay or clays commonly refer to either rocks that may be consolidated or unconsolidated, or a group of minerals having unique properties. Traditionally, clays (rocks) are distinctive in at least two properties which render them technologically useful: plasticity and composition. Plasticity signifies the property of the clay when wetted that permits deformation by application of relatively slight pressure and retention of the deformed shape after release of the pressure. This property distinguishes clay from hard rocks. Clays are predominantly composed of hydrous phyllosilicates, referred to as clay minerals. These are hydrous silicates of Al, Mg, K, and Fe, and other less abundant elements. Clay minerals are extremely fine crystals or particles, often colloidal in size and usually platelike in shape. The nonclay mineral portion of clays (rocks) may consist of other minerals, portions of rocks, and organic compounds.

The very fine particles yield very large specific surface areas that are physically sorptive and chemically surface reactive. Many clay mineral crystals carry an excess negative electric charge owing to internal substitution by lower valent cations, and thereby increase internal reactivity in chemical combination and ion exchange. Clays, which may have served as substrates selectively absorbing and catalyzing amino acids in the origin of life, apparently catalyze petroleum formation in rocks (see PETROLEUM).

Because clays (rocks) usually contain more than one mineral and the various clay minerals differ in chemical and physical properties, the term clay may signify entirely different things to different clay users. Whereas the geologist views clay as a raw material for shale, the pedologist as a dynamic system to support plant life, and the ceramist as a body to be processed in preparation for vitrification, the chemist and technologist view clay as a catalyst, adsorbent, filler, coater, or source of aluminum or lithium compounds, etc.

For geologists, clay refers to sediments or sedimentary rock particles having a diameter of 3.9 μm or less. Soil scientists define clays as disperse systems of the colloidal products of weathering in which secondary mineral particles of dimensions smaller than 2 μm predominate. Ceramists, who probably process the greatest quantity of clay, usually emphasize aluminosilicate content and plasticity (1). Plasticity is defined as resulting from colloids of organic or mineral nature (2) but it may also arise from other causes (3). Even the origin of the clay has been included in one definition (4). Clay has been recognized as being a product of deep-seated alteration of silicate minerals by hydrothermal solutions rising from an igneous source as well as a product of surface weathering (5,6). Although clay minerals are usually considered breakdown products of silicates, largely by hydrolysis, clays may be built up from hydrates of silica and alumina (7).

A clay deposit usually contains nonclaylike minerals as impurities and these impurities may actually be essential in determining the unique and specially desired properties of the clay. Both crystalline and amorphous minerals and compounds may be present in a clay deposit (8).

A broad definition of clays includes the following properties:

(*1*) Crystalline hydrated silicates of aluminum, iron, and magnesium comprise the majority of clay minerals, however, amorphous hydrated aluminum compounds are also included. Distinctions among clay minerals are made by chemical and structural parameters. The chemical variations range from kaolinite [*1318-74-2*], which is relatively uniform in chemical composition, to smectite minerals, which vary widely in chemical composition, base exchange properties, and expanding crystal lattice. The illite group, typically a component of sediment (9), includes micas, although illite [*1273-60-3*] differs from muscovite [*1318-94-1*], $K_2O \cdot 3Al_2O_3 \cdot 6SiO_2 \cdot 2H_2O$, because illite contains less potassium, and more water. The chlorite clay minerals resemble metamorphic chlorite [*1318-59-8*] (10), but aluminous or at least aluminum-rich chlorites have also been found in soils (4,11). Vermiculite [*1318-00-9*], characterized by a highly expanding crystal lattice, and sepiolite [*15501-74-3*], $Mg_8Si_{12}O_{34}H_4$, and palygorskite [*12174-11-7*], which possess chain or fiber structures, must also be included. Moreover, clay minerals are excellent examples of mixed layering, both random and regular, in layer-structure silicates. These mixed-layer clays are among the most ubiquitous of the various clay minerals. The structural differences among clay minerals are related to the arrangement of tetrahedral (T) and octahedral (O) layers, and the manner in which electrostatic charge imbalances, created by chemical substitution, are neutralized. Figure 1 shows several examples.

(*2*) The possible content of hydrated alumina and iron. Hydrated alumina minerals like gibbsite [*14762-49-3*], $Al(OH)_3$, boehmite [*1318-23-6*], $AlOOH$, and diaspore [*14457-84-2*], $AlOOH$, occur in bauxitic clays. Bauxites grade chemically into hydrated ferruginous and manganiferous laterites. Hence, finely divided M_2O_3, usually hydrated, may be a significant constituent of a clay where M may be Al or Fe. Hydrated colloidal silica may play a role in the slippery and sticky properties of certain clays.

(*3*) The extreme fineness of individual clay particles, which may be of colloidal size in at least one dimension. Clay minerals are usually platy in shape, and less often lathlike and tubular or scroll shaped (13). Because of this fineness clays exhibit the surface chemical properties of colloids (qv) (14). Some clays possess relatively open crystal lattices and show internal surface colloidal effects. Other minerals and rock particles, which are not hydrous aluminosilicates but which also show colloidal dimensions and charactristics, may occur intimately intermixed with the clay minerals and play an essential role.

(*4*) The property of thixotropy in various degrees of complexity (3). Thixotropic properties may lead to loss of stability as shown by the sometimes catastrophic flow of quick clays, especially in Norway, Sweden, and Canada (15).

(*5*) The possible content of quartz [*14808-60-7*], SiO_2, sand and silt, feldspars, mica [*12001-26-2*], chlorite, opal [*14639-88-4*], volcanic dust, fossil fragments, high density so-called heavy minerals, sulfates, sulfides, carbonate minerals, zeolites, and many other rock and mineral particles ranging upward in size from colloids to pebbles. An extreme example is that of a clay from western Texas composed of 98.5% dolomite [*16389-88-1*], $CaMg(CO_3)_2$, and 1.5% iron oxide and

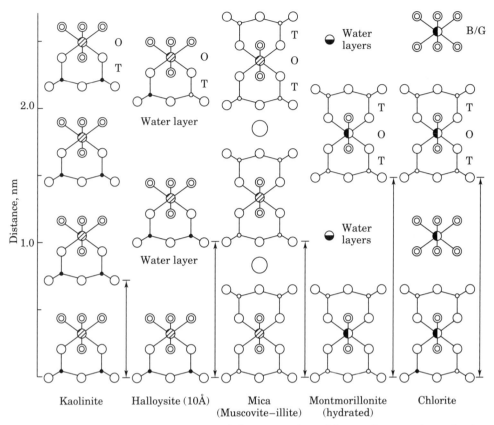

Fig. 1. Diagrammatic representation of the succession of layers in some layer lattice silicates (12) where O is oxygen; ◎, hydroxyl; •, silicon; ○, Si–Al; ⊘, aluminum; ◑, Al–Mg; ◯, potassium; ◑, Na–Ca. Sample layers are designated as O, octahedral; T, tetrahedral; and B/G, brucite- or gibbsitelike. The distance depicted by arrows between repeating layers in nm are 0.72, kaolinite; 1.01, halloysite (10 Å); 1.00, mica; ca 1.5, montmorillonite; and 1.41, chlorite.

alumina; it occurs in rhombic particles averaging 0.008 mm in diameter, and possesses sufficient plasticity to be molded into bricklets (16).

The synthesis of clay minerals has been extensively studied in order to understand the genesis of clay minerals and more recently to calibrate analytical techniques (17–25). Many experiments were performed at high temperatures, some using synthetic chemicals, others using part naturally occurring minerals. Organic compounds were found to facilitate the synthesis of kaolinite at low temperature by condensing aluminum hydroxide into octahedrally coordinated sheets (18).

Geology and Occurrence

Clays may originate through several processes: (*1*) hydrolysis and hydration of a silicate, ie, alkali silicate + water → hydrated aluminosilicate clay + alkali hydroxide; (2) solution of a limestone or other soluble rock containing relatively insoluble clay impurities that are left behind; (3) slaking and weathering of shales (clay-rich sedimentary rocks); (4) replacement of a preexisting host rock by invading guest clay where the constituents are carried in part or wholly by solution; (5) deposition of clay in cavities or veins from solution; (6) bacterial and other organic activity, including the extraction of metal cations as nutrients by plants; (7) action of acid clays, humus, and inorganic acids on primary silicates; (8) alteration of parent material or diagenetic processes following sedimentation in marine and freshwater environments (26,27); and (9) resilication of high alumina minerals.

Clays or shales that may be utilized in the manufacture of bricks, tiles, and other heavy clay products exist in every state in the United States. (See BUILDING MATERIALS, SURVEY). Some blending of materials is often necessary to control shrinkage of the product, and the economics of manufacture are governed by the demands of fuel, labor, transportation costs, and the market.

Glacial clays, as unassorted glacial till or secondarily deposited melt water, are abundant in the United States north of the Missouri and Ohio Rivers. Quartz-rich sand, silt, or pebbles, especially limestone, may occur mixed with the clay. Kaolins are plentiful in North Carolina, South Carolina, Georgia (28), Florida, and Vermont. Certain flint clays and other clays having a kaolinitic composition which can replace kaolins in some uses (29), occur in Missouri, Arkansas, Colorado, Texas, Ohio, Indiana, Oregon, and Pennsylvania. Ball clays, ie, clays having high plasticity and strong bonding power, are obtained primarily from western Tennessee and Kentucky, though some are found in New Jersey. Other plastic clays, especially plastic fire clay, are extensively produced in Missouri, Illinois, Ohio, Kentucky, Mississippi, Alabama, and Arkansas.

Fire clays are those that resist fusion at a relatively high temperature, usually around 1600°C. Missouri, Pennsylvania, Ohio, Kentucky, Georgia, Colorado, New Jersey, Texas, Arkansas, Illinois, and Maryland are large producers of fire clays. Loess is a quartz-rich, clayey silt, windblown in origin but in some cases reworked by water. It is prevalent along the Missouri, Mississippi, and Ohio Rivers and their tributaries. Loess has been used primarily for brick making. Adobe, a calcareous, sandy to silty clay used extensively for making sundried brick, is available in the more arid southwestern states. Slip clay for glazing pottery is produced near Albany, New York.

Bentonite [1302-78-9], widely distributed geographically and geologically, also varies widely in properties. Swelling bentonite occurs in Wyoming, South Dakota, Montana, Utah, Nevada, and California. Bentonite that swells little or not at all occurs in large quantities in Texas, Arkansas, Mississippi, Kentucky, and Tennessee. Fuller's earth and bleaching clays are found chiefly in Georgia and Florida. Many of these fuller's earths contain palygorskite (locally called attapulgite) as the most important clay mineral constituent.

High alumina clays refer in the ceramic industries to nodular clays, burley-flint clay, burley and diaspore, gibbsitic or bauxitic kaolins (clays), abrasive clays, and others. Since the depletion of diaspore varieties in Missouri and Pennsylvania, most bauxitic kaolin and clay is produced in Alabama and Arkansas.

Each continent has clays of almost every type, however, certain deposits are outstanding (28–31). There are tremendous reserves of white kaolin in Brazil, and deposits of bauxitic clay shared with countries across Brazil's northern borders. England has the famous kaolins of the Cornwall district. Refractory clays in Scotland provide the raw materials for a refractories (qv) industry. Similar deposits are found in Brittany and neighboring areas in France. Czechoslovakia and Germany have large reserves of kaolin accompanied by quartz and mica. There are large deposits of bauxitic clays (with bauxite) in Hungary and the former republics of Yugoslavia. Both flint clay and white kaolin are found in South Africa. Kaolins are found in the People's Republic of China. Japan has notable hydrothermal kaolins, sedimentary kaolin, smectite, and flint clay. Hydrothermal kaolins are also widely distributed in Mexico. Australia has large deposits of flint clay and kaolin clay naturally calcined by the burning of coal beds. New Zealand has hydrothermal clays. India has flint and lateritic clays. Sepiolite is found near Madrid, Spain. Localities producing bauxite almost always have a potential for producing associated high alumina clays.

The commercial value of a clay deposit depends on market trends, competitive materials, transportation facilities, new machinery and processes, and labor and fuel costs. Naturally exposed outcrops, geological area and structure maps, aerial photographs, hand and power auger drills, core drills, earth resistivity, and shallow seismic methods are used in exploration for clays (32). Clays are mined primarily by open-pit operation, including hydraulic extraction; however, underground mining is also practiced.

Specific information concerning the geology or occurrence of a particular deposit or variety of clay may be obtained from a state geological survey; general information may be obtained from the U.S. Geological Survey or the U.S. Bureau of Mines. Similar agencies may be contacted in foreign countries, or literature references may be consulted, eg, References 33–35, and the general references.

Technological Classification

The technological classification of a clay (rock, desposit) should take into account the following factors: (1) the dominant clay mineral type including breakdown into its polymorphs, the sites and amount of charge on it, and shape of clay crystal and particle; (2) the clay minerals present in minor quantities, but perhaps coating the surface of the primary constituent; (3) the particle-size distribution of the clay and other minerals; (4) ion-exchange capacity and neutral molecule sorption; (5) the type of exchangeable ions present on the clay and degree of saturation of exchange sites; (6) hygroscopicity of the clay; (7) reactivity of the clay with organic compounds; (8) expansion potentialities of the clay mineral lattice; (9) electrolytes and solutions in association with the clay deposit; (10) the accessory minerals, or mineral impurities, their sizes, homogeneity of mixture, and ion-exchange capacity; (11) content of organic matter and especially its occurrence, size and discreteness of particles, its adsorption on and/or within the clay crystal units, and protective colloidal action; (12) presence or absence of bacteria or other living organisms because organic growth may rapidly change the pH and other properties of a clay deposit; (13) content of hydrated alumina and/or silica, which are relatively soluble in ground water or in dilute acid or alkali; (14) the structure and texture of the clay deposit, such as lamination, orientation of mineral particles, and other gross features; (15) the rheological properties of both the natural and processed clay; and (16) the engineering strength and sensitivity to moisture, desalination, and shock, eg, the quick clays.

Mineralogy

The development of apparatus and techniques, such as x-ray diffraction, contributed greatly to research on clay minerals. Crystalline clay minerals are identified and classified (36) primarily on the basis of crystal structure and the amount and locations of charge (deficit or excess) with respect to the basic lattice. Amorphous (to x-ray) clay minerals are poorly organized analogues of crystalline counterparts.

The structural variations among the clay minerals can be understood by considering various physical combinations of tetrahedral and octahedral sheets and the electrostatic effect chemical substitution has on the structural units. The tetrahedral sheets are composed primarily of Si^{4+} and oxygen, but minor amounts of Al^{3+} or Fe^{3+} may substitute for Si^{4+}. The substitution of M^{3+} for Si^{4+} leaves the tetrahedral sheet negatively charged. The cations of the octahedral sheet are composed primarily of Al^{3+}, Fe^{3+}, Mg^{2+}, and Fe^{2+}, but all other transition elements, except Sc, may be included. The anions of the octahedral sheet are O^{2-}, OH^-, and F^-. The smallest unit of the octahedral sheet contains three octahedra having an ideal net charge of negative six, ie, three O^{2-}. If the negative charge is balanced by two trivalent cations, the layer is referred to as a dioctahedral layer; if balanced by three bivalent cations, the layer is referred to as a trioctahedral layer. Substitution of bivalent cations for trivalent cations, univalent cations (Li^+) for bivalent cations or unfilled octahedral sites, leaves the ocahedral layer a net negative charge. The tetrahedral apical oxygen is shared with the ocahedral layer to join the two types of layers (37,38).

The least complicated clay minerals are the 1:1 clay minerals composed of one tetrahedral (T) layer and one octahedral (O) layer (see Fig. 1). These 1:1 clay minerals are also referred to as TO minerals. The TO package has a basal spacing (nominal thickness) of 0.7 nm (7 Å) and they are commonly referred to as 7 Å minerals. Kaolinite, the dioctahedral 1:1 mineral, has Al^{3+} filling two of three octahedral sites, and serpentine [12168-92-2], $(Mg)_3Si_2O_5(OH)_4$, the trioctahedral 1:1 mineral has Mg^{2+} filling all three octahedral sites. The kaolin minerals have limited substitution in the octahedral layer and have negligible layer charge, as do most of the serpentine minerals. The net electrostatic charge, referred to as x is approximately zero for kaolin and serpentine minerals. A group of serpentine minerals have significantly paired substitution, eg, Al^{3+} in both the tetrahedral and octahedral layers, but still maintain electrostatic neutrality.

Clay minerals that are composed of two tetrahedral layers and one octahedral layer are referred to as 2:1 clay minerals or TOT minerals. The apical oxygens of the two tetrahedral sheets project into the octahedral sheet. The 2:1 structure has a basal spacing (nominal thickness) of 1.0 nm (10 Å). Pyrophyllite [12269-78-2], $Al_2Si_4O_{10}(OH)_2$, is the dioctahedral mineral, ie, Al^{3+} in the octahedral sites, and talc [14807-96-6], $Mg_3Si_4O_{10}(OH)_2$, is the trioctahedral, ie, Mg^{2+} in the octahedral sites. Both these minerals are essentially free of substitution in the octahedral site and therefore do not have a net charge deficit $(x = 0)$ in the TOT layer.

The multitude of variation in clay minerals is caused by substitution in the octahedral and tetrahedral layers resulting in charge deficits. The manner in which the charge deficit is balanced leads to many of the useful and unique properties of clay minerals.

The smectite minerals, where the charge deficit is balanced by loosely held hydrated cations in the interlayer space between 2:1 layers, have the smallest charge deficit, $x = 0.2$–0.6. Smectite minerals may be dioctahedral or trioctahedral depending on the predominant cation in the octahedral layer. The amount of water in the interlayer space varies with the cation content and the basal spacing may vary between 1.0–1.8 nm (10–18 Å). Vermiculite has a charge deficit > 0.6. The deficit is balanced by cations with two water layers and has a basal spacing of 1.4 nm (14 Å). Illite has a net negative charge of ≈ 0.9 and the charge deficit is balanced by K^+. The K^+ ion fits nearly perfectly into the basal tetrahedral plane and the resulting structure has a basal spacing of approximately 1.0 nm (10 Å). The basal spacing cannot be easily altered by common chemical treatments designed to identify clay minerals in x-ray analysis. Glauconite [1317-57-3] is an Fe-rich dioctahedral mica similar to illite. The mica minerals, eg, muscovite, biotite, etc, have $x \approx 1.0$ and balance the charge deficit using K^+ in a manner similar to illite. Chlorite minerals also have ca $x = 1$, and the charge deficit is balanced by the inclusion of a Mg or Al hydroxide layer that has a net positive charge of approximately 1.0. The chlorite minerals are composed of 2:1 layers plus an interlayer hydroxide sheet yielding a 1.4 nm (14 Å) basal spacing.

Palygorskite and sepiolite minerals are 2:1 layered phyllosilicates that differ from the above mentioned clays because the octahedral sheets have significant intracrystalline void space caused by discontinuous octahedral layers. The basal tetrahedral unit is connected to an adjacent inverted basal tetrahedral creating a void space or channel. Charge deficits are balanced by hydrated cations in the intracrystalline space.

Mixed-layer clays are a combination of minerals, with the exception of palygorskite and sepiolite. Mixed-layer minerals may contain equal proportions of or may be dominated by one mineral. The arrangement of the component minerals may be random, short-range ordered, or long-range ordered. In the case of illite–smectite many studies have demonstrated a complete range of compositions varying from dominantly smectite through equal amounts of smectite and illite components, to dominantly illite, with increasing temperature and/or depth of the sediment. A more detailed discussion of the structure of clay minerals is available (37–40).

Various techniques are used to study crystalline clay minerals. These include: the polarizing microscope (41), chemical analysis and computation of the mineral formula taking into account the substitution of atoms (37,42,43), staining (44–46), density, possible electrical double refraction (47,48), dehydration, base exchange (49,50), electron micrographs, transmission electron microscopy (tem) (51–53) scanning electron microscopy (sem) (54–57), energy-dispersive x-ray spectral (edxs) microanalysis (58), scanning tunneling microscopy (stm) (59), atomic force microscopy (afm) (59–61), analytical transmission electron microscopy (atem) (62–64), x-ray or electron powder diffraction patterns (xrd)

(36,43,65,66), differential thermal analysis (dta) and thermal balance analysis (tga), imbibition (67–69), high pressure dta (70), infrared absorption (49,51,71–74), magic-angle spin nuclear magnetic resonance (mas/nmr) spectroscopy and solid-state nmr (75–77), electron paramagnetic resonance (78), Mössbauer spectroscopy (78–80), and field appearance, especially responses to weathering (81, and the general references) (see ANALYTICAL METHODS).

Clay minerals are divided into crystalline and paracrystalline groups and a group amorphous to x-rays. Although the clays of the different groups are similar, they show vastly different mineralogical, physical, thermal, and technological properties (3). Chemical analysis alone has limited value in revealing the clay's identity or usefulness. The mineral composition, which reveals the organization of the constituent elements is most important.

Crystalline and Paracrystalline Groups

Kaolins. The kaolin minerals include kaolinite [1318-74-7], dickite [1318-45-2], and nacrite [12279-65-1] which all have composition $Al_2O_3 \cdot 2SiO_2 \cdot 2H_2O$; halloysite (7 Å) [12069-16-1], $Al_2O_3 \cdot 2SiO_2 \cdot 2H_2O$, and halloysite (10 Å) [12244-16-5], $Al_2O_3 \cdot 2SiO_2 \cdot 4H_2O$ (82–85). The structural formulas for kaolinite and halloysite (10 Å), which are shown in Figure 1, are $Al_4Si_4O_{10}(OH)_8$ and $Al_4Si_4O_{10}(OH)_8 \cdot 4H_2O$, respectively. The so-called fire clay mineral is a b-axis disordered kaolinite (12); halloysite (7 Å) and halloysite (10 Å) are disordered along both the a- and b-axes. Indeed, most variations in the kaolin group originate as structural polymorphs, related to variations in layer stacks. Representative analyses (82–84) of the kaolin minerals are given in Table 1.

Kaolinite and dickite are easily distinguished where they occur in recognizable crystals (87). Nacrite is relatively rare. Halloysite (7 Å) is usually exceedingly fine-grained, showing a mean index of refraction of about 1.546. The index of refraction for halloysite (10 Å) varies somewhat with the immersion liquid used; it ranges from 1.540 to 1.552 (83).

X-ray studies show that kaolin minerals have two-layer crystal structures: a sheet of silica tetrahedra and an alumina–gibbsitelike sheet. Adjacent cells are spaced about 0.71 nm across the (001) plane. The interplanar spacings normal to the (001) cleavage are the most significant criteria used in x-ray differentiation between the clay mineral groups. Within the kaolin group other x-ray structural differences are used to distinguish the members (12). The kaolin minerals are dioctahedral minerals and therefore only two of the three octahedral sites are filled. In kaolinite the vacant octahedral site is either always the B site or the C site giving kaolinite a one-layer triclinic structure (88,89). In dickite, the vacancy alternates regularly between the B and C sites giving dickite a two-layer monoclinic structure (88). Nacrite is an uncommon mineral that has a six-layer monoclinic structure with an inclined z-axis.

Halloysite, a mineral in the kaolin family, has a chemical composition similar to, but physical properties that differ greatly from, kaolinite. Halloysite differs from kaolinite in tetrahedral Al content, layer stacking sequence, and configuration of the six-fold rings (90). Earlier work referred to the hydrated form of halloysite as endellite [11244-16-1] (83). This is a naturally occurring 1.0 nm min-

Table 1. Chemical Analyses of the Kaolin Minerals, wt %

Component	Sample					
	1[a]	2[b]	3[c]	4[d]	5[e]	6[f]
SiO_2	45.44	40.26	46.5	45.78	42.68	44.90
Al_2O_3	38.52	37.95	39.5	36.46	38.49	38.35
Fe_2O_3	0.80	0.30		0.28	1.55	0.43
FeO			1.08			
MgO	0.08			0.04	0.08	trace
CaO	0.08	0.22		0.50		trace
K_2O	0.14 ⎱	0.74		⎱ 0.25	0.49	0.28
Na_2O	0.66 ⎰			⎰	0.28	0.14
TiO_2	0.16				2.90	1.80
H_2O loss						
at 105°C	0.60	4.45		2.05		
above 105°C	13.60	15.94	14.0	13.40	14.07	14.20
Total	*100.08*	*99.86*	*100.0*	*99.84*	*100.54*	*100.10*

[a]Kaolinite, Roseland, Va. (84).
[b]Halloysite, Huron Co., Ind. (82).
[c]Theoretical kaolinite.
[d]Washed kaolin, Webster, N.C. (6).
[e]Flint fire clay, near Owensville, Mo. (86)
[f]Typical sedimentary kaolin, S.C. Ga., Ala. Courtesy S. C. Lyons.

eral having loosely bound water responsible for expansion to 1.0 nm along the c-axis. This mineral, referred to as halloysite (10 Å) (91), can also be expanded to 1.0 nm when solvated in ethylene glycol. If halloysite (10 Å) is dehydrated its c-axis dimension is reduced to 0.7 nm. It then becomes halloysite (7 Å) and its c-axis dimension is not able to rehydrate to 1.0 nm.

Four basic morphologies of halloysite are recognized: tubular (long and short), spheroidal, platy, and prismatic. Tubular halloysite is formed in order to compensate for the lateral misfit between the larger tetrahedral and the smaller octahedral sheets (13,90). An alternative interpretation is that certain crystals have an inherent roundish, elongate morphology (92). In platy halloysite the lateral misfit is relieved by the substitution of the larger Fe^{3+} for octahedral Al^{3+}. Spheroidal halloysite is a result of the dissolution–precipitation of amorphous particles, commonly volcanic glass and pumice. There is no consensus concerning the origin of prismatic halloysite (90). The substitution of Al in the tetrahedral layer is a function of the pH of the liquid medium during formation (93). Aluminum, bound by water in octahedral coordination at relatively low pH levels, is readily incorporated into the octahedral sheets of 1:1 minerals and as a result kaolinite is formed. In a pH range from 5.3 to 6 at 25°C, or 4 to 5 at 100°C, aluminum coordination changes from octahedral to tetrahedral. The change in aluminum coordination enables Al^{3+} to substitute for Si^{4+} in the tetrahedral sheet, leading to a charge imbalance in the crystal structure that is satisfied by cations in bound water. The substitution of Al^{3+} for Si^{4+} also leads to a misfit between octahedral and tetrahedral layers. In other kaolin group minerals, any misfit between octahedral and tetrahedral layers is compensated by tetrahedral rotation; however, the cations in the bound water of halloysite prevent tetrahedral rotation, and the misfit is compensated by the rolling of the tetrahedral sheet to the octahedral sheet. Rolling brings tetrahedral and octahedral cations into close proximity and water helps to reduce cation–cation Coulomb repulsion forces (90).

Halloysite may be differentiated from kaolinite and dickite by treatment with potassium acetate, ethylene glycol, formamide, and hydrazine (54,94). Staining kaolin minerals with aniline dyes produces varied artificial pleochroism which may be sufficiently selective to identify particular mineral species (45). In differential thermal analysis, kaolinite shows a strong endothermic peak at about 620°C and a strong exothermic peak at about 980°C, which sharply differentiates it from other clay mineral groups. Electron micrographs show kaolinite in roughly equidimensional pseudohexagonal plates and halloysite most commonly as lath-shaped tubes. Halloysite may also have platy, prismatic, or spherical shapes (90). Kaolinite, the most abundant, and halloysite, the second most abundant, of the kaolin group minerals are the most used by industry. Problems of clay mineral differentiation for the kaolin group rarely arise; however, adequate means for identifying and distinguishing various species are available (34,45,94,95). The cationic base–exchange capacity (CEC) of kaolinite is low, less than 10 meq/100 g of dry clay; the CEC of halloysite has been reported to be as high as 60 meq/100 g (96).

Kaolin most commonly originates by the alteration of feldspar or other aluminum silicates via an intermediate solution phase (97,98) usually by surface weathering (26,99) or by rising warm (hydrothermal) waters. A mica, or hydrated alumina solid may form as an intermediate phase during the alteration from parent material to kaolin minerals.

Large deposits of relatively pure kaolinite have developed from parent, feldspar-rich pegmatites, whereas others are secondarily deposited in sedimentary beds after transportation. Colloidal fractions of geologically ancient soils were presumably concentrated in old swamps and leached to develop kaolinitic clay deposits (6,29). Kaolinite is formed by weathering in an oxidizing environment under acid conditions, and in a reducing environment where bases such as calcium, magnesium, alkalies, and iron(II) are removed. Removal of the bases is essential in kaolin formation (100). With more intense leaching of silica from the clay an aluminous hydrate remains and the clay becomes bauxitic. Kaolinite may develop from the silication of gibbsite (7). Halloysite may be formed either by hydrothermal or weathering processes. Allophane [12172-71-3] may have led to halloysite and thence to kaolinite as crystallization became more highly ordered following weathering (101; however, this sequence does not always prevail. Some indications point toward a possible association of acid sulfate waters and mobile potassium for the origin of halloysite (99). The relationship between pH and aluminum coordination and the formation of either kaolinite or halloysite (93) has been demonstrated. The fact that aluminum may be either tetrahedrally coordinated and a precursor to halloysite, or octahedrally coordinated and a precursor to kaolinite, makes it possible that one may form from the other.

The textures of kaolin (rock) include varieties similar to examples observed in igneous and metamorphic as well as sedimentary rocks (98). Kaolin grains and crystals may be straight or curved, sheaves, flakes, face-to-face or edge-to-edge floccules, interlocking crystals, tubes, scrolls, fibers, or spheres. (98). The kaolin group is transformed at high temperatures to a silica–alumina spinel structure (102), thence to mullite [1302-93-8], $Al_6Si_2O_{13}$, with or without accompanying cristobalite [14464-46-1], SiO_2, (103,104). The alkali metal, flux, and content of kaolin clay strongly influence the phases formed upon heating.

Serpentines. Substituting 3 Mg^{2+} for the 2 Al^{3+} in the kaolin structure results in the serpentine minerals, $Mg_3Si_2O_5(OH)_4$. In serpentines all three possible octahedral cation sites are filled. Most serpentine minerals are tubular to fibrous in structure presumably because of misfit between Mg octahedral and tetrahedral layers. Different varieties of magnesium serpentine are recognized as resulting from the different structures that have evolved to accommodate the misfit between the size of the octahedral and tetrahedral sheets. Lizardite [12161-84-1] results from a 1:1 planar layer structure; chrysotile [12001-27-5] from a cylindrical curvature of the layers; antigorite [66076-98-0] from a curvature of alternating wave modulation; and carlosturanite [98443-44-8] from ordered vacancies and modifications of the tetrahedral sheet within a planar structure (105).

Chrysotile (serpentine) occurs in both clino and ortho structures. Both one-layer ortho and clino, and six-layer ortho (as in nacrite) structures have been observed. Chrysotile transforms at high temperature to forsterite [15118-03-3] and silica. Particularly fibrous varieties are called asbestos (qv).

A number of serpentine group minerals have substitutions in both the tetrahedral and octahedral layer, but they still maintain electrostatic neutrality. Amestite [12413-27-5], which approximates $(Mg_2Al)(SiAl)O_5(OH)_4$ in composition, cronstedite [61104-43-3], $(Fe_2^{2+}, Fe^{3+})(SiFe^{3+})O_5(OH)_4$, chamosite, [12173-07-2], $(Fe^{2+}, Fe^{3+})_{2.3}(Fe^{2+}, Al)_{0.7}(Si_{1.14}Al_{0.86})O_5(OH)_4$, and berthierine [12178-37-9], $(Fe^{2+}, Fe^{3+}, Mg)_{2-3}(Si, Al)_2O_5(OH)_4$, are examples of such minerals (105). Garnierite [12198-10-6], a general term for hydrous nickel silicates, probably falls in this class of serpentine group minerals. Also, a cobalt serpentine has been synthesized (12).

Talc and Pyrophyllite. Talc (qv) and pyrophyllite are 2:1 layer clay minerals having no substitution in either the tetrahedral or octahedral layer. These are electrostatically neutral particles ($x = 0$) and may be considered ideal 2:1 layer hydrous phyllosilicates. The structural formula of talc, the trioctahedral form, is $Mg_3Si_4O_{10}(OH)_2$ and the structural formula of pyrophyllite, the dioctahedral form, is $Al_2Si_4O_{10}(OH)_2$ (106). Ferripyrophyllite has the same structure as pyrophyllite, but has ferric iron instead of aluminum in the octahedral layer. Because these are electrostatically neutral they do not contain interlayer materials. These minerals are important in clay mineralogy because they can be thought of as pure 2:1 layer minerals (106).

Talc and pyrophyllite are found in metamorphic rocks that are rich in Mg and Al, respectively. Talc is most common in metamorphosed ultrabasic rocks and in metamorphosed siliceous dolomite. Pyrophyllite is found in metapelites, including metabauxites and metakaolinites, and in rocks enriched in Al by hydrothermal processes (106).

Smectites (Montmorillonites). Smectites are the 2:1 clay minerals that carry a lattice charge and characteristically expand when solvated with water and alcohols, notably ethylene glycol and glycerol. In earlier literature, the term montmorillonite was used for both the group (now smectite) and the particular member of the group in which Mg is a significant substituent for Al in the octahedral layer. Typical formulas are shown in Table 2. Less common smectites include volkhonskoite [12286-87-2] which contains Cr^{2+}; medmontite [12419-74-8], Cu^{2+}; and pimelite [12420-74-5], Ni^{2+} (12).

Smectites are structurally similar to pyrophyllite [12269-78-2] or talc [14807-96-6], but differ by substitutions mainly in the octahedral layers. Some substitution may occur for Si in the tetrahedral layer, and by F for OH in the structure. Deficit charges in smectite are compensated by cations (usually Na, Ca, K) sorbed between the three-layer (two tetrahedral and one octahedral, hence 2:1) clay mineral sandwiches. These are held relatively loosely, although stoichiometrically, and give rise to the significant cation exchange properties of the smectite. Representative analyses of smectite minerals are given in Table 3. The determination of a complete set of optical constants of the smectite group is usually not possible because the individual crystals are too small. Representative optical measurements may, however, be found in the literature (42,107).

X-ray diffraction patterns yield typical 1.2–1.4 nm basal spacings for smectite partially hydrated in an ordinary laboratory atmosphere. Solvating smectite in ethylene glycol expands the spacing to 1.7 nm, and heating to 550°C collapses it to 1.0 nm. Certain micaceous clay minerals from which part of the metallic interlayer cations of the smectites has been stripped or degraded, and replaced by H_3O^+, expand similarly. Treatment with strong solutions of potassium salts may permit differentiation of these expanding clays (108).

Smectite [12199-37-0] from an oxidized outcrop is stained light blue by a dilute solution of benzidine hydrochloride. The color does not arise from smectite specifically, but from reaction of a high oxidation state of elements such as Fe^{3+} or Mn^{4+} (46).

Transmission electron micrographs show hectorite and nontronite as elongated, lath-shaped units, whereas the other smectite clays appear more nearly equidimensional. A broken surface of smectite clays typically shows a "corn flakes" or "oak leaf" surface texture (54). High temperature minerals formed upon heating smectites vary considerably with the compositions of the clays. Spinels commonly appear at 800–1000°C, and dissolve at higher temperatures. Quartz, especially cristobalite, appears and mullite forms if the content of aluminum is adequate (38).

Table 2. Formulas of Smectite Minerals[a]

Mineral	CAS Registry Number	Formula[b]
montmorillonite	[1318-93-01]	$[Al_{1.67}Mg_{0.33}(Na_{0.33})]Si_4O_{10}(OH)_2$
beidellite	[12172-85-9]	$Al_{2.17}[Al_{0.33}(Na_{0.33})Si_{3.17}]O_{10}(OH)_2$
nontronite	[12174-06-0]	$Fe(III)\,[Al_{0.33}(Na_{0.33})Si_{3.67}]O_{10}(OH)_2$
hectorite	[12173-47-6]	$[Mg_{2.67}Li_{0.33}(Na_{0.33})]Si_4O_{10}(OH,F)_2$
saponite	[1319-41-1]	$Mg_{3.00}[Al_{0.33}(Na_{0.33})Si_{3.67}]O_{10}(OH)_2$
sauconite	[12424-32-7]	$[Zn_{1.48}Mg_{0.14}Al_{0.74}Fe(III)_{0.40}][Al_{0.99}Si_{3.01}]O_{10}(OH)_2X_{0.33}$

[a]Ref. 42.

[b]More substitution takes place than shown; $Na_{0.33}$ or $X_{0.33}$ refers to the exchangeable base (cation) of which 0.33 equivalent is a typical value.

Table 3. Chemical Analyses of the Smectite Minerals, wt %

Component	1[a]	2[b]	3[c]	4[d]	5[e]	6[f]
				Sample		
SiO_2	51.14	47.28	43.54	55.86	42.99	34.46
Al_2O_3	19.76	20.27	2.94	0.13	6.26	16.95
Fe_2O_3	0.83	8.68	28.62	0.03	1.83	6.21
FeO				0.99	2.57	
MnO	trace			none	0.11	
ZnO	0.10					23.10
MgO	3.22	0.70	0.05	25.03	22.96	1.11
CaO	1.62	2.75	2.22	trace	2.03	
K_2O	0.11	trace		0.10	trace	0.49
Na_2O	0.04	0.97		2.68	1.04	
Li_2O				1.05		
TiO_2	none			none		0.24
P_2O_6						
F				5.96		
H_2O removed						
at 150°C	14.81	19.72	14.05	9.90	13.65	6.72
above 150°C	7.99		6.62	2.24	6.85	10.67
Total	*99.75*	*100.37*	*100.02*	*102.98*	*100.29*	*99.95*

[a]Montmorillonite, Montmorillon, France (42).
[b]Beidellite, Beidell, Colo. (42).
[c]Nontronite, Woody, Calif. (42).
[d]Hectorite, Hector, Calif. (42).
[e]Saponite, Ahmeek Mine, Mich. (107).
[f]Sauconite, Friedensville, Pa. (107).

The cation-exchange capacity of smectite minerals is notably high, 80–90 meq or higher per 100 g of air-dried clay, and affords a diagnostic criterion of the group. The crystal structure is obviously weakly bonded. Moreover, the structure of smectite is expandable between the silicate layers so that when the clay is soaked in water it may swell to several times its dry volume (eg, bentonite clays). Soil colloids having high cation-exchange capacity facilitate the transfer of plant nutrients to absorbing plant rootlets.

The minerals of the smectite group have been formed by surface weathering, low temperature hydrothermal processes, alteration of volcanic dust in stratified beds (109), action of circulating water of uncertain source along fractures and in veins, and laboratory synthesis. The optimum weathering environment for smectite genesis is one in which calcium, iron(II), and especially magnesium are present in significantly high concentrations. Potassium should be low or low in relation to magnesium, calcium, and ferrous iron. Organic matter that exerts reducing action is usually concomitant, and a neutral to slightly alkaline medium generally prevails under conditions where the alkali and alkaline-earth metals are not readily removed. The weathering environment for smectite is different from that in which kaolinite is formed (100). If the system permits effective leaching and H^+ ions become available in sufficient quantity to cause the metallic cations to be easily leached away, kaolinite tends to form. The reverse reaction rarely, if ever, takes place.

Bentonite is a rock rich in montmorillonite that has usually resulted from the alteration of volcanic dust (ash) of the intermediate (latitic) siliceous types. In general, relicts of partially unaltered feldspar, quartz, or volcanic glass shards offer evidence of the parent rock. Most adsorbent clays, bleaching clays, and many clay catalysts are smectites, although some are palygorskite [1337-76-4].

Pillared clays are smectite minerals or illite–smectite minerals that have been structurally modified to contain pillars of stable inorganic oxide. The pillars prop open the smectite structure so they have a basal space of approximately 3.0 nm. Typical metals in the pillars include Al, Zr, Ti, Ce, and Fe, and these materials are used in catalytic processes to crack heavy crude oils (110–112).

The original pillared clays were made by: (1) mixing smectite with a polymeric cationic hydroxy metal complex such as aluminum chlorhydrol; (2) allowing a minimal amount of time for the cationic hydroxy metal complex to exchange with the interlayer cations; and (3) calcining the resulting material to decompose the hydroxy metal complex (110). A number of newer methods have been developed to make pillared clays (111–117).

Illite. Illite is a general term for the clay mineral constituents of argillaceous sediments that strongly resemble mica minerals (9,118). Other names that have been used for illite include: bravaisite [12197-39-6], degraded mica, hydromica, hydromuscovite [12173-60-3], hydrous illite, hydrous mica, K-mica, micaceous clay, and sericite [12174-53-7] (37). Illite and the mica minerals have a 2:1 sheet structure similar to the smectite minerals except that the maximum charge deficit in mica is typically in the tetrahedral layers and contains potassium held tenaciously in the interlayer space, which contributes to a 1.0 nm basal spacing. Illite differs from mica minerals in the following ways: (1) one-sixth of the Si^{4+} tetrahedral sites of illite are replaced by Al^{3+} compared to one-fourth in mica minerals; (2) illite has a net charge deficiency of 1.3 per unit cell compared to 2.0 for mica minerals; (3) illite has a higher silica to alumina ratio than mica minerals; (4) in illite, the potassium ions between the unit layers may be partially replaced by other cations, possibly Ca^{2+}, Mg^{2+}, H^+, whereas, in the mica minerals the interlayer ion is almost exclusively potassium; (5) illite crystals contain some randomness in the stacking of the layers in the c direction; and (6) the size of the illite particles occurring naturally is very small, of the order of 1 to 2 μm or less (38). Because the mica minerals and illite in argillaceous sediments may be widely diverse in origin, considerable variations exist in the composition and polymorphism of the illite minerals including trioctahedral illites, sodium illite (brammallite [12197-36-3]), and ammonium illite (118).

The formula of illite can be expressed as $2K_2O \cdot 3MO \cdot 8R_2O_3 \cdot 24SiO_2 \cdot 12H_2O$ (9), and the crystal structure (119) by the formula $K_y[Al_{4-x}(Fe,Mg)_x]$ $(Si_{(8-y)+x}Al_y)O_{20}(OH)_4$ where y refers to the K^+ ions that satisfy the excess charges resulting when about 15% of the Si^{4+} positions are replaced by Al^{3+}. A representative chemical analysis of illite is found in Table 4. Optical constants of illite minerals are difficult to obtain because of the small size of the available crystals. The highest (γ) index of refraction ranges from about 1.588 to 1.610, the

Table 4. Chemical Analysis of Illite, Glauconite, and Attapulgite, wt %

Component	Illite	Glauconite	Attapulgite
SiO_2	51.22	48.66	55.03
Al_2O_3	25.91	8.46	10.24
Fe_2O_3	4.59	18.8	3.53
FeO	1.70	3.98	
MgO	2.84	3.56	10.49
CaO	0.16	0.62	
K_2O	6.09	8.31	0.47
Na_2O	0.17		
TiO_2	0.53		
H_2O removed			
at 110°C	7.49		
at 150°C			9.73
above 150°C			10.13
Total	*100.7*	*99.8*	*99.62*

birefringence is about 0.033, the optical character is negative, and the axial angle, $2V$, is small, on the order of 5°.

A 1.0 nm basal spacing exhibited in a diffractogram peak that is somewhat broad and diffuse and skewed toward wider spacings characterizes the x-ray diffraction pattern of illite. Polymorphs may be present (120). Muscovite derivatives are typically dioctahedral; phlogopite derivatives are trioctahedral.

Differential thermal analysis curves of illite show three endothermic peaks in the ranges 100–150, 500–650, and at about 900°C, and an exothermic peak at about 940°C, or immediately following the highest endothermic peak. Minerals formed from illite at high temperature vary somewhat with the composition of the clay, but usually a spinel-structure mineral followed by mullite at still higher temperatures is observed (42). The cation-exchange capacity of illite is 20–30 meq/ 100 g of dry clay. The interlayer potassium exerts a strong bond between adjacent clay structures. Illite that has lost part of its original potassium by weathering processes may be reconstituted with the sorption and incorporation of transient dissolved potassium (108).

Illite was defined as the most abundant clay mineral in Paleozoic shale and is widespread in many other sedimentary rocks; it is common in soils, slates, certain alteration products of igneous rocks, and recent sediments. Its origin has been attributed to alteration of silicate minerals by weathering and hydrothermal solutions, reconstitution, wetting and drying of soil clays, and diagenesis involving other three-layer minerals and potassium during geologic time and pressure under deep burial (27,121,122). Illitization of smectite via illite–smectite mixed-layer intermediates is a very common and an important reaction in the formation of shales during burial diagenesis.

Glauconite. Glauconite [*1317-57-3*] (123–126) is a green, dioctahedral, micaceous clay rich in ferric iron and potassium. The generally accepted formula for glauconite is $(Na,K)_{0.78}(Fe^{3+}_{1.01}Al_{0.45}Mg_{0.39}Fe^{2+})_{2.05}(Si_{3.65}Al_{0.35})O_{10}(OH)_2$ (39). Glauconite has many characteristics common to illite, but much glauconite contains random mixed expanding layers, and can be referred to as interstratified glauconite–smectite minerals (127). In addition, glauconite found in Late Cenozoic rocks tends to have less crystallographic order than older glauconite; therefore, the modifiers ordered (well crystalline) and disordered (poorly crystalline) are commonly used (39,127). Prior to the 1950s the term glauconite was used for any type of green-colored sedimentary rocks or for green-colored grains, therefore the term glauconite in early literature differs in usage (127).

Glauconite occurs abundantly in sand-size or bigger pellets, or in pellets within fossils, notably foraminifera, giving it an organic connotation (12). Occurrences as replacements, and matrix and flakes in sandstone as a product of diagenesis (121,128), indicate other possible origins. Glauconite is typically formed in a marine environment (129), but glauconitic mica has been reported from nonmarine rocks (81,130).

The chemical analysis of glauconite (Bonneterre, Missouri), is given in Table 4. Powder x-ray diffraction patterns resemble those of illite in which intensities of even-numbered basal spacings are minimal. The glauconitic green sands of New Jersey have been used in ion-exchange, water-softening installations (see ION EXCHANGE), and as a source of slowly released potassium in soil amendments.

Celadonite [71606-04-7] is an iron-rich dioctahedral micaceous mineral that is similar to glauconite. Celadonite has a composition of: $(Na,K)_{0.83}$ $(Fe^{3+}_{0.72}Al_{0.49}Mg_{0.63}Fe^{2+}_{0.20})_{2.05}(Si_{3.81}Al_{0.19})O_{10}(OH)_2$ (39) and, like glauconite, has well crystalline, poorly crystalline, and interstratified varieties (127). Distinctions between glauconite and celadonite may be vague in the area of approximately 0.2% tetrahedral Al. Celadonite is found as an alteration mineral in mafic volcanic rocks (127).

Chlorite and Vermiculite. Chlorite is a 1.4 nm (14 Å) clay mineral that cannot be expanded or collapsed by traditional laboratory procedures. Chlorite occurs commonly in argillaceous sedimentary rocks and in certain soils. Samples of ocean sediments show that chlorite is most common in the high latitudes (131,132), and that chlorite is also common in sedimentary rocks that have undergone low temperature metamorphism (133) and burial diagenesis (134,135). Chlorite is also commonly found rimming quartz grains, filling pores in sandstones, and replacing grains in limestones (136). In early clay mineral studies chlorite was less well explored than the other clay minerals or the chlorite of igneous and metamorphic occurrences (10,137).

Structurally, the unit layer of chlorite is composed of a 2:1 layer combined with a 0.4 nm Mg or Al interlayer or hydroxide sheet. The 2:1 layer structure has a composition of $[(R^{2+},R^{3+})_3(Si_{4-x}R^{3+}_x)O_{10}(OH)_2]^-$ and the octahedral interlayer has a composition of $[(R^{2+},R^{3+})_3(OH)_6]^+$. The tetrahedral cations are typically Si^{4+} or Al^{3+}, the three-layer (TOT) octahedral cations are typically Mg^{2+}, Fe^{3+}, Fe^{2+}, Mn^{2+}, Cr^{3+}, and Ti^{4+} and the interlayer cations are typically Mg^{2+} (brucitelike layer) or Al^{3+} (gibbsitelike layer) (37,38,138–140). Clinochlore [12252-52-7], $Mg_3(Si_3Al)O_{10}(OH)_2 \cdot (Mg_2Al)(OH)_2$, chamosite [12173-01-2], Fe^{2+}_3 $(Si_3Al)O_{10}(OH)_2 \cdot (Fe^{2+}_2Al)(OH)_2$, nimite [71618-47-8], $Ni_3(Si_3Al)O_{10}(OH)_2 \cdot$ $(Ni_2Al)(OH)_2$, baileychlore [14705-13-4], $Zn_3(Si_3Al)O_{10}(OH)_2 \cdot (Zn_2Al)(OH)_2$, donbassite [12415-16-6], $Al_2(Si_3Al)O_{10}(OH)_2 \cdot (Al_2)(OH)_2$, cookite [1302-92-7], Al_2-

$(Si_3Al)O_{10}(OH)_2 \cdot (LiAl_2)(OH)_2$, sudoite [12211-44-8], $Al_2(Si_3Al)O_{10}(OH)_2 \cdot (AlMg_2)$ $(OH)_2$, and franklinfurnaceite [110778-47-7], $Mn_2^{2+}Mn_2^{3+},(Zn_2Si_2)O_{10}(OH)_2 \cdot$ $(Ca_2Mn^{2+}Fe^{3+})(OH)_2$, are also chlorites.

Regularly interstratified (1:1) chlorite and vermiculite has been attributed to the mineral corrensite [12173-14-7] (141). Chlorite mixed layers have been documented with talc, vermiculite, smectite, illite, biotite, kaolinite, serpentine, and muscovite. The mixed-layer mineral is named after the components, eg, talc–chlorite. The earlier literature, however, has reference to specific minerals such as kulkeite [77113-95-2] (talc–chlorite) and tosudite [12424-41-1] (chlorite–dioctahedral smectite) (142).

Aluminum chlorite, $(Al,Fe)_4(Si,Al)_4O_{10}(OH)_8$, in which a gibbsitelike interlayer proxies in part for the brucitelike interlayer, is being discovered in increasing occurrences and abundance (11,141). Chloritelike structures have been synthesized by precipitation of Mg and Al between montmorillonite sheets (143). Cookite [1302-92-7], an aluminous chlorite containing lithium, has been found in high alumina refractory clays and bauxite [1318-16-7] (139).

Vermiculite is an expandable 2:1 mineral like smectite, but vermiculite has a negative charge imbalance of 0.6–0.9 per $O_{10}(OH)_2$ compared to smectite which has ca 0.3–0.6 per $O_{10}(OH)_2$. The charge imbalance of vermiculite is satisfied by incorporating cations in two water layers as part of its crystal structure (144). Vermiculite, which can be either trioctahedral or dioctahedral, often forms from alteration of mica and can be viewed as an intermediate between illite and smectite. Also, vermiculite is an end member in a compositional sequence involving chlorite (37). Vermiculite may be viewed as a mica that has lost part of its K^+, or a chlorite that has lost its interlayer, and must balance its charge with hydrated cations.

The chemical composition of vermiculite can be quite variable (145). The megascopic varieties are generally trioctahedral, and the clay-size varieties contain both dioctahedral and trioctahedral varieties (144). Smectite minerals do not commonly occur as macroscopic single crystals.

Palygorskite and Sepiolite. Palygorskite (attapulgite) and sepiolite are clay minerals in which the 2:1 layers are linked together in chainlike or a combination of chain–sheet structures (12). Because of commercial applications, attapulgite [1337-76-4], named for its occurrence near Attapulgus, Georgia, is a commonly used name for palygorskite. However, it has been determined that the short crystals, called attapulgite, and the longer crystals, palygorskite, have the same crystalline structure (146) and the mineral name palygorskite has precedence (147,148).

Palygorskite and sepiolite are different from other clay minerals in the manner in which the 2:1 layers are joined. Rather than being joined in a continuous manner, the tetrahedral sheets are joined to an adjacent inverted tetrahedral layer, making the octahedral layers noncontinuous and leaving an open channel in the mineral structure (37,38,148). The b-dimension of palygorskite is ≈ 1.8 nm (18 Å); the b-dimension of sepiolite is ≈ 2.7 nm (27 Å) (37).

Palygorskite has an ideal formula that approximates $MgAl_3Si_8O_{20}(OH)_3$ $(OH_2)_4 \cdot x[R^{2+}(H_2O)_4]$; the ideal formula for sepiolite is $Mg_8Si_{12}O_{30}(OH)_4$ $(OH_2)_4 \cdot x[R^{2+}(H_2O)_8]$ (37). The chemical composition of a specific sample may vary widely because there is substitution of Na, Fe, Mn, Al, and Ni in the octahedral sheet of sepiolite, and substitution of Na, Fe, and Mn in palygorskite (37) giving rise to varieties modified by their substitutional component such as: Mn–sepiolite, Mn–palygorskite, Mn–ferrisepiolite, Mn–ferripalygorskite, (149), falcondoite [62996-88-7], Ni–sepiolite with Ni > Mg (150), Na–sepiolite (loughlinite [22830-49-5]), $Na_2Mg_3Si_6O_{16}\cdot 8H_2O$ (151), and Al–sepiolite (152).

These clays have distinctive uses and properties not shown by platy clay minerals. The Georgia–Florida deposits originated from evaporating sea water (153). Palygorskites sorb both cations and neutral molecules. Typical cation-exchange capacities are in the order of 20 meq/100 g dry clay. For chemical analysis of attapulgite see Table 4. Sepiolite is used in drilling muds where resistance to flocculation in briny water is desired (see PETROLEUM). Sepiolite and attapulgite are best identified by their 110 reflections, 1.21 and 1.05 nm, respectively, in x-ray powder diffraction (154,155).

Mixed-Layer Minerals.   In addition to polymorphism resulting from the disordering and proxying of one element for another, clay minerals exhibit ordered and random intercalation sandwiches with one another (12). For example, in mixed-layer clay minerals, sheets of illite may be interspersed with montmorillonite either randomly or regularly (156). Mixed-layer minerals having three components are rare but include illite–chlorite–smectite and illite–smectite–vermiculite (37,157). The accepted names for the identified mixed-layer minerals having components present in fixed percentages are available (158). The concept of how the mineral components are placed in consecutive locations is commonly referred to as Reichweite, or ordering (37,159). A random mixed-layer clay has an ordering of R = 0 (R0) which means that the first mineral (designated A) may be followed by either a second mineral (designated B) or itself, in a random manner. A perfectly ordered mixed-layer that is a 50:50 mixture of two minerals has an ordering of R = 1 (R1) indicating an ABAB order (37). Long-range ABAA order has an order of R = 3 (R3) and is found in illite–smectite and glauconite–smectite mixed-layer minerals (157). The determination of the ordering and the pecentage of the mineral components is commonly obtained by x-ray data (37).

Mixed-layer clays, particularly illite–smectite, are very common minerals and illustrate the transitional nature of the 2:1 layered silicates. The transition from smectite to illite occurs when smectite, in the presence of potassium from another mineral such as potassium feldspar, or from thermal fluids, is heated and/or buried. With increasing temperature smectite plus potassium is converted to illite (37,39).

The physical structure of mixed-layer minerals is open to question. In the traditional view, the MacEwan crystallite is a combination of 1.0 nm (10 Å) nonexpandable units (illite) that forms as an epitaxial growth on 1.7 nm expandable units (smectite) that yield a coherent diffraction pattern (37). This view is challenged by the fundamental particle hypothesis which is based on the existence of fundamental particles of different thickness (160–162).

The fundamental particle hypothesis assumes that random illite–smectite mixed-layer minerals are actually discrete particles of illite and smectite. During burial diagenesis, or as a response to thermal activity, smectite dissolves and produces fine illite particles. Fine illite crystals stacked upon other fine illite crystals expand upon ethylene–glycol treatment, and can be identified by x-ray techniques as smectite and the entire diffraction package is identified as regular (R1) illite–smectite. Subsequently, as illite fuses to form thicker particles there are not as many interparticle surfaces to adsorb ethylene–glycol; therefore, the larger crystals appear to be illite-rich illite–smectite mixed-layer minerals (R3). Therefore, in the view of many clay researchers, random illite–smectite may be a physical mixture of discrete illite and smectite particles, whereas regular illite–smectite may be only illite.

An alternative description of illite–smectite mixed-layer clays begins with megacrystals of smectite that incorporate smaller packets of illite (163). These constituents are observed as mixed-layer minerals in x-ray analysis. Diagenesis increases the percentage of illite layer and with increasing alteration the mixed-layer mineral takes on the characteristics of an illite dominated illite–smectite.

Chlorite is another mineral that is commonly associated with mixed-layered clays. Complete solid solutions of chlorite mixed-layer minerals have not been identified. In contrast to illite–smectite mixed-layer minerals, chlorite mixed-layer minerals occur either as nearly equal proportions of end-member minerals (R1) or dominated by one end member (R0) (142). Mixed-layer chlorite may consist of any of the di–tri combinations of chlorite and chlorite mixed-layering occurs with serpentine, kaolinite, talc, vermiculite, smectite, and mica. References of specific chlorite mixed-layer minerals of varied chemical compositions are available (142,156).

Amorphous and Miscellaneous Groups

Allophane and Imogolite.   Allophane is an amorphous clay that is essentially an amorphous solid solution of silica, alumina, and water (82). In allophane less than one-half of the aluminum is held in tetrahedral coordinations and the SiO_2 to Al_2O_3 ratio typically varies between 1.3 and 2.0, but values as low as 0.83 have been reported. The typical morphology of allophane is cylindrical (37). Allophane may be associated with halloysite, smectite minerals, or it may occur as a homogeneous mixture with evansite, an amorphous solid solution of phosphorus, alumina, and water. Its composition, hydration, and properties vary. Chemical analyses of two allophane samples are given in Table 5.

The index of refraction of allophane ranges from below 1.470 to over 1.510, with a modal value about 1.485. The lack of characteristic lines given by crystals in x-ray diffraction patterns and the gradual loss of water during heating confirm the amorphous character of allophane. Allophane has been found most abundantly in soils and altered volcanic ash (101,164,165). It usually occurs in spherical form but has also been observed in fibers.

Imogolite [12263-43-3] is an uncommon paracrystalline clay mineral assigned the formula $1.1\ SiO_2 \cdot Al_2O_3 \cdot 2.3 - 2.8H_2O$ (166). All the aluminum is held in octahedral coordination and the SiO_2 to Al_2O_3 ratio typically varies between 1.5–1.15. The morphology of imogolite has been reported as thread-shaped and as hollow spheres. Imogolite is generally viewed as an intermediate between allophane and kaolinite. In modern environments both allophane and imogolite are associated with volcanic material in areas of high rainfall.

High Alumina Clay Minerals.   Several hydrated alumina minerals should be grouped with the clay minerals because the two types may occur so intimately

Table 5. Chemical Analyses of Allophane, wt %

Component	1^a	2^b
SiO_2	32.30	4.34
Al_2O_3	30.41	41.41
Fe_2O_3	0.23	0.86
MgO	0.29	0.22
CaO	0.02	0.20
$K_2O + Na_2O$	0.10	0.10
TiO	none	none
CuO	1.60	1.80
ZnO	4.06	4.30
CO_2	0.65	2.07
P_2O_5	0.02	9.23
SO_3	0.21	0.08
H_2O removed		
at 105°C	16.38	20.92
above 105°C	14.43	14.43
Total	*100.70*	*99.96*

^{a}Allophane, Monte Vecchio, Sardinia (82).
^{b}Allophane–evansite, Freienstein, Styria (82).

associated as to be almost inseparable. Diaspore (α-AlO(OH)) and boehmite (γ-AlO(OH)), both $Al_2O_3 \cdot H_2O$ (Al_2O_3, 85%; H_2O, 15%) are the chief constituents of diaspore clay, which may contain over 75% Al_2O_3 on the raw basis (27). Gibbsite, $Al_2O_3 \cdot 3H_2O$ (Al_2O_3, 65.4%; H_2O, 34.6%), and cliachite [12197-64-7], the so-called amorphous alumina hydrate (much cliachite is probably cryptocrystalline), as well as the monohydrates, occur in bauxite (33,35,167), bauxitic kaolin, and bauxitic clays (168,169).

The hydrated alumina minerals usually occur in oolitic structures (small spherical to ellipsoidal bodies the size of BB shot, about 2 mm in diameter) and also in larger and smaller structures. They impart harshness and resist fusion or fuse with difficulty in sodium carbonate, and may be suspected if the raw clay analyzes at more than 40% Al_2O_3. Optical properties are radically different from those of common clay minerals, and x-ray diffraction patterns and differential thermal analysis curves are distinctive.

High alumina minerals are found where intense weatheirng and leaching has dissolved the silica. It is generally believed that a very humid, subtropical climate is required for this (lateritic) stage of weathering.

BIBLIOGRAPHY

"General Survey" under "Clays" in *ECT* 1st ed., Vol. 4 pp. 24–38, by W. D. Keller, University of Missouri; "Survey" under "Clays" in *ECT* 2nd ed., Vol. 5, pp. 541–560, by W. D. Keller, University of Missouri; in *ECT* 3rd ed., Vol. 6, pp. 190–206, by W. D. Keller, University of Missouri.

1. ASTM Committee on Standards *J. Am. Ceram. Soc.* **11,** 347 (1928).
2. F. H. Norton, *Refractories*, 2nd ed., McGraw-Hill Book Co., Inc., New York, 1962.

3. R. E. Grim, *Applied Clay Mineralogy*, McGraw-Hill Book Co., Inc., New York, 1962.
4. H. Wilson, *Ceramics–Clay Technology*, McGraw-Hill Book Co., Inc., New York, 1927.
5. H. Ries, *Clays, Their Occurrences, Properties, and Uses*, John Wiley & Sons, Inc., New York, 1927, p. 1.
6. H. Ries, "Clay," in *Industrial Minerals and Rocks*, American Institute of Mining and Metallurgical Engineers, Washington, D.C., 1937, pp. 207–242.
7. M. Goldman and J. I. Tracey, Jr., *Econ. Geol.* **41,** 567 (1946).
8. T. Sudo, *Mineralogical Study on Clays of Japan*, Maruzen Co., Ltd., Tokyo, 1959.
9. R. E. Grim, R. H. Bray, and W. F. Bradley, *Am. Mineral,* **22,** 813 (1937).
10. M. D. Foster, *U.S. Geol. Surv. Prof. Pap.* **414-A** (1962).
11. J. E. Brydon, J. S. Clark, and V. Osborne, *Can. Mineral.* **6,** 595 (1961).
12. G. Brown, Ed., *The X-ray Identification and Crystal Structures of Clay Minerals*, Mineralogical Society, London, 1961.
13. T. F. Bates, F. A. Hildebrand, and A. Swineford, *Am. Mineral.* **35,** 463 (1959).
14. C. E. Marshall, *The Colloid Chemistry of the Silicate Minerals*, Academic Press, Inc., New York, 1949.
15. I. T. Rosenqvist, "Marine Clays and Quick Clay Slides in South and Central Norway," *Guide to Exc. No. C-13, 21st International Geological Congress, Oslo*, 1960.
16. H. Ries, *Am. J. Sci.* **44**(4), 316 (1917).
17. C. DeKimpe, M. C. Gastuche, and G. W. Brindley, *Am. Mineral.* **46,** 1370 (1961).
18. J. Linarea and F. Huertas, *Science* **171,** 896 (1971).
19. R. Roy, *C.N.R.S. Groupe Fr. Argiles C.R. Reun. Etud.* **105,** 83 (1962).
20. B. Velde, *Clays and Clay Minerals In Natural and Synthetic Systems*, Elsevier, New York, 1977.
21. C. E. Weaver and L. D. Pollard, *The Chemistry of Clay Minerals*, Elsevier, New York, 1973.
22. M. A. Wilson, K. Wada, S. I. Wada, and Y. Kakuto. *Clay Minerals,* **23,** 161–174 (1988).
23. T. Mizutani, Y. Fukushima, and T. Kobayashi, *Clays and Clay Miner.* **39,** 381–387 (1991).
24. H. Tateyama and co-workers, *Clays Clay Miner.* **10,** 180–185 (1992).
25. H. Nakazawa, H. Yamada, and T. Fujita, *Appl. Cly Sci.* **6,** 359–401 (1992).
26. W. D. Keller, *Principles of Chemical Weathering*, Lucas Bros., Columbia, Mo., 1957.
27. W. D. Keller, "Processes of Origin of the Clay Minerals," *Proceedings of the Soil Clay Mineral Institute*, Virginia Polytechnic Institute, Blacksburg, Va., 1962.
28. S. H. Patterson and B. F. Buie, "Field Conference on Kaolin and Fuller's Earth, Nov. 14–16, 1974" *Guidebook 14, Georgia Dept. Natl. Resources*, Atlanta, Ga., 1974.
29. W. D. Keller, J. F. Westcott, and A. O. Bledsoe, *Proceedings of the 2nd Conference of Clays and Clay Minerals, National Acdemy of Science-National Research Council Publication 327*, 1954, pp. 7–46.
30. S. H. Patterson and H. W. Murray, "Clays," *Industrial Minerals and Rocks*, 4th ed., AIME, 1975, pp. 519–585.
31. J. Vachtl, *Proc. XXIII Int. Geol. Cong.* **15,** 13 (1968).
32. M. Kuzvart and M. Bohmer, *Prospecting and Exploration of Mineral Deposits*, Academia Press, Prague, Czechoslovakia, 1978.
33. Gy. Bardossy, *Acta Geol. Acad. Sci. Hung.* **6**(1–2), I (1959).
34. R. C. Mackenzie, ed., *The Differential Thermal Investigation of Clays*, Mineralogical Society, London, 1957.
35. S. H. Patterson, *U.S. Geol. Surv. Bull.* **1228** (1967).
36. C. M. Warshaw and R. Roy, *Bull. Geol. Soc. Am.* **72,** 1455 (1961).
37. D. M. Moore and R. C. Reynolds, Jr., *X-ray Diffraction and the Identification and Analysis of Clay Minerals*, Oxford University Press, Oxford, UK, 1989.
38. R. E. Grim. *Clay Mineralogy*, McGraw-Hill Book Co., Inc., New York, 1968.

39. H. Chamley, *Clay Sedimentology*, Springer-Verlag, Berlin, 1989.
40. S. W. Bailey, ed., *Micas, Rev. Mineral.* **13**, (1984).
41. T. R. P. Gibb, Jr., *Optical Methods of Chemical Analyses*, McGraw-Hill Book Co., Inc., New York, 1942, pp. 243–319.
42. C. S. Ross and S. B. Hendricks, *U.S. Geol. Surv. Prof. Pap.* **205-B**, 23 (1945).
43. D. L. Bish and J. E. Post, eds. *Modern Powder Diffraction, Rev. Mineral.* **20**, (1989).
44. G. T. Faust, *U.S. Bur. Mines Rep. Invest.* **3522**, (1942).
45. E. A. Hauser and M. B. Leggett, *J. Am. Chem. Soc.* **62**, 1811 (1940).
46. J. B. Page, *Soil Sci.* **51**, 133 (1941).
47. C. E. Marshall, *Z. Kristallogr. Mineral.* **90**, 8 (1935).
48. G. D. Brunton, *Clays Clay Miner.* **36**, 94 (1988).
49. P. F. Kerr, ed., "Reference Clay Minerals," *American Petroleum Institute Research Project 49*, 1950.
50. C. S. Piper, *Soil and Plant Analysis*, Interscience Publishers, Inc., New York, 1944.
51. H. Beutelspacher and H. Van der Marel, *Atlas of Electron Microscopy of Clay Minerals and Their Admixtures*, Elsevier, New York, 1968.
52. J. F. Banfield and R. A. Eggleton, *Clays Clay Miner.* **36** (1988).
53. B. Güvan and W.-L. Huang, *Clays Clay Miner.* **39**, 387–399 (1991).
54. R. L. Borst and W. D. Keller, *Proc. Int. Clay Conf. Tokyo* **871** (1969).
55. W. D. Keller and R. F. Hanson, *Clays Clay Miner.* **23**, 201 (1975).
56. W. D. Keller, *Clays Clay Miner.* **25**, 311 (1977).
57. W. D. Keller, *Clays Clay Miner.* **26**, 1–20 (1978).
58. B. Singh and R. J. Gilkes, *Clays Clay Miner.* **39**, 571–579 (1991).
59. H. Lingreen and co-workers, *Am. Mineral.* **76**, 1218–1222 (1991).
60. F. J. Wicks, K. Kjoller, and G. S. Henderson, *Can. Mineral.* **30**, 83–91 (1992).
61. H. Hartman and co-workers, *Clays Clay Miner.* **38**, 337–342 (1990).
62. W. D. Huff, J. A. Whiteman, and C. D. Curtis *Clays Clay Miner.* **36**, 86–93 (1988).
63. K. Tazaki, *Proc. Int. Clay Conf. Italy*, 573–584 (1981).
64. Y.-C. Yua, D. R. Peacor, and S. D. McDowell, *J. Sed.Pet.* **57**, 335–342 (1987).
65. P. J. Heaney, J. E. Post, and H. T. Evans, Jr., *Clays Clay Miner.* **40**, 129–144 (1992).
66. C. de la Calle, H. Suquet, and C.-H. Pons, *Clays Clay Miner.* **36**, 481–490 (1988).
67. J. Konta. *Am. Mineral.* **46**, 289 (1961).
68. R. E. Grim, *Clays Clay Miner.* **36**, 97–101 (1988).
69. M. Kawano and K. Tomita, *Clays Clay Miner.* **39**, 597–608 (1991).
70. A. F. Foster van Gross, *Proceedings of the 9th International Clay Conference, Strasbourg, Sciences Gelogiques Memoir*, Vol. 5, 123–132.
71. J. E. Amonette and Dh. Rai, *Clays Clay Miner.* **38**, 129–136 (1990).
72. C. Blanco and coworkers, *Clays Clay Miner.* **36**, 364–368 (1988).
73. P. F. McMillian and A. M. Hofmeister, *Spectroscopic Methods Mineral. Geol., Rev. Mineral.* **18**, 99–160 (1988).
74. J. K. Crowley and N. Vergo, *Clays Clay Miner.* **36**, 310–316 (1988).
75. R. J. Kirkpatrick in Ref. 73, pp. 341–404.
76. L. Huve and co-workers, *Clays Clay Miner.* **40**, 186–191 (1992).
77. S. P. Altaner, R. J. Kirkpatrick, and C. A. Weiss, Jr., in Ref. 70, pp. 161–170.
78. G. Calas, in Ref. 73, pp. 513–572; T. Mizutani, Y. Fukushima, and T. Kobayashi, *Clays Clay Miner.* **39**, 381–386 (1991).
79. P. R. Lear, P. Komadel, and J. W. Stucki, *Clays Clay Miner.* **36**, 376–378 (1988).
80. D. Tichit and co-workers, *Clays Clay Miner.* **36**, 369–375 (1988).
81. W. D. Keller, *U.S. Geol. Surv. Bull.* **1150** (1962).
82. C. S. Ross and P. F. Kerr, *U.S. Geol. Surv. Prof. Pap.* **185-G**, 135 (1934).
83. L. T. Alexander and co-workers, *Am. Mineral.* **28**, 1 (1943).
84. C. S. Ross and P. F. Kerr, *U.S. Geol. Surv. Prof. Pap.* **165-E**, 151 (1930).

85. G. T. Faust, *Am. Mineral.* **40,** 1110 (1955).

86. M. H. Thornberry, *Mo. Univ. Sch. Mines Metall. Bull.* **8**(2), 34 (1925).

87. C. S. Ross, *Proc. U.S. Nat. Mus.* **69,** 1 (1926).

88. S. W. Bailey, *Miner. Soc. Am. Rev. Mineral.* **19,** 1–8 (1988).

89. R. F. Giese, *Hydrous Phyllosilicates (exclusive of micas), Miner. Soc. Am. Rev. Mineral.* **19,** 29–62 (1988).

90. S. W. Bailey, in Ref. 70, Vol. 2, pp. 89–98.

91. G. W. Brindley and G. Pedro, *AIPEA Newsletter* **12,** 5–6 (1976).

92. F. V. Chukhrov, and B. B. Zvyagin, *Proc. Int. Clay Conf., Jerusalem* **1,** 11 (1966).

93. E. Merino, C. Harvey, and H. H. Murray, *Clays Clay Miner.* **37,** 135–142 (1989).

94. B. K. G. Theng, G. J. Churchman, J. S. Whitton, and G. G. C. Claridge, *Clays Clay Miner.* **32,** 249–258 (1984).

95. R. L. Parfitt and A. D. Wilson, "Estimation of Allophane and Halloysite in Three Sequences of Volcanic Soils, New Zealand," *Volcanic soils, Catena suppl.,* **7** (1985).

96. G. W. Kunze and W. F. Bradley, *Clays Clay Miner.* **12,** 523–527 (1964).

97. L. B. Sand, *Am. Mineral.* **41,** 28 (1956).

98. W. D. Keller, *Clays Clay Miner.* **26,** 1 (1978).

99. W. D. Keller, *Tenth National Conference of Clays and Clay Minerals,* Pergamon Press, New York, 1963, pp. 333–343.

100. W. D. Keller, *Bull. Am. Assoc. Petrol. Geologists* **40,** 2689 (1956).

101. M. Fieldes, *N. Z. J. Sci. Technol.* **37,** 336 (1955).

102. G. W. Brindley and M. Nakahira. *J. Am. Ceram. Soc.* **42,** 311 (1959).

103. M. Slaughter and W. D. Keller, *Am. Ceram. Soc. Bull.* **38,** 703 (1959).

104. F. M. Wahl, R. E. Grim, and R. B. Graf, *Am. Mineral.* **46,** 1064 (1961).

105. F. J. Wicks and D. S. O'Hanley, in Ref. 89, pp. 91–159.

106. B. W. Evans and S. Guggenheim, in Ref. 89, pp. 225–294.

107. C. S. Ross, *Am. Mineral.* **31,** 411 (1946).

108. C. E. Weaver, *Am. Mineral.* **43,** 839 (1958).

109. M. Slaughter and J. W. Earley, *Geol. Soc. Am. Spec. Pap.* **83,** (1965), 95 pp.

110. U.S. Pat. 4,176,190 (Sept. 1979), D. E. W. Vaughan, R. J. Lassier, and J. S. Magee, Jr. (to W. R. Grace and Co.).

111. U.S. Pat. 4,248,739 (June 1981), D. E. W. Vaughan, R. J. Lassier, and J. S. Magee, Jr. (to W. R. Grace and Co.).

112. R. Burch, ed., *Catal. Today* **2,** 188–366 (1988). Two issues dedicated to pillared clays with 14 chapters.

113. J. M. Adams, *Appl. Clay Sci.* **2,** 309–342 (1987).

114. W. Y. Lee, R. H. Raythatha, and B. J. Tatarchuk, *J. Catal.* **115,** 159–179 (1989).

115. G. W. Brindley and R. E. Sempels, *Clay Miner.* **12,** 229 (1977).

116. J. Sterte and J. Shabtai, *Clays Clay Miner.* **35,** 429–439 (1987).

117. J. Sterte, *Clays Clay Miner.* **38,** 609–616 (1990).

118. J. Srodon and D. D. Eberl, in Ref. 40, pp. 495–539.

119. R. E. Grim. *Bull. Am. Assoc. Petrol. Geologists* **31,** 1491 (1947).

120. A. A. Levinson, *Am. Mineral.* **40,** 41 (1955).

121. J. F. Burst, Jr., *Proceedings of the 6th National Conference of Clays and Clay Minerals,* Pergamon Press, Inc., New York, 1959, pp. 327–341.

122. W. D. Keller, "Diagenesis of Clay Minerals—A Review," *Proceedings of the 11th National Conference of Clays and Clay Minerals, 1962,* Pergamon Press, Inc., New York, 1963.

123. J. F. Burst, Jr., *Bull. Am. Assoc. Petrol. Geologists* **42,** 310 (1958).

124. J. F. Burst, Jr., *Am. Mineral.* **43,** 481 (1958).

125. J. Hower, *Am. Mineral.* **46,** 313 (1961).

126. E. G. Wermund, *Bull. Am. Assoc. Petrol. Geologists* **45,** 1667 (1961).

127. I. E. Odom, in Ref. 40, pp. 545–571.
128. P. M. Hurley and co-workers, *Bull. Am. Assoc. Petrol. Geologists* **44,** 1793 (1960).
129. P. E. Cloud, *Bull. Am. Assoc. Petrol. Geologists* **39,** 484 (1955).
130. W. D. Keller, *Fifth National Conference of Clays and Clay Minerals, National Academy of Science-National Research Council Publication 566,* 1958, pp. 120–129.
131. G. M. Griffin and B. S. Parrot, *Am. Assoc. Petrol. Geol. Bull.* **48,** 57–69 (1964).
132. H. L. Windom, "Lithogenous Material in Marine Sediments," *Chemical Oceanography,* Academic Press, Inc., New York, 1976.
133. F. B. Van Houten and M. E. Purucker, *Earth Sci. Rev.* **20,** 211–243 (1984).
134. J. Hower and co-workers, *Geol. Soc. Am. Bull.* **87,** 725–737 (1976).
135. S. Ferry, P. Cotillon, and M. Rio, *C. R. Acad. Sci., Paris,* **297,** 51–56 (1983).
136. J. E. Welton, *SEM Petrology Atlas,* American Association of Petroleum Geologists, Tulsa, Okla., 1984.
137. A. L. Albee, *Am. Mineral.* **47,** 851 (1962).
138. S. W. Bailey and B. E. Brown, *Am. Mineral.* **47,** 819 (1962).
139. W. F. Bradley, *Second National Conference on Clays and Clay Minerals, National Academy of Science-National Research Council Publication 327,* 1954, pp. 324–334.
140. G. W. Brindley and F. H. Gillery, *Am. Mineral.* **41,** 169 (1956).
141. W. F. Bradley and C. E. Weaver, *Am. Mineral.* **41,** 497 (1956).
142. R. C. Reynolds, Jr., in Ref. 89, pp. 601–626.
143. M. Slaughter and I. Milne, eds., *Proceedings of the 7th National Conference of Clays and Clay Minerals,* Pergamon Press, Inc., New York, 1960, pp. 114–124.
144. C. de la Calle and H. Suquet, in Ref. 89, pp. 455–496.
145. W. A. Deer, R. A. Howie, and J. Zussman, *An Introduction to the Rock Forming Minerals,* Longman, Bungay, Suffolk, UK, 1977.
146. C. W. Huggins, M. V. Denny, and H. R. Shell, *U.S. Bureau of Mines, Report, Investigation,* Vol. 6017, 1964.
147. S. W. Bailey, G. W. Brindley, W. D. Johns, R. T. Martin, and M. Ross, *Clays Clay Miner.* **19,** 132–133 (1971).
148. B. T. Jones and E. Galan, in Ref. 89, pp. 631–667.
149. E. I. Semenov, *Inst. Mineral. Geokhim. Krystallokhim. Redk Elementov, Izdat. "Nauka"* (1969).
150. G. Springer, *Can. Mineral.* **14,** 407–409 (1976).
151. J. J. Fahey, M. Ross, and D. J. Axelrod *Am. Mineral.* **45,** 270–281 (1960).
152. L. E. Roger, J. Quirk, and K. Norrish, *J. Soil Sco.* **7,** 177–184 (1956).
153. S. H. Patterson, *U.S. Geol. Surv. Prof. Pap.* **828** (1974).
154. S. Caillere and S. Henin, "The X-ray Identification and Crystal Structures of Clay Minerals." *Mineralogical Society Great Britain Monograph, 325–342,* 1961, Chapt. VIII.
155. W. F. Bradley, *Am. Mineral.* **25,** 405 (1940).
156. F. M. Allen, "Mineral Definition by HRTEM: Problems and Opportunities," *Minerals and Reactions at the Atomic Scale, Minerals Society of America Reviews in Mineralogy,* Nov. 1992, Chap. 8.
157. E. Eslinger and D. Pevear, "Clay Minerals for Petroleum Geologist and Engineers," *SEPM Short Coarse Notes No. 22.* (1988).
158. S. W. Bailey, *Am. Mineralogist,* **67,** 394–398 (1982).
159. H. Jadgozinski, *Acta Crystallogr.* **2,** 201–207 (1949).
160. P. H. Nadeau, M. J. Wilson, W. J. McHardy, and J. M. Tait, *Science* **225,** 923–925 (1984).
161. P. H. Nadeau, J. M. Tait, W. J. McHardy, and M. J. Wilson, *Clay Min.* **19,** 67–76 (1984).
162. P. H. Nadeau, M. J. Wilson, W. J. McHardy, and J. M. Tait, *Mineral. Mag.* **49,** (1985).

163. J. H. Ahn and D. R. Peacor, *Clays Clay Miner.* **34,** 164–179 (1984).

164. K. S. Birrell and M. Fieldes, *N.Z. J. Soil Sci.* **3,** 156 (1952).

165. W. A. White, *Am. Mineral.* **38,** 634 (1953).

166. N. Yoshinga and S. Aomine, *Soil Sci. Plant Nutr. Tokyo* **8,** 22 (1962).

167. I. Valeton, *Bauxites,* Elsevier, New York, 1972.

168. A. F. Frederickson, ed., "Problems of Clay and Laterite Genesis," *AIME Symposium Volume,* American Institute Mining and Mechanical Engineers, New York, 1952.

169. M. Gordon, Jr., J. I. Tracey, Jr., and M. W. Ellis, *U.S. Geol. Surv. Prof. Pap.* **299** (1958).

General References

References 3, 12, 14, 21, 26, 28, 80, 34, 35, 37, 39, 40, 43, 51, 73, 75, 77, 78, 89, 90, 112, 136, 153, 156, 167, and 168 of the numbered bibliography may also be considered general reference works.

T. DOMBROWSKI
Engelhard Corporation

USES

Clays are composed of extremely fine particles of clay minerals which are layer-type aluminum silicates containing structural hydroxyl groups. In some clays, iron or magnesium substitutes for aluminum in the lattice, and alkalies and alkaline earths may be essential constituents in others. Clays may also contain varying amounts of nonclay minerals such as quartz [*14808-60-7*], calcite [*13397-26-7*], feldspar [*68476-25-5*], and pyrite [*1309-36-0*]. Clay particles generally give well-defined x-ray diffraction patterns from which the mineral composition can readily be determined.

Clay particles are so finely divided that clay properties are often controlled by the surface properties of the minerals rather than by bulk chemical composition. Particle size, size distribution, and shape; the nature and amount of both mineral and organic impurities; soluble materials, nature, and amount of exchangeable ions; and degree of crystal perfection are all known to affect the properties of clays profoundly.

Clays are classified into six groups by the U.S. Bureau of Mines (1): kaolin, ball clay, fire clay, bentonite, fuller's earth, and common clay and shale. About half the tonnage of clays produced in the United States is in the last category. In terms of monetary value, however, kaolin accounts for about two-thirds of the dollar volume.

Ceramic Products

A large proportion of the annual production of ball clay, fire clay, and common clay and shale are used for ceramics (qv). Ceramic products are generally considered to be products made from fine-grained oxides, silicates, and many other naturally occurring materials through the application of high temperature. The resulting fired products may be polycrystalline, as are whiteware, bricks, etc, or a vitreous glassy material, eg, glass (qv), or combinations of both types of materials, such as porcelain enamel (see ENAMELS, VITREOUS OR PORCELAIN). Increasingly ceramics are being produced from synthetic materials of high purity because of the unique properties that may be obtained (see ADVANCED CERAMICS). The clays used in conventional ceramics are far from being pure compounds.

In general, ceramic ware is produced by plasticizing the clay by the addition of water so that it may be shaped or formed into the desired object. Ceramic products may also be formed by dispersing the clay in water to form a slip which is then cast in a plaster mold. After being shaped, the object is dried to increase its strength so that it may be handled, and is then fired at elevated temperatures until there has been some vitrification or fusion of the components to form a glassy bond that makes the shape permanent and strong so that the object does not disintegrate in water. In the case of porcelain enamel the slip is sprayed on a metal surface and then fired.

Properties. *Plasticity.* Plasticity may be defined as the property of a material that permits it to be deformed under stress without rupturing and to retain the shape produced after the stress is removed. When water is added to dry clay in successive increments, the clay becomes workable, that is, readily shaped without rupturing. The workability and retention of shape develop within a very narrow moisture range.

Plasticity may be measured by determination of: (1) the water of plasticity defined as the amount of water necessary to develop maximum plasticity, a subjective judgment, or the range of water content in which plasticity is demonstrated; (2) the amount of penetration of an object, frequently a needle or some type of plunger, into a plastic mass of clay under a given load or rate of loading and at varying moisture contents; and (3) the stress necessary to deform the clay and the maximum deformation the clay undergoes before rupture at different moisture contents and with varying rates of stress application.

In ceramics, plasticity is usually evaluated by means of the water of plasticity. Values for the common clay minerals are given in Table 1. Each clay mineral can be expected to show a range of values because particle size, exchangeable ion composition, and crystallinity of the clay mineral also exert an influence. Nonclay mineral components, soluble salts, organic compounds, and texture can also affect the water of plasticity.

In general, a relatively low value for water of plasticity is desired in ceramics and hence kaolinite, illite, and chlorite [14998-27-7] clays have better plasticity characteristics than attapulgite or montmorillonite. The plasticity values of the first group are changed only slightly by variations in the exchangeable cation composition. However, sodium gives lower values than calcium, magnesium, potassium, and hydrogen. In the case of montmorillonite, the water of plasticity varies considerably with the nature of the exchangeable cations, sodium giving higher values than the others.

Clays composed only of clay minerals may have higher water of plasticity values than desired. Consequently, the presence of substantial amounts of nonclay minerals or the addition of materials that reduce the water of plasticity may improve the working characteristics of a clay.

Table 1. Ceramic Properties of Clay Minerals

Property	Mineral				
	Kaolinite	Illite	Halloysite	Montmoril-lonite	Attapulgite
CAS Registry Number	[1318-74-7]	[12173-60-3]	[12244-16-5]	[1318-93-0]	[1337-76-4]
water of plasticity, %[a]	8.9–56.3	17–38.5	33–50	83–250	93
strength, kg/cm^2					
green[b]	0.34–3.2	3.2	5	5[c]	
dry[a]	69–4840	1490–7420	1965	1896–5723	4482
linear shrinkage, %[a]					
drying[d]	3–10	4–11	7–15	12–23	15
firing[e]	2–17	9–15	20	11	23

[a]Ref. 2.
[b]Ref. 3.
[c]Calcium montmorillonite.
[d]Percentage of dry length: 5 h at 105°C.
[e]Allophane [12172-71-3] has a linear shrinkage value for firing of 50%.

Plasticity in clay–water systems is caused by a bonding force between the particles and water which acts as a lubricant and permits some movement between the particles under the application of a deforming force. The bonding force is in part a result of the charges on the particles (see CLAYS, SURVEY).

Green Strength. Green strength is the transverse breaking strength measured while the plasticizing water is still present. As water is continuously added to a dry clay, strength increases to a maximum and then decreases. The strength at water of plasticity is, in general, lower than the maximum strength. Values for the common clay minerals are given in Table 1.

As in the case of plasticity, green strength values would be expected to vary with exchangeable cation composition to only a slight degree for kaolinite, illite, and chlorite, and to a considerable degree for montmorillonite. In the last, sodium would be expected to provide higher maximum green strength than other common cations. Poorly crystallized varieties of kaolinite and illite yield higher green strength than well-crystallized varieties. The presence of large quantities of non-clay minerals reduces the green strength, whereas small amounts may actually increase the strength because these permit the development of a more uniform clay body. Green strength is also related to the particle size such that smaller particles provide higher strength. If the clay mineral particles develop preferred orientation in certain directions during formation of the ware, the breaking strength is somewhat greater in the transverse direction to the preferred orientation.

Drying Properties. Drying Shrinkage. The reduction in length or volume that takes place on drying is termed drying shrinkage. As a rule, drying skrinkage increases as the water of plasticity increases and, for a particular clay mineral, it increases as the particle size decreases. In addition, drying shrinkage varies with the degree of crystallinity. Ball clay, which contains relatively poorly ordered kaolinite, shows values at the high end of the range of typical values shown in

Table 1. The nature of the adsorbed cation causes variations in the amount of drying shrinkage only as it affects the water of plasticity.

The presence of nonclay minerals tends to reduce drying shrinkage depending on mineral shape, particle size distribution, and abundance. Granular particles having a wide distribution of sizes are most effective. The presence of nonclay minerals at about 25% of the ceramic body composition is generally desirable for minimizing shrinkage. Drying shrinkage is also related to texture. For example, if the clay mass shows parallel orientation of the basal plane surfaces of the clay minerals, shrinkage in the direction at right angles to the basal planes is substantially greater than in the direction parallel to them (4).

In the initial drying phase of a clay body the volume shrinkage is about equal to the volume of water evaporated. Beyond a given moisture content there is either no further shrinkage or only a very small amount of water is lost. The water lost during the shrinkage interval is called shrinkage water and is that which separates the component particles. The critical point at which shrinkage stops is reached when the moisture film around the particle becomes so thin that the particles touch one another and shrinkage can go no further. The water loss following the shrinkage period is called pore water.

In the production of ceramic ware the shape of the ware must be retained after drying and the ware must be free from cracks and other defects. Controlled drying helps to minimize defects. In general, clays containing moderate amounts of nonclay minerals are easier to dry than those composed wholly of clay minerals. Furthermore, clays composed of illite, chlorite, and kaolinite are relatively easier to dry than those composed of montmorillonite.

Dry Strength. Dry strength is measured as the transverse breaking strength of a test piece after drying long enough, usually at 105°C, to remove almost all the pore and adsorbed water. Values, given in Table 1, usually show a large range because of variations in particle size distribution, crystallinity, and, especially for montmorillonite, the nature of the exchangeable ions.

Large amounts of nonclay mineral components, especially if the particles are well sorted, tend to reduce the dry strength. In general, the dry strength is higher when sodium is the adsorbed cation. The presence of organic matter in some clays increases dry strength and this appears partly to be the explanation for the high dry strength for some ball clays. A principal factor in determining dry strength is the particle size of the clay mineral component. The maximum strength increases rapidly as the particle size decreases.

Firing Properties. Heating clay materials to a sufficiently high temperature results in fusion of the material. In the 100–150°C range, the shrinkage and pore water are lost with the attendant dimensional changes. In general, the rate of oxidation increases with increasing temperature. The oxidation of sulfides, present in many clays, frequently in the form of pyrite, FeS_2, begins between 400 and 500°C. Beginning at about 500°C and in some cases continuing to 900°C, the hydroxyl groups of the clay minerals condense and are driven off as water vapor. The exact temperature, rate, and abruptness of the loss of hydroxyls depend on the nature of the clay minerals and the particle sizes. Reduction of particle size, particularly if accompanied by poor crystallinity, tends to reduce the temperature interval. Kaolinite and halloysite minerals lose hydroxyls abruptly at 450–600°C. The loss of hydroxyls from montmorillonite minerals varies greatly with structure and composition but is generally slower and more gradual than that for kaolinite and halloysite.

The loss of hydroxyls is usually accompanied by a modification, not a complete destruction of the structure. In the montmorillonite-type clay minerals, hydroxyl loss is not accompanied by shrinkage, whereas in kaolinite and halloysite the loss is accompanied by shrinkage, which continues up into the vitrification range. In the range of 800 to 900°C, the structure of the clay mineral is destroyed and significant firing shrinkage develops. Values for firing shrinkage are also given in Table 1. The range of shrinkage values results from variations in size and shape of the clay mineral particles, the degree of crystallinity, and in the case of the montmorillonite-type of minerals, variations in composition.

At temperatures above about 900°C new crystalline phases develop from all the clay minerals except those containing large amounts of iron, alkalies, or alkaline earths. In these latter cases fusion may result after the loss of structure without any intervening crystalline phase. Frequently there is a series of new high temperature phases developing in an overlapping sequence as the mineral is heated to successively higher temperatures. This is followed by complete fusion of the mineral, which, in the case of kaolinite, takes place at 1650–1775°C. For the montmorillonite-type minerals, the fusion temperature varies from about 1000 to 1550°C, the lower values being found in minerals relatively rich in iron, alkalies, and alkaline earths.

The initial high temperature phases are frequently related to the structure of the original clay mineral, whereas the later phases developing at higher temperatures are related to the overall composition. In the development of high temperature phases, nucleation of the new lattice configuration takes place first, followed by a gradual growth of the new structure and an increase in its perfection as the temperature is raised. Traces of various elements cause substantial changes in the temperature and the rate of formation of the high temperature phases.

Miscellaneous. Other important properties are resistance to thermal shock, attack by slag, and, in the case of refractories (qv), thermal expansion. For whiteware, translucency, acceptance of glazes, etc, may be extremely important. These properties depend on the clay mineral composition, the method of manufacture and impurity content.

Raw Materials. Raw material requirements vary widely, depending on use.

Brick. Almost any clay composition is satisfactory for the manufacture of brick unless the clay contains a large percentage of coarse material that cannot be eliminated or ground to adequate fineness. A high concentration of nonclay material in a silt-size range may cause difficulties by greatly reducing the green and firing strength of the brick. Montmorillonite should be absent, or present only in very small amounts, or the shrinkage may be excessive. Clays composed of mixtures of clay minerals having from 20–50% of unsorted fine-grain nonclay materials are most satisfactory. Large amounts of iron, alkalies, and alkaline earths, either in the clay minerals or as other constituents, cause too much shrinkage and greatly reduce the vitrification range; thus, a clay with a substantial amount of calcareous material is not desirable. Face bricks, which are of superior quality, are made from similar materials but it is even more desirable to avoid these detrimental components (see BUILDING MATERIALS, SURVEY).

Tile. Roofing and structural tiles are usually made from the same material as face brick. Drain tiles have a high porosity, which is frequently obtained by firing at a low temperature. Drain tiles are often made from clays having about 75% of fine-grained nonclay mineral material in addition to components that provide a high green and dry strength and a low fusion point. Wall and floor tiles are frequently made of mixtures where talc and kaolin are the primary components.

Terra-Cotta, Stoneware, Sewer Pipe, and Paving Brick. Clays composed of mixtures of clay minerals containing 25–50% fine-grained unsorted quartz are well suited for the manufacture of terra-cotta, stoneware, sewer pipe, and paving brick. A small amount of montmorillonite can be tolerated, but a large amount gives undesirable shrinkage and drying properties. In general, clays having low shrinkage, good plastic properties, and a long vitrification range should be used.

Whiteware. Porcelain and dinnerware are made up of about equal amounts of kaolin, ball clay, flint (ground quartz), feldspar, or some other white-burning fluxing material such as talc [14807-96-6] and nepheline. The kaolin clay is composed of well-crystallized particles of kaolinite. Ball clays are white-burning, highly plastic, and easily dispersible. They provide the plasticity necessary in the forming of the ware and adequate green and dry strength for handling. The chief component of most ball clays is extremely fine-grained and poorly organized kaolinite. However, some ball clays are known, for example, those in south Devonshire in Great Britain, that contain remarkably well-ordered kaolinite. Some ball clays also contain small amounts of illite and/or small amounts of montmorillonite which may add to desired properties. Many ball clays also contain a small but significant amount of organic material that also appears to enhance the desired properties. Small amounts of bentonites and, in some cases, halloysite, are also used in whiteware bodies as replacements of ball clay to increase dry strength.

Porcelain Enamel. The slurry used in enameling is commonly composed of ball clay, frits, and coloring pigments (qv). The frits are finely ground particles of glass with a low fusion temperature.

Refractories. Refractory products are prepared from a wide variety of naturally occurring materials such as chromite [1308-31-2] and magnesite [546-93-0] or from clays predominantly composed of kaolinite. Increasingly, higher purity synthetic materials are being used to obtain special properties. On the other hand, for many refractory uses, a somewhat lower fusion point than that provided by kaolinite may be adequate, so that clay materials having a moderate amount of other components as, for example, illite, may be satisfactory. High alumina clays are also used extensively for the manufacture of special types of refractories.

An interesting type of clay used widely in the manufacture of refractories is so-called flint clay, which is very hard and has very slight plasticity even when finely ground. Flint clays are essentially pure, extremely fine-grained kaolinite. In some cases the hardness appears to result from the presence of a small amount of free silica acting as a cement, whereas in other cases it is the result of an intergrowth of extremely small kaolinite particles.

Paper

The paper (qv) industry is the largest consumer of processed clays, nearly all of which is kaolin (5,6). Kaolin has two main uses in paper: as a filler where kaolin is mixed with pulp (qv) fibers (see FILLERS); and as a coating where kaolin is combined with water, binders, and various additives and coated onto the surface of the paper sheet (see PAPERMAKING MATERIALS AND ADDITIVES). Its wide-spread use results from the fine particle size, chemical inertness, insolubility over a wide range of pH, white or nearly white color, and low cost. As a filler, kaolin improves opacity of the sheet, imparts smoothness to the surface, and replaces some of the more expensive pulp fibers. In coatings (qv) the kaolin imparts opacity, brightness, a glossy finish, and greatly improved printing quality over that of the uncoated sheet.

Types of Kaolin. Kaolins for use in paper are generally classified into three groups according to processing: air-floated, water-washed, or calcined. Kaolins that are air-floated are processed by selecting appropriate crudes, drying, crush-ing, and pulverizing. In some cases oversize particles are removed by air classi-fication. Some air-floated clays may be slurried in water at 70% solids for conven-ience in handling and shipping. The addition of a deflocculating agent is required in order to obtain a low viscosity.

Water-washed kaolins are of higher added value than air-floated clays be-cause the more elaborate processing that they receive gives more uniform prod-ucts. Wet processing of kaolin consists of mining selected crudes, dispersing in water, degritting to remove oversize particles (screening, sedimentation, cy-clones), centrifugal fractionation into different size fractions, chemical bleaching (oxidative, reductive or both), filtration, redispersion, and drying. Many variations of this scheme are used to obtain kaolin products having different properties. High intensity magnetic separation of weakly magnetic impurities (see SEPARATIONS, MAGNETIC), froth flotation (qv), and selective flocculation are used to remove col-ored impurities and improve product brightness. Delamination, a selective grind-ing process that gives platelike particles of high aspect ratio, may also be used to give products that provide improved opacity and print quality (7). Figure 1 gives a general outline of wet processing of kaolin.

Following wet processing, fine particle size kaolins may be calcined, ie, heat treated at about 1000°C. This treatment converts the kaolin to an amorphous pigment of significantly higher brightness and opacity (8). Properties of the var-ious types of kaolins used in paper are shown in Table 2.

Sources. The largest sources of kaolin for the paper industry are Cornwall in southwestern England and the middle Georgia area in the United States. Smaller, but important, sources of production are located in Australia and Brazil. Many other sources of kaolins for paper are located in various countries but pro-duction is small and frequently the product characteristics are such that use is limited to filler applications. Kaolins from each area have different characteristics that can be traced to the geologic origin of the crude material (9). For example, English kaolins tend to be coarser and slightly whiter than standard brightness United States kaolins, whereas commercially produced Australian and Brazilian kaolins are very fine in particle size.

Properties. The properties of kaolin that make it useful in the paper in-dustry are brightness, viscosity, and particle size and shape.

Paper is usually white, thus it is important that materials used in making it are also white. Brightness, the percent reflectance of blue light, is the commonly used measure of kaolin whiteness. Higher brightness is usually more desirable, thus brighter kaolins command higher prices. Water-washed kaolins are gener-

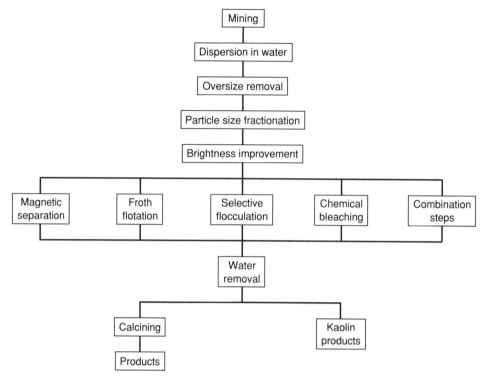

Fig. 1. Schematic for the wet processing of kaolin.

ally higher in brightness and of more uniform color than the air-floated products. Calcined products are of still higher brightness.

For paper coating applications it is desirable that kaolins give fluid aqueous slurries at high (typically 70%) solids, because coating is carried out at high solids to minimize costly drying on the paper machine. A considerable amount of kaolin is shipped as a slurry for convenience in handling so that higher solids gives lower shipping costs. Except when shipped as a slurry, low viscosity at high solids is less important in filler applications because the clays are mixed with pulp fibers in a dilute suspension.

Particle size or particle size distribution is important because properties of the kaolin pigments are very dependent on the size of the individual particles. In general, in paper coatings, finer particle size kaolins give higher brightness, better opacity, and higher gloss to the finished sheet. All coating kaolins must be virtually free of oversize particles, generally defined as 44 μm (325 mesh) sieve residue. These larger pieces cause scratches on the paper in high speed coating operations and give visible imperfections in the sheet surface.

The finer particle fractions of kaolin are typically small platelets having a roughly hexagonal shape. The platey nature of the particles is one of the characteristics of kaolin that makes it so valuable as a paper coating pigment because in the coating operation the particles orient themselves parallel to the fiber substrate to give the desired smooth surface. The coarser particles, at least from the middle Georgia deposits, consist of stacks of loosely bound platelets, commonly

Table 2. Kaolin Grades for the Paper Industry

Kaolin type[a]	Brightness, %	Maximum solids, %	Median particle size, μm	Uses	Relative cost[b]
		Air-floated			
	81–85	c	0.30–0.35	filling	L
		Water-washed			
standard brightness					
coarse particle (no. 3)	82–85	c	1.0–2.0	filling	L–M
intermediate (no. 2)	86	70	0.40–0.55	filling, coating	M
fine (no. 1)	87	70	0.30–0.40	coating	M
ultrafine (ultrafine no. 1)	87	70	0.25–0.30	high gloss coating	M
high brightness					
intermediate	90	70	0.40–0.55	filling, coating	M–H
fine	90	70	0.30–0.40	coating	M–H
ultrafine	90	70	0.25–0.30	high gloss coating	M–H
		Calcined			
	93	50c	0.70	filling, coating	Ht

[a]Usual paper-grade designation is given in parentheses.
[b]L = low; M = moderate; H = high; Ht = highest.
[c]Filler applications do not require low viscosity at high solids.

described as booklets (10). These can be selectively ground to give delaminated kaolin products of quite different characteristics than those of the naturally occurring material (11). The larger diameter but thin platelets that are obtained by delamination are especially valuable in obtaining good coated paper quality even at very light coating weights. Delamination also improves brightness because the resulting particles are better light scatterers and do a better job at obscuring discolored impurities. When used as filler pigments, the large diameter platey delaminated particles are retained well in the sheet and improve its optical properties.

Filling. Changes in paper sheet brightness and opacity resulting from addition of various levels of some kaolin clay products are given in Figure 2. It is evident that the improvements in both brightness and opacity are greatest for the fine-particle calcined kaolin. All of the uncalcined kaolins give a lesser brightness and opacity improvement.

Coating. Table 3 shows how coated paper properties change with particle size. As a general rule, finer particle kaolins give improved gloss, opacity, and brightness. For extremely fine kaolins, however, opacity and brightness may decrease as a result of the loss of light scattering power of the very fine fractions and an increase in colored impurities.

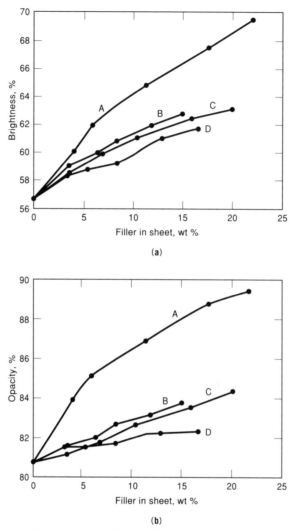

Fig. 2. (a) Brightness improvement obtained by the use of kaolin as a filler in paper and (b), opacity improvement obtained as pulp fibers are replaced with various kaolins. In both cases, A represents fine-particle calcined clay; B, high brightness No. 1 kaolin; C, coarse-particle water-washed kaolin; and D, air-floated kaolin.

Table 3. Effect of Particle Size on Coated Sheet Properties

Particle size		Coated sheet properties		
% < 2 μm	Median, μm	Gloss	Brightness, %	Opacity, %
35	3.80	25	71.6	84.2
54	1.80	31	72.2	84.6
78	0.75	33	72.6	84.6
85	0.68	39	72.7	84.7
96	0.46	45	72.6	84.7

Testing. The Technical Association of the Pulp and Paper Industry (TAPPI) gives test methods that are widely used by kaolin suppliers to the paper industry. These include tests for viscosity, viscosity stability, brightness, pH, particle size distribution, moisture content, and screen residue (12). Other tests on kaolins that relate to use properties are carried out by kaolin suppliers.

Molding Sands

Molding sands, composed essentially of sand and clay, are used extensively in the metallurgical industry for the shaping of metal by the casting process (see METALLURGY). Using a pattern, a cavity of the desired shape is formed in the sand into which molten metal is poured and then allowed to cool.

The molding sand may be a natural sand containing clay or a synthetically prepared mixture of clean quartz sand and clay. Synthetic sands are widely used because they can be prepared to meet property specifications and properties are more easily controlled when used. A small amount of water must be added to the molding sand to impart plasticity to develop cohesive strength so that the sand can be molded around the pattern, and to give the sand sufficient strength to maintain the cavity after the pattern is removed and while the metal is poured into it. These properties vary greatly with the amount of tempering water as well as the nature of the clay.

In foundry practice the same sand is used repeatedly. Because the high temperature of the metal dehydrates and vitrifies some of the clay, fresh clay must be added continuously as the sand is used. The only completely adequate test for the satisfactory use of a clay in bonding molding sands is the result obtained by actual use in foundry practice.

Raw Materials. The bentonites, composed essentially of montmorillonite and used extensively in bonding molding sands, are of two types. The type carrying sodium as a principal exchangeable cation is produced largely in Wyoming. The calcium carrying type is produced in Mississippi and in many countries outside the United States, such as England, Germany, Switzerland, Italy, the former USSR, South Africa, India, and Japan. Natural calcium montmorillonites are occasionally treated with various sodium compounds so that the properties are similar to the naturally occurring sodium bentonites of Wyoming.

Plastic clays composed largely of poorly crystallized kaolinite but having small amounts of illite, and at times montmorillonite, are widely used in bonding molding sands, especially in the United States.

A third type of clay used in foundries is composed essentially of illite. Most illite clays have a bonding strength and plasticity too low for bonding use, but there are some varieties that have properties approaching those of montmorillonite. The illite in such clays is fine-grained, poorly organized, and frequently associated with mixed layer assemblages containing montmorillonite. Illite bonding clays are produced extensively in Illinois.

Properties. The value of a clay for bonding molding sand is usually determined by the green and dry compression strengths of mixtures with varying amounts of the clay and to which varying amounts of tempering water have been added. Other properties such as the bulk density, flowability, permeability, and hot strength may be important. Standard procedures for determining the properties of bonding clays have been published (13).

Green compression strengths in the range from about 35 to 75 kPa (5–11 psi) are desired in actual practice. Green compression strength is highly dependent on the amount of clay present as well as its type. Typically calcium montmorillonite gives the highest green compression strength, whereas kaolinite and illite have about the same strength but considerably lower values. Sodium montmorillonite and halloysite give a value intermediate between these two extremes (14).

Plastics and Rubber

Clays used in plastics and rubber have historically been divided into two categories: fillers that extend a polymer or fillers that reinforce a polymer. Extenders are generally classified as such because they are low in cost compared to the polymer into which they are incorporated. A preferred term for clays used as extenders might be functional fillers because addition to a polymer almost always alters its properties to some extent (15).

Kaolin is by far the primary clay mineral filler used in polymers and various grades are produced by several manufacturers specifically for use in this industry. Table 4 lists the most important kaolin grades used as components in plastics and rubber. More than half the tonnage in polymer applications is accounted for by air-floated kaolin, primarily because of its extensive use in rubber. Attapulgite is used for some applications. The use of attapulgite as a replacement for asbestos (qv) in phenolic-based brake linings is increasing (see BRAKE LININGS AND CLUTCH FACINGS).

Table 4. Kaolin Grades for Polymer Applications

Kaolin type	Brightness, %	Median particle size, μm	Uses	Relative cost[a]
		Air-floated		
regular	73–76	0.2–1.0	rubber	L
surface treated	73–76	0.2–1.0	rubber	M
		Water-washed		
coarse particle	79–82	4.8	thermosets	M
intermediate	85–87	0.6	PVC	M
fine	86–88	0.4	PVC	M
		Calcined		
meta kaolin	84–86	1.4	PVC insulation	H
high temperature	90–92	1.4	engineering plastics, PVC	H
surface treated	90–92	1.4	engineering plastics, PVC	Ht

[a]L = low; M = moderate; H = high; Ht = highest.

Types of Kaolin. As can be seen from Table 4, kaolins for use in polymers may be subdivided into the categories of air-floated, water-washed, and calcined. The processing of each of these types is similar to that for use in the paper industry.

The calcined or meta kaolin is used almost exclusively in poly(vinyl chloride) wire insulation because it enhances the electrical resistivity of the compound.

An important group of clays used in polymers are those labeled "surface treated" in Table 4. These products are made from a fully calcined kaolin, an air-floated kaolin, or a water-washed kaolin. Several types of chemical compounds are used to convert the somewhat hydrophilic surface into a more oleophilic surface that is more compatible with polymers (15). Of special utility are calcined kaolins that have been treated with silanes. Although silane treatment raises the cost of kaolin-based fillers significantly, the improvement in polymer properties justifies the extra cost. The data of Table 5, for example, show the effect on physical properties of adding 40% of an aminosilane-treated high temperature calcined kaolin to a nylon compound. The improvement in strength properties and the marked increase in heat deflection temperature are noteworthy. Custom processors can treat clay minerals using a wide variety of materials to alter surface properties.

Sources. Kaolin for use in polymers is obtained from the same sources as those for paper clays. Attapulgite is obtained from producers in southern Georgia and northwest Florida.

Properties. Properties of clays that make them useful in the plastics and rubber industries are color, particle size and shape, and viscosity. Except for applications in black compounds, clays used in the polymers industry are required to be white or nearly so. Because the refractive index of clays quite closely matches that of most polymer compounds, clays contribute little to opacity and color. In many cases the color of the compound is detrimentally affected by clays because high refractive index iron and titanium impurities become more visible. Frequently a titanium dioxide [13463-67-7] pigment is used along with the clay because the high refractive index gives good opacity and obscures any discoloring materials present in the clay.

Particle size and shape are important because large particles may give a rougher than desired surface whereas small particles are more effective in

Table 5. Effect of Clay Filler on Nylon Properties[a]

Property	Unfilled	Filled
tensile strength, MPa[b]	82	97
tensile elongation, %	60	8
tensile modulus, MPa[b]	2900	6200
flexural strength, MPa[b]	97	159
flexural modulus, MPa[b]	2900	6200
izod impact, J/m	0.020	0.017
deflection temp. at 1.82 MPa, °C	77	204

[a]Data for 40% loading of aminosilane-treated calcined kaolin in nylon-6,6.
[b]To convert MPa to psi, multiply by 145.

increasing the polymer hardness. Particles having a high aspect ratio, ie, "platey" particles, can give greater reinforcement than more isometric particles.

The incorporation of a clay filler into a polymer may strongly affect the viscosity of the compound, especially when finer clays are used. In general, finer particle clays increase viscosity more than coarser particle ones. Depending on the polymer and its application, increasing the viscosity may be desirable or undesirable. For example, a polymer that is too fluid to handle by some processing equipment may be rendered sufficiently viscous to be utilized by the incorporation of a clay filler. Surface treated clays, because of the more oleophilic surface, give lower viscosity compounds than do untreated counterparts.

Testing.   Various test methods are provided by ASTM (16). These include pigment tests of importance such as chemical analysis, presence of oversize particles, oil absorption, particle size distribution, degree of dispersion, presence of soluble components, etc. Numerous tests are also given by ASTM for the properties of filled and unfilled polymers. These include, for example, such properties as impact resistance, stiffness, viscosity, tear resistance, hardness, color, and electrical resistivity.

Drilling Fluids

In oil well drilling a fluid is pumped down the well through a hollow drill string and through nozzles in the bit in the bottom. The fluid returns to the surface through the annulus between the drill string and the wall of the bore. This fluid is usually water-based although oil-based fluids are sometimes used. The use of a drilling fluid has several purposes: it cools and lubricates the drill string, it suspends the cuttings removed by the bit, its density helps to suspend the drill string, and it keeps out formation fluids. Once the drilling fluid reaches the surface, cuttings are removed and the fluid is recycled down the bore (see PETROLEUM, DRILLING FLUIDS AND OTHER OIL RECOVERY CHEMICALS).

Clays are an important ingredient in most water-based drilling fluids. Specifically montmorillonite (in the form of bentonite), attapulgite, and sepiolite [15501-74-3] are used. The primary function is to control viscosity and the suspension properties of the fluid. In a drilling fluid, non-Newtonian viscosity is desired. At the high shear rates encountered at the drill bit, the fluid should be of low viscosity yet under lower shear rates it should be sufficiently viscous to suspend the cuttings removed. Under static conditions it should give sufficient gel strength to prevent settling of the cuttings. Clays are not usually used as the sole viscosity control agent and typically organic water-soluble polymers (qv) such as starch [9005-225-8] (qv), carboxymethylcellulose (CMC) [9004-32-4], natural gums, or synthetic organic polymers are used in conjunction with them. These polymers also aid the clay component of the drilling mud to serve as a filtration aid, ie, prevent the drilling fluid in the bore hole from being lost into the formation.

In addition to clays, weighting agents are added to increase the drilling fluid density so as to prevent blowouts from high pressure in the formation. Dispersants (qv) are added to control viscosity. Weighting agents are usually finely ground natural minerals of high density such as barite [13462-86-7] or hematite [1317-60-8]. Dispersants used are those commonly used for deflocculating clay such as tetrasodium pyrophosphate [7722-88-5], sodium hexamethaphosphate [10124-56-8], and lignin sulfonates [8061-51-6].

Bentonite is the principal clay used in drilling fluids and most domestic production is in the Wyoming–Montana–South Dakota area. This bentonite is largely in the sodium form and gives the desired high viscosity at low concentrations. Attapulgite from northern Florida and southern Georgia is preferred in those cases where saltwater is used in the drilling fluid or encountered in the borehole. Attapulgite, although more costly than bentonite, maintains its high viscosity better than bentonite in the presence of flocculating salts (17). Sepiolite is used where the highest temperature stability is needed. Bentonite that has been treated to render it oleophilic is used in oil-based muds. Test methods for clay properties of value in drilling fluids are provided by the American Petroleum Institute (18).

Paint

Clays are widely used in both oil-based and water-based paints (see PAINT). In this application, clays perform several important functions: they extend the much higher cost titanium dioxide opacifying pigment, control viscosity so as to prevent pigments from settling during storage, provide thixotropy so that the paint is easily applied yet does not sag after application, improve gloss retention, promote film integrity, and aid in tint retention.

Kaolins for use in paints are similar to those used in both plastics and paper but are processed somewhat differently so as to enhance the properties of the paint. Both calcined and uncalcined kaolins are used. Calcined kaolins are of higher cost than uncalcined kaolin, but considerably less costly than titanium dioxide. The calcined kaolins contribute significantly to opacity so that they can be used to replace part of the TiO_2 with no loss of hiding power. Although kaolins are easily wetted by both oil and water, surface treatment may sometimes be used to enhance the oleophilic properties. The kaolin in paint may represent 20–30% of the pigment.

Bentonite and attapulgite are also used but as a viscosity control and suspending agent rather than as an extender pigment because their very fine particle size contributes little to opacity. Bentonites and attapulgites for paint use must be processed to remove oversize particles. Organic treated bentonites are widely used as suspending and antisag agents in oil-based paints. Whiter grades are preferable so as to contribute as little as possible to the color of the finished paint. Procedures for paint raw materials and finished paint are available (16).

Miscellaneous

Clays are used in a vast number of products. In a few cases, clays are used as a chemical raw material as, for example, for synthetic zeolite production or for production of aluminum.

Adsorbents. Acid activated clays have been widely used to treat mineral, vegetable, and animal oils (see VEGETABLE OILS). The primary objective of such treatment is decolorization and, at least in the case of edible oil, to remove components that contribute to off-tastes. Typically the oil is filtered through a granular clay product or treated with finely ground clay and subsequently filtered.

A wide range of clay materials have been used for decolorizing. These may be substantially crude clay such as fuller's earth, which largely contains montmorillonite as the active clay ingredient, or specially treated attapulgites, montmorillonites, and kaolinites. Proprietary acid activation processes are frequently used for production of clay-derived materials of superior performance.

To be of value in an oil decolorizing operation, a relatively small amount of clay must reduce color substantially. Oil retention must be low, ie, only a small amount of oil is retained in the clay during the decolorization process. Further, the clay must be readily removed by filtration.

Because of the increasing popularity of cats as pets, a significant quantity of clay is being used in pet litter. Generally this material is attapulgite or montmorillonite based because these minerals have a high absorptive capacity for liquids and are readily granulated. Halloysite has also been used (19). Other applications for absorptive clays include cleaning up chemical spills and oil and grease removal from garage floors.

Adhesives. Clays, especially kaolin and attapulgite, are widely used in various adhesive formulations. Adhesives (qv) containing clays can be derived from natural products such as starch or protein, or be wholly synthetic, eg, latex, hot melt, emulsion, etc.

In addition to serving as an extender for the adhesive, the use of clay can also improve the properties of the system. The addition of clay to the adhesive can increase viscosity by reducing dripping and sagging, improve smoothness of the surface, and slow the penetration of the adhesive into the substrate thus reducing cost by lowering the adhesive requirement. Both air-floated and water-washed kaolins are used. The former is less costly, the latter is of better color and more uniform. Attapulgites are typically used to control viscosity and provide a thixotropic system for ease in handling.

Catalysts. Historically, crude clays have been used to some extent in petroleum refining (20). More recently, however, processed clays are increasingly used as raw materials and converted to more reactive catalyst products. Various proprietary processes are used and numerous patents have been issued.

Frequently kaolin clay is used for a raw material. In one of the older processes kaolin is treated with sulfuric acid at elevated temperature and subsequently washed to remove liberated alumina and alkalies. Kaolin can also serve as a raw material for preparing a number of synthetic zeolites (21).

Cement. Portland cement, a mixture of calcium silicate and calcium aluminate minerals, is produced by the calcination of argillaceous limestone or mixtures of limestone and clay (see CEMENT). Although other clays can be used, kaolin is preferred because of its alumina and silica content and low level of impurities. It is especially desirable in the manufacture of white cement and other types requiring careful control of chemical composition. Air-floated kaolin, because of its low cost, is usually used.

Chemical Raw Material. In addition to use as a catalyst raw material, clays are used or have been extensively studied as chemical raw material. For example, kaolin has been investigated as a raw material for aluminum metal production.

Kaolin has a 38 to 40% alumina content and is available in the United States in large quantities whereas the higher alumina bauxite reserves are very limited. The Bureau of Mines has actively carried out research in the aluminum from kaolin area for many years. Activity increases whenever imports of bauxite are threatened by war or other trade interruptions (1,22,23).

Kaolin, usually air-floated, is an essential ingredient in continuous filament fiber glass and significant quantities are used in this application (24).

Inks. Refined kaolin is a common ingredient in a large variety of printing inks (qv). In addition to extending the more expensive polymers present, kaolin also contributes to improved color strength, limits the penetration of the ink into the paper, controls rheology, and improves adhesion. Kaolin for this application must usually be as white as possible and free from oversize particles. Surface treated clays are used to improve compatibility with oil-based ink. Clays can also be an ingredient in the newer water-based or uv-cured inks.

Pelletizing. In many industries it is common practice to agglomerate fine-particle materials such as iron ores or fertilizers into larger, more easily handled aggregates. Bentonites of high dry strength are best suited for pelletizing because they provide a strong bond. Attapulgite can also be used.

Pesticides. Many pesticides are highly concentrated and are in a physical form requiring further treatment to permit effective application. Typically carriers or diluents are used (see INSECT CONTROL TECHNOLOGY). Although these materials are usually considered inert, they have a vital bearing on the potency and efficiency of the dust or spray because the carrier may consist of up to 99% of the final formulation. The physical properties of the carrier or diluent are of great importance in the uniform dispersion, the retention of pesticide by the plant, and in the preservation of the toxicity of the pesticide. The carrier must not, for example, serve as a catalyst for any reaction of the pesticide that would alter its potency.

Clays composed of attapulgite, montmorillonite, and kaolinite are used for pesticides in finely pulverized or granular form. Granular formulations are reportedly less expensive, more easily handled, reduce loss caused by wind drift, and produce a more effective coverage.

Other Clay Uses. Other applications for clays include use as a suspending agent, eg, montmorillonite and attapulgite in liquid fertilizers and dishwasher detergents; in pharmaceuticals (qv), eg, kaolinite and attapulgite for diarrhea control; in cosmetics, montmorillonite and attapulgite; and in water impedence where bentonite linings are used for reservoirs and waste disposal areas.

Economic Aspects

Clays vary in price from only a few dollars per ton for common clay to >$0.25/kg for some of the specialty surface treated clays. For clays that are used in large quantities such as kaolins for paper coatings, transportation to the point of use may be the primary cost component.

In the United States, kaolin is the principal clay product and about 9 million metric tons were reported mined and processed during 1991 (25). Bentonite production was reported as being 2 million tons for the same year. World production data for clays, often as a function of geographical location, use, or specific producer, are available (26).

BIBLIOGRAPHY

"Clays" in *ECT* 1st ed., Vol. 4: "Ceramic Clays," pp. 38–49, by W. W. Kriegel, North Carolina State College; "Fuller's Earth," pp. 49–53, by W. A. Johnston, Attapulgus Clay Co.; "Activated Clays," pp. 53–57, by G. A. Mickelson and R. B. Secor, Filtrol Corp.; "Papermaking, Paint, and Filler Clays," pp. 57–71, by S. C. Lyons, Georgia Kaolin Co.; "Rubbermaking Clays," pp. 71–80, by C. A. Carlton, J. M. Huber Corp.; and "Clays (Uses)" in *ECT* 2nd ed., Vol. 5, pp. 560–586, by R. E. Grim, University of Illinois; in *ECT* 3rd ed., Vol. 6, pp. 207–223, by R. E. Grim, University of Illinois.

1. *Potential Sources of Aluminum,* U.S. Bureau of Mines, Washington, D.C., IC8335, 1967.
2. W. A. White, *The Properties of Clays,* Ph.D. dissertation, University of Illinois, Urbana, Ill., 1947.
3. U. Hofmann, *Rapport Europees Cong. Electronenmicroscopic,* Ghent, 1954, pp. 161–172.
4. W. O. Williamson, *Trans. Brit. Ceram. Soc.* **40,** 225 (1941).
5. *Mineral Industries Surveys, Clays in 1990,* U.S. Dept. of the Interior, Bureau of Mines, Washington, D.C., 1991.
6. *Mineral Facts and Problems, Clays,* U.S. Dept. of the Interior, Bureau of Mines, Washington, D.C., 1985.
7. U.S. Pat. 3,171,718 (Mar. 2, 1965), F. A. Gunn and H. H. Morris (to Freeport Kaolin Co.).
8. P. Sennett, in V. C. Farmer and Y. Tardy, eds., *Proceedings of the 9th International Clay Conference, Strasbourg, Sci. Geol., Mem.* **89,** 71–79 (1990).
9. H. H. Murray, *Rev. Mineral.* **19,** 67 (1988).
10. H. H. Morris, P. Sennett, and R. J. Drexel, *Tappi* **48**(12), 92A (1965).
11. P. Sennett, R. J. Drexel, and H. H. Morris, *Tappi* **50,** 560 (1967).
12. *Tappi Test Methods,* Technical Association of the Pulp and Paper Industry, Atlanta, Ga., 1991.
13. *Testing and Grading of Foundry Clays,* 6th ed., American Foundry Society, Chicago, Ill., 1952.
14. R. E. Grim and F. L. Cuthbert, *Ill. State Geol. Surv. Rep. Invest.,* 102 (1945).
15. D. G. Sekutowski, in J. D. Edenbaum, ed., *Plastics Additives and Modifiers Handbook,* Van Nostrand Reinhold, New York, 1992.
16. *ASTM Handbook,* Philadelphia, Pa., 1992.
17. W. L. Haden and I. A. Schwint, *Ind. Eng. Chem.* **59,** 58 (1967).
18. *API Specifications on Oil-Well Drilling Fluid Materials,* API Spec. 13A (and supplements), 7th ed., American Petroleum Institute, Dallas, Tex., 1979.
19. *Ind. Miner.,* 49 (June 1991).
20. L. B. Ryland, M. W. Tanele, and J. N. Wilson, in P. H. Emmett, ed., *Catalysis,* Vol. VII, Reinhold Publishing Co., New York, 1960, p. 1.
21. D. W. Breck, *Zeolite Molecular Sieves; Structure, Chemistry and Use,* John Wiley & Sons, Inc., New York, 1974.
22. F. A. Peters, P. W. Johnson, and R. C. Kirby, *U.S. Bur. Mines Rept. Invest.* No. 6229 (1963).
23. F. A. Peters, P. W. Johnson, and R. C. Kirby, *U.S. Bur. Mines Rept. Invest.* No. 6133 (1963).
24. *Ind. Miner.,* 38 (Nov. 1991).

25. *Mining Eng.* **44**(6), 555 (1992).
26. *Ind. Miner.,* a monthly publication of Industrial Minerals Division, Metal Bulletin plc, London.

General References

H. H. Murray, ed., *Applied Clay Science,* Vol. 5, No. 5 and 6, Mar. 1991. Special issue on some applications of selected clay minerals.
R. E. Grim, *Applied Clay Mineralogy,* McGraw-Hill Book Co., Inc., New York, 1962.
Mineral Facts and Problems, Clays, U.S. Dept. of the Interior, Bureau of Mines, Washington, D.C., 1985.

PAUL SENNETT
Engelhard Corporation

CLINICAL CHEMISTRY. See AUTOMATED INSTRUMENTATION,
CLINICAL CHEMISTRY.

CLUSTERS. See SUPPLEMENT.

CLUTCH FACINGS. See BRAKE LININGS AND CLUTCH FACINGS.

COAGULANTS AND ANTICOAGULANTS. See BLOOD,
COAGULANTS AND ANTICOAGULANTS.

COAL

The use of coal, known as the rock that burns, was recorded in China, Greece, and Italy over 2000 years ago. Coal mining began in Germany around the tenth century AD and enough coal was mined in England for export in the thirteenth century. Coal mining began in the United States in about 1700.

Coal is usually a dark black color, although geologically younger deposits of brown coal have a brownish red color (see LIGNITE AND BROWN COAL). The color, luster, texture, and fracture vary with rank, type, and grade. Coal is the result of combined biological, chemical, and physical degradation of accumulated plant matter over geological ages. The relative amounts of remaining plant parts leads to different types of coal, which are sometimes termed banded, splint, nonbanded (cannel and boghead); or hard or soft; or lignite, subbituminous, bituminous, or anthracite. In Europe the banded and splint types are generally referred to as ulmic or humic coals. Still other terms refer to the origins of the plant parts through maceral names such as vitrinite, liptinite, and inertinite. The degree of conversion of plant matter or coalification is referred to as rank. Brown coal and lignite, subbituminous coal, bituminous coal, and anthracite make up the rank series with increasing carbon content. The impurities in these coals cause differences in grade.

Coal consists primarily of carbon, hydrogen, and oxygen, and contains lesser amounts of nitrogen and sulfur and varying amounts of moisture and mineral matter. The mode of formation of coal, the variation in plant composition, the microstructure, and the variety of mineral matter indicate that there is a mixture of materials in coal. The nature of the organic species present depends on the degree of biochemical change of the original plant material, on the historic pressures and temperatures after the initial biochemical degradation, and on the finely divided mineral matter deposited either at the same time as the plant material or later. The principal types of organic compounds have resulted from the formation and condensation of polynuclear and heterocyclic ring compounds containing carbon, hydrogen, nitrogen, oxygen, and sulfur. The fraction of carbon in aromatic ring structures increases with rank.

Nearly all coal is used in combustion and coking (see COAL CONVERSION PROCESSES). At least 80% is burned directly in boilers for generation of electricity (see MAGNETOHYDRODYNAMICS; POWER GENERATION) or steam for industrial purposes. Small amounts are used for transportation, space heating, firing of ceramic products, etc. The rest is essentially pyrolyzed to produce coke, coal gas, ammonia (qv), coal tar, and light oil products from which many chemicals are produced (see FEEDSTOCKS, COAL CHEMICALS AND FEEDSTOCKS). Combustible gases and chemical intermediates are also produced by the gasification of coal (see FUELS, SYNTHETIC), and different carbon (qv) products are produced by various heat treatments. A small amount of coal is used in miscellaneous applications such as fillers (qv), pigments (qv), foundry material, and water (qv) filtration (qv).

In 1991 the annual coal production averaged ca 900×10^6 t in the United States and 4.7×10^9 t, of which 1.2×10^9 t were brown coal and lignite, worldwide (1,2). World reserves of bituminous coal and anthracite are ca 5.6×10^{12} t of coal equivalent, ie, 29.3 GJ/t (12.6×10^3 Btu/lb), and subbituminous and lignite are 2.9×10^{12} t of coal equivalent (see FUEL RESOURCES). For economic and environmental reasons coal consumption has been cyclic.

Origin of Coal

Coal evolved from partially decomposed plants in a shallow-water environment. Various chemical and physical changes occurred in two distinct stages: one biochemical and the other physicochemical (geochemical) (3–7). Because some parts of plant material are more resistant to biochemical degradation than others, optical variations in petrologically distinguishable coals resulted. The terms vitrain and clarain refer to bright coals; durain is a dull coal, and fusain is structured fossil charcoal. Exposure to pressure and heat during the geochemical stage caused the differences in degree of coalification or rank that are observable in the continuous series: peat, brown coal and lignite, subbituminous coal, bituminous coal, and anthracite. The carbon containing deposits in which the inorganic material predominates, such as in oil shale (qv) and bituminous shale, are not classified as coal.

Complete decay of plant material by oxidation and oxygen-based bacteria and fungi is prevented only in water-logged environments such as swamps in regions where there is rapid and plentiful plant growth. Peat is formed in such swamps from plant debris such as branches and twigs, bark, leaves, spores and pollen, and even tree trunks that are rapidly submerged in the swamp water. A series of coal seams have been formed from peat swamps growing in an area that has undergone repeated subsidence followed by deposition of lacustrine or marine intrusion material. Periods during which vegetation flourished and peat accumulated were followed by rapid subsidence resulting in submergence of the peat swamp and covering of the deposit with silt and sand. It has been suggested that in the United States the Dismal Swamp of Virginia and North Carolina, which is gradually being flooded by Lake Drummond, is an area undergoing active subsidence (8).

According to the autochthonous, *in situ*, theory of coal formation, peat beds and subsequently coal were formed from the accumulation of plants and plant debris in place. According to the allochthonous theory, the coal-producing peat bogs or swamps were formed from plant debris that had been transported, usually by streams or coastal currents, to the observed burial sites.

Biochemical Stage. The initial biochemical decomposition of plant matter depends on two factors: the ability of the different plant parts to resist attack and the existing conditions of the swamp water. Fungi and bacteria can cause complete decay of plant matter that is exposed to aerated water or to the atmosphere. The decay is less complete if the vegetation is immersed in water containing anaerobic bacteria. Under these latter conditions, the plant protoplasm, proteins, starches, and to a lesser extent the cellulose (qv) are easily digested. Lignin (qv) is more resistant. The most decay-resistant plant parts, for both anaerobic and aerobic decomposition, are the waxy protective layers, ie, cuticles, spore, and pollen walls, and the resins. Vitrain results from the partial decay of lignin and cellulose in stagnant water. The original cell structure of the parent plant tissue can be recognized in many samples.

The clarain (9) and bright attritus (8) are finely banded bright parts of coal that evolved from the residues of fine woody material such as branches, twigs, leaves, spores, bark, and pollen. In aerated waters, the plant parts were more decomposed and show a higher concentration of resins, spores, and cuticles. Dull coal, called durain (9), was formed under these conditions and occurs commonly in Europe. It is not as widely found in the United States where it is known as splint or block coal. More selective chemical and biochemical activity, probably in a drier environment, led to the formation of soft, charcoal-like fusain from woody plant material. The conversion was rapid and probably complete by the end of the peat formation stage. Cannel coal is believed to have formed in aerated water, which decomposed all but the spores and pollen. The name is derived from its quality of burning in splints with a candlelike flame. Boghead coal closely resembles cannel coal but was derived from algae instead of plant spores.

Geochemical Stage. The conversion of peat to bituminous coal is the result of the cumulative effects of temperature and pressure over a long time. The sediment covering the peat provides the pressure and insulation so that the earth's internal heat can be applied to the conversion. The temperature increase is about 4 to 8°C for each 100 m of depth. The changes in plant matter are termed normal coalification.

Moisture is lost and the chemical composition changes during coalification. Oxygen and hydrogen decrease and carbon increases. These compositional changes are accompanied by decreases in volatile matter and increases in calorific value. The volatile matter and calorific content are the main criteria used for commercial classification in the United States and for the International Classification.

The change in rank from bituminous coal to anthracite involves the application of significantly higher pressures, ie, as in mountain building activity, and temperatures, ie, as in volcanic activity. The more distant the coal from the disruption, the less proportionate the alteration. Tectonic plate movements involved in mountain building provide pressure for some changes to anthracite. As a general rule, the older the coal deposit, the more complete the coalification and the higher the rank of coal. Most commercial bituminous coal fields were deposited during the Pennsylvanian (ca 285–320 million years ago), Upper Cretaceous (ca 65–100 million years ago), and early Tertiary (ca 20–65 million years ago) ages. The lower rank coals come primarily from the Tertiary and Upper Cretaceous ages, and peat deposits are relatively recent, less than one million years old. However, age alone does not determine rank. The brown coal of the Moscow basin is not buried deeply, and although it was deposited during the Lower Carboniferous or Mississippian age (ca 320–360 million years ago), there was not enough heat and pressure to convert it further.

Coal Petrography

Careful examination of a piece of coal shows that it is usually made up of layers or bands of different materials which upon microscopic examination are distinct entities distinguishable by optical characteristics (10–12). The study of the origin, composition, and technological application of these materials is called coal petrology, whereas coal petrography involves the systematic quantification of the amounts and characteristics by microscopic study. The petrology of coal may involve either a macroscopic or microscopic scale.

On the macroscopic scale, two coal classifications have been used: humic or banded coals and sapropelic or nonbanded coals. Stratification in the banded coals, which result from plant parts, is quite obvious; the nonbanded coals, which derive from algal materials and spores, are much more uniform. The physical and chemical properties of the different layers in a piece of coal or a seam can vary significantly. Therefore the relative amounts of the layers are important in determining the overall characteristics of the mined product. Coal petrography has been widely applied in cokemaking and is important in coal liquefaction programs.

If the mineral matter in the coal exceeds about 40%, then the material is referred to as a coaly or carbonaceous shale. If the mineral matter is a finely divided clay, well dispersed in the coal, then the material may be described as a stony coal or bone coal.

Macerals. Coal parts derived from different plant parts, are referred to as macerals (13). The maceral names end in "-inite" as do the mineral forms of rocks. The most abundant (about 85%) maceral in U.S. coal is vitrinite, derived from the woody tissues of plants. Another maceral, called liptinite, is derived from the waxy parts of spores and pollen, or algal remains. The liptinite macerals fluoresce under blue light permitting a subdivision based on fluorescence. A third maceral, inertinite, is thought to be derived from oxidized material or fossilized charcoal remnants of early forest fires.

A number of subdivisions of the maceral groups have been developed and documented by the International Commission on Coal Petrology (14). Table 1 lists the Stopes-Heerlen classification of higher rank coals. Periodic revisions include descriptions of the macerals, submacerals, morphology, physical properties, and chemical characteristics. Theories on the mode of formation of the macerals and their significance in commercial applications are also included of Reference 14.

The macerals in lower rank coals, eg, lignite and subbituminous coal, are more complex and have been given a special classification. The term huminite has been applied to the macerals derived from the humification of lignocellulosic tissues. Huminite is the precursor to the vitrinite observed in higher rank coals.

The elemental composition of the three maceral groups varies. The vitrinite, which frequently is about 85% of the sample in the United States, is similar to

Table 1. Stopes-Heerlen Classification of Maceral Groups, Macerals, and Submacerals of Higher Rank Coals[a]

Maceral group	Maceral	Submaceral
vitrinite	telinite	telocollinite
	collinite	gelocollinite
		desmocollinte
		corpocollinite
liptinite	sporinite	
	cutinite	
	suberinite	
	resinite	
	alginite	
	liptodetrinite	
	fluorinite	
	bituminite	
	exudatinite	
inertinite	fusinite	
	semifusinite	
	macrinite	
	micrinite	
	sclerotinite	
	inertodetrinite	

[a]Ref. 9.

the parent coal. The liptinites are richer in hydrogen, whereas the inertinites are relatively deficient in hydrogen and richer in carbon. The liptinites also contain more aliphatic materials; the inertinites are richer in aromatics. The term inertinite refers to the relative chemical inertness of this material, making it especially undesirable for liquefaction processes because it tends to accumulate in recycled feedstock streams.

Vitrinite Reflectance.   The amount of light reflected from a polished plane surface of a coal particle under specified illumination conditions increases with the aromaticity of the sample and the rank of the coal or maceral. Precise measurements of reflectance, usually expressed as a percentage, are used as an indication of coal rank.

Precise reflectance measurements are carried out using incident light having wavelength of 546 nm (green mercury line), and carefully polished coal specimens immersed in an oil having a refractive index of 1.518 at 23°C. Comparison is made to a calibrated standard, using a photomultiplier system. Coal is an anisotropic material, so the reflectance varies according to the orientation of the particle. A typical procedure involves making many measurements on the vitrinite particles in a coarsely ground sample to obtain a range of values that are then used to determine a maximum vitrinite reflectance that correlates with coal rank. Minimum values can also be correlated. Figure 1 illustrates the relationship between reflectances and the carbon content (12). The reflectance of liptinite macerals is less than that for vitrinite, and the petrographer can distinguish the two for low rank coals. However as measurements are made on progressively higher rank coals, the reflectivities of the liptinites and vitrinites become similar and are the same for medium volatile bituminous coals. For inertinites, the distinctions

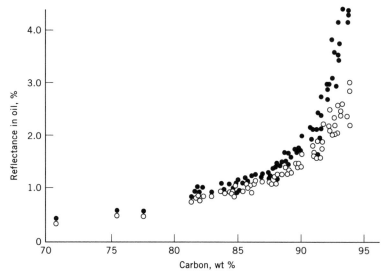

Fig. 1.   The relationship between carbon content and maximum (•) and minimum (○) reflectances of vitrinite (11).

between reflectivities for vitrinite persist into the anthracite range. Table 2 indicates the vitrinite reflectances for the various coal ranks (12).

Application of Coal Petrology and Petrography. Petrographic analysis is frequently carried out for economic evaluation or to obtain geologic information. Samples are usually lumps or more coarsely ground material that have been mounted in resins and polished. Maceral analysis involves the examination of a large number (usually > 500) of particles during a traverse of a polished surface to identify the macerals at specified intervals. A volume percentage of each of the macerals present in a sample is calculated.

Seam correlations, measurements of rank and geologic history, interpretation of petroleum (qv) formation with coal deposits, prediction of coke properties, and detection of coal oxidation can be determined from petrographic analysis. Constituents of seams can be observed over considerable distances, permitting the correlation of seam profiles in coal basins. Measurements of vitrinite reflectance within a seam permit mapping of variations in thermal and tectonic histories. Figure 2 indicates the relationship of vitrinite reflectance to maximum temperatures and effective heating time in the seam (11,15).

The coking behavior of coal depends on the rank of the coal, the properties of the individual constituents, and their relative amounts (3–7,10). For some purposes, a blend of coals can be selected to achieve desired coking properties. The maceral groups behave differently on heating: vitrinite from most medium rank coal (9–33% volatile matter) has good plasticity and swelling properties and produces an excellent coke; inertinite is almost inert and does not soften on heating, and exinite becomes extremely plastic and is almost completely distilled as tar. By careful control of the petrological composition and the rank of a coal blend, behavior during carbonization can be controlled. Additionally, coking behavior can be reasonably predicted using petrography and maceral breakage (10). Oxidation reduces the coke forming properties of a given coal and can also be detected by petrograhic techniques (16).

Table 2. Vitrinite Reflectance Limits, Taken in Oil, and ASTM Coal Rank Classes[a]

Coal rank	Maximum reflectance, %
subbituminous	<0.47
high volatile bituminous	
C	0.47–0.57
B	0.57–0.71
A	0.71–1.10
medium volatile bituminous	1.10–1.50
low volatile bituminous	1.50–2.05
semianthracite	2.05–3.00[b]
anthracite	>3.00[b]

[a]Ref. 11.
[b]Approximate value.

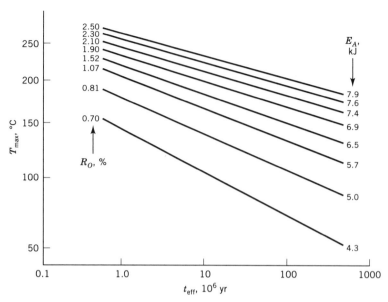

Fig. 2. Relation of vitrinite reflectance (R_o) in percent to maximum temperature (T_{max}) and effective heating time (t_{eff}) where E_A = activation energy in kJ and t_{eff} is within 15°C of T_{max}. To convert kJ to kcal, divide by 4.184 (11).

Classification Systems

Prior to the nineteenth century, coal was classified according to appearance, eg, bright coal, black coal, or brown coal. A number of classification systems have since been developed. These may be divided into two types, which are complementary: scientific and commercial. Both are used in research, whereas the commercial classification is essential industrially. In the scientific category, the Seyler chart has considerable value.

The Seyler Classification. The Seyler chart, shown in Figure 3, is based on the carbon and hydrogen content of coals determined on a dry mineral-matter-free basis (17). Points representing different coal samples lie along a broad band. The center band on the chart shows the properties of coal rich in vitrinite. The location of the band indicates the range and interrelationship of the properties. Coals above the band are richer in hydrogen, eg, cannel and boghead coals, and the liptinite macerals in the usual coals. Coals below the band are represented by the maceral inertinite. Other properties, such as moisture and swelling indexes, also fit into specific areas on this chart. The curve in the solid band represents a composition range where the properties of the coal change rapidly. Swelling indexes, coking power, and calorific values are maximized, and moisture is minimized. The lowest rank coals lie on the right side of this curve; the highest rank coals are on the left at the lower part of the band.

The ASTM Classification. The ASTM classification system was adopted in 1938 as a standard means of specification. This system is used in the United States and in many other parts of the world, and is designated D388 in the ASTM

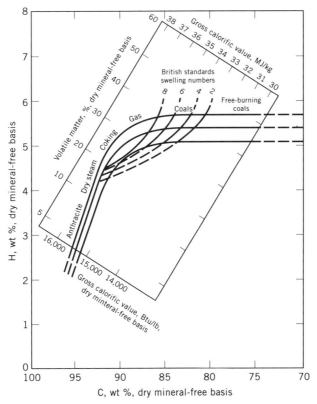

Fig. 3. Simplified form of Seyler's coal classification chart (17). An updated version of Seyler's coal classification is described in Reference 6. Note that ASTM uses the free-swelling index (18).

Standards (18). The higher rank coals are specified by fixed carbon ≥ 69%, or for volatile matter ≤ 31%, on a dry, mineral-free basis. Lower rank coals are classified by calorific value on the moist, mineral-matter-free basis. These parameters are given in Table 3. Calorific value depends on two properties: moisture absorbing capacity and the calorific value of the pure coal matter. When some overlap between bituminous and subbituminous coals occurs, it is resolved on the basis of the agglomerating properties.

National Coal Board Classification for British Coals. The classification proposed in 1946 by the UK Department of Scientific and Industrial Research led to the system in use by the National Coal Board for coals in the United Kingdom. There are two parameters: the quantity of volatile matter determined on a dry, mineral-matter-free basis, and the Gray-King coke-type assay, a measure of coking power as designated in the British Standards (18). This latter assay is used as a primary means of classification for lower rank coals. The classification applies to coals having less than 10% ash. High ash coals are cleaned before analysis by a float–sink separation to reduce the ash content below 10%.

International Classification. *Hard Coal.* The amount of coal in international commerce since ca 1945 necessitated an international system of coal clas-

Table 3. Classification of Coals by Rank

Coals	Fixed carbon, %[a] ≥	<	Volatile matter, %[a] >	≤	Gross calorific value, kJ/kg[b] ≥	<	Agglomerating character
Anthracitic							
meta-anthracite	98			2			
anthracite	92	98	2	8			nonagglomerating
semianthracite[c]	86	92	8	14			
Bituminous							
low volatile	78	86	14	22			
medium volatile	69	78	22	31			
high volatile							commonly
A		69	31		32,500[d]		agglomerating[e]
B					30,200[d]	32,500	
C					26,700	30,200	
					24,400	26,700	agglomerating
Subbituminous							
A					24,400	26,700	
B					22,100	24,400	
C					19,300	22,100	
Lignitic							
A					14,600	19,300	nonagglomerating
B						14,600	

[a]Dry, mineral-matter-free basis.

[b]To convert from kJ/kg to Btu/lb, multiply by 0.4302; moist mineral-matter-free basis, ie, contains inherent moisture but not water visible on the surface.

[c]If agglomerating, classify in low volatile group of the bituminous class.

[d]Coals having 69% or more fixed carbon on the dry, mineral-matter-free basis are classified according to fixed carbon, regardless of gross calorific value.

[e]There may be nonagglomerating varieties in the groups of the bituminous class, and there are notable exceptions in high volatile C bituminous group.

sification and in 1956 the Coal Committee of the European Economic Community agreed on a system designated the International Classification of Hard Coal by Type (3). Volatile matter and gross calorific value on a moist, ash-free basis are among the parameters considered. Table 4 shows the various classes of the international system and gives the corresponding national names used for these coals.

A three-digit classification is employed in the international system (10) where the first digit indicates the class or rank, such that higher digits correspond to lower ranks; the second digit indicates the group indicated by caking properties such as the free-swelling index or the Roga index; and the third digit defines a subgroup based on coking properties as measured using a dilatometer or the Gray-King assay. Coals having volatile matter up to 33% are divided into classes 1–5; coals having volatile matter greater than 33% are divided into classes 6–9. The

Table 4. The International and Corresponding National Systems of Coal Classes[a]

	International system			National classifications							
Class no.	Volatile matter, %	Calorific value, kJ/g[b,c]	Belgium	Germany	France	Italy	The Netherlands	Poland	United Kingdom	United States
0	0–3					antraciti speciali		meta-antracyt		*meta-*anthracite
1A	3–6.5		maigre	Anthrazit	anthracite	antraciti communi	anthraciet	antracyt	anthracite	anthracite
1B	6.5–10							polantracyt		
2	10–14		¼ gras	Mager-kohle	maigre	carboni magri	mager	chudy	dry steam	semianthracite
3	14–20		½ gras	Esskohle	demigras	carboni semigrassi	esskool	polkoksowy meta-koksowy	coking steam	low volatile bituminous
4	20–28		¾ gras	Fettkohle	gras à courte flamme	carboni grassi corta fiamma	vetkool	orto-koksowy	medium volatile coking	medium volatile bituminous
5	28–33		gras	Gaskohle	gras proprement dit	carboni grassi media fiamma		gazowo koksowy		high volatile bituminous A
6	>33 (33–40)	32.4–35.4				carboni da gas	gaskool			
7	>33 (32–44)	30.1–32.4			flambant gras	carboni grassi da vapore	gasvlam-kool	gazowy	high volatile	high volatile bituminous B
8	>33 (34–46)	25.6–30.1		Gas flamm-kohle	flambant sec	carboni secchi	vlamkool	gazowo-plomienny		high volatile bituminous C
9	>33 (36–48)	<25.6						plomienny		subbituminous

[a]Ref. 3.
[b]Calculated to standard moisture content.
[c]To convert kJ/g to Btu/lb, multiply by 430.2.

433

calorific values are given for a moisture content obtained after equilibrating at 30°C and 96% rh. The nine classes are then divided into four groups as measured through either the free-swelling index (17) or the Roga index. These tests indicate properties observed when the coal is heated rapidly.

Brown Coal and Lignite. The brown coals and lignites, defined as coals having heating values that are less than 23,860 kJ/kg (10,260 Btu/lb, 5700 kcal/kg), are classified separately (see LIGNITE AND BROWN COAL). A four-digit code is used for classification. The first two digits (class parameter) are defined by total moisture content of freshly mined coal on an ash-free basis. The third and fourth digit are defined by the tar yield on a dry, ash-free basis.

Composition and Structure

The constitution of a coal involves both the elemental composition and the functional groups that are derived therefrom. The structure of the coal solid depends to a significant extent on the arrangement of the functional groups within the material.

Composition. The functional groups within coal contain the elements C, H, O, N, or S (3,4,5,19). The significant oxygen-containing groups found in coals are carbonyl, hydroxyl, carboxylic acid, and methoxy. The nitrogen-containing groups include aromatic nitriles, pyridines, carbazoles, quinolines, and pyrroles (20). Sulfur is primarily found in thiols, dialkyl and aryl–alkyl thioethers, thiophene groups, and disulfides. Elemental sulfur is observed in oxidized coal (20).

The relative and absolute amounts of the various groups vary with coal rank and maceral type. The principal oxygen-containing functional groups in vitrinites of mature coals are phenolic hydroxyl and conjugated carbonyls as in quinones. Spectroscopic evidence exists for hydrogen bonding of hydroxyl and carbonyl groups. There are unconjugated carbonyl groups such as ketones in exinites. The infrared absorption bands are displaced from the normal carbonyl range for simple ketones by the conjugation in vitrinites. Interactions between the carbonyl and hydroxyl groups affect the normal reactions.

A range of quantitative organic analytical techniques may be used to determine functional group concentrations. Acetylation and *O*-alkylation are used to determine hydroxyl groups, whereas carbonyl groups are difficult to quantify using simple procedures. A variety of instrumental techniques has also been used to aid in the understanding of coal structure and constitution. Magnetic resonance techniques have been particularly helpful in determining relative amounts of different carbon species within a coal. Table 5 contains data obtained using these techniques (21).

Aromaticity of coal molecules increases with coal rank. Calculations based on several models indicate that the number of aromatic carbons per cluster varies from nine for lignite to 20 for low volatile bituminous coal, and the number of attachments per cluster varies from three for lignite to five for subbituminous through medium bituminous coal. The value is four for low volatile bituminous (21).

Reaction of coals and mild selective oxidizing agents such as benzoquinone (20,22) causes the coals to lose much of the hydrogen content. Similarly, a palla-

Table 5. Carbon Structural Distribution of the Argonne Premium Coals Based on Nmr Measurements[a]

Coal[b]	Fraction of carbon type[c]											
	f_a	f_a'	f_a^C	f_a^H	f_a^N	f_a^P	f_a^S	f_a^B	f_{al}	f_{al}^H	$f_{al}*$	f_{al}^O
North Dakota (L)	0.61	0.54	0.07	0.26	0.28	0.06	0.13	0.09	0.39	0.25	0.14	0.12
Wyodak (SB)	0.63	0.55	0.08	0.17	0.38	0.08	0.14	0.16	0.37	0.27	0.10	0.10
Blind Canyon (HVB)	0.65	0.64	0.01	0.22	0.42	0.07	0.15	0.20	0.35	0.22	0.13	0.04
Illinois #6 (HVB)	0.72	0.72	0.00	0.26	0.46	0.06	0.18	0.22	0.28	0.19	0.09	0.05
Pittsburgh (HVB)	0.72	0.72	0.00	0.27	0.45	0.06	0.17	0.22	0.28	0.13	0.15	0.03
Lewiston-Stockton (HVB)	0.75	0.75	0.00	0.27	0.48	0.05	0.21	0.22	0.25	0.14	0.11	0.04
Upper Freeport (MVB)	0.81	0.81	0.00	0.28	0.53	0.04	0.20	0.29	0.19	0.09	0.10	0.02
Pocahontas (LVB)	0.86	0.86	0.00	0.33	0.53	0.02	0.17	0.34	0.14	0.08	0.06	0.01

[a] Ref. 21.
[b] L = lignite; SB = subbituminous; HVB = high volatile bituminous; MVB = medium volatile bituminous; and LVB = low volatile bituminous.
[c] The symbols f_a and f_{al} correspond to total fraction of sp^2 and sp^3 hybridized carbon, respectively. f_a' represents the fraction of sp^2 carbon in aromatic rings; f_a^C, the fraction in carbonyls, $\delta > 165$ ppm; f_a^H, the aromatic fraction that is protonated; f_a^N, the aromatic fraction that is nonprotonated; f_a^P, the phenolic or phenolic ether carbon, $\delta = 150$–165 ppm; f_a^S, the alkylated aromatic carbon, $\delta = 135$–150 ppm; f_a^B, the aromatic bridgehead carbon; f_{al} represents the fraction of CH or CH_2 aliphatic carbon; $f_{al}*$, the CH_3 or nonprotonated aliphatic carbon; and f_{al}^O, the aliphatic carbon bound to oxygen, $\delta = 50$–90 ppm.

435

dium catalyst can cause the evolution of molecular hydrogen (23,24). These methods may give an indication of the minimum amount of hydrogen in the coal that is involved in hydroaromatic rings. This amount is close to the total nonaromatic hydrogen determined for lower rank coals. Other hydrogen determining methods involve dehydrogenation using sulfur (25) and using halogens (26). The values obtained by these last methods are somewhat lower than that of benzoquinone.

Hydrogen can be added to the aromatic structures converting them to hydroaromatic rings. The hydrogen addition and removal is generally but not entirely reversible (24).

High resolution mass spectrometry (qv) has been used with extracts of a series of coals to indicate the association of different heteroatoms (27). Various types of chromatography (qv) have also been used to identify the smaller species that can be extracted from coal.

Coal Structure.   *Bonding in Macromolecules.*   Conclusions regarding the chemical structure of the macromolecules within coal are generally based on experimental measurements and an understanding of structural organic chemistry (3,4,20,28). The description given herein refers to vitrinites.

Several requirements must be met in developing a structure. Not only must elementary analysis and other physical measurements be consistent, but limitations of structural organic chemistry and stereochemistry must also be satisfied. Mathematical expressions have been developed to test the consistency of any given set of parameters used to describe the molecular structure of coal and analyses of this type have been reported (4,6,19,20,29,30).

Evidence suggests that the structure for vitrinites in bituminous coals and anthracite has the following characteristics: (*1*) the molecule contains a number of small aromatic nuclei or clusters, each usually having from one to four fused benzene rings. The average number of clusters in the molecule is characteristic of the coal rank. This average increases slightly to 90% carbon and then increases rapidly; (*2*) the aromatic clusters are partly linked together by hydroaromatic, including alicyclic, ring structures. These latter rings also contain six carbon atoms and thus, upon loss of hydrogen can become part of the cluster, increasing the average cluster size; (*3*) other linkages between clusters involve short groupings such as methylene, ethylene, and ether oxygen; (*4*) a significant amount of hydrogen sites on the aromatic and hydroaromatic rings have been substituted by methyl and sometimes larger aliphatic groups; (*5*) the oxygen in vitrinites usually occurs in phenolic hydroxyl groups and ethers (19), substituting for hydrogen on the aromatic structures; (*6*) a small fraction of the rings, both aromatic and hydroaromatic, have oxygen substituted for a carbon atom. Some of these heterocyclic rings may be five-membered; (*7*) a lesser amount of oxygen than that occurring in phenolic groups appears in the carbonyl moieties, ie, in quinone form on aromatic rings or as ketones attached to hydroaromatic ones. In both cases the oxygen is apparently hydrogen bonded to adjacent hydroxyl groups: the reactivity and peak location for the characteristic infrared absorptions are different from those in the typical quinones and ketones; (*8*) nitrogen is less abundant in vitrinites than oxygen. It is usually present as a heteroatom in a ring structure or as a nitrile (19); (*9*) most of the sulfur in as-mined coal, especially if the S content exceeds 2 wt %, is associated with inorganic material. However, for very low sulfur coals the organic sulfur which occurs as both aliphatic and aromatic usually exceeds the inorganic. Sulfide and disulfide groups may also link clusters (19,20); (*10*) a given piece of coal contains a variety of molecules having different proportions of a given structural feature and varying molecular weight. These molecules are composed of planar fused ring clusters linked to nonplanar hydroaromatic structures. The overall structure is irregular, open, and complex. Entanglement between molecules occurs, as do cross-links of hydrogen bonded species, leading to difficulties in molecular-weight determinations for extracts and to changes in properties on heating. The different molecular shapes and sizes in a piece of coal lead to irregularities in packing and hence the amorphous nature and the extensive ultrafine porosity; and (*11*) some of the evidence suggests that the molecular weights of extracts representing 5–50 wt % of the coal average 1000–3000. Larger transient units also exist owing to aggregation and it is probable that unextracted coal contains even larger molecular units (19).

Figure 4 gives a representation of the coal molecule (28) that correlates with products obtained from liquefaction. Heating above 400°C or mild chemical oxi-

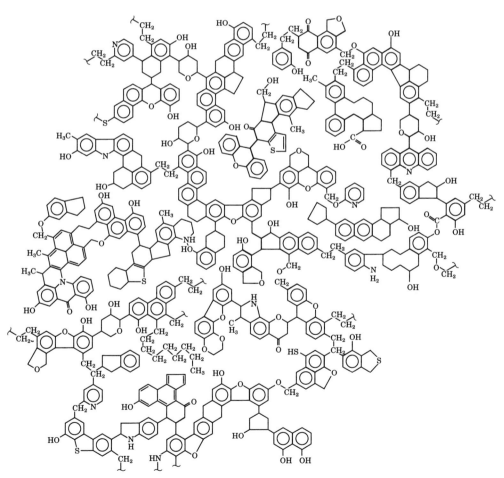

Fig. 4. Model of bituminous coal structure (28).

dation changes some hydroaromatic into aromatic structures. Hydroaromatic structures appear to be involved in tar formation (31); however, the tar probably contains some of the same smaller aromatic structures also found in the coal. Tar and gas production, including changes with rank, is consistent (3,4) with the structural model described. Most of the larger units remain in the char or coke produced by heating. The average size of the molecular units is increased by polymerization and conversion of hydroaromatic links to aromatic rings.

Bonding between Macromolecules. The macromolecules that make up the coal structure are held together by a variety of forces and bonds (32). The coal network model is one approach to describing the three-dimensional structure of the solid. Aromatic clusters are linked by a variety of connecting bonds, through oxygen, methylene or longer aliphatic groups, or disulfide bridges, and the proportions of the different functional groups change as the rank of the coal is progressively increased. For example, oxygen is diminished. Acid groups decrease early in the rank progression and other groups follow leaving ether moieties that act as cross-links between the larger clusters in the high rank coals. Another type of linkage involves hydrogen bonds which, for example, hold hydroxy and keto-groups together in the solid. A review of coal models, the mechanical properties of the network, and the glass-transition temperature corresponding to the change from a fluid to a rigid amorphous solid are available (32).

A model for coal fluidity based on a macromolecular network pyrolysis model has been developed (33). In that model, bond breaking is described as a first-order reaction having a range of activation energies. A variety of lattices have also been used to describe the bonding in coal. In turn these structures have been used to describe devolatilization, combustion, and char formation (34). The form of the macromolecule in a liquid extract tends to be spherical as a result of surface energy considerations, hydrogen bonds, and van der Waals forces (32).

Coal Constitution. Chemical composition studies (35,36) indicate that brown coals have a relatively high oxygen content. About two-thirds of the oxygen is bonded carboxyl, acetylatable hydroxyl, and methoxy groups. Additionally, unlike in bituminous coals, some alcoholic hydroxyl groups are believed to exist.

Anthracites. The anthracites, which approach graphite in composition (see CARBON, GRAPHITE), are classified higher in rank, have less oxygen and hydrogen, and are less reactive than bituminous coals. Anthracites are also insoluble in organic solvents. These characteristics become more pronounced as rank increases within the anthracite group. The aromatic carbon fraction of anthracites is at least 0.9, and the number of aromatic rings per cluster is greater than that for the low volatile bituminous coals, with a value of about 10 for anthracite having 95 wt % C. There is x-ray diffraction evidence (37) to indicate that the aromatic rings are more loosely and variably assembled than those in bituminous coal clusters. The anthracites have greater optical and mechanical anisotropy than lower rank coals, and the internal pore volume and surface increase with rank after the minimum below about 90 wt % C.

Mineral Matter in Coal. The mineral matter (7,38) in coal results from several separate processes. Some comes from the material inherent in all living matter; some from the detrital minerals deposited during the time of peat formation; and a third type from secondary minerals that crystallized from water which has percolated through the coal seams.

A variety of instrumental techniques may be used to determine mineral content. Typically the coal sample is prepared by low temperature ashing to remove the organic material. Then one or more of the techniques of x-ray diffraction, infrared spectroscopy, differential thermal analysis, electron microscopy, and petrographic analysis may be employed (7).

The various clay minerals are the most common detrital mineral (see CLAYS); however, other common ones include quartz, feldspar, garnet, apatite, zircon, muscovite, epidote, biotite, augite, kyanite, rutile, staurolite, topaz, and tourmaline. The secondary minerals are generally kaolinite, calcite, and pyrite. Analyses have shown the presence of almost all elements in at least trace quantities in the mineral matter (39). Certain elements, ie, germanium, beryllium, boron, and antimony, are found primarily with the organic matter in coal, whereas zinc, cadmium, manganese, arsenic, molybdenum, and iron are found with the inorganic material. The primary elemental constituents of mineral matter in coal are aluminum, silicon, iron, calcium, magnesium, sodium, and sulfur. The relative concentrations depend primarily on the geographical location of the coal seam, and vary from place to place within a given field. In the eastern United States the most abundant mineral elements are silicon, aluminum, and iron and there are much lower amounts of alkali and alkaline-earth elements. West of the Mississippi River the relative amounts of silicon, aluminum, and iron are much less and the alkaline-earth and alkali elements are much greater.

Properties

Pieces of coal are mixtures of materials somewhat randomly distributed in differing amounts. The mineral matter can be readily distinguished from the organic, which is itself a mixture. Coal properties reflect the individual constituents and the relative proportions. By analogy to geologic formations, the macerals are the constituents that correspond to minerals that make up individual rocks. For coals, macerals, which tend to be consistent in their properties, represent particular classes of plant parts that have been transformed into coal (40). Most detailed chemical and physical studies of coal have been made on macerals or samples rich in a particular maceral, because maceral separation is time consuming.

The most predominant maceral group in U.S. coals is vitrinite. The other important maceral groups include inertinite consisting of micrinite, a dull black amorphous material, fusinite, a dull fibrous material similar to charcoal, and the liptinite group including sporinite, which is relatively fusible and volatile. Differences in macerals are evident over the range of coal rank, ie, from brown coal or lignite to anthracite. The definition of rank is that generally accepted as the wt % C, on a dry, mineral-free basis, in the vitrinite associated with the given coal in the seam. The range of ranks in which differences between macerals are most significant is 75–92 wt % C content of the vitrinite. These coals are bituminous.

In the United States the commercial classification of coals is based on the fixed carbon (or volatile matter) content and the moist heating value. One correlation is made by plotting the hydrogen content, on a mineral-free basis, against the corresponding carbon content. A similar plot, made using the commercial criteria of volatile matter and heating placed on axes at an appropriate angle to the % C, % H axes, forms the Seyler coal classification chart (17). Both are illustrated in Figure 3. Table 6 indicates the usual range of composition of commercial coals of increasing rank.

Table 6. Composition of Humic Coals

Type of coal	Composition, wt %[a]						Calorific value, kJ/g[b]
	C	H	O	N	Moisture as found	Volatile matter	
peat	45–60	3.5–6.8	20–45	0.75–3.0	70–90	45–75	17–22
brown coals and lignites	60–75	4.5–5.5	17–35	0.75–2.1	30–50	45–60	28–30
bituminous coals	75–92	4.0–5.6	3.0–20	0.75–2.0	1.0–20	11–50	29–37
anthracites	92–95	2.9–4.0	2.0–3.0	0.5–2.0	1.5–3.5	3.5–10	36–37

[a] Dry, mineral-matter-free basis except for moisture value.
[b] To convert kJ/g to Btu/lb, multiply by 430.2.

Physical Methods of Examination. Physical methods used to examine coals can be divided into two classes which, in the one case, yield information of a structural nature such as the size of the aromatic nuclei, ie, methods such as x-ray diffraction, molar refraction, and calorific value as a function of composition; and in the other case indicate the fraction of carbon present in aromatic form, ie, methods such as ir and nuclear magnetic resonance spectroscopies, and density as a function of composition. Some methods used and types of information obtained from them are (41):

Measurement	Yields information on
optical properties	true rank criterion
	size and ordering of aromatic layers
	graphite structure and anisotropy
density and pore structure	reactivity
	diffusion of gases
	absorbing power
	molecular sieve material
electron microscopy	ultrafine granular structure
	pore structure
	graphitic crystallization and layering
	element distribution in fine components
x-ray diffraction	size distribution of aromatic ring systems
	diameter and thickness of lamellae
	mean bond length
electrical properties	free-radical concentrations
	ring structure and sizes
	energy gap semiconductor properties
	graphitic structure and anisotropy

thermal analysis	rate of change of physical state
coal liquids–sample	solvent separation
separations	polarity
	functionality
mass spectrometry	molecular weight and formulas
	hydrocarbon-type analysis
	carbon number distribution
	trace elements
nuclear magnetic resonance	ratios of aromatic carbon and hydrogen to total
	distribution of isotropic tracers
electron spin resonance	free radicals, charge carriers
ir absorption	functional groups, eg, OH, aromatic and aliphatic CH, CH_3
	minerals
uv/vis absorption	aromaticity
	aromatic ring size
optical fluorescence	identification of coal-derived components
electron spectroscopy for chemical analysis (esca)	elemental distribution and chemical state of surfaces
emission spectroscopy	quantitative and qualitative elemental analysis
	single element and survey analysis
	primary, minor, and trace element analysis for metals and semimetals
x-ray fluorescence	inorganic element analysis
atomic absorption	precise quantitative determination for metals and semimetals
	single-element analysis for minor and trace constituents
Möessbauer spectroscopy	iron compounds and association with sulfur

The scattering of x-rays (6,37,42) gives information on the average distances between the carbon atoms in coal and insight into the bonding between these atoms. Because x-ray scattering depends on the number of protons in the nucleus, carbon is much more effective in scattering x-rays than hydrogen (see X-RAY TECHNOLOGY). The ultraviolet and visible spectra (6) of coal and various solvent extracts show decreasing absorption with increasing wavelength and lack features to aid in interpreting structure except for one peak around 270 nm, which is believed to result from superposition of effects from many similar species. In studies of specific features, comparisons are usually made between coal or coal-derived samples, and pure, usually aromatic, compounds indicating probable presence of particular structures or functional groups. Similar statements can be made concerning reflectance and refractive index (3,4). The derived optical anisotropy is especially evident in coals having carbon contents that exceed 80–85 wt %. Measurements perpendicular and parallel to the bedding plane give different results for optical and some other characteristics (see SPECTROSCOPY).

A significantly greater amount of information concerning functional groups such as hydroxyl can be obtained from infrared absorption (3,4,6); however, this is less specific than the information obtained from an individual organic compound (Fig. 5) (6). An estimate of the relative amount of hydrogen attached to aromatic and nonaromatic structures can, however, be made by using this method (see INFRARED AND RAMAN SPECTROSCOPY). Studies may be carried out on raw coal or products derived from the coal. Physical separation is used to separate fractions of extract and aid in the deduction of the parent coal structure. A method of characterizing coal liquids in terms of ten fractions of different functionality has been described (43).

Magnetic resonance (nmr) spectra (^{1}H and ^{13}C) (6,21) also yield information on bonding for hydrogen and carbon, including the distribution between aromatic and nonaromatic structures, as well as bonding to various heteroatoms. Additional estimates may be made of hydrogen in CH, CH_2, and CH_3 groups. Developments in solid-state ^{13}C nmr spectroscopy, coupled with cross-polarization (cp), magic angle spinning (mas), and dipolar-decoupling techniques have made these estimates somewhat more quantitative than those from ir measurements (see MAGNETIC SPIN RESONANCE). Quantitation has been somewhat limited because of a fraction of the atoms that are not observed owing to paramagnetic centers or spin dynamics of the system (21) (see Table 5).

Electron spin resonance (esr) (6,44) has had more limited use in coal studies. A rough estimate of the free-radical concentration or unsatisfied chemical bonds in the coal structure has been obtained as a function of coal rank and heat treatment. For example, the concentration increases from 2×10^{18} radicals/g at 80 wt % carbon to a sharp peak of about 50×10^{18} radicals/g at 95 wt % carbon content and drops almost to zero at 97 wt % carbon. The concentration of these radicals is less than that of the common functional groups such as hydroxyl. However, radical existence seems to be intrinsic to the coal molecule and may affect the reactivity of the coal as well as its absorption of ultraviolet radiation. Measurements from room temperature to 900 K indicate that the number of electron spins/g increases sharply above about 600 K and peaks from 773 to 850 K, with increasing values for higher rank coals. Oxidation increases the number of radicals by a factor of three over eight days (44).

The other physical measurements (4,6), except for diamagnetic susceptibility (4) and possibly density (4), are primarily of interest for determining chemical structural properties of coal.

Physical Properties. Most of the physical properties discussed herein depend on the direction of measurement as compared to the bedding plane of the coal. Additionally, these properties vary according to the history of the piece of coal. Properties also vary between pieces because of coal's brittle nature and the crack and pore structure. One example concerns electrical conductivity. Absolute values of coal sample specific conductivity are not easy to determine. A more characteristic value is the energy gap for transfer of electrons between molecules, which is determined by a series of measurements over a range of temperatures and is unaffected by the presence of cracks. The velocity of sound is also dependent on continuity in the coal.

The specific electrical conductivity of dry coals is very low, specific resistance 10^{10}–10^{14} ohm·cm, although it increases with rank. Coal has semiconducting

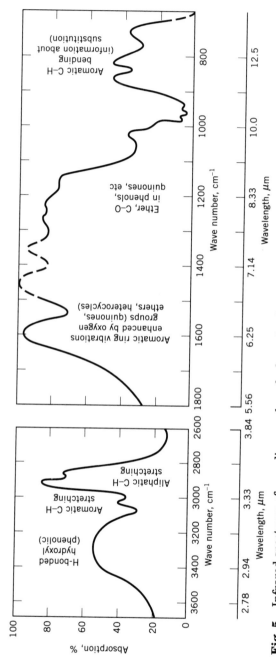

Fig. 5. Infrared spectrum of a medium rank coal where the dashed lines represent variations that occur as a result of differing maceral content or because of functional group conjugation.

properties. The conductivity tends to increase exponentially with increasing temperatures (4,6). As coals are heated to above ca 600°C the conductivity rises especially rapidly owing to rearrangements in the carbon structure, although thermal decomposition contributes somewhat below this temperature. Moisture increases conductivity of coal samples through the water film.

The dielectric constant is also affected by structural changes on strong heating. Also the value is very rank dependent, exhibiting a minimum at about 88 wt % C and rising rapidly for carbon contents over 90 wt % (4,6,45). Polar functional groups are primarily responsible for the dielectric of lower ranks. For higher ranks the dielectric constant arises from the increase in electrical conductivity. Information on the freedom of motion of the different water molecules in the particles can be obtained from dielectric constant studies (45).

Magnetic susceptibility measurements indicate that the organic part of the coal is diamagnetic, having traces of paramagnetic behavior resulting from free radicals or unpaired electrons (6).

Density values (4,6) of coals differ considerably, even after correcting for the mineral matter, depending on the method of determination. The true density of coal matter is most accurately obtained from measuring the displacement of helium after the absorbed gases have been removed from the coal sample. Density values increase with carbon content or rank for vitrinites. Values are 1.4–1.6 g/mL above 85 wt % carbon where there is a shallow minimum. A plot of density versus hydrogen content gives almost a straight-line relationship, and if the reciprocal of density is plotted, the linear relationship is improved. Values for different macerals as well as for a given maceral of different ranks are almost on the same line.

Thermal conductivity and thermal diffusivity are also dependent on pore and crack structure. Thermal conductivities for coals of different ranks at room temperature are in the range of 0.23–0.35 W/(m·K). The range includes the spread owing to crack variations and thermal diffusivities of $(1-2) \times 10^{-3}$ cm^2/s. At 800°C these ranges increase to 1–2 W/(m·K) and $(1-5) \times 10^{-2}$ cm^2/s, respectively. The increase is mainly caused by radiation across pores and cracks.

The specific heat of coal can be determined by direct measurement or from the ratio of the thermal conductivity and thermal diffusivity. The latter method gives values decreasing from 1.25 J/(g·K) (0.3 cal/(g·K)) at 20°C to 0.4 J/(g·K) (0.1 cal/(g·K)) at 800°C. The specific heat is affected by the oxidation of the coal (46).

Ultrafine Structure.   Coal contains an extensive network of ultrafine capillaries (3,4,6,47) that pass in all directions through any particle. The smallest and most extensive passages are caused by the voids from imperfect packing of the large organic molecules. Vapors pass through these passages during adsorption, chemical reaction, or thermal decomposition. The rates of these processes depend on the diameters of the capillaries and any restrictions in them. Most of the inherent moisture in the coal is contained in these capillaries. The porous structure of the coal and products derived from it have a significant effect on the absorptive properties of these materials.

A range of approaches has been developed for studying the pore structure. For example, heat of wetting by organic liquids is one measure of the accessible surface. The use of liquids having different molecular sizes gives information about restrictions in the pores. Measurements of the apparent density in these liquids give corresponding information about the volume of capillaries. Measurement of the adsorption of gases and vapors provide information about internal volume and surface area. Pores have been classified into three size ranges: (1) micropores (<0.4–2.0 nm) measured by CO_2 adsorption at 298 K; (2) mesopores (2.0–50 nm) from N_2 adsorption at 77 K; and (3) macropores (>50 nm) from mercury porosimetry. For coals having <75 wt % C, macropores primarily determine porosity. For 76–84 wt % C about 80% of the pore volume primarily results from micro and transitional pores. For the higher rank coals porosity is caused primarily by micropores (48).

Bituminous coals appear to have specific internal surfaces in the range of 30–100 m²/g arising almost entirely from ultrafine capillaries of <4 nm diameters. The surface area of the very fine capillaries can be measured accurately by using methods not too far below room temperature, depending on the gas or vapor used (49). Diffusion into the particle is very slow at low temperatures. Therefore, measurements at liquid nitrogen temperature (77 K) relate to the external surfaces and macro- as well as mesopores, and may yield areas that are lower than ambient temperature measurements by factors of 100. Sorption by neon or krypton near room temperature and heat of wetting in methanol have given surface area values. The methanol method is affected, up to a factor of 4, by polar groups, but it is faster. Pore characteristics may also be determined by nmr measurements, as with xenon (50). Measurements of the change in the nmr chemical shift with varying Xe pressures permit the calculation of pore diameters. Total porosity volumes of bituminous and anthracite coal particles are about 10–20%, and about 3–10% are in the micro-phase range. There are shallow minima in plots of internal area as can be seen from Figure 6 and in plots of internal porosity against coal rank in the range of vitrinite carbon content of 86–90% (6). It is possible to use low angle scattering of x-rays to obtain a value of internal surface, but this does not distinguish between accessible capillaries and closed pores (51).

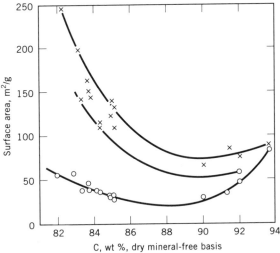

Fig. 6. Surface area of coals as estimated from (○) neon sorption at 25°C and (x) methanol heat of wetting.

Different coals have been observed in the electron microscope when two pore-size ranges appear, one of >20 nm and the other <10 nm (52). Fine pores from 1–10 nm across have been observed using a lead impregnation procedure (53). Effectiveness of coal conversion processes depends on rapid contact of gases with the surface. Large internal surfaces are required for satisfactory rates (54).

Mechanical Properties. Mechanical properties (4,6,55) are important for a number of steps in coal preparation from mining through handling, crushing, and grinding. The properties include elasticity and strength as measured by standard laboratory tests and empirical tests for grindability and friability, and indirect measurements based on particle size distributions.

Deformation Under Load. The mechanical behavior of coal is strongly affected by the presence of cracks, as shown by the lack of proportionality between stress and strain in compression tests or between strength and rank. However, tests in triaxial compression indicate that as the confirming pressure is increased different coals tend to exhibit similar values of compressive strength perpendicular to the directions of these confining pressures. Except for anthracites, different coals exhibit small amounts of recoverable and irrecoverable strain under load.

Dynamic tests have been used to measure the variation of elastic properties with coal rank. Tests using vitrains suggested that coals were mechanically isotropic up to 92 wt % C, with anisotropy increasing above this value. Dynamic tests were used to measure internal energy losses in vibration and to study the fluidity changes of coking coals on oxidation. The Young's modulus for median rank coals has been found to be about 4 GPa (4×10^{10} dyn/cm^2) (6). Sharp increases in the Young's and shear moduli have been found in vitrinites having increase in carbon content over 92 wt %.

Strength. The strength of a coal as measured in the laboratory may not be relevant to mining or size reduction problems where the applied forces are much more complex. There are indications that compressive strength, measured by compression of a disk, may give useful correlations to the ease of cutting for different kinds of equipment. Studies of the probability of survival of pieces of different size suggest that the breaking stress S should be most closely related to the linear size x rather than the area or volume of the piece. The results of a number of studies (6) indicate that S is proportional to x^{-r} where r frequently has the value 1/2.

The effects of rank on both compressive and impact strength have been studied, and usual minima were found at 20–25% dry, ash-free volatile matter (88–90 wt % carbon). Accordingly, the Hardgrove grindability index exhibits maximum values in this area.

Size Distribution Relationships. Different models have been used to describe the size distribution of particles experiencing single and multiple fractures. A model based on fracture at the site of the weakest link and a distribution of weakest links in the system gave results that could be described as well by the Rosin-Rammler relation (56). The latter is based on the concept that fracture takes place at pre-existing flaws that are distributed randomly throughout the particle.

Comminution. The size reduction of coal during handling of comminution results from many ill-defined forces. Grindability and friability tests are useful indicators of size reduction for any given coal having a specific energy consumption. The Hardgrove test yields an index that varies with coal rank, moisture content, and ash and maceral distribution. The higher the grindability index, the lower the energy requirements to achieve a given size reduction. These indexes are useful in establishing capacity factors for pulverizers. Grinding is easiest for coals having 75–80% dry, ash-free fixed carbon. Optimum moisture contents have been observed for the younger coals. High moisture contents lead to difficulty in grinding, and excessive drying causes the coal particles to be tougher than the optimum dryness.

A relationship between energy consumption and size reduction would be helpful for comminution processes, but none of the many attempts to develop this have been broadly applicable. One reason is that generation of new surface is only one of many phenomena in the size reduction (qv) process. The energy requirements of a comminution system may, however, be estimated from laboratory tests for given amounts of size reduction. For pulverized coal-fired boilers 70% <74 μm size ($-$200 mesh) is frequently used. Product size distributions of reproducible forms are obtained from a range of graded coal input sizes and careful control of crushing conditions in the laboratory. The Hardgrove test gives data for comparing ease of grinding different coals. The efficiency of pulverizers can then be calculated from the energy requirements for each product size from a series of tests. The relationship between a particle size and energy consumption obtained from plant data is frequently expressed in terms of Kick's or Rittinger's laws (57) or some modification such as Bond's law (58). These empirical relationships do not provide much insight into the mechanism of the grinding process.

The development of a continuous grinding index was the focus of work in the late 1970s (59). The laboratory test equipment used is similar to that for the Hardgrove test but permits classifying the product and recycling the oversize material. An improved correlation is obtained that may, however, need to be corrected for the relative sizes of the test grinding balls versus those used in commercial-scale equipment. The continuous grinding index is especially useful for lower rank coals.

Properties Involving Utilization. Coal rank is the most important single property for application of coal. Rank sets limits on many properties such as volatile matter, calorific value, and swelling and coking characteristics. Other properties of significance include grindability, ash content and composition, and sulfur content.

Combustion. Most of the mined coal is burned to produce steam for electric power generation (qv). The calorific value determines the amount of steam that can be generated. However, the design and operation of a boiler requires consideration of a number of other properties (see FURNACES, FUEL-FIRED).

In general, high rank coals (high calorific value) are more difficult to ignite, requiring supplemental oil firing and slower burning in large furnaces to reach complete combustion. Greater reactivity makes lower rank coals better suited for cyclone burners, which carry out rapid, intense combustion to maximize carbon utilization and minimize smoke emission. The burning profile, a derivative thermogravimetric analysis of oxidation, is used to characterize coal for oxidation or combustion behavior (see COMBUSTION TECHNOLOGY; THERMAL, GRAVIMETRIC, AND VOLUMETRIC ANALYSES).

Volatile matter is important for ease of coal ignition. Because high rank coals have low volatile matter contents, they burn more slowly and with a short flame.

They are primarily used for domestic heating, where heat is transferred directly from the fuel bed. For kilns, long hot flames are preferred and the coal should have medium to high volatile matter. The heating value released with the volatile matter for the various coals is given in Table 7 (6).

The swelling and caking properties of coal are not important for most boiler firing, such as pulverized coal-fired use. Some units, however, such as retort stokers, form coke in their normal operation. The smaller domestic heating units also require noncaking coal for satisfactory operation. For pulverized coal firing, a high Hardgrove index or grindability index is desired. The easier the coal is to grind, the lower the energy cost for pulverizing. The abrasiveness of the coal is also important because this determines the wear rate on pulverizer elements.

Coal moisture content, which affects both handling characteristics and freight costs, is most important for fine (smaller than 0.5 mm) particles. The lower rank coals have higher moisture contents. The moisture acts as a diluent, lowering flame temperatures and carrying sensible heat out with the flue gases. For pulverized coal firing the moisture content must be low to maximize grindability and avoid clogging. Thus dry run-of-mine coal having up to 30% ash may be more desirable than cleaning and drying coal.

The moisture content of peat or brown coal that is briquetted for fuel must be reduced to about 15% for satisfactory briquetting. Mechanical or natural means are used because of the cost of thermal drying. Moisture is sometimes desirable. About 8% is necessary for prevention of combustible loss from a chain-grate stoker.

Ash content is also important. Ash discharge at high temperature, as molten ash from a slagging boiler, involves substantial amounts of sensible heat. However, the higher cost of washed coal of lower (about 10%) ash content does not always merit its use. Ash disposal and extra freight costs for high ash coals also affect the coal selection. The use of continuous mining equipment produces coal having about 25% ash content. The average ash content of steam coal burned in the United States is about 15%. For some applications, such as chain-grate stokers, a minimum ash content of about 7–10% is needed to protect the metal parts.

Ash fusion characteristics are important in ash deposition in boilers. Ash deposition occurring on the furnace walls is termed slagging, whereas accumulation on the superheater and other tubes is termed fouling. A variety of empirical indexes have been developed (60,61) to relate fouling and slagging to the ash

Table 7. Rank and Heating Value in Volatile Matter

Rank (ASTM)	Volatile matter, %	Total heat energy liberated in volatile matter, %
anthracite	<8	5–14
semianthracite	8–14	14–21
low volatile bituminous	14–22	21–28
medium volatile bituminous	22–31	28–36
high volatile A bituminous	>31	36–47

chemical composition through parameters such as acidic and basic oxides content, sodium, calcium and magnesium, and sulfur.

A related property is the viscosity of coal ash. Ash viscosity affects the rate at which ash deposits may flow from the walls, and thus the requirements for ash removal equipment such as wall blowers and soot blowers. The preferred coal ash has a narrow temperature range through which it passes the plastic range, ca 25,000–1,000,000 mPa·s($=$cP) (62).

Some minor constituents can interfere in firing. High ($>$0.6%) chlorine is associated with high sodium and complex sulfate deposits that appear to be required to initiate deposition on superheater tubes, as well as initiate stress corrosion cracking of superheater tubes. Phosphorus ($>$0.03% of the coal) contributes to phosphate deposits where high firing temperatures are used. Sulfur forms complex sulfates, however, its most damaging effect is corrosion of the boiler's coolest parts through condensation of sulfur oxides as sulfuric acid. Control is achieved by setting flue gas temperatures above the acid dew point in the boiler areas of concern.

Sulfur content plays an important role in meeting air quality standards. In the United States, the EPA has set an emission limit for SO_2 of 516 g/10^6 kJ (490 g/10^6 Btu) of coal burned. To meet this, steam coals have to contain less than 1% sulfur. Regulations resulting from the Clean Air Act of 1991 call for reduction of the total amount of sulfur oxide emissions by 8–9 million tons annually. Half of this reduction is required by 1995, the remainder by the year 2000. A cap on the total emissions is given, and reductions in NO_x and particulate emissions are also mandated. Credits given for reductions of emissions that exceed the amount indicated for a given plant may be sold to other facilities unable to meet the requirements. This requirement is expected to force the addition of SO_2, NO_x, and particulate removal equipment to all boilers. Technology is being developed to control SO_x and NO_x through a combination of sorbent injection into the furnace and scrubbing and/or baghouse treatment to neutralize the acid gases and catalytically convert the NO_x to nitrogen (63).

Fluidized-bed boilers have been built with sizes up to 150 megawatts for commercial power generation and cogeneration units. This type of technology is displacing some stoker fired units at the low capacity (less than 200,000 kg/h of steam) boilers and smaller pulverized coal units at the large size of the fluidized-bed range. Bubbling and circulating bed designs are used, with operating temperatures in the 816–899°C range. Sulfur oxides are controlled using dolomite or limestone injection in the bed. Higher (2:1–5:1) Ca:S ratios are needed for fluid bed units than for wet scrubbing (1:1) or spray dryers (1.2–1.5:1). Nitrous oxide emissions are higher for fluid-bed units than for the other methods, possibly because of formation and oxidation of hydrogen cyanide (64,65) (see REACTOR TECHNOLOGY).

Coke Production. Coking coals are mainly selected on the basis of the quality and amount of coke that they produce, although gas yield is also considered. About 65–70% of the coal charged is produced as coke. The gas quality depends on the coal rank and is a maximum, measured in energy in gas per mass of coal, for coals of about 89 wt % carbon on a dry, mineral matter-free basis, or 30% volatile matter.

Coals having 18–32% volatile matter are used to produce hard metallurgical coke. Methods have been developed to blend coals having properties outside this range to produce coke. Several coals are frequently blended to improve the quality of the coke (6,66). Blending also affects the shrinkage required to remove the coke from the ovens after initial swelling. Lower rank coals having up to 40% volatile matter may be used alone or in blends at a gasmaking plant. This coke, which need not be as strong as metallurgical coke, is more reactive, and is used in the domestic market.

Coking coal is cleaned so that the coke ash content is not over 10%. An upper limit of 1–2 wt % sulfur is recommended for blast furnace coke. A high sulfur content causes steel (qv) to be brittle and difficult to roll. Some coal seams have coking properties suitable for metallurgical coke, but the high sulfur prevents that application. Small amounts of phosphorus also make steel brittle, thus low phosphorus coals are needed for coke production, especially if the iron (qv) ore contains phosphorus.

Solvent Extraction.   Coal partially dissolves in a number of solvents and this property has been used to aid in characterization of coal material, because the composition of extracts is sometimes similar to the coal. A wide range of organic solvents can be used (6,67), but dissolution is never complete and usually requires heating to temperatures sufficient for some thermal degradation or solvent reaction to take place, eg, ca 400°C. Dissolution of up to 40 wt % can be achieved near room temperature and up to 90% near 400°C. At room temperature the best solvents are primary aliphatic amines, pyridine, and some higher ketones, especially when used with dimethylformamide. Above 300°C large amounts can be dissolved using phenanthrene, 1-naphthol, and some coal-derived high boiling fractions. Coals having 80–85% carbon in the vitrinite give the largest yields of extract. Very little coal having >90 wt % C dissolves. Ultrasonic enhancement of extraction increases the yield of product by about 2.5 times the nonirradiated material (68) (see ULTRASONICS). The increase occurs only in solvent mixtures that significantly swell coals that range in rank from lignite to high volatile bituminous coal.

When the concentration of dissolved coal exceeds about 5% of the solution by weight, the extracted material resembles the parent coal in composition and some properties. The extract consists of the smaller molecules within the range of the parent coal. Recovered extract is relatively nonvolatile and high melting. A kinetic study of coal dissolution indicated increasing heats of activation for increasing amounts of dissolved coal (69).

Gasification.   Many of the coal selection criteria for combustion apply to gasification, which is typically a form of partial oxidation. Gasifiers are primarily described as fixed bed, fluidized bed, entrained, or rotating bed (70). The fixed bed involves an upward flow of reaction gas through a relatively stationary bed of hot coal. The gas velocity is slow enough to avoid blowing the coal out of the bed. The fluidized bed operates at higher gas velocities than the fixed bed and utilizes somewhat smaller particles. The entrained bed operates with parallel flows of reaction gas and finely pulverized coal particles to minimize reaction time and maximize throughput of product. The rotating bed is similar to a kiln that operates with the coal entering at the upper part of the inclined kiln. Rotation avoids clinkering and exposes fresh surfaces to enhance completion of the reaction. The range of coals that may be used vary from one gasifier type to another. Entrained flow gasifiers are able to handle the widest range of raw coals. Fixed-bed gasifiers require mildly caking or noncaking feedstocks for normal operation.

The Lurgi fixed-bed gasifier operates using lump coal of a noncaking type having an ash composition chosen to avoid a sticky, partly fused ash in the reactor. A slagging version of this gasifier has been tested in Westfield, Scotland. Other fixed-bed gasifiers have similar coal requirements.

The Shell-Koppers-Totzek gasifier is an entrained-bed type. It can gasify lignite and subbituminous or bituminous coal. The coal is fed as a pulverized fuel, usually ground to 70% <74 μm (−200 mesh) as used for pulverized coal fired boilers. Residence times are only a few seconds, therefore coal reactivity is important. The gasifier operates at >1650°C, so that coal ash flows out of the gasifier as a molten slag. Coal ash composition must permit continuing molten ash flow.

Fluidized-bed gasifiers typically require a coal feed of particles near 2–3 mm dia. Caking coals are to be avoided because they usually agglomerate in the bed. This can be avoided using a pretreatment consisting of a surface oxidation with air in a fluidized bed. A useful flue gas is produced. Examples of this type include the commercially available Winkler, and the U-Gas technology developed at the Institute of Gas Technology in Chicago.

Chemistry

Coal reactions, which on heating are important to the production of coke and synthetic fuels, are complicated by its structure.

Mature (>75 wt % C) coals are built of assemblages of polynuclear ring systems connected by a variety of functional groups and hydrogen-bonded cross-links (Fig. 4) (3,4,7,21). The ring systems themselves contain many functional groups. These polynuclear coal molecules differ one from another to some extent in the coal matter. For bituminous coal, a tarlike material occupies some of the interstices between the molecules. Generally coal materials are nonvolatile except for some moisture, light hydrocarbons, and contained carbon dioxide. The volatile matter produced on carbonization reflects decomposition of parts of the molecule and the release of moisture. Rate of heating affects the volatile matter content such that faster rates give higher volatile matter yields.

Coal composition is denoted by rank. Rank increases with the carbon content and decreases with increasing oxygen content. Table 8 gives a listing of the empirical formulas in terms of hydrogen, oxygen, nitrogen, and organic sulfur per 100 carbon atoms for a set of eight premium coal samples.

The surface of coal particles undergoes air oxidation, a process that may initiate spontaneous combustion in storage piles or weathering with a loss of heating and coking value during storage. Combustion produces oxides of sulfur and nitrogen as well as carbon dioxide and water vapor. The SO_x results from oxidation of both organic sulfur and inorganic forms such as pyrite. Nitrogen oxides are formed primarily from the nitrogen in the coal during high temperature combustion, rather than from the air used for combustion.

Partial oxidation as carried out in gasification produces carbon monoxide, hydrogen gas, carbon dioxide, and water vapor. The carbon dioxide reacts with

Table 8. Empirical Composition of Argonne Premium Coal Samples[a]

Coal		Composition, atoms/100 atoms C			
Type	From	H	O	N	S
Nonbituminous coals					
lignite	Beulah-Zap	79.5	20.9	1.4	0.4
subbituminous	Wyodak-Anderson	85.6	18.0	1.3	0.2
Bituminous coals					
high volatile 1[b]		77.3	13.1	1.5	1.2
high volatile 2[c]		85.7	10.8	1.7	0.2
high volatile 3[d]		76.3	8.9	1.6	0.3
high volatile 4[e]		76.7	8.0	1.7	0.4
medium volatile	Upper Freeport	66.0	6.6	1.6	0.3
low volatile[f]		58.5	2.0	1.2	0.2

[a]Ref. 71.
[b]Illinois No. 6 coal.
[c]Blind Canyon coal.
[d]Lewiston-Stockton coal.
[e]Pittsburgh coal.
[f]Pocahontas No. 3 coal.

hot carbon from the coal to produce carbon monoxide, and steam reacts with the carbon to produce carbon monoxide and hydrogen. The hydrogen can react with carbon through direct hydrogen gasification:

$$C + 2\,H_2 \rightarrow CH_4$$

at high hydrogen pressure, frequently 6.9 MPa (1000 psi) and moderate temperatures of 650–700°C. Methane may also be produced from

$$CO + 3\,H_2 \rightarrow CH_4 + H_2O$$

in a nickel-catalyzed reactor. This latter reaction is highly exothermic and is used to provide steam for the process. The correct 3:1 ratio of hydrogen to carbon monoxide is achieved using the water gas shift reaction:

$$CO + H_2O \rightarrow H_2 + CO_2$$

A mixture of CO and H_2, called synthesis gas, may also be used in other catalytic reactors to make methanol (qv) or hydrocarbons (qv):

$$CO + 2\,H_2 \rightarrow CH_3OH$$

or

$$n \text{ CO} + 2n \text{ H}_2 \rightarrow (\text{CH}_2)_n + n \text{ H}_2\text{O}$$

Surface oxidation short of combustion, or using nitric acid or potassium permanganate solutions, produces regenerated humic acids similar to those extracted from peat or soil. Further oxidation produces aromatic acids and oxalic acid, but at least half of the carbon forms carbon dioxide.

Treatment with hydrogen at 400°C and 12.4 MPa (1800 psi) increases the coking power of some coal and produces a change that resembles an increase in rank. Hydrogenation using an appropriate solvent liquefies coal. Noncatalyzed processes primarily produce a tarlike solvent-refined-coal used as a boiler fuel. Catalysts and additional hydrogen were used in the H-Coal process developed by Hydrocarbon Research, Inc. to produce a higher quality liquid product. A 450 t/d plant was built in Catlettsburg, Kentucky, to demonstrate this process by making a coal-derived refinery feedstock. The reactor used a catalyst suspended in a process derived liquid or ebulated bed. Hydrogen reactions over short (0.1–2 s) times with very rapid heating produce a range of liquids such as benzene, toluene, xylene, and phenol. A less rapid heating and lower maximum temperatures permit removal of some sulfur and nitrogen from the coal (72). These efforts have not been commercialized.

Treatment of coal with chlorine or bromine results in addition and substitution reactions. At temperatures up to 600°C chlorinolysis produces carbon tetrachloride, phosgene, and thionyl chloride (73). Treatment with fluorine or chlorine trifluoride at atmospheric pressure and 300°C can produce large yields of liquid products.

Hydrolysis using aqueous alkali has been found to remove ash material including pyrite. A small pilot plant for studying this process was built at the Battelle Memorial Institute in Columbus, Ohio (74) and subsequently discontinued. Other studies have produced a variety of gases and organic compounds such as phenols, nitrogen bases, liquid hydrocarbons, and fatty acids totaling as much as 13 wt % of the coal. The products indicate that oxidation and other reactions as well as hydrolysis take place.

The pyritic sulfur in coal can undergo reaction with sulfate solutions to release elemental sulfur (see SULFUR REMOVAL AND RECOVERY). Processes to reduce the sulfur content of coal have been sought (75). The reaction of coal and sulfuric acid has been used to produce cation exchangers, but it was not very efficient and is no longer employed. Efforts have turned to the use of hot concentrated alkali in a process called Gravimelt.

Many of the products made by hydrogenation, oxidation, hydrolysis, or fluorination are of industrial importance. Concern about stable, low cost petroleum and natural gas supplies is increasing the interest in some of the coal products as upgraded fuels to meet air pollution control requirements as well as to take advantage of the greater ease of handling of the liquid or gaseous material and to utilize existing facilities such as pipelines (qv) and furnaces. A demonstration plant was built in North Dakota for conversion of coal to methane, also known as substitute natural gas (SNG) production. This plant, operated by Great Plains Gasification Associates and in use at this writing, may be converted to produce methanol instead of methane (see GAS, NATURAL). A chemistry based on the conversion of synthesis gas has been developed and applied extensively in South

Africa to the production of liquid fuels and many other products. A small-scale production is used in the manufacture of photographic film materials from coal-derived synthesis gas in the Eastman Kodak plant in Kingsport, Tennessee. However, the principal production of chemicals from coal involves the by-products of coke manufacturing.

Reactions of Coal Ash. Mineral matter impurities have an important effect on the utilization of a coal. One of the constituents of greatest concern is pyrite because of the potential for sulfur oxide generation on combustion. The highest concentrations of pyrite are associated with coal deposition under marine environments, as typified by the Illinois Basin, including parts of Illinois, Indiana, and Kentucky. Additionally, the mineral matter has a tendency to form sticky deposits in a boiler. This tendency is most pronounced using mixtures that are rich in water-soluble alkalies such as are found in the Western Plains states. Coals from North Dakota, South Dakota, Wyoming, and Montana are typically low in the sulfur bearing constituents and therefore otherwise desirable as fuels.

Coal deposits from east of the Mississippi River generally have acidic mineral constituents, ie, they are richer in silica and alumina and tend to produce higher melting ash mixtures. These materials do not soften until above 1000°C and have limited problems with deposition on the inside walls of the boiler (slagging) or on the superheater tubes inside the boiler (fouling).

Coal ash passes through many reactors without significant chemical change. High temperature, exceeding the ash-softening temperature for the coal, permits reactions of the simpler ash constituents to form more complex species. Molten ash behavior affects slagging and ash removal. Correlations of viscosity have been made with a variety of chemical parameters, and descriptions based on acid–base chemistry appear to correlate with observed effects (74). Iron may be interconverted between the Fe(II) and Fe(III) states. Significant reduction in viscosity occurs as ferrous concentrations increase.

Corrosion of boiler tubes appears to be initiated in some cases with the formation of a white layer of general composition $(Na,K)_3Al(SO_4)_3$. Conditions for initiation of the deposit are favored by coals having high alkali and sulfur contents. The white layer bonds to the tubes and permits growth of ash deposits that insulate the layer and permit further corrosion.

Plasticity of Heated Coals. Coals having a certain range of composition associated with the bend in the Seyler diagram (Fig. 3) and having 88–90 wt % carbon soften to a liquid condition when heated (4,6). These materials are known as prime coking coals. The soft condition is somewhat reversible for a time, but does not persist for many hours at 400°C, and is not observed above ca 550°C if the sample is continuously heated as in a coking process. Continuous or lengthy heating result in degradation of coal matter, releasing vapors and resulting in polymerization of the remaining material. The coal does not behave like a Newtonian fluid and only empirical measurements of plasticity can be made (see RHEOLOGICAL MEASUREMENTS). About 10–30% of the coal becomes liquid, having a melting point below 200°C, and this molten material plasticizes the solid matrix remaining.

The molten part of a vitrinite is similar to the gross maceral, and a part of the maceral is converted to a form that can be melted after heating to 300–400°C. The molten material is unstable and forms a solid product (coke) above 350°C at rates that increase with temperature. The decomposition of the liquid phase is rapid for lower rank noncoking coals, and less rapid for prime coking coals. The material that melts resembles coal rather than tar and, depending on rank, only a slight or moderate amount is volatile.

The fluidity of coal increases and then decreases at a given temperature. This has been interpreted in terms of reaction sequence of coal → fluid coal → semicoke. In the initial step, a part of the coal is decomposed to add to that which normally becomes fluid. In the second step, the fluid phase decomposes to volatile matter and a solid semicoke. The semicoke later fuses accompanied by evolution of additional volatile matter to form a high temperature coke.

Formation of a true coke requires that the fluid phase persist long enough during heating for the coal pieces to form a compact mass before solidification occurs from the decomposition. Too much fluidity leads to an expanded froth owing to formation of dispersed bubbles from gas evolution in the fluid coal. Excess bubble formation results in a weak coke. The porous nature of true coke is caused by the bubble formation during the fluid phase. The strength of semicoke is set by the degree of fusion during the fluid stage and the thickness of the bubble walls formed during the frothing. In the final conversion to a hard high temperature coke, additional gas evolution occurs while the solid shrinks and is subjected to thermal stresses. The strength of the resultant coke and the size of the coke pieces are strongly affected by the crack structure produced as a result of the thermal stresses. Strong large pieces of coke are desirable to support the ore burden in blast furnaces.

Several laboratory tests (3,6) are used to determine the desirability of a coal or blend of coals for making coke. These are empirical and are carried out under conditions that approach the coking process. The three properties that have been studied are swelling, plasticity, and agglomeration.

Several dilatometers have been developed to determine the swelling characteristics of coals. The sample is placed in a cylindrical chamber with a piston resting on the coal surface. The piston motion reflects coal volume changes and is recorded as a function of temperature with a constant heating rate. When the coal first softens, contraction is caused by the weight of the piston on softened coal particles that deform to fill void spaces. Swelling then takes place when the particles are fused sufficiently to resist the flow of the evolving gases. The degree of swelling depends on the rate of release of volatile matter and the plasticity of the coal. The mass stabilizes at 450–500°C as the semicoke hardens. The shapes of the curves depend on the dilatometers. Curves obtained using the Hofmann apparatus have been classified into four main types that permit distinguishing coals having the optimum softening and swelling properties for production of a strong coke (76). Types A and C, and to some extent type D, can soften so that curves like those shown in Fig. 7 can be obtained.

Free-swelling tests are commonly used to measure a coal's caking characteristics. A sample of coal is packed in a crucible or tube, without compaction, and heated at a fixed rate to about 800°C. Infusible coals distill without changing appearance or state of agglomeration. The fusible coals soften, fuse, and usually swell. The profile of the resultant coke is compared to a series of reference profiles so that a swelling index can be assigned. The profiles represent indexes between 0 and 9. The best cokes come from coals having indexes between 4 and 9.

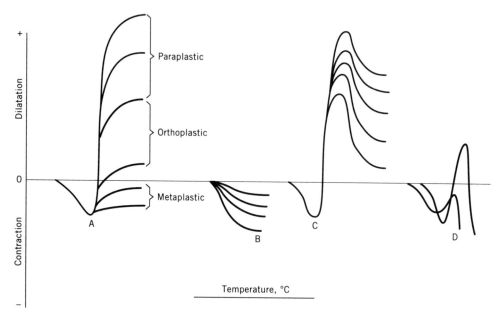

Fig. 7. Coal classification system according to Hofmann where A = Eu plastic; B = Sub plastic; C = Per plastic; and D = Fluido plastic. Courtesy of Centre d'Etudes et Recherches des Charbonnages de France and Brennstoff-Chemie.

The Gray-King assay, primarily carried out in Europe and the UK, is obtained from a similar test. The coal is heated to 600°C in a horizontal tube. Standard photographs are used to compare general appearance, profile, and size of the coke mass. Before testing, the more fusible coals are mixed with varying amounts of a standard electrode carbon of carefully selected size. A nonuniform scale termed A-F and G-G9 has been developed from the coke appearance for low swelling coals or from the amount of carbon required to give a standard appearance for the high swelling coals. The UK National Coal Board Rank Code Numbers are partly assigned on the Gray-King assay and partly on the volatile matter. The Gray-King assay procedure can also permit evaluation of yields of tar, gas, and liquor.

Plasticity can be studied using a device known as the Gieseler plastometer. A constant torque is applied to a shaft with rabble arms imbedded in coal in a crucible heated at a fixed rate. The rate of rotation of the shaft indicates the fluidity of the coal and is plotted as a function of the coal temperature. These curves, as shown in Figure 8, have a well-defined peak for coking coals usually near 450°C. Softening occurs at 350–400°C. At a normal heating rate of 3°C/min, the fluid hardening may be complete by 500°C.

Several agglutinating and agglomerating tests that indicate the bonding ability of the fusible components and depend on the crushing strength of a coke button produced, in some cases with addition of inert material, are also used. The Roga agglutinating test, developed in Poland, provides one of the criteria of the Geneva International Classification System. The coal sample is mixed with carefully sized anthracite, compacted, and heated to 850°C in 15 min. The part of

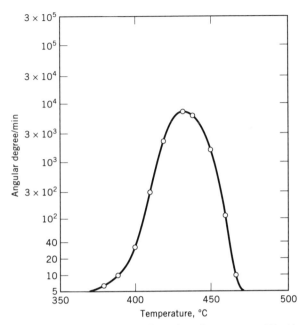

Fig. 8. Plasticity curve obtained using the Gieseler plastometer. Heating rate is 2°C/min. Courtesy of Centre d'Etudes et Recherches des Charbonnages de France.

the product that passes through a 1 mm (ca 18 mesh) screen is weighed, and a rotating drum further degrades the product. Roga indexes from 0–70 have been determined.

A coherent plastic layer from a few mm to 2–3 cm thick separates the semi-coke and coke from the unfused coal in the coke oven. Coking properties are assessed in Russia and some other countries by a measurement of the thickness of this plastic layer. A standardized test widely used in eastern Europe is the best known of this type (6) and involves a penetrometer used to measure the thickness of the plastic layer in a column of coal heated from the bottom. The various standard tests give results that are similar but do not give close correlations with each other.

The behavior of different polymerizing and gas-relating materials has been used to relate the plastic behavior of coal with known kinds of chemical change (3). The plastic nature of coal matter is determined by the competition between the reactions that generate the liquid phase, and those that convert it to semicoke. In general for vitrinites the greater the heating rate, the greater the fluidity or plasticity and the dilatation. Inertinite essentially does not contribute to the plastic properties of the coal. Exinite becomes fluid when heated, but also rapidly devolatilizes instead of forming semicoke, has little value as a binder, and can increase the fluidity to an undesirable extent.

Pyrolysis of Coal. Most coals decompose below temperatures of about 400°C (5,6), characteristic of the onset of plasticity. Moisture is released near 100°C, and traces of oil and gases appear between 100–400°C, depending on the coal rank. As the temperature is raised in an inert atmosphere at a rate of

1–2°C/min, the evolution of decomposition products reaches a maximum rate near 450°C, and most of the tar is produced in the range of 400–500°C. Gas evolution begins in the same range but most evolves above 500°C. If the coal temperature in a single reactor exceeds 900°C, the tars can be cracked, the yields are reduced, and the products are more aromatic. Heating beyond 900°C results in minor additional weight losses but the solid matter changes its structure. The tests for volatile matter indicate loss in weight at a specified temperature in the range of 875–1050°C from a covered crucible. This weight loss represents the loss of volatile decomposition products rather than volatile components.

A predictive macromolecular network decomposition model for coal conversion based on results of analytical measurements has been developed called the functional group, depolymerization, vaporization, cross-linking (FG-DVC) model (77). Data are obtained on weight loss on heating (thermogravimetry) and analysis of the evolved species by Fourier transform infrared spectrometry. Separate experimental data on solvent swelling, solvent extraction, and Gieseler plastometry are also used in the model.

Six factors form the basis of this model: (*1*) the decomposition of functional group sources in the coal yield the light gas species in thermal decomposition. The amount and evolution kinetics can be measured by thermogravimetry/Fourier transform infrared spectrometry (tg/ftir) and the functional group changes by ftir and nmr; (*2*) the decomposition of a macromolecular network yields tar and metaplast. The amount and kinetics of the tar evolution can be measured by tg/ftir and the molecular weight by field ionization mass spectrometry (fims). The kinetics of metaplast formation and destruction can be measured by solvent extraction, by Gieseler plastometry, and by proton magnetic resonance thermal analysis (pmrta); (*3*) the molecular-weight distribution of the metaplast depends on the network coordination number, ie, the average number of attachments on aromatic ring clusters. The coordination number can be determined by solvent swelling and nmr; (*4*) the network decomposition is controlled by bridge breaking. The number of bridges broken is limited by the available donatable hydrogen; (*5*) the network solidification is controlled by cross-linking. The changing cross-link density can be measured by solvent swelling and nmr. Cross-linking appears to occur with evolution of both CO_2 prior to bridge breaking and CH_4 after bridge breaking. Thus low rank coals, which form a lot of CO_2, cross-link prior to bridge breaking and are thermosetting. High volatile bituminous coals, which form little CO_2, undergo significant bridge breaking prior to cross-linking and become highly fluid. Weathering, which increases the CO_2 yield, causes increased cross-linking and lowers fluidity; and (*6*) the evolution of tar is controlled by mass transport in which the tar molecules evaporate into the light gas species and are carried out of the coal at rates proportional to their vapor pressure and the volume of light gases. High pressures reduce the volume of light gases and hence reduces the yield of heavy molecules having low vapor pressures. These changes can be studied using field ionization mass spectrometry.

Nature and Origin of Products.   Volatile matter yields decrease with increasing coal rank. For slow heating the final weight loss depends on the maximum temperature (Fig. 9). A variety of reactions take place and increasing temperatures provide the thermal energy required to break the stronger chemical bonds. Much decomposition takes place in a short time (apparently <1 s), but detection

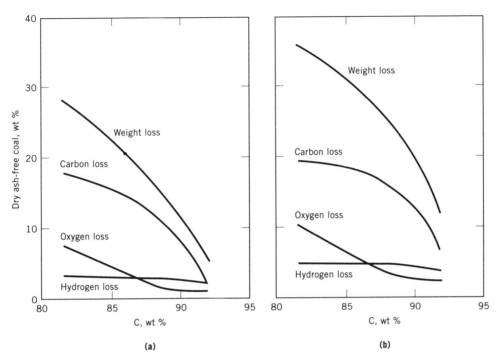

Fig. 9. Composition of volatile matter as a function of rank (bright coals) at (**a**) 500°C and (**b**) 900°C. The wt % of C is on a dry ash-free basis of unheated coal. Courtesy of Institute of Fuel.

is limited by the rate of diffusion of the volatile products through the solid. The liquids result from initial decomposition and gases from decomposition of liquid material. Very rapid heating rates produce weight losses as high as 72% at 1900°C, suggesting that the intrinsic volatile matter is limited only by the vapor pressure of the initial pyrolysis fragments, and would be expected to increase with temperature and decreasing coal rank (78).

The residual solid or char heated to 500°C contains 3–3.5 wt % H and up to 5 wt % O. On further heating to 900°C the solid contains only 0.8 wt % H and up to 0.3 wt % O. An aqueous liquor is produced that comes from the moisture in the coal as well as hydroxyl and possibly other oxygen-containing groups. Phenols in the tar are probably derived from hydroxyl aromatic groups in the coal. The total tar yield appears to be proportional to the fraction of aromatic carbon in the coal (see TAR AND PITCH). Coke oven gas is obtained from a variety of reactions that include cracking some of the tar. The hydrogen in the gas is generated after the char is heated to 400°C, but most is evolved in the conversion of the fluid coal to semicoke or coke at 550–900°C. The steam in the ovens can also produce hydrogen on reaction with hot coke.

Pyrolysis Reaction Mechanisms. An overall picture of the pyrolysis process is generally accepted but the detailed mechanism is controversial. Information has been obtained from: the sequence of volatile material appearing in a coke plant as determined by gas chromatography; laboratory work simulating coking and minimizing secondary reactions by working in vacuum or sweeping with inert gas (79); laboratory studies using model organic compounds to determine the mechanism by which these materials are converted to coke, liquid, and gaseous products; and laboratory work with more complex materials, including specially synthesized polymers, to better provide a model of coal (4,80). Radioactive tracers have been used in the last two studies to follow the transformation to materials in the products (4). In the last study, gas generating materials were added to aid in simulating the swelling process. The dehydrogenation of coal, which can alter the distribution of products, also provides information regarding the formation mechanism (81). The mechanism of formation of metallurgical coke and its effect on coke properties has also been described (82).

The mechanism of coal pyrolysis has been discussed (77,79,82) and a table summarizing the various changes has been prepared (79). The early stages involve formation of a fluid through depolymerization and decomposition of coal organic matter containing hydrogen. Around 400–550°C aromatic and nonaromatic groups may condense after releasing hydroxyl groups. The highest yields of methane and hydrogen come from coals having 89–92 wt % C. Light hydrocarbons other than methane are released most readily below 500°C; methane is released at 500°C. The highest rate for hydrogen occurs above 700°C (77,79).

Resources

World Reserves. Amounts of coal of some specified minimum deposit thickness and some specified maximum overburden thickness existing in the ground are termed resources. There is no economic consideration for resources, but reserves represent the portion of the resources that may be recovered economically

Table 9. Estimated Total Original Coal Resources of the World by Continents[a]

Continent	Identified resources,[a] 10^9 t	Hypothetical[b] resources,[c] 10^9 t	Estimated total resources,[d] 10^9 t
Asia[c]	3,635[d]	6,362	9,997[e]
North America	1,727	2,272	3,999
Europe[f]	273	454	727
Africa	82	145	227
Oceania[g]	64	55	118
Central and South America	27	9	36
Totals	*5,808*	*9,297*	*15,104*

[a]Ref. 84.
[b]Original resources in the ground in beds 30 cm or more thick and generally less than 1299 m below surface but includes small amount between 1200 m and 1800 m.
[c]Includes European Russia.
[d]Includes about 2090 × 10^9 metric tons in Russia.
[e]Includes about 8600 × 10^9 metric tons in Russia.
[f]Includes Turkey.
[g]Australia, New Zealand, and New Caledonia.

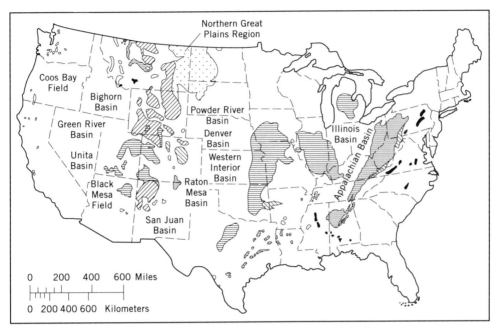

Fig. 10. Coal fields of the conterminous United States where ■ represents anthracite and semianthracite; ▨, low volatile bituminous coal; ▤, medium and high volatile bituminous coal; ▩, subbituminous coal; and ⬚, lignite (84).

using conventional mining equipment. The first inventory of world coal resources was made during the Twelfth International Geological Congress in Toronto in 1913. An example of the changes since 1913 can be seen from an examination of the coal resources for Canada. These were estimated to have been 1217×10^9 metric tons in 1913, based on a few observations and statistical allowance for all possible coalbeds to a minimum thickness of 0.3 m and to a maximum depth of 1220 m below the surface. In 1974, however, the estimate of solid fossil fuel resources (excluding peat) from the World Energy Conference gave the total resources as only 109×10^9 metric tons, and in 1986 the proven recoverable resources and estimated additional amount in place was given as 50×10^9 metric tons (83), less than 5% of the earliest figures (see FUEL RESOURCES).

Comprehensive reviews of energy sources are published by the World Energy Conference, formerly the World Power Conference at six-year intervals (83). The 1986 survey includes reserves and also gives total resources. In 1986 the total proven reserves of recoverable solid fuels were given as 6×10^{11} metric tons. One metric ton is defined as 29.2×10^3 MJ (27.7×10^6 Btu) to provide for the variation of calorific value in different coals. The total estimated additional reserves recoverable and total estimated additional amount in place are 2.2×10^{12} and 7.7×10^{12} metric tons, respectively. These figures are about double the 1913 estimates, primarily because significantly increased reserves have been indicated for Russia.

The part of the resource that is economically recoverable varies by country. The estimates made in the survey show that the proven recoverable reserves

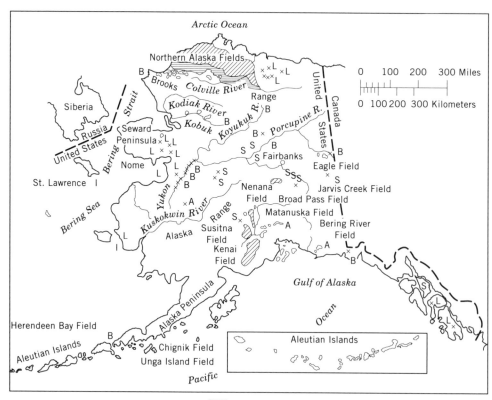

Fig. 11. Coal fields of Alaska where ▤ represents bituminous coal; ▨, subbituminous coal and lignite; and X is an isolated occurrence of coal of unknown extent. A = anthracite; B = bituminous; S = subbituminous; and L = lignite (84).

would last about 1200 years at the 1988 annual rate of production and that the estimated additional amount in place represent almost 1700 years at 1988 annual consumption.

In Table 9 (84), a somewhat different basis is used. The estimated total original coal resources of the world include beds 30-cm thick, and generally <1220 m below surface but also include small amounts between 1229 m and 1830 m. The data from column 1 are from earlier World Power Conference Surveys, whereas the figures for hypothetical resources (col. 2) and total estimated resources (col. 3) may be less reliable. This estimate represents about one-third more than the World Energy Conference Survey.

Reserves in the United States.   Coal is widely distributed and abundant in the United States as indicated in Figures 10 and 11. A large portion of the coal fields contain lignite and subbituminous coal, however, and another portion of the coal is contained either in thin or deep beds that can be mined only with difficulty or great cost. Reserve estimates for the United States as of 1974 were on the order of 1.6 × 10^{12} t (84). Little mapping and prospecting has been done to change the 1974 estimate. This information is useful for showing the quantitative distribution of reserves, selecting appropriate areas for further exploration of development, and in planning coal-based industrial activity.

The reserves of 21 states have been classified by overburden thickness, reliability of estimates, and bed thickness. This coal represents about 60% of the total identified tonnage. Of this, 91% is less than 305 meters from the surface, 43% is bituminous, and 58% is in beds thick enough to be mined economically.

Table 10. U.S. Coal Production by State, 1984–1991, t × 10[3a]

State	1984	1985	1986	1987	1988	1989	1990[b]	1991[c]
				East				
Alabama	24,596	25,240	23,450	23,190	24,078	25,417	26,359	24,778
Georgia	114							
Illinois	57,902	53,755	56,174	53,713	53,203	53,814	54,837	53,231
Indiana	34,100	30,251	29,830	31,061	28,394	30,546	32,604	28,544
Kentucky								
eastern	106,485	102,832	102,361	108,875	106,712	114,171	116,584	105,954
western	38,378	35,431	37,411	41,119	36,618	37,817	40,793	37,443
Maryland	3,726	2,710	3,547	3,597	2,944	3,065	3,166	3,552
Ohio	35,644	32,327	33,088	32,496	30,911	30,590	32,009	27,219
Pennsylvania								
bituminous	66,585	60,563	61,159	60,712	60,919	61,061	60,843	56,983
anthracite	3,779	4,275	3,897	3,232	3,228	3,040	3,183	2,883
Tennessee	6,640	6,761	6,238	5,849	5,911	5,884	5,623	3,860
Virginia	36,654	37,174	37,390	40,445	41,664	39,049	42,601	39,682
West Virginia	118,955	116,010	117,956	124,102	131,665	139,451	153,638	151,447
Total East	*533,559*	*507,327*	*512,500*	*528,391*	*526,246*	*543,907*	*572,240*	*535,577*
				West				
Alaska	780	1,301	1,426	1,355	1,584	1,436	1,549	1,302
Arizona	10,462	8,740	10,493	10,332	11,257	10,837	10,264	11,988
Arkansas	74	73	152	76	251	64	54	41
California		64		42	49	37	55	46
Colorado	16,314	15,657	13,835	13,093	14,448	15,548	17,170	15,802
Iowa	479	537	439	425	310	390	346	318
Kansas	1,206	903	1,349	1,835	669	777	655	378
Louisiana		188	2,047	2,498	2,623	2,709	2,893	2,853
Missouri	6,114	5,058	4,256	3,897	3,785	3,067	3,166	2,000
Montana	29,964	30,227	30,852	31,234	35,304	34,270	34,155	34,610
New Mexico	19,321	20,160	19,518	17,371	19,797	21,521	22,057	19,747
North Dakota	20,078	24,401	23,281	22,829	26,996	26,846	26,525	26,813
Oklahoma	4,213	3,030	2,768	2,606	1,939	1,592	1,542	1,685
Texas	37,360	41,277	44,120	45,880	47,471	48,899	50,626	48,770
Utah	11,189	11,604	12,956	14,989	16,492	18,253	20,029	19,783
Washington	3,516	4,030	4,178	4,040	4,694	4,575	4,541	4,674
Wyoming	118,870	127,768	124,238	133,340	148,925	155,775	167,298	176,299
Total West	*279,939*	*295,017*	*295,907*	*305,843*	*336,596*	*346,596*	*362,925*	*367,110*
Grand Total	*813,498*	*802,344*	*808,407*	*834,233*	*862,842*	*890,503*	*934,401*	*902,686*

[a]Ref. 2. Data includes mines producing less than 9000 t/yr. Courtesy of U.S. Dept. of Energy/Energy Information Administration.
[b]Revised.
[c]Preliminary.

On a uniform calorific value basis, coal constitutes 69% of the total estimated recoverable resources of fossil fuel in the United States. Petroleum and natural gas are about 7% and oil in oil shale, which is not as of this writing used as a fuel, is about 23%. The 1989 total recoverable reserves of coal are about 500 times the 1989 annual production (2), whereas the reserves of oil and gas are smaller, the production and consumption rate of oil and gas in the United States is three times that of coal.

Coal Production.   In 1860 world coal production was 122×10^6 t/yr. Production increased to 1140×10^6 t in 1913, giving a 4.2% annual average rate of increase. The rate has slowed and been erratic since that time. Statistical data on world coal production from 1860 to 1960 is given in Reference 81. World coal and lignite production rose to about 4.7×10^9 t in 1988 (1).

United States Coal Production.   Coal production in the United States, which dates back to 1702, started in earnest about 1820. It has increased with fluctuations to 617×10^6 t in 1976, 755×10^6 t in 1982, and 878×10^6 t in 1989 (2). In 1988 the United States produced 24.6% of the world's bituminous coal supply (1). U.S. coal production by state for the years 1984–1989 is given in Table 10 (2). Bituminous coal shipments in 1988 totaled 795×10^6 t to United States destinations. Of these, electric utilities received 81.5% (682×10^6 t), coke and gas plants 8.3% (38×10^6 t), and industrial, commercial, retail, and transportation 10.5% (75×10^6 t) (2).

Anthracite production, which takes place in Pennsylvania, fell from 90×10^6 t in 1918 to 42×10^6 t in 1938; increased in 1944 to 58×10^6 t but has declined steadily to 23×10^6 t in 1957, 15×10^6 t in 1961, 6×10^6 t in 1974, and 3.2×10^6 t in 1989. Of the 6 million tons produced in 1974, half was mined and the other half was recovered from culm banks (mine waste piles).

The demand for energy is continually increasing and the highest energy consumption in the world occurs in the United States. In 1989 consumption totaled 8.6×10^{13} MJ (81.3×10^{15} Btu) or 11.7 metric tons of coal-equivalent per capita (85). World recoverable reserves were about 120 times the annual coal production in 1988 and about 10 times that for the additional reserves believed to be in place (1). Estimated coal consumption reduces the known recoverable reserves at about 1%/yr. Whereas the use of bituminous coal is expected to continue to increase in terms of tonnage, the percentage of coal used in the United States has stabilized as shown in Table 11.

Table 11. Energy Source Usage in the United States, % Energy Basis[a]

Source	1942	1952	1962	1972	1988	1991
coal,	68.7	45.2	32.5	25.6	23.5	23.1
petroleum	17.1	25.4	23.6	22.8	42.7	40.1
natural gas	9.0	23.3	38.3	44.4	23.1	24.7
hydroelectric	5.2	6.1	5.5	6.0	3.3	3.8
nuclear			0.1	1.2	7.1	8.0

[a] Ref. 2.

Sample Sources

Basic coal research requires a variety of coal samples of different ranks that workers may access using a minimum of effort. Coal sample banks fill this need. Moreover, over the past decades it has become evident that the quality of samples degrades from atmospheric oxidation and the degradation has limited the ability of researchers to compare results. The U.S. Department of Energy Office of Basic Energy Sciences has sponsored the Argonne Premium Coal Sample Program to permit the acquisition of ton-sized samples of each of eight different coals representing a range of coal ranks, chemical composition, geography, and maceral content (71).

The samples were collected and processed in a manner to avoid exposure to oxygen and control loss of moisture; then they were pulverized to convenient sizes for research, ie, <149 μm (-100 mesh) and <840 μm (-20 mesh); the entire ton was thoroughly mixed; and packaged in sealed glass ampules under nitrogen. These samples have been well characterized and are available in quantities that are expected to last for decades (Table 8).

Other coal sample banks are also in existence. The Penn State Sample Bank at Pennsylvania Sate University has the most diverse collection of samples (86). The Illinois Basin Coal Sample Program at the Illinois State Geological Survey specializes in samples from the Illinois Basin (89). The European Center for Coal Specimens has a significant collection of samples from the entire world and is located in Eygelshoven in The Netherlands (88). Each makes samples available in kilogram quantities.

Mining and Preparation

Mining. Coal is obtained by either surface mining of outcrops or seams near the surface or by underground mining depending on geological conditions, which may vary from thick, flat seams to thin, inclined seams that are folded and need special mining methods. Coal mining has changed from a labor intensive activity to one that has become highly mechanized. In 1988 the average output per person per day in underground mines in the United States was 17.7 metric tons. For surface mines the output was 42 metric tons per person per day (2).

Strip or open-pit mining involves removal of overburden from shallow seams, breaking of coal by blasting or mechanical means, and loading of the coal. The development of very large excavating equipment, including draglines, bulldozers, front-end loaders, and trucks, has been responsible for significantly increased production from strip mining.

The two methods of underground mining commonly used are room-and-pillar and longwall. In room-and-pillar mining the coal is removed from two sets of corridors that advance through the mine at right angles to each other. Regularly spaced pillars, constituting about half of the coal seam, are left behind to support the overhead layers in the mined areas. This method is used extensively in the United States and other nations having large reserves. The pillars may later be removed, leading to probable subsidence of the surface. Longwall mining is used to permit recovery of as much of the coal as possible (89). Two parallel headings are made 100–200 m apart and at right angles to the main heading. The longwall between the two headings is then mined away from the main heading. The equipment provides a movable roof support system that advances as the coal is mined and allows the roof to collapse in a controlled manner behind it. This method also leads to subsidence of the overhead layers.

Another method used in Europe for steeply inclined seams is called horizontal mining. Horizontal shafts are cut through rock below the coal seams. Vertical connections are then made to the seam to permit coal removal.

The mechanical equipment used in room-and-pillar underground mining usually involves a series of specific operations with continuous mining equipment. Continuous miners use rotating heads equipped with bits to pick or cut through the coal without blasting and load it into a shuttle car for conveying to a belt system.

Preparation.   Coal preparation is of significant importance to the coal industry and to consumers (6,55,90). Preparation normally involves some size reduction of the mined coal and the systematic removal of some ash-forming material and very fine coal. The percentage of mined coal that is mechanically cleaned in the principal coal producing countries has risen during the past 30 years. There are a number of reasons for this. The most important in the United States is the increased use of continuous mining equipment. The nature of this operation tends to include inorganic foreign matter from the floor and ceiling of the seams, thus run-of-mine coal includes about 25% mineral matter. The size consist of the mined coal is also smaller when produced using continuous mining equipment. The depletion of the better quality seams, which are low in ash and sulfur, in many coal fields necessitates cleaning of the remainder. Moreover, the economic need to recover the maximum amount of salable coal has led to cleaning of the finer sizes which had previously been discarded. Stringent customer demands for coal meeting definite specifications, regulations requiring the removal of pyrites to control air pollution (qv), increased freight rates, and ash disposal costs all contribute to the upsurge in coal cleaning (see COAL CONVERSION PROCESS, CLEANING AND DESULFURIZATION).

In earlier times the coal was hand-picked for removal of larger pieces of refuse, but higher labor costs have made this method uneconomical. Mechanical separation methods are used, most of which depend on the difference in density between the coal and refuse. The washability characteristics of a coal determine the extent to which the refuse may be removed. The laboratory float-and-sink analysis gives information on the percentages and quality of the coal material occupying different density ranges. From this information, graphs are constructed showing the composite quantity and coal quality that can be obtained by cleaning at different specific gravities. This information is considered with the economic factors involved in the sale of the washed coal to choose an optimum method of cleaning. Cleaning plants are usually designed to handle the output of specific mines and to clean for a specific market. The plant includes various types of cleaning methods designed to move the different fractions through several cleaning circuits to maximize recovery.

In some areas, run-of-mine coal is separated into three products: a low gravity, premium-priced coal for metallurgical or other special use, a middlings product for possible boiler firing, and a high ash refuse. The complete preparation of coal usually requires several processes.

Cleaning Methods. Jig washing is the most widely used of all cleaning methods. A bed of coal particles is subjected to alternate upward and downward currents of water causing a moving bed of particles to stratify. The lighter clean coal particles go to the top and the heavier refuse particles to the bottom (see MINERAL RECOVERY AND PROCESSING). The heavy medium process is a simple float-and-sink one that is widely used for coarse coal cleaning. The medium is usually a suspension of pulverized magnetite, which is mixed to the desired specific gravity. This method is also used in cyclones for a wide range of coal cleaning.

Using trough washers, the coal is fed to a trough in a stream of water that carries the coal particles forward but allows heavier particles to sink and be removed. Washing tables are used for cleaning fine coal. A coal suspension in water flows across a slanted table that oscillates at right angles to the direction of flow. The heavier refuse particles settle onto the table and are trapped by riffles or bars, while the lighter coal particles are carried over the riffles in the current. If dry cleaning is used, the coal passes over a perforated, oscillating table through which air is blown. This method creates a dust problem, although it eliminates the need for drying the coal. In countries using hydraulic mining or underground dust suppression with water, there is limited opportunity for dry cleaning.

Froth flotation (qv) is the most important method for cleaning fine coal because very small particles cannot be separated by settling methods. Air is passed through a suspension of coal in water to which conditioning reagents, usually special oils, have been added. The oils are selected so that the coal particles preferentially attach themselves to the bubbles and separate from the refuse that remains in suspension.

Dewatering. The coal leaving the cleaning plant is very wet and must be at least partially dried to reduce freight charges, meet customer requirements, and avoid freezing. Draining on screens removes substantial amounts of water from larger coal, but other dewatering (qv) methods are required for smaller sizes having larger surface areas.

Vibrating screens and centrifuges are used for dewatering. For very fine coal, such as that obtained from flotation, vacuum filtration with a disk or drum-type filter may be used. Flocculants may be added to aid filtration (see FLOCCULATING AGENTS). They are also used for cleaning wastewater and pollution control. If very low ($\leq 2-3\%$) moisture contents are required, thermal drying must be used. A number of dryer types are available including fluidized bed; suspension; and rotary- and cascade-dryers. All of these are expensive to operate, however (see DRYING).

Storage. Storage of the coal may be necessary at any of the various steps in production or consumption, ie, at the mine, preparation plant, or consumer location. Electric utilities have the largest amounts of coal in storage, having stockpiles that frequently are able to meet 60–100 day normal demand thus protecting against delays, shortages, price changes, or seasonal demands. Stockpiles have tended toward smaller sizes; a 65-day supply was reported for 1989 (2).

For utilities, two types of storage are used. A small amount of coal in storage meets daily needs and is continually turned over. This coal is loaded into storage bins or bunkers. However, long-term reserves are carefully piled and left undisturbed except as necessary to sustain production.

Coal storage results in some deterioration of the fuel owing to air oxidation. Moreover, if inadequate care is taken, spontaneous heating and combustion may result. As the rank of coal decreases, it oxidizes more easily and must be piled more carefully. Anthracite does not usually present a problem.

The surface of the coal particles oxidizes or weathers resulting in cracks, finer particles, and reduced agglomeration all of which may destroy coking properties. If spontaneous heating takes place, the calorific value of the coal is reduced. Hot spots must be carefully dug out and used as quickly as possible. Without spontaneous heating and with good compaction, calorific value losses below 1%/yr have been recorded.

Coal piles are carefully constructed to exclude air or to allow adequate ventilation. The latter requires larger sizes, graded as 4 cm + without fines, for avoiding heating by ventilation. For exclusion of air, mixed sizes provide fines to fill the gaps between larger pieces. Pockets of large sizes must not be allowed because these provide access for air. The coal should be compacted to maximize the bulk density of the coal pile.

Several approaches have been effective for storage: (1) large compacted layered piles where sides and top are sealed using an oil or asphalt (qv) emulsion. Four liters of oil seals one square meter of coal; (2) large compacted layered piles where sides and top are covered first with fines to seal the pile and then with coarse coal to protect fines from wind and weather. The sides may slope at angles ≥30°; (3) piles of compacted layers in open pits having tight sides so that the air has access only at the top; (4) sealed bins or bunkers in which airtight storage can be provided for smaller amounts of coal for long times; and (5) underwater storage in concrete pits. This is expensive and rarely used but effectively prevents deterioration although it introduces other problems related to handling wet coal.

Large compacted storage piles should be located on hard surfaces and not subject to flooding. A layer of fines may be put down first to facilitate recovery. Each layer of coal should be compacted after it is deposited. The top of the pile should have a slight crown to avoid water accumulation. Excessive heights should be avoided to prevent air infiltration caused by wind. Coal removal should be done in layers followed by compacting and smoothing the surface. Piles should be limited to the same rank of coal depending on the intended use.

Transportation

The usual means of transporting coal are railroad, barge, truck, conveyer belt from mine to plant, and slurry pipelines (2,4) (see TRANSPORTATION). In 1988 769 × 10⁶ t of coal was transported to United States destinations; of this, 57.5% was shipped by railroad, 16.0% by barge, 12.3% by truck, and 14.2% by conveyer, slurry pipeline, and other methods (2). Electric utilities consumed 85.83% of the coal transported in the United States in 1988 (2).

The unit train handles about half of rail transportation (71). Most unit trains consist of about 100 rail cars and are dedicated to coal haulage from the mine to the consuming plant. Almost no time is wasted at either the loading or unloading site as a result of efficient loading and unloading equipment. Diesel and diesel-electric trucks having capacities up to 320 t (2) handled the off-highway transport of coal.

A 437 km slurry pipeline, 46 cm in diameter, was started in 1970 to move coal from Arizona to southern Nevada. The coal is crushed and ground to the fineness needed for proper viscosity and settling. About 18–20% is -325 mesh (<44 μm), 35–45% is -100 mesh (<149 μm), and 0–2% is $+14$ (< 1070 μm) mesh. The solids content of the slurries has approached 70% using additives to stabilize the mixture. The slurry is dewatered with centrifuges before combustion of the coal.

For shipment in cold climates, a freezeproofing treatment using inorganic chemicals or oil spray is used. An oil spray on the inside of the coal cars is also effective at 3.8–5.7 L/car for four round trips of 1280 km each. Oil treatment has also been used for dustproofing; wind loss can be prevented by use of an asphalt emulsion on the top of rail cars (6).

Coal pipelines have been built in countries such as France (8.8 km), and Russia (61 km), and pipelines are also used for transporting limestone, copper concentrates, magnetite, and gilsonite in other parts of the world. The first coal pipeline, built in Ohio, led to freight rate reductions. The pipeline stopped operation after introduction of the unit train, used exclusively to transport coal from the mine to an electric power generation station.

Hydraulic transport is used in mines and for lifting coals to the surface in Russia, Poland, and France. Pneumatic transport of coal is used over short distances in power plants and steel mills. The longest (14.5 km) single flight conveyer belt in the world near Uniontown, Kentucky, has a capacity of 1360 t/h.

Economic Aspects

Table 12 gives the estimated destinations for United States bituminous coal exports in 1976 and 1989. The volume of exports almost doubled during this period. About one-third of the bituminous coal exported from the United States in 1976 went to Europe, another third to Asia, and the remainder to North and South America. The pattern shifted to about 24% to North and South America, about 50% to Europe, 24% to Asia, and the balance to Africa (1).

Of the 1989 total, 65,128,000 t were metallurgical coal and 34,910,000 t were steam coal. Exports of coke from the United States in 1989 were 1,169,120 t, whereas anthracite exports were 745,749 t. Lignite exports were 163,628 t. In 1989 Canada produced 77,727,000 t, imported 16,160,000 t and exported 36,094,000 t. Japan is the principal recipient of Canadian exports. Selected coal exports and imports are given in Table 13.

The weighted average values fob underground and surface mines decreased from $28.24/t in 1984 to $23.99/t in 1990 (2). Underground mine prices decreased from $36.66/t to $31.51/t, whereas surface mine prices dropped from $22.70/t to $18.72/t.

Table 12. U.S. Bituminous Coal Exports for 1976 and 1989, 10^3 t[a]

Country of destination	1976	1989
North America		
Canada	14,966	16,270
Mexico	228	10
other		98
Total	*15,194*	*16,378*
South America		
Argentina	477	741
Brazil	2,033	5,675
Chile	132	943
other		
Total	*2,642*	*7,359*
Europe		
Belgium/Luxembourg	1,998	7,052
Denmark		3,178
France	3,109	6,517
Germany	902	734
Greece	422	213
Ireland		1,248
Italy	3,820	11,237
The Netherlands	3,126	6,074
Portugal	234	1,424
Spain	2,280	3,316
United Kingdom	765	4,457
Total EEC	*13,760*	*45,449*
Albania		120
Bulgaria		96
Finland		103
Iceland		64
Norway	112	105
Rumania	192	1,558
Sweden	740	734
Switzerland	13	117
Yugoslavia	167	1,509
Total Europe	*17,920*	*49,740*
Asia, Mideast		
Israel		487
Japan	17,058	13,845
Korea	425	3,836
Taiwan		4,497
Turkey	217	1,686
Total Asia, Mideast	*17,700*	*24,351*
Africa		
Algeria		740
Egypt	292	585
Morocco		824
South Africa	114	0
other		1
Total Africa	*406*	*2,150*
Grand Total	*53,862*	*100,038*

[a]Ref. 1.

Table 13. Coal Exports and Imports, t $\times$ 10^{3a}

Country	Exports[b]	Imports[b]
Australia	99,391	
Belgium[c]	853	11,057
Canada	32,773	14,673
Colombia	13,842	
France[c]	1,535	12,025
Germany[c]	5,025	7,516
Indonesia[c]	658	
Italy		20,809
Japan		101,467
Korea[c]		22,087
The Netherlands[c]	1,748	13,584
Poland[c]	32,370	1,086
Republic of South Africa	46,724	
Spain[c]	7	8,776
Sweden[c]	35	3,846
United Kingdom[c]	1,739	12,004
United States	90,835	2,570[d]
Russia	37,534	12,011

[a]Ref. 1.
[b]Data for 1989 unless otherwise indicated.
[c]Data for 1988.
[d]Bituminous coal.

Analysis

Most countries have an official national organization, which is responsible for developing and maintaining standards for testing and analysis. The ASTM serves this purpose in the United States as does the British Standards Organization (BS) in the United Kingdom. In Geneva, the International Organization for Standardization (ISO) formed a committee (T.C.27), which is responsible for developing international standards. Each organization issues periodic updates of their standards. ASTM does this on an annual basis. The ASTM coal standard methods are each assigned a number, preceded by a D. The methods are periodically reviewed and revised. A two-digit number may follow the method number to indicate the year of the revision. BS standards of similar type also exist for most methods described. Details of an individual test may be found in the compilation for the respective source organization.

Sampling.   The procedures for taking a sample, reducing the particle size of the sample, and separation of a smaller portion for later analysis are given in ASTM D2234 and D2013 (18) and BS1017. The procedures describe the minimum amount of sample needed to maintain a representative sample for analysis.

Size Analysis.   ASTM and BS (18) provide a number of methods dealing with the size specifications and size analysis procedures including D197, D410, D311, and D431.

Moisture Holding Capacity.   The bed or equilibrium moisture is the amount of moisture retained after equilibration at 96–97% rh at 30°C (D1412) (18). Total moisture is determined by air drying, crushing to smaller particle sizes and heating at 107°C to constant weight (D3302).

Analysis.   The proximate analysis is based on determinations of volatile matter, moisture, and ash for a coal sample. Fixed carbon is then calculated by difference. Volatile matter is determined empirically by measuring the weight loss when coal is heated in a covered crucible at either 950°C (ASTM D3175 or D5142) or 900°C (BS).

The ultimate analysis gives the elemental composition in terms of C, H, N, S, and O (D3176, D3177 or D4239, and D3178) (18). C and H analyses are based on oxidation of the sample in a tube and reaction of the gaseous products with absorbents to permit calculation of C and H content. N involves a Kjeldahl determination. S involves oxidation and detection of SO_2 by infrared detection, titration with base, or conversion to an insoluble sulfate to determine total S. Oxygen is determined by difference. Oxygen in the organic material is calculated by subtracting the dry, ash-free percentage of C, H, N, and organic S from 100. Sulfur forms are determined in a separate procedure. Sulfate and pyritic sulfur are determined separately and subtracted from total S to give organic S.

Other Elements.   To determine chlorine, the sample is mixed with Eschka's mixture and burned to convert the chlorine to chloride or decomposed in an oxygen bomb (18). Chloride is determined by titration (D2361) or using a chloride selective electrode (D4208) (18) (see ELECTROANALYTICAL TECHNIQUES).

Phosphorus determination involves the conversion of phosphorus to soluble phosphate by digesting the coal ash with a mixture of sulfuric, nitric, and hydrofluoric acids (18). Phosphate is precipitated as ammonium phosphomolybdate, which may be reduced to give a blue solution that is determined colorimetrically or volumetrically (D2795) (18).

Calorific Value.   To determine calorific value, a sample is placed in a bomb, pressurized with oxygen, and ignited. The temperature rise in the water bath of the calorimeter (see THERMAL, GRAVIMETRIC, AND VOLUMETRIC ANALYSIS) surrounding the bomb is used to determine the calorific value (D2015, D3286, or D1989) (18).

Ash Fusibility.   A molded cone of ash is heated in a mildly reducing atmosphere and observed using an optical pyrometer during heating. The initial deformation temperature is reached when the cone tip becomes rounded; the softening temperature is evidenced when the height of the cone is equal to twice its width; the hemispherical temperature occurs when the cone becomes a hemispherical lump; and the fluid temperature is reached when no lump remains (D1857) (18).

Swelling and Coking Tests

For the free swelling index which is also known as the crucible swelling number (ASTM D720), a coal sample is rapidly heated to 820°C in a covered crucible. Then the profile of the resulting char is compared to a series of standard numbered profiles (18). For the Roga index weighed amounts of coal and standard anthracite are mixed and carbonized, and the product coke is tested in a Roga drum for its resistance to abrasion (89).

For the Gray-King coke-type assay test (91,92) coal is heated in a retort tube to 600°C and the product coke is compared to a series of standard cokes. For a strongly swelling coal, enough anthracite or electrode carbon is added to the coal to suppress the swelling. This method is primarily used in Europe.

In the Audibert-Arnu dilatometer test (91), a thin cylinder of compressed powdered coal contacting a steel piston is heated at a rate not over 5°C/min. The piston movement is used to calculate the percent dilation.

Hardgrove Grindability Index and Strength Tests. A specially-sized coal sample is ground in a specifically-designed ball and race grinding mill (D409). The index is determined from the amount of coal remaining on a 74 μm (200 mesh) screen (18). The higher the index, the easier it is to grind the coal.

The drop shatter test indicates the resistance of a coal or coke to breakage on impact (see D440). A sample is dropped in a standard way a number of times from a specified height. For the tumbler test or abrasion index (ASTM D441), the coal or coke is rotated in a drum to determine the resistance to breakage by abrasion (18).

Health and Safety Factors

Coal mining has been a relatively dangerous occupation (2,91–93). During the period from 1961–1967 the average fatality rate in the United States for each million person hours worked was 1.05. In the seven years after the passage of the Federal Coal Mine Health and Safety Act of 1969, the average fatality rate decreased to 0.58, and by 1989 the rate was 0.25 (2).

The rates of occupational injuries are reported per 200,000 employee-hours which correspond to about 100 employee-years. In 1989, the total for all mines was 11.84 or 11.84%. Over the three-year periods before and after passage of the 1969 act the rates for underground mining were 48.60% and 40.07%, respectively. The principal causes of fatalities are falling rock from mine roofs and faces, haulage, surface accidents, machinery, and explosions. For disabling injuries the primary causes are slips and falls, handling of materials, use of hand tools, lifting and pulling, falls of roof rock, and haulage and machinery (2).

Gases and Coal Dust Explosions. Gases can be hazardous in coal mines. Methane is of greatest concern, although other gases including carbon monoxide and hydrogen sulfide may be found in some mines. Methane must be detected and controlled because mixtures of air and 5–15% of methane are explosive.

The U.S. Mine Health and Safety Act of 1969 requires that a mine be closed if there is 1.5% or more methane in the air. The use of an electrical methane detection device is required. High capacity ventilation systems are designed to sweep gases from the cutting face and out of the mine. These systems remove all gases before they become harmful.

Whereas an explosion from methane tends to be localized, it may start coal dust explosions resulting in more widespread injury and loss of life. All coal breaking operations result in formation of fine coal particles; some are controlled with water during the mining operation. Breakage associated with hauling disperses dust, and dust accumulations can be made safe by rockdusting. Powdered limestone is spread over the mine surfaces to cover the dust.

Drainage. Some mines are located beneath subsurface streams, or the coal seams may be aquifers. These mines may become flooded if not continually pumped. In Pennsylvania anthracite mines as much as 30 tons of water may be pumped for each ton of coal mined (94).

Air or biological oxidation of pyrite leads to sulfate formation and dilute sulfuric acid in the mine drainage. This pollutes streams and the water supplies into which the mine water is drained. Means of controlling this problem are under study.

Other Hazards. Rocks falling from the roofs of mines used to cause the largest number of accidents. Roof bolts are placed in holes drilled into the roofs of working areas to tie the layers of rock together and thus prevent rock falls. A disease called pneumoconiosis, also called black lung, results from breathing coal dust over prolonged periods of time. The coal particles coat the lungs and prevent proper breathing.

Regulations. The U.S. Bureau of Mines, Mining Enforcement and Safety Administration (MESA) studies hazards and advises on accident prevention. MESA also administers laws dealing with safety in mines. Individual states may also have departments of mines to administer state standards.

The Federal Coal Mine Health and Safety Act set standards for mine ventilation, roof support, coal dust concentrations levels, mine inspections, and equipment. As a part of this comprehensive act, miners must receive medical examinations at employer expense, and payments are made from the U.S. government to miners who cannot work because of black lung disease.

Uses

Coal As Fuel. Coal is used as a fuel for electric power generation, industrial heating and steam generation, domestic heating, railroads, and coal processing. About 87% of the world's coal production is burned to produce heat and derived forms of energy. The balance is practically all processed thermally to make coke, fuel gas, and liquid by-products. Other uses of coke and fuel gas also contribute to coal consumption for heat. In the United States coal use for power generation has increased to 86.1% in 1988, whereas coking coal use has dropped to 4.7% and the industrial/retail market declined to 9.2% (2).

Electric Power Generation. Coal is the primary fuel for thermal electric power generation. Since 1940 the quantity of bituminous coal consumed by electric utilities has grown substantially in each successive decade, and this growth is expected to continue for many years. Coal consumed by electric utilities increased from about 536×10^6 t in 1981 to 689×10^6 t in 1989 (2). The reasons for increased coal demand include availability, relative stability of decreasing coal prices, and lack of problems with spent fuel disposal as experienced in nuclear power plants (see NUCLEAR REACTORS).

The overall efficiency of electric power plants consisting of coal-fired boilers and steam turbines has plateaued at about 39%. The addition of pollutant control equipment has increased the internal power use on the stations and lowered the effective efficiency of the plant. The increased efficiencies have been achieved through use of larger units (up to 1500 MW) and higher pressures to 24.1 MPa

(3500 psi) and reheat, but concerns about reliability and ability to match power generation and demand have kept plant sizes below these values. Maximum temperatures have not been increased because of the difficulties of corrosion owing to coal ash constituents, materials properties, and costs of better alloys. The advent of any future increases in efficiency depends on development of new systems of power generation, which might include fluid-bed boilers, gasification of coal to power a gas turbine having hot exhaust directed to a waste heat boiler in a combined cycle (gas turbine and steam turbine), or use of magnetohydrodynamics (qv) (see FURNACES, FUEL-FIRED).

Almost all modern large coal-fired boilers for electric power generation use pulverized coal. The cyclone furnace, built mainly for use in Germany and the United States, uses coarser pulverized coal. The ash is removed primarily as a molten slag from the combustor. Apparently this design is no longer offered in the United States. This method of firing has not been accepted in the United Kingdom because of the higher softening temperature of the ash of the British coals. Stoker firing is generally limited to the smaller obsolete stand-by utility plants and generation plants used by industrial companies.

One significant advantage of pulverized coal boilers is the ability to use any kind of coal, including run-of-mine or uncleaned coals. However, with the advent of continuous mining equipment, the ash content frequently is ca 25%, and some preparation is frequently practiced. There were 931 coal preparation plants in the United States in 1988, mainly in Kentucky, West Virginia, and Pennsylvania.

The advent of fluidized-bed boilers has enabled the size of units to go to 150 megawatts for commercial power generation and cogeneration in the last decade. This technology is displacing some stoker fired units at the low capacity applications and smaller pulverized coal units at the large size of the fluidized-bed range. Bubbling and circulating bed designs are used, and operating temperatures are in the 815–900°C range. Sulfur oxides are controlled using dolomite or limestone injection in the bed. Higher (2:1–5:1) calcium to sulfur ratios are needed for fluid-bed units than for wet scrubbing (1.0:–1) or spray dryers (1.2–1.5:1). Nitrous oxide emissions are higher than for other methods, possibly as a result of formation and oxidation of hydrogen cyanide (64). Several processes are being developed to reduce nitrogen oxides emissions. For example, three pressurized fluid-bed combustors are being demonstrated under the U.S. Department of Energy's Clean Coal Technology Program (15).

Integration of coal gasification and a combination of a gas turbine for power generation and a waste heat boiler for power generation is termed integrated gasification combined cycle (IGCC). Efficiencies are currently about 42% and promise to be higher as gas turbine technology improves (63). As of 1992 five plants using this technology have been announced in the United States. The IGCC technology uses sulfur gas removal techniques that result in higher removal rates than conventional scrubbers, in part because of the improved efficiency of scrubbing the more concentrated gases.

A primary concern in coal-fired power generation is the release of air pollutants. Limits on SO_2 output, 0.52 g/MJ equivalent of coal input to a new plant, have been established. For a bituminous coal of 27.9 MJ/kg there is thus an upper limit of 0.72% sulfur content. Relatively few coals can meet this requirement. The U.S. Department of Energy indicated recoverable reserves of 420×10^9 t in 1987

(2) that were categorized by sulfur content: 33.5% had 0.6% S or less, 15.4% had between 0.61% and 0.83% S, 16.1% had between 0.84 and 1.67% S, 12.4% had between 1.68 and 2.50% S, and 22.6% had more than 2.5% S. The lowest sulfur coal, ~86%, is found west of the Mississippi River, mainly in Montana and Wyoming, quite distant from the electric power demand centers in the East. A trend to utilization of the western coals has developed.

Industrial Heating and Steam Generation. The principal industrial users of coal include the iron (qv) and steel (qv) industry and the food, chemicals, paper (qv), engineering, bricks, and other clay products, and cement (qv) industries, and a group of miscellaneous consumers such as federal and local government installations, the armed services, and small industrial concerns. Most of the coal is burned directly for process heat, ie, for drying and firing kilns and furnaces, or indirectly for steam generation for process needs or for space heating, and for a small amount of electric power generation. Industrial coal usage in the United States has diminished significantly in past decades, especially among small users, because of the greater convenience in storing and handling gaseous and liquid fuels and the higher initial cost of coal-fired equipment.

Several developments are being pursued to utilize coal directly, ie, automation of controls, coal and ash handling equipment for smaller stoker and pulverized coal-fired units, design of packaged boiler units, and pollution control equipment. In the cement industry coal firing has been used, because the sulfur oxides react with some of the lime to make calcium sulfate in an acceptable amount.

Coal Processing to Synthetic Fuels and Other Products. The primary approaches to coal processing or coal conversion are thermal decomposition, including pyrolysis or carbonization (5,6), gasification (6), and liquefaction by hydrogenation (6). The hydrogenation of coal is not currently practiced commercially.

In the United States the Clean Coal Technology program was created to develop and demonstrate the technology needed to use coal in a more environmentally acceptable manner. Activities range from basic research and establishing integrated operation of new processes in pilot plants through demonstration with commercial-scale equipment.

High Temperature Carbonization. High temperatures and long processing times are used in carbonizing coking coals in coke ovens or gas retorts. Besides metallurgical or gas coke the products include fuel gas, crude tar, light oils (benzene, toluene and xylene, referred to as BTX, and solvent naphtha), and ammonia gas (see COAL CONVERSION PROCESSES, CARBONIZATION).

Most coal chemicals are obtained from high temperature tar with an average yield over 5% of the coal which is carbonized. The yields in coking are about 70% of the weight of feed coal. Tars obtained from vertical gas retorts have a much more uniform chemical composition than those from coke ovens. Two or more coals are usually blended. The conditions of carbonization vary depending on the coals used and affect the tar composition. Coal-tar chemicals include phenols, cresols, xylenols, benzene, toluene, naphthalene, and anthracene.

The largest consumer of coke is the iron and steel industry. In the United States, ca 600 kg of coke is used to produce a metric ton of steel. Japanese equipment and practice reduce the requirement to 400–450 kg. Coke is also used to make calcium carbide (see CARBIDES), from which acetylene is made. Synthesis gas for methanol and ammonia production is also made from gasification of coke.

Considerable research has been carried out to produce metallurgical-grade coke from low rank bituminous and subbituminous coal. This is especially true in areas where coking coal reserves are becoming significantly depleted or are unavailable. The leading countries in this area of research are the United States (FMC Formcoke), Japan (Itoh process), and Germany (BFL process). These processes generally involve carbonization of crushed coal in fluidized beds, agglomerating the semicoke into conveniently-sized balls with a binder, and calcining. The advantages of this technology include better heat transfer, shorter carbonizing time, continuous operation, and utilization of a much broader range of coals.

Low Temperature Carbonization.   Lower temperature carbonization of lump coal at ca 700°C, primarily used for production of solid smokeless fuel, gives a quantitatively and qualitatively different yield of solid, liquid, and gaseous products than does the high temperature processes.

Although a number of low temperature processes have been studied, only a few have been used commercially. These have been limited in the types of coal that are acceptable, and the by-products are less valuable than those obtained from high temperature processing. The Disco process is used in the United States to supply a limited amount of fuel to meet requirements of smoke ordinances. The British Coalite and Rexco processes produced substantial amounts of domestic smokeless fuel. Development of fluid-bed methods of carbonizing finer coal at ca 400°C has been studied in the United Kingdom. A reactive char is briquetted without a binder to produce a premium open-fire smokeless fuel.

Gasification.   Gasification of coal is used to provide gaseous fuels by surface and underground applications, liquid fuels by indirect liquefaction, ie, catalytic conversion of synthesis gas, and chemicals from conversion of synthesis gas. There are also applications in steelmaking (see COAL CONVERSION PROCESSES, GASIFICATION).

Gasifier Designs.   A number of gasifiers are either available commercially or in various stages of development. These are described as fixed bed, fluidized bed, and entrained or rotating bed. The fixed bed involves an upward flow of reaction gas through a relatively stationary bed of hot coal. The gas velocity is slow enough to avoid blowing the coal out of the bed. The fluidized bed operates at higher gas velocities than the fixed bed and utilizes somewhat smaller particles. The entrained bed operates with parallel flows of reaction gas and finely pulverized coal particles to minimize reaction time and maximize throughput of product. The rotating bed is similar to a kiln, which operates with the coal entering at the upper part of the inclined kiln. Rotation avoids clinkering and exposes fresh surfaces to enhance completion of the reaction. The range of coals that may be used vary from one gasifier type to another with entrained flow gasifiers able to handle the widest range of raw coals. Fixed-bed gasifiers require mildly caking or noncaking feedstocks for normal operation.

The Lurgi fixed-bed gasifier operates using lump coal of a noncaking type with an ash composition chosen to avoid a sticky, partly fused ash in the reactor. A slagging version of this gasifier has been tested in Westfield, Scotland. Other fixed-bed gasifiers have similar coal requirements.

Fluidized-bed gasifiers typically require a coal feed of particles near 2–3 mm in diameter. Caking coals are to be avoided, because they usually agglomerate in the bed. This can be avoided using a pretreatment consisting of a surface oxidation with air in a fluidized bed. A useful fuel gas is produced. Examples of this type include the commercially available Winkler, and the U-Gas technology developed at the Institute of Gas Technology in Chicago. The latter is offered by a joint venture of Stone & Webster and Tampella Keeler. This system uses air-blown gasification and hot gas cleanup.

The Texaco gasifier and a similar unit developed by The Dow Chemical Company are pressurized entrained gasifiers. At the top pulverized coal is mixed with reaction gas and is blown down into the gasifier. The reaction products leave from the side, and ash is blown down to a water pool where it is quenched. These units have operated at an Eastman Kodak facility in Kingsport, Tennessee and at the Coolwater power station in California for an integrated combined cycle power plant.

Pulverized coal is used in several entrained gasifiers and was studied in Germany before World War II. The Koppers-Totzek gasifier has been used commercially in different parts of the world. The original design used multiple (2 or 4) heads to feed coal, air or oxygen, and steam into an entrained atmospheric pressure reactor. Molten slag is discharged. The Babcock and Wilcox company also built an entrained bed gasifier for the DuPont Company at Belle, West Virginia, for chemical feedstock.

The Shell-Koppers-Totzek gasifier is also an entrained type. It can gasify lignite and subbituminous or bituminous coal. The coal is fed as a pulverized fuel, usually ground to 70% <74 μm (-200 mesh) as used for pulverized coal fired boilers. Residence times are only a few seconds, therefore coal reactivity is important. The gasifier operates at $>1650°C$, at 2.2 MPa (22 atm) so that coal ash flows out of the gasifier as a molten slag. Coal ash composition must permit continuing molten ash flow. The joint development of the Shell Oil Co. and Koppers-Totzek led to a demonstration plant in The Netherlands having a gasifier for a 250 MW (50 cycle) integrated gasification combined cycle scheduled to begin operation in 1993. This is to be one of the first of the new generation of these plants to operate (95).

Surface Gasification Technology.   Gasification of coal for fuel gas and chemical intermediate production has been developed commercially, and improvements in technology are being studied in a number of facilities. In the United States, the purpose of a number of programs has shifted from production of a substitute natural gas (methane) to electric power generation (qv) through the integrated gasification-combined-cycle (IGCC) plants. The interest in this use of coal results from the low emission levels that can be achieved and the potential for higher power generation efficiency.

Efficiencies of about 42% from natural gas to electricity have been indicated and can improve as the high temperature capabilities of turbines improve. Coal gasification would lower the overall energy efficiency but still give efficiencies greater than those with conventional coal-fired plants having typical emission control systems. Conventional power plants are able to produce electricity having heat rates of about 10 MJ/kWh and 90% SO_2 removal. The heat rates for IGCC plants are expected to be from 8 to 9.5 MJ/kWh having 99% SO_2 removal (96).

The Lurgi process (6) is the most successful complete gasification process for converting weakly caking coals as well as noncaking ones. The gasification takes place with steam and oxygen at 2–3 MPa (20–30 atm) to produce a 13.0–14.9 MJ/m^3 (350–400 Btu/ft^3) gas, which may be enriched with hydrocarbons to meet town gas specifications. The reactor is a slowly moving bed and is fed lump coal. Fine coal particles are usually removed before feeding to the gasifier.

The first commercial operation of the Lurgi process was in Germany in 1936 using brown coal. The reactor was modified to stir the coal bed to permit utilization of bituminous coal. One plant was built at the Dorsten Works of Steinkohlengas AG, and the Sasol plants were built in South Africa to provide synthesis gas for liquid fuels.

The gasifier for the 250 MW IGCC project in The Netherlands, scheduled to begin operation in 1993, is a 55 MW gas turbine with the balance of the power from a steam turbine. An Australian coal is to be used, and sulfur removal is expected to be 98.5% (95).

In the 1970s a combined U.S. Federal government–American Gas Association program supported the development of second generation processes for making pipeline quality gas. In these processes coal is prepared, gasified, the gas is cooled, shifted if necessary to adjust the H_2:CO ratio to about 3:1, the acid gases (H_2S and CO_2) removed, and then catalytic conversion to methane is carried out. Under this program the Institute of Gas Technology in Chicago developed the Hygas process in a 68 t/d pilot plant in which the gasifier at 6.9 MPa (1000 psi) accepts a coal slurry, dries it, goes through the first stage hydrogenation at 650–730°C, and second stage at 815–930°C before steam–oxygen gasification of the char to obtain high carbon utilization. The process also produces some benzene, toluene, and xylene, which were used in the pilot plant to make up the slurry. This process has been operated successfully using lignite, subbituminous, and bituminous coals.

The CO_2 Acceptor Process was also developed under this program by Consolidation Coal Co. in a 36 t/d pilot plant at Rapid City, South Dakota. Heat to drive the gasification process was provided by the reaction of calcined dolomite (MgO–CaO) and CO_2 produced in gasification of lignite or subbituminous coal using steam at 1 MPa (10 atm) and 815°C. The spent dolomite is regenerated at 1010°C in a separate vessel and returned to the gasifier. The process has operated successfully using lignite and subbituminous coal.

Still another process, called BI-GAS, was developed by Bituminous Coal Research in a 73 t/d pilot plant in Homer City, Pennsylvania. In this entrained-bed process, pulverized coal slurry was dried and blown into the second stage of the gasifier to contact 1205°C gases at ca 6.9 MPa (1000 psi) for a few seconds residence time. Unreacted char is separated and recycled to the first stage to react with oxygen and steam at ca 1650°C to produce hot gas and molten slag that is tapped.

The Synthane process was developed by the DOE at the Pittsburgh Energy Research Center. This fluidized-bed process operated at ca 6.9 MPa (1000 psi) and 980°C to gasify coal and produce some char. It used subbituminous coal. A third-generation process called Steam-Iron was also developed by the Institute of Gas Technology at a pilot plant in Chicago. This plant generates hydrogen from char produced in any gasification process. A gas producer uses air to make a reducing gas from the char in one vessel. The reducing gas converts iron oxide to iron in the upper two stages of a second vessel. Steam is converted to hydrogen and reoxidizes the iron in two stages in the lower half of the vessel.

None of these second- or third-generation processes has been commercialized, largely because of the relatively low price of available liquid and gaseous fuels.

A large commercial plant was completed in 1981 by a consortium of American Natural Gas and Peoples Gas, Light and Coke, and others for Mercer County, North Dakota. This plant has a design capacity of 3.7×10^6 m^3 (137.5×10^6 standard cubic feet (SCF)) of methane per day. The plant uses 14 Lurgi gasifiers and 12,700 t/d of lignite, 2,585 t/d oxygen, and 12,383 t/d of steam. The air separation plant is the largest in the hemisphere (see CYROGENICS; NITROGEN). The Phosam process is used for recovery of 113 t/d of ammonia, and the Stretford process was initially used for the recovery of 106 t/d of sulfur. The other products are used primarily as boiler fuels and include tar oil, naphtha, and crude phenol. The coal supply is Beulah-Zap lignite produced at an adjacent mine. The fine coal is removed before gasification and is sold to the neighboring electric utility (Basin Electric Power Cooperative) for use in the adjacent power plant and another plant about 48 km away. The cost of the gas was subsidized in the initial decade of operation. Future plans include the production of substitute natural gas (SNG) and the use of the site for demonstration of a coal to methanol plant (97). In 1988 the ownership of the plant was transferred to the Dakota Gasification Co. of Bismarck. The agreement calls for operation of the plant until 2009 as long as revenues exceed expenses.

Processes for intermediate-Btu gas, ie, 9.3–18.6 MJ/m^3 (250–500 Btu/ft^3), or synthesis gas production were also developed. In the IGT U-Gas, or a similar Westinghouse process, crushed coal is fed into a fluidized-bed gasifier. Steam and oxygen enter the base of the bed. A part of these gases carry unreacted fines into a hot spout, which accelerates gasification and permits the ash to soften and particles to agglomerate. Ash agglomerates discharge below the spout. Product gases can be cleaned and pipelined as industrial fuel gas near 11.2 MJ/m^3 (300 Btu/ft^3). This technology has been offered by a consortium of Stone & Webster and Tampella Keeler. This group uses air-blown gasification and hot gas cleanup to lower the capital costs. The modification provides a lower heating value product.

Several plants use the Texaco partial oxidation gasifier developed as a modification of Texaco's oil consuming partial oxidation process. Pulverized coal falls through the reactor at high pressure and temperature to produce the gas which is then cleaned. The ratio of carbon monoxide to hydrogen can be adjusted by the water gas shift reaction as needed for a variety of chemical intermediates. This design was used in a plant to make chemical intermediates for the Eastman-Kodak Co. in Kingsport, Tennessee. The Coolwater IGCC demonstration plant in Southern California used this gasifier to provide fuel gas for boilers for electric power generation or for gas turbines for combined cycle power generation. The plant was technically successful but not able to compete economically. The scrubbing system removes a very high amount of the sulfur in the coal (95,96).

The high capital cost, about $1500/kW, is the principal deterrent to growth of the IGCC concept. The ability to remove up to 99% of the sulfur species from the combustion products make the IGCC an environmentally desirable option as the Clean Air Act Amendment of 1991 phases in and the increased efficiency reduces the carbon dioxide emissions.

A similar design has been developed using a 161 MW plant by The Dow Chemical Company in its Plaquemine, Louisiana location. Destec, Inc. is a power subsidiary of The Dow Chemical Company and has joined with Public Service Of Indiana to build a new 230 MW plant near Terre Haute, Indiana. Operation is projected for 1995 (95).

Future large gasification plants, intended to produce ca 7×10^6 m^3 standard (250 million SCF) of methane per day, are expected to be sited near a coal field having an adequate water supply. It is cheaper to transport energy in the form of gas through a pipeline than coal by either rail or pipeline. The process chosen is expected to utilize available coal in the most economical manner.

Underground Coal Gasification (UCG). Underground coal gasification is intended to gasify a coal seam *in situ*, converting the coal into gas and leaving the ash underground. This approach avoids the need for mining and reactors for gasification. UCG is presently considered most interesting for deep coal or steeply sloping seams. This approach involves drilling holes to provide air or oxygen for gasification and removal of product gases and liquids (98).

A low calorific value gas, which includes nitrogen from air, could be produced for boiler or turbine use in electric power production, or an intermediate calorific value gas containing no nitrogen for an industrial fuel gas, or synthesis gas for chemical and methane production could be provided. This approach which has been studied in Russia, Europe, Japan, and the United States, is still noncommercial in part because it is not economically competitive.

Although many environmental and safety problems can be avoided using UCG, there is some concern about groundwater contamination as a result of the process (see GROUNDWATER MONITORING).

In the United States a program, carried out near Hanna, Wyoming for the Department of Energy, examined different approaches to gasification, including use of air and oxygen. Other programs under government sponsorship included use of a longwall generator at the Morgantown, West Virginia Energy Technology Center.

Industrial testing programs have been carried out by Gulf Research and Development Co. in Western Kentucky (99) on a coal seam at a depth of 32.6 m and a thickness of 2.7 m. The coal seam was excavated for study after the gasification program. Another program using Russian technology is being carried out by Texas Utilities Services in an East Texas lignite deposit.

A joint Belgian–West German program is aimed at gasifying seams ≥ 1000 m underground and using the gases for combined cycle electrical generating plants. If initial efforts are effective, hydrogen is to be pumped into hot coal seams to make substitute natural gas from the exothermic carbon–hydrogen reaction. Work using underground coal gasification has been most extensive in Russia where the technology has been applied to produce gas for four or five electrical generation stations. An institute was established in 1933 to study this process and has primarily studied air-blown gasification that produced a gas of about 3.35–4.20 MJ/m^3 (90–113 Btu/ft^3) heat content. Other work has produced synthesis gas suitable for chemical production.

The chemistry of underground gasification has much in common with surface gasification; however, many of the parameters cannot be controlled because the reaction occurs in a remote site. Heat energy to drive the gasification comes primarily from carbon combustion to produce CO and then CO_2. Because many coal seams are also aquifers there is a considerable amount of water intrusion, which leads to steam generation at the expense of the reaction energy. As a result the rate of air or oxygen passage through the injection wells and seam are adjusted to maintain a low level of moisture in the product gas. The steam is beneficial for char gasification and some is consumed in the water gas shift reaction to produce H_2 and CO_2 from H_2O and CO. Some H_2 reacts with C to produce CH_4, which enhances the calorific value of the gas.

UCG is started by drilling wells to serve as injection points for oxidant and steam as well as collection points for product gases. Permeability of the coal seam is achieved by directional drilling, countercurrent combustion, electro-linking or hydraulic fracturing. Permeability is needed to provide a high rate of production with a minimum of pressure drop through the reaction zone. Low rank, ie, lignite and subbituminous, coals crack and shrink during gasification, rendering the seams more permeable. The bituminous coals swell and plug gas channels unless carefully preconditioned with preliminary oxidation to avoid this.

Liquid Fuels and Chemicals from Gasification of Coal. Gasification of coal using steam and oxygen in different gasifiers provides varying proportions of carbon monoxide and hydrogen. Operations at increasing pressures increases the formation of methane. Because mixtures of CO and H_2 are used as the start of chemical synthesis and methane is not wanted or needed for chemical processes, the conditions favoring its formation are avoided. The product gases may then be passed over catalysts to obtain specific products. Iron-based catalysts are used to produce hydrocarbons in the Fischer-Tropsch process, or zinc or copper catalysts are used to make methyl alcohol.

The Fischer-Tropsch process has not been economical in competitive markets. The South African Sasol plant (100) has operated successfully using the Kellogg and German Arge (Ruhr Chemie Lurgi) modification of the Fischer-Tropsch process. The original plant was designed to produce 227,000 t/yr of gasoline, diesel oil, solvents, and chemicals from 907,000 metric tons of noncaking high ash subbituminous coal. The Lurgi gasification process is used to make the synthesis gas. The capacity of this plant was expanded substantially with Sasol II and Sasol III commissioned in 1980 and 1983 to meet transportation fuel needs for South Africa. The combined annual production capacity of the three Sasol facilities is 8×10^6 m^3 (50 million barrels) of liquid products (97).

The success of the Sasol project is attributed to the availability of cheap coal and the reliability of the selected components. Plants using Lurgi or Koppers-Totzek gasifiers for making chemicals are located in Australia, Turkey, Greece, India, and Yugoslavia, among other countries.

A variety of pilot plants using fluid-bed gasifiers have been built in the United States, Germany, and elsewhere. The Winkler process is the only one that has been used on a large scale. It was developed in Germany in the 1920s to make synthesis gas at atmospheric pressure. Plans were being made to develop a pressurized version. Plants using bituminous coal have been built in Spain and Japan with the atmospheric pressure gasifier.

Gasification and Metallurgy. Some interesting combinations of these technologies include direct reduction of iron ore and direct injection of coal into the blast furnace. In direct reduction, a reducing gas mixture of methane or carbon monoxide and hydrogen reduces iron ore pellets into elemental iron by reaction at 1000–1200°C. These pellets may later be used to feed steelmaking processes. In 1983, 45 plants having a capacity of 15×10^6 metric tons were in operation. Pulverized coal has been successfully injected into the tuyeres of a blast furnace of the Armco Co. in Middletown, Ohio to supplement coke (see IRON BY DIRECT REDUCTION).

Liquefaction. Liquefaction of coal to oil was first accomplished in 1914. Hydrogen was placed with a paste of coal, heavy oil, and a small amount of iron oxide catalyst at 450° and 20 MPa (200 atm) in stirred autoclaves. This process was developed by the I. G. Farbenindustrie AG to give commercial quality gasoline as the principal product. Twelve hydrogenation plants were operated during World War II to make liquid fuels (see COAL CONVERSION PROCESSES, LIQUEFACTION).

Imperial Chemical Industries in Great Britain hydrogenated coal to produce gasoline until the start of World War II. The process then operated on creosote middle oil until 1958. As of this writing none of these plants is being used to make liquid fuels for economic reasons. The present prices of coal and hydrogen from coal have not made synthetic liquid fuels competitive. Exceptions are those cases, as in South Africa, where there is availability of cheap coal, and fuel liquids are very important.

The Pott-Broche process (101) was best known as an early industrial use of solvent extraction of coal but was ended owing to war damage. The coal was extracted at about 400°C for 1–1.5 h under a hydrogen pressure of 10–15 MPa (100–150 atm) using a coal-derived solvent. Plant capacity was only 5 t/h with an 80% yield of extract. The product contained less than 0.05% mineral matter and had limited use, mainly in electrodes.

Solvent extraction work was carried out by a number of organizations in the United States. Pilot plants for producing solvent refined coal (SRC) were built and initially sponsored by the Southern Company Services and Electric Power Research Institute in Wilsonville, Alabama in 1973 and built with Department of Energy sponsorship near Tacoma, Washington in 1974 having capacities of 5 and 45 t/d of coal input, respectively. The Wilsonville plant was closed in 1992 after many modifications from the initial design; the Tacoma plant is closed.

In the SRC work, coal was slurried with a process-derived anthracene oil and heated to 400–455°C at 12.4–13.8 MPa (1800–2000 psi) of hydrogen for 0–1 h. A viscous liquid was extracted. The product stream contains some hydrocarbon gases, and H_2S. The residue is gasified to generate hydrogen for the process. The remaining filtrate is separated into solvent, which is recycled, and SRC, a low ash, tarlike boiler fuel.

Heating value of the product (SRC) is ca 37 MJ/kg (16,000 Btu/lb). Sulfur contents have been reduced from 2–7% initially to 0.9% and possibly less. Ash contents have been reduced from 8–20% to 0.17% (102). These properties permit compliance with EPA requirements for SO_2 and particulate emissons. The SRC is primarily intended to be used as a boiler fuel in either a solid or molten form (heated to ca 315°C). The solid has a Hardgrove index of 150 (103). Boiler tests have been successfully carried out using a utility boiler.

A series of process improvements have been developed at Wilsonville to produce high quality transportation fuels. Two integrated stages of liquefaction separated the initial coal dissolution from the hydrogenation to upgrade the product. This was known as SRC-II. An intermediate step, critical solvent deashing, was added to remove mineral matter to extend the life of catalysts used in hydrogenation. Later efforts involved the use of an ebulated bed developed by Hydrocarbon Research, Inc. (HRI) and eliminated the mineral matter removal between stages. Temperatures were lowered to reduce contamination of catalysts, which were also added to the first stage. This approach has been called the integrated catalytic two-stage liquefaction process (104).

Several processes progressed to demonstration scales but have not been commercialized, primarily because of economic inability to compete with available petroleum products. The H-Coal process developed by Hydrocarbon Research, Inc. was demonstrated at Catlettsburg, Kentucky using a 545 t/d plant and DOE support. The Exxon donor solvent liquefaction process was not commercialized either.

Processes for hydrogen gasification, hydrogen pyrolysis, or coking of coal usually produce liquid co-products. The Hygas process produces about 6% liquids as benzene, toluene, and xylene. Substitution of petroleum residuum for the coal-derived process oil has been used in studies of coal liquefaction and offers promise as a lower cost technology (104).

Bioprocessing and Biotreatment of Coal. The use of biotechnology to process coal to make gaseous and liquid fuels is an emerging field (105). Bacteria and enzymes have been studied to establish the technical feasibility of conversion. The earliest work was done on microbial decomposition of German hard coals (106). Reactors have been designed to use a variety of bacteria and fungi to break down the large molecular structure into smaller units that may be useful as intermediates (solubilization) or as liquid and gaseous fuels (conversion). Efforts have focused on lower rank coals, lignite or brown coal and subbituminous coal, because of greater reactivity. The conversion processes frequently introduce chemically combined oxygen through hydrolysis or related reactions to make the solid soluble in the reaction mixture as an initial step. Further reaction involves biological degradation of the resulting material to form gases or liquids.

The large-scale processing of coal is expected to involve plants similar to sewage treatment facilities in the handling of liquid and solid materials (see WATER, SEWAGE). The reaction rates are substantially lower than those achieved in high temperature gasifiers and liquefaction reactors requiring much larger systems to achieve comparable coal throughput.

Biological processes are also being studied to investigate ability to remove sulfur species in order to remove potential contributors to acid rain (see AIR POLLUTION). These species include benzothiophene-type materials, which are the most difficult to remove chemically, as well as pyritic material. The pyrite may be treated to enhance the ability of flotation processes to separate the mineral from the combustible parts of the coal. Genetic engineering (qv) techniques are being applied to develop more effective species.

Other Uses. The quantity of coal used for purposes other than combustion or processing is quite small (2,6). Coal, especially anthracite, has established markets for use as purifying and filtering agents in either the natural form or converted to activated carbon (see CARBON). The latter can be prepared from bituminous coal or coke, and is used in sewage treatment, water purification, respirator absorbers, solvent recovery, and in the food industry. Some of these markets are quite profitable and new uses are continually being sought for this material.

Carbon black from oil is the main competition for the product from coal, which is used in filters. Carbon for electrodes is primarily made from petroleum coke, although pitch coke is used in Germany for this product. The pitch binder used for electrodes and other carbon products is almost always a selected coal tar pitch.

The preparation of pelletized iron ore represents a substantial market for coke and anthracite for sintering. Direct injection into the blast furnace of an auxiliary fuel, coal, or oil is now practiced to provide heat for the reduction and some of the reducing agent in place of the more expensive coke that serves these purposes. Some minor uses of coal include the use of fly ash, cinders, or even coal as a building material (see BUILDING MATERIALS, SURVEY); soil conditioners from coal by oxidizing it to humates (see SOIL STABILIZATION); and a variety of carbon and graphite products for the electrical industry, and possibly the nuclear energy program. The growth of synthetic fuels from coal should also provide substantial quantities of by-products including elemental sulfur, fertilizer as ammonia or its salts, and a range of liquid products. The availability of ammonia and straight-chain paraffins may permit future production of food from fossil fuels.

BIBLIOGRAPHY

"Coal" in *ECT* 1st ed., Vol. 4, pp. 86–134, by H. J. Rose, Bituminous Coal Research, Inc.; in *ECT* 2nd ed., Vol. 5, pp. 606–678, by I. G. C. Dryden, British Coal Utilisation Research Association; in *ECT* 3rd ed., Vol. 6, pp. 224–283, by K. S. Vorres, Institute of Gas Technology.

1. *International Coal 1990 Edition*, National Coal Association, Washington, D.C., 1990.
2. *Coal Data 1990 Edition*, National Coal Association, Washington, D.C., 1990.
3. D. W. van Krevelen, *Coal*, Elsevier Scientific Publishing Co., Amsterdam, The Netherlands, 1961.
4. H. H. Lowry, ed., *Chemistry of Coal Utilization*, Vols. 1 and 2, John Wiley & Sons, Inc., New York, 1945.
5. *Ibid.*, Suppl. Vol., 1963.
6. M. Elliott, in Ref. 4, Second Supplementary Vol., 1981; this is an exceptionally extensive source.
7. H. J. Gluskoter, N. F. Shimp, and R. R. Ruch, in Ref. 6, Chapt. 7.
8. R. Thiessen, *U.S. Bur. Mines Inform. Circ.*, 7397 (1947).
9. M. C. Stopes, *Proc. R. Soc. London Ser. B.* **90** 470 (1919); *Fuel* **14** 4 (1935).
10. R. C. Neavel, in Ref. 4, Chapt. 3.
11. C. R. Ward, ed., *Coal Geology and Coal Technology*, Blackwell Scientific Publications, London, 1984. This is an excellent text on this area.
12. A. Davis, in Ref. 11, Chapt. 3.
13. M. C. Stopes, *Fuel* **14**, 4–13 (1935).
14. *International Handbook of Coal Petrography*, 2nd ed., International Committee for Coal Petrology, Centre National de la Recherche Scientifique, Paris, France, 1963, 252 pp.; 2nd suppl. to 2nd ed., 1976.

15. A. Hood, C. C. M. Gutjahr, and R. L. Heacock, *Bull. Am. Assn. Petrol. Geol.* **59,** 986–996 (1975).
16. R. J. Gray, A. H. Rhoades, and D. T. King, *Trans. Soc. Min. Engrs. AIME* **260,** 334–341 (1976).
17. C. A. Seyler, *Fuel* **3,** 15, 41, 79 (1924); *Proc. S. Wales Inst. Eng.* **53,** 254, 396 (1938).
18. "Gaseous Fuels, Coal and Coke," *Annual Book of ASTM Standards*, Vol. 5.05, American Society for Testing and Materials, Philadelphia, Pa., published annually; *British Standards 1016,* parts 1–16, British Standards Institute, London, published annually.
19. I. Wender and co-authors, in Ref. 6, Chapt. 8.
20. L. M. Stock, R. Wolny, and B. Bal, *Energy Fuels* **3,** 651 (1989).
21. M. S. Solum, R. J. Pugmire, and D. M. Grant, *Energy Fuels,* **3**(2), 187–193 (1989).
22. M. E. Peover, *J. Chem. Soc.,* 5020 (1960).
23. R. Raymond, I. Wender, and L. Reggel, *Science* **137,** 681 (1962).
24. L. Reggel, I. Wender, and R. Raymond, *Fuel* **43,** 75 (1964).
25. B. K. Mazumdar and co-workers, *Fuel* **41,** 121 (1962).
26. B. K. Mazumdar, S. S. Choudhury, and A. Lahiri, *Fuel* **39,** 179 (1960).
27. R. E. Winans and P. H. Neill, in W. L. Orr and C. M. White, eds., *Geochemistry of Sulfur in Fossil Fuels*, ACS Symposium Series, No. 429, American Chemical Society, Washington, D.C., 1990, p. 249.
28. J. H. Shinn, *Fuel,* 1187 (1984).
29. I. G. C. Dryden, *Fuel* **37,** 444 (1958).
30. I. G. C. Dryden, *Fuel* **41,** 55 and 301 (1962).
31. A. C. Bhattacharya, B. K. Mazumdar, and A. Lahiri, *Fuel* **41,** 181 (1962); S. Ganguly and B. K. Mazumdar, *Fuel* **43,** 281 (1964).
32. T. Green, J. Kovac, D. Brenner, and J. W. Larsen, in R. A. Meyers, ed., *Coal Structure,* Academic Press, Inc., New York, 1982, p. 199.
33. P. R. Solomon, P. E. Best, Z. Z. Yu, and S. Charpenay, *Energy Fuels* **6,** 143 (1992); see also P. R. Solomon, D. G. Hamblen, R. M. Carangelo, M. A. Serio, and G. V. Deshpande, *Ibid.* **2,** 405 (1988).
34. D. M. Grant, R. J. Pugmire, T. H. Fletcher, and A. R. Kerstein, *Energy Fuels* **3,** 175 (1989).
35. I. Wender, *Chem. Rev.-Cat. Sci.* **14,** 97 (1976).
36. A. L. Chaffee G. J. Perry, R. B. Johns, and A. M. George, *Am. Chem. Soc. Adv. Chem. Ser.* **192,** Chapt. 8 (1981).
37. L. Cartz and P. B. Hirsch, *Phil. Trans. R. Soc. London Ser. A* **252,** 557 (1960).
38. R. D. Harvey and R. R. Ruch, in K. S. Vorres, ed., *Mineral Matter and Ash in Coal,* ACS Symposium Series, No. 301, American Chemical Society, Washington, D.C., 1986, Chapt. 2.
39. H. J. Gluskoter and co-workers, *Ill. State Geol. Survey Circ.,* 499 (1977).
40. W. Spackman, "What Is Coal?", *Short Course on Coal Characteristics and Coal Conversion Processes,* Pennsylvania State University, University Park, Pa., Oct. 1973, 48 pp.
41. A. G. Sharkey, Jr. and J. T. McCartney, in Ref. 6, Chapt. 4.
42. D. L. Wertz, *Energy Fuels* **4**(5), 442–447 (1990).
43. M. Farcasiu, *Fuel* **56,** 9 (1977).
44. M. Bakr, T. Yokono, and Y. Sanada, *Proceedings of the 1989 International Conference on Coal Science,* Oct. 23–27, Vol. 1, Tokyo, Japan, p. 217.
45. I. Chatterjee and M. Misra, *J. Microwave Power Electromagnet. Energy* **25**(4), 224–229 (1990).
46. R. A. MacDonald, J. E. Callanan, and K. M. McDermott, *Energy Fuels,* **1**(6), 535 (1987).

47. O. P. Mahajan, *Coal Porosity in Coal Structure*, Academic Press, Inc., New York, 1982, p. 51.
48. H. Gan, S. P. Nandi, and P. L. Walker, *Fuel* **51,** 272 (1972).
49. J. W. Larsen and P. Wernett, *Energy Fuels*, **2**(5), 719 (1988).
50. P. C. Wernett, J. W. Larsen, O. Yamada, and H. J. Yue, *Energy Fuels* 4(4), 412 (1990).
51. Z. Spitzer and L. Ulicky, *Fuel* **55,** 212 (1976).
52. J. T. McCartney, H. J. O'Donnell, and S. Ergun, *Coal Science, Advances in Chemistry Series*, Vol. 55, American Chemical Society, Washington, D.C., 1966, p. 261.
53. G. H. Taylor, in Ref. 6, p. 274.
54. W. H. Wiser, "Some Chemical Aspects of Coal Liquefaction," in Ref. 40.
55. H. F. Yancey and M. R. Geer, in J. W. Leonard and D. R. Mitchell, eds., *Coal Preparation*, 3rd ed., American Institute of Mining, Metallurgical and Petroleum Engineers, Inc., New York, 1968, pp. 3–56.
56. P. Rosin and E. Rammler, *J. Inst. Fuel* **7,** 29 (1933); J. G. Bennett, *J. Inst. Fuel* **10,** 22 (1936).
57. F. Kick, *Dinglers Polytech. J.* **247,** 1 (1883); P. von Rittinger, *Lehrbuch der Aufbereitungskunde*, Ernst and Korn, Berlin, Germany, 1867, p. 595.
58. F. C. Bond, *Min. Eng.* **4,** 484 (1952).
59. S. J. Vecci and G. F. Moore, *Power* 74 (1978).
60. R. C. Attig and A. F. Duzy, *Coal Ash Deposition Studies and Application to Boiler Design*, American Power Conference, Chicago, Ill., 1969.
61. E. C. Winegartner and B. T. Rhodes, *J. Eng. Power* **97,** 395 (1975).
62. *Steam, Its Generation and Use*, The Babcock & Wilcox Co., New York, 1972, p. 15–4.
63. R. Smock, *Power Eng.* **95**(2), 32 (1991).
64. J. Makansi, *Power* **135**(3), 15 (1991).
65. G. A. Nelkin and R. J. Dellefield, *Mech. Eng.* **112**(9), 58 (1990).
66. J. A. Harrison, H. W. Jackman, and J. A. Simon, *Ill. State Geol. Survey Circ.*, 366 (1964).
67. T. Takanohashi and M. Iino, *Energy Fuels* 4(5), 452–455 (1990).
68. M. G. Matturro, R. Liotta, and R. P. Reynolds, *Energy Fuels*, 4(4), 346 (1990).
69. G. R. Hill and co-workers in *Advances in Chemistry Series*, Vol. 55, American Chemical Society, Washington, D.C., 1966, p. 427.
70. Ref. 6, Chapt. 24.
71. K. S. Vorres, *Energy & Fuels* 4(5), 420–426 (1990).
72. D. K. Fleming, R. D. Smith, and M. R. Y. Aquino, *Preprints, Fuel Chem. Div., Am. Chem. Soc.* **22**(2), 45 (1977).
73. S. C. Spalding, Jr., J. O. Burckle, and W. L. Teiser, in Ref. 69, p. 677.
74. E. P. Stambaugh in Coal Desulfurization, *Chemical and Physical Methods*, ACS Symposium Series 64, American Chemical Society, Washington, D.C., 1977, p. 198.
75. J. W. Hamersma, M. L. Kraft, and R. A. Meyers, in Ref. 72, pp. 73, 84.
76. H. Hofmann and K. Hoehne, *Brennstoff Chemie* **35,** 202, 236, 269, 298 (1954).
77. P. R. Solomon and co-workers, *Preprints, Fuel Chem Div., Am. Chem. Soc.* **36**(1), 267 (1991).
78. M. D. Kimber and M. D. Gray, *Combust. Flame* **11,** 360 (1967).
79. D. Fitzgerald and D. W. van Krevelen, *Fuel* **38,** 17 (1959).
80. K. Ouchi and H. Honda, *Fuel* **38,** 429 (1959).
81. B. K. Mazumdar, S. K. Chakrabartty, and A. Lahiri, *Proceedings of the Symposium on the Nature of Coal*, Central Fuel Resource Institute, Jealgora, India, 1959, p. 253; S. C. Biswas and co-workers, *Ibid.* p. 261.
82. H. Marsh and D. E. Clark, *Erdol und Kohle* **39,** 113 (1986); see also *Proceedings of the Iron & Steel Society*, meeting in Toronto, Apr. 1992, and *Iron & Steel Society AIME*, in Aug. 1992.

83. *Surveys of Energy Resources 1986*, World Energy Conference, Central Office, London, 1986.
84. P. Averitt, "Coal Resources of the United States, Jan. 1, 1974," *U.S. Geological Survey Bulletin*, 1975, p. 1412.
85. U.S. Dept. of Energy, Energy Information Agency, *Monthly Energy Review*, Mar. 1990.
86. *The Penn State Coal Sample Bank and Data Base*, Energy and Fuels Research Center, Pennsylvania State University, University Park, Pa., Apr. 1988.
87. R. D. Harvey and C. W. Kruse, *J. Coal Quality* **7**(4), 109–113 (1988).
88. *SBN Sample Catalogue*, European Center for Coal Specimens, Eygelshoven, The Netherlands, revisions issued periodically.
89. *Coal Age* **82,** 59 (1977).
90. *Coal Age* **68,** 226 (1963).
91. *International Classification of Hard Coals by Type*, United Nations Publication No. 1956 II, E.4 E/ECE/247; E/ECE/Coal/100, 1956.
92. *Analysis and Testing of Coal And Coke*, British Standards, parts 1–16, 1957–1964, p. 1016.
93. *Coal Age* **68,** 62 (1963).
94. *World Book Encyclopedia*, Field Enterprises Educational Corp., Chicago, Ill., 1975, p. 566.
95. M. Valenti, *Mechanical Engineering* **114**(1), 39 (1992).
96. R. Smock, *Power Engineering* **95**(2), 32 (1991).
97. R. D. Doctor and K. E. Wilzbach, *J. Energy Resources Technol.* **111,** 160 (1989).
98. T. F. Edgar and D. W. Gregg, "Underground Gasification of Coal," in Ref. 6.
99. D. Raemondi, P. L. Terwilliger, and L. A.Wilson, Jr., *J. Petrol. Tech.* **27,** 35 (1975).
100. *PETC Review*, Issue 4, Fall 1991, p. 16.
101. A. Pott and co-workers, *Fuel* **13,** 91, 125, 154 (1934).
102. *Environ. Sci. Technol.* **8,** 510 (1974).
103. W. Downs, C. L. Wagoner, and R. C. Carr, *Preparation and Burning of Solvent Refined Coal*, presented at American Power Conference, Chicago, Ill., Apr. 1969.
104. *PETC Review*, Issue 3, Pittsburgh Energy Technology Center, Pittsburgh, Pa., Mar. 1991.
105. D. L. Wise, *Bioprocessing and Biotreatment of Coal*, Marcel Dekker, Inc. New York, 1990, 744 pp.
106. R. M. Fakoussa, translation of the *Investigations of the Microbial Decomposition of Untreated Hard Coals*; *Coal as a Substrate for Microorganisms*: doctoral dissertation of R. Fakoussa, Bonn, 1981; prepared for the U.S. Department of Energy, Pittsburgh Energy Technology Center; translated by the Language Center Pittsburgh under Burns and Roe Services Corp., Pittsburgh, Pa., June 1987.

General References

D. L. Crawford, ed., *Biotransformations of Low Rank Coals*, CRC Press Inc., Boca Raton, Fla., 1992.
D. L. Wise, ed., *Bioprocessing and Biotreatment of Coal*, Marcel Dekker, New York, 1990, 744 pp.
Fuel **70**(3), (1991) contains a series of papers by leading researchers presented at the International Conference on Coal Structure and Reactivity; chemical, physical, and petrographic aspects, Cambridge, UK, Sept. 5–7, 1990.
Fuel **70**(5), (1991) contains a series of papers presented at "Biotechnology for the Production of Clean Fuels", Aug. 27–28, 1990, Washington, D.C., pp. 569–620.
W. Francis, *Coal*, 2nd ed., Edward Arnold & Co., London, 1961.

E. Stach and co-workers, *Stach's Textbook of Coal Petrology*, 3rd ed., Gebruder Borntraeger, Berlin, Germany, 1982, 535 pp. Excellent text on coal petrography.

KARL S. VORRES
Argonne National Laboratory

COAL CHEMICALS. See FEEDSTOCKS, COAL CHEMICALS AND FEEDSTOCKS.

COAL CONVERSION PROCESSES

CARBONIZATION

Coal carbonization is the process for producing metallurgical coke for use in iron-making blast furnaces and other metal smelting processes. Carbonization of coal (qv) entails heating coal to temperatures as high as 1100°C in the absence of oxygen in order to distill out tars and light oils (see TAR AND PITCH). A gaseous by-product referred to as coke oven gas (COG) along with ammonia, water, and sulfur compounds are also thermally removed from the coal. The coke that remains after this distillation largely consists of carbon (qv), in various crystallographic forms, but also contains the thermally modified remains of various minerals that were in the original coal. These mineral remains, commonly referred to as coke ash, do not combust and are left as a residue after the coke is burned. Coke also contains a portion of the sulfur from the coal. Coke is principally used as a fuel, reductant, and support for other raw materials in ironmaking blast furnaces (see FURNACES, FUEL-FIRED; IRON). A much smaller tonnage of coke is similarly used in cupola furnaces in the foundry industry. The carbonization by-products are usually refined, within the coke plant, into commodity chemicals such as elemental sulfur (qv), ammonium sulfate, benzene, toluene, xylene, and naphthalene (qv) (see also AMMONIUM COMPOUNDS; BTX PROCESSING). Subsequent processing of these chemicals produces a host of other chemicals and materials. The COG is a valuable heating fuel used mainly within steel (qv) plants for such purposes as firing blast furnace stoves, soaking furnaces for semifinished steel, annealing furnaces, and lime kilns as well as heating the coke ovens themselves.

Cokemaking dates to seventeenth century England where it was discovered that interrupting the burning of coal heaps produced solid blocks of carbon from the bottom of the heap (1). This carbon quickly supplanted wood charcoal as the main blast furnace fuel. The first commercially successful coal carbonization plant was developed in 1709 (2). Subsequent generations of cokemaking facilities proceeded to ever more effectively exclude air (oxygen) from contact with the carbonizing coal. These facilities evolved from the initial coal heaps first to pits, then to masonry-walled nonroofed ovens. Dome-shaped mud-walled ovens, and then domed refractory brick ovens, commonly called beehive ovens, appeared by 1840 (1). At about this same time, rectangular-shaped ovens having arched roofs and removable doors on one or both ends of the oven appeared. This latter type allowed for pushing the coke out of the oven so that the coke could be quenched with water. Earlier ovens were designed for quenching the coke within the oven, necessitating subsequent reheating of the oven as well as repair of damage caused by thermal shock to the oven structure.

Within 15 years, the enclosing of the coal during carbonization allowed for the first attempts at profitably recovering the off-gases from the coking process. Commercial success in this endeavor is generally credited to Germany's by-product coke ovens of the early 1880s (1) which led to rapid growth in the steel industry as well as to development of chemical industry for use of the cokemaking by-products. Carbonization facilities at different locations in various countries aimed at different products depending on local needs. For some facilities, COG (also called coal gas) for street lights, etc was the prime product and the coke produced was a troublesome by-product. Production of COG or illuminating gas began about 1800 and was accomplished by heating coal in iron or steel retorts. Other facilities concentrated on producing tar and oil for use as rope preservatives. But a growing proportion of cokemaking facilities concentrated on quality coke for blast furnaces and used sale of the other carbonization products to offset the coke costs.

Developments in cokemaking technology in the United States closely followed those elsewhere. By the mid-1800s, coke had displaced charcoal as the principal smelting fuel (1) coinciding with application of hot blast principles to iron smelting. The first U.S. use of beehive ovens occurred in 1833 in Pennsylvania and in 1893 Semet-Solvay constructed the first narrow vertical slot by-product ovens in the United States (1). These coke ovens were built in Syracuse, New York and were designed for recovering ammonia for the Solvay soda-ash process. United States Steel Corp. shortly thereafter built the first vertical coke ovens for production of blast furnace coke near Joliet, Illinois. Annual construction of beehive coke ovens in the United States continued to grow through 1910, but most coke in the United States was produced via vertical by-product coke ovens by about 1920 (1).

Lack of availability of coal-derived chemicals from Germany during World War I led to rapid development of the by-product cokemaking industry in the United States during this period. This included production of blast furnace coke to support steelmaking, recovery of numerous organic liquids to support the chemical self-sufficiency effort, and recovery of by-product gas for industrial and home lighting. These products were in high demand until the availability of large quantities of low cost natural gas (see GAS, NATURAL) and petroleum (qv) products from the southwestern United States became available in the mid-1940s.

As of this writing, coke oven batteries are almost entirely of the vertical by-product oven type. Numerous alternative cokemaking processes, such as formcoke processes, have had some success, but widespread adoption has not occurred.

Supply and Demand

The vast majority of coke is produced from slot-type by-product coke ovens. Total coke production worldwide was about 378×10^6 t in 1990 (3). As shown in Table 1, this tonnage has remained relatively stable for the last two decades. In 1990, the former USSR (CIS) was the largest coke producer, producing 80×10^6 t, followed closely by the People's Republic of China, producing 73×10^6 t. Japan produced 53×10^6 t and the United States produced about 27×10^6. Since 1970, CIS production has remained in the 75–85 million metric ton range, but massive shifts in production have occurred in the United States, Japan, and the People's Republic of China. Since 1970, United States production has decreased by more than 50%; Japanese production has increased by 50%. During the same time period, the People's Republic of China increased coke production by over 300%. Thus the United States dropped from being the No. 2 producer in 1970 to being the No. 4 producer in 1990.

Worldwide demand for blast furnace coke has decreased over the past decade. Although, as shown in Figure 1, blast furnace hot metal production (pig iron) increased by about 4% from 1980 to 1990, coke production decreased by about 2%

Table 1. Production of Blast Furnace Coke, t $\times 10^{6}$[a]

Country	1970	1980	1985	1987	1988	1989	1990
North America	69.7	47.2	30.7	30.0	30.3	34.7	30.9
Western Europe	98.6	78.1	70.7	63.5	62.6	60.1	58.5
Japan	36.4	54.4	51.7	46.4	50.6	51.6	53.0
others[b]	10.1	9.6	5.9	7.5	8.4	8.5	8.5
Total industrial countries	*214.8*	*189.3*	*159.0*	*147.4*	*151.9*	*154.9*	*150.9*
Latin America	4.4	7.7	10.6	9.8	11.7	11.6	11.4
Asia	8.9	17.0	19.9	21.5	23.2	23.6	25.0
Africa and Middle East	0.6	2.2	2.2	2.1	2.1	2.2	2.2
Total developing countries	*13.9*	*26.9*	*32.7*	*33.4*	*37.0*	*37.4*	*38.6*
Total Western World	*228.7*	*216.2*	*191.7*	*180.8*	*188.9*	*192.3*	*189.5*
China, etc	25.5	46.4	51.4	61.4	64.6	69.7	76.7
Russia and Eastern Europe	102.8	122.5	119.5	120.9	120.8	115.2	111.7
Total Eastern countries	*128.3*	*168.9*	*170.9*	*182.3*	*185.4*	*184.9*	*188.4*
World Total	*357.0*	*385.1*	*362.6*	*363.3*	*374.3*	*377.2*	*377.9*

[a]Ref. 4.
[b]Mainly Australia and South Africa.

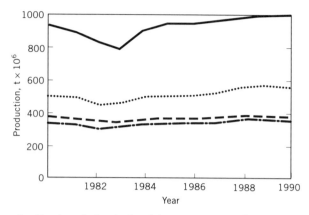

Fig. 1. World production trends in steelmaking raw materials where (—) represents iron ore production; (·····), pig iron (hot metal) production; (- - -), coke production; and (—·—·), scrap consumption (3).

over the same time period (3). This discrepancy of increased hot metal and decreased coke production is accounted for by steady improvement in the amounts of coke required to produce pig iron. Increased technical capabilities, although not universally implemented, have allowed for about a 10% decrease in coke rate, ie, coke consumed per pig iron produced, because of better specification of coke quality and improvements in blast furnace instrumentation, understanding, and operation methods (4). As more blast furnaces implement injection of coal into blast furnaces, additional reduction in coke rate is expected. In some countries that have aggressively adopted coal injection techniques, coke rates have been lowered by 25% (4).

Production of coke is expected to continue to decrease unless new cokemaking facilities are constructed, because the effective production capacities of coke plants decrease as the coke plants age. This situation is particularly acute in North America where the majority of coke plants are over 25 years in age (5) and the economic life spans of conventional coke batteries average about 20 years.

Coals for Cokemaking

Known world coal reserves in 1990 were estimated to be about 1000–1600 billion metric tons (4). The geographic distribution of these reserves is widespread, but about two-thirds of this coal resides in the United States, People's Republic of China, and the Commonwealth of Independent States. Some 637–1075 billion metric tons is classified as anthracite and bituminous coals of which 10% is estimated to be suitable for cokemaking. Thus this 60–108 billion metric tons of coking coal, if recovered in a fully useful form, represents enough coal to supply coke plants at 1990 consumption rates for about 100–200 years. North America is estimated to possess about 130 billion metric tons of bituminous coal of which, perhaps, one-tenth would be classified as coking coal, most of which resides in the United States (4).

For by-product coke ovens, it is general practice to blend two or more types of coals that have complimentary technical as well as economic characteristics. Because most by-product coke plants are located near the large industrial users of the coke and by-products, coals usually have to be transported from the coal mines to the coke plants. Thus coal blends are designed on integration of coke quality needs, by-product quality needs, coal costs, transportation costs, impacts of productivity, and impacts on the coke ovens themselves. The physical behavior of coal blends during coking can damage coke ovens.

The United States possesses a wealth of good quality coking coals in the Appalachian states as well as in locations in some southern and western states (4) (see COAL). Coal blends normally consist of higher rank (more metamorphosed) coals in minor proportion relative to certain lower rank coals. The higher rank coals are referred to as medium volatile and low volatile according to the classification system shown in Table 2. Similarly, the lower rank coking coals are referred to as high volatile. The reference to volatility reflects the relative amounts

Table 2. Classification of Coals by Rank[a]

Class	Fixed carbon limits, %[b]	Volatile matter limits range, %[b]	Calorific value limits, J/kg[c,d]
Anthracitic coals[e]			
meta-anthracite	98[f]	2[g]	
anthracite	92–98	2–8	
semianthracite	86–92	8–14	
Bituminous coals[h]			
low volatile	78–86	14–22	
medium volatile	69–78	22–31	
high volatile A	69[g]	31[f]	14000[f]
high volatile B			13000–14000
high volatile C			11500–13000, 10500–11500
Subbituminous coals[e]			
subbituminous A			10500–11500
subbituminous B			9500–10500
subbituminous C			8300–9500
Lignitic coals[e]			
lignite A			6300–8300
lignite B			6300[g]

[a]Ref. 7.
[b]Dry, mineral-matter-free basis.
[c]Moist, mineral-matter-free basis.
[d]To convert J to cal, divide by 4.184.
[e]Nonagglomerating.
[f]Value is equal or greater than.
[g]Value is less than.
[h]Commonly agglomerating.

of by-products derived from the coals. High volatile coals generate more gas and tar during coking than do the medium and low volatile coals. Coals having either very low or very high volatile contents are not extensively used in cokemaking for technical reasons.

Medium and low volatile coals contribute great strengthening components to the resultant coke. High volatile coals provide the background network of coke structure because of the high propensity to become fluid and to intermix with other coals during coking including the reactives–inerts theory in which coking coals are thought of as contributing aggregate and cementlike materials that, if properly balanced, produce a strong concretelike product. However, as other coke properties besides cold strength have gained importance, basic research on coal carbonization behavior has shown the carbonization mechanisms to be much more complex than is considered in the current views of coal blending technology (6).

From a bulk chemical standpoint, coking coals may be considered to consist of irregular macromolecules that are nearly two-dimensional and are held together by noncovalent bonds. Up to 80% of the carbon is hydroaromatic or aromatic. This percentage of aromatic molecules, versus aliphatic ones, increases with coal rank. The aromatics are of a condensed angular ring structure and are often connected together via aliphatic or ether bridges. These linked rings form chains of molecules along which exist short methyl, ethyl, or hydroxyl side chains. These chains may also contain other elements such as oxygen and sulfur. Coals having lower rank than about high volatile A are generally considered noncoking in that the individual coal particles only weakly, or not at all, fuse together under standard coking conditions. Lower rank coals have a lower aromaticity to the point that most of the carbonization decomposition products consist of low molecular-weight hydrocarbons that are gaseous, thus little or no plastic material is generated. Coals higher in rank than low volatile are similarly viewed because their highly aromatic molecules are strongly cross-linked, as a result of the coalification process, which prevents generation of much plastic behavior during coking.

The smallest semihomogeneous components of coal identified are a group of materials of microscopic size called macerals. Coal, formed from geological processes acting for hundreds of millions of years on remains of ancient plant life, is a complex, heterogeneous material. Indeed, microscopically, coal can be seen to be composed of different groups of materials that, within a group, have uniform and unique combinations of color, texture, and structure. These macerals, the organic analogues to the inorganic minerals, are routinely quantified when coals are characterized in support of determining coking abilities and suitability for blending with other coals. For simplicity, macerals are discussed in terms of the three main groups, although it is clear that different macerals within a group may actually possess quite distinct behaviors during carbonization (6). Petrographic characterization of coking coal is the main science used worldwide to design coal blends for metallurgical cokemaking.

In most coals, the predominant group maceral is vitrinite, largely derived from the woody parts of the plants from which coal originates. Under standard microscope techniques, vitrinite appears as an intermediate gray material of relatively uniform mixture. Owing to its molecular structure, vitrinite is semistable when subjected, in the absence of oxygen, to the conditions of cokemaking. As the less stable portions of the heated vitrinite decompose to form volatile compounds that are eventually recovered as part of the cokemaking by-products, the particles soften. Vitrinite particles can become fluid and act as a solvent relative to other coal components. This mobility and ability to interact with other coal constituents results in fusing of adjacent coal particles to form coherent large carbon-rich structures that account for most of the carbon in coke. In the process, small optically active liquid crystal spheres of aromatic molecules, called mesophase, promote crystallographic alignment of the carbon atoms into locally oriented regions known as mosaics. These mosaic structures have unique optical, chemical, and structural properties. Among these properties is anisotropy, the ability to reflect light to different degrees depending on orientation of the material. This crystallinity has a profound impact on the subsequent behavior of coke in the blast furnace.

Another group maceral is inertinite which results from coal-forming dehydrogenation processes, whether chemical or thermal, acting on virtually any type of plant material. As the name implies, inertinite generally is thermally stable at cokemaking temperature and does not become fluid or even appreciably soften. Inertinite macerals may originate from any part of the original plants, but these plant parts were chemically altered during the coalification process. Any structure, such as plant cell walls, present in the archetypical inertinite can still be present in the material after it has been coked. Inertinite macerals, however, do evolve gaseous hydrocarbon compounds during carbonization. The remaining molecules are not normally mobile in inertinite and the heterogeneous carbon arrangement results in an isotropic type of carbon that performs quite differently in the blast furnace than does anisotropic carbon from other macerals. Microstructural analysis of coke is widely used to characterize carbon forms and spacial distribution, in order to better understand the suitability of the coke for blast furnace use.

Yet another group maceral, exinite (also called liptinite), represents the waxy and resinous plant components and is largely aliphatic and very unstable at coking temperatures. These macerals, in general, almost totally decompose into gases during carbonization and very little, if any, of a given exinite maceral remains in the product coke. Other types of maceral behaviors, which are not easily classified, have also been found. All ranges of behavior intermediate to those of the vitrinites, inertinites, and exinites exist and some macerals can influence the carbonization behavior of others. For example, some inertinite particles have been found to soften and fuse if located adjacent to some exinite particles. The resulting carbon from these inertinites can become aligned in the process and be optically anisotropic, similar to the carbon derived from some vitrinites. A host of other material interactions, including those that inhibit coalescence of adjacent softened particles, are also known (6).

Coal arrives at the coke plant by ship, rail, conveyor, or truck. Each type of coal is unloaded into a separate stockpile in the coal field. Reclaiming of coal from the stockpile can be accomplished using mobile equipment or bridge-mounted hoists. Coal of each type is moved to its coal bunker. In some plants, it is possible to crush each coal independently prior to it reaching the coal bunker. This crushing is often done in two stages. The first stage may be to simply ensure that no large, ie, usually no larger than about 25 mm, coal lumps remain and to ensure that large pieces of rock or foreign material are removed. The second stage of crushing takes place in a hammermill or similar facility that is equipped to pulverize the coal. It is common practice to pulverize coal for cokemaking to more than 80%, being less than 3 mm in size. In other plants, crushing of the coal takes place after the various coals have been blended. This blending is accomplished by discharging measured, either mass or volumetric, amounts of each coal into mixing bins. This may be accomplished through use of intermediate conveyor belt systems as well. The coal blend is then withdrawn from the mix bins and is transported to the coal blend silo at the coke battery. This material flow is depicted in Figure 2.

Coals are not usually stored at coke plants for lengthy time periods. Besides the costs to maintain such inventories, coal undergoes low temperature oxidation that can adversely impact its coking behavior (8). Oxygen can form cross-links in the coal molecules and inhibit the softening character needed for coke formation. Also, coke shrinkage can be significantly destroyed, leading to coke oven damage as the coke is pushed after the end of the carbonization cycle.

Coking Mechanism

There are several necessary conditions for coal to be transformed into coke. These include a heat supply, enclosure, or blanketing to prevent oxygen contact with the coal, and close contact between the coal particles during the carbonization process. In conventional vertical coke ovens these conditions are readily met. The heat is supplied from gas-burning flues located within the walls of each oven. Loading, called charging, of the coal into the oven, and retrieving, called pushing, of the product coke from the oven are accomplished via openings in the oven that can be sealed to prevent incursion of air. Finally, the coal particles are charged into the oven by being dropped from a height so that the particles are packed together between the opposite walls of the oven, causing contact between particles. In this configuration, even for cubic particles and excluding contact with gas pockets, each particle, on average, comes into contact with six other particles.

Upon heating, coal molecules undergo many reactions. The primary reactions involve pyrolysis and formation of radicals having lower molecular weight than in the original coal. Some of the radicals, enriched with hydrogen, form liquid and gaseous products. In other reactions some radicals form more stable substances of higher molecular weight and less hydrogen content. Surface tension on the liquid components promotes additional contact between these components to further facilitate fusing the coal particles as they are heated.

When coal is charged into the hot coke oven, the coal particles adjacent to the coke oven walls and doors begin to devolatilize and soften immediately. The softened, or fluidized, particles sinter into each other and are further devolatilized to form a layer of fully sintered coal particles, ie, coke against the oven walls and doors. Away from these hot surfaces are successive layers of coal in various stages of softening, melting, fusing, and resolidifying. This arrangement results in the existence of an envelope of "plastic" coal that continues to move inward, away from the heated surfaces, until the plastic envelope converges at the center of the coal charge. What results are three separate regimes of coal/coke transformation

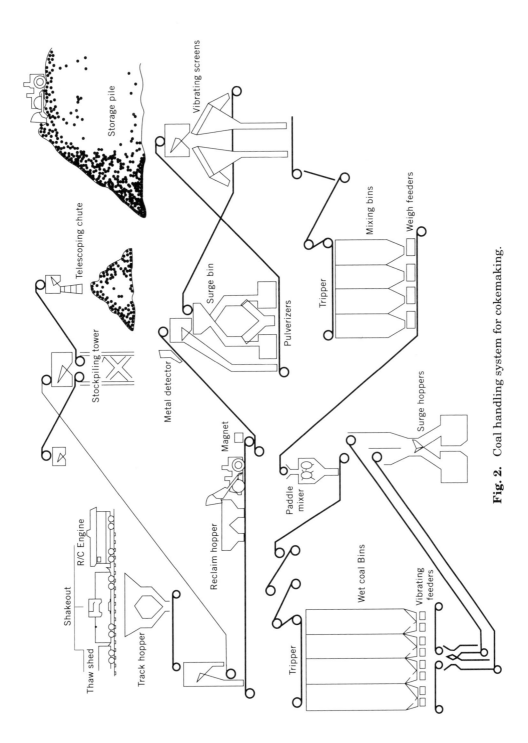

Fig. 2. Coal handling system for cokemaking.

497

that occur simultaneously in a given coke oven until the plastic envelope is consumed. Formation and movement of the plastic envelope is shown in Figure 3.

The plastic envelope represents a barrier to gas movement. Although gases generated in the envelope move out in both directions, gases generated on either side of the envelope do not generally cross this barrier. On the hotter side of the envelope, toward the heating surfaces, decomposition of the envelope leaves a porous solid carbon often referred to as semicoke. The semicoke adjacent to the envelope still contains much volatile gas which continues to be driven off as it reaches higher temperatures from heat transfer from the oven walls and doors. The remaining volatiles in this semicoke decrease to nearly zero as proximity to the hot surfaces increases. During this heating, the carbon crystal structure grows and becomes increasingly oriented, and the porosity of the carbon decreases as the pore walls thicken and densify. This process continues even to very high temperatures, but conventional cooking is normally considered to have concluded by the time the carbon reaches about 1000°C.

On the cooler side of the plastic envelope, toward the center of the coal charge, the coal remains at least as cool as about 100°C until all of the water in the coal, both chemically bound and on the surface of the coal particles, is converted to steam and flows out of the coal mass. This process also occurs progressively in layers away from the plastic layer. Immediately adjacent to the colder side of the plastic layer, the coal particles are completely dry and elevated to a temperature just below that needed to cause the particles to soften. At the center of the coal charge, temperatures may not rise above that of the boiling point of

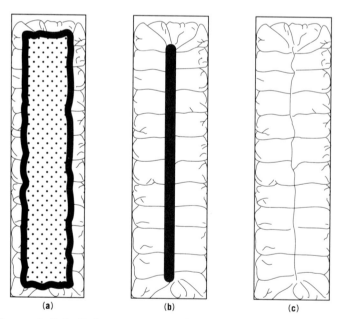

Fig. 3. Development of plastic layer movement during coking: (**a**), about midway in coking cycle; (**b**), convergence of plastic envelope; and (**c**), end of coking cycle. The thick, dark, solid line represents the plastic envelope, defining the boundary between coal and semicoke; and ⬚ represents the plasticized coal.

water for several hours after the coal is charged into the oven. For coke oven wall temperatures of about 1200°C, the complete coking cycle requires 18 hours or more. This cycle starts when coal is charged into the hot oven and ends when the last of the volatiles are degassed from the coke at the center of the charge, which is the coolest part of the charge.

After semicoke has been formed and continues to degas, it tends to physically shrink. The volume of this material is lowered as the carbon atoms align into more compact forms. This effect of itself would tend to cause the coke mass to contract away from the oven walls and doors. However, buildup of gas pressure within the plastic layer and thermal expansion of the coal particles produce forces that tend to continue pressing the coke in contact with the oven walls and doors, as shown in Figure 4. The force keeping the coke in contact with the walls and doors acts to maintain good heat transfer from the oven surfaces into the coke mass. However, if this pressure is high, the oven surfaces, in particular the walls, can themselves be deflected. The walls are constructed of brittle refractory materials (see REFRACTORIES) and such deflection leads to a gradual deterioration of the strength of the oven structure. Moreover, high instantaneous wall pressures can cause catastrophic failure of the oven walls in a matter of hours. Thus coal blends and coking conditions must be carefully matched to ensure that the pressures attained are not excessive. Additionally, if the coke mass remains in intimate contact with the coke oven walls through the coking cycle, friction between the coke and walls as the coke is pushed from the oven can also damage the walls. Thus the final balance of pressure and contraction should favor contraction so that the coke surfaces have moved away from the oven surfaces prior to pushing of the coke from the oven. In actual practice, wall pressure and coke shrinkage requirements place a significant restriction on usable coal blends for vertical oven cokemaking.

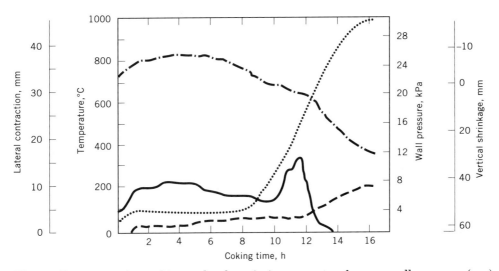

Fig. 4. Parameters in a coking cycle where (—) represents coke oven wall pressure; (····) coal charge center temperature; (- - -) coke mass lateral contraction; and (–·–·), coke mass vertical shrinkage. To convert kPa to psi, multiply by 0.145.

Coke Ovens and Battery Operation

Figure 5 shows a schematic of a modern coke oven. Individual coke ovens are constructed of interlocking silica bricks that are produced in numerous shapes for special purposes. It is not uncommon for modern coke oven batteries to contain 2000 different shapes and sizes of brick. Typical coke ovens are 12–14 m in length, 4–6 m in internal height, and have less than 0.5 m internal width. On each side of the oven are heating flues also constructed of silica brick. Batteries of adjacent ovens, where ovens share heating flues, contain as many as about 85 ovens. At each end of each oven, refractory-lined steel doors are removed and reseated for each oven charge and push. Beneath each oven is a refractory substructure of the heating system regenerator checkers and sole flues for heating the oven floors. Combustion air, and sometimes combustion gases, are preheated in the regenerator. Reversing equipment periodically, perhaps twice per hour, reverses gas and air flow to the oven flues in order to maintain uniform temperature distributions in all of the flues. A concrete basement in which various battery equipment and portions of the heating system are contained is also standard. Coke batteries are generally heated with part of the coke oven gas generated in the cokemaking process, however, they can be heated using blast furnace gas and natural gas also. Once heated, the battery generally remains hot for its entire life because cooling of the silica brick causes a mineralogical change in the silica that lowers the strength of the brick. This effect is not reversed upon reheating, thus unexpected, sustained loss of heat can be catastrophic to the ovens.

Above the ovens is a roof system capable of supporting the moving Larry car from which coal is discharged into each oven through 3–5 charging holes in the roof of each oven. The Larry car itself is filled, for each oven charge, from a large blended coal silo that is constructed above the travel of the Larry car, usually at one end of the coke battery. Modern Larry car technology includes telescopic charging chutes for minimizing dust emissions during charging. Many facilities also include automatic charging hole lid removal and reseating. Considerable attention is given to the order in which coal is charged through each charging hole, overlapping of charging through the different holes, etc in order to produce as uniform a charge as possible and minimal emissions. After completion of charging and reseating of the charging hole lids, a small subdoor at the top of one of the oven doors is opened and a steel leveling bar is inserted along the length of the oven at the top of the coal charge. The leveling bar is moved back and forth over the coal charge to produce a level coal charge having sufficient free space above the charge. This free space is important in ensuring balanced heating of the coal and is needed for conveying the carbonization volatiles out of the oven. Most coke batteries charge wet coal into the ovens, however, a few facilities are equipped with coal preheaters that not only remove all moisture from the coal, but preheat it to 150–200°C in order to quicken the carbonization process. The preheat charge facilities function very similarly to wet charge facilities with the exception of more attention being paid to potentially higher charging emissions because of the dryness of the coal.

On the battery top, at either one or both ends of each oven, are refractory-lined standpipes mounted on additional roof openings into each oven. The volatile

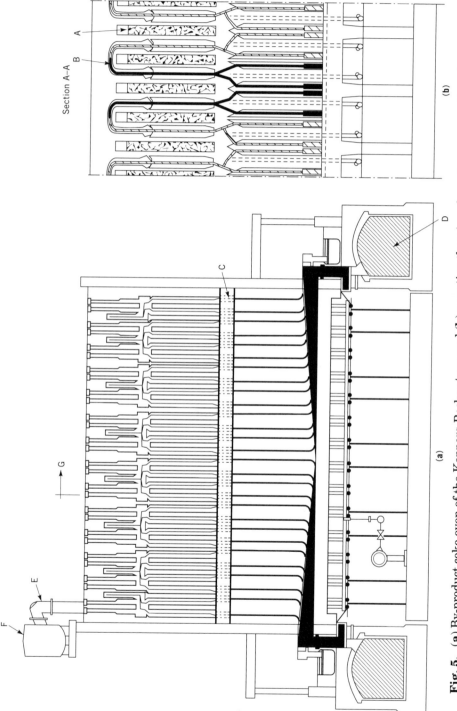

Fig. 5. (a) By-product coke oven of the Koppers-Becker type and (b) a section therein, where A represents the coke oven chamber containing coal; B, oven wall heating flue; C, sole heating flue; D, air preheating; E, standpipe; F, collecting main; and G, relationship between the two sections.

501

gases generated from the coal during carbonization flow to the top of the oven, into the free space, and out the standpipes. The standpipes are all connected to huge collecting mains that run the length of the battery. These mains transport the gases to the by-product plant in which the gases are processed into various materials. Cooling water is sprayed into the mains in order to cool the gases and to condense some of the tar out of the gas. A typical flow diagram for a modern by-product coke plant is shown in Figure 6.

During charging, standpipe valves are positioned so that all gases pass directly into the collecting main. This includes the volume of air displaced from the oven by the coal, coal dust generated during charging, and steam and other gases generated during initial contact of the coal with the oven surfaces. To prevent gas pressure in the oven from building up during charging to the point that gas is forced past the doors, aspiration nozzles in the standpipes use steam to maintain a slight suction on the oven during charging. After pushing, standpipe valves are generally positioned to allow escape to the atmosphere of carbon dioxide gas produced from combustion of wall carbon while the oven is empty.

At the end of each oven's coking cycle, which ranges from about 16 to 24 hours depending on production needs and battery condition, the doors are removed from the oven. A pusher machine equipped with a large water-cooled ram then pushes the coke from the oven into a hot or quench car. After the coke is pushed from the oven, the doors are replaced to maintain oven heat and for maintenance of oven carbon. Oven carbon refers to carbon from the coal that coats the internal surfaces of the oven. Because the refractory bricks are essentially mortarless, the carbon acts to seal the oven wall and prevent cross leakage of gases between the ovens and the flues. Too much carbon buildup, however, can interfere with heat transfer, particularly at the oven top, and reduces the oven volume. Adjustments of the time during which each oven is empty, and the internal conditions of the oven, are made to maintain a balance between carbon burnoff in the empty oven and carbon buildup in the charged oven. The hot car may or may not be constructed with a moveable roof or partial roof to minimize gaseous and particulate emissions. The car moves on rails and positions the hot coke beneath a large water tank equipped with nozzles on its underside. Water flow is timed to quench the coke with a minimal amount of excess water remaining on the cooled coke. After quenching, the hot car again moves to dump the coke onto a refractory covered coke wharf sloped away from the hot car. The coke flows to the bottom of the wharf at which point it drops onto a conveyor system for transportation to blast furnace, storage pile, or for further transportation out of the plant.

Coke ovens in a battery are charged with coal and pushed according to planned schedules. These schedules attempt to ensure that wall pressures generated in adjacent ovens are balanced to prevent wall movement, and that heat utilization and movement of charging and pushing equipment are optimized. To balance heat utilization from the flues, it is undesirable to charge ovens sequentially because each flue would be trying to heat ambient, wet coal in both of its ovens. Therefore, ovens are pushed in any one of a number of pushing series which may consist of pushing every tenth oven, every seventh oven, odd ovens, etc depending on productivity, battery life, and heating impacts on the facility. In modern coke plants, computerized control systems manage oven scheduling and heating.

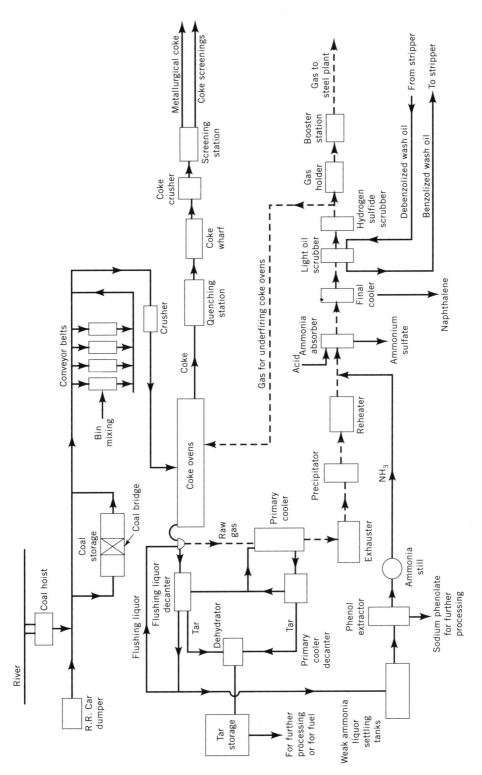

Fig. 6. Schematic for a by-product coke plant (9).

503

Coke Properties and Use

Coke, used in ironmaking blast furnaces, provides three primary functions. First, coke is the fuel that is burned in the blast furnace. The heat generated by combustion of coke provides the energy needed for the various reactions in the process including the melting of iron raw materials and the elevation of liquid iron to the temperatures required for downstream processing into steel. Second, the gases produced from combustion and gasification of coke are the reducing agents to remove oxygen from the iron raw materials in the blast furnace. Third, coke is the only blast furnace material that remains solid at the high temperatures that exist in the lower portions of the furnaces. This means that the coke can support the rest of the materials and provide passageways for the ascending gases and descending iron droplets to meet and interact. The ability to maintain permeability in the blast furnace is perhaps the most important function of coke. These functions and interactions are shown in Figure 7.

The combustion and reduction functions of coke are enhanced by maximization of the carbon in the coke. This means that the other chemical constituents of coke derived from the coal sulfur and coal minerals should be as low as possible

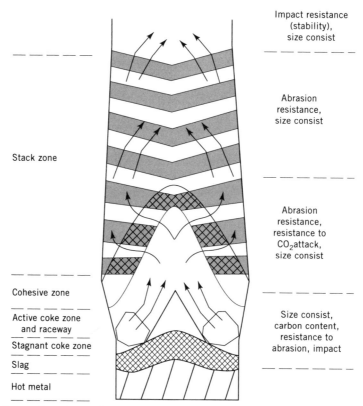

Fig. 7. Primary functions of coke in a blast furnace showing the various furnace zones where arrows indicate direction of gas flow. The term size consist indicates the presence of a distribution of sizes.

as the presence of these impurities dilutes the amount of carbon in coke. Additionally, these impurities must be melted, prevented from entering the molten product iron, and must then be removed from the blast furnace. This is accomplished by maintaining a molten slag layer that floats on the molten iron and periodically "tapping" this slag from the furnace. The slag chemistry and properties are continually adjusted to drive the coke impurities, as well as impurities from other raw materials, into the slag rather than into the pig iron. The materials added to the blast furnace in order to effect this slag volume, chemistry, and properties must also be melted. Slag components arising from coke impurities create a thermal drain on the blast furance and occupy furnace volume that could have otherwise contributed to productivity. Thus coke impurities have a twofold negative effect on the blast furnace: dilution of the useful coke, and increased consumption of the useful coke in order to melt the impurities and associated slag materials. Of special concern are phosphorus and alkalies. The boiling points of most alkali compounds result in recirculation of some alkali in the blast furnace. The refractory lining materials used in blast furnaces can be attacked, consumed, and weakened by these alkalies. Phosphorus, which does not readily combine with blast furnace slag, largely exits the blast furnace in the hot metal. The phosphorus content of steel has a significant impact on the properties of the steel, so hot metal phosphorus must be controlled through blast furnace input levels or through expensive hot metal or liquid steel dephosphorization processes.

The burden support and permeability function of coke is enhanced by the coke degrading in size as little as possible as it progresses downward through the blast furnace. This means that, in addition to minimum breakage as the coke is transported from coke plant to blast furnace, the coke should be as resistant as possible to all of the blast furnace conditions that act to degrade the coke.

When coke is first charged into the blast furnace, it falls through some distance before coming to rest on top of other burden materials in the blast furnace. This acts to break the coke at the relatively cool temperatures that exist at the furnace top. Various laboratory tests exist that attempt to gauge the ability of coke to resist this breakage. In North America the most widely used of these tests is the Tumbler Test, which measures coke stability, standardized by ASTM (7). Elsewhere in the world, similar tests such as the Micum, JIS, and Irsid tests gauge coke resistance to breakage in tumbler-type machines (8). The results of these tests have been statistically correlated to blast furnace performance, ie, to coke consumption per unit of pig iron and blast furnace productivity, even though the breakage phenomena are known to only partially reflect the actual performance. Most cokes presently measure between 55 and 65 stability. Values of over 60 are usually desired for use in large modern blast furnaces.

Coke degradation upon charging into the blast furnace is known to be only a minor effect compared to other degradation (10). These other mechanisms include thermal shock of the coke, gasification of the coke, attack by molten iron and slag, intercalation of alkalies into the coke crystal structure, carbon attack via oxygen liberated from coke impurities, and abrasion at various temperatures and atmospheres in the blast furnace (11). Only the effect of gasification on coke degradation has been studied to the point that a widely accepted laboratory test of this coke characteristic has been developed. The coke strength after reaction (CSR) test, developed by Japanese steel companies and building on earlier reactivity tests, subjects coke to a prescribed gasification and subsequent measurement of abrasion resistance (12). The higher the CSR, the better resistance to degradation stemming from gasification. Typical CSR values range from about 50 to 65. In large, modern blast furnaces, CSR values in excess of 60 are generally desired. All of the coke degradation mechanisms and ways to meaningfully characterize coke are under continued investigation. These studies are expected to generate additional coke quality measurement techniques (of an empirical nature), but are ultimately aimed at fundamental understanding.

Other Cokemaking Technology

Owing to the importance of coke to the steel industry, means for improving the quality of coke, lowering its cost of production, and developing cleaner cokemaking processes are always under investigation. Alternative technologies include modifications to by-product vertical coke oven processes, completely new process designs, and modifications to cokemaking processes used in the past.

One modification to the vertical coke oven process, coal preheating, gained attention and acceptance in the late 1960s and early 1970s. Several coke plants were built around the world in which the coal blend was thoroughly dried and preheated before it was charged into the oven via special Larry cars, enclosed conveyors, or pneumatic transport. The stated benefits of this technology were to shorten coking times to increase productivity, and to allow for use of marginal quality coals. By preheating the coal outside of the coke oven, coking cycles could be shortened by six hours (35%) or more and substantial reductions in coke battery operating costs could be realized. This preheating also changed carbonization conditions in the coke ovens such that use of lower rank and lower cost coals was possible while maintaining coke quality. Over the years, however, most of these facilities shut down as it was realized that costs of operating and maintaining the preheaters and additional battery maintenance negated the benefits of shortened coking cycles.

Another modification to the vertical coke oven process, dry quenching of the coke as first developed in the CIS (13), started becoming of interest in the 1960s and has continued to grow steadily. Using dry quenching, the coke is discharged into a hot car and is quickly transported and dumped into a sealed vessel. Contact with cooling gases, of various sorts, accomplishes slow cooling of the coke rather than the more conventional rapid quenching with water. In this way, the coke is dry when used at the blast furnace. Also, there are some claims of coke strength improvements resulting from dry quenching. Some facilities recover the heat from the cooling gases and make use of this heat for steam raising or electricity generation. Besides plants in the CIS, Japanese coke plants have extensively adopted this technology as have some European ones.

Extensions to the size of vertical coke ovens have been developed and this idea continues to be evaluated. The largest ovens actually implemented are in Germany. Rather than the conventional limit of 6-m oven heights, the Huckingen Plant contains a battery of ovens that are nearly 8 m in height (14). Also, beyond the conventional 460-mm mean oven width, the Prosper plant contains batteries of ovens tht are 590-mm in width (14). Such large ovens can hold more than twice the coal of conventional coke ovens and offer an advantage of fewer openings that can leak emissions per ton of production. These facilities were quite expensive to build, however, and wear and tear of operation over the years remains to be evaluated. Further extension of oven dimensions is planned for a jumbo reactor in Germany (15).

Formcoke Processes.   A completely different approach to making coke is embodied in the various types of formcoke processes which produce coke briquettes in a series of reactors and vessels. The formcoke process, such as that shown in Figure 8, entails heating the coal, in a fluidized bed for example, to drive off some of the volatile matter. This volatile matter is collected and modified for future use. The remaining coal, referred to as char, can be prepared by crushing and screening and then mixed with an organic binder such as a petroleum product, a product of the volatile matter removed earlier in the formcoke process, or even a special type of highly fluid coal. The char/binder mixture is briquetted and the briquettes are further heated in a moving bed furnace to finish the carbonization cycle. The briquettes can be either dry or wet quenched. Alternatively, coals with or without additives may be briquetted in a hot briquetting press without need for the separate devolatilizing step. Formcoke offers the advantages of using a wide range of coals, minimizing emissions because of the closed vessels, and producing a uniform size coke product. In practice, the formcoke has not been considered suitable for general blast furnace use, however, because of its high degradation at higher temperatures. A few operating plants exist, but these mainly produce formcoke for nonblast furnace uses. Additional study may produce commercial formcoke plants for support of ironmaking blast furnaces (16).

Nonrecovery Cokemaking.   Another cokemaking technology that is being practiced in various forms in several countries is nonrecovery cokemaking (17). This technology evolves around horizontal ovens somewhat similar to those that were used historically along with beehive ovens. In the People's Republic of China and other developing countries, numerous coke plants exist based on oven designs that have changed little since the early 1800s. These consist of rectangular ovens constructed of simple clay bricks. The ovens can be on the order of 1–1.5 m high, 1.5–2.5 m in width, and 5–10 m in length. Coal is loaded into cold ovens and is covered with movable roofing material and/or mud and straw mixtures. The coal is then ignited through openings in the walls or roof. The coal in the vicinity of the openings burns and generates heat that is conducted into the rest of the oven to carbonize that coal. At an appropriate time, usually after several days, the roof materials are removed and the coke is water quenched while still in the horizontal ovens. After a few days of drying, the coke is then removed from the oven and the cycle is restarted. In these simple coke plants, most movement of coal and coke is done using manual labor, beasts of burden, and simple mechanized equipment.

Updated versions of horizontal coke ovens exist in developed countries including Australia and the United States (18). These nonrecovery ovens are constructed of silica brick and have permanent roofs. Doors at each end of each oven are removed for pushing, and, for some facilities, charging of coal into the ovens. The ovens remain hot at all times and are heated by a combination of heat from combustion of coal volatiles and coal in the oven itself and coal volatiles that are drafted through flue systems in the oven walls and floors. The existence of flues in the oven floors, also called soles, gives rise to another name for these ovens,

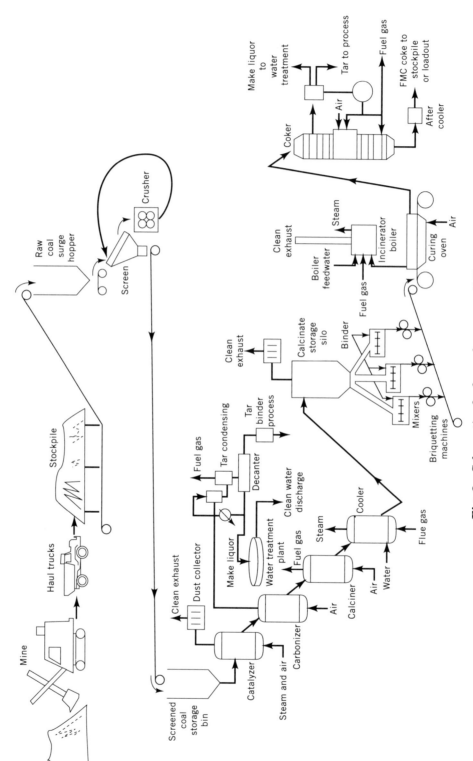

Fig. 8. Schematic of a formcoke process (16).

508

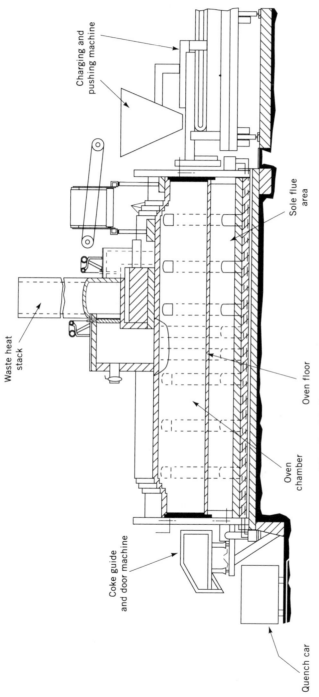

Fig. 9. Modern nonrecovery coke oven (19).

Charging and pushing machine

Waste heat stack

Coke guide and door machine

Quench car

Sole flue area

Oven floor

Oven chamber

509

sole flue ovens. These types of nonrecovery ovens have about the same dimensions as those of past generations, but the coke is pushed and water quenched outside of the oven as in modern vertical coke ovens. Though this cokemaking technology produces coke suitable for use in blast furnaces, it is not very prominent as of this writing. However, low costs and relative low emission of pollutants may stimulate future growth in use of the most technologically advanced versions of nonrecovery cokemaking. Figure 9 shows one of these nonrecovery coke ovens.

Other types of cokemaking technology include both batch and continuous processes, and processes that use electrical induction as the heat-transfer mechanism. Processes under development are further described in Reference 16.

BIBLIOGRAPHY

"Carbonization" in *ECT* 1st ed., Vol. 3, pp. 156–178, by W. O. Keeling, Koppers Co., Inc. and F. W. Jung, Research Consultant; in *ECT* 2nd ed., Vol. 4, pp. 400–423, by C. C. Russell, Koppers Co., Inc.; "Carbonization" under "Coal Conversion Processes" in *ECT* 3rd ed., pp. 284–306, by M. Perch, Koppers Co., Inc.; "Carbonization and Coking" under "Coal Chemicals and Feedstocks" in *ECT* 3rd ed., Supplement, pp. 228–234, by G. Collin, Rütgerswerke AG, and G. Löhnert, Ruetgers-Nease Chemical Co.

1. C. D. King, *Seventy-Five Years of Progress in Iron and Steel,* AIME, New York, 1948, Chapt. 1.
2. R. Baker and F. B. Traice, *AIME Ironmaking Proceedings,* 1991, p. 31.
3. IISI, *Steel Statistical Yearbook 1991,* Brussels, 1991.
4. IISI, *Future Supplies of Coking Coal,* Brussels, 1992.
5. IISI, *Western World Cokemaking Capacity,* Brussels, 1989.
6. D. D. Kaegi, H. S. Valia, and C. H. Harrison, *AIME Ironmaking Proceedings,* Pittsburgh, Pa., 1988, pp. 339–349.
7. *Annual Book of ASTM Standards,* Part 26, ASTM, Philadelphia, Pa., 1978, pp. 430–432.
8. R. Loison, R. Foch, and A. Boyer, *Coke Quality and Production,* Butterworth's, London, 1989.
9. R. A. Meyers, ed., *Coal Handbook,* Marcel Dekker, New York, 1981.
10. R. R. Willmers, J. R. Monson, and H. C. Wilkinson, *The Degradation of Coke in the Blast Furnace,* EUR 8703, Commission of the European Communities, 1984, p. 83.
11. P. M. Fellows and R. R. Willmers, *AIME Ironmaking Proceedings,* Pittsburgh Pa., 1985, pp. 239–251.
12. M. Sakawa and co-workers, *J. Fuel Soc. Jpn.,* 397 (1984).
13. S. S. Eluind, *Coke Chem. (USSR)* (5), (1960).
14. D. Rreidenback and co-workers, *1st International Cokemaking Congress,* F3, Essen, Germany, 1987.
15. W. Eisenhut and co-workers, *AIME Ironmaking Proceedings,* Pittsburgh, Pa., 1991, pp. 15–25.
16. AISI, *Alternative Cokemaking Technologies,* Pittsburgh, Pa., 1991.
17. R. G. Sandercock, *Australian IMM Symposium Proceedings,* Sydney, Australia, 1967, pp. 224–231.
18. J. J. Knoerzer, C. E. Ellis, and C. W. Pruitt, *AIME Ironmaking Proceedings,* Pittsburgh, Pa., 1991, pp. 191–198.
19. A. J. Buonicore and W. T. Davis, eds., *Air Pollution Engineering Manual,* Van Nostrand Reinhold, New York, 1992.

General References

H. C. Porter, *Coal Carbonization,* The Chemical Catalog Co., New York, 1924.
L. Grainger and J. Gibson, *Coal Utilization: Technology, Economics & Policy,* King's English Bookprinters, Leeds, UK, 1981.
C. R. Ward, *Coal Geology and Coal Technology,* Blackwell Scientific Publications, Victoria, Australia, 1984.

DENNIS KAEGI
VALERY ADDES
HARDARSHAN VALIA
MICHAEL GRANT
Inland Steel Company

CLEANING AND DESULFURIZATION

Coal (qv) is a primary source of energy for the United States and is expected to continue to be so into the twenty-first century (see FUEL RESOURCES). However, combustion of raw coal directly in the furnace of an electric power generating plant yields flue gases containing sulfur oxides (SO_x), nitrogen oxides (NO_x), and compounds of toxic metals (see POWER GENERATION). These materials, which are principally derived from impurities in the coal feed, may be reduced below permissible emission levels by various processes (1): the coal can be cleaned before it is fed to the furnace; the contaminants can be captured in a solid sorbent during, or immediately following, combustion as the hot product gases pass through the boiler; or the cool flue gases can be cleaned after leaving the heat exchange region (see AIR POLLUTION CONTROL METHODS; EXHAUST CONTROL, INDUSTRIAL; FUELS, SYNTHETIC, GASEOUS FUELS; SULFUR REMOVAL AND RECOVERY).

Coal Cleaning

In 1990 coal production in the United States reached 0.9 billion metric tons (2) and worldwide production was estimated to be over four billion metric tons. In 1982 it was estimated that at least 50% of the world coal production was cleaned in some manner before use (3). As higher quality coal reserves are depleted and more stringent environmental regulations on pollutants, particularly sulfur oxides, are enacted, this percentage is expected to increase.

Impurities. The three categories of potential pollutants in coal are sulfur, nitrogen, and ash. Sulfur and ash are associated with both the mineral and organic portions of coal, whereas nitrogen is mainly associated with the organic matter (4).

Most commercial coals of the eastern United States contain 0.5–4.0 wt % sulfur; most western coals contain less than 1 wt % sulfur. Sulfur is present in coal as sulfate, pyrite, and organic sulfur. Sulfate sulfur is of minor concern as its concentration in coal is much less than 1 wt %. Furthermore, sulfate compounds can be easily removed by washing because of their high solubility in water. No definite relationship between the organic and pyritic sulfur coal contents has been established. In the United States, both the organic and pyritic sulfur content in raw coal may vary from 20–80% of the total sulfur. Theoretically, organic sulfur cannot be removed from coal unless chemical bonds are broken or the organic sulfur compound is extracted. Thus the amount of organic sulfur present sets the lowest limit to which a coal can be cleaned by physical methods. The pyritic particles may be macroscopic or microscopic in size. For some coals, pyritic sulfur is finely dispersed in the coal matrix and removal of this form of sulfur by physical cleaning methods can be difficult.

Nitrogen, unlike pyritic sulfur, is mostly chemically bound in organic molecules in the coal and therefore not removable by physical cleaning methods. The nitrogen content in most U.S. coals ranges from 0.5–2.0 wt %.

Coal ash is derived from the mineral content of coal upon combustion or utilization. The minerals are present as discrete particles, cavity fillings, and aggregates of sulfides, sulfates, chlorides, carbonates, hydrates, and/or oxides. The key ash-forming elements and compounds are (4,5):

Minor elements	Trace elements
Pollutant	*Named as hazardous*
sulfur	
nitrogen	beryllium
	fluorine
Ash-forming	arsenic
	selenium
sodium	cadmium
potassium	mercury
iron	lead
calcium	manganese
magnesium	copper
silica	chromium
alumina	
titania	

Minor elements contribute ≥ 1 wt % to the ash; trace elements contribute ≤ 0.1 wt %. The degree of de-ashing achievable by physical cleaning depends on the distribution of mineral matter in the coal. In some cases, a considerable amount of the mineral matter can be removed; in other cases, especially where the mineral matter is distributed throughout the coal as microscopic particles, de-ashing by physical cleaning is not practical.

Concern over the release of hazardous trace elements from the burning of coal has been highlighted by the 1990 Clean Air Act Amendments. Most toxic elements are associated with ash-forming minerals in coal (5). As shown in Table 1, levels of many of these toxic metals can be significantly reduced by physical coal cleaning (6).

Conventional Coal Preparation Plants. Coal cleaning (preparation) is based principally on size and density differences, with the exception of flotation (qv). In this manner physical impurities, ie, ash and pyrite, may be removed from coal. Four general categories of coal preparation plants can be defined based on levels or degrees of cleaning (7): level 1 involves crushing and screening only; level 2, coarse coal cleaning only; level 3, coarse coal and partial fine coal cleaning; and

Table 1. Effect of Coal Cleaning on Trace Elements[a]

Coal	Trace element content, ppm							
	Cd	Cr	Cu	F	Hg	Mn	Ni	Pb
feed	3.15	55	25	156	0.20	53	26	18
product[b]	0.05	28	10	71	0.09	7.9	11	3.0
reduction, %	98	49	60	54	55	85	58	83

[a]Upper Freeport Coal, W.Va., 0.075 mm top size, ie, all particles ≤ 0.075 mm.
[b]Float at 1.40 specific gravity.

level 4, total cleaning, ie, all size fractions are cleaned. At each successive level, the process design becomes increasingly more sophisticated.

In a typical modern coal preparation plant, shown in Figure 1, coal is subjected to (1) size reduction (qv) and screening; (2) separation of impurities; and (3) dewatering (qv) and drying (qv). Size reduction is accomplished in rotary or roll crushers. More impurities are liberated as the coal size is reduced. Then, coal is screened, either wet or dry, to separate the various size fractions. Before treatment, the crushed raw coal is divided into coarse (>10 mm), intermediate (0.6–10 mm), and fine (<0.6 mm) sizes. Coarse coal is cleaned using one or more pieces of equipment based on gravity separation, such as jigs, or dense-medium baths. The intermediate size coals are usually cleaned using dense-medium baths/cyclones, jigs, concentrating tables, or spirals. Fine size coals can only be effectively treated by nongravimetric washing methods, such as froth flotation.

The product of any wet-separation process must be dewatered or dried depending on the mode of transportation and use. Coarse coal can be easily dewatered by natural drainage using screens. Intermediate size coal is dewatered using sieve bends or centrifuges. For fine coal, dewatering may require not only more complicated mechanical devices, such as centrifuges and vacuum filters, but also thermal drying, to achieve an acceptable moisture content.

In 1992 there were more than 400 physical coal cleaning plants throughout the United States having a total capacity of over 400 million metric tons of raw coal per year. Table 2 shows the types of coal cleaning equipment used in these plants. Historically, jigs are the equipment of choice and remain the most popular device for cleaning coal. The use of dense-medium baths and cyclones has been increasing steadily however, particularly where difficult-to-clean coals are involved and where the relative density differences between coal and refuse are small. Concentrating tables, eg, Deister shaking tables, have been employed by many plants. These provide good separation, especially in removing pyrite from coal. Pneumatic-dry-separation processes are less likely to be used because of the inability to achieve sharp separations, partly because of changes in mining laws requiring large quantities of water to be sprayed on the coal during mining and handling to suppress dust. The efficiency of pneumatic processes is severely impeded by added moisture.

Conventional coal cleaning processes can remove about 50% of pyritic sulfur and 30% of total sulfur. For northern Appalachian region coals it has been shown that a greater sulfur reduction can be achieved by applying physical coal cleaning to finer size coals (Table 3) (8).

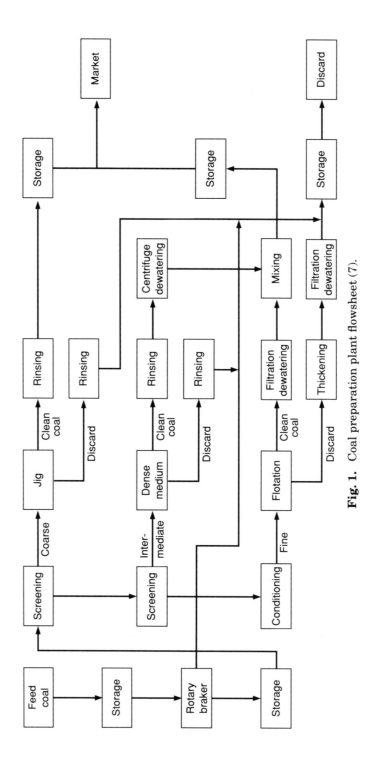

Fig. 1. Coal preparation plant flowsheet (7).

Table 2. Mechanical Cleaning of U.S. Coal by Equipment Type

Equipment	Annual percentage						
	1940	1950	1960	1970	1975	1978	1983[a]
jigs	46.0	47.4	50.0	43.0	46.6	46.6	39.3
dense medium vessels	6.5	14.6	24.3	31.4	32.6	33.2	37.6
concentrating tables	2.3	2.4	11.3	13.6	10.7	10.5	10.5
classifiers[b]	7.6	9.1	4.0	1.1	2.3	2.7	5.7
launders	15.9	5.8	2.8	1.6	1.0	0.6	0.8
others[c]	21.7	20.7	7.0	5.5	2.5	1.9	1.3
flotation			0.6	3.3	4.3	4.5	4.8
Total cleaned, %[d]	*22.2*	*38.5*	*65.7*	*53.6*	*41.2*	*33.8*	*42.6*

[a]1983 is the last year for which these data were collected.
[b]Includes cyclones.
[c]Includes pneumatic separation.
[d]As percentage of total production.

Table 3. Effects of Physical Cleaning on Sulfur Reduction in Coal[a]

Cleaning level	Products, 10^6 t[b]		Sulfur reduction	
	Coal	Sulfur	10^6 t	%
no cleaning	95	2.69		
nominal cleaning, 100% particle top size	89	1.82	0.87	32
3.8 cm[c]	88	1.73	0.96	36
1.0 cm[c]	87	1.50	1.19	44
0.14 cm[d]	85	1.15	1.54	57

[a]Northern Appalachian Region Coals.
[b]Tonnage necessary to produce a heating value of 2.87×10^{12} MJ.
[c]At specific gravity = 1.40.
[d]At specific gravity = 1.3.

Advanced Coal Cleaning Technologies. As the easy-to-remove relatively clean coals are gradually mined out and as fuel specifications become more stringent in order to meet environmental regulations, the need for advanced fine coal cleaning processes has grown. For any fine coal cleaning process, two characteristics tend to dominate. As coal particles are crushed into finer size, the specific surface area increases and the mass of each particle becomes smaller. This leads to the development of surface force-controlled processes or advanced density-based processes that are quite different from the specific gravity-controlled processes found in coarse coal cleaning. In general, advanced processes are capable of producing a deep-cleaned coal product having low ash and low sulfur content. However, as of this writing, most of these processes are in the small-scale demonstration stage and have not been tested on a commercial scale.

Flotation. The application of flotation (qv) to coal cleaning is a relatively new development in the United States. In 1960, only 0.6% of the clean coal produced came from flotation. However, by 1983 flotation accounted for about 5% of the clean coal production (Table 2). Utilization of the flotation process is expected to grow rapidly because more fine size coal is produced as a result of beneficiation schemes that require significant size reduction of the raw coal prior to cleaning to enhance the liberation of pyrite and ash minerals.

The flotation process usually involves three steps: (1) the conditioning of the coal surface in a slurry with reagents, (2) adhesion of hydrophobic coal particles to gas bubbles, and (3) the separation of the coal-laden bubbles from the slurry. In the conventional flotation process, when the coal particles become attached to air bubbles, the particles are allowed to rise to the top of the flotation cell and form a stable froth layer (9). A mechanical scraper is used to remove the froth layer and separate the clean coal product from the refuse-laden slurry.

Reverse flotation is a two-stage process (10). The first stage is a conventional froth flotation in which most of the high ash refuse and some of the coarser or liberated pyrite are rejected as tailings. The coal froth concentrate and some dilution water is fed to a second stage flotation where a hydrophilic colloid, such as starch or dextrine, is added to depress the coal, followed by a xanthate collector to float the pyrite. This process has been successfully tested at 90 kg/h. The process is specifically designed for pyritic sulfur removal. When tested on a Ohio No. 9 seam coal sample, this process achieved a 93.8% reduction of pyritic sulfur.

Although froth flotation is recognized as the best available fine coal cleaning technique, it becomes ineffective when the particle size is much smaller than 0.1 mm or when the feed contains a large amount of clay, resulting in low coal recovery or poor selectivity. A solution to these problems is the use of modified flotation devices.

The KEN-FLOTE column (11) is one of several column flotation processes based on a countercurrent principle. The feed slurry containing reagents is introduced into the column just below the froth zone. Air is injected at the bottom of the column via an air sparger. Wash water is sprayed within the froth zone to reject the entrained impurities from the froth. Test results on this column indicate that a 6% ash product coal having a combustible-recovery of 75–80% can be obtained. A 70–80% pyrite reduction is also claimed. Figure 2 shows the operation of such a column.

The packed-bed flotation column (12) utilizes a stack of corrugated plates as the packing elements arranged in blocks positioned at right angles to each other. These stacked corrugated plates provide a tortuous flow path to attain intimate particles–bubbles contact and limit impurity entrainment. It also features countercurrent flow of air and pulp in the column. It is reported that less than 1% ash in product coal has been attained for a two-stage cleaning.

Another modification is the use of microbubble column flotation (13). In this process, smaller bubbles are generated to enhance the recovery of micrometer-sized particles. A countercurrent flow of feed slurry is also used to further enhance the bubble–particle attachment. The process is capable of producing ultraclean coals containing less than 0.8% ash.

Similarly, small (0.2–0.6 mm) air bubbles are introduced into a 2.6-m Deister Flotaire column at an intermediate level allowing rapid flotation of readily floatable material in the upper recovery zone. The bottom air permits longer retention time of the harder-to-float particles in the presence of micrometer-sized

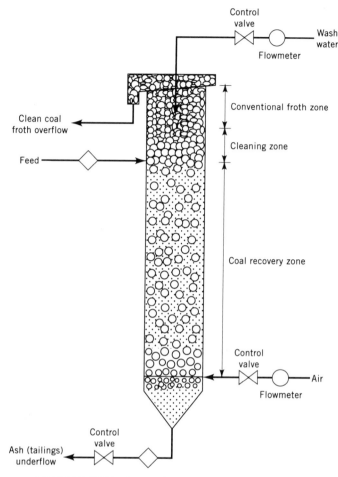

Fig. 2. Diagram of KEN-FLOTE column flotation cell. Courtesy of CAER, University of Kentucky (14).

bubbles at a reduced downward velocity. The first commercial unit went on stream in 1986. It was used to improve the recovery of <0.6 mm (−28 mesh) coal in the plant's tailings. An average of 5.5% increase in coal recovery resulted from its use (14). The second commercial use processed <0.15 mm (−100 mesh) coal feed.

A new flotation cell developed by AFT, Inc. (15) is designed to promote a "skin" flotation process for deep cleaning of fine size coal. In this process plant fines are slurried and then treated with flotation reagents to enhance the natural hydrophobicity of the coal. The conditioned feed slurry is sprayed through nozzles onto the surface of the Sprayflot flotation cell. This operation provides excellent opportunity for air bubble attachment by the air-avid coal particles, which remain on the surface of the cell. The associated hydrophilic impurities remain dispersed in the slurry throughout the cell. The feed to flotation cell contains flocs that form during the conditioning of the feed slurry prior to beneficiation. Such flocs are normally contaminated with entrained mineral matter. The shearing action of the slurry being induced through the spray nozzles is designed to eliminate such flocs and thus provide an improved clean coal product. The froth generated is shown to be drier than a normal flotation froth, thus providing a significant cost advantage during the clean coal dewatering operation.

In 1981, a novel flotation device known as the air-sparged hydrocyclone, shown in Figure 3, was developed (16). In this equipment, a thin film and swirl flotation is accomplished in a centrifugal field, where air sparges through a porous wall. Because of the enhanced hydrodynamic condition, separation of fine hydrophobic particles can be readily accomplished. Also, retention times can be reduced to a matter of seconds. Thus, this device provides up to 200 times the throughput of conventional flotation cells at similar yields and product qualities.

Agglomeration-Based Fine Coal Cleaning. Most recently a search for nonaqueous collectors or reagents for fine coal cleaning has been undertaken. A number of liquids have been tested and found to be suitable as agglomeration agents. These include heavy oil, Freon, pentane, hexane, heptane, 2-methylbutane, methyl chloride, and liquid carbon dioxide.

The use of a water-immiscible liquid to separate coal from impurities is based on the principle that the coal surface is hydrophobic and preferentially wetted by the nonaqueous medium whereas the minerals, being hydrophilic, remain suspended in water. Hence, separation of two phases produces a clean coal containing a small amount of a nonaqueous liquid, eg, oil, and an aqueous suspension of the refuse. This process is generally referred to as selective agglomeration.

One of the best known examples is the spherical agglomeration process, developed by the National Research Council of Canada (17,18). This is a two-stage fine coal cleaning process using No. 2 fuel oil as the agglomerant at a rate of about 6% on a dry solid basis. A typical flow diagram is shown in Figure 4. In the first stage, fine coal–oil agglomerates are formed under high shear agitation condition. The formed agglomerates are then fed to the second-stage low shear contactor where the agglomerate size increases. The agglomerated product is in turn separated from fine clay and other high ash material on a Vor-Siv screen. The screen oversize material is further dewatered in high speed screen-bowl centrifuges to about 12% moisture. The product coal is pelletized using a lignin binder prior to storage.

The Otisca-T process (19) is a three-step selective agglomeration process. First, the coal is ground to minus 0.002 mm in a controlled environment. Then, the finely ground coal is agglomerated using a low molecular-weight hydrocarbon such as pentane, leaving the associated pyrite and mineral matter in the water phase. Finally, the agglomerant (pentane) is recovered for reuse. In this process, heating value recovery has been achieved in the range of 93–98% with pyritic sulfur reductions up to 90%. Product coal ash contents less than 1% are routinely obtained.

The LICADO process based on *l*iquid *ca*rbon *d*ioxide (20,21), is a novel nonaqueous process for fine coal cleaning. Liquid CO_2 is used both as an agglomerant and a transport agent. The process relies on selective agglomeration of coal particles and transport of coal–liquid CO_2 agglomerates from the aqueous phase to the CO_2 phase to achieve the desired separation. It was reported that >90% pyritic sulfur rejection and over 85% clean coal recovery have been achieved. A

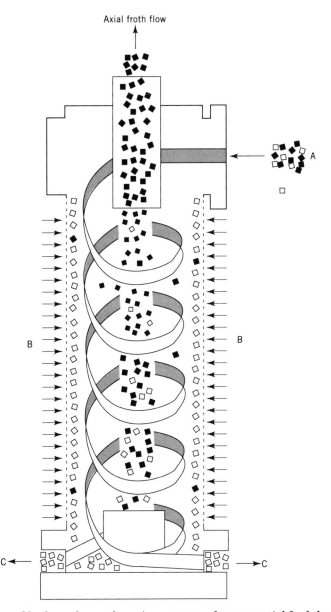

Axial froth flow

A

B   B

C   C

Fig. 3. Air-sparged hydrocyclone, where A represents the tangential feed that establishes swirl flow; B, the area of small bubbles formed by high shear at the porous wall; and C, the outlet for the (□) hydrophilic particles rejected by the swirl flow. The (■) hydrophobic particles are in the axial froth flow. Courtesy of Professor J. D. Miller, University of Utah (16).

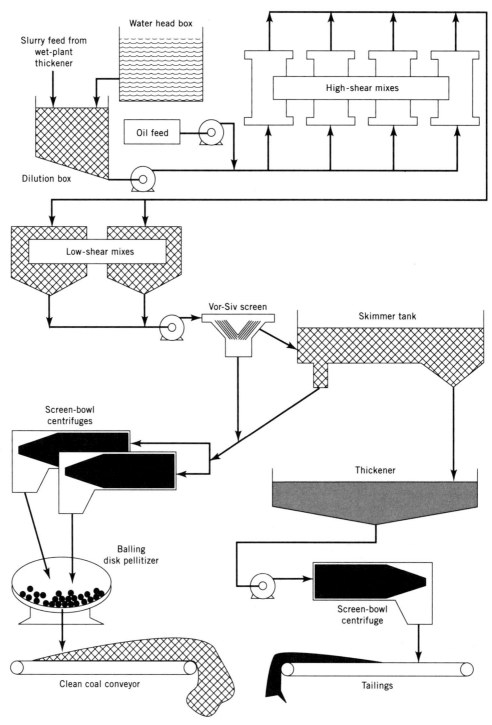

Fig. 4. Flowsheet of Florence Mining Co. oil agglomeration process (7).

principal advantage of the LICADO process is that the product coal contains very little moisture and requires no further dewatering. A continuous research unit of the LICADO process is shown in Figure 5.

Heavy-Medium and Heavy-Liquid Cycloning Processes. Heavy-medium cyclones are widely used for cleaning intermediate size coals and application has been extended to fine coal cleaning by using very fine magnetite particles to provide the desired specific gravity of the processing medium. For example, the Micro-Mag process (22) uses a magnetite medium ground to less than 0.01 mm top size. An alternative method to handle very fine size coal particles is to use heavy-liquids as media for the cyclone separators. Examples of these heavy-liquids include: Freon-113, methylene chloride, sulfuric acid, zinc chloride, calcium chloride, and sugar solutions. All heavy-liquid cycloning processes are capable of effecting separation down to about 0.04 mm (325 mesh) coal particles. Unfortunately, most of these liquids cannot be used in commercial applications because of the potential hazards to workers or the environment and high cost.

Dry Coal Cleaning. Developments in the areas of magnetic and electrostatic separation as a means of cleaning coals in the dry state include high gradient magnetic separation (HGMS), triboelectrostatic separation (TESS), and dry coal purifier (D-CoP).

Several coal cleaning processes have been developed based on the differences in magnetic properties of coal, which is diamagnetic, and pyrite, which is paramagnetic, and some of the ash-forming minerals which are weakly magnetic. The best known example is the HGMS technique which can be used for both dry and

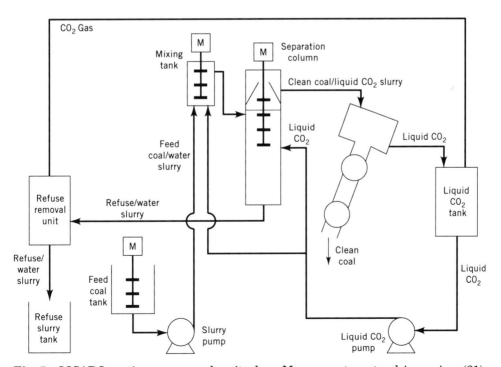

Fig. 5. LICADO continuous research unit where M represents motor-driven mixer (21).

wet coal cleaning (23). A magnetic field as strong as 2 Tesla has been employed to capture fine-sized feebly magnetic particles more efficiently. The separation efficiency can be further increased by using a matrix column filled with fine filaments made of ferromagnetic materials, such as stainless steel wool. The HGMS process is usually operated in a cyclic mode. In the first cycle, pyrite and other impurities are captured in the filament matrix. The trapped particles are then flushed out by high pressure air in the next cycle before returning to the first cycle. This technique has been tested on a variety of coals having ash contents in the range of 10–28% and sulfur contents up to 6.5%. The pyrite rejection ranged from 14–94%. The high rejections were usually associated with low clean coal recoveries (see SEPARATION, MAGNETIC).

In the TESS process, dry pulverized coal is blown rapidly past a copper baffling device that imparts a positive charge on the coal particles and a negative charge on the pyrite and mineral matter particles (24). The charged particles are immediately introduced into an electrostatic separator where negatively charged plates attract the positively charged coal and positively charged plates attract the pyrite and mineral matter. Ash reductions of up to 93% and pyritic sulfur reductions greater than 95% were obtained in processing minus 0.037 mm coal in a two-stage configuration. However, the yield was as low as 40%.

Another dry coal cleaning process is the dry coal purifier (D-CoP) (25). This technique processes crushed coal in an air-fluidized bed with magnetite particles. When a fluidized bed contains particles of different densities and sizes, there is a tendency at near minimum bubbling conditions for the solids to stratify in the vertical direction according to density, and to a lesser extent, size. In a case of a bed consisting of magnetite particles and crushed coal, the clean fraction of the coal segregates at the top of the bed, the liberated minerals settling toward the bottom. As a consequence, the ash content of the coal at the top of the bed is lowered, thereby permitting recovery of coal having significantly reduced amounts of pyrite and other materials. Because it is a dry process and involves fine coal particles, D-CoP can be integrated directly into a pulverized coal power plant.

Chemical and Biological Coal Cleaning. Whereas physical coal cleaning is capable of removing most of the ash (mineral matters) and inorganic (pyritic) sulfur, it cannot be used to remove organic sulfur. For most bituminous coals in the United States, 40 to 60% of the sulfur content is bound in organic matter. Organic sulfur can be classified into four types (26): thiols, sulfides, disulfides, and heterocyclic thiophenes, such as dibenzothiophene [132-65-0], $C_{12}H_8S$. The sulfur from these compounds can only be removed using chemical or biological methods. Some of these processes have progressed to the miniplant stage. However, most are still at laboratory scale.

Oxidative Desulfurization Process. Oxidative desulfurization of finely ground coal, originally developed by The Chemical Construction Co. (27,28), is achieved by converting the sulfur to a water-soluble form with air oxidation at 150–220°C under 1.5–10.3 MPa (220–1500 psi) pressure. More than 95% of the pyritic sulfur and up to 40% of the organic sulfur can be removed by this process.

The applicability of a nitrogen dioxide oxidative cleaning process for 10 coals on the laboratory scale has been examined via a pilot-plant program (29). In this process, dry pulverized coal is treated with gaseous NO_2, to convert the sulfur to an alkali-soluble form at about 120°C and 342 kPa (496 psi). The coal is then

washed using aqueous sodium hydroxide solution to dissolve the sulfur compounds. The waste effluent is treated with lime to regenerate the caustics for reuse
and the gypsum formed in the process goes to disposal. The generated gaseous
sulfur dioxide and sulfur trioxide are removed by a gas scrubbing system. The
process is capable of removing essentially all of the pyrite sulfur and up to 40%
of the organic sulfur from a Lower Kittanning coal (29).

Coal Cleaning by Reactive Leaching. During World War II, Germany developed the first chemical coal cleaning technique to produce ultraclean coal (30). A
run-of-mine bituminous coal (26 t/h) was cleaned by multistage flotation to less
than 1% ash, and then leached using a mixture of 0.4% hydrofluoric acid and 1.4%
hydrochloric acid to obtain 0.5% ash coal. This development reached the pilot-
plant stage. A flotation product assaying 0.8% ash was cleaned to 0.28% ash and
10% moisture.

More recently, the molten caustic leaching (MCL) process developed by TRW,
Inc. has received attention (28,31,32). This process is illustrated in Figure 6. A
coal is fed to a rotary kiln to convert both the mineral matter and the sulfur into

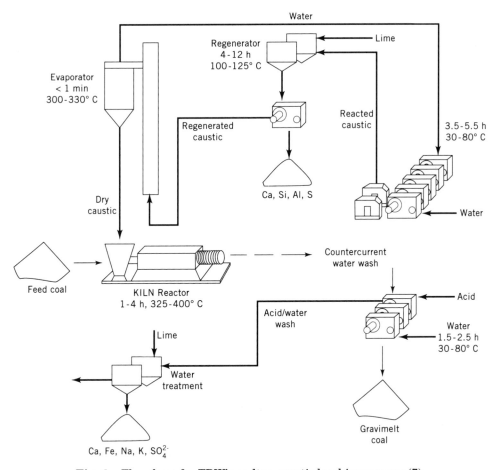

Fig. 6. Flowsheet for TRW's molten caustic leaching process (7).

water- or acid-soluble compounds. The coal cake discharged from the kiln is washed first with water and then with dilute sulfuric acid solution countercurrently. The effluent is treated with lime to precipitate out calcium sulfate, iron hydroxide, and sodium–iron hydroxy sulfate. The MCL process can typically produce ultraclean coal having 0.4 to 0.7% sulfur, 0.1 to 0.65% ash, and 25.5 to 14.8 MJ/kg (6100–3500 kcal/kg) from a high sulfur, ie, 4 wt % sulfur and ca 11 wt % ash, coal. The moisture content of the product coal varies from 10 to 50%.

Based on the same principle, several other chemical leaching processes, including the promoted oxidative leaching, wet-oxygen leaching, hydrothermal leaching, and hydrogen peroxide–sulfuric acid leaching processes have been developed and exhibited promising desulfurization characteristics (28).

Microwave Desulfurization.   Microwave desulfurization of coal is another modification of the alkali leaching method. A mixture of coal and sodium hydroxide is heated at about 250°C to promote reaction of the sodium hydroxide with the sulfur contained in the coal. The wet-coal cake is then irradiated with microwaves for 25 to 45 seconds under an inert atmosphere. The coal is washed with water and acid to remove soluble sulfide, usually Na_2S, and other solubilized mineral matter. The entire process can be repeated several times to obtain a desired level of sulfur and ash removal. In tests using an Illinois No. 6 coal containing 15.4 wt % ash and 3.4 wt % sulfur, nearly all of the pyritic sulfur, 75% of the organic sulfur, and 87% of the ash were removed, resulting in a clean coal product having less than 2% ash and 0.7% sulfur by weight (33) (see MICROWAVE TECHNOLOGY).

Chlorinalysis.   Chlorine can be used to remove the sulfur from coal. The coal is contacted with chlorine gas in methylchloroform at about 75°C (34). After separation of the coal from the slurry, the solvent is recovered by distillation. The chlorinated coal is washed with water and finally dechlorinated by heating at about 300–350°C. The process is capable of extracting about 90% of the pyritic sulfur and up to 70% of the organic sulfur from some coals. This process could thus expand the availability of low sulfur solid fuels by making a greater portion of the high organic sulfur coals available as a clean source of energy. However, much research and development is needed to overcome several technical problems, including chlorine retention by the coal, chlorine regeneration, and recycling.

Self-Scrubbing Coal.   A novel coal cleaning process marketed by Custom Coal International (35) is designed to produce a self-scrubbing coal. In this process, crushed run-of-mine coal is first cleaned by using a heavy-medium bath to remove noncombustible material, including 90% of the pyritic sulfur content of the coal. Limestone-based additives then are mixed with the beneficiated coal to produce a clean coal product. These additives react with the remaining organic sulfur, which is released during combustion, to remove an additional 70–80% of resulting SO_2 from the coal, effecting a reduction of 80–90% of total sulfur.

Microbial Coal Cleaning.   Some of the organic sulfur compounds, such as thiophene and dibenzothiophene (DBT), can be degraded by a variety of microorganisms that often proliferate in petroleum-saturated soil. A strain of *Pseudomonas,* named CB-1, that can convert thiophenic sulfur to sulfate has been isolated (36), as have a number of other microbes including CB-2, which is an *Acinetobacter.* Both CB-1 and CB-2 have been tested in a continuous bench-scale

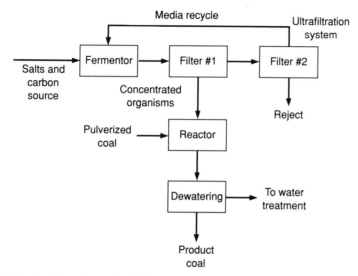

Fig. 7. Flowsheet for ARC's microbial coal cleaning process (7).

unit. The coal slurry can contain as high as 26 wt % solids. The organisms are grown in continuous fermenters at 25 to 35°C and then fed to the reactor as a thick broth. A process flowsheet is depicted in Figure 7. Organic sulfur removal is in the neighborhood of 25% and the combined use of microbes, either simultaneously or sequentially, could potentially improve organic sulfur rejection. The limiting factors appear to be those of accessibility and residence time. Therefore, finer size coal should be used not only to improve accessibility of microbes to coal particle surfaces but also to reduce the overall retention time in the bioreactor. Additional information and a detailed review of conventional and advanced coal cleaning technologies may be found in Chapters 7 and 14 of Reference 7.

Combustion of Synthetic Fuels. Sulfur may also be removed from coal before combusting the coal for energy production by converting to a synthetic gaseous fuel (syngas) (see COAL CONVERSION PROCESSES, GASIFICATION; FUELS, SYNTHETIC, GASEOUS FUELS), then removing the H_2S from the syngas (see SULFUR REMOVAL AND RECOVERY). Rather than simply combusting the syngas in a boiler, this clean, hot-burning fuel can first be burned in a gas turbine. The hot exit gases from the gas turbine can be fed to a conventional boiler and steam turbine. This combined cycle technique, ie, gas turbine cycle plus steam cycle, integrated with coal gasification, is denoted as gasification combined cycle (GCC) (37). The GCC process provides a significant boost in overall thermal efficiency over other commercially available systems for deriving power from coal. GCC projects include an early conceptual one using the Westinghouse coal-gasification technology (38), and one being planned to meet the requirements of the 1990 Clean Air Act is the repowering by Public Service of Indiana of Unit 1 at its Wabash River Station (39).

Coal Desulfurization in the Furnace and Ductwork

Sulfur dioxide in the hot gases from a coal-fired combustor may be transferred to a solid reaction product that usually contains the calcium ion, at any point in the furnace. In a fluidized-bed combustor the SO_2-sorbent is intimately mixed with the coal throughout the region where combustion occurs. In entrained-bed systems the sorbent is introduced in finely divided form into the hot exhaust gases as they pass through the heat exchange system inside the boiler or through the ductwork leading to the particulate removal equipment following the boiler.

Fluidized-Bed Combustion. Fluidized-bed combustors are able to burn coal particles effectively in the range of 1.5 mm to 6 mm in size, which are floating in place in an expanded bed (40). Coal and limestone for SO_2 capture can be fed to the combustion zone, and ash can be removed from it, by pneumatic transfer. Very little precombustion processing is needed to prepare either the coal or the sorbent for entry into the furnace (41).

In the 1970s commercial fluidized-bed combustors were limited to the atmospheric, bubbling-bed system, called the atmospheric fluidized-bed combustor (AFBC). In the late 1970s the circulating fluidized combustor (CFC) was introduced commercially, and in the 1980s the new commercial unit was the pressurized fluidized-bed combustor (PFBC).

Atmospheric Fluidized-Bed Combustors. By the late 1970s coal-fired AFBCs such as those shown in Figure 8 were offered having single units up to 230×10^3 t/h of steam (60 megawatts-electric and denoted as MWe) and fluidizing velocities up to 3 m/s (41). By 1991 the largest fluidized-bed unit in operation had grown in size to 160 MWe at the Shawnee Station of the Tennessee Valley Authority (43). A brief report of a 1990 survey of seven manufacturers of fluidized-bed combustors with regard to boiler statistics and problems generally common to the systems may be found in Reference 43. Three projects of particular interest are the fluidized-bed cogeneration system at the Europoort Tank Farm of Shell Nederland Raffinadeij (SNR) (44); the retrofit conversion at the Black Dog Station of Northern States Power (NSP) Co. (45); and the relocation, repowering, and reconfiguration of the power plant of Florida Crushed Stone (FCS) Corp. at its quarry near Brooksville, Florida (46).

SNR's fluidized-bed cogeneration system is an early example of the commercial development of AFBC technology. Foster Wheeler designed, fabricated, and erected the coal-fired AFBC/boiler, which generates 6.6 MWe and 37 MW thermal (also denoted as MWt) of heat energy. The thermal energy is transferred via medium-pressure hot water to satisfy the heat demand of the tank farm. The unit burns 6.4 t/h of coal and uses a calcium to sulfur mole ratio of 3 to set the limestone feed rate. The spent bed material may be reinjected into the bed as needed to maintain or build bed inventory. The fly ash, collected in two multicyclone mechanical collectors, may also be transferred pneumatically back to the combustor to increase the carbon burnup efficiency from 93%, without fly ash reinjection, to 98%.

NSP's retrofit conversion is the electric utility industry's demonstration of the use of AFBC to repower an aging pulverized-coal furnace using a clean combustor, capable of burning fuel of lower quality. Partly funded by the Electric

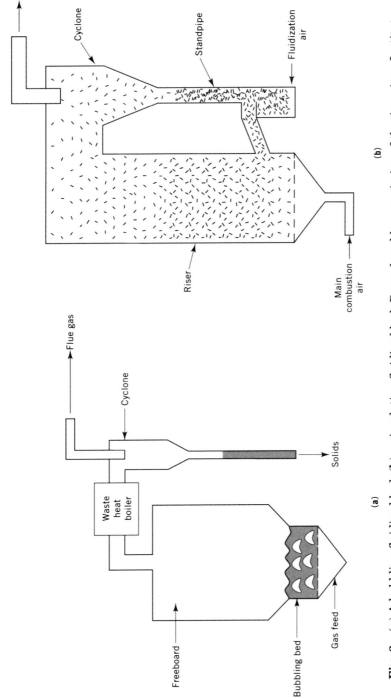

Fig. 8. (a) A bubbling fluidized bed; (b) a circulating fluidized bed. Reproduced by permission of the American Institute of Chemical Engineers, 1990 (42).

527

Power Research Institute (EPRI), the 130 MWe FBC unit was constructed in 1985 by Foster Wheeler Energy Corp. The new unit is expected to provide a 25-year unit life extension and it has already reduced emissions per unit of electric power produced. Details of startup and equipment performance during the first eight months of operation are provided in Reference 45. Another smaller retrofit project, in which two bubbling-bed units were installed in 1990 by Energy Products of Idaho in the 25 MWe Stream Plant No. 2 of Tacoma City Light, can burn coal, wood (qv), refuse-derived fuel (see FUELS FROM WASTE), or a mixture of all three fuels in a cost effective, efficient, and environmentally clean manner (43).

In FCS's 1986 repowering project Babcock and Wilcox (B&W) constructed a bubbling-bed section to FCS's existing 125 MWe pulverized-coal furnace to produce 31.3 t/h of lime, using crushed coal as the source of heat to calcine limestone in the fluidized bed. A portion of the lime is drawn from the bed as bottom ash and a portion is collected as fly ash. Both portions are transferred to a cement (qv) plant adjacent to the boiler. The hot flue gas from the FBC flows into the existing main pulverized-coal furnace, in which a B&W LIMB system was also installed to absorb sulfur dioxide during those times when the FBC is not operating.

Circulating Fluidized-Bed Combustors. Commercialization of the circulating fluidized-bed combustor (CFBC) began in the late 1970s (42). CFBCs operate with fluidizing velocities up to 10 m/s, which greatly increases elutriation from the bed within the main vessel, or riser (Fig. 8**b**). The effluent from the riser passes to a cyclone, where particulates are separated from the combustion gases and returned to the riser through an enlarged dipleg, or standpipe. The CFBC can handle feeds having many more fine particles than can an AFBC. This leads to more rapid burnout of the carbon content of the feed than for AFBCs and allows the firing of a larger size range of fuels.

By the late 1980s six principal commercial CFBC technologies were available (42). In 1993 the largest CFBC in operation is expected to be the Pyropower Corporation's 165 MWe reheat coal-fired unit, under construction since 1991 at the Point Aconi Station of Nova Scotia Power Corp. (43). Combustion and SO_2 control in this unit is to be carried out in the water-cooled riser. The unit is expected to operate at 870°C to optimize sulfur capture. The cyclone separators are refractory-lined and are supported approximately 30 m above grade.

An earlier Pyropower CFBC installation was completed in 1987 at the 110 MWe Nucla Station of the Colorado-Ute Electric Association (47). During shakedown, ie, the first year of its operation, beginning July 1, 1987, coal was fed to the unit for 37% of the time. A number of significant problems were identified and solved during this period, including several associated with a significant overheating incident, which shut the unit down for 74 days during Fall 1987. Following acceptance of the unit in Fall 1988, an extensive two-year test, sponsored by EPRI, provided a successful demonstration of the Pyropower technology (43).

Another of the six CFBC technologies, the multisolid fluidized-bed combustor (MSFBC), has been under development by Battelle Memorial Institute since 1974 (48). In an MSFBC a CFBC is superimposed on an AFBC in the combustor section. An early 15 MWt commercial version of MSFBC was designed and constructed by Struthers Thermo-Flood Corp. for Conoco.

Pressurized Fluidized-Bed Combustors. By 1983 the pressurized fluidized-bed combustor (PFBC) had been demonstrated to have capacities up to 80 MWt (49). PFBCs operate at pressures of up to 1500 kPa (220 psi) and fluidization velocities of 1–2 m/s. Compared to an AFBC of the same capacity, a PFBC is smaller, exhibits higher combustion efficiencies with less elutriation of fine particles, and utilizes dolomite, $CaCO_3 \cdot MgCO_3$, rather than limestone to capture SO_2.

Considerable development work on PFBCs was carried out in the United States, the United Kingdom, and Sweden through the 1980s (50). By the beginning of the 1990s two PFBCs were commercially available: ASEA Babcock's bubbling-bed technology and Pyropower Corp.'s circulating-bed system (51). A 70-MWe version of the ASEA Babcock technology has been installed at the Tidd Station of Ohio Power Co. and a 40-MWe version of the Pyropower system is planned for the Alma Station of Dairyland Power Cooperative.

The PFBC at the Tidd Station operates at 1200 kPa (170 psi) and a bed temperature of 860°C (51). A pressure vessel, 13.4 m in diameter by 20.7 m high, houses the combustor and its ancillaries. Coals, which contain ash contents less than about 25%, are blended with dolomite and pumped to the combustor as a paste having a total water content of 20–25%. Coals, which contain ash contents higher than 25%, and dolomite are individually fed pneumatically via separate lock hoppers. Both coal and dolomite are crushed to 3-mm top size before being fed to the unit.

Second-generation PFBC, currently being developed by Foster Wheeler Development Corp. places a pressurized carbonizer ahead of a circulating-bed PFBC (CPFBC) (52). The carbonizer produces a low heating-value fuel gas and a char, which are burned separately in a topping combustor and in the CPFBC respectively. This technology promises higher electrical power generating efficiencies (45%) than is obtained with first-generation PFBC systems (40%) (51).

Entrained-Bed Combustion. Entrained-bed combustors burn coal particles finer than 100 μm, which are carried rapidly through the flame by the combustion gases (40). To obtain particles of this size, larger pieces of coal are fed to pulverizers and thence to the burners, giving rise to the designation of these devices as pulverized coal (PC) furnaces. Flame temperatures are frequently above the ash fusion temperature of the coal, ie, >1200°C. In contrast to FBCs, little significant sulfur capture is possible by cofeeding calcium-based sorbents into the combustion zone, because calcium sulfite and calcium sulfate are unstable above 1200°C. Thus sorbents for sulfur removal from PC furnaces are injected as the combustion gases are being cooled. This can be accomplished in the upper region of the furnace or in the ductwork leading from the furnace to the particulate removal system (53).

Sulfur Removal Above the Furnace Combustion Zone. The first full-scale demonstration of injecting sorbent into a furnace above the combustion zone came in 1970, when fine limestone and pulverized coal were cofed to one of the units at the Shawnee Station of the Tennessee Valley Authority (54). Whereas this test disappointingly provided less than 30% removal of SO_2, the research that emerged showed that relocating the limestone feed point could significantly increase SO_2 capture. As a result, in the early 1980s the U.S. Environmental Protection Agency (EPA) initiated a program for limestone injection with multistage burners (LIMB) to develop and demonstrate processes to remove SO_2 from the upper region of PC furnaces (55). In the LIMB process finely-divided limestone is injected into the furnace at the 1300°C level, which is just above the furnace outlet plane and ahead of the pendent superheaters as shown in Figure 9. There, after being flash-calcined to lime, the sorbent reacts with SO_2 in the presence of O_2 to form $CaSO_4$. The temperature regime, where the sulfate is both stable and formed at reasonable rates, lies between 1200 and 900°C.

Three demonstrations of the LIMB technology have been carried out. The first was a privately funded project in the 75 MWt Boiler 405 at the No. 4 AC Station of Inland Steel Industries, Inc. (56). By injecting 70 wt % minus 200 mesh (74 μm) limestone, approximately 40% SO_2 removal was achieved at a Ca:S ratio of 3. This rose to 50% removal when the Ca:S ratio was increased to 4. The second LIMB demonstration was the backup desulfurization system installed by B&W as part of the relocation, repowering, and reconfiguration of the FCS power plant (46).

The third, and most significant LIMB, was the installation at the 105 MWe wall-fired Unit 4 Boiler at Ohio Edison's Edgewater Station (57). Three rows of eight sorbent injector nozzles each were installed on the front wall at elevations of 55.2, 57.0, and 58.2 m (54). The row at the 57.0 m level extends for two additional nozzles on each of the two adjacent side walls. All nozzles can be tilted through a 30° arc and are designed to achieve a momentum flux, which permit the pneumatically-transported sorbent to penetrate the flue gas from the wall. Problems associated with sorbent feeding, increased soot blowing, electrostatic precipitator (ESP) operation, and handling of increased ash have been corrected since the LIMB system was started up at the Edgewater Station in July 1987 (58). The problem associated with the ESP was the production of a back corona, caused by the high electrical resistivity of the sulfate-containing fly ash. The solution was simply to install a 18.3-m horizontal duct having a 4.3-m by 4.3-m cross section, containing an array of 100 water-spray nozzles, to humidify the flue gas before it passed to the ESP. The B&W LIMB process has proven technically very successful: SO_2 removals of 55 to 60% were obtained at a Ca:S stoichiometry of 2 and the goal of 50% removal at a stoichiometry of 1.6 was anticipated before the conclusion of the demonstration program (54). As of this writing sorbent injection in the upper portion of three additional commercial boilers in Illinois is being carried out in conjunction with gas reburning tests for NO_x reduction (59).

Removal in the Ductwork. For that portion of the furnace where temperatures drop from 900 to 150°C, both thermodynamics and kinetics prevent effective desulfurization using calcium-based sorbents. However, below 150°C humidification of the flue gas reestablishes a supportive environment for SO_2 removal, just as it improves ESP performance. At least six processes, based on the addition of a sorbent in conjunction with humidification of the flue gas downstream of the preheater, are being developed as of this writing (60). Three are funded as part of the U.S. Department of Energy's program to develop duct injection technology, and a fourth is an extension of the LIMB demonstration at the Edgewater Station.

The Bechtel confined zone dispersion (BCZ) process involves the injection of a fine slurry mist of pressure hydrated dolomitic lime or calcitic lime, using two-fluid atomizing nozzles. A demonstration at the 70 MWe Seward Station of the Pennsylvania Electric Co., performed in 15.2 m of ductwork with a 2.4-m by 3.4-m cross section, achieved a 50% removal of SO_2 at a Ca:S ratio around 1.1.

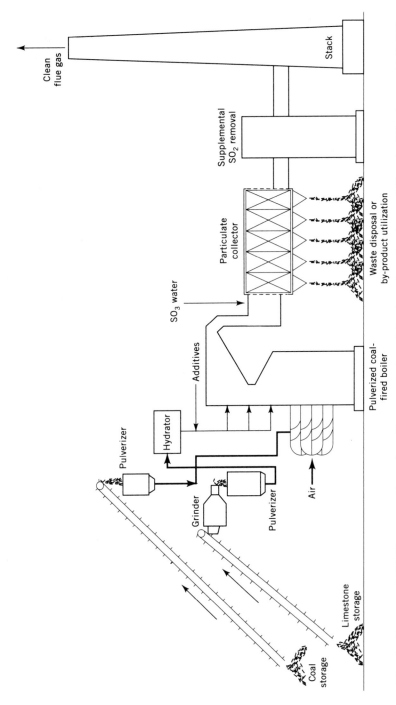

Fig. 9. Integrated LIMB system. Reproduced by permission of the American Institute of Chemical Engineers, 1985 (55).

531

The General Electric in-duct scrubbing (IDS) process involves the atomization of a slaked lime slurry, using a rotary disk atomizer. A test at the 12 MWe scale at the Muskingum River Station of Ohio Power, performed in a duct with a 4.3-m^2 cross section, achieved 50% SO$_2$ removal with good lime utilization.

The Dravo hydrate addition at low temperature process involves a two-step injection of water and dry sorbent in a rectangular 19.8-m duct having a cross section of 2 m^2. In one step water is injected through atomization nozzles to cool the flue gas from 150°C to approximately a 15°C approach to adiabatic saturation. The other step involves the dry injection of hydrated lime, either downstream or upstream of the humidification nozzles. Typical SO$_2$ removals were 50–60% at a Ca:S ratio of 2.

The Coolside process of Consolidation Coal Co. (Consol) (Fig. 10) involves injecting hydrated lime upstream of flue gas humidification. A second sorbent, initially sodium hydroxide, may be added to the water spray, using two-fluid atomizing nozzles. Being tested in a 17-m duct having a 19.8-m^2 cross section at the Edgewater Station of Ohio Edison, the process is projected to remove 70% SO$_2$ at a Ca:S ratio of 2 and a Na:Ca ratio of 0.185, based on pilot-plant tests (61). Considerable work has been carried out by Consol (62–64) and a team from Acurex Corp. and EPA (65) to seek process improvements by studying the characteristics of sorbents, developing sorbent modifications, and examining new substances for this purpose.

Sulfur Removal from Flue Gases

In 1983 there were 116 flue-gas desulfurization (FGD) systems in service, representing 47 gigawatts-electric of power generation capacity (66). As of 1992, more

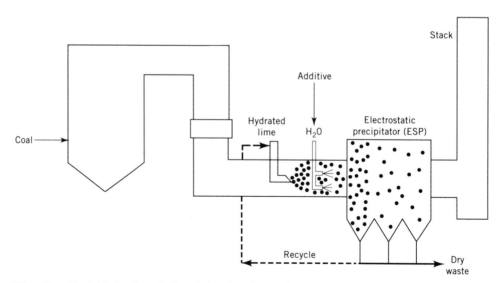

Fig. 10.   Coolside hydrated lime injection. Reproduced by permission of the American Institute of Chemical Engineers (61).

than 150 coal-fired boilers in the United States operated with FGD systems. The total electrical generating capacity of these plants has risen to 72 gigawatts (67). FGD processes are classified into (1) wet-throwaway, (2) dry-throwaway, (3) wet-regenerative, and (4) dry-regenerative processes (68).

   Wet-Throwaway Processes. By 1978, three wet-throwaway systems were in commercial operation: lime scrubbing, limestone slurry scrubbing, and dual alkali (1). Lime/limestone wet scrubbing (Fig. 11) remains the most common post-combustion control technique applied to utility boilers (67). The waste product

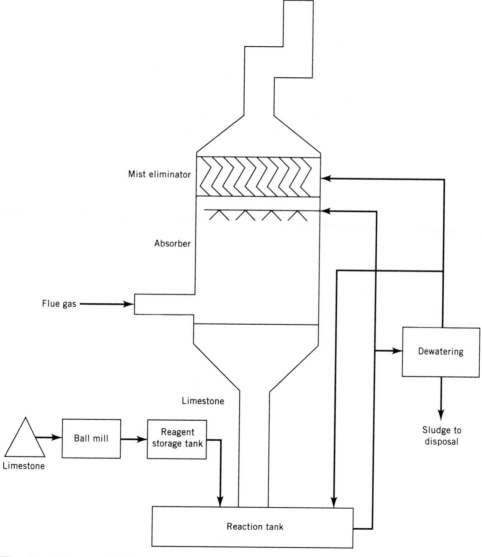

Fig. 11. Limestone FGD system. Reproduced by permission of the American Institute of Chemical Engineers (67).

from the scrubbers can either be sent to a landfill or be upgraded by oxidation to become saleable gypsum.

Whereas it is not precisely a lime/limestone wet scrubber because it uses alkali already present in the coal ash, the Colstrip FGD system has been most expansively described (69). Three scrubber modules, operating in parallel, are used on two 360 MWe coal-fired boilers at the Colstrip Station of the Montana Power Co. and the Puget Sound Power and Light Co. Each module consists of a downflow venturi scrubber, centered within an upflow spray tower contactor, and is designed to clean 120 MWe of equivalent gas flow under normal conditions and 144 MWe under emergency conditions. Thus, when one module is off line, the boilers can still operate at 80% capacity, using the remaining two modules to clean the generated flue gas. Test data show that the levels of pollutants in the plant emissions are well below the vendor guarantee and the applicable federal standards. Scrubber availability and plant load for a 22-month period shortly after startup and a number of operating details for the FGD system may also be found in Reference 69.

In the design of a lime/limestone scrubber, there are numerous considerations to be evaluated, including particulate removal (if any), ash removal, scrubber type and configuration, scaling prevention, absorbent feed control, water balance, mist elimination, reheat, and sludge disposal (70). In addition, the size of the boiler is a factor. FGD systems for smaller industrial coal-fired boilers can be purchased as packages, as opposed to the large specially engineered systems for field construction at utility stations (71). These latter systems are typically designed as vertical towers and use either lime or limestone as the sorbing agent. If the sulfur content is high or liquid waste is not permitted, the double alkali process may be used, where the sodium ion is replaced by a calcium ion to form an insoluble precipitate of the sulfur compounds. The precipitate is then filtered from the liquid stream and the regenerated liquid, containing the original sodium ions, is returned to the scrubber.

Dry-Throwaway Processes.   Dry-throwaway systems were the precursor of processes that removed SO_2 in the ductwork, eg, the BCZ and IDS processes. Here, however, the device is a spray chamber similar to the wet scrubbers such as the three modules of the Colstrip installation (Fig. 12). Into the upper portion of the chamber a slurry or clear solution containing sorbent is sprayed. Water evaporates from the droplets, the sorbent reacts with SO_2 both before and after drying, and the dry product is removed in a downstream baghouse or ESP (72). Unfortunately, dry scrubbing is much less efficient than wet scrubbing and lime, instead of the much less expensive limestone, is required to remove SO_2 effectively. Consequently, a search has been conducted for more reactive sorbents (72–75).

One commercial dry-scrubbing process is the system designed and installed by MikroPul Corp. at Strathmore Paper Company's 3.2 MWe PC cogeneration boiler in Woronoco, Massachusetts (76). The system consists of a slaked-lime spray drier reactor and a fabric filter. The stainless-steel spray drier is 4.4 m in diameter and 8.5 m tall. Four flue gas/sorbent slurry diffusers, containing two-fluid, external-mix-type nozzles, are installed at the top of the chamber. Early modifications to the chamber included changes in its aerodynamics to improve mixing and changes in the slurry feed and distribution system. The unit generally removes 80% SO_2 at a Ca:S ratio of 1.8.

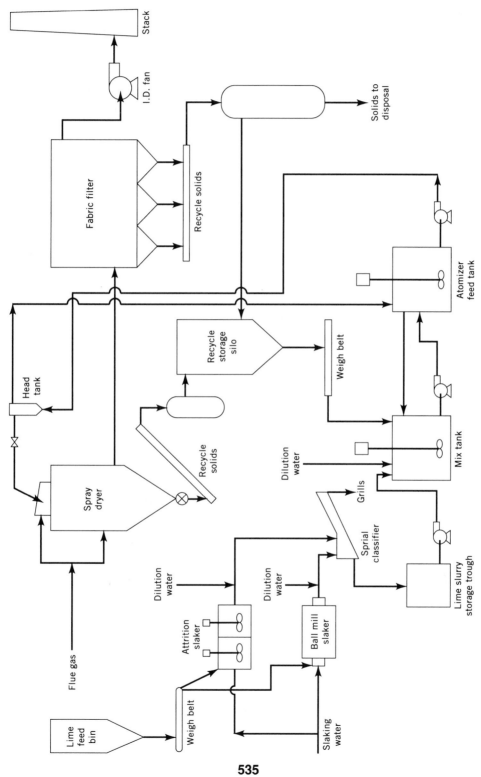

Fig. 12. Lime spray dryer process flow diagram. Reproduced by permission of the American Institute of Chemical Engineers, 1991 (67).

535

Regenerative Processes. By 1984 two commercial regenerative processes, the Davy McKee Wellman-Lord system and the United Engineers & Constructors Magnesium Oxide system were in operation in five generating stations, for a total of eleven units, in the United States (77). The Wellman-Lord process, of which only one unit is still in operation and which is no longer offered commercially, uses an aqueous solution of sodium sulfite to absorb SO_2 and form sodium bisulfite. The sodium bisulfite is thermally decomposed in the regenerator to re-form sodium sulfite and release SO_2 and the sulfite is returned to the absorber. The SO_2 is converted to either sulfur or sulfuric acid.

The MgO system shown in Fig. 13 uses an aqueous solution of $Mg(OH)_2$ to absorb SO_2, forming crystalline $MgSO_3 \cdot 3H_2O$. After removal by centrifugation the $MgSO_3 \cdot 3H_2O$ is dried in a direct-fired rotary kiln to produce anhydrous magnesium sulfite [7757-88-2], $MgSO_3$, which may then be shipped to a fluidized-bed calciner. In the calciner $MgSO_3$ is decomposed to magnesium oxide [1309-48-4], MgO, and SO_2. MgO absorbers were installed at Units 1 and 2 of the Eddystone Station and Unit 1 of the Cromby Station of Philadelphia Electric Co., and regenerators were built at sulfuric acid plants in Delaware and New Jersey by Allied Corp. and Essex Chemical, respectively. By February 1984 about 23,800 t of sulfuric acid had been made from by-product SO_2 and the scrubbers had consistently reached 96–98% removal of sulfur dioxide. The MgO system continues to operate well, but the regenerative process market has moved toward other systems, particularly the NOXSO process, which provides combined NO_x/SO_2 removal.

The NOXSO process has many of the elements of a traditional dry scrubber (78–80). The sorbent for both NO_x and SO_2 is sodium carbonate impregnated on a high surface-area gamma alumina. From a fluidized-bed absorber the sorbent is first heated from 120 to 600°C, driving off the NO_x and loosely-bound SO_2. The sorbent then passes to the first chamber of a regenerator where it is contacted with natural gas to remove most of the sulfur (about 70%), which it carries into that chamber. In the second chamber of the regenerator steam drives off the remaining sulfur. The sulfur-laden gases from both chambers are combined and processed to yield products for sale. A contract to construct a 5 MWe proof-of-concept plant for the NOXSO process at either Boiler 10 or Boiler 11 of Ohio Edison's Toronto Station was signed in 1989. A full-scale demonstration is in operation at Ohio Edison's Niles Station.

Development efforts regarding regenerative processes have also focused on higher temperature sorption using fluidized beds of metals supported on porous solids. The most advanced of these developments is the WSA-SNOX cleaning technology, offered by Haldor Topsoe, Inc. (81). After NO_x removal by selective catalyst reduction (SCR), the SO_2 in the flue gas is oxidized to SO_3 over a conventional sulfuric acid catalyst. Upon cooling, the SO_3 is hydrated to sulfuric acid, which is condensed, concentrated, and stored for sale. In 1988 several WSA-SNOX units were in operation in Europe and the first one in the United States was being planned by Ohio Edison for Boiler No. 2 of the Niles Station.

At a much earlier stage in the research and development cycle, fluidized-bed processes use porous sorbents containing copper oxide (82), cerium oxide (83), and other metal oxides (84).

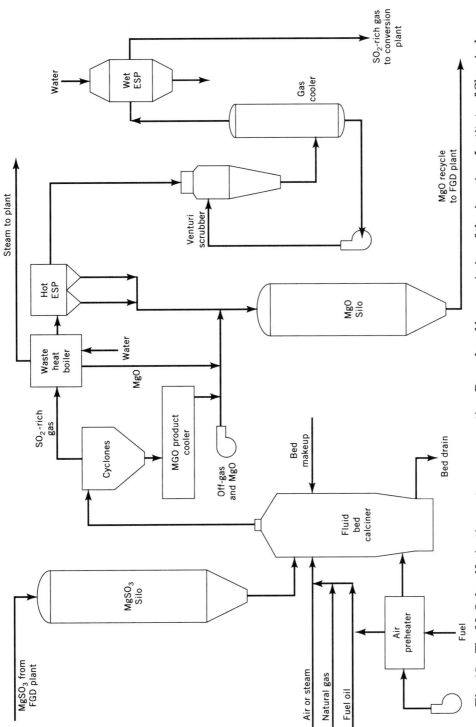

Fig. 13. The MgO desulfurization system: regeneration. Reproduced by permission of the American Institute of Chemical Engineers, 1984 (77).

537

Clean Coal Choices

When the Clean Air Act of 1990 was signed into law, electric utilities were required to establish plans and initiate projects to comply with that Act's Title IV. Each utility had to evaluate how the various commercial and emerging clean coal systems fit into the utility's technical and business environment resulting in strategies to utilize fuel switching and wet throwaway FGD processes almost exclusively (38,85,86).

BIBLIOGRAPHY

"Coal Conversion Processes, Desulfurization" in *ECT* 3rd ed., Vol. 6, pp. 306–324, by E. Stambaugh, Battelle Memorial Institute.

1. F. T. Princiotta, *Chem. Eng. Prog.* **74**(2), 58–64 (Feb. 1978).
2. *Coal*, National Coal Association, Mar. 1992, p. 9.
3. E. Zimmerman, *9th International Coal Preparation Congress*, New Delhi, India, Nov. 29–Dec. 4, 1982, pp. 32–44.
4. H. H. Lowery, *Chemistry of Coal Utilization*, Supplementary Vol. I, pp. 119–149, 1963.
5. "Trace Elements in Coal—An Interlaboratory Study of Analytical Techniques," *Technical Bulletin*, Consolidation Coal Co., Oct. 1987.
6. C. T. Ford and A. A. Price, "Evaluation of the Effect of Coal Cleaning on Fugitive Elements," Phase II, Part I, *Bituminous Coal Research*, BCR Report L-1082, 1980.
7. J. W. Leonard and B. C. Hardinge, ed., *Coal Preparation*, 5th ed., Society for Mining, Metallurgy and Exploration, Inc., Littleton, Colo., Chapts. 7, pp. 271–496, and 14, pp. 966–1005, 1991.
8. J. A. Cavallaro and co-workers, *Sulfur and Ash Reduction Potential and Selected Chemical and Physical Properties of United States Coals, DOE/PETC-91/2*, Jan. 1990.
9. D. W. Fuerstenau, in P. Somasundaran, ed., *Fine Particles Processing*, Vol. 1, AIME, New York, 1980, p. 669.
10. K. J. Miller and A. W. Deurbrouck, *Physical Cleaning of Coal: Present and Developing Methods*, Marcel Dekker, Inc., 1982, pp. 255–291.
11. B. K. Parekh and co-workers, *Column Flotation '88*, AIME, pp. 227–233, 1980.
12. A. F. Taggart, *Elements of Ore Dressing*, John Wiley & Sons, Inc., New York, 1951.
13. R. H. Yoon and co-workers, *2nd Int. Conf. on Processing and Utilization of High Sulfur Coals*, Carbondale, Ill., Sept. 27–Oct. 1, 1987, pp. 533–543.
14. D. E. Zipperian and U. Svensson, *Column Flotation '88*, 43–54, 1988.
15. L. E. Burgess and co-workers, *5th International Symposium on Coal Slurry Combustion and Technology*, Tampa, Fla., 1983, pp. 255–268.
16. J. D. Miller and M. C. Van Camp, *Tran. AIME* **272,** 1575 (1982).
17. C. E. Capes and K. Darcovich, *Powder Technol.* **40,** 43–52 (1984).
18. J. C. Knight, *Coal*, 58–60 (May 1989).
19. D. V. Keller, Jr. and W. M. Burry, *Coal Prep.* **8,** 1 (1990).
20. U.S. Pat. 4,613,429 (1986), S. H. Chiang and G. E. Klinzing (to University of Pittsburgh).
21. G. Araujo and co-workers, *Energy Prog.* **7**(2), 72 (1987).
22. R. D. Stoessner and co-workers, *Coal Prep '86*, 1986, pp. 5–34.
23. R. E. Hucko and C. P. Maronde, *9th International Coal Preparation Congress*, F2, 1982, pp. F2-1–F2-16.

24. R. E. Hucko and co-workers, "Status of DOE-sponsored Advanced Coal Cleaning Processes," in R. R. Klimpel and P. T. Luckie, eds., *Industrial Practice of Fine Coal Processing*, SME, Inc., Littleton, Colo., 1989.

25. E. Levy and co-workers, *8th Pittsburgh Coal Conference*, Pittsburgh, Pa., Oct. 14–18, 1991, pp. 1021–1026.

26. R. Markuszewski and co-workers, *Div. Fuel Chem.* **49,** 187–194 (1980).

27. S. Friedman, R. B. LaCount, and R. P. Warzinski, *173rd National Meeting of ACS*, New Orleans, La., Mar. 21–25, 1977, pp. 100–105.

28. R. R. Oder and co-workers, *Coal Conference*, American Mining Congress, Pittsburgh, Pa., May 1977.

29. U.S. Pat. 3,909,211 (1975), A. F. Diaz and E. D. Guth (to KVB Engineering Inc.).

30. A. Crawford, *Trans.* **3**(4), 204–219 (Jan. 1952).

31. J. L. Anastasi and co-workers, "Molten-Caustic-Leaching (Gravimelt Process) Integrated Test Circuit Operation Results," Report to the Gravimelt Process Advisory Board, Summer 1989.

32. J. L. Anastasi and co-workers, *5th Annual Coal Preparation, Utilization and Environmental Control Contractors Conference*, U.S. Department of Energy, Pittsburgh, Pa., July 31–Aug. 3, 1989.

33. C. K. Richardson and co-workers, *3rd Annual Pittsburgh Coal Conference*, Pittsburgh, Pa., Sept. 8–12, 1986, pp. 130–141.

34. P. S. Ganguli and co-workers, *Div. Fuel Chem. Prepr., ACS,* **21**(7), 118 (1976).

35. J. K. Kindig, *8th Pittsburgh Coal Conference*, Pittsburgh, Pa., Oct. 14–18, 1991, pp. 231–236.

36. U.S. Pat. 4,562,156 (1985), J.D. Isbister (to Atlantic Research Corp.).

37. J. Yasin and J. C. Gwozdz, *Power Eng.* **92**(1), 34–36 (Jan. 1988).

38. C. W. Schwartz, *Energy Prog.* **2**(4), 207–212 (Dec. 1982).

39. R. Smock, *Power Eng.* **95**(8), 17–22 (Aug. 1991).

40. R. H. Essenhigh, in C. Y. Wen and E. S. Lee, eds., *Coal Conversion Technology*, Addison-Wesley Publishing Co., Inc., New York, 1979, pp. 171–312.

41. E. C. McKenzie, *Chem. Eng.* **85**(18), 116–127 (Aug. 14, 1978).

42. R. J. Dry and R. D. LaNauze, *Chem. Eng. Prog.* **86**(7), 31–47 (July 1990).

43. D. J. Smith, *Power Eng.* **95**(12), 18–23 (Dec. 1991).

44. W. R. Kelly and co-workers, *Chem. Eng. Prog.* **80**(1), 35–40 (Jan. 1984).

45. W. J. Larva and S. Moore, *5th Pittsburgh Coal Conference*, Pittsburgh, Pa., Sept. 12–16, 1988, pp. 131–146.

46. J. A. Barsin and co-workers, *4th Pittsburgh Coal Conference*, Pittsburgh, Pa., Sept. 28–Oct. 2, 1987, pp. 629–640.

47. K. J. Heinschel, *5th Pittsburgh Coal Conference*, Pittsburgh, Pa., Sept. 12–16, 1988, pp. 368–376.

48. H. Nack and co-workers, *Chem. Eng. Prog.* **80**(1), 41–47 (Jan. 1984).

49. K. K. Pillai, *Proc. Instn. Mech. Engrs.* **197C,** C72/83 (Nov. 1983).

50. M. Marrocco, *5th Pittsburgh Coal Conference*, Pittsburgh, Pa., Sept. 12–16, 1988, pp. 235–244.

51. L. K. Carpenter and co-workers, *7th Pittsburgh Coal Conference*, Pittsburgh, Pa., Sept. 10–14, 1990, pp. 931–940.

52. A. Robertson and co-workers, *8th Pittsburgh Coal Conference*, Pittsburgh, Pa., Oct. 14–18, 1991, pp. 1036–1045.

53. B. K. Gullett and J. C. Kramlich, *4th Pittsburgh Coal Conference*, Pittsburgh, Pa., Sept. 28–Oct. 2, 1987, pp. 219–230.

54. P. S. Nolan and co-workers, *5th Pittsburgh Coal Conference*, Pittsburgh, Pa., Sept. 12–16, 1988, pp. 224–234.

55. D. G. Lachapelle, *Chem. Eng. Prog.* **81**(5), 56–62 (May 1985).

56. R. R. Landreth, *5th Pittsburgh Coal Conference*, Pittsburgh, Pa., Sept. 12–16, 1988, pp. 160–174.

57. R. V. Hendriks and P. S. Nolan, *J. Air Pollut. Control Assn.* **36**(4), 432–438 (Apr. 1986).

58. J. L. Hoffmann and T. R. Goots, *8th Pittsburgh Coal Conference*, Pittsburgh, Pa., Oct. 14–18, 1991, pp. 444–449.

59. W. Bartok and co-workers, *Environ. Prog.* **9**(1), 19–23 (Feb. 1990).

60. R. M. Statnick and co-workers, *4th Pittsburgh Coal Conference*, Pittsburgh, Pa., Sept. 28–Oct. 2, 1987, pp. 250–266; J. P. Gooch and co-workers, *6th Pittsburgh Coal Conference*, Pittsburgh, Pa., Sept. 25–29, 1989, pp. 289–298.

61. H. Yoon and co-workers, *Environ. Prog.* **7**(2), 104–111 (May 1988).

62. M. R. Stouffer and co-workers, *Ind. Eng. Chem. Res.* **28**(1), 20–27 (Jan. 1989).

63. J. A. Withum and H. Yoon, *Environ. Sci. Technol.* **23**(7), 821–827 (July 1989).

64. M. R. Stouffer and H. Yoon, *AIChE J.* **35**(8), 1253–1262 (Aug. 1989).

65. W. Jozewicz and co-workers, *J. Air Pollut. Control Assoc.* **38**(8), 1027–1034 (Aug. 1988).

66. K. E. Yeager, *Pollut. Eng.*, 24–28 (June 1984).

67. M. Maibodi, *Environ. Prog.* **10**(4), 307–313 (Nov. 1991).

68. R. McInnes and R. Van Royen, *Chem. Eng.* **97**(9), 124–127 (Sept. 1990).

69. C. Grimm and co-workers, *Chem. Eng. Prog.* **74**(2), 51–57 (Feb. 1978).

70. A. V. Slack, *Chem. Eng. Prog.* **74**(2), 71–75 (Feb. 1978).

71. J. D. Brady, *Chem. Eng. Prog.* **80**(9), 59–62 (Sept. 1984).

72. H. T. Karlsson and co-workers, *J. Air Pollut. Control Assoc.* **33**(1), 23–28 (Jan. 1983).

73. C. Jorgensen and co-workers, *Environ. Prog.* **6**(2), 26–32 (Feb. 1987).

74. J. M. Markussen and H. W. Pennline, *6th Pittsburgh Coal Conference*, Pittsburgh, Pa., Sept. 25–29, 1989, pp. 299–308.

75. A. Pakrasi and co-workers, *J. Air Waste Manage. Assoc.* **40**(7), 987–992 (July 1990).

76. T. V. Reinauer and co-workers, *Chem. Eng. Prog.* **79**(3), 74–81 (Mar. 1983).

77. C. Murawczyk and J. S. MacKenzie, *Chem. Eng. Prog.* **80**(9), 62–68 (Sept. 1984).

78. J. L. Haslbeck and co-workers, *6th Pittsburgh Coal Conference*, Pittsburgh, Pa., Sept. 25–29, 1989, pp. 319–329.

79. J. L. Haslbeck and co-workers, *7th Pittsburgh Coal Conference*, Pittsburgh, Pa., Sept. 10–14, 1990, pp. 330–339.

80. J. L. Haslbeck and co-workers, *8th Pittsburgh Coal Conference*, Pittsburgh, Pa., Oct. 14–18, 1991, pp. 479–484.

81. D. R. Juist and co-workers, *5th Pittsburgh Coal Conference*, Pittsburgh, Pa., Sept. 12–16, 1988, pp. 1165–1176.

82. J. T. Yeh and co-workers, *Environ. Prog.* **4**(4), 223–228 (Nov. 1985).

83. W. G. Wilson and co-workers, *8th Pittsburgh Coal Conference*, Pittsburgh, Pa., Oct. 14–18, 1991, pp. 457–463.

84. O. Faltsi-Saravelou and J. A. Vasalos, *Ind. Eng. Chem. Res.* **29**(2), 251–258 (Feb. 1990).

85. M. M. Peplowski, *Coal* **97**(2), 39–41 (Feb. 1992).

86. S. A. Mitnick, *Electr. J.*, 44–49 (Jan./Feb. 1992).

Shiao-Hung Chiang
James T. Cobb, Jr.
University of Pittsburgh

GASIFICATION

The gasification of coal (qv) to produce coal gas dates back to 1792. Coal gas, first produced in the United States in Baltimore in 1819, was widely used during the mid-1800s in urban areas for lighting and heating. By 1930 there were over 11,000 coal gasifiers operating in the United States and in the early 1930s over 11 million metric tons of coal were gasified annually.

Early gasifiers were air-blown, low pressure units that produced gas of low heating value, typically less than $5600 kJ/m^3$ (150 Btu/ft^3). The development of large-scale processes began in the late 1930s, with much of the work carried out in Germany. The first improvements were higher pressure, oxygen-blown gasifiers that resulted in a higher quality coal gas and higher efficiencies. Following World War II, interest in coal gasification waned, particularly in the United States, because of the increasing availability of inexpensive oil and natural gas (see GAS, NATURAL; PETROLEUM). Interest in coal gasification was renewed in 1973 when international oil and gas prices increased sharply.

Extensive development efforts over the past 20 years by companies such as Shell, Texaco, British Gas, and Lurgi have led to high temperature, high pressure slagging processes that offer high efficiencies, improved economics, and excellent environmental performance. Simultaneously, the development of high firing-temperature gas turbines has created a new and potentially very large market for coal gas as a fuel for combined cycle power generation (1,2) (see also COMBUSTION TECHNOLOGY; FUELS, SYNTHETIC-GASEOUS FUELS; POWER GENERATION).

Coal Gasification Combined Cycle Power Generation

Coal, the primary fuel for electricity generation in the United States and other countries, is expected to have an increasing role in the future. Conventional coal-fired electricity generation has resulted in numerous environmental problems, notably emissions of sulfur and nitrogen compounds, both of which have been linked to acid rain, and emissions of particulates (see AIR POLLUTION). Conventional coal-firing technologies only partially solve these problems. Modern coal gasification combined cycle (CGCC) power generation technologies, also known as integrated gasification combined cycle systems (IGCC), present electric power producers with important options and opportunities to improve efficiency, environmental performance, and overall cost effectiveness (1,3).

Electricity Demand. Although energy conservation efforts by electric utilities and customers are expected to help temper growth in electricity demand into the year 2010, electricity use in all sectors is expected to show continued growth. The Department of Energy (DOE) forecasts that, even when comprehensive energy conservation programs are taken into account, electricity consumption in the United States by the year 2000 is expected to be more than 20% above 1990 levels. The DOE expects the demand for electricity to grow at almost twice the rate of total energy demand; electricity demand in 2010 is predicted to be almost 50% greater than the demand in 1990.

Electric utilities are therefore expected to build new power plants or to extend the lives of existing, older ones. Ready availability, secure supply, and low price are expected to make coal the fuel of choice for most of the new baseload generating capacity.

Coal Gasification Combined Cycle. Coal gasification combined cycle (CGCC) integrates two commercially proven technologies: the manufacture of a clean-burning fuel gas from coal and the highly efficient use of that gas to produce electricity in a combined cycle power generation system. The combined cycle system has two basic components: (1) high efficiency gas turbines, which burn the clean fuel gas to produce electricity (4,5), and (2) exhaust heat, which is recovered to power traditional high efficiency steam turbines to generate additional electricity. The overall system is shown in Figure 1. The combination of the gas turbine and steam turbine cycles gives CGCC systems a coal-to-power efficiency of 41–43%, based on coal higher heating value (HHV), compared with about 34–35% achieved by conventional coal combustion steam cycle power plants. Additional efficiency gains are being pursued in CGCC systems using innovations such as hot-gas cleanup in the gasification island and improved gas turbines in the power block.

Demonstration Projects. The principal developers of advanced coal gasification technologies constructed and successfully operated large-scale demonstration plants during the 1980s. The British Gas Corporation constructed a 550 t/d slagging Lurgi gasifier in 1982. This unit, located in Westfield, Scotland, operated periodically until 1992, gasifying nine different coals.

In 1982, Texaco started up a 900 t/d gasifier at Southern California Edison's Cool Water facility. This was the first coal gasification plant to operate in an electric utility environment, providing coal gas as fuel to a GE-frame 7E combustion turbine. The Cool Water gasification plant operated for over 25,000 h on four different bituminous coals before it was shut down in 1989.

As of this writing, Destec Energy, Inc. has been operating a CGCC power plant using its coal gasification technology at a Dow Chemical Company plant in Plaquemine, Louisiana, since April 1987. This plant is designed for 1435 t/d of subbituminous coal on a dry basis or 1835 t/d of low quality lignites on a dry basis (see LIGNITE AND BROWN COAL). The syngas is fed to two Westinghouse WD 501

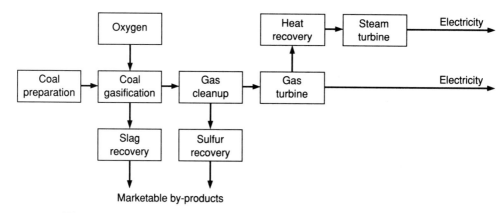

Fig. 1. Schematic of coal gasification combined cycle power generation.

gas turbines that in conjunction with the steam cycle, produce a net power output of 161 MW.

Shell's demonstration unit, called SCGP-1, was placed in service in 1987 at Shell's Deer Park manufacturing complex near Houston, Texas. SCGP-1 has a capacity ranging from 227 t/d on bituminous coal to 364 t/d on high moisture, high ash lignite. During four years of operation, SCGP-1 logged over 14,000 h while providing engineering and environmental data on 18 significantly different feedstocks, including Texas lignite, a subbituminous Powder River Basin coal, a wide variety of bituminous coals, and petroleum coke.

Commercialization of Coal Gasification Technologies. The successful operation of the demonstration plants is expected to lead to widespread commercialization of coal gasification technologies in the 1990s (Table 1). The Texaco coal gasification process (TCGP) has been licensed several times for use in chemical manufacture. The first license was issued to Tennessee Eastman for an 800 t/d plant, which was started up in 1983 and is used in the production of acetic anhydride [108-24-7]. In 1984, a second Texaco coal gasification plant having a coal capacity of 1650 t/d was placed in commercial service by Ube Industries in Japan to produce ammonia. A third commercial plant, having a coal capacity of 800 t/d, has been operated since 1986 by SAR in Oberhausen, Germany, for use in oxo-chemicals manufacture. Texaco has subsequently executed licensing agreements with the People's Republic of China for two relatively small facilities to start up in the mid-1990s.

The Shell coal gasification process (SCGP) has been licensed to Demkolec B.V., a subsidiary of the Dutch Electricity Generating Board, for use in a 253-MW

Table 1. Coal Gasification Combined Cycle Projects

Location	Capacity, MW	Design[a]	Estimated start-up
	Europe		
Buggenum (Demkolec)	2500	Shell	1993
Berrenrath (Rheinbraun)	300	HTW	1995
Spain	250	undecided	1996
Italy	350	undecided	1996
	North America		
Tennessee (TVA)	265[b]	Shell	1998
Wabash River	265	Destec	1995
Cool Water, Fla.	100[c]	Texaco	1992
Lakeland, Fla.	260	Texaco	1996
Delaware City, Del.	125	Texaco	1995
Springfield, Ill.	60	ABB CE	1996
Reno, Nev.	80	KRW	1997
Coeburn, Va. (Tamco)	107	U-Gas	1996
Canada	250	undecided	1996

[a]Terms are defined in text.
[b]Urea is also to be produced.
[c]Methanol is also a product.

integrated coal gasification combined cycle (CGCC) power plant being built in Buggenum, The Netherlands. The Demkolec project, the largest and most integrated CGCC plant in the world, is a significant step for CGCC commercialization (4). Construction was started in 1990 for operation in 1993 at an initial investment of $450 million. The net capacity at full load should be 253 MW and net efficiency at full load approximately 43% basis lowest heating value (LHV). The desulfurization level is expected to be 97.85% min (75 mg/m^3), and a maximum NO$_x$ emission from the gas turbine of 95 g/GJ. By-products are to be slag, fly ash, sulfur, and salt. The coal consumption at full load is expected to be ca 2000 t/d (585 MW) using dry ground coal. The gasification pressure is ca 2.8 MPa (28 bar); reaction temperature, ca 1500°C; steam pressures of 12.5, 4, and 0.8 MPa (125, 40, and 8 bar); and sulfur production, approximately 5000 t/yr.

Destec and PSI Energy, Inc. are sponsoring a 265-MW Wabash River Coal Gasification Repowering Project in Indiana. Selected for funding under the DOE's Clean Coal Technology Program, the project is expected to demonstrate, in a commercial setting, advancements in Destec's coal gasification technology to process high sulfur bituminous coal. Lurgi is involved in engineering several advanced combined cycle coal gasification projects in Europe and the United States, including a 300-MW CGCC plant in Berrenrath near Cologne, Germany. This plant, scheduled for start-up in late 1995, is designed to gasify low rank brown coal from Germany in a high temperature Winkler (HTW) fluidized bed gasifier; the syngas is expected to fuel a Siemens 94.3 gas turbine.

Types of Gasifiers

There are essentially three types of coal gasifiers: moving-bed or countercurrent reactors; fluidized-bed or back-mixed reactors; and entrained-flow or plug-flow reactors. The three types are shown schematically in Figure 2.

Moving-Bed Gasifier. The moving-bed gasifier is also called a fixed-bed gasifier. It involves a series of countercurrent reactions in which large (5 cm ×

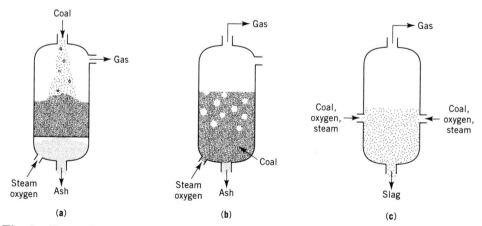

Fig. 2. Types of gasifiers: (**a**) moving-bed (dry-ash), (**b**) fluidized-bed, and (**c**) entrained-flow.

0.6 cm) particles of coal move slowly down the bed and react with gases moving up through the bed. The Lurgi gasifier is a prime example. At the top of the gasifier, the entering coal is heated and dried in the drying zone while cooling the product gas as it exits the reactor. The gas exit temperature ranges from 315°C for high moisture lignites to 550°C for bituminous coals. The coal is further heated and devolatilized by hotter gas as it descends through the carbonization zone. Below this zone, the devolatilized coal is gasified by reaction with steam and carbon dioxide in the gasification zone. The highest temperatures are reached in the combustion zone near the bottom of the gasifier, where the oxygen reacts with the char, which, together with ash, is all that remains of the original coal. Reaction of the char and steam, together with the presence of excess steam, moderates the temperature below the ash slagging temperature in this combustion zone. The whole bed is supported by a grate below the combustion zone where the ash is cooled by releasing heat to the entering steam and oxygen.

Characteristics of moving-bed gasifiers are low gasification temperatures, relatively low oxygen requirements, relatively high methane content in syngas produced, relatively low product gas temperature, production of hydrocarbon liquids such as tars and oils, and limited ability to handle fines. A slagging version of the fixed-bed Lurgi gasifier is the British Gas/Lurgi slagging gasifier (6), which offers many improvements, including the ability to handle caking coals and coal fines.

Fluidized-Bed Gasifier. A fluidized-bed gasifier consists of a back-mixed gasifier where feed coal particles are well mixed with coal and char particles already undergoing gasification. This gasifier is operated at a constant temperature below the initial ash fusion temperature in order to avoid sticky molten slag formation. Some coal particles are reduced in size during gasification and are entrained with the hot raw gas as it leaves the reactor. These char particles are recovered and recycled to the reactor. Ash particles are removed from below the bed and are cooled by heating the incoming steam and recycle gas. Examples of fluidized-bed gasifiers are the high temperature Winkler (HTW) and the Kellogg Rust Westinghouse (KRW) gasifiers. Fluidized-bed gasifiers typically utilize significant fly-ash recycle to capture unconverted carbon, have uniform and moderate temperature throughout the gasifier, and are limited in ability to convert high rank coals.

Agglomerated ash operation improves the ability of fluidized-bed processes to gasify unreactive high rank coals and caking coals efficiently. In the KRW gasifier, the base of the bed provides the hot ash agglomerating zone. Unconverted char is recycled to this zone, which is hot enough to gasify the char and soften the ash. The ash particles stick together, growing in size and density, until separated from the char and removed by dry lockhoppers.

Entrained-Flow Gasifier. The entrained-flow gasifier consists of a plug-flow system in which the fine coal particles concurrently react with steam and oxygen. Residence time is a few seconds. These systems operate at high temperatures, well above ash slagging conditions, in order to assure good carbon conversion and provide a mechanism for removal of ash as molten slag. Entrained-flow gasifiers are utilized in the Shell coal gasification process, Texaco coal gasification process, Dow coal gasification process by Destec, and Prenflo by Krupp-Koppers (7). The short residence time required in entrained gasifiers can result in potentially high throughputs at elevated pressures. Entrained gasifiers have high feedstock flexibility. The agglomerating tendency and fines content of feed coal that greatly limit the operation of the moving-bed and fluid-bed gasifiers are not a problem in entrained gasification. Entrained gasifiers, which can use 100% of the mine output, have a small coal inventory that results in rapid start-up, shutdown, and load-following characteristics. Entrained gasifiers have greater turndown capacity than do fluid-bed gasifiers, and the product gases contain no tars and light oils, thus facilitating heat recovery and requiring less gas cleaning and purification. Also, the product gas contains much lower quantities of other impurities, such as mercaptans, ammonia (qv), carbon disulfide (qv), carbonyl sulfide, and thiophene (qv), than does that of other types of gasifiers. Treatment of the wastewater from the gas cleaning operation is therefore simpler for entrained-flow gasifiers.

Commercial Processes

Shell Coal Gasification Process. The SCGP is based on a dry feed, entrained-bed, high pressure, high temperature slagging design (3,8,9). The process can handle a wide variety of coals, ranging from bituminous to lignite, in an environmentally acceptable way, and produces a high purity, medium heating value gas that is attractive for use in power generation. A typical configuration for a combined cycle power plant incorporating the SCGP for its fuel supply is shown in Figure 3.

Raw coal is crushed and fed to a pulverizer similar to those used in a pulverized coal boiler. This mill grinds the coal to a size range suitable for efficient gasification, that is, to 90 wt % $\leq$ 100 μm. As the coal is being ground, it is simultaneously dried by means of a heated inert gas stream that carries the evaporated water from the system as it sweeps the pulverized coal through an internal classifier to collection in a baghouse. The dried, milled coal is delivered to the gasifier feed system using a pneumatic conveying system (see CONVEYING).

The oxygen required in the gasification step is supplied by an air separation plant (see CRYOGENICS). Nitrogen (qv) from the air separation unit is compressed for use in the gasification plant, for example, for the makeup of inert gas to coal milling and drying and for transporting coal in the feed system.

Pressurized coal, oxygen, and, if necessary, steam enter the gasifier through opposing burners. The gasifier consists of an outer pressure vessel and an inner gasification chamber with a water-cooled membrane wall. The inner gasifier wall temperature is controlled by circulating water through the membrane wall to generate saturated steam. The membrane wall encloses the gasification zone, from which two outlets are provided. One opening at the bottom of the gasifier is used for the removal of slag; the other outlet allows hot raw gas and fly slag to exit from the top of the gasifier.

Most of the mineral content of the feed coal leaves the gasification zone in the form of molten slag. The high (over 1500°C) gasifier temperature ensures that the molten slag flows freely down the membrane wall into a water-filled compartment at the bottom of the gasifier. High (above 99%) carbon conversions are obtained, and the high temperature ensures that essentially no organic compo-

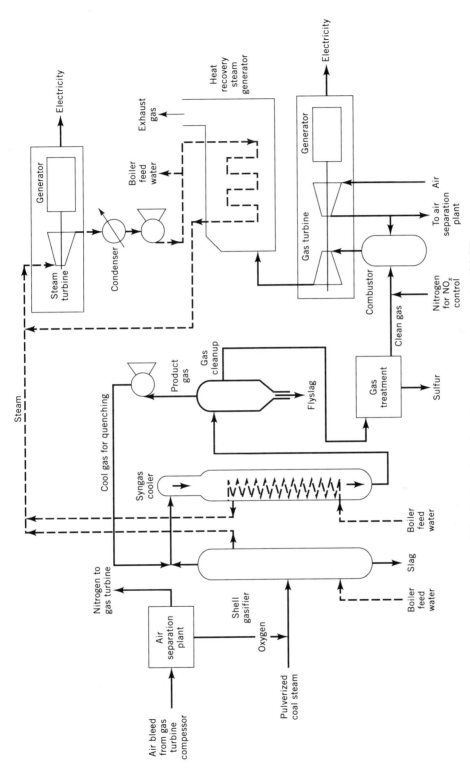

Fig. 3. Combined cycle power generation using SCGP.

nents heavier than methane are formed. The insulation provided by the slag layer in the gasifier minimizes heat losses, so that cold gas efficiencies are high and CO_2 levels in the syngas are low. The recycle of fly slag enhances gasification efficiency.

Flux may be added to the coal feed to promote the appropriate slag flow from the gasifier at the preferred operating temperature. As the molten slag contacts the water bath, the slag solidifies into dense, glassy granules. These slag granules fall into a collecting vessel located beneath the slag bath and are transferred to a lockhopper that operates on a timed cycle to receive the slag. After the lockhopper is filled, the slag is washed with clean makeup water to remove entrained gas and any surface impurities. After washing, the lockhopper is depressurized and the slag is fed to intermediate storage.

The hot raw product gas leaving the gasification zone is quenched with cooled, recycled product gas to convert any entrained molten slag to a hardened solid material prior to entering the syngas cooler. The syngas cooler recovers high level heat from the quenched raw gas by generating high pressure steam. The gasifier and syngas cooler included in the SCGP plant are similar to the water wall boilers widely used in other utility processes.

The bulk of the fly slag contained in the raw gas leaving the syngas cooler is removed from the gas using commercially demonstrated equipment such as filters or cyclones. The recovered fly slag can be recycled back to the gasifier. The syngas then goes to a scrubbing system, where the remaining traces of solids and water-soluble contaminants are removed, and thereafter to an acid gas removal system, where an amine-based solvent, such as Sulfinol, typically removes more than 99% of the sulfur species (see SULFUR REMOVAL AND RECOVERY).

A bleed from the scrubbing system is sent to a sour slurry stripper. The water is then clarified and can be recycled to minimize the volume of effluent to be biotreated and discharged or evaporated. The acid gas from the acid gas removal system and from the sour slurry stripper is fed to a Claus plant, where salable elemental sulfur (qv) is produced. For maximum sulfur recovery and minimal sulfur emissions, the Shell Claus off-gas treating process (SCOT) is used.

Texaco Coal Gasification Process. The TCGP incorporates a single-stage, slagging, pressurized, entrained-bed downflow gasifier. Rather than using a dry coal feed system, the Texaco gasifier uses a concentrated water slurry of coal ground to a carefully controlled size distribution (10–14).

The TCGP begins by wet grinding coal using a controlled amount of water in conventional rod or ball mills. The coal–water slurry, having a controlled solids concentration, is pumped, along with oxygen, to the refractory-lined gasifier through a specially designed burner. Gasification occurs at temperatures in the 1200–1500°C range, which destroys most organic species more complex than methane. Oxygen input to the gasifier is tightly controlled to maintain a reducing atmosphere. The predominant gasification products in the resulting syngas are carbon monoxide and hydrogen. Other components include carbon dioxide, nitrogen, hydrogen sulfide, and small amounts of ammonia and carbonyl sulfide.

The TCGP allows three possible configurations that differ in the amount of high level heat that is recovered from the gasifier exit gas. As the quantity of high level heat recovery increases, overall plant efficiency increases. This efficiency gain is accompanied by increased capital cost. In the maximum heat recovery configuration, the hot gas exiting the gasifier passes through two waste heat boilers, called radiant and convective syngas coolers for the predominant mode of heat transfer used to cool the gas. High pressure steam is generated in these boilers. This configuration is commonly known as the radiant plus convective or RC mode. In the total quench or TQ mode of the TCGP, the gas from the reactor passes immediately into a water quench. No attempt is made to recover the high level syngas heat. Lower level heat is recovered from the quenched syngas and from the hot quench water. The radiant only or RO mode of the TCGP is a compromise between the RC and TQ extremes. Radiant heat is recovered to generate high pressure steam, and then the gas is water-quenched. Lower level heat is recovered as in the TQ mode.

The cooled syngas from any of the TCGP modes is then routed to a water scrubber to remove particulates. The gas is cooled further and is passed through an absorber that removes nearly all the sulfur, mainly H_2S, in the gas. Valuable elemental sulfur is produced from the H_2S stream. Normally 97–99% of the sulfur entering the TCGP in the coal is recovered as salable elemental sulfur (12).

Most of the solid waste, or glassy slag, from the coal drops through the radiant boiler or quench chamber and is collected in the slag handling section. The fine soot carried in the syngas is removed in the water scrubber and collected by the black water handling section. These two solids streams are dewatered (see DEWATERING) and often combined. The soot and the glasslike slag are composed of the feed coal ash mineral matter and a small amount of unconverted carbon.

Dow Coal Gasification Process. The Dow coal gasification process (Destec) is a two-stage, slurry feed, entrained-flow, slagging gasifier. The first stage assures high carbon conversion and optimum slag removal. The second stage reduces the raw product gas temperature to about 1000°C. This latter step helps to improve cold gas efficiency relative to other slurry fed processes and to lower waste heat recovery costs (15–18).

Subbituminous coal or lignite is received in railroad hopper cars, unloaded, and then ground and slurried with water recycled from the syngas cleanup process. The slurry is then transferred to a slurry storage tank to provide hold-up for transfer to the gasification area. The ground coal slurry is pumped to a slurry feed tank in the gasification area. Provisions are made for back flushing lines and recirculating coal slurry to prevent plugging. The slurry is pumped to the slurry preheaters, where it is heated to within 25 to 50°C of the boiling point of the slurry at the reactor pressure. Positive displacement pumps capable of handling liquid–solid suspensions at high pressure are used to control the slurry feed rate to the preheaters. After the preheating, the slurry is fed to the reactor, where it is mixed with oxygen in the burner nozzles. The feed rate of oxygen is controlled to maintain the reactor temperatures in a specific range that depends on the properties of the coal. Under these conditions, the coal is almost totally gasified by partial combustion to CO, H_2, CO_2, and H_2O.

The sulfur in the subbituminous coal or lignite is converted mostly to hydrogen sulfide. Small amounts of carbonyl sulfide are also produced. Molten slag is drained from the bottom of the gasifier into a water quench. The slag is withdrawn continuously as a slurry through grinders and a pressure letdown system. The hot syngas is cooled in an integral heat recovery system to about 1000°C. The second stage is unique to the Dow coal gasification process. Hot gases leaving the first stage are cooled by additional slurry introduced into the second stage. The raw gas is then sent to the heat recovery system.

The high temperature heat recovery train, consisting of a steam boiler and a steam superheater, generates steam for use in the Dow steam system. The syngas is cooled to within 30°C of the condensation temperature to prevent condensation in the economizer. After the syngas is cooled to near its condensation point, it is fed to a wet particulate scrubber that is operated at the boiling point of the recirculating water. The dilute slurry produced is concentrated and blended with the reactor feed stream. Water collected from other parts of the process is used in the wet scrubber. The scrubbed syngas is then cooled through a series of heat exchangers to about 50°C prior to H_2S removal. Water condensed from the syngas, as it is cooled, is recycled to the process and slurry unit after removal of NH_3, H_2S, and other soluble gases.

H_2S is removed from the syngas in the GAS/SPEC ST-1 process licensed from Dow. The sweetened syngas is suitable for fuel for the gas turbine power generation system. The separated acid gas, consisting of H_2S, CO_2, and water, is fed to the sulfur recovery unit. Sulfur is recovered from the acid gas produced in the H_2S removal section using the Selectox Process, licensed from Union Oil Company of California through the Ralph M. Parsons Company. The acid gas is preheated, mixed with a controlled flow of air, and fed to the catalytic reactor, which partially oxidizes the H_2S to sulfur and water. The effluent gas from the reactor is cooled to condense the sulfur. The tail gas is fed to an incinerator, which burns the remaining H_2S to SO_2 and vents it to the atmosphere. The sulfur removal and recovery processes are designed to meet existing environmental regulations.

High Temperature Winkler Process. The high temperature Winkler (HTW) process developed by Rheinbraun is especially targeted for the gasification of brown and hard coals, peat, and biomasses in a fluidized-bed gasifier (19–25) (see FUELS FROM BIOMASS). The raw brown coal containing 50–60% moisture is dried down to 12% using Rheinbraun's drying process (19), which involves a fluidized-bed dryer using immersed heating surfaces for internal waste heat utilization. The vapor produced in this process is cleaned in an electrostatic precipitator that removes the fine coal particles and is then used either for fluidization or, after recompression, for drying the coal. The condensate serves to saturate the cleaned fuel gas before its entry into the gas turbine, thus further increasing the overall plant efficiency.

Dried coal is pneumatically conveyed to feed bins, pressurized through lockhoppers, and fed to the refractory-lined fluidized-bed gasifier vessel by variable-speed screws. The gasifying agent, mixed oxygen and steam, is fed near the bottom of the gasifier. The bed operates at 2.5 MPa (25 bar) and 750–800°C. Oxygen and steam are also injected above the bed to increase carbon conversion and reduce yields of methane and other hydrocarbons. The freeboard zone above the bed operates as much as 150 to 230°C above the bed temperatures (24).

A portion of the ash and some char are entrained in the reactor effluent. Rheinbraun's design removes coarse particles in a primary hot cyclone and returns them to the bed through a hot dip-leg. Fine particles are separated in a secondary hot cyclone and removed from the system through a lockhopper. Bottom ash is removed through another lockhopper. The process provides the option, depending on carbon conversion, to mix the bottom ash and secondary cyclone with brown coal for burning in existing brown coal-fired boiler power stations (25). After the hot cyclones or ceramic candle filters, the gas is cooled in a horizontal fire-tube type boiler, which generates medium-pressure steam. The gas is then further cooled and fine dust removed by water quench by means of a venturi scrubber.

British Gas/Lurgi Slagging Gasifier. The technology developed by the British Gas Corporation (BGC) and Lurgi started with the dry ash Lurgi gasifier and incorporated enhancements such as operation at a higher temperature that melts the coal ash to slag. A significant efficiency advantage is gained by reducing the steam requirement to only about 15% of that required by the dry-ash Lurgi gasifier.

Coal having a top size of 5 cm and a fines content up to 35 wt % < 0.6 cm is fed at the top of the gasifier through a lockhopper. The coal reacts while moving down through the gasifier. The coal ash is removed from the bottom of the gasifier as molten slag through a slag tap, then quenched in water, and removed by a lockhopper. Steam and oxygen, injected through tuyeres at the bottom of the bed, react with the coal as the gases move up through the bed. This countercurrent action results in a wide temperature difference between the top and bottom of the gasifier. Reaction zones are similar to those in the conventional Lurgi gasifier. The solids entrained in the raw product gas and hydrocarbon by-products such as tars and oils, naphtha, and phenols can be recycled to the top of the gasifier or reinjected into the gasifier at the tuyeres, where they are gasified. Additional coal fines can also be fed through the tuyeres.

Compared with raw gas from the dry-ash Lurgi gasifier, the raw gas from the slagging gasifier has lower H_2O, CO_2, and CH_4 and higher CO content, primarily because of the lower steam consumption. Recycle of the tar and oil in the slagging gasifier increases the gas yield by reducing the net hydrocarbon liquid production to only naphtha and phenols. The slagging gasifier offers additional advantages over the dry-bed gasifier in terms of feed flexibility because it can handle caking coals and a significant amount of fines. The slagging gasifier technology was developed at BGC's Westfield facility in Scotland, initially on a 275 t/d pilot plant, and subsequently on a 550 t/d demonstration unit.

Coal Gasification Chemistry

During combustion, oxygen in the air converts the combustible matter in coal, consisting of complex molecules mostly made up of carbon and hydrogen, to carbon dioxide and water. Excess air is supplied to ensure complete combustion. In contrast, gasification involves incomplete combustion in an oxygen-deficient or reducing atmosphere. Gasification uses 20–30% of the oxygen theoretically required for complete combustion to carbon dioxide and water. Carbon monoxide and hydrogen are the principal products, and only a fraction of the carbon in the coal is oxidized completely to carbon dioxide. The heat released by partial combustion provides the bulk of the energy necessary to drive the gasification reactions.

In fixed-bed gasifiers the coal is first dried by evaporation of the surface and inherent moisture, then devolatilized, which breaks the weaker chemical bonds, forming tars, oils, phenols, and hydrocarbon gases. These devolatilization products exit the gasifier with the syngas, because of low temperatures and lack of oxygen in fixed-bed gasifiers. Fluidized-bed gasifiers provide better mixing and uniform temperatures that allow oxygen to react with the devolatilization products. These products also undergo thermal cracking, primarily on hot char surfaces, reacting with steam and H_2. In dry fluidized-bed gasifiers, temperatures have to be maintained below the ash melting point, which leads to incomplete carbon conversion for unreactive coals. Agglomerating ash gasifiers operate at higher temperatures, near the ash softening point, which provides improved carbon conversion. Entrained-bed slagging gasifiers provide uniform high temperatures, resulting in complete conversion of all coals to hydrogen, carbon monoxide, and carbon dioxide, and producing no tars, oils, or phenols. Thus the principal gasification reactions are (26,27)

$$C + O_2 \rightarrow CO_2 \quad \text{(exothermic)}$$

$$C + \tfrac{1}{2}O_2 \rightarrow CO \quad \text{(exothermic)}$$

$$CO + H_2O \rightarrow CO_2 + H_2 \quad \text{(exothermic—shift reaction)}$$

$$C + H_2O \rightarrow CO + H_2 \quad \text{(endothermic)}$$

$$C + CO_2 \rightarrow 2\,CO \quad \text{(endothermic)}$$

$$CO + 3\,H_2 \rightarrow CH_4 + H_2O \quad \text{(exothermic—methanation)}$$

$$C + 2\,H_2 \rightarrow CH_4 \quad \text{(exothermic—methanation)}$$

Trace elements such as sulfur and nitrogen are also involved in the gasification reactions. Sulfur in coal is converted primarily to H_2S under the reducing conditions of gasification. Approximately 5 to 15% of the sulfur is converted to COS, whereas the coal nitrogen is converted primarily to N_2; trace amounts of NH_3 and HCN are also formed.

Coal Properties

Developers of coal gasification technology have studied the impact of key coal properties on different parts of the gasification process. These tests have provided a good understanding of the influence of coal properties and have led to the development of process and equipment options. For example, a comprehensive demonstration program conducted on the SCGP at the demonstration plant near Houston (SCGP-1) included 18 different feeds converting a very broad range of coals (28). The coal property extremes are shown in Table 2.

   Ash Content.   Ash content affects gasifier performance, especially for most slagging gasifiers, because molten ash in the form of slag provides an insulating coverage on the wall of the gasifier, which reduces the heat transferred during the gasification reaction (29). Ash content also influences the requirements of the slag tap and the slag handling system. A related parameter is the slagging efficiency, which is the percentage of mineral solids recovered as slag out of the bottom of the gasifier relative to the total mineral solids produced by the process. As shown in Table 2, the feeds gasified at SCGP-1 had ash contents ranging from 0.5% moisture free (MF) for petroleum coke to 24.5% MF for Texas lignite.

Table 2. SCGP-1 Feed Property Extremes

Constituent	Composition, wt %	
	High	Low
ash	24.5[a] (up to 35)	0.5[b]
oxygen	16.3[c]	0.1[b]
sulfur	5.2[b]	0.3[d]
chlorine	0.41[e]	
moisture	30.7[a]	
Na_2O	3.1[d]	
K_2O	3.3[f]	
CaO	23.7[c]	0.8[b]
$CaO + MgO + Fe_2O_3$		10.2[g]
Fe_2O_3	27.8[h]	
SiO_2	58.9[i]	
Al_2O_3	32.6[j]	

[a]Texas lignite.
[b]Petroleum coke.
[c]Buckskin, Powder River Basin, Wyo.
[d]SUFCo.
[e]Pyro No. 9.
[f]Pike County.
[g]Newlands.
[h]R&F.
[i]El Cerrejon.
[j]Skyline.

Ash Melting Point/Slag Viscosity. For gasification technologies utilizing a slagging gasifier, slag flow behavior is an important parameter. To determine slag viscosity, the viscosity of coal ash is measured in a reducing atmosphere. Coals having a wide range of ash fusion temperatures were tested at fluid temperatures ranging from 1190°C for Illinois No. 5 coal to 1500°C and higher for several Appalachian coals, such as Skyline, Robinson Creek, and Pike County. Figure 4 shows coal ash viscosity plots versus temperature for a variety of coals. Slag viscosity varies over several orders of magnitude for the different coals at representative gasifier temperatures. For instance, Buckskin, which is a subbituminous coal from the Powder River Basin in Wyoming, has a much lower viscosity than an Appalachian coal such as Blacksville No. 2. For coals having high slag viscosities, slag behavior can be modified by the addition of a flux such as limestone (calcium carbonate).

Fouling Characteristics. Fouling of heat transfer surfaces can result from constituents such as chlorine, sodium, potassium, and calcium. Most fouling indexes are based on experience with pulverized coal boilers. This information is often used to select conventionally high fouling coals to obtain fouling data on gasification units. In order to establish fouling indexes, the coal selection criteria in demonstration programs has included a wide range of fouling agents. As shown in Table 2, the calcium oxide content ranged from 0.8% for petroleum coke to 23.7% for Buckskin coal.

Reactivity. Reactivity is used to describe the relative degree of ease with which a coal undergoes gasification reactions. The primary property affecting the ease of conversion is the oxygen content of the coal, which in turn reflects its age or rank. Other factors that have impact are maceral distribution, volatile content,

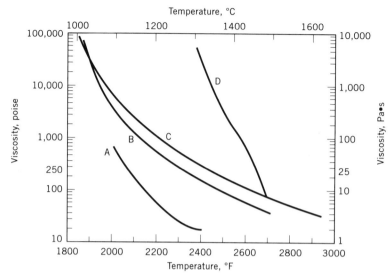

Fig. 4. Viscosity–temperature profiles of coal ash samples, where A represents Buckskin; B, Pyro No. 9; C, Blacksville (Appalachia) No. 2; and D, Drayton.

and, to a lesser degree, some mineral components. From Table 2 it can be seen that petroleum coke has the lowest reactivity (an oxygen content of 0.1% MF), whereas Buckskin has the highest reactivity (an oxygen content of 16.3% MF).

Corrosion. The primary coal properties affecting corrosion are sulfur and chlorine levels. The range of sulfur levels in Table 2 shows a low of 0.3% for SUFCo and a high of 5.2% for petroleum coke. R&F coal has 4.2% sulfur. The highest chlorine level in this group of coals is 0.41% for Pyro No. 9 coal.

Gasifier Performance

Entrained-flow gasifiers can process a wide variety of coals, and data for several SCGP-1 coals are shown. Table 3 gives compositional analyses; Table 4 shows ash minerals; and Table 5 presents a summary of gasification parameters for the different feeds. These feeds range from high rank bituminous coals from Appalachia, for example, Pike County coal, to low rank Texas lignite and petroleum coke (30–37).

Operating Parameters. The primary gasifier operating parameters are coal composition, coal throughput, oxygen/coal ratio and steam/oxygen ratio. The amount of oxygen and steam fed to the gasifier depends on the coal composition. In general, low rank coals are very reactive and require less oxygen and little to no steam, whereas high rank coals are relatively unreactive, requiring more oxygen and a moderate amount of steam. Steam provides an alternative source of oxygen for the gasification reaction and helps to moderate the gasification temperature. As a source of hydrogen, steam also helps to balance the H_2/CO ratio, giving a constant syngas composition for all coals including petroleum coke. Gas-

Table 3. Analyses of SCGP-1 Feedstocks[a]

Component	Texas lignite	Pike County[b]	Pike County[c]	Dotiki	Newlands	El Cerrejon	Skyline	Robinson Creek	R&F	Pocahontas No. 3	Petroleum coke
					Samples as received						
moisture	30.70	8.55	6.04	6.71	7.01	11.86	7.95	5.61	5.70	7.52	9.31
ash	16.96	6.87	10.71	8.36	14.14	7.75	8.80	7.23	13.02	4.68	0.45
volatile	30.19	32.13	30.80	34.61	25.30	33.30	33.36	32.63	35.05	16.22	9.62
fixed carbon	22.15	52.45	52.45	50.32	53.55	47.09	49.89	54.53	46.23	71.58	80.62
Total	*100.00*	*100.00*	*100.00*	*100.00*	*100.00*	*100.00*	*100.00*	*100.00*	*100.00*	*100.00*	*100.00*
					Dry basis samples						
ash	24.48	7.51	11.40	8.96	15.20	8.79	9.56	7.66	13.81	5.06	0.50
carbon	56.22	79.40	75.24	74.47	71.49	74.71	74.44	78.31	69.64	86.51	89.23
hydrogen	4.36	5.18	4.67	5.23	4.28	4.99	4.83	5.09	4.55	4.29	3.59
nitrogen	1.13	1.59	1.49	1.58	1.61	1.53	1.51	1.43	1.41	1.20	1.35
chlorine	0.08	0.20	0.12	0.23	0.09	0.02	0.02	0.16	0.17	0.11	0.03
sulfur	1.67	0.67	0.79	3.10	0.63	0.98	1.11	1.15	4.15	0.88	5.22
oxygen	12.06	5.46	6.29	6.44	6.70	8.99	8.52	6.20	6.28	1.96	0.10
Total	*100.00*	*100.01*	*100.01*	*100.00*	*100.00*	*100.00*	*100.00*	*100.00*	*100.01*	*100.01*	*100.02*
					Other properties						
HHV[d], kJ/kg[e]	22839	32586	30749	31439	29225	30918	30523	32567	29344	34879	35679
Hardgrove grindability	63	47	49	55	57	53	36	43	60	91	61

[a]Composition is given in units of wt % unless other units are indicated.
[b]Washed samples.
[c]Run-of-mine samples.
[d]HHV = higher heating value, on dry basis.
[e]To convert kJ to kcal, divide by 4.184.

555

Table 4. Ash Minerals Variability of SCGP-1 Feedstocks[a]

Component	Texas lignite	Pike County[b]	Pike County[c]	Dotiki	Newlands	El Cerrejon	Skyline	Robinson Creek	R&F	Pocahontas No. 3	Petroleum coke
					Ash mineral						
P_2O_5	0.16	0.06	0.31	0.25	1.36	0.20	0.30	0.28	0.38	0.22	0.2
SiO_2	48.82	52.30	53.22	51.10	50.89	58.94	51.20	50.92	47.14	38.64	6.8
Fe_2O_3	7.26	5.89	7.38	13.31	7.59	9.04	9.09	11.05	27.76	17.28	6.3
Al_2O_3	15.12	31.00	28.70	21.50	31.62	17.45	32.61	29.06	19.08	26.05	1.9
TiO_2	1.05	1.33	1.49	1.30	1.61	0.78	1.59	1.38	0.86	1.33	0.8
CaO	11.67	4.30	1.23	4.46	2.18	4.16	1.19	1.63	1.18	6.57	0.8
MgO	2.05	1.04	1.10	0.69	0.43	2.45	0.70	1.04	0.66	1.63	0.6
SO_3	10.89	1.18	0.75	4.05	0.94	3.97	0.94	1.05	0.97	4.41	0.3
K_2O	1.06	2.24	3.32	2.34	0.51	1.83	1.85	2.74	1.64	1.72	0.1
Na_2O	0.42	0.27	0.36	0.44	0.12	0.53	0.21	0.52	0.16	1.05	1.2
Total	98.49	99.61	97.86	99.44	97.25	99.36	99.66	99.67	99.83	98.90	98.14[d]
					Ash content						
ash (MF), mean %	24.48	7.50	11.40	8.96	15.20	8.79	9.56	7.66	13.81	5.06	0.50
standard deviation	3.31	0.72	1.09	0.42	0.99	1.18	2.97	2.17	1.99	0.35	0.39
standard deviation, %	13.52	9.60	9.56	4.69	6.51	13.42	31.07	28.33	14.41	6.92	78.00

[a]Composition is given in units of wt %.
[b]Washed sample.
[c]Run-of-mine sample.
[d]Sample also contains 71.77 wt % V_2O_5 and 7.37 wt % NiO.

Table 5. Summary of SCGP-1 Gasification Performance

Parameter	Texas lignite[a]	Texas lignite[b]	Pike County[c]	Pike County[d]	Dotiki	New-lands	El Cerrejon	Skyline	Robinson Creek	R&F	Pocahontas No. 3	Petroleum coke
coal to plant, t/d	335	248	154	175	166	171	246	179	171	197	156	156
oxygen/MAF–coal ratio	0.877	0.865	1.006	0.974	0.970	0.986	0.922	0.955	0.985	0.966	1.021	1.038
burner steam/oxygen ratio			0.141	0.108	0.128	0.089	0.141	0.122	0.107	0.053	0.217	0.231
gasifier off-gas, vol %[e]												
CO	60.59	61.82	63.08	64.43	62.05	65.30	63.41	63.17	64.22	64.26	62.80	65.23
H_2	28.20	28.01	29.81	30.14	30.33	26.90	30.78	29.24	29.42	28.46	28.39	25.90
CO_2	5.38	4.47	2.59	0.68	2.47	2.26	1.68	1.99	0.81	1.24	1.86	2.17
$H_2S + COS$	0.71	0.80	0.24	0.36	0.90	0.28	0.27	0.34	0.47	1.01	0.23	1.58
$N_2 + Ar + CH_4$	5.08	4.83	4.18	4.34	4.18	5.19	3.83	5.21	5.01	4.97	6.68	5.07
sweet syngas, kg/h	12062	10381	12213	13230	12275	12116	16707	13437	13971	13716	12973	13325
HHV[f] energy basis, GJ/h[g]	155.2	130.9	158.6	173.0	161.7	153.2	227.2	172.0	179.4	178.3	168.6	166.5
sulfur removal, %[h]	99.1	99.8	99.7	99.8	99.8	99.5	98.6	99.5	99.5	99.7	99.6	99.8
carbon conversion, %	99.7	99.4	99.9	99.1	99.9	99.7	99.6	99.9	99.7	99.5	99.3	99.5
cold gas efficiency, % HHV[f] (sweet gas basis)	78.8	80.3	80.9	83.0[i]	80.1	80.3	83.4	82.4	82.2	79.6	92.4	78.9

[a]High ash content.
[b]Low ash content.
[c]Washed samples.
[d]Run-of-mine samples.
[e]Dry gas.
[f]HHV = higher heating value.
[g]To convert J to cal, divide by 4.184.
[h]From syngas.
[i]See Figure 5.

557

ifier performance is evaluated in terms of syngas production and composition, carbon conversion, and cold gas efficiency (see Table 5).

Cold Gas Efficiency (CGE). Cold gas efficiency, a key measure of the efficiency of coal gasification, represents the chemical energy in the syngas relative to the chemical energy in the incoming coal. Cold gas efficiency on a sweet gas basis is calculated as the percentage of the heating value in coal that is converted to clean product syngas after removal of H_2S and COS.

Carbon Conversion. Carbon conversion on a once-through basis is a function of the coal composition and is strongly influenced by the oxygen/coal ratio. For some coals, the conversion pattern is also affected by the level of steam in the blast. Another factor is fly slag recycle, which raises the carbon conversion by recycling the unconverted carbon, most of which resides on the fly slag. This results in an overall carbon conversion greater than 99%.

Gas Composition and Heating Value. Table 5 shows syngas composition for a number of feedstocks. These numbers reflect the composition of the gasifier off-gas on a dry basis. The primary gas components are CO and H_2, ranging from 59 to 67% and from 25 to 31%, respectively. Generally the gas composition is constant within a fairly narrow band for all coals including petroleum coke. The moderate variation is primarily because of variation in the CO_2 concentration caused by different steam/oxygen levels in the blast and oxygen moisture and ash-free (MAF) coal ratios. The HHV of the product syngas after removal of H_2S, COS, and CO_2 is typically 12 MJ/m^3 (300 Btu/ft^3) and does not change significantly with changes in feedstock or gasifier conditions. The product syngas, also called medium-Btu gas (MBG), makes an excellent fuel for commercial gas turbines because of the constant CO and H_2 concentrations.

Heat Balance. Mass and heat balances are calculated around the gasification block, which includes the gasifier, quench, syngas cooler, and solids removal systems. A typical heat balance for Pike County coal is shown in Figure 5. Input streams are HHV of the incoming coal and sensible heat of the coal–oxygen–steam blast. The output streams are HHV of the sour syngas obtained

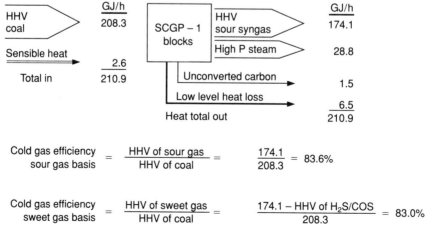

Fig. 5. SCGP-1 heat balance for run-of-mine Pike County coal. To convert GJ/h to Btu/h, multiply by 9.48×10^5.

from gas analysis, HHV of unconverted carbon obtained from analysis and weights of solids, heat recovered in the steam system in the gasifier and syngas cooler, and low level heat representing unrecovered sensible heat in the syngas. The low level heat is calculated by difference, thus forcing the heat balance to 100%. This low level heat is typically 3 to 4% for all feedstocks. Thus in most cases at least 95% of the energy of the feed streams, mostly heating value of the coal, is converted to usable energy in the form of syngas and high pressure steam.

Environmental Performance of Coal Gasification Technology

One advantage of modern coal gasification combined cycle systems is excellent environmental performance. Not only are regulatory standards met, but emissions and effluents are well below accepted levels (13,16,28).

Acid Rain Emissions.   Coal gasification combined cycle (CGCC) represents a superior technology for controlling SO_2 and NO_x emissions. Emissions are much lower than those from traditional coal combustion technologies (38). During gasification, the sulfur in the coal is converted to reduced sulfur compounds, primarily H_2S and a small amount of carbonyl sulfide, COS. Because the sulfur is gasified to H_2S and COS in a high pressure concentrated stream, rather than fully combusted to SO_2 in a dilute-phase flue gas stream, the sulfur content of the coal gas can be reduced to an extremely low level using well-established acid gas treating technology. The sulfur is recovered from the gasification plant as salable, elemental sulfur. A small quantity of sulfur can also be captured in the slag as sulfates.

The gas treating and sulfur recovery processes employed in coal gasification have been broadly applied and operated for decades in refinery and petrochemical facilities and in natural gas sweetening plants. Operating experience from SCGP-1 (28,36) has confirmed that overall sulfur removal efficiencies of 99.4% from the raw syngas are achievable, independent of coal sulfur content for a variety of coals (Fig. 6). Modern pulverized coal (PC) plants generally have flue gas desulfurization (FGD) units capable of 95% sulfur capture. See Figure 7 for a comparison of

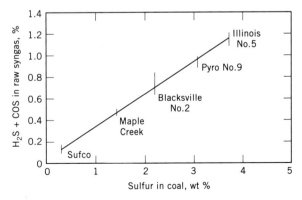

Fig. 6. Sulfur profile for SCGP-1 feedstocks. Overall sulfur removal is >99.4% and the sulfur in the sweet syngas is <20 ppm.

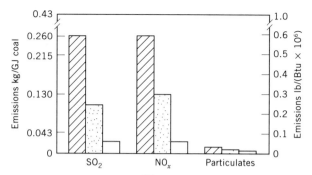

Fig. 7. Environmental emissions, where ▨ represents new source performance standards (NSPS) requirements; ▢ represents a pulverized coal (PC) plant; and ▢ represents SCGP-1.

emissions. New technologies are being developed for removing sulfur and other contaminants at high temperature. One hot-gas cleanup process uses metal oxide sorbents to remove $H_2S + COS$ from raw gas at high (>500°C) temperature and system pressure.

During coal gasification the nitrogen content of coal is converted to molecular nitrogen, N_2, ammonia, NH_3, and a small amount of hydrogen cyanide, HCN. In moving-bed gasifiers, some of the nitrogen also goes into tars and oils. The NH_3 and HCN can also be removed from the coal gas using conventional (cold) gas treating processes. Other techniques are being investigated in hot-gas cleanup technologies. After removal of HCN and NH_3, combustion of the coal gas in the gas turbine produces no fuel-based NO_x. Only a small amount of thermal NO_x is formed, and this can be controlled to low levels through turbine combustor design and, if necessary, steam or nitrogen addition. Based on tests using SCGP-type coal gas fired in a full-scale GE-frame 7F combustor (39), a NO_x concentration of no more than 10 ppm in the gas turbine flue gas is attainable. See Figure 7 for a comparison of NO_x emissions from a PC plant equipped with low NO_x burners.

Criteria Air Pollutants.   Moving-bed gasifiers produce tars, oils, phenols, and heavy hydrocarbons (qv), the concentrations of which are controlled by quenching and water scrubbing. Fluidized-bed gasifiers produce significantly lower amounts of these compounds because of higher operating temperatures. Entrained-flow gasifiers operate at even higher temperatures, typically in excess of 1650°C. SCGP-1 experience has confirmed that carbon conversions of greater than 99.5% are easily attainable for any coal and that essentially no organic compounds heavier than methane are produced (40). Emissions of volatile organic compounds (VOC) from a CGCC plant are expected to be approximately 300 times lower than those from a similarly sized coal-fired steam plant equipped with low NO_x burners and an FGD unit.

The product gas after cleanup consists of primarily CO and H_2. Combustion of coal gas in high firing-temperature gas turbines converts virtually all of the CO to CO_2, and gas turbine exhaust is expected to contain no more than 10 ppm CO when operating at design conditions. Carbon monoxide emissions from a CGCC plant are thus expected to be around one-tenth those of a modern coal-fired plant equipped with low NO_x burners.

Particulate removal from the coal gas is effected either through a series of dry-solid and wet-solid removal steps or through the use of dry solids filters, so that the gas fed to the combustion turbine is essentially free of suspended particulates. The emissions of total suspended particulates (TSP) from a CGCC plant are about one-third those from a comparable pulverized coal plant equipped with a fabric filter and FGD unit.

Hazardous Air Pollutants. A number of the metals present in coal have the potential to be released as toxic air emissions. In moving-bed and fluidized-bed gasifiers, these metals are captured in the water. In entrained-bed gasifiers, a majority of these metals are captured in the slag. Because the coal ash in entrained gasifiers becomes vitrified at the high gasifier temperatures, the resultant glasslike slag encapsulates the metals in nonleachable form. In conventional treating systems using cold-gas cleanup, the small fraction of metals released to the gas phase is captured effectively in the gas cooling and gas treating steps. The combination of gas cooling and multistage gas–liquid contacting reduces very substantially the potential for airborne emissions of volatile metals such as lead, beryllium, mercury, or arsenic.

The total emissions of hazardous air pollutants from a CGCC plant having wet cleanup are expected to be at least an order of magnitude lower than those achievable from a modern coal-fired steam plant (41). Metals removal in hot-gas cleanup systems is still under development.

Water Consumption and Effluent Characterization. Another advantage of CGCC power generation is derived from lower water requirements. Because more than half of the power generated in a CGCC plant comes from the gas turbine, the water requirement is only 70–80% of that required for a coal-fired power plant, where all of the power is generated from steam turbines.

Whereas moving-bed gasifiers require complex water-treatment systems to address tars, phenols, and metals, this complexity is mostly alleviated for fluidized-bed gasifiers and is eliminated for entrained-flow gasifiers. The exiting water streams of SCGP-1 contain no detectable amounts of volatile or semivolatile organics. The effluent from a CGCC facility can be biotreated to meet National Pollutant Discharge Elimination System (NPDES) standards (42). Biological treatment provides oxidation for the small amounts of inorganic nitrogen and sulfur species that remain in the water. Effluent from SCGP-1 has pH 7.8 and contains:

Chemical analysis	ppmwt
oil and grease	<1
phenols	<0.1
ammonia	<0.5
nitrite	<0.5
nitrate	50
total cyanide	<0.27
thiocyanate	0.1
formate	<0.1
thiosulfate	<0.1
sulfate	109
sodium	470
chloride	510
pH	7.8

The biotreated SCGP-1 effluent contains fully oxidized products and very low concentrations of trace metals:

Metal	ppmwt
antimony	0.01
arsenic	0.02
beryllium	<0.01
cadmium	<0.03
chromium	<0.01
copper	<0.01
lead	<0.03
mercury	<0.001
nickel	<0.02
selenium	0.35
silver	0.03
thallium	<0.001
zinc	0.03

Both acute and chronic toxicity testing of the treated effluent on daphnia shrimp and fathead minnows have indicated that the effluent is completely suitable for discharge into receiving waters with no adverse impact (42).

Solid By-Products. Coal gasification power generation systems do not produce any scrubber sludge, a significant advantage over both direct coal combustion processes that use limestone-stack gas scrubbers and fluidized-bed combustion processes that use solid absorbents for sulfur capture. In coal gasification, the sulfur in the coal is recovered as bright yellow elemental sulfur for which there are several commercial applications, the largest being in the phosphate fertilizer industry (see FERTILIZERS). Elemental sulfur is a commodity traded worldwide, with 1990's prices in excess of $100/t.

The ash in the coal is converted to slag, fly slag, or fly ash. Moving-bed and fluidized-bed gasifiers produce fly ash, which may be disposed of in a manner similar to that used for conventional power plant fly ash. In slagging gasifiers, the coal ash is mostly converted to a glasslike slag that has very low leachability. Environmental characterization of SCGP-1 slag and fly slag was performed for several coals using the extraction procedure (EP) toxicity tests and the toxicity characteristic leaching procedure test (TCLP), confirming that toxic trace metal concentrations in the leachate were well below Resource Conservation and Recovery Act (RCRA) requirements (43). Many of the elements, if present, were even below the detection limits. Additionally, the runoff from the slag storage area was collected and analyzed for comparison with the National Interim Primary Drinking Water Standards. The results of this comparison show that the measured values are typically much less than those allowed by the stringent national standard (43).

As part of a solids utilization program at SCGP-1, gasifier slag has been used as a principal component in concrete mixtures (Slagcrete) to make roads, pads, and storage bins. Other applications of gasifier slag and fly slag that are expected to be promising are in asphalt (qv) aggregate, Portland cement kiln feed, and lightweight aggregate (see CEMENT) (44). Compressive strength and dynamic creep tests have shown that both slag and fly slag have excellent construction properties.

CO$_2$ Emissions and Global Warming. The high coal-to-busbar efficiency of a CGCC system provides a significant advantage in responding to CO$_2$ emissions and thus to global warming concerns. High efficiency translates to lower coal consumption and lower CO$_2$ production per unit of electricity generated. The average existing PC unit has a heat rate of more than 10,550 kJ/kWh (10,000 Btu/kWh) on a higher heating value basis, which means that associated CO$_2$ emissions for a nominal 450-MW plant are well over 100,000 t/y. The most efficient CGCC units offer heat rates of 8650 kJ/kWh (8200 Btu/kWh) and reduce CO$_2$ emissions by about 15–20% relative to the emissions from a PC unit.

Syngas Chemistry

Whereas near-term application of coal gasification is expected to be in the production of electricity through combined cycle power generation systems, longer term applications show considerable potential for producing chemicals from coal using syngas chemistry (45). Products could include ammonia, methanol, synthetic natural gas, and conventional transportation fuels.

The economics of coal gasification are influenced by the availability of oil and natural gas, but coal is expected to continue to play an ever-increasing role as a significant resource base for both energy and chemicals.

Ammonia. Ammonia is produced through the reaction of hydrogen and nitrogen. In a coal-to-ammonia facility, coal gasification produces the hydrogen and an air separation plant, which also provides oxygen for coal gasification, supplies the nitrogen. Because coal gasification produces a mixture of hydrogen and carbon monoxide, the CO is combined with steam in a water gas shift reactor to produce carbon dioxide and H$_2$. Following CO$_2$ removal, the hydrogen stream is fed to an ammonia synthesis reactor where it reacts with molecular nitrogen to produce ammonia.

The water gas shift reaction is an exothermic reaction:

$$CO + H_2O \longrightarrow CO_2 + H_2 \qquad \Delta H_R = -39.8 \text{ kJ } (-9.5 \text{ kcal})$$

For shifting coal-derived gas, conventional iron–chromium catalysts can be used. Because coal gas has a significantly higher concentration of carbon monoxide than is found in gas streams in conventional refineries, the catalyst must be able to withstand high thermal loads. However, potential catalyst poisons such as phenol and other hydrocarbons are not a concern in entrained-bed gasifiers.

Methanol. Methanol is produced by stoichiometric reaction of CO and H$_2$. The syngas produced by coal gasification contains insufficient hydrogen for complete conversion to methanol, and partial CO shifting is required to obtain the desired concentrations of H$_2$, CO, and CO$_2$. These concentrations are expressed in terms of a stoichiometric number, $(H_2 - CO)/(H_2 + CO_2)$, which has a desired value of 2. In some cases CO$_2$ removal is required to achieve the stoichiometric number target. CO and H_2 are then reacted to form methanol in a catalytic methanol synthesis reactor.

The exothermic reaction

$$CO + 2\,H_2 \longrightarrow CH_3OH \qquad \Delta H_R = -109 \text{ kJ } (-26 \text{ kcal})$$

is enhanced by high pressures and low temperatures. Catalysts used in the reactor are based on copper, zinc, or chromium oxides (46), and reactors are designed to remove the exothermic heat of reaction effectively.

Mobil Oil Corporation has developed a process on a pilot scale that can successfully convert methanol into 96 octane gasoline. Although methanol can be used directly as a transportation fuel, conversion to gasoline would eliminate the need to modify engines and would also eliminate some of the problems encountered using gasoline–methanol blends (see ALCOHOL FUELS; GASOLINE AND OTHER MOTOR FUELS).

Synthetic Natural Gas.   Another potentially very large application of coal gasification is the production of synthetic natural gas (SNG). The syngas produced from coal gasification is shifted to produce a H_2-to-CO ratio of approximately 3 to 1. The carbon dioxide produced during shifting is removed, and CO and H_2 react to produce methane (CH_4), or SNG, and water in a methanation reactor.

The following reactions can occur simultaneously within a methanation reactor.

$$CO + 3\,H_2 \longrightarrow CH_4 + H_2O$$

$$CO_2 + 4\,H_2 \longrightarrow CH_4 + 2H_2O$$

$$CO + H_2O \longrightarrow CO_2 + H_2$$

$$2\,CO \longrightarrow CO_2 + C$$

The heat released from the CO–H_2 reaction must be removed from the system to prevent excessive temperatures, catalyst deactivation by sintering, and carbon deposition. Several reactor configurations have been developed to achieve this (47).

The tube wall reactor (TWR) system features the use of catalyst-coated tubes. The Raney nickel catalyst is flame-sprayed onto the inside surface of the tubes, and the tubes are immersed in a liquid, such as Dowtherm, which conducts the heat away. Some quantity of recycle gas, in the ratio from zero to five, may also be used. In fluidized-bed catalyst systems, Raney nickel or thorium nickel catalysts operate under moderate pressure, and heat is quickly removed from the system by the off-gas stream. In the liquid-phase methanation system developed by Chem Systems, an inert liquid is pumped upward through the reactor, operating at 2–7 MPa (300–1000 psi) and 300–350°C, at a velocity sufficient to fluidize the catalyst and remove process heat. At the same time, the coal gas is passed up through the reactor, where methanation occurs in the presence of the catalyst. It has been found that catalyst attrition is substantially reduced over that in gas-fluidized beds because of the cushioning effect of the liquid. Processes have also been developed for hydrogasification that maximize direct conversion of coal to methane. A good example is the HYGAS process, which involves the direct hydrogenation of coal in the presence of hydrogen and steam, under pressure, in two fluidized-bed stages. Additional developments have been pursued with catalysts, such as Exxon's catalytic gasification process, but these processes have not been commercialized.

A coal-to-SNG facility can be built at a coal mine-mouth location, taking advantage of low cost coal. SNG can then be pipelined to local distribution companies and distributed through the existing infrastructure. This approach is used in the Great Plains Coal Gasification Project in Beulah, North Dakota, which employs Lurgi gasifiers followed by shift and methanation steps. SNG has the advantage that it can directly displace natural gas to serve residential, industrial, and utility customers reliably.

Another technology that is being pursued for fuel utilization of coal is mild gasification. Similar to pyrolysis, mild gasification is performed at atmospheric pressure at temperatures below 600°C. By drying and heating under controlled conditions, the coal is partially devolatilized and converted to gases and a solid residue. The gases can be used as fuel and partially condensed to produce a liquid fuel similar to residual fuel oil. The solid product is similar to low moisture, high heating value coal. A demonstration project for Powder River Basin coal is being implemented by ENCOAL in Wyoming.

Conventional Transportation Fuels. Synthesis gas produced from coal gasification or from natural gas by partial oxidation or steam reforming can be converted into a variety of transportation fuels, such as gasoline, aviation turbine fuel (see AVIATION AND OTHER GAS TURBINE FUELS), and diesel fuel. A widely known process used for this application is the Fischer-Tropsch process which converts synthesis gas into largely aliphatic hydrocarbons over an iron or cobalt catalyst. The process was operated successfully in Germany during World War II and is being used commercially at the Sasol plants in South Africa.

More recently, Shell has developed proprietary technology for converting syngas into liquid hydrocarbons (48). This technology is particularly well suited for producing high quality distillate fractions and is therefore referred to as the Shell middle distillate synthesis (SMDS) process. This is a modernized version of the classical Fischer-Tropsch technique. In the first step, the synthesis gas components, hydrogen and carbon monoxide, react to form predominantly long-chain paraffins that extend well into the wax range.

Underlying the Fischer-Tropsch reaction is a chain-growth mechanism. The product distribution is in accordance with Schultz-Flory polymerization kinetics and can be characterized by the probability of chain growth. The higher the probability of chain growth, the heavier the waxy product. In the development of the SMDS process, proprietary catalysts have been developed with a high selectivity toward heavier products and, therefore, with a low yield of products in the gas and gasoline range. Much attention has been paid to the selection of a reactor for this very highly exothermic process. In principle, three different types of reactors can be used for the synthesis: a fixed-bed reactor, an ebulliating or fluidized-bed reactor, and a slurry reactor.

The use of a fluidized-bed reactor is possible only when the reactants are essentially in the gaseous phase. Fluidized-beds are not suitable for middle distillate synthesis, where a heavy wax is formed. For gasoline synthesis processes like the Mobil MTG process and the Synthol process, such reactors are especially suitable when frequent or continuous regeneration of the catalyst is required. Slurry reactors and ebulliating-bed reactors comprising a three-phase system with very fine catalyst are, in principle, suitable for middle distillate and wax synthesis, but have not been applied on a commercial scale.

For the Fischer-Tropsch reaction in the first stage (heavy paraffin synthesis, or HPS) of the SMDS process, a tubular fixed-bed reactor has been chosen for its inherent simplicity in design and operation and also for its proven technology in other processes, such as methanol synthesis. The catalyst is located in the tubes, which are cooled by boiling water around them, and considerable heat can thus be removed by boiling heat transfer. The good stability of the SMDS catalyst makes it possible to use a fixed-bed reactor. In the next step, heavy paraffin cracking (HPC), the long-chain waxy paraffins are cracked to desired size under mild hydrocracking conditions using a commercial Shell catalyst. In the final step, by selection of the corresponding cut points, the product stream is split into fractions of the required specification. The products manufactured in the SMDS process are predominantly paraffinic and free of impurities such as nitrogen and sulfur.

BIBLIOGRAPHY

"Gasification" under "Coal Chemicals and Feedstocks" in *ECT* 3rd ed., Supplement pp. 194–215, by J. Falbe, D. C. Frohning, and B. Cornils, Ruhrchemie AG.

1. Staff Report, *Modern Power Systems*, **10** (Nov. 1990).
2. D. R. Simbeck, R. L. Dickenson, and E. D. Oliver, *EPRI Report AP-3109*, June 1983.
3. R. N. Franklin, R. P. Jensen, and R. T. Perry, "SCGP—Efficient Clean Coal Power for Today and Tomorrow," *EPRI Conference on Technologies for Producing Electricity in the Twenty-First Century*, San Francisco, Calif., Oct. 30–Nov. 2, 1989.
4. B. Becker, "Gas Turbine Design for Buggenum, the First Modern European ICGCC," *Ninth Annual EPRI Conference on Gasification Power Plants*, Palo Alto, Calif., Oct. 16–19, 1990.
5. S. V. D. Linden, "ABB's Gas Turbine Activities and Plans for Future Integrated Gasification Power Plants," *Ninth Annual EPRI Conference on Gasification Power Plants*, Palo Alto, Calif., Oct. 16–19, 1990.
6. J. A. Lacey and co-workers, "An Update of the BGL Gasifier," *Tenth Annual EPRI Conference on Coal Gasification Power Plants*, San Francisco, Calif., Oct. 1991.
7. W. Schellberg and E. Kuske "Status of Prenflo Technology," *Tenth Annual EPRI Conference on Coal Gasification Power Plants*, San Francisco, Calif., Oct. 1991.
8. Shell Coal Gasification Project, "SCGP-1 Design and Construction," *EPRI Report No. GS-6372*, Interim Report, Aug. 1988.
9. Shell Coal Gasification Project, "Shakedown and Demonstration Phases," *EPRI Report No. GS-6373*, Interim Report, Sept. 1989.
10. J. H. Kolaian and W. G. Schlinger, *Energy Progress* **2**(4), 228 (Dec. 1982).
11. V. R. Shorter and P. A. Smith, paper presented at *California Clean Air and New Technologies Conference*, Los Angeles, Calif., 1990.
12. "Cool Water Coal Gasification Program," *Fifth Progress Report, EPRI AP-5931*, Interim Report, Oct. 1988.
13. "Cool Water Coal Gasification Program," *Final Report EPRI GS-6806*, Dec. 1990.
14. P. A. Smith and P. F. Curran, "Commercial Scale Power Generation Using Texaco Coal Gasification," *International Conference for the Power Generation Industries*, Dec. 1991.
15. D. R. Simbeck, R. L. Dickenson, A. J. Moll, E. D. Oliver, and F. E. Biasca, *SFA Pacific Quarterly Report*, Apr. 1987.

16. D. G. Sundstrom and J. U. Bott, paper presented at the *Tenth Annual EPRI Conference on Gasification Power Plants*, San Francisco, Calif., Oct. 1991.

17. R. H. Fisackerly and D. G. Sundstrom, "The Dow Syngas Project—Project Overview and Status Report," presented at the *Sixth Electric Power Research Institute Gasification Contractors Conference*, Palo Alto, Calif., Oct. 1986.

18. J. P. Henly, *Development and Commercialization of the Dow Coal Gasification Technology*, presented in Japan and China by Dow Chemical, Sept. 1986.

19. W. Adlhoch, J. Keller, and P. K. Herbert, Presented at the *Tenth Annual EPRI Conference on Gasification Power Plants*, San Francisco, Calif., Oct. 1991.

20. K. A. Theis and U. Femmer, *Braunkohle* **4**, 120–124 (Apr. 1982).

21. *Plants for Coal Technology: Rheinbraun HTW Process*, Uhde brochure Ro 1 5 19 2000 81, Dortmund, West Germany, 1981.

22. W. Adlhoch and K. A. Theis, "The Rheinbraun High Temperature Winkler (HTW) Process," paper presented at the *EPRI Workshop on Synthetic Fuels: Status and Future Direction*, San Francisco, Calif., 1980.

23. *High Temperature Winkler Gasification: Gas From Lignite*, Rheinbraun brochure, Cologne, West Germany, 1981.

24. H. Teggers and co-workers, "Latest Status of the Rheinbraun High-Temperature Winkler (HTW) Process," paper presented at the *Fourth International Coal Utilization Exhibition and Conference, Coal Technology '81*, Houston, Tex., Nov. 1981.

25. K. A. Theis and E. Nitschke, *Hydrocarbon Process.* 233–237 (Sept. 1982).

26. D. Hebden and H. J. F. Stroud, "Coal Gasification Processes," in M. A. Elliott, ed., *Chemistry of Coal Utilization*, second supplementary volume, Wiley-Interscience Publications, New York, 1981, Chapt. 24.

27. G. L. Baughman, *Synthetic Fuels Data Handbook*, 2nd ed., Cameron Engineers Inc., 1978.

28. J. N. Phillips, M. B. Kiszka, U. Mahagaokar, and A. B. Krewinghaus, "Shell Coal Gasification Project Final Report on Eighteen Diverse Feeds," *EPRI TR-100687*, 1992.

29. P. C. Richards and A. B. Krewinghaus, "Coal Flexibility of the Shell Coal Gasification Process," *Sixth Annual International Pittsburgh Coal Conference*, Sept. 1989.

30. U. Mahagaokar and A. B. Krewinghaus, "Shell Coal Gasification Project, Gasification of SUFCo Coal at SCGP-1, *EPRI Report No. GS-6824*, Interim Report, May 1990.

31. U. Mahagaokar, A. B. Krewinghaus, and M. B. Kiszka, "Shell Coal Gasification Project: Gasification of Six Diverse Coals," *EPRI Report GS-7051*, Interim Report, Nov. 1990.

32. J. N. Phillips, U. Mahagaokar, and A. B. Krewinghaus, "Shell Coal Gasification Project: Gasification of Eleven Diverse Feeds," *EPRI Report GS-7531*, Interim Report, May 1992.

33. R. P. Jensen, U. Mahagaokar, and A. B. Krewinghaus, "SCGP—Progress in a Proven, Versatile, and Robust Technology," *Ninth EPRI Conference on Coal Gasification Power Plants*, Palo Alto, Calif., Oct. 16–19, 1990.

34. U. Mahagaokar and A. B. Krewinghaus, "Shell Coal Gasification Plant No. 1—Recent Results on Domestic Coals," *Power-Gen '90 Conference*, Orlando, Fla., Dec. 4–6, 1990.

35. U. Mahagaokar and A. B. Krewinghaus, "Shell Coal Gasification Process—Recent Performance Results on Drayton, Buckskin, Blacksville No. 2, and Pyro No. 9 Coals," *1990 International Joint Power Conference*, Boston, Mass., Oct. 1990.

36. U. Mahagaokar and co-workers, "Shell's SCGP-1 Test Program—Final Overall Results," *Tenth Annual EPRI Conference on Gasification Power Plants*, San Francisco, Calif., Oct. 1991.

37. J. N. Phillips and co-workers, "SCGP Recent Results on Low Ash Feedstocks Including Petroleum Coke," *Eighth Annual Pittsburgh Coal Conference*, Pittsburgh, Pa., Oct. 1991.

38. G. A. Cremer and C. A. Bayens, "Shell GCC Sets New Standards For Clean Power From Coal," *Alternate Energy 1991*, Scottsdale, Ariz., Apr. 1991.
39. R. P. Allen, R. A. Battista, and T. E. Ekstrom, "Characteristics of an Advanced Gas Turbine with Coal Derived Fuel Gases," *Ninth Annual EPRI Conference on Gasification Power Plants*, Palo Alto, Calif., Oct. 1990.
40. W. V. Bush, K. R. Loos, and P. F. Russell, "Environmental Characterization of the Shell Coal Gasification Process. I. Gaseous Effluent Streams," *Fifteenth Biennial Low-Rank Fuels Symposium*, St. Paul, Minn., May 22–25, 1989.
41. D. C. Baker, W. V. Bush, and K. R. Loos, "Determination of the Level of Hazardous Air Pollutants and Other Trace Constituents in the Syngas from the Shell Coal Gasification Process," *Conference on Managing Hazardous Air Pollutants—State of the Art*, Washington, D.C., Nov. 4–6, 1991.
42. D. C. Baker, W. V. Bush, K. R. Loos, M. W. Potter, and P. F. Russell, "Environmental Characterization of the Shell Coal Gasification Process. II. Aqueous Effluent," *Sixth Annual Pittsburgh Coal Conference*, Pittsburgh, Pa., Sept., 1989.
43. R. T. Perry and co-workers, "Environmental Characterization of the Shell Coal Gasification Process. III. Solid By-Products," *Seventh Annual International Pittsburgh Coal Conference*, Pittsburgh, Pa., 1990.
44. J. A. Salter and co-workers, "Shell Coal Gasification Process: By-Product Utilization," *Ninth International Coal Ash Utilization Symposium*, Orlando, Fla., Jan. 1991.
45. G. A. Cremer, N. Hauser, and C. A. Bayens, paper presented at the *AICHE Spring National Meeting*, Mar. 1990.
46. *Lignite to Methanol: An Engineering Evaluation of Winkler Gasification and ICI Methanol Gasification Route, AP-1592*, Electric Power Research Institute, Palo Alto, Calif., Oct. 1980.
47. P. F. H. Rudolph, "The Lurgi Process—The Route to SNG From Coal," *Fourth Synthetic Pipeline Gas Symposium*, Chicago, Ill., 1972.
48. J. R. Williams and G. A. Bekker, paper presented at the *Thirteenth International LNG/LPG Conference*, Kuala Lumpur, Malaysia, Oct. 1988.

UDAY MAHAGAOKAR
A. B. KREWINGHAUS
Shell Development Company

LIQUEFACTION

Coal (qv) can be converted to liquid and gaseous fuels by two different processing routes normally termed "direct" and "indirect" (see also FUELS, SYNTHETIC-GASEOUS FUELS; FUELS, SYNTHETIC-LIQUID FUELS). Direct liquefaction processes include those that normally proceed to liquids in a single processing sequence, using solid coal as the primary reactant. Some direct liquefaction schemes also involve chemical pretreatment steps. Indirect liquefaction processes involve gasification of coal as the first conversion step, followed by catalytic recombination of the resulting synthesis gas mixture ($CO + H_2$) to form hydrocarbons and oxygenates.

Review of coal liquefaction research may be found in References 1–3. Herein, those processing schemes for coal liquefaction that, since the 1970s, have received attention beyond the laboratory to pilot plants or process development units are presented.

Direct Liquefaction

Coal liquefaction involves raising the atomic hydrogen-to-carbon ratio from approximately 0.8/1.0 for a typical bituminous coal, to 2/1 for liquid transportation fuels or 4/1 for methane (4). In this process, molecular weight reduction and removal of mineral matter and heteroatoms such as sulfur, oxygen, and nitrogen may need to be effected.

Hydrogenation or hydroliquefaction and pyrolysis are the two means used for direct liquefaction. In direct hydrogenation the primary reactions are a combination of homogeneous thermal cracking, ie, free-radical generation, and heterogeneous hydrogenation involving hydroaromatics in the slurry vehicle or the coal itself as hydrogen transfer agents. Rapid and efficient capping of the primary free radicals generated by thermolysis is thought to be necessary in order to prevent regressive reactions leading to formation of char (5). Other theories of coal liquefaction suggest that hydrogen can engender reactions involving scission of strong bonds in the coal macromolecule, and hence can act as an active bond cleaving agent rather than simply a passive radical quencher (6).

Process schemes that apply pyrolysis chemistry normally involve thermolysis in an inert or reducing atmosphere and produce two principal products from coal: a tar and char. The relative proportion of char to tar can be quite high, hence the rationale for liquefaction by pyrolysis is often not production of coal-derived distillate materials. Hydropyrolysis processing schemes involving thermolysis in the presence of hydrogen and/or pyrolysis under conditions of rapid heating can, however, generate yields of distillate products significantly in excess of the volatile matter content of the starting coal.

Bergius was awarded the Nobel Prize for Chemistry in 1931 based on his pioneering work on coal liquefaction (7). The work of I. G. Farben on the Bergius process led to the development of a two-stage direct hydrogenation liquefaction process, in which primary coal solubilization was carried out in the first stage using added disposable catalysts in bubble-column reactors. Distillate materials were subsequently catalytically upgraded to liquid transportation fuels using supported hydrotreating/hydrocracking catalysts in fixed-bed reactors (8).

Hydrogenation. *Solvent Refined Coal Process.* Work in the mid-1960s by the Spencer Chemical Co. (9) and during the 1970s by the Gulf Chemical Corp. led to two solvent refined coal (SRC) processing schemes: SRC-I for production of low ash solid boiler fuels and SRC-II for distillates, eg, "syn-crude."

A schematic flow diagram for the SRC-I process is shown in Figure 1. There are essential features of this process where coal is first slurried in a recycle solvent, then preheated and finally reacted in a bubble column-type reactor at 450°C in the presence of gaseous hydrogen. Because of the high reactivities of the coals tested, primarily eastern U.S. high and medium volatile bituminous coals, no catalysts were added and the reaction was carried out at pressures as low as 6.9 MPa (1000 psig). Mean residence time in the reactor was reported to be on the order of 30 minutes. Solids were removed by use of either rotary pressure precoat filters or hydroclones. SRC yields of approximately 60%, computed as the mass of SRC-I per mass of moisture and ash free (MAF) coal exclusive of light hydrocarbon gas make, were achieved. Ash removal was quite high by this process, and

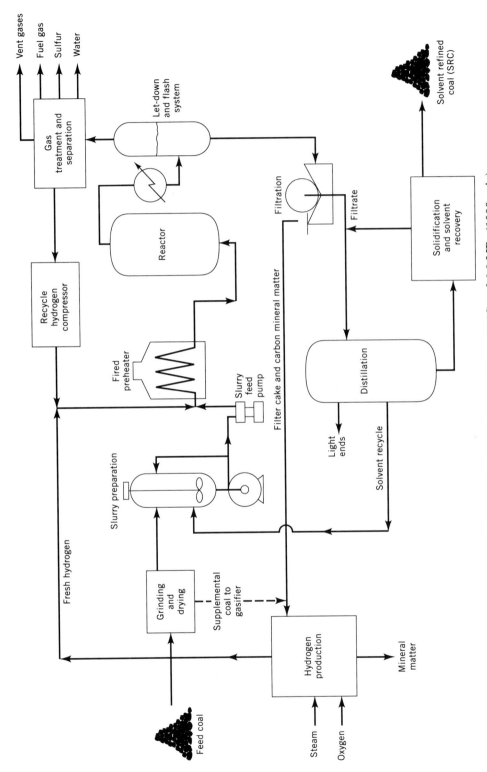

Fig. 1. SRC-I process. The reactor operated at 450°C and 6.9 MPa (1000 psig).

depending on the forms of sulfur in the parent coal (inorganic/organic), sulfur reduction was also substantial. Data for a Kentucky No. 9 high pyrite eastern U.S. coal showed SRC-I/parent coal weight percent ratios for total sulfur as 0.22 and for mineral matter as 0.0095. The heating value ratio was 1.2.

The SRC-I technology was tested at a large scale demonstration plant in Fort Lewis, Washington, commissioned in 1974. This plant operated for several years, but severe problems were encountered, primarily with solvent balance and operation and reliability of the solids separation portion of the facility (10). The design for this facility differs in that an expanded bed hydrocracking unit was added as a second-stage reactor to increase the yield of distillate material (11). Total solids (SRC plus two-stage liquefaction or TSL solids) were reduced to approximately 27% by this modification, resulting in an increase in distillate materials (naphtha plus middle distillate). Bench-scale testing of the hydrocracking step indicated that the coal-derived naphtha, ie, the C_5 or 200°C boiling fraction, would be low (0.01 wt %) in sulfur, but high (approximately 0.1%) in nitrogen. Severe hydrotreating of this material would be required before refining into gasoline via catalytic reforming. Overall thermal efficiency of this proposed facility was calculated to be 70%.

The SRC-II process, shown in Figure 2, was developed in order to minimize the production of solids from the SRC-I coal processing scheme. The principal variation of the SRC-II process relative to SRC-I was incorporation of a recycle loop for the heavy ends of the primary liquefaction process. It was quickly realized that minerals which were concentrated in this recycle stream served as hetero-

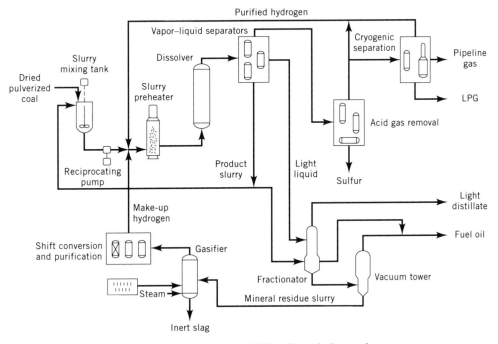

Fig. 2. SRC-II process where LPG is liquefied petroleum gas.

geneous hydrogenation catalysts which aided in the distillate production reactions. In particular, pyrrhotites, Fe_xS_y or nonstoichiometric iron sulfides, produced by reduction of iron pyrite were identified as being especially important as heterogeneous catalysts and pyrite was subsequently added for cases where the inherent pyrite content of the coal was low (12,13). Results on the yield of some of the primary liquefaction products, based on wt % of MAF coal, for addition of pyrite to the parent coal slurry for a moderately reactive but relatively low (0.9 wt %) pyrite Pittsburgh seam bituminous coal from West Virginia include:

Product	Pyrite addition, wt %		
	0.0	3.0	7.5
light hydrocarbon gases	16.6	17.1	17.6
naphtha	7.3	9.4	11.4
total oil	37.5	40.9	44.7
SRC	29.8	27.5	23.5
insoluble organic matter	5.9	5.3	5.2

where light hydrocarbon gases have carbon content up to C_4.

A yield comparison between the products of the SRC-I and SRC-II processes operating on a high volatile Kentucky bituminous coal is

Product	Product yield, wt %	
	SRC-I	SRC-II
C_1–C_4	10.5	16.1
total oil	25.9	38.9
SRC solids	42.7	21.0
insoluble organic matter	4.1	5.1
H_2	−2.4	−5.6

where the negative sign indicates that hydrogen is being consumed.

Changing process configuration to SRC-II was successful in producing about 50% additional oil. However, a large increase in light hydrocarbon gas make accompanied this increase with an attendant reduction in hydrogen utilization efficiency, and problems persisted using coals other than Kentucky 9/14.

Exxon Donor Solvent (EDS) Process. A schematic flow diagram for the Exxon donor solvent (EDS) process is shown in Figure 3. The principal modification in this technology was the incorporation of a fixed-bed catalytic hydrotreating unit for the recycle solvent stream. This additional unit was required to keep the hydrogen donating/shuttling capacity of the recycle solvent oil at an acceptably high value (14). The use of bottoms slurry recycle to increase the solvent "make" fraction by taking advantage of the catalytic properties of minerals was also investigated and improved yields in the bottoms recycle mode were generally reported. Recycle of this fraction was also reported to dramatically improve operability of the process, especially using low rank coals where viscosity of the bottoms stream was a significant problem (15). The primary liquefaction part of

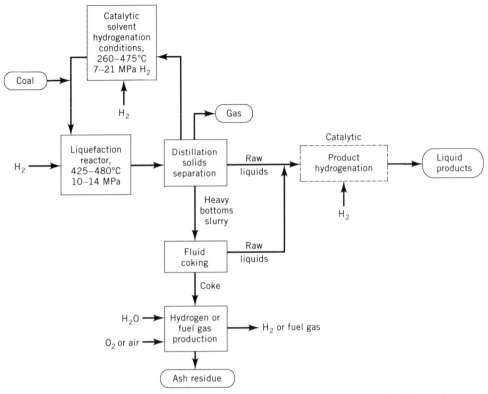

Fig. 3. Exxon donor solvent process. To convert MPa to psi, multiply by 145.

the reaction system operated at temperatures from 425 to 480°C and pressures from 10 to 14 MPa (1450–2030 psi) using mean residence times in the range of 15 minutes to 2 hours depending on coal reactivity and process configuration.

Exxon was the first to investigate the suitability of a wide range of different U.S. coals for conversion. Operation of the EDS process was demonstrated in a 230 t/d unit in Baytown, Texas that had a start-up in May of 1980. Data on the response of a variety of coals to once-through and bottoms recycle operations are shown in Figure 4. Figure 5 presents typical liquefaction product distributions for the system operated both with and without the Flexicoking (fluidized-bed coking) option.

H-Coal. A significantly different scheme for direct coal liquefaction, developed by Hydrocarbon Research Inc. (HRI), was based on research and development on the H-Oil ebullated bed catalytic reactor for hydrotreating and hydrocracking heavy oil. The heart of this process is the reactor, where coal, catalyst, solvent, and hydrogen are all present in the same vessel (Fig. 6). The reactor is maintained in a "bubbling" or ebullated, ie, well-mixed, state by internal agitation coupled with the action of the gas bubbling through the fluid. This process was piloted by HRI and Ashland Synthetic Fuels, Inc. in a 600 t/day pilot plant adjacent to Ashland's refinery in Catlettsburg, Kentucky (10). The process consists

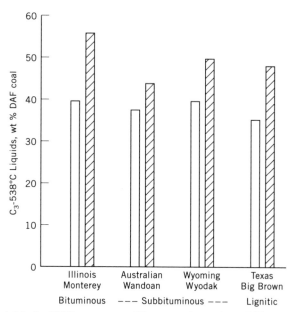

Fig. 4. Product yields for EDS process on □ a once-through and ▨ a bottoms recycle basis for various typs of coal. DAF = dry ash free. C_3-538°C = a boiling fraction.

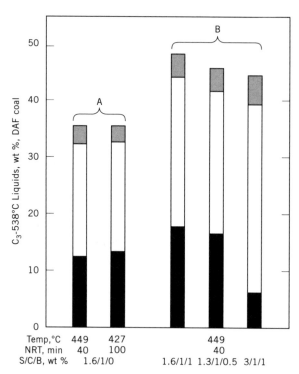

Fig. 5. Product yields for the EDS process in a 34-kg/d pilot plant A, with, and B, without the Flexicoking option where ▦ represents gas; □, naphtha; and ■, oil. The pressure for the once-through process is 10.3 MPa, for the recycle, 13.8 MPa. NRT, nominal residence time; S/C/B, solvent/coal/bottoms. To convert MPa to psi, multiply by 145.

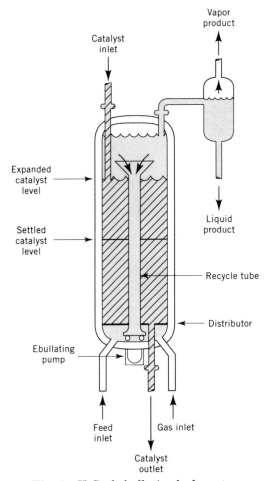

Fig. 6. H-Coal ebullating bed reactor.

of slurry preparation followed by catalytic hydrogenation/hydrocracking at 450°C and 15 MPa (2200 psi) in the ebullated bed reactor.

A principal focus of this project was research and development for catalysts that were tolerant of the coal-derived mineral matter in the reactor. Typical early catalysts showed rapid deactivation because of coking and loss of surface area, presumably from pore-mouth blockage by coke and metals laydown. Coke build-ups of 20 to 25 wt % and surface area reductions from 300 m^2/g for the fresh catalyst to 25 m^2/g for the aged catalyst were reported after only five days on-stream (16). Although one of the primary advantages of the H-Coal processing scheme was the ability to continuously add and withdraw catalyst from the reactor in order to maintain a stable level of activity, catalyst replacement and consumption rates were unacceptably high under these conditions. More recent developments in catalyst design have led to reduced catalyst consumption rates. Catalyst consumptions on the order of 0.50 to 0.75 kg of catalyst per ton of coal have been reported (17). Research on other processing options, including two-stage liquefaction, may be effective in reducing this value even further (18).

The H-Coal process could operate in one of two modes, depending on the desired product slate. In the "syn-crude" mode, a fluid-bed coking unit was employed to maximize recovery of distillate from the liquefaction product (Fig. 7a). When operated in the fuel oil mode (Fig. 7b), no coker was used and the primary product was a coal-derived low sulfur fuel oil. Total hydrogen demand on the process was also reduced in the latter mode of operation.

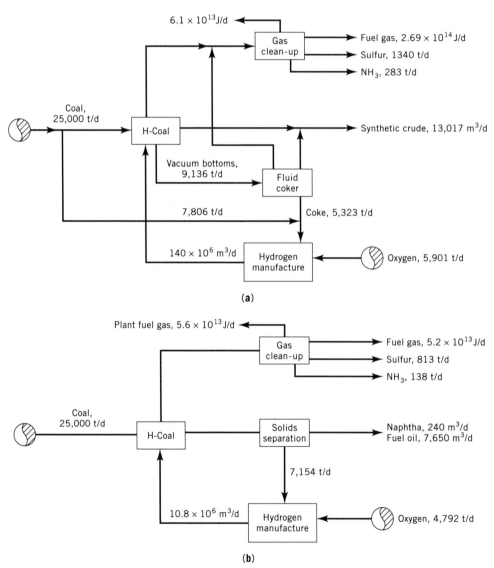

Fig. 7. H-Coal process using Illinois No. 6 coal: (**a**) in syncrude mode, and (**b**) in low sulfur fuel oil mode. To convert J to Btu, multiply by 6.48×10^{-4}. To convert m^3 to bbl, multiply by 6.29. To convert m^3 to standard cubic feet (SCF), multiply by 35.3

Wilsonville Coal Liquefaction Facility. Beginning in 1973, a 6 t/d coal lique-faction pilot plant was built by the Edison Electric Institute (EEI) and Southern Company Services in Wilsonville, Alabama. The Electric Power Research Institute (EPRI) assumed project sponsorship in late 1973, and the U.S. Department of Energy became the primary sponsor in 1976. Amoco Oil Co. joined the project in 1984. The purpose of the Wilsonville Advanced Coal Liquefaction R&D facility was to provide a flexible but reasonably large-scale pilot plant where effects of coal type and processing, ie, reactor configuration could be tested and evaluated. Research on the Kerr-McGee critical solvent de-ashing technology was also carried out, resulting in the development of alternative methods for solids removal from primary liquefaction products.

The plant began operation in 1974 in the SRC-I mode, but evolved to a two-stage operation utilizing two ebullating bed catalytic reactors (19). Initial efforts in two-stage liquefaction focused on catalytic upgrading of the thermal products, or nonintegrated two-stage liquefaction (NTSL). This configuration, termed non-integrated because the coal-derived resid hydrocracking step did not interact with the primary thermal part of the plant, was excessively inefficient because of high hydrogen consumptions associated with the thermal part of the operation. In the integrated two-stage approach (ITSL), a short contact-time thermal reactor was close coupled to an ebullated bed catalytic reactor and process solvent was generated by distillation of the hydrocracked products. The thermal resid produced in the ITSL under short contact time conditions was more reactive toward expanded bed hydrocracking, thus permitting operation of the ebullated bed reactor at lower severity and minimizing gas make (20). Results on liquefaction of an Illinois No. 6 high volatile bituminous coal at Wilsonville using both the NTSL and ITSL modes are shown in Table 1 (21).

Coal throughput, ie, space velocity per unit reactor volume, was substantially improved in going to the ITSL mode. The higher reactivity of the coal-derived resid permitted operation of the hydrocracker at lower temperature; this would be expected to reduce the rate of coke lay-down on catalyst, and improve hydrogen utilization efficiency by minimizing formation of light hydrocarbon gases (higher distillate selectivity). A 35% increased yield of C_4^+ distillate was obtained.

Reconfigured integrated two-stage liquefaction (RITSL) where solvent de-ashing was practiced after the hydrocracking step, and close coupled integrated two-stage liquefaction (CC-ITSL) where the two reactors (thermal/catalytic) were linked directly without any intervening processing steps (22,23) were also explored. Typical results for these processes are also shown in Table 1. Incremental improvements in distillate yield and selectivity were realized by changing the process configuration, but at the expense of increased hydrogen consumption.

From 1985 to 1992, process development at Wilsonville focused on development of a catalytic/catalytic two-stage liquefaction scheme (CTSL) utilizing ebullating bed catalytic reactors in both stages. Initial work (24) indicated that distillate yields as high as 78% could be obtained by operating the first stage at low severity (399°C) and by using a large pore bi-modal NiMo catalyst having mean micropore diameter in the 11.5–12.5 nm range. Results in the CTSL mode for three different coals are shown in Table 2. These data show the significant improvement in distillate production that can be achieved by use of catalyst in both

Table 1. Operating Conditions and Yields at Wilsonville Plant[a]

Parameter	Mode of operation[b]				
	NTSL	ITSL	RITSL	CC-ITSL	CC-ITS
run number	241CD	7242BC; 243JK/244B	247D	250D	250G(a)
catalyst	Armak	Shell 324M	Shell 324M	Amocat IC	Amocat
Thermal stage					
average reactor temperature, °C	429	460;432	432	440	443
coal space velocity at temp > 371°C, kg/m^3	320	690;450	430	320	320
pressure, MPa[c]	15	17;10–17	17	17	17
Catalytic stage					
average reactor temperature, °C	416	382	377	399	399
space velocity catalyst[d], h^{-1}	1.7	1.0	0.9	2.08	2.23
catalyst age, resid/catalyst	260–387	278–441; 380–850	446–671	697–786	346–439
Yields[e]					
C_1–C_3 gas	7	4;6	6	7	8
C_4^+ distillate	40	54;59	62	64	63
resid	23	8;6	3	2	5
hydrogen consumption	4.2	4.9;5.1	6.1	6.1	6.4
Other					
hydrogen efficiency, C_4^+ distillate/H_2 consumed	9.5	11;11.5	10.2	10.5	9.8
distillate selectivity, C_2–C_3/C_4^+ distillate	0.18	0.07;0.10	0.10	0.11	0.12
energy content of feed coal reject to ash concentrate, %	20	24;20–23	22	23	16

[a]Illinois No. 6 coal.
[b]See text for term definitions.
[c]To convert MPa to psi, multiply by 145.
[d]On a wt of feed per wt of catalyst basis.
[e]Wt % on a MAF coal basis.

Table 2. Operating Conditions and Yields at Wilsonville Plant for CTSL Mode[a]

Parameter	Coal type		
	Illinois No. 6	Ohio 6[b]	Wyodak
run number	253A	254G	251-IIIB
catalyst	Shell 317	Shell 317	Shell 324
First stage			
average reactor temperature, °C	432	433	441
inlet hydrogen partial pressure, MPa[c]	14.1	15.0	17.3
feed space velocity, h^{-1}	4.8	4.3	3.5
pressure, MPa[c]	17.9	18.8	17.9
catalyst age, resid/catalyst	150–350	1003–1124	760–1040
Second stage			
average reactor temperature, °C	404	421	382
space velocity, feed/catalyst, h^{-1}	4.3	4.2	2.3
catalyst age, resid/catalyst	100–250	1166–1334	371–510
Yield[d]			
C_1–C_3 gas	6	8	11
C_4^+ distillate	70	78	60
resid	−1	−1	+2
hydrogen consumption	6.8	6.9	7.7
Other			
hydrogen efficiency, C_4^+ distillate/H_2 consumed	10.3	11.3	7.8
distillate selectivity, C_1–C_3/C_4^+ distillate	0.08	0.11	0.18
energy content of feed coal rejected to ash concentrate, %	20	10	15

[a]CTSL = catalytic/catalytic two-stage liquefaction.
[b]Approximately 6% ash.
[c]To convert MPa to psi, multiply by 145.
[d]Wt % on a MAF coal basis.

stages, but at the cost of increasing levels of hydrogen consumption. The Wilsonville Advanced Liquefaction R&D facility was shut down in early 1992 and is scheduled to be decommissioned.

Other Processes. In the period from roughly 1970 through 1990, several other liquefaction processes were developed. For example, direct hydrocracking of coal using molten zinc chloride as the catalyst was investigated (25). This reaction was carried out at temperatures ranging from 385 to 440°C, and pressures from 13.8 to 20.3 MPa (2001–2900 psi). Mean residence time in the stirred reactor was between 30 and 190 minutes, depending on coal type and reaction conditions. Catalyst to coal ratios were varied in the range of 0.8 to 2.0 (mass catalyst/mass

MAF coal). Because unreacted organic matter and coal mineral matter were retained in the molten catalyst, a key to the process configuration was catalyst recovery and regeneration which was carried out by high temperature (800–1040°C) distillation. Yields of low sulfur and low nitrogen coal-derived C_5^+ distillate from 55 to 65 wt % were reported for bench-scale runs where bituminous coal was processed. The catalyst also served as an effective sulfur and nitrogen scavenger. These species were retained in the melt as $ZnCl_2 \cdot NH_3 \cdot NH_4Cl$ and ZnS. Of this distillate material, 75 to 80% was 90 research octane number (RON) gasoline indicating that a significant fraction of the coal's original aromaticity was preserved during the hydrocracking step. Catalyst recovery and corrosion in both the reactor and regenerator vessels were significant problems that hampered further development of this processing scheme.

Research on catalytic coal liquefaction was also carried out using an emulsified molybdenum catalyst added to the slurry medium to enhance rates of coal conversion to distillate (26). Reaction at 460°C, 13.7 MPa (1980 psi) in the presence of the dispersed catalyst was sufficient to greatly enhance conversion of a Pittsburgh No. 8 bituminous coal to hexane-soluble oils:

	Mo added, ppm		
Product	216	108	0
gases and light oil, wt %	33.3	32.6	17.6
hexane-soluble oil, wt %	22.2	26.6	5.4
asphaltenes	23.4	21.1	50.3
hydrogen consumed	6.1	6.1	4.4

where asphaltenes are defined as toluene-soluble, hexane-insoluble and hydrogen consumed is kg H_2/100 kg MAF coal.

Other variations of catalytic and noncatalytic coal liquefaction schemes were also developed (27,28). Additionally, bench-scale and semiworks systems have been operated in Germany by researchers at Bergbau-Forschung in Essen (29). A 2.5 ton per day pilot plant is being operated by the National Coal Board in the United Kingdom at Point of Ayr in Wales (30). This facility is notable for the use of semibatch or candle filters for removal of mineral matter and unreacted coal from the primary liquefaction products.

Pyrolysis and Hydropyrolysis.   The second category of direct liquefaction aimed at producing distillate materials from coal is pyrolysis and hydropyrolysis. Pyrolysis, sometimes called destructive distillation, essentially involves heating the coal in an inert atmosphere, followed by recovery of coal-derived tars and distillates in the off-gas stream (31). Depending on the coal type and processing conditions, yields of condensables can equal or even exceed the volatile matter content of the parent coal (32,33). Pyrolysis processes are, however, usually burdened with poor liquid yield, relative to direct hydrogenation, and the coal-derived liquids are high in heteroatoms and both organic and inorganic fine particulate matter. When high yields of coal-derived distillates are desired, pyrolysis is usually carried out in a hydrogen atmosphere, eg, hydropyrolysis, or at extremely rapid heating rates, eg, flash pyrolysis.

Liquefaction/Pyrolysis.   Large-scale research and development on coal pyrolysis was carried out on the char oil energy development (COED) process (34).
This scheme involved temperature staged pyrolysis in three interacting fluidized
beds, as shown in Figure 8, and was tested in a 36 t/d process demonstration unit
during the early 1970s. Pyrolysis temperatures ranged from 450 to 540°C in the
COED process, and the long residence times associated with the fluid beds mandated low yields of liquid products. Typical product yields for four different U.S.
coals are shown in Table 3. The yield structure is heavily weighted toward production of char and gas. Production of coal-derived liquids ranged from 0.04 to
0.21 m³/t of coal as compared to 0.61–0.79 m³/t for direct hydrogenation. Further,

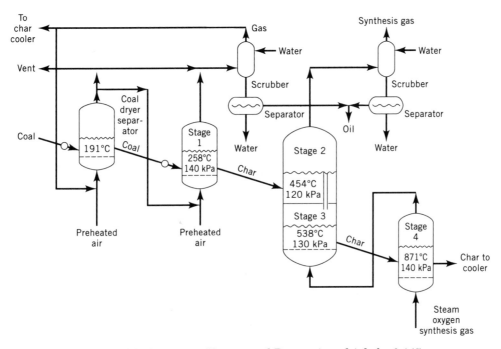

Fig. 8.   COED process. To convert kPa to psi, multiply by 0.145.

Table 3. FMC/COED Process Product Distribution

	Coal composition, wt %			
Yields, dry coal basis	N.D. lignite	Utah	Illinois	West Kentucky
char	55.8	54.5	59.5	63.0
tar	5.3	21.5	19.3	17.3
gas	37.6	18.3	15.1	13.0
liquor (aq)[a]	1.3	5.7	6.1	6.7

[a]Water containing water-soluble organics produced during pyrolysis.

the liquids produced were high in heteroatoms (especially nitrogen) and required extensive hydrotreating before use as a synthetic crude oil.

Process development on fluidized-bed pyrolysis was also carried out by the Consolidation Coal Co., culminating in operation of a 32 t/d pilot plant (35). The CONSOL pyrolysis process incorporated a novel stirred carbonizer as the pyrolysis reactor, which made operation of the system feasible even using strongly agglomerating eastern U.S. bituminous coals. This allowed the process to bypass the normal pre-oxidation step that is often used with caking coals, and resulted in a nearly 50% increase in tar yield. Use of a sweep gas to rapidly remove volatiles from the pyrolysis reactor gave overall tar yields of nearly 25% for a coal that had Fischer assay tar yields of only 15%.

Other large-scale coal pyrolysis process developments were carried out by the Tosco Corp., with its TOSCOAL process (36). Essentially a direct copy of Tosco's rotating kiln technology that was developed for pyrolysis of oil shale, this slow heating scheme achieved tar yields at maximum temperatures of 482–521°C that were essentially identical to those obtained by a Fischer assay.

Process development of the use of hydrogen as a radical quenching agent for the primary pyrolysis was conducted (37). This process was carried out in a fluidized-bed reactor at pressures from 3.7 to 6.9 MPa (540–1000 psi), and a temperature of 566°C. The pyrolysis reactor was designed to minimize vapor residence time in order to prevent cracking of coal volatiles, thus maximizing yield of tars. Average residence times for gas and solids were quoted as 25 seconds and 5–10 minutes. A typical yield structure for hydropyrolysis of a subbituminous coal at 6.9 MPa (1000 psi) total pressure was char 38.4, oil 29.0, water 19.2, and gas 16.2, on a wt % MAF coal basis.

Tar yields of approximately 0.32 m^3/t were quoted. Because the scheme used hydrogen, the pyrolysis liquids generally exhibited lower heteroatom contents than conventional tars derived from coal pyrolysis in an inert atmosphere. Process development proceeded through a 270 t/d semiworks plant which was operated successfully on noncaking coals. Operability for caking coals was difficult, however.

Flash Pyrolysis.   Development of a rapid, ie, flash, pyrolysis process was carried out in the late 1960s and early 1970s (38). The process was designed to heat coal at rates in excess of 5000 K/s. Process development proceeded through to a 2.7 t/d process development unit (PDU), in which operation was proven on a variety of caking and noncaking coals. The reactor section facilitated rapid heating by direct contact with hot char from the char burner. Gas residence times (< 2 sec) were carefully controlled in order to minimize secondary cracking reactions and maximize the yield of coal-derived liquids. Typical yield structures for pyrolysis at 580°C for two coals are shown in Table 4. Whereas rapid heating and hence high tar yields could be obtained with this system, rapid quenching of reaction products proved to be a significant problem especially as the process was scaled up from the laboratory.

Development of a flash coal pyrolysis reaction system was also carried out by Lurgi-Rhurgas (39). During the time period from 1940 to 1960 units processing 10 t/h were operated, and a small commercial plant was built and has operated in the former Yugoslavia since 1963. As shown in Figure 9, coal is rapidly heated by mixing with hot recycled char in a screw conveyor-type reactor. Volatiles re-

Table 4. Product Distribution for the Occidental Flash Pyrolysis Process

	Coal	
Yield, wt %	Western Kentucky bituminous	Wyoming subbituminous
tar	35	27
char	56	52
gas	7	13

covery is completed at 750°C in vessel number 4. A typical product distribution for this system operating on a high volatile West Virginia bituminous coal gave a tar yield of 28 wt %, char of 58 wt %, and gas + liquor of 14 wt % on a basis of MAF coal.

A novel high pressure flash hydropyrolysis reaction system was designed and operated by Rockwell Corp. during the mid-1970s (40). The reactor was designed to mix hot high pressure hydrogen and coal in a highly turbulent zone such that extremely rapid (> 10,000 K/sec) heating rates could be obtained. A schematic of the reactor is shown in Figure 10. In this system the energy required to heat coal to temperatures between 871 and 1038°C was generated by combustion of a portion of the hydrogen feed to the reactor. Rapid heating then was facilitated by direct contact with hot hydrogen and the combustion gases. The rapid heating

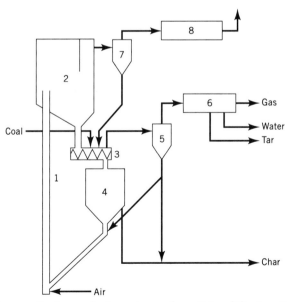

Fig. 9. Lurgi-Rhurgas flash pyrolysis system, where 1 is a lift pipe; 2, primary pyrolysis reactor; 3, screw feeder; 4, secondary pyrolysis reactor; 5 and 7, cyclones; and 6 and 8, product recovery and tailgas cleaning.

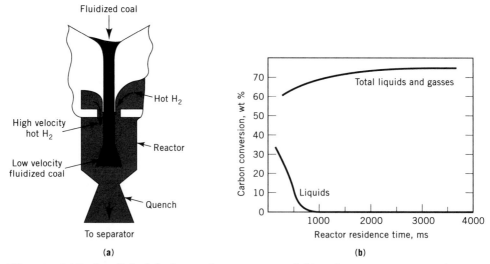

Fig. 10. (a) Rockwell flash hydropyrolysis reactor and (b) carbon conversion as a function of reactor residence time. The reaction was run at 1038°C and 10.3 MPa (1500 psig).

coupled with extremely fast transition through the coal's plastic regime obviated problems associated with operation using agglomerating coals. Further, the extremely short residence times for coal-derived volatiles and the activity of hydrogen as a radical scavenger helped minimize secondary cracking reactions thus permitting yields of coal-derived tars to greatly exceed that predicted by the Fischer Assay. Whereas total coal conversion was relatively insensitive to reactor residence time, the yield of liquid or oil was a maximum at ~0.1 s for a U.S. bituminous coal. Longer residence times favored formation of gases. Operating pressure also had an effect on coal conversion and product distribution. Higher pressures favored production of liquids. This process was operated in a 1 t/h pilot plant, where the technology was successfully demonstrated for a variety of different feed coals.

Indirect Liquefaction

The second category of coal liquefaction involves those processes which first generate synthesis gas, a mixture of CO and H_2, by steam gasification of coal

$$C \ (s) \ + \ H_2O \longrightarrow CO \ + \ H_2$$

followed by production of solid, liquid, and gaseous hydrocarbons and oxygenates via catalytic reduction of CO in subsequent stages of the process (41). Whereas coal is usually the preferred feedstock, other carbon-containing materials such as coke, biomass, or natural gas can also be used (see FUELS FROM BIOMASS; GAS, NATURAL).

Synthesis gas from the gasifier is first cleaned to remove gasifier tars and organic sulfur, and the composition of the gas is adjusted in a catalytic shift converter to raise the hydrogen content

$$CO + H_2O \xrightarrow[\text{catalyst}]{} CO_2 + H_2$$

This clean and shifted gas is then converted to hydrocarbons and other products in a series of catalytic reactors. The synthesis reaction is usually carried out using two or three reactors in series because of the highly exothermic nature of the overall reaction.

The first demonstration of catalytic conversion of synthesis gas to hydrocarbons was accomplished in 1902 using a nickel catalyst (42). The fundamental research and process development on the catalytic reduction of carbon monoxide was carried out by Fischer, Tropsch, and Pichler (43). Whereas the chemistry of the Fischer-Tropsch synthesis is complex, generalized stoichiometric relationships are often used to represent the fundamental aspects:

$$n\, CO + 2n\, H_2 \longrightarrow (-CH_2-)_n + n\, H_2O$$

$$2n\, CO + n\, H_2 \longrightarrow (-CH_2-)_n + n\, CO_2$$

$$n\, CO + 2n\, H_2 \longrightarrow H(-CH_2-)_n OH + (n-1)\, H_2O$$

By proper selection of catalyst and reaction conditions, hydrocarbons and oxygenates ranging from methane and methanol through high ($> 10,000$) molecular weight paraffin waxes can be synthesized as indicated in Figure 11 (44).

Low Pressure Synthesis.   Processes which operated at relatively low pressures, in the range of 100–200 kPa (1–2 atm) dominated commercial applications of the Fischer-Tropsch process in Germany prior to 1939 (45). Catalysts were primarily cobalt based. Two compositions were typical: 100 Co:18 ThO$_2$:100 kieselguhr and 100 Co:5 ThO$_2$:7.5 MgO:200 kieselguhr where kieselguhr [61790-53-2] is a type of diatomaceous earth (see DIATOMITE). Catalyst lives of one to two months were normally experienced (46). Catalyst deactivation was caused primarily by buildup of high molecular-weight waxes, and regeneration was accomplished simply by solvent extraction of the catalyst using gasoline and/or reactivation with hydrogen (see CATALYSTS, REGENERATION).

Medium Pressure Synthesis.   Pressures of 500–2000 kPa (5–20 atm) were typical for the medium pressure Fischer-Tropsch process. Cobalt catalysts similar to those used for the normal pressure synthesis were typically used at temperatures ranging from 170 to 200°C in tubular "heat exchanger" type reactors.

Both normal and medium pressure Fischer-Tropsch syntheses produced a hydrocarbon product that was highly paraffinic in nature. The olefin content (primarily monoolefins) of the liquid products could also be quite high depending on operating conditions. Selectivity of the normal pressure process was higher for gasoline, whereas the medium pressure process produced more diesel fuel and paraffin waxes. The gasoline had a very low (generally <50) octane number but

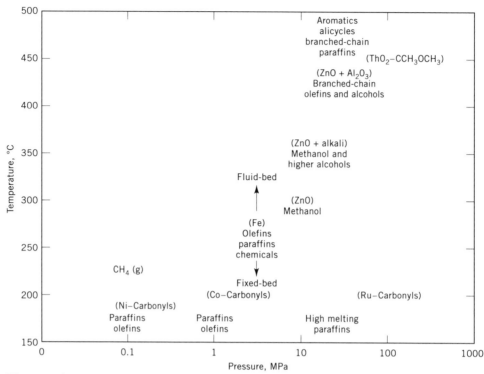

Fig. 11. Optimum pressure/temperature ranges for Fischer-Tropsch synthesis processes showing the various catalysts in parentheses. To convert MPa to psi, multiply by 145.

the cetane number of the diesel fraction was excellent. Further, the products were essentially sulfur-free because of the extensive gas cleaning practiced prior to the synthesis step. The primary differences between normal and medium pressure synthesis were the 10 to 15% higher hydrocarbon yield and increased catalyst life for the higher pressure process.

High Pressure Synthesis. Reaction at pressures of 10 to 20 MPa (100–200 atm) and temperatures in the 400°C range is known as the high pressure process.

Development of SASOL. Over 70% of South Africa's needs for transportation fuels are being supplied by indirect liquefaction of coal. The medium pressure Fischer-Tropsch process was put into operation at Sasolburgh, South Africa in 1955 (47). An overall flow schematic for SASOL I is shown in Figure 12. The product slate from this facility is amazingly complex. Materials ranging from hydrocarbons through oxygenates, alcohols, and acids are all produced.

The plant utilizes iron catalysts, and both fixed and fluidized reactor schemes. The fixed-bed reactors, designed by Lurgi, contained approximately 40 m^3 of catalyst in over 2000 vertical tubes having diameters of 4.5 cm OD. Catalyst was manufactured by precipitation from an iron nitrate solution using sodium carbonate. Copper and potassium were added as promoters, and the final material was pelletized and reduced with hydrogen prior to use. The fixed-bed part of SASOL I contains five reactors in parallel, each reactor processing 30,000 m^3 of feed at 220 to 255°C and 2.5 MPa (25 atm) and producing 87.4 m^3 (550

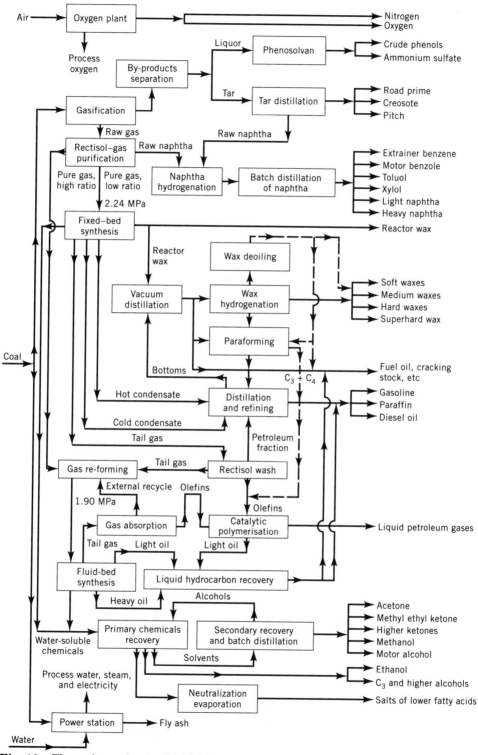

Fig. 12. Flow scheme for the SASOL I Fischer-Tropsch process. To convert MPa to psig, multiply by 145.

barrels) of product per day (48). A flowsheet showing one fixed-bed reactor train is given in Figure 13. Catalyst life was reported to be approximately six months.

The principal advance in technology for SASOL I relative to the German Fischer-Tropsch plants was the development of a fluidized-bed reactor/regenerator system designed by M. W. Kellogg for the synthesis reaction. The reactor consists of an entrained-flow reactor in series with a fluidized-bed regenerator (Fig. 14). Each fluidized-bed reactor processes 80,000 m^3/h of feed at a temperature of 320 to 330°C and 2.2 MPa (22 atm), and produces approximately 300 m^3 (2000 barrels) per day of liquid hydrocarbon product with a catalyst circulation rate of over 6000 t/h (49).

The overall processing scheme utilized at SASOL I involves steam-oxygen gasification of coal using high pressure 3 MPa (30 atm) Lurgi gasifiers producing 22,500 m^3 each of raw gas having a hydrogen-to-CO ratio of 1.7-to-1. Coal fed to the plant is high ash (35 wt %), low energy content (23 MJ/kg) from coal fields near Sasolburgh. SASOL I consumes approximately 5.5 million tons per year of coal, with 60% going for gasification and synthesis and 40% for generation of on-site power. The raw gas is purified using Rectisol (chilled methanol) technology for removal of gasification tars, H_2S, CO_2, and some methane. The purified gas is then sent to either the fixed-bed reactors (Arge synthesis) or fluid-bed reactors (Synthol synthesis).

A comparison of the selectivities and distribution patterns for the fixed-bed and fluid-bed parts of SASOL I is given in Table 5 (50). As shown, the fixed-bed system gave higher yields of liquid products (C_5^+), whereas the fluidized bed system was considerably more selective for formation of products in the gasoline (C_5–C_{11}) boiling range. Hydrocarbons from the Synthol reactors were substan-

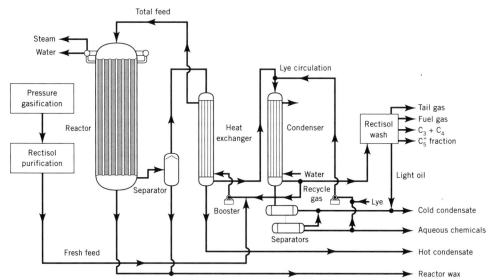

Fig. 13. Flowsheet of medium pressure synthesis, fixed-bed reactor (Lurgi-Ruhrchemie-Sasol) having process conditions for SASOL I of an alkaline, precipitated-iron catalyst, reduction degree 20–25%; having a catalyst charge of 32–36 t, at 220–255°C and 2.48 MPa (360 psig) at a fresh feed rate of 20,000–22,000 m^3/h in the reactor.

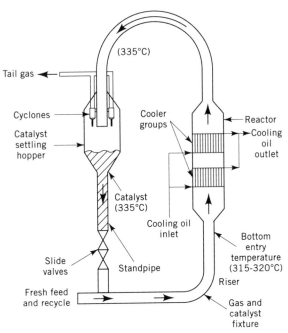

Fig. 14. Fluidized-bed reactor for SASOL I (M. W. Kellogg) having process conditions of an alkaline, reduced high grade magnetite, reduction degree 95%; 75% < 150 μm, 25% < 49 μm catalyst at a catalyst charge of 100–400 t, at 320–330°C and 2.3 MPa (330 psig); fresh feed, 80,000–110,000 m³/h in the reactor.

tially more olefinic than those produced in the Arge system, and the fluid-bed reactors produced considerably more oxygenated species, with ethanol being the primary product.

SASOL II and III. Two additional plants were built and are in operation in South Africa near Secunda. The combined annual coal consumption for SASOL II, commissioned in 1980, and SASOL III, in 1983, is 25×10^6 t, and these plants together produce approximately 1.3×10^4 m³ (80,000 barrels) per day of transportation fuels. A block flow diagram for these processes is shown in Figure 15. The product distribution for SASOL II and III is much narrower in comparison to SASOL I. The later plants use only fluid-bed reactor technology, and extensive use of secondary catalytic processing of intermediates (alkylation, polymerization, etc) is practiced to maximize the production of transportation fuels.

Developments in Indirect Liquefaction. Much of the research and process development on indirect liquefaction of coal in the 1990s is aimed at matching the synthesis conditions with modern, efficient coal gasifiers such as those developed by Texaco, Dow, and Shell (see COAL CONVERSION PROCESSES, GASIFICATION). A comparison of the gas product mix from a Shell gasifier with that from the older standard Lurgi system is shown in Table 6. Whereas the newer gasifiers are considerably more efficient than the older design, there is the drawback that the gas produced is much lower in hydrogen content. This problem may be solved by shift conversion of the raw synthesis gas, a process that is expensive, or it may be

Table 5. Selectivities and Product Distributions for Fixed- and Fluidized-Bed Units at SASOL I

Product	Fixed-bed process			Fluid-bed process	
Selectivity over average catalyst life[a]					
CH_4	7.8			13.1	
C_2H_4	0.6			4.4	
C_2H_6	2.6			5.8	
C_3H_6	3.9			12.8	
C_3H_5	2.2			3.4	
C_4H_6	2.5			10.0	
C_4H_{10}	2.4			3.2	
C_5 and above	75.7			39.0	
nonacid chemicals	2.3			7.3	
acids				1.0	
Liquid product composition					
liquefied petroleum gas, C_2–C_4	5.6			7.7	
petrol, C_5–C_{11}	33.4			72.3	
middle oils[b]	16.6			3.4	
waxy oil or Gatsch	10.3			3.0	
medium wax, mp 57–60°C	11.8				
hard wax, mp 95–97°C	18.0				
alcohols and ketones	4.3			12.6	
organic acids	traces			1.0	
	C_5–C_{10}	C_{11}–C_{15}	C_5–C_{18}	C_{11}–C_{14}	
paraffins, vol %	45	55	13	15	
olefins	50	40	70	60	
aromatics	0	0	5	15	
alcohols	5	5	6	5	
carbonyls	traces	traces	6	5	

[a]Wt % basis.
[b]Diesel, furnace oil, etc.

obviated by design of synthesis reactors and catalysts that can utilize the low H_2/CO gas directly. Because of the exothermic nature of the synthesis reactions, design efforts have focused on the development of slurry-phase reactors to replace conventional fixed-bed and fluid-bed systems (50).

Research on slurry-phase reactors and catalysts is being carried out by the U.S. Department of Energy at the Pittsburgh Energy Technology Center (PETC). In such a reactor, a schematic of which is shown in Figure 16, the catalyst is suspended in a slurry medium which is comprised of the waxy portion of the Fischer-Tropsch products. The catalyst is finely divided, thus reducing diffusional mass and heat-transfer limitations and the reactor is agitated by bubbling the reactant gas through the catalyst slurry. Use of internal heat exchangers directly in the bed and the inherently high heat-transfer rates that are typical in this type of well-mixed configuration provides for much higher rates of heat removal from the bed. The better temperature control resulting from this design reduces cata-

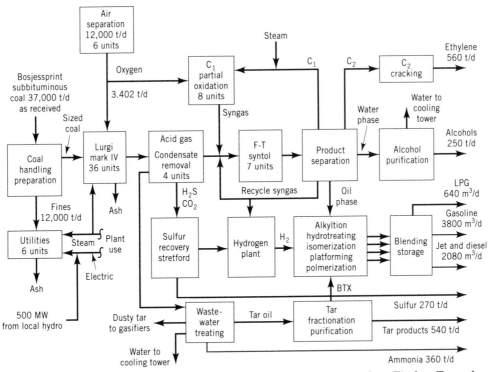

Fig. 15. Block flow diagram, SASOL II and III, where F-T corresponds to Fischer-Tropsch, and BTX is benzene, toluene, xylene. To convert m^3 to barrels, multiply by 6.29.

Table 6. Synthesis Gas Compositions from Gasifiers Operating on Western U.S. Coals

Material	Dry gasifier product, vol %	
	Shell	Lurgi
H_2	30	50
CO	66	25
CO_2	3	10
CH_4 + higher hydrocarbons		15
inerts	1	
H_2/CO ratio, vol basis	0.45	2
net efficiency to synthesis gas, %[a]	78–80	61

[a]Net efficiency includes thermal losses in reforming methane to synthesis gas.

591

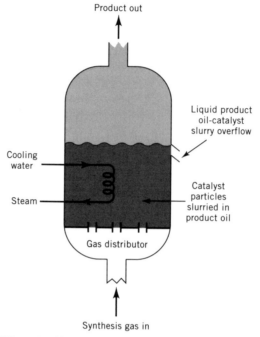

Fig. 16. Slurry-phase Fischer-Tropsch reactor.

lyst deactivation when low H_2-to-CO gases are used. Use of slurry reactors for the Fischer-Tropsch synthesis is being investigated at SASOL in a small (1 m dia) pilot reactor.

BIBLIOGRAPHY

"Hydrogenation" under "Coal Chemicals and Feedstocks" in *ECT* 3rd ed., Supplement, by J. Langhoff, pp. 216–228.
1. R. F. Probstein and R. E. Hicks, *Synthetic Fuels*, McGraw-Hill, New York, 1976.
2. D. D. Whitehurst, T. O. Mitchell, and M. Farcasiu, *Coal Liquefaction*, Academic Press, Inc., New York, 1980.
3. G. P. Curren, R. T. Struck, and E. Gorin, *Ind. Eng. Chem. Proc. Des. Devel.* **6,** 173 (1967).
4. L. W. Vernon, *FUEL* **59,** 102 (1980).
5. S-C. Shin, R. M. Baldwin, and R. L. Miller, *Energy Fuels* **3,** 71 (1989).
6. D. F. McMillen, R. Malhotra, S. J. Chang, and E. S. Nigenda, *Energy Fuels* **1,** 193 (1987).
7. Ger. Pat. 301,231 (Nov. 26, 1919), F. Bergius and J. Billwiller.
8. E. E. Donath, "Hydrogenation of Coal and Tar," in H. H. Lowry, ed., *Chemistry of Coal Utilization, Supplementary Volume*, John Wiley & Sons, Inc., New York, 1963.
9. D. L. Kloepper, T. F. Rogers, C. H. Wright, and W. C. Bull, *Office of Coal Research R&D Report No. 9*, U.S. Dept. of the Interior, Washington, D.C., 1965.
10. S. B. Alpert and R. H. Wolk, in M. A. Elliott, ed., *Chemistry of Coal Utilization, Second Supplementary Volume*, John Wiley & Sons, Inc., New York, 1981.

11. J. C. Tao, *Proceedings of the 3rd International Coal Utilization Conference*, Houston, Tex., 1980.
12. C. H. Wright, and D. E. Severson, *ACS Div. Fuel Chem. Preprints* **16**(2), 68 (1972).
13. B. F. Alexander and R. P. Anderson, *ACS Div. Fuel Chem. Preprints* **27**(2), 18 (1982).
14. W. R. Epperly and J. W. Taunton, *Sixth International Conference of Coal Gasification, Liquefaction, and Conversion to Electricity*, Pittsburgh, Pa., 1979.
15. S. J. Hsia, *Proceedings of the Eleventh Biennial Lignite Symposium*, Grand Forks, N.D., 1981.
16. C. C. Kang and E. S. Johanson, in R. T. Ellington, ed., *Liquid Fuels from Coal* Academic Press, Inc., New York, 1977.
17. A. G. Comolli, E. S. Johanson, S. V. Panvelker, and R. H. Stalzer, *Proceedings of the DOE Liquefaction Contractors' Review Meeting*, Pittsburgh, Pa., 1992.
18. R. E. Lumpkin, *Science* **239**, 873 (1988).
19. E. L. Huffman; *Proceedings of the Third Annual International Conference on Coal Gasification and Liquefaction*, Pittsburgh, Pa., 1976.
20. H. D. Schindler, J. M. Chen, and J. D. Potts, *Final Technical Report on DOE Contract No. DE-AC22-79ET14804*, 1983.
21. H. D. Schindler; "Coal Liquefaction: A Research Needs Assessment," Vol. 2, Technical Background, *Final Report on DOE Contract No. DE-AC01-87ER30110*, 1989.
22. S. R. Hart, Jr. and E. L. Huffman, *Electric Power Research Institute Report No. AP-4257-SR* Vol. 2, paper 34, 1985.
23. S. R. Hart, Jr. and E. L. Huffman, *Proceedings of the Joint Conference on Coal Gasification and Synthetic Fuels for Power Generation*, San Francisco, Calif., 1985.
24. A. G. Comolli, E. S. Johanson, J. B. McLean, and T. O. Smith, *Proceedings of the DOE Direct Liquefaction Contractors' Review Meeting*, Pittsburgh, Pa., 1986.
25. C. W. Zielke, R. T. Struck, J. M. Evans, C. P. Costanza, and E. Gorin, *Ind. Eng. Chem. Proc. Des. Devel.* **5**(2), 158 (1966).
26. N. B. Moll and G. J. Quarderer, *Chem. Eng. Prog.* **75**, 46 (1979).
27. P. M. Yavorsky, S. Akhtar, and S. Friedman, *Chem. Eng. Prog.* **69**(3), 31 (1973).
28. J. D. Potts, K. E. Hastings, R. S. Chillingworth, and K. Unger, *U.S. DOE Interim Report FE-2038-42*, 1980.
29. B. O. Strobel, *Proc. Int. Conf. Coal Sci.* **2**, 735 (1989).
30. *Gasoline from Coal*, National Coal Board, Coal House, Harrow, Middlesex, UK, 1986.
31. C. Y. Wen and S. Dutta, in C. Y. Wen and E. S. Lee, eds., *Coal Conversion Technology*, Addison Wesley, Inc., Reading, Mass., 1979.
32. J. B. Howard, in Ref. 31, Chapt. 12.
33. P. Arendt and K. H. van Heek, *Fuel* **60**, 779 (1981).
34. *Final ERDA Report*, Report No. FE-1212T9, Vol. 1, Char Oil Energy Development, 1975.
35. L. Seglin and S. A. Bresler, in Ref. 31, Chapt. 13.
36. F. B. Carlson, *Proceedings of the 101st Annual AIME Meeting*, San Francisco, Calif., 1972.
37. E. T. Coles, *ERDA Report*, Report No. PER(A)-0, 1975.
38. A. Sass, *Chem. Eng. Prog.* **70**(1), 72 (1974).
39. W. Peters, *Chem. Ing. Tech.* **32**(3), 178 (1960).
40. *Chem. Eng. News*, 27 (Nov. 20, 1978).
41. H. Juntgen, J. Klein, K. Knoblauch, H-J. Schroter, and J. Schulze, in Ref. 31, Chapt. 30.
42. P. Sabatier and J. B. Senderens, *C.R. Acad. Sci. (Paris)* **134**, 514 (1902).
43. F. Fischer and H. Tropsch, *Chem. Ber.* **59**, 830 (1926).
44. J. Schulze, *Chem.-Ing.-Tech.* **46**, 976 (1974).

45. S. S. Penner, ed., *U.S. DOE Working Group on Research Needs for Advanced Coal Gasification Techniques (COGARN)*, DOE Contract No. DE-AC01-85ER30076, Washington, D.C., 1987.
46. J. R. Katzer and B. C. Gates, "Catalytic Processing in Fossil Fuel Conversion," *AIChE Today Series*, American Institute of Chemical Engineers, New York, 1975.
47. J. C. Hoogendoorn, in *Clean Fuels from Coal, Institute of Gas Technology Symposium Series*, IGT, Chicago, Ill., 1973, p. 353.
48. P. E. Rosseau, J. W. van der Merwe, and J. D. Louw, *Brennstoff Chem.* **44,** 162 (1963).
49. J. C. Hoogendoorn, "Gas from Coal for Synthesis of Hydrocarbons," paper presented at *AIME 23rd Annual Meeting*, 1974.
50. G. J. Steigel, *PETC Rev.* **4,** 14 (1991).

ROBERT M. BALDWIN
Colorado School of Mines

COAL CONVERSION PROCESSES, MAGNETOHYDRODYNAMICS. See MAGNETOHYDRODYNAMICS.

COAL GAS. See COAL CONVERSION PROCESSES; FUELS, SYNTHETIC.

COAL, GASIFICATION OF COAL. See COAL CONVERSION PROCESSES, GASIFICATION; FUELS, SYNTHETIC.

COAL, POWER FROM COAL BY GASIFICATION AND MAGNETOHYDRODYNAMICS. See COAL CONVERSION PROCESSES, GASIFICATION; MAGNETOHYDRODYNAMICS; POWER GENERATION.

COAL, SYNTHETIC CRUDE OIL FROM COAL. See FUELS, SYNTHETIC.

COAL TAR. See COAL CONVERSION PROCESSES; TAR AND PITCH.

COATED FABRICS

A coated fabric is an engineered product derived by a combination of textile and polymer coating technology. Each type of coated fabric is designed to fulfill the requirements of a specific application. The textile substrate is chosen based on a knowledge of the strength and dimensional control requirements of the finished product. The coating, a product of polymer compounding knowledge, is chosen for its specific properties such as abrasion resistance, oil resistance, resistance to leaching, etc.

Textile Component

Industrial-grade fabrics, as opposed to apparel-grade fabrics, typically constitute the substrate classification from which the coated fabrics producer selects a construction to fulfill specific end use requirements. Because of increasing imports of apparel-grade textiles, the profitability of industrial-grade textiles has drawn resources away from other segments of the textile industry. By the mid-1980s industrial coated fabrics accounted for over 10% of all fiber consumption (1). Most textile manufacturers have divisions that specialize in the development and production of various industrial fabrics (see GEOTEXTILES). In partnership with the coating company, the textile company can aid in the selection of the proper fabric and provide hand samples, pilot yardage, and the ultimate production quantities. A comprehensive listing of industrial fabric suppliers has been compiled (2).

Fibers. For many years cotton (qv) and wool (qv) were the primary textile components, contributing the properties of strength, elongation control, and aesthetics to the finished product. Although the modern coated fabrics industry began by coating wool to make boots, cotton has been used more extensively. Up until the 1980s cotton constructions, including sheetings, drills, sateens, and knits, had commanded a significant share of the market. This is because cotton is easily dyed, absorbs moisture, withstands high temperature without damage, and is stronger wet than dry. As cotton again became an important fiber for the apparel industry during the 1980s, the supply and pricing situation changed significantly and polyester–cotton blends and 100% polyesters became the fabrics of choice. Polyester is now the most widely used industrial coating and laminating fiber (3). When moisture absorbance and glueability are the most important properties, polyester–cotton blends and in some cases 100% cotton textiles still see significant use, ie, wallcovering and case goods covering.

Polyester has numerous advantages since its fibers are smooth, crisp, and resilient. Since moisture does not penetrate polyester, it does not affect the size and shape of the fiber. Polyester is also resistant to chemical and biological attack. Because it is thermoplastic, the heat required for good adhesion to the polyester substrate can also create a shrinkage problem during the coating process. Because of this, 100% polyester textiles are often heat set to relatively high temperatures in a post operation after weaving (see FIBERS, POLYESTER).

Nylon is the strongest of the commonly used fibers. Since it is both elastic and resilient, articles made with nylon will return to their original shape. Nylon has a degree of thermoplasticity so that when articles are formed and heated to a temperature above its heat set temperature, it will retain that shape. Nylon fibers are also smooth, very nonabsorbent, do not soil easily, and resist most chemical and biological action. Nylon substrates are used in applications where very high strength is required. One newly developed use for nylon that is predicted to grow to $11-14 \times 10^6$ m by the year 2000 is the use of Neoprene (polychloroprene) coated nylon in the production of driver's side automotive air bags. Lightweight nylon knits and taffetas that are thinly coated with polyurethane or poly(vinyl chloride) are used extensively in apparel. PVC coatings do not adhere well to nylon, so in recent years there has been an effort to develop improved adhesives and bonding agents that improve the coating adhesion to nylon textile (see FIBERS, POLYAMIDES).

Rayon and glass fibers are the least used because of their poor qualities. Rayon's strength approaches that of cotton but its smoother fibers make adhesion more difficult (see FIBERS, REGENERATED CELLULOSICS). Rayon also has a tendency to shrink more than cotton which makes processing more difficult. *Glass fibers* offer very low elongation, very high strength, and have a tendency to break under compression. Because of this, glass fibers are typically used where support with low stretch is required and the product will be subject to minimal flexing or where flame retardant properties are required. These include applications such as ducting and duct tape, protective clothing, and insulation. With increasing emphasis on the fire retardant properties of upholstery constructions, the textile industry has recently developed knit constructions based on glass fibers overspun with polyester. These have acceptable flex properties and the glass fibers serve to prevent the burning upholstery covering from penetrating into the urethane foam cushioning (see GLASS).

Textile Construction.   There are many choices in textile construction. The original and still the most commonly used is the woven fabric. Woven fabrics have four basic constructions: the plain weave, the drill weave, the satin weave, and the twill weave. The plain and the drill weave are the strongest constructions because they have the tightest interlacing of fibers. Twill weaving produces distinct surface appearances and is used for styling effects. Satin weave is used primarily for high style applications because it is the weakest of the wovens. Woven nylon and polyester have displaced heavy cotton in most tarpaulin applications. Also, as the technology for producing weft inserted fabrics improves, wefts are also finding increasing use in the production of tarpaulins. For shoe uppers and other applications where strength is important, woven cotton fabrics are used.

Knitted fabrics are used where moderate strength and considerable elongation are required. Whereas cotton yarns formerly dominated the knit market, they were first displaced by polyester–cotton knits and later by 100% polyester knits because the polyester knits have a higher strength to weight ratio; therefore, lighter weight backings can be used to achieve similar properties. When a polymeric coating is applied to a knit fabric, the stretch properties are somewhat reduced from those of the fabric. The stretch and set properties of the final construction are important for upholstering and forming. The main use of knit fabrics is in apparel, heavyweight transportation and furniture upholstery, shoe liners, boot shanks, and any product where elongation is required.

Many types of nonwoven fabrics (qv) are utilized as substrates for coated fabrics, including products made by the wet web method, saturated nonwovens, spunbonded nonwovens, and needled nonwovens. Today's nonwovens are engineered fabrics designed for specific end uses. The wet web process gives a nonwoven fabric with paperlike properties and poor drape. Most often wet web nonwovens are used in the production of coated fabrics such as wallcoverings. These webs are often treated with latex polymers to improve the strength and stripability properties of the finished wallcovering. Saturated nonwovens are prepared by laying dry webs, tightly compressing by needle punching, and then impregnating them with 50–100% by weight with a soft latex. The finished product often resembles a split leather; in fact, it is often split on a leather skiving machine to produce thinner substrates. Most often these impregnated nonwoven fabrics are used for shoe liners. It is difficult to achieve uniformity of stretch and strength in two directions as well as a smooth surface; therefore a high quality nonwoven of this type is very expensive. Spunbonded nonwovens are available in both polyester and nylon in a wide range of weights. They are often lower priced than other nonwovens. Their strength properties are very high and elongation is low. Since they are quite stiff, these materials are used where strength and price are the primary considerations.

The most common nonwovens used for coated fabric substrates are the lightly needled, low density nonwovens which are typically prepared with either polypropylene or polyester fibers. Specialty fibers such as Kevlar or Nomex are also used when high strength or fire retardant properties are desired (see HIGH PERFORMANCE FIBERS). By varying the amount of needling combined with careful orientation of the fibers and selection of the fiber length, the nonwoven manufacturer can obtain very good strength and balanced stretch. Optionally, a very thin layer of polyester-based polyurethane foam can be needled into the nonwoven to improve the drape and surface coating process. Upholstery produced on polyester and polypropylene substrates has now replaced most expanded PVC on knit fabric constructions in automotive, furniture, and marine upholstery applications. This is because a product that has many of the properties of the expanded products, such as plushness, tailorability, and stretch can be produced at a much lower cost.

Post Finishes of the Textile Component. The construction that results from either weaving or knitting is called greige good. In many cases, other steps are required before the fabric can be coated. This often includes scouring to remove surface impurities and finishes added to the yarns to improve weaving, and heat setting to correct the width, stabilize the textile, and minimize shrinkage during coating. Optional treatments include: dyeing if a colored substrate is required; napping of cotton and polyester–cotton blends to add bulk, impart a softer hand, or increase adhesion; and mildew and/or antiwicking treatments for textiles that will be used to produce products for outdoor applications.

Polymeric Coating Component

Rubber and Synthetic Elastomers. For many years nondecorative coated fabrics consisted of natural rubber on cotton cloth. Natural rubber is possibly the best all-purpose rubber but some characteristics, such as poor resistance to oxygen and ozone attack, reversion and poor weathering, and low oil and heat resistance, limit its use to special application areas (see ELASTOMERS, SYNTHETIC; RUBBER, NATURAL).

Polychloroprene (Neoprene), introduced in 1933, rapidly gained prominence as a general purpose synthetic elastomer having oil, weather, and flame resistance. The introduction of new elastomers in solid or latex form was accelerated by World War II. In addition to natural rubber and polychloroprene, other rubbery polymers in use include: styrene–butadiene (SBR), polyisoprene, polyisobutylene (Vistanex), isobutylene–isoprene copolymer (Butyl), polysulfides (Thiokol), poly-acrylonitrile (Paracril), silicones, chlorosulfonated polyethylene (Hypalon), poly(vinyl butyral), acrylic polymers, ethylene–propylene–diene monomer (EPDM), fluorocarbons (Viton), polybutadiene, polyolefins, and many more. Co-polymerization makes the number of variations available staggering (see ACRYLIC ESTER POLYMERS; ACRYLONITRILE POLYMERS; COPOLYMERS; FLUORINE COMPOUNDS; OLEFIN POLYMERS; POLYMERS CONTAINING SULFUR; SILICON COMPOUNDS; URETHANE POLYMERS). The number of commercially available elastomers is large with many producers of those polymers offering several variations that provide a wide range of properties.

Most elastomers are vulcanizable; they are processed in a plastic state and later cross-linked in a heating process to provide elasticity in their final form. With the number of elastomeric coatings available, almost any use requirement can be met. Also, by compounding these various elastomers, it is possible to develop products that meet particular area requirements and specific environmental conditions. Additional information about compounding and processing of elastomeric coatings is available (4). The only limitations to possible constructions that can be developed are in the areas of processability and cost. These elastomers are applied to the textile by either calender or solution coating. Thin coatings are typically applied from solution and thicker coatings by direct calendering to the textile. A typical natural rubber-based formulation is shown in Table 1.

SBR (styrene–butadiene rubber) has replaced natural rubber in many applications because of price and availability. It has good aging properties, abrasion resistance, and flexibility at low temperatures. A typical SBR-based formulation is shown in Table 2.

Table 1. Natural Rubber Formulation

Component	Parts
smoked sheet	100.00
stearic acid	1.00
ZnO	3.00
Vanplast R	3.00
Agerite White	0.50
SRF Black (N774)	4.00
calcium carbonate	75.00
clay	50.00
sulfur	0.75
miscellaneous accelerators	1.35
Total	*238.60*

Neoprene offers resistance to oil, weathering, is inherently nonflammable, and is processable on either coaters or calenders. Other elastomers such as EPDM can often be formulated for equivalent performance and unlike Neoprene, they do not discolor and can be formulated into light-colored products. EPDM is most often used to produce roofing membranes because of its excellent outdoor aging. Some Neoprene is also used in the uncured state for forming flashings on single-ply rubber roofing systems. A typical Neoprene based formulation is shown in Table 3. This mixture can be calendered or dissolved in toluene to 25–60% solids for application as a thin coating.

Isobutylene–isoprene elastomer (Butyl) has high resistance to oxidation, resists chemical attack, and is the elastomer most impervious to air. Because of these properties it is often used for protective garments, inflated air structures, cold air balloons, and fumigation covers.

Chlorsulfonated polyethylene (Hypalon) resists ozone, oxygen, and oxidizing agents, and is the coating of choice for roofing applications where oil resistance is needed, ie, fast food restaurants, etc. In addition, it has nonchalking weathering properties and does not discolor, permitting pigmentation for decorative effects.

Table 2. SBR Compounding

Component	Parts
SBR	100.00
stearic acid	2.50
ZnO	3.00
Vanplast R	7.00
Agerite White	0.50
SRF Black (N774)	4.00
Cumar MH	20.00
calcium carbonate	75.00
clay	75.00
sulfur	2.50
miscellaneous accelerators	2.35
Total	*291.85*

Table 3. Neoprene Compounding

Component	Parts
Neoprene W	100.00
magnesium oxide	4.00
Agerite Stalite S	2.00
Vanwax H	3.00
Thermax (N990)	60.00
Vanplast PL	5.00
sulfur	0.50
miscellaneous accelerators	2.00
Total	*176.50*

Polyurethane.   Urethanes have a number of important applications in coated fabrics. The most important is the production of fabrics for apparel, tenting, life vests, evacuation slides, flexible fuel storage tanks, and other industrial-grade coated fabrics. This is because polyurethanes are lighter weight than vinyl polymers and have better abrasion resistance and strength. Polyurethane fabrics can also be easily decorated to look like leather. Many urethane-coated fabrics are used in women's footwear and apparel where styling is important and light weight is desirable. These products usually consist of 0.05 mm of polyurethane on a napped woven cotton fabric. The result is a lightweight product that has good abrasion and scuff resistance. Urethane coated fabrics have not been successful in high quality shoes for either men or children because they do not have the long-term durability of a natural leather shoe.

Low weight coatings of polyurethane on 22, 44, and 89 tex (200, 400, and 800 den) woven nylon fabrics produce products that are suitable for apparel, luggage, and athletic bags. The lightest weight products are used for windbreakers and industrial clothing whereas the heavier weight fabrics are used for luggage and athletic bags. The coatings that are typically on the back side of the textile provide water repellency to the product and help prevent fraying when the fabrics are die cut when manufacturing the finished products. Polyurethanes have found only limited application in the upholstery market because they are often subject to hydrolysis with long-term exposure to body oils. Also because of their light weight, they do not offer the plushness and stretch that is normally required for upholstery applications.

Poly(vinyl chloride).   By far the most important polymer used in coated fabrics is poly(vinyl chloride). This relatively inexpensive polymer resists aging processes readily, resists burning, and is durable. It can be compounded readily to improve processing, aging, burning properties, softness, etc. In addition, it can be decorated to fit nearly any required use including leather prints, textile looks, or detailed patterns. PVC-coated fabrics are used for furniture, marine and automotive upholstery, window shades, automotive trim, wallcoverings, book covers, convertible topping, shoe uppers and liners, and many other uses. Two of the largest uses of PVC-coated fabrics are in vinyl wallcoverings and upholstery. Fabric backed wallcoverings range from very lightweight products with 0.08 mm of PVC on lightweight nonwoven backings for residential use to heavyweight expanded wallcovering for commercial applications with up to 0.50 mm of PVC that is later expanded to approximately 1.00 mm. Vinyl upholstery is most popular in commercial applications where durability and cleanability are important. Even the recent trend of increasing use of natural leather in upscale residential upholstery and automotive end uses has created a market for matching PVC-coated fabrics that are used for trim and backs and other nonseating surfaces. Every year millions of meters of PVC-coated fabrics are produced for these uses. Tables 4 and 5 show typical PVC formulations (see VINYL POLYMERS, VINYL CHLORIDE AND POLY(VINYL CHLORIDE).

Processing

Coated fabrics can be prepared by lamination, direct calendering, direct coating, or transfer coating (see COATING PROCESSES). The basic problem is to bring the

Table 4. Formulation for Calendering PVC

Component	Parts
poly(vinyl chloride) resin (calender-grade)	98.50
acrylic processing aid	1.50
epoxy plasticizer	5.00
phthalate plasticizer	65.00
BaZn stabilizer	3.00
TiO$_2$ (and other pigments)	15.00
calcium carbonate (filler)	25.00
stearic acid (lubricant)	0.25
Total	*213.25*

Table 5. Plastisol PVC Formulation

Component	Parts
poly(vinyl chloride) resin (dispersion-grade)	100.00
epoxy plasticizer	4.00
phthalate plasticizer	70.00
BaZn stabilizer	2.50
TiO$_2$ (and other pigments)	10.00
calcium carbonate (filler)	25.00
dispersant (wetting agent)	1.00
Total	*212.50*

polymer and the textile together without altering undesirably the properties of the textile. Almost all of the coating processes require that the polymer be in a fluid or semifluid condition during lamination and often heat is required to obtain adhesion between the polymer and the textile. Because of this, it is important that the heat during lamination not damage sensitive synthetic or thermoplastic fabrics.

Calendering. The base polymer must first be mixed with other compounding ingredients such as stabilizers, extenders, fillers, plasticizers, colorants, lubricants, etc, before it can be transferred to the calender (see PLASTICS PROCESSING). With PVC compounds, the first step is to prepare a dry blend and then transfer it to a Banbury mixer for fluxing. With other elastomeric polymers, the ingredients are typically added directly to the Banbury. During fluxing the temperature of the polymer compound is raised to 150–170°C through the internal heat of mixing. This fluxed mixture is then transferred to a series of mills where it is further mixed before transfer to a calender. On the calender it passes through a series of heated rolls where the smoothness and gauge is adjusted to the desired conditions before being married directly to the preheated textile fabric on the bottom roll. The object is to get the required amount of adhesion without driving the compound into the fabric excessively, which would cause a clothy appearance and lower the stretch and tear properties of the finished coated fabric.

Coating. Coating operations require a more fluid compound. Rubbers and other elastomers are often dissolved in solvents before being coated on textiles.

In the case of PVC, fluidity is achieved by dispersing the resin system in plasticizers and making a plastisol. If lower viscosity is required, an organosol can be made by adding solvent to the plastisol. This plastisol or organosol can be applied to the textile by various methods. In the past the most common coating method was to use a knife over roll or reverse roll coater. The knife over roll or knife over plate is most common in the production of polyurethane-coated fabrics. It usually has significant speed limitations and is best suited for applying thin coatings. The reverse roll coater is probably the most versatile coating technique. It is not without limitations, but it can typically handle the widest variety of coating viscosities, speeds, and application rates (5). Today, a rotary screen coater is often used to apply coatings to the textile. Unless the fabric is very dense, knife over roll or reverse roll coaters often give too much penetration of the coating. An advantage of a rotary screen coater is that it can be used to apply a metered thickness of coating to a textile with minimum penetration. There is even technology available that allows the fabric to pass between two rotary screen heads so that different compounds or colors can be applied to both sides of a textile at the same time (see COATING PROCESSES).

Another common coating method is a transfer coating method. A release paper, most often with an inverse leatherlike grain, is coated by either a knife over roll coater or a reverse roll coater. Many polyurethane coated fabrics and most expanded PVC-coated fabrics are transfer coated. The expanded products consist of a wear layer, an expanded layer, and the textile substrate. The wear layer is coated on release paper and gelled. Then a layer of vinyl-based compound containing a chemical blowing agent such as azodicarbonamide is applied. The fabric is placed on top of the second layer and sufficient heat is applied to decompose the blowing agent causing the expansion.

Lamination.   In lamination a film is prepared by calendering or extrusion (see LAMINATES). It is then adhered to a textile at a laminator by either an adhesive or sufficient heat which partially melts the film to obtain a mechanical bond. There are a variety of adhesives available for lamination, including solvent systems, water-base latex systems, and various forms of hot melt adhesives (qv).

Post Treatment.   Coated fabrics can be decorated and protected by applying inks and coatings to the surface. Often the finished product is an attempt to simulate the look of leather. This is most common in upholstery, luggage, and athletic bag constructions where natural leather is the main competitor, although any number of decorative effects can be created and are used in producing products such as wallcovering.

The typical method for applying these print inks is a rotogravure method. In the past these inks were usually made up of vinyl copolymer resins and pigments dissolved in ketones and other solvents. These print inks are usually applied to a flat surface and often multiple print patterns are applied to obtain a highly decorative effect. There has been a gradual movement to the use of water-base inks since the 1980s and the passage of the Clean Air Act of 1990 has targeted a reduction of air toxics emissions by 75% through the year 2000 (6). This target is to be met by applying MACT (maximum achievable control technology) standards. Although this reduction can be achieved by emissions control devices, most manufacturers still using solvent-based inks and coatings will choose to accelerate their conversion from solvent to water-base inks rather than purchase and operate control devices.

Most decorative coated fabrics, especially those used in upholstery, apparel, or luggage applications, also have a clear final finish. Again this can be either a solvent or water-based system and typically is based on a hard vinyl or vinyl acrylic polymer or a urethane. This finish usually serves several purposes. Because most coated fabrics, particularly vinyl-coated fabrics, are slightly tacky, the coating serves to dry the surface and provide slip. Also the finished luster of the coating can be adjusted with ultrafine silica to anything from very dull to full bright. Finally, the coating often imparts improved stain resistance to the product. This is particularly true of some of the special stain resistant coatings being offered by several manufacturers of PVC upholstery and wallcovering.

If a textured surface is desired, the coated fabric is heated to soften it and pressure is applied by an engraved embossing roll. Most often embossing is the final step and comes after the printing and finishing operations are complete; however, sometimes the coated fabric is embossed first and then printed or given a wash coat of ink (spanishing) to achieve a unique decorative effect.

A typical solvent-based finish coating for PVC is shown in Table 6.

Table 6. Vinyl-Based Topfinish for PVC Coated Fabrics

Component	Parts
vinyl chloride–vinyl acetate copolymer	85.50
polymethacrylate resin	14.50
BaZn vinyl stabilizer	1.75
silica gel	14.00
methyl ethyl ketone	950.00
toluene	50.00
Total	*1115.75*

Economic Aspects

In the total market for coated fabrics, the first consideration is the required performance of the finished product, followed by cost considerations. Because of their higher cost, the rubber and specialty elastomers are only used for producing coated fabrics that require oil or chemical resistance, low air permeability, or other unique properties. Likewise, urethane-coated fabrics are often used for the production of style and design oriented coated fabrics for the apparel, shoe, and handbag market because it is easy to produce small run sizes using transfer coating methods. In addition, the finished products have light weight and good abrasion and scuff resistance and are significantly less expensive than the natural leather hides that they replace.

Because of performance and cost considerations, PVC-coated fabrics are the workhorse coated fabrics. They are easily processed by either calender coating or transfer coating, are easily decorated by a combination of printing and embossing, and depending on the choice of compound formulation and textile backing can be designed for many different uses. The largest uses of PVC coated fabrics are in upholstery applications for marine, contract, and automotive applications, and for the production of fabric backed vinyl wallcovering. These products offer the combination of high style decorative effects with the benefits of performance and long-term durability.

Testing

Depending on the application for a coated fabric there are many different specifications that the products must meet. These specifications refer to specific tensile and tear values for the construction, abrasion specifications, chemical resistance, air permeability, fire resistance, and toxicity, and many other properties. Most often the specifications are set by the individual customers of these products, however, there are also various industry and trade groups and the federal government that set specifications for families of products. Several of these are published by the Chemical Fabrics & Film Association, Inc. (CFFA), and the federal government, such as the Federal Specification–Wallcovering, Vinyl Coated CCC-W-408C. Most of the test methods used for measuring these properties are standard ASTM methods or methods suggested by the customer or industry trade groups.

Health and Safety Factors

Some materials used in coating operations have been identified by the federal government as being hazardous to workers' health. Because of this, manufacturers of coated fabrics are subject to OSHA standards relating to acceptable exposure to these chemicals. In most cases, depending on the individual chemical, this required engineering changes to the process, protective equipment, personnel monitoring, and extensive record keeping. One change that many manufacturers have made or are presently working on is the development of formulations for coatings, inks, and finishes that do not contain heavy metals. This is particularly true in the vinyl industry where there is a movement away from stabilizer systems containing cadmium, mildewcides that contain arsenic, and pigment systems containing lead chromates and lead molybdates.

Most exposure problems are related to solvents and dusts, so particular attention should be given to raw material mixing areas and solvent exposure during coating and post-printing and finishing operations. In particular, emptying bags and bulk transfer of raw materials should be monitored. Even when calender coating textiles, consideration should be given to exposure to the gases that are given off during heating. Under federal "Right to Know" laws, training is required for all employees exposed to these chemicals and the MSDS for these chemicals must be readily available in the workplace. No coating operation should be initiated without making a full review of applicable federal and state OSHA standards for the raw materials that will be used in the processes.

The actual coated fabrics themselves are not subject to these federal regulations. They are classified as "Articles of Commerce." However, in recent years most manufacturers have, for the convenience of their customers, generated an MSDS for their products. Their customers, the manufacturers, and contractors using these coated fabrics have requested this information because they are also covered under federal and state "Right to Know" laws and want to ensure the safety of their employees.

BIBLIOGRAPHY

"Coated Fabrics" in *ECT* 1st ed., Vol. 4, pp. 134–144, H. B. Gausebeck, Armour Research Foundation of Illinois Institute of Technology; in *ECT* 2nd ed., Vol. 5, pp. 679–690, by D. G. Higgins, Waldron-Hartig Division of Midland-Ross Corp., in *ECT* 3rd ed., Vol. 6, pp. 377–386, by F. N. Teumac, Uniroyal, Inc.

1. R. P. Antoshak, *J. Coated Fabrics* **15,** 239 (Apr. 1986).
2. *Industrial Fabric Products Review, 1990 Buyer's Guide,* Industrial Fabrics Association International, St. Paul, Minn., 1990.
3. W. C. Smith, *J. Coated Fabrics* **15,** 180 (Jan. 1986).
4. *The Vanderbuilt Rubber Handbook,* R. T. Vanderbuilt Co., Inc., Norwalk, Conn., 1990.
5. J. A. Pasquale, III, *J. Coated Fabrics* **15,** 271 (Apr. 1986).
6. *Clean Air Act Law and Explanation,* Commerce Clearing House, Chicago, Ill., 1991.

General References

R. M. Murray and D. C. Thompson, *The Neoprenes,* E. I. du Pont de Nemours & Co., Inc., Wilmington, Del., 1963.
H. L. Weiss, *Coating and Laminating Machines,* Converting Technology Co., Milwaukee, Wis., 1977.
1990 Rubber Red Book, Communication Channels, Inc., Atlanta, Ga., 1990.
Modern Plastics Encyclopedia 1990, McGraw-Hill Inc., New York, 1990.
1990 Plastics Directory, The Cahners Publishing Co., Newton, Mass., 1990.
ASTM Annual Book of Standards, American Society of Testing and Materials, Philadelphia, Pa., 1990.
CFFA Standard Test Methods, Chemical Films & Fabrics Association, Inc. (CFFA), Cleveland, Ohio, 1984.
CFFA-W-101-A, CFFA, Cleveland, Ohio, 1984.
CFFA-M-101A, CFFA, Cleveland, Ohio, 1991.
CFFA-U-101A, Cleveland, Ohio, 1984.

BRUCE BARDEN
GenCorp Polymer Products

COATING PROCESSES

SURVEY

Coatings technology covers a wide variety of products and processes (1–4). Typical are paints (see PAINT) and the diverse surface coatings (qv) used to protect houses, bridges, appliances, and automobiles. These coatings provide functional needs such as waterproofing, flameproofing, and corrosion protection, as well as having decorative aesthetic qualities (see CORROSION AND CORROSION CONTROL). Coating processes are also important to the production of such coated products as photographic films for medical, industrial, graphic arts, and consumer use (see COLOR PHOTOGRAPHY; PHOTOGRAPHY); magnetic media for audio and visual use and for data storage (see INFORMATION STORAGE MATERIALS; MAGNETIC MATE-RIALS); adhesives (qv); printing plates; and paper (qv).

Whole industries are based on coatings and coating processes. For example, in publishing coatings are used to give paper its gloss, strength, and ink acceptability (see INKS). Lithographic printing plates for printing presses are photosensitive coatings on aluminum (see LITHOGRAPHY). Photographic film, itself a coated product, is used to expose the plates, set the type, and prepare the printed pictures. Other industries based on coatings are the entertainment industry, which uses magnetic tape (qv) and film, and compact disks, and the computer industry where information is stored on magnetic coated structures such as hard drives and floppy disks (see COMPUTER TECHNOLOGY). Coated products such as photoresist films are used to fabricate circuit boards, which have coatings to connect components (see ELECTRICAL CONNECTORS; PHOTORESISTS).

Coating is defined herein as replacing air at a substrate interface with a new material. The replacing is the coating process and includes application techniques, the importance of solution parameters, the selection of coating method, and the mechanisms involved. Because most coating solutions are applied from some solvent that must be removed for the coating to be functional, the drying step is also an integral part of the coating process.

Coating Methods Selection

The application of a liquid to a traveling web or substrate is accomplished via a large number of diverse coating methods. Steps in the coating process involve: preparing the solution to be coated; metering the coating to the desired coating weight; applying the coating to the support; followed by removing the solvent. The coating method selected depends on several factors which include: the nature of the support to be coated; the coating composition rheology; coating solvent; wet-coating weight or coverage desired; coating uniformity; the desired coating speed; the number of layers; and whether the coating is to be continuous or intermittent.

The primary substrates or support include many types of paper and paperboard, polymer films such as polyethylene terephthalate, metal foils, woven and nonwoven fabrics, fibers, and metal coils. Although the coating process is better suited to continuous webs than to short individual sheets, it does work very well for intermittent coating, such as in the printing process. In general, there is an ideal coater arrangement for any given product. However, most coating machines are required to produce many different products and coating thickness and the machine is therefore usually a compromise for several applications. Table 1 describes the capabilities of some of the principal coating processes.

Rheology. Rheology is the science of deformation and flow of matter (see RHEOLOGICAL MEASUREMENTS). Because the coating process imparts shear and extensional stresses to the solution, the rheological properties of coating liquids are important factors in the selection and successful running of a coating operation.

The shear or dynamic viscosity, ie, the ratio of the shear stress to the shear rate, is a measure of the resistance of the solution to flow under mechanical stress. A high viscosity solution requires a high level of stress to change shape. Typically, the shear rate, γ, where γ is the velocity divided by the thickness or gap width versus shear-stress, or force per unit area, curves at a given temperature or over a range of temperatures measured using a viscometer. The slope of this curve is the viscosity. Kinematic viscosity, υ, is the shear viscosity divided by the density, ρ, ie, $\upsilon = \mu/\rho$. For extensional flows, the extensional or elongational viscosity is the ratio of tensile stress to the rate of extension. Extensional properties are much

Table 1. Summary of Coating Methods[a,b]

Process	Viscosity, Pa·s[c]	Wet thickness, μm[d]	Coating accuracy, %	Speed max, m/min	Effect of web roughness
		Single layer			
rod, wire wound	0.02–1	5–50	10	250	large
reverse roll	0.1–50	12–1200	5	300	slight
forward roll	0.02–1	10–200	8	150	
air knife	0.005–0.5	2–40	5	500	large
knife over roll	0.1–50	25–750	10	150	large
blade	0.5–40	1–30		1500	large
gravure	0.001–5	1–50	2	700	
slot	0.005–20	15–250	2	400	slight
extrusion	50–5000	15–750	5	700	
		Multilayer			
slide	0.005–0.5	15–250	2	300	slight
curtain, precision	0.005–0.5	2–500	2	300	slight

[a]Ref. 1.
[b]Values given are meant to be guideline values.
[c]1 Pa·s = 1000 cP.
[d]1 μm = 1 cm^3/m^2 of wet coating, and 1 μm = 1 g/m^2 for a density of 1 g/cm^3.

more difficult to measure than shear properties. Although often ignored, extensional properties can be important, especially in the coating of polymeric liquids.

A Newtonian fluid has a constant viscosity independent of shear rate. The shear stress is a linear function of shear rate and the stress is zero at zero shear rate. In practice most coating liquids exhibit non-Newtonian behavior over some range of shear rate. Typical shear stress-shear rate behaviors include: pseudo plastic or shear thinning, where the viscosity decreases with increasing shear rate; dilatant or shear thickening, where the viscosity increases with increasing shear rate; and Bingham or yield-stress behavior, where the stress must exceed some finite value before the fluid flows. Combinations of these behaviors, such as shear thinning with a yield stress, are also common. Time-dependent behavior is also possible and the terms used are thixotropic, where the viscosity decreases with time, or rheopectic, where the viscosity increases with time.

Elasticity is another manifestation of non-Newtonian behavior. Elastic liquids resist stress and deform reversibly provided that the strain is not too large. The elastic modulus is the ratio of the stress to the strain. Elasticity can be characterized using transient measurements such as recoil when a spinning bob stops rotating, or by steady-state measurements such as normal stress in rotating plates.

Coating solutions often exhibit a mixture of viscous and elastic behavior, with the response of a particular system depending on the structure of the material and the extent of deformation. For example, polymer melts can be highly elastic if a polymer chain can stretch when subjected to deformation. The behavior of colloidal suspensions is controlled by interparticle forces, the range of which rarely extends more than a particle diameter (see COLLOIDS). Consequently suspensions tend to behave like viscous liquids except at very high particle concentrations when the particles are forced into close proximity. Because many coating solutions consist of complex mixtures of polymer and colloidal material, a thorough characterization of the bulk rheology requires a number of different measurements.

Because the coating process involves the formation and maintenance of interfaces, interfacial or surface rheology must also be considered. The surface tension is a measure of the surface energy per unit area, which depends on the strength of the intermolecular attractive forces. High surface tension fluids have high attractive forces and tend to bead on a surface so as to minimize interfacial energy. In the coating process the solution should flow out and thus the surface tension should be low. Surfactants (qv) are used in aqueous coating solutions to control this phenomenon. Time-dependent or dynamic surface tension reflects the variation in surface tension because of the finite rate of diffusion of the surfactant to the surface. The dynamic surface tension is relevant when the diffusion time scale is comparable to the age of the coating surface under consideration.

An analogous property is the wetting behavior of the solution on a substrate. A drop of liquid when placed on a surface may either wet and flow out or contract to form a stationary drop. The behavior depends on the balance of surface tension, the interfacial tension, and the surface forces of the substrate so that the coating fluid and substrate must be considered as a unit. Even a low surface tension fluid does not coat on a nonwetting substrate such as Teflon. The wetting tendency is measured by the contact angle, where a zero contact angle corresponds to wetting.

To improve wettability a variety of surface treatments, such as flame treatment or corona discharge, are applied to substrates to increase surface energy. Wetting behavior is complicated by the fact that liquids wet stationary and moving surfaces differently. That is, the contact angle measured on a stationary surface is less than from the contact angle measured on a moving surface, and the contact angle increases with surface velocity. Thus experimental techniques are needed to characterize wetting under the conditions of the coating process.

Coating Processes

Many different coating processes are used in the coating industry. Processes herein are classified by the number of layers that the method can apply and whether the method coats a continuous web or discrete surface. Continuous web coating methods include both single layers and multilayers. Dip, rod, knife, blade, air knife, gravure, forward and reverse roll, and extrusion coating methods are all single layer; slide coating methods are multilayer; and slot and curtain coating methods may be either single or multilayer. For discrete surface coating, spray, dip, spin, vacuum, or curtain coating methods can be used.

The control of the coverage or coating weight of the coating and its resulting uniformity is an important characteristic of the coating method. There are two basic classes: premetered, which deliver a set flow rate of solution per unit width to the coated web, and postmetered, in which the coverage is a function of the liquid properties, the system geometry, and the web and roll speeds.

In most coating operations a single layer is coated. When more than one layer must be applied one can make multiple passes, or use tandem coaters where the next layer is applied at another coating station immediately following the dryer section for the previous layer, or a multilayer coating station can be used. Slot, extrusion, slide, and curtain coaters are used to apply multiple layers simultaneously. Slide and curtain coaters can apply an unlimited number of layers simultaneously, whereas slot and extrusion coaters are limited by the complexity of the die internals.

The precision or uniformity of the coating is very important for some products such as photographic or magnetic coatings. Some processes are better suited for precise control of coverage. When properly designed, slot, slide, curtain, gravure, and reverse-roll coaters are able to maintain coverage uniformity to within 2%. In many of the other coating processes the coverage control may be only 10%. Table 1 lists attainable control.

Limits of Coatability. In any coating process there is a maximum coating speed above which coating does not occur. Above this coating speed air is entrained resulting in many bubbles in the coating, or in ribs and finally rivulets, or wet and dry patches. In slot coatings, to coat thinner, below a critical speed often means to coat slower. Above the critical speed the minimum thickness depends only on the gap. Above some higher speed a coating cannot be made (5). Using bead vacuum thinner coats can be obtained. Similar effects were found in slide coating except the critical speed is never reached (6). The maximum coating speed in slide coating for thick coatings where no bead vacuum is used, is identical to the velocity of air entrainment for a tape plunging into a pool of coating fluid.

Lower viscosity liquids can be coated faster and thinner. Polymer solutions can be coated at higher speeds than Newtonian liquids. These phenomena have been explained in terms of a balance of forces acting on the coating bead, ie, the coating liquid in the region where it first makes contact with the web (7). Stabilizing forces are mainly bead vacuum or electrostatic assist, if used. The destabilizing forces are primarily the drag force on the coating liquid and the momentum of the air film carried along by the web. Thus there is a net destabilizing force which is balanced by the cohesive strength of the liquid. Limits of coatability occur in all coating operations but under different conditions in each process.

The air entrainment velocity for plunging tapes does not depend on the wettability of the surface but does increase with surface roughness (8). Presumably the rough surface lets air that would otherwise be entrained escape in the valleys between the peaks that are covered with coating liquid (9). In the converting industry, which involves coatings on rough and porous paper surfaces, much higher (up to 25 m/s) coating speeds can be attained than in photographic coatings on smooth plastic films. A good description of the window of coatability in slot coating can be found in Reference 10.

Knife and Blade Coating.   Knife and blade coating are in many ways similar. In both cases the knife or the blade doctors off excess coating that has been picked up in the applicator pan. Knives are usually held perpendicular to the web, whereas blades are usually tilted toward the incoming web. Typically blades are thin, only 0.2–0.5 mm thick, and can be rigid or flexible (of spring steel). Knives are thicker and are always rigid. Blades, being thinner, wear faster, and have to be changed relatively often, perhaps 2–4 times a day. Blades are always pressed against the web, which is supported by a backing roll either made of chrome plated steel or rubber covered. Knives may also be pressed against unsupported web which is held taut by the drive tension. Knives may also be held at a fixed gap from the supported web on a backing roll.

The ends of the knives can be square, beveled, or rounded. If the end is square and parallel to the web, if the upstream face is perpendicular to the web, and if there is a fixed gap between the end of the knife and the web, then the wet coverage is exactly one-half the gap. On the other hand, if there is a low angle in a converging section of the knife or of the blade, leading up to a tight gap, as there is for many knives and for all bent blades, then strong hydrodynamic forces build up and tend to lift the knife or blade away from the web. This forces more fluid under the knife or blade, so that the coated thickness is greater than half the gap.

In all cases of knife and blade coating, except in knife coating at a fixed gap, a rigid member and a flexible member are pressed together. The flexible member can deflect, to allow for nonuniformities in the web. In knife coating and beveled blade coating, the knife or blade is rigid, and the unsupported web or the web on a rubber-covered roll is flexible. In bent blade coating the blade is flexible, and the web on the roll is rigid or relatively rigid as in the case of a rubber-covered roll. Knife coating against unsupported web is more difficult to control than the other knife and blade-coating techniques, because here the web tension is very important.

The simplest and least expensive, but still effective, coating method is knife coating, either against a backing roll or on unsupported web. Coating against a backing roll is more accurate, as it is independent of web tension. The knife, held perpendicular to the web, acts as a doctor blade and removes excess coating liquid. The coating can be applied by any convenient method, such as with an applicator roll, or by pumping the fluid into a pool formed by the web, the knife, and two end dams. The control of the coverage is by proper positioning of the knife. The unsupported knife shown in Figure 1a is used for coating open fabric webs where coating penetration is desired or cannot be prevented. A full width endless belt can be used to support a weak web and pull it through the knife area without tearing, to overcome the drag of the knife.

The knife-over roll coater, Figure 1b, is probably the most common of the knife coaters. It is simple and compact. The driven back-up roll may be precision made and chrome-plated, having a controlled gap between the web and the knife. The backing roll may also be rubber-covered, the knife pressing against the web. Here the coating weight is determined by the pressure against the knife. Higher pressures give lower coating weights.

Knife coaters can apply high coverages, up to 2.5 mm wet, and can handle high viscosities, up to 100,000 mPa·s (= cP). These tend to level rough surfaces rather than give uniform coverage, a characteristic that can be desirable or not depending on the needs of the finished coating. Streaks and scratches are hard to avoid, especially using high viscosity liquids.

Blade Coating.   Flexible blade coaters can be used with either a downward moving web, as shown in Figure 2a, or with an upward moving web, as shown in Figure 2b, and as with knife coaters, there are many ways of feeding the metering blade. A puddle behind the blade is shown in Figure 2a, a forward turning applicator roll in Figure 2b, and a slot applicator or die fountain in Figure 2c. Jet fountains, where the coating liquid spurts out to the web 25–50 mm away, are occasionally used.

Blade coaters are commonly used on pigmented coatings. They have the unique feature of troweling in the low areas in a paper web, thus producing coated surface that has excellent smoothness and printing qualities. The backing roll is usually covered with resilient material and is driven at the same speed as the web to stabilize the web and draw it past the blade. A replaceable blade is rigidly clamped at one end, and the unsupported end is forced against the substrate. The wet coverage is adjusted by varying blade thickness, the blade angle, and the force

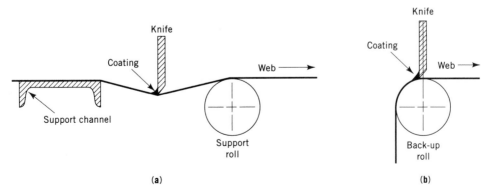

Fig. 1. (**a**) Unsupported knife; (**b**) knife over roll.

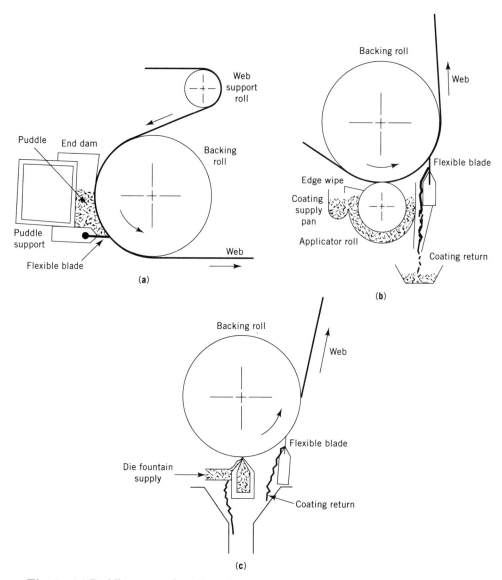

Fig. 2. (**a**) Puddle coater; (**b**) roll applicator blade coater; (**c**) fountain blade coater.

pushing the blade against the substrate. The force on the blade can be obtained by various means: a rubber tube between the blade and a rigid member can be filled with varying air pressures; or the blade holder can be rotated so as to apply a greater or lesser force at the tip, while keeping constant the angle the blade tip makes with the web as in Figure 3.

As the force on the blade increases, and the force is concentrated at the blade tip, the wet coverage decreases rapidly. However, further increases in force bend the blade, and a larger area of the blade presses the liquid against the web. Increasing the loading of the blade causes the tip to lift up and the coverage is then increased. At further increases in load the coverage again decreases. This behavior is shown in Figure 4.

The beveled blade coater uses a rigid blade held at an angle of 40–55° to the web. The end of the blade is parallel to the substrate and pressed against it. If initially the end of the blade is not parallel to the web, it soon is as a result of abrasion by the pigmented fluid. When the loading on the blade increases, the wet coverage decreases. With the same force but using a thicker blade, the pressure, or force per unit area, on the coating fluid between the blade and the web decreases, and the coverage increases.

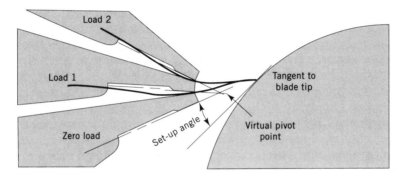

Fig. 3. The Beloit S-matic coating head (2).

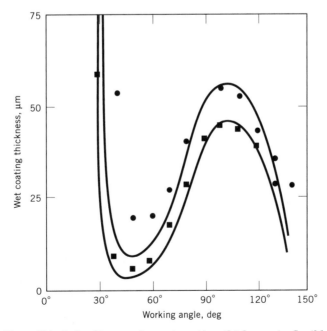

Fig. 4. The effect of blade loading on the wet-coating thickness in flexible blade coating where (●) represents a less stiff blade; (■) represents a stiffer blade. The working angle represents the blade loading (11).

In the rod-blade coater unit a rod is mounted at the end of the blade. This coater behaves more like the beveled blade coater than a flexible blade coater.

Two-Blade Coaters.   In order to coat both sides of a web simultaneously, two flexible blade coaters can be used back-to-back, ie, with both blades pressing against each other and the web between them. The web usually travels vertically upward. Different coatings can be applied on each side of the web. The blades tend to be thinner and more flexible than the standard blades and the angle to the web is lower. The web has to have sufficient tensile strength to be pulled through the nip.

Simultaneous coatings can also be made with one flexible blade against a roll, where the web moves downward and supplies the coating fluid with puddles and edge dams between the web and the blade and between the web and the roll. The roll may rotate faster than the web. Figure 5 shows a version where the fluid on the roll side is supplied by a transfer roll.

Air-Knife Coater.   The air-knife coater is a versatile coating process in use for a wide range of products. A coating pan and roll are used to apply the coating solution and then an air knife is positioned after the pan to regulate the final wet-coating weight by applying a focused jet of air to the web. The excess solution is collected in an overflow pan and can either be recirculated and used again or scrapped. The air knife can function either in the precision or the squeegee mode and give very different types of coating and performance characteristics, although the same name is used for all of these processes.

In the precision mode the air knife uses low pressures and doctors off some of the coating to control coating weight and levels, or maintains the surface to give a uniform coating of reasonable quality. The coating weight is a function of web speed, viscosity of solution, surface tension, and air-knife pressure. This precision mode has been used to coat photographic films where the air velocity is at 13–130 m/min and air-knife pressures are 50–2500 Pa (0.2–10 in. of water) to give 1–200 μm wet thickness.

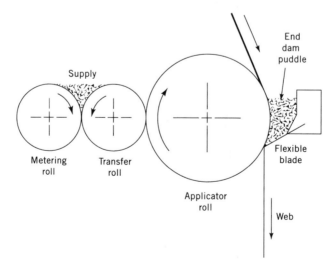

Fig. 5. Billblade with transfer rolls.

In the squeegee mode, the air knife operates at much higher pressures and coating speeds than in the precision mode and effectively doctors off the majority of the coating. This process is used for porous supports such as paper where the coating is absorbed into the voids. After the squeegee the coating solids remain in the voids and in a thin surface layer, effectively functioning as a leveling device.

The advantage of the air-knife processes are low initial cost, versatility for coating a variety of webs and solutions, ease of changing and maintaining coating, and the coating quality. The disadvantages are the noise and contamination problems created by the air stream, solution viscosity limitations, a somewhat restricted coating weight range, and a high operating cost because of the energy to operate the air blowers.

Wire-Wound Rod Coating. The wire-wound rod coater shown in Figure 6 uses a wire-wound rotating rod, called a Mayer rod, to meter off excess applied coating solution. The rod is rotated for two reasons: to increase life by inducing even wear, and to prevent particles from getting caught in the rod and causing streaks. Normal rotation is in the reverse direction to the web travel. The rod wire size controls the coating weight. As the rod has an undulating surface because of the wire, the coating has a similar unevenness, and if the solution does not self level, a smoothing rod must be used to smooth out the surface. Thus rod coaters are best used with low viscosity liquids.

Rod coaters are commonly used for low solids, low viscosity coatings such as those used in poly(vinylidene chloride), carbon paper, and silicone release papers. Coating weights range from 1.5–10 g/m^2, and speeds are as high as 300 m/min. The wire-wound rod can be held against unsupported web, as shown in Figure 6, or against a backing roll. When used against unsupported web the web tension affects the coverage. Coating rods are compact, simple, and inexpensive, but wear rapidly when used with abrasive fluids.

Roll Coating. *Meniscus or Bead-Roll Coater.* One of the simplest coating machines used on coatings that have stringent optical requirements is the meniscus coater. Figure 7 shows an early design. The web is supported by an idler-type back-up roll so that the web is a few micrometers above the level of coating in a constant-level pan. To start coating, the pan is usually raised to make contact with the web and then retracted to form the meniscus, or coating bead. The pan design is very critical. The coating speed is very slow, only about 10 m/min on watery, thin liquids.

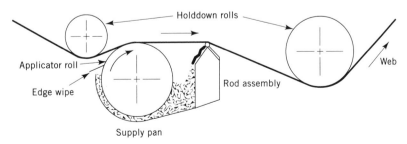

Fig. 6. Metering-rod coater.

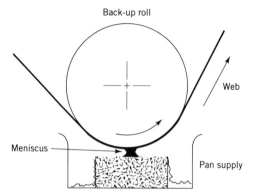

Fig. 7. Pan-type meniscus coater.

Kiss-Roll Coaters. In kiss-roll coating, the web passes over a roll wet with the coating fluid, and has no backing roll. There are many types of kiss-roll coaters. The kiss roll can turn in the direction of the web or in the reverse direction but usually operates in the web direction. Kiss-roll coaters are tension sensitive and are often used to apply excess coating prior to a metering device.

Forward-Roll Coaters. In roll coating the web passes between two rolls rotating in the same direction, one of which applies the coating fluid, as shown in Figure 8 (11). The applicator roll drags fluid into the nip. The fluid exiting the nip splits in two. Some adheres to the web and some to the applicator roll. The flow rate through the nip q is

$$q = \gamma\, GU$$

where G is the gap between the rolls, U is the average surface speed of the two rolls, and γ is approximately equal to 1.3, but varies between 1.29 and 1.33 for equal roll speeds. γ drops to 1.23 when one roll is stationary.

Each roll carries away some of the flow. The ratio of film thickness on the two rolls, t_2/t_1, depends on the speed ratio, and is defined as

$$t_2/t_1 = (U_2/U_1)^{0.65}$$

where the subscript 1 corresponds to the web and the subscript 2 to the applicator roll. The total flow through the gap q is equal to the sum of the flow on the web, $t_1 U_1$, and that on the exit side of the applicator roll, $t_2 U_2$. Thus the wet thickness on the web can be easily calculated.

In forward-roll coating it is fairly common to have an instability called ribbing, where the coating thickness varies sinusoidally across the web, and the coating looks as if a giant comb were dragged down the wet coating. Ribbing occurs when the capillary number, Ca, exceeds a certain value depending on the gap to diameter ratio. The capillary number is defined as

$$Ca = \eta U / \sigma$$

where U is the average surface speed of the two rolls, η is the viscosity of the

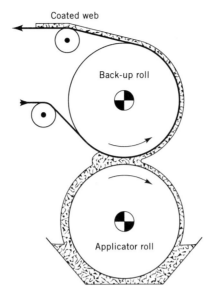

Fig. 8. Two-roll forward roll coater (11).

coating fluid, and σ is the surface tension. It is very difficult to avoid ribbing in forward-roll coating, and if the fluid is not self-leveling a smoothing bar should be used when a smooth coating is required.

Reverse-Roll Coating. Reverse-roll coating is an extremely versatile coating method and can give a very uniform, defect-free coating from 12–1200 μm thick at a very wide range of coating speeds, using coating fluids having from low to extremely high viscosities. In reverse-roll coating, the coating fluid is applied to the applicator roll by any of a number of techniques, such as having the applicator roll sit in a pan of fluid, using a fountain roll, or a fountain or slot die. The excess fluid is then metered off by a reverse-turning metering roll and the remaining fluid is completely transferred to the web traveling in the reverse direction. Two of the many possible configurations are shown in Figure 9.

All the flow remains on the applicator roll after the metering roll is transferred to the web; thus it is important to know what this flow is. The thickness of the metered coating on the applicator roll, t_a, is found to be a function of the gap, of the ratio of the speed of the metering roll to that of the applicator roll, and of the capillary number based on the roll speed (Fig. 10).

In reverse-roll coating, as in forward-roll coating, instabilities can form. However, it is possible to obtain defect-free coatings at high speed and sometimes increasing the speed can lead to a smooth coating when a ribbing condition is present.

Another defect, called cascade or seashore, can also form in reverse-roll coating. This defect is caused by the entrapment of air under certain conditions and can appear in the metered flow on the applicator roll. An operability diagram, showing the region of stable flow as well as the regions where these defects form, is given in Figure 11 for two gaps. The region in the speed ratio where stable coatings can be made is at high capillary numbers, ie, at high speeds. There is

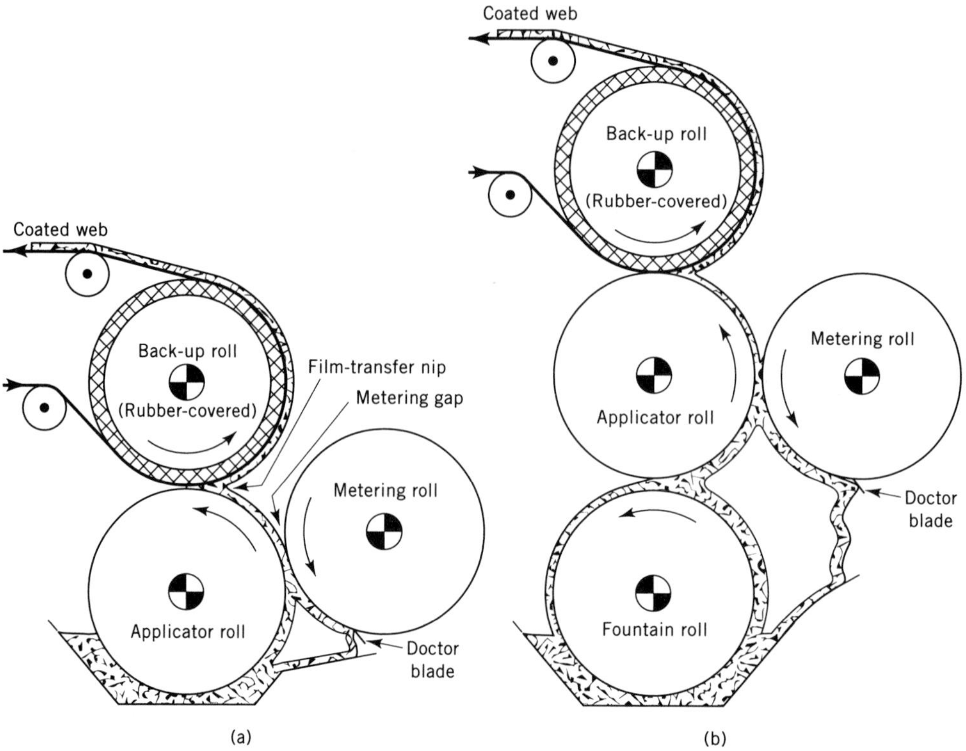

Fig. 9. Pan-fed reverse-roll coaters: (**a**) three roll; (**b**) four roll (11).

also a stable region at very low speeds, but low speeds are not usually desirable. The principal advantage of reverse-roll coating is that conditions can be adjusted to give a stable, defect-free coating at high coating speeds. Using precision bearings, reverse-roll coaters can lay down as uniform a coating as any coating process, about ±2%.

Gravure Coating. Gravure coating is an accurate way of coating thin (1–25 μm wet coverage) layers of low (10–5000 mPa·s (= cP)) viscosity liquids. The coating liquid is picked up by a patterned chrome-plated roll, the excess doctored off, and the liquid transferred from the filled cells to the web. Figure 12 illustrates two types of gravure coaters. In offset gravure the liquid in the cells is transferred to a rubber-covered offset roll before the final transfer to the web. In reverse gravure the gravure roll turns in the reverse direction with respect to the web. In differential gravure the forward rotating gravure cylinder runs at a different speed than the backing or impression roll. The coating liquid can be applied to the gravure roll by a number of methods, not just by the pan-fed system illustrated.

The three common cell patterns for the gravure cylinder are illustrated in Figure 13. The pyramidal and quadrangular cells are similar, except that the quadrangular has a flat, not a pointed, bottom. The trihelical pattern consists of continuous grooves spiraling around the roll, usually at a 45° angle. The volume factor relates to the cell volume per unit area, and has units of height, typically

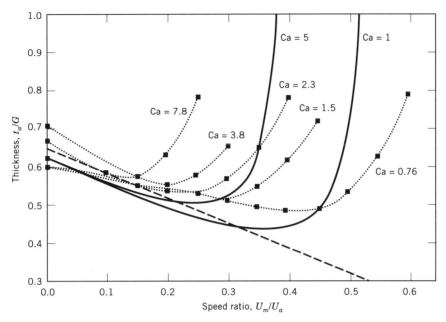

Fig. 10. Reverse-roll, metered film thickness on the applicator roll divided by gap, t_a/G, as a function of the ratio of the metering roll speed, U_n, to applicator roll speed, U_a, for various capillary numbers. (—) represents theoretical values; ($\cdot\cdot\cdot$) experimental ones; and (- - -) is the lubrication model (11).

ranging from 4 to 300 μm. The cell pitch or count is the number of cells per centimeter measured perpendicular to the pattern and usually ranges from 4–160/cm. The pattern is made by mechanical engraving, chemical etching, or electromechanical engraving.

After the gravure cylinder is covered with coating liquid, the excess is doctored off, normally using a 0.1–0.4 mm spring steel blade. Usually the doctor blade makes a 55–70° angle with the incoming gravure roll surface and is oscillated 6–50 mm to give even wear and to dislodge dirt that could cause streaks. A reverse-angle doctor blade can also be used. It often makes an angle of 65–90° with the exiting surface. This blade does not have to be loaded against the cylinder face because fluid forces press the blade against the surface, and so the reverse blade can be made of softer materials, such as bronze or plastic. There is no need to oscillate this blade because in this position it cannot trap dirt; however, the standard blade is felt to do a better job of doctoring.

As for the flexible-blade coater, a softer doctor blade or one having a lower loading and an almost smooth cylinder with a shallow pattern allows excess liquid to pass through. A stiff, highly loaded blade and a cylinder having a large volume factor wipes the surface clean. The gravure roll has to be heavily loaded against the back-up or impression roll in order to achieve good transfer to the web. The usual force is about 2000–20,000 N/m.

The most important factor in determining the transfer or web pickup is the gravure pattern design. The volume factor controls the average coating thickness

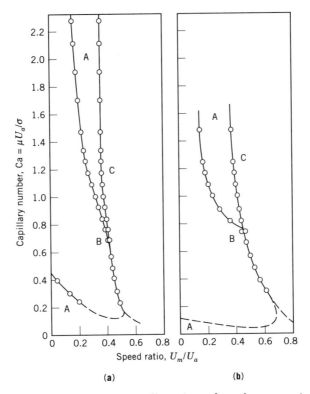

Fig. 11. Operability diagram for reverse-roll coating, where A represents a stable coatings area; B, ribbing; and C, cascade; for (**a**) a gap, $G = 750$ μm and (**b**) a gap, $G = 250$ μm (11).

and normally about 58% of the cell volume transfers. The cell pitch controls the stability of web pickup. The leveling of the coating can be a problem. Large spacing between cells often results in printing of the cell pattern, rather than a uniform coating. Reverse and differential gravure tend to give better leveling. A smoothing bar can also be used.

Dip Coating. Dip coating is one of the oldest coating methods in use. Using continuous webs, the web passes under an applicator roll partially submerged in a pan of the coating fluid. The web is thus actually dipped into the coating solution, and a doctor blade is sometimes used to remove excess fluid, if reduction of the wet coating weight is desired. Otherwise the coverage is determined by the pickup characteristics of the liquid, ie, viscosity and surface tension, the coating speed, and the characteristics of the support.

Dip coating is very commonly used for coating continuous objects that are not flat, such as fibers, and for irregularly shaped discrete objects. Tears or drops of coating at the bottom of dip coated articles may be removed by electrostatic attraction as the article is moved along a conveyor.

Extrusion. Extrusion coating and slot coating are in principle very similar. In extrusion coating a high viscosity material, often a polymer melt, is forced out of the slot of the coating die unto a substrate, where it is cooled to form a contin-

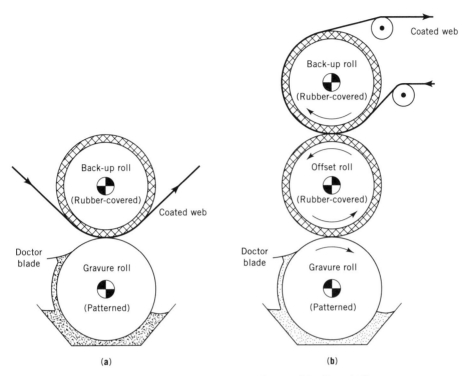

Fig. 12.   Gravure coaters (**a**) direct; (**b**) offset (11).

uous coating. As can be seen in Figure 14**a**, the highly viscous liquid does not wet the lips of the die. Similarly in slot coating, a relatively low viscosity liquid, usually < several thousand mPa·s(= cP), which is often a polymer solution, is forced out of the slot and onto the web. In slot coating the coating liquid does wet the lips of the die, as shown in Figure 14**b**.

Extrusion coating is often used in food packaging (qv) where vapor and oxygen barriers are required and heat sealability is desired. The expanding food packaging industry is the direct result of packaging improvements that can be attained from improving the surface and physical characteristics of a flexible web by extrusion coating (see FILM AND SHEETING MATERIALS).

Because of the high viscosities involved in extrusion coating, the coating die and the auxiliary equipment are massive. An extruder is needed to heat and melt the thermoplastic polymer, the die is heated by electric heaters, and the die also contains adjusting bolts that are sometimes computer controlled, every few centimeters across the width, to try and obtain a uniform cross-web coverage. There is usually also a laminating station to combine the plastic sheet with the substrate and to cool the laminate. The plastic leaves the die at approximately 175°C and is approximately 0.5 mm thick. It is then elongated owing to the pulling effect caused by the pressure nip in the laminating section, and the substrate which is moving at the higher velocity. The elongation effect also reduces the width of the extruded film approximately 2–6 cm and reduces film thickness to approximately 12–25 μm before it makes contact with the substrate.

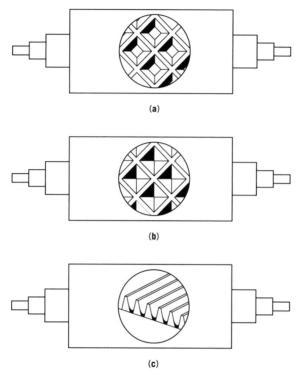

(a)

(b)

(c)

Fig. 13. Common cell patterns in gravure coating: (**a**) quadrangle; (**b**) pyramid; and (**c**) trihelical (2).

Good temperature control of the plastic and pressure control ahead of the coating die is important to the success of the coating. Variations in temperature lead to irregularities in the coating thickness both in the machine direction and across the web. Variations in the cross-web direction can be reduced by adjusting the slot opening via the adjusting bolts. The extruded film width is adjustable by external deckles to block off the exit of the die. In the laminator the nip helps to promote bonding, before chilling the molten plastic. The driven chill roll is chromium or nickel plated and can have a mirror, matte, or an embossed surface. Once the extruded film passes through the laminating nip, it takes on the finish of the chill roll. The chill roll is 60–90 cm in diameter and utilizes refrigerated water to reduce the film temperature to ca 65°C in the 120° wrap of the chill roll before the film is stripped. To improve adhesion of the extruded film to the substrate, adhesion-promoting primers are usually applied to the web before the laminator. Priming can be electrostatic, chemical, or in the form of ozone treatment. Coating weights are controlled by the line and extruder speeds. However, in many cases the chill roll capacity limits the maximum thickness to be obtained.

Extrusion coating lines operate at speeds up to 1000 m/min and can apply 10–30 g/m² of coating. Additionally, multiple layers can be extruded from one die. Multiple extruders feeding one die provide the unique capability of producing layers of different resins to give superior functional properties. For example, an

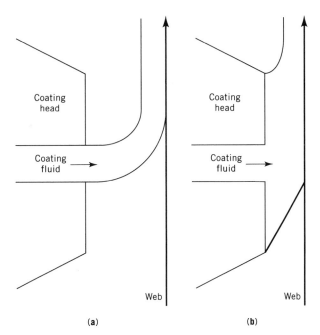

Fig. 14. Comparison of (**a**) extrusion coating and (**b**) slot coating.

inexpensive resin can be used as the core of a three-ply extrudate, the outer plies being more expensive but also much thinner than if extruded alone.

Slot Coating.   Slot coating (Fig. 14**b**), which involves a relatively low viscosity fluid, uses much simpler equipment than extrusion coating. An ordinary pump or a pressurized vessel feeds the fluid through a flow meter and control valve to the coating die, which often operates at room temperature. If heating is required, water flowing through internal channels is usually adequate, and electrical heaters are not needed. Because of the simple rheologies of the fluids, the die can be designed to give uniform flow across the width with no adjustments, and so the die is much less massive. In fact, adjusting the slot opening should be avoided. It is not difficult to design the die to give uniform flow, but it is very difficult to make the exact correct adjustments. Because the viscosity is relatively low, the pressures within the die are also, and so the die can be much less massive and still withstand the spreading forces.

Normally the web is supported by the backing roll in slot coating. However, for very thin (< ca 15 μm) coatings, the gap between the coating lips and the web becomes very tight, ie, under about 100 μm, and the system becomes difficult to control and operate. The run out of the bearings can become a significant fraction of the gap. Dirt can hang up in the gap to cause streaks. If the web contacts the coating die the web can tear, causing a shutdown of the operation. For thin coatings unsupported web can be used, however, then web tension becomes an important variable.

Multiple layers can be coated simultaneously from one slot coating die, with the layers coming together internally and flowing out through a single slot. Multiple slots can also exit onto one set of lips, but this arrangement does not seem as satisfactory as using a single slot. Multiple layers in slot coating works well, but the die internals are complicated.

In slot coating, bead vacuum is often used to increase the window of coatability, that is to allow thinner coatings and also to allow coatings at higher speeds. A vacuum box is placed under the coating die, and a vacuum of up to about 1000 Pa (4 in. of water) is pulled by a vacuum fan. There should be a tight vacuum seal against the sides or ends of the rotating backing roll, but no rubbing contact where the web enters the vacuum chamber in order to prevent scratches. The air in the vacuum chamber can resonate as in a musical instrument, so that the air leakage should be kept as small as possible thus reducing the amplitude of the resulting pressure fluctuations. Otherwise chatter occurs at wider gaps.

Multilayer Methods

Slide Coating. A slide coater, illustrated in Figure 15, can coat an unlimited number of layers simultaneously. Each layer flows out onto the slide yet does not mix with the other layers as they all flow together down the slide, across the gap, and onto the web. Slide coating is extensively used in coating photographic films and papers, both color and black and white. In color films, nine or more layers are coated simultaneously.

Instabilities can form on the slide in the form of interfacial waves, which would disturb the desired laminar flow and cause mixing. The closer the physical properties of all the layers are, the closer the system is to being single layer and

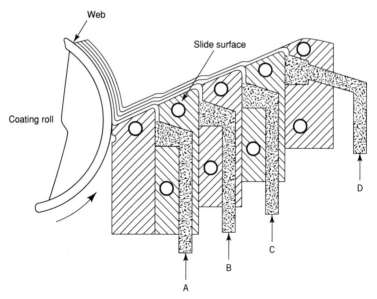

Fig. 15. A slide coater where A, B, C, and D correspond to the inlets for the liquids for layer 1, 2, 3, and the top layer, respectively (7,12).

internal waves do not form in a single-layer system. The layer densities are always reasonably close to each other, although densities are usually not subject to control. The viscosities of adjacent layers should generally not vary by more than a factor of three, except for the bottom layer, which should be of a low viscosity to reduce drag forces and allow higher coating speeds.

As for slot coating, a slight vacuum (up to 1 kPa) under the coating bead aids in coating by allowing thinner coatings and higher coating speeds. The bottom edge of the slide should be sharp and have a tight radius of curvature of no more than about 50 μm, to pin the bottom meniscus and reduce the chance of cross-web barring or chatter.

Curtain Coating. In the converting industry curtain coating is used to deliver coating liquid in a falling sheet or curtain to the substrate, which moves through the curtain at the coating speed. In one version a slot coating head is aimed downward and the coating emerges as a falling film or sheet as seen in Figure 16. The curtain thickness is controlled by the feed rate and by precise adjustments of the slot opening. The vertical distance of the coating head above the substrate can be adjusted. The falling curtain is protected from stray air movements by transparent enclosure sheets. Coating thicknesses as low as 12 μm are possible when coating with lacquers or with low viscosity wax melts, and are as low as 20 μm with hot-melt compositions of higher viscosity. There is no problem in obtaining heavier coating coverages.

Air-bubble entrapment may occur in the case of a gravity-applied continuous coating over an impermeable substrate. Bubbles may also be caused by moisture vaporization from the substrate. Remelting of the coating may minimize the bubble defects. Curtain coating equipment of this design is capable of operation at substrate speeds up to 500 m/min.

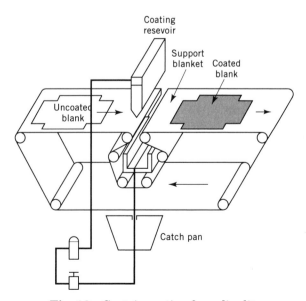

Fig. 16. Curtain coating from die slit.

Curtain-coating equipment is also available in which the falling curtain is generated by overflow from an open weir. The coating is delivered to the open weir uniformly across its width by a pipe having diffuser jet openings. As the coating overflows the low side of the weir, it travels down a short flat skirt before dropping. The thickness of the falling curtain is adjusted by precise control of the rate of delivery of coating to the weir. Hot-melt coatings can also be applied by the open weir, as for the slot die. Because there is no close restriction to flow as in the slot die, the open weir does not tend to form scratches or coating streaks because of crusting or coating hang-up in the slot opening. When applying hot-melt coating formulations the coating supply is held in a reservoir at a temperature that does not thermally degrade the material during its residence. The coating is brought to temperature using heat exchangers as it is pumped to the weir. Weir-type equipment is recommended for operation at substrate speeds up to 400 m/min. The coating fluid not carried away on the coated surface falls into a collection trough for recirculation.

Curtain coating is adaptable for coating irregularly sized sheets such as slotted cut-out corrugated carton blanks or sheets of plywood, as well as for continuous substrates. Coatings may also be applied to uneven geometric shapes such as blocks. The principal limitation of curtain coating is that a high flow rate of about 0.5 cm^3/(s·cm) width is needed to maintain an intact curtain. Usually about double this minimum is desirable. Thus to obtain a thin coating, high coating speeds are required. Curtain coating is inherently a high speed process.

For precision coating the curtain has to be completely uniform across the width. Precision multilayer curtain coating is used to coat color photographic materials. This is illustrated in Figure 17, where the liquid layers flow down a slide over the edge to form the curtain. In most precision coatings the curtain is narrower than the web. Edge guides are used to prevent the curtain from necking in.

Discrete Surface-Coating Methods

Coatings may be applied by spraying on irregularly shaped and compound curved or sharp-edged surfaces (see COATING PROCESSES, SPRAY COATINGS). Many coat-

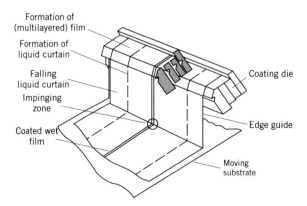

Fig. 17.   Curtain coating apparatus (13).

ing materials of suitable dielectric constant may be electrically polarized so that the powders are attracted to a grounded or oppositely charged surface (see COATING PROCESSES, POWDER TECHNOLOGY).

Dip and Spin Coating. The dip coating technique described for webs can also be used to coat discrete surfaces such as toys and automotive parts. The surface to be coated is suspended on a conveyor and the part dipped into the coating solution. The surface is then removed, the coating drains, then levels to give the desired coverage. The object is then dried or cured in an oven.

The spin coating process is used to produce a thin uniform coating on discrete supports. In this process a colloidal suspension is placed on a substrate, which rests on a driven platform. The speed of the platform, which can be as high as 10,000 rpm, is increased to the desired coating speed. Centrifugal force then forces much of the coating off the support leaving a thin, uniform film behind. In addition, the coating is drying during the process and as a result the viscosity increases, resistance to flow occurs, and a level thin coating is left. The coating chamber can also be conditioned to provide hot air to the coating to dry or cure the remaining film. Subsequent coatings of different coating materials can then be applied to develop a multilayer structure. This process is used to coat structures such as photomasks, magnetic disks, optical coatings, and a variety of layered products in the microelectronics industry.

Vacuum Deposition Techniques. Thin coatings are applied to a variety of substrates for use on semiconductors, ceramics, and electrooptical devices using a wide variety of vacuum deposition techniques. Vacuum deposition is a rapidly advancing area of coating technology. In these processes the support to be coated is placed in a vacuum chamber, which contains the coating material. Typically the source is a metal such as gold or tungsten. A vacuum is then drawn, electrical energy is applied to the source, and the metal evaporates or sputters off in the vacuum. The metal coats or plates out on the surface creating a coated support. Individual supports such as a target to be examined in a scanning electron microscope (see MICROSCOPY), or a continuous system on a running web, such as in metallized polyethylene terephthalate, can be used. The coatings can be continuous or circuits can be made if the support is masked.

There are several vacuum processes such as physical vapor deposition (PVD) and chemical vapor deposition (CVD), sputtering, and anodic vacuum arc deposition. Materials other than metals, ie, tetraethylorthosilicate, silane, and titanium aluminum nitride, can also be applied.

Coating Process Mechanisms

One of the principal advances in the coating process area in the 1980s was the development of techniques to understand and define basic coatings mechanisms. This has led to improved quality and a wider range of utility for most coating techniques. Advances have been in the computer modeling of the coating process and the development of visualization techniques to actually see the flows in the coating process. The flow patterns predicted by the computer models can be verified by the visualization techniques.

Free surfaces and interfaces make the physics of coating flow systems extremely difficult to handle by classical mathematical methods. As a result, coater designs and parameter ranges of defect-free coating have traditionally been determined through expensive and time-consuming statistical experimentation. Therefore, coating developed largely as an art rather than a science and is based on pieces of information. The ability to model the coating process by using modern methods of numerical and functional analysis, and to explain many of the complex mechanisms of coating instabilities and the resulting defects, is thus refreshing.

The most successful models are based on the finite element method. The flow is discretized into small subregions (elements) and mass and force balances are applied in each. The result is a large system of equations, the solution of which usually gives the speed of the coating liquid in each element, pressure, and the location of the unknown free surfaces. The smaller the elements, the more the equations which are often in the range of 10,000 to upward of 100,000.

It is now possible to simulate steady transversely uniform flows of Newtonian or non-Newtonian liquids by using commercially available software packages such as FIDAP, FLUENT, PHOENICS, and POLYFLOW. Using these codes, it is possible to locate regions of flow recirculation that may cause coating defects as a result of the increased residence time of solution. The free-surface handling capabilities of currently available commercial codes are limited to relatively simple steady flows and transient response to specified transversely uniform disturbances. For a steady uniform flow to exist in nature, however, it should have the tendency to recover from all small disturbances, eg, building vibrations and even molecular fluctuations, that are always present; ie, the flow should be stable. Flow instabilities resulting in defects such as vacuum streaks cannot be predicted by commercial software. Using more advanced methods developed first at the University of Minnesota and more recently at Du Pont (14,15), it is now possible to predict most coating flow instabilities including bead break-up, flooding, crossweb barring (chatter), down-web ribbing (vacuum streaks), and diagonal chatter. It is also possible to follow the long time development of the resulting defects and to explore parameter ranges of stable, defect-free operation (coating windows) at a fraction of the cost of the actual physical experiment. In a computer model the geometry can be quickly changed without having to construct expensive new parts. Therefore, considerable time and cost savings can be realized by optimizing coating systems computationally.

Experimental techniques to visualize flows have been extensively used to define fluid flow in pipes and air flow over lift and control surface of airplanes. More recently this technology has been applied to the coating process and it is now possible to visualize the flow patterns (16,17). The dimensions of the flow field are small, and the flow patterns both along the flow and inside the flow are important. Specialized techniques such as utilizing small hydrogen bubbles, dye injection, and optional sectioning, are required to visualize these flows.

A stereo zoom microscope having a very small depth of field and a clear window on the side of the applicator is employed. The hydrogen bubbles are generated by fine electrodes in the fluid and the dye is injected at an appropriate point near the region of interest. Accurate control of the light level and position is needed to avoid reflections that mask the details of the flow field. The microscope is focused at varying points and flow nonuniformities are recorded. Using this technique, the air entrainment, flow recirculation in the bead, curtain coating formation, and the teapot effect have all been visualized for the many types of slide and curtain coating. This information leads to an improved understanding of the coating process. The same technology can also be applied to many other types of coating.

Drying

The drying (qv) process after the application of the coating is as important as the coating process itself. Coatings applied in the fluid state are not complete until dried. The coated film or web is transported through the dryer where the properties of the coating can either be enhanced or deteriorated by the drying process. Drying of coatings involves the removal of the inert inactive solvent (vehicle) used to suspend, dissolve, or disperse the active ingredients of the coating, ie, the polymer, binder, pigments, dyes, slip agents, hardener, coating aids, etc. After drying, only the desired coating on the substrate should be retained. Coating solvents can range from the easy to handle water to flammable and toxic organic materials. The drying process is a thermal one in which the solvent evaporates. Drying must occur without adversely effecting the coating formulation, maintaining the desired physical uniformity of the coating.

Whereas drying can be a physical process involving only solvent removal, chemical reactions can also be used to help solidify the coating. Solidification, or cross-linking can be accelerated by catalysts or be accomplished by an electron beam or uv radiation. When this happens, the process is called curing. Practically, most coatings undergo both drying and some form of curing in the dryer and the distinction is in the relative degree. An example is aqueous gelatin coatings, which cure through aldehyde cross-linkers. The cross-linking starts in the dryer.

The functions of the dryer are to provide a source of heat to volatilize the solvent and a means to carry the solvent away from the coating. Efficient hardware to minimize energy costs that are equipped with the appropriate pollution abatement devices to meet OSHA standards are used. Dryers that use hot air to provide heat and carry away the solvent are the most common. Heat sources for drying are generally steam, oil, radiant electric, and gas. Selection depends on availability of supply, the temperature range desired, and costs. Dryers can also use other sources of energy such as microwaves or radio-frequency waves. However, a medium such as air is still needed to carry off the evaporating solvent and the other sources of heat tend to be more expensive and are only used in special circumstances. If the coating can react with oxygen or if the solvents are flammable, inert gases can be used in place of the air.

Because the evaporation of the solvent is an endothermic process, heat must be supplied to the system, either through conduction, convection, radiation, or a combination of these methods. The total energy flux into a unit area of coating, q_t, is the sum of the fluxes resulting from conduction, convection, and radiation (see HEAT EXCHANGE TECHNOLOGY, HEAT TRANSFER).

$$q_t = q_{\text{Conduction}} + q_{\text{Convection}} + q_{\text{Radiation}} \tag{1}$$

Whereas conductive and radiative heating are useful techniques for some applications, convective heating is by far the most common means of supplying the energy needed to evaporate the solvent, because convection is the only means of heating that also provides a means of transporting solvent vapor away from the surface of the coating.

The rate of convective heating can be estimated

$$q_{\text{Convection}} = q_t = hA\Delta t \tag{2}$$

and the solvent evaporation rate is q_t/λ, where q_t = rate of heat transfer in W; λ = heat of evaporation of solvent in J/kg; h = heat-transfer coefficient in W/(m²·K); and $\Delta t = t_1 - t_2$, ie, the temperature difference between web and air supply. This is the driving force for evaporation and t_1 = drying air temperature in K; t_2 = temperature of material being dried in K; and A = area in m².

Equation 2 defines the basic mode of operation of the dryer. The heat-transfer coefficient is a key property of the dryer configuration.

Air Impingement Dryers. Air impingement dryers, the most widely used for coated webs, basically consist of a heat source and heat exchangers, fans to move the air, nozzles or air delivery devices positioned close to the web, and solvent removal ducts. If the air is recirculated, a means to remove the solvent from the air is also provided. Figure 18 shows a typical dryer. In addition, there are devices to control the temperature and velocity of air being delivered and in some cases to limit the solvent concentration well below the lower explosive unit. The dryer can have separate sections or zones so that varied time-temperature histories for a coating can be obtained.

The dryer must also transport the web through the dryer. This can be accomplished by combinations of driven and idler rolls. The web path can be either horizontal or vertical, or, with the appropriate web-turning devices, fold back upon itself to conserve space. The idler rolls in single-sided dryers are spaced so that there is enough wrap for the rolls to turn, and so as to keep the coating within the effective area of the nozzles. The rolls can also be driven using tendency drives, which have two sets of bearings, with the axle being driven and the roll idling on the axle; thus very little force is needed to keep the roll at the speed of the web. Tendency driers are needed for light webs which do not generate enough wrap to turn idler rolls. Typically the web should be within 6–7 nozzle slot widths of the impingement nozzles.

In these single-sided dryers, the air impinges only one of the coated sides, heating and drying from the coated side only. The air can be delivered to the web from plenums with slots, from plenums with holes, or from specially designed nozzles. There are many possible configurations which can easily be adapted to particular process requirements. In the most basic of the configurations, the nozzles and idler rolls are contained in insulated boxes to minimize heat losses, solvent escape, and noise. The efficiency of the dryers depends on the air velocity, the temperatures used, the solvent level in the air, and the heat-transfer coefficient which is controlled by the nozzle geometry, the distance from the web, and the velocity of the air. Wet coatings are fluid, however, and the air should not disturb the coating.

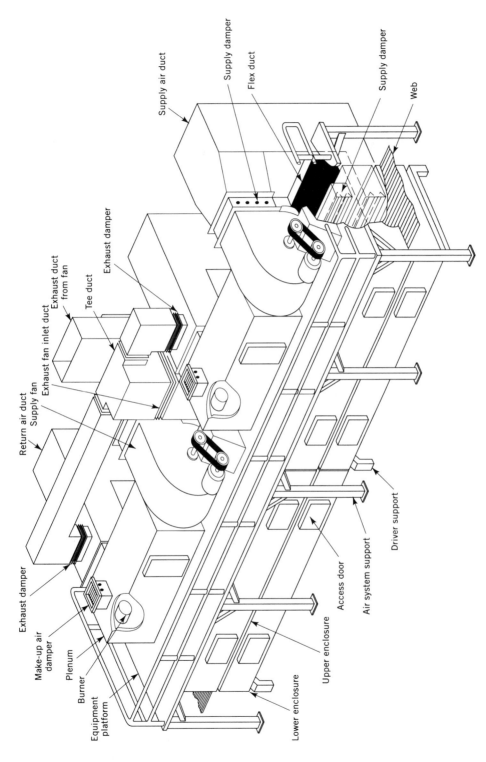

Fig. 18. Dryer components of a top mounted air system TEC Systems F00056 dryer. Courtesy of TEC Systems, W. R. Grace and Co.

631

The two-sided or floater dryer is the most modern dryer design. In this configuration the roll transport system in the dryer is replaced with air nozzles on the rear of the web so that the air transports and supports the web as well as heating and drying the web from both sides. When impervious webs are used, the drying is only from the coated side. The two-sided heat transfer results in higher drying rates and permits higher coating speeds using the same dryer length, while eliminating problems from rolls. This process does, however, use more air.

Two types of floater nozzles are currently in use and they are based on two different principles. The Bernoulli principle is used in the airfoil flotation nozzles, in which the air flows from the nozzle parallel to the web and the high velocities create a reduced pressure, which attracts the web while keeping the web from touching the nozzles. The Coanda effect is used to create a flotation nozzle when the air is focused and thus a pressure pad is created to support the web as shown in Figure 19.

Air flotation dryers have excellent heat-transfer coefficients, give very uniform drying across the web, and give excellent web stability. These dryers, which can be used for a wide range of web types and tensions, also tend to be quieter and thus pose less noise problems than higher velocity single-sided dryers. Floater dryers are totally enclosed and compact so that they are clean and cause less dirt defects in the coating. Air dryers can also be coupled with radio frequency or infrared units to give even higher drying rates. Often these units are added to existing dryers because no additional dryer length is required.

Single-sided dryer air velocities can run from 150–600 m/s and have heat-transfer coefficients from 30–140 W/(m²·K). Floater dryers operate at slightly

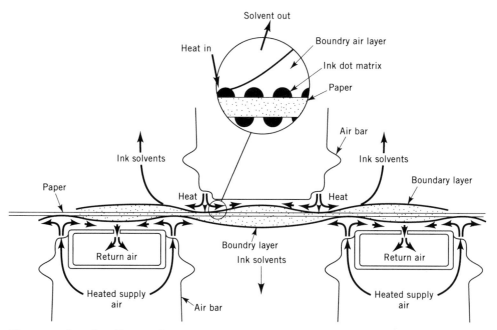

Fig. 19. Coanda effect air bars in the TEC F00029 system. Courtesy of TEC Systems, W. R. Grace and Co.

lower (150–500 m/s) slot velocities and have higher (50–275 W/(m².K) heat-transfer coefficients.

Contact Dryers.   Coatings on webs as well as sheets of newly formed paper can be dried by contacting the web around the surface of a heated drum. Thus conduction is used to transfer the heat. Drums can also be used to rapidly cool warm extruded films, to increase the viscosity, and solidify the film.

Convection Drying Modeling.   Models of the drying process have been developed to estimate whether a particular coating can dry under the conditions of an available dryer. These models can be run on desktop personal computers (see COMPUTER TECHNOLOGY).

To model convection drying both the heat transfer to the coated web and the mass transfer (qv) from the coating must be considered. The heat-transfer coefficient can be taken as proportional to the 0.78 power of the air velocity or to the 0.39 power of the pressure difference between the air in the plenum and the ambient pressure at the coating. The improvement in heat-transfer coefficients in dryers since the 1900s is shown in Figure 20. The mass-transfer coefficient for solvent to the air stream is proportional to the heat-transfer coefficient and is related to it by the Chilton-Colburn analogy

$$\frac{h}{\rho C_p k_m} = N_{\mathrm{Le}}^{2/3} \tag{3}$$

where h is the heat transfer-coefficient in W/(m².K); k_m is the mass-transfer coefficient, kg/(s·m²·(kg/m³)); C_p is the heat capacity of the air, J/(kg·K); ρ is the density of the air, kg/m³; and N_{Le} is the Lewis number, equal to the thermal diffusivity of the air divided by the mass diffusivity of solvent vapors in the air.

Using this information, the heat transfer to the coated web can be equated to the heat consumed by the evaporating solvent, allowing calculation of the tem-

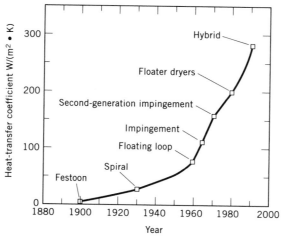

Fig. 20.   Dryer efficiency improvement over time (19). To convert W/(m².K) to Btu/(h·ft².°F), multiply by 0.176.

perature of the coated web (18). These equations can also be used to calculate the psychrometric chart for water or for any solvent. For single-sided drying, the web temperature is the wet-bulb temperature of the air for that particular solvent. This relationship holds for the constant rate period of drying, where the coating behaves as if it were a pool of solvent. When dry patches appear on the surface, the rate of drying decreases, and the falling rate period begins.

Modeling the falling rate period is more difficult, because the drying rate then depends on the mechanisms occurring within the coating. In coating on impervious webs the rate-limiting process is diffusion; in coatings on porous paper it may be capillary action. In aqueous coatings most of the drying occurs in the constant rate period; for organic solvent systems most of the drying occurs in the falling rate period, to the extent that in some cases the constant rate period is over before the coated web enters the dryer.

The higher the air temperature the more rapid the drying. However, there are temperature limitations both for the web and for the coating. Plastic films should not be heated above their glass-transition temperature (the softening point) to prevent distortion and stretching. Many photographic coatings, for example, should not be heated above about 50°C.

Commercial Availability

All of the many types of coaters and dryers discussed herein are commercially available from many different vendors. These vendors usually have pilot facilities so that new coating and drying techniques can be easily tested. Contract coating companies, specializing only in coating, also exist.

BIBLIOGRAPHY

"Coating Processes," in *ECT* 3rd ed., Vol. 6, pp. 386–426, by S. C. Zink.

1. E. D. Cohen and E. B. Gutoff, eds., *Modern Coating and Drying Technology*, VCH Publishers, New York, 1992, pp. 64, 67, 83, 89, 92, 105, 120.
2. D. Satas, ed., *Web Processing and Converting Technology and Equipment*, Van Nostrand Reinhold, New York, 1984, pp. 32, 44.
3. D. Satas, ed., *Coatings Technology Handbook*, Marcel Dekker, Inc., New York, 1991.
4. H. L. Weiss, *Coating and Laminating Machines*, Converting Technology Co., Milwaukee, Wis., 1977.
5. K. Y. Lee, L. D. Lu, and J. T. Liu, *Chem. Eng. Sci.* **47,** 1703–1713 (1992).
6. E. B. Gutoff and C. E. Kendrick, *AIChE J.* **33,** 141–145 (1987).
7. E. B. Gutoff, "Premetered Coating," in Ref. 1.
8. R. A. Buonopane, E. B. Gutoff, and M. M. I. Rimore. *AIChE J.* **32,** 682–82 (1986).
9. L. E. Scriven, "Air Entrainment," presented at *AIChE Spring Meeting,* Atlanta, Ga., Mar. 1984.
10. L. E. Sartor, "Slot Coating: Fluid Mechanics and Die Design," Ph.D. dissertation, University of Minnesota, 1990.
11. D. J. Coyle, "Roll Coating," in Ref. 1.
12. U.S. Pat. 2,761,410 (Sept. 4, 1956), J. A. Mercier and co-workers (to Eastman Kodak).

13. S. F. Kistler and L. E. Scriven, "Finite Element Analysis of Dynamic Wetting for Curtain Coating at High Capillary Numbers," presented at *AIChE Spring Meeting*, Orlando, Fla., Mar. 1982.
14. K. N. Christodoulou and L. E. Scriven, *J. Sci. Comput.* **3**, 355–406 (1988).
15. K. N. Christodoulou, D. F. Scofield, and L. E. Scriven, "Stability of Multilayer Coating Flows to Three-Dimensional Disturbances," paper 96e, presented at the *AIChE Spring National Meeting*, Orlando, Fla., Mar. 18–22, 1990.
16. P. M. Schweizer, *J. Fluid Mech.* **193**, 285–302 (1988).
17. L. E. Scriven and W. J. Suzynski, *Chem. Eng. Prog.* **86**(9), 24–29 (Sept. 1990).
18. E. B. Gutoff, *Chem. Eng. Prog.* **87**(1), 73–79 (Feb. 1991).
19. E. D. Cohen, "Thin Film Drying," in Ref. 1.

EDWARD D. COHEN
E. I. du Pont de Nemours & Co., Inc.
EDGAR B. GUTOFF
Consultant

POWDER TECHNOLOGY

The fluidized-bed coating process was discovered and developed in the mid-1950s (1). The first basic U.S. patent was issued in 1958 (2) and exclusive rights to the process in North America were licensed to The Polymer Corp., Reading, Pennsylvania. Polymer's coating business was subsequently acquired in 1986 by Morton International. Acceptance of this coating process was rather slow. In 1960, the annual sales of coating powders in the United States was below 450 metric tons in part because of a lack of expertise in the methodology. In addition, the available powder coating materials were expensive, efficient production techniques had not been worked out, and volume of production was low.

In the fluidized-bed coating process, the coating powder is placed in a container having a porous plate as its base. Air is passed through the plate causing the powder to expand in volume and fluidize. In this state, the powder possesses some of the characteristics of a fluid. The part to be coated, which is usually metallic, is heated in an oven to a temperature above the melting point of the powder and dipped into the fluidized bed where the particles melt on the surface of the hot metal to form a coating. Using this process, it is possible to apply coatings ranging in thickness from ca 250 to 2500 μm. It is difficult to obtain coatings thinner than 250 μm and, therefore, fluidized-bed applied coatings are generally referred to as thick-film coatings, differentiating them from most conventional thin-film coatings applied from solution or as a powder at thicknesses of 20–75 μm (see FILM DEPOSITION TECHNIQUES; THIN FILMS).

In the electrostatic spray process, the coating powder is dispersed in an air stream and passed through a corona discharge field where the particles acquire an electrostatic charge. The charged particles are attracted to and deposited on the grounded object to be coated. The object, usually metallic and at room temperature, is then placed in an oven where the powder melts and forms a coating. Using this process it is possible to apply thin-film coatings comparable in thickness to conventional solution coatings, ie, 20–75 μm. A hybrid process based on a combination of high voltage electrostatic charging and fluidized-bed application techniques (electrostatic fluidized bed) has evolved as well as triboelectric spray application methods (see COATING PROCESSES, SPRAY COATINGS).

Powder coating processes are considered to be fusion-coating processes, that is, at some time in the coating process the powder particles must be fused or melted. Although this is usually carried out in a convection oven, infrared, resistance, and induction heating methods also have been used. Therefore, with minor exceptions, fixed coating installations are required, excluding the use of powder coatings in maintenance applications. Additionally the substrate must be able to withstand the temperatures required for melting and curing the polymeric powder, limiting powder coating methods to metal, ceramic, and glass (qv) substrates for the most part, although some plastics have been powder coated successfully.

Compared to other coating methods, powder technology offers a number of significant advantages. These coatings are essentially 100% nonvolatile, ie, no solvents or other pollutants are given off during application or curing. They are ready to use, ie, no thinning or dilution is required. Additionally, they are easily applied by unskilled operators and automatic systems because they do not run, drip, or sag as do liquid coatings. The reject rate is low, the finish tougher and more abrasion resistant, than that of most conventional paints (see PAINT). Thicker films provide electrical insulation, corrosion protection, and other functional properties. Powder coatings cover sharp edges for better corrosion protection. The coatings material is well utilized: overspray can be collected and reapplied. No solvent storage, solvent dry off oven, or mixing room are required. Air from spray booths is filtered and returned to the room rather than exhausted to the outside. Moreover, less air from the baking oven is exhausted to the outside thus saving energy. Finally, there is no disposal problem because there is no sludge from the spray booth wash system. Whereas the terms coating powder and powder coating are sometimes used interchangeably, herein the term coating powder refers to the coating material and powder coating to the process and the applied film.

Coating powders are frequently separated into decorative and functional grades. Decorative grades are generally finer in particle size, and color and appearance are important. They are applied to a cold substrate using electrostatic techniques at a relatively low film thickness, eg, 20–75 μm. Functional grades are usually applied in heavier films, eg, 250–2500 μm using fluidized-bed, flocking, or electrostatic spray coating techniques to preheated parts. Corrosion resistance and electrical, mechanical, and other functional properties are more important in functional coatings (see CORROSION AND CORROSION CONTROL).

Coating powders are based on both thermoplastic and thermosetting resins. For use as a powder coating, a resin should possess: low melt viscosity, which affords a smooth continuous film; good adhesion to the substrate; good physical properties when properly cured, eg, high toughness and impact resistance; light color, which permits pigmentation in white and pastel shades; good heat and chemical resistance; and good weathering resistance. The resin should remain stable on storage at 25°C for at least six months and should possess a sufficiently high glass-transition temperature, T_g, so as to resist sintering on storage.

The volume of thermosetting powders sold exceeds that of thermoplastics by a wide margin. Thermoplastic resins are almost synonymous with fluidized-bed applied thick-film functional coatings whereas thermosetting powders are used almost exclusively in electrostatic spray processes and applied as thin-film decorative coatings.

Thermoplastic resins have a melt viscosity that is at least an order of magnitude higher than thermosetting resins at normal application temperatures. It is, therefore, difficult to pigment thermoplastic resins sufficiently to obtain complete hiding in thin films, yet have sufficient flow to give a smooth coating. In addition, thermoplastic resins are much more difficult to grind than thermosetting resins and grinding is usually carried out under cryogenic conditions. Because powders designed for electrostatic spraying generally have a maximum particle size of about 75 μm (200 mesh), the thermoplastic powders are predominant in the fluidized-bed coating process where heavier coatings are applied and the particle size of the powder is coarser. Fluidized-bed powders typically contain only about 10–15% of particles below 44 μm (325 mesh), whereas the high end of the particle-size distribution ranges up to about 200 μm (70 mesh). Most thermoplastic coating powders require a primer to obtain good adhesion and priming is a separate operation that requires time, labor, and equipment. In automotive applications, some parts are primed by electrocoating. Primers are not usually required for thermosetting powder coatings.

Thermoplastic resins have one advantage over thermosetting resins: they do not require a cure and the only heating necessary is that required to complete melting or fusion of the powder particles. Thermoplastic resins have applications in coating wire, fencing, and other applications where the process involves continuous coating at high line speeds.

Thermoplastic Coating Powders

As a coating powder, a thermoplastic resin must melt and flow at the application temperature without significant degradation. Attempts to improve the melt flow characteristics of a polymer by lowering the molecular weight and plasticizing or blending with a compatible resin of lower molecular weight can result in poor impact resistance or a soft film in the applied coating. Attempts to improve the melt flow by increasing the application temperature are limited by the heat stability of the polymer. If the application temperature is too high, the coating shows a significant color change or evidence of heat degradation. Most thermoplastic powder coatings are applied between 200 and 300°C, well above the generally considered upper temperature limits, but the application time is usually ≤5 min. The principal polymer types are based on plasticized poly(vinyl chloride) (PVC) [9002-86-27], polyamides (qv), and other specialty thermoplastics. Typical properties of coating powders are given in Table 1.

PVC Coatings.   All PVC powder coatings are plasticized formulations (see VINYL POLYMERS, VINYL CHLORIDE AND POLY(VINYL CHLORIDE)). Without plasticizers (qv), PVC resin is too high in melt viscosity and does not flow sufficiently under the influence of heat to form a continuous film. Suspension and bulk polymerized PVC homopolymer resins are used almost exclusively because vinyl chloride–vinyl acetate and other copolymer resins have insufficient heat stability. A typical melt-mixed PVC power coating formulation is given in Table 2. The

Table 1. Physical and Coating Properties of Thermoplastic Powders[a]

Property	Vinyls	Poly-amides	Poly-ethylene	Polypro-pylene	PVDF[b]
melting point, °C	130–150	186	120–130	165–170	170
preheat/postheat temperatures, °C[c]	290–230	310–250	230–200	250–220	230–250
specific gravity	1.20–1.35	1.01–1.15	0.91–1.00	0.90–1.02	1.75–1.90
adhesion[d]	G–E	E	G	G–E	G
surface appearance[e]	smooth	smooth	OP	smooth	sl OP
gloss, Gardner 60° meter	40–90	20–95	60–80	60–80	60–80
hardness, Shore D	30–55	70–80	30–50	40–60	70–80
resistance[d,f]					
impact	E	E	G–E	G	G
salt spray	G	E	F–G	G	G
weathering	G	G	P	P	E
humidity	E	E	G	E	G
acid[g]	E	F	E	E	E
alkali[g]	E	E	E	E	G
solvent[g]	F	E	G	E	G–E

[a]All powders require a primer and pass the flexibility test, which means no cracking under a 3-mm dia mandrel bend.
[b]Poly(vinylidene fluoride).
[c]Typical ranges.
[d]E = excellent; G = good.
[e]OP = orange-peel effect; sl OP = slight orange-peel effect.
[f]F = fair; P = poor.
[g]Inorganic; dilute.

Table 2. Melt-Mixed PVC Coating Powder Formulation[a]

Ingredient	Parts by weight	Wt %
PVC homopolymer resin	100	47.8
DNODP[b]	65	31.1
epoxidized soya oil	5	2.4
$TiO_2/CaCO_3$[c]	25	12.0
stabilizers		
barium–cadmium	6	2.9
organic phosphite	3	1.4
dispersion-grade PVC	5	2.4
Total	*209*	*100.0*

[a]Ref. 3.
[b]*n*-Octyl *n*-decyl phthalate [*119-07-3*], $C_{26}H_{42}O_4$.
[c]Pigment/extender.

dispersion-grade PVC resin is added in a postblending operation to give good fluidizing characteristics (4). Whereas most PVC powder coatings are made by the dry-blend process, melt-mixed formulations are used where superior performance, such as in outdoor weathering applications and electrical insulation, is required.

Almost all PVC powder coatings are applied by the fluidized-bed coating process. Although some electrostatic spray-grade formulations are available, they are very erratic in their application characteristics. The resistivity of plasticized PVC powders is low compared to other powder coating materials and the applied powder quickly loses its electrostatic charge. For the same reason, PVC powders show poor application characteristics in an electrostatic fluidized bed and are seldom used in this process.

Dishwasher baskets are coated with fluidized-bed PVC powder. Other applications are washing machine retainer rings and various types of wire mesh and chain-link fencing. PVC coatings have a cost/performance balance that is difficult to match with any of the other thermoplastic materials. Properly formulated PVC powders have good outdoor weathering resistance and are used in many applications where good corrosion resistance is required (see CORROSION AND CORROSION CONTROL). These coatings are also resistant to attack by most dilute chemicals except solvents. In addition, PVC coatings possess excellent edge coverage.

Powder coatings as a class are superior to liquid coatings in ability to coat sharp edges and isolate the substrate from contact with corrosive environments. PVC coatings are softer and more flexible than any of the other powder coating materials. Primers used for PVC plastisols have been found generally suitable for powder coatings as well (5).

Polyamides. Coating powders based on polyamide resins have been used in fusion-coating processes for a long time (1). Nylon-11 [25587-80-8] has been used almost exclusively; however, more recently, coating powders also have been sold based on nylon-12 [24937-16-4]. The properties of these two resins are quite similar. Nylon-6 [25038-54-4] and nylon-6,6 [32131-17-2] are not used because the melt viscosities are too high.

Polyamide powders are prepared by both the melt-mixed and dry-blend process. In the latter, the resin is ground to a fine powder and the pigments are mixed in with a high intensity mixer. Melt-mixed powders have a higher gloss, eg, 70–90% on the Gardner 60° glossmeter, whereas dry-blended powders have a gloss in the range of 40–70%. Because the pigment is not dispersed in the resin in the dry-blend process, it must be used at very low concentrations, usually less than 5%. Even in melt-mixed formulations, the concentration of pigment and fillers (qv) seldom exceeds about 30%.

Nylon coating powders are available for electrostatic spray and fluidized-bed application. Nylon coatings are very tough, resistant to scratching and marring, have a pleasing appearance, and are suitable for food contact applications when properly formulated. These coatings are used for chair bases, hospital furniture, office equipment, knobs, handles, and other hardware. Because of expense, nylon is generally applied only to premium items. Nylon coatings have good solvent and chemical resistance and are used for dishwasher baskets, food trays, hot water heaters, plating and chemical-etching racks, and large diameter water pipes in power-generating stations.

For maximum performance, a primer is used. Most nylon primers are un-pigmented resin solutions applied at only ca 10% solids. This gives a dry film thickness of about 3–5 μm. If nylon is applied by electrostatic spray techniques, it is necessary to prime the parts. Although the exact reason is not known, it is speculated that either the primer partially insulates the coating from the substrate and prevents the charge in the electrostatically applied powder layer from leaking off too quickly, or the primer is tacky at the fusion temperature and causes the coating to adhere to the substrate. Nylon coating powders are discussed in more detail in Reference 6.

Other Thermoplastic Coating Powders.   Coating powders based on poly-ethylene [9002-88-4] and polypropylene [9003-07-0] have been available for many years but have achieved limited commercial success (see OLEFIN POLYMERS). A primary problem in using polyolefin-based coating powders is poor adhesion to metal. Modified polyolefin powders have been disclosed in which an unsaturated carboxylic acid or anhydride is grafted onto the polyolefin resin backbone, along with other modifiers to improve adhesion (7). Also, multilayer coatings having an intermediate layer of epoxy resin and modified polyolefins have also been developed (8).

Thermoplastic polyesters achieved some commercial success during the mid-1980s; however, these were eventually replaced by nylon coating powders in functional coatings and thermosetting polyester powders in decorative applications because of lack of any unique characteristics or price advantages (see POLYESTERS, THERMOPLASTIC).

Architectural coatings (qv) based on poly(vinylidene fluoride) [25101-45-5] (PVDF) and applied as dispersions in organic solvents have been available and used successfully for many years (see VINYLIDENE CHLORIDE AND POLY(VINYLIDENE CHLORIDE)). Because of the significant reduction in volatile organic compounds (VOCs) mandated by 1991 legislation, efforts have been made to develop coating powders based on PVDF resins (9). Thermoplastic acrylic polymers are used as a modifying resin to improve flow, pigment wetting, and adhesion (10). Thermosetting coating powders based on hydroxy functional fluoropolymer resins cross-linked with blocked isocyanates have also been disclosed (11).

Several other coating powders are available. These powders, which sell in the range of $11–$22/kg and have very limited application, are based on specialty polymers such as ethylene–chlorotrifluoroethylene [25101-45-5] (E-CTFE), poly(phenylene sulfide) (PPS) [25212-74-2], and tetrafluoroethylene–ethylene co-polymers [68258-85-5] (see FLUORINE COMPOUNDS, ORGANIC; POLYMERS CONTAINING SULFUR). Such powders are used in functional applications where resistance to corrosion and elevated temperatures are required. They are usually applied by fluidized-bed coating techniques but can also be applied by electrostatic techniques to a heated substrate. Extremely high application temperatures in the range of 250–350°C are required for these polymers because of high melting point and high melt viscosity.

Thermosetting Coating Powders

Thermosetting coating powders, with minor exceptions, are based on resins that cure by addition reactions. Thermosetting resins are more versatile than thermoplastic resins in that: many types are available both in varying molecular-weight ranges and having different functional groups; numerous cross-linking agents are available, thus the properties of the applied film can be modified; the resins possess a low melt viscosity during application allowing application of thin films (qv), and addition of pigments and fillers required to achieve opacity in the thin films can be incorporated without adversely affecting flow; gloss, textures, and special effects can be produced by modifying the curing mechanism or through the use of additives; and manufacturing costs are lower because compounding is carried out at lower temperatures and the resins are friable and can be ground to a fine powder without using cryogenic techniques.

The properties of thermosetting coating powders are given in Table 3. The molecular weight, or the glass-transition temperature, T_g, of coating powder thermosetting resins must be high enough to prevent the individual particles from sintering or fusing during transportation and storage. The minimum T_g required is in the range of 40–50°C and preferably above 50°C. Epoxy resins (qv), because of the aromatic backbone, have the required T_g at a relatively low molecular weight and corresponding low melt viscosity. Other thermosetting resins, however, require some linear comonomers to achieve flexibility. This results in a lower T_g and higher molecular-weight resins must be used. Thus, at an equivalent range of T_g, polyester resins have a melt viscosity of about 4000–9000 mPa·s(= cP) at 200° or about 2–10 times that of an epoxy resin. As pigment and filler loadings are increased, the difference in flow becomes even more pronounced. Therefore,

Table 3. **Physical and Coating Properties of Thermosetting Powder**

Property	Epoxy	Polyurethane[a]	Polyester[b]	Hybrid	Acrylic[a]
fusion range, °C	120–200	160–220	160–220	140–210	120–200
cure time[c], min	1–30[d]	5–15	5–15	5–15	5–15
storage temp, °C[e]	30	30	30	30	30
adhesion[f]	E	G–E	G–E	G–E	G
gloss, Gardner 60° meter	5–95	5–95	40–95	20–95	80–95
pencil hardness[g]	H-4H	H-2H	H-4H	H-2H	H-2H
flexibility[f]	E	E	E	E	F–P
resistance[f]					
impact	E	G–E	G–E	G–E	F
overbake	F–P	G–E	E	G–E	G
weathering	P	G–E	G–E	P–F	G–E
acid[h]	G–E	F	G	G	F
alkali[h]	G–E	P	F	G	P
solvent	G–E	F	F–G	F	F

[a]Hydroxy function-blocked isocyanate cure.
[b]TGIC (triglycidyl isocyanurate) cure.
[c]Value is given at 160–200°C, unless otherwise indicated.
[d]At 240–135°C.
[e]Maximum value is given.
[f]E = excellent; G = good; F = fair; P = poor.
[g]Refers to highest degree of lead hardness at which coating can be marred.
[h]Inorganic; dilute.

considerable efforts have been made to develop resins that give a smooth finish, and good storage stability, physical properties, and cure response (12).

The ability of a coating powder to form a smooth film with a low degree of orange-peel effect, ie, good appearance, is dependent on a number of factors other than melt viscosity. Flow control additives, necessary to prevent cratering, have a profound effect on surface characteristics (13). The geometry of the electrostatically deposited powder layer prior to fusion has an effect on the smoothness of the cured coating (14). The particle size, shape, and distribution of the coating powder are also significant because the geometry of the electrostatically applied powder layer is influenced by these factors (15). The rate of the reaction between the resin and curing agent affects the rheology of the molten coating layer prior to gelation and is also an important factor in developing optimum surface smoothness (16). This, in turn, is influenced by the rate of temperature increase of the resinous film, ie, the cure schedule. Optimum curing conditions can be predicted based on the reaction kinetics of the coating powder system (17).

The relative reactivity of thermosetting powders can be easily determined by the gel time or stroke-cure test. A small amount of powder is placed on a hot plate, usually at 200°C, and the time until the coating composition gels, or no longer forms fibers, is determined. Powders are characterized by relative gel times (cure rate) as shown in Figure 1.

Formulation. In addition to the resin and curing agents or hardeners, a variety of other ingredients are normally present in coating powder formulations. Catalysts and accelerators are used to modify the reaction rate and curing characteristics (Fig. 2). Flow control additives are employed to prevent cratering and

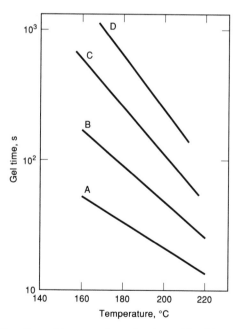

Fig. 1. Gel time as a function of temperature for A, residual heat curing; B, fast curing; C, medium-fast curing; and D, slow curing powders.

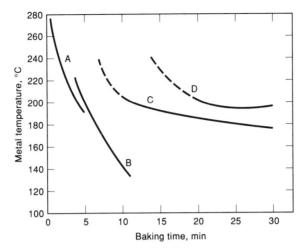

Fig. 2. Baking time as a function of temperature for A, residual heat curing; B, fast curing; C, medium-fast curing; D, slow curing powders where (---) represents undercured and (—) cured material.

promote leveling of the molten polymer film and wetting of the substrate. The most widely used types are low molecular-weight polymers of butyl acrylate [9003-49-0] and copolymers of ethyl acrylate and 2-ethylhexyl acrylate [26376-86-3] (see ACRYLIC ESTER POLYMERS). Thermoplastic resins such as benzoin [119-53-9] are sometimes used to aid the release of bubbles from the molten film.

Matting or flattening agents are employed to control gloss, which is dependent on microscopic surface smoothness. Thus, materials that disrupt surface smoothness through incompatibility, such as polypropylene (18), can be used to control gloss. Certain fillers such as coarse grades of calcium carbonate [1317-65-3] and various silicates are also used to modify gloss, usually in combination with other techniques. Curing agents having widely different reactivities cause a two-stage polymerization to occur, resulting in reduced gloss (19). Mixing powders of varying reactivities (gel time) also result in a reduction in the gloss of the final film, although this effect is not always desired.

Colorants are used in most powder coating formulations. Carbon black, titanium dioxide, iron oxides, and other inorganic pigments (qv) are widely used. Most organic pigments also are used in powder coatings; however, because of high surface area and thixotropic characteristics organic pigments cannot be used at high, eg, >10%, levels without giving excessive orange peel. Fillers such as calcium carbonate, blanc fixe [07727-43-7], barium sulfate [07727-43-7], and various silicates are used to modify gloss, hardness, permeability, and other coated film characteristics. Polyolefins, fluorocarbons, and waxes (qv) are used to modify the slip and mar characteristics of the film.

Specialty Coatings. Clear coatings are formulated using curing agents and flow control additives, which have a high degree of compatibility with the resin. Conventional uv stabilizers (qv) can be added to improve exterior durability and metallic finishes can also be prepared. In order to obtain an optimum metallic effect, aluminum flake must be added as a post-blending operation, after the powder is ground to final particle size (20). The addition of silicones or other incompatible agents that cause cratering, produce special effects. Textured coatings are produced by adding thixotropes or high oil absorption fillers to the coating composition and by modifying the particle size. A wrinkle finish can be obtained using selected curing agents and catalysts (21).

Epoxy Coating Powders. Thermosetting coating powders based on epoxy resins [25068-38-6], $C_{15}H_{16}O_2 \cdot (C_3H_5ClO)_x$, have been used longer in powder coating processes and sold in greater volume than any other resin class. The earliest epoxy powders, based on latent curing agents such as dicyandiamide [461-58-5], $C_2H_4N_4$, frequently accelerated using tertiary amines or imidazoles, were slow to medium-fast curing having a typical cure cycle of 20–30 min at 200°C (Fig. 1). With minor modificatins, some of these powders are used in such applications as buss bar and motor core insulation and corrosion-resistant coatings where their excellent long-term performance has been well documented. However, the development of electrostatic spray powders led to faster curing coating powders having better flow and designed primarily for decorative applications (22). During the 1970s, epoxy resins designed specifically for decorative powder coatings and characterized by very good color, low melt viscosity, uniform properties, and a low incidence of gels and other insoluble contaminants were developed. These were filtered through a 10–25 μm filter and packaged in multiwall foil bags to prevent moisture absorption. A special class of coatings was developed for application on pipe and reinforcing bar (rebar) in which the powder is applied to the preheated substrate and the powder particles melt, fuse, flow, and cure within 10–20 seconds from the residual heat in the pipe or rebar (23). These materials are referred to as residual heat cure powders (Fig. 2). Epoxy powder coated rebar, first developed in the United States because of the nationwide problem of rebar corrosion, primarily in bridge decks, have now been adopted worldwide (24).

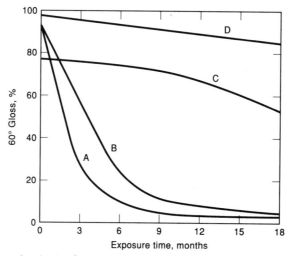

Fig. 3. Gloss retention in outdoor exposure in Florida for A, epoxyresin; B, epoxy–polyester hybrid resin; C, brown; and D, white powder coatings. White coatings are acrylics, polyester–urethane, or polyester-TGIC based (see text).

Decorative epoxy powders are used in a wide variety of applications, eg, for light fixtures, garden equipment, floor cleaning machines, motor control cabinets, and for automotive under-the-hood items. Resins of the "3" type, ie, epoxide equivalent weight (EEW) ca 650–750 are frequently used. Resins in this molecular weight range have the lowest melt viscosity (high flow) commensurate with a T_g high enough to prevent powder sintering under most conditions of storage. Type 4 epoxy resins (EEW 750–900) are more often used in functional applications such as for electrical insulation where thicker coatings are used and a higher degree of edge coverage is necessary. Even higher equivalent weight resins are sometimes used to enhance flexibility or other special requirements. In outdoor applications, epoxy coatings chalk readily; however, they protect the substrate for many years. Figure 3 compares the gloss retention of epoxy coatings with other thermoset types.

Epoxy–Polyester Hybrids. Epoxy resins react with acid functional curing agents to yield a coating having many desirable properties. For many years, coating powders based on epoxy resins cured using trimellitic anhydride (TMA) [552-30-7] were used for both decorative and functional applications (25). However, as the irritant and sensitizing characteristics of TMA have become more widely recognized, this system became less popular.

A series of acid functional saturated polyester resins specifically designed for curing using bisphenol A epoxy resins was developed. These resins have an acid number between about 35 and 75 (acid equivalent weight ca 750–1600), are relatively linear, and are used at about a 1:1 epoxy:polyester stoichiometric ratio. Because the polyester resins are somewhat less expensive than epoxy resins, especially in Europe, higher ratios of polyester to epoxy are sometimes used. Properties of these polyester–epoxy hybrids are similar to those of a straight epoxy, but differ in several respects. The overbake resistance (resistance to color change after extending curing) and resistance to discoloration on exposure to sunlight is superior. Most epoxy curing agents contain nitrogen compounds that discolor on extended heating. Because the cross-link density for hybrid coatings is generally less than for straight epoxies, hybrid coating powders are inferior in solvent resistance and hardness. They are also somewhat inferior in salt spray and corrosion resistance. Polyester resins, having a higher melt viscosity than epoxy resins, cause the hybrids to have more orange peel than epoxy formulations, especially at the higher polyester–epoxy ratios.

The reaction rate between carboxyl end groups of the polyester and the epoxide groups of the epoxy resin is generally quite slow requiring a catalyst to obtain a practical baking time. Catalysts are frequently mixed with the polyester resin by the resin manufacturer (26). The ideal catalyst should exhibit good reactivity at the desired baking temperature, eg, 140°C, while providing good flow and shelf stability (27). Tertiary amines, amic acids, and quaternary phosphonium compounds are effective catalysts for the epoxy–carboxyl reaction (28,29).

Epoxy–polyester hybrids are marginally better than straight epoxies in gloss retention on exterior exposure (Fig. 3) but are not recommended for exterior applications in most cases. For the most part, applications for the hybrids are the same as those for decorative epoxies. These latter are gradually being replaced by the hybrids, which are gaining in market share in the United States. In Europe, hybrids are the most widely used powder coating type.

Polyester–TGIC Cured. A principal class of exterior durable powder coatings is based on acid functional saturated polyester resins cured using triglycidylisocyanurate (TGIC) [02451-62-9]. This system was first developed in Europe in the early 1970s. The acid functional polyester resins used in TGIC-cured coating powders are similar to those used in epoxy polyester hybrids. However, the TGIC-cured resins have a higher equivalent weight, typically in the range of 1600 to 1900. Thus most resins are used at a 93:7 ratio of resin to TGIC. Acid functional resins are normally prepared by a two-step process: the reaction of excess polyol and dibasic acids followed by esterification of the hydroxyl terminated resin using dibasic acids or anhydrides (30). This technique yields a resin where the functional groups are at the end of the molecule, rather than occurring randomly along the polymer chain. It is also possible to prepare carboxyl group terminated polyesters in a one-step process (31). The exterior durability of polyester resins cured using TGIC is significantly affected by the monomer composition, and it has also been observed that resins having a higher T_g give improved weatherability on Florida exposure. The high T_g resins appear to be more resistant to hydrolysis, which also contributes to film degradation on exterior exposure (26). Reactivity, flow, and physical properties are addressed through the use of specific catalysts (32) and control of molecular structure (33).

TGIC-cured polyester resins and coating powders have gained a significant market share position in the exterior durable market in both Europe and North America. European usage has led to satisfactory performance for over 20 years (34). Powder coatings based on TGIC-cured polyesters have excellent physical properties and flexibility, good resistance to color change on overbake, and a generally good overall balance of properties. However, TGIC has shown mutagenic activity in several tests and other curing agents for acid functional resins are being investigated. Among the most promising are the β-hydroxy-alkylamides (35). Accelerated testing and early Florida exposure results indicate the β-hydroxy-alkylamide-cured polyesters have equivalent gloss retention to TGIC-cured systems (36). Because β-hydroxyalkylamide curing agents and acid functional resins react via a condensation reaction, problems of surface imperfections resulting from the volatiles generated have been observed in coatings where the thickness exceeds about 75–80 μm. Coating powders based on this technology as of this writing have not gained significant commercial success.

Urethane Polyesters. In the United States the search for exterior durable coating powders led to technology based on hydroxyl functional polyester resins. The earliest curing agents evaluated were based on melamine–formaldehyde resins, such as hexa(methoxymethyl) melamine (HMMM) [68002-20-0], which are widely utilized as curing agents in conventional paint systems (see AMINO RESINS). Coating powders based on this chemistry suffer limitations: the melamine resin depresses the T_g of the coating powder to the point where the powder sinters during storage, especially at elevated temperatures; and the methanol generated during the curing process becomes trapped in the film, especially at thicknesses above about 50 μm, resulting in a frosty or visually nonuniform surface. Curing agents based on polyisocyanates blocked with caprolactam [00105-60-2] (qv) give an outstanding combination of properties in the final film (see Table 3). Because the unblocking reaction does not start to occur until about 160°C, the powder has a chance to flow out and give a smooth uniform film prior to any substantial cross-linking. In addition, not all of the caprolactam present evolves during the curing reaction. Some remains in the film, apparently acting as a plasticizer so that the urethane polyesters yield a much smoother, more orange peel-free film than the TGIC polyesters. Urethane polyesters are preferred in the United States though not in Europe (37).

The polyester resins used in this technology are typically based on terephthalic acid [100-21-0], $C_8H_6O_4$, or isophthalic acid [121-91-5], $C_8H_6O_4$, neopentyl glycol [126-30-7], $C_5H_{12}O_2$, and branched using trimellitic anhydride (38). The most commonly used curing agents are adducts of isophorone diisocyanate (IPDI) [4098-71-9], $C_{12}H_{18}N_2O_2$, and low molecular-weight polyols, blocked with caprolactam (39). Other aliphatic or cycloaliphatic isocyanate adducts blocked with caprolactam can also be used (40) (see ISOCYANATES, ORGANIC). Caprolactam blocked aromatic isocyanates such as toluene diisocyanate [584-84-9], $C_9H_6N_2O_2$, are also used, but result in coatings having limited exterior durability. Polyester resins having an acid number of 40–50 (equivalent weight 1100–1400) are mostly used, but highly branched resins of equivalent weights as low as 200 are available, which can be mixed with these resins to produce coatings having exceptionally high hardness and outstanding stain, solvent, and chemical resistance. A wide variety of polyester resins can be synthesized yielding a diversity of properties in the final coating (41,42).

Environmental concern about the caprolactam evolved from urethane polyester coating powders during curing has led to research into using higher T_g derivatives such as tetramethoxyglycoluril [17464-88-9] (43), and higher molecular-weight condensates (44). The aim is to reduce the degree of functionality of the alkoxy methyl derivatives and thus reduce the condensation volatiles, and to increase the T_g to improve storage and handling characteristics.

The retention of gloss on exterior exposure for the urethane polyesters is quite similar to coatings based on TGIC-cured polyesters (Fig. 3). In addition to the polymeric backbone of the resin itself, exterior durability is influenced by formulating variables, such as the gloss control system used and color. Darker colors lose gloss more rapidly on Florida exposure than lighter colors. This may well be because of the higher surface temperatures generated by the dark colors which leads to hydrolytic instability and more rapid degradation.

Unsaturated Polyester Resins. A special class of coating powders is based on unsaturated polyester resins. They are utilized in matched metal die molding operations such as sheet molding compounds (SMC) and bulk molding compounds (BMC) molding, where the mold is coated with the powder prior to placing the resin charge in the mold (see POLYESTERS, UNSATURATED). The powder melts and flows on the mold surface, and when the mold is closed, the powder reacts with the molding compound forming a coating on the molded part. This process is known as in-mold coating (45). Unsaturated polyester resin powders can provide a colored and finished exterior molded surface or a finish ready for painting. Normally, a primer/sealer must be applied to molded articles prior to painting. In addition to the unsaturated polyester resin, multifunctional unsaturated monomers such as triallyl cyanurate (TAC) [101-37-1] or diallyl phthalate (DAP) [131-17-9], suitable peroxide initiators (qv) or mixtures thereof, and mold release agents (qv) are used to formulate the coating powder (46).

Acrylic Resins. Coating powders based on acrylic resins (47,48) have been available in both Europe and the United States since the early 1970s, but have never achieved significant commercial success. The development of acrylic coating powders for automotive topcoats, where acrylic resins have a long history of successful performance, is being pursued. Among the obstacles to the acceptance of powder coatings for automotive topcoats is the inability to change colors rapidly and economically. Acrylic-based clearcoats are being developed, however.

Acrylic coating powders have achieved some success in Japan utilizing resins having glicydyl methylacrylate functionality cured with C_{10}–C_{12} dicarboxylic acids (49). Hybrid polyester–acrylic coating powders have also been reported in which an acid functional polyester resin coreacts with a glycidyl-containing acrylic polymer (50). Hydroxyl functional acrylic resins cured with blocked isocyanates have also been available for many years in the United States and achieved some commercial success as appliance finishes.

Manufacture

Coating powders are either melt-mixed or dry-blended as shown in Figure 4. Production methods based on spray drying from solution have been investigated (51) but it is extremely difficult to remove the last traces of solvent that volatilizes during the curing operation resulting in imperfections in the film. Other wet methods, such as precipitating the powder from solution with water have also been evaluated (52), but it is difficult to dry the powder completely, the process is inefficient for small lots, and equipment cleaning is time consuming.

Melt-Mixing. Dry ingredients are weighed into a batch mixer such as the high intensity impeller mixers, but medium intensity horizontal plow mixers such as the Littleford are also used. Tumble mixers can be employed if a thorough distribution of all ingredients is ensured. Mixing times range from 30–60 minutes for tumble mixers to 1–2 minutes for the high intensity mixers. The latter have the advantage that small amounts of liquids can be added to the premix; however, the batch size is frequently smaller than for a tumble mixer and more batches have to be weighed (see MIXING AND BLENDING).

The premix is then melt compounded in a high shear extruder where the ingredients are compacted, the resin melts, and individual components are thoroughly dispersed in molten resin. These compounding machines generate sufficient heat through mechanical shear so that after start-up, little external heat needs to be supplied. Both single-screw machines such as the Buss Ko-kneader and twin-screw extruders such as the Werner Pfleiderer or Baker Perkins are used (see PLASTICS PROCESSING). Residence time in the extruder is in the range of 0.5–1.0 minutes, and melt temperatures for thermosetting materials normally run from 60–140°C, slightly above the melting temperature of the resin. Because of the low temperatures and short residence time, little if any reaction between the resin and curing agents occurs. The molten compound is cooled rapidly by passing it through the nip of a water-cooled roll and subsequently onto a continuous stainless steel water-cooled belt or drum. The cooled compound is broken into small chips, about 10 × 12 mm, suitable for fine grinding. Thermosetting resins are quite friable and are usually ground to final particle size in an air

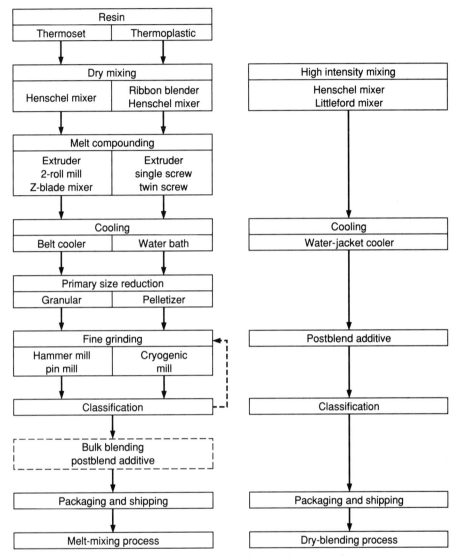

Fig. 4. Flow diagram for powder coating manufacture.

classifying mill. In this grinder, a blower generates an air stream in which the product is entrained. A variable speed separator reduces air flow through the mill so that the larger particles are returned to the grinding chamber. The fine powder is separated from the air stream by a cyclone or dust collector. Finished powder can be packaged directly from the cyclone, or screened to be sure there are no large particles in the final product. Frequently, a colloidal silica or alumina is mixed with the powder to improve handling and dry free-flow characteristics (53).

Dry-Blending. Most plasticized PVC powders are prepared by a dry-blend process in which the plasticizers, stabilizers, pigments, and additives are absorbed on the porous PVC particles at elevated temperatures while they are being agitated in a high speed mixer. Thermosetting powders are almost never prepared by this process.

Application Methods

Fluidized-Bed Coating. Fluidized-bed coating, the first commercial process for applying 100% nonvolatile polymeric materials to a substrate to form a uniform coating, is the method of choice for many applications where a heavy functional coating is required. The U.S. market for fluidized-bed powders is estimated to be in the range of ~10,000 t/yr. The process is relatively simple. The main variables are the temperature of the part as it enters the fluidized bed, the mass of the part being coated, dip time, and post-heat temperature. Other variables, such as motion of the part on the bed and density and temperature of the powder in the bed, also effect the quality of the coating (54–56). The process is especially useful in coating objects having a high surface to mass ratio such as expanded metal and fabricated wire goods. Sharp edges and intersections are well covered because of the heavy film thickness applied. The size of parts that can be coated is limited because the fluidized bed must be large enough to accommodate them.

Electrostatic Fluidized-Bed Coating. In an electrostatic fluidized bed, the fluidizing container and the porous plate must be constructed of a nonconductive material, usually plastic. Ionized air is used to fluidize and charge the powder. The parts to be coated are passed over the bed and charged powder is attracted to the grounded substrate. The rate of powder deposition varies significantly based on the distance of the part from the fluidizing powder. Therefore, this process is usually utilized only when the object to be coated is essentially planar or symmetrical and can be rotated above the charged powder. Electrostatic fluidized-bed coating is an ideal method for continuously coating webs, wires, fencing, and other articles that are normally fabricated in continuous lengths and are essentially two-dimensional. In a variation of this process, two electrostatic fluid beds are arranged back to back and the continuous web of material is passed between them, coating both sides simultaneously. Millions of lineal meters of window screen have been coated using this technique (57) and variations of this process have been proposed for powder coating magnet wire (58).

Electrostatic Spray Coating. Electrostatic spray coating is the most widely utilized method for the application of powder coatings (see COATING PROCESSES, SPRAY COATINGS). In a typical high voltage system, powder is maintained in a fluidized-bed reservoir, injected into an air stream, and carried to the gun where it is charged by passing through a corona discharge field. The charged powder is transported to the grounded part to be coated through a combination of electrostatic and aerodynamic forces. Ideally, the powder should be projected toward the substrate by aerodynamic forces so as to bring the powder particles close to the substrate where electrostatic forces then predominate and cause the particles to be deposited (59). The powder is held by electrostatic forces to the surface of the substrate, which is subsequently heated in an oven where the particles fuse and form a continuous film. The processes involved are powder charging, powder transport, adhesion mechanisms, back ionization, and self-limitation (60–62). The thickness that can be applied is self-limiting. As charged powder particles and free ions generated by the high voltage corona discharge approach the powder layer already deposited, the point is reached where the charge on the layer exceeds its dielectric strength and back ionization occurs. At this point, any oncoming powder is rejected and loosely adhering powder on the surface falls off. It has been demonstrated that some imperfections in the final coating are a result of defects in the powder layer, eg, from back ionization (63).

The characteristic of the electrostatic spray process to form limiting films enables operation of equipment after only brief training and instruction. It is almost impossible to create runs, drips, or sags characteristic of spray-applied liquid finishes. Furthermore, the practical design of automatic spray installations is possible. Multiple electrostatic guns mounted on reciprocators are positioned in opposition to each other in an enclosed spray booth and parts to be coated are moved between the two banks of guns where a uniform coating of powder is applied. Oversprayed powder is captured in the reclaim system and reused. Powder coating booths have been designed with interchangeable filter units to facilitate change from one powder type of color to another (64). A typical automatic powder spray installation having replaceable cartridge filters for color changes is shown in Figure 5. One disadvantage of the electrostatic powder spray process using corona discharge guns is that a high voltage field is set up between the gun and the parts to be coated. Parts having deep angles or recesses are sometimes difficult to coat because of the Faraday Cage effect (65).

Another method of overcoming the Faraday Cage effect is by the use of triboelectric guns in which powder charging occurs by the frictional contact of the powder particles and the interior surface of the gun. Electrons are separated from the powder particles, which become positively charged and attracted to the substrate (66). Because there is no electrostatic field between the gun and the article being coated, a Faraday Cage is not developed and particles are able to penetrate into recessed areas of the substrate. There are a number of commercial powder coating systems using triboelectric charging technology, but this number is quite small compared to that of the more conventional corona guns. The powder application rate of triboelectric guns is significantly lower than for corona guns; additionally, whereas powders based on a wide variety of resins and formula types charge and apply readily with corona guns, only certain resin systems charge well in triboelectric guns (67). Most recently electrostatic guns, which utilize both triboelectric and corona charging of the powder, have been developed (68).

Stricter control of particle size distribution improves powder handling (qv) and transport through the system, and application efficiency. It also reduces blinding of the final filter (69). Additives have been developed which improve the triboelectric charging characteristics (70).

Hot Flocking. Several nonfluidized-bed coating methods are based on contacting a preheated substrate with powder to form a film. Although these techniques are not widely used, for certain parts they are the preferred method of application. For example, the coating of motor stators or rotors using a thermosetting powder provides primary insulation between the core and windings. The part is preheated to about 200°C and the powder is directed from a fluidized bed using an air venturi pump, similar to the devices used to supply powder from the reservoir to electrostatic guns, through flexible tubes and directed at the preheated part. Multiple tubes, usually in pairs in opposition to each other, are nor-

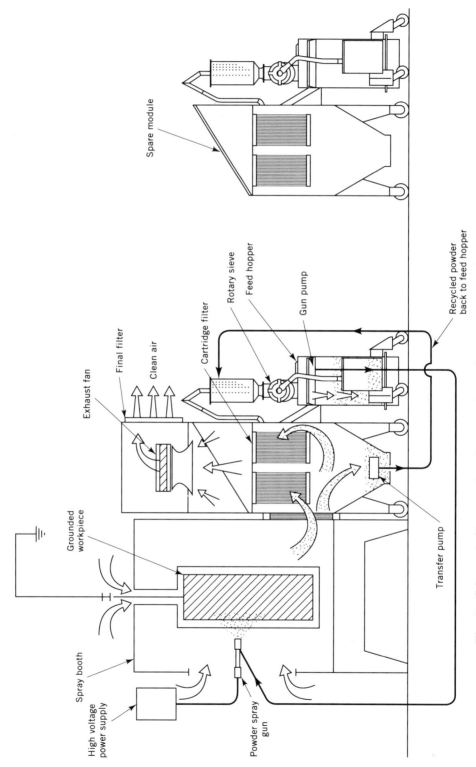

Fig. 5. Schematic diagram of an electrostatic powder spray system. Courtesy of Nordson Corp.

652

mally used. In a similar fashion, the inside diameter (ID) of pipe can be coated by entraining powder in an airstream and blowing it through a preheated section of pipe (71). Very uniform coatings in the range of 200 to 300 μm can be applied by this process to provide corrosion protection for drill pipe, and water injection and gathering pipe in diameters up to ~50 cm.

Metal Cleaning and Preparation. As in any finishing operation, the surface of the object to be coated must be clean, dry, and free from rust, mill scale, grease, oil, drawing compounds, rust inhibitors, or any soil that might prevent good wetting of the surface by the coating powder (72). Steel should be sandblasted or centrifugally blast-cleaned to give a near white finish (73) (see METAL SURFACE TREATMENTS). Phosphate and chromate conversion coatings improve performance in harsh environments (74).

Economic Aspects

The worldwide market for coating powders increased at an annualized growth rate of 12–13% in the 1980s. Moreover, worldwide coating powder usage for thermosetting decorative powders is expected to increase to 332,000 t by 1992, a 264% increase over the quantity sold in 1984 (75). A significant factor in the growth of coating powders has been the increasingly stringent environmental regulations governing volatile organic compounds (VOCs) and disposal of hazardous waste (see HAZARDOUS WASTE TREATMENT). A breakdown of worldwide coating powder consumption and estimated growth rates is given in Table 4. The European market is larger and more mature than the North American market. However, it is estimated that in 1989 only about 15% of industrial coatings as a percentage of dry film on the surface of a component were powder coatings. This value is expected to grow to 18% by 1994. In North America, powder coatings as a percentage of dry film on original equipment manufacturers (OEM) components was only 7% in 1989. This number is expected to be 12% by 1994 (76).

The distribution of powder coating sales by resin type for various geographical areas is given in Table 5. The polyester–epoxy hybrids account for 50% of the European market, yet have only market share ≤25% in the rest of the world. Similarly, the polyester urethanes, which account for only a small market share in Europe, find significant usage in the United States and Japan. Acrylics hold a significant share only in the Japanese market.

Table 4. 1989 World Market Distribution of Thermoset Decorative Powder Coatings[a]

	t × 10³	%	Growth rate, %
Europe	127.5	54	11
North America	52.0	22	10
Far East	40.0	17	15
other	16.5	7	12
World Total	*236.0*	*100*	*12–13*

[a]Ref. 75.

Table 5. 1989 World Resin Distribution in Thermoset Decorative Powder Coatings[a]

Resin	Distribution, %			
	United States	Canada	Europe	Japan
urethane	33	14	6	35
epoxy	27	36	26	22
TGIC	23	24	18	5
hybrid	17	26	50	20
acrylic				18
Total	*100*	*100*	*100*	*100*

[a]Ref. 75.

Thermoset decorative coatings are by far the largest powder coatings market segment amounting to 48,500 t worth $279,000,000. However, in the United States in 1989, the market for functional thermoset powders (all epoxy) was estimated to be about 12,000 t, at a value of $67,000,000, primarily going into the pipe and rebar markets. The market for functional thermoplastic coatings was about 7700 t, worth $42 million. This last is primarily PVC powder going into the dishwasher and chain-link fence markets. The total powder coating market in the United States in 1989 amounted to some 68,000 having a dollar value of almost $400 million (77). Table 6 details the markets for powder coating by use.

Another indication of the growth of the powder coating market is reflected in the membership statistics of the Powder Coating Institute (PCI). From 1987 to 1991 the number of members associated with powder coating manufacturers increased from 5 to 22; equipment suppliers from 3 to 9; custom applicators from 1 to 29; and suppliers to the powder coating industry from 5 to 38 (78).

Table 6. 1989 U.S. Powder Coating Consumption By Use[a]

Industry	Quantity		Value	
	$t \times 10^3$	%	$\$ \times 10^6$	%
automotive	8.5	12.5	50.3	13.0
appliance	10.5	15.4	56.5	14.6
furniture and fixtures	11.6	17.0	59.4	15.3
electrical/electronic	3.3	4.8	20.2	5.2
general metal	16.9	24.8	113.3	29.2
machinery and equipment	2.5	8.0	34.1	8.8
pipe/rebar	8.7	12.8	44.7	11.5
wire goods/fencing	3.2	4.9	9.5	2.4
Total (rounded)	*68.2*	*100*	*388*	*100*

[a]Ref. 75–77.

Test Methods

Methods for evaluating the performance of powder coatings are the same as those used for conventional coatings (qv) in which test coatings are compared to standards. Test methods for coating powders have been reviewed in detail and reported in the literature (79). In addition, the American Society for Testing Materials has issued a comprehensive standard that covers the most important testing methods for the characterization and evaluation of powder coatings (80).

Environmental and Energy Considerations

One of the most significant factors contributing to the growth of powder coating processes has been the increase in federal and state environmental regulations. The 1991 Clean Air Act amendments drastically control the release of VOCs in manufacturing operations, maintenance, and home painting and decorating. Powder coating processes offer a method to eliminate VOCs. Whereas trace quantities of volatiles are released by some powder coatings during application and cure, these volatiles do not meet the definition for VOCs (81,82). Eg, caprolactam, which is given off during the curing of urethane polyester powder coatings at a level of about 4–6%, is not considered a VOC according to the 1991 regulations (83). High volume usage plants may have restrictions on caprolactam emissions, however, depending on the reportable quantities (RQs) established in the Clean Air Act amendments.

In addition to the environmental advantages, the low volatile emissions of powder coatings during the baking operation has economic and energy saving advantages. Fewer air changes per hour in the baking oven are required for powder coatings than for solvent-based coatings, which saves fuel. Further, in the coating operation almost all powder is recovered and reused, resulting in higher material utilization, and waste minimization. The air used in the coating booths during application is filtered and returned to the workplace atmosphere, reducing heating and cooling demands. Additionally because of the need for more sophisticated devices to control emission of VOCs in liquid systems, the capital investment to install a new powder coating line is becoming increasingly more economically favorable. The savings in material and energy costs of powder systems has been documented in several studies (84,85).

The only components in a coating powder which might cause the waste to be classified as hazardous are certain heavy-metal pigments sometimes used as colorants. Lead- (qv) and cadmium-based pigments (qv) are seldom used, however, and other potentially hazardous elements such as barium, nickel, and chromium are usually in the form of highly insoluble materials that seldom cause the spent powder to be characterized as a hazardous waste (86).

Health and Safety Factors

Any finely divided organic material can form ignitable mixtures with air at specified concentrations. The most significant hazard in the manufacture and application of coating powders is the potential of a dust explosion (see POWDERS HANDLING). The severity of a dust explosion is related primarily to the material involved, its particle size, and concentration in air at time of ignition. The lower explosive limit (LEL), the lowest concentration of a material dispersed in air that explodes when ignited (87), is essentially the same as the minimum explosive concentration (MEC) (88). The LEL values for a number of coating powders are given in Table 7.

The energy required to initiate an explosion and the maximum explosive pressure developed by a number of polyester–epoxy powder coatings has been studied in some detail (89). The variables studied included composition, level and type of pigmentation, particle size, and concentration in air. The lowest MEC for unfilled and unpigmented powders was 33–35 g/m^3.

In powder coating installations, the design of the spray booth and duct work, if any, should be such that the powder concentration in the duct is always kept below the LEL employing a wide margin of safety. General safety considerations are detailed (90). The use of flame detection systems in all automatic powder coating installations is required. Most spray booths being installed, such as that illustrated in Figure 5, do not utilize baghouses or cyclones for dust collection and, therefore, because no enclosed or confined space for a dust cloud is present, pose little hazard for a dust explosion. If powder ignition should occur, flame detection sensors shut down the power to the coating system with the exception of overhead lights.

If a cyclone or dust collector is utilized in the coating system, either a pressure relief or explosion suppression system is necessary whenever the baghouse cannot be located outside the building (see AIR POLLUTION CONTROL METHODS). If located inside the building, dust collectors should be near an outside wall and duct work from the explosion vents should be directed to the outside through short runs, not exceeding 3 m. Required explosion vent areas and other design considerations can be found in the literature (91). The spray guns, spray booth, duct work, dust collection, and powder reclaim system, as well as the workpiece, must all be properly grounded (92).

There were over 3000 electrostatic powder spray lines and about 800 other powder coating installations operating in the United States as of 1991. For additional general information on the safety aspects of powder coatings, see References 90, 93, and 94.

The health hazards and risk associated with the use of powder coatings must also be considered (see HAZARD ANALYSIS AND RISK ASSESSMENT). Practical

Table 7. Lower Explosive Limits Range for Coating Powders

Resin type	g/m^3	Reference
epoxy	45–78	87
epoxy–polyester	33–35[a]	89
polyester–urethane	65–71	87
polyester–TGIC	40–70	87
polypropylene	32	87

[a]Unfilled–organic binder only.

methods to reduce employee exposure to powder such as the use of long-sleeved shirts to prevent skin contact should be observed. Furthermore, exposure can be minimized by good maintenance procedures to monitor and confirm that the spray booths and dust collection systems are operating as designed. Ovens should also be properly vented so that any volatiles released during curing, such as caprolactam in the case of urethane polyesters, do not enter the workplace atmosphere. Whereas the vast majority of powder coatings do not give off any volatiles during the cure, in addition to caprolactam, small amounts of methanol are released during the cure of powders using melamine derivatives and, in some cases, low molecular-weight polymers, oligimers, or additives may become volatilized during the curing operation.

In general, the raw materials used in the manufacture of powder coatings are relatively low in degree of hazard. None of the epoxy, polyester, or acrylic resins normally used in the manufacture of thermoset powder coatings are defined as hazardous materials by the OSHA Hazard Communication Standard. Most pigments and fillers used in powder coatings generally have no hazards other than those associated with particulate nature. Some epoxy curing agents are skin irritants, however, most of these characteristics are greatly diminished when these materials are compounded into the powder coating. In addition to being diluted, the materials are dispersed in a resinous matrix having a low degree of water solubility which appears to make them less biologically accessible. For example, anhydrides and anhydride adducts generally elicit a strong respiratory or eye irritant response. However, when powder coatings containing anhydride-based curing agents were tested in animal exposures, the coatings were found to be nonirritating to the skin, eye, and respiratory tract (95,96). Similarly, TGIC is a skin irritant, but TGIC-containing powders are not (97). The reproductive genetic toxicity of TGIC alone and on a coating powder containing 10% TGIC, established the no-observable-effect level (NOEL) at 2.5 mg/m^3 for neat TGIC, and at >100 mg/m^3 for the 10% TGIC containing powder (98). These tests were considered to be especially significant because inhalation is the most likely route of exposure for coating powders and the tests clearly indicated that coating powders containing hazardous ingredients such as TGIC can be significantly less hazardous than would be expected based only on the dilution of the active ingredient. The threshold limit value–time-weighted average (TLV–TWA) for coating powders containing TGIC has been recommended not to exceed temporary concentrations of 2–5 mg/m^3 air. The lower limit is more appropriate for unpigmented high TGIC percentage powders (99).

Although these data indicate that coating powders do not appear to pose significant hazards to the health of personnel working with the powders, worker exposure should nevertheless be minimized. Coating powders should be treated as nuisance dusts and the concentration in air kept below 10 mg/m^3 TLV–TWA (100), primarily through environmental controls. Hoods and proper ventilation should be provided during handling and application of powders; ovens should be properly vented to the outside atmosphere and provided with sufficient forced air circulation to assure that they are operating under negative pressure with regard to the workplace environment. When environmental control of dust cannot be designed to keep concentrations below the TLV, protective equipment such as dust and fume masks or externally supplied air respirators should be used (101,102).

BIBLIOGRAPHY

1. Ger. Pat. 933,019 (Sept. 15, 1955), E. Gemmer (to Knapsack-Griesheim, AG).
2. U.S. Pat. 2,844,489 (July 22, 1958), E. Gemmer (to Knapsack-Griesheim, AG).
3. U.S. Pat. 3,640,747 (Feb. 8, 1972), D. S. Richart (to the Polymer Corp.).
4. U.S. Pat. 3,264,271 (Aug. 2, 1966), H. M. Gruber and L. Haag (to the Polymer Corp.).
5. U.S. Pat. 3,008,848 (Nov. 14, 1961), R. W. Annonio (to Union Carbide Corp.).
6. D. S. Richart, in M. I Kohan, ed., *Nylon Plastics,* John Wiley & Sons, Inc., New York, 1973, Chapt. 14.
7. U.S. Pat. 4,946,895 (Aug. 7, 1990), T. Ohmae and co-workers (to Somitomo Chemical Co. Ltd.).
8. U.S. Pat. 4,345,004 (Aug. 17, 1982), N. Miyata and H. Murase (to Hercules, Inc.).
9. U.S. Pat. 4,770,939 (Sept. 13, 1988), W. Sietsess and co-workers (to Labofina SA).
10. *Powder Coatings from Kynar 500PC,* Technical Data Bulletin, ATOCHEM North America, Philadelphia, Pa., June 1990.
11. U.S. Pat. 4,916,188 (Apr. 10, 1990), J. C. Reising (to Glidden Co.).
12. S. Harris, *Polym. Paint Colour J.* **180**(4273), 708 (Nov. 28, 1990).
13. S. A. Stachowiak, in G. D. Parfitt and A. V. Patsis, eds., *Organic Coatings Science and Technology,* Vol. 5, Marcel Dekker, Inc., New York, 1983, pp. 67–89.
14. U. Zorll, *Polym. Paint Colour J.* **179**(4241), (July 12, 1989).
15. S. Wu, *Polym. Plast. Technol. Eng.* **7**(2), 1976).
16. P. G. deLange, *J. Coat. Tech.* **56**(717), (Oct. 1984).
17. R. P. Franiau, *Paintindia* **37**(9), 33 (Sept. 1987).
18. U.S. Pat. 4,242,253 (Dec. 30, 1980), M. D. Yallourakis (to E. I. du Pont de Nemours & Co., Inc.).
19. U.S. Pat. 3,947,384 (Mar. 30, 1976), F. Schülde and co-workers (to Veba-Chemie AG).
20. D. P. Chapman, "Aluminum Pigment Technology for Water-borne and Powder Coatings," *Proceedings of the 18th Annual Water-Borne, Higher-Solids* and *Powder Coatings Symposium,* New Orleans, La., sponsored by University of Southern Mississippi, Feb. 1991.
21. U.S. Pat. 4,341,819 (July 27, 1982), D. A. Schreffler and C. M. Noonan (to the Polymer Corp.).
22. U.S. Pat. 3,631,149 (Dec. 28, 1971), H. Gempeler and P. Zuppinger (to Ciba Ltd.).
23. U.S. Pat. 4,855,358 (Aug. 8, 1989), S. C. Hart (to Morton Thiokol, Inc.).
24. K. McLeod, *Polymers Paint Colours J.* **179**(4250), 835 (Nov. 29, 1989).
25. U.S. Pat. 3,477,971 (Nov. 11, 1969), R. A. Allen and W. L. Lantz (to Shell Oil Co.).
26. Y. Merck, *Corrosion Prev. Control* **35**(2), 36–40 (Apr. 1988).
27. R. Van der Linde and E. G. Belder, in Ref. 13, pp. 55–66.
28. S. P. Pappas, V. D. Kunz, and B. C. Pappas, *J. Coat. Tech.* **63**(796), 39–46 (May 1991).
29. U.S. Pat. 3,792,011 (Feb. 12, 1974), J. D. B. Smith and R. N. Kauffman (to Westinghouse Elect. Corp.).
30. U.S. Pat. 4,147,737 (Apr. 3, 1979), A. J. Sein and co-workers (to DSM).
31. U.S. Pat. 4,740,580 (Apr. 26, 1988), Y. Merck and co-workers (to UCB, SA).
32. U.S. Pat. 4,910,287 (Mar. 20, 1990), J. J. McLafferty and L. Wang (to Ruco Polymer Corp.).
33. M. Hoppe, *J. Coat. Tech.* **60**(763), (Aug. 1988).
34. E. Bodnar, *State of the Arts of Weathering Resistant Powder Coatings for Outdoor Architectural Applications,* Paper FC 89-604, Finishing '89 Conference Papers, S.M.E., Dearborn, Mich., 1989.
35. U.S. Pat. 4,076,917 (Feb. 28, 1978), G. Swift and H. J. Cemci (to Rohm and Haas Co.).
36. K. Wood and D. Hammerton, "Hydroxyalkylamide Cross-linkers for Powder Coatings," in Ref. 20.

37. R. Dhein, H. K. Kreuder, and H. Rudolph, "Masked Isocyanates as Raw Materials for Powder Coatings," *Paint Mfg.* **16,** (July/Aug. 1973).
38. U.S. Pat. 4,264,751 (Apr. 28, 1981), A. S. Schiebelhoffer (to Goodyear Tire and Rubber Co.).
39. U.S. Pat. 4,354,014 (Oct. 12, 1982), E. Wolf and R. Gras (to Chemische Werke Hüls, AG).
40. U.S. Pat. 4,295,529 (July 26, 1983), K. A. Pai Panandiker and C. Danick (to Cargill Inc.).
41. U.S. Pat. 4,275,189 (June 23, 1981), C. Danick and co-workers (to Cargill Inc.).
42. U.S. Pat. 4,859,760 (Aug. 22, 1989), F. W. Light, Jr. and J. D. Hood (to Eastman Kodak Co.).
43. U.S. Pat. 4,118,437 (Oct. 3, 1978), G. Parekh (to American Cyanamid Co.).
44. U.S. Pat. 4,102,943 (July 25, 1978), R. A. Isaksen and F. J. Locke (to Monsanto Co.).
45. V. Reddy, "In-Mold Powder Coating," *Prod. Finish.* **53**(10), (July 1989).
46. U.S. Pat. 4,873,274 (Oct. 10, 1989), F. L. Cummings and G. D. Correll (to Morton Thiokol, Inc.).
47. U.S. Pat. 3,790,513 (Feb. 5, 1974), C. Victorius (to E. I. du Pont de Nemours & Co., Inc.).
48. U.S. Pat. 4,788,255 (Nov. 29, 1988), P. H. Pettit and M. L. Kaufman (to PPG Industries, Inc.).
49. U.S. Pat. 4,042,645 (Aug. 16, 1977), K. Hirota and co-workers (to Mitsui Toatsu).
50. U.S. Pat. 4,499,239 (Feb. 12, 1985), Y. Murakami and co-workers (to Dainippon Ink and Chemicals Inc.).
51. U.S. Pat. 3,561,003 (Feb. 2, 1971), B. J. Lanham and V. G. Hykel (to Magnavox Co.).
52. U.S. Pat. 3,737,401 (June 5, 1973), I. H. Tsou and J. W. Garner (to Grow Chemical Corp.).
53. M. Ettlinger, *Metalloberfläche* **41,** 9 (1987).
54. R. R. Sharetts and D. S. Richart, *Plas. Des. Process,* **10** (June 1962); **26** (July 1962).
55. C. K. Pettigrew, *Mod. Plas.* **43**(12), III (Aug. 1966); **44**(1), 150 (Sept. 1966).
56. A. H. Landrock, *Chem. Eng. Prog.* **63**(2), 86 (1967).
57. G. T. Robinson, *Prod. Finish.,* 16 (Sept. 1976).
58. U.S. Pat. 4,188,413 (Feb. 12, 1980), J. H. Lupinski and B. Gorowitz (to General Electric Co.).
59. M. L. Lang and P. J. Lloyd, *Int. J. Multiphase Flow* **13**(6), 823 (1987).
60. J. F. Hughes, *Inst. Met. Fin.* **64,** Pt. 2, 10 (May 1986).
61. A. W. Bright, *J. Oil Colour Chem. Assoc.* **60,** 23 (1977).
62. J. F. Hughes and Y. C. Ting, *IEEE/IAS 12th Annual Meeting,* Los Angeles, Calif., 1977, pp. 906–909.
63. V. G. Nix and J. S. Dodge, *J. Paint Tech.* **45**(586), 59 (Nov. 1973).
64. U.S. Pat. 4,498,913 (Feb. 12, 1985), G. F. Tank and S. O. Dawson (to Nordson Corp.).
65. U.S. Pat. 4,811,898 (Mar. 14, 1989), J. C. Murphy (to Nordson Corp.).
66. B. D. Moyle and J. F. Hughes, *J. Electrostatics* **16,** 277–286 (1985).
67. R. Lehmann, *Poly. Paint Colours J.* **179**(4246), (Oct. 4, 1989).
68. U.S. Pat. 4,747,546 (May 31, 1988), R. Talacko (to Ransburg-Gema AG).
69. "A Fully Classified Powder," *Paint Resin* **59,** 5 (Oct. 1989).
70. P. Binda, *Polymers Paint Colours J.* **180**(4), 273,710 (Nov. 28, 1990).
71. U.S. Pat. 4,243,699 (Jan. 6, 1981), J. E. Gibson.
72. "Surface Cleaning Finishing and Coating," in T. Lyman, ed., *Metals Handbook,* 9th ed., Vol. 5, American Society for Metals, Metals Park, Ohio, 1989.
73. *Near White Blast Cleaning,* Specification SSPC-SP10 Steel Structures Painting Council, Pittsburgh, Pa., 1963.
74. B. M. Perfetti *J. Coat. Tech.* **63**(795), 43 (Apr. 1991).

75. G. Bocchi, "North American Market Overview," *Powder Coating '90,* Conference Papers, Gardner Publications, Cincinnati, Ohio, Oct. 1990.
76. M. S. Reisch *Chem. Eng. News* **68**(38), 58 (Sept. 17, 1990).
77. I. Skiest and co-workers, *Powder Coatings II Multiple-Client Survey,* Skiest Inc., Whippany, N.J., Sept. 1990.
78. Data supplied by the Powder Coatings Institute, Alexandria, Va., July 1991.
79. C. H. J. Klaren, *J. Oil Colour Chem. Assoc.* **60,** 205 (1977).
80. "Paint—Tests for Formulated Products and Applied Coatings" and "Standard Recommended Practices for Testing Polymer Powders in Powder Coatings," ASTM 3451, *1991 Annual Book of ASTM Standards* Vo. 06.01, American Society for Testing and Materials, Philadelphia, Pa.
81. J. J. Brezinski, ed., *Manual on Determination of Volatile Organic Compounds in Paints, Inks, and Related Coating Products,* ASTM, Philadelphia, Pa., 1989.
82. J. Berry, "Control of Volatile Organic Compound (VOC) Emissions from Painting Operations in the United States," *Polymers Paint Colour J.* **181**(4281), (Apr. 17, 1991).
83. D. Richart, *Update on Health and Safety Aspects of Powder Coatings,* Powder Coating '90 Conference Papers, Gardner Publications, Cincinnati, Ohio, Oct. 1990.
84. I. Yeboah, *Conversion to Powder Coating Case Study,* Paper FC 89-627, Finishing '89 Conference Papers S.M.E., Dearborn, Mich., Oct. 1989.
85. R. Henshaw, *Conversion from Liquid to Powder—Cost Justification and Payback Analysis,* Powder Coating '90 Conference Papers, Gardner Publications, Cincinnati, Ohio, Oct. 1990.
86. "Toxicity Characteristic Leaching Procedure," *Code of Federal Regulations,* 40 CFR, Pt. 261, Appendix II, Method 1311.
87. P. H. Dobson, *Ind. Fin.* **77,** (Sept. 1974).
88. *Spray Finishing Using Flammable and Combustible Materials,* Code No. 33, National Fire Protection Association (NFPA), Quincy, Mass., 1989.
89. R. K. Eckhoff and G. H. Pedersen, *J. Hazard Materials,* **19**(1), (1988).
90. D. R. Scarbrough, in *Fire Protection Handbook,* 17th ed., NFPA, 1991, Chapt. 02-12.
91. *Guide for Explosion Venting,* Code No. 68, NFPA, 1988.
92. *Static Electricity,* Code No. 77, NFPA, 1988.
93. *Dust Explosion Prevention-Plastics Industry,* Code No. 654, NFPA, 1988.
94. *OSHA Regulations CFR,* Part 1910, Sections 1910.93, 1910.94, and 1910.107, 1974, et seq.
95. J. F. Fabries and co-workers, *Toxicity of Powder Paints by Inhalation,* Report No. 1092/RI, Institut National De Recherche Et De Sécurité (INRS), Dept. of Occupational Pathology, France, 1982.
96. *Power of Six Powder Paints to Cause Irritations and Allergies of the Skin INRS,* Dept. of Occupational Pathology, France, May 1979.
97. *Toxicological Studies of Uralac Powder Coating Resin and Powders,* Scado B.V., Zwolle, The Netherlands, 1979.
98. A. Wiedow, *Results of Inhalation Studies on PT810 and PT810 Containing Powders,* paper presented at CIBA GEIGY PT810 Health & Safety Seminar, Greenville, S.C., Sept. 23–24, 1991.
99. *Safe Handling of PT810 Based Powder Coatings,* CIBA GEIGY GG, Basle, Switzerland, Nov. 7, 1991.
100. *Threshold Limit Values for Chemical Substances and Physical Agents in the Workroom Environment,* American Conference of Governmental Industrial Hygienists, Cincinnati, Ohio, 1990–1991.
101. *OSHA Regulations,* 29 CFR, Section 1910.34.
102. *American National Standard Practices for Respiratory Protection,* National Standards Institute, Inc., New York, 1989.

General References

D. S. Richart, "Coating Methods, Powder Coating," in J. I. Kroschwitz, ed., *Encyclopedia of Polymer Science and Engineering,* 2nd ed., Vol. 3, John Wiley & Sons, Inc., New York, 1985, pp. 575–601.

M. T. Gillies, ed., *Powder Coatings—Recent Developments,* Noyes Data Corp., Park Ridge, N.J., 1981.

D. A. Bate, *The Science of Powder Coatings—Chemistry Formulation and Application,* Vol. 1, Scholicem Int., Port Washington, N.Y., 1990.

T. Misev, *Powder Coatings in Chemistry and Technology,* John Wiley & Sons, Inc., New York, 1991.

E. Miller, ed., *Users Guide to Powder Coatings,* Society of Mechanical Engineers, Dearborn, Mich., 1985.

DOUGLAS S. RICHART
Morton International, Inc.

SPRAY COATING

A coating may be applied to articles, ie, workpieces, by spraying. This application method is especially attractive when the articles have been previously assembled and have irregularly shaped and curved surfaces. The material applied is frequently a paint (qv), ie, a combination of resin, solvent, diluent, additives, and pigment. The material can also be a hot thermoplastic, an oil, or a polymer dissolved in a solvent. Many types of spray equipment are available. Methods can be used in combinations, and most of the techniques can be used for simple one-applicator manual systems or in highly complex computer-controlled automatic systems having hundreds of applicators. In an automatic installation, the applicators can be mounted on fixed stands, reciprocating or rotating machines, or even robots (Fig. 1).

Atomization

Airless Atomization. In airless- or pressure-atomizing systems, the coating is atomized by forcing the coating (or the liquid) through a small-diameter nozzle under high pressure. The fluid pressure is typically between 5 and 35 MPa (700–5000 psi); fluid flow rates are between 150–1500 cm³/min. In most commercial applications, a pump designed for the type of material sprayed is used to develop the high pressure. The pump can be mechanically, electrically, pneumatically, or hydraulically driven and the nozzle apertures have diameters ranging from 0.2 to 2.0 mm. The more viscous the fluid and the higher the desired liquid-flow rate, the larger the nozzle. As the fluid is forced through the nozzle, it accelerates to a high velocity and leaves the nozzle in a thin sheet or jet of liquid in the relatively motionless ambient air, producing a shear force between the fluid and air. The fluid is atomized by turbulent or aerodynamic disintegration, depending on the specific conditions (Fig. 2). The most common nozzles produce a long, narrow fan-shaped pattern of various sizes; others produce a solid or hollow

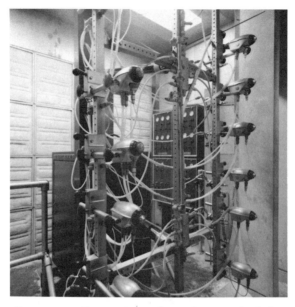

Fig. 1. An automatic spray-coating system, where air-automized electrostatic spray guns are mounted on reciprocators.

Fig. 2. Airless atomization process.

cone. The pattern of a fan nozzle can split and form "fingers" if the pressure is not sufficient. Because the coating material is often abrasive, the nozzle is typically made from tungsten carbide (see CARBIDES).

Airless atomization generally produces a medium-to-coarse particle size. Using a given nozzle, the higher the fluid pressure, the finer the atomization. Airless atomization can be used to atomize a large amount of material, or can atomize at high flow rates that can be rapidly deposited on the workpiece with minimal overspray or misting and excellent penetration into recessed areas. The flow rate is controlled by the nozzle size and the fluid pressure. The minimum nozzle size is determined by the size required to prevent plugging under operating conditions; minimum application pressure is determined by the required degree of atomization and the elimination of fingers.

A variation of airless atomization is called air-assisted airless. A small amount of compressed air at 35–170 kPa (5–25 psi) is introduced adjacent to the airless nozzle and impinges upon the thin sheet of fluid as it exits from the nozzle. This air aggravates the turbulence in the fluid and results in improved atomization at lower fluid pressures. Often, material that cannot be properly atomized using straight airless atomization can be using the air-assisted airless method. In some cases, the introduction of the air allows some control of the fan size.

Supercritical Atomization. Atomization can be obtained by mixing a supercritical fluid (SCF) with the material to be atomized. This process reduces volatile organic compound (VOC) emissions as the SCF acts as a solvent and replaces some of the hydrocarbon solvents in the material (see SUPERCRITICAL FLUIDS).

The material sprayed is generally a very high solids or viscous material that is thinned with the SCF to a spray viscosity in a special mixing/metering system. This mixture is then sprayed in an airless-type spray gun specifically designed for the process. In addition to the pressure or airless-type atomization, there is secondary atomization that results when the SCF dissolved in the spray material changes to a gas and rapidly expands (Fig. 3). Thus atomization quality is excellent, and materials that cannot be atomized using other methods can be used.

Fig. 3. Supercritical fluid atomization process.

An aerosol container can be considered a special application of airless atomization (see AEROSOLS). The pressure is usually supplied by a liquefied gas in the container at its equilibrium pressure. The material being sprayed has a very low viscosity to provide easy material flow through the feed tube and to permit fine atomization.

Air Atomization. In an air atomizer, an external source of compressed air, usually supplied at pressures of 70–700 kPa (10–100 psi), is used to atomize the liquid. Air atomization is perhaps the most versatile of all the atomization methods. It is used with liquids of low to medium viscosity, and flow rates of 50–1000 cm^3/min are common. Medium-to-fine particle sizes are produced, and the resulting surface finish is very good. It is sometimes difficult to penetrate small recessed areas, however, because the atomization air forms a barrier in the recess that the coating particles must then penetrate. When higher air pressures are used, air atomization produces considerable misting and overspray, which can be a disadvantage under some conditions. Air-atomizing devices can be of internal- or external-mix design.

The most common type of air atomizer is an external-mix where the coating material and atomization air are mixed in the space in front of the nozzle and air cap. An annulus of air generally surrounds the fluid as it leaves the tip, and the shear stress between the fluid and the air causes the initial atomization. In addition to the annulus, numerous (in some cases, as many as 10) air holes are placed in the air cap to direct air jets that continue the atomization process, assist in keeping the cap clean, and help shape the spray. Air jets coming from holes located in two diametrically opposed "horns" or ears produce a fan-shaped pattern that is oriented 90° from the horns. Where a fan pattern is not required, special nozzles expel the coating from a circular annulus about 0.6 mm thick surrounded by an air annulus. A swirl imparted to the atomization air in the annulus results in a very efficient atomization process and a spray having low momentum and misting. This is especially useful when coating long, narrow objects or when the atomizers are reciprocated or rotated to blend the patterns from several atomizers. It is difficult to coat a flat surface manually using this type of spray pattern.

In an internal-mix nozzle, the atomization air is mixed with the coating material before being forced through a nozzle or tip. As the mixture of air and coating material passes through the nozzle, its pressure is significantly reduced, and the resulting expansion produces the atomization of the coating material. This is a very efficient method of atomization, but usually coating material builds up around the fluid tip where the atomization occurs. This material is then torn away without being properly atomized, producing slugs that may be deposited on the workpiece and blemish the finish. It is therefore necessary to clean the tip of an internal-mix air atomizer frequently.

A special case of air atomization is high volume low pressure (hvlp) spray. In this case the air pressure at the spray gun is less than 70 kPa (10 psig) and there are relatively large (up to 0.32 cm) holes in the air cap to easily pass the low pressure air. This type of atomizer produces a soft or slow moving spray and is generally considered to be rather efficient in depositing the material on the workpiece. However, the use of low pressure air for atomization usually limits the viscosity and/or flow rate of the material that can be atomized.

The fluid delivery in an air-spray system can be pressure or suction fed. In a pressure-fed system, the fluid is brought to the atomizer under positive pressure generated with an external pump, a gas pressure over the coating material in a tank, or an elevation head. In a suction system, the annular flow of air around the fluid tip generates sufficient vacuum to aspirate the coating material from a container through a fluid tube and into the air stream. In this case, the paint supply is normally located in a small cup attached to the spray device to keep the elevation differential and frictional pressure drop in the fluid-supply tube small.

Most industrial production systems use a pressure-feed system, whereas many touch-up and recoating operations use suction feed. In a pressure-feed system, the fluid-flow rate and the atomization air are controlled independently. This permits the fluid-flow rate to be set to the desired value within a very large range. The pressure of the automization air is then matched with the fluid-flow rate to give the desired fineness of atomization. In a suction-feed system, the coating flow rate is determined by the flow of the atomization air and the size of the fluid orifice. Generally, it is not possible to suction feed more material than can be atomized by the quantity of air being used. A suction-feed system works best with low viscosity fluids, as the pressure differential available to transport the fluid is small, and higher viscosity liquids generally do not have sufficient flow rate for practical applications.

Electrostatic Atomization. The atomization of the coating material by electrostatic forces occurs when an electrical charge is placed on a filament or thin sheet of coating, and the mutual repulsion of the charges tears the coating material apart. For this process to produce acceptable atomization, the physical properties of the coating material must be within a relatively narrow range. The material is charged by an external source of high voltage, either prior to or as it is forced to flow over a knife edge, through a thin slot, or orifice, or it is discharged from the edge of a slowly rotating disk or bell (cup). The thin sheet or small diameter stringers or cusps of coating material are torn apart by the mutual repulsion of the charges on the material.

When a disk or bell is used, the coating material is fed near the center, and the rotation, generally 900–3600 rpm, provides a means of distributing the coating material to the edge of the device. At the edge, fluid surface tension or mechanical features of the rotating surface become the controlling factor, and the coating material comes off the surface in cusps. At higher rotation speeds, mechanical forces become significant and the stringers can break off at their base, resulting in larger particles being formed. Electrostatic atomization is limited to about 4 cm^3/min per centimeter of discharge length; voltages of 100–150 kV are used. This can be a very efficient coating method, but because of the required combination of low surface tension, low viscosity, and proper balance of electrical characteristics, this method of atomization has not been very successful for high solids or waterborne coatings (see COATINGS). Penetration into recessed areas is generally fair to poor, and excessive buildup of material on edges of the workpiece is possible.

Rotary Atomization. In rotary atomization, a bell (cup) or disk rotates at a speed of 10,000–40,000 rpm. In contrast to electrostatic atomization, mechanical forces dominate. The coating material is introduced near the center of the rotating device, and centrifugal force distributes it to the edge, where the material has an angular velocity close to that of the rotating member. As the coating material leaves the surface, its main velocity component is tangential, and it is spun off in the form of a thin sheet or small cusps as illustrated in Figure 4. The material is then atomized by turbulent or aerodynamic disintegration, depending on exact conditions. The diameter of a typical bell is 4–10 cm, whereas that of a disk is typically 10–25 cm. When a bell is used, the part to be coated is transported across the open face of the bell. A disk is generally used in an omega-shaped loop with the centerline of the disk on the centerline of the loop. The disk has a 360° spray pattern and is reciprocated up and down to cover the length of the part.

Rotary atomization produces the most uniform atomization of any of the aforementioned techniques, and produces the smallest maximum particle size. It is almost always used with electrostatics and at lower rotational speeds the electrostatics assist the atomization. At higher rotational speeds the atomization is principally mechanical in nature and does not depend on the electrical properties of the coating material. If the viscosity of a coating material is sufficiently low that it can be delivered to a rotary atomizer, the material can generally be atomized. The prime mover is usually an air-driven turbine and, provided that the turbine has the required power to accelerate the material to the angular velocity, liquid-flow rates of up to 1000 cm^3/min can be atomized using an 8-cm diameter bell.

Rotary atomization produces an excellent surface finish. The spray has low velocity, which allows the electrostatic forces attracting the paint particles to the

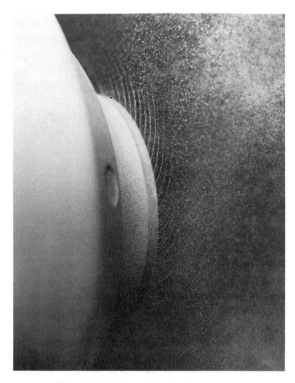

Fig. 4.   Rotary atomization process.

ground workpiece to dominate, and results in transfer efficiencies of 85–99%. The pattern is very large and partially controlled and directed by shaping air jets. The spray when using a metallic cup has relatively poor penetration into recessed areas. Excessive material deposited on the edges of the workpiece can also be a problem.

Recent developments in rotary atomization include the use of semiconductive composites (qv) for the rotary cup permitting the construction of a unit that does not produce an ignition spark when brought close to a grounded workpiece yet has the transfer efficiencies associated with a rotary atomizer. In addition, the use of the semiconductive material softens the electrostatic field and results in less edge buildup and better penetration into recess areas. Other systems use electronic means to effectively prevent arcing to grounded surfaces.

Electrostatic Spraying

Use of electrostatic spraying or electrostatic deposition increases the efficiency of material transfer to the workpiece (see COATING PROCESSES, POWDER TECHNOLOGY). The cost of solvents and coating materials and the emphasis on reducing emissions to the atmosphere have both increased dramatically since the late 1970s. These factors have effected an emphasis on increased transfer efficiency, ie, the fraction of the material removed from the coating bucket that is placed on the workpiece. The transfer efficiency is affected by the painting technique, workpiece geometry, the coating material, how the workpiece is presented to the atomizer, the ambient air movement, and other variables.

Electrostatic forces can be very effective in increasing the transfer efficiency. An electrical charge, usually negative, is placed on the coating material before atomization or as the coating particles are being formed. This is accomplished either by direct charging, where the coating material comes in contact with a conductor at high voltage, or by an indirect method, where the air in the vicinity of the coating particles is ionized and these ions then attach themselves to the coating particles. An external voltage source of 60–125 kV is normally used. A voltage gradient is established between the vicinity of the atomizer and the grounded workpiece by using the charged coating particles, charged metal atomizer, or an electrode near the atomizer as a local source of a high voltage field. An electrostatic force is exerted on each coating particle equal to the product of the charge it carries and the field gradient. The trajectory of the particle is determined by all the forces exerted on the particle. These forces include momentum, drag, gravity, and electrostatics. The field lines influencing the coating particles are very similar in arrangement to the alignment of iron particles when placed between two magnets. Using this method, coating particles that would normally pass alongside the workpiece are attracted to it, and it is possible to coat part or all of the back side of the workpiece.

Electrostatic spray atomizers are constructed from metal or nonconductive materials. A metal atomizer has sufficient electrical capacitance that when it is approached rapidly by a grounded object, eg, workpiece, an electrical arc may occur that can have sufficient energy to ignite certain solvent-air mixtures. A metal atomizer offers maximum ruggedness and efficiency but may present a fire hazard if not electronically protected. Thus this type of system often employs an electronic feedback system to reduce voltage and prevent arcing under these conditions. Most nonconductive atomizers are of a nonincendiary design; the rate of energy discharge has been specifically limited in such a way not to cause ignition. However, this type of atomizer is generally not as rugged as a metal atomizer, and in operation, the working voltages decrease, resulting in somewhat lower transfer efficiency.

All of the atomization techniques that produce a spray can be used with electrostatic spraying. Electrostatic atomization by definition uses electrostatic deposition. Furthermore, in rotary atomization the momentum toward the workpiece is relatively low and the transfer efficiency very poor if it is used without electrostatic deposition. Air and airless sprays also benefit from electrostatics in transfer efficiency. One problem for electrostatic spraying is that penetration into recessed areas is more difficult because the coating particles are attracted to the edges of the workpiece. The edges are closer to the high voltage source and therefore concentrate the field gradients. This problem can be overcome by reducing the voltage level or by using atomizers having high particle momentum, such as air or airless atomizers.

Economic Aspects

Spray equipment is marketed in a variety of ways in the United States. Several large manufacturers have broad product lines and sell equipment both directly to the user and through distributors. These companies can also provide custom engineered automatic systems for specific application. The largest U.S. manufacturers include Binks, Graco, ITW (DeVilbiss and Ransburg), and Nordson. There is also a multitude of smaller companies that provide a limited product line. Automatic spray systems can also be purchased through system houses that engineer an entire system, which might include the pretreatment system, ovens, and conveyor line.

Both individual components and small prepackaged systems are usually available from the many general distributors as are manual systems. Some manual systems are purchased in conjunction with automatic systems. Some of the larger distributors also custom design small automatic systems using standard components.

General References

K. J. Coeling and T. Bublick, in J. I. Kroschwitz, ed., *Encyclopedia of Polymer Science and Engineering,* 2nd ed., Vol. 3, John Wiley & Sons, Inc., New York, pp. 567–575.
R. P. Fraser, N. Dombrowski, and J. H. Routley, *Chem. Eng. Sci.* **18,** 339 (1969).
R. Ingebo, M. M. Elkotb, M. A. El-Sayed Mandy, and M. E. Montaser, *Proceedings of the 2nd International Conference on Liquid Atomization and Spray Systems,* Madison, Wis., 1982, pp. 9–17, 107–115.
N. Dombrowski and W. R. Johns, *Chem. Eng. Sci.* **18,** 203 (1963).
D. Beadley, *J. Phys. D.* **6,** 1724 (1973).
J. Gretzinger and W. R. Marshall, Jr., *AIChE J.* **7**(2), 312 (June 1961).
R. Tholome and G. Sorcinelli, *Indust. Finish.* (Nov. 1977).
J. O. Hinte and H. Milborn, *J. Appl. Mech.,* 145 (June 1950).

N. Dombrowski and T. L. Lloyd, *Chem. Eng. J.,* 63 (1974).

K. J. Coeling, *Operating Characteristics of High Speed Centrifugal Atomizers,* The DeVilbiss Co., Toledo, Ohio, 1981.

F. A. Robinson, Jr., G. Pickering, and J. Scharfenberger, *Paint Con. '84,* Hitchcock Publishing Co., Carol Stream, Ill., 1984.

J. M. Lipscomb, *Surface Coating '83,* Chemical Coaters Association, *Finishing '83 Conference Proceedings,* Association for Finishing Processes of SME, pp. 10-1, 10-10.

C. Dumouchel and M. Ledoux, *Proceeding of the 5th International Conference on Liquid Atomization and Spray Systems,* Garthersberg, Md., 1991, pp. 157–164.

D. C. Busby and co-workers, *Supercritical Fluid Spray Application Technology: A Pollution Prevention Technology for the Future,* pp. 218–239; *Proceedings of the 17th Water-Borne and High-Solid Coating Symposium,* New Orleans, La., 1990.

J. Schrantz and J. M. Bailey, *Indust. Finish.* (June 1989).

F. Robinson and D. Stephens, *Indust. Finish.* (Sept. 1990).

J. Wesson, *Prod. Finish.* (Aug. 1990).

K. J. COELING
Nordson Corporation

COATINGS

Coatings are ubiquitous in an industrialized society. U.S. shipments of coatings in 1990 were about 3.8×10^6 m³ having a value of \$11.6 billion (1). Coatings are used for decorative, protective, and functional treatments of many kinds of surfaces (see also PAINTS). Coatings such as interior flat wall paints are used for decoration but incidentally serve to reflect light diffusely to provide more uniform illumination. Some coatings, such as those on undersea pipelines, are only for protective purposes (see also COATINGS, MARINE). Others, such as exterior automobile coatings, fulfill both decorative and protective functions. Still others provide friction control on boat decks or car seats. Some coatings control the fouling of ship bottoms, others protect food and beverages in cans (see FOOD PACKAGING). Silicon chips, printed circuit panels, coatings on waveguide fibers for signal transmission, and magnetic coatings on videotapes and computer disks are among many so-called high-tech applications for coatings (see also ELECTRONIC COATINGS; INFORMATION STORAGE MATERIALS; MAGNETIC TAPE).

Each year tens of thousands of coating types are manufactured. In general, these are composed of one or more resins or polymers, a mixture of solvents (except in powder coatings), commonly one or more pigments (qv), and frequently one or more additives. Coatings can be classified into thermoplastic and thermosetting coatings. Thermoplastic coatings contain at least one polymer having sufficiently high molecular weight to provide the required mechanical strength properties without further polymerization. Thermosetting coatings contain lower molecular-weight polymers that are further polymerized after application in order to achieve desired properties. The term thermosetting is used not only when the coating is cured or baked in an oven after application, but also when it is dried or cured at ambient temperature (see also COATING PROCESSES).

Film Formation

Most coatings are manufactured and applied as liquids and are converted to solid films after application to the substrate. In the case of powder coatings, the solid powder is converted after application to a liquid, which in turn forms a solid film (see COATING PROCESSES, POWDER TECHNOLOGY). The polymer systems used in coatings are amorphous materials and therefore the term solid does not have an absolute meaning, especially in thermoplastic systems such as lacquers, most plastisols, and most latex-based coatings. A useful definition of a solid film is that it does not flow significantly under the pressures to which it is subjected during testing or use. Thus a film can be defined as a solid under a set of conditions by stating the minimum viscosity at which flow is not observable in a specified time interval. For example, it is reported that a film is dry-to-touch if the viscosity is greater than about 10^6 mPa·s($=$ cP) (2). However, if the definition of dry is that the film resists blocking, that is, sticking together, when two coated surfaces are put against each other for two seconds under a mass per unit area of 1.4 kg/cm^3 (20 psi), the viscosity of the film has to be greater than 10^{10} mPa·s($=$ cP).

The viscosity of amorphous systems is a function of free-volume availability. The free volume of a material is the summation of the spaces or holes that exist between molecules of a material resulting from the impact of one molecule or molecular segment striking another. Such holes open and close as the molecules vibrate. Above the glass-transition temperature (T_g) the holes are large enough and last long enough for molecules or molecular segments to move into them. Free volume increases as temperature increases and the rate of volume increase is higher above T_g. An important factor affecting the free volume of a system is thus the difference between the temperature, T, and the T_g. The relationship between viscosity, η, and T_g is expressed in the Williams-Landel-Ferry (WLF) equation. Using so-called universal constants and assuming that viscosity at T_g is 10^{15} mPa·s($=$ cP) the WLF equation, when η is in units of Pa·s, becomes:

$$\ln \eta = 27.6 - \frac{40.2(T - T_g)}{51.6 + (T - T_g)} \qquad (1)$$

Using this equation, the approximate $(T - T_g)$ value required for a film of a thermoplastic copolymer to be dry-to-touch, ie, to have a viscosity of 10^6

mPa·s($=$cP), can be estimated (3,4). The calculated $(T - T_g)$ for this viscosity is 54°C, which, for a film to be dry-to-touch at 25°C, corresponds to a T_g value of -29°C. The calculated T_g necessary for block resistance at 1.4 kg/cm³ for two seconds and 25°C, ie, $\eta = 10^{10}$ mPa($=$cP), is 4°C. Because the universal constants in the WLF equation are only approximations, the T_g values are estimates of the T_g required. However, if parameters such as the mass per area applied for blocking were larger, the time longer, or the test temperature higher, the T_g of the coating would also have to be higher.

For practical coatings, it is not sufficient just to form a film; the film must also have a minimum level of strength depending on product use. Film strength depends on many variables, but one critical factor is molecular weight. For example, the acrylic polymers used in lacquers for refinishing automobiles must have a weight average molecular weight $(\overline{M}_w)$ above 75,000. This required molecular weight varies according to the chemical composition of the polymer and the mechanical properties required for a particular application.

Solvent Evaporation from Solutions of Thermoplastic Polymers. A solution of a copolymer of vinyl chloride (chloroethene) [75-01-4], C_2H_3Cl, vinyl acetate (acetic acid ethenyl ester) [108-05-4], $C_4H_6O_2$, and a hydroxy-functional vinyl monomer having a number average molecular weight $(\overline{M}_n)$ of 23,000 and a T_g of 79°C, gives coatings having good mechanical properties without cross-linking (5). A simple coating having only the resin and 2-butanone (methyl ethyl ketone, MEK) [78-93-3], C_4H_8O, as the sole solvent would give a polymer concentration of about 19 wt % solids or approximately 12 vol % in order to have a viscosity of about 100 mPa·s($=$cP) for spray application (see COATING PROCESSES, SPRAY COATINGS). Because of the relatively high vapor pressure under application conditions, MEK evaporates rapidly and a substantial fraction of the solvent evaporates in the time interval between the coating leaving the orifice of the spray gun and arrival on the surface being coated. As the solvent evaporates, the viscosity increases and the coating reaches the dry-to-touch stage very rapidly after application and does not block under the conditions discussed. However, if the film is formed at 25°C, the dry film contains several percent retained solvent.

In the first stages of solvent evaporation from such a film the rate of evaporation depends on the vapor pressure at the temperatures encountered during the evaporation, the ratio of surface area to volume of the film, and the rate of air flow over the surface (6), and is essentially independent of the presence of polymer. However, as the solvent evaporates the T_g increases, free volume decreases, and the rate of loss of solvent from the film, at some point, becomes dependent not on how fast the solvent evaporates, but on how rapidly the solvent molecules can diffuse to the surface of the film. In this diffusion-control stage, solvent molecules must jump from free-volume hole to free-volume hole to reach the surface where evaporation can occur. As solvent loss continues, T_g increases, and free volume decreases. When the T_g of the remaining polymer solution approaches the temperature at which the film is being formed, the rate of solvent loss becomes very slow. If the film is being formed at 25°C from a solution of a resin having a higher T_g, eg, 79°C, loss of solvent becomes very slow when the T_g of the film exceeds 25°C and a significant amount of MEK remains in the hard, dry film indefinitely, acting as a plasticizer. In order to remove the last of the MEK in a short time, it is necessary to heat the film to a temperature significantly above the T_g of the solvent-free polymer.

The rate of solvent diffusion through the film depends not only on the temperature and the T_g of the film but also on the solvent structure and solvent-polymer interactions. The solvent molecules move through free-volume holes in the films and the rate of movement is more rapid for small molecules than for large ones. Additionally, linear molecules may diffuse more rapidly because their cross-sectional area is smaller than that of branched-chain isomers. For example, although isobutyl acetate (IBAc) [105-46-4], $C_6H_{12}O_2$, has a higher relative evaporation rate than n-butyl acetate (BAc) [123-86-4], $C_6H_{12}O_2$, IBAc diffuses more slowly out of a film of a nitrocellulose lacquer than BAc during the second stage of drying (7). Similarly, n-octane [111-65-9], C_8H_{18}, diffuses more rapidly out of alkyd films than isooctane (2,2,4-trimethylpentane) [540-84-1], C_8H_{18}, although isooctane has the higher relative evaporation rate (8).

Film thickness is an important factor in solvent loss and film formation. In the first stage of solvent evaporation, the rate of solvent loss depends on the first power of film thickness. However, in the second stage when the solvent loss is diffusion rate controlled, it depends on the square of the film thickness. Although thin films lose solvent more rapidly than thick films, if the T_g of the drying film increases to ambient temperature during the evaporation of the solvent, then, even in thin films, solvent loss is extremely slow. Models have been developed that predict the rate of solvent loss from films as functions of the evaporation rate, thickness, temperature, and concentration of solvent in the film (9).

Thermoplastic polymer-based coatings have low solids contents because the relatively high molecular weight requires large amounts of solvent to reduce the viscosity to that required for application. Air pollution (qv) regulations limiting the emission of volatile organic compounds (VOC) and the increasing cost of solvents has led increasingly to replacement of such coatings with types that require less solvent for application (see COATING PROCESSES, SURVEY).

Film Formation from Solutions of Thermosetting Polymers. Substantially less solvent is required in formulating a coating from a low molecular-weight resin that can be further polymerized to a higher molecular weight after application to the substrate and evaporation of the solvent. Theoretically, difunctional reactants could be used. However, in contrast to polymers for fibers and some plastics, this is not feasible for coatings where the close control of stoichiometric ratio and purity required to achieve a desired molecular weight reproducibly with difunctional reactants is impractical. Therefore, the average functionality must be over two in order to ensure that the molecular weight of the final cured film is high enough for good properties. Not only should the average functionality be over two, it is usually preferable for the number of monofunctional molecules to be at a minimum because these terminate polymerization. If any of the resin molecules have no functional groups, they cannot react and remain in the film as a plasticizer (see PLASTICIZERS). The reactions are commonly called cross-linking reactions. A cross-linked film not only has very high molecular weight it is also insoluble in solvents. For many but not all applications this solvent resistance is an advantage of thermosetting coatings over thermoplastic coatings.

The mechanical properties of the cross-linked film depend strongly on many factors; two of the most important are the lengths of the segments between cross-links and the T_g of the cross-linked resin. Segment length depends on the average equivalent weight and the average functionality of the components and the fraction of cross-linking sites actually reacted. The size of the segments between cross-links is often expressed as cross-link density (XLD): the closer the cross-links, the higher the XLD. Everything else being equal, the higher the XLD, the higher the modulus, that is, the harder the film. The T_g of cross-linked polymers is controlled by four factors and corresponding interactions: T_g of the segments of polymer between cross-links, the cross-link density, the presence of dangling ends, and the possible presence of cyclic segments (10). The T_g of the polymer segments between cross-links is governed by the chemical structures of the resin and cross-linking agent and by the ratio of these components. Because cross-links restrict segmental mobility, T_g increases as XLD increases. T_g also increases as the fraction of dangling ends decreases, that is, as cross-linking reactions proceed. The effect of cyclization has not been directly measured, but would be expected to restrict chain mobility and hence to increase T_g.

In the initial steps of cross-linking low molecular-weight resins, the molecular weight and XLD increase whereas the fraction of dangling ends decreases resulting in an increase in T_g. As the cross-linking reaction continues, a gel forms. Reaction does not stop at gelation but continues as long as there are functional groups to react and there is sufficient mobility in the matrix to permit the reactive groups to move into position for reaction. As the reaction continues, modulus above T_g increases and the film becomes insoluble in solvent. Solvent can still dissolve in a cross-linked film leading to swelling; the extent of swelling decreases as XLD increases.

A problem in thermosetting systems is the relationship between stability of the coating during storage before application and the time and temperature required to cure the film after application. It is generally desirable to be able to store a coating for many months or even years without significant increase in viscosity that would result from cross-linking during storage. On the other hand, after application, the cross-linking reaction should proceed in a short time at as low a temperature as possible. Because reaction rates depend on the concentration of functional groups, storage life can be increased by using more dilute coatings, ie, adding more solvent increases storage life. When the solvent evaporates after application, the reaction rate increases initially. However, VOC regulations are forcing the use of less and less solvent, increasing the problem of storage stability.

The reaction rates are also dependent on the rate constants for the reactions at the temperatures of storage and curing.

$$\ln k = \ln A - E_a/RT \qquad (2)$$

In general terms, the rate constant k changes most rapidly with temperature if the activation energy, E_a, of the reaction is high. However, the reaction rate is slow unless the preexponential term A is also large. Under the assumptions of required ratios of rate constants, rate equations and Arrhenius equations have been used to calculate what orders of magnitude of E_a and A are required to permit various combinations of storage times and curing temperatures (11). Such calculations show that to formulate a coating stable for six months at 30°C, the calculated kinetic parameters become unreasonable if cure is desired in 30 minutes below about 120°C. No known chemical reactions, whether useful in cross-linking coatings or not, have a combination of E_a and A that would permit design of such a system.

More reactive combinations can be used in so-called two-package coatings where one package contains a resin with one of the reactive groups and the second package contains the component with the other reactive group. The packages are mixed shortly before use. Two-package coatings are used on a large scale commercially but are not generally desirable. They take extra time and are generally more expensive. Material is usually wasted and there is a chance of error in mixing. Two-package coatings are increasingly being referred to as "2K coatings" and single-package coatings as "1K coatings." The K stands for the German word for component.

The design of stable coatings that cure at lower temperatures or shorter times must be based on factors other than the kinetics of the cross-linking reaction. To avoid the kinetic limitations on the ratio of storage stability to cure schedule several techniques are used. These include uv-curing systems; use of systems requiring an atmospheric component as a catalyst or reactant, eg, oxygen or moisture; use of a volatile inhibitor; use of a cross-linking reaction that is a reversible condensation reaction involving the loss of a volatile reaction product that includes some of this monofunctional volatile by-product as solvent in the coating; and use of catalysts or reactants that change phase over a narrow temperature range.

The problems of cross-linking reaction rates are not just a question of kinetics of the reaction of pairs of functional groups. Before two functional groups can react, they must be in close proximity. If there is no free volume, the functional groups cannot reach each other to react. If the free volume is sufficiently large, the functional groups have easy access to each other and the rate of reaction is governed by the kinetic parameters of the reaction. At intermediate levels of free volume, the reaction rate is controlled by the rate of diffusion of the reactants through the reaction matrix and the principal factor controlling availability of free volume is $(T - T_g)$. As low molecular-weight resins and cross-linking agents react, T_g increases and free volume decreases. If the cure temperature is at least somewhat higher than the T_g of the fully reacted system, the reaction can go to completion at rates governed by kinetic parameters. If, however, the cure temperature is significantly below the T_g of the fully reacted system, the reaction slows to a rate controlled by diffusion and stops before completion. As the reaction proceeds, T_g increases and in some cases it has been suggested that reactions stop when T_g has increased to the reaction temperature (12,13). Other work indicates that when the T_g equals T, the reaction rate diminishes by two to three orders of magnitude but continues slowly until the T_g has increased to a temperature 50°C above the reaction temperature (14). Examination of the WLF equation indicates that the universal constant in the denominator is 51.6, indicating that at 51.6°C below T_g, viscosity becomes infinite, that is, free volume approaches zero. Further studies are required, but coatings formulators can make practical use of the principles involved.

An example of the importance of free-volume availability on cross-linking has been reported in the evaluation of a trifunctional derivative of an aliphatic isocyanate which contains an aromatic ring, m-tetramethylxylidene diisocyanate

(TMXDI) [2778-42-9], $C_{14}H_{16}N_2O_2$, as a cross-linking agent for hydroxy-functional resins (15).

CH3—C—N=C=O with CH3 on C; benzene ring; C—N=C=O with CH3 groups

TMXDI

$$O=C=N-(CH_2)_6-N=C=O$$

HDI

Because of steric effects, TMXDI is less reactive than the trifunctional derivatives of hexamethylene diisocyanate (HDI) (1,6-diisocyanatohexane) [822-06-0], $C_8H_{12}N_2O_2$, but this can be overcome by catalyst composition and concentration. Although essentially complete reactions were obtained for films cured at elevated temperatures, when cured at 21°C the reaction was slow and essentially stopped at about 50% completion. The acrylic resin being cross-linked had been designed for use with a more flexible triisocyanate cross-linking agent derived from hexamethylene diisocyanate. In a series of acrylic resins having lower T_g, as the T_g decreased, the cross-linking reaction with TMXDI at 21°C was faster and went more nearly to completion. Using an appropriately designed acrylic resin and catalyst, the new isocyanate gave films which cured at room temperature at rates, extent reaction, and film hardness comparable to the results obtained using the linear, more flexible isocyanate and an acrylic resin designed for use with it. Resins and cross-linking agents must be selected or designed for use with each other, especially for curing systems used at ambient temperature.

Film Formation by Coalescence of Polymer Particles. Latex paints have low solvent emissions as well as many other advantages. A latex is a dispersion of high molecular-weight polymer particles in water (see LATEX TECHNOLOGY). The dispersion is stabilized by charge repulsion and entropic repulsion (16). (The terms steric repulsion or osmotic repulsion are sometimes used instead of entropic repulsion). Because the latex polymer is not in solution, the rate of water loss by evaporation is almost independent of concentration until near the end of the evaporation process (17). When a dry film is prepared from a latex, the forces that stabilize the dispersion of latex particles must be overcome and the particles must coalesce into a continuous film. As the water evaporates, the particles come closer and closer together. As they approach each other, they can be thought of as forming the walls of capillary tubes in which surface tension leads to a force striving to collapse the tube. The smaller the diameter of the tube, the greater the force. When the particles get close enough together so that that force pushing them together exceeds the repulsive forces holding them apart, coalescence is possible. A surface tension driving force also promotes coalescence because of the decrease in surface area when the particles coalesce to form a film. Both factors have been shown to be important in film formation from latexes (18). Coalescence, however, also requires that the polymer molecules in the particles be free to intermingle with those from adjoining particles. This movement can occur only if there are sufficient number and size of free-volume holes in the polymer particles into which the polymer molecules from other particles can move. In other words, the T_g of the latex particles must be lower than the temperature at which film formation is being attempted (19).

The rate of coalescence is controlled by the free-volume availability, which in turn depends mainly on $(T - T_g)$. The viscosity of the coalesced film is also dependent on free volume. If a film is to resist the mild blocking test described earlier, $(T - T_g)$ would have to be on the order of 21°C. If the film is to resist blocking at 40°C, the T_g should have to be above 19°C. However, in many cases the paint must be formulated in such a way that it can be applied at a temperature as low as 5°C, and therefore, the T_g of the latex particles would have to be below 5°C. (In this oversimplified example, the possible effects of pigments on film formation are ignored (3)). This problem is usually solved by adding a coalescing agent, such as 2,2,4-trimethyl-1,3-pentanediol monoisobutyrate (2-methylpropanoic acid 2,2,4-trimethyl-1,3-pentanediyl ester) [*132503-14-1*], $C_{12}H_{24}O_3$, tributyl phosphate [*126-73-8*], $C_{12}H_{27}O_4P$, and 2-butoxyethanol [*111-76-2*], $C_6H_{14}O_2$, among others. The coalescing agent dissolves in the latex particles, acts as a plasticizer, increases free-volume availability, and hence permits film formation at a lower temperature. After the film has formed, the coalescing agent diffuses slowly to the surface and evaporates. Because the free volume in the film is relatively low, the last traces of coalescing solvent evaporate slowly. Even though the films feel dry, they block and pick up dirt for a long time after application.

In another approach for forming a film at low temperatures and resisting blocking at higher temperatures, latexes are used in which the particles have outer shells of low T_g and interiors of higher T_g. The composition of the polymer in the particles is a gradient such that there is a gradient of T_g from the center of the particle to the outer shell. The T_g of the outer shell is low, permitting coalescence at low temperatures, but in time, after application, intermixing occurs, leading to a higher average T_g and permitting block resistance at higher temperatures (16).

Powder coatings form films by coalescence. Because the powder must not fuse or sinter during storage, the free volume at storage temperature must be sufficiently low to avoid coalescence at this stage. T_gs of powder coating particles are commonly of the order of 50 to 55°C. Higher T_gs must be avoided because rapid coalescence after application requires that the $(T - T_g)$ under baking conditions be as large as possible without requiring excessively high baking temperatures. The problem is even more complex in thermosetting powder coatings where overly rapid cross-linking can impede coalescence and leveling. Other examples of coalescing systems include nonaqueous dispersion, plastisol, "water-reducible" resin, and electrodeposition coatings (3).

Resins

Alkyds. Although no longer the principal class of resins used in coatings in the United States, alkyds are still very important and a wide range of types of alkyds are manufactured (see ALKYD RESINS). Whereas some nonoxidizing alkyds are used as plasticizers in lacquers and cross-linked with melamine–formaldehyde resins in baking enamels, the majority are oxidizing alkyds for use in coatings for air-dry and force-dry applications. The principal advantages of alkyds are low cost and relatively foolproof application characteristics, resulting primarily from low surface tensions. The principal shortcomings of these resins are embrittlement and discoloration upon aging and relatively poor hydrolytic stability.

Emphasis is being placed on developing high solids alkyd systems, for example, by using longer-oil alkyds. However, as oil length increases, air drying slows down for two reasons: the average number of functional groups, ie, activated methylene groups between two double bonds, eg, $-CH=CHCH_2CH=CH-$, per molecule decreases because the molecular weight decreases, and the ratio of aromatic rings to long aliphatic chains is reduced, resulting in lower T_g (see DRYING AGENTS, PAINTS; DRYING OILS). Solids can be increased at the same $\overline{M}_n$ by reducing polydispersity, that is, the breadth of molecular weight distribution. However, the properties of films from such narrow molecular-weight distribution alkyds tend to be inferior to those of broader molecular-weight distribution resins of similar composition (20). Earlier work on high solids baking alkyds showed that high functionality polyols such as tripentaerythritol (2,2-bis[[3-hydroxy-2,2-bis(hydroxymethyl)propoxy]methyl]]-1,3-propanediol) [78-24-0], $C_{15}H_{32}O_{10}$, produce high performance, high solids oxidizing alkyds for baking applications (21). For air-drying purposes, a high solids alkyd might be made using tripentaerythritol, phthalic anhydride (1,3-isobenzofurandione) [85-44-9], $C_8H_4O_3$, highly unsaturated fatty acids, and sufficient benzoic acid [65-85-0], $C_7H_6O_2$, to give an appropriate ratio of aromatic to aliphatic moieties.

Another approach to high solids alkyd coatings is the partial replacement of solvent in a conventional long-oil alkyd with a reactive diluent. For example, dicyclopentenyloxyethyl methacrylate [70191-60-5], $C_{16}H_{22}O_3$ (22), mixed amides from the condensation of acrylamide (qv) (2-propeneamide) [79-06-1], C_3H_5NO, and drying oil acid amides with hexamethoxymethylmelamine [3089-11-0], $C_{15}H_{30}N_6O_6$, (23,24), and trimethylolpropane triacrylate [37275-47-1], $C_{12}H_{18}O_5$ (25), have been recommended. Such reactive diluents give solutions of higher viscosity, ie, lower alkyd concentration at the same viscosity, but do not evaporate after application; they coreact with the alkyd resin and become part of the final film. Because molecular weight and resultant contribution to viscosity are lower than that of the alkyd, VOC emissions of the final coatings are reduced.

dicyclopentadienyloxyethyl methacrylate

A water-reducible coating or resin is one that is diluted with water before use. Water-reducible alkyds give comparable drying performance to solvent-borne alkyds. However, they are not widely used because film properties tend to be poorer than those of solvent-borne alkyds, especially in air-dry systems (26). This is partly because of alcoholysis of the alkyd by primary alcohols such as 1-butanol [71-36-3], $C_4H_{10}O$, a common solvent in water-reducible alkyds (27,28); secondary alcohols such as 2-butanol [*78-92-2*], $C_{10}H_4O$, minimize this problem (27). In any case, the slow loss of amine or ammonia leads to short-term high sensitivity to water. Even in the fully dry films, the presence of unreacted carboxylic acid groups leads to films having comparatively poor water resistance limiting their usefulness.

Polyesters. The term polyester is used in the coatings field almost entirely for low molecular-weight hydroxy, or sometimes carboxylic acid, terminated oil-free polyesters (see POLYESTERS, THERMOPLASTIC). Polyesters have been one important class of replacements for alkyd resins in melamine–formaldehyde cross-linked baking enamels. Hydroxy-terminated polyesters are also used with polyfunctional isocyanates in making air-dry and force-dry coatings as well as with blocked isocyanates in coatings for higher baking temperature and in powder coatings (see ISOCYANATES, ORGANIC). Carboxylic acid-terminated polyesters are used predominantly with epoxy cross-linkers in powder coatings (see EPOXY RESINS). When adhesion directly to metal is required, polyesters are generally preferred over acrylic resins. On the other hand, when highest exterior durability is needed, especially over primers, acrylic resins are generally preferred over polyesters. This is probably primarily a result of the greater hydrolytic stability of acrylic resin coatings. In order to maximize the hydrolytic stability of polyesters for exterior durability, isophthalic acid (IPA) (1,3-benzenedicarboxylic acid) [*121-91-5*], $C_8H_6O_4$, is commonly used instead of phthalic anhydride (PA) in making these resins (see PHTHALIC ACID AND OTHER BENEZENE POLYCARBOXYLIC ACIDS). An aliphatic acid, such as adipic acid (AA) (1,6-hexanedioic acid) [*124-04-9*], $C_6H_{10}O_4$, is commonly used or dimer acids (qv), such as that shown may be employed.

$$(CH_2)_7COOH$$
$$(CH_2)_7COOH$$
$$CH_2CH{=}CH(CH_2)_4CH_3$$
$$(CH_2)_3CH_3$$

dimer fatty acid
(one isomer of mixture)

The flexibility of the final film is partly controlled by the ratio of aromatic to aliphatic dibasic acids. The esters of highly substituted polyols such as neopentyl glycol (NPG) (2,2-dimethyl-1,3-propanediol) [*126-30-7*], $C_5H_{12}O_2$, 1,4-dimethylolcyclohexane (cyclohexanedimethanol (CHDM)) [*27193-25-5*], $C_8H_{16}O_2$, and trimethylolpropane (TMP) (2-ethyl-2-(hydroxymethyl)-1,3-propanediol) [*77-99-6*], $C_6H_{14}O_3$, are more hydrolytically stable than esters of simple polyols such as ethylene glycol (1,2-ethanediol) [*107-21-1*], $C_2H_6O_2$ (see GLYCOLS). Mixtures of diol and triol are used to give an average functionality greater than two. The cross-link density of the final film is controlled by the equivalent weight, which is related to the average functionality and the number average molecular weight.

There has been significant progress in the development of polyesters for high solids coatings. In contrast to acrylic resins, the preparation of low molecular-weight and hence high solids polyesters, where substantially all of the molecules have a minimum of two hydroxyl groups is straightforward. The lower limit of the average molecular weight that is useful in baking systems is controlled by the volatility of the lowest molecular weight fractions. For conventionally prepared polyesters, the optimum number-average molecular weight for baking enamels has been reported to be 800–1000 (29). Further improvements are achieved by synthetic techniques that give narrower molecular-weight distributions (30,31). Very narrow molecular-weight, linear, hydroxy-terminated polyesters are commercially available, for example, K-Flex (King Industries) (32), presumably prepared by removal of low molecular-weight fractions using thin-film vacuum evaporation. Low molecular-weight, hydroxy-terminated polyester diols and triols derived from the reaction of caprolactone (2-oxepanone) [502-44-3], $C_6H_{10}O_2$, and a diol or triol are also commercially available, for example, Tone Polyols (Union Carbide Corp.) (33). These latter polyesters are used predominantly in mixtures having hydroxy-functional acrylic resins or somewhat higher molecular-weight polyesters to increase solids without as much sacrifice in physical properties as would result from their use as the sole polyester component. Because viscosity is in general increased by increasing the average number of functional groups per molecule, high solids polyesters are usually made with an average number of hydroxyl groups only a little over two. Although hindered polyols such as neopentyl glycol provide polyester coatings having hydrolytic stability superior to those of most less hindered polyols, the terminal hydroxyl groups on the resins seem to be less reactive toward a cross-linking resin such as melamine–formaldehyde (30). Coatings based on polyesters of dimethylolcyclohexane cross-linked with melamine–formaldehyde resins are reported to have superior test results for properties related to hydrolytic stability but tend to have higher viscosities than neopentyl glycol-based polyesters (34).

Water-reducible polyester resins have terminal hydroxyl and carboxylic acid groups and an acid number of 40–60. The acid number is the number of milligrams of KOH required to neutralize one gram of resin solids. In order to make such a resin reproducibly with a minimum risk of gelation, the reactivity of the different carboxylic acid groups must vary significantly. The use of trimellitic anhydride (1,3-dihydro-1,3-dioxo-5-isobenzofurancarboxylic acid) [552-30-7], $C_9H_4O_5$, near the end of the reaction at a lower temperature takes advantage of the higher reactivity of the anhydride group (35). Another method uses dimethylolpropionic acid (3-hydroxy-2(hydroxymethyl)-2-methylpropanoic acid) [4767-03-7], $C_5H_{10}O_4$, as part of the diol in making a polyester. The highly hindered carboxylic acid esterfies more slowly than the carboxylic acid groups from isophthalic and adipic acids.

trimellitic anhydride   dimethylolpropionic acid

After carrying out the reaction to the appropriate acid number, the water-reducible polyester is dissolved in an ether-alcohol such as 1-propoxy-2-propanol

[*1569-01-3*], $C_6H_{14}O_2$. A secondary alcohol should be used because primary alcohols lead to more rapid transesterification than secondary alcohols (27). An amine such as 2-(dimethylamino)ethanol [*108-01-0*], $C_4H_{11}NO$, is used to neutralize the acid, pigment is dispersed in the resin solution, melamine–formaldehyde resin is added as are additives, and the coating is reduced to application viscosity with water. The resin in the diluted coating is not in solution but forms aggregates swollen with solvent and water, which are present as a reasonably stable dispersion. The storage life of such systems is limited by their relatively facile hydrolysis (36). They can be used, but coatings should be stored only for a few weeks after the addition of water. Another potential problem with such polyesters is formation of some low molecular-weight nonfunctional cyclic molecules. In baking ovens, small amounts of such polyesters gradually accumulate by condensation in cool spots in the oven and eventually sufficient resin can accumulate to drip on products passing through the ovens. Water-reducible acrylic resins are used in much greater volumes because of the problems with water-reducible polyesters.

Amino Resins. Melamine–formaldehyde (MF) resins are the most widely used cross-linking agents for baking enamels (see AMINO RESINS AND PLASTICS). Unlike the MF resins used in plastics, textile, and paper (qv) treatment, the MF resins used in coatings are ether derivatives. The resins are made by reacting melamine (1,3,5-triazine-2,4,6-triamine) [*108-78-1*], $C_3H_6N_6$, and formaldehyde [*50-00-0*], CH_2O, followed by etherification of the methylol groups using an alcohol. Two classes of MF resins are widely used (37). Class I resins are made using excess formaldehyde and a high fraction of the methylol groups are etherified with alcohol. Commercial resins contain a range of compounds having a large fraction of the amine groups substituted with two alkoxymethyl groups; some oligomers have two or more triazine rings coupled with methylene ether groups. Various alcohols can be used. The largest volume Class I resins contain a high proportion of hexamethoxymethylmelamine (HMMM).

Class II MF resins are made using a lower ratio of formaldehyde to melamine and a significant fraction of the nitrogens have one alkoxymethyl group and a hydrogen. An example of one compound present is *N,N',N''*-trimethoxymethyl-melamine (TMMM). Class II resins have a larger fraction of oligomers.

The MF resins react with both hydroxyl and carboxylic acid groups and are used to cross-link any resins having such substituents. They also undergo self-condensation reactions, the Class II more rapidly than Class I. Class I resins tend to give cross-linked films having greater toughness and flexibility but Class II resins tend to cure at lower temperatures. Class I resins require strong acid catalysts such as sulfonic acids whereas Class II resin reactions are catalyzed by weaker acids such as carboxylic acids.

As with other resin systems, work has been concentrated on increasing the solids content of coatings. Class I resins have lower molecular weights than Class II resins and provide higher solids. Lowest molecular-weight resins are produced under conditions that minimize self-condensation and maximize HMMM content. Mixed methyl butyl ether Class I MF resins have lower T_g and viscosity and lead to coatings having somewhat lower VOC emissions (38). A further emphasis in research is to minimize emission of free formaldehyde and generally Class I resins are superior in this respect. Attempts to further limit free formaldehyde are being actively pursued.

In high solids coatings, especially those made with low average functionality polyester resins, good film properties require closer control of curing time and temperature, catalyst concentration, and ratio of reactants than do the older, lower solids, higher average functionality systems (39). The potential problem is especially notable using Class II MF resins; there is greater latitude using the higher functionality Class I MF resins. Although it had been believed that steric hindrance limited the reaction of Class I resins so that only about half of the potential functional groups reacted (37), complete reaction has been demonstrated (40). Stoichiometric amounts of Class I MF resin and hydroxyl groups react, but it is frequently desirable to use a higher ratio of MF resin. The excess MF resin gives coatings having a higher modulus above T_g, apparently by increasing the extent of self-condensation.

In water-borne coatings, methyl ether MF resins are preferred because of greater water solubility compared to that of higher alcohol derivatives. Although either Class I or II MF resins can be used in water-borne coatings, Class I resins are generally preferred because of better storage stability. In the case of anionic electrodeposition coatings, mixed methyl ethyl ether resins are commonly used for cross-linking the epoxy ester or other binder resin. The water solubility of these mixed ether MF resins is high enough for them to be incorporated into the system but sufficiently limited that essentially all of the MF resin dissolves in the epoxy ester aggregates rather than staying in solution in the continuous phase of the system. This behavior is essential to maintain balanced composition during application by electrodeposition.

Because of the high content of nitrogens having two alkoxymethyl groups, Class I MF resins require strong acid catalysts such as p-toluenesulfonic acid (pTSA) (4-methylbenzenesulfonic acid) [104-15-4], $C_7H_8O_3S$. The higher the concentration of catalyst the lower the curing temperature of the coating. However, the catalyst remains in the film and can catalyze hydrolysis of the cross-link bonds on exterior exposure, leading to film failure. Addition of sulfonic acid decreases storage stability. In order to avoid this problem, it is common to use blocked catalysts which are generally salts of a sulfonic acid and a volatile amine. It has been found that some grades of titanium dioxide pigment neutralize the catalyst and slow down curing as the coating is stored for longer times.

The choice of sulfonic acid can be critical in affecting film performance. Dinonylnapthalenedisulfonic acid [60223-95-2], $C_{28}H_{44}O_6S_2$, catalyst commonly gives superior film properties as compared to pTSA, perhaps because of greater solubility in the coating (41). On the other hand dodecylbenzenesulfonic acid [27176-87-0], $C_{18}H_{30}O_6S$, which is also more soluble, gives superior properties when the coating is applied over primer, but poor adhesion results when applied directly to steel. Apparently the sulfonic acid is strongly adsorbed on the steel surface and the adhesion of the coating to this monolayer of catalyst is poor.

Other amino resins besides MF resins are used to a lesser degree in coatings. Urea–formaldehyde resins are used in some coatings for wood furniture because these resins cross-link at lower temperatures than MF resins and the higher water resistance and exterior durability that can be obtained using MF resins are not needed. Ethers of formaldehyde derivatives of 6-phenyl-1,3,5-triazine-2,4-diamine [91-76-9], $C_9H_9N_5$ (benzoguanamine resins) give coatings having superior resistance to alkalies and detergents and are widely used as cross-linkers for coatings used on home laundry machines and dishwashers. Ethers of formaldehyde derivatives of glycoluril (tetrahydroimidazo[4,5-*d*]imidazole-2,5-(1*H*,3*H*)-dione) [496-46-8], $C_4H_6N_4O_2$, are especially useful for coatings requiring outstanding extensibility during forming (42). Amides react with formaldehyde to give methylol derivatives that in turn react with alcohols to give reactive ethers. Thermosetting acrylics having such reactive groups derived from acrylamide were among the first thermosetting acrylic resins.

benzoguanamine   glycoluril

Acrylic Resins.   Acrylic resins are the largest volume class of coatings resins. The term acrylic is used in a general sense to mean resins where a significant fraction of the comonomers are acrylic or methacrylic esters (see ACRYLIC ESTER POLYMERS; METHACRYLIC POLYMERS). Other monomers such as styrene (qv) (ethenylbenzene) [100-42-5], C_8H_8, vinyl acetate, and others, may be included in the copolymer. High molecular-weight thermoplastic acrylics, which were widely used in acrylic lacquers for new automobiles, are now restricted to repair and aftermarket lacquers because of the need to control VOC emissions from automobile assembly plants. As VOC controls become stricter, use of acrylic lacquers in repair shops can be expected to be eliminated. Use of thermosetting acrylic resins continues to expand with increasing emphasis on high solids and water-reducible types. In architectural coatings, acrylic latexes and styrene or vinyl acetate acrylic copolymer latex-based paints have replaced drying oil and alkyd-based paints almost entirely except in gloss enamels. The consumption of acrylic latexes in industrial coatings applications is small but growing. The main advantages of acrylic coatings involve the high degree of resistance to thermal and photoxidation and to hydrolysis, giving coatings that have superior color retention, resistance to embrittlement, and exterior durability.

Hydroxy-functional thermosetting acrylics are widely used in baking enamels for automobile and appliance top coats, exterior can coatings, and coil coating. Research efforts have been directed at increasing the solids content of such coatings while maintaining the excellent properties. In contrast to polyesters, where virtually all molecules have at least two hydroxyl groups, synthesis of very low molecular-weight acrylic resins having an average functionality of two to three and containing few molecules that are nonfunctional or only monofunctional is difficult. Free-radical polymerization, the usual method for synthesizing thermosetting acrylics, results in a random distribution of the 2-hydroxyethyl methacrylate (2-methyl-2-propenoic acid 2-hydroxyethyl ester) [868-77-9], $C_6H_{10}O_3$, comonomer in the oligomer chains and hence significant fractions of nonfunctional and monofunctional molecules unless the number average molecular weight is on the order of 3500 or higher and the average functional monomers per molecule is on the order of three or higher (43).

Various techniques have been studied to increase solids content. Hydroxy-functional chain-transfer agents, such as 2-mercaptoethanol [60-24-2], C_2H_6OS, reduce the probability of nonfunctional or monofunctional molecules, permitting lower molecular-weight and functional monomer ratios (44). Making low viscosity acrylic resins by free-radical initiated polymerization requires the narrowest possible molecular-weight distribution. This requires careful control of temperature, initiator concentration, and monomer concentrations during polymerization.

The initiator structure may be critical: it should yield radicals that are least active in hydrogen-abstraction reactions. Azo initiators have been widely used. More recently, initiators containing the *t*-amyl peroxide group have been recommended (45). Most so-called high solids acrylic resin coatings contain only 45–50 vol % solids. However, there are proprietary acrylic resins available that can be used to make coatings having as high as 70 vol % solids (46). Introduction of functional groups onto methacrylate oligomers synthesized by alkoxide initiated anionic polymerization can lead to narrow molecular-weight distribution and, therefore, higher solids (47). Group transfer polymerization is another approach to narrow distribution, low molecular-weight hydroxy-functional acrylic resins (48).

In contrast to the situation with high solids coatings, acrylic resins are, in general, more appropriate than polyesters for water-reducible baking coatings. Acrylic copolymers using acrylic acid (2-propenoic acid) [79-10-7], $C_3H_4O_2$ (see ACRYLIC ACID AND DERIVATIVES) and 2-hydroxyethyl methacrylate as functional comonomers, prepared in an ether–alcohol solvent and partially neutralized with an amine such as 2-(dimethylamino)ethanol, are stable against both hydrolysis and alcoholysis. After pigmenting and adding MF resin, the coatings can be diluted with water and applied with relatively low VOC emissions. Class I, high methoxymethyl content, MF resins are used as cross-linking agents. Although such systems are commonly called water-soluble, these acrylic resins are not truly soluble in water. On dilution with water, adequately stable dispersions of aggregates swollen with solvent and water are formed. As a result of the formation of these aggregates, the change in viscosity on dilution with water is abnormal. Initially the viscosity decreases; however, on further dilution, the viscosity increases. On still further dilution, the viscosity decreases rapidly. The pH after dilution is also abnormal, being on the order of 9 even with only 75% of the amount of amine required to neutralize the carboxylic acid groups on the resin. The viscosity at application solids is essentially independent of molecular weight. The morphology of such systems and the effects of variables on the abnormal viscosity and pH profiles obtained on dilution with water have been studied (49,50).

Another example of water-reducible acrylic resins is used in the interior linings of two-piece beverage cans. A graft copolymer of styrene, ethyl acrylate (2-propenoic acid ethyl ester) [140-88-5], $C_5H_8O_2$, and acrylic acid on a bisphenol A epoxy resin is prepared in an ether–alcohol solvent (51,52). Benzoyl peroxide (dibenzoyl peroxide) [94-36-0], $C_{14}H_{10}O_4$, is used as the initiator at a temperature high enough so the benzoyl free radicals generated rapidly lose carbon dioxide to form phenyl free radicals. The phenyl free radicals are good hydrogen abstractors, maximizing the opportunity for grafting. The product is, of course, a mixture of graft copolymer, unreacted epoxy resin, and ungrafted acrylic copolymer. The resin is partially neutralized with 2-(dimethylamino)ethanol and diluted with water to spray application viscosity. The resin is again not truly soluble in water but dilution using water gives a stable dispersion of resin aggregates swollen with solvent and water. The epoxy groups are probably hydrolyzed.

Latexes are used as binders in a majority of architectural paints and a small but increasing fraction of industrial coatings. A significant advantage of latex-based systems is the reduced VOC emissions. For household paints the ease of cleanup with water and low odor as compared with alkyd paints are primary advantages especially for interior paints. Because the molecular weight of the polymer in latexes is very high, cross-linking is not needed in most cases to obtain excellent film properties. Most performance properties of paint films made from latex paints are far superior to the performance obtained using alkyd paints, particularly desirable is the exterior durability and low probability of embrittlement with age and blistering. The principal limitations of latex-based coatings are the relatively poorer leveling as compared to solvent-borne coatings and the inability to make very high gloss latex coatings.

So-called pure acrylic latexes are employed for maximum durability as required, for example, in high performance exterior latex paints. On the other hand, interior flat wall latex paints do not need the high resistance to exterior exposure and hydrolysis. The most widely used latexes for this application are vinyl acetate copolymer latexes such as vinyl acetate/butyl acrylate (2-propenoic acid butyl ester) [141-32-2], $C_7H_{12}O_2$, copolymers having just sufficient butyl acrylate to provide the needed reduction in T_g to permit coalescence. Latexes made from these comonomers, having widely different reactivity ratios, require preparation under monomer starved polymerization conditions (53) (see POLYMERIZATION MECHANISMS AND PROCESSES).

A wide range of latexes is available for various performance, cost, and viscosity requirements. In some cases, such as for air-dry gloss enamels, uniform, small particle size latexes are required. For a high solids latex, a broad or bimodal particle size distribution is needed. Differences in composition from the interior core of the polymer particles to the other shell can be desirable (16). A small quantity of acrylic acid as a comonomer permits stabilization of latex particles without a high requirement for surfactants (16). Specialty comonomers, such as 2-(dimethylamino)ethyl methacrylate (2-methyl-2-propenoic acid 2-(dimethylamino)ethyl ester) [2867-47-2], $C_8H_{15}NO_2$, permit preparation of latexes having superior adhesion after exposure of films on gloss paint or steel substrates to water.

Epoxy Resins.   Epoxy resins (qv) are used to cross-link other resins with amine, hydroxyl, and carboxylic acid (or anhydride) groups. The epoxy group, properly called an oxirane, is a cyclic three-membered ether group. By far the most widely used epoxy resins in coatings are bisphenol A (BPA) (4,4'-(1-methylethylidene)bisphenol) [80-05-7], $C_{15}H_{16}O_2$, epoxy resins.

bisphenol A (BPA) epoxy resin

A principal use for epoxy resins is as a component in two-package primers for steel. One package contains the epoxy resin and the other a polyfunctional amine. In coatings, generally low molecular-weight polyamines are not useful because the equivalent weight is so low that the ratio of the two components would be very high, increasing the probability of mixing ratio errors; furthermore, low molecular-weight amines tend to have greater toxic hazards. Amine-terminated polyamides are widely used, they are sometimes called amido-amines but frequently just polyamides. Amide groups do not react readily with epoxy groups. Polyamides are made from long-chain dibasic acids, such as dimer acids, and a stoichiometric excess of a polyamine, such as diethylenetriamine (DETA) (*N*-(2-aminoethyl)-1,2-ethanediamine) [*111-40-0*], $C_4H_{13}N_3$. The long aliphatic chain contributes flexibility to the cross-linked film. If more rigid films are needed, so-called amine adducts are used. Amine adducts are the reaction products of a low molecular-weight BPA epoxy resin and a large excess of polyamine from which any unreacted polyamine is removed by vacuum thin-film evaporation. The product is a polyamine having substantially higher molecular and equivalent weight and rigid aromatic rings. In designing combinations of epoxy resin and amine for ambient temperature cure coatings, care must be taken so that the T_g of the fully cured coating is not too much over the temperature at which the film is to be cured, otherwise, the cross-linking stops because of restricted mobility in the film. Epoxy/amine coatings are particularly effective as corrosion protective primers for steel because the amine groups resulting from the cross-linking reaction promote adhesion to the steel in the presence of water and because the cross-linked resins are completely resistant to hydrolysis (see also CORROSION AND CORROSION CONTROL).

Epoxy resins also are used to cross-link phenolic resins (qv). Such coatings are widely used in interior can linings. The hydrolytic stability and adhesion of the coatings are critical in terms of selection for this application (54). Adhesion is still further improved by the incorporation of a small amount of epoxy phosphate in the coatings. Epoxy phosphates are made by reacting BPA epoxy resins and small amounts of phosphoric acid and water. Complex reactions occur including formation of partial phosphate esters of a primary alcohol from a ring opening reaction with an epoxy group (55).

Epoxy resins are widely used in powder coatings. Probably the largest volume usage is of BPA epoxy resins cross-linked with dicyanodiamide (cyanoguanidine) [*461-58-5*], $C_2H_4N_4$. Because BPA epoxy resins are easily photoxidized, they are not useful in coatings requiring exterior exposure. Triglycidylisocyanurate (TGIC) (1,3,5-tris(oxiranylmethyl)-1,3,5-triazine-2,4,6(1*H*,3*H*,5*H*)-trione) [*2451-62-9*], $C_{12}H_{15}N_3O_6$, is widely used in powder coatings that require exterior durability. It is most commonly used to cross-link carboxylic acid-terminated polyesters.

triglycidylisocyanurate

Another use for epoxy resins is as a raw material to make epoxy esters. Epoxy esters are made by reaction of BPA epoxy resins and drying oil fatty acids; each epoxy ring can potentially react with two fatty acid molecules and furthermore hydroxyl groups on the backbone of the epoxy resin can also esterify with the fatty acids. Generally a low molecular-weight BPA resin initially reacts with BPA to make a higher molecular-weight BPA resin and then the fatty acid is added directly to the reactor for the esterification step. Lower molecular-weight resins give lower viscosity epoxy esters but the functionality is also low so that more rapid cross-linking is obtained for epoxy esters based on higher molecular-weight BPA resins. These synthetic drying oils provide coatings having better adhesion and substantially better saponification resistance than alkyd resins, because backbone linkages are stable ether groups rather than ester groups. Exterior durability of these materials is, however, poor, and the principal use is in primers for steel. Epoxy esters of fatty acids having some conjugated double bonds can be converted to water-reducible resins by reaction with maleic anhydride (2,5-furandione) [108-31-6], $C_4H_2O_3$ (see MALEIC ANHYDRIDE, MALEIC ACID, AND FUMARIC ACID), followed by amine neutralization. Baking primers made with such resins provide the same corrosion protection as conventional solvent-borne epoxy ester primers.

For some years, this type of vehicle was used in anionic electrodeposition primers. However for automobiles cationic electrodeposition primers are now preferred. The vehicles for cationic primers are proprietary but probably consist of the reaction product of epoxy resins and polyfunctional amines solubilized by the conversion of amine groups into salts using an acid such as lactic acid (2-hydroxypropanoic acid) [50-21-5], $C_3H_6O_3$. Alcohol-blocked isocyanates are used as the cross-linking agents.

Urethane Systems.   Isocyanates react with a wide variety of functional groups to give cross-links. The most widely used coreactants are hydroxy-functional polyester and acrylic resins. The isocyanate group reacts with the hydroxyl group to generate a urethane cross-link. The reaction proceeds relatively rapidly at ambient or modestly elevated temperatures. In addition to the low curing temperatures, significant advantages of urethane coatings are generally excellent abrasion and impact resistance combined with solvent resistance.

A variety of polyisocyanates are commercially available; those most widely used in coatings are toluene diisocyanate (TDI) (2,4-diisocyanato-1-methylbenzene) [584-84-9], $C_9H_6N_2O_2$; hexamethylene diisocyanate (HDI) (1,6-diisocyanatohexane); isophorone diisocyanate (IPDI) (5-isocyanato-1-(isocyanatomethyl)-1,3,3-trimethylcyclohexane) [4098-71-9]; and tetramethylxylidene diisocyanate (TMXDI). Diisocyanates are generally relatively highly toxic materials and are not, in general, desirable as components of coatings. The diisocyanates are converted to some higher molecular-weight, higher functionality reactant.

TDI   IPDI

TDI and TMXDI are available as prepolymers derived by reaction of excess TDI or TMXDI with trimethylolpropane and removing the excess diisocyanate by vacuum thin-film evaporation. The average NCO functionality is a little over three. TDI, HDI, and IPDI are available as trimers having an isocyanurate ring, ie, substituted 1,3,5-triazine-2,4,6-($1H,3H,5H$)-triones. The average functionality is somewhat over three. HDI is also available as a biuret derivative having an average functionality of about three.

HDI-isocyanurate   HDI-biuret

Aromatic isocyanates such as TDI give films that discolor rapidly on exterior exposure. Aliphatic isocyanates, TMXDI is classified as an aliphatic isocyanate because the NCO group is not directly on an aromatic ring, give coatings having excellent exterior durability. Polyisocyanates can also be made by the copolymerization of m-isopropenyl-α,α-dimethylbenzyl isocyanate (TMI) (1-(1-isocyanto-1-methylethyl)-3-(1-methylethenyl)benzene) [2094-99-7], $C_{13}H_{15}NO$, and acrylic esters (56). The lower vapor pressure and skin permeability of these trifunctional and polyfunctional isocyanates reduces the toxic hazard but, as for any other reactive coating, caution should be exercized, especially when coatings are applied by spray techniques.

TMI

The reaction of isocyanates and alcohols is too rapid to permit formulation of one-package stable coatings from polyols and polyisocyanates. Thus two-package coating systems are used on a large scale: one package contains the polyisocyanate and the other a polyhydroxy-resin and the pigment(s). Reaction rates are controlled by catalysts. The most widely used catalysts are organotin compounds, most commonly dibutyltin dilaurate (DBTDL) (dibutyl bis[1-(oxododecyl)oxy]-stannate) [77-58-7], $C_{32}H_{64}O_4Sn$. Highly catalyzed systems cure rapidly at moderate temperature but have pot lives so short that they must be used in proportioning mixing spray guns so that the two packages are mixed just before spraying (see COATING PROCESSES, SPRAY COATING). Slower cure systems may have pot lives of several hours. Any polyhydroxy-resin can be used; the most widely used are polyesters and acrylics.

$$
\begin{array}{c}
\text{C}_4\text{H}_9 \quad \diagdown \quad \diagup \quad \text{O--O--C}\left(\text{CH}_2\right)_{10}\text{CH}_3 \\
\text{Sn} \\
\text{C}_4\text{H}_9 \quad \diagup \quad \diagdown \quad \text{O--O--C}\left(\text{CH}_2\right)_{10}\text{CH}_3
\end{array}
$$

DBTDL

A principal challenge is formulating to maximize pot life while still curing in a short time at low temperature. This is an increasing problem as coatings are formulated to higher and higher solids. The difficulty can be severe because in uncatalyzed systems, reaction rates are second order in respect to alcohol, so that the reaction rate slows down dramatically as the concentration of alcohol groups decreases. Use of tin catalysts like DBTDL is especially desirable because the catalyzed reaction rate with respect to alcohol has been shown to be one-half (57). Using DBTDL, pot life can be extended by including a volatile carboxylic acid such as acetic acid [64-19-7], $C_2H_4O_2$, in the formula.

The reaction rate of isocyanates and alcohols is affected by the media. Rates are slowest in strong hydrogen bond-acceptor solvents and most rapid in hydrocarbon, especially aliphatic hydrocarbon, solvents. Therefore, hydrogen-bonding solvents and resins having as low hydrogen bond potential as possible are used. In formulating air-dry coatings, one should select combinations of polyol and polyisocyanate such that the T_g permits complete reaction at curing temperature. Usually some excess isocyanate is used because some isocyanate also reacts with water from the atmosphere. In aircraft coatings, it is common to use a 2:1 ratio of isocyanate to hydroxyl. The excess isocyanate reacts with water to yield an amine which reacts very rapidly with another molecule of isocyanate to give a urea cross-link.

Another class of urethane coatings is known as moisture cure coatings. These are stable one-package coatings based on isocyanate-terminated resins as the sole vehicle. The cross-links formed are urea groups. Because pigments have water adsorbed on the surface of the particles, these coatings are almost always clears, ie, unpigmented, because the cost of drying pigment is usually prohibitive. Moisture cure clear coatings are employed where high abrasion resistance is needed, such as on floors, bowling alleys and pins, and similar installations.

Still another important class of urethane coatings is based on blocked isocyanates (58–60). Blocked isocyanates are made by reaction of a diisocyanate and a blocking agent giving a product that is quite stable in the presence of alcohols and water at ambient temperature, but which reacts with hydroxyl or amine groups at elevated temperatures. Using blocked isocyanates and hydroxy-functional acrylic or polyester resins, stable one-package coatings can be formulated without need for moisture-free pigments and solvents and with substantially reduced toxic hazard. However, the coatings require relatively high temperature cures and, in some cases, the blocking agent that evolves during cure presents a pollution or toxic hazard problem. The earliest blocking agents were phenols, which are still used in wire coatings. Oximes, especially MEK oxime (2-butanone oxime) [96-29-7], C_4H_9NO, give blocked isocyanates that react with alcohols at

130 to 140°C in the presence of a catalyst, but still give reasonable storage stability.

$$R-N=C=O \; + \; \begin{array}{c} R' \\ \diagdown \\ C-N-OH \\ \diagup \\ R'' \end{array} \; \rightleftarrows \; \begin{array}{c} \; \; H \;\; O \;\;\;\;\;\; R' \\ \; \; | \;\;\; \| \;\;\;\;\;\; \diagup \\ R-N-C-O-N-C \\ \;\;\;\;\;\;\;\;\;\;\;\;\;\;\;\; \diagdown \\ \;\;\;\;\;\;\;\;\;\;\;\;\;\;\;\; R'' \end{array}$$

These are used in coil coatings and other industrial coatings. Caprolactam (qv) (hexahydro-2H-azepin-2-one) [105-60-2], $C_6H_{11}NO$, is the blocking agent most widely used in urethane powder coatings. The reaction products of diisocyanates and diethyl malonate (propanedioic acid diethyl ester) [105-53-3], $C_7H_{12}O_4$, act as cross-linking agents for polyols at the lowest temperature of any commercial blocked isocyanates but they are not truly blocked isocyanates. Rather, they react with alcohols by transesterification to yield films cross-linked with ester and amide bonds instead of urethane cross-linked coatings (60,61).

Polyamines react with blocked isocyanates at lower temperatures than polyols. Oximes are used as blocking agents with polyamines in applications such as magnetic tape coatings where the curing temperature must be kept low to avoid heat distortion of the plastic tape substrate. The reactivity is too high to permit one-package coatings but the combination of amine and blocked isocyanate provides adequate pot life to permit application by roll coating. Alcohol-blocked isocyanates are sufficiently stable at ambient temperatures in the presence of amines to be used in one-package coatings. They are used in cationic electrodeposition coatings where there is also need for long-term stability in the presence of large amounts of water. Cures at reasonable temperatures are possible because the blocked isocyanates react with amine groups on the resins to give urea cross-links.

Isocyanates are also used to make coating resins which do not cross-link through reactions of the isocyanate groups. These have the advantage that the toxic hazards associated with isocyanates are handled in the resin factory rather than by the coatings applicator. One class of such products is urethane oils, also called urethane alkyds and uralkyds. A diisocyanate such as TDI reacts with partial glycerol esters of drying oils to yield polyurethanes analogous to alkyd resins. Like alkyds, urethane oils form dry films by autoxidation. The dry films exhibit superior resistance to hydrolysis and abrasion as compared to alkyd films. Many coating products for the do-it-yourself market labeled as varnish are urethane oils. Their properties are generally superior to varnishes as well as alkyds.

High molecular-weight, linear, hydroxy-terminated polyurethanes are used in lacquers for top coats for coated fabrics (qv). Low molecular-weight, hydroxy-terminated polyurethanes can be used with melamine–formaldehyde resins to replace polyesters to take advantage of the greater hydrolytic stability of the urethane groups as compared to ester groups and the greater abrasion resistance of the urethane coatings. These coatings are more expensive than polyesters, especially when made with aliphatic diisocyanates, and tend to give higher viscosity solutions than polyesters. Water-reducible polyurethanes can be made by using dimethylolpropionic acid as one of the polyols in the reaction. The carboxylic acid is so hindered that it reacts very slowly with the isocyanate, resulting in a carboxylic acid-substituted hydroxy-terminated polyurethane. Melamine–formaldehyde resins are used as cross-linking agents. Such polyurethanes have excellent hydrolytic stability compared to water-reducible polyesters and superior abrasion resistance. In view of the importance of developing low solvent emission coatings, considerable effort is being devoted to new types of water-borne urethane resins (62,63).

Other Coatings Resins. A wide variety of other resin types are used in coatings. Phenolic resins, ie, resins based on reaction of phenols and formalde-hyde, have been used in coatings for many years. Use has been declining but there are still significant applications, particularly with epoxy resins in interior can coatings.

Silicone resins provide coatings having outstanding heat resistance and ex-terior durability. The cost is relatively high but for specialized applications these resins are important binders (64). Silicone-modified acrylic and polyester resins provide binders having intermediate durability properties and cost. Fluorinated polymers also exhibit outstanding durability properties. High cost, however, lim-its applicability.

Polyfunctional 2-hydroxyalkylamides can serve as cross-linkers for carbox-ylic acid-terminated polyester or acrylic resins (65). The hydroxyl group is acti-vated by the neighboring amide linkage (66). Solid grades of hydroxyamides are finding use as cross-linkers for powder coatings (67).

$$\left(\begin{array}{c} CH_3 \\ | \\ HOCHCH_2 \end{array}\right)_2 NC(CH_2)_4 CN \left(\begin{array}{c} CH_3 \\ | \\ CH_2CHOH \end{array}\right)_2$$

hydroxyalkylamide cross-linker

A range of acetoacetylated resins has been introduced (68,69). The acetoace-toxy functionality can be cross-linked with melamine–formaldehyde resins, iso-cyanates, polyacrylates, and polyamines. There is particular interest for possible corrosion protection on steel because the acetoacetoxy group can form coordina-tion compounds (qv) with iron, perhaps enhancing the adhesion to steel surfaces (see CHELATING AGENTS).

Volatile Components

In most coatings, solvents are used to adjust the viscosity to the level required for the application process and to provide for proper flow after application (70). Most methods of application require coating viscosities of $50–1000$ mPa·s($=$cP). Many factors must be considered in the selection of the solvent or, more commonly, solvent mixtures. Except for water-borne systems, solvents are usually chosen that dissolve the resins in the coating formulation. Solubility parameters have been recommended as a tool for selecting solvents that can dissolve the resins. However, the concept of solubility parameters for the prediction of polymer sol-ubility is a gross oversimplification (71,72) and the old principle that like dissolves like is the most useful selection criterion. For resins that are soluble in esters (see ESTERS, ORGANIC) and ketones (qv), costs can generally be decreased by using mixtures that contain hydrocarbons (qv) and alcohols (see ALCOHOLS, HIGHER ALIPHATIC; ALCOHOLS, POLYHYDRIC) to replace part of the esters and ketones.

The problem of solvent selection is most difficult for high molecular-weight polymers such as thermoplastic acrylics and nitrocellulose in lacquers. As molecular weight decreases, the range of solvents in which resins are soluble broadens. Even though solubility parameters are inadequate for predicting all solubilities, they can be useful in performing computer calculations to determine possible solvent mixtures as replacements for a solvent mixture that is known to be satisfactory for a formulation.

An important characteristic of solvents is rate of evaporation. Rates of solvent loss are controlled by the vapor pressure of the solvent(s) and temperature, partial pressure of the solvent over the surface, and thus the air-flow rate over the surface, and the ratio of surface area to volume. Tables of relative evaporation rates, in which n-butyl acetate is the standard, are widely used in selecting solvents. These relative rates are determined experimentally by comparing the times required to evaporate 90% of a weighed amount of solvent from filter paper under standard conditions as compared to the time for n-butyl acetate. The rates are dependent on the standard conditions selected (6). Most tables of relative evaporation rates are said to be at 25°C. This, however, means that the air temperature was 25°C, not that the temperature of the evaporating solvent was 25°C. As solvents evaporate, temperature drops and the drop in temperature is greatest for solvents that evaporate most rapidly.

The higher the ratio of surface area to volume, the more rapid is the evaporation of solvent. When coatings are applied by spray gun the atomized particles have a very high ratio of surface area to volume and hence solvent evaporation is much more rapid during the time when the atomized particles are traveling from the orifice of the spray gun to the surface being coated than from the film of the coating on the surface. Except in the case of some high solids coatings, this rapid evaporation of solvent during the atomized stage permits formulation and spray application of coatings that level well yet do not sag on vertical surfaces. By adjusting the combination of solvents to the particle size of the droplets obtained with a particular spray gun and the distance between the spray gun and the object being sprayed, the viscosity of the coating arriving at the surface can be adjusted to stay low enough to permit leveling, yet increase rapidly enough to avoid, or at least minimize, sagging. The control of sagging and leveling is also affected by film thickness. Because a thin film has a higher surface area to volume ratio than a thicker film, the concentration of solvent left in a thin film decreases more rapidly with time than the concentration in a thick film. The variations in solvent evaporation are so dependent on the particular application conditions that tables of relative evaporation rate are only useful as general guidelines.

Final adjustment of solvent selection must be done under actual field conditions. In many baking coatings a significant fraction of the solvent is lost in the baking oven, yet the tables of relative evaporation rates are based on 25°C air. Information on evaporation rates for a small number of solvents as a function of temperature up to 150°C has been published (73).

The loss of solvent during spray application of high solids coatings is usually less than that from conventional lower solids coatings (74,75). As a result, it is difficult to control sagging during spray application of many high solids coatings.

The lower rate of solvent loss from high solids coatings has not been fully explained. Because of the lower molecular weight and higher concentrations of resins in high solids coatings, the ratio of numbers of solvent molecules to resin molecules is lower than in lower solids coatings. Hence, vapor pressure depression is greater in the case of high solids coatings. However, this difference would not seem to be large enough to account for the very large differences in solvent losses that have been reported. Another factor may be that high solids coatings may reach a stage where solvent loss is controlled by diffusion rate much earlier than is the case in low solids coatings (8,43).

Toxic hazards, environmental considerations, flammability, odor, surface tension, and viscosity also affect solvent selection. In high solids coatings, the effect of solvent choice on viscosity can be critical. Because the resins used in high solids coatings tend to have polar functional groups, it is usually desirable to use hydrogen bond-acceptor solvents so as to minimize hydrogen bonding between resin molecules, which tends to increase viscosity. Because VOC emission regulations are based on mass of solvent per unit volume of coating, it is important to take solvent density into consideration in comparing the effect of solvent selection on viscosity and choice of solvents.

Pigments

Pigments (qv) in coatings provide opacity and color. Pigment content governs the gloss of the final films and can have important effects on mechanical properties. Some pigments inhibit corrosion. Pigmentation affects the viscosity and hence the application properties of coatings. An important variable determining the properties of pigments is particle size and particle size distribution. Pigment manufacturing processes are designed to afford the particle size and particle size distribution that provide the best compromise of properties for that pigment. In the process of drying the pigment, the particles generally aggregate. The coatings manufacturer must disperse these dry pigment aggregates in such a way as to achieve a stable dispersion where most, if not all, of the pigment is present as individual particles. An excellent treatise on the chemistry, properties, and uses of pigments is available (76).

Pigments can be divided into four broad classes: white, color, inert, and functional pigments. The ideal white pigment, when dispersed in the coating, would absorb no visible light and would efficiently scatter light entering the film. Scattering leads to the reflection of diffuse light back out of the film. The opacity of the film increases as light scattering increases. Light scattering increases rapidly as the difference in refractive index between the pigment particles and the binder increases. Light scattering is also affected by the particle size of the pigment and its concentration. Efficient scattering of light permits hiding the substrate under a film of coating by the thinnest film. Hiding is also increased by light absorption so that the combination of light scattering and light absorption permits covering with the thinnest film.

Because of its high refractive index and relatively low absorption of light, rutile [1317-80-2], a form of titanium dioxide [13463-67-7], TiO_2, is the most widely used white pigment. The optimum particle size of rutile TiO_2 for scattering visible light is 0.19 µm at 560 nm. Because light scattering drops off faster on the lower side of the optimum than on the higher side, commercial rutile TiO_2 is made with an average particle size of 0.22–0.25 µm. If aggregates of TiO_2 particles are not separated and stabilized as a dispersion of individual particles, light scattering is low. In the case of rutile TiO_2, hiding increases linearly with concentration in a dry coating film until the pigment volume concentration (PVC) exceeds about 10%. For a further increase in concentration, hiding increases less than proportionally. Finally above about 22% PVC, hiding actually decreases because of less light scattering. Whereas the exact pigment content is system dependent, the most cost-efficient level of pigmentation using TiO_2 for white coatings is about 18%. Another crystal type of TiO_2, anatase [1317-70-0], is available; it absorbs even less visible light than rutile and hence gives whiter films but has a lower refractive index so that it does not scatter light as efficiently. TiO_2 can promote the photodegradation of coatings on exterior exposure, leading to erosion of binder from the surface exposing loose pigment particles. This phenomenon is called chalking. Surface treatment using alumina and silica gives pigments which lead to minimal chalking.

Small air bubbles also scatter light because the refractive index of air is about 1.0, whereas the refractive index of most polymers is approximately 1.5. Air bubbles in films are sometimes useful in increasing opacity but the efficiency in scattering light is much less than for rutile TiO_2 because the difference in refractive index is much smaller.

Color pigments selectively absorb some wavelengths of light more strongly than others. A wide variety of color pigments are used in coatings. A discussion of the different pigments and the advantages and disadvantages thereof is available (76). The selection of pigments for a coating formulation is based among other factors on color, cost, transparency or opacity, durability, and resistance to heat, chemicals, and bleeding, ie, solubility in solvents. In contrast to white pigments, where there is an optimum particle size for scattering, in the case of color pigments, smaller particle size leads to stronger light absorption. However, other properties are commonly adversely affected if the particle size is too small. For example, exterior durability commonly decreases and solubility increases as particle size is decreased. Pigments are produced having average particle sizes that provide the best compromise of color strength and other properties.

Aluminum-flake pigments are also widely used. There are two important classes: leafing and nonleafing aluminum pigments. Leafing aluminum pigments are surface treated so that when a coating is applied, the platelets come to the surface of the drying film, resulting in a barrier to permeation of water and oxygen and giving high light reflectance so that the appearance approaches that of aluminum metal. The nonleafing grade does not behave in this manner but tends to orient within the film parallel to the surface. Nonleafing aluminum pigments are widely used in automobile top coats to give the so-called metallic effect.

Inert pigments, also called extenders and fillers (qv), do not exhibit significant absorption or scattering of light when incorporated into coatings. In most cases, inert pigments are used to occupy volume in the coating composition. The viscosity and flow properties of coatings can be substantially affected by increasing the volume of pigments dispersed in a coating. Many properties of the dried film depend on the pigment volume concentration in the film. Although such effects could be obtained using higher volumes of white and color pigments than needed for opacity and color, the volume effects can generally be obtained by using less expensive inert pigment and just sufficient white and color pigments required for the opacity and color. A wide variety of clays (qv), silica, and carbonates are used as inert pigments (76).

The most widely used functional pigments are the corrosion-control pigments. These pigments inhibit the corrosion of steel.

Pigment Dispersion.   The dispersion of pigments involves wetting, separation, and stabilization. Wetting, that is displacement of air and water from the surface of the pigment by the vehicle, requires that the surface tension of the dispersion medium be lower than that of the pigment surface. Except when the vehicle is water, wetting is rarely a problem. Wetting agents are required for dispersion of low polarity surface pigments such as organics, in water, and occasionally for dispersion of pigments in organic media. The rate of wetting of pigment is fastest when the viscosity of the vehicle is low.

Pigment aggregates are separated into individual particles by a variety of dispersion equipment (77,78), which transmit shear stress of sufficient magnitude to break up the aggregates. If little force is required, low viscosity dispersions can be made using equipment that provides low shear stresses. The most widely used equipment of this type is a high speed impeller in a tank. Pigments where the aggregates are not easily separated require equipment that exerts a higher shear stress on the aggregates, such as ball mills, sand mills, shot mills, attritor mills, extruders, and others. For pigment aggregates that are even more difficult to separate, equipment imparting even higher shear stress is needed, eg, dough mixers and two-roll mills. Because this last type of equipment is expensive and labor costs are high, use is limited to the preparation of dispersions from expensive pigments where ultimate color strength is important. This class of dispersion equipment is especially useful when the aggregates must be reduced to the ultimate particle size in order to minimize light scattering, such as in transparent pigment dispersions for use in metallic coatings (qv) for automobiles.

Stabilization of the pigment dispersion is usually the most critical aspect of the process. If the dispersion is not fully stabilized, the separated pigment particles flocculate, ie, reagglomerate. Although flocculates are easily separated again by low shear stresses, flocculated dispersions are not desirable because the hiding and color strength are lower, reflecting the larger effective particle size of the flocculates in the film. Flocculation of the pigment in gloss coatings generally leads to reduction in gloss. Furthermore, at low shear rates the viscosity of flocculated systems is much higher than that of nonflocculated systems.

Pigment dispersions are stabilized by charge repulsion and entropic, ie, steric or osmotic, repulsion. Although both types of stabilization force may be present in most cases, for pigment dispersions in solvent-borne coatings entropic repulsion is usually the most important mechanism for stabilization. In solvent-borne coatings, entropic stabilization can generally be achieved by adsorption of resin on the pigment surfaces. The adsorbed layer of resin swells with solvent. It has been shown for a wide variety of pigments and binders, that a stable dispersion is obtained if the thickness of the adsorbed layer is at least 8 to 10 nm. If the thickness is less, flocculation occurs (79,80). The reduction in entropy involved in the compression of an adsorbed layer having an average layer thickness > 10 nm requires a force greater than that imparted by the Brownian motion of the pigment particles. The principal factors controlling adsorbed layer thickness are molecular weight of the resin and the presence of multiple, but limited numbers of adsorbable groups scattered along the resin chain.

Solvent selection can sometimes affect the stability of a pigment. A change in solvent may swell the adsorbed layer further and promote stabilization. On the other hand, some solvents might be so strongly adsorbed on the pigment surface that the resin could not successfully compete with it despite having multiple interaction sites. Such a system would not lead to a stable dispersion because the adsorbed layer thickness of a solvent layer is < 1 nm. Most conventional coatings are stabilized using the same resin system that has been selected for the coating binder. In some cases, a related resin having more functional groups or higher molecular weight has to be designed. However, in high solids coatings, low molecular-weight resins having a limited number of functional groups must be used, leading to thinner adsorbed layers. Therefore, flocculation is a more common problem than for conventional coatings (81). Lower molecular-weight stabilizing resins having multiple functional groups are being designed and used as stabilizers (82).

Another problem with high solids coatings is that the adsorbed layer thickness can be a limiting factor in the pigment loading. The adsorbed layer increases the volume fraction of internal phase that, in turn, is important in affecting the viscosity of highly pigmented high solids coatings (43). Stabilization of thinner adsorbed layers (4–5 nm) is possible using surfactants, presumably because the adsorbed layer thickness is uniform, whereas the minimum average 8–10 nm layer required for resins is not of uniform thickness. However, high concentrations of conventional surfactants can be required so that at equilibrium with surfactant in solution, the pigment surface is still completely covered by adsorbed surfactant.

Some surfactants have been designed that are sufficiently strongly adsorbed on a pigment surface that little excess is required (83). Such systems provide stable dispersions for high solids coatings and increased pigment loading, but a system must be developed for each individual pigment, at least in the case of organic pigments. Pigment manufacturers commonly treat the pigment surfaces during the manufacturing process to reduce the sensitivity to flocculation.

Experimental determination of adsorbed layer thickness requires a laboratory effort appropriate for research studies but not for day-to-day formulation development. The Daniels flow point method (77) is a simple procedure that permits determination of whether a resin-solvent-pigment combination can give a stable dispersion and, if so, approximately what concentration of resin in the solvent is required to prevent flocculation. The procedure permits use of the lowest possible viscosity vehicle for pigment dispersion. This permits more rapid wetting by the vehicle and also permits higher pigment loading at the same viscosity. The most expensive equipment in a coatings factory is the dispersion equipment. Therefore, it is desirable to disperse as high a volume of pigment per unit time as possible. Thus high pigment loading of mill bases provides the most efficient operations. Use of the Daniels flow point method provides the data required for calculating the highest pigment loading formulas for dispersing pigment using equipment such as ball mills and sand mills.

The determination of the extent of dispersion is not a simple task. Color strength of colored pigment dispersions and scattering efficiency of white pigment dispersions are determined by comparing the tinting strength to a standard. Flocculation can be detected by color changes on application of low shear stress, rapid settling or centrifugation to a bulky sediment, and by the presence of shear thinning flow. In the coatings industry, a steel bar having a tapered groove called a Hegman gauge is widely used for testing the degree of dispersion. A sample is placed in front of the deep end of the groove and the dispersion is drawn down to see at what gauge reading particles can be detected. The scale is 0–8 for a groove depth varying from 100 μm to 0. The diameter of most white and color pigments is less than 1 μm. The procedure is of little practical value because it detects only the presence of relatively large-size aggregates and does not give any indication of whether most particles are present in their ultimate particle size. Furthermore, the Hegman gauge cannot detect the presence of flocculation. For research purposes, the most accurate technique available for quantifying pigment dispersion is infrared back scattering (see INFRARED AND RAMAN SPECTROSCOPY).

Dispersion of pigments for latex paints is the principal application of aqueous dispersions in coatings. Commonly, three surfactants are used in preparing the dispersion of the white and inert pigments: potassium tripolyphosphate, an anionic surfactant, and a nonionic surfactant. The final colored latex paint formulations, which are very complex, commonly contain seven pigments, some having high surface energies (inorganic pigments) and some having low surface energies (organic pigments). The latex polymer is present as a dispersion that must be stabilized against coalescence and flocculation, and latexes themselves commonly contain two or more surfactants and a water-soluble polymer. Generally, but not always, the latex particles have a low surface tension. Complexity is increased by the use of at least one and sometimes two water-soluble polymers in the latex paint formula that can adsorb on the surface of some of the pigment particles and, in some cases, the latex particles. Although wetting inorganic pigments with water usually presents no problem, many organic pigments require a wetting agent to displace the air from the surface of the pigment particles. All dispersions must be stabilized against flocculation (85). The appropriate surfactants, wetting agents, and water-soluble polymers are selected largely by trial and error. Accumulation of data banks of successful and unsuccessful combinations provides a valuable tool for more efficient formulation.

Pigment Volume Relationships.   Pigmentation can have profound effects on the properties of coating films depending on the level of pigmentation of these films. Variations in these effects are best interpreted in terms of volume relationships rather than weight relationships. Pigment volume concentration (PVC) is defined as the volume of pigment in a dry film divided by the total volume of the dry film, commonly expressed as a percentage. The term pigment volume content is sometimes used for pigment volume concentration.

As the volume of pigment in a series of formulations is increased, properties change, and at some PVC there is a fairly drastic change in a series of properties that is a function of PVC. This PVC is defined as the critical pigment volume concentration (CPVC) for that system (86). The CPVC is the maximum PVC that can be present in a dry film of that system, having sufficient solvent-free resin to adsorb on all the pigment surfaces and fill all the interstices between the pigment particles. In other words, when the PVC is above the CPVC, there are voids in the film. Gloss decreases as PVC increases. Hiding generally increases as PVC increases but above CPVC there is a rapid increase in the rate of increase of hiding because of the presence of voids. In the unique case of rutile TiO_2, hiding passes through a maximum at about 22%. If the PVC of rutile pigmented coatings were increased above CPVC, the hiding would again start to increase rapidly. Tinting strength of white paints behaves like hiding, increasing rapidly above CPVC. Tensile strength of films increases with increasing PVC, passing through a maximum at CPVC. Stain resistance is poorer and the ease of removing stains becomes more difficult for coatings above CPVC, compared to those with PVC below CPVC. Blistering of films on wood is less likely to occur above CPVC. The intercoat adhesion to a primer is improved when the primer has a PVC greater than CPVC. The CPVC effects result from the voids in dry films having a PVC greater than CPVC.

CPVC depends on the pigments and pigment combinations. Pigment density is an important variable in comparing weight data with volume data. CPVC of dense pigments tends to be lower than might be expected looking at weight data. CPVC is generally lower for smaller particle size pigments, because these have a higher volume proportion of adsorbed resin than larger particle size pigments. Particle size distribution is very important. Broad distribution systems pack more efficiently and have higher CPVC values. Flocculation results in lower CPVC in a dry film as compared to the same pigment combination having a well-stabilized dispersion. Each coating application has a PVC:CPVC ratio that provides the best compromise of properties for the use involved. By determining the appropriate PVC:CPVC ratio for some application, this ratio can be used as a starting point for formulation of other pigment combinations for that application (86).

Experimental determination of CPVC is time consuming. A reasonable approximation can be obtained by determining the pigment loading that gives a viscosity approaching infinity in a 100% solids system, such as linseed oil with a pigment or combination of pigments. When expressed as grams of linseed oil required per 100 g of pigment, this infinite viscosity value is called the oil absorption of the pigment. Oil absorption values can be converted to CPVC values using the densities of the pigment(s) and the oil. Because different pigments have different particle size distributions, mixtures give CPVC values that are higher than the average based on the CPVC values of the individual pigments. Equations have been developed to calculate CPVC of pigment mixtures from a combination of oil absorption values and particle size distributions of the individual pigments (87). A spatula rub-up procedure has been used in the determination of oil absorption (77), but a Brabender plastograph gives much greater accuracy (88). Even if CPVC data are not available, a formulator can still make use of the concept by thinking in terms of volume rather than weight relationships. Volume relationships are particularly useful in maximizing the volume of inert pigment that can be used in a coating, which in turn minimizes cost while retaining the required physical properties.

Application of CPVC concepts to latex paints is controversial (89). However, most workers seem to agree that for a particular pigment combination, the CPVC of a latex paint is always lower than for the same pigment combination in a solution-based paint (77). This may result from the alignment of pigment and latex particles as the water evaporates in forming a film. Latexes of low T_g and small particle size give coatings having higher CPVCs than when higher T_g and/or larger particle size latexes are used. The addition of a coalescing agent to the formula tends to increase CPVC. To reduce cost of low gloss latex paints, the formula having the highest possible CPVC permits the incorporation of the largest volumes of low cost inert pigments.

Color Matching. Most pigmented coatings must be color matched to a certain standard. In many cases, a significant part of the time involved in formulating a new coating is in establishing the color match and a significant part of the manufacturing cost is the color matching operation. Poor color matching is a common source of customer complaints and problems and costs can be minimized if established specifications for the initial color match and for judging the acceptability of production batches are made.

Acceptance of a color recommendation made by the coatings supplier effectively eliminates the time and cost involved in an initial color match and ensures selection of a pigment combination appropriate to the coating use. If, however, a coatings user provides a sample or standard for a color match, formulators need the following information.

(1) *Possibility of a Spectral (Nonmetameric) Match.* Only if exactly the same pigments, including white and black ones if needed, can be used in establishing the new coating standard as were used in the customer's sample, can the fraction of light absorbed at each wave length be identical to the sample. Although a color can usually be matched under one light source using different pigment combinations, it is only possible to match the color under all light sources if chemical compositions are identical. If this is not possible, the user must accept a metameric match; that is, the colors match under some light source but not under others. For example, because the colorants cannot be identical if the sample is a dyed fabric, the color of a textile cannot be matched under all light sources by any coating. If the coatings user has been using a coating that has lead-containing pigments and wishes to have a lead-free paint, only a metameric match is possible.

(2) *Light Sources.* If the match is to be metameric, the coatings user and the supplier must agree on the light source(s) under which the color is to be evaluated. Furthermore, a decision should be made whether it is more desirable to have a close match under one light source without regard to how far off that match might be under other light sources, or to have only a fair match under several light sources.

(3) *Gloss and Texture.* The color of a coating depends in part on gloss and surface texture. The light reaching the eye of an observer has been reflected both from the inside and the surface of the coating film. The light from within the film is "colored light," whereas the light reflected from the surface of the film is "white light." The color seen by the observer varies with the ratio of light from within and that from the surface of the film. At most angles of viewing, more light is reflected from a low gloss surface than from a high gloss surface. Therefore, at most angles of viewing, a low gloss coating having exactly the same colorant combination as a high gloss coating has a lighter color than the high gloss coating. It is impossible to match the colors of high gloss and low gloss coatings from all angles of viewing. Therefore, there must be agreement as to the gloss, and if the gloss of the color standard is different from the gloss desired for the new coating, the angles of illumination and viewing must be agreed upon. It is not possible to make even a metameric match of the color of some fabric samples with a coating

under all angles of viewing because the colorants as well as the surface texture must be different. Claims to the contrary are misleading.

(4) *Color Properties Required.* Colorants must be chosen to permit formulation of a coating that can meet performance requirements such as exterior durability and resistance to solvents, chemicals, and heat. Health and safety regulations may also affect colorant choice.

(5) *Film Thickness and Substrate.* In most cases, a coating does not completely hide the substrate, and the color of the substrate affects the color of the applied coating. The extent of substrate effect depends on film thickness and is particularly important in applications such as can and coil coatings that are commonly less than 25 μm thick. A thin coating where the color was established over a gray primer does not match a standard applied over a red primer; a coating designed for a one-coat application on aluminum does not match the color standard when applied on steel.

(6) *Baking Schedule.* Because resin color can be affected by heating, the color of the coating is affected by the time and temperature of baking, and the baking schedule must be specified. Color requirements for overbaking must also be established. The combination of a short time, high temperature baking schedule, which cannot be duplicated in the laboratory, and the thin-film thickness used makes matching of colors for coil coating application particularly difficult.

(7) *Cost.* The color matcher should know any cost limitations. An important element affecting cost is the tolerance limits permitted in production.

(8) *Tolerances.* The closeness of the color match required is of great importance. In some cases, such as coatings for exterior siding or automotive top coats, very close matches are required. In many other applications, users set tight tolerance limits even if they are not needed. Overly tight tolerance requirements raise costs without performance benefits. For coatings that are produced over time and have many repeat batches, the most appropriate way to set tolerances is by a series of limit panels. For example, for a deep yellow coating, the greenness and redness limits and limits of brightness and darkness are needed. Because colors may change on aging, spectrophotometric measurements need to be made of both the standard and limit panels, and Commission Internationale d'Eclairage (CIE) tristimulus values should be calculated (see COLOR). Any coating giving a color that fell within this color space should thus be acceptable. Attempts have been made to assign numerical values to color differences thus permitting specifications that would give a numerical definition of the allowable tolerance in a single number. This objective has not been attained, although progress has been made in developing equations that permit calculation of color difference numbers ΔE in which the scale is equal for all colors (90). The use of ΔE specifications is not desirable except for single colors, because a ΔE value of 1 is a tighter specification for some part of the color range than for other parts, even with 1976 CIE color difference equations (90). Furthermore ΔE color difference specifications permit variation around the standard equally in any direction. In practice, coatings users are more concerned about color variation in one specific direction. For example, whites that are too blue are more acceptable, in general, than whites that are too yellow.

Once the laboratory establishes a color match, the factory should match the color in spite of the fact that batch-to-batch color variations are to be expected in the pigments as manufactured. Whenever possible, the color match should be made using at least four colorants, including black and/or white, if they are needed. Four colorants give the four degrees of freedom necessary to move in any direction in three-dimensional color space. Sometimes this is not possible. For example, because the type of color mixing involved in using mixtures of pigments is subtractive color mixing, if a bright color is desired that can be made only from a single pigment, eg, phthalocyanine green, there is no pigment available that could make a batch of phthalocyanine green that was grayer than the standard less gray. A large fraction of color matching, particularly when repeated production batches are made of the same color, is now performed using instrumental measurements and color matching computer programs (90).

Metallic colors for automotive top coats are an example of critical formulation and color matching problems. These colors owe their popularity to the "color flop" as the viewing angle is changed. The change is in the opposite direction and larger than is the case when viewing ordinary high gloss coatings. The depth of the color is light when viewed from near normal, ie, perpendicular to the surface (face color), and darker when viewed from a wide angle from normal (flop color). A high degree of flop, that is a large color change with a change in viewing angle, requires that three conditions be met: the gloss of the coating must be high, the coating matrix in which the aluminum pigment is suspended must be essentially transparent, and the nonleafing aluminum pigment used must be oriented parallel or nearly parallel to the surface of the film. In order to have a transparent matrix, the pigment-free dry film must be completely transparent and the color pigment selection and dispersion must be such that there is no light scattering from the pigment. The aluminum pigment is oriented in the film as it is forming by film shrinkage because of evaporation of volatiles in the applied coating. The upper layers of the film increase in viscosity more rapidly than the lower layers. The extent of shrinkage and, hence, the degree of orientation decreases as the solids of the coating increase. The highest degree of orientation has been achieved for lacquers having volume solids on the order of 10%. Similar results were obtained using water-reducible acrylic coatings applied at around 18% volume solids. Conventional thermosetting acrylic enamels having a volume solids of about 30% give noticeably less flop. About 45% volume solids is the highest solids content that gives a significant degree of orientation of the aluminum pigment particles. The considerations in obtaining a pleasing appearance using the increasingly popular pearlescent pigments in automotive coatings are similar, because pearlescent pigments are flake pigments that must be oriented approximately parallel to the surface of the film.

Flow

Rheological properties, that is, flow and deformation, of coatings have significant impacts on application and performance properties. The application and film formation of liquid coatings require control of the flow properties at all stages. The mechanical properties of the applied coating films are controlled by the viscoelastic responses of the films to stress and strain. A good overview of the field of rheology in coatings is available (91); a more extensive but somewhat dated discussion of the flow aspect of the rheology of coatings is also available (77).

Viscosity of Resin Solutions. The viscosity of coatings must be adjusted to the application method to be used. It is usually between 50 and 1000 mPa·s(=cP), at the shear rate involved in the application method used. The viscosity of the coating is controlled by the viscosity of the resin solution, which is in turn controlled mainly by the free volume (4). The factors controlling free volume are temperature, resin structure, solvent structure, concentration, and solvent-resin interactions.

The temperature dependence of viscosity of resin solutions can be expressed by the WLF equation (eq. 3) where the reference temperature T_r is taken as the lowest temperature for which data are available (92,93).

$$\ln \eta = \ln \eta_{T_r} - \frac{c_1(T - T_r)}{c_2 + (T - T_r)} \qquad (3)$$

This relationship has been shown to hold for a wide variety of coating resins and resin solutions over a wide range of concentrations. A simplification of equation 3 where T_g is the reference temperature is given in equation 2, which assumes that the viscosity at T_g is 10^{12} Pa·s.

$$\ln \eta = 27.6 - \frac{A(T - T_g)}{B + (T - T_g)} \qquad (4)$$

Equation 4, which is a rewrite of equation 1, does not model the relationship of viscosity with temperature quite as well as equation 3, but it is useful because it shows the relationship to T_g of the resin solution. The T_g of the resin solution is an important, but not singular, factor in controlling viscosity. The T_g of the resin solution depends on the T_g of the resin, the T_g of the solvent, the concentration of the resin solution, and the effects of resin-solvent interactions (92). The T_g of the resin depends on the molecular weight and the structure of the resin. The flow of solutions of coatings resins in good solvents in the viscosity range of 50 to 10,000 mPa·s(=cP) is Newtonian and log viscosity increases as the square root of the molecular weight (92,94). Much further work is needed to elucidate the solvent-resin interaction effects, but it appears that low viscosity hydrogen bond-acceptor solvents give the greatest reduction in viscosity for polar substituted low molecular-weight resins (43,93,95).

The viscosity of high solids coatings in the range of application varies more rapidly with temperature than is the case for conventional lower solids coatings (43). Hence the viscosity is reduced more by using hot-spray systems that permit a further increase in solids. Within the range of 0.05 to 10 Pa·s, log viscosity varies approximately directly with concentration. The viscosity of higher viscosity solutions varies more steeply with concentration (92). It is also generally true that the viscosity of high solids coatings varies somewhat more steeply with concentration than the simple log relationship.

Viscosity of Systems with Dispersed Phases. A large proportion of coatings are pigmented and, therefore, have dispersed phases. In latex paints, both the pigments and the principal polymer are in dispersed phases. The viscosity of a coating having dispersed phases is a function of the volume concentration of the dispersed phase and can be expressed mathematically by the Mooney equation (96), a convenient form of which is

$$\ln \eta = \ln \eta_e + K_E V_i/(1 - V_i/\phi) \tag{5}$$

where η_e is the viscosity of the external phase, K_E is a shape constant (2.5 for spheres), V_i is the volume fraction of internal phase, and ϕ is the packing factor, ie, the volume fraction of internal phase when the V_i is at the maximum close-packed state possible for the system. The Mooney equation assumes rigid particles having no particle–particle interaction. It fits pigment dispersions and latexes that exhibit Newtonian flow.

If there is particle–particle interaction, as is the case for flocculated systems, the viscosity is higher than in the absence of flocculation. Furthermore, a floccu-lated dispersion is shear thinning and possibly thixotropic because the floccules break down to the individual particles when shear stress is applied. Considered in terms of the Mooney equation, at low shear rates in a flocculated system some continuous phase is trapped between the particles in the floccules. This effectively increases the internal phase volume and hence the viscosity of the system. Under sufficiently high stress, the floccules break up, reducing the effective internal phase volume and the viscosity. If, as is commonly the case, the extent of floccule separation increases with shearing time, the system is thixotropic as well as shear thinning.

Shear thinning systems that are generally also thixotropic, also result if the disperse phase particles are not rigid. In the shear field, the nonrigid particles are distorted, resulting in less crowding and therefore lower viscosity. In terms of equation 5, K_E becomes smaller and the packing factor becomes larger. Hence, the viscosity of the system is lower as the shear rate increases and, generally, as the time at a given higher shear rate is extended. Emulsions (qv), ie, dispersions of liquids in liquids, show thixotropic flow because the dispersed phase particles are fluid and can be distorted. The degree of thixotropy is related to the difference in viscosity between the internal and external phase. An internal phase of higher viscosity distorts more slowly under a given shear stress and hence the thixotropic response is slower. The rate of viscosity reduction during the application of stress is slower as the viscosity of the internal phase increases; the rate of recovery of the higher viscosity after application of the shear stress has been stopped is also slower.

Flow During Application. Many methods of film application to a substrate give an uneven surface. For example, application by brush gives brush marks that are not related to the bristles of the brush, but rather to the fact that the film of wet paint is split between the substrate and the brush. Similarly, when paint is applied by a roller, the film is split such that part of the paint remains on the roller and part transfers to the substrate. Film splitting gives an uneven surface. It is generally desirable for these nonuniformities to level out. Leveling occurs most rapidly if the viscosity of the coating on the substrate is low. In order to keep the viscosity low for a longer period, it is common to formulate brush-applied coatings using relatively slow evaporating solvents. However, low viscosity coatings applied to a vertical wall sag, that is, they run down the wall unevenly. The viscosity requirements to promote leveling and to minimize sagging are in conflict with each other. Sometimes solvents can be chosen that evaporate slowly enough to permit reasonable leveling, but evaporate rapidly enough so that the viscosity increases before serious sagging occurs.

In brush or roller applied paints, solvent selection alone seldom provides an adequate compromise between leveling and sagging. Leveling with minimum sagging can frequently be achieved by formulating the coating with thixotropic flow properties. As a coating is applied by brush or roller, a high shear rate is exerted on it; in a thixotropic system, this results in low viscosity. If viscosity recovers slowly enough, leveling can occur, but, then the viscosity can increase again before the advent of serious sagging. Thixotropic coatings can be achieved by flocculation of pigment particles; but this also reduces hiding, color strength, and CPVC. Thixotropic behavior is better achieved by adding a material that produces a nonrigid dispersed phase component. Gel particles, for example, swell with solvent but do not dissolve. Very fine particle size inert pigments can adsorb a layer of resin and solvent on the surface; if the particle size is small enough, most of the volume of the particles after being added to a coating is adsorbed layer which can distort in a shear field. Some pigments, such as treated bentonite clays, absorb solvent and give thixotropy (97). Attapulgite clay can be used in water systems; it absorbs large amounts of water giving a distortable dispersed phase (see CLAYS).

It has been suggested that the driving force for leveling is surface tension (98). However, flow resulting from surface tension differentials has been shown in some cases to be more important (99). When a coating is applied by brush or roller, film thickness is uneven. Differentials in concentration of the resin solution in the film develop as solvent evaporates. The thinnest sections increase in resin concentration most rapidly. As resin concentration increases, surface tension of the solution increases. Liquids of low surface tension flow to cover adjoining liquid surfaces having higher surface tension. Because the thick sections have lower surface tensions, coatings from those areas flow over the higher surface tension thin areas. The surface tension differential induced flow thus reduces differences in film thickness and promotes leveling.

Leveling in the case of spray application can frequently be controlled by adjusting the rate of solvent evaporation in such a way that it is slow enough to permit leveling but rapid enough that the viscosity builds up before sagging becomes a serious problem. However, solvent loss from high solids coatings during spraying is generally slower and it is more difficult to control sagging of high solids coatings; the addition of thixotropic agents is frequently necessary (43,100).

If the evaporation rate of solvents is too high during spray application, areas of different surface tension can develop leading to surface irregularities called orange peel as a result of surface tension differential driven flows (101). The last spray droplets to arrive on the surface of the wet film travel for a longer time after leaving the spray gun and, hence, lose more solvent. As a result, these particles have a higher surface tension; when the material of lower surface tension flows to cover these high surface tension areas, bumps develop, ie, an orange-peel effect is obtained. The effect can be minimized by using slower evaporating solvents. Growth of orange peel of this type can be avoided by adding a very small amount of a solution of a low molecular-weight poly(dimethylsiloxane) fluid before spraying. The methyl groups of the silicone orient rapidly at the surface, imparting a uniform low surface tension to the entire surface. Because there is no surface tension differential, orange peel does not grow. It is said that the silicone fluid promotes leveling, although it would be more precise to say that it prevents the growth of the surface irregularities. In fact, when leveling is driven by surface tension, low surface tension resulting from the addition of silicone fluid slows down leveling.

Another type of flow defect that can occur during application is crawling. In some cases, after application of a smooth film, the liquid coating partially dewets and attempts to form balls on the substrate, leading to a highly irregular film with very thin or even bare areas. Crawling results from applying a coating having high surface tension over a substrate of lower surface tension. On a metal substrate such as steel, it is usually the result of incomplete removal of oil from the surface before coating. On plastic substrates, it can result from failure to remove mold release compounds from the surface of the plastic article before coating. In the case of plastics having inherent low surface tensions such as polyethylene, it may be difficult to make a coating having lower surface tension than that of the plastic. To avoid crawling, it may be necessary to oxidize the surface of the plastic to increase its surface tension. Because the surface tension of many high solids coatings is higher than that of conventional coatings, crawling occurs on a wider range of plastic substrates as the transition is made to higher and higher solids coatings.

Cratering is another common flow defect encountered in the application of coatings. Craters appearing on the surface are not a result of the eruption of bubbles from the inside of the drying film, but are commonly caused by small particles of contaminants of low surface tension being deposited on the surface of the wet, freshly applied film. Because liquids of low surface tension flow to try to cover liquids of high surface tension, the coating near the contaminating particle flows away over the surrounding area. As the coating flows, solvent evaporates, further increasing the surface tension differential and accelerating the flow. As the liquid flows further away from the contaminating particle, the viscosity increases, and the flow slows to a stop, resulting in the craterlike appearance (101). Again, cratering can commonly be eliminated by the addition of a small amount of a silicone fluid to the coating. However, caution is needed not to add excess silicone fluid, which can cause crawling if an insoluble droplet reaches the substrate surface. Such insoluble droplets have been attributed to the presence of some higher molecular-weight fraction of silicone fluid that is insoluble in many coating formulations (102). Modified silicone fluids, such as polysiloxane–polyether block copolymers have been developed which are compatible with a wider variety of coatings and are less likely to cause undesirable side effects. Low molecular-weight poly(n-octyl acrylate) derivatives can also be used to minimize cratering with less danger of side effects.

Coatings having vehicles of low surface tension such as alkyds are less likely to show imperfections such as crawling and cratering, and generally level well. On the other hand, high solids coatings, because these generally have higher surface tensions, are more likely to show film defects such as cratering than conventional coatings (81). A discussion of coating defects is available (103).

Film Properties

The film properties required for some application can only be determined by the performance of the applied coating in practice. Because requirements and exposure conditions vary widely, devising laboratory tests to predict film performance is difficult and frequently not possible. Data banks of actual field performance as functions of coating compositions, application variables, and environmental factors can be very useful.

For coatings where performance requirements are high, such as for automotive exteriors, cans, marine applications, and corrosion protection, coating suppliers and users have accumulated such data banks and use them as guides in formulating new coatings or devising new application procedures. Data on films that fail are even more useful than data on successful films. As the variables in the data bank expand, it becomes increasingly possible to explain correlations between composition and performance and thus predict the performance of new coatings. Data banks are also very useful for checking the relationship between field performance and empirical tests.

Because following actual field performance often requires observations over many years, it is desirable to accelerate the testing process. A widely used test method is field exposure under conditions that are more drastic than commonly encountered in actual use. For example, test panels are exposed at a 5° angle facing south in Florida for accelerated exterior durability testing. This does not provide a complete basis for predicting actual performance because it is only one environment. The change in temperature from day to night is not as great as in other locations; the uv content of the sunlight is not as high as at higher elevations; the pH of the rain is not as acidic as in some other parts of the country; and so on. Therefore, it is desirable to expose panels at several different geographic locations. In some cases, laboratory tests can be devised simulating actual performance conditions. For example, a laboratory shaker test using six packs of beverage cans was devised to simulate the abrasion of can coatings during shipment of six packs of beer (qv) stacked two pallets deep in a railcar (104). This test was carefully validated by comparison with real field performance and was found to give reasonably reliable and predictive results.

Numerous paint tests are widely used (105). Some are suitable quality control tests. Others are guides to be used along with a data bank of known behavior in considering the effect of changes in composition on properties such as impact resistance and ability to withstand deformation without cracking. In general, however, these tests are not satisfactory for predicting field performance of a new coating, although they are widely used in this way. It is not uncommon for a specification based on quality control tests to be used as a design criterion for a new coating.

Adhesion

In most cases, it is desirable to have a coating that is difficult to remove from the substrate to which it has been applied. An important factor controlling this property is the adhesion between the substrate and the coating (see ADHESIVES). In formulating a coating, it is important to remember that difficulty in removing a coating also can be affected by how difficult it is to penetrate through the coating and how much force is required to push the coating out of the way as the coating is being removed from the substrate as well as the actual force holding the coating onto the substrate. Furthermore, the difficulty of removing the coating can be strongly affected by the roughness of the substrate. If the substrate has undercut areas that are filled with cured coating, a mechanical component makes removal of the coating more difficult. This is analogous to holding two dovetailed pieces of wood together.

Surface roughness affects the interfacial area between the coating and the substrate. The force required to remove a coating is related to the geometric surface area, whereas the forces holding the coating onto the substrate are related to the actual interfacial contact area. Thus the difficulty of removing a coating can be increased by increasing the surface roughness. However, greater surface roughness is only of advantage if the coating penetrates completely into all irregularities, pores, and crevices of the surface. Failure to penetrate completely can lead to less coating to substrate interface contact than the corresponding geometric area and leave voids between the coating and the substrate, which can cause problems. The physical chemistry of adhesion has been reviewed (106).

Adhesion to Metals. For interaction between coating and substrate to occur, it is necessary for the coating to wet the substrate (107). Somewhat oversimplified, the surface tension of the coating must be lower than the surface tension of the substrate. In the case of metal substrates, clean metal surfaces have very high surface tensions and any coating wets a clean metal substrate.

Penetration of the vehicle of the coating as completely as possible into all surface pores and crevices is critical to achieving good adhesion. This requires that the surface tension of the coating be low enough for wetting, but the extent of penetration is controlled by the viscosity of the continuous phase of the coating. Although broad rigorous scientific studies of the relationship between continuous-phase viscosity and adhesion have not been published, the importance of this relationship is evident from the formulating decisions made over many years in the manufacture of coatings having good adhesion performance.

The critical viscosity is that of the continuous phase because many of the crevices in the surface of metal are small compared to the size of pigment particles. Because penetration takes time, the initial viscosity of the external phase should be low and the viscosity should be kept as low as possible for as long as possible. Slow evaporating solvents are best for coatings that are to be applied directly on metal. Systems that cross-link slowly minimize the increase of viscosity of the continuous phase. Because viscosity of the vehicle drops with increasing temperature, baking coatings can be expected to provide better adhesion than a similar composition coating applied and cured at room temperature. Work on epoxy/phenolic coatings for cans shows results consistent with this discussion of factors affecting adhesion (54).

Adhesion is strongly affected by the interaction between coating and substrate. On a clean steel substrate, hydrogen bond or weak acid–weak base interactions between the surface layer of hydrated iron oxide that is present on any clean steel surface and polar groups on the resin of the coating provide such interaction. It has been suggested that several polar groups spaced along a resin backbone, having some flexible units to permit facile orientation of the groups to the interface during coating application, and some rigid segments to promote only partial adsorption of the groups so that there can be interaction with the balance of the coating resin, can provide cooperative interactions that enhance adhesion (108). Conversely adhesion can be adversely affected if additives having single-polar groups and long nonpolar tails are present in a coating. For example, do-decylbenzenesulfonic acid is a catalyst for melamine–formaldehyde resins. However, its use in coatings directly applied on steel can adversely affect adhesion (41). Presumably, the sulfonic acid groups interact strongly with the steel surface leading to a monolayer having a surface of long hydrocarbon chains. The effect is similar to trying to achieve adhesion to oily steel or polyethylene.

Fracture mechanics (qv) affect adhesion. Fractures can result from imperfections in a coating film which act to concentrate stresses. In some cases, stress concentration results in the propagation of a crack through the film, leading to cohesive failure with less total stress application. Propagating cracks can proceed to the coating/substrate interface, then the coating may peel off the interface, which may require much less force than a normal force pull would require.

Adhesion of coatings is also affected by the development of stresses as a result of shrinkage during drying of the film. For example, in the case of uv-cure coatings, where curing is achieved by photoinitiated free-radical polymerization of acrylic double bonds, a substantial volume reduction occurs in the fraction of a second required. This loss in volume leads to stresses in the film that, in effect, partly supply the force needed to pull the film from the substrate. Hence, less external force must be applied to remove the film, and adhesion is poorer. Sometimes such stresses can be relieved by heating the coating to anneal it. Internal stresses can also result from solvent loss from a film and other polymerization reactions, as well as changes in temperature and relative humidity, particularly at temperatures below T_g. These stresses can affect coating durability, especially adhesion (109).

The formation of covalent bonds between resin molecules in a coating and the surface of the substrate can enhance adhesion. Thus, adhesion to glass is promoted by reactive silanes having a trimethoxysilyl group on one end that reacts with a hydroxyl group on the glass surface. The silanes have various functional groups that react with the cross-linking agent in the coating on the other end of the molecule (110). Although there has been some indication that such interactions occur with hydroxyl groups from hydrated iron oxide on the surface of steel, the effect apparently is small because the use of such additives to enhance adhesion to steel is not as widespread as in the case of glass. Acetoacetoxylated resins appear to enhance the adhesion of coatings to steel in the presence of water (69).

Adhesion to Plastics and Coatings. In contrast to the application of coatings on clean steel, wetting can be a serious problem for adhesion of coatings to plastics. Some plastic substrates have such low surface tension that it may be difficult to formulate coatings having a sufficiently low surface tension to wet the substrate. Polyolefin plastics, in particular, are difficult to wet. Frequently the surface of the polyolefin plastic must be oxidized to increase surface tension and provide groups to interact with polar groups on the coating resin. The surface can be treated using an oxidizing solution, by flame treatment, or by exposure to a corona discharge. The treated surface should be coated soon after this treatment because the surface is not stable. Unoxidized molecules having low surface tension can migrate to the surface of the uncoated surface treated plastic. Difficulties in wetting plastics can also result from residual mold release agents on the surface of the plastic. Mold releases must be completely removed before coating.

Adhesion to plastics can be enhanced if resin molecules from the coating can penetrate into the surface layers of the plastic. Penetration would take place through free-volume holes; hence, raising the temperature above the T_g of the plastic substrate generally promotes adhesion. The T_g of the plastic can be lowered by penetration of solvent from the coating into the surface of the plastic. This, in turn, may permit penetration of resin molecules from the coating into the surface of the plastic. Hence solvent selection in formulating a coating can be critical in achieving adhesion to plastics. In selecting coatings solvents for application to articles fabricated from high T_g thermoplastics such as polystyrene and poly(methyl methacrylate), a solvent system having too high a rate of evaporation should not be used in order to avoid crazing, that is, cracking of the surface.

Intercoat adhesion to other coatings is a specialized case of adhesion to plastics. The same considerations apply. A further design parameter is available in formulating primers. Because adhesion to rough surfaces is better than to gloss surfaces, primers are usually highly pigmented and hence have a rough surface to promote adhesion of the top coat to the primer surface. Adhesion of top coats can be further enhanced if the primer is formulated using a PVC higher than CPVC. A film from such a primer has voids into which resin solution from the top coat can flow, providing a mechanical anchor between the top coat and the primer. Because the voids in the primer coat are thus filled with vehicle from the top coat, the effective PVC of the final primer layer of the overall coating is approximately equal to CPVC. The PVC of the primer should be only a little above CPVC. Loss of significant amounts of vehicle from the top coat into the pores of the primer can increase the PVC of the topcoat and hence reduce its gloss and affect other properties.

Testing for Adhesion.   Because of the wide range of exposures to stresses in actual use, there is no really satisfactory laboratory test for the adhesion of a coating film to the substrate during use. Probably the most useful guide for an experienced coatings formulator is the use of a penknife to see how difficult it is to remove the coating and to observe its mode of failure. Many tests for adhesion have been devised (105,111). From the standpoint of obtaining a measurement related to the work required to separate the coating from the substrate, the direct pull test is probably the most widely used. The accuracy of the test is subject to considerable doubt and the precision is not very good. Even for experienced personnel, reproducibility variations of 15% or more must be expected. The most common specification test, cross-hatch adhesion, is of little value beyond separating systems having very poor adhesion from others.

Exterior Durability

In many cases, an important performance requirement for coatings is exterior durability. There are many potential modes of failure when coatings are exposed outdoors. Commonly, the first indication of failure is reduction in gloss resulting from surface embrittlement and erosion leading to the development of roughness and cracks in the surface of the coating. In some cases, the next step is "chalking," that is, the erosion of resin from the surface of the coating leaving loose pigment particles on the surface. Chalking reduces gloss and, as chalking proceeds, can lead to complete film erosion and also to drastic color change caused by the increased surface reflectance, which makes colors shift to light shades. Colors can also change if the pigment or pigment-resin combination undergoes photochemical degradation on exterior exposure. Photochemical oxidation and hydrolysis of the resins in coatings are common causes of failure. On exterior exposure, films may crack or check as a result of embrittlement, reducing the elongation-to-break. Such films crack as they undergo expansion and contraction. These changes may be caused by temperature changes, especially when the substrate and the coating have different coefficients of thermal expansion. Similar stresses on coatings can be caused by expansion and contraction of wood substrates with changes in moisture content of the wood.

Various kinds of chemical attack, such as those resulting from acid rain and bird droppings, can result in film degradation and discoloration. Retention of dirt from the atmosphere on the coating surface can lead to drastic color changes and blotchy appearance. Mildew can grow on the surface of many coating films, again leading to blotches of gray discoloration. In view of the large variety of exposure conditions and possible modes of failure, no laboratory test has been devised to predict field performance of coatings on exterior exposure. However, careful accumulation of actual field use results correlated with environmental, compositional, and application variables is the most useful way of understanding the causes of failures and, hence, of being able to forecast the possible performance of a new coating material.

Although many failure mechanisms are involved, the two most common modes are hydrolysis and photochemical oxidation by free-radical chain reactions. In general, resins that have backbone linkages that cannot hydrolyze provide better exterior durability than systems having, for example, ester groups in the backbone. Some esters are more resistant to hydrolysis than others. In general, esters of highly hindered alcohols such as neopentyl glycol are less easily hydrolyzed than those of less hindered alcohols such as ethylene glycol (112). Esters of isophthalic acid are more resistant to hydrolysis in the range of pH 4 to 8 than esters of phthalic acid. Ester groups on acrylic polymers are highly resistant to hydrolysis. Even if some of the ester groups do hydrolyze, it would not result in breaking the polymer backbone.

Susceptibility to free-radical induced photoxidation varies with structure of the resins and pigments and, in some cases, with the interactions between pigment and resin. In general terms, resistance of resins to photochemical failure is related to the ease of abstraction of hydrogens from the resin molecules by free radicals. The greatest resistance is shown by fluorinated resins and silicone resins, especially methyl substituted silicone resins. The greatest sensitivity to degradation is shown by resins having methylene groups between two double bonds; methylene groups adjacent to amine nitrogens, ether oxygens, or double bonds; and methine groups. Resins containing aromatic rings substituted with a hetero atom directly on the ring, such as bisphenol A epoxy resins and urethane resins based on aromatic isocyanates, are very susceptible to photochemical failure.

Pigment selection can also be critical in formulating for high exterior durability. Some pigments are more susceptible to color change on exposure than others. Some pigments act as photosensitizers to accelerate degradation of resins in the presence of uv and water. For example, anatase TiO_2 accelerates the chalking of coatings during exterior exposure. Rutile TiO_2 pigments with appropriate surface treatments, on the other hand, lead to little, if any, increase in chalking. It cannot be concluded that because one colored pigment gives greater exterior life in one resin system compared to another colored pigment the order of durability is the same in other resins systems. Reversals are fairly common.

The exterior durability of relatively stable coatings can be enhanced by use of additives. Ultraviolet absorbers reduce the absorption of uv by the resins and hence decrease the rate of photodegradation. Further improvements can be gained by also adding free-radical trap antioxidants (qv) such as hindered phenols and especially hindered amine light stabilizers (HALS). A discussion of various types of additives is available (113).

It has also been found that there can be interactions between hydrolytic degradation and photochemical degradation. Especially in the case of melamine–formaldehyde cross-linked systems, photochemical effects on hydrolysis have been observed.

Although many variations in methods have been tried, no reliable laboratory test is available to predict exterior durability (115). Exterior exposure of panels in various locations such as Florida is widely used to forecast performance with reasonable success. However, coatings that are sensitive to acid rain or to cracking on rapid temperature change might not perform as well in actual use as predicted by Florida exposure results. In the EMMAQUA tests (DSET Laboratories, Inc.), panels are exposed in Arizona on a machine that turns to keep mirrors that reflect the sunlight to the panel surface approximately normal to the sun. The machine increases the radiation intensity shining on the panels about sevenfold over direct exposure of panels in the same location. To simulate the effect of rain, the panels are sprayed with water each night. The test can provide useful guidance in a few weeks, especially if comparisons are based on exposure to equal intensities of uv radiation rather than for equal periods of time. Many other accelerated laboratory tests have been devised. However, reversals in performance between pairs of coatings in the field as compared with the laboratory results indicate that frequently these tests are not reliable predictors. A review article is available which discusses the difficulties of evaluating durability of coatings (116).

The most promising approach to laboratory techniques for predicting performance is to understand the mechanism of failure and then use instrumental methods to study the susceptibility of a coating to failure. The most powerful tool available now is the use of esr spectrometry to monitor the rate of free-radical appearance and disappearance (117–119) (see MAGNETIC SPIN RESONANCE).

Corrosion Control

An important function of many coatings is to protect metals, especially steel, against corrosion. Corrosion protection is required in two different situations: in one case, the steel is protected against corrosion with intact coating films; in the other case, the objective is to protect the steel against corrosion even when the film has been ruptured.

Protection by Intact Films. In the case of intact films, the key factors responsible for corrosion protection are adhesion of the coating to the steel in the presence of water, oxygen and water permeability, and the resistance of the resins in the film to saponification (120). If the resins in the coating are adsorbed to cover the steel surface completely and if the adsorbed groups cannot be displaced by water, oxygen and water permeating through the film cannot contact the steel and corrosion does not occur. In effect, only a monolayer is required to protect against corrosion, if the monolayer stays in place over the period of exposure.

The factors affecting adhesion are critical in controlling corrosion. Clean surfaces are critical. Sandblasting to white metal provides a good surface for application of corrosion-protective coatings if the coating is applied over the sandblasted surface before it can become contaminated. Conversion coating having insoluble phosphate crystals improves adhesion and hence corrosion protection. Assuming high performance coatings are being used, it has been found that in automobiles the quality of the conversion coating, which is affected by the quality of the steel, is the most important variable in corrosion control (121,122). Viscosity of the external phase of the coating applied to the steel surface can also be a critical factor. Because viscosity drops as temperature increases, greater penetration of vehicles may help account for the generally superior corrosion protection observed with baking coatings. In the case of corrosion control, penetration into the micropores and crevices is especially critical because if those surfaces do not have resin adsorbed on them when the water and oxygen permeate through the film, corrosion begins. Corrosion leads to a solution of ionic materials in water under the coating film, establishing an osmotic cell. The osmotic driving force accelerates the permeation of water and the osmotic pressure provides a further force to displace the film from the surface to form blisters.

Because water permeates through any coating film, it is desirable to have groups adsorbed on the surface that cannot be displaced by water, or at least where the equilibrium strongly favors maintaining the adsorbed layer. Although there is some controversy as to the mode of action, it is a common observation that resins having multiple amine groups along the chain give better corrosion protection than those having multiple hydroxy groups, perhaps because amines are less readily displaced from the surface of steel by water. Even though a particular interaction between a polar group on a resin molecule and the surface of the steel or the phosphate groups of the conversion coating is displaced by water, if there are multiple groups from the same resin molecule adsorbed, there can be a cooperative enhancement of adhesion (108). If one group desorbs, other groups that can do so readsorb the molecule onto the substrate surface. In addition to amine groups, increasing evidence is being reported that phosphoric acid partial esters, as in epoxy phosphates, give enhanced adhesion and resistance to displacement by water (55). Resins that have saponifiable groups in the backbones are particularly likely to show poor wet adhesion.

If a coating had perfect wet adhesion, no other factors would affect corrosion protection, but frequently in practice such a high degree of wet adhesion cannot be attained. Thus it has been shown that the corrosion protection is also enhanced if the permeability of the film to oxygen and water is low (123). A significant factor controlling these permeabilities is the free-volume availability in the film, which is in turn related to T_g. From the point of view of corrosion protection, it is desirable to have the T_g of films at least 50°C higher than the service temperatures of the coatings. This condition is difficult, if not impossible, to achieve using ambient cure coatings and for the highest corrosion protection, baking coatings are desirable whenever it is feasible to use them.

Permeability is controlled not only by the diffusion rate but also by the solubility of the diffusing molecules. There can be significant variations of the solubility of water in various films. Films containing highly polar groups have the highest tendency to dissolve water. Films made with chlorinated polymers such as chlorinated rubber, vinyl chloride copolymers, and vinylidene chloride copolymers dissolve very small amounts of water and are widely used as vehicles for top coats for corrosion protection coating systems. Permeability of coatings to oxygen and water is also reduced by pigmentation. Lowest permeability results from having a PVC close to but not above CPVC. Platelet pigments such as leafing aluminum pigments and mica can orient parallel to the surface of a film during solvent evaporation and therefore act to improve the barrier properties of coatings. The mechanical integrity of films is important to maintain the coatings intact.

Protection by Nonintact Films.   It is also possible to achieve corrosion control by coatings after a film has been ruptured. Because the coatings used to achieve this control generally give poorer protection when their films are not ruptured, such systems should be used only when film rupture must be anticipated or when complete coverage of the steel interface cannot be achieved. There are two techniques used on a large scale: primers containing corrosion inhibiting pigments and zinc-rich primers.

The mechanism of action of corrosion inhibiting pigments is not completely understood, but it is generally agreed that they promote oxidation at the surface of anodic areas of the steel to form a barrier layer. This action is called passivation (120). For the pigments to be effective, they must have a minimum solubility in water. If the solubility is too high, however, they would be rapidly lost from the film by leaching with water and thus provide only short-term protection. Because the pigments are somewhat water-soluble, their presence in an intact film can lead to blistering and loss of adhesion. Therefore, it is undesirable to use such pigments except when the need for protection by nonintact films is most important.

The oldest corrosion inhibiting pigments are red lead [1314-41-6], Pb_3O_4, containing approximately 15% PbO, and zinc yellow [85497-55-8], $3ZnCrO_4 \cdot K_2CrO_4 \cdot Zn(OH)_2 \cdot 2H_2O$, commonly, but mistakenly, called zinc chromate. There is concern about toxic hazard with red lead. Even more serious, soluble chromates, such as are present in zinc yellow, are carcinogenic. Because there are no satisfactory laboratory tests to predict the effectiveness of coatings containing corrosion-inhibiting pigments, extended field experience is necessary to determine the utility of new pigments. Zinc-calcium molybdates, surface-treated barium metaborate, zinc phosphate, zinc salts of nitrophthalic acid, and calcium-barium phosphosilicates and phosphoborates are pigments which have been recommended as corrosion inhibitors.

Zinc-rich primers can be very effective in protecting steel with nonintact films against corrosion. High contents of zinc metal powder are required in the primers; PVC of the zinc powder must exceed CPVC to permit the necessary electrical contact between the zinc particles and between zinc and steel. Also having the PVC above CPVC makes the coating porous so that water can enter the film permitting completion of the electrical circuit. The zinc becomes the anode and the steel the cathode of an electrolytic cell, hence the zinc acts as a sacrificial metal to protect the steel. Because a base, $Zn(OH)_2$, is generated, saponification-resistant binders are required. The most widely used systems are inorganic zinc-rich primers, also called zinc silicate primers. The vehicle is an alcoholic solution of partially polymermized tetraethyl orthosilicate (silicic acid tetraethyl ester) [78-10-4], $C_8H_{20}O_4Si$. After application, atmospheric moisture continues the hydrolysis and completes the polymerization of the tetraethyl orthosilicate and zinc salts of the polysilicic acid form. The primer must be protected by a top coating. In applying top coats, penetration into the pores of the primer must be avoided (120). Increasingly, latex paints are being used as top coats (124) because they do not permit vehicle penetration into the pores and they have the important advantage of lower VOC emissions (120).

It is highly desirable to clean a steel surface thoroughly before applying a corrosion-protecting coating system, but it is not always possible. Sometimes it is necessary to apply coatings over oily, rusty steel. In this case, it is essential to have a vehicle that can displace and dissolve the oily contamination from the surface and have sufficiently low viscosity for a sufficiently long time after application for penetration through the rusty areas down to the surface of the steel. Drying oil primers pigmented with red lead are still widely used for this purpose. They have the very low surface tension necessary for wetting, and can dissolve the oil and penetrate through the rust. Although their wet adhesion to steel and saponification resistance are inferior to those of many other primers, it is better to have some vehicle on the surface of the steel rather than having no vehicle penetrate down through the rust to the steel surface. This inadequacy is partly offset by the use of red lead as a corrosion-inhibiting pigment.

Formulation of effective corrosion-resistant coatings is made difficult by the lack of a laboratory test that can provide reliable predictions of field performance. The most widely used test is exposure in a salt fog chamber. It has been shown repeatedly, however, that the results of such tests do not correlate with actual performance (125). Outdoor exposure of panels can provide useful data, especially in locations where salt spray occurs, but predictions of performance are not always satisfactory (126).

Useful guidance in evaluating wet adhesion can be obtained by checking adhesion after exposure in a humidity chamber. In some cases, cathodic disbonding tests may provide useful data (ASTM Standards G8-79, G19-83, and G42-80). Another approach to testing for delamination is the use of electrochemical impedance spectroscopy (eis) (127,128). Impedance is the apparent opposition to flow of an alternating electrical current, and is the inverse of apparent coating capacitance. When a film begins to delaminate there is an increase in apparent capacitance. The rate of increase of capacitance is proportional to the amount of surface area delaminated by loss of wet adhesion. High performance systems show slow rates of increase of capacitance so tests must be continued for long time periods. Eis can be a powerful tool for study of the effect of variables on delamination.

An extensive survey of accelerated test methods for anticorrosive coating performance which emphasizes the need to develop more meaningful methods of testing has been published (129). The most powerful tool available is the accumulated material in data banks correlating substrate, composition, application conditions, and specifics of exposure environments with performance.

Mechanical Properties

Hundreds of tests of the mechanical properties of coatings have been developed. Many are suitable only for quality control (qv) work, but a few have been sufficiently correlated with actual field performance and hence have value for predicting performance. A complete listing and discussion of paint testing is available (105). The ASTM provides detailed procedures for many paint tests (130). A monograph on mechanical properties of coatings has been published that gives an excellent presentation of a limited range of tests (131). In this monograph and an earlier review paper (132), the advantages of determining basic mechanical properties is emphasized. Basic mechanical properties such as tensile strength-to-break, elongation-to-break, work-to-break, loss and storage modulus, and loss tangent (tan delta) can be more readily related to structure and correlated with actual performance than most paint tests. Dynamic mechanical thermal analysis has become a powerful tool for the assessment of cure and the study of the effect of extent of sure on film properties and performance (133,134).

Because of the much larger volume of single products, there have been more intensive studies of the mechanical properties of rubber, plastics, and fibers in the past than of coating where the volume of individual products is generally relatively small. However, the mechanical property requirements of films for various coating applications and how to vary composition in order to achieve these needs is under investigation.

Coating films are viscoelastic, therefore mechanical properties depend on temperature and the rate of application of stress. Behavior is shifted toward the elastic mode with increased tensile strength and decreased elongation-to-break and with a more nearly constant modulus as a function of stress as temperature is decreased or as the rate of application of stress is increased. The shifts can be particularly large if results are compared above and below T_g. Below T_g, coatings show dominantly elastic response and, therefore, are brittle and break if the relatively low elongation-to-break is exceeded. Above T_g, the viscous component of deformation is larger and films are softer (lower modulus) and less likely to break during forming. Caution is necessary in considering the relationship of T_g to formability because some materials, such as poly(methyl methacrylate) and especially bisphenol A polycarbonate, are ductile at temperatures substantially below T_g (135). Modulus above T_g is controlled primarily by the cross-link density. In fact, it has now been shown that the cross-link density of melamine–formaldehyde cross-linked coatings can be calculated from the storage modulus above T_g (40).

In designing coatings that must withstand large deformations without cracking, such as in container and coil coating applications, the T_g, as determined at a rate of application of stress comparable to that which is experienced in the forming operation, should be lower than the temperature at which forming takes place. If the deformation is large, the cross-link density must be low or the elongation-to-break even above T_g is too small to withstand the deformation. In general, thin films resist deformation better than thicker films. In applications such as can coatings and coil coatings, the film thicknesses applied are kept to the minimum necessary to achieve the required protection and hiding. Formability is also affected by adhesion. If the adhesion between the substrate and the coating is excellent, the stresses can be dissipated and failure is less likely. Impact resistance, such as is encountered when gravel hits the coating on a speeding automobile, is related to formability of coatings. The rates of application of stresses encountered can vary widely. In coatings, as in plastics, impact resistance is enhanced in coatings that exhibit broad tan delta peaks in comparison with coatings having relatively narrow tan delta peaks (40,131). Complex phenomena are involved in the deformation of multiple coats of paint when struck by gravel. In addition to impact resistance, effects of fracture mechanical properties are being studied (136).

Abrasion resistance of floor coatings has been related to work-to-break of coatings. There is often no correlation to the most widely used specification test for abrasion, the Tabor Abraser test (137). Polyurethane coatings, in general, provide superior abrasion resistance. This may result from the presence of what might be thought of as two types of cross-links, the covalent cross-links and cross-links resulting from intermolecular hydrogen bonds between urethane groups on different molecules. At low levels of stress, the hydrogen bonds remain intact and the coating shows adequate hardness and resistance to swelling with solvents, whereas at higher levels of stress the hydrogen bonds can be separated, permitting extension without rupture of the covalent cross-links. When the stress is removed, hydrogen bonds can reform. An excellent example of the effect of these phenomena on performance can be seen in aircraft finishes. Polyurethane aircraft coatings combine high abrasion resistance with resistance to the effects of the strong solvents used as hydraulic fluids in aircraft.

In other cases, abrasion resistance is related to the coefficient of friction between the surfaces in contact with each other. For example, the abrasion resistance of coatings on can exteriors that rub on other cans can be enhanced by adding waxes or fluorine-substituted surfactants that reduce surface tension and, hence, coefficient of friction. In some applications, abrasion resistance can be increased by incorporating a relatively large particle size inert pigment. For example, the burnish resistance of flat wall paint can be improved by incorporating large particle size silica, SiO_2. Presumably, the particles reduce abrasive wear by reducing contact areas and, therefore, reduce the chance of transmission of force from one surface to the other. Abrasion resistance is also related to the tendency for stresses to be concentrated or dissipated within the film. If there are imperfections that concentrate the stress, the film can tear more easily and abrasion resistance is poorer. On the other hand, if there are particles within the film that serve to dissipate stress, the film is less likely to tear and abrasion resistance is improved. Abrasion is one of several types of wear in a broad approach to fracture energetics and surface energetics of polymer wear (138).

Architectural Coatings

Roughly half the volume of all coatings sold in the United States are classified as architectural coatings (1). In 1990 that amounted to 1.9×10^6 m^3 of coatings having a value of \$4.91 billion. These coatings are designed to be applied to residences and offices, and for other light-duty building purposes. In contrast to most product coatings, they are designed to be applied in the field, in some cases by contractors but in large measure by do-it-yourself consumers. A wide range of products is involved.

Flat Wall Paint.   The largest volume of architectural coatings is flat wall paint. In the United States, virtually all flat wall paint is latex-based. Latex paints have the advantages of low odor, fast drying, easy clean-up when wet, durability of color and film properties, and lower VOC emissions as compared to oil- or alkyd-based paints. They are manufactured primarily as white base paints, which are tinted by the retailer to the color selected by the customer from large collections of color chips. Two or three base whites are made: one for pastel shades, one for deep shades, and sometimes one for medium color shades. Bases for pastels require a relatively high content of TiO$_2$ to provide hiding. The deeper color paints cannot be matched using the large amount of TiO$_2$ in the white base. Even if the color could be matched, the cost would be excessive because there would be more hiding than needed and large amounts of extra color pigments would be required to match the color. This method of marketing has the enormous advantage of being able to supply a very wide choice of colors to the consumer carrying a relatively low inventory stock. However, it is critical to maintain the same level of hiding when formulating new white base paints. Otherwise all color mixing formulas and color chips would have to be replaced. The tinting colors used by retail stores are designed to be used for a wide variety of formulas. Commonly, the tint colors are not manufactured by the paint manufacturer but by companies specializing in making these color dispersions.

The performance of quality grades of latex flat wall paints is excellent in most respects, thus the emphasis in technical efforts is on cost reduction. In most cases, vinyl acetate copolymer latexes are used; a comonomer such as butyl acrylate provides the required reduction in T_g for film formation. Only when high scrub resistance is required would a higher cost latex having a high content of acrylate ester monomers be used. The least expensive component in the paint, on a volume cost basis, is the inert pigment. Therefore, the inert pigment content is maximized. In order to have low gloss, flat paints have a PVC near the CPVC. In order to have the highest inert pigment content, formulas are developed having CPVC as high as possible by using a broad distribution of particle sizes. As described earlier, CPVC of latex paints is also affected by latex composition. Smaller particle size and lower T_g give higher CPVC, flocculation of latex or pigment reduces CPVC.

In some applications, such as ceiling paint, paints having PVC higher than CPVC may give excellent performance. The cost of such paints is lower because of the higher inert pigment content and, at the same time, hiding is better because the final film incorporates air voids, which increase light scattering. For ceiling paints one coat hiding is particularly important. On a ceiling the reduction in stain and scrub resistances resulting from having the PVC higher than CPVC is not a significant disadvantage. Quality flat wall paints are formulated with PVC slightly less than CPVC and therefore provide good stain and scrub resistances.

On a volume basis, the most expensive component of flat wall paints is the TiO_2. Therefore, significant efforts are applied to reach a standard hiding with the minimum possible level of TiO_2. The efficiency of hiding by TiO_2 decreases as the TiO_2 concentration is increased above about 10 vol %; it becomes economically inefficient at roughly 18 vol % in the dry film. Above a PVC of about 22, hiding actually decreases with increasing TiO_2 concentration. Although there is some controversy, it is felt that in the critical range of 15 to 18 vol % of TiO_2, the efficiency of hiding can be increased by using as part of the inert pigment some material having a particle size in the same range as that of the TiO_2 (139). This permits substituting inexpensive inert pigment for some of the TiO_2 while maintaining the same level of hiding and tinting strength.

Another approach to minimizing TiO_2 content is the use of Spindrift pigment (Dulux, Ltd., Australia), which consists of resin particles that contain air bubbles. The air bubbles contain TiO_2 particles (140). The larger refractive index differences and the larger number of interfaces give substantial increases in light scattering and hiding. Still another approach is the use of a high T_g latex as part of the pigment volume; paints can be formulated using PVC somewhat higher than CPVC without giving a porous surface (141). These paints provide equal hiding at lower TiO_2 content, while retaining good enamel hold out and stain resistance. Yet another method of minimizing TiO_2 requirement is the use as pigments of high T_g latexes, the particles of which contain air voids, for example, Ropaque (Rohm and Haas Co.) (142,143). When used in paints having a PVC slightly lower than CPVC, the air voids inside the high T_g polymer particles in the film provide hiding and permit lower TiO_2 concentrations. Such paints can have stain and scrub resistances equal to other latex paints.

Most wall paints are applied by roller. During roller application, some latex paints give excess spatter, that is, small particles of paint are thrown into the air when the paint is applied. This is the result of the growth of the filaments that are produced by film splitting in roller application to the length where they break in two places rather than in one. Paints having high extensional viscosity exhibit this behavior (144). Extensional viscosity increases when high molecular-weight water-soluble polymers with very flexible backbones are used as thickeners, leading to increased spattering. Minimum spattering is obtained with low molecular-weight water-soluble thickeners having rigid segments in the polymer backbone, such as low molecular-weight hydroxyethylcellulose (145).

A continuing challenge in latex paints of all kinds is to protect against the effects of bacteria in the can and mildew growth after application. Excellent housekeeping in a latex paint plant is essential to minimize the introduction of bacteria into the paint. Bacteria can generate foul odors or, in extreme cases, sufficient gas pressure inside the cans to blow the can lids off. The more common problem is that bacteria feed on many water-soluble cellulosic thickeners used in latex paints. The enzymes split the cellulose molecules reducing the molecular weight, leading to a sizeable reduction in viscosity. It is also essential to avoid introducing enzymes that might have been generated through bacterial growth. A bacteriocide incorporated in the paint can kill the bacteria without destroying the enzyme, and viscosity can drop even using adequate bacteriocide in the paint. Mildew can grow on almost any paint surface but latex paints are particularly susceptible, and therefore contain a fungicide.

Many fungicides and bacteriocides are available. Testing is difficult because fungal and bacterial growth are so dependent on ambient conditions. Additionally, many fungicides and bacteriocides are effective against only a limited number of organisms. The broadest spectrum biocides available that protect against both bacterial and mildew growth are aromatic organomercury compounds such as phenylmercuric acetate [62-38-4], $C_8H_8HgO_2$. However as of August 1990, use of mercury compounds in interior paint in the United States has been banned.

In order to be effective, fungicides must be low molecular weight and slightly soluble in water which means that these compounds may be lost from the film by leaching or volatilization. The problem has been approached by developing polymers having pendent fungicide groups which can slowly hydrolyze off over time (146). A review of biocides for latex paints is available (147).

Exterior House Paints.   Latex paints dominate the exterior house paint market in the United States because of superior performance in almost all cases. As compared to oil or alkyd paints, the exterior durability of latex paints, that is, resistance to chalking, checking, and cracking, particularly when using latexes having high methyl methacrylate (2-methyl-2-propenoic acid methyl ester) [96-33-3], $C_4H_6O_2$, content. Methyl methacrylate has excellent resistance to photodegradation and hydrolysis. Another advantage of latex paints used on wood (qv) surfaces is that because of high moisture vapor permeability, they are much less likely to blister than oil or alkyd paints.

An application where latex paints show outstanding performance is over masonry such as stucco or cinder block construction. This performance results from saponification resistance in the presence of the alkali from the cement. Furthermore because masonry surfaces are porous, having both small and large pores, the low viscosity external phase of a latex paint can penetrate rapidly into the small pores, causing a rapid increase in the viscosity of the remaining paint. The bulk paint, in turn, sinks into the larger holes more slowly than a solution-based paint. Thus less latex paint is required to cover the same surface area as compared to alkyd paints.

There are limitations to the applicability of exterior latex house paints providing a small continuing market for oil or alkyd exterior house paints. Because film formation from latex paints occurs by coalescence, there is a temperature limit, below which the paint should not be applied. This temperature can be varied by choice of the T_g of the latex polymer and the amount of coalescing agent in the formula. In the United States, most latex paints are formulated for application at temperatures above 5–7°C. If painting must be done when the temperature is below 5–7°C, oil or alkyd paint is preferable.

Another limitation is that latex paints do not give good adhesion when applied over a chalky surface such as weathered oil or alkyd paint. The latex particles are large compared to the pores between the "chalk" particles on the surface of the old paint. Thus, when a latex paint is applied, no vehicle penetrates between the chalk particles to the surface beneath and there is nothing to provide adhesion to the underlying substrate. This problem can be minimized by replacing part of the latex polymer solids with a modified drying oil emulsified into the latex paint. When the film dries, the emulsion breaks and the modified drying oil can penetrate between the chalk particles to the substrate. In order to have package stability, high resistance to saponification is required. For example, drying oil fatty acid esters of low molecular-weight copolymers of styrene and allyl alcohol (2-propen-1-ol) [107-18-6], C_3H_6O, are used. The outstanding exterior durability of the exterior latex paint is reduced by incorporating modified drying oil. It is common to use an oil- or alkyd-based primer over a chalky surface then apply a latex top coat. The problem of adhesion to chalky surfaces is becoming less important with time because, as latex paints are more commonly used, there are fewer chalky surfaces to repaint.

Cost is an important factor in exterior house paints which are designed with low gloss because this permits higher pigment loading especially in high CPVC paints. Inert pigments should be used that give a broad distribution of particle size because this leads to high CPVC, which in turn permits incorporation of greater quantities of low cost inert pigment. Low gloss also reduces dirt pickup, which is a greater problem with latex paints than alkyd paints because the former remain permanently thermoplastic. Calcium carbonate pigments, however, should be avoided. The latex paint film is permeable to water and carbon dioxide. Calcium carbonate can dissolve in the solution of carbonic acid and the soluble calcium bicarbonate leaches out to the surface of the film. The reaction is reversed when water evaporates, and the calcium bicarbonate decomposes depositing a frost of calcium carbonate on the surface of the film.

Gloss Enamels.   In contrast to exterior and flat wall paint, about half of the gloss paint or enamels sold are based on alkyd resins. Professional painters particularly favor the continued use of alkyd gloss paints. The need for reduction of VOC emission levels, especially in California, has led to efforts to increase the solids content of alkyd paints or overcome the disadvantages of latex gloss paints.

It is not possible to make latex enamels that have as high a gloss as solution-based coatings. In solution-based coatings, gloss is enhanced during the solvent evaporation by the formation of a thin surface layer that has a lower pigment content than the average PVC of the coating as a whole. In many cases, there is a layer of roughly 1 μm on the surface of the gloss paint film that contains essentially no pigment particles. In latex paints, formation of such a clear surface layer is not possible, and therefore the gloss is lower. The ratio of pigment to binder at the surface of a latex paint film can be decreased somewhat by using a finer particle size latex. Although gloss is enhanced in this way, it is still low compared to high gloss alkyd paints.

Also reducing the gloss of latex paints is the haze resulting from the incompatibility of surfactants with the latex polymer film and blooming of surfactant to the surface. Whereas this can be ameliorated by making latexes with as low a surfactant concentration as possible, it probably can never be completely eliminated. On the other hand, the far superior resistance of the latex polymer to photoxidation as compared to any alkyd leads to superior gloss retention by latex paints. The difference in gloss retention is particularly large in exterior applications. Furthermore, the latex paint films are more resistant to cracking, checking, and blistering. The gloss of a latex coating can also be considerably improved by including in the formula a water-soluble polymer such as an amine salt of a co-polymer of acrylic esters, styrene, and a substantial amount of acrylic acid. Such coatings are widely used as floor waxes. When the wax coating is dirty, it can be removed by washing with ammonia water. However, in most cases such an approach would not be appropriate for paints. Proprietary resins are on the market that are reported to enhance gloss without sacrificing film properties.

The principal limitation of gloss latex paints is not gloss, rather it is the greater difficulty of getting adequate hiding from one coat. In professionally applied paint the cost of labor is higher than the cost of the paint. Alkyd paints, which provide hiding in one coat are favored by painting contractors in many cases over latex paints which commonly require two coats. There are several factors involved in the hiding differences that exist between gloss latex and gloss alkyd paints.

The volume solids of latex gloss paints are substantially lower than the volume solids of alkyd gloss paints. The solids of latex paints are commonly around 33% compared to around 67% for an alkyd paint. In order to apply the same dry film thickness, twice as much wet paint has to be applied. The main factor controlling the wet film thickness initially applied is the viscosity of the paint at the high shear rates experienced during brush application. In order to apply a thicker wet film, latex paint should have a higher viscosity at high shear rate than is appropriate for an alkyd paint. For many years the high shear viscosity of most latex gloss paints was lower than that of a corresponding alkyd paint. The film thickness applied is also affected by the judgment of the painter and the fact that the wet hiding of latex paints is substantially higher than the dry hiding. When the water (refractive index 1.33) evaporates from the film, the latex particles (refractive index approximately 1.5) coalesce. The number of interfaces for scattering light drops and, therefore, hiding decreases. This effect is augmented by the larger difference in refractive index between rutile TiO_2 (refractive index 2.76) and water, which is the interface in the wet film, as compared to TiO_2 and the latex polymer which is the interface in the dry film. The effect is large because a small change in a large refractive index difference makes a larger difference in light scattering than the same change where there is a smaller refractive index difference. Furthermore, in the wet paint the TiO_2 scatters light more efficiently because of the lower volume concentration as compared to the dry film. There is a similar difference in alkyd paints but the effect is smaller because the volume change during drying is less.

Probably the largest factor affecting hiding by gloss latex paints is poor leveling. Assuming that a uniform dry film of 50 μm of a paint provides satisfactory hiding in some application, if the film has thinner areas of, for example, 35 μm, and thicker areas, for example, 65 μm, the hiding power of the uneven film is poor. Actually, the uneven film is even worse than a uniform 35 μm film, because the thick and thin film areas are right next to each other and the contrast in hiding is thus emphasized.

The poorer leveling generally encountered for latex paints has several causes. First, in alkyd paints for brush application the solvent is very slow evaporating mineral spirits, whereas in latex paints the water evaporates more rapidly, unless the relative humidity is very high. As a result, the viscosity of the latex paint increases more rapidly after application. Second, the volume fraction of internal phase in latex paints is much higher than in alkyd paints because for a latex paint both the binder and the pigment are in the dispersed phase. Therefore, the viscosity changes more rapidly with loss of volatiles than is the case for alkyd paints. Another possible factor is that, in the case of latex paint, leveling may occur only by surface tension driven flow, whereas in the case of solvent-based brush applied paints the principal driving force is surface tension differential driven flow, which tends to lead to a more uniform film thickness (98). Whereas the surface tension of water is high, the surface tension of latex paints is low because of the presence of surfactants. Furthermore, the surface tension of latex paint probably does not change during the early stages of water loss, whereas the surface tension of a solvent–solution paint does change in the early stages of volatile loss. Therefore, surface tension differentials are less likely to develop in latex paint films than in films from solvent-based paints.

The rheological properties of gloss latex paints greatly influence leveling, and therefore, hiding (see RHEOLOGICAL MEASUREMENTS). Latex paints have exhibited a much higher degree of shear thinning than alkyd gloss paints, leading to paints having viscosity that is too low at high shear rate, and a subsequent applied film thickness that is too thin. At low shear rates, the viscosity is too high to permit adequate leveling. The problem is particularly severe for latex paints because the rate of viscosity recovery after stopping a high shear rate tends to be very fast. The viscometer used for architectural paints, the Stormer viscometer, measures the viscosity related to a midrange of shear rates giving no information about the viscosity in the most critical high or low regions. Thus the viscosity problems associated with latex paints went undetected for some time.

The reasons for the greater dependency of viscosity on shear rate in latex paints have not been fully elucidated. It appears, however, that it may at least partly result from flocculation of latex particles in the paint. Progress in minimizing this problem has been made by using associative thickeners in formulating latex paints. Many kinds of associative thickeners have been made that are all moderately low molecular-weight, water-soluble polymers having occasional long-chain nonpolar hydrocarbon groups spaced along the backbone. Such thickeners reduce shear thinning of latex paints, perhaps by stabilizing the latex particles against flocculation.

Latex paints formulated with associative thickeners have increased high shear viscosity allowing the application of thicker wet films. Furthermore, the thickeners afford reduced low shear rate viscosity and a slower rate of recovery of the low shear rate viscosity which improves leveling at the same time (148). The thicker wet film in itself also promotes leveling, because the leveling rate increases with the cube of wet film thickness (98). Leveling is better, and gloss is somewhat higher, but on the other hand, more difficulty is experienced in controlling sagging (149). Whereas sag control of the latex paints is still superior to that of most alkyd paints, the leveling and gloss even of the best latex paints still do not equal that of alkyd paints. Unfortunately, no adequate laboratory tests are available to measure gloss (150) or absolute hiding (151) of paints.

Another shortcoming of latex, which is particularly evident in gloss paints, is the time required to develop full film properties. Latex paints dry to touch and even to handling much more rapidly than do alkyd paints, but the latex requires a much longer time to reach the full dry properties. For example, in blocking situations such as placing a heavy object on a newly painted shelf, or the problem of sticking windows and doors, latex paints require more time to develop block resistance than do alkyd paints. The initial coalescence of latex particles is rapid but the rate of coalescence is limited by free-volume availability. Because $(T - T_g)$ must be small, the free volume is small. This situation is helped by using coalescing solvents, but the loss of these solvents is diffusion rate controlled and the rate is affected strongly by free-volume availability. It has been reported that by having the core of the latex particles having a high T_g with a gradient to low T_g on the outer shell of the particles, it is possible to achieve film formation at low temperatures yet fairly quickly achieve blocking resistance (16).

Another important potential problem for use of gloss latex paints is adhesion to an old gloss paint surface when water is applied to the new dry paint film. Adequate adhesion to an old gloss paint surface is always a problem for any new coat of paint. It is essential to wash any grease off the surface to be painted and to roughen the surface by sanding. But after wetting with water, some latex paint films can be peeled of the old paint surface in sheets. The resistance to adhesion loss by wetting with water improves as the system ages, but for several weeks or even months wet adhesion can be a serious problem. Proprietary latexes have been developed that minimize this problem. It has been said that amine-substituted latex polymers exhibit greater resistance to loss of adhesion. Such polymers can be prepared using an amine-substituted acrylic monomer, such as 2-(dimethylamino)ethyl methacrylate, as a comonomer in preparing the latex or reacting the carboxylic acid groups on a latex polymer and hydroxyethylethyleneimine (1-aziridinoethanol) [1072-52-2].

Product Coatings

About 32% of the total volume of coatings in 1990 (1.2×10^6 m^3) having a value of $4.08 billion are applied in factories to a very large variety of products ranging from automobiles to toys (1). These are called product or, commonly, industrial coatings, or industrial finishes. These coatings are all proprietary. Additionally, because government regulations of air quality, effluent, waste disposal, and toxic hazards are continually changing, any specific information that is available as of this writing may already be outdated.

A large fraction of the product coatings are applied to metal. The essential first step in metal coating is the preparation of the metal. Oil and related contaminants must be removed by detergent or solvent washing. Solvent degreasing is the most effective. In detergent washing, the last trace of detergent must be rinsed off before drying preparatory to painting. For best adhesion and corrosion resistance, the surface of the metal should be treated (see METAL SURFACE TREATMENTS). In the case of steel, phosphate conversion coating treatments are used: for aluminum, chromate conversion treatments are available. The surface should be carefully rinsed before applying paint. Surface contaminants can result in crawling and/or blistering of the coating. Treated surfaces should not be touched. Water-soluble contaminants such as salts, can lead to blistering after the coated product is put into service. Oils can lead to crawling of the wet coating film applied on top of the oil.

Primers for Metal. If reasonably high performance is required in the end product and unless cost is of paramount importance, a minimum of two coats, usually a primer and a top coat, should be applied to metal. For highest performance, primer vehicles should provide good wet adhesion, be saponification resistant, and have low viscosity to permit penetration of the vehicle into microsurface irregularities in the substrate. Color, color retention, exterior durability, and other such properties are generally not important in primers. Resin systems such as those including bisphenol A epoxy resins which provide superior wet adhesion can thus be used in spite of their poor exterior durability.

In order to provide a suitable surface for adhesion of the top coat, the gloss of the primer should be low and the cross-link density should be as low as handling characteristics permit. In some cases, it is desirable to have the PVC greater than CPVC permitting good intercoat adhesion with the top coat and relatively easy sanding. After penetration by the top coat vehicle, the PVC becomes approximately equal to CPVC. High pigmentation is desirable because the least expensive component is inert pigment. Furthermore, high pigmentation reduces oxygen and water permeability of the final combined film as long as highly hydrophilic pigments are avoided. One drawback for high solids is that highly pigmented, high solids coatings have higher viscosities.

A primer should be designed for baking at as high a temperature as possible because this promotes penetration into the conversion coating and micropores on the surface of the metal. The primer vehicle should be designed to provide maximum resistance to displacement by water, that is, to have good wet adhesion. Multiple polar groups spaced along the polymer backbone tend to promote wet adhesion and resins substituted with amine groups or phosphoric acid partial esters show enhanced wet adhesion. Cost is sometimes the dominant factor on composition and alkyds are still widely used as primer vehicles even though epoxy esters, epoxy/amine, or epoxy/phenolic-based primers generally provide better performance.

Electrodeposition. Primers for automobiles and other products and a significant part of primers for household appliances are applied by electrodeposition (152,153). Almost all electrodeposition primers are now cationic. The compositions are proprietary, but the vehicles in some primers are epoxy/amine resins neutralized with volatile organic acids such as lactic acid; an alcohol-blocked isocyanate is used as a cross-linking agent. The pigments are dispersed in the resin system and the coating is reduced with water. The amount of amine salt is such that a stable dispersion, not a solution, results in the aqueous phase. All of the resin, cross-linking agents, and pigments must be located in the dispersed phase aggregates, so that all components deposit at equal rates.

In application, the automobile or other article to be coated is made the cathode in an electrodeposition system. A current differential on the order of 250 to 400 V is applied, which attracts the positively charged coating aggregates to the cathode. At the cathode, hydroxide ions from the electrolysis of water precipitate the aggregates on the surface of the metal. As the conveyor removes the coated product from the bath, residual liquid is rinsed off with water and the article is conveyed into a baking oven for a high temperature bake.

As the coating operation continues, acid accumulates in the electrodeposition bath. Thus pH must be controlled by the addition of acid-deficient make-up coating and by the removal of excess acid from the bath. Other soluble materials also tend to accumulate in the bath. These must be removed by continuous passage through ultrafiltration units. Additionally, the temperature must be closely controlled. Highly automated electrodeposition tanks are used. The high voltage application is required in order to have high throw power, that is, deposition on internal surfaces of the metal article in the short time that it is in the tank. High throw power without film rupture at the readily accessible coated surfaces requires a limited conductivity of the applied film and low conductivity of the continuous water phase. The low conductivity requirement is another reason that the efficient operation of the ultrafiltration process is essential. The voltage that can be used varies with the type of metal being coated. Whereas film rupture has been attributed to gas evolution under the film from the electrolysis of water, it appears that electric discharges through the film may be more important (154).

Cationic electrodeposition primers show substantially better corrosion protection than anionic ones. In anionic electrodeposition, the phosphate conversion coating on the steel partially dissolves with the hydrogen ions generated at the anode surface. However, in the case of cationic primers, if zinc–iron phosphate conversion coatings are used on steel and zinc–manganese–nickel conversion coatings are used on zinc-coated steels, the conversion coatings do not dissolve (155). Furthermore, the resin used in the cationic electrodeposition primer is generally saponification resistant and promotes good wet adhesion, critical requirements for corrosion protection.

Electrodeposition primers offer substantial advantages in some applications over conventional primers: the highly automated lines permit low operating costs; the VOC emissions are lower than from other primer systems; and areas not reached by spray application are coated, giving superior corrosion protection. There is also the advantage of essentially 100% utilization of the coating and no waste from overspray. Furthermore, electrodeposition gives uniform film thickness to the coating on all areas of the article, avoiding thin spots, fatty edges, sagging, and so forth.

The capital cost of application facilities is high for electrodeposition, however, effectively limiting this technique to high production articles. Furthermore, it is more difficult to obtain adequate intercoat adhesion of a top coat to an electrodeposited primer than to a spray-applied primer. This adhesion problem is partially because the PVC must be lower than CPVC. The lower pigment content gives a higher gloss surface, which reduces the opportunity for intercoat adhesion. Commonly, this problem is overcome by applying a thin coat of sealer, that is, a thin coating of a high solvent content coating, before application of the top coat. Because the film thickness of the electrodeposited primer is uniform, the surface of the coating in effect replicates any irregularities in the surface of the metal. If high gloss top coats are to be applied over the primer, care must be taken in selecting steel and steel treatment procedures that permit adequate smoothness or a coat of primer-surfacer over the electrodeposited primer may be required.

Top Coats.   The selection of a top coat depends on cost, method of application, and product use and performance requirements, among other factors. As a result of increasingly stringent air quality standards and increased solvent costs, approaches to reduction of solvent emissions are being sought.

High Solids. There is no agreement on a definition of high solids coatings. For any particular use, this term means higher solids than previously used in that application. In automotive metallic coatings, high solids usually is taken to be about 45 vol %. This limit results from the solvent level required to give sufficient shrinkage during film formation to orient the aluminum pigment to give the desired color effect. Furthermore, thermosetting acrylic resins made by free-radical initiated polymerization give films having satisfactory performance for automotive top coats only when the molecular weight and degree of functionality are such that solids for spray application are less than 50 vol % solids.

In the case of clear or high gloss polyester coatings, volume solids of over 70% can be achieved with reasonable film properties. These do not have the exacting requirements of automotive top coats. Because of the effect of pigments on viscosity, highly pigmented high solids coatings (over about 75% volume solids) cannot be used. Low gloss clear coatings become more and more difficult to formulate as solids increase because the flatting agent concentration near the film surface must be increased by transporting fine particle size pigments to that surface by convection currents within the drying film caused by solvent loss.

Whereas the main driving force behind the development of higher and higher solids coatings has been the reduction of VOC emissions, solvent cost is also a factor. A further advantage is that the same dry film thickness can be applied in less time. High solids coatings are made using lower molecular-weight resins having fewer average functional groups per molecule as compared to conventional coatings. As a result, more complete reaction of the functional groups is necessary to achieve good film properties. Greater care is necessary for high solids coatings manufacturers to maintain close adherence to quality control standards. Additionally, application techniques, baking times, and temperatures for which the coating is designed, must be carefully adhered to. There is a relatively small window of cure. Shorter times or lower temperatures can result in undercure and the properties are more affected by overcure, in general, than conventional coatings (39). High solids coatings are likely to have high surface tensions and hence are more likely to give film defects such as crawling and cratering than conventional coatings. High solids coatings are more subject to oven sagging than are conventional coatings because of the greater temperature dependency of their viscosities (43).

Sagging of spray-applied high solids coatings is more difficult to control than for conventional coatings. Little solvent is lost between the spray gun and the surface being coated during the application of high solids coatings (74). Therefore, there is little increase in viscosity before the coating droplets arrive on the object being sprayed, thus increasing the probability of sagging. In many cases, the only available method of controlling sagging is the incorporation of an additive that imparts thixotropic flow properties. In many applications, fine particle size silicon dioxide provides the desired flow properties. However, for automotive metallic coatings, the difference in refractive index between the SiO_2 and the polymer in the dried film is sufficient to lead to some light scattering and hence reduce the flop effect in the metallic color. In order to minimize this problem, microgel additives have been developed. The acrylic gel particles are highly swollen by solvent and hence occupy a substantial volume in the liquid coating, but they are easily distorted, conferring thixotropic flow properties to the coating. After the solvent has evaporated, the refractive index is essentially identical to that of the acrylic coating polymer. No light scattering interferes with the color flop.

Microgel additives also minimize another problem in using high solids automotive top coats where base coat/clear coat systems are being increasingly used. A metallic coating is applied over the primer and, after enough solvent has evaporated to increase the viscosity of the layer, a clear top coat is applied over the wet base coat film. In the case of high solids base coats, the solvent loss and viscosity build up are so slow that the application of the top coat seriously roughens the base coat surface when it is applied. The presence of a microgel additive in a high solids metallic base coat gives the higher viscosity needed to permit top coat application without disturbing the base coat surface.

Water-Borne Systems.   In industrial coatings two classes of water-borne systems are used: water-reducible and latex systems. As of this writing water-reducible systems are used on a much larger scale but the consumption of latex product coatings is increasing. Water-reducible systems are sometimes called water-soluble systems but the resins are not truly soluble in water. When the coating is reduced with water, the polymer separates into relatively stable aggregates containing the pigment and cross-linking agent and these are swollen by solvent and water. In anionic systems, the resin is sufficiently substituted with COOH groups to give the desired water dilutability after partially neutralizing with an amine. Acid numbers are in the range of 40–80 mg KOH/g resin solids (49,50). The largest volume coatings of this type contain acrylic resins cross-linked with melamine–formaldehyde resins or blocked isocyanates. Polyester–melamine and alkyd–melamine systems are less important because of the difficulty of achieving adequate hydrolytic stability for many applications. Water-reducible urethane resins can be used with melamine–formaldehyde cross-linkers to give coatings having superior abrasion resistance, but at a higher cost.

Because the molecular weight and average functionality of the resins used in water-reducible coatings are comparable to those of conventional solution thermosetting coatings, film properties obtained after curing are fully equivalent. Similarly, and in contrast to high solids coatings, they have tolerances to variations in cure conditions comparable to those of conventional solution coatings. Surface tensions are relatively low, and few problems with crawling and cratering are encountered. Because the solids are low, good orientation of aluminum pigment can be obtained and such water-reducible coatings are being used increasingly in base coats for automobiles along with a high solids clear coat (156).

There are, of course, disadvantages to the water-reducible coatings. Solids contents at application viscosities are relatively low, 18–25 vol %. However, because most of the volatile material is water, VOC emissions are still lower than from most high solids coatings. The evaporation rate of water depends on the relative humidity, which can lead to inconsistent flow behavior during flash off. Above a critical relative humidity, water evaporates more slowly than the solvent; below this critical value the solvent evaporates more slowly than water. The viscosity of the coating depends strongly on the ratio of solvent to water as well as on the concentration, thus abnormal viscosity changes can occur when flash off occurs above the critical relative humidity (157).

The extent to which water is lost before the applied coating enters the baking oven can also be a factor in the possible "popping" of the films. If the viscosity of the surface of a coating film becomes high while there is still a significant amount of volatile material in the lower levels of the film, then as the coating film is heated in the oven the volatile material can vaporize under the viscous top layer. A bubble is created that frequently pops, leaving a blown-out crater. Popping is more difficult to control in water-reducible baking enamels than in solvent-based counterparts, even when the relative humidity in the flash off zone is well controlled.

Variations in coating composition can affect the probability of popping. The T_g of the resin in the coating has been shown to affect the film thickness of coating that can be applied without popping (158). The lower the T_g of the resin, the thicker the film that can be applied without popping. Slower evaporating solvents keep the surface of the film open longer and hence minimize popping. Popping is more likely to occur as film thickness increases. In applications such as exterior can coatings where thin films are applied, problems with popping are seldom, if ever, encountered.

Acrylic coatings of fairly similar composition can also be applied as top coats directly on metal by anionic electrodeposition. Some iron is dissolved at the anode however, and anionic electrodeposition coatings tend to become discolored. Cationic electrodeposition top coats can be made, using for example, 2-(dimethyl-amino)ethyl acrylate as a comonomer and a blocked aliphatic diisocyanate as the cross-linking agent for hydroxy-functional groups. Electrodeposition top coats are particularly useful where it is difficult to achieve full, uniform coverage by other means such as on articles having sharp edges, eg, finned heat exchange units. The highly automated electrodeposition process can lead to significant cost reduction. It has been reported for cationic electrodeposition coating of air conditioners that replacing a former system of flow-coated primer and spray-applied top coat required only one operator. In the former system, 50 people, including those doing the required touch up and repair were needed (159).

Until the 1980s, latex-based coatings were infrequently used in product coating applications. High gloss coatings cannot be made using latexes and transparent coatings are difficult if not impossible to make. Furthermore, the rheological properties of the coatings limit utility. Flow problems are least severe when coatings are applied by curtain coating or reverse roll coating because these two methods do not involve film splitting. The availability of associative thickeners that minimize flocculation of latex particles permits formulation of industrial coatings that are less thixotropic and hence level better after application. Popping can cause difficult problems in applying thick coats of latex-based coatings. However, in contrast to the situation with water-reducible coatings, latexes having low T_g are more likely to give popping problems. When coalescence is slower because of a higher T_g, there is more time for the water to escape from the film before the surface layer coalesces. Low VOC emissions and excellent film properties resulting from the high molecular weight of the polymer are expected to lead to increases in the use of latexes in product coatings.

Coil Coating. An important segment of the product coatings market is sold for application to coiled metal, both steel and aluminum (160). In this process, the metal is first coated and then fabricated into the final product rather than fabricating first into product. The method offers considerable economic advantages because coatings are applied by direct or reverse roll-coating at high (up to 400m/min) speeds to wide (up to 3 m) coils in a continuous process. The metal is cleaned, conversion coated, and coated with primer (if desired) and top coat all in an in-line operation. In many cases, coatings are applied to both sides of the metal during the same run through the coil coating line. The labor cost of coating application is much less than for application to a previously fabricated product. There is essentially 100% effective utilization of coating, and the loss by overspray involved in coating fabricated products is avoided.

Use of coil-coated stock reduces fire risk and hence insurance costs for the metal fabricator. The problem of controlling VOC emissions is also avoided because no coating is done in the factory. VOC emission control problems for the coil coater are minimal. The oven exhaust is used for the air needed for the gas heaters for the oven. In other words, the organic solvents are used as fuel rather than being allowed to escape into the atmosphere. Film thickness of the coatings is more uniform than can be applied by spray, dip, or brush coating of the final product. The coatings are all baked coatings, and when coil-coated metal is used, substantially greater exterior durability and corrosion protection can be achieved as compared with field-applied air-dry coatings.

There are, of course, limitations to the applicability of the process. The metal must be flat. Coil coating is economical only for long runs of the same color and quality of coated metal. Offsetting the economic advantages of coil coating is the higher cost of maintaining inventories of coated metal coils. Additionally, because the films must sometimes withstand extreme extensions in fabricating a final product, they must generally have long elongations-to-break which may limit the range of other film properties that can be designed into the coating. For example, it may be difficult to achieve a high modulus at a low rate of application of stress while still having adequate elongation-to-break to permit fabrication without film cracking. Although some coating may be smeared over the cut edge of the metal when the coil is cut, the cut edge of the metal usually has little, if any, coating on it. These cut edges can be a weak spot for corrosion protection and can present an appearance problem if the edges are visible in the fabricated product. Although coatings can be made that can be welded through, they char from the intense heat and have, therefore, an unacceptable appearance if the weld marks are visible. Design of a final fabricated product must take the cut edge and weld marks into account, hiding and/or locating them in positions least likely to cause corrosion problems.

In order to have reasonable oven lengths for curing at high speeds, the coatings must cure in short times, less than 1 min. Therefore, the ovens are operated at high temperatures, above 290°C. The short time, high temperature cure schedules are difficult to duplicate in a coatings development laboratory. Substantial experience is required to make informed estimates of how to translate laboratory cure response to coating line cure response. This problem can be particularly acute for color matching. The ability to do satisfactory color matching for coil lines is a prerequisite to commercial success in supplying coil coatings. Because during start up, and sometimes in splicing a new coil onto the previous coil, the line must be slowed down, it is essential that there be a minimum change in color and film properties when the standard curing time is exceeded.

A wide range of resin compositions are used in coil coatings, depending on the product performance requirements and cost limitations. The lowest cost coatings are generally alkyd coatings. They are appropriate for indoor applications where color requirements are not stringent. Alkyd–melamine coatings are sometimes used outdoors when long-term durability is not needed. Greater durability and better color retention on overbaking are obtained using polyester–melamine systems. Polyesters usually provide better adhesion to unprimed metal than acrylic coatings. For two-coat systems, epoxy ester primers are widely used along with acrylic–melamine top coats. Blocked isocyanates can also be used for cross-linking the polyesters and acrylics but care is needed because urethanes decompose thermally and loss of film properties and discoloration on overbake may be severe. For this reason, it is necessary to use blocked diisocyanates that react at relatively low temperatures. Oximes are the most widely used blocking agents for coil-coating systems. Thermoplastic coatings based on vinyl plastisols and acrylic latexes resist severe deformation during fabrication and exhibit good exterior durability.

For greater exterior durability, silicone-modified polyesters or silicone-modified acrylic resins are used. Hydroxy-functional polyester or acrylic resins are partially reacted with a silicone intermediate having methoxy substituents on the siloxane resin chain. Cross-linking is completed during curing by reaction of the remaining methoxy substituents and hydroxyl groups on the polyester or acrylic resin. In order to avoid softening during prolonged exposure to high humidity, which can result in reversible hydrolysis of cross-links with groups having three oxygens on a silicone atom, it is common to use a small amount of a melamine–formaldehyde resin as a supplementary cross-linker. Silicone-modified coatings, especially having 50% or more silicone modification, show substantially longer exterior life times than the corresponding unmodified polyester or acrylic coating. Silicone-modified coatings are, however, more expensive. They are particularly useful in colored coatings where the color change can be substantial when chalking occurs. The white coated product in the same quality line may well be made without silicone modification to reduce cost, because color change resulting from the chalking of the white coating after exposure for a long period is less obvious. For still greater exterior durability, fluorinated polymer systems are used.

In the can industry, large-volume three-piece cans, used for packing many fruits and vegetables, are made from coil-coated stock. In this case, generally, oil-modified phenolic resins are used for coating tinplated steel coils on the side that becomes the interior of the can. Coil stock for making ends for two-piece beverage cans is coated with epoxy–melamine, epoxy–phenolic, or a cationic uv-cure epoxy coating.

Radiation-Cure Coatings. Radiation has several significant advantages over heat as the source of energy to carry out cross-linking reactions (161,162). Coatings can be designed that cure rapidly, in a second or less, at room temperature yet have relatively long storage lives. The energy requirements for curing are much lower than for thermal baking systems. Rather than volatile solvents, reactive monomers can be used to provide for the low viscosities needed for application. Thus VOC emissions can be very low. In some cases, the energy source is high energy electron, but more commonly ultraviolet, radiation curing systems are used.

In uv-curing coatings, a photoinitiator is required in the formula, whereas in electron-cured coatings the energy is directly absorbed by the reacting molecules. Both free-radical and cationic initiated polymerization systems are employed for uv cures. Benzoin derivatives, benzil ketals, acetophenone derivatives, α-hydroxyalkylphenones, O-acyl-α-oximinoketones, and benzophenone (diphenylmethanone) [119-61-9], $C_{13}H_{10}O$, and 2-(dimethylamino)ethanol combinations are examples of photoinitiators for free-radical polymerization (163). The most widely used vehicle systems are oligomers substituted with several acrylic ester groups mixed with low molecular-weight difunctional or trifunctional acrylates and a monofunctional acrylate. Acrylic ester systems provide a much more reactive cure than methacrylate systems. Styrene is sometimes used as a monomer. However, cure is slower than for acrylic esters and significant amounts of styrene evaporate. Many pigments absorb uv radiation, thus most commercial uv-cure coatings are unpigmented, clear coatings. Printing inks (qv), which are applied as very thin films, and particle board filler, where only inert pigments which absorb little uv are used, are examples of uv-cure pigmented systems.

Photoinitiators for cationic polymerization generate reactive electrophiles (acids). Three classes of cationic photoinitiators have been used commercially: triarysulfonium, diaryliodonium, and ferrocenium salts of very strong acids, such as hexafluorophosphoric acid (164). When the uv absorption of a potential photoinitiator is low, a strongly absorbing photosensitizer can also be used (165). The most widely used vehicles are epoxy resins diluted with monofunctional and difunctional epoxy compounds to adjust the viscosity. An important advantage of this type of system is that shrinkage is substantially less than in systems where acrylic double bonds are polymerized. Furthermore, the protons generated by the photoinitiator are more stable than the free radicals generated in radical systems. Hence curing can continue after the article being coated has passed beyond the uv source. However, reaction rates are slower than for free-radical polymerized acrylate coatings. High speed curing can be carried out by combining uv exposure to release the acid catalyst and baking to give a rapid cross-linking reaction.

There are many advantages to uv curing, but there are also limitations. Only flat surfaces that can be passed under the focused uv source, or cylindrical surfaces that can be rotated under the source, can be practically cured by this method. Although it is desirable in many cases to make coatings that cure at low temperatures, in uv-cure systems, as in all other coatings, the T_g of the final cured film is limited to temperatures a little above the temperature of the film during curing by the limitation of available free volume on the cross-linking reactions. Radical initiated uv-cure systems are poor candidates for coatings requiring good exterior durability, because the residual photoinitiator is also a photoinitiator for photodegradation reactions. Solvent-free pigmented uv-cure coatings are limited not only by the depth of penetration of the uv radiation but also by the effect of the pigmentation on flow properties. Even for gloss coatings, such as exterior white can coatings, the amount of pigmentation is sufficient to increase the viscosity enough to affect leveling adversely. As a result of the substantial shrinkage which occurs in a very short time in the uv curing of acrylate systems using radical photoinitiators, adhesion to smooth surfaces such as metals is generally reduced by the stresses in the film. Such stresses can frequently be relieved by heat treatment after curing. Also costs tend to be relatively high.

Cationic systems have fewer limitations. They are not air inhibited, the photoinitiators do not initiate photodegradation reactions, there is less shrinkage during curing, and the stability of the acids generated permits migration if initiator through a pigmented film, allowing cure of somewhat thicker pigmented films. However, commercial adoption has been slow, possibly because of higher costs and slower cure rates at ambient temperatures.

Uses that have developed for uv curing reflect the special advantages of the system rather than replacement to reduce VOC emissions or energy consumption. Clear coatings on heat-sensitive flat plastic substrates, where rapid cure is needed, is an area where uv curing has found application, for example, uv-cure abrasion-resistant top coats for vinyl plastic flooring. These coatings are made using acrylic ester-terminated polyurethanes as the oligomers. Other examples of application to heat-sensitive substrates are coatings for thin wood veneers used for door skins and coatings for paper.

The high solids of uv-cure systems can also be important in some cases. For example, uv curing fillers for particleboard for use in paneling and furniture has the advantage over conventional coatings that the high solids permit filling of the rough surface in one coat rather than using multiple coats of solvent-based systems. In printed circuit boards, uv-cure systems are widely used. Another application, which illustrates the advantages of high cure rates, is the use of uv-cure clear coatings of glass optical fibers in waveguides. This application also illustrates that it is sometimes desirable to use solvents in uv-cure coatings, and not restrict the formulation to 100% solids (see also ELECTRONIC COATINGS).

Powder Coatings. Another coating technique that substantially reduces VOC emissions is powder coating. A wide variety of coating types and application methods are used (166–168) and powder coatings have been the most rapidly growing part of the coatings industry. The two most widely used methods of application involve: passing heated objects to be coated into a fluidized bed of powder particles suspended in air; and electrostatically spraying grounded articles at ambient temperature with powder, then baking. Triboelectric charging of the powder particles has improved spray application of powders (168) (see COATING PROCESSES, POWDER TECHNOLOGY).

Thermoplastic powders are generally applied by a fluidized-bed process, which results in relatively thick films. As the coating builds up, the temperature at the surface drops. The last particles picked up stick to the surface but do not completely fuse into the film. A conveyor transfers the article into an oven when fusion and leveling are completed. The polymers are usually vinyl chloride copolymers, polyolefins, or polyesters, especially scrap poly(ethylene) terephthalate. The growth products are thermosetting powders which are generally applied by electrostatic spray. The sprayed parts are carried into an oven for fusion, leveling, and cross-linking.

Powder coatings have many advantages. VOC emissions approach zero because no solvents are used. Compared to many solvent-borne baking coatings, fuel cost for heating ovens are low, even though the baking temperature may be higher. This results from the substantial increase in recirculation of the hot air in the ovens, because there is less need to exhaust air to keep the solvent concentration below the lower explosive limit. Lack of solvent permits even application of coatings to the inside of pipes without the solvent wash encountered when solvent coatings are applied to an enclosed area. As compared to electrostatic spray applied solvent-borne or water-borne coatings, utilization efficiency of powder coatings can be much higher because the oversprayed powder can be collected and reused. On the other hand, the overspray from liquid coatings must be caught in a water wash spray booth. This overspray cannot be directly recycled and generally the resulting sludge becomes a solid waste disposal problem.

There are disadvantages or limitations to powder coatings as well. Only substrates which can withstand the relatively high baking temperatures can be coated. Color changeover in the application line requires shutting down the line and cleaning the booth. Therefore, powder coating is most applicable to product lines where long runs of the same color are processed. Color matching is more difficult than with conventional coatings. Because the final coatings cannot be blended or shaded with tinting colors, the whole batch must be reworked if the color match is not satisfactory. By careful control of incoming raw materials and processing variables, satisfactory reproducibility of color matching for many applications is possible. Metallic colors showing the color change with angle of viewing such as are widely used in automotive top coats cannot be made in powder coatings although other metallic colors can be obtained.

The T_g of the powder coating before curing must be above the storage temperature in order to avoid sintering. As a result, the range of physical properties of the final films is limited. The high T_g of the powder before application requires high baking temperatures. In order to coalesce after application, the temperature must be sufficiently above the T_g for the free volume to be adequate for ready coalescence. The T_g of the resins used in powder coatings is controlled by the monomers and molecular weight. It has been reported that, at least in one case, it is advantageous to use higher molecular-weight, more flexible resins because these can have adequate package stability but flow more easily at higher temperatures than a lower molecular-weight resin of similar T_g with more rigid chains (169). The problem of coalescence and flow is further compounded by the decrease in free volume and increase in viscosity resulting from the cross-linking reaction that is simultaneously proceeding. Although leveling is adequate for many applications, it can be a limitation for powder coatings (170).

The range of suitable resin compositions is much narrower than in liquid coatings. The T_g of the powder must be sufficiently high to prevent sintering during storage. The components are blended and the pigments are dispersed in the vehicle by passing through a heated extruder that subjects the material to high shear stress. The cross-linking reaction must proceed slowly enough under the conditions encountered in the extruder that very little polymerization occurs during the process. Because some cross-linking does occur, the number of times that the same material passes through the extruder must be limited as well as the fraction of recycled material that is incorporated in any batch. These considerations place a premium on raw material control for resins as well as pigments. In solvent coatings, the final viscosity and application properties can be adjusted by the additions of solvents and varying their evaporation rates. In powder coatings, on the other hand, the application properties are governed by the resins alone.

Bisphenol A epoxy resins with dicyandiamide as the cross-linking agent are widely used in coatings where exterior durability is not required. Phenolic resins are used as cross-linkers for epoxy resins when greater chemical resistance is required. So-called hybrid polyester powder coatings are based on carboxylic acid-terminated polyesters and BPA epoxy resins. These have superior color retention at lower cost but still inferior exterior durability. For applications where exterior durability is required, triglycidyl isocyanurate is used along with carboxylic acid-functional polyesters. Hydroxy-functional polyester or acrylic resins cross-linked with blocked isocyanates give powder coatings having superior exterior durability combined with abrasion resistance. Caprolactam (qv) blocked isophorone diisocyanate is the most widely used blocked isocyanate. Blocked isocyanate powder coatings tend to give better leveling than most other powder coatings perhaps because the caprolactam released by the unblocking reaction volatilizes only slowly. While the caprolactam is still present in the film, it can lower viscosity and may help promote coalescence and leveling. Tetramethoxymethylglycoluril is also used as a cross-linker for hydroxy-functional polyesters (171). Tetra-(2-hydroxyalkyl)bisamides are used as cross-linkers for carboxylic acid-functional resins (67).

Examples of applications for powder coatings include pipe lining, coatings for rebars, underbody automotive parts, wheels, garden tractors, home appliances, playground equipment, metal furniture, fire extinguishers, and many others. As VOC restrictions become more restrictive, powder coatings can be expected to increase in usage.

Wood Product Coatings. Furniture is one important class of wood (qv) products that is industrially coated. In most cases, the appearance standards are set by the fine furniture industry where the flat areas are composed of plywood having high quality top veneer; the legs, rails, and so forth are solid wood; and there are frequently carved wood decorative additions (see also WOOD-BASED COMPOSITES AND LAMINATES). The finishing process for fine furniture is long and requires significant artistic skill. If the final overall color of the furniture is lighter than the color of any part of the wood, the first step is bleaching using a solution of hydrogen peroxide in methanol [67-56-1], CH_4O. The bleached wood, or unbleached for darker color finishes, is given a wash coat of size to stiffen the fibrils so they can be cleanly removed by sanding. The wood is then coated with stain, that is, a solution of substantive acid dyes in methanol, to give a desired overall color tone to the piece of furniture. A wash coat, generally a low solids vinyl chloride copolymer lacquer, is applied over the stain, partially to seal the stain in place but also to prepare the surface for the next operation, filling. The filler, a dilute dispersion of pigments, usually in linseed oil with mineral spirits solvent, is sprayed over the whole piece of furniture. It is then wiped off, leaving filler only in the pores of the wood. The colors of fillers are commonly dark brown; the filler serves to emphasize the grain pattern in the hardwood veneer. Next, shading stains are selectively sprayed on the wood to give different colors to various sections.

It is also common to distress the surface for resemblance to antique furniture. For example, spots of black pigment stain can simulate India ink spots dropped from quill pens, and dark stains carefully applied into corners simulate dirt accumulated over many years. When this "art work" is completed, a sanding sealer is applied to immobilize the lower layers of the finish and to provide a surface that can be sanded smooth. Finally a top coat is applied and polished smooth.

The primary binder used in the sanding sealer and the top coat is nitrocellulose. Nitrocellulose lacquers supply a coating that brings out the natural beauty of the wood and gives a depth of finish that has not been matched with other systems. The film properties of the lacquers improve with increasing molecular weight of the nitrocellulose, whereas the solids of the lacquers decrease. Lacquers suitable for high quality furniture have volume solids less than 20%, although this can be increased somewhat by using hot spray application. Other components of the lacquer are plasticizers and hard resins to provide a combination of flexibility and hardness for sanding and polishing while retaining flexibility to withstand cracking as the wood in the furniture expands and contracts. Zinc stearate [557-05-1], $C_{36}H_{70}O_4Zn$, is incorporated in the sanding sealer to reduce clogging of sand paper.

Top coats generally include fine particle size silicon dioxide to reduce the gloss. In making a low gloss top coat, it is essential to retain the transparency. Low gloss can be achieved using a minimum amount of fine particle size SiO_2 pigment. When the solvent evaporates from the low solids lacquer, convection currents carry the fine particle size SiO_2 to the surface. Hence the PVC of the surface layer is relatively high, giving the low gloss, whereas the PVC of the entire film is low (2–4%) in order to maintain transparency. As higher and higher solids lacquers while maintaining clarity are attempted, low gloss becomes more and more difficult. The convection currents are not as strong using the lesser amounts of solvent.

Nitrocellulose lacquers offer other important advantages for furniture. There is latitude as to when a complete piece of furniture is rubbed after finishing, whereas thermosetting systems must be rubbed after some cross-linking but before cross-linking has advanced to the point that it cannot be rubbed without leaving permanent scratches. Using lacquers, the production line can be shut down at the end of a shift and pieces on the line can be rubbed the next day or after a weekend; using thermoset systems, the line must be stopped early or an extra shift must be brought in just for the rubbing. Because of warehousing space problems for furniture, factories frequently prefer to load furniture onto trucks almost directly off the finishing line. The finish must therefore withstand print tests equivalent to the pressures that are going to be encountered in shipping very soon after they are applied. This is a problem for water-based top coats because the rate of loss of water depends on humidity. The loss of the last of the water from an almost dry film can be slow, resulting in delay before the furniture can be wrapped and shipped. The ease of repair of the thermoplastic nitrocellulose lacquers is another factor strongly in their favor.

This finished furniture is a beautiful product attained at a high cost. Most good furniture is made at substantially lower cost with as little sacrifice in appearance as possible. The expensive hardwood veneer plywood is replaced by printed, coated particleboard. The artists' work of filling, shading, distressing, glazing, and so on needs only to be done once on a carefully selected beautiful wood laid up meticulously into fine patterns. The finished panels are then photographed to make a series of three gravure printing plates of the finest through the boldest parts of the pattern. The particleboard is coated with a uv-cure filler and sanded smooth. A lacquer base coat is sprayed on, in a color corresponding to the first coat of stain in fine furniture finishing. Then three prints are applied by offset gravure printing using three carefully selected color inks. The panel is assembled into the furniture and finished with lacquers as in the standard process.

Only experts can distinguish the prints from real wood by just looking at the surface. Except for the most expensive furniture, high density polyurethane foam parts formed in flexible molds made from master wood carvings are used. The interior of the mold is sprayed with a base coat lacquer before charging with the urethane components of the foam. This lacquer coat serves as a mold release agent and imparts the color to the surface that matches the stain color of the furniture. The part is attached to the furniture and finished as if it were wood. The strength of these parts is actually superior to that of wood carvings. For low cost furniture, impact-resistant polystyrene injection molded parts are used. Although complete pieces of furniture can be made this way, they can easily be distinguished from real wood.

Many alternatives to nitrocellulose lacquers for top coats have been investigated and some are used commercially. Especially for lower cost furniture exposed to hard use such as motel and institutional furniture, alkyd–urea top coats are used. Urea–formaldehyde resins, in contrast to melamine–formaldehyde resins are used because the former can be cured using butyl acid phosphate [1623-15-0], $C_4H_{11}O_4P$, catalyst at room temperature or force dried at 60–70°C. These resins are applied at around 35 vol % solids and require about half the spraying time of a nitrocellulose lacquer. They are also more difficult to repair and do not provide the depth of appearance of nitrocellulose lacquers.

Conventional nitrocellulose lacquer finishing leads to the emission of large quantities of solvents into the atmosphere. An ingeneous approach to reducing VOC emissions is the use of supercritical carbon dioxide as a component of the solvent mixture (172). The critical temperature and pressure of CO_2 are 31.3°C and 7.4 MPa (72.9 atm), respectively. Below that temperature and above that pressure, CO_2 is a supercritical fluid. It has been found that under these conditions, the solvency properties of CO_2 are similar to aromatic hydrocarbons (see SUPERCRITICAL FLUIDS). The coating is shipped in a concentrated form, then metered with supercritical CO_2 into a proportioning airless spray gun system in such a ratio as to reduce the viscosity to the level needed for proper atomization. VOC emission reductions of 50% or more are projected.

Considerable efforts have been invested in developing water-borne coatings for wood furniture. The two most promising approaches seem to be water-reducible acrylic resins with urea–formaldehyde cross-linkers and hydroxy-functional vinyl acetate copolymer latexes again with urea–formaldehyde cross-linking. Water-borne systems would be used more widely if it were not for the effect of relative humidity on drying and the relatively long time for the coatings to develop print resistance. Because of grain raising caused by direct contact of water with wood, water systems are not suitable for the first layer of coating on the wood.

Uv-cure coatings are widely used in European furniture manufacture but have found more limited applications in the United States. Most United States furniture has a relatively low gloss finish, frequently has curved surfaces, and is finished after assembly; most European furniture has a relatively higher gloss finish with primarily flat surfaces, and the furniture is generally assembled after coating. Uv-cure coatings are more easily adapted to coating before assembly. Because of the pressure on the furniture industry to decrease VOC emissions, uv-cure coatings may become more widely adopted in the United States. The use of high pressure laminated plastics having wood grain reproductions can be expected to take a larger share of the furniture top market in response to VOC regulations.

The other principal component of industrial wood finishing is the panel industry. The highest cost segment of this industry, fine hardwood veneer paneling for executive office walls, is comparable to the fine furniture industry, similar nitrocellulose lacquer finishing systems are used. However, the bulk of the industry requires less expensive finishing operations. Some plywood is stained followed by roll coating with a low gloss nitrocellulose lacquer top coat. It is common to print grain patterns from woods such as walnut onto inexpensive, relatively featureless veneers such as luan before applying a lacquer top coat. Lacquer topcoats are being replaced by alkyd–urea coatings to reduce VOC.

Large volumes of wood grain paneling are made from hardboard paneling. The coating systems are usually a base coat, three prints, and a clear top coat. In higher style paneling, the hardboard is embossed in the pressing step involved in making the hardboard. For example, if a pecky cypress paneling is desired, the hardboard would be embossed with holes resembling those of pecky cypress. The first step in finishing such a board is to apply a so-called hole color with a brush roll roller coater. Then the flat surfaces are precision coated with a base coat. Precision coating is applied using an overall gravure roller, it can be thought of as an overall print. Because the holes do not contact the print roller, they are not coated with base coat and the color of the holes is not covered. The base coat color is the equivalent of the stain color if real wood were being used. Then three prints are applied from gravure rollers made by photographing carefully selected and finished real cypress. Finally an overall low gloss top coat is applied. Because hardboard can withstand contact with water and relatively high temperature baking, the coatings are most commonly water-reducible acrylic–melamine systems. Using this method, imitations can be made of any wood and also of brick paneling, marble paneling, porcelain tile, and so forth.

Hardboard paneling is not restricted to interior wall paneling. Factory coated hardboard is also widely used for exterior siding for homes. The largest volume of such hardboard is so-called preprimed board. A primer is applied in the factory that is suitable for exterior exposures up to six months before painting with exterior house paint. While only one coat of latex paint need be applied, it has been reported that greater durability is obtained if a primer is applied after house construction using an acrylic latex top coat (173). Fully prefinished hardboard is also sold commercially. The exterior durability of this paneling is superior to that of the field painted paneling, but the styling is more limited; it is used primarily in commercial buildings (see also BUILDING MATERIALS, SURVEY).

Special Purpose Coatings

Special purpose coatings include those coatings that do not fit under the definition of architectural or product coatings. About 16% of the volume (6.1×10^5 m^3) of United States coatings, value $2.62 billion in 1990, falls into this category (1).

Heavy-Duty Maintenance and Marine Coatings.   Heavy-duty maintenance coatings are applied to bridges, off-shore drilling rigs, chemical or petroleum refinery tanks, and similar structures. The applications are usually over steel at ambient temperatures and the objective is corrosion protection. In choosing a system for an application, many factors must be taken into consideration: cleaning of the steel surface, film integrity, temperature of the painting operation, environment, materials, costs, etc.

For applications where the surface can be thoroughly cleaned and where maintenance of film integrity can be reasonably expected, it is most appropriate to use a primer without passivating pigments but having excellent wet adhesion. For example, a two-package epoxy–amine coating would serve. This combination of epoxy resin and polyamine should be designed in such a way that the T_g of the fully reacted coating permits the reaction to go to completion in a reasonable time period (commonly a week) at the temperatures that prevail during the curing period. Epoxy–amine coatings have the further advantage that they can cure underwater and hence are suitable for piers and off-shore installations. The top coats should be designed to minimize oxygen and water permeability. Chlorinated polymers such as vinyl chloride copolymers, vinylidene chloride (1,1-dichloroethene) [75-35-4], $C_2H_2Cl_2$, copolymers, or chlorinated rubber are widely used in top coats. Urethane coatings based on aliphatic isocyanates are widely used because of good abrasion resistance. Where applicable, use of leafing aluminum pigment in the final coat is desirable to provide a further barrier to water and oxygen permeation.

In situations where complete cleaning of the steel is not possible and where film ruptures are to be expected, primers having passivating pigments or zinc-rich primers are used. Cost is generally lower and mechanical integrity of the films is generally greater using passivating pigment primers. On the other hand, corrosion protection can generally be extended for longer time periods using zinc-rich primers. The substrate should be cleaned as well as possible before painting, generally by sandblasting, and primer should be applied as soon as possible after the surface has been sandblasted. Top coats for passivating pigment primers are the same as for epoxy–amine primers. When using zinc-rich primers the effectiveness of the primer depends on maintaining its porous structure, therefore, the next coating should not penetrate down into the pores of the primer coat. For solvent-based top coats, this is generally accomplished by applying a very thin film on the primer surface. The viscosity of a thin film increases rapidly as a result of fast solvent evaporation, minimizing penetration into the primer pores. The application requires highly skilled workmanship and careful inspection. Further top coat can then be applied without concern for penetration into the primer.

Application of latex paints directly over a zinc-rich primer essentially eliminates the penetration problem. The latex polymer particles are large compared to the primer pores, and after coalescence the viscosity of the polymer is so high that it does not penetrate into the pores. The water and oxygen permeability of latex paints is generally higher than that of solvent-based paints. Thus it is especially desirable to incorporate platelet pigments such as mica and leafing aluminum. Use of vinylidene chloride–acrylate ester copolymer latexes has been recommended because of the lower water permeability (174). Field performance using latex paints over zinc-rich primers in such applications as highway bridges has been reported to equal solvent-borne paint performance (124,125).

Not only are latex paints being used over zinc-rich primer, they are also increasingly being used directly on steel for corrosion protection. The steel should be thoroughly cleaned, eg, by sandblasting, and the first coat of paint applied quickly. A sandblasted steel surface is highly activated and can flash rust when water contacts the surface, ie, a layer of hydrated iron oxide forms almost instantly. This can be avoided when a latex paint is applied by incorporating a low volatility amine such as 2-amino-2-methyl-1-propanol [124-68-5], $C_4H_{11}NO$, in the formula. Latexes have been designed that have superior wet adhesion to metal, for example, by using a small amount of 2-(dimethylamino)ethyl methacrylate as a comonomer. Because the size of the latex particles is large compared to the cross-section of surface pores in steel and the viscosity of the polymer after coalescence is high, the vehicle of latex paints cannot fully penetrate the micropores on the steel surface. Therefore, passivating pigments should be incorporated in the primer formula to provide corrosion protection in these areas. Passivating pigments for latex paints must be carefully selected. Solubility of the pigment must be high enough to provide the desired passivating action, but low enough that the ions do not destabilize the latex dispersion. Recommendations are that latex paints should not be applied on highway bridges when the temperature is less than 10°C and the relative humidity above 75% (124).

Many marine coatings play a key role in extending the life of ships by corrosion protection (175). Products are similar to those used for other heavy-duty maintenance applications. Ship bottom coatings, designed to delay the growth of barnacles, algae, and other marine life on underwater hulls, are widely called antifouling paints (176). Cuprous oxide has been used as a toxicant in antifouling coatings, but it was replaced in large measure by toxicant pigments based on tributyltin derivatives because of the longer service life of the latter. The service life of antifouling paints was substantially extended by the development of polymers having tributyltin groups covalently bonded to the polymer. The polymers were designed so that organotin compounds were slowly released by hydrolysis (176).

All toxicants used to control fouling are also toxic to marine life in harbors, and although there is considerable controversy with regard to environmental risk, regulations are expected to become more restrictive. Efforts are now concentrated on means other than toxicants to control marine growths on ship bottoms. Some progress has been reported using silicone coatings that have such low surface tensions that marine growth has difficulty adhering to the surface (177) (see COATINGS, MARINE).

Automotive Refinish Paints. Paint for application to automobiles after they have left the assembly plant is a significant market. Although some of this paint is used for full repainting, especially of commercial trucks, most is used for repairs after accidents, commonly just one door or part of a fender, and so forth. In order to be able to serve this market, it is necessary to supply paints that match the colors of all cars and trucks, both domestic and imported, that have been manufactured over the previous ten years or so. Repair paints for the larger volume car colors are manufactured and stocked, but for the smaller volume colors formulas are supplied by the coatings manufacturer to the paint distributor that permit a reasonable color match for any car by mixing standard bases.

The nitrocellulose lacquers used for primers are being replaced with latex primers and there are two broad classes of top coats: lacquers and enamels. Acrylic lacquers are made from high molecular-weight thermoplastic acrylic polymers and cellulose acetobutyrate and plasticizer. They are applied at solids contents as low as 10–12 vol % and have many advantages: all colors can be matched, including bright metallics that were applied to cars in previous years as lacquers; gloss retention approaches that of factory applied coatings; and the lacquer films reach the "dust-free" state rapidly. The paint shops have dirt particles and spray dust in the air. If drying the coating takes a long time, to the point where dust particles do not adhere to the surface, the final appearance is poor.

Acrylic lacquers also have disadvantages: the gloss of the air-dry film is too low and the surface must be rubbed to achieve the high gloss necessary for automotive top coats; and most importantly, VOC emissions are very high. Acrylic lacquers are expected to be phased out when controls become more restrictive.

The other broad class of refinish coatings is cross-linking coatings, called enamels in the trade. In the United States, most shops cure the enamels at room temperature; in Europe, cure is commonly carried out at temperatures of 60–75°C. Traditionally, the vehicle was an oxidizing medium oil alkyd. These enamels have some advantages over lacquers: the solids are higher, 35–40 vol %, resulting in lower VOC emissions; the same dry film thickness can be sprayed more rapidly; and the films dry to a high gloss so they do not require polishing. On the other hand, there are disadvantages: only a limited range of metallic colors can be matched; the films dry more slowly and can pick up dust for an hour or more; and the exterior durability, both gloss and color retention, are noticeably inferior to the original finish. Many systems have been devised to overcome the deficiencies of alkyd enamels while retaining the advantages. Acrylic–alkyd graft copolymers dry to a dust-free stage more rapidly than conventional alkyds. Another type of acrylic refinish enamel is based on acrylic copolymers with glycidyl acrylate (2-propenoic acid oxiranylmethyl ester) [106-90-1], $C_6H_8O_3$, as a comonomer. The epoxy groups and drying oil fatty acids react to give a so-called acrylic–alkyd. These afford both faster drying and better durability. The durability is particularly enhanced because the drying is rapid enough, because of the acrylic backbone, that metal driers are not needed. The metal driers used in regular alkyds not only accelerate the drying, but also accelerate the film degradation.

In another approach, the out-of-dust time of enamels was reduced by using a polyfunctional isocyanate, such as isophorone diisocyanate trimer, as an additive to an alkyd enamel just before spraying. The isocyanates react with the hydroxyl groups of an alkyd and give faster drying. Two-package urethane coatings having hydroxy-functional acrylics and polyisocyanates have been widely used. Many painters have decided that they do not want to use isocyanate systems because some have experienced respiration difficulties after spraying. All manufacturers are therefore developing nonisocyanate cross-linking systems. Systems having toxic hazards as low as possible are the goal. Painters also need to be trained to handle them. The toxic hazard of reactive cross-linking agents can be minimized by having a molecular weight high enough for the vapor pressure to approach zero and for a low rate of permeation through skin and body membranes.

Recommendations for ventilation, protective clothing, and masks should be provided from the supplier.

Other Special Purpose Coatings. Large volumes of traffic paint are used to mark the center lines and edges of highways. The white paints are pigmented with TiO_2 and the yellow paints have been pigmented with chrome yellow (lead chromate). Concern about the toxic hazard of using chrome yellow is leading to its replacement with organic yellow pigments. The paint must dry fast enough after application so that a car can drive over it within minutes. The most widely used vehicles are alkyds with chlorinated rubber and fast evaporating solvents. Immediately after application, glass reflector beads are applied. To an increasing degree, hot melt coatings are being applied and, in high traffic areas where the lifetime of paint is short, paints are being replaced with thermoplastic tapes. Other examples of special purpose coatings include aerosol-packaged coatings, swimming pool paints, and nonstick coatings for cooking utensils.

Economic Aspects

The value of United States coatings consumption in 1991 reached a record $13.6 billion, continuing the slow but steady growth in dollar value that has occurred over many years. For comparison, the value in 1986 was $9.6 billion (1).

The volume of coatings declined more than 6% from 1989 to 1990 to about 3.8×10^6 m^3 after remaining approximately level for the preceding three years. However, in 1991, consumption increased to 4.5×10^6 m^3. Because there are very large differences in the prices per unit volume, changes in product mix can give distorted comparisons of value/volume totals. A primary factor affecting both price and volume has been changes resulting largely from the effect of regulations reducing permissible VOC emissions. For any particular use there is usually a required dry film thickness for the coating after application. High solids coatings require lesser volumes of coating to achieve the same film thickness than conventional solvent content coatings. Therefore, for the same amount of applied coating the volume of coating sold decreases. Because solvents are usually the lowest cost components of the coating, cost per unit volume increases.

Coating volume is also affected by other technologies which can substantially reduce the need for coatings. For example, in many cases coatings are not required on molded plastics products which have replaced coated metal products. High pressure laminates are increasingly used as furniture tops. In some cases, the replacements have been from one kind of coating to another. For example, recoated siding for residential housing, both metal and hardboard, have substantially reduced the potential market for exterior house paint for two reasons. The initial coating is sold industrially rather than as architectural coating and also the durability of the coatings can be much greater than that of field applied paint increasing the time interval before repainting. Such coated siding has to a degree been replaced in turn by uncoated vinyl siding.

The effect of environmental regulation and the increasing recognition of immediate or potential toxicity hazards of coating components has led to a technical revolution in the coatings field in the 1970s and 1980s. Further drastic changes are expected in the 1990s.

In 1990 there were 1065 paint companies operating in the United States (178). This is a marked decrease in numbers as compared to 1980. A substantial fraction of the industry is concentrated in chemical companies and a few large independent companies. However, there are a large number of small and medium-sized companies that generally serve specialized segments of the business or restricted geographical areas. It is estimated that 45% of the companies have fewer than 20 employees. The number of production and nonproduction paint industry workers in the United States was 61,800 in 1990.

Whereas the larger companies are international in scope, imports and exports are relatively small. In general, requirements in different countries are quite different and there is generally a need for relatively close contact between consumer and supplier so that the United States industry faces little competition from imported coatings and, conversely, exports play a minor role in the field.

BIBLIOGRAPHY

"Coatings (Industrial)" in *ECT* 1st ed., Vol. 4, pp. 145–189, by H. C. Payne, American Cyanamid Co.; "Coatings, Industrial" in *ECT* 2nd ed., Vol. 5, pp. 690–716, by W. von Fischer, Consultant, and E. G. Bobaleck, University of Maine; in *ECT* 3rd ed., Vol. 6, pp. 427–445, by S. Hochberg, E. I. du Pont de Nemours & Co., Inc.; "Coatings, Resistant" in *ECT* 3rd ed., Vol. 6, pp. 455–481, by C. G. Munger, Consultant.

1. M. S. Reisch, *Chem. Eng. News* **69**(49), 29 (1991).
2. H. Burrell, *Off. Dig. Fed. Soc. Paint Technol.* **34**(445), 131 (1962).
3. Z. W. Wicks, Jr., *Film Formation*, Federation of Societies for Coatings Technology, Blue Bell, Pa., 1986.
4. Z. W. Wicks, Jr., *J. Coat. Technol.* **55**(743), 23 (1986).
5. W. P. Mayer and L. G. Kaufman, *FATIPEC Fed. Assoc. Tech. Ind. Paint Vernis Emaux Encres Impr. Eur. Coat. Cong. XVII* **I**, 110 (1984).
6. A. Rocklin, *J. Coat. Technol.* **48**(622), 45 (1976).
7. D. J. Newman and C. J. Nunn, *Prog. Org. Coat.* **3**, 221 (1973).
8. W. H. Ellis, *J. Coat. Technol.* **55**(696), 63 (1983).
9. H. P. Blandin, J. C. David, J. M. Vergnaud, J. P. Illien, and M. Malizewicz, *Prog. Org. Coat.* **15**, 163 (1987).
10. H. Stutz, K.-H. Illers, and J. Mertes, *J. Polym. Sci.: B. Polym. Phys.* **28**, 1483 (1990).
11. S. P. Pappas and L. W. Hill, *J. Coat. Technol.* **53**(675), 43 (1981).
12. K. Horie, I. Mita, and H. Kambe, *J. Polym. Sci.: A. Polym. Chem.* **6**, 2663 (1968).
13. M. T. Aronhime and J. K. Gilham, *J. Coat. Technol.* **56**(718), 35 (1984).
14. H. E. Bair, *Polym. Prepr. Am. Chem. Soc. Div. Polym. Chem.* **26**(1), 10 (1986).
15. D. E. Fiori and R. W. Dexter, *Proceedings of the Water-Borne Higher-Solids Coatings Symposium*, New Orleans, 1986, p. 186.
16. K. L. Hoy, *J. Coat. Technol.* **51**(651), 27 (1979).
17. S. G. Croll, *J. Coat. Technol.* **59**(751), 81 (1987).
18. S. T. Eckersley and A. Rudin, *J. Coat. Technol.* **62**(780), 89 (1990).
19. J. W. Vanderhoff, E. B. Bradford, and W. K. Carrington, *J. Polym. Sci. Polym. Symp.* **41**, 155 (1973).
20. S. L. Kangas and F. N. Jones, *J. Coat. Technol.* **59**(744), 99 (1987).
21. U.S. Pat. 2,577,770 (Dec. 11, 1951), P. Kass and Z. W. Wicks, Jr. (to Interchemical Corp.).
22. D. B. Larson and W. D. Emmons, *J. Coat. Technol.* **55**(702), 49 (1983).

23. Resimene AM-300 and AM-325, technical bulletin, Monsanto Chemical Co., St. Louis, Mo., Jan. 1986.

24. U.S. Pat. 4,293,461 (Oct. 6, 1981), W. F. Strazik, J. O. Santer, and J. R. LeBlanc (to Monsanto Chemical Co.).

25. E. Levine, *Proceedings of the Water-Borne Higher-Solids Coatings Symposium*, New Orleans, 1977, p. 155.

26. R. Hurley and F. Buona, *J. Coat. Technol.* **54**(694), 55 (1982).

27. C. J. Bouboulis, *Proceedings of the Water-Borne Higher-Solids Coatings Symposium*, New Orleans, 1982, p. 18.

28. J. J. Engel and T. J. Byerly, *J. Coat. Technol.* **57**(723), 29 (1985).

29. S. N. Belote and W. W. Blount, *J. Coat. Technol.* **53**(681), 33 (1981).

30. F. N. Jones and D. D.-L. Lu, *J. Coat. Technol.* **59**(751), 73 (1987).

31. J. D. Hood, W. W. Blount, and W. T. Sade, *J. Coat. Technol.* **58**(739), 49 (1986).

32. L. J. Calbo, in Ref. 15, p. 356.

33. *TONE Polyols*, technical bulletin, Specialty Polymers and Composites Division, Union Carbide Corp., Stamford, Conn., 1986.

34. D. J. Golob, T. A. Odom, Jr., and R. W. Whitson, *Polym. Mater. Sci. Eng.* **63,** 826 (1990).

35. *TMA-109e*, technical bulletin, Amoco Chemicals Corp., Chicago, Ill., 1984.

36. K. L. Payne, F. N. Jones, and L. B. Brandenburger, *J. Coat. Technol.* **57**(723), 35 (1985).

37. J. O. Santer, *Prog. Org. Coat.* **12,** 309 (1984).

38. N. Albrecht, in Ref. 15, p. 200.

39. D. R. Bauer and R. A. Dickie, *J. Coat. Technol.* **54**(685), 57 (1982).

40. L. W. Hill and K. Kozlowski, *J. Coat. Technol.* **59**(751), 63 (1987).

41. L. Calbo, *J. Coat. Technol.* **52**(660), 75 (1980).

42. G. G. Parekh, *J. Coat. Technol.* **51**(658), 101 (1979).

43. L. W. Hill and Z. W. Wicks, Jr., *Prog. Org. Coat.* **10,** 55 (1982).

44. R. A. Gray, *J. Coat. Technol.* **57**(728), 83 (1985).

45. V. R. Kamath and J. D. Sargent, Jr., *J. Coat. Technol.* **59**(746), 51 (1987).

46. *Joncryl 500*, technical bulletin, PSN 0-104012/83, S.C. Johnson & Sons, Inc., Racine, Wis., 1983.

47. R. A. Haggard and S. N. Lewis, *Prog. Org. Coat.* **12,** 1 (1984).

48. J. A. Simms and H. J. Spinelli, *J. Coat. Technol.* **59**(752), 125 (1987).

49. L. W. Hill and Z. W. Wicks, Jr., *Prog. Org., Coat.* **8,** 161 (1980).

50. Z. W. Wicks, Jr., E. A. Anderson, and W. J. Culhane, *J. Coat. Technol.* **54**(668), 57 (1982).

51. J. T. K. Woo and co-workers, *J. Coat. Technol.* **54**(689), 41 (1982).

52. J. T. K. Woo and R. R. Eley, in Ref. 15, p. 432.

53. S. C. Misra, C. Pichot, M. S. El-Aasser, and J. W. Vanderhoff, *J. Polym. Sci.: A. Polym. Chem.* **21,** 2383 (1983).

54. P. S. Sheih and P. L. Massingill, *J. Coat. Technol.* **62**(781), 25 (1990).

55. J. L. Massingill, *J. Coat. Technol.* **63**(797), 47 (1991).

56. R. W. Dexter, R. Saxon, and D. E. Fiori, *J. Coat. Technol.* **58**(737), 43 (1986).

57. F. W. van der Weij, *J. Polym. Sci.: A. Polym. Chem.* **19,** 381 (1981).

58. Z. W. Wicks, Jr., *Prog. Org. Coat.* **3,** 73 (1975).

59. Z. W. Wicks, Jr., *Prog. Org. Coat.* **9,** 3 (1981).

60. T. A. Potter, J. W. Rosthauser, and H. G. Schmelzer, in Ref. 15, p. 162.

61. Z. W. Wicks, Jr. and B. W. Kostyk, *J. Coat. Technol.* **49**(634), 77 (1977).

62. J. W. Rosthauser and K. Nachtkamp, in K. C. Frisch and D. Klempner, eds., *Advances in Urethane Science and Technology*, Vol. 10, Technomics, Westport, Conn., 1987, p. 121.

63. P. L. Jansse, *J. Oil Colour Chem. Assoc.* **89,** 478 (1989).

64. L. H. Brown, in R. R. Myers and J. S. Long, eds., *Treatise on Coatings*, Vol. I, Part III, Marcel Dekker, New York, 1972, pp. 513–563.

65. J. Lomax and G. F. Swift, *J. Coat. Technol.* **50**(643), 49 (1978).

66. Z. W. Wicks, Jr., M. R. Appelt, and J. C. Soleim, *J. Coat. Technol.* **57**(726), 51 (1985).

67. A. Mercurio, *Proceedings of the XVIth International Conference of Organic Coatings Science and Technology*, Athens, Greece, 1990, p. 235.

68. R. J. Clemens and F. D. Rector, *J. Coat. Technol.* **61**(770), 83 (1989).

69. F. D. Rector, W. W. Blount, and D. R. Leonard, *J. Coat. Technol.* **61**(771), 31 (1989).

70. W. H. Ellis, *Solvents*, Federation of Societies for Coatings Technology, Blue Bell, Pa., 1986.

71. P. L. Huyskens, M. C. Haulait-Pirson, L. D. Brandts Buys, and X. M. van der Berght, *J. Coat. Technol.* **57**(724), 57 (1985).

72. N. J. Kamlet, R. M. Doherty, J.-L. M. Abboud, M. H. Abraham, and R. W. Taft, *Chemtech* **16,** 586 (1986).

73. H. L. Jackson, *J. Coat. Technol.* **58**(741), 87 (1986).

74. S. H. Wu, *J. Appl. Polym. Sci.* **22,** 2769 (1978).

75. D. R. Bauer and L. M. Briggs, *J. Coat. Technol.* **56**(716), 87 (1984).

76. T. C. Patton, ed., *Pigment Handbook*, 3 Vol., Wiley-Interscience, New York, 1973; P. A. Lewis, ed., 2nd ed., Vol. 1, Wiley-Interscience, New York, 1989.

77. T. C. Patton, *Paint Flow and Pigment Dispersion*, 2nd ed., Wiley-Interscience, New York, 1979.

78. J. Winkler, E. Klinke, and L. Dulog, *J. Coat. Technol.* **59**(754), 35 (1987); J. Winkler, E. Klinke, M. N. Sathyanarayana, and L. Dulog, *J. Coat. Technol.* **59**(754), 45 (1987); J. Winkler and L. Dulog, *J. Coat. Technol.* **59**(754), 55 (1987).

79. K. Rehacek, *Ind,. Eng. Chem. Prod. Res. Dev.* **15,** 75 (1976).

80. A. Saarnak, *J. Oil Colour Chem. Assoc.,* **62,** 455 (1979).

81. S. Hochberg, in Ref. 27, p. 143.

82. H. L. Jakubaukas, *J. Coat. Technol.* **58**(736), 71 (1986).

83. R. B. McKay, *Proceedings of the VIth International Conference of Organic Coatings Science and Technology*, Athens, Greece, 1980, p. 499.

84. J. E. Hall, R. Bordeleau, and A. Brisson, *J. Coat. Technol.* **61**(770), 73 (1989).

85. W. H. Morrison, Jr., *J. Coat. Technol.* **57**(721), 55 (1985).

86. G. P. Bierwagen and T. K. Hay, *Prog. Org. Coat.* **3,** 281 (1975).

87. G. P. Bierwagen, *J. Paint Technol.* **44**(574), 46 (1972).

88. T. K. Hay, *J. Paint Technol.* **46**(591), 44 (1974).

89. G. P. Bierwagen and D. C. Rich, *Prog. Org. Coat.* **11,** 339 (1983).

90 F. W. Billmeyer and M. Saltzman, *Principles of Color Technology*, 2nd ed., Wiley-Interscience, New York, 1981.

91. C. K. Schoff, *Rheology*, Federation of Societies for Coatings Technology, Blue Bell, Pa., 1991.

92. Z. W. Wicks, Jr., G. F. Jacobs, I. C. Lin, E. H. Urruti, and L. G. Fitzgerald, *J. Coat. Technol.* **57**(725), 51 (1985).

93. A. Toussaint and I. Szigetvari, *J. Coat. Technol.* **59**(750), 49 (1987).

94. P. R. Sperry and A. Mercurio, *Am. Chem. Soc. Div. Org. Coat. Plast. Chem. Prepr.* **43,** 427 (1978).

95. Z. W. Wicks, Jr. and L. G. Fitzgerald, *J. Coat. Technol.* **57**(730), 45 (1985).

96. L. F. Nielsen, *Polymer Rheology*, Marcel Dekker, Inc., New York, 1977, p. 133.

97. S. J. Kemnetz, A. L. Still, C. A. Cody, and R. Schwindt, *J. Coat. Technol.* **61**(776), 47 (1989).

98. S. E. Orchard, *Appl. Sci. Res., Sect. A* **11,** 452 (1962).

99. W. S. Ovediep, *Prog. Org. Coat.* **14,** 159 (1986).

100. *Ibid.*, p. 1.
101. F. J. Hahn, *J. Paint Technol.* **43**(562), 58 (1971).
102. F. Fink, W. Heilen, R. Berger, and J. Adams, *J. Coat. Technol.* **62**(791), 47 (1990).
103. P. E. Pierce and C. F. Schoff, *Coating Film Defects*, Federation of Societies for Coatings Technology, Blue Bell, Pa., 1988.
104. G. A. Vandermeerssche, *Closed Loop* **3**, 1 (Apr. 1981).
105. E. M. Corcoran, A. G. Roberts, and G. G. Schurr, *Paint Testing Manual*, 13th ed., American Society for Testing and Materials, Philadelphia, Pa., 1972. A new edition is in preparation, 1992.
106. D. H. Kaeble, *Physical Chemistry of Adhesion*, Wiley-Interscience, New York, 1971.
107. W. A. Zisman, *J. Paint Technol.* **44**(564), 42 (1972).
108. W. Funke, *J. Coat. Technol.* **55**(705), 31 (1983).
109. D. Y. Perara and D. Van der Eynde, *J. Coat. Technol.* **59**(748), 55 (1987).
110. E. F. Plueddemann, *Prog. Org. Coat.* **11**, 297 (1983).
111. T. R. Bullett and J. L. Prosser, *Prog. Org. Coat.* **1,** 45 (1972).
112. E. T. Turpin, *J. Paint Technol.* **47**(602), 40 (1975).
113. P. J. Schirmann and M. Dexter, in L. Calbo, ed., *Handbook of Coatings Additives*, Marcel Dekker, Inc., New York, 1986, pp. 225–269.
114. D. R. Bauer, *Prog. Org. Coat.* **14**, 193 (1986).
115. Association of the Automotive Industry Working Group on Test Methods for Paints, *J. Coat. Technol.* **58**(734), 57 (1986).
116. R. R. Blakey, *Prog. Org. Coat.* **13**, 279 (1985).
117. J. L. Gerlock, D. R. Bauer, L. M. Briggs, and R. A. Dickie, *J. Coat. Technol.* **57**(722), 37 (1985).
118. J. L. Gerlock, D. R. Bauer, and L. M. Briggs, *Prog. Org. Coat.* **15**, 197 (1987).
119. A. Sommer, E. Zirngiebl, L. Kahl, and M. Schonfelder, *Prog. Org. Coat.* **19**, 79 (1991).
120. Z. W. Wicks, Jr., *Corrosion Protection by Coatings*, Federation of Societies for Coatings Technology, Blue Bell, Pa., 1987.
121. J. J. Wojttowiak and H. S. Bender, *J. Coat. Technol.* **50**(642), 86 (1978).
122. S. Maeda, *J. Coat. Technol.* **55**(707), 43 (1983).
123. W. Funke, *J. Oil Colour Chem. Assoc.* **68**, 229 (1985).
124. R. Warness, *Low-Solvent Primer and Finish Coats for Use on Steel Structures*, Tech. Rep. FHWA/CA/TL-85/02, National Technical Information Service, Springfield, Va., 1985.
125. J. Mazia, *Met. Finish.* **75**(5), 77 (1977); R. D. Wyvill, *Met. Finish,* **80**(1), 21 (1982); R. Athey and co-workers, *J. Coat. Technol.* **57**(726), 71 (1985).
126. T. S. Lee and K. L. Money, *Mater. Perform.* **23**, 28 (1984).
127. J. R. Scully, *Electrochemical Impedance Spectroscopy for Evaluation of Organic Coating Deterioration and Under Film Corrosion—A State of the Art Review*, Report No. DTNSRDC/SME-86/006, D. W. Taylor Naval Ship Research and Development Center, Bethesda, Md., 1986.
128. W. S. Tait, *J. Coat. Technol.* **61**(768), 57 (1989).
129. B. R. Appleman, *J. Coat. Technol.* **62**(787), 57 (1990).
130. *Annual Book of Standards*, Vols. 06.01, 06.02, 06.03, American Society for Testing and Materials, Philadelphia, Pa., new editions annually.
131. L. W. Hill, *Mechanical Properties of Coatings*, Federation of Societies for Coatings Technology, Blue Bell, Pa., 1987.
132. L. W. Hill, *Prog. Org. Coat.* **5,** 277 (1977).
133. D. J. Skrovanek and C. K. Schoff, *Prog. Org. Coat.* **16,** 135 (1988).
134. D. J. Skrovanek, *Prog. Org. Coat.* **18,** 89 (1990).
135. S. H. Wu, *J. Appl. Polym. Sci.* **20,** 327 (1976).

136. E. Ladstater and W. Gessner, *Proceedings of the International Conference of Organic Coatings Science and Technology*, Athens, Greece, 1986, p. 203.

137. R. M. Evans, in Ref. 64, Vol. 2, Part I, 1969, pp. 13–190; R. M. Evans and J. Fogel, *J. Coat. Technol.* **47**(639), 50 (1977).

138. L. H. Lee, *Am. Chem. Soc. Polym. Mat. Sci. Eng. Prepr.* **50,** 65 (1984).

139. J. H. Braun, *J. Coat. Technol.* **60**(758), 67 (1988).

140. R. W. Hislop and P. L. McGinley, *J. Coat. Technol.* **50**(642), 69 (1978).

141. A. Ramig, Jr. and F. L. Floyd, *J. Coat. Technol.* **51**(658), 63, 75 (1979).

142. D. M. Fasano, *J. Coat. Technol.* **59**(752), 109 (1987).

143. J. W. Vanderhoff, J. M. Park, and M. S. El-Aasser, *Polym. Mat. Sci. Eng. Prepr.* **64,** 345 (1991).

144. D. B. Massouda, *J. Coat. Technol.* **57**(722), 27 (1985).

145. J. E. Glass, *J. Coat. Technol.* **50**(640) 53, 61, (641), 56 (1978).

146. C. U. Pittman, Jr., G. A. Stahl, and H. Winters, *J. Coat. Technol.* **50**(636), 49 (1978).

147. W. B. Woods, *Paint Coat. Ind.* **38**(5), 25 (1987).

148. R. H. Fernando, W. F. McDonald, and J. E. Glass, *J. Oil Colour Chem. Assoc.* **69,** 263 (1986).

149. J. E. Hall, P. Hodgdson, L. Krivanek, and P. Malizia, *J. Coat. Technol.* **58**(738), 65 (1986).

150. U. Zorll, *Prog. Org. Coat.* **1,** 113 (1972).

151. E. Cremer, *Prog. Org. Coat.* **9,** 241 (1981).

152. F. Beck, *Prog. Org. Coat.* **4,** 1 (1976).

153. M. Wismer and co-workers, *J. Coat. Technol.* **54**(688), 35 (1982).

154. R. E. Smith and D. W. Boyd, *J. Coat. Technol.* **60**(756), 77 (1988).

155. C. K. Schoff, *J. Coat. Technol.* **62**(789), 115 (1990).

156. A. J. Backhouse, *J. Coat. Technol.* **54**(688), 35 (1982).

157. L. B. Brandenburger and L. W. Hill, *J. Coat. Technol.* **51**(659), 57 (1979).

158. B. C. Watson and Z. W. Wicks, Jr., *J. Coat. Technol.* **55**(698), 59 (1983).

159. T. J. Miranda, *J. Coat. Technol.* **60**(760), 47 (1988).

160. J. E. Gaske, *Coil Coatings*, Federation of Societies for Coatings Technology, Blue Bell, Pa., 1987.

161. S. P. Pappas, ed., *Radiation Curing: Science and Technology*, Plenum Press, New York, 1992.

162. C. Deckler, *J. Coat. Technol.* **59**(751), 97 (1987).

163. H. J. Hageman, *Prog. Org. Coat.* **13,** 123 (1985).

164. J. V. Crivello, *J. Coat. Technol.* **63**(793), 35 (1991).

165. S. P. Pappas, *Prog. Org. Coat.* **13,** 35 (1986).

166. T. E. Misev, *Powder Coatings, Chemistry and Technology*, John Wiley & Sons, Inc., New York, 1991.

167. L. Kapilow and R. Samuel, *J. Coat. Technol.* **59**(750), 39 (1987).

168. J. H. Jilek, *Powder Coatings*, Federation of Societies for Coatings Technology, Blue Bell, Pa., 1991.

169. M. J. Hannon, D. Rhom, and K. J. Wissrun, *J. Coat. Technol.* **48**(675), 42 (1976).

170. P. G. de Lange, *J. Coat. Technol.* **56**(717), 23 (1984).

171. W. Jacobs, P. W. Lucis, G. G. Parekh, and R. G. Lees, in Ref. 67, p. 509.

172. K. A. Nielsen, *Federation of Societies for Coatings Technology Symposium*, Louisville, Ky., 1990.

173. W. Bailey and co-workers, *J. Coat. Technol.* **62**(789), 133 (1990).

174. H. R. Friedli and C. M. Keillor, *J. Coat. Technol.* **59**(748), 65 (1987).

175. H. R. Bleile and S. Rodgers, *Marine Coatings*, Federation of Societies for Coatings Technology, Blue Bell, Pa., 1989.

176. C. M. Sghibartz, *FATIPEC Fed. Assoc. Tech. Ind. Paint Vernis Emaux Encres Impr. Eur. Cont. Cong. XVII* **IV,** 145 (1982).
177. A. Milne and M. E. Callow, *Trans. Inst. Marine Eng.* **97,** 229 (1984).
178. *The Rauch Guide to the U.S. Paint Industry*, Rauch Associates, Bridgewater, N.J., 1990.

General References

Monograph series published by the Federation of Societies for Coatings Technology, Blue Bell, Pa. Titles issued thus far: *Introduction to Coatings Technology, Introduction to Polymers and Resins, Film Formation, Organic Pigments, Inorganic Primer Pigments, Solvents, Mechanical Properties of Coatings, Corrosion Protection by Coatings, Application of Paints and Coatings, Coating Film Defects, Rheology, Aerospace and Aircraft Coatings, Marine Coatings, Coil Coatings, Automotive Coatings, Radiation Cure Coatings, Cationic Radiation Curing, Sealants and Caulks,* and *Powder Coatings.*
Z. W. Wicks, Jr., F. N. Jones, and S. P. Pappas, *Organic Coatings: Science and Technology,* John Wiley & Sons, Inc., New York, Vol. I, 1992; Vol. II, 1993.

ZENO W. WICKS, JR.
Consultant

COATINGS, MARINE

The marine environment is highly aggressive. Materials in marine service are constantly exposed to water, corrosive salts, strong sunlight, extremes in temperature, mechanical abuse, and chemical pollution in ports. This climate is very severe on ships, buoys, and navigational aids, offshore structures such as drilling platforms, and facilities near the shore such as piers, locks, and bridges.

Marine coatings are the most important, cost-effective means to preserve steel (qv) and other metals in the marine environment. These coatings impart physical and chemical properties, eg, antifouling, color, and slip resistance, to surfaces that cannot be obtained in any other way. Modern coatings are sophisticated mixtures of polymers and other chemicals, and they require careful control of surface preparation and application conditions. These materials are not as adaptable as earlier coatings to the variety of metals, design features, and surface conditions encountered in marine construction, but they provide long-lasting cost-effective protection when used as part of a corrosion control program (see CORROSION AND CORROSION CONTROL). In 1989 a total of 38×10^6 L of marine coatings were sold in the United States. Shipbuilding consumed 4×10^6 L, construction of pleasure boats required 3×10^6 L, and 31×10^6 L were employed for maintenance and repair. The selection and application of marine coatings has become a highly specialized discipline in which governmental regulations are a dominant influence. Many paints (qv) and painting procedures used in the past are no longer permitted or are extremely costly to use.

Corrosion Control Plan. A corrosion control plan for each ship or structure, which is designed to control deterioration in the most economical and practical manner and to include all appropriate mechanisms for corrosion control, must be developed before construction begins. For steel in the marine environment, the chief methods available are protective coatings and cathodic protection. Cathodic protection is an electrical method of preventing metal corrosion in a conductive medium by placing a negative charge on the item to be protected. This protection mechanism is specifically designed as part of the total corrosion control system and protects only the submerged portions of steel ships and structures.

Materials most resistant to deterioration are chosen, the strength and available shapes of which have a critical influence upon design. Crevices, areas where water can collect, and sharp edges are to be avoided, and dissimilar metals must be electrically isolated from one another in order to prevent corrosion of the more anodic metal. Sharp edges cause paint to draw thin and should be removed by grinding or sanding. Welds have sharp projections that should be removed by grinding. Weld spatter should be scraped or ground from metal surfaces. Outside corners should be rounded, and inside corners should be filled because they provide a collection site for excess paint that may not fully cure. Crevices and pits should be filled with weld metal or caulking because they collect corrosive agents and accelerate deterioration.

Protective Coatings. Each coating in the protective coating system is designed to perform a specific function and to be compatible with the total system. Selection of a coating system is influenced by the chemical nature of the coating and by the conditions the coating is designed to resist. The identity of each material, the number of coats and the dry film thickness of each, the maximum times between blasting and coating and between coats, and the minimum time between application of the last coat and commencement of service need to be defined as do suitable surface preparation techniques, proper methods for the application of paint, and effective quality control procedures. The effectiveness of a coating is directly related to its ability to maintain adhesion to the substrate, its integrity, and its thickness. Areas that cannot be easily or safely repaired, especially those which require drydocking for repair, need to be given the best available coating.

Fouling organisms attach themselves to the underwater portions of ships and have a severe impact on operating costs. They can increase fuel consumption and decrease ship speed by more than 20%. Warships are particularly concerned about the loss of speed and maneuverability caused by fouling. Because fouling is controlled best by use of antifouling paints, it is important that these paints be compatible with the system used for corrosion control and become a part of the total corrosion control strategy.

Environmental Concerns

Local environmental regulations have significantly affected the production, transportation, use, and disposal of coatings.

Volatile Organic Compounds. As coatings dry, solvents are released into the atmosphere, where they undergo chemical reactions in sunlight and produce photochemical smog and other air pollutants (see AIR POLLUTION). As a general rule, the volatile organic compound (VOC) content of marine coatings is restricted to 340 g/L. In the locations where ozone (qv) levels do not conform to the levels established by the Environmental Protection Agency, regulations require an inventory of all coatings and thinners from the time they are purchased until they are used.

The VOC regulations have been the driving force behind the development of entirely new coatings technologies, the reformulation of coatings, and the creation of new surface preparation and paint application methods. High solids coatings, eg, those of epoxy and urethane, have displaced alkyds and are now the principal marine coatings, but some have short pot lives. Alkyd coatings have been extensively reformulated and are still important but dry more slowly than their forebears and are not used as extensively as in the 1970s. Coatings that contain high levels of solvents, such as vinyl and chlorinated rubber coatings, are disappearing from the marine industry.

VOC-conforming paints are more demanding than their predecessors, and attention to application conditions and techniques is imperative. The coatings can be more viscous and harder to apply, wet films have poorer leveling, and coatings of uniform thickness are more difficult to achieve.

Heavy-Metal Pigments.   Lead (qv) and chromate pigments (qv), used for many years as corrosion inhibitors in metal primers and topcoats for marine coating systems, have been linked to adverse health and environmental effects (see CHROMIUM COMPOUNDS). Inhaling or ingesting droplets of lead- or chromate-containing coatings is a potential source of poisoning of paint applicators. Regulations concerning the removal of lead-based coatings require that existing coatings be analyzed before removal for toxic metals and that all debris be contained during removal. The air in the vicinity must also be monitored during removal to ensure safe conditions, and old paint and blasting abrasive must be disposed of as toxic waste.

Because of these concerns, lead- and chromate-containing pigments are not used in marine coatings. Chromate pigments, which contain the metal in the $+6$ valence state, are proscribed, but pigments containing chromium in the $+3$ valence state, such as chromium oxide, Cr_2O_3, are unregulated and continue to be used. The corrosion inhibitive pigments that were commonly used and those which have replaced them are listed in Table 1.

Organotins.   In the mid-1970s compounds based on derivatives of triphenyl- or tributyltin (see TIN COMPOUNDS) known generically as organotins, were found to be much more effective than cuprous oxide paints in controlling fouling, and numerous products were introduced. These fell into two classes. Free-association coatings contained a tributyltin salt, eg, acetate, chloride, fluoride, or oxide, physically mixing into the coating. These were available in a variety of resins and characterized by a leach rate of organotin which is quite high when the coating is new, and which falls off rapidly until insufficient to prevent fouling. In contrast, copolymer coatings contain organotin which is covalently bound to the resin of the coating and is not released until a tin–oxygen bond hydrolyzes in seawater (1). This controlled hydrolysis produces a low and steady leach rate of organotin and creates hydrophilic sites on the binder resin. This layer of resin subsequently washes away and exposes a new layer of bound organotin. These coatings are also

Table 1. Active Pigments in Anticorrosive Coatings

Pigment	CAS Registry Number	Molecular formula
	Prohibited	
red lead	[1314-41-6]	Pb_3O_4
white lead	[1344-36-1]	$PbCO_3$
lead chromate	[7758-97-6]	$PbCrO_4$
zinc chromate	[13530-65-9]	$ZnCrO_4$
strontium chromate	[7789-06-2]	$SrCrO_4$
basic lead silicochromate	[11113-70-5]	$PbCrO_4 \cdot nSiO_2$
	Allowed	
zinc oxide	[1314-13-2]	ZnO
zinc phosphate	[7779-90-0]	$Zn_3(PO_4)_2$
zinc phosphosilicate		$Zn_3(PO_4)_2 \cdot nSiO_2$
zinc molybdate	[13767-32-3]	$ZnMoO_4$
calcium borosilicate		$CaO \cdot B_2O_3 \cdot nSiO_2$
calcium phosphosilicate		$Ca_3(PO_4)_2 \cdot nSiO_2$

known as controlled release, self-polishing, or ablative coatings, and last for five years when applied at a dry-film thickness of 375 μm.

However, there is now considerable evidence that sufficiently high concentrations of organotins kill many species of marine life and affect the growth and reproduction of others. Thus many nations restrict organotin coatings to vessels greater than 25 meters in length. In the United States, laws prohibit the retail sale of copolymer paints containing greater than 7.5% (dry weight) of tin, and of free-association paints containing greater than 2.5% (dry weight) of tin, but do not restrict the size of the ship to which the paints may be applied (2). Organotins must be used on vessels with aluminum hulls, because copper is cathodic to aluminum and causes rapid pitting and perforation when used on an aluminum hull. Regulations to minimze the exposure of workers and protect the environment during the application and removal of organotin coatings also exist.

Abrasive Blast Cleaning. Removal of paint by abrasive blasting may lead to adverse health effects for workers who breathe dust formed during the operation. Regulations restrict blasting operations to such procedures as blasting within enclosures, using approved mineral abrasives, using a spray of water to reduce dust, and blasting with alternative materials such as ice, plastic beads, or solid carbon dioxide. A military specification (3) describes abrasives that are approved for use in U.S. naval shipyards. Limits are placed on carbonates, gypsum, and free silica, all of which are not abrasive but only contribute to dust, and on the amount of arsenic, beryllium, cadmium, lead, and 13 other toxic materials in blasting abrasives.

Debris from the removal of paint may contain lead, chromium, or other heavy metals. Collection of such debris is required to prevent release of these metals into the environment and to avoid exposure and contamination of workers. Blasting debris is contained in two ways: by use of containment systems, eg, screens, panels, tarpaulins, and shrouds, which enclose the removal area, and by paint removing machines equipped with vacuum collection devices. The latter include both powered mechanical tools, eg, grinders, brushes, and sanders, and self-contained abrasive blasting equipment. Blasting enclosures that draw in air help to contain particles of paint but do not ensure worker safety, and these are seldom more than 85% effective in preventing dust and debris from escaping from the system. In order to evaluate the efficiency of the containment method, periodic medical examinations of workers for respired air contaminants are required.

Hazardous waste generated by removal of toxic paints may be stored for only a limited time and must be disposed of in conformance with prevailing regulations. The amount of hazardous waste can be greatly reduced by cleaning and recycling the abrasive.

Reactive Coatings.   Coatings that cure by chemical reaction of two component parts are the most widely used in marine applications and protection of workers from the reactive ingredients is required. For example, urethane coatings may contain isocyanates that may cause respiratory difficulties, and epoxy coatings may contain glycidyl ethers which are skin sensitizers. This danger is diminished in modern coatings when the reactive groups are bound to oligomers having low vapor pressures and the likelihood of exposure to vapors is considerably reduced.

Surface Preparation for Marine Painting

Surface preparation, always important in obtaining optimal coatings performance, is critical for marine coatings (see METAL SURFACE TREATMENTS). Surface preparation usually comprises about half of the total coating costs, and if inadequate may be responsible for early coating failure. Proper surface preparation includes cleaning to remove contaminants and roughening the surface to facilitate adhesion.

Standards for Cleaned Steel Surfaces.   The most important standards (4) used to specify and evaluate cleaned steel surfaces are summarized in Table 2 in order of increasing cost. Photographic standards consistent with the written standards in Table 2 are also available (5).

Abrasive Blasting.   Blast cleaning using mineral abrasives (qv) is the preferred method for cleaning steel prior to applying marine coatings. Blasting not only provides the highest level of cleanliness but also roughens the surface to provide for good adhesion of the primer. As much blasting as possible is done in purpose-built enclosures to minimize the amount of particulates produced and to provide better and less costly cleaning. Shop blasting is accomplished by equipment having high speed rotating wheels that propel shot or grit abrasive onto steel. Portable closed-cycle vacuum blasting equipment is available for field use. Special machines have been made for steel decks and hulls which recycle the abrasive several times, saving costly abrasive and reducing the amount of blasting waste.

Abrasive blasting of steel ships using conventional equipment is usually accomplished at a nozzle pressure of about 700 kPa. Abrasives can completely remove rust, scale, dirt, and old coatings, but grease and oil are smeared and driven into the surface. Abrasive blasting must, therefore, be preceded by solvent cleaning if any grease or oil is present.

Table 2. Steel Surface Preparation Standards[a]

Number[b]	Title	Intended use
SSPC-SP-1	Solvent Cleaning	removal of oil and grease prior to further cleaning by another method
SSPC-SP-2	Hand Tool Cleaning	removal of loose surface contaminants before spot repair
SSPC-SP-3	Power Tool Cleaning	removal of loose surface contaminants before spot repair
SSPC-SP-7	Brush-Off Blast	removal of loose surface contaminants before spot repair
SSPC-SP-6	Commercial Blast	for interior steel to be coated with alkyd paint
SSPC-SP-10	Near-White Metal Blast	for most exterior surfaces, decks, water, fuel tanks, etc
SSPC-SP-5	White Metal Blast	for the most demanding coating conditions and for those products, ie, inorganic zinc primers, thermal sprayed metals, and powder coatings, which require an uncontaminated surface

[a]Ref. 4.
[b]SSPC = Steel Structures Painting Council, Pittsburgh, Pa.

To reduce the amount of dust produced, water can be added to the abrasive from a circular water sprayer around the nozzle. Chemical corrosion inhibitors must be dissolved in the water to prevent flash rusting of the steel. Newer methods to reduce dust include the use of ice, solid carbon dioxide (dry ice), or plastic beads as abrasives. Blasting with dry ice is inexpensive and effective, but the accumulation of carbon dioxide must be avoided in enclosures. Plastic beads are inexpensive, but the cutting efficiency is low and paint removal is slow; the beads can be cleaned of paint particles and reused.

Softer metals such as aluminum and its alloys can be blast cleaned using abrasives that are not as hard as those used on steel. Garnet, walnut shells, corncobs, peach pits, glass or plastic beads, and solid carbon dioxide have been used successfully.

Other Cleaning Methods. Solvent cleaning, ie, degreasing, is chiefly used to remove grease and oil. Solvent is applied to rags which are replaced when they become contaminated. The final rinse is always made using fresh solvent. Individual ship components can be solvent-cleaned by dipping in tanks of solvent.

Hand and power tool cleaning is used on ships mostly for spot repair of damaged areas. Hand tools include scrapers, wire brushes, and sanders. Electric and pneumatic power tools, which include grinders and needle guns, clean faster and more thoroughly than hand tools. Most power tools have vacuum lines connected to collect paint debris.

Blast cleaning with water, sometimes called hydroblasting, is used to remove marine fouling and sometimes to clean metal surfaces for coating. The water may be heated and detergent may be added to facilitate removal of oil, dirt, and marine slimes. Cleaning bare steel for coating may be achieved using pressures over 200 MPa (30,000 psi) and water volumes of only 8 to 56 L/min. Abrasive may be injected into the stream of water or used in a second operation to produce the rough surface needed for adhesion. Corrosion inhibitors must be used in the water, and extreme caution must be maintained using these high pressures.

Steam cleaning may also be used to remove grease and oil. On large surface areas such as the hulls of ships, steam cleaning is usually more economical and efficient than solvent cleaning. Detergents are sometimes brushed onto the hull before steam cleaning.

Types of Coatings

Coatings ingredients fall into four principal classes. Resins form a continuous solid film after curing, bind all ingredients within the film, and provide adhesion to the substrate (6). The properties of a coating are determined principally by the resins it contains. Pigments are metals or nearly-insoluble salts that impart opacity, color, and chemical activity to the coating. Solvents are used primarily to facilitate manufacture and application, but are lost by evaporation after application and are not a permanent part of the coating. Additives used in small (1–50 ppt) amounts give the coating such desirable additional properties as ease of manufacture, stability in shipment and storage, ease of application, or increased performance of the dried film.

These ingredients may be formulated to give coatings that protect against corrosion in different ways (7). Barrier coatings physically separate oxygen, water, ions, and other corrosive agents from the steel surface. Inhibitive coatings prevent corrosion by absorbing or neutralizing corrosive agents, or by slowly releasing protective ions. Sacrificial coatings contain a metal (usually zinc) that is oxidized more rapidly than steel, thereby providing protection for the substrate by electrochemical action. Conversion coatings chemically oxidize the surface of the substrate to a depth of 7–10 μm, producing a passive layer which resists corrosion better than the metal itself.

Modern marine coatings fall into eight generic categories, each named for the principal resin it contains. Significant variation in each category is achieved by varying pigments and other ingredients. The categories are discussed in rough order of importance.

Epoxies. Epoxy coatings are the workhorse materials for premium marine applications (8). High performance primers and anticorrosive coatings that conform to VOC regulations are widely available (9). The cured coatings are durable, tough, and smooth, and demonstrate excellent resistance to solvents, alkalies, and abrasion. Multiple-coat systems are applied before a preceding coat cures completely, in order to obtain chemical reaction between coats. Fully-cured epoxies have a hard finish that is difficult to topcoat. Epoxy resins (qv) photolyze in sunlight, leaving a dust of unbound pigment known as chalk. Thus they are always topcoated, usually using urethanes, alkyds, or vinyls, for exterior use.

The coatings are usually formulated using an epoxy resin in a first component and a polyamide, amine adduct, or polyamine curing agent in a second component. The coatings cure by a chemical reaction between the components, and curing time depends primarily on temperature. Epoxy–polyamide coatings can tolerate some surface dampness and contamination during application and a near white blast is satisfactory, although the best possible surface preparation is always desirable. Organic zinc primers containing about 30% epoxy and polyamide resins and about 69% zinc metal provide long-lasting corrosion protection but are not suitable for immersion service.

Urethanes. Urethane coatings are comparatively expensive and are used almost exclusively as topcoats over epoxy or inorganic zinc primers. Urethanes containing aliphatic polyisocyanates as curing agents are the best choice for excellent high gloss cosmetic topcoats where prolonged resistance to ultraviolet radiation and retention of appearance are important. Coatings containing aromatic polyisocyanates lose gloss and become yellow in sunlight, but give excellent service as tank linings and in other interior applications. Both types of curing agents produce tough, durable, smooth coatings with excellent resistance to chemicals and abrasion, and both can be formulated to give highly flexible elastomeric coatings if desired.

Urethane coatings are formulated in two components (10). The first contains a polyester polyol, pigments, additives, and solvents, and the second contains a polyisocyanate curing agent. Modern curing agents have low vapor pressures, which minimize worker exposure to isocyanate fumes. Yacht finishes that may be applied by brush are available for the individual user not wishing to apply the coating by spray. Use of urethanes requires careful attention to worker safety and application procedures but, when properly applied, they are the best finishes available for most exterior marine surfaces (see URETHANE POLYMERS).

Alkyds. Alkyd resins (qv) are polyesters formed by the reaction of polybasic acids, unsaturated fatty acids, and polyhydric alcohols (see ALCOHOLS, POLYHYDRIC). Modified alkyds are made when epoxy, silicone, urethane, or vinyl resins take part in this reaction. The resins cross-link by reaction with oxygen in the air, and carboxylate salts of cobalt, chromium, manganese, zinc, or zirconium are included in the formulation to catalyze drying.

Alkyd coatings were the standard products in the marine industry for atmospheric service until they were superseded by epoxies in the early 1970s. Alkyds dry reasonably fast, are easy to apply, and demonstrate good weathering in mild environments but are not suitable for immersion service. They have a high moisture vapor transmission rate and to be effective must contain inhibitive pigments (Table 1). The elimination of lead and chromate pigments has made the formulation of an effective alkyd coating challenging but not impossible.

Inorganic Silicate Coatings. Inorganic silicate coatings are available for marine use in diverse formulations, which cure by different mechanisms. A silicate binder is formed when sodium, potassium, or lithium silicates in alkaline aqueous solution polymerize. Partially hydrolyzed ethyl silicate in an alcohol–water solution is also used in these coatings. Water is necessary for curing to take place but must evaporate before a film can form. High humidity, low temperature, or poor air circulation may retard evaporation and film formation, but very low humidity retards curing.

The coatings demonstrate excellent abrasion resistance, hardness, and toughness, but they are not flexible. Inorganic zinc-rich coatings containing more than 80 wt % metallic zinc in a silicate binder are used in automated blasting and priming operations as preconstruction primers for steel plate. They are also used near the seashore as primer coatings on bridges, electric power transmission towers, and structural steel but are not suitable for immersion service. A single coat of 75–125 μm provides galvanic protection to steel. These coatings are almost always topcoated. Vinyls, epoxies, and urethanes are suitable, but alkyd coatings are not stable to the alkaline surface of zinc and should not be used as topcoats.

Vinyls. Vinyl resins are thermoplastic polymers made principally from vinyl chloride; other monomers such as vinyl acetate or maleic anhydride are co-polymerized to add solubility, adhesion, or other desirable properties (see MALEIC ANHYDRIDE, MALEIC ACID, AND FUMARIC ACID). Because of the high, from 4,000 to 35,000, molecular weights large proportions of strong solvents are needed to achieve application viscosities. Whereas vinyls are one of the finest high performance systems for steel, many vinyl coatings do not conform to VOC requirements (see VINYL POLYMERS).

Vinyl coatings are lacquers, that is, they form films solely by the evaporation of solvent. Thus throughout their lives they are soluble in the solvents used to apply them, allowing for good intercoat adhesion when solvents in a later coat soften the resins in an earlier coat. Vinyl coatings have been widely used on bridges, locks, ships, dams, and on- and off-shore steel structures. They are tough, flexible, adherent coatings. Excellent primer, anticorrosive, and antifouling formulations are available. The coatings have a low moisture vapor transmission rate, and inhibitive pigments are rarely used. These coatings are particularly useful where fast drying at low (0–10°C) temperatures is required, and they require only a short curing time before being placed in service, but they do not tolerate surface moisture. Vinyls are excellent for immersion service only if applied over a near white or white blast. They have good weather resistance but are softened by heat and are not suitable for prolonged use above 65°C.

The wash primer is a special type of vinyl coating. This material contains a poly(vinyl butyral) resin, zinc chromate, and phosphoric acid in an alcohol–water solvent. The coating is so thin it is literally washed onto a freshly blasted steel surface, where it passivates the metal surface by converting it to a thin iron phosphate–chromate coating. The alcohol solvent makes it possible to apply the coating over damp surfaces. The coating forms the first coat of an all-vinyl system and can also be used to preserve a freshly-cleaned steel surface until an epoxy or other primer coat can be applied. These coatings contain chromates and are very high in VOC.

Chlorinated Rubber. Chlorinated rubber coatings are lacquers. These thermoplastic materials have low moisture vapor transmission and excellent acid, water, salt, and alkali resistance. They dry under very cold conditions but do not tolerate surface moisture. They are useful at extreme service temperatures (− 35 to 120°C) and are easily repaired. Toughness and high chemical resistance are similar to vinyls. Chlorinated rubber coatings are widely used in Europe and are common in the United Kingdom, where coatings pigmented with micaceous iron oxide are used by British Railways. They find fewer applications in the United States, primarily as topcoats on exterior steel exposed to high humidity.

Coal Tar. Coal-tar resins are made from processed coal-tar pitch (see TAR AND PITCH). They undergo rapid and severe cracking in sunlight and thus are suitable only for underground use on steel and saltwater immersion. Coal-tar resins are frequently combined with epoxy resins to add the chemical and abrasion resistance of the latter. Coal-tar epoxy coatings cured with polyamides are widely used on marine structures because of low water permeability. All of these are cost-effective coatings but are available only in black or dark shades and demand a white or near white blast for long life. The U.S. Navy does not use coal tar or coal-tar epoxy coatings because low levels of carcinogens may be present in processed coal tar.

Powder Coatings. Coating films can be formed from dry thermoplastic powder (see COATING PROCESSES, INDUSTRIAL-POWDER TECHNOLOGY). Small objects may be dipped in a fluidized bed of powder or grounded and coated by electrostatic spray; the object is then heated to fuse the particles and form a film. The powder may also be melted as it is sprayed; molten droplets coalesce and form a film on the object before cooling. Epoxy, polyester, and vinyl resins are widely used, and the particular properties of the coating depend on the type of resin. These coatings are used on electrical junction boxes, motor housings, hatches, and other small pieces of equipment, which are usually removed from the ship and coated in a shop.

Application Methods

The application of marine coatings is a critical factor in achieving maximum performance. Protective clothing and breathing equipment should be worn during application. Because of large surface areas, ships are usually spray painted (see COATING PROCESSES, SPRAY COATINGS). Three techniques are widely used: air, airless, and electrostatic spraying. In air spraying, paint is forced by 200–400 kPa (30–60 psi) of compressed air into a spray gun where a second stream of air atomizes the paint and carries it onto the surface. Paint losses can be as high as 40% because some paint misses the surface (overspray) and some rebounds from the surface. Nearby objects must be protected from inadvertent painting.

Airless spray uses hydraulic pressure to deliver the paint. Paint is brought to the spray gun under 7–40 mPa (1000–6000 psi), where it is divided into small separate streams and forced through a very small orifice to produce the spray. Airless spray is faster, cleaner, and less wasteful than air atomization, but demands good technique because it delivers paint very quickly.

Electrostatic spraying is used in shops to coat conductive objects. It is very useful for odd-shaped objects such as wire fence, cables, and piping. An electrostatic potential of 60,000 volts on the object attracts oppositely-charged paint particles; the spray can wrap around and coat the side of the object opposite to the spray gun. This technique produces very uniform finishes and has the least paint loss of the three methods. However, it is slow, requires expensive equipment, produces only thin coats, and is sensitive to wind currents.

Manual painting occurs mostly during touch-up or repair, and is best suited for piping, railings, and other hard to spray places. The conventional tools for manual application are brushes, rollers, paint pads, and paint mitts. These methods are very slow but are suitable for unskilled applicators and allow the painter to work the coating deeply into the surface being painted.

Transfer Efficiency. Many components of ships and marine structures are now coated in the shop under controlled conditions to reduce the amount of solvents released into the atmosphere, improve the quality of work, and reduce cost.

Regulations designed to limit the release of volatile organic compounds into the air confine methods of shop application to those having transfer efficiencies of 65%. Transfer efficiency is defined as the percent of the mass or volume of solid coating that is actually deposited on the item being coated, and is calculated as

$$\text{transfer efficiency (\%)} = \frac{\text{mass of solid coating on item} \times 100}{\text{mass of solid coating consumed}}$$

or

$$= \frac{\text{volume of solid coating on item} \times 100}{\text{volume of solid coating consumed}}$$

The principal factors affecting transfer efficiency are the size and shape of the object, the type of application equipment, the air pressure to the spray gun, and the distance of the spray gun from the object. The transfer efficiency becomes lower as the object becomes smaller or more complex. The transfer efficiency increases when the spray gun is brought closer to the object and when the atomizing pressure is reduced. The transfer efficiency of different types of application equipment in descending relative order is manual > electrostatic spray > airless spray > conventional atomized air spray.

Selection of Coatings

Underwater Hull. Hull coatings consist of two layers of an anticorrosive coating topped with one layer of an antifouling coating. The coating system must resist marine fouling, severe corrosion, the cavitation action of high speed propellers, and the high current densities near the anodes of the cathodic protection system. Epoxy and coal-tar epoxy systems are commonly used as anticorrosive coatings. The U.S. Navy uses two coats of epoxy polyamide paint (11), each 75 μm thick when dry. Epoxy and coal-tar epoxy systems are used extensively on commercial ships. Coal-tar epoxy systems (12) are usually applied in two coats to give a total dry film thickness of 200 μm.

Anticorrosive systems require an antifouling topcoat. Marine antifouling coatings (13) contain materials that are toxic to fouling organisms and are the only effective way to prevent the growth of marine organisms on the hull. The nature of the toxic substance is heavily regulated in Europe, Japan, and the United States. Arsenic, cadmium, and mercury are proscribed and organotins are severely restricted. Cuprous oxide has always been and remains the most widely used toxicant.

The system of hull coatings, including antifouling paint, must be compatible with the cathodic protection system. Thus the coating system must have good dielectric properties to minimize cathodic protection current requirements and must be resistant to the alkalinity produced by the electric current. The cathodic protection system should prevent corrosion undercutting of coatings that become damaged, and the current density should be able to be increased easily to meet the increased electrical current needed as the coating deteriorates.

Antifouling paints containing a vinyl-rosin base and cuprous oxide (14) were used beginning in the 1940s but are being discontinued because of their high VOC levels. They provided about two years of protection against fouling but needed to be cleaned about every three months, depending on operational schedules and the waters in which the vessel operated. Organotin antifouling paints effectively prevent fouling for much longer periods, and the copolymer paints containing covalently-bound tin furnish about five years of protection when applied at a dry film thickness of 375 μm (15 mils).

Hull coatings having low surface energies, known as fouling release coatings, provide fouling protection without the use of toxins (15). These coatings form only weak bonds with fouling organisms, and the fouling loses adhesion by its own weight or by the motion of the ship through the water. Heavily fluorinated urethane coatings were tested for some years, but toughened silicone coatings are now providing superior performance. Foulant release coatings are environmentally benign and promise extended service lives.

Epoxy and polyester systems filled with flake glass provide a finish that is tough and resistant to abrasion. One commercial system is filled with copper flakes to provide intrinsic antifouling action. These systems are applied at a total dry film thickness of about 625 μm and are used on pleasure boats.

Boottop and Freeboard Areas. The boottop is that part of the hull that is immersed when the ship is loaded and exposed when the ship is empty. The freeboard is the area from the upper limit of the boottop to the main deck. The boottop suffers mechanical damage from tugs, piers, and ice, and experiences intermittent wet and dry periods with nearly constant exposure to sunlight. Thus coatings for this area require resistance to sunlight and mechanical damage, good adhesion, and flexibility. Frequently the hull coating system is used in the boottop area, and one or two extra topcoats are applied for added strength.

In the freeboard areas, commercial ships use organic zinc-rich primers extensively and usually topcoat them with a two- or three-coat epoxy system. U.S. Navy ships use an organic zinc-rich primer, two to three coats of an epoxy-polyamide coatings, and a silicone-alkyd topcoat (16); the entire dry system is 150–225 μm thick.

Weather Decks. Coatings for decks must be resistant to abrasion by pedestrians and small vehicles, and must be slip-resistant. Inorganic zinc primers overcoated with epoxy coatings for additional corrosion protection perform well on steel decks, or a multiple-coat all-epoxy system can be used. Nonskid coatings are used on aircraft carrier landing and hangar decks and in passageways of all ships to maintain traction during wet and slippery conditions. The coatings contain epoxy resins and a coarse grit and are applied using a roller over epoxy primers to produce a textured finish. Nonskid coatings are 6 to 10 mm thick when dry.

Aluminum and galvanized steel are widely used in equipment on weather decks. Historically, galvanized steel and aluminum surfaces have been treated using a thin coat of wash primer after cleaning and before coating. For environmental reasons, the wash primer is omitted and these metals are coated directly with an epoxy primer. An aliphatic polyurethane, alkyd or silicone-alkyd enamel provides improved weather resistance. Galvanizing can best be cleaned by water blasting if no rusting is present or a light brushoff blast can be used if rusting exists. The same blast technique can be used to clean aluminum surfaces. Alkyds must not be used on galvanized steel because zinc and moisture rapidly hydrolyze the resin, producing zinc soaps that destroy adhesion.

Superstructure. Coatings for superstructures must have good resistance to sun, salt, and corrosion, and good gloss and appearance retention properties. In addition, coatings on antennas and superstructures must be resistant to acidic exhaust fumes and high temperatures. Deck hardware and machinery, masts, and booms are coated with an inorganic zinc primer, an intermediate coat of epoxy, and a finish coat of aliphatic polyurethane or silicone-alkyd enamel. Powder coatings are used effectively on antennas and other equipment on the superstructure. This equipment, as well as exhaust stacks, steam riser valves and piping, and other hot surfaces, can also be coated using thermal-sprayed aluminum. Powder coatings and sprayed metallic coatings (qv) are applied in shops under controlled conditions. Inorganic zinc primers are becoming widespread in new construction.

Tanks. Coatings for liquid cargo tanks are selected according to the materials that the tanks (qv) are to contain. Tank coatings protect the cargo from contamination and must be compatible with the material carried. Epoxy systems are most frequently selected because they perform well with both aqueous and organic products. A carefully applied three-coat epoxy system having a dry-film thickness of 225–300 μm can be expected to last for 12 years.

Coatings for potable water tanks must not impart taste or odor to the water and must not allow corrosion products to enter the water. Epoxy coatings are usually used. In the United States these coatings must be approved by the National Sanitation Foundation, acting as agent for the U.S. Environmental Protection Administration. Complete cure is very important and up to two weeks at 20°C may be necessary.

Petroleum products far exceed all other substances carried aboard ship, as cargoes or as fuel. Fuel tanks have specialized requirements because they may be filled with seawater ballast after the fuel is consumed. The coating must resist attack by both fluids, and seawater is much more corrosive than hydrocarbons. A three-coat epoxy system totalling 250–300 μm dry thickness gives good service in U.S. Navy ships. Zinc primers are not permitted in fuel tanks because, in addition to being unsuitable for seawater immersion, zinc may dissolve in automotive or aviation fuels causing damage to the engines in which the fuel is subsequently used. The same three-coat epoxy system used in fuel tanks is used in a variety of other tanks, including ballast tanks, sanitary holding tanks, and hydraulic fluid reservoirs. Fluorinated polyurethane coatings pigmented with 24% of poly(tetrafluoroethylene) give exceptionally long service in fuel tanks (17).

Machinery Spaces, Bilges, and Holds. Machinery spaces and bilges are so inaccessible that surface preparation is a significant problem and damage to machinery that cannot be removed must be avoided. Chemical cleaning by aqueous citric acid solutions, followed by degreasing using a nonflammable solvent, is widely used. Surfaces are best protected using a two- or three-coat epoxy–polyamide system having a total thickness of 250–300 μm. Alkyd enamels perform well in dry machinery spaces. Holds for carrying cargo may be painted with either of these systems, but the epoxy system is preferred for chemical and abrasion resistance.

The interior of piping has been protected from corrosion and abrasion by aromatic amine-cured epoxy coatings (18). These coatings are forced through intact piping systems by compressed air and cure within 10 minutes, forming a hard impervious lining. They have been used to protect 70:30 and 90:10 copper:nickel pipes in sanitary systems from sulfide corrosion, and are also suitable for use in potable water piping systems.

Living Areas. Coatings for living areas must be easily cleaned and resistant to bacteria, soiling, and fire. Living areas are generally painted with nonflaming coatings, or with intumescent coatings which foam when heated and produce a thick char that lessens damage to the substrate. For ceilings and walls in living and sleeping areas the U.S. Navy uses coatings based on highly chlorinated alkyd resins (19) or on aqueous emulsions of vinylidene chloride (20). Epoxy systems are generally used in damp areas such as galleys, washrooms, and showers where moisture deteriorates enamels.

BIBLIOGRAPHY

"Coatings, Marine" in *ECT* 3rd ed., Vol. 6, pp. 445–454, by R. W. Drisko, U.S. Naval Civil Engineering Laboratory.

1. D. Atherton, J. Verborgt, and M. A. M. Winkeler, *J. Coatings Technol.* **51**(657), 88 (1979).
2. R. Abel, N. J. King, J. L. Vosser, and T. G. Wilkinson, in M. A. Champ, ed., *Proceedings of Oceans '86,* Marine Technology Society, Washington, D.C., 1986, p. 1314.
3. U.S. Military Specification MIL-A-22262A, *Abrasive Blasting Media, Ship Hull Blast Cleaning,* Feb. 6, 1987.
4. J. D. Keane and co-edits., *Systems and Specifications,* 6th ed., Steel Structures Painting Council, Pittsburgh, Pa., 1991, pp. 9–48.
5. SSPC-VIS-1-89, *Visual Standard for Abrasive Blast Cleaned Steel,* Steel Structures Painting Council, Pittsburgh, Pa., 1989.
6. R. F. Brady, Jr., *J. Protective Coatings Linings* **4**(7), 42 (1987).
7. R. F. Brady, Jr., H. G. Lasser, and F. Pearlstein, in R. S. Shane and R. Young, eds., *Materials and Processes,* 3rd ed., Marcel Dekker, Inc., New York, 1985, pp. 1267–1319.
8. R. F. Brady, Jr., *J. Protective Coatings Linings* **2**(11), 24 (1985).
9. R. F. Brady, Jr. and C. H. Hare, *J. Protective Coatings Linings* **6**(4), 49–60 (1989).
10. K. B. Tator, *J. Protective Coatings Linings* **2**(2), 22 (1985).
11. U.S. Military Specification MIL-P-24441A, *Paint, Epoxy Polyamide, General Specification For,* July 15, 1980.
12. Ref. 4, pp. 233–239.
13. U.S. Military Specification DOD-P-24647, *Paint, Antifouling, Ship Hull,* Mar. 22, 1985.
14. U.S. Military Specification MIL-P-15931D, *Paint, Antifouling, Vinyl,* May 27, 1980.
15. R. F. Brady, Jr., J. R. Griffith, K. S. Love, and D. E. Field, *J. Coatings Tech.* **59**(755), 113 (1987).
16. U.S. Military Specification DOD-E-24635, *Enamel, Gray, Silicone Alkyd Copolymer, for Exterior Use,* Sept. 13, 1984.
17. J. R. Griffith and R. F. Brady, Jr., *Chemtech* **19**(6), 370 (1989).
18. R. F. Brady, Jr., *Surface Coatings Australia* **28**(5), 12 (1991).
19. U.S. Military Specification DOD-E-24607, *Enamel, Interior, Nonflaming, Chlorinated Alkyd Resin, Semigloss,* Oct. 13, 1981.
20. U.S. Military Specification DOD-C-24596, *Coating Compounds, Nonflaming, Fire-Protective,* Nov. 6, 1979.

General References

H. R. Bleile and S. D. Rogers, *Marine Coatings,* Federation of Societies for Coatings Technology, Blue Bell, Pa., 1989.

R. F. Brady, Jr., "Marine Applications," in J. I. Kroschwitz, ed., *Encyclopedia of Polymer Science and Engineering,* Vol. 9, John Wiley & Sons, Inc., New York, 1988, pp. 295–300.

J. D. Costlow and R. D. Tipper, eds., *Marine Biodeterioration: An Interdisciplinary Study,* Naval Institute Press, Annapolis, Md., 1984.

C. H. Hare, *The Painting of Steel Bridges,* Reichhold, New York, 1988.

J. D. Keane and co-eds., *Good Painting Practice,* 2nd ed., Steel Structures Painting Council, Pittsburgh, Pa., 1982.

J. D. Keane and co-eds., *Systems and Specifications,* 6th ed., Steel Structures Painting Council, Pittsburgh, Pa., 1991.

C. G. Munger, *Corrosion Protection by Protective Coatings,* National Association of Corrosion Engineers, Houston, Tex., 1984.

Z. W. Wicks, Jr., *Corrosion Protection by Coatings,* Federation of Societies for Coatings Technology, Blue Bell, Pa., 1987.

Journal of Coatings Technology, published monthly by the Federation of Societies for Coatings Technology, Blue Bell, Pa.

Journal of Protective Coatings and Linings, published monthly by the Steel Structures Painting Council, Pittsburgh, Pa.

ROBERT F. BRADY, JR.
U.S. Naval Research Laboratory

RICHARD W. DRISKO
U.S. Naval Civil Engineering Laboratory

COATINGS, RESISTANT.   See COATINGS; CORROSION AND CORROSION CONTROL; REFRACTORY COATINGS.

COBALT AND COBALT ALLOYS

COBALT

Cobalt [7440-48-4], a transition series metallic element having atomic number 27, is similar to silver in appearance. Cobalt was used as a coloring agent by Egyptian artisans as early as 2000 BC and cobalt-colored lapis or lapis lazuli was used as an item of trade between the Assyrians and Egyptians. In the Greco-Roman period cobalt compounds were used as ground coat frit and coloring agents for glasses. The common use of cobalt compounds in coloring glass and pottery led to their import to China during the Ming Dynasty under the name of Mohammedan blue.

The ancient techniques for mining cobalt and the use of cobalt compounds were lost during the Dark Ages. However, in the sixteenth century mining techniques became widely known through the works of Georgius Agricola, the German mineralogist. At that time cobalt was supplied as smalt or zaffre, the latter being a cobalt arsenide of sulfide ore that was roasted to yield a cobalt oxide. When fused with potassium carbonate to form a type of glass, zaffre became smalt. In the sixteenth century the ability of cobalt to color glass (qv) blue was rediscovered. Metallic cobalt, isolated in 1735 by a Swedish scientist, was established as an element in 1780.

Cobalt and cobalt compounds (qv) have expanded from use as colorants in glasses and ground coat frits for pottery (see COLORANTS FOR CERAMICS) to drying agents in paints and lacquers (see DRYING AGENTS, PAINTS), animal and human nutrients (see MINERAL NUTRIENTS), electroplating (qv) materials, high temperature alloys (qv), hardfacing alloys, high speed tools (see TOOL MATERIALS), magnetic alloys (see MAGNETIC MATERIALS, BULK; MAGNETIC MATERIALS, THIN-FILM), alloys used for prosthetics (see PROSTHETIC AND BIOMEDICAL DEVICES), and uses in radiology (see RADIOACTIVE TRACERS). Cobalt is also used as a catalyst for hydrocarbon refining from crude oil for the synthesis of heating fuels (1,2) (see FUELS, SYNTHETIC; PETROLEUM).

Occurrence

Cobalt is the thirtieth most abundant element on earth and comprises approximately 0.0025% of the earth's crust (3). It occurs in mineral form as arsenides, sulfides, and oxides; trace amounts are also found in other minerals of nickel and iron as substitute ions (4). Cobalt minerals are commonly associated with ores of nickel, iron, silver, bismuth, copper, manganese, antimony, and zinc. Table 1 lists the principal cobalt minerals and some corresponding properties. A complete listing of cobalt minerals is given in Reference 4.

The world's largest cobalt reserves are in Zaire, Zambia, Morocco, Canada, and Australia. Together the ores of these countries contain well over one-half of the world cobalt supply. The richest deposits are in Zaire and Zambia. The reserves of Canada and Australia comprise approximately one-fourth of the world supply. Smaller but commercially practical ore bodies also exist in Russia, Finland, Uganda, and the Philippines.

The largest cobalt producing mine in the world is a national company in Zaire, La Générale des Carriers et des Mines du Congolaise des Mines (Gecomines). In addition to cobalt, this mining complex also produces copper and zinc (6). Zaire has been producing cobalt since 1914 and has been the world leader of cobalt ore production since 1940.

Cobalt deposits in Zaire are either cobaltiferous or cobalt–copper ores. A small amount of these deposits occur as high grade oxidized ore. As much as 0.4 wt % cobalt is contained in malachite, $CuCO_3 \cdot Cu(OH)_2$, deposits as $CuO \cdot 2Co_2O_3 \cdot 6H_2O$. The oxidized cobalt minerals that have been found in these

Table 1. Important Cobalt Minerals and Corresponding Properties[a]

Mineral	CAS Registry Number	Chemical formula	Crystalline form	Approximate hardness, Mohs'	Density, kg/m³	Cobalt, wt %	Location
			Arsenides				
smaltite	[12044-42-1]	$CoAs_2$	cubic	6.0	6.5	23.2	United States, Canada, Morocco
safflorite	[12044-43-8]	$CoAs_2$	orthogonal	5.0	7.2	28.2	Morocco, Canada
skutterudite	[12196-91-7]	$CoAs_3$[b,c]	cubic	6.0	6.5	20.8	Ontario, Morocco
			Sulfides				
carrollite	[12285-42-6]	$CuCo_2S_4$,$CuS \cdot Co_2S_3$, Co_3S_4	cubic	5.5	4.85	38.7	Zaire, Zambia
linnaeite	[1308-08-3]	Co_3S_4	cubic	5	4.5	48.7	Zaire
siegenite	[12174-56-0]	$(Co,Ni)_3S_4$				26.0	United States
cattierite	[12017-06-0]	CoS_2[b]					Zaire
			Arsenide–sulfide				
cobaltite	[1303-15-7]	$CoAsS$	cubic	6	6.5	35.5	United States, Canada, Australia
			Oxide				
asbolite	[12413-71-7]	$CoO \cdot 2MnO_2 \cdot 4H_2O$	ore	1–2	1.1		Zaire, Zambia
erythrite	[149-32-6]	$3CoO \cdot As_2O_5 \cdot 8H_2O$	ore	2	3	29.5	
heterogenite	[12323-83-0]	$CuO \cdot 2Co_2O_3 \cdot 6H_2O$, $Co_2O_3 \cdot H_2O$[d]	ore	4	3.5	57	Zaire
sphaerocobaltite	[14476-13-2]	$CoCO_3$	ore			49.6	Zaire, Zambia

[a]Refs. 5 and 6.
[b]May also include some nickel.
[c]May also include some nickel and some iron.
[d]$CuO \cdot 2Co_2O_3 \cdot 6H_2O$ and $CoO \cdot 3Co_2O_3 \cdot CuO \cdot 7H_2O$ are also present.

762

deposits are heterogenite, stainierite, asbolite, sphaerocobaltite, carrollite, co-baltite, cattiertite, cobalto-vaesite, $NiCoS_2$, siegenite, and selenio–seigenite, $(Co,Ni)_2(Si,Se)_4$.

In Zambia, having a production second only to Zaire, cobalt is mined by Nchanga Consolidated Copper Mines Ltd. (NCOM) and Roan Consolidated Ltd. (RCM). The cobalt minerals found there include minnaeite, carrollite, and cobaltiferrous pyrite.

In Moroccan deposits, cobalt occurs with nickel in the forms of smaltite, skutterudite, and safflorite. In Canadian deposits, cobalt occurs with silver and bismuth. Smaltite, cobaltite, erythrite, safflorite, linnaeite, and skutterudite have been identified as occurring in these deposits. Australian deposits are associated with nickel, copper, manganese, silver, bismuth, chromium, and tungsten. In these reserves, cobalt occurs as sulfides, arsenides, and oxides.

In the United States, cobalt is found at Blackbird, Idaho, and the Grace and Cornwall mines in Pennsylvania. At the Blackbird mine, cobalt occurs with chal-copyrite [1308-56-1], gold, silver, and nickel. The Pennsylvania deposits occur with magnetite as cobaltiferrous pyrite. As of this writing, the Colombian Gov-ernment is studying apparently large deposits of cobalt ore in Cerro Matoso. There are also large deposits in Cuba.

Future Sources. Lateritic ores (7) are becoming increasingly important as a source of nickel, and cobalt is a by-product. In the United States, laterites are found in Minnesota, California, Oregon, and Washington. Deposits also occur in Cuba, Indonesia, New Caledonia, the Philippines, Venezuela, Guatemala, Aus-tralia, Canada, and Russia (see NICKEL AND NICKEL ALLOYS).

The laterites can be divided into three general classifications: (1) iron nick-eliferous limonite which contains approximately 0.8–1.5 wt % nickel. The nickel to cobalt ratios for these ores are typically 10:1; (2) high silicon serpentinous ores that contain more than 1.5 wt % nickel; and (3) a transition ore between type 1 and type 2 containing about 0.7–0.2 wt % nickel and a nickel to cobalt ratio of approximately 50:1. Laterites found in the United States (8) contain 0.5–1.2 wt % nickel and the nickel occurs as the mineral goethite. Cobalt occurs in the la-teritic ore with manganese oxide at an estimated wt % of 0.06 to 0.25 (9).

Properties

The electronic structure of cobalt is [Ar] $3d^74s^2$. At room temperature the crys-talline structure of the α (or ϵ) form, is close-packed hexagonal (cph) and lattice parameters are $a = 0.2501$ nm and $c = 0.4066$ nm. Above approximately 417°C, a face-centered cubic (fcc) allotrope, the γ (or β) form, having a lattice parameter $a = 0.3544$ nm, becomes the stable crystalline form. The mechanism of the allo-tropic transformation has been well described (5,10–12). Cobalt is magnetic up to 1123°C and at room temperature the magnetic moment is parallel to the c-direction. Physical properties are listed in Table 2.

Many different values for room temperature mechanical properties can be found in the literature. The lack of agreement depends, no doubt, on the different mixtures of α and γ phases of cobalt present in the material. This, on the other

Table 2. Properties of Cobalt

Property	Value		
at wt	58.93		
transformation temperature, °C	417		
heat of transformation, J/g[a]	251		
mp, °C	1493		
latent heat of fusion, ΔH_{fus}, J/g[a]	259.4		
bp, °C	3100		
latent heat of vaporization, ΔH_{vap}, J/g[a]	6276		
specific heat, J/(g·°C)[a]			
15–100°C	0.442		
molten metal	0.560		
coefficient of thermal expansion, °C^{-1}			
cph at RT	12.5		
fcc at 417°C	14.2		
thermal conductivity at RT, W/(m·K)	69.16		
thermal neutron absorption, Bohr atom	34.8		
resistivity, at 20°C[b], 10^{-8} Ω·m	6.24		
Curie temperature, °C	1121		
saturation induction, $4\pi I_S$, T[c]	1.870		
permeability, μ			
initial	68		
max	245		
residual induction, T[c]	0.490		
coercive force, A/m	708		
Young's modulus, GPa[d]	211		
Poisson's ratio	0.32		
hardness[f], diamond pyramid, of % Co	99.9	99.98[e]	
at 20°C	225	253	
at 300°C	141	145	
at 600°C	62	43	
at 900°C	22	17	
strength of 99.9% cobalt, MPa[g]	as cast	annealed	sintered
tensile	237	588	679
tensile yield	138	193	302
compressive	841	808	
compressive yield	291	387	

[a] To convert J to cal, divide by 4.184.
[b] Conductivity = 27.6% of International Annealed Copper Standard.
[c] To convert T to gauss, multiply by 10^4.
[d] To convert GPa to psi, multiply by 145,000.
[e] Zone refined.
[f] Vickers.
[g] To convert MPa to psi, multiply by 145.

hand, depends on the impurities present, the method of production of the cobalt, and the treatment.

The hardness on the basal plane of the cobalt depends on the orientation and extends between 70 and 250 HK. Cobalt is used in high temperature alloys of the superalloy type because of its resistance to loss of properties when heated to fairly high temperatures. Cobalt also has good work-hardening characteristics, which contribute to the interest in its use in wear alloys.

Whereas finely divided cobalt is pyrophoric, the metal in massive form is not readily attacked by air or water or temperatures below approximately 300°C. Above 300°C, cobalt is oxidized by air. Cobalt combines readily with the halogens to form halides and with most of the other nonmetals when heated or in the molten state. Although it does not combine directly with nitrogen, cobalt decomposes ammonia at elevated temperatures to form a nitride, and reacts with carbon monoxide above 225°C to form the carbide Co_2C. Cobalt forms intermetallic compounds with many metals, such as Al, Cr, Mo, Sn, V, W, and Zn.

Metallic cobalt dissolves readily in dilute H_2SO_4, HCl, or HNO_3 to form cobaltous salts (see also COBALT COMPOUNDS). Like iron, cobalt is passivated by strong oxidizing agents, such as dichromates and HNO_3, and cobalt is slowly attacked by NH_4OH and NaOH.

Cobalt cannot be classified as an oxidation-resistant metal. Scaling and oxidation rates of unalloyed cobalt in air are 25 times those of nickel. The oxidation resistance of Co has been compared with that of Zr, Ti, Fe, and Be. Cobalt in the hexagonal form (cold-worked specimens) oxidizes more rapidly than in the cubic form (annealed specimens) (3).

The scale formed on unalloyed cobalt during exposure to air or oxygen at high temperature is double-layered. In the range of 300 to 900°C, the scale consists of a thin layer of the mixed cobalt oxide [1308-06-1], Co_3O_4, on the outside and a cobalt(II) oxide [1307-96-6], CoO, layer next to the metal. Cobalt(III) oxide [1308-04-9], Co_2O_3, may be formed at temperatures below 300°C. Above 900°C, Co_3O_4 decomposes and both layers, although of different appearance, are composed of CoO only. Scales formed below 600°C and above 750°C appear to be stable to cracking on cooling, whereas those produced at 600–750°C crack and flake off the surface.

Processing

Sulfide Ores. In the Zairian ores, cobalt sulfide as carrollite is mixed with chalcopyrite and chalcocite [21112-20-9]. For processing, the ore is finely ground and the sulfides are separated by flotation (qv) using frothers. The resulting products are leached with dilute sulfuric acid to give a copper–cobalt concentrate that is then used as a charge in an electrolytic cell to remove the copper. Because the electrolyte becomes enriched with cobalt, solution from the copper circuit is added to maintain a desirable copper concentration level. After several more steps to remove copper, iron, and aluminum, the solution is treated with milk of lime to precipitate the cobalt as the hydroxide.

Zambian copper sulfide ores are leaner in cobalt and are, therefore, concentrated twice. On the first stage, the bulk of the copper ore is floated off in a high lime circuit. The carrollite is not carried by the lime, but is recovered in a second pass using different flotation methods. The second concentration, which contains 25% Cu, 17% Fe, and 3.5–4% Co, undergoes a sulfatizing roasting to convert the cobalt to water-soluble cobalt sulfate. The iron and copper in the concentrate form insoluble oxides and sulfates. After roasting, the matte is leached with water and filtered. After removing the last of the copper, milk of lime is added to precipitate the cobalt hydroxide which is filtered and dissolved in sulfuric acid. The resulting solution is then used as an electrolyte from which cobalt is electrodeposited. The cobalt thus produced is marketed either in a granulated form or still attached to the cathode (see METALLURGY, EXTRACTIVE).

Arsenic-Free Cobalt–Copper Ores.   The arsenic-free cobalt ores of Zaire are treated by smelting. Ores such as heterogenite which have high cobalt content can be sent directly to an electric furnace in lump form. The fines must be sintered before being charged. Smelting the cobalt–copper feed along with lime and coke produces a slag and two alloys.

Arsenic Sulfide Ores.   The high grade ores of Morocco are magnetically separated to give arsenides and oxides (see SEPARATION, MAGNETIC). Ores that are most concentrated in the arsenides are subjected to an oxidizing roast to remove the arsenic. The concentrates are used as feed in a blast furnace, resulting in speiss, matte, and perhaps boullion. The cobalt-containing speiss is then crushed and roasted and the roasted speiss is treated with sulfuric acid and the solids removed and roasted. After the iron compounds are precipitated with sodium chlorate and lime, the cobalt–nickel solution is treated with sodium hypochlorite to precipitate a cobalt hydrate.

Pressure-acid leaching was used to extract cobalt from Blackbird mine ores before its closing in 1974. The result was a very fine cobalt powder which was subjected to a seeding process to produce cobalt granules. Leaching methods are also used in the refinement of lateritic ores.

Lateritic Ores.   The process used at the Nicaro plant in Cuba requires that the dried ore be roasted in a reducing atmosphere of carbon monoxide at 760°C for 90 minutes. The reduced ore is cooled and discharged into an ammoniacal leaching solution. Nickel and cobalt are held in solution until the solids are precipitated. The solution is then thickened, filtered, and steam heated to eliminate the ammonia. Nickel and cobalt are precipitated from solution as carbonates and sulfates. This method (8) has several disadvantages: (1) a relatively high reduction temperature and a long reaction time; (2) formation of nickel oxides; (3) a low recovery of nickel and the contamination of nickel with cobalt; and (4) low cobalt recovery. Modifications to this process have been proposed but all include the undesirable high 760°C reduction temperature (9).

A similar process has been devised by the U.S. Bureau of Mines (8) for extraction of nickel and cobalt from United States laterites. The reduction temperature is lowered to 525°C and the holding time for the reaction is 15 minutes. An ammoniacal leach is also employed, but oxidation is controlled, resulting in high extraction of nickel and cobalt into solution. Mixers and settlers are added to separate and concentrate the metals in solution. Organic strippers are used to selectively remove the metals from the solution. The metals are then removed from the strippers. In the case of cobalt, spent cobalt electrolyte is used to separate the metal-containing solution and the stripper. Metallic cobalt is then recovered by electrolysis from the solution. Using this method, 92.7 wt % nickel and 91.4 wt % cobalt have been economically extracted from domestic laterites containing 0.73 wt % nickel and 0.2 wt % cobalt (8).

Deep Sea Nodules.   Metal prices influence the type of extraction process used for sea nodules. Whereas there are those typically rich in manganese, most of the mining and refining expenses must be met by the price of commodity metals such as copper and cobalt rather than by an abundant metal like manganese. It has been suggested that a method be used that would selectively remove cobalt, nickel, and copper, leaving manganese stored in the tailings for future use (13,14). This method involves smelting of the reduced new nodules to produce a manganiferous slag and an alloy containing the metals. The alloy is converted into a matte by oxidation and sulfidation. An oxidative pressure leach is used to produce a purified solution from which the metal is obtained (13). Research is continuing in the areas of pyrometallurgy, hydrometallurgy, and electrometallurgy to find more efficient extraction techniques (15–18) (see OCEAN RAW MATERIALS).

Analysis

The detection and determination of traces of cobalt is of concern in such diverse areas as solids, plants, fertilizers (qv), stainless and other steels for nuclear energy equipment (see STEEL), high purity fissile materials (U, Th), refractory metals (Ta, Nb, Mo, and W), and semiconductors (qv). Useful techniques are spectrophotometry, polarography, emission spectrography, flame photometry, x-ray fluorescence, activation analysis, tracers, and mass spectrography, chromatography, and ion exchange (19) (see ANALYTICAL METHODS; SPECTROSCOPY; TRACE AND RESIDUE ANALYSIS).

For colorimetric or gravimetric determination 1-nitroso-2-naphthol can be used. For chromatographic ion exchange (qv), cobalt is isolated as the nitroso-(R)-salt complex. The cyanate complex is used for photometric determination and the thiocyanate for colorimetry. A rapid chemical analysis of alloys, powders, and liquids, employing x-ray spectrography has been developed. With this method, cobalt, in a concentration of 10–60%, may be analyzed in a few minutes with accuracies on the order of ± 1%.

Economic Aspects

Table 3 gives the U.S. consumption of cobalt between 1985 and 1990 by form; Table 4 gives consumption according to products. During the period of 1978–1981 the price and availability of cobalt was erratic. In May 1978, the production of cobalt in Zaire came to a temporary standstill and the world price of the metal immediately increased at least fourfold. Many users sought alternative materials. For example, magnetic materials manufacturers moved in the direction of using cobalt-free rare-earth materials and ferrites, whereas wear alloy and superalloy producers worked to develop nickel alloy alternatives. The supply and price have stabilized somewhat since 1978. However, in the first eight months of 1992 cobalt

Table 3. U.S. Consumption of Cobalt, t[a]

Form	1986	1987	1988	1989	1990
metal	3507	3819	4247	3901	4070
scrap	1196	1025	1018	1184	1225
chemical compounds[b]	1778	1802	2020	2067	2177
Total	*6483*	*6645*	*7286*	*7152*	*7472*

[a]Ref. 20.
[b]Includes oxide.

Table 4. U.S. Reported Consumption of Cobalt by Products, t[a]

Application	1985	1986	1987	1988	1989	1990
superalloys	2418	2924	2873	2926	2898	3391
hardfacing alloys	566	518	654	704	654	677
magnetic alloys	660	699	666	869	861	700
other alloys	175	204	243	231	337	251
ceramics and chemicals	1797	2139	2210	2556	2402	1453
Total	*5616*	*6483*	*6645*	*7286*	*7152*	*6472*

[a]Based on cobalt content (20).

price fluctuations ranged from $77/kg to $26/kg and as of the beginning of September 1992, cobalt was priced at ~ $43/kg.

Uses

As Metal. The largest consumption of cobalt is in metallic form in magnetic alloys, cutting and wear-resistant alloys, and superalloys. Alloys in this last group are used for components requiring high strength as well as corrosion and oxidation resistance, usually at high temperatures.

During World War II German scientists developed a method of hydrogenating solid fuels to remove the sulfur by using a cobalt catalyst (see COAL CONVERSION PROCESSES). Subsequently, various American oil refining companies used the proces in the hydrocracking of crude fuels (see CATALYSIS; SULFUR REMOVAL AND RECOVERY). Cobalt catalysts are also used in the Fisher-Tropsch method of synthesizing liquid fuels (21–23) (see FUELS, SYNTHETIC).

Cobalt–molybdenum alloys are used for the desulfurization of high sulfur bituminous coal, and cobalt–iron alloys in the hydrocracking of crude oil shale (qv) and in coal liquefaction (6).

As Salts. The second largest use of cobalt is in the form of salts (see COBALT COMPOUNDS), which have the largest application as raw material for electroplating (qv) baths and as highly effective driers for lacquers, enamels, and varnishes. Addition of cobalt salts to paint greatly increases the rate at which paint (qv) hardens. Cobalt oxide colors glass pink or blue depending on the environment of the CoO_x molecule within the glass. The pink colors are formed in boric oxide or alkali-borate glasses. The cobalt concentration determines the intensity of the color obtained, eg, glass used in making foundryman's goggles requires 4.5 kg of cobalt for every ton of glass. By contrast, the glass used in making decorative bottles requires only about 280 g/t. Cobalt is also used to decolorize soda–lime–silica glass. Pottery enamels react very similarly because the fusible enamels are forms of fusible glasses (see ENAMELS, PORCELAIN OR VITREOUS). Colors varying from blue to black can be obtained, depending on the oxides added to the frit to improve the adherence of porcelain enamel to steel sheet metal. Combinations of cobalt compounds are also used as ceramic pigments (qv) ranging in colors of violet, blue, green, and pink (23–25).

Radioactive cobalt, ^{60}Co, produced by bombarding stable ^{59}Co with low energy neutrons, has application in radiochemistry, radiography, and food sterilization (26–28) (see FOOD PROCESSING; RADIOISOTOPES; STERILIZATION TECHNIQUES).

Cobalt is an essential ingredient in animal nutrition. It has been shown that animals deprived of cobalt show signs of retarded growth, anemia, loss of appetite, and decreased lactation. Dressing the top soil of pastures with cobalt increases the cobalt content of the vegetation. Cobalt is known to be necessary to the synthesis of vitamin B_{12} [68-19-9] (see VITAMINS), a lack of which has been linked to pernicious anemia in humans. Cobalt is also used as the target material in electrical x-ray generators (see X-RAY TECHNOLOGY).

COBALT ALLOYS

Pure metallic cobalt has a solid-state transition from cph (lower temperatures) to fcc (higher temperatures) at approximately 417°C. However, when certain elements such as Ni, Mn, or Ti are added, the fcc phase is stabilized. On the other hand, adding Cr, Mo, Si, or W stabilizes the cph phase. Upon fcc-phase stabilization, the energy of crystallographic stacking faults, ie, single-unit cph inclusions that impede mechanical slip within the fcc matrix, is high. Stabilizing the cph phase, however, produces a low stacking fault energy. In the case of high stacking fault energy, fcc stabilization, only a few stacking faults occur, and the ductility of the alloy is high. When foreign elements are dissolved throughout the matrix lattice, the mechanical slip is generally impeded. This results in increased hardness and strength, a metallurgical phenomenon known as solid-solution hardening.

Mechanical properties depend on the alloying elements. Addition of carbon to the cobalt base metal is the most effective. The carbon forms various carbide phases with the cobalt and the other alloying elements (see CARBIDES). The presence of carbide particles is controlled in part by such alloying elements such as chromium, nickel, titanium, manganese, tungsten, and molybdenum that are added during melting. The distribution of the carbide particles is controlled by heat treatment of the solidified alloy.

Cobalt alloys are strengthened by solid-solution hardening and by the solid-state precipitation of various carbides and other intermetallic compounds. Minor phase compounds, when precipitated at grain boundaries, tend to prevent slippage at those boundaries thereby increasing creep strength at high temperatures. Aging and service under stress at elevated temperature induce some of the carbides to precipitate at slip planes and at stacking faults thereby providing barriers to slip. If carbides are allowed to precipitate to the point of becoming continuous along the grain boundaries, they often initiate fracture (see FRACTURE MECHANICS). A thorough discussion of the mechanical properties of cobalt alloys is given in References 29 and 30 (see also REFRACTORIES).

Cobalt-Base Alloys

As a group, the cobalt-base alloys may generally be described as wear-resistant, corrosion-resistant, and heat-resistant, ie, strong even at high temperatures. Ta-

Table 5. Compositions of Cobalt-Base Wear-Resistant Alloys, wt %[a]

Alloy trade name	Cr	W	Mo	C	Fe[b]	Ni	Si	Mn
Cobalt-base wear-resistant alloys[c,d]								
Stellite 1	31	12.5	1[b]	2.4	3	3[b]	2[b]	1[b]
Stellite 6	28	4.5	1[b]	1.2	3	3[b]	2[b]	1[b]
Stellite 12	30	8.3	1[b]	1.4	3	3[b]	2[b]	1[b]
Stellite 21	28		5.5	0.25	2	2.5	2[b]	1[b]
Haynes alloy 6B	30	4	1	1.1	3	2.5	0.7	1.5
Tribaloy T-800	17.5		29	0.08[b]				
Stellite F	25	12.3	1[b]	1.75	3	22	2[b]	1[b]
Stellite 4	30	14.0	1[b]	0.57	3	3[b]	2[b]	1[b]
Stellite 190	26	14.5	1[b]	3.3	3	3[b]	2[b]	1[b]
Stellite 306[e]	25	2.0		0.4		5		
Stellite 6K	31	4.5	1.5[b]	1.6	3	3[b]	2[b]	2[b]
Cobalt-base high temperature alloys[d,f]								
Haynes alloy 25 (L605)	20	15		0.10	3	10	1[b]	1.5
Haynes alloy 188	22	14		0.10	3	22	0.35	1.25
MAR-M alloy 509[g]	22.5	7		0.60	1.5	10	0.4[b]	0.1[b]
Cobalt-base corrosion-resistant alloys[h]								
MP35N, multiphase alloy	20		10			35		
Ultimet[i]	25.5	2	5	0.08[b]	3[j]	9		

[a]Where the balance of the alloy consists of cobalt.
[b]Value given is maximum value.
[c]Stellite and Tribaloy are registered trademarks of Stoody Deloro Stellite.
[d]Haynes is a registered trademark of Haynes International, Inc.
[e]Also contains 6 wt % niobium.
[f]MAR-M-Alloy is a registered trademark of Martin-Marietta.
[g]Also contains 3.5 wt % tantalum, 0.2 wt % titanium, and 0.5 wt % zirconium.
[h]MP35N is a registered trademark of Standard Pressed Steel Co.
[i]Also contains 0.1 wt % nitrogen.
[j]Value given is not necessarily maximum.

ble 5 lists typical compositions of cobalt-base alloys in these application areas. Many of the alloy properties arise from the crystallographic nature of cobalt, in particular its response to stress; the solid-solution-strengthening effects of chromium, tungsten, and molybdenum; the formation of metal carbides; and the corrosion resistance imparted by chromium. Generally, the softer and tougher compositions are used for high temperature applications such as gas-turbine vanes and buckets. The harder grades are used for resistance to wear.

Historically, many of the commercial cobalt-base alloys are derived from the cobalt–chromium–tungsten and cobalt–chromium–molybdenum ternaries first investigated at the turn of the twentieth century. The high strength and stainless nature of the binary cobalt–chromium alloy, and powerful strengthening agents, tungsten and molybdenum, were identified early on. These alloys were named Stellite alloys after the Latin *stella* for star because of their starlike luster. Stellite alloys were first used as cutting tools and wear-resistant materials.

Following the success of cobalt-base tool materials during World War I, these alloys were used from about 1922 in weld overlay form to protect surfaces from wear. These early cobalt-base hardfacing alloys were used on plowshares, oil well drilling bits, dredging cutters, hot trimming dies, and internal combustion engine valves and valve seats. In 1982, approximately 1500 metric tons of cobalt-base alloys were sold for the purpose of hardfacing, one-third of this quantity being used to protect valve seating surfaces for both fluid control and engine valves.

In the 1930s and early 1940s, cobalt-base alloys for corrosion and high temperature applications were developed (31). Of the corrosion-resistant alloys, a cobalt–chromium–molybdenum alloy having a moderately low carbon content was developed to satisfy the need for a suitable investment cast dental material (see DENTAL MATERIALS). This biocompatible material, which has the tradename Vitallium, is used for surgical implants. In the 1940s this same alloy underwent investment casting trials for World War II aircraft turbocharger blades, and, with modifications to enhance structural stability, was used successfully for many years in this and other elevated-temperature applications. This early high temperature material, Stellite alloy 21, is in use in the 1990s predominantly as an alloy for wear resistance.

Cobalt-Base Wear-Resistant Alloys

The main differences in the Stellite alloy grades of the 1990s versus those of the 1930s are carbon and tungsten contents, and hence the amount and type of carbide formation in the microstructure during solidification. Carbon content influences hardness, ductility, and resistance to abrasive wear. Tungsten also plays an important role in these properties.

Types of Wear. There are several distinct types of wear that can be divided into three main categories: abrasive wear, sliding wear, and erosive wear. The type of wear encountered in a particular application is an important factor influencing the selection of a wear-resistant material.

Abrasive wear is encountered when hard particles, or hard projections on a counter-face, are forced against and moved relative to a surface. In alloys such as the cobalt-base wear alloys which contain a hard phase, the abrasion resistance generally increases as the volume fraction of the hard phase increases. Abrasion resistance is, however, strongly influenced by the size and shape of the hard-phase precipitates within the microstructure, and the size and shape of the abrading species (see ABRASIVES).

Sliding wear is perhaps the most complex in the way different materials respond to sliding conditions. The metallic materials that perform best under sliding conditions are cobalt-based either by virtue of oxidation behavior or ability to resist deformation and fracture. Little is known of the influence of metal-to-metal bond strength during cold welding. For materials such as the cobalt-base wear alloys having a hard phase dispersed throughout a softer matrix, the sliding-wear properties are controlled predominantly by the matrix. Indeed, within the cobalt alloy family, resistance to galling is generally independent of hard particle volume fraction and overall hardness.

Four distinct forms of erosive wear have been identified: solid-particle erosion, liquid-droplet erosion, cavitation erosion, and slurry erosion.

The abrasion resistance of cobalt-base alloys generally depends on the hardness of the carbide phases and/or the metal matrix. For the complex mechanisms of solid-particle and slurry erosion, however, generalizations cannot be made, although for the solid-particle erosion, ductility may be a factor. For liquid-droplet or cavitation erosion the performance of a material is largely dependent on ability to absorb the shock (stress) waves without microscopic fracture occurring. In cobalt-base wear alloys, it has been found that carbide volume fraction, hence, bulk hardness, has little effect on resistance to liquid-droplet and cavitation erosion (32). Much more important are the properties of the matrix.

Alloy Compositions and Product Forms.   The nominal compositions of various cobalt-base wear-resistant alloys are listed in Table 5. The six most popular cobalt-base wear alloys are listed first. Stellite alloys 1, 6, and 12, derivatives of the original cobalt–chromium–tungsten alloys, are characterized by their carbon and tungsten contents. Stellite alloy 1 is the hardest, most abrasion resistant, and least ductile.

Stellite alloy 21 differs from the first three alloys in that molybdenum rather than tungsten is used to strengthen the solid solution. Stellite alloy 21 also contains considerable less carbon. Each of the first four alloys is generally used in the form of castings and weld overlays. Haynes alloy 6B differs in that it is a wrought product available in plate, sheet, and bar form. Subtle compositional differences between alloy 6B and Stellite alloy 6, eg, such as silicon control, facilitate processing. The advantages of wrought processing include greatly enhanced ductility, chemical homogeneity, and resistance to abrasion.

The Tribaloy alloy T-800, is from an alloy family developed by DuPont in the early 1970s, in the search for resistance to abrasion and corrosion. Excessive amounts of molybdenum and silicon were alloyed to induce the formation during solidification of hard and corrosion-resistant intermetallic compounds, known as Laves phase. The Laves precipitates confer outstanding resistance to abrasion, but limit ductility. As a result of this limited ductility the alloy is not generally used in the form of plasma-sprayed coatings.

The physical and mechanical properties of the six commonly used cobalt wear alloys are presented in Table 6. In the case of the Stellite and Tribaloy alloys this information pertains to sand castings. Notable are the moderately high yield

Table 6. Mechanical and Physical Properties of Cobalt-Base Wear-Resistant Alloys

Property	Alloy[a]					
	1	6	12	21	6B	T-800
hardness, Rockwell	55	40	48	32	37[b]	58
yield strength, MPa[c]		541	649	494	619[b]	
ultimate tensile strength, MPa[c]	618	896	834	694	998[b]	
elongation, %	<1	1	<1	9	11	
thermal expansion coeff, μm/(m·°C)						
20–100°C	10.5	11.4	11.5	11.0	13.9[d]	
20–500°C	12.5	14.2	13.3	13.1	15.0[d]	12.6
20–1000°C	14.8		15.6		17.4[d]	15.1
thermal conductivity, W/(m·K)					14.8	14.3
specific gravity	8.69	8.46	8.56	8.34	8.39	8.64
electrical resistivity, $\mu\Omega$·m	0.94	0.84	0.88		0.91	
melting range, °C						
solidus	1255	1285	1280	1186	1265	1288
liquidus	1290	1395	1315	1383	1354	1352

[a]See Table 5.
[b]3.2 mm (⅛ in.) thick sheet.
[c]To convert MPa to psi, multiply by 145.
[d]Starting temperature of 0°C.

strengths and hardnesses of the alloys, the inverse relationship between carbon content and ductility in the case of the Stellite alloys, and the enhanced ductility imparted to alloy 6B by wrought processing. Typical applications of the cobalt wear-resistant alloys are given in Table 7. Generally, the alloys are used in moderately corrosive and/or elevated-temperature environments.

Cobalt-Base High Temperature Alloys

For many years, the predominant user of high temperature alloys was the gas-turbine industry. In the case of aircraft gas-turbine power plants, the chief material requirements were elevated-temperature strength, resistance to thermal fatigue, and oxidation resistance. For land-base gas turbines, which typically burn lower-grade fuels and operate at lower temperatures, sulfidation resistance was the primary concern. The use of high temperature alloys (qv) in the 1990s is more diversified, as more efficiency is sought from the burning of fossil fuels and waste, and as new chemical processing techniques are developed.

Cobalt-base alloys are not as widely used as nickel and nickel–iron alloys in high temperature applications. Nevertheless, cobalt-base high temperature alloys play an important role because of excellent resistance to sulfidation and strength at temperatures exceeding those at which the γ'- and γ''-precipitates in the nickel and nickel–iron alloys dissolve. Cobalt is also used as an alloying element in many nickel-base high temperature alloys. The various types of iron-base, nickel-base, and cobalt-base alloys for high temperature application are discussed

Table 7. Applications of Cobalt-Base Wear-Resistant Alloys

Applications	Stellite alloys[a]	Forms	Mode of degradation
Automotive industry			
diesel engine valve seating surface	6, F	weld overlay	solid particle erosion, hot corrosion
Power industry			
control valve seating surfaces	6, 21	weld overlay	sliding wear, cavitation erosion
steam turbine erosion shields	6B	wrought sheet	liquid droplet erosion
Marine industry			
rudder bearings	306	weld overlay	sliding wear
Steel industry			
hot shear edges	6	weld overlay	sliding wear, impact, abrasion
bar mill guide rolls	12	weld overlay	sliding wear, impact, abrasion
Chemical processing industry			
control valve seating surfaces	6	weld overlay	sliding wear, cavitation erosion
plastic extrusion screw flights	1, 6, 12	weld overlay	sliding wear, abrasion
pump seal rings	6, 12	weld overlay	sliding wear
dry battery molds	4	casting	abrasion
Pulp and paper industry			
chainsaw noses	6B, 6	wrought sheet weld overlay	sliding wear, abrasion
Textile industry			
carpet knives wool spinning	6K, 12	wrought sheet weld overlay	abrasion
Oil and gas industry			
rotary drill bearings	190	weld overlay	abrasion, sliding wear
Aircraft industry			
helicopter rotor blade erosion shields	6B	sheet	abrasion

[a]See Table 5.

in Reference 33. Nickel-base and cobalt-base castings for high temperature service are also covered.

Alloy Compositions and Product Forms. Stellite 21, an early type of cobalt-base high temperature alloy, is used primarily for wear resistance. The use of tungsten rather than molybdenum, moderate nickel contents, lower carbon contents, and rare-earth additions typify cobalt-base high temperature alloys of the 1990s as can be seen from Table 5.

Haynes alloys 25, also known as L605 and 188, are wrought alloys available in the form of sheets, plates, bars, pipes, and tubes together with a range of matching welding products for joining purposes. MAR-M alloy 509 is an alloy designed for vacuum investment casting. Selected mechanical properties of these three alloys are given in Table 8.

Table 8. Properties of Cobalt-Base High Temperature Alloys

Property	Alloy[a]		
	25	188	MAR-M509
yield strength, MPa[b]			
at 21°C	445[c]	464[d]	585[e]
at 540°C		305[f]	400[d]
tensile strength, MPa[b]			
at 21°C	970[c]	945[d]	780[e]
at 540°C	800[g]	740[f]	570[e]
1000-h rupture strength, MPa[b]			
at 870°C	75	70	140
at 980°C	30	30	90
elongation, %	62	53[c]	3.5[e]
thermal expansion coeff, μm/(m·K)			
from 21–93°C	12.3	11.9	
from 21–540°C	14.4	14.8	
from 21–1090°C	17.7	18.5	
thermal conductivity, W/(m·K)			
at 20°C	9.8[h]	10.8	
at 500°C	18.5[i]	19.9	
at 900°C	26.5[j]	25.1	
specific gravity	9.13	8.98	8.86
electrical resistivity, μΩ·m	0.89	1.01	
melting range, °C			
solidus	1329	1302	1290
liquidus	1410	1330	1400

[a]See Table 5.
[b]To convert MPa to psi, multiply by 145.
[c]3.2mm (⅛ in.) thick.
[d]Sheet 0.75–1.3 mm (0.03–0.05 in.) thick.
[e]As cast.
[f]Sheet, heat treated at 1175°C for 1 h with rapid air cool.
[g]Sheet, heat treated at 1230°C for 1 h with rapid air cool.
[h]At 38°C.
[i]At 540°C.
[j]At 815°C.

Cobalt-Base Corrosion-Resistant Alloys

Although the cobalt-base wear-resistant alloys possess some resistance to aqueous corrosion, they are limited by grain boundary carbide precipitation, the lack of vital alloying elements in the matrix after formation of the carbides or Laves precipitates, and, for the case and weld overlay materials, by chemical segregation in the microstructure. By virtue of homogeneous microstructures and lower carbon contents, the wrought cobalt-base high temperature alloys, which typically contain tungsten rather than molybdenum, are even more resistant to aqueous corrosion, but still fall well short of the nickel–chromium–molybdenum alloys in corrosion performance. To satisfy the industrial need for alloys which exhibit resistance to aqueous corrosion yet share the attributes of cobalt as an alloy base, ie, resistance to various forms of wear and high strength over a wide range of temperatures, several low carbon, wrought cobalt–nickel–chromium–molybdenum alloys are produced. The compositions of two of these are presented in Table 5. In addition, the cobalt–chromium–molybdenum alloy Vitallium is widely used for prosthetic devices and implants owing to excellent compatibility with body fluids and tissues.

The two corrosion-resistant alloys presented in Table 5 rely on chromium and molybdenum for their corrosion resistance. The corrosion properties of Ultimet are also enhanced by tungsten. Both alloys are available in a variety of wrought product forms: plates, sheets, bars, tubes, etc. They are also available in the form of welding (qv) consumables for joining purposes.

Mechanical Properties. An advantage of the two corrosion-resistant alloys is that they may be strengthened considerably by cold working. MP35N alloy is intended for use in the work-hardened or work-hardened and aged condition, and the manufacturers have supplied considerable data concerning the mechanical properties of the alloy at different levels of cold work. Some of these data are given in Table 8.

Uses. Applications of both these alloys include pump and valve components and spray nozzles. MP35N alloy is also popular for fasteners, cables, and marine hardware.

Economic Aspects

With cobalt historically being approximately twice the cost of nickel, cobalt-base alloys for both high temperature and corrosion service tend to be much more expensive than competitive alloys. In some cases of severe service their performance increase is, however, commensurate with the cost increase and they are a cost-effective choice. For hardfacing or wear applications, cobalt alloys typically compete with iron-base alloys and are at a significant cost disadvantage.

BIBLIOGRAPHY

"Cobalt and Cobalt Alloys" in *ECT* 1st ed., Vol. 4, pp. 189–199, by G. A. Roush, Mineral Industry; in *ECT* 2nd ed., Vol. 5, pp. 716–736, by F. R. Morral, Cobalt Information Center,

Battelle Memorial Institute; in *ECT* 3rd ed., Vol. 6, pp. 481–494, by F. Planinsak and J. B. Newkirk, University of Denver.

1. R. S. Young, *Cobalt,* American Chemical Monograph Series, No. 149, American Chemical Society, Rhinehold, New York, 1960.
2. *Cobalt Monograph,* Cobalt Information Center, Brussels, 1960.
3. A. H. Hurlich, *Met. Progr.* **112**(5), 67 (Oct. 1977).
4. C. S. Hurlbut Jr., *Dana's Manual of Mineralogy,* 17th ed., John Wiley & Sons, Inc., New York, 1966.
5. J. W. Christian, *Proc. R. Soc.* **206**A, 51 (1951).
6. *U.S. Bureau of Mines Minerals Yearbook,* Vol. III, U.S. Bureau of Mines, Washington, D.C., 1971, 1974.
7. C. Chandra, *Characterization of Lateritic Nickel Ores by Electron-Optical and X-Ray Techniques,* Ph.D. dissertation, University of Denver, Denver, Colo., 1976.
8. R. E. Siemens, *Process for Recovery of Nickel from Domestic Laterites,* presented at the 1976 Mining Convention, U.S. Bureau of Mines, 1976.
9. L. F. Power and G. H. Geiger, *Miner. Sci. Eng.* **9**(1), 32 (1977).
10. J. B. Hess and C. S. Barrett, *Trans. Am. Inst. Min. Met. Eng.* **194,** 645 (1952).
11. H. Bibring and F. Sebilleall, *Rev. Met.* **52,** 569 (1955).
12. A. Seeger, *Z. Metallkunde* **47,** 653 (1956).
13. R. Sridhar, W. E. Jones, and J. S. Warner, *J. Met.* **28**(4), 32 (1976).
14. J. C. Agarwal and co-workers, in Ref. 13, p. 24.
15. M. Wadsworth, *J. Met.* **28**(3), 4 (1976).
16. P. Duby, in Ref. 15, p. 8.
17. P. Tarassoff, in Ref. 15, p. 11.
18. M. G. Manzone, in Ref. 15, p. 16.
19. C. Tombu, *Cobalt* **20,** 103; **21,** 185 (1963).
20. H. R. Millie, ed., *Minerals and Materials,* U.S. Bureau of Mines, Washington, D.C., Mar. 1978, p. 28.
21. Ger. Pat. 1,012,124 (July 11, 1957), B. Lopmann.
22. U.S. Pat. 3,576,734 (Apr. 27, 1971), H. L. Bennett (to Bennett Engineering Co.).
23. M. A. Aglan and H. Moore, *J. Soc. Glass Tech.* **39,** 351 (1955).
24. J. Berk and J. de Jong, *J. Am. Ceram. Soc.* **41,** 287 (1958).
25. J. C. Richmond and co-workers, *J. Am. Ceram. Soc.* **36,** 410 (1953).
26. A. Charlesby, *Nucleonics* **14,** 82 (Sept. 1956).
27. E. B. Darden, E. Maeyens, and R. C. Bushland, *Nucleonics* **12**(10), 60 (1954).
28. A. E. Berkowitz, F. E. Joumot, and F. C. Nix, *Phys. Rev.* **98,** 1185 (1954).
29. C. T. Sims, N. S. Stoloff, and W. G. Hagel, eds., *Superalloys II,* John Wiley & Sons, Inc., New York, 1987, p. 135.
30. N. J. Grant and J. R. Lane, *Trans. ASM* **41,** 95 (1949).
31. R. D. Gray, *A History of the Haynes Stellite Company,* Cabot Corp., Kokomo, Ind., 1974.
32. K. C. Antony and W. L. Silence, *ELSI-5 Proceedings,* University of Cambridge, UK, 1979, p. 67.
33. *ASM Metals Handbook,* Vol. 1, 10th ed., ASM International, Materials Park, Ohio, 1990.

F. Galen Hodge
Haynes International, Inc.,

COBALT COMPOUNDS

Cobalt [7440-48-4] forms numerous compounds and complexes of industrial importance, and although total cobalt consumption has remained largely constant since the 1970s, nonmetallic cobalt usage has increased from 30% in 1970 to about 40% in 1990 (1). Cobalt, at wt 58.933, is one of the three members of the first transition series of Group 9 (VIIIB). The electronic configuration is $[Ar]3d^7 4s^2$. There are thirteen known isotopes, but only three are significant: ^{59}Co is the only stable and naturally occurring isotope; ^{60}Co has a half-life of 5.3 years and is a common source of γ-radioactivity; and ^{57}Co has a 270-d half-life and provides the γ-source for Mössbauer spectroscopy (see COBALT AND COBALT ALLOYS).

Cobalt exists in the $+2$ or $+3$ valence states for the majority of its compounds and complexes. A multitude of complexes of the cobalt(III) ion [22541-63-5] exist, but few stable simple salts are known (2). Werner's discovery and detailed studies of the cobalt(III) ammine complexes contributed greatly to modern coordination chemistry and understanding of ligand exchange (3). Octahedral stereochemistries are the most common for the cobalt(II) ion [22541-53-3] as well as for cobalt(III). Cobalt(II) forms numerous simple compounds and complexes, most of which are octahedral or tetrahedral in nature; cobalt(II) forms more tetrahedral complexes than other transition-metal ions. Because of the small stability difference between octahedral and tetrahedral complexes of cobalt(II), both can be found in equilibrium for a number of complexes. Typically, octahedral cobalt(II) salts and complexes are pink to brownish red; most of the tetrahedral Co(II) species are blue (see COORDINATION COMPOUNDS).

Cobalt metal is significantly less reactive than iron and exhibits limited reactivity with molecular oxygen in air at room temperature. Upon heating, the black, mixed valence cobalt oxide [1308-06-1], Co_3O_4, forms; at temperatures above 900°C the olive green simple cobalt(II) oxide [1307-96-6], CoO, is obtained. Cobalt metal reacts with carbon dioxide at temperatures greater than 700°C to give cobalt(II) oxide and carbon monoxide.

In the absence of complexing agents and in acidic solution the cobalt(II) hexaaquo ion [15276-47-8] oxidizes with difficulty.

$$Co(H_2O)_6^{3+} + e^- \rightarrow Co(H_2O)_6^{2+} \qquad E^0 = 1.86 \text{ V}$$

Indeed the cobalt(III) ion is sufficiently unstable in water to result in release of oxygen and formation of cobalt(II) ion. Under alkaline conditions the oxidation is much more facile and in the presence of complexing agents, eg, ammonia or cyanide, the oxidation may occur with ease or even spontaneously.

$$CoO(OH)(s) + H_2O + e^- \rightarrow Co(OH)_2(s) + OH^- \qquad E^0 = 0.17 \text{ V}$$

$$Co(NH_3)_6^{3+} + e^- \rightarrow Co(NH_3)_6^{2+} \qquad E^0 = 0.1 \text{ V}$$

$$Co(CN)_6^{3-} + H_2O + e^- \rightarrow [Co(CN)_5(H_2O)]^{3-} + CN^- \qquad E^0 = -0.80 \text{ V}$$

Preparation and Properties

Cobalt(II) Salts. Cobalt(II) acetate tetrahydrate [71-48-7], $Co(C_2H_3O_2)_2 \cdot 4H_2O$, occurs as pink, deliquescent, monoclinic crystals. It can be prepared by reaction of cobalt carbonate or hydroxide and solutions of acetic acid, by reflux of acetic acid solutions in the presence of cobalt(II) oxide, or by oxygenation of hot acetic acid solutions over cobalt metal. The tetrahydrate is soluble in water, alcohol, and acidic solutions. Dehydration of the crystals occurs at about 140°C. It is used as a bleaching agent (see BLEACHING AGENTS) and drier in inks (qv) and varnishes, and in pigments (qv), catalysis (qv), agriculture, and the anodizing industries (see DRYING AGENTS, PAINTS).

The mauve colored cobalt(II) carbonate [7542-09-8] of commerce is a basic material of indeterminate stoichiometry, $(CoCO_3)_x \cdot (Co(OH)_2)_y \cdot zH_2O$, that contains 45–47% cobalt. It is prepared by adding a hot solution of cobalt salts to a hot sodium carbonate or sodium bicarbonate solution. Precipitation from cold solutions gives a light blue unstable product. Dissolution of cobalt metal in ammonium carbonate solution followed by thermal decomposition of the solution gives a relatively dense carbonate. Basic cobalt carbonate is virtually insoluble in water, but dissolves in acids and ammonia solutions. It is used in the preparation of pigments and as a starting material in the preparation of cobalt compounds.

Cobalt(II) acetylacetonate [14024-48-7], cobalt(II) ethylhexanoate [136-52-7], cobalt(II) oleate [14666-94-5], cobalt(II) linoleate [14666-96-7], cobalt(II) formate [6424-20-0], and cobalt(II) resinate can be produced by metathesis reaction of cobalt salt solutions and the sodium salt of the organic acid, by oxidation of cobalt metal in the presence of the acid, and by neutralization of the acid using cobalt carbonate or cobalt hydroxide.

Cobalt(II) chloride hexahydrate [7791-13-1], $CoCl_2 \cdot 6H_2O$, is a deep red monoclinic crystalline material that deliquesces. It is prepared by reaction of hydrochloric acid with the metal, simple oxide, mixed valence oxides, carbonate, or hydroxide. A high purity cobalt chloride has also been prepared electrolytically (4). The chloride is very soluble in water and alcohols. The dehydration of the hexahydrate occurs stepwise:

$$CoCl_2 \cdot 6H_2O \xrightarrow[-4H_2O]{50°C} CoCl_2 \cdot 2H_2O \xrightarrow[-H_2O]{90°C} CoCl_2 \cdot H_2O \xrightarrow[-H_2O]{140°C} CoCl_2$$

The anhydrous chloride is blue, ie, tetrahedral cobalt, and commonly used as a humidity indicator in desiccants (qv).

Cobalt(II) hydroxide [1307-86-4], $Co(OH)_2$, is a pink, rhombic crystalline material containing about 61% cobalt. It is insoluble in water, but dissolves in acids and ammonium salt solutions. The material is prepared by mixing a cobalt salt solution and a sodium hydroxide solution. Because of the tendency of the cobalt(II) to oxidize, antioxidants (qv) are generally added. Dehydration occurs above 150°C. The hydroxide is a common starting material for the preparation of cobalt compounds. It is also used in paints and lithographic printing inks and as a catalyst (see PAINT).

Cobalt(II) nitrate hexahydrate [*10026-22-9*], $Co(NO_3)_2 \cdot 6H_2O$, is a dark reddish to reddish brown, monoclinic crystalline material containing about 20% cobalt. It has a high solubility in water and solutions containing 14 or 15% cobalt are commonly used in commerce. Cobalt nitrate can be prepared by dissolution of the simple oxide or carbonate in nitric acid, but more often it is produced by direct oxidation of the metal with nitric acid. Dissolution of cobalt(III) and mixed valence oxides in nitric acid occurs in the presence of formic acid (5). The trihydrate forms at 55°C from a melt of the hexahydrate. The nitrate is used in electronics as an additive in nickel–cadmium batteries (qv), in ceramics (qv), and in the production of vitamin B_{12} [*68-19-9*] (see VITAMINS, VITAMIN B_{12}).

Cobalt(II) oxalate [*814-89-1*], CoC_2O_4, is a pink to white crystalline material that absorbs moisture to form the dihydrate. It precipitates as the tetrahydrate on reaction of cobalt salt solutions and oxalic acid or alkaline oxalates. The material is insoluble in water, but dissolves in acid, ammonium salt solutions, and ammonia solution. It is used in the production of cobalt powders for metallurgy and catalysis, and is a stabilizer for hydrogen cyanide.

Cobalt(II) phosphate octahydrate [*10294-50-5*], $Co_3(PO_4)_2 \cdot 8H_2O$, is a red to purple amorphous powder. The product is obtained by reaction of an alkaline phosphate and solutions of cobalt salts. The material is insoluble in water or alkali, but dissolves in mineral acids. The phosphate is used in glazes, enamels, pigments (qv) and plastic resins, and in certain steel (qv) phosphating operations (see ENAMELS, PORCELAIN OR VITREOUS).

Cobalt(II) sulfamate [*14017-41-5*], $Co(NH_2SO_3)_2$, is generally produced and sold as a solution containing about 10% cobalt. The product is formed by reaction of sulfamic acid and cobalt(II) carbonate or cobalt(II) hydroxide, or by the aeration of sulfamic acid slurries over cobalt metal. Cobalt(II) sulfamate is used in the electroplating (qv) industry and in the manufacture of precision molds for record and compact discs (see INFORMATION STORAGE MATERIALS).

Cobalt(II) sulfate heptahydrate [*10026-24-1*], $CoSO_4 \cdot 7H_2O$, is a reddish pink monoclinic crystalline material that effloresces in dry air to form the hexahydrate. The dehydration–decomposition occurs according to the following:

$$CoSO_4 \cdot 7H_2O \xrightarrow[-H_2O]{41°C} CoSO_4 \cdot 6H_2O \xrightarrow[-5H_2O]{71°C} CoSO_4 \cdot H_2O \xrightarrow[-H_2O]{250°C} CoSO_4$$

$$3\,CoSO_4 \xrightarrow{710°C} Co_3O_4 + 3\,SO_2 + O_2$$

Cobalt(II) sulfate can be prepared by solution of cobalt(II) carbonate, cobalt(II) hydroxide, or cobalt(II) oxide in sulfuric acid. The digestion of the metal in sulfuric acid solution is assisted by air sparging. Also, cobalt(III) and mixed valence oxides in the presence of formic acid can be dissolved in sulfuric acid solution (5). High concentration (6) and high purity (7) cobalt sulfate solutions have been prepared electrolytically. Cobalt sulfate heptahydrate and cobalt(II) sulfate monohydrate [*10124-43-3*] are the most economical sources of cobalt ion and are used in feed supplements (see FEEDS AND FEED ADDITIVES) as well as in the electroplating industry, in storage batteries, in porcelain pigments, glazes, and as a drier for inks.

Cobalt Oxides. Cobalt(II) oxide [1307-96-6], CoO, is an olive green, cubic crystalline material. The product of commerce is usually dark gray and contains 75–78 wt % cobalt. The simple oxide is most often produced by oxidation of the metal at temperatures above 900°C. The product must be cooled in the absence of oxygen to prevent formation of Co_3O_4. Cobalt(II) oxide is insoluble in water, ammonia solutions, and organic solvents, but dissolves in strong mineral acids. It is used in glass (qv) decorating and coloring and is a precursor for the production of cobalt chemicals.

Cobalt(II) dicobalt(III) tetroxide [1308-06-1], Co_3O_4, is a black cubic crystalline material containing about 72% cobalt. It is prepared by oxidation of cobalt metal at temperatures below 900°C or by pyrolysis in air of cobalt salts, usually the nitrate or chloride. The mixed valence oxide is insoluble in water and organic solvents and only partially soluble in mineral acids. Complete solubility can be effected by dissolution in acids under reducing conditions. It is used in enamels, semiconductors, and grinding wheels. Both oxides adsorb molecular oxygen at room temperatures.

Cobalt Carbonyls. Dicobalt octacarbonyl [15226-74-1], $Co_2(CO)_8$, is an orange-red solid that decomposes in air. It is prepared by heating cobalt metal to 300°C under 20–30,000 kPa (3–4000 psi) of carbon monoxide, by reduction of cobalt(II) carbonate with hydrogen under pressure at high temperatures, or by heating a mixture of cobalt(II) acetate and cyclohexane to 160°C in the presence of carbon monoxide and hydrogen at 30,000 kPa (4000 psi) pressure. The octacarbonyl is reduced with sodium amalgam and acidified to yield tetracarbonylhydridocobalt [16842-03-8], $HCo(CO)_4$, a yellow liquid that is an active oxo catalyst (see CARBONYLS; OXO PROCESS).

Economic Aspects

Prices of cobalt compounds are directly related to the cost of cobalt metal which fluctuates widely. Zaire is the primary cobalt supplier. The price of cobalt metal was $46.30/kg in July 1992, down from $76.75/kg in December 1991, but up from the $27.56/kg July 1991 price. Annual usage of cobalt in the western world averaged 15,950 t in 1980–1984. The nonmetallic uses of cobalt were about 35% in 1984 and estimated to be 40% in 1990 (1,8,9).

The high cost of cobalt metal has led to substitution such that alloys containing low or no cobalt are produced where possible. Because the cost of cobalt compounds in specialty applications is usually much less significant than in alloys, there is less incentive to develop alternatives.

Analysis

Typical analyses of selected cobalt compounds are given in Table 1.

Separation. 1-Nitroso-2-naphthol [131-91-9], $C_{10}H_7NO_2$, can be used for the preliminary separation of cobalt from other metals by extraction into chloroform (10,11). Cobalt can be separated from cadmium, lead, and zinc by extraction using dithizone [60-10-6], $C_{13}H_{12}N_4S$, at pH 6–10 followed by hydrolysis in dilute

Table 1. Analysis of Cobalt Compounds[a]

Assay, wt %	Acetate tetrahydrate	Carbonate	Chloride hexahydrate	Hydroxide	Nitrate	Mixed oxide[b]	Sulfate hexahydrate	Sulfate feed grade
Co	23.5	46.0	24.8	61.0	20.3	71.0	21.0	21.0
Ni	0.04–0.10	0.1–0.3	0.07–0.1	0.1–0.2	0.03–0.10	0.15–0.35	0.05	0.10
Fe	0.005–0.015	0.1–0.6	0.005	0.01–0.10	0.001–0.002	0.05–0.20	0.001–0.002	0.01
Cu	0.0005		0.002	0.01	0.004	0.008–0.02	0.001–0.004	0.005
Mn	0.001–0.01	0.006		0.015	0.005	0.03–0.3	0.002	0.005
Pb		0.01	0.002	0.005	0.001	0.02	0.001	0.001
HCl insols[c]				0.02		0.01	0.05	0.10
H$_2$O insols[c]	0.007–0.015				0.01			
ABD[d], kg/m^3	940	660	1000	350	950	1500	1000	1000

[a]All materials are technical-grade cobalt(II) compounds unless indicated.
[b]Material is Co$_3$O$_4$.
[c]Insols = insoluable material.
[d]ABD = apparent bulk density.

HCl. The more stable cobalt complex remains in the organic extract. The formation of stable anionic halide complexes allow for the separation from nickel. Extraction of the complexes is effected by amines such as triisooctylamine, trioctylmethylammonium salts, or strongly basic anionic resins. Nickel and chromium do not form anionic chloride complexes and are not extracted under conditions of the experiment. Copper, zinc, and iron can be eluted from basic anionic resins using 0.01 M HCl whereas cobalt is eluted with 4 M HCl. Cobalt can also be separated from nickel as the sulfate salt by extraction with bis(2,4,4-trimethylpentyl)phosphinic acid. The blue thiocyanate complex of cobalt can be extracted into a mixture of ether and amyl alcohol to bring about separation from nickel and iron(III) in the presence of citrate ion (12). Manganese can be separated from cobalt by chlorination at a pH of 2.0. Cobalt sulfate is separated from a variety of metals using a chelating ion-exchange (qv) resin (13).

Cobalt(II) can be separated from cobalt(III) as the acetylacetonate (acac) compounds by extraction of the benzene soluble cobalt(III) salt (14). Magnesium hydroxide has been used to selectively adsorb cobalt(II) from an ammonia solution containing cobalt(II) and cobalt(III) (15).

Determination. Pure cobalt compounds can be assayed by EDTA titration at 40°C using hexamine [100-97-0], $C_6H_{12}N_4$, buffer (pH = 6) to a xylenol orange endpoint. Cobalt and nickel can be determined by cyanometry (16) or potentiometric titration using ferricyanide (17) can be used in the presence of large amounts of nickel, zinc, or copper. Colorimetry can be used to determine large amounts of cobalt by formation of the blue thiocyanate complex, $Co(SCN)_4^{2-}$. An excellent discussion of interferences and their elimination is available (18). Colorimetric methods using nitroso-naphthols (19,20) or nitroso-R salt [525-05-3], $C_{10}H_5NNa_2O_8S_2$ (21,22) are quite specific for cobalt, but not particularly sensitive. 2-(5-Bromo-2-pyridylazo)-5-diethylaminophenol (5-Br-PADAP) is a very sensitive reagent for certain metals and methods for cobalt have been developed (23). Nitroso-naphthol is an effective precipitant for cobalt(III) and is used in its gravimetric determination (24,25). Atomic absorption spectroscopy (26,27), x-ray fluorescence, polarography, and atomic emission spectroscopy are specific and sensitive methods for trace level cobalt analysis (see SPECTROSCOPY; TRACE AND RESIDUE ANALYSIS).

Health and Safety

Cobalt is one of twenty-seven known elements essential to humans (28) (see MINERAL NUTRIENTS). It is an integral part of the cyanocobalamin [68-19-9] molecule, ie, vitamin B_{12}, the only documented biochemically active cobalt component in humans (29,30) (see VITAMINS, VITAMIN B_{12}). Vitamin B_{12} is not synthesized by animals or higher plants, rather the primary source is bacterial flora in the digestive system of sheep and cattle (8). Except for humans, nonruminants do not appear to require cobalt. Humans have between 2 and 5 mg of vitamin B_{12}, and deficiency results in the development of pernicious anemia. The wasting disease in sheep and cattle is known as bush sickness in New Zealand, salt sickness in Florida, pine sickness in Scotland, and coast disease in Australia. These are essentially the same symptomatically, and are caused by cobalt deficiency. Symptoms include initial lack of appetite followed by scaliness of skin, lack of coordination, loss of flesh, pale mucous membranes, and retarded growth. The total laboratory synthesis of vitamin B_{12} was completed in 65–70 steps over a period of eleven years (31). The complex structure was reported by Dorothy Crowfoot-Hodgkin in 1961 (32) for which she was awarded a Nobel prize in 1964.

Cobalt compounds can be classified as relatively nontoxic (33). There have been few health problems associated with workplace exposure to cobalt. The primary workplace problems from cobalt exposure are fibrosis, also known as hard metal disease (34,35), asthma, and dermatitis (36). Finely powdered cobalt can cause silicosis. There is little evidence to suggest that cobalt is a carcinogen in animals and no epidemiological evidence of carcinogenesis in humans. The LD_{50} (rat) for cobalt powder is 1500 mg/kg. The oral LD_{50} (rat) for cobalt(II) acetate, chloride, nitrate, oxide, and sulfate are 194, 133, 198, 1700, 5000, and 279 mg/kg, respectively; the intraperitoneal LD_{50} (rat) for cobalt(III) oxide is 5000 mg/kg (37).

Several nonoccupational health problems have been traced to cobalt compounds. Cobalt compounds were used as foam stabilizers in many breweries throughout the world in the mid to late 1960s, and over 100 cases of cardiomyopathy, several followed by death, occurred in heavy beer drinkers (38,39). Those affected consumed as much as 6 L/d of beer (qv) and chronic alcoholism and poor diet may well have contributed to this disease. Some patients treated with cobalt(II) chloride for anemia have developed goiters and polycythemia (40). The impact of cobalt on the thyroid gland and blood has been observed (41).

The exposure level established by NIOSH for the workplace is 0.1 mg/m^3 (42). ACGIH has recommended a TLV of 0.05 mg/m^3 for cobalt. At the ACGIH worker exposure levels it has been suggested that occupational health problems would not occur (41).

Uses

The uses of cobalt compounds are summarized in Table 2.

Cobalt in Catalysis. Over 40% of the cobalt in nonmetallic applications is used in catalysis. About 80% of those catalysts are employed in three areas: (1) hydrotreating/desulfurization in combination with molybdenum for the oil and gas industry (see SULFUR REMOVAL AND RECOVERY); (2) homogeneous catalysts used in the production of terphthalic acid or dimethylterphthalate (see PHTHALIC ACID AND OTHER BENZENE POLYCARBOXYLIC ACIDS); and (3) the high pressure oxo process for the production of aldehydes (qv) and alcohols (see ALCOHOLS, HIGHER ALIPHATIC; ALCOHOLS, POLYHYDRIC). There are also several smaller scale uses of cobalt as oxidation and polymerization catalysts (44–46).

Cobalt's ability to actively catalyze reactions comes from one or more of the following properties (43): (1) the facile redox properties of the element, the ability to form stable species in multiple oxidation states, and the ease of electron transfer between the oxidation states via unstable intermediates or free radicals; (2) the ability to form complexes with suitable donor groups and the relative stabilities of those complexes; (3) the ability of cobalt to form complexes of varying coordination number and to exist in equilibrium in more than one stereochemical form; (4) the ability to undergo equilibrium decomposition reactions such as

Table 2. Uses of Cobalt Compounds[a]

Compound	CAS Registry Number	Molecular formula	Molecular weight	Uses
cobalt(III) acetate	[917-69-1]	$Co(C_2H_3O_2)_3$	235.9	catalyst, powder
cobalt(II) acetate tetrahydrate	[71-48-7]	$Co(C_2H_3O_2)_2 \cdot 4H_2O$	248.9	drier for lacquers and varnishes, sympathetic inks, catalyst, pigment for oil cloth, mineral supplement, anodizing agent
cobalt(II) acetylacetonate	[14024-48-7]	$Co(C_5H_7O_2)_3$	355.9	vaporplating of cobalt, catalyst synthesis enamels
cobalt(II) aminobenzoate		$Co(C_7H_6NO_2)_2$	194.9	tire cord adhesion
cobalt(II) ammonium sulfate	[13586-38-4]	$CoSO_4 \cdot (NH_4)_2SO_4 \cdot 6H_2O$	394.9	catalyst, plating solutions
cobalt(II) bromide	[7789-43-7]	$CoBr_2$	218.7	catalyst, hydrometers
cobalt(II) carbonate	[513-79-1]	$CoCO_3 \cdot Co(OH)_2$	118.9	pigment, ceramics, feed supplement, catalyst, fodder fat stabilizer
cobalt(II) carbonate (basic)	[7542-09-8]	$2CoCO_3 \cdot Co(OH)_2 \cdot H_2O$	348.7	chemicals
dicobalt octacarbonyl	[15226-74-1]	$Co_2(CO)_8$	341.8	catalyst, powder
cobalt(II) chloride	[7791-13-1]	$CoCl_2 \cdot 6H_2O$	237.9	chemicals, sympathetic inks, hydrometers, plating baths, metal refining, pigment, catalyst, dyestuffs, magnetic recording materials, moisture indicators
cobalt(II) citrate	[18727-04-3]	$Co_3(C_6H_5O_7)_2 \cdot 2H_2O$	590.7	therapeutic agents, vitamin preparations, plating baths
cobalt(II) fluoride	[10026-17-2]	CoF_2	96.7	fluorinating agent
cobalt(II) fluoride tetrahydrate	[13817-37-3]	$CoF_2 \cdot 4H_2O$	168.7	catalyst
cobalt(III) fluoride	[10026-18-3]	CoF_3	115.6	fluorinating agent
cobalt(II) fluorosilicate hexahydrate	[15415-49-3]	$CoSiF_6 \cdot 6H_2O$	308.3	ceramics
cobalt(II) formate	[6424-20-0]	$Co(CHO_2)_2 \cdot 2H_2O$	184.9	catalyst
cobalt(II) hydroxide	[1307-86-4]	$Co(OH)_2$	92.9	paints, chemicals, catalysts, printing inks, battery materials, powder
cobalt(II) iodide	[15238-00-3]	CoI_2	312.7	moisture indicator
cobalt(II) linoleate	[14666-96-7]	$Co(C_{18}H_{31}O_2)_2$	616.9	paint and varnish drier
cobalt(II) naphthenate		$Co(C_{11}H_{10}O_2)_2$	406.9	catalyst, paint and varnish drier, tire cord adhesive, antistatic adhesive
cobalt(II) nitrate hexahydrate	[10026-22-9]	$Co(NO_3)_2 \cdot 6H_2O$	290.9	pigments, chemicals, ceramics, feed supplements, catalysts, battery materials
cobalt(II) 2-ethylhexanoate	[136-52-7]	$Co(C_{18}H_{15}O_2)_2$	344.9	paint and varnish drier, adhesion additive
cobalt(II) oleate	[14666-94-5]	$Co(C_{18}H_{33}O_2)_2$	620.9	paint and varnish drier, antistatic adhesive

(Continued)

Table 2. *(Continued)*

Compound	CAS Registry Number	Molecular formula	Molecular weight	Uses
cobalt(II) oxalate	[814-89-1]	CoC_2O_4	146.9	catalysts, cobalt powders
cobalt phthalocyanine		$Co(pc)$	572.9	coating, conducting polymer
cobalt(II) potassium nitrite	[17120-39-7]	$K_3Co(NO_2)_6 \cdot 1.5H_2O$	478.9	pigment
cobalt(II) resinate		$Co(C_{44}H_{62}O_4)_2$	1366.9	paint and varnish drier, catalyst
cobalt Schiff-base complexes				oxygen-sensing and oxygen-indicating agents
cobalt(II) stearate	[13586-84-0]	$Co(C_{18}H_{35}O_2)_2$	624.9	paint and varnish drier, tire cord adhesive
cobalt(II) succinate trihydrate	[3267-76-3]	$Co(C_4H_4O_4) \cdot 3H_2O$	228.9	therapeutic agents, vitamin preparations
cobalt(II) sulfamate		$Co(NH_2SO_3) \cdot 3H_2O$	208.9	plating baths
cobalt(II) sulfate	[10026-24-1]	$CoSO_4 \cdot xH_2O$	154.9	chemicals, ceramics, pigments, plating baths, dyestuffs, magnetic recording materials, anodizing agents, corrosion protection agent
cobalt(II) sulfide	[1317-42-6]	CoS	90.0	catalysts
cobalt(II) thiocyanate	[3017-60-5]	$Co(CNS)_2 \cdot 3H_2O$	174.9	humidity indicator, drug testing
cobalt(II) aluminate	[13820-62-7]	$CoAl_2O_4$	176.7	pigment, catalysts, grain refining
cobalt(II) arsenate	[24719-19-5]	$Co_3(AsO_4)_2 \cdot 8H_2O$	598.5	pigment for paint, glass, and porcelain
cobalt(II) chromate	[24613-38-5]	$CoCrO_4$	174.8	pigment
cobalt(II) ferrate	[12052-28-7]	$CoFe_2O_4$	234.5	catalyst, pigment
cobalt(II) manganate	[12139-69-4]	$CoMn_2O_4$	232.7	catalyst, electrocatalyst
cobalt(II) oxide	[1307-96-6]	CoO	74.9	chemicals, catalysts, pigments, ceramic gas sensors, thermistors
cobalt oxide	[1308-06-1]	Co_3O_4	240.7	enamels, semiconductors, solar collectors pigments, magnetic recording materials
cobalt dilanthanum tetroxide	[39449-41-7]	La_2CoO_4	400.7	anode, catalyst
tricobalt tetralanthanum decaoxide	[60241-06-7]	$La_4Co_3O_{10}$	892.3	catalyst
lithium cobalt dioxide	[12190-79-3]	$LiCoO_2$	97.8	battery electrode
sodium cobalt dioxide	[37216-69-6]	$NaCoO_2$	113.8	battery electrode
dicobalt manganese tetroxide	[12139-92-3]	$MnCo_2O_4$	236.7	catalyst
dicobalt nickel tetroxide	[12017-35-5]	$NiCo_2O_4$	240.5	anode, catalyst
lanthanum cobalt trioxide	[12016-86-3]	$LaCoO_3$	245.8	oxygen electrode
cobalt(II) phosphate	[10294-50-5]	$Co_3(PO_4)_2 \cdot 8H_2O$	510.9	glazes, enamels, pigments, steel pretreatment
cobalt(II) tungstate	[12640-47-0]	$CoWO_4$	305.7	paint and varnish drier

$Co(RCOO)_2 \rightleftarrows Co^{2+} + 2\ RCOO^-$ in polymerization systems. The decomposition products may take part in the catalysis; (5) cobalt complexes are able to undergo ligand exchange reactions which can be important in assisting catalysis. This is an example of vitamin B_{12}'s activity as a catalyst; and (6) cobalt oxides and sulfides, because of lattice vacancies, bond energies, or electronic effects, are active heterogeneous catalysts.

The breadth of reactions catalyzed by cobalt compounds is large. Some types of reactions are hydrotreating petroleum (qv), hydrogenation, dehydrogenation, hydrodenitrification, hydrodesulfurization, selective oxidations, ammonoxidations, complete oxidations, hydroformylations, polymerizations, selective decompositions, ammonia (qv) synthesis, and fluorocarbon synthesis (see FLUORINE COMPOUNDS, ORGANIC).

Hydrotreating Catalysts. A preliminary step in refining of crude oil feedstocks is the removal of metals, significant reduction of the organic sulfur and nitrogen constituents, and reduction of the molecular weight of the high molecular-weight fraction. This step is called hydrotreating and it typically employs a catalyst of cobalt and molybdenum on a high surface area alumina or silica support. Hydrotreating of a nominal 4000 cubic meters (25,000 barrel) per day operation produces 200 tons per day of sulfur and deposits ca 450 kg of metal, primarily vanadium and nickel, on the catalyst (47) which must be sulfur tolerant.

The primary reactions occurring during hydrotreating are (43) desulfurization of sulfides, polysulfides, mercaptans, and thiophene as exemplified by

$$R—S—R' + 2\ H_2 \rightarrow RH + R'H + H_2S$$

denitrification of pyridine and pyrrole resulting in the corresponding alkane and ammonia; deoxygenation of phenol; hydrogenation of alkenes and aromatics, eg, of butadiene and benzene; and hydrocracking of alkanes such as *n*-decane

$$C_{10}H_{22} + H_2 \rightarrow 2C_5H_{12}$$

Hydrodesulfurization, by far the most common hydrotreating reaction, usually occurs in the presence of a cobalt–molybdenum catalyst support on alumina. Hydrodesulfurization constitutes the largest single use of cobalt in catalysis (qv). These catalysts are not easily poisoned, are regenerable numerous times, and usable for several years (see CATALYSTS, REGENERATION). The catalysts are prepared by precipitation of cobalt and molybdate solutions in the presence of an alumina support, by mixing the dried components in powder form, by precipitation of the three components, or by impregnation of the support with solutions of cobalt salts and molybdic acid or ammonium molybdate. The material is calcined at about 600°C to produce oxides of the cobalt–molybdenum–alumina followed by activation by reduction–sulfidation to yield the cobalt–molybdenum sulfide catalyst. The activity of the catalyst depends on several factors: type of support, ratio of cobalt to molybdenum, content of cobalt and molybdenum, impregnation procedure, calcination conditions, and reduction–sulfidation (48,49). The cobalt content is commonly between 2.5 and 3.7 wt %.

Homogeneous Oxidation Catalysts. Cobalt(II) carboxylates, such as the oleate, acetate, and naphthenate, are used in the liquid-phase oxidations of *p*-xylene to terephthalic acid, cyclohexane to adipic acid, acetaldehyde (qv) to acetic acid, and cumene (qv) to cumene hydroperoxide. These reactions each involve a free-radical mechanism that for the cyclohexane oxidation can be written as

initiation $C_6H_{12} + R\cdot \rightarrow C_6H_{11} + RH$

chain $C_6H_{11} + O_2 \rightarrow C_6H_{11}OO\cdot$

                          $C_6H_{11}OO\cdot + C_6H_{12} \rightarrow C_6H_{11}OOH + C_6H_{11}$

decomposition $C_6H_{11}OOH \rightarrow C_6H_{10}O + H_2O$

                          $Co^{2+} + C_6H_{11}OOH \rightarrow Co^{3+} + C_6H_{11}O\cdot + OH^- \rightarrow C_6H_{12}$

                          $Co^{3+} + C_6H_{11}OOH \rightarrow Co^{2+} + C_6H_{11}OO\cdot + H^+ \rightarrow chain$

Cobalt-catalyzed oxidations form the largest group of homogeneous liquid-phase oxidations in the chemical industry.

Oxo or Hydroformylation and Hydroesterification. Reactions of alkenes with hydrogen and formyl groups are catalyzed by $HCo(CO)_4$

$$HCo(CO)_4 + RCH=CH_2 \rightarrow RCH_2CH_2Co(CO)_4$$

$$RCH_2CH_2Co(CO)_4 + CO \rightarrow RCH_2CH_2COCo(CO)_4$$

$$RCH_2CH_2COCo(CO)_4 + H_2 \rightarrow RCH_2CH_2COH + HCo(CO)_4$$

and hydroesterification occurs according to

$$R-CH=CH_2 + R'OH + CO \rightarrow RCH_2CH_2COOR'$$

The oxo reactions occur under 10,000–30,000 kPa (1450–4000 psi) of synthesis gas at temperatures of 75–200°C. Oxo syntheses account for the third largest use of cobalt in catalysis.

Cobalt compounds are used as catalysts in several processes of lesser industrial importance. Fischer-Tropsch catalysts produced from cobalt salts produce less methane than the corresponding iron catalysts. The chemistry of carboxylation reactions catalyzed in the presence of cobalt is also rich (50). Cobalt cyanide can be used to hydrogenate a broad range of organic and inorganic materials. Exhaust gas purification including oxidation of carbon monoxide and nitrogen monoxide is effected in the presence of Co_3O_4 (see EXHAUST CONTROL, INDUSTRIAL). The oxide is also effective in the production of nitric acid (qv) from ammonia. Cobalt carboxylates are used in the adhesive industry as cross-linking and polymerization catalysts (see ADHESIVES).

Cobalt in Driers for Paints, Inks, and Varnishes. The cobalt soaps, eg, the oleate, naphthenate, resinate, linoleate, ethylhexanoate, synthetic tertiary neodecanoate, and tall oils, are used to accelerate the natural drying process of unsaturated oils such as linseed oil and soybean oil. These oils are esters of unsaturated fatty acids and contain acids such as oleic, linoleic, and eleostearic (see VEGETABLE OILS). On exposure to air for several days a film of the acids convert from liquid to solid form by oxidative polymerization. The incorporation of oil-soluble cobalt salts effects this drying process in hours instead of days. Soaps of manganese, lead, cerium, and vanadium are also used as driers, but none are as effective as cobalt (see DRYING AGENTS, PAINTS-DRIERS AND METALLIC SOAPS).

A film of drying oils undergoes a series of reactions, but the primary reaction is autooxidation polymerization where oxygen attacks a carbon adjacent to a double bond forming a hydroperoxide. Once peroxide formation begins, dissociation into free radicals occurs giving polymers having carbon–carbon, carbon–oxygen–carbon, alcohol, ketone, aldehyde, and carboxylic acid linkages (43). The reactions depend on drying conditions and properties of the dried film may change.

The concentration of drier also effects the rate and physical characteristics of the dried film. In dryer combinations cobalt is considered a surface oxidation catalyst, whereas driers such as lead are considered a through drier or polymerization catalyst. Lead, cerium, manganese, and iron soaps are often used in combination with cobalt to slow down the rate of drying. Zinc and calcium soaps are not active as driers, but can also be used as cobalt diluents to slow down the drying rate. The mixture of metal soaps determines the rate of drying as well as the properties of the final dried film. Coordination driers such as aluminum and zirconium soaps aid the cross-linking polymerization process and are always used along with an oxidative polymerization catalyst.

For water-based alkyd paints, greater (0.2% cobalt on a resin basis) concentrations of drier are required than for other systems because the reaction of the drier with water decreases the activity of the catalyst. The cobalt content of oil-based paint formulations is usually 0.01–0.05% cobalt. Although the concentration of cobalt in the formulations is small, the large volume of paints, inks, and varnishes constitute a significant use for cobalt chemicals.

Cobalt as a Colorant in Ceramics, Glasses, and Paints. Cobalt(II) ion displays a variety of colors in solid form or solution ranging from pinks and reds to blues or greens. It has been used for hundreds of years to impart color to glasses and ceramics (qv) or as a pigment in paints and inks (see COLORANTS FOR CERAMICS). The pink or red colors are generally associated with cobalt(II) ion in an octahedral environment and the chromophore is typically $Co-O_6$. The tetrahedral cobalt ion, $Co-O_4$ chromophore, is sometimes green, but usually blue in color.

Cobalt pigments are usually produced by mixing salts or oxides and calcining at temperatures of 1100–1300°C. The calcined product is then milled to a fine powder. In ceramics, the final color of the pigment may be quite different after the clay is fired. The materials used for the production of ceramic pigments are

Color	Ingredients
purple	Co_3O_4–Al_2O_3–MgO
blues	Co_3O_4–SiO_2–CaO
Thenard blue	Co_3O_4–Al_2O_3
blue-greens	Co_3O_4–Cr_2O_3–Al_2O_3
blue-black	Co_3O_4–Fe_2O_3–Cr_2O_3
black	Co_3O_4–Cr_2O_3–Fe_2O_3–MnO_2–Ni_2O_3

Paint pigments do not change colors on application. Other common colors are violet from cobalt(II) phosphate [18475-47-3], pink from cobalt and magnesium oxides, aureolin yellow from potassium cobalt(III) nitrite [13782-01-9], $KCo(NO_2)_4$, and cerulean blue from cobalt stannate [6546-12-5]. Large quantities of cobalt are used at levels of a few ppm to decolorize or whiten glass and ceramics. Iron oxide or titanium dioxide often impart a yellow tint to various domestic ware. The cobalt blue tends to neutralize the effect of the yellow.

Cobalt is used in ceramic pigments and designated as underglaze stains, glaze stains, body stains, overglaze colors, and ceramic colors. The underglaze is applied to the surface of the article prior to glazing. The glaze stain uses cobalt colorants in the glaze. A body stain is mixed throughout the body of the ceramic. Overglaze colors are applied to the surface and fired at low temperatures. Ceramic colors are pigments used in a fusible glass or enamel and are one of the more common sources of the blue coloration in ceramics, china, and enamel ware.

Cobalt oxides are used in the glass (qv) industry to color or decolorize. Usually the cobalt is tetrahedrally coordinated and the color that it imparts to the glass is blue. Some special low temperature glasses contain pink, octahedrally, coordinated cobalt. These pink glasses turn blue on heating. The blue color from cobalt is very stable and its intensity easily controlled by doping rates. A discernable tint can be noticed for glass containing only a few parts per million of cobalt ion. The more typical blue cobalt glass usually contains closer to 200 ppm cobalt.

Cobalt is used as a blue phosphor in cathode ray tubes for television, in the coloration of polymers and leather goods, and as a pigment for oil and watercolor paints. Organic cobalt compounds that are used as colorants usually contain the azo (51) or formazon (52) chromophores.

Miscellaneous Uses.   *Adhesives in the Tire Industry.*   Cobalt salts are used to improve the adhesion of rubber to steel. The steel cord must be coated with a layer of brass. During the vulcanization of the rubber, sulfur species react with the copper and zinc in the brass and the process of copper sulfide formation helps to bond the steel to the rubber. This adhesion may be further improved by the incorporation of cobalt soaps into the rubber prior to vulcanization (53,54) (see TIRE CORDS).

Adhesion of Enamel to Steel.   Cobalt compounds are used both to color and to enhance adhesion of enamels to steel (55). Cobalt oxide is often incorporated into the ground frit at rates of 0.5–0.6 wt %, although levels from 0.2 to 3 wt % have been used. The frit is fired for ten minutes at 850°C to give a blue enamel that is later coated with a white cover coat.

Agriculture and Nutrition.   Cobalt salts, soluble in water or stomach acid, are added to soils and animal feeds to correct cobalt deficiencies. In soil application the cobalt is readily assimilated into the plants and subsequently made available to the animals (56). Plants do not seem to be affected by the cobalt uptake from the soil. Cobalt salts are also added to salt blocks or pellets (see FEEDS AND FEED ADDITIVES).

Electroplating.   Cobalt is plated from chloride, sulfate, fluoborate, sulfamate, and mixed anionic baths (57). Cobalt alloyed with nickel, tungsten, iron, molybdenum, chromium, zinc, and precious metals are plated from mixed metal baths (58,59). A cobalt phosphorus alloy is commonly plated from electroless baths. Cobalt tungsten and cobalt molybdenum alloys are produced for their excellent high temperature hardness. Magnetic recording materials are produced by electroplating cobalt from sulfamate baths (60) and phosphorus-containing baths or by electroless plating of cobalt from baths containing sodium hypophosphite as the reducing agent. Cobalt is added to nickel electroplating baths to enhance hardness and brightness or for the production of record and compact discs (61).

Miscellaneous. Small quantities of cobalt compounds are used in the production of electronic devices such as thermistors, varistors, piezoelectrics (qv), and solar collectors. Cobalt salts are useful indicators for humidity. The blue anhydrous form becomes pink (hydrated) on exposure to high humidity. Cobalt pyridine thiocyanate is a useful temperature indicating salt. A conductive paste for painting on ceramics and glass is composed of cobalt oxide (62).

Cobalt salts are used as activators for catalysts, fuel cells (qv), and batteries. Thermal decomposition of cobalt oxalate is used in the production of cobalt powder. Cobalt compounds have been used as selective absorbers for oxygen, in electrostatographic toners, as fluoridating agents, and in molecular sieves. Cobalt ethylhexanoate and cobalt naphthenate are used as accelerators with methyl ethyl ketone peroxide for the room temperature cure of polyester resins.

BIBLIOGRAPHY

"Cobalt Compounds" in *ECT* 1st ed., Vol. 4, pp. 199–214, by S. B. Elliott, Ferro Chemical Corp. and C. Mueller, General Aniline & Film Corp.; in *ECT* 2nd ed., Vol. 5, pp. 737–748, by F. R. Morral, Battelle Memorial Institute; in *ECT* 3rd ed., Vol. 6, pp. 495–510, by F. R. Morral.

1. M. Jones, *Proceedings of the 2nd International Conference, Cobalt Metallurgy and Uses, Venice, 1985,* Cobalt Development Institute, London, 1986.
2. W. Levason and C. A. McAuliffe, *Coord. Chem. Rev.* **12,** 151 (1974).
3. A. Werner, *Ber.* **46,** 3674 (1913); **47,** 1964, 1978 (1914); *Z. Anorg. Chem.* **3,** 267 (1893); *Alfred Werner and Cobalt Complexes, Werner Centennial ACS Monograph Series, Advances in Chemistry Series,* Vol. 62, Washington, D.C., 1966.
4. Jpn. Kokai Tokkyo Koho, JP 55 158280 (1980), (Sumitomo Metal Mining).
5. Czech. CS 140,580 (1971), V. Stastny and J. Sedlacek.
6. Jpn. Kokai Tokkyo Koho, JP 59 83785 (1984), (Nippon Soda).
7. Jpn. Kokai Tokkyo Koho, JP 57 57876 (1982), (Sumitomo Metal Mining).
8. J. D. Donaldson, S. J. Clark, and S. M. Grimes, *Cobalt in Medicine, Agriculture and the Environment,* monograph series, Cobalt Development Institute, London, 1986.
9. I. DuBois, *Cobalt News* **7,** (1978).
10. Z. Marczenko, *Separation and Spectrophotometric Determination of Elements,* Ellis Horwood Ltd., West Sussex, UK, 1986.
11. E. B. Sandell, *Colorimetric Determination of Traces of Metals*, Interscience Publishers, New York, 1959.
12. N. S. Bayliss and R. W. Pickering. *Ind. Eng. Chem., Anal. Ed.* **18,** 446 (1946).
13. T. Jeffers and M. Harvey, Report No. 8927, U.S. Bureau of Mines, Washington, D.C., 1985.
14. J. E. Hicks, *Anal. Chem. Acta* **45,** 101 (1969).
15. I. V. Melikhov, M. Belousova, and V. Peshkova, *Zh. Analit. Khim.* **25,** 1144 (1970).
16. B. S. Evans, *Analyst* **62,** 363 (1937).
17. B. Bagshawe and J. Hobson, *Analyst* **73,** 152 (1948).
18. G. Charlot and D. Bezier, *Quantitative Inorganic Analysis,* Methuen and Co., Ltd., London, 1957, p. 410.
19. E. Boyland, *Analyst* **71,** 230 (1946).
20. J. Yoe and C. Barton, *Ind. Eng. Chem., Anal. Ed.* **12,** 405 (1940).

21. H. Willard and S. Kaufmann, *Anal. Chem.* **19,** 505 (1947).
22. T. Ovenston and C. Parker, *Anal. Chem. Acta* **4,** 142 (1950).
23. J. Zbiral and L. Sommer, *Z. Anal. Chem.* **306,** 129 (1981).
24. C. Mayr and W. Prodinger, *Z. Anal. Chem.* **117,** 334 (1939).
25. H. Brintzinger and R. Hesse, *Z. Anal. Chem.* **122,** 241 (1941).
26. A. Varma, *CRC Handbook of Atomic Absorption Analysis,* Vol. 1, CRC Press, Boca Raton, Fla., 1984, pp. 435–449.
27. W. Slavin, *Graphite Furance AAS, A Source Book*, Perkin-Elmer, Ridgefield, Conn., 1984.
28. W. Mertz, *Nutrition Today* **18,** 26 (1983).
29. D. Dolphin, ed., B_{12}, Vols. 1 and 2, John Wiley & Sons, Inc., New York, 1982.
30. R. S. Young, *Cobalt in Biology and Biochemistry,* Academic Press, Inc., London, 1979.
31. J. N. Krieger, *Chem. Eng. News* (Mar. 12, 1973).
32. P. Senhard and D. Hodgkin, *Nature* **192,** 937 (1961); D. Crowfoot-Hodgkin, *Proc. Royal Soc. (London)* **A288,** 294 (1965).
33. E. Matromatteo, *Am. Ind. Hyg. Assoc. J.,* 29 (1986).
34. H. Jobs and C. Ballhausen, *Vertrauensartz Krankenkasse* **8,** 142 (1940).
35. A. P. Wehner, R. H. Busch, R. J. Olson, and D. K. Craig, *Am. Ind. Hyg. Assoc. J.* **38,** 338 (1977).
36. D. Munro-Ashman and A. J. Miller, *Contact Dermatitis* **2,** 65 (1976).
37. G. Speijers, E. Krajnc, J. Berkvens, and M. Van Logten, *Food Chem. Toxicol.* **20,** 311 (1982); J. Llobet and J. Domingo, *Rev. Esp. Fisiol.* **39,** (1983); W. Frederick and W. Bradley, *Ind. Med.* **14,** 482 (1946).
38. Y. L. Morin, A. R. Foley, G. Martineau, and J. Roussel, *Can. Med. Assoc.* **97,** 881 (1967).
39. Y. L. Morin and P. Daniel, *Can. Med. Assoc. J.* **97,** 926 (1967).
40. R. T. Gross, J. P. Kriss, and T. H. Spaet, *Pediatrics* **15,** 284 (1955).
41. A. G. Cecutti, in G. Tyroler and C. Landolt, eds., *Extractive Metallurgy of Nickel and Cobalt,* The Metallurgical Society, Warrendale, Pa., 1988.
42. *NIOSH Occupational Hazard Assessment,* DHHS Publication No. 82-107, U.S. Dept. of Labor, Washington, D.C., 1981.
43. J. D. Donaldson, S. J. Clark, and S. M. Grimes, *Cobalt in Chemicals, The Monograph Series,* Cobalt Development Institute, London, 1986.
44. B. Delmon, *Proceedings of the International Conference on Cobalt Metallurgy and Uses,* ATB Metallurgy, Brussels, Belgium, 1981.
45. E. de Bie and P. Doyen, *Cobalt* (1962).
46. P. Granger, in Ref. 1.
47. B. G. Silbernagel, R. R. Mohan, and G. H. Singhal, in T. Whyte, R. Dalla Betta, E. Derouane, and R. Baker, eds., *Catalytic Materials: Relationship Between Structure and Reactivity*, ACS Symposium Series No. 248, ACS, Washington, D.C., 1984, pp. 91–99.
48. C. Wivel, B. Clausen, R. Candia, S. Moerup, and H. Topsoe, *J. Catal.* **87,** 497 (1984).
49. L. Petrov. C. Vladdy, D. Shopov, V. Friebova, and L. Beranek, *Coll. Czech. Chem. Commun.* **48,** 691 (1983).
50. H. Alper, *Adv. Organomet. Chem.* **19,** 190 (1981).
51. R. Price, in K. Ventkataraman, ed., *Chemistry of Synthetic Dyes*, Academic Press, Inc., New York, 1970, p. 303.
52. A. Nineham, *Chem. Rev.* **55,** 355 (1955).
53. W. Van Ooij, *Rubber Chem. Technol.* **52,** 605 (1979).
54. G. Haemers, *Rubber World* **182,** 26 (1980).
55. K. Bates, *Enameling, Principles & Practice,* Funk & Wagnalls, Ramsey, N.J., 1974.
56. E. Underwood, *Trace Elements in Humans and Animal Nutrition,* Academic Press, Inc., New York, 1971.

57. W. H. Safranek, *The Properties of Electrodeposited Metals and Alloys*, 2nd ed., Elsevier, New York, 1986.
58. W. Betteridge, *Cobalt and Its Alloys*, Ellis Horwood, Ltd., Chichester, UK, 1982.
59. R. Brugger, *Nickel Plating: A Comprehensive Review of Theory, Practice, Properties, and Applications Including Cobalt Plating*, Draper, Teddington, UK, 1970.
60. T. Chen and P. Caralloti, *Appl. Phys. Lett.* **41,** 206 (1982).
61. Jpn. Pat. 62083486 (1987), S. Takei (to Seiko Epson Corp.).
62. U.S. Pat. 4,317,749 (1982), J. Provance and K. Allison (to Ferro Corp.).

H. WAYNE RICHARDSON
CP Chemicals, Inc.

COCAINE. See ALKALOIDS; ANESTHETICS; PSYCHOPHARMACOLOGICAL AGENTS.

COCHINEAL. See COLORANTS FOR FOODS, DRUGS, AND COSMETICS; DYES, NATURAL.

COCOA. See CHOCOLATE AND COCOA.

COCONUT OIL. See FATS AND FATTY OILS; VEGETABLE OILS.

COFFEE

Coffee was originally consumed as a food in ancient Abyssinia and was presumably first cultivated by the Arabians in about 575 AD (1). By the sixteenth century it had become a popular drink in Egypt, Syria, and Turkey. The name coffee is derived from the Turkish pronunciation, kahveh, of the Arabian word *gahweh*, signifying an infusion of the bean. Coffee was introduced as a beverage in Europe early in the seventeenth century and its use spread quickly. In 1725, the first coffee plant in the western hemisphere was planted on Martinique, West Indies. Its cultivation expanded rapidly and its consumption soon gained wide acceptance.

Commercial coffees are grown in tropical and subtropical climates at altitudes up to ca 1800 meters; the best grades are grown at high elevations. Most individual coffees from different producing areas possess characteristic flavors. Commercial roasters obtain preferred flavors by blending or mixing the varieties before or after roasting. Colombian and washed Central American coffees are generally characterized as mild, winey-acid, and aromatic; Brazilian coffees as heavy body, moderately acid, and aromatic; and African robusta coffees as heavy body, neutral, slightly acid, and slightly aromatic. Premium coffee blends contain higher percentages of Colombian and Central American coffees.

Green Coffee Processing

The coffee plant is a relatively small tree or shrub belonging to the family Rubiaceae. It is often controlled to a height of 3 to 5 meters. *Coffea arabica* (milds) accounts for 69% of world production; *Coffea canephora* (robustas), 30%; and *Coffea liberica* and others, 1%. Each of these species includes several varieties. After the spring rains the plant produces white flowers. About six months later the flowers are replaced by fruit approximately the size of a small cherry, hence they are called cherry. The fruit on a tree can include underripe, ripe (red, yellow, and purple color), and overripe cherries. It can be selectively picked (ripe only) or strip picked (predominantly ripe plus some underripe and overripe).

Green coffee processing is effected by either the dry or wet method. The dry method produces so-called natural coffees. The wet method usually produces the more uniform and higher quality washed coffees.

The dry method is used in most of Brazil and in other countries where water is scarce in the harvesting season. The cherries from strip picking are spread on open drying ground and turned frequently to permit thorough drying by the sun and wind. Sun-drying usually takes two to three weeks depending on weather conditions. Some producing areas use hot air, indirect steam, and other machine-drying devices. When the coffee cherries are thoroughly dry, they are transferred to hulling machines which remove the skin, pulp, parchment shell, and silver skin in a single operation.

In the wet method, as practiced in Colombia, freshly picked ripe coffee cherries are fed into a tank for initial washing. Stones and other foreign material are removed. The cherries are then transferred to depulping machines which remove the outer skin and most of the pulp. However, some pulp mucilage clings to the parchment shells that encase the coffee beans. Fermentation tanks, usually containing water, remove the last portions of the pulp. Fermentation may last from twelve hours to several days. Because prolonged fermentation may cause development of undesirable flavors and odors in the beans, some operators use enzymes to accelerate the process.

The beans are subsequently dried either in the sun or in mechanical dryers. Machine-drying continues to gain popularity in spite of higher costs because it is faster and independent of weather conditions. When the coffee is thoroughly dried, the parchment is broken by rollers and removed by winnowing. Further rubbing and winnowing removes the silver skin to produce ordinary green unroasted coffee, containing about 10–12% moisture.

Coffee prepared by either the wet or dry method is machine-graded by sieves, oscillating tables, and airveyors into large, medium, and small beans. Damaged beans and foreign matter are removed by handpicking, machine separators, electronic sorters, or a combination of these techniques. Commercial coffee is graded according to the number of imperfections present, such as black beans, damaged beans, stones, pieces of hull, or other foreign matter. Processors also grade coffee by color, roasting characteristics, and cup quality of the beverage. After all processing, coffee will maintain acceptable quality for approximately one year.

Coffee Chemistry

Chemical Composition of Green Coffee. The chemical composition of green coffee can affect roasted flavor quality. The chemistry can vary according to species, variety, growing environment, post-harvest handling including wet or dry processing, and storage time, temperature, and humidity. The composition data in Table 1 are given as a range compiled from literature (2–4).

Robustas generally have lower amounts of lipid, trigonelline [535-83-1], and sucrose [57-50-1], and higher levels of caffeine [58-08-2] and chlorogenic acid (3-caffeoylquinic acid) [327-97-9] when compared to arabicas. While the protein level is considered similar between robustas and arabicas, protein structure and function, including enzymatic activities, is under genetic control and can vary. Some literature indicates slightly higher levels of free amino acids (nonprotein) in green robustas compared to arabicas, eg, 0.8 vs 0.5%, respectively (5). Wet processed beans tend to have slightly lower mineral content compared to dry processed beans.

Table 1. Typical Analyses of Green Coffee, %[a]

	Type[b]	
Constituent[c]	Robusta	Arabica
moisture	11(10–13)	12.5(10–13)
lipids	10(7–11)	15(14–17)
ash	4.2(3.9–4.5)	4.0(3.5–4.5)
caffeine	2.0(1.5–2.6)	1.3(1.1–1.4)
chlorogenic acid	9(7–10)	7(5–8)
carboxylic acids	2(1–3)	2.5(1.5–3.5)
trigonelline	0.7(0.3–0.9)	1.1(0.9–1.2)
protein	11(9–13)	11(9–13)
free amino acids	0.8	0.5
sucrose	4(3–6)	8(5–9)
reducing sugars	0.5(0.4–0.6)	0.1(0.1–0.2)
others[d]	8.8(5–10)	5.5(3–8)
polymeric carbohydrate		
mannan	22	22
arabinogalactan	17(16–18)	15(14–16)
cellulose	8(7–9)	7(7–8)

[a]Refs. 2–4.
[b]Typical value and (range).
[c]Dry basis (db) (ex moisture; as is basis).
[d]Others by difference.

Numerous organic acids in coffee include acids of metabolic origin, eg, acetic, lactic, citric, malic, and oxalic; free quinic acid [77-95-2]; and various chlorogenic acid (CGA) isomers that appear to be species specific.

quinic acid

Of these, the CGA isomers are the principal acids and result from the esterification of the 3, 4, and 5 position hydroxyls of quinic acid with the carboxyls of several phenolic acids, including caffeic acid (C) [331-39-5], ferulic acid (F) [1135-24-6], and p-coumaric acid (Cm) [501-98-4].

caffeic	R = OH	
ferulic	R = OCH$_3$	
p-coumaric	R = H	

The principal CGA isomers identified in green coffee include three caffeoylquinic acid isomers, 3-CQA [327-97-9], 4-CQA [905-99-7], and 5-CQA [906-33-2]; three dicaffeoylquinic acid isomers, 3,4-diCQA [14534-61-3], 3,5-diCQA [2450-53-5], and 4,5-diCQA [57378-72-0]; and three feruloylquinic acid isomers, 3-FQA [1899-29-2], 4-FQA, and 5-FQA [40242-06-6]. The total CGA level is somewhat higher in robustas compared to arabicas. The 5-CQA is the predominant isomer both in arabicas, ie, 4–5% dry basis (db), and in robustas, 5–6% db, and is known to form in vitro and possibly in vivo complexes with caffeine [58-08-2]. Greater compositional differences between robustas and arabicas are found in the minor CGA isomers, eg, 3,4-diCQA, 5-FQA, CmQA, etc, which appear correlated to genetic origin (6). The greater level of certain CGA isomers in robustas, compared to arabicas, may explain its greater production of spicy, medicinal, phenolic flavors upon roasting. Analysis has demonstrated greater levels of compounds of phenolic origin in roasted robustas, eg, guiacol derivatives (7). There are indications that the undesirable brown and black pigments in green coffee are formed by polyphenol oxidase [9002-10-2] polymerization of CQA isomers.

The cell wall complex of both robustas and arabicas contain a mannan [9052-06-6], an arabinogalactan [9036-66-2], and cellulose [9004-34-6]. The mannan is a low molecular-weight linear polymer of about 20–40 mannose sugars with β-1,4-linkages. It functions as a reserve polysaccharide providing energy to the sprouting coffee seed. The arabinogalactan is a high molecular-weight and highly branched polymer consisting of a β-1,3-linked galactose [59-23-4] polymer backbone with frequent side chains containing arabinose [147-81-9] and mannose [3458-28-4]. The cellulose appears similar to other plant celluloses (4).

The surface of the green coffee contains a cuticular wax layer (0.2–0.3% db) for both varieties. The wax contains insoluble hydroxytryptamides derived from 5-hydroxytryptamine [61-47-2] and saturated C18–C22 fatty acids.

$$HO-\underset{\underset{H}{\overset{|}{N}}}{\overbrace{}}-CH_2CH_2NH-\overset{\overset{O}{\parallel}}{C}(CH_2)_nCH_3$$

About 75% of the interior oil consists of triglycerides containing about 35–41% palmitic (C16, n-14), 37–46% linoleic (C18:2), 8–12% oleic (C18:1), and 7–11% stearic (C18, n-16) [57-11-4] as the principal fatty acids on an oil basis. The remaining 20% of the lipid is distinctive and consists of fatty acid esters of the diterpene alcohols, cafestol [469-83-0] and kahweol [6894-43-5] (8). Free cafestol (0.3–0.7% db coffee) and kahweol (0.1–0.2% db coffee) are found in the oil of arabicas. Smaller amounts of cafestol (0.2–0.4% db coffee), trace levels of kahweol, and a new diterpene, 16-O-methylcafestol [108214-28-4] (0.06–0.1% db coffee), are found in the oil fraction of robustas; the latter is proposed as an indicator for the presence of robusta in a blend (9).

The characteristic note of green beans is attributed to methoxypyrazines which have very low threshold values (10). The characteristic musty earthy note of robustas was recently identified as 2-methylisoborneol [2371-42-8] (11).

Green bean storage for extended periods, or under abusive conditions of elevated temperature and humidity, causes fading of color, and changes in roast flavor. The loss of reducing sugars and free amino acids upon storage at 60°C have been noted and suggest the Maillard reaction. Enzymatic or oxidative mechanisms in the development of off flavors are also possible (12).

Effects of Roasting on Components.   Green coffee has no desirable taste or aroma; these are developed upon roasting. Many complex physical and chemical changes occur during roasting including the obvious change in color from green to brown. In the first stage of roasting, loss of free water (typically 11% of the bean) occurs. In the second stage, chemical dehydration, fragmentation, recombination, and polymerization reactions occur. Many of these are associated with the Maillard reaction and lead to the formation of lower molecular-weight compounds such as carbon dioxide, free water, and those associated with flavor and aroma; and higher molecular-weight colored materials, both soluble and insoluble in water. Some of the reactions are exothermic and produce a rapid rise in temperature, usually accompanied by a sudden expansion or puffing of the beans, with a volume increase of 50–100% depending on variety and roasting parameters. The loss of carbon dioxide and other volatile substances, as well as the water loss produced by chemical dehydration during the latter stage, accounts for most of the 2–5% dry weight roasting loss. Most of the lipid, caffeine, inorganic salts, and polymeric carbohydrate survive the roasting process (13).

Table 2 indicates some of the chemical changes that occur in green coffee as a result of roasting. The data are presented as a range and typical value (when appropriate) compiled from the literature and other sources and reflect the fact

Table 2. Approximate Analyses of Roasted, Brewed, and Instant Coffee, %[a]

Constituent[b]	Roasted	Brewed[c]	Instant
moisture	4 (1–5)		3.5 (2–5)
oil	17 (16–20)	0.8 (0.2–1.0)	(0.1–0.6)
ash	4.5 (4–5)	14	9 (7–11)
caffeine	1.2 (1.0–1.6)	4.8	3.5 (2–5)
chlorogenic acid	1.5 (1–3.5)	14.8	(3–9)
carboxylic acids	3 (2–4)	3.0	5.5 (4–8)
trigonelline	0.5–1.0	1.6	(0.5–2)
protein	8–10	6	(1–6)
reducing sugar	0.2	0.4	(1–5)
sucrose	0.2 (0–0.5)	0.8	(0.6–0)
aroma compounds	~0.1	~0.1–2	~0.05
browning products and others	(22–11)	29.4	(20–35)
polymeric carbohydrate	38	24	(30–50)
mannan	20		
arabinogalactan	12		
cellulose	6		

[a]Typical value and (range).
[b]Dry basis (ex moisture; as is basis).
[c]Ref. 14.

that the chemistry is highly dependent on degree of roast and starting material. The principal water-soluble constituents, about 25% of the green coffee, are involved. These include some of the protein, the free amino acids, trigonelline, reducing sugars, sucrose, chlorogenic acids, organic acids, and inorganic salts. Most sucrose disappears early in the roast. The reducing sugars that form react rapidly with the free amino acids via the Maillard reaction to form aromatic compounds and color.

Roasting denatures and insolubilizes much of the protein. Some of the constituent amino acids in the protein are destroyed during the latter stages of roasting and thus contribute to aroma and flavor development. Analysis of the protein amino acids in green and corresponding roasted coffee show marked decreases in arginine, cysteine, lysine, serine, and threonine in both arabica and robusta types after roasting. Alanine, glycine, leucine, glutamic acid, and phenylalanine increase upon roasting and are relatively stable (15). Cysteine [52-90-4] and methionine [63-68-3] are the probable sources of the many potent sulfur compounds found in coffee aroma, eg, mercaptans, organic sulfides, thiazoles, etc. Other amino acids are capable of generating aromatic compounds such as pyrazines, pyrroles, etc.

About 50–80% of the trigonelline is decomposed during roasting. Trigonelline is a probable source for niacin [59-67-6] but also a source of some of the aromatic nitrogen compounds such as pyridines, pyrroles, and bicyclic compounds found in coffee aroma (16). Certain acids, such as acetic, formic, propionic, quinic, and glycolic, are formed or increase upon roasting, while other acids present initially in the green coffee, such as chlorogenic acid isomers, citric, and malic, disappear with increasing degree of roast. The composite of acids in brew of a lightly roasted coffee contribute to the taste quality termed "fine acidity." Brews of darkly roasted coffees are less acid. Slight cleavage of the triglycerides and the diterpene and sterol esters occurs during roasting. It is likely that some oxidation of lipid components is initiated during the roasting stage. However, some of the Maillard products generated upon roasting may act as antioxidants (qv), slowing down the deterioration of the lipids upon storage. The aromatic, oil-soluble aroma compounds slowly partition into the oil phase after the roasted bean is allowed to cool and equilibrate.

Aroma. The chemistry for the formation of aroma compounds during roasting is complex (17,18). A significant portion of these compounds is derived from lipids, organic acids, eg, chlorogenic acids, carbohydrates, eg, sugars, and proteinaceous material, eg, amino acids, either by degradation to reactive products, which include saturated and unsaturated aldehydes, ketones, dicarbonyls, amines, and hydrogen sulfide, or by interaction via the Maillard nonenzymatic browning reaction. The latter stages of the Maillard reaction result in the formation of structurally more complicated heterocyclic compounds, many containing sulfur, oxygen, and nitrogen. The number of volatile flavor compounds identified in coffee has increased to about 800 (19) and may ultimately exceed 1000 as a result of a combination of advanced analytical techniques including gas chromatography (gc), mass spectrometry (ms), fourier transform infrared spectrometry (ftir), and nuclear magnetic resonance spectrometry (nmr). Although present in minute quantities as low as ppb or ppt, volatile flavor compounds are extremely important to the aroma of freshly roasted and ground coffee and to the flavor balance in a cup of coffee. Table 3 summarizes the volatile components by chemical class of roasted coffee (19,20).

A relatively new methodology called aroma dilution analysis (ada), which combines aroma dilution and gas chromatography-olfactometry to gain a better understanding of the relative importance of aroma compounds, was recently done for coffee. In a roasted Colombian coffee brew, 41 impact compounds were found with flavor dilution threshold factors (FD) greater than 25, and 26 compounds had FD factors of 100 or above. While the technique permits assessment of the impact of individual compounds, it does not evaluate synergistic effects among compounds (13).

Chemistry of Brewed Coffee. The chemistry of brewed coffee is dependent on the extraction of water-soluble and hydrophobic aromatic components from the coffee cells and lipid phase, respectively. Factors that affect extraction and flavor quality of brewed coffee are degree of roast; blend composition; grinding technique; particle size and density; water quality; water to coffee ratio; and brewing technique or device, such as drip filter, percolator, or espresso, which defines the water temperature, steam pressure, brewing time, water recycle, etc. Extraction yields upon home brewing range from about 9 to 28% and typically about 23% dry basis roasted and ground (R&G) (21). The trend in the United States with some notable exceptions has been toward weaker brew strengths, with typical recipes of about 5 g of R&G coffee per 6 oz cup (brew solids concentration about 0.7%), compared to about 10 g a generation ago (brew solids about 1.2%). As a comparison, espresso typically uses about 8 to 12 g of coffee per 2 oz of beverage (brew solids ranging from about 3 to 5%). Espresso being brewed rapidly under steam pressure (brew time ranging from about 15 to 35 s) contains a relatively large

Table 3. Aromatic Components of Roasted Coffee[a]

Chemical class	Number of compounds
hydrocarbons	74
alcohols	20
aldehydes	30
ketones	73
acids	25
esters	31
lactones	3
phenols (and ethers)	48
furans	127
thiophenes	26
pyrroles	71
oxazoles	35
thiazoles	27
pyridines	19
pyrazines	86
amines and misc. nitrogen compounds	32
sulfur compounds	47
miscellaneous	17
Total	*791*

[a]Ref. 19,20.

amount of oil droplets (about 0.1–0.2% basis brew) and suspended colloidal solids (about 0.3% basis brew) which contribute to the greater turbidity and mouthfeel of this beverage compared to filter brewed coffee (22).

Instant Coffee. The chemistry of instant or soluble coffee is dependent on the R&G blend and processing conditions. This is indicated in Table 2 by the wide range of constituents. In addition to the atmospherically extractable solids found in brewed coffee, commercial percolation generates water-soluble carbohydrate by hydrolysis which contributes to the yield. This additional carbohydrate includes the sugars, arabinose, mannose, and galactose; oligosaccharides derived from mannan and arbinogalactan; and the partially hydrolyzed polysaccharides, mannan and arabinogalactan. It improves the drying properties and retention of volatiles by the extract, and reduces hygroscopicity. These water-soluble carbohydrates formed the basis for the first 100% pure instant coffee developed by General Foods Corp. in the late 1940s.

Roasted and Ground Coffee Processing and Packaging

The main processing steps in the manufacture of roast and ground coffee products are blending, roasting, grinding, and packaging. Green coffee is shipped in burlap bags (60–70 kg) or in bulk containers (16,500–18,000 kg). Prior to processing, the green coffee is cleaned to remove string, lint, dust, husk, and other foreign matter. Coffee of different varieties or from different sources may be blended before or after roasting at the option of the manufacturer.

Roasting Technology.   Roasting is usually by hot combustion gases in rotating cylinders or fluidized-bed systems; infrared and microwave roasted coffees have recently come to market in Japan. Roasting times in batch cylinders were traditionally 8–15 minutes, whereas much of the coffee produced today is roasted in continuous fluidized beds in only 0.5–4 minutes (23–25). In either case, the initial step of roasting is a moisture elimination and uniform heating step. When the bean temperature exceeds about 165°C, the reactions have switched from endothermic to exothermic and the roast has begun to develop. This stage is generally accompanied by a noticeable crackling sound like that of corn popping and the beans swell to as much as twice their unroasted volume. The final bean temperature of 185–250°C is determined by the flavor development desired for the finished goods, whether a blend or individual varieties are roasted. A water or air quench terminates the roasting reaction. Most, but not all, of any water added is evaporated from the heat of the beans. Theoretically, about 700 kJ is needed to roast one kg of coffee beans. A roaster efficiency of 75% or more (933 kJ/kg) is possible with recirculation of the roaster gas, whereas older, nonrecirculating units operate with an efficiency as low as 25% (2800 kJ/kg). The acceptability of the roast is judged by either a photometric reflectance measurement on the roasted bean or a ground sample of the bean, and adjustments of the temperature controls which initiate the quenching end point. The faster fluidized-bed roasting processes (batch or continuous) are the basis of high yield coffee products. These units are generally operated at lower air temperature, ie, 185–400°C vs 425–490°C, resulting in a more uniform roast throughout the bean, an increase in extractable soluble solids of 20% or more, and higher aroma retention. The higher circulation rate of roaster gas required for fluidization increases heat transfer to allow a faster roast at lower temperature. Exhausted roaster gas often must be incinerated for environmental pollution purposes. The use of infrared heat (gas-fired ceramics) or microwave energy to speed up the roasting process while providing a more even roast has also been patented (26,27).

The roasted and quenched beans are air cooled and conveyed to storage bins for moisture and temperature equilibration before grinding. Residual foreign matter (mostly stones), which may have passed through the initial green cleaning step, is removed in transit to the storage bins by means of a high velocity air lift which leaves the heavier debris behind. The roasted beans may flow by gravity or be airveyed to the grinders.

Grinding.   Grinding of the roasted coffee beans is tailored to the intended method of beverage preparation. Average particle size distributions range from very fine (500 μm or less) to very coarse (1100 μm). A finer grind will allow greater solids extraction, but may slow the brewing process because of increased flow resistance and reduced wettability. Most coffee is ground in multistep steel roll mills in order to produce the most desirable particle size distribution. After passing through cracking rolls, the broken beans are fed between two more rolls, one of which is cut or scored longitudinally; the other, circumferentially. The paired rolls operate at speeds designed to cut rather than crush the cracked particles. For finer grinds, a second pair of more finely scored rolls running at higher speed is positioned below the first set. Some coffee is flaked, to increase extractability without slowing brewing speed, by passing through closely spaced smooth rolls after grinding (28). This is a crushing step which disrupts the cellular structure.

A normalizer mixing section after grinding provides a uniform distribution and may be used to adjust density before packaging.

Packaging.   Most roasted and ground coffee sold directly to consumers in the United States is vacuum-packed in 0.37, 0.74, or 1.1 kg metal cans. After roasting and grinding, the coffee is conveyed, usually by gravity, to weighing-and-filling machines that achieve the proper fill by tapping or vibrating. A loosely set cover is partially crimped. The can then passes into the vacuum chamber maintained at 3.3 kPa (25 mm Hg) absolute pressure or less. The cover is clinched to the can cylinder wall and the can moves through an exit valve or chamber. This process removes 95% or more of the oxygen from the can. Polyethylene snap caps for reclosure are placed on the cans before they are stacked in cardboard cartons for shipping. A case usually contains 8.8 kg of coffee, and a production packing line usually operates at a rate of 250–350 0.37 kg cans per minute.

Vacuum-packed coffee retains a high quality rating for at least one year. The slight loss in fresh roasted character that occurs is because of chemical reactions with the residual oxygen in the can and previous exposure to oxygen prior to packing (29).

Coffee vacuum-packed in flexible, bag-in-box packages has gained wide acceptance in Europe. The inner liner, usually a plastic-laminated foil, is formed into a hard brick shape during the vacuum process (30). In the United States, a printed multilaminated flexible structure is used to form the brick pack which is sold as is at retail. These types of packages provide a barrier to moisture and oxygen similar to that of a metal can.

Inert gas flush packing in plastic-laminated pouches, although less effective than vacuum packing, can remove or displace 80–90% of the oxygen in the package. These packages offer satisfactory shelf life and are sold primarily to institutions.

Some coffee in the United States, and an appreciable amount in Europe, is distributed as whole beans, which are ground in stores or by consumers in their homes. Whole-bean roasted coffee remains fresh longer than ground coffee. The specialty gourmet shop trade based on this system has grown significantly in the United States since the early 1980s. If the coffee is freshly roasted an excellent product is provided; if the roasted beans are allowed to sit in an unprotected bin or bag for more than a few days, oxidation reactions will cause the product to degrade.

Optimal packaging of roasted whole beans in foil laminate bags requires the use of a one-way valve which allows carbon dioxide gas released from the beans to escape and prevent air from entering the package. This permits packing the coffee soon after roasting. A specialty mail-order trade for roasted whole beans packaged in this form, as well as roasted and ground in vacuum-packed foil laminate pouches, has also grown since the early 1980s.

Instant Coffee Processing and Packaging

Instant coffee is the dried water-extract of ground, roasted coffee. Although used in Army rations as early as the U.S. Civil War, the popularity of instant coffee as a grocery product grew only after World War II, coincident with improvements in manufacturing methods and consumer trends toward convenience. Extensive patent literature dates back to 1865. Instant coffee products represented 15% of the coffee consumed in the United States in 1991 (31).

Green beans for instant coffee are blended, roasted, and ground similarly to those for roasted and ground products. A concentrated coffee extract is normally produced by pumping hot water through the coffee in a series of cylindrical percolator columns. The extracts are further concentrated prior to a spray- or freeze-drying step, and the final powder is packaged in glass or other suitable material. Some soluble coffees, both spray- and freeze-dried, are manufactured in producing countries for export.

Blend/Roast/Grind.   Blends of Brazilian, Central American, and Colombian milds as well as African, Asian, and Brazilian robustas are prepared to achieve desired flavor characteristics. The batch- or continuous-type roasters used for roasted and ground coffee also are used for instant coffee. Grinding of roasted beans for an instant coffee process is adjusted to suit the type of commercial percolation system to be used. The average particle size is generally larger than that used for domestic brewing to avoid excessive pressure drops across the percolator columns. Similarly, very fine particles are avoided.

Extraction.   Commercial extraction equipment and conditions have been designed to obtain the maximum yield of soluble solids with the desired flavor character. Conceptually, most commercial systems can be represented by a series of countercurrent batch extractors. The freshwater feed, at pressures well above one atmosphere and temperatures high enough to hydrolyze the coffees' polysaccharides to oligosaccharides, contacts the most spent coffee grounds. During the final extraction stage, fresh ground coffee is contacted with an extract of these oligosaccharides at temperatures near the atmospheric boiling point.

Significant factors influencing extraction efficiency and product quality are grind size, feed water temperature and temperature profile through the system, percolation time, ratio of coffee to water, premoistening or wetting of the ground coffee, design of extraction equipment, and flow rate of extract through the percolation columns.

Percolation trains consisting of 5–10 columns are the norm. Height to diameter ratios usually range from 4:1 to 7:1. To improve extraction, the ground coffee may be steamed or wetted. Feed water temperatures ranging from 154 to 182°C are common and the final extract exits at 60–82°C. To minimize flavor and aroma loss prior to drying, the effluent extract may be cooled in a plate heat exchanger. The yield, a function of the properties of the particular blend and roast, the operating temperatures, and the percolation time, is generally controlled through adjustment of the soluble solids drawn off from the final stage. Extraction yield is calculated from both the weight of extract collected and the soluble solids concentration as measured by specific gravity or refractive index. Soluble yields of 24–48% or higher on a roasted coffee basis are possible. Robusta coffees give yields about 10% higher than arabica because of a higher level of available polysaccharides and caffeine. The latest technology in thermal extraction of spent grounds provides roasted yields in excess of 60% (32).

Extract is stored in insulated tanks prior to drying. Because high soluble solids concentration is desirable to reduce aroma loss and evaporative load in the driers, most processors concentrate the 15–30% extract to 35–55% prior to drying

(33). This may be accomplished by vacuum evaporation or freeze concentration. Clarification of the extract, normally by centrifugation, may be used to assure the absence of insoluble fine particles.

The flavor of instant coffee can be enhanced by recovering and returning to the extract or finished dry product some of the natural aroma lost in processing. The aroma constituents from the grinders, percolation vents, and evaporators may be added directly or in concentrated or fractionated form to achieve the desirable product attributes.

Drying. The criteria for good instant coffee drying processes include minimization of loss or degradation of flavor and aroma, uniformity of size and shape in a free-flowing form, acceptability of the bulk density for packaging, product color acceptability, and moisture content below the level required to maintain shelf stability (less than 5%). Operating costs, product losses, and capital investment are also considerations in the selection of a drying process.

Spray Drying and Agglomeration. Most instant coffee products are spray-dried. Stainless steel towers with a concurrent flow of hot air and atomized extract droplets are utilized for this purpose. Atomization, through pressure nozzles, is controlled based on selection of the nozzles, properties of the extract, pressures used, bulk density, and capacity requirements. Low inlet air temperatures (200–280°C) are preferred for best flavor quality. The spray towers must be provided with adequate dust collection systems such as cyclones or bag filters. The dried particles are collected from the conical bottom of the spray drier through a rotary valve and conveyed to bulk storage bins or packaging lines. Processors may screen the dry product to assure a uniform particle size distribution.

Most spray-dried instant coffees have been marketed in a granular form, rather than the small spherical spray-dried form, since the mid-1960s. The granular appearance is achieved by steam fusing the spray-dried material in towers similar to the spray drier. Belt agglomerators are also common.

Freeze Drying. Commercial freeze drying of instant coffee has been a common practice in the United States since the mid-1960s. The freeze-drying process provides the opportunity to minimize flavor degradation due to heat (34).

Sublimation of ice crystals to water vapor under a very high vacuum, about 67 Pa (0.5 mm Hg) or lower, removes the majority of the moisture from the granulated frozen extract particles. Heat input is controlled to assure a maximum product end point temperature below 49°C. Freeze drying takes significantly longer than spray drying and requires a greater capital investment.

Packaging. In the United States, instant coffee for the consumer market is usually packaged in glass jars containing from 56 to 340 g of coffee. Larger units for institutional, hotel, restaurant, and vending machine use are packaged in bags and pouches of plastic or laminated foil. In Europe, instant coffee is packaged in glass jars or foil-laminated packages.

Protective packaging is primarily required to prevent moisture pickup. The flavor quality of regular instant coffee changes very little during storage. However, the powder is hygroscopic and moisture pickup can cause caking and flavor impairment. Moisture content should be kept below 5%.

Many instant coffee producers in the United States incorporate natural coffee aroma in coffee oil into the powder. These highly volatile and chemically unstable flavor components necessitate inert-gas packing to prevent aroma deterioration and staling from exposure to oxygen.

Decaffeinated Coffee Processing

Decaffeinated coffee products represented 18% of the coffee consumed in 1991 in the United States (31). Decaffeinated coffee was first developed commercially in Europe about 1900. The process as described in a 1908 patent (35) consists of first, moisturizing green coffee to at least 20% to facilitate transport of caffeine through the cell wall, and then contacting the moistened beans with solvents.

Until the 1980s, synthetic organic solvents commonly were used in the United States to extract the caffeine, either by direct contact as above or by an indirect secondary water-based system (36). In each case, steaming or stripping was used to remove residual solvent from the beans and the beans were dried to their original moisture content (10–12%) prior to roasting.

In the 1980s, manufacturers' commercialized processes which utilized either naturally occurring solvents or solvents derived from natural substances to position their products as naturally decaffeinated. The three most common systems use carbon dioxide under supercritical conditions (37), oil extracted from roasted coffee (38), or ethyl acetate, an edible ester naturally present in coffee (39). Specificity for caffeine and caffeine solubility is key to selection and system design. Because caffeine can be selectively removed from water extracts of green beans by activated charcoal, several processes which utilize water or recycled green coffee extract have been described and are also considered natural decaffeination. If water is used, the green coffee extract produced is externally decaffeinated and the noncaffeine solids containing important flavor precursors are reabsorbed before drying and roasting. The use of recycled green extract obviates the need for a separate reabsorption step as the caffeine deficient green extract selectively leaches caffeine. Pre-absorbing sugar on activated charcoal to improve its specificity also has been commercialized (40). The degree of decaffeination, based on comparison to the starting material, is controlled using known time-temperature relationships for the particular process for each bean type.

In all the above mentioned processes of coffee decaffeination, changes occur that affect the roast flavor development. These changes are caused by the pre-wetting step, the effects of extended (four hours plus) exposure at elevated temperature as required to economically extract the caffeine from whole green beans, and the post-decaffeination drying step.

To make an instant decaffeinated coffee product, the decaffeinated roast and ground coffee is extracted in a manner similar to nondecaffeinated coffee. Alternatively, the caffeine from the extract of untreated roasted coffee is removed by using the solvents described previously.

Economic Importance of Coffee

Coffee has been a significant factor in international trade since the early 1800s. It is among the leading agricultural products in international trade along with wheat, corn, and soybeans. The total world production of green coffee in the 1989–1990 growing season was 97.1 million bags; exportable production was 73.4 million bags with an export value of $6.7 billion (Table 4) (41).

In 1990, the United States import from producing countries totaled 21 million bags of green coffee equivalent (Table 5). This includes 19.6 million bags of green coffee, 0.2 million bags of roasted coffee, and 1.2 million bags of soluble coffee with a total value of $1.9 billion (42). More than 79% of this import came from countries in the western hemisphere.

Historically, any factors that affect the balance of supply and demand, eg, political, climatic, etc, have contributed to the high volatility of green coffee prices. The International Coffee Organization (ICO), consisting of seventy-two producer and consumer nations, developed the International Coffee Agreement (ICA) in 1962 to achieve stable prices through export quotas adjusted by indicator price change. The breakdown of the ICA in 1989 led to the historically low prices of the early 1990s. In September 1990, an extension of the 1983 International Coffee Agreement was approved until the end of September 1992 (43) to provide more time for the seventy-two national council of the ICO to renegotiate a new accord.

Table 4. World Production of Green Coffee, 1989–1990[a]

Country	Exportable production[b]	%
Brazil	15,500	21.1
Colombia	11,538	15.7
Indonesia	5,575	7.6
Ivory Coast	4,700	6.4
Mexico	3,350	4.6
Guatemala	3,162	4.3
Uganda	3,045	4.1
El Salvador	2,607	3.6
Costa Rica	2,198	3.0
Ecuador	1,849	2.5
Zaire	1,780	2.4
Honduras	1,722	2.3
Kenya	1,665	2.3
Cameroon	1,393	1.9
Peru	1,219	1.7
Ethiopia	1,200	1.6
Papua New Guinea	1,081	1.5
India	1,000	1.4
Vietnam	950	1.3
Madagascar	870	1.2
Tanzania	844	1.1
others	6,137	8.4
Total	*73,385*	*100.0*

[a]Ref. 41.
[b]Thousands of 60-kg bags.

Table 5. World Imports of Green Coffee in 1990[a]

Country	Imports[b]	%
United States	21,009	27.5
Germany	13,012	17.0
France	6,301	8.2
Japan	5,506	7.2
Italy	5,242	6.9
The Netherlands	3,199	4.2
Spain	3,017	3.9
United Kingdom	2,898	3.8
Canada	2,253	3.0
Belgium/Luxembourg	2,149	2.8
Eastern Europe	1,604	2.1
other Western Europe	9,006	11.8
other Asia and Oceania	1,216	1.6
Total	*76,412*	*100.0*

[a]Ref. 42.
[b]Thousands of 60-kg bags.

Coffee Regulations and Standards

Various standards and regulations for green coffee are set either by legislation in the producing and consuming countries, eg, the United States, France, or the European Economic Community, or by voluntary standards set by various trade organizations and associations in the consuming countries, eg, National Coffee Association or London Terminal Market. These standards and regulations define what is being purchased by contract and protect consumers against fraud and potential health risks. They include defining designations as species, geographical origin, processing, and crop year; and quality grades/types including level and type of defects, bean size, extraneous matter, agricultural residues, flavor, color, bulk density, and percentage of maximum moisture. Legislation in the consuming countries also covers processed coffee for fraud or potential health risks. International organizations such as the International Coffee Organization (ICO) based in London, the International Organization for Standardization (ISO), the Codex Alimentarius Commission of the United Nations (FAO/WHO) based in Rome, and the National Coffee Association (NCA) based in New York help achieve standardization in international trade, support research, and provide information about methods of analysis, guide to storage and transport, publications, meetings, etc (44,45).

Decaffeination Regulations. For decaffeinated roasted coffee, EEC standards indicate the maximum content of caffeine as 0.1% db; for decaffeinated instant coffee it is 0.3% db. In the United States, decaffeination usually signifies that 97% of the caffeine has been removed. Permissible solvents for decaffeination processes are defined by national legislation, eg, FDA or EEC directive. The maximum residual solvent content after decaffeination, roasting, or instant coffee processing is to be kept within good manufacturing practice, ie, very low ppm levels or below at point of sale (46).

Caffeine and Health.   Caffeine is listed in the U.S. Code of Federal Regulations as a multipurpose, generally recognized as safe (GRAS) substance and in *Food Chemicals Codex*, published by the U.S. National Academy of Sciences/ National Research Council, as a flavoring agent and stimulant. Consumption of caffeine in coffee, tea, and soft drinks has been studied for health concerns with regard to every organ system. In the early 1990s, numerous publications continue to appear and much active research continues to be done, but a consensus has developed among knowledgeable scientists that consumption of caffeine in moderate amounts is not associated with any serious health consequences (47–53).

Caffeine is considered by pharmacologists to be a mild stimulant of the central nervous system. It has been shown to promote feelings of well being and increased ability to perform certain mental tasks efficiently. There are people who are oversensitive to the effects of caffeine; overindulgence by these individuals, eg, intake of more than 600 mg caffeine/d, can bring unwanted effects such as anxiety, restlessness, sleeping difficulties, headache, or palpitations of the heart (54).

Coffee Substitutes

Coffee substitutes, which include roasted chicory, chick peas, cereal, fruit, and vegetable products, have been used in all coffee consuming countries. Although consumers in some locations prefer the noncoffee beverages, they are generally used as lower cost beverage sources. Additionally, it is not unusual for consumers in some of the coffee producing countries to blend coffee with noncoffee materials.

Chicory is harvested as fleshy roots which are dried, cut to uniform size, and roasted. Chicory contains no caffeine, and on roasting develops an aroma compatible with that of coffee. It gives a high yield, about 70%, of water-soluble solids with boiling water and can also be extracted and dried in an instant form. Chicory extract has a darker color than does normal coffee brew (55).

New Technology

Coffee Biotechnology.   A number of advanced biotechnology techniques are being investigated to improve and/or develop coffee varieties as well as to better utilize roasted coffee (56).

In South and Central America, breeding selection programs are in progress to improve coffee disease resistance, productivity, and cup quality. Micropropagation methods based on multiplying cells in a bioreactor have also been reported (57) and will allow for the propagation of identically cloned plants, thus eliminating nonuniformity in coffee plant material. Coffee plant regeneration studies from protoplasts, ie, coffee cells without cell walls, have been reported (58) and may serve as a key part of the methodology involved in transforming coffee plants via recombinant DNA techniques. The first successful results in transforming coffee cells with marker genes have been described (59). The potential applications of recombinant DNA to coffee plants are immense.

Coffee bioconversions through enzymatic hydrolysis have been used to modify green coffee and improve the finished product (60). Similarly, enzymes have been reported which increase yield and improve flavor of instant coffee (61). Fermentation of green coffee extracts to produce diacetyl [431-03-8], a coffee flavor compound, has also been demonstrated (62).

Potential consumer benefits from biotechnology (56) are cost and quality. The use of biotech means to increase the level of various sulfur-containing amino acids in coffee proteins, and to enhance sucrose and oil levels, could have an impact on the flavor and aroma of the finished ground coffee product. Also, caffeine level modification/elimination through genetic manipulations of the coffee plant could yield low caffeine coffee without additional processing by the manufacturer.

Liquid Coffee Products. Liquid coffee concentrates in frozen form have been available for many years for the food service or catering businesses, but they have not made a significant impact in either this market segment or in the grocery market, mainly because of inconvenience. Liquid coffee products, generally presweetened and ready to drink with milk added, represent a significant part of the Japanese and Korean coffee markets. Much of this product is sold in vending machines, hot and cold, for immediate consumption. The beverage is prepared by mixing the diluted coffee extract from a commercial percolator with the desired additives such as milk, sugar, and flavorants. It is then packaged in a container (generally a can) suitable for retort processing to provide a product that can be distributed without refrigeration. Recently the U.S. market has seen entry of several similarly processed products.

BIBLIOGRAPHY

"Coffee" in *ECT* 1st ed., Vol. 4, pp. 215–223, by L. W. Elder, General Foods Corp.; "Coffee, Instant" in *ECT* 1st ed., Suppl. 2, pp. 230–234, by H. S. Levenson, Maxwell House Division of General Foods Corp.; "Coffee" in *ECT* 2nd ed., Vol. 5, pp. 748–763, by R. G. Moores and A. Stefanucci, General Foods Corp.; in *ECT* 3rd ed., Vol. 6, pp. 511–522, by A. Stefanucci, W. P. Clinton, and M. Hamell, General Foods Corp.

1. W. A. Ukers, *All About Coffee*, 2nd ed., Tea & Coffee Trade Journal, New York, 1935, pp. 1–3.
2. R. J. Clarke and R. Macrae, eds., *Coffee*, Vol. 1, *Chemistry*, Elsevier Applied Science Publishers, Ltd., Barking, UK, 1985, Chapts. 2–8.
3. M. N. Clifford and K. C. Wilson, eds., *Botany, Biochemistry and Production of Beans and Beverage*, The Avi Publishing Company, Inc., Westport, Conn., 1985, Chapt. 13.
4. A. G. W. Bradbury and D. J. Halliday, "Polysaccharides in Green Coffee Beans," *Proceedings of the 12th Colloquium of ASIC*, Montreux, 1987, pp. 265–269.
5. R. Tressl, M. Holzer, and H. Kamperschroer, *Proceedings of the 10th Colloquium of ASIC*, Salvador-Bahia, Brazil, 1982, pp. 279–292.
6. M. N. Clifford, in Ref. 2, pp. 183–189.
7. R. Tressl, in T. Parliment, R. McGorrin, and C. T. Ho, eds., *Thermal Generation of Aromas*, ACS Symposium Series 409, Washington, D.C., 1989, pp. 286–287.
8. P. Folstar, in Ref. 2, pp. 203–220.
9. K. Speer and A. Montag, *D. Leben. Rundschau* **85**(12), 381–384 (1989).
10. S. Dart and H. E. Nursten, in Ref. 2, p. 237.
11. O. G. Vitzthum, C. Weisemann, R. Becker, and H. S. Kohler, *Cafe Cacoa* **XXXIV**(1), 27–35 (1990).

12. Ref. 2, pp. 84, 189, 207, 271.
13. W. Holscher, O. G. Vitzthum, and H. Steinhart, *Cafe Cacoa* **XXXIV**(3), 205–212 (1990).
14. O. G. Vitzthum, in O. Eichler, ed., *Kaffee und Coffein*, Springer, Berlin, 1976, pp. 30–31.
15. Ref. 2, pp. 141–142.
16. Ref. 2, pp. 127–137.
17. W. Baltes, in P. A. Pinot, H. V. Aeschbacher, R. F. Hurrell, and R. Liardon, eds., *The Maillard Reaction in Food Processing, Human Nutrition and Physiology*, Birkhäuser Verlag, Boston, Mass., 1990.
18. Ref. 7, pp. 285–301.
19. H. Marse, C. Visscher, L. Willemsens, and M. H. Boelens, *Volatile Compounds in Food: Qualitative and Quantitative Data*, Vol. II, TNO-CIVO, Food Analysis Institute, A. J. Zeist, The Netherlands, 1989, pp. 661–679.
20. I. Flament, *Food Rev. Int.* **5**(3), 317–414 (1989).
21. G. Pictet, in Ref. 2, Vol. 2, *Technology*, 1987, pp. 221–255.
22. M. Petracco, *Proceedings of the 13th Colloquium of ASIC*, Paipa, Colombia, 1989, pp. 246–260.
23. U.S. Pat. 4,322,447 (Mar. 20, 1982), M. Hubbard (to Hills Bros. Coffee Co.).
24. U.S. Pat. 4,737,376 (Apr. 12, 1988), L. Brandlein and co-workers (to General Foods Corp.).
25. U.S. Pat. 4,988,590 (Jan. 29, 1990), S. E. Price and co-workers (to Procter & Gamble Co.).
26. U.S. Pat. 4,860,461 (Aug. 29, 1989), Y. Tamaki and co-workers (to Pokka Corp. and NGK Insulators Ltd.).
27. U.S. Pat. 4,780,586 (Oct. 25, 1988), T. Le Viet and T. Bernard (to Nestec S.A.).
28. U.S. Pat. 3,615,667 (Oct. 26, 1971), F. M. Joffe (to Procter & Gamble Co.).
29. W. Clinton, "Evaluation of Stored Coffee Products," *Proceedings of the 9th Colloquium of ASIC*, London, 1980, p. 273.
30. A. L. Brody, *Food and Flavor Section*, Arthur D. Little, Inc., Cambridge, Mass.; *Flexible Packaging of Foods*, CRC Press, Division of the Chemical Rubber Co., Cleveland, Ohio, 1970, pp. 41–42.
31. *Coffee Drinking Study*, National Coffee Association, New York, 1991.
32. Eur. Pat. Appl. 0363529A3 (June 10, 1988), H. D. Stahl and co-workers (to General Foods Corp.).
33. U.S. Pat. 4,107,339 (Aug. 15, 1978), B. Shrimpton (to General Foods Corp.).
34. U.S. Pat. 3,438,784 (Apr. 15, 1969), W. P. Clinton and co-workers (to General Foods Corp.).
35. U.S. Pat. 897,763 (Sept. 1, 1908), J. F. Meyer (to Kaffee-Hag).
36. U.S. Pat. 2,309,092 (Jan. 26, 1943), N. E. Berry and R. H. Walters (to General Foods Corp.).
37. U.S. Pat. 4,820,537 (Apr. 8, 1989), S. Katz (to General Foods Corp.).
38. U.S. Pat. 4,465,699 (Aug. 14, 1964), F. A. Pagliaro and co-workers (to Nestlé SA).
39. U.S. Pat. 4,409,253 (Oct. 11, 1983), L. R. Morrison, Jr. (to Procter & Gamble Co.).
40. W. Heilmann, in *Proceedings of the 14th Colloquium of ASIC*, San Francisco, Calif., 1991, pp. 349–356.
41. *World Coffee Situation*, Foreign Agricultural Service, USDA, Washington, D.C., June 1991.
42. Documents, International Coffee Organization, London, EB3225/90, EB3271/91.
43. Newsletters, National Coffee Association, New York, Sept. 24, 1990; Oct. 8, 1990.
44. R. Clarke, in Ref. 2, Vol. 6, *Commercial and Techno-Legal Aspects*, 1988, pp. 105–142, 211–215.
45. R. Clarke, in Ref. 21, pp. 35–38.
46. Ref. 45, pp. 137–149, 214.

47. *Evaluation of Caffeine Safety*, Institute of Food Technologists' Expert Panel on Food Safety and Nutrition, Institute of Food Technologists, Chicago, Ill.; also published in *Food Tech. (Aust.)* **40**(3), 106–115 (Mar. 1988).

48. G. B. Schreiber, C. E. Maffeo, M. Robbins, M. N. Masters, and A. P. Bond, *Prev. Med.* **17**, 280–294 (1988).

49. G. B. Schreiber and co-workers, *Prev. Med.* **17**, 295–309 (1988).

50. G. E. Boecklin, *J. Am. Diet. Assoc.* **88**(3), 366–370 (Mar. 1988).

51. S. Heyden, in Ref. 2, Vol. 3, *Physiology*, 1988, pp. 57–80.

52. *What You Should Know About Caffeine*, International Food Information Council (IFIC), Washington, D.C., June 1990. This educational material was also reviewed favorably by the American Academy of Family Physicians.

53. "Caffeine Update: The News Is Mostly Good," *Wellness Letter*, Vol. 4, Issue 10, University of California, Berkeley, July 1988, pp. 4–5.

54. Ref. 47, pp. 111, 107.

55. Ref. 2, Vol. 5, *Related Beverages*, 1987.

56. M. Sondahl, in Ref. 22, pp. 407–419.

57. A. Zamarripa and co-workers, "Mass Propagation of Coffea spp. by Somatic Embryogenesis in Liquid Medium," in Ref. 40.

58. J. Spiral and co-workers, "Protoplast Culture and Regeneration in Coffea Species," in Ref. 40.

59. C. R. Barton and co-workers, "Stable Transformation of Foreign DNA into Coffea Arabica Plants," in Ref. 40.

60. U.S. Pat. 4,904,484 (Feb. 27, 1990), L. E. Smith and T. N. Asquith (to Procter & Gamble Co.).

61. U.S. Pat. 4,983,408 (Jan. 8, 1991), R. Colton.

62. U.S. Pat. 4,867,992 (Sept. 19, 1989), B. Boniello and co-workers (to General Foods Corp.).

General References

R. J. Clarke and R. Macrae, eds., *Coffee*, Vol. 1, *Chemistry*, 1985; Vol. 2, *Technology*, 1987; Vol. 3, *Physiology*, 1988; Vol. 4, *Agronomy*, 1988; Vol. 5, *Related Beverages*, 1987; Vol. 6, *Commercial and Technico-Legal Aspects*, 1988; Elsevier Applied Science Publishers, Ltd., Barking, UK. An excellent reference series.

M. Sivetz and N. Desrosier, *Coffee Technology*, AVI Publishing Co., Inc., Westport, Conn., 1979. Somewhat dated but good reference for the basic coffee processing technology.

GERALD WASSERMAN
HOWARD D. STAHL
WARREN REHMAN
PETER WHITMAN
Kraft General Foods Corporation

COGENERATION. See ENERGY MANAGEMENT; POWER GENERATION.

COKE. See COAL; COAL CONVERSION PROCESSES, CARBONIZATION.

COKE OVEN GAS.   See COAL; COAL CONVERSION PROCESSES,
CARBONIZATION.

COLLIDINES (TRIMETHYLPYRIDINES, ETC)   See PYRIDINE AND
PYRIDINE BASES.

COLLOIDS

A colloid is a material that exists in a finely dispersed state. It is usually a solid particle, but it may be a liquid droplet or a gas bubble. Typically, colloids have high surface-area-to-volume ratios, characteristic of matter in the submicrometer-size range. Matter of this size, from approximately 100 nm to 5 nm, just above atomic dimensions, exhibits physicochemical properties that differ from those of both the constituent atoms or molecules and the macroscopic material. The differences in composition, structure, and interactions between the surface atoms or molecules and those on the interior of the colloidal particle lead to the unique character of finely divided material, specifics of which can be quite diverse (see FLOCCULATING AGENTS).

Colloids and their singular and often curious properties were known to early civilizations, though their diversity and common traits were probably not appreciated (1–6). Domestic examples include clays (qv), soils, the swelling behavior of previously dry wood when wet, milk, butter and cheese, dough and pastry, inks and paints, latex, and silk, wool, and cotton fibers. Terrestrial colloidal phenomena, such as fogs, mists, smoke, and clouds, were also observed but must have seemed unrelated to other colloids. Applications of chemistry to the extraction and working of metals, the manufacture of pottery, glass, dyes, fabrics, and papyrus and paper, and the practice of medicine, among other tasks, clearly demanded the manipulation of colloids. Yet, concepts about the existence of this state of matter and, moreover, that its properties often govern macroscopic events seem to have not existed before the nineteenth century.

Pioneering investigations during the nineteenth century helped to set a scientific foundation for the understanding of the widely varied features of colloids (1–7). Notable research in this period addressed electrophoresis and electro-osmosis (8), Brownian movement (9), protein denaturation (10), coagulation of inorganic colloids (11), the optical properties of gold sols (12), dialysis (13), and the polymerization and purification of silica sol (14,15), to name just a few. In turn, these and other studies led to pivotal practical and theoretical developments, including the ultramicroscope (16), the molecular kinetic theory (17–19), the ultracentrifuge (20), a quantitative explanation of suspension viscosity (17), and experimental evaluation of Avogadro's number (17,19).

Today, colloids are encountered in a wide variety of industries, and many important technological problems relate to the behavior of dispersions, the mixtures obtained when colloids are distributed, ie, dispersed, in a medium, such as a liquid or a gaseous fluid (see DISPERSANTS). Commercial examples include metals, magnetic powders, catalysts, ceramics, minerals, oil recovery, technical glasses, paints and pigments, polymers, pulp and paper (qv), prepared foods, pharmaceuticals, fibers, detergents, and purified water (see LATEX TECHNOLOGY; PLASTICIZERS; PRINTING PROCESSES). Natural and biological systems may also depend, to some extent, on the behavior of colloids. Phenomenological examples are found in soil science and plant nutrition, meteorology, hematology, membrane science, and antigen–antibody reactions in medical technology (see MEMBRANE TECHNOLOGY). In view of the ubiquity and importance of colloids, the ability to control their formation, destruction, or stabilization is often a critical technological problem. This is true whether the specific colloidal material is deemed desirable, as in the reinforcement of plastics, metals, and ceramics, or undesirable, as is often true in water-treatment, meteorology, and other environmentally important systems.

General Properties

Dimensions. Most colloids have all three dimensions within the size range ~100 nm to 5 nm. If only two dimensions (fibrillar geometry) or one dimension (laminar geometry) exist in this range, unique properties of the high surface area portion of the material may still be observed and even dominate the overall character of a system (21). The non-Newtonian rheological behavior of fibrillar and laminar clay suspensions, the reactivity of catalysts, and the critical magnetic properties of multifilamentary superconductors are examples of the numerous systems that are ultimately controlled by such colloidal materials.

A dispersion factor, defined as the ratio of the number of surface atoms to the total number of atoms in the particle, is commonly used to describe highly dispersed systems that do not exhibit a particularly high surface-area-to-volume ratio (22). Representative values for 10-, 100-, and 1000-nm particles are, respectively, on the order of 0.15–0.30, 0.40, and 0.003–0.02, depending on the specific dimensions of the atoms or molecules that comprise the particles. Other quantities can be used to describe the degree of dispersion (6,7), but these tend to assume, at least, quasi-equilibrium conditions that are not always met (7,23).

In view of the facts that three-dimensional colloids are common and that Brownian motion and gravity nearly always operate on them and the dispersing medium, a comparison of the effects of particle size on the distance over which a particle translationally diffuses and that over which it settles elucidates the colloidal size range. The distances traversed in 1 h by spherical particles with specific gravity 2.0, and suspended in a fluid with specific gravity 1.0, each at 293 K, are given in Table 1. The dashed lines are arbitrary boundaries between which the particles are usually deemed colloidal because the effect of Brownian diffusion or sedimentation, relative to the other, is on the order of 1% in this region.

Nomenclature. Colloidal systems necessarily consist of at least two phases, the colloid and the continuous medium or environment in which it resides, and their properties greatly depend on the composition and structure of each phase. Therefore, it is useful to classify colloids according to their states of subdivision and agglomeration, and with respect to the dispersing medium. The possible classifications of colloidal systems are given in Table 2. The variety of systems represented in this table underscores the idea that the problems associated with colloids are usually interdisciplinary in nature and that a broad scientific base is required to understand them completely.

These traits have led to a wide diversity of theoretical and analytical advances needed to interpret colloidal behavior properly. Various traits that relate to the properties and behavior of colloids have developed since the middle of the

Table 1. Effect of Spherical Particle Size on the Relative Brownian Diffusion and the Sedimentation Distances After 1 h in Water at 293 K

Radius, a, nm	D_{tr}/D_{ions} [a,b]	Distance diffused, x, nm	Distance settled, Δh [c]	$x/\Delta h$ [d]
0.1	1	3.9×10^6	0.08 nm	4.9×10^7
1 (solvent)	0.1	1.2×10^6	8 nm	1.5×10^5
---------				---------
10	10^{-2}	3.9×10^5	0.8 μm	4.9×10^2
100	10^{-3}	1.2×10^5	80 μm	1.5
1,000	10^{-4}	3.9×10^4	8 mm	4.9×10^{-3} ($<1\%$)
---------				---------
10,000	10^{-5}	1.2×10^4	0.8 m	1.5×10^{-5}

[a]D_{tr} is the translational diffusion coefficient and is given by $kT/6\pi\eta a$ for which k, T, and η are Boltzmann's constant, Kelvin temperature, and fluid viscosity, respectively.
[b]D_{tr}-values are scaled to the D_{tr}-value for ions, D_{ions}.
[c]Densities, ρ_p and ρ_{fluid}, are 2.0 and 1.0 g/cm^3, respectively.
[d]Dashed lines indicate the boundaries between which colloids typically exist.

Table 2. Classifications of Two-Phase Colloidal Systems[a]

Dispersing medium	Dispersed (colloidal) matter	Names[b]
gas	liquid	aerosol, fog, mist
	solid	aerosol, smoke
liquid	gas	foam
	liquid	emulsion
	solid	sol[c], gel, dispersion, suspension
solid	gas	solid foam
	liquid	gel, solid emulsion
	solid	solid sol, alloy

[a]Adapted from References 6, 7, 21, and 24–26.
[b]The appropriate name depends on the specific properties of the system.
[c]Aqueous sols are commonly called hydrosols.

twentieth century (21,25,26). Notable examples include experimental use of the Mie and inelastic theories of light scattering, photon correlation spectroscopy, neutron and x-ray scattering, electrical double-layer models, quantitative formulation of van der Waals interactions, gas adsorption theory, electron microscopy, ultracentrifugation, various spectroscopic techniques based on electron, ion, and photon probes, pulsed nuclear magnetic resonance methods, and direct measurement of interparticle forces. Many of these can be used, either directly or indirectly, to study the changes that a colloidal system undergoes, with the appearance or disappearance of the colloidal state being the most critical event and a change in the tendency to agglomerate (solids) or to coalesce (liquids, gases) being the most common industrial problems.

Behavior. Diffusion, Brownian motion, electrophoresis, osmosis, rheology, mechanics, and optical and electrical properties are among the general physical properties and phenomena that are primarily important in colloidal systems (21,24–27). Of course, chemical reactivity and adsorption often play important, if not dominant, roles. Any physical and chemical feature may ultimately govern a specific industrial process and determine final product characteristics.

One common situation, the required pumping for the transport of fluid systems with a high concentration of suspended solids, eg, those exceeding 10 vol %, exemplifies the interplay of the phenomena just mentioned in conjunction with colloidal behavior. The pumping operation is needed for many processes. Extrusion in the polymer industry, the processing of gelatinous foods and cosmetic items, the fabrication of various high performance materials in the ceramic and metallurgical industries, the preparation and handling of pigment slurries in the paint industry, and the treatment of waste by-products in the minerals and water industries are a few examples in which the transport behavior of colloids is very important. The prediction and control of suspension rheology, especially the thixotropic and dilatant tendencies, is primarily important for these and other uses. These rheological properties depend, in turn, on the specific interactions among the colloidal particles, the dispersing medium, and the solute additives, ie, salts, surfactants, and polymers. Wetting, electrical double-layer, and van der Waals forces nearly always effect the interactions that determine the colloidal behavior of a given system.

Monitoring Techniques. The best physical technique for investigating or monitoring colloids is usually identified by considering the nature of the colloidal material and the dispersing medium. The purpose for which the information is sought, viz, for fundamental knowledge or technological use, must also be taken into account. In principle, any distinctive physical property of the colloidal system in question can be used, at least empirically, to monitor changes in the dispersed state. The more complex a system is chemically or with respect to its particulate heterogeneity, the less likely it is that a single property uniquely and completely describes changes in the colloidal state.

The preparation and description of colloids is often studied using electron and optical microscopy, scattering of light, x-rays, neutrons, and rheology and surface tensiometry. Agglomeration or coalescence of the colloidal material can be monitored by various techniques. Light scattering, neutron scattering, microscopy, rheology, conductivity, filtration, sedimentation, and electrokinetics are among those successfully used to follow colloidal changes.

Physical and Chemical Properties

Formation. Colloid formation involves either nucleation and growth phenomena or subdivision processes (6,21,24–29). The former case requires a phase change; the latter one pertains to the comminution or atomization of coarse particles (solids) or droplets (liquids). Three nucleation and growth processes by which colloids have been synthesized in the laboratory or in nature are condensation of vapor to yield liquid or solid directly (27), condensation of vapor to form colloidal liquid droplets (aerosols, emulsions) that may subsequently solidify (30), and precipitation from liquid or solid solutions (31). Chemical reactions often induce these phase changes; however, they are not essential for producing either homogeneously or heterogeneously nucleated colloids. Certainly, growth of a colloid is controlled kinetically and depends fundamentally on the rate at which fresh solute material in the correct orientation approaches a growing nucleus and the bond energy requirements at the surface of the nucleus. This last consideration suggests that dislocations in solid particles critically influence particle growth. The various mechanisms of colloid formation and the resultant powder properties of industrially produced single- and multi-component colloidal solids are listed in Table 3.

Processes based on subdivision contrast these chemical processes and are widely used (27,32–38). Though it is difficult to reconcile experimental and practical observations of comminution into a single coherent theory, such a theory would clearly have to consider a variety of phenomena including mechanics, particle fracture, and agglomeration. Consequently, subdivision of solids on an industrial scale should be considered more of an art than a science at the present time. Nonetheless, a complex technology, if not a specialized one, has developed to conduct and to control comminution and size-fractionation processes. Mathematical models are available to describe changes in particle size distribution during comminution, but these are generally restricted to specific processes. Comprehensive reviews of the developments in preparing colloidal solids by subdivision should be consulted for further details (27).

Specific advancements in the chemical synthesis of colloidal materials are noteworthy. Many types of generating devices have been used to produce colloidal liquid aerosols (qv) and emulsions (qv) (39–43); among them are atomizers and

Table 3. Industrially Produced Colloidal Materials and Related Processes

Mechanism	Examples[a]
vapor → liquid → solid ↓ → → → ↑	oxides, carbides via high intensity arc; metallic powders via vacuum or catalytic reactions
vapor + vapor → solid	chemical vapor deposition, radio frequency-induced plasma, laser-induced precipitation
liquid → solid	ferrites, titanates, aluminates, zirconates, molybdates via precipitation
solid → solid	oxides, carbides via thermal decomposition

[a]Ref. 27.

nebulizers of various designs (30,44–50). A unique feature of producing liquid or solid colloids via aerosol processes (Table 3) is that material with a relatively narrow size distribution can be routinely prepared. These monosized colloids are often produced by relying on an electrostatic classifier to select desired particle sizes in the final stage of aerosol production. Another significant development is the increasing availability of colloidal powders prepared as liquid suspensions. These powders have uniform chemical and phase composition, particle size, and shape (21). Such colloids of certain elements, eg, sulfur, gold, selenium, and silver, have been obtainable for a long time (6,51–55), whereas uniform colloids of other elements, eg, carbon, cobalt, and nickel, have been generated more recently (56–58). The laboratory and industrial preparation of uniform, monosized colloids of complex compositions has advanced rapidly and has the potential for becoming routine, along with the development of aerosol, hydrosol, and dispersion techniques. Examples include many inorganic compounds, such as NaCl, AgCl, AgBr, $BaSO_4$, $FePO_4$, β-FeOOH, $Cr(OH)_3$, SiO_2, ZrO_2, V_2O_5, Al_2O_3, α-Fe_2O_3, Fe_3O_4, TiO_2, La_2O_3, Ga_2O_3, ZnO, Co-ferrite, Ni-ferrite, Cr-ferrite, ZnS, PbS, and CdS (21,30,58–67), and organic latices of poly(vinyl acetate), polystyrene, poly(vinyl chloride), styrene–butadiene rubber, poly(acrylic acid), polyurea, poly(styrene)–poly(acrylate), and poly(methacrylate)–poly(acrylate) (21,61,62,66–67). Industrial synthesis of these and related colloidal materials can be limited by problems of scale-up. Still, monosized powders and monodispersed colloidal sols are frequently used in many products, eg, pigments, coatings, and pharmaceuticals (21,61).

Characterization. The proper characterization of colloids depends on the purposes for which the information is sought because the total description would be an enormous task (27). The following physical traits are among those to be considered: size, shape, and morphology of the primary particles; surface area; number and size distribution of pores; degree of crystallinity and polycrystallinity; defect concentration; nature of internal and surface stresses; and state of agglomeration (27). Chemical and phase composition are needed for complete characterization, including data on the purity of the bulk phase and the nature and quality of adsorbed surface films or impurities.

Particle Size. The most direct methods for determining the particle size, shape, and morphology of colloids are scanning and transmission electron microscopy. Resolution of optical microscopes is usually insufficient to determine primary particle size, although agglomerate size can be effectively evaluated using conventional optical microscopes. Indirect techniques for determining particle size or particle-size distribution include sedimentation and centrifugation, conductometric techniques, light scattering, x-ray diffraction, gas and solute adsorption, ultrafiltration, and diffusiometric methods (24,73). Care must be taken in selecting an indirect method since these require assumptions about either the real size distribution or the process on which the analysis is based. For instance, commercial conductometric equipment, eg, Electrozone and Coulter Counters, relies on the sphericity of particles; light-scattering techniques are reliable only if the particle shape is known or assumed (73); and adsorption analyses rely on model adsorption isotherms, the uniformity of particle size and porosity, and the orientation of adsorbed species. Thus the best analytical choice also depends, to some extent, on the physical properties of the colloid and on the dispersing medium described in Table 2. Typically, more than one method is needed to evaluate the particle size suitably.

Surface Area and Permeability or Porosity.   Gas or solute adsorption is typically used to evaluate surface area (74,75), and mercury porosimetry is used, in conjunction with at least one other particle-size analysis, eg, electron microscopy, to assess permeability (76). Experimental techniques and theoretical models have been developed to elucidate the nature and quantity of pores (74,77). These include the kinetic approach to gas adsorption of Brunauer, Emmett, and Teller (78), known as the BET method and which is based on Langmuir's adsorption model (79), the potential theory of Polanyi (25,80) for gas adsorption, the experimental aspects of solute adsorption (25,81), and the principles of mercury porosimetry, based on the Young-Dupré expression (24,25).

Solid-State Structure.   The degree of crystallinity and polycrystallinity of colloids is obtained via x-ray diffraction and the interpretation of Bragg reflections (82). Laue techniques (82) are not routinely employed for studying colloids because they require single crystals of larger size. The structure and concentration of defects in single-crystal colloids, down to a particle-size range of ca 1–10 μm, can be determined by conventional methods using x-rays and measurement of physical properties. In principle, inherent properties can be studied on finer material. However, in practice, it is often necessary first to conduct analyses of defects on samples of the same material in particles larger than the colloidal size range and then to assume that the determined defect structure also applies to the corresponding colloids. Neutron diffraction is expected to be increasingly used (83), particularly with regard to magnetic structure (84), to complement x-ray studies of colloidal powders.

Internal and Surface Stresses.   These stresses in colloidal particulates can be evaluated using the procedures developed for macroscopic samples, namely, electron microscopy, x-ray and neutron diffraction, and physical property measurements (27,82–84). Since many equilibrium crystals are encountered and apparently equilibrated surfaces are not smooth but rather saw-toothed owing to entropic factors (25), the actual planes that predominate at surfaces in practical systems may be determined by experimental or operational conditions. The detailed study of surface stresses in solid colloids is therefore difficult because actual crystal planes at powder surfaces are usually incomplete and imperfect; moreover, these imperfections may actually represent unequilibrated surface stresses (25). Adsorbed gaseous species or the dispersing medium with its solutes probably affects the energy and atomic configurations responsible for surface stresses. All these factors compound the problems associated with stress analysis. However, some progress in unravelling this problem has been made with the experimental confirmation of surface distortions that result from an imbalance of surface forces, eg, those reported in studies on elemental semiconductors of silicon, germanium, and diamond (85–88). Thus the practical consequences of internal and surface stresses are most important in the formation of solid colloids, particularly during their growth when surface structure changes most radically, and in the dispersibility of solids in liquids.

As an example, the following discussion illustrates the impact of surface states on the dispersion of colloidal silicon powder. The process consists of three stages: the wetting of powder, the breakdown of large agglomerates into colloidal particles or small clusters of primary particles, and the stabilization of the particulates against agglomeration (29). The first two involve the replacement of silicon–vapor (air) or silicon–silicon interfaces by a silicon–liquid one and, possibly, the mechanical disintegration of agglomerates. These steps constitute complete wetting, for which the change in free energy per unit area, ΔG_d, is

$$\Delta G_d = \gamma_{SL} - \gamma_{SV} = -\gamma_{LV}\cos\theta_e$$

The terms γ_{SL}, γ_{SV}, and γ_{LV} represent the surface tensions or free energies of the solid–liquid, solid–vapor, and liquid–vapor interfacial regions, respectively, when the three phases coexist and establish an equilibrium contact angle, θ_e, measured within the liquid phase. Available theories of wetting (88,89) adequately describe various experimental dispersibility phenomena exhibited by silicon powder (90,91). Thus, ΔG_d, θ_e, and various contributions to the surface free energy of the powder (γ_{Si}) can be estimated using macroscopic contact angles and the fundamental physical properties of silicon and common liquids. Values for γ_{Si} and its dispersion force component are 39.9 mJ/m^2 and 38.0 mJ/m^2 (9×10^{-7} cal/cm^2), respectively (53); the difference can be attributed to the electronic band structure of silicon. This approach can be extended to more complicated solids; but, the surface structure is often unknown, limiting its success.

State of Agglomeration. Nearly all colloidal systems undergo some agglomeration. This leads to a distribution of aggregate size for solid colloids and of droplet size for liquid colloids. Consequently, either the rate or the extent of agglomeration is of practical concern whenever colloidal material is encountered. Ultramicroscopy (UM) is the most desired method for evaluating the kinetic or equilibrated aspects of agglomeration; unfortunately, it is often not feasible or too intricate and time consuming to use. Although UM is the only direct method widely available, this technique will be fully exploited once procedures for wet or environmental cells are developed. The most common indirect methods for evaluating agglomeration consist of optical techniques (24,25,28), ie, measurements of light scattered or optical density and photon correlation spectroscopy; here, the magnitude of the optical effect increases as the agglomerates become larger via aggregation or coalescence. Other methods have been devised which are based on sedimentation, rheology, and electrical and thermal conductivity. These techniques tend to be empirical, qualitative, and specific to a given colloidal system. More than one technique is required to assess the state of agglomeration when a wide range of colloidal dimensions exists. If agglomeration is severe and produces clusters exceeding approximately 5 μm, the aggregate-size distribution can be evaluated using a variety of classification techniques developed for macroscopic solids, eg, sieving (27,92), provided such methods do not themselves induce or break agglomerates.

The state of agglomeration is intimately related to colloidal stability or the resistance to agglomerate or coalesce. Wetting phenomena play a significant role here, as do various nonwetting, colloidal factors. The stabilizing role of second solvents and surfactants can be exploited if a colloidal solid is not completely wetted by the dispersing fluid and if the interfacial energy in fluid–fluid dispersions such as emulsions and foams (qv) is initially unfavorable. This step requires an understanding of the energetics of wetting and adsorption (25,29,93,94) which

can be obtained via calorimetry (29) and adsorption isotherms (29,95), or by studying the specific micellization behavior and hemimicelle formation of surface-active
agents (95,96). Nonwetting factors in colloidal stability usually pertain to the
resistance to agglomeration. The agglomerating process is induced by particulate
collisions arising from diffusion, as in Brownian motion, velocity or shear gradients in a liquid-dispersion medium, and gravitational settling (97). The colli-
sional frequency as a result of Brownian diffusion depends on the viscosity of the
liquid, concentration of particles, temperature, and fraction of collisions that lead
to irreversible agglomeration.

This last consideration can be quantified using various models for repulsive
or attractive electrostatic (24,25,28,29), London-van der Waals (24,98–101), steric
(102), and hydrodynamic (103) factors that affect the stabilization of aqueous and
nonaqueous colloidal systems. These forces can be evaluated separately because
they originate from different sources. A comprehensive model of colloid stability,
the Derjaguin-Landau-Verwey-Overbeek or DLVO model (28) has proved to be
essentially correct, regarding the roles of electrolytes, dielectric constant, and
other physical quantities in colloidal systems, and is regarded as a cornerstone of
modern colloid science (6,7,24). The DLVO model describes the interplay of electrostatic and London-van der Waals interactions in the stability of hydrophobic
suspensions. It treats the electrostatic interactions between two identically
charged, suspended particles to be repulsive and to arise from the overlap of the
electrical double layers associated with each particle. London-van der Waals
forces are described in this model by Hamaker's formulas (99), although more
precise treatments now exist (100,101). The total interaction for two mutually
approaching particles is viewed, therefore, to be the combined effect of surface
charges and the London-van der Waals forces. These interactions are portrayed
in Figure 1 as functions of particle separation. If the particles are electrostatically
charged and have an interfacial potential whose magnitude exceeds 25 mV, the
DLVO model predicts for binary particle interactions that a substantial, repulsive
potential energy barrier will inhibit the close approach of the particles, thereby
stabilizing them against agglomeration (viz, V_{max} in Figure 1). The critical potential depends, in part, on the particle size and the Hamaker constant (99). The
value of 25 mV given usually applies to particles smaller than ~200 nm. A value
approaching $|50|$ mV may be required for larger particles and those with a large
Hamaker constant, eg, ceramic and metal particles (105,106). The model has been
extended to systems with dissimilar particles that differ in size, shape, and chemical composition (107), and to those with particles that have an adsorbed layer of
ions (6,7,24,25,28,108), as depicted in Figure 2.

Quantitative understanding of the interplay of steric or solvation factors
with either the electrostatic or London-van der Waals forces is not part of the
DLVO theory, but it has developed since the early 1970s (101). Long-chain surfactants and high molecular-weight polymers have commonly been added to colloidal suspensions to ensure long-term stability. However, the mechanisms of
steric stabilization have only recently been elucidated so that these relatively
short-range interactions can be used industrially to produce systems with the
desired behavior at an economical cost. Molecular architecture, thickness and the
hydrodynamic volume of the adsorbed layer, temperature, and chain or segment
solvation are critical parameters in determining the effectiveness of a dispersing

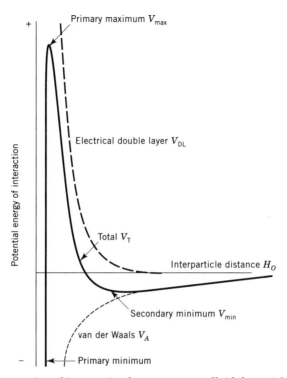

Fig. 1. Potential energies of interaction between two colloidal particles as a function of their surface–surface separation, H_O, for electrical double layers due to surface charge (V_{DL}), London-van der Waals forces (V_A), and the total interaction (V_T). The primary maximum usually ensures stability, if its magnitude (V_{max}) exceeds the range 10 to 15 kT; smaller barriers lead to irreversible agglomeration in the primary minimum (6,7,24,28). The secondary minimum (V_{min}) can promote weak, reversible agglomeration of large particles, if its magnitude is on the order of 10 kT or more (101,104–106). The overall energy barrier to coagulation in the primary minimum ($V_{barrier}$) is given by the expression, $V_{max} - V_{min}$, where the primary minimum represents the potential energy at contact ($H_O \rightarrow 0$); stability ensues if the magnitude of $V_{barrier}$ exceeds 10 to 15 kT. Note that the primary minimum has a finite depth because of the contributions of the Born repulsion and the structure of liquid at the particle surface (not shown) for separations very close to particle contact (7,101).

agent to provide sufficient steric stabilization (99,102,109). Depletion stabilization is related to steric stabilization, in that added soluble polymer ensures stability; however, this mechanism relies on the ability of free polymer in the bulk solution to sustain a thermodynamically metastable barrier to the close approach of particles (110). Above a threshold polymer concentration, the local concentration of free-polymer segments between the particles is depleted on the close mutual approach of the particles, inducing an unfavorable free energy that provides colloidal stability.

If velocity or shear gradients are present and are sufficiently large, the frequency of collisions depends on the volume fraction of solids and the mean velocity gradient, as well as on the system parameters listed earlier. Several research groups have initiated fundamental studies dealing with colloidal systems in which

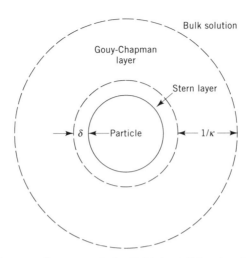

Fig. 2. Schematic diagram of a suspended colloidal particle, showing relative locations of the Stern layer (thickness, δ) that consists of adsorbed ions and the Gouy-Chapman layer (1/κ) which dissipates the excess charge, not screened by the Stern layer, to zero in the bulk solution (108). In the absence of a Stern layer, the Gouy-Chapman layer dissipates the surface charge.

shear is important (111–117). The rheological behavior of dilute colloidal systems can be Newtonian, for which the viscosity is independent of shear rate and the systems do not exhibit a nonzero yield stress (7). Often the behavior of dilute Newtonian systems can be considered Einsteinian, where the ratio of the viscosity of the colloidal system to that of the continuous phase linearly depends on the volume fraction of colloid and the colloidal particles do not interact with one another (17,113,118). Sometimes, the interpretation of data is complicated by non-Newtonian flow, stemming from either polydispersity or high solids content. For these systems, various mathematical and rheological techniques (114,119,120) help to describe the rheology (see RHEOLOGICAL MEASUREMENTS).

Assuming that sedimentation is slow compared to the first two collision mechanisms, the overall agglomeration rate, $-dN/dt$, is

$$-dN/dt = k_d N^2 + k_s N,$$

where N is the particle-number concentration, k_d and k_s are the respective rate constants corresponding to diffusion-controlled and shear-induced collision processes, and the minus sign denotes that the number concentration decreases with time, t. The constants, k_d and k_s, depend on particle properties such as chemical composition of bulk and surface phases, dielectric constant, dipole moment, size, size distribution, shape, surface charge, solid-phase distribution within particles, and particle anisotropy. Properties of the liquid-dispersing medium that contribute significantly to the values of these rate constants are dielectric constant, dipole moment, and the ability to dissolve electrolytes and polymers, in addition to those properties cited earlier. The k_d term usually dominates in quiescent systems containing submicrometer particles. The full expression for $(-dN/dt)$ and its use are treated in more detail elsewhere (24,97).

Chemical reactions can also affect the k_d and k_s terms and thereby influence or control colloidal stability (21,121). Pertinent examples are dissolution, reprecipitation, hydrolysis, precipitation, and chemical complexing. The last reaction may involve either simple species, eg,

$$Al^{3+} + SO_4^{2-} \rightleftharpoons AlSO_4^+$$

or complicated solutes such as $Al_8(OH)_{20}^{4+}$, chelated metals (97), synthetic and natural polymers (29,98,102), and a variety of surfactants and dispersants (95,122). Many of the possible bulk solution chemical reactions that influence colloidal stability, along with specific sample reactions and their general interfacial analogues, are listed in Table 4. The use of natural and synthetic polymers (dispersants) to stabilize aqueous and nonaqueous colloidal suspensions is technologically important, and research in this area has focused on polymer adsorption and steric stabilization (102). Of course, colloidal destabilization by electrolytes and bridging flocculation by polymers have been addressed experimentally and theoretically. Comprehensive reviews on the relevant phenomena that derive from soluble and adsorbed polymers are available (102).

Chemical Properties. Any classical wet-chemical analyses or instrumental techniques that are routinely used to analyze the bulk composition of solids and liquids are, in principle, also suitable for colloids. The available instrumental methods range, for example, from spectrographic analysis for chemical composition, which is limited to crude estimates for impurities, to Raman spectroscopic identification of chemical functionalities, which provides accuracy to within 1%, depending on the desired accuracy of stoichiometric determinations. Surface-chemical analyses that cannot be conducted using conventional methods designed for bulk materials are usually accomplished by optical, diffraction, and spectroscopic techniques; these are often applied under conditions of an ultrahigh vac-

Table 4. Representative Solution and Surface Equilibria Influencing Colloidal Stability

Solution	Surface analogue
Hydrolysis	
$CH_3(CO)OCH_3 + H_2O \rightleftharpoons CH_3(CO)OH + CH_3OH$	$M_2O + H_2O \rightleftharpoons 2\,MOH$
$PO_4^{3-} + H_2O \rightleftharpoons HPO_4^{2-} + OH^-$	$MOH + H_2O \rightleftharpoons MOH_2^+ + OH^-$
Dissociation	
$Al(OH)_2^+ \rightleftharpoons Al^{3+} + 2\,OH^-$	$MOH_2^+ \rightleftharpoons MO^- + 2\,H^+$
$C_6H_5(CO)OH \rightleftharpoons C_6H_5(CO)O^- + H^+$	$MOH \rightleftharpoons MO^- + H^+$
Dissolution	
$ZnC_2O_4(s) \rightleftharpoons Zn^{2+} + C_2O_4^{2-}$	
$Al(OH)_3(s) + OH^- \rightleftharpoons AlO_2^- + 2\,H_2O$	
Complexation	
$Cu^{2+} + 4\,OH^- \rightleftharpoons Cu(OH)_4^{2-}$	$MO^- + Na^+ \leftrightharpoons MO^-Na^+$
$n\text{-}C_{12}H_{25}N(CH_3)_2 + HCl \rightleftharpoons n\text{-}C_{12}H_{25}NH^+(CH_3)_2Cl^-$	$MOH + HCl \rightleftharpoons MOH_2^+Cl^-$

uum. The spectroscopies measure the responses of solid surfaces to beams of electrons, ions, neutral species, and photons. Each spectroscopy has unique attributes (25,123), making it suitable for certain colloids but unsuitable for others. Many of the techniques developed for microchemical analysis of surfaces or thin films are given in Table 5. Classical analytical techniques and instrumental methods can also be used for surface-chemical analyses, if the outermost layers of a solid colloid or the adsorbed layers can be quantitatively desorbed and studied. Visible, ultraviolet, and, especially, infrared spectroscopy (125) can be used to research adsorbed material. Lastly, surface conductance methods provide qualitative data on the surface-chemical aspects of colloids (6,126).

Table 5. Diffraction and Scattering Methods for Surface Structure and Composition[a,b]

Emitted (diffracted or scattered) probe	Incident probe		
	Electrons	Ions	Photons
electrons	AES	EID	PES
	CEELS	INS	UPS
	HEED		XES
	LEED		XPS
ions	ESD	ISS	LEIS
		NIRMS	
		SIMS	
photons	APS	IILE	ellipsometry
	ESCA		IRS
			OSEE
			PhD

[a]Adapted from Refs. 25, 123, 124.
[b]AES = auger electron spectroscopy; APS = appearance potential spectroscopy; CEELS = characteristic electron energy-loss spectroscopy; EID = electron impact desorption; ESCA = electron spectroscopy for chemical analysis; ESD = electron-stimulated desorption; HEED = high energy electron diffraction; IILE = ion-induced light emission; INS = ion-neutralization spectroscopy; IRS = infrared spectroscopy; ISS = ion-scattering spectroscopy; LEED = low energy electron diffraction; LEIS = low energy ion scattering; NIRMS = noble-gas-ion reflection mass spectrometry; OSEE = optically stimulated exoelectron emission; PES = photoelectron spectroscopy; PhD = photoelectron diffraction; SIMS = secondary ion mass spectroscopy; UPS = ultraviolet photoelectron spectroscopy; XES = x-ray emission spectroscopy; and XPS = x-ray photoelectron spectroscopy.

Hazards of Colloidal Systems

Two situations exist for which the chemical reactivity of colloidal materials may be detrimental or even physically harmful. The first one is a result of their high specific surface area. Colloidal chemical reactivity may differ considerably from that of the identical macroscopic material with less surface area. This is particularly important if the colloidal surface is easily and rapidly oxidized. Dust explosions and spontaneous combustion are potential dangers whenever certain materials exist as finely divided dry matter exposed to oxidizing environments (see POWDERS, HANDLING). Many of the metals in Table 6 exhibit this property. The

Table 6. Exposure Limits of Selected Colloidal Materials[a]

Substance	Limit mg/m^3	Substance	Limit mg/m^3
asbestos	70.6[b]	nicotine	0.5
boron-oxide	15	oil mist	5
cadmium dust	0.2[c]	osmium tetroxide	0.002
calcium oxide	7	quartz dust	30
carbaryl	5	paraffin (fume)	2
carbon black	3.5	phosphoric acid	1
chromium (metal, insoluble salt)	1	phosphorus	0.1
coal dust	24	plaster of Paris	15[c,d]
coal-tar pitch volatiles	0.2	rhodium dust	0.1
cobalt dust	1	silver metal	0.01
copper dust	1	sulfuric acid	1
cotton dust	1	talc	706[b]
diatomaceous earth	80	tantalum	5
DDT	1	tellurium	5
ferrovanadium dust	1	thallium	0.2
fluoride dust	2.5[c]	soluble compounds	0.1
hafnium	0.5	tin compounds[e]	
iron oxide (fume)	10[d]	inorganic	2
kaolin	10	organic	0.1
magnesium oxide (fume)	10	tungsten	0.1
mica	0.05	insoluble compounds	5
mercury	0.1[c]	vanadium oxide	0.3[f]
molybdenum (insoluble compounds)	15	zinc chloride	1
nickel	1	zinc oxide fume	5

[a]Adapted from Ref. 127.
[b]10^6 particles per m^3.
[c]8-h, time-weighted average.
[d]Total dust.
[e]Except SnH_4 and SnO_2.
[f]0.1–0.5 mg/m^3.

second hazard is that many colloidal substances, including fibrillar matter, are often inhaled to the extent that physiological problems may arise if they are retained by bodily tissues such as the lungs. This hazard may be quite acute if the particulates are ca 1 μm; exceedingly small particulates may be exhaled and particulates that are significantly larger than 1 μm settle and are not often inhaled into the lungs. Short-term allergic conditions are common, eg, those induced by pollen and household dust, asthma and hay fever. Long-term effects may be fatal, as is the case with silicosis, asbestosis, and black lung disease.

Specific potential hazards have been associated with a diverse spectrum of colloidal materials: chemicals, coals, minerals, metals, pharmaceuticals, plastics, and wood pulp, to name a few. Limits for human exposure for many particulate, hazardous materials are published (127) and are given in Table 6. If the material is a colloidal solid or liquid, the exposure limits, expressed in mg/m^3, may be misleading. Large fractions of readily hydrolyzable metals exist as adsorbed species on suspended (colloidal) solids in fresh and marine water systems (128) and can also be anticipated in industrial wastewater. Elements such as lead, zinc, and vanadium that are released into the atmosphere as vapors subsequently condense or are removed as solid particulates by rain (98,128). Liquid droplets may also constitute a hazard; for instance, smog often contains sulfuric acid aerosols. Trace metals that are commonly found as suspended matter in the chemical form of hydrous oxides and other insoluble matter are given in Table 7.

Some of the industries that produce waste containing significant amounts of suspended matter are listed in Table 8. The cost to protect the general population from these solids is, of course, highly variable, owing to transient economic issues and governmental restrictions (129). Although the latter constraints set a rigorous schedule for technologically isolating and containing waterborne and airborne pollutants (130,131), they are largely unenforced. Yet, economic and governmental forces have helped to foster quite large water- and air-treatment industries (131,132). The increasing cost of research on the particulate pollutants listed in Tables 6 to 8 and related hazardous materials is possibly as great as 60% of the increasing annual cost of research (131,133) and acutely affects the progress made in controlling colloidal pollutants.

Until the economic and governmental forces surmount present limitations, the greatest cost will continue to be linked to the treatment of health problems caused by solids and liquids suspended in air or water. Presently, the effects of the colloidal solids and liquids that comprise smog are widespread and well

Table 7. Metals Commonly Found in Freshwater and Dust Systems[a]

Metal	Aqueous concentration, ppm[b]	Dust concentration, ppm	
		Laboratory dust	Furnace filters
Ag		5	
Al	310		
B	31		
Ba	90		
Ca		70	10,000
Co		19	
Cu	2	38	82
Fe	740	140	5,300
K			3
Mg		720	8,200
Mn	200	140	160
Ni	8	89	160
Pb		14,620	7,400
Sn		110	66
Sr	67		
Ti	7		
V		20	70
Zn			900

[a]Adapted from Ref. 128.
[b]Ecological regulation of the metals is treated in Ref. 97.

Table 8. Industrial Sources That Produce Suspended, Colloidal Waste[a]

Industry	Origin	Principal disposal method
Food and drug		
canned goods	fruits, vegetables	filtration, lagooning
meat, poultry	bones, grease, fatty residue	filtration, settling, flotation
beet sugar	lime sludge	coagulation, lagooning
yeast	residue	anaerobic digestion, filtration
coffee	pulp	settling, filtration
rice	extractables (starch)	coagulation
soft drinks	equipment washing	municipal sewer
Apparel		
textiles	fibers	precipitation, filtration
leather goods	precipitated lime	sedimentation
laundry trades	fabric washing	flotation
Chemicals		
detergents	saponified soaps	flotation, precipitation
explosives	metals, oils, soaps	flotation, precipitation
phosphate/phosphorus	clays, slimes	lagooning, coagulation
Other materials		
pulp and paper	paper pulp washing	settling, lagooning
steel	coal coking, oils, scale	coagulation
iron-foundry	clay, coal	filtration
oil	drilling muds, sludge	burning
rubber	latex washing	aeration
glass	polishing, cleaning	precipitation
Energy		
coal processing	cleaning	flotation

[a]Adapted from Ref. 129.

known. But, owing to the chemical and physical complexity of smog, exposure to it often causes health problems that are far worse than those caused by exposure to its constituents. Thus synergistically detrimental effects can be anticipated, and these are difficult to treat. Similarly, exposure to airborne pollutants found indoors and in confined spaces, many of which are particulates or microbes of colloidal size, can lead to complex physiological responses; these reactions range from minor irritation of the eyes, nose, skin, and upper airways to chronic disease, such as pneumoconiosis, ie, fibrosis because of the inhalation of dust particulates over long periods of time, asthma, and cancer (134). Although indoor pollutant sources may emit low levels of particulates, the related health effects are not correspondingly low. Thus fundamental research is needed to characterize indoor materials, products, and activities, in terms of the health hazards linked to colloidal pollutants (135).

Applications of Colloids

General Uses. Although colloids may be undesirable components in industrial systems, particularly as waste or by-products and, in nature, in the forms of fog and mist, they are desirable in many technologically important processes such as mineral beneficiation and the preparation of ceramics, polymers, composite materials, paper, foods, textiles, photographic materials, cosmetics, and detergents. Colloids are deemed either desirable or undesirable, based on their unique physiochemical properties of diffusion, chemical reactivity with their environments' components, particulate interactions, adhesion, deposition, and electrical, thermal, and magnetic behavior. These properties underpin the wide use of colloids. Sample uses are as reinforcement aids in metals, ceramics, and plastics; as adhesion promoters in paints and thermoplastics; as nucleating agents in cloud seeding; as activated powder catalysts; as thickening agents in gels and slurries; and as abrasives in toothpastes (21,24,27,29,92,136). This list embodies the broad applications of (solid) colloids.

The remainder of this section specifies some applications for colloidal solids, liquids, and gases. Whereas some applications of colloidal materials have been mentioned earlier regarding specific phenomena, other examples will illustrate how colloids affect many technologically important systems in a positive manner and demonstrate the broad range of applications that permeates current synthetic materials.

Colloidal Solids. Used as reinforcement agents in metals, ceramics, and polymers (136,137), particles of colloidal solids may be spherical, angular, fibrillar, or flake-shaped. Examples include alumina and thoria to reinforce aluminum and nickel, respectively, by providing obstacles to the movement of dislocations in the metals, and zirconia and silicon carbide to reinforce a variety of ceramics, eg, alumina, silicon nitride, and glass, by inhibiting the propagation and opening of cracks in the matrix (see COMPOSITES). Asbestos, crystalline silicas, and organic solids are added to concrete to improve its strength by providing an interlocking particulate structure within the concrete matrix (139); asbestos (qv) (140), various oxides (141), and carbon black (27,141) are added to reinforce polymers by inducing a stiffened or high yield matrix. For instance, maximum strength of natural rubber may be achieved with ~10 vol % ZnO or ~22 vol % carbon black (qv) and that of TD (thoria-dispersed) nickel by ~3 vol % ThO_2 (136).

The magnitude of the strengthening often depends on particle shape. Fibrillar fillers are commonly used as discontinuous fibers in metals and plastics, eg, for 65 vol % E-glass in epoxy resin (136) and as whiskers in ceramics (138). Glass and aluminum oxide are common fillers that are occasionally pretreated with a polymeric or metallic coating; silica and various clays are also used. Unidirectionally oriented continuous fibers are less commonly employed, but they are successfully incorporated into laminated structures so that the fibers in successive layers are orthogonal or otherwise specifically oriented, yielding an alternating fiber structure (136).

Lastly, regarding reinforcement, the mechanism by which ceramics and metals are reinforced often involves precipitation of colloidal material during thermal treatment of the matrix composition (142–145). This approach is used for the TiO_2-precipitation in borosilicate glasses designed for enamels (142,145), the precipitation in AgCl–NaCl alloy systems (143), the $Ni_3(Al,Ti)$-precipitation in Inconel x-750 matrix (144), and the Fe_3C-precipitation in tempered martensite (144). Other examples of fillers are listed in Table 9. Strengthening mechanisms (27) include precipitation and dispersion when the reinforcing phase is metallic and the toughened materials are metals or ceramics. When inorganic, nonmetallic ceramics strengthen metals or polymers (146,147), the mechanism may be dispersion or reinforcement, for example, by cross-linking. Reinforcement implies a higher volume fraction than in dispersion hardening.

Fillers (qv) can be added not only to improve mechanical properties such as impact strength, fracture toughness, and tensile strength of structural ceramics, as indicated for concrete, but also to enhance optical properties, as is done for colored glasses containing colloidal gold or crystalline, chromium-based oxides. Other applications of colloidal solids include the preparation of rigid, elastic and thixotropic gels (136,148,149), aerogel-based thermal insulators (148,150), and surface coatings (27,29,148). Commercial uses of silica gel and sol-gel processing often focus on rigid gels having 20–30 vol % SiO_2. The principal interparticulate forces in a rigid gel are chemical and irreversible, and the colloid improves the gel's mechanical strength. Elastic gels are commonly associated with cellophane,

Table 9. Fillers for Reinforcement

Fillers	Matrix substance
Nonmetallic	
asbestos	mica
boron on tungsten	nylon-6,6
calcium carbonate	silica
fused silica	tungsten carbide
glass	titanium carbide
glass (SiO_2, Al_2O_3, B_2O_3); E-glass	titanium dioxide
graphite	zinc oxide
kaolin and other clays	zirconium dioxide
Metallic	
copper	titanium
gold	tungsten
nickel	zinc, cermets
Whiskers	
alkali halides	iron
Al_2O_3	α-SiC
B_4C	sulfides
graphite	graphite

[a]After Refs. 27, 136, 140–146.

rubber, cellulosic fibers, leather, and certain soaps; weak van der Waals forces between the particles in these gels render them reversible or elastic. Lastly, many thixotropic gels and surface coatings contain colloidal solids, eg, clays, alumina, ferric oxide, titania, silica, and zinc oxide. Consumer and industrial pastes belong to this category; putty, dough, drilling mud, lubricating grease, toothpaste, and paint are some examples.

Colloidal Liquids.   These fluids are commonly used in the form of emulsions by many industries. Permanent and transient antifoams consisting of an organic material, eg, polyglycol, oils, fatty materials, or silicone oil dispersed in water, is one application (25,26,151–153) that is important to a variety of products and processes: foods, cosmetics, pharmaceuticals, pulp and paper, water treatment, and minerals beneficiation. Other outlets for emulsions are paints (qv), lacquers, varnishes, and electrically and thermally insulating materials. The lowering of the interfacial tension, usually upon the adsorption of surfactant molecules, is a prerequisite for the formation of emulsions (qv). But other factors, such as interfacial hydration, droplet (interfacial) charge, the presence of either an adsorbed steric-stabilizing layer or a solid that is wetted differently by the bulk fluid and the colloidal liquid, and the self-association of amphiphilic molecules (those with hydrophilic and hydrophobic entities) may enhance the stability of an emulsion. Similarly, the ability of biological amphiphilic molecules to aggregate into spherical and nonspherical clusters, ie, vesicles, may have been important for the development of early living cells (101). Cellular biological membranes in plants and animals share features with these colloidal systems, although the molecular and hierarchical membrane structures, their hydration, and their dynamic properties are complex (101,154,155). Lastly, the macroscopic nature of concentrated gels, such as lubricating greases formed by dispersing short-chain surfactants (qv), eg, lithium 12-hydroxystearate, in mineral oil (156), is akin to the behavior of biological amphiphiles, being also dependent on self-assembly mechanisms. The associations between fibrous clusters, the length of threadlike surfactant strands, and the density of their contact points (cross-links) govern the grease's shear-resistance (156).

De-emulsification, ie, the breaking of foams or emulsions, is an important process, with the oil industry being a common one in which the process is often critical. Chemical and particulate agents that displace the surfactant and permit an unstabilized interface to form are used for this purpose.

Colloidal Gases.   Fluid foams are commonplace in foods, shaving cream, fire-fighting foam, mineral flotation, and detergents (25,26,152,153,157). Solid foams, eg, polyurethane foam and natural pumice, also contain dispersed gas bubbles that are often produced via viscoelastic polymer melts within which gas, eg, carbon dioxide, bubbles are nucleated (158). Thus, in view of the fact that the concentration of bubbles greatly affects the properties of foams, the production, dispersion, and maintenance of colloidal gas bubbles are basic to foams (qv) and related materials. Often, natural and synthetic soaps and surfactants are used to make fluid foams containing colloidal gas bubbles. These agents reduce the interfacial tension and, perhaps, the viscosity at the gas–liquid interface, making the foam stable. Also, some soluble proteins that denature upon adsorption or with agitation of the liquid phase can stabilize foams by forming insoluble, rigid layers at the gas–liquid interface (26).

New Developments

Important problems in colloid science remain to be addressed if the potential of colloids is to be fully exploited, among them, extension of understanding to more concentrated suspensions, testing of predictions using model powders, and examination of relaxation phenomena in ordered colloids. Much is known about colloids and their formation and behavior, but considerably more remains unknown. Thus the full potential to control colloids is not presently realized.

Recent work on the structure of regularly arranged colloidal suspensions indicates that ordering of particles in a fluid occurs under restricted solution conditions and solids concentrations exceeding approximately 50 vol % (159–162). Experimental advances have spurred theoretical modeling of the liquidlike-to-solidlike phase transition in suspensions (163,164) and the formation of hierarchical colloidal structures in the dry state that resemble grains in polycrystalline materials (165). Regular patterns that correspond to hexagonal or cubic packing, along with associated lattice defects, dislocations, grain boundaries, and segregation phenomena, have been identified in suspensions of solid colloids (160,165), and novel processing techniques that exploit these periodic structures are needed (166). An interesting feature of these suspensions is that, although liquid separates the particles and there are virtually no particle–particle contacts, the dynamical aspects of the thermodynamic structures and phase transitions in these systems resemble those found in atomic assemblies (164,167). Accordingly, as one example, these stable but ordered suspensions can be regarded as precursory systems for ordered, prefired, ceramic components. Outlets for such systems include various processing techniques, eg, slip, tape, freeze, pressure, centrifugal, and ultrasonic casting and isostatic, and hot pressing (168).

Whether or not such structures are successfully fabricated on a large scale, the ability to control concentrated colloidal suspensions, their rheological properties, and the microscopic packing of their particulates continue to be emphasized in research and play a central role in the production of advanced ceramics. In some instances, the ability to prevent particle agglomeration is paramount, whereas in other cases, the prediction of the non-Newtonian rheology of concentrated colloidal systems is most critical. Often, the two needs cannot be easily separated. The properties of mixed suspensions, in which more than one type of colloidal particle is dispersed, is an example that typifies this situation. Although the general properties of these systems resemble those of simpler colloidal systems, the different colloids present can interact with one another, increasing the complexity of the behavior (107,169). For instance, the minor component in a colloidal mixture can control the overall stability of the suspension (105,106,170), the suspension rheology (106,171), and even the suitability of a processing method (171). A few of the important areas for which concepts about the formation and stability of colloids need to be further developed are the fabrication of composites and other materials, the prevention of corrosion, the extraction of low-grade ores, and the treatment of complex, polluted environmental systems, including those with radionuclides formed by the weathering of high level nuclear waste glass (172) or the deterioration of aging hazardous waste sites (173). Many applications increasingly demand an improved ability to control the rheology of concentrated suspensions; ferrofluids, electrorheological fluids, and the storage and transport of blood belong to this group.

Another arena of maturing research that is likely to be greatly expanded is the use of modern and developing surface-science techniques to understand better the colloidal phenomena encountered in various disciplines, eg, heterogeneous catalysis, prevention of corrosion, processing of semiconductors and development of new materials, tribology, and adhesion (25,174,175). A particularly interesting surface-science utility is the atomic force microscopy (25,175), which can be a tool to study two-dimensional colloidal films. This technique, unlike scanning tunneling microscopy, which detects a current of tunneling electrons between a scanning tip and a surface (176), does not require the surface to be electrically conducting (175,177), though it can be used in such cases (178). Instead, the tip-surface interactions derived from van der Waals forces, electrostatic, frictional, and magnetic forces are measured on a localized scale, often over subnanometer dimensions (173). Being compatible with a variety of liquid environments (179) and capable of imaging a force as low as 10^{-7} N, atomic force microscopy has also been used to study many organic materials (175,179), including proteins, DNA, and various lipids (180).

Advances in the capability to measure surface and intermolecular forces directly, to a surface separation resolution of 0.1 nm and a force sensitivity of $\sim 10^{-8}$ N, greatly affect the ability to prepare or destroy, to stabilize, and to control colloidal systems (101,181). Besides the testing of quantitative models for the interaction force between surfaces, direct force measurements help in the understanding of a variety of complex colloidal phenomena. Some of these are phase diagrams for certain surfactant–water mixtures (182), the tolerance of some dispersions to high concentrations of salt (101), and the role of hydration in colloidal systems (101,183). The combination of measurements of the forces between surfaces and the insights offered by other evolving approaches, eg, ultrasonic studies of emulsions and suspensions (184,185), electroacoustics (186), and small-angle x-ray and neutron scattering (187), will hasten the knowledge and the control of phase transitions, compressibility, creaming, and sedimentation in concentrated colloidal systems.

BIBLIOGRAPHY

"Colloids" in *ECT* 3rd ed., Suppl. Vol., pp. 241–259, by Alan Bleier, Massachusetts Institute of Technology.

1. A. E. Alexander and P. Johnson, *Colloid Science,* Vol. 1, Oxford University Press, London, 1949, Chapt. 1, p. 1.
2. H. Freundlich, *Colloid & Capillary Chemistry,* H. S. Hatfield, Transl., Methuen, London, 1926.
3. J. R. Partington, *A Short History of Chemistry,* 3rd ed., Dover, New York, 1989.
4. E. Farber, *The Evolution of Chemistry,* 2nd ed., The Ronald Press, New York, 1969, Chapt. 20, p. 312.
5. S. Voyutsky, *Colloid Chemistry,* MIR Publishers, Moskow, Russia, 1978.
6. H. R. Kruyt, *Colloid Science, Volume I, Irreversible Systems,* Elsevier, Amsterdam, The Netherlands, 1952.

7. R. J. Hunter, *Foundations of Colloid Science,* Vol. 1, Clarendon Press, Oxford, UK, 1987.

8. F. F. Reuss, *Mémoires de la Société Imperiale des Naturalistes de Moscou* **2,** 327 (1809), cited in Refs. 1, 5, and 6.

9. R. Brown, *Phil. Mag.* **4,** 161 (1828); R. Brown, *Ann. du Phys. u Chem.* **14,** 294 (1828); M. Gouy, *J. Phys.* **7,** 561 (1988); F. M. Exner, *Ann. du Phys.* **2,** 843 (1900).

10. Ascherson, *Archiv. Anat. Physiol. Lpz.,* 44 (1840), cited in Ref. 1.

11. F. Selmi, *Nuovi Ann. d. Scienze Naturali di Bologna Serie II* **IV,** 145 (1845), cited in Ref. 1.

12. M. Faraday, *Phil. Trans.* **147,** 154 (1857), cited in Refs. 1 and 2.

13. T. Graham, *Phil. Trans. Roy. Soc.* **151,** 183 (1861); *J. Chem. Soc.,* 618 (1864), cited in Refs. 1 and 2.

14. T. Graham, *Lieb. Ann. Chem.* **121,** 36 (1862); *Ibid.* **135,** 65 (1865); *J. Chem. Soc.* **15,** 216 (1862); *Ibid.* **17,** 318 (1864); *Ann. Phys. Leipzig* **190,** 187 (1861); *Phil. Trans. R. Soc. London* **151,** 204 F(1861); *J. Chem. Soc.* **15,** 216 (1862).

15. Ref. 2, pp. 625–628.

16. R. Zsigmondy, *Z. Anal. Chem.,* **40,** 697 (1901); *Colloids and the Ultramicroscope,* J. Alexander, Transl., John Wiley & Sons, Inc., New York, 1909.

17. A. Einstein, *Ann. Physik* **17,** 549 (1905); *Ibid.* **19,** 371 (1906); *Ibid.* **34,** 591 (1911); *Investigations on the Theory of Brownian Movement,* Dover, New York, 1956.

18. M. von Smoluchowski, *Ann. Physik* **21,** 756 (1906).

19. J. Perrin, *Ann. Chim. Phys.* **18,** 5 (1909); in *Le Conseil Solvay,* 1911, cited in Ref. 4; *Abh. der Bunsen-Gesellschaft,* (7), Wilhelm Knapp, Halle, 1914, p. 205, cited and quoted in Ref. 4.

20. T. Svedberg, *Colloid Chemistry,* ACS Monograph 16, Chemical Catalog Co., New York, 1924; T. Svedberg and K. O. Pederson, *The Ultracentrifuge,* Clarendon, Oxford, UK, 1940.

21. E. Matijević, *Chem. Technol.* **3,** 656 (1973); *Ibid.* **21,** 176 (1991); *Ann. Rev. Mater. Sci.* **15,** 483 (1985).

22. R. Van Hardevald and F. Hartog, *Surf. Sci.* **15,** 189 (1969).

23. J. Th. G. Overbeek, *Adv. Colloid Interface Sci.* **15,** 251 (1982).

24. P. C. Hiemenz, *Principles of Colloid and Surface Chemistry,* 2nd ed., Marcel Dekker, Inc., New York, 1986; R. D. Vold and M. J. Vold, *Colloid and Interface Chemistry,* Addison-Wesley, Reading, Mass., 1983; H. Sonntag and K. Strenge, *Coagulation and Stability of Disperse Systems,* Halsted, New York, 1972; D. J. Shaw, *Introduction to Colloid and Surface Chemistry,* 3rd ed., Butterworth, London, 1980.

25. A. W. Adamson, *Physical Chemistry of Surfaces,* 5th ed., John Wiley & Sons, Inc., New York, 1990.

26. D. H. Everett, *Basic Principles of Colloid Science,* Royal Society of Chemistry, London, 1988.

27. C. R. Veale, *Fine Powders,* John Wiley & Sons, Inc., New York, 1972; J. K. Beddow, *Particulate Science and Technology,* Chemical Publishing, New York, 1980.

28. B. V. Derjaguin and L. D. Landau, *Acta Physiochem. URSS* **14,** 633 (1941); E. J. Verwey and J. Th. G. Overbeek, *Theory of the Stability of Lyophobic Colloids,* Elsevier, Amsterdam, The Netherlands, 1948.

29. G. D. Parfitt, in G. D. Parfitt, ed., *Dispersion of Powders in Liquids,* 3rd ed., Applied Science, London, 1981, Chapt. 1, pp. 1–50.

30. B. J. Ingebrethsen, Ph.D. dissertation, Clarkson University, Potsdam, N.Y., 1982.

31. H. Füredi-Milhofer and A. G. Walton, in Ref. 29, Chapt. 5, p. 203.

32. G. C. Lowrison, *Crushing and Grinding,* Chemical Rubber Co., Boca Raton, Fla., 1974.

33. G. E. Agar and P. Somasundaran, paper presented at *10th International Mineral Processing Congress,* London, UK, 1973.

34. A. Z. Frangiskos, Ph.D. dissertation, University of Leeds, Leeds, UK, 1956.

35. A. G. Evans, *J. Mater. Sci.* **7**, 1137 (1972).

36. J. A. Holmes, *Trans. Inst. Chem. Eng.* **35**, 125 (1957).

37. N. Arbiter and U. N. Bhrany, *Trans. Am. Inst. Min. Metall. Pet. Eng.* **217**, 245 (1960).

38. D. W. Fuerstenau and P. Somasundaran, *Proceedings of the 6th International Mineral Processing Congress, Cannes 1963,* Pergamon Press, Oxford, UK, 1965, p. 25.

39. O. G. Raabe, in B. Y. H. Liu, ed., *Fine Particles,* Academic Press, Inc., New York, 1975, p. 57.

40. M. Kerker, *Adv. Colloid Interface Sci.,* **5**, 105 (1975).

41. N. A. Fuchs and A. G. Sutugin, in C. N. Davies, ed., *Aerosol Science,* Academic Press, Inc., New York, 1966, Chapt. 1, p. 1.

42. H. Willeke, *Generation of Aerosols and Facilities for Exposure,* Ann Arbor Press, New York, 1980.

43. M. I. Tillery, G. O. Wood, and H. J. Ettinger, *Environ. Health Perspect.* **16**, 25 (1976).

44. T. T. Mercer, M. I. Tillery, and H. Y. Chow, *Am. Ind. Hyg. Assoc. J.* **29**, 66 (1968).

45. M. B. Denson and D. B. Swartz, *Rev. Sci. Instrum.* **45**, 81 (1974).

46. R. N. Bergland and B. Y. H. Liu, *Environ. Sci. Technol.* **7**, 147 (1973).

47. W. H. Walton and W. C. Prewett, *Proc. Phys. Soc. London* **62**, 341 (1949).

48. B. Vonnegut and R. Neubauer, *J. Colloid Sci.* **7**, 616 (1952).

49. A. L. Heubner, *Science* **168**, 118 (1970).

50. E. P. Knutson and K. T. Whitby, *Aerosol Sci.* **6**, 443 (1975); J. K. Agarwal and G. J. Sem, *TSI Quarterly* **6**, 3 (1978).

51. R. Zsigmondy, *Z. Phys. Chem. (Leipzig)* **56**, 65 (1906).

52. E. Wiegel, *Kolloidchem. Beih.* **25**, 176 (1927).

53. H. R. Kruyt and A. E. van Arkel, *Recl. Trav. Chim. Pay-Bas* **39**, 656 (1920).

54. V. K. LaMer and M. D. Barnes, *J. Colloid Sci.* **1**, 71 (1946).

55. V. K. LaMer and R. Dinegar, *J. Am. Chem. Soc.* **72**, 4847 (1950).

56. P. Pendleton, B. Vincent, and M. L. Hair, *J. Colloid Interface Sci.* **80**, 512 (1981).

57. R. S. Sapieszko and E. Matijević, *Corrosion* **36**, 522 (1980).

58. E. Matijević, in L. L. Hench and D. R. Ulrich, eds., *Ultrastructure Processing of Ceramics, Glasses, and Composites,* Wiley-Interscience, New York, 1984, Chapt. 27, pp. 334–352.

59. W. Stöber, A. Fink, and E. Bohn, *J. Colloid Interface Sci.* **26**, 62 (1968).

60. D. Sinclair and V. K. LaMer, *Chem. Rev.* **44**, 245 (1949).

61. E. Matijević, *Acc. Chem. Res.* **14**, 22 (1981); in L. L. Hench and D. R. Ulrich, eds., *Science of Ceramic Chemical Processing,* Wiley-Interscience, New York, 1986, Chapt. 50, pp. 463–481.

62. T. Sugimoto, *Adv. Colloid Interface Sci.* **28**, 65–108 (1987).

63. R. H. Ottewill and R. F. Woodbridge, *J. Colloid Sci.* **16**, 581 (1961).

64. G. Chiu, *J. Colloid Interface Sci.* **83**, 309 (1981).

65. A. Bleier and R. M. Cannon, *Am. Ceram. Soc. Bull.* **61**, 336 (1982); C. J. Brinker, D. E. Clark, and D. R. Ulrich, eds., *Better Ceramics Through Chemistry II,* Materials Research Society, Pittsburgh, Pa., 1986, pp. 71–78.

66. J. H. L. Watson, W. Heller, and W. Wojtowicz, *J. Chem. Phys.* **16**, 997 (1948).

67. S. Hamada and E. Matijević, *J. Colloid Interface Sci.,* **84**, 274 (1982); D. Murphy and E. Matijević, *J. Chem. Soc., Faraday Trans. I* **80**, 563 (1984); T. Sugimoto and E. Matijević, *J. Colloid Interface Sci.* **74**, 227 (1980); G. J. Muench, S. Arajs, and E. Matijević, *J. Appl.* **52**, 2493 (1981).

68. J. W. Vanderhoff and co-workers, in G. Goldfinger, ed., *Clean Surfaces,* Marcel Dekker, Inc., New York, 1970, Chapt. 2, p. 15; J. W. Vanderhoff, *Pure Appl. Chem.* **52,** 1263 (1980).

69. A. Homola and R. O. James, *J. Colloid Interface Sci.* **59,** 123 (1977); Y. Chung-li, J. W. Goodwin, and R. H. Ottewill, *Progr. Colloid Polymer Sci.* **60,** 163 (1976).

70. V. I. Eliseeva, S. S. Ivanchev, S. I. Kuchanov, and A. V. Lebedev, *Emulsion Polymerization and Its Applications,* Consultants Bureau, New York, 1976.

71. R. Partch, E. Matijević, A. W. Hodgson, and B. E. Aiken, *J. Polymer Sci., Polym. Chem. Ed.* **21,** 961 (1983); R. E. Partch, K. Nakamura, K. J. Wolfe, and E. Matijević, *J. Colloid Interface Sci.* **105,** 560 (1985).

72. J. Ugelstad, P. C. Mork, K. H. Kaggerud, T. Ellingsen, and A. Berge, *Adv. Colloid Interface Sci.* **13,** 101 (1980); A. T. Skjeltorp, J. Ugelstad, and T. Ellingsen, *J. Colloid Interface Sci.* **113,** 577 (1986).

73. B. H. Kaye, *Direct Characterization of Fineparticles,* Wiley-Interscience, New York, 1981; M. Kerker, *The Scattering of Light and Other Electromagnetic Radiation,* Academic Press, Inc., New York, 1969; J. D. Stockham and E. G. Fochtman, eds., *Particle Size Analysis,* Ann Arbor Sci., Ann Arbor, Mich., 1979; D. W. Schuerman, ed., *Light Scattering by Irregularly Shaped Particles,* Plenum Press, New York, 1980.

74. M. J. Jaycock and G. D. Parfitt, *Chemistry of Interfaces,* Ellis Horwood Ltd., Chichester, UK, 1981.

75. G. D. Parfitt and K. S. W. Sing, eds., *Characterization of Powder Surfaces,* Academic Press, Inc., New York, 1976.

76. D. H. Everett and J. M. Haynes, in D. H. Everett, ed., *Colloid Science,* Vol. 1, The Chemical Society, London, 1973, Chapt. 4, p. 123; E. A. Boucher, *J. Mater. Sci.* **11,** 1734 (1976); F. A. L. Dullien and V. K. Batra, *Ind. Eng. Chem.* **62,** 25 (1970); A. A. Liabastre and C. Orr, *J. Colloid Interface Sci.* **64,** 1 (1978).

77. S. J. Gregg and K. S. W. Sing, *Adsorption, Surface Area and Porosity,* Academic Press, Inc., London, 1967.

78. S. Brunauer, P. H. Emmett, and E. Teller, *J. Am. Chem. Soc.* **60,** 309 (1938); S. Brunauer, *The Adsorption of Gases and Vapors,* Vol. 1, Princeton University Press, Princeton, N.J., 1945.

79. I. Langmuir, *J. Am. Chem. Soc.* **40,** 1361 (1918).

80. M. Polanyi, *Verh. Deut. Physik Ges.* **16,** 1012 (1914).

81. D. Graham, *J. Phys. Chem.* **59,** 896 (1955); H. J. van den Hul and L. Lyklema, *J. Colloid Interface Sci.* **23,** 500 (1967); J. F. Padday, *Pure and Applied Chemistry, Surface Area Determination,* Butterworths, London, 1969; G. Schay and L. G. Nagy, *J. Colloid Interface Sci.* **38,** 302 (1973); D. H. Everett and co-workers, International Union of Pure and Applied Chemistry, Division of Physical Chemistry, *Manual of Symbols and Terminology for Physical Quantities and Units,* Appendix II, Part I; *Pure and Appl. Chem.* **31,** 579 (1972); D. H. Everett, in D. H. Everett, ed., *Colloid Science,* Vol. 1, The Chemical Society, London, 1973, Chapt. 2, p. 49; M. A. Rahman and A. K. Ghosh, *J. Colloid Interface Sci.* **77,** 50 (1980); D. H. Everett, in J. W. Goodwin, ed., *Colloidal Dispersions,* The Royal Society of Chemical Society, London, 1982, Chapt. 4, p. 71.

82. B. D. Cullity, *Elments of X-Ray Diffraction,* 2nd ed., Addison-Wesley, Reading, Mass., 1978.

83. G. E. Bacon, *Neutron Diffraction,* 3rd ed., Oxford University Press, London, 1975.

84. J. M. Ziman, *Principles of the Theory of Solids,* 2nd ed., Cambridge University Press, Cambridge, UK, 1972, Chapt. 10, p. 329; M. J. Schmank and A. D. Krawitz, *Metall. Trans. A* **13,** 1069 (1982); A. D. Krawitz and co-workers in *Proceedings of the International Conference on the Science of Hard Materials,* Jackson, Wyo., Aug. 1981.

85. J. A. Appelbaum, G. A. Baraff, and D. R. Hamann, *Phys. Rev. Lett.* **35,** 729 (1975).

86. J. J. Lander and J. Morrison, *J. Appl. Phys.* **34,** 1411 (1963).

87. J. J. Lander and J. Morrison, *Surf. Sci.* **4,** 241 (1966).

88. L. A. Girifalco and R. J. Good, *J. Phys. Chem.* **61,** 904 (1957); R. J. Good and L. A. Girifalco, *J. Phys. Chem.* **64,** 561 (1960); R. J. Good, in R. J. Good and R. R. Stromberg, eds., *Surface and Colloid Science,* Vol. 11, Plenum Press, New York, 1979, Chapt. 1, p. 1.

89. F. M. Fowkes, *Ind. Eng. Chem.* **56,** 40 (1964).

90. S. Mizuta, W. R. Cannon, A. Bleier, and J. S. Haggerty, *Am. Ceram. Soc. Bull.* **61,** 872 (1982).

91. A. Bleier, *J. Phys. Chem.* **87,** 3493–3500 (1983); *J. Am. Ceram. Soc.* **66,** C-79 (1983).

92. C. Orr, *Particulate Technology,* Macmillan, New York, 1966.

93. G. C. Benson, P. Scheiber, and F. van Zeggeren, *Can. J. Chem.* **34,** 1553 (1956).

94. A. J. Tyler, J. A. G. Taylor, B. A. Pethica, and J. A. Hockey, *Trans. Faraday Soc.* **67,** 483 (1971).

95. M. J. Rosen, *Surfactants and Interfacial Phenomena,* John Wiley & Sons, Inc., New York, 1978.

96. W. C. Preston, *J. Phys. Colloid Chem.* **52,** 84 (1948).

97. W. Stumm and J. J. Morgan, *Aquatic Chemistry,* 2nd ed., John Wiley & Sons, Inc., New York, 1981.

98. M. J. Vold, *J. Colloid Sci.* **16,** 1 (1961).

99. H. G. Hamaker, *Physica (Utrecht)* **4,** 1058 (1937).

100. I. E. Dzyaloshinskii, E. M. Lifshitz, and L. P. Pitaevskii, *Adv. Phys.* **10,** 165 (1961).

101. J. N. Israelachvili, *Intermolecular and Surface Forces,* 2nd ed., Academic Press, Inc., New York, 1992.

102. T. Sato and R. Ruch, *Stabilization of Colloidal Dispersions by Polymer Adsorption,* Marcel Dekker, Inc., New York, 1980; Th. F. Tadros, ed., *The Effect of Polymers on Dispersion Properties,* Academic Press, Inc., London, 1982; J. Lyklema, *Adv. Colloid Interface Sci.* **2,** 65 (1968); Y. S. Lipatov and L. M. Sergeeva, *Adsorption of Polymers,* Halsted, New York, 1974; B. Vincent, *Adv. Colloid Interface Sci.* **4,** 193 (1974); C. A. Finch, ed., *Chemistry and Technology of Water-Soluble Polymers,* Plenum Press, Inc., New York, 1983; B. Vincent, in Th. F. Tadros, ed., *Solid/Liquid Dispersions,* Academic Press, Inc., 1987, pp. 147–162; H. J. Ploehn and W. B. Russel, *Adv. Chem. Eng.* **15,** 137 (1990).

103. B. V. Derjaguin, *Discuss. Faraday Soc.* **42,** 317 (1966); J. W. Th. Lichtenbelt, H. J. M. C. Ras, and P. H. Wiersema, *J. Colloid Interface Sci.* **46,** 522 (1974); J. W. Th. Lichtenbelt, C. Pathmamanoharan, and P. H. Wiersema, *J. Colloid Interface Sci.* **49,** 281 (1974); J. Th. G. Overbeek, *J. Colloid Interface Sci.* **58,** 408–422 (1977).

104. R. J. Pugh and J. A. Kitchener, *J. Colloid Interface Sci.* **35,** 656 (1971).

105. A. Bleier, *Colloids Surfaces,* **66,** 157–179 (1992); A. Bleier and C. G. Westmoreland, *J. Am. Ceram. Soc.* **74** (12), 3100–3111 (1991); S. Baik, A. Bleier, and P. F. Becher, in C. J. Brinker, D. E. Clark, and D. R. Ulrich, eds., *Better Ceramics through Chemistry II,* Materials Research Society, Pittsburgh, Pa., 1986, pp. 791–800; A. Bleier and C. G. Westmoreland, in C. J. Brinker, D. E. Clark, and D. R. Ulrich, eds., *Better Ceramics through Chemistry III,* Materials Research Society, Pittsburgh, Pa., 1988, pp. 145–154.

106. A. Bleier and C. G. Westmoreland, in B. J. J. Zelinski, C. J. Brinker, D. E. Clark, and D. R. Ulrich, eds., *Better Ceramics through Chemistry IV,* Materials Research Society, Pittsburgh, Pa., 1990, pp. 185–190.

107. B. V. Derjaguin, *Discuss. Faraday Soc.* **18,** 85 (1954); B. V. Derjaguin, N. V. Churaev, and V. M. Muller, *Surface Forces,* V. I. Kisin, transl., J. A. Kitchener, ed. (Transl.), Consultants Bureau, New York, 1987, Chapt. 9, pp. 311–326.

108. O. Stern, *Z. Elektrochem.* **30,** 508 (1924); D. C. Grahame, *Chem. Rev.* **41,** 441 (1947).

109. M. J. Garvey, Th. F. Tadros, and B. J. Vincent, *J. Colloid Interface Sci.* **55,** 440 (1976); F. Lafuma, K. Wong, and B. Cabane, *J. Colloid Interface Sci.* **143,** 9 (1976).

110. R. I. Feigin and D. H. Napper, *J. Colloid Interface Sci.* **74,** 567 (1980); *Ibid.* **75,** 525 (1980).

111. A. Okagawa, G. J. Ennis, and S. G. Mason, *Can. J. Chem.* **56,** 2815, 12824 (1978); M. Zuzousky, Z. Priel, and S. G. Mason, *J. Colloid Interface Sci.* **75,** 230 (1980).

112. T. G. M. Van den Ven and R. J. Hunter, *J. Colloid Interface Sci.* **69,** 135 (1979); R. J. Hunter and J. Frayne, *J. Colloid Interface Sci.* **71,** 30 (1979); R. J. Hunter and J. Frayne, *J. Colloid Interface Sci.* **76,** 107 (1980).

113. Th. G. M. van de Ven, *Colloidal Hydrodynamics,* Academic Press, Inc., London, 1989.

114. I. M. Krieger, *Adv. Colloid Interface Sci.* **3,** 111 (1972).

115. B. A. Firth, *J. Colloid Interface Sci.* **57,** 257 (1976).

116. K. Higashitani, S. Miyafusa, T. Matsuda, and Y. Matsuno, *J. Colloid Interface Sci.* **77,** 21 (1980).

117. C. D. Han and R. G. King, *J. Rheol.* **24,** 213 (1980).

118. R. J. Hunter, *Foundations of Colloid Science,* Vol. 2, Clarendon Press, Oxford, UK, 1989, Chapt. 14, p. 992; J. Happel and H. Brenner, *Low Reynolds Number Hydrodynamics,* Martinus Nijhoff Publishers, The Hague, The Netherlands, 1983, pp. 431–473.

119. C. C. Mill, ed., *Rheology of Disperse Systems,* Pergamon Press, Inc., Elmsford, N.Y., 1959.

120. M. R. Rosen, *Polym.-Plast. Technol. Eng.* **12,** 1 (1979).

121. E. Matijević, in K. J. Mysels, C. M. Samour, and J. H. Hollister, eds., *Twenty Years of Colloid and Surface Chemistry,* American Chemical Society, Washington, D.C., 1973, p. 283; *J. Colloid Interface Sci.* **43,** 217 (1973).

122. *McCutcheon's Detergents and Emulsifiers,* MC Publ. Co., Glen Rock, N.J., 1980.

123. P. F. Kane and G. B. Larrabee, *Anal. Chem.* **49,** 221R (1977); P. F. Kane and G. B. Larrabee, *Anal. Chem.* **51,** 308R (1979); G. B. Larrabee and T. J. Shaffner, *Anal. Chem.* **53,** 163R (1981); S. N. K. Chaudhari and K. L. Cheng, *Appl. Spectrosc. Rev.* **16,** 187 (1980); E. N. Sickafus, *Ind. R&D,* June 1980; M. H. Koppelman and J. G. Dillard, in M. M. Mortland and V. C. Farmer, eds., *International Clay Conference, 1978,* Elsevier, Amsterdam, The Netherlands, 1979, p. 153; R. J. Blattner and C. A. Evans, Jr., in W. Bardsley, D. T. J. Hurle, and J. B. Mullin, eds., *Crystal Growth: A Tutorial Approach,* North-Holland, The Netherlands, 1979, p. 269; M. Beer, R. W. Carpenter, L. Eyring, C. E. Lyman, and J. M. Thomas, *C&EN* **59,** 40 (1981); H. H. Brongersma, F. Meijer, and H. W. Werner, *Phillips Tech. Rev.* **54** (11/12), 357 (1974); W. A. Beers, *Res./Dev.,* 18 (Nov. 1975); C. A. Evans, Jr., *Anal. Chem.* **47,** 818A (1975); J. W. Coburn, *Thin Solid Films* **64,** 371 (1982); T. Cosgrove, in D. H. Everett, ed., *Colloid Science,* Vol. 3, The Chemical Society, London, 1979, Chapt. 7, p. 293; G. A. Somorjai, in T. Fort and K. H. Mysels, eds., *Eighteen Years of Colloid and Surface Chemistry, The Kendall Award Addresses, 1973–1990,* American Chemical Society, Washington, D.C., 1991, p. 161.

124. G. A. Somorjai, *Chemistry in Two Dimensions: Surfaces,* Cornell University Press, Ithaca, N.Y., 1981; G. A. Somorjai and M. A. Van Hove, *Adsorbed Monolayers on Solid Surfaces,* Springer-Verlag, New York, 1979, p. 121.

125. H. L. Little, *Infrared Spectra of Adsorbed Species,* Academic Press, Inc., London, 1966; M. L. Hair, *Infrared Spectroscopy in Surface Chemistry,* Marcel Dekker, Inc., New York, 1967; P. P. Yaney and R. J. Becker, *Appl. Surf. Sci.* **4,** 356 (1980); M. R. Basila, *Appl. Spectrosc. Rev.* **1,** 289 (1969); R. P. Eischens, *Acc. Chem. Res.* **5,** 74 (1972); T. A. Egerton and A. H. Hardin, *Catal. Rev.* **11,** 71 (1975); G. H. Rochester, *Powder Technol.* **13,** 157 (1976).

126. R. W. O'Brien, *J. Colloid Interface Sci.* **81,** 234 (1981).

127. R. C. Weast and M. J. Astle, eds., *Handbook of Chemistry and Physics,* 62nd ed., Chemical Rubber Company, Boca Raton, Fla., 1981, p. D-101; S. Budavari and co-eds., *The Merck Index,* 11th ed., Merck, Rahway, N.J., 1989, p. MISC-53.

128. A. E. Martell, *Pure Appl. Chem.* **44,** 81 (1975); F. T. Mackenzie and R. Wollast, in E. D. Goldberg, ed., *The Sea,* Vol. 6, Wiley-Interscience, New York, 1977, p. 739, cited in Ref. 97.

129. W. F. Echelberger, *Water Pollution Control Technology,* Course Syllabus and Study Materials, Center for Professional Advancement, East Brunswick, N.J., July 1977.

130. *Amendment of the Federal Water Pollution Control Act,* U.S. Government, Public Law 92-500, 92nd Congress, 5.2770, Washington, D.C., Oct. 18, 1972.

131. J. Wei, T. W. F. Russel, and M. W. Swartzlander, *The Structure of the Chemical Processing Industries,* McGraw-Hill Book Co., Inc., New York, 1979.

132. M. P. Freeman and J. A. Fitzpatrick, eds., *Physical Separations,* Engineering Foundation, New York, 1980; A. J. Rubin, ed., *Chemistry of Wastewater Technology,* Ann Arbor Sci., Ann Arbor, Mich., 1978.

133. H. W. Zussman, *Adv. Chem. Ser.* **83,** 116 (1968); *C&EN* **60,** 41 (1982).

134. R. E. Glenn, *Ceram. Eng. Sci. Proc.* **13,** 153 (1992); W. G. Tucker, *Ann. N.Y. Acad. Sci.* **641,** 1 (1992); B. A. Tichenor, *Ann. N.Y. Acad. Sci.* **641,** 63 (1992); D. J. Moschandreas, *Ann. N.Y. Acad. Sci.* **641,** 87 (1992).

135. W. G. Tucker, *Ann. N.Y. Acad. Sci.* **641,** 322–327 (1992).

136. Z. D. Jastrzebski, *The Nature and Properties of Engineering Materials,* 2nd ed., John Wiley & Sons, Inc., New York, 1977; J. E. Gordon, *The New Science of Strong Materials,* 2nd ed., Princeton University Press, Princeton, N.J., 1976; A. G. Guy, *Essentials of Materials Science,* McGraw-Hill Book Co., Inc., New York, 1976.

137. S. J. Lefond, ed., *Industrial Minerals and Rocks,* 4th ed., American Institute of Minerology and Metallurgy, Petroleum Engineering, Inc., New York, 1975.

138. P. F. Becher, *J. Am. Ceram. Soc.* **74,** 255–269 (1991); A. Kelly and R. B. Nicholson, eds., *Strengthening Methods in Crystals,* Applied Science Publishing, London, 1971.

139. A. A. Ames, in Ref. 137, p. 129.

140. N. Severinghaus, in Ref. 137, p. 235.

141. L. Mitchell, in Ref. 137, p. 33.

142. W. D. Kingery, H. K. Bowen, and D. R. Uhlmann, *Introduction to Ceramics,* 2nd ed., Wiley-Interscience, New York, 1976.

143. R. J. Stokes and C. H. Li, *Acta Metall.* **10,** 535 (1962).

144. K. M. Ralls, T. H. Courtney, and J. Wulff, *Introduction to Materials Science and Engineering,* John Wiley & Sons, Inc., New York, 1976.

145. R. J. Charles, in G. Piel and co-eds., *Materials,* Scientific American, Inc., W. H. Freeman and Co., San Francisco, Calif., 1967; A. Kelly, in G. Piel and co-eds., *Materials,* Scientific American, Inc., W. H. Freeman and Co., San Francisco, Calif., 1967.

146. L. E. Murr, *Interfacial Phenomena in Metals and Alloys,* Addison Wesley, London, 1975; S. J. Burden, *Ceram. Eng. Sci. Proc.,* **3,** 1 (1982).

147. Ref. 125, pp. C740 and C-747.

148. S. J. Teichner, *Chemtech* **21,** 372–377 (1991).

149. E. Dickinson, *Chemtech* **21,** 665–669 (1991).

150. H. D. Gesser and C. Goswami, *Chem. Rev.* **89,** 765 (1989); R. J. Ayen and P. A. Iacobucci, *Rev. Chem. Eng.* **5,** 157 (1988).

151. S. Ross, *Chem. Eng. Progr.* **63,** 41 (1967); S. Ross and J. N. Butler, *J. Phys. Chem.* **60,** 1255 (1956); S. Ross and R. M. Haak, *J. Phys. Chem.* **62,** 1260 (1958); J. G. Hawke and A. E. Alexander, *J. Colloid Sci.* **11,** 419 (1956); R. E. Prattle, *J. Soc. Chem. Ind. (London)* **69,** 363, 368 (1950); P. Becher, *Emulsions,* Reinhold Publishing Corp., New York, 1965.

152. J. J. Bikerman, *Foams, Theory and Industrial Applications,* Reinhold Publishing Corp., New York, 1952; F. Sebba, *J. Colloid Interface Sci.* **35,** 643 (1971); J. A. Kitchener, in J. F. Danielli, K. G. A. Pankhurst, and A. C. Riddlford, eds., *Recent Progress in Surface Science,* Vol. I, Academic Press, Inc., New York, 1964; J. J. Bikerman, *Foams,* Springer-Verlag, New York, 1973; M. C. Phillips, in F. Franks, ed., *Water, A Comprehensive Treatise,* Vol. 5, Plenum Publishing Corp., New York, 1981, pp. 133–172.

153. E. Dickinson, ed., *Food Emulsions and Foams,* Royal Society of Chemistry, London, 1987.

154. G. Cevc and D. Marsh, *Phospholipid Bilayers,* John Wiley & Sons, Inc., New York, 1987; D. Marsh, *Handbook of Lipid Bilayers,* CRC Press, Boca Raton, Fla., 1990; F. Franks, *Water,* Royal Society of Chemistry, London, 1984, pp. 69–78.

155. R. Miller and G. Kretzschmar, *Adv. Colloid Interface Sci.* **37,** 97–121 (1991).

156. J. Prost and F. Rondelez, *Suppl. Nature* **350,** 11–23 (1991).

157. D. F. Darling and R. J. Birkett, in Ref. 151, pp. 1–29.

158. F. W. Billmeyer, Jr., *Textbook of Polymer Science,* 2nd ed., Wiley-Interscience, New York, 1971.

159. P. A. Hiltner and I. M. Krieger, *J. Phys. Chem.* **73,** 2386 (1969); A. Kose and S. Hachisu, *J. Colloid Interface Sci.* **46,** 460 (1974).

160. S. Okamuto and S. Hachisu, *J. Colloid Interface Sci.* **61,** 172 (1977).

161. K. Takano and S. Hachisu, *J. Colloid Interface Sci.* **66,** 124 (1978).

162. I. F. Efremov, in E. Matijecić, ed., *Colloid and Surface Science,* Vol. 8, Plenum Publishing Corp., New York, 1976, Chapt. 2, p. 85.

163. W. van Meegan and I. Snook, *Faraday Disc.* **65,** 92 (1978); E. Dickenson, *J. Chem. Soc., Faraday Trans. II* **75,** 466 (1979); E. Dickinson, in D. H. Everett, ed., *Colloid Science,* Vol. 4, London, 1983, p. 150.

164. P. N. Pusey, W. van Megen, S. M. Underwood, P. Bartlett, and R. H. Ottewill, *J. Phys.: Condens. Matter* **2,** SA373-SA377 (1990); *Physica A* **176,** 16–27 (1991).

165. I. F. Efremov, G. M. Lúkashenko, and O. G. Us'yarov, in B. V. Derjaguin, ed., *Research in Surface Forces,* Vol. 4, R. K. Johnston, Transl., Consultants Bureau, New York, 1975, pp. 32–38; I. A. Aksay and R. Kikuchi, in L. L. Hench and D. R. Ulrich, eds., *Science of Ceramic Chemical Processing,* Wiley-Interscience, New York, 1986, Chapt. 54, pp. 513–521.

166. A. Bleier, in R. F. Davis, H. Palmour III, and R. L. Porter, eds., *Emergent Process Methods for High Technology Ceramics,* Plenum Publishing Corp., New York, 1984, p. 71.

167. P. N. Pusey, in D. Levesque, J.-P. Hansen, and J. Zinn-Justin, eds., *Liquids, Freezing and the Glass Transition,* Les Houches Session LI, Elsevier, Amsterdam, The Netherlands, 1991, cited in Ref. 162.

168. F. Y. Wang, *Ceramic Fabrication Processes,* Academic Press, Inc., New York, 1976; J. S. Reed, *Introduction to the Principles of Ceramic Processing,* Wiley-Interscience, New York, 1988, Chapt. 22, p. 380.

169. P. M. Adler, A. Nadim, and H. Brenner, *Adv. Chem. Eng.* **15,** 1–72 (1990).

170. A. Bleier and E. Matijević, *J. Colloid Interface Sci.* **55,** 510 (1976); *J. Chem. Soc., Faraday Trans. I.* **74,** 1346 (1978); in A. J. Rubin, ed., *Chemistry of Waste Water Technology,* Ann Arbor Science, Ann Arbor, Mich., 1978, p. 81.

171. O. O. Omatete, A. Bleier, C. G. Westmoreland, and A. C. Young, *Ceram. Eng. Sci. Proc.* **12,** 2084–2094 (1991); A. Bleier, O. O. Omatete, and C. G. Westmoreland, in M. J. Hampden-Smith, W. G. Klemperer, and C. J. Brinker, eds., *Better Ceramics through Chemistry V,* Vol. 271, Materials Research Society, Pittsburgh, Pa., 1992, pp. 269–275; A. Bleier and O. O. Omatete, "Rheology and Microstructure of Concentrated Zirconia—Alumina Suspensions for Gelcasting Composites," submitted for publica-

tion in L. J. Struble, C. F. Zukoski, and G. Maitland, eds., *Flows and Microstructure of Dense Suspensions*, Materials Research Society, Pittsburgh, Pa., in press.

172. J. K. Bates, J. P. Bradley, A. Teetsov, C. R. Bradley, and M. B. ten Brink, *Science* **256**, 649–651 (1992).

173. J. F. McCarthy and J. M. Zachara, *Environ. Sci. Technol.* **23**, 496 (1989).

174. J. T. Yates, Jr., *C&EN* **70**, 22–35 (1992).

175. G. Binnig, C. F. Quate, and Ch. Gerber, *Phys. Rev. Lett.* **56**, 930 (1986); P. K. Hansma and co-workers, *Science* **242**, 209 (1988); J. Frommer and E. Meyer, *J. Phys.: Condens. Matter* **3**, S1–S9 (1991); H. G. Hansma and coworkers, *Langmuir* **7**, 1051 (1991).

176. G. Binnig and co-workers, *Phys. Rev. Lett.* **49**, 57 (1982); C. F. Quate, *Physics Today* **39**, 26 (1986); P. K. Hansma and J. Tersoff, *J. Appl. Phys.* **61**, R1 (1987); J. Schneir and P. K. Hansma, *Langmuir* **3**, 1025 (1987).

177. A. L. Weisenhorn and co-workers, *Scanning Microsc.* **4**, 511 (1990).

178. S. Manne and co-workers, *Science* **251**, 183 (1991).

179. B. Drake and co-workers, *Science* **243**, 1586 (1989).

180. E. Meyer and co-workers, *Nature* **349**, 398 (1991); A. L. Weisenhorn and co-workers, *Science* **247**, 1330 (1990); H. G. Hansma and co-workers, *J. Vac. Sci. Technol. B* **9**, 1282 (1991); M. Egger and co-workers, *J. Struct. Biol.* **103**, 89 (1990); B. Drake and co-workers, *Science* **243**, 1586 (1989); J. A. Zasadzinski and co-workers, *Biophys. J.* **59**, 755 (1991); J. Frommer, *Angew. Chem.* in press (1992).

181. D. Tabor and R. H. S. Winterton, *Proc. Roy. Soc. London A* **312**, 435–450 (1969); J. N. Israelachvili and D. Tabor, *Proc. Roy. Soc. London A* **331**, 19–38 (1972); J. N. Israelachvili and D. Tabor, in J. F. Danielli, M. D. Rosenberg, and D. A. Cadenhead, eds., *Recent Progress in Surface Science*, Vol. 7, Academic Press, Inc., New York, 1973, pp. 1–55; J. N. Israelachvili and G. E. Adams, *J. Chem. Soc. Faraday Trans. I* **74**, 975–1001 (1978); J. N. Israelachvilli, *Acc. Chem. Res.* **20**, 415–421 (1987).

182. P. M. Claesson, C. E. Blom, P. C. Herder, and B. W. Ninham, *J. Chem. Soc. Faraday Trans. I* **82**, 2735–2746 (1986).

183. B. V. Velamakanni, J. C. Chang, F. F. Lange, and D. S. Pearson, *Langmuir* **6**, 1323–1325 (1990).

184. D. J. McClements, *Adv. Colloid Interface Sci.* **37**, 33–72 (1991).

185. A. J. Babchin, R. S. Chow, and R. P. Sawatzky, *Adv. Colloid Interface Sci.* **20**, 111 (1989).

186. B. J. Marlow, D. Fairhurst, and H. P. Pendse, *Langmuir* **4**, 611–626 (1988); R. W. O'Brien, *J. Fluid Mech.* **190**, 71–86 (1988); R. W. O'Brien, *J. Fluid Mech.* **212**, 81 (1990); N. P. Miller and J. C. Berg, *Colloids Surfaces* **59**, 119–128 (1991); U.S. Pat. 4,497,207 (1985), T. A. Oja, G. L. Peterson, and D. W. Cannon.

187. D. W. Schaefer and K. D. Keefer, in L. Pietronero and E. Tosatti, eds., *Fractals in Physics*, Elsevier, Amsterdam, The Netherlands, 1986, pp. 39–45; C. J. Brinker and G. W. Scherer, *Sol-Gel Science: They Physics and Chemistry of Sol-Gel Processing*, Academic Press, Inc., Boston, Mass., 1990; Ref. 35, Chapt. 14, p. 827; J. B. Hayter, *Mater. Sci. Forum* **27/28**, 345 (1988); J. B. Hayter, in V. Degiorgio and M. Corti, eds., *Physics of Amphiphiles: Micelles, Vesicles, and Microemulsions*, North Holland, Amsterdam, The Netherlands, 1985, p. 59.

ALAN BLEIER
Oak Ridge National Laboratory

COLOR

Color is the part of perception that is carried to the eye from our surroundings by differences in the wavelengths of light. This involves, first, the nature and spectral power distribution in the light from the illuminating light sources. Next, there are several often interrelated processes derived from the interaction of the illumination with matter, including absorption, reflection, refraction, diffraction, scattering, and fluorescence. Finally, there is the perception system, involving the eye and the transmission system from eye to brain, leading to the final interpretation reached in the brain. This last is a complex process, involving psychological as well as physiological factors. As one example, a specific shade of green may have a quite different meaning for a jungle resident than it has for a desert dweller. One cannot know precisely what another sees as color, but the development of an agreed terminology based on common experience has led to a satisfactory science of color description and measurement.

The precision measurement of color is of significance in many branches of science and technology. It serves as a record for archival description, for standardization purposes, and for matching and controlling the many colorful products of commerce. In a field that has changed significantly even in the last decade, five current books can be particularly recommended for further details (1–5). Additional books for a well-rounded basic library on color might include References 6 to 18. The latest Commission Internationale de l'Éclairage (CIE) publications (19) should be consulted for the current definitive word.

Color Fundamentals

An immediate complexity is illustrated in the two early and apparently incompatible theories of color vision. Trichromatic theory, first proposed in 1801 by Thomas Young and later refined by Hermann von Helmholtz, postulated three types of color receptors in the eye. This explained many phenomena, such as various forms of color blindness, and was confirmed when three types of blue-, green-, and red-sensitive cones were reported to be present in the retina in 1964. Yet Ewald Hering's 1878 opponent theory, which used three pairs of opposites, light-dark, red-green, and blue-yellow, also offered much insight, including the explanation of contrast and afterimage effects and the absence of some color combinations such as reddish greens and bluish yellows. In the modern zone theories it is now recognized that the data from three trichromatic detectors in the eye are processed on their way to the brain into opponent signals, thus removing the apparent incompatibility.

In color technology and measurement, both types of approaches are used. Color printing, for example, generally employs three colors (usually plus black), and the ever useful CIE system was founded on experiments in which colors were matched by mixtures of three primary colors, often blue, green, and red. Yet transmitted television signals are based on the opponent system, with one intensity and two color-balance signals, as are the modern representations of color, such as the CIELAB and related color spaces based on red-green and yellow-blue opponent axes.

Light and Color. Visible light is that part of the electromagnetic spectrum, shown in Figure 1, with wavelengths between the red limit at about 700 nm and the violet limit of 400 nm. Depending on the observer, light intensity, etc, typical values for the spectral colors are red, 650 nm; orange, 600; yellow, 580; green, 550 and 500; and blue, 450.

In 1666, Sir Isaac Newton first split white light with a prism into its component colors, the spectrum, and he assigned the seven colors red, orange, yellow, green, blue, indigo, and violet, using just seven colors, possibly by analogy with the seven notes of the musical scale. The eye, however, functions quite differently from the ear, which is able to perceive the sound from individual instruments sounding together, while the eye always perceives only a single color, whether this be spectrally pure orange, an equivalent mixture of yellow and red, an equivalent mixture of green and red, and so on. Newton did recognize that his "rays . . . are not coloured. In them there is nothing else than a certain power . . . to stir up a sensation of this or that colour." Nevertheless, terms such as "orange light" are commonly used and need not produce any confusion if this qualification is kept in mind.

The spectral color sequence, joined by some nonspectral colors such as purple and magenta, is one of several attributes used in descriptions of color, variously designated hue, chromatic color, dominant wavelength, or simply, but imprecisely and quite unsuitably for the present purposes, color. A second attribute is saturation, chroma, tone, or purity, which gives a measure of how little or much gray (or white or black) is present. Thus a mixture of pure spectral orange with gray gives an unsaturated orange, transforming into gray (or white or black) as the amounts are altered. For any given hue and saturation, there can be different levels of brightness, lightness, luminance, or value, completing the three parameters normally required to specify color.

Color, in the broadest colloquial sense, is taken to mean the single dimension of the various hues of the spectrum, joined by a few nonspectral hues such as purple. Color in the full technical sense encompasses a multidimensional space defined at a minimum by the foregoing three parameters hue, saturation, and brightness. An example that clarifies this point is that a strong orange becomes, when the brightness is reduced but with constant hue and saturation, not a weak

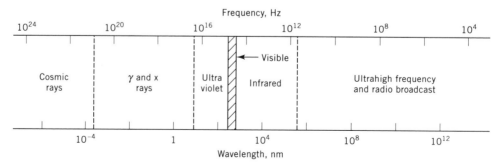

Fig. 1. The electromagnetic spectrum (5).

orange but a brown. It has been estimated that, in this full sense, an individual with normal color vision can distinguish a total of seven million different colors (20).

The appearance of color depends significantly on the exact circumstances. Normally one thinks of viewing a colored object under some type of illumination, the object mode. Viewing a light source there is the illuminant mode. Finally, in viewing through a hole in a screen there is the aperture mode. Perception differs significantly in these modes. In the object mode the eye/brain has the ability to compensate for a wide range of illuminants (white sun, blue sunless sky, reddish incandescent lamp or candle) and infer something very close to the true color. At the same time, various surface effects enter, as shown in Table 1. Even a non-metallic object with a deeply colored surface will reflect almost pure illuminant in the glare of a glancing angle if it is glossy. The perceived color is influenced by the presence of adjacent colors in the object mode, and several additional color-appearance phenomena also can influence color perception. One or two additional parameters (3,4) may be required in addition to the customary three for a full specification of the perceived color, but fortunately this is unusual.

Table 1. Object Mode Perceptions[a]

Object	Dominant perception	Dominant attributes[b]
opaque metal, polished	specular reflection	reflectivity, gloss, hue
opaque metal, matte	diffuse reflection	hue, saturation, brightness, gloss
opaque nonmetal, glossy	diffuse and specular reflections	hue, saturation, gloss, brightness
opaque nonmetal, matte	diffuse reflection	hue, saturation, brightness
translucent nonmetal	diffuse transmission	translucency, hue, saturation
transparent nonmetal	transmission	hue, saturation, clarity

[a]Ref. 3.
[b]In approximate sequence of importance.

Interactions of Matter with Light. In the most generalized interaction of light with matter the many phenomena of Figure 2 are possible. In absorption, electrons are excited by the absorbed photons and their energy may subsequently appear as heat or as fluorescence, an additional emitted light at a lower energy, ie, longer wavelength. This effect is utilized in fluorescent whitening agents used in detergents, paper, textiles, etc. Scattering may derive from irregularities of the surface, diffuse or matte as distinguished from glossy or specular reflection. Interference, ie, diffraction grating effect, derives from regularly repeated patterns. Irregularities in the interior, depending on the size and geometry, may involve Rayleigh or Mie-type scattering and then lead to translucency or opacity if there is pronounced multiple scattering.

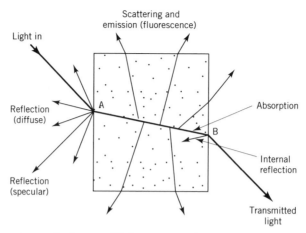

Fig. 2. The adventures of a beam of light passing through a block of partly transparent substance (5).

Color Vision

The Eye. Light passes through the cornea, the transparent outer layer of the eye, through the lens and the aqueous and vitreous humors, and is focused on to the retina. The iris, forming the pupil, acts as a variable aperture to control the amount of light that enters the eye, varying from about f/2.5 to f/13 with a 30:1 light intensity ratio. The two humors serve merely as neutral transmission media and to keep the eyeball distended. The retina is a layer about 0.1 mm thick that contains the light-sensitive rods and cones. Only the rods function in low levels of illumination of about less than 1 lux, providing an achromatic, noncolor image. The rod spectral response is shown in Figure 3**a**. The cones are of three types, designated B, G, and R in this figure. Although these are often spoken of as blue-, green-, and red-detecting cones, such designations are incorrect. As can be seen, each set of cones is sensitive to a wide range of wavelengths with extensive overlap. Appropriate designations are short-, medium-, and long-wavelength sensitive cones, but the B, G, and R labels provide a convenient mental picture useful as long as the actual functioning is kept in mind.

The distribution of rods and cones is shown in Figure 3**b** centered about the fovea, the area of the retina that has the highest concentration of cones with essentially no rods and also has the best resolving capability, with a resolution about one minute of arc. The fovea is nominally taken as a 5° zone, with its central 1° zone designated the foveola. There are about 40 R and 20 G cones for each B cone in the eye as a whole, whereas in the fovea there are almost no B cones. A result of this is that color perception depends on the angle of the cone of light received by the eye. The extremely complex chemistry involved in the stimulation of opsin molecules, such as the rhodopsin of the rods, and the neural connections in the retinal pathway are well covered in Reference 21.

The trichromatic theory, subsequently confirmed by the existence of the three sets of cones, must be combined with the opponent theory, which is involved

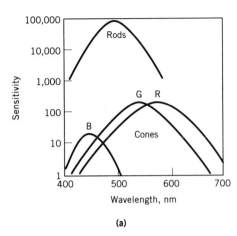 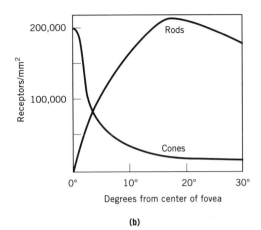

Fig. 3. (**a**) The relative sensitivities of the rods and the three sets of cones in the human eye (5). (**b**) The distribution of rods and cones in the central part of the human retina. The x-axis is marked in degrees from the center of the fovea.

in the retinal pathway. A third approach, the appearance theory (2) or the retinex theory, must be added to explain color constancy and other effects. As one example of this last, consider an area perceived as red in a multicolored object such as a Mondrian painting when illuminated with white light. If the illumination is changed so that energy reflected by this same area is greater at shorter wavelengths than the energy reflected at longer wavelengths, this area is still perceived as red within the overall visual context. If all other colors are now covered so that only our "red" area is visible, corresponding to the aperture mode, then this area is perceived as green. Clearly, all three approaches must be melded to give a full description of color perception, a process that is not yet complete.

In addition to this color constancy phenomenon, there are several other well-known effects that can influence the perception of color. In simultaneous contrast phenomena, there are effects from both luminosity differences and color differences across a boundary. The Bezold-Brücke effect involves a change in hue with luminance, a shift toward the blue end of the spectrum as the luminance is increased. Colors also appear more desaturated at both very high and very low luminances than at intermediate values. In the Helmholtz-Kohlrausch effect there is an increase in the apparent luminance as reflected light increases in spectral purity at constant luminosity. Finally, there is the well-known aftereffect, when prolonged viewing of a color distorts the next-viewed color in the direction of the complementary color of the first.

Color Vision Defects. Anomalous color vision is present, eg, if one of the three sets of cones is inoperative (dichromacy) or defective (anomalous trichromacy). This affects 2–3% of the population with males more prone because these defects reside on the X-chromosome, with one present in males but two in females. Eye specialists have standard tests for detecting these and other defects. Summaries of this whole field are available (6,9,22).

Color Order Systems

Many one-, two-, and three-dimensional systems have been developed over the years to order colors in a systematic way and provide specimen colors for visual comparison. Coordination has now been achieved with computer programs between essentially all of these systems and the CIE systems described below and conversions can easily be made between them.

The Munsell System. The best known and most widely used is the Munsell system (15,23), developed by the artist A. H. Munsell in 1905 and modified over the years. This is a three-dimensional space shown in Figure 4, and Figure 5 on the color plate. Munsell value V (or lightness) is used as the vertical axis with 0/ for black at the bottom and 10/ for white at the top. Radially there is the Munsell chroma C with /0 at the center and maximum /10 or higher at the periphery. The Munsell hue H at the periphery uses the five principal hues: red, yellow, green, blue, and purple and the five adjacent binary combinations such as BG, with ten steps within each, for a total of 100 hue steps as in Figures 4 and 5. These 100 steps can be further subdivided. A full designation for an orange school bus in the usual HVC sequence and using interpolation might then be 9.5YR 7/9.25. The Munsell system was renotated in 1943 to make it more uniform and consistent with the CIE system. The *Munsell Book of Color* (23) consists of about 1500 painted paper chips available in both glossy and matte versions; there are also textile color collections (23), now discontinued. With interpolations some 100,000 colors can be distinguished within the Munsell system by visual comparison under carefully standardized viewing conditions.

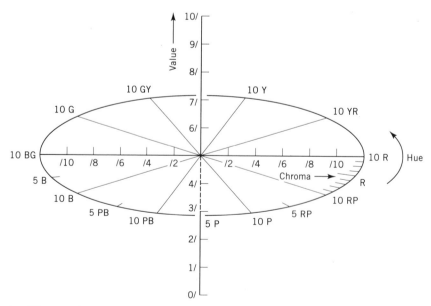

Fig. 4. The coordinate system of the Munsell color order system: R, red; Y, yellow; G, green; B, blue; and P, purple (3).

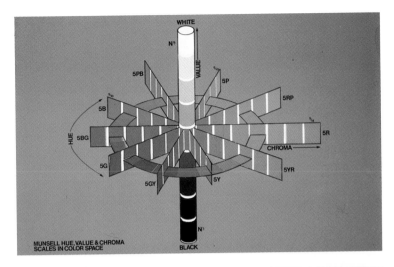

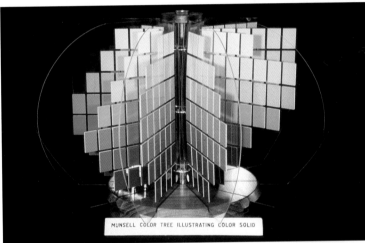

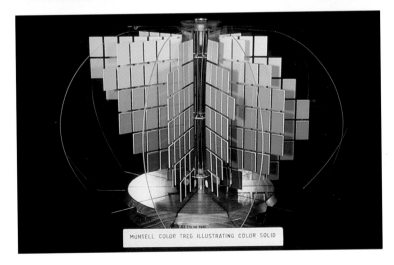

Fig. 5. Munsell hue, value, and chroma scale arrangement. Courtesy of Munsell Color, Newburgh, New York.

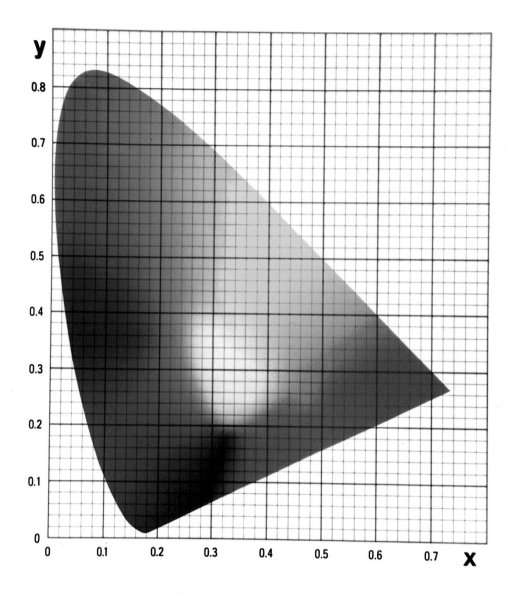

Fig. 9. The 1931 CIE x, y chromaticity diagram. Courtesy of Minolta.

Other Color Order Systems. The Natural Color System (24), abbreviated NCS, developed in Sweden is an outgrowth of the Hesselgren Color Atlas, and uses the opponent color approach. Here colors are described on the basis of their resemblances to the basic color pairs red-green and blue-yellow, and the amounts of black and white present, all evaluated as percentages. Consider a color that has 10% whiteness, 50% blackness, 20% yellowness, and 20% redness; note that the sum is 100%. The overall NCS designation of this color is 50, 40, Y50R indicating in sequence the blackness, the chromaticness (20 + 20), and the hue (50% on the way from yellow to red; the sequence used is Y, R, G, B, Y).

The Ostwald Color System (25) is a nonequally spaced system and is no longer published, as is also true of the *Maerz and Paul Dictionary of Color* (26). There are also the Colorcurve, the Coloroid, and the German DIN systems, among others. The Optical Society of America has published the OSA Uniform Color Scale System (27) with 558 equally spaced color chips. This uses a designation such as 2:5:3, with the first number (ranging from -7 to $+5$) specifying the lightness, the second (-6 to $+15$) giving the blue-yellow content, and the third (-10 to $+6$) for red-green. For many purposes a much simpler set of 267 color regions as provided by the ISCC-NBS Centroid System, a joint project of the Inter-Society Color Council and the USA National Bureau of Standards, now NIST (12,28), is convenient. This uses color names with adjectival modifications of the form vivid blue, brilliant yellow-green, light yellowish brown, and so on. The Universal Color Language (12) was devised during this development, with six levels of color discrimination varying from 13 in the simplest system to five million in the most sophisticated.

Many systems of limited range or of lower dimensionality are used in industry for the designation and control of color circumstances where a limited scale is suitable. Examples are the systems for describing plant tissues; soils; and skin, hair, and eye colors (23). There are no less than 19 color scales for yellow in oil, fat, varnish, etc, under names such as Gardner, Saybolt, Lovibond, and Hellige (1). Then there is the D, E, ... scale used for near-colorless diamonds devised by the Gemological Institute of America. These systems are well known within highly specialized industries and many have been standardized by industry organizations such as ASTM, TAPPI, and Federal Tests. A summary is given in Reference 1.

Basic Colorimetry

The International Commission on Illumination (abbreviated CIE from the French expression) over the years has recommended a series of methods and standards in the field of color; for a history of this process see Reference 8.

When considering light of a certain spectral energy distribution falling on an object with a given spectral reflectance and perceived by an eye with its own spectral response, to obtain the perceived color stimulus it is necessary to multiply these factors together as in Figure 6. Standards are clearly required for both the observer and the illuminant.

The CIE Standard Observer. The CIE standard observer is a set of curves giving the tristimulus responses of an imaginary observer representing an aver-

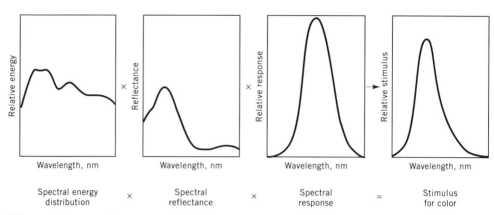

Fig. 6. The stimulus perceived as color is made up of the spectral power (or, as here, energy) curve of a source times the spectral reflectance (or transmittance) curve of an object times the appropriate spectral response curves (one shown here) of the eye (3).

age population for three primary colors arbitrarily chosen for convenience. The 1931 CIE standard observer was determined for 2° foveal vision, while the later 1964 CIE supplementary standard observer applies to a 10° vision; a subscript 10 is usually used for the latter. The curves for both are given in Figure 7 and the differences between the two observers can be seen in Table 2. The standard observers were defined in such a way that of the three primary responses $\bar{x}(\lambda)$, $\bar{y}(\lambda)$, and $\bar{z}(\lambda)$, the value of $\bar{y}(\lambda)$ corresponds to the spectral photopic luminous efficiency, ie, to the perceived overall lightness of an object.

CIE used the 1931 CIE standard observer to establish a color representation system in which the hue and saturation could be represented on a two-dimensional diagram. Three tristimulus values X, Y, and Z are first obtained, based on the standard observer, so that the hue and saturation of two objects match if they have equal values of these three parameters. Each of these is defined, following the concept of Figure 6, in the form:

$$X = k\int S(\lambda)R(\lambda)\bar{x}(\lambda)d\lambda$$

where $S(\lambda)$ is the spectral power distribution of the illuminant, $R(\lambda)$ is the spectral reflectance factor of the object, and $\bar{x}(\lambda)$, $\bar{y}(\lambda)$, and $\bar{z}(\lambda)$ are the color matching functions of one of the standard observers of Figure 7. The constant k is defined in terms of $S(\lambda)$, the spectral power distributions of the illuminant as:

$$k = 100/\int S(\lambda)\bar{y}(\lambda)d\lambda$$

so that y for a perfectly reflecting diffuser is 100.

Chromaticity Diagrams. The CIE 1931 chromaticity diagram uses the chromaticity coordinates:

$$x = X/(X + Y + Z); y = Y/(X + Y + Z); z = Z/(X + Y + Z).$$

It is not actually necessary to specify z, since $x + y + z = 1$. The two-dimensional

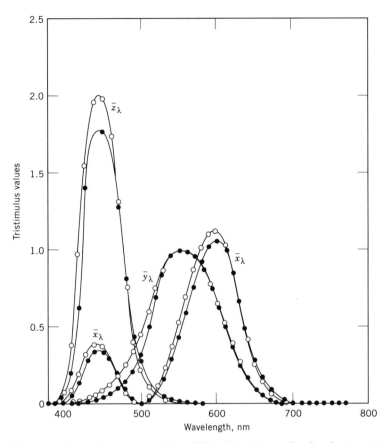

Fig. 7. Color matching functions of the CIE 1931 2° standard colorimetric observer (—●—), and the CIE 1964 10° supplementary standard colorimetric observer (—○—) (3).

Table 2. Spectral Chromaticity Coordinates[a]

Wave-length, nm	x	y	x_{10}	y_{10}	u'	v'	u'_{10}	v'_{10}
400	0.1733	0.0048	0.1784	0.0187	0.2558	0.0159	0.2488	0.0587
450	0.1566	0.0177	0.1510	0.0364	0.2161	0.0550	0.1926	0.1046
500	0.0082	0.5384	0.0056	0.6745	0.0035	0.5131	0.0020	0.5477
550	0.3016	0.6923	0.3473	0.6501	0.1127	0.5821	0.1375	0.5789
600	0.6270	0.3725	0.6306	0.3694	0.4035	0.5393	0.4088	0.5387
650	0.7260	0.2740	0.7137	0.2863	0.6005	0.5099	0.5700	0.5145
700	0.7347	0.2653	0.7204	0.2796	0.6234	0.5065	0.5863	0.5121

[a]Values for the 2° CIE chromaticity coordinates x, y and the 10° coordinates x_{10}, y_{10}; the 2° and 10° red-green metric chromaticity coordinates u' and u'_{10} and the 2° and 10° yellow-blue metric coordinates v' and v'_{10}.

presentation of x and y is shown in Figure 8, and Figure 9 on the color plate. Here Newton's pure spectral colors in fully saturated form follow the horseshoe-shaped curve with wavelengths from 400 nm violet to 700 nm red. Coordinates for these spectral colors at 50-nm intervals are given in Table 2 for both 2° and 10° observers. Closing the curves in these figures is the dashed line from red to blue, including the saturated nonspectral purples and magentas.

   Any straight line passing through the central achromatic point marked W for white (standard daylight D_{65} in this instance) connects complementary colors, such as the line connecting 480 nm blue and 580 nm yellow shown. These two colors will together give white in amounts given by the law of the lever lengths (length WY of the blue and length BW of the yellow). Saturated orange at 620 nm becomes unsaturated when mixed with white as at D. The color of D can be specified by its X, Y, and Z tristimulus values, by the chromaticity coordinates $x = 0.4$, $y = 0.3$ together with the lightness Y in the frequently used x, y, Y designation, or it could be described by Y together with the dominant wavelength of 620 nm, with 25% purity, since D corresponds to three parts of W and one part of 620 nm orange. If a point is in the lower right part of Figure 8, where the dominant hue is on the nonspectral blue to red join, then there is no dominant wavelength; instead the line is extended upward through the W point to reach the spectral curve at the complementary dominant wavelength.

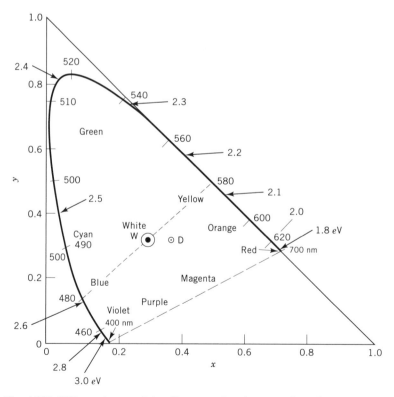

Fig. 8. The 1931 CIE x,y chromaticity diagram showing wavelengths in nm and energies in eV. The central point W (for white) corresponds to standard daylight D_{65} (5).

One of several defects of the chromaticity diagram of Figures 8 and 9 is that the minimum distinguishable colors are not equally spaced, that is, that equal changes in x, y, and Y do not correspond to equally perceived color differences. Very obviously, the greens occupy a disproportionately large area in these figures. Many transformations have been studied to adjust this, but none is perfect. Probably, the most useful employs the CIE 1976 Chromaticity Coordinates u' and v' as shown in Figure 10 obtained from:

$$u' = 4X/(X + 15\,Y + 3\,Z) = 4x(-2\,x + 12\,y + 3)$$

$$v' = 9Y/(X + 15\,Y + 7\,Z) = 9y(-2\,x + 12\,y + 3).$$

Standard Illuminants. Three of many sources with quite different energy distributions, which the eye nevertheless accepts as white, are the daylight, incandescent light, and fluorescent lamp light shown in Figure 11. The chromaticity diagram is convenient for representing light sources. One particular use is to describe ideal black body colors, the sequence black, red, orange, yellow, white, and blue exhibited as any object is heated up under idealized conditions. For real incandescent objects, ie, nonideal black bodies, that ideal black body temperature is used which gives the closest visual match.

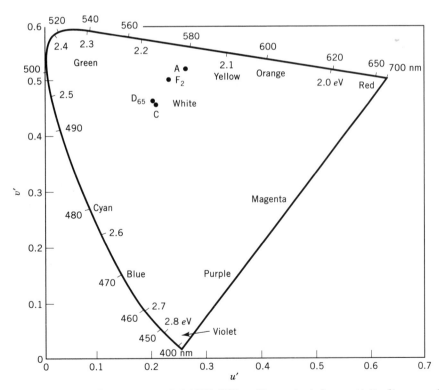

Fig. 10. The currently recommended 1976 CIE uniform u', v' chromaticity diagram showing wavelength in nm and energies in eV; A, C, and D_{65} are standard illuminants and F_2 is a typical fluorescent lamp.

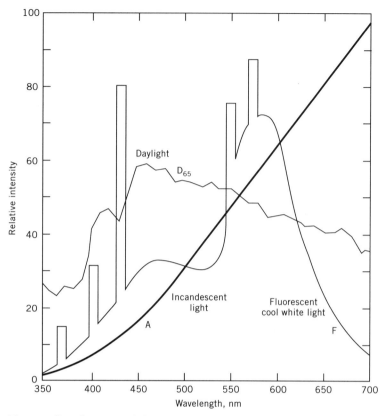

Fig. 11. Energy distributions of CIE standard illuminant A, a tungsten incandescent lamp; a cool white fluorescent lamp F; and CIE standard illuminant D_{65} which approximates average daylight.

Clearly, standardized light sources are desirable for color matching, particularly in view of the phenomenon of illuminant metamerism described below. Over the years CIE has defined several standard illuminants, some of which can be closely approximated by practical sources. In 1931 there was Source A, defined as a tungsten filament incandescent lamp at a color temperature of 2854 K. Sources B and C used filtering of A to simulate noon sunlight and north sky daylight, respectively. Subsequently a series of D illuminants was established to better represent natural daylight. Of these the most important is Illuminant D_{65}.

Gloss and Opacity.   Attributes such as gloss, transparency, translucency, opacity, haze, and luster may apply to some materials (Table 1), and these are relevant in that they may influence the judgment of color differences. As one example, gloss can produce veiling reflections that change the apparent contrast. When present, these attributes can be measured using specialized approaches. Six types of gloss are distinguished (1): specular gloss, sheen, contrast gloss or luster, absence-of-bloom gloss, distinctness-of-image gloss, and surface uniformity gloss. Opacity can be measured by the contrast ratio method, using the reflectance with both a white and a black backing. Standardized procedures for some of these methods have been established by organizations such as ASTM and TAPPI (1).

Light Mixing.   Light or additive mixing applies to light beams. White results when any suitable set of three-color beams of the appropriate intensity are mixed. On the chromaticity diagram of Figures 8 and 9, the condition for equal intensity beams is that W lies at the center of gravity of the triangle formed by the three sources. A suitable set of primary light beams is red, blue, and green, each being near a corner of Figures 8 and 9. Red and green by themselves add to give yellow, red and blue give purple and magenta, and blue and green give blue-green and cyan, as can be established by tie lines on Figure 8. It is important to distinguish magenta from red and cyan from blue to avoid confusing the additive from the subtractive system described next.

Colorant Mixing.   A colorant, whether a dye dissolved in a medium or pigment particles dispersed in it, produces color by absorbing and/or scattering part of the transmitted light. If only absorption is present, the Beer-Lambert law applies:

$$A(\lambda) = \log 1/T(\lambda) = \sum_i a_i(\lambda)bc_i$$

where A is the absorbance, T is the transmittance, $a_i(\lambda)$ is the absorptivity or specific absorbance of absorber i at wavelength λ, b is the length of the absorbing path, and c_i the concentration of the absorber. When colorants are mixed, they function by each independently absorbing light and the subtractive mixing rules merely specify this additivity. Here the result of mixing three primary colorants is to absorb all light and produce black if sufficiently concentrated. In actual practice a fourth pigment, usually a white or black opacifier, needs to be added to ensure opacity. The preferred primary colorants are the complementary colors of the corners of Figures 8 and 9, namely cyan, yellow, and magenta. Combining yellow and cyan colorants then produces green, yellow and magenta give red, and cyan plus magenta give blue.

When both absorption and scattering are present, the Beer-Lambert law must be replaced by the Kubelka-Munk equation employing the absorption and scattering coefficients K and S, respectively. This gives the reflectivity R_∞

$$(1 - R_\infty)^2/2R_\infty = K/S = \sum_i c_i K_i(\lambda)/\sum_i c_i S_i(\lambda)$$

where c_i is the concentration and $K_i(\lambda)$ and $S_i(\lambda)$ are the specific absorbance and scattering parameters, respectively, of absorber and scatterer i at wavelength λ.

The color of an opaque paint depends both on the size of the pigment particles and on refractive index considerations. The scattering is maximum, as shown in Curve A of Figure 12, when the particle size is about one-half of the wavelength of light, ie, in the 200 nm to 350 nm range; this is desired for opacity in typical inorganic pigments with refractive index much larger than that of the medium. With organic pigments, where the refractive index is not very different, scattering is replaced by absorption as the principal interaction; here the color strength or absorptive power increases as the particle size is decreased to a much smaller size, as shown in Curve B of Figure 12. Unfortunately, too small a particle size may lead to a reduction of the lightfastness, requiring compromises.

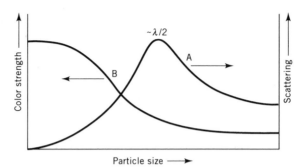

Fig. 12. A, the schematic variation of scattering, and B, color strength for pigment particles of various sizes (5).

The ready availability of computers has led to the detailed analysis of the colorant formulation problems faced every day by the textile, coatings, ceramics, polymer, and related industries. The resulting computer match prediction has produced improved color matching and reductions in the amounts of colorants required to achieve a specific result with accompanying reductions of cost. Detailed treatments have been given for dyes and for pigments (13,29,30).

Metamerism. There are several types of metamerism, the phenomenon in which two objects perceived as having a perfect color match under one set of conditions are found to differ in color under other conditions. Most common is illuminant metamerism that occurs when a change in illuminant is the cause. This originates from the above described situation that a given visually identified hue can be caused by many different stimuli, for example by an object reflecting, say, only pure spectral orange, or red plus yellow, and so on. A change in illuminant to one having more energy in the red, for example, would leave the perceived color of the pure spectral orange unchanged, but would make the red plus yellow combination appear to be a more reddish orange. Another related cause would be the presence of an ultraviolet component in one of the sources (usually actual daylight) causing a fluorescent object to emit light in addition to that reflected compared to a visible-light-only source such as an incandescent lamp. Note that the standard illuminant D_{65} specifies significant intensity at wavelengths less than 400 nm in the ultraviolet.

Observer metamerism derives from the significant differences in spectral response found among persons with normal color vision. The variation in cone concentration and cone-type distribution discussed above results in field size metamerism. Here an example is the difference between the 2° and 10° response curves of the standard observers of Figure 7. Finally, there is geometric metamerism observed with a change in apparent color and with a change in viewing angle, as with some metallic paints.

Quantification of metamerism is difficult and its avoidance is a principal aim in color technology. The car where body color, upholstery, and plastic parts match in the lit showroom should not show clashing colors in daylight. The obvious but rarely applicable solution is to use the same pigments in all parts. Some illuminant metamerism is almost unavoidable, the aim of the color expert being to keep it within acceptable limits, ie, to achieve adequate color consistency.

Advanced Colorimetry

The 1931 CIE system with its x, y, Y chromaticity diagram system is useful for general colorimetry and color matching, but it suffers several drawbacks. These were partly corrected in the 1976 CIE chromaticity diagram; u', v' achieve a more uniform color space. Both the x, y and u', v' systems are still basically two-dimensional, with the lightness Y added. A series of transformations led to modified systems that are fully three-dimensional with the length of any lines, however oriented within this space, more closely representing perceived color differences. Some of these are opponent related. Two systems have been ultimately agreed upon, designated the CIELUV and CIELAB color spaces, and have come into widespread use.

The 1976 CIELUV Color Space. Properly designated CIE $L^*u^*v^*$, this uses a white object or light source designated by the subscript n as the reference standard and employs the transformations:

$$L^* = (116Y/Y_n)^{1/3} - 16; \ Y/Y_n > 0.008856$$

$$L^* = 903.3Y/Y_n); \ Y/Y_n \le 0.008856$$

$$u^* = 13L^*(u' - u'_n)$$

$$v^* = 13L^*(v' - v'_n)$$

The CIELUV space preserves a property of the CIE 1931 chromaticity space which is important in the field of color reproduction, eg, in the television, photographic, and the graphic arts industries. This is the characteristic that the chromaticities of additive mixtures of color stimuli lie on the straight line connecting the chromaticities of the component stimuli; this is true of the 1976 metric chromaticity diagram but not of the CIELAB space that follows.

The 1976 CIELAB Color Space. Defined at the same time as the CIELUV space, the CIELAB space, properly designated CIE $L^*a^*b^*$, is a nonlinear transformation of the 1931 CIE X, Y, Z space. It also uses the metric lightness coordinate L^*, together with:

$$a^* = 50[(X/X_n)^{1/3} - (Y/Y_n)^{1/3}]$$

$$b^* = 200[(Y/Y_n)^{1/3} - (Z/Z_n)^{1/3}].$$

These equations apply for X/X_n, Y/Y_n, and Z/Z_n all > 0.008856. For $X/X_n \le 0.008856$, the term $[(X/X_n)^{1/3}]$ is replaced by $[7.787(X/X_n) + 16/116]$ and similarly for Y and for Z in these two equations.

This transformation results in a three-dimensional space that follows the opponent color system with $+a^*$ as red, $-a^*$ as green, $+b^*$ as yellow, and $-b^*$ as blue. CIELAB is closely related to the older Adams-Nickerson, modified Adams-Nickerson, and other spaces of the L,a,b type, which it replaced (1,3).

The CIELAB coordinates L^*,a^*,b^*, either in that form or in the L^*,C^*_{ab},h_{ab} form discussed below, are the most commonly used color descriptors in the field of paints, pigments, textiles, paper, ceramics, polymers, and most other opaque to transparent substances.

The 1976 CIE Metric Color Spaces. Both the CIELUV and CIELAB spaces can have their Cartesian coordinates converted to cylindrical coordinates, called metric or hue-angle coordinates, with L^* unchanged. These coordinates are designated CIE $L^*C_{uv}^*h_{uv}$ and CIE $L^*C_{ab}^*h_{ab}$, respectively. The metric hue-angles are given as:

$$h_{uv} = \tan^{-1}(v^*/u^*) \qquad h_{ab} = \tan^{-1}(b^*/a^*)$$

and the metric chromas:

$$C_{uv}^* = [(u^*)^2 + (v^*)^2]^{1/2} \qquad C_{ab}^* = [(a^*)^2 + (b^*)^2]^{1/2}.$$

The close analogy of h_{ab} to the Munsell hue and of C_{ab}^* to the Munsell chroma of Figure 4 is evident; L^* is also closely related to the Munsell value.

In the system derived from CIELUV it is also possible to specify a metric saturation with reference to a standard white designated by subscript n:

$$S_{uv} = 13[(u^* - u_n^*)^2 + (v^* - v_n^*)^2]^{1/2}.$$

A saturation correlate cannot be given for the CIE $L^*C_{ab}^*h_{ab}$ space.

Hunter *L,a,b* and Other Color Spaces. The CIELAB and CIELUV color spaces were the outgrowth of a large and complex group of interrelated early systems and have replaced essentially all of them except for the 1942 Hunter L,a,b group of color spaces (3). This was the earliest practical opponent-based system which is still widely used. In this system, for illuminant C and the 2° standard observer:

$$L = 10Y^{1/2} \qquad \text{(lightness coordinate)}$$

$$a = 17.5(1.02X - Y)/Y^{1/2} \qquad \text{(red-green coordinate)}$$

$$b = 7.0(Y - 0.847Z)/Y^{1/2}. \qquad \text{(yellow-blue coordinate)}$$

There are other equations for other illuminants and other observers (3) and also various modifications for special conditions (31).

Color Difference Assessment. Color difference scales include those of Judd-Hunter, Macadam, Adams-Nickerson, ANLAB, and ANLAB40. All of these have limitations in some way or another; they are described in most texts (1–4). Each applies only to the precise conditions used in their determination and interconversion is not possible. Modifications of CIELAB in the metric form such as the *CMC(1:c)* system (14) promise improved performance for the future.

In the CIELAB and CIELUV color spaces, the difference between a batch sample and a reference standard designated with a subscript s, can be designated by its components, eg, $\Delta AL^* = L^* - L_s^*$. The three-dimensional total color differences are given by Euclidian geometry as the 1976 CIE $L^*a^*b^*$ and 1976 CIE $L^*u^*v^*$ color difference formulas:

$$\Delta E_{ab}^* = [(\Delta L^*)^2 + (\Delta a^*)^2 + (\Delta b^*)^2]^{1/2}$$

$$\Delta E_{uv}^* = [(\Delta L^*)^2 + (\Delta u^*)^2 + (\Delta v^*)^2]^{1/2}$$

In CIE metric coordinates, either for CIELAB or CIELUV, ΔL^* and ΔE^* are the same, $\Delta C^* = C^* - C_s^*$, $\Delta h = h - h_s$ and $\Delta H^* = [(\Delta E^*)^2 - (\Delta L^*)^2 - (\Delta C^*)^2]^{1/2}$. The last of these, the metric hue difference ΔH^*, is preferred to Δh, since the latter is in degrees rather than in units compatible with ΔL^* and ΔC^*, as in ΔH^*.

An example may clarify this system. Consider a red apple with CIELAB coordinates measured as $L^* = 41.75$, $a^* = 45.49$, $b^* = 9.61$. This converts to metric as $L^* = 41.75$, $C^* = 46.49$, $h = 13.25$ and, incidentally, to Munsell 10RP 4/10 (recall that the sequence is hue, value, chroma). A second apple has $L^* = 49.23$, $a^* = 40.13$, $b^* = 12.20$, $C^* = 41.94$, $h = 18.79$, and Munsell 2.5R 5/9. Taking the differences $\Delta L^* = +7.48$, ie, the second apple is brighter (note Munsell values are 4 and 5); $\Delta a^* = -5.36$, ie, more green (or less red); $\Delta b^* = +2.59$, ie, more yellow (or less blue); $\Delta C^* = -4.55$, ie, less saturation (note Munsell chromas are 10 and 9); $\Delta h = +5.54$, ie, larger hue angle which means more yellow in this instance (note Munsell hue 2.5R is 2½ steps from 10RP as in Figure 4 or $2.5 \times 360/100 = 9°$ away). The total three-dimensional color difference $\Delta E^* = 9.56$. The metric hue difference can also now be calculated as $\Delta H^* = 3.84$. The lightness difference $\Delta L^* = +7.48$ is the dominant component in the total color difference $\Delta E^* = 9.56$ and the hue difference $\Delta H^* = 3.84$ is less important than the chroma difference $\Delta C^* = -4.55$.

The color difference magnitudes derived from the CIE as well as from other color-space and color-ordering systems do not agree as well as could be desired with one another or with the visually perceived differences; they cannot be inter-converted by a constant factor in general. This is probably the least satisfactory part of colorimetry theory. Many factors contribute to this. In an ideal three-dimensional color space, the region that is not distinguishably different from a given point would be a sphere. In actual practice this is not achieved fully in any color space; these regions are the so-called MacAdam ellipses (1–4) on two-dimensional chromaticity diagrams; they are ellipsoids in the three-dimensional color spaces. An example of the direction of more recent research for improvement is the *CMC(1:c)* color difference formula (14).

A perceived color difference varies with the mode (object, illuminant, aperture); the texture (glossy, rough, metallic, etc); size, flatness, and transparency characteristics of objects; the level, color, and geometry (point-source versus diffuseness) of the illumination; the presence of ultraviolet light, fluorescence, and polarized light; and the nature and color of the surroundings and background. Finally, the various metamerisms discussed above are at work as well as observer experience and adaptation to the observing situation. Nevertheless the system does work and color difference measurements can be used successfully in actual practice if all these parameters are controlled.

Finally, it cannot be overemphasized that despite instrumental measurements and data manipulations, it is the perception of the eye that still is the final arbiter as to whether or to what degree two colors match. Instrumental methods do serve well for the typical industrial task of maintaining consistency under sufficiently well-standardized conditions; however, a specific technique may not serve in extreme or unusual conditions for which it was not designed.

Color Measuring Instruments

There has been a tremendous change in the last two decades as computers have taken over the tedious calculations involved in color measurement. Indeed, microprocessors either are built into or are connected to all modern instruments, so that the operator may merely need to specify, for example, x, y, Y or L^*, a^*, b^* or $L^*, C^*, h,$ either for the 2° or the 10° observer, and for a specific standard illuminant, to obtain the desired color coordinates or color differences, all of which can be stored for later reference or computation. The use of high intensity filtered Xenon flash lamps and array detectors combined with computers has resulted in almost instantaneous measurement in many instances.

Measurement Conditions.   In 1968, the CIE recommended the four geometries of Figure 13 for reflectance measurements. In the first 0°/45° or normal/

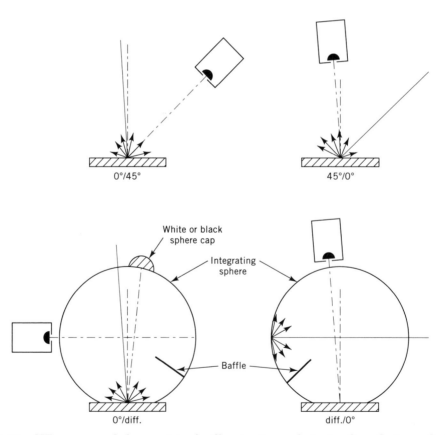

Fig. 13. CIE recommended geometries for illumination and viewing for reflectance-factor measurements (11).

45° geometry, the illumination is normal to the sample and the detector is at 45°. In the second 45°/0° geometry the two conditions are reversed. These techniques usually give the same results, except when polarized light interacts, for example, with oriented metallic flakes. In the third 0°/diffuse or normal/diffuse geometry the incident light beam impinges normally onto the sample and the reflected and scattered light is collected by an integrating sphere. In the last diffuse/0° geometry the incident light is first diffused by the integrating sphere before interacting with the sample, with the detector normal to the sample.

The illumination used is usually filtered white light to approximate daylight. The detector may respond to all light, may use a monochromator, a diode array, a rotating wedge interference filter, or band pass filters. Often several photodetectors are covered with filters approximating the CIE color matching functions $\bar{x}$, $\bar{y}$, $\bar{z}$ or a suitable linear combination of these. There can be considerable differences when fluorescence is present. For example, a monochromator detector system would be unaffected except at the fluorescence wavelength, but a broad band detector would show a strongly enhanced spurious response whenever fluorescence is stimulated.

For transmittance measurements there are two principal geometries, one with a full collection of all transmitted light by an integrating sphere, the other with both illuminating and detector beams collimated. Translucency is measured, for example, by two reflectance measurements, one with a white background, the other with black. Goniophotometric instruments are used to measure specular reflections, important in glossy and metallic objects and coatings.

Spectrophotometers and Spectrocolorimeters. Spectrophotometers are the most sophisticated color measuring instruments and provide the most detailed and accurate information. They may provide continuous spectral data of reflectance and transmittance against wavelength or use up to 20-nm wavelength steps, with high precision in reflectance (down to 0.01%) and tristimulus values (down to 0.01). There may be dual beams with one for the sample and a second as reference for the most stable and precise operation. Many spectrophotometers use an integrating sphere, although some use other geometries or permit alternative ones.

Examples of spectrophotometers are the ACS (Applied Color Systems Inc., Princeton, New Jersey) Chroma Sensor series; the BYK-Gardner (Silver Springs, Maryland) Color Sphere and Color Machine units; the Hunterlab (Reston, Virginia) UltraScan, ColorQUEST, and LabScan units; the Macbeth Division of Kollmorgen Instruments (Newburgh, New York) Color-Eye series; the Milton Roy (Rochester, New York) Diano Match-Scan, Color Scan, and Color Graph units; the Minolta Corp. (Rahway, New Jersey) CM1000; and the X-Rite Inc. (Grandville, Michigan) 968. Some of these have additional capabilities such as limited gloss measurements, while some are compact, battery-powered, portable units with built-in microprocessors, such as the ACS PCS-500, the Minolta CM-1000, and the X-Rite 968.

Slightly less sophisticated are spectrocolorimeters that determine spectral response curves for further computation but from which the spectral curve itself is not available. An example is the X-Rite 948.

Colorimeters. Also known as tristimulus colorimeters, these are instruments that do not measure spectral data but typically use four broad-band filters to approximate the $\bar{y}$, $\bar{z}$, and the two peaks of the $\bar{x}$ color-matching functions of the standard observer curves of Figure 7. They may have lower accuracy and be less expensive, but they can serve adequately for most industrial color control functions. Examples of colorimeters are the BYK-Gardner Co. XL-835; the Hunter Lab D25 series; the Minolta CA, CL, CS, CT, and CR series (the last of these is portable with an interface); and the portable X-Rite 918.

Other Instruments and Supplies. A goniospectrophotometer, such as the Macbeth Color-Eye 5010, measures the angularly dependent spectral variation of light reflection and scattering, and goniophotometers determine this in a nonspectral manner. Such measurements are important in characterizing metallic and pearlescent finishes, for example. A variety of goniophotometers, reflectometers, gloss meters, haze meters, etc, are available from most of the sources listed above, as are computers and software for acquiring and manipulating data from various instruments and for converting among the various color-order and color-measuring systems.

Calibration with standard reflectance and transmittance samples should be routinely used for optimum results in spectrophotometry and colorimetry. Calibration of the wavelength (32) and photometric (33) scales is also advisable. The calibration of a white reflectance standard in terms of the perfect reflecting diffuse, τ, has been discussed (34), as have diagnostic tiles for tristimulus colorimetry (35). A collaborative reference program is available on instrument performance (36).

Color-order systems, such as the many Munsell collections available from Macbeth, have been described previously. Essential for visual color matching is a color-matching booth. A typical one, such as the Macbeth Spectralite, may have available a filtered 7500 K incandescent source equivalent to north-sky daylight, 2300 K incandescent illumination as horizon sunlight, a cool-white fluorescent lamp at 4150 K, and an ultraviolet lamp. By using the various illuminants, singly or in combination, the effects of metamerism and fluorescence can readily be demonstrated and measured. Every user should be checked for color vision deficiencies.

The Fifteen Causes of Color

No less than 15 distinct chemical and physical mechanisms explain the various causes of color, ordered into five groups as in Table 3. In the first group, covered by quantum theory, there are incandescence, simple electronic excitations, and vibrational and rotational excitations. Most chemical compounds contain only paired electrons that require very high energies to become unpaired and form excited energy levels; this requires ultraviolet, hence there is no visible absorption and no color. Absorption color can, however, be derived from the easier excitation of unpaired electrons in transition-metal compounds and impurities, covered by ligand field theory in the second group. Absorptions from paired electrons can be shifted into the visible by increasing the size of the region over which the electrons are localized, as in organic compounds covered by molecular orbital theory in the

Table 3. Fifteen Causes of Color

Cause	Examples
Vibrations and simple excitations	
incandescence	flames, lamps, carbon arc, limelight
gas excitations	vapor lamps, flame tests, lightning, auroras, some lasers
vibrations and rotations	water, ice, iodine, bromine, chlorine, blue gas flame
Transitions involving ligand field effects	
transition-metal compounds	turquoise, chrome green, rhodonite, azurite, copper patina
transition-metal impurities	ruby, emerald, aquamarine, red iron ore, some fluorescence and lasers
Transitions between molecular orbitals	
organic compounds	most dyes, most biological colorations, some fluorescence and lasers
charge transfer	blue sapphire, magnetite, lapis lazuli, ultramarine, chrome yellow, Prussian blue
Transitions involving energy bands	
metals	copper, silver, gold, iron, brass, pyrite, ruby glass, polychromatic glass, photochromic glass
pure semiconductors	silicon, galena, cinnabar, vermillion, cadmium yellow and orange, diamond
doped semiconductors	blue and yellow diamond, light-emitting diodes, some lasers and phosphors
color centers	amethyst, smoky quartz, desert amethyst glass, some fluorescence and lasers
Geometrical and physical optics	
dispersive refraction	prism spectrum, rainbow, halos, sun dogs, green flash, fire in gemstones
scattering	blue sky, moon, eyes, skin, butterflies, bird feathers, red sunset, Raman scattering
interference	oil slick on water, soap bubbles, coating on camera lenses, some biological colors
diffraction	diffraction gratings, opal, aureole, glory, some biological colors, most liquid crystals

third group; this also explains various forms of charge transfer. In the fourth group there is color in metals and in semiconductors such as yellow cadmium sulfide, both pure and doped, covered by band theory; this also covers color centers. In the final group there are four color-causing mechanisms explained by geometrical and physical optics.

Color from Incandescence.   Any object emits light when heated, with the sequence of blackbody colors, black, red, orange, yellow, white, and bluish-white as the temperature increases. The locus of this sequence is shown on a chromaticity diagram in Figure 14.

The distribution of energy under idealized conditions is given by Planck's equation in which the energy E_R radiated into a hemisphere in W/cm^2 in wavelength interval $d\lambda$ at wavelength λ at T in K is

$$E_R = 37415 \ d\lambda/\lambda^5[e^{(14338/\lambda T)} - 1]$$

The total energy emitted E_T in W/cm^2 is given by Stefan's law as:

$$E_T = 5.670 \times 10^{-12} \ T^4$$

and Wien's law gives the peak wavelength λ_m in nm as

$$\lambda_m = 2,897,000/T$$

Real or gray bodies deviate from these ideal blackbody values by the λ-dependent emissivity, but the color sequence remains essentially the same. This mechanism explains the color of incandescent light sources such as flames in a candle, tungsten filament light bulb, flash bulb, carbon arc, limelight, lightning in part, and the incandescent part of pyrotechnics (qv).

Color from Gas Excitation.   When the atoms in a gas or vapor are excited, eg, by an electric discharge, electrons on the atoms are elevated to higher energy states. On their way back to the ground state they can emit light, the energies being limited by the quantum states available and by the selection rules that control the probabilities of the various possible transitions. Examples are efficient

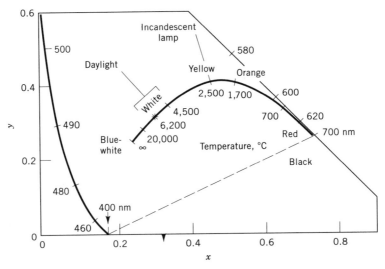

Fig. 14.   Blackbody colors shown on a chromaticity diagram; the asterisk is standard illuminant D$_{65}$ at 6500 K (5).

gas discharge light sources such as sodium and mercury vapor lamps, the latter emission also being present in fluorescent lamps in part; lightning in part; auroras, excited by energetic particles from the sun, in part; the Bunsen-burner flame tests for Na, Li, Sr, etc, where the excitation is thermal; and the electrically excited gas lasers such as the helium–neon, carbon dioxide, and argon-ion lasers as well as metal vapor lasers such as the helium–cadmium laser.

Color from Vibrations and Rotations. Vibrational excitation states occur in H_2O molecules in water. The three fundamental frequencies occur in the infrared at more than 2500 nm, but combinations and overtones of these extend with very weak intensities just into the red end of the visible and cause the blue color of water and of ice when viewed in bulk (any green component present derives from algae, etc). This phenomenon is normally seen only in H_2O, where the lightest atom H and very strong hydrogen bonding combine to move the fundamental vibrations closer to the visible than in any other material.

Electronic energy levels in molecules are often modified by vibrational and rotational excitations. In iodine this mechanism is involved in the intense purple color, as well as in the much weaker colors of bromine and chlorine, both in the condensed and gaseous states. Blue and green emissions from such energy levels in the unstable CH and C_2 molecules occur in the premixed region of candle and gas flames. Such emissions also occur in some auroras, particularly in the broad bands from the excited N_2^* molecule. There are some laser transitions from vibrational states such as those of CO_2, but these occur in the infrared since the energy spacings are small. Finally, the slow phosphorescence from triplet states in organic molecules usually involves vibrational states.

Color from Transition-Metal Compounds and Impurities. The energy levels of the excited states of the unpaired electrons of transition-metal ions in crystals are controlled by the field of the surrounding cations or cationic groups. From a purely ionic point of view, this is explained by the electrostatic interactions of crystal field theory; ligand field theory is a more advanced approach also incorporating molecular orbital concepts.

Consider a crystal of corundum [1302-74-5], pure Al_2O_3. Each Al is surrounded by six O ligands in the form of a slightly distorted octahedron. All electrons are paired and there are no absorptions in the visible region and hence no color. If a few percent of the aluminum atoms are replaced with chromium, the result is the red mineral, gem, laser, and maser crystal ruby [12174-49-1]. The term diagram, giving the effect of different strength ligand fields on the ^2E, $^4T_1$, and $^4T_2$ excited energy levels with respect to the ground state $^4A_2$ level of the three unpaired electrons in the $3d$ orbitals of Cr^{3+}, is shown at (**a**) in Figure 15. The 2.23 eV ligand field in ruby results in the energy level scheme at (**b**) in this figure, with absorption of light in the violet and green as at (**c**), leading to the intense red color with a weak blue component in transmission. There is also the well-known red fluorescence of ruby as the system returns to the ground state after light absorption and heat emission, as shown at (**b**) and (**c**). The change in chemistry from Al_2O_3 to the $Be_3Al_2Si_6O_{18}$ of beryl [1302-52-9] does not change the nature or the symmetry of the Al-surrounding ligands but reduces the ligand field some 8% to 2.05 eV. In Cr^{3+}-containing beryl the result is the scheme at (**d**) in Figure 15, with both ^4T levels shifted to slightly lower energy levels. The result is the intense green color of emerald [12415-33-7], with essentially the same red fluorescence as in ruby.

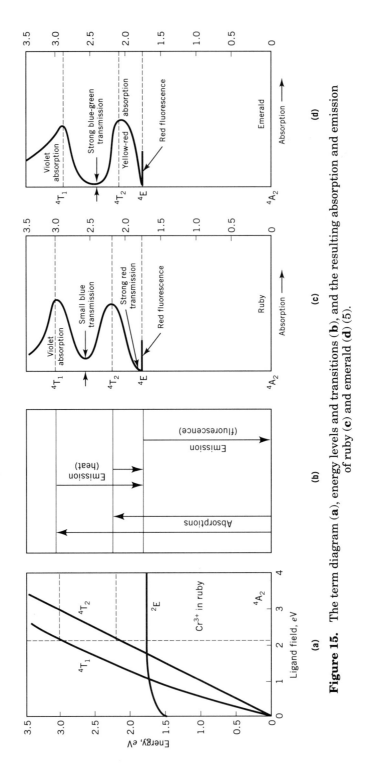

Figure 15. The term diagram (**a**), energy levels and transitions (**b**), and the resulting absorption and emission of ruby (**c**) and emerald (**d**) (5).

864

The unexpected situation intermediate between red ruby and green emerald is chrysoberyl [1304-50-3], BeAl$_2$O$_4$, called alexandrite [12252-02-7] when containing Cr^{3+}. Here an intermediate crystal field of 2.17 eV produces evenly balanced red and green transmissions, resulting in the alexandrite effect with a red color in red-rich incandescent light and a green color in blue- and green-rich daylight and fluorescent light. This same change occurs as the chromium content of ruby is increased in the solid-solution series Al$_2$O$_3$:Cr$_2$O$_3$, the Cr$_2$O$_3$ end member being the pigment chrome green [1308-38-9] with a ligand field near 2.07 eV. At about 25% Cr$_2$O$_3$ the color changes from red via gray to green. In the intermediate region, pressure shifts the color to red by shortening the bonds and increasing the ligand field in piezochromism, while temperature has the reverse effect, shifting the color to green in thermochromism (see CHROMOGENIC MATERIALS).

The ligand field decreases in the spectrochemical sequence CN$^-$ > NH$_3$ > O^{2-} > H$_2$O > F$^-$ > Cl$^-$ > Br$^-$ > I$^-$ and increases in isoelectronic sequences such as V(II) < Cr(III) < Mn(IV). Whereas in the 3d transition elements the color varies strongly with the symmetry and magnitude of the ligand field, in the lanthanides and actinides the unpaired electrons in the 4f and 5f shells, respectively, are shielded by the outer electrons so that there are only minor energy level and, therefore, color changes with the ligand field. Most ligand-field transitions are formally forbidden by the selection rules, hence transitions have low oscillator strengths and colors tend to be weak.

Idiochromatic (self-colored) transition-metal compounds, occur where the transition-metal ions are an essential part of the structure and contribute to the nature of the ligand field. Examples in addition to the chrome green Cr$_2$O$_3$ are purple chrome alum KCr(SO$_4$)$_2$·12H$_2$O; pink rhodonite [14567-57-8], MnSiO$_3$; green vitriol, FeSO$_4$·7H$_2$O, and yellow geothite [1310-14-1], FeO(OH); blue CoO [1307-96-6] and pink sphaerocobaltite [14476-13-2], CoCO$_3$; green bunsenite [1313-99-1], NiO; and yellow tenorite [1317-92-6], CuO, blue azurite [1319-45-5], Cu$_3$(CO$_3$)$_2$(OH)$_2$, and turquoise [1319-32-0], CuAl$_6$(PO$_4$)$_4$(OH)$_8$·5H$_2$O, and the various green salts present in copper patina. Allochromatic impurities may modify these colors.

Allochromatic (other-colored) transition-metal compounds, involve small amounts of these same transition elements but in the ligand field of the host lattice. Examples in addition to the above discussed chromium-containing ruby, emerald, and alexandrite are red beryl containing manganese; the iron-containing green or blue beryl aquamarine [1327-51-1] and many brown and red iron-containing minerals such as sandstone and red iron ore; the intense blue cobalt glass; a green vanadium-containing form of emerald; and purple neodymium-containing yttrium aluminum garnet YAG [12005-21-9], Y$_3$Al$_5$O$_{12}$. Some of these, such as ruby and Nd:YAG, serve as the active media of optically pumped crystal lasers. The absorptions and fluorescence emissions from ligand field energy levels tend to be relatively narrow in crystals; in glasses, where the disorder leads to a range of ligand fields, these absorptions and emissions are much broader, as in the Nd:glass used in lasers such as the NOVA thermonuclear fusion lasers. In the decolorizing of glass, the greenish color caused by iron impurities is removed by adding Mn^{IV}O$_2$ [1313-13-9]. This acts in two ways: it reduces some of the green-producing Fe(II) to the yellow-producing but weaker colorant Fe(III) while forming some Mn(III). This latter also produces a purple color; since this is complementary to the green of Fe(II), it results in an inconspicuous very pale grey.

Color in Organic Compounds. In organic molecules, particularly those containing conjugated chains of alternating single and double bonds, eg, as in the polyenes $H_3C\!(\!-CH\!=\!CH)_{\overline{n}}CH_3$, the double bonding p_z orbitals have a variety of excited states. For $n = 2$ in 2,4-hexadiene [592-46-1], the transition between the HOMO (highest occupied molecular orbital) and the LUMO (lowest unoccupied MO) is in the ultraviolet, hence no color. With $n = 8$ in 2,4,6,8-deca-tetraene [2423-96-3] this transition has shifted down into the violet end of the visible part of the spectrum, thus producing a yellow color. The electronic energy levels are also modified by vibrational effects, resulting in transitions to triplet states and leading to phosphorescence from the very slow forbidden transition back to the singlet ground state.

Examples of polyene-type colorants are the orange β-carotene pigment [7235-40-7] of carrots, the pink carotenoids of flamingos, and the yellow crocin [42553-65-1] present in saffron. Cyclic but nonbenzenoid conjugated colored systems include the green chlorophyll [1406-65-1] of the vegetable kingdom, the red to brown hemoglobins and porphyrins of the animal kingdom, and the related blue dye copper phthalocyanin [147-14-8].

The conjugated chromophore (color-causing) system can be extended by electron-donor groups such as $-NH_2$ and $-OH$ and by electron-acceptor groups such as $-NO_2$ and $-COOH$, often used at opposite ends of the molecule. An example is the aromatic compound alizarin [72-48-0], also known since antiquity as the red dye madder.

alizarin   indigo

Another ancient dye is the deep blue indigo [482-89-3]; the presence of two bromine atoms at positions * gives the dye Tyrian purple [19201-53-7] once laboriously extracted from certain sea shells and worn by Roman emperors.

Organic colors caused by this mechanism are present in most biological colorations and in the triumphs of the dye industry (see AZINE DYES; AZO DYES; FLUORESCENT WHITENING AGENTS; CYANINE DYES; DYE CARRIERS; DYES AND DYE INTERMEDIATES; DYES, ANTHRAQUINONE; DYES, APPLICATION AND EVALUATION; DYES, NATURAL; DYES, REACTIVE; POLYMETHINE DYES; STILBENE DYES; and XANTHENE DYES). Both fluorescence and phosphorescence occur widely and many organic compounds are used in tunable dye lasers such as rhodamine B [81-88-9], which operates from 580 to 655 nm.

rhodamine B

Color from Charge Transfer. This mechanism is best approached from MO theory, although ligand field theory can also be used. There are several types of color-producing charge-transfer (CT) processes.

Consider corundum, Al_2O_3 containing both Fe^{2+} and Ti^{4+} substituting on adjacent Al sites. The transfer of one electron can occur with the absorption of about 2.2 eV, resulting in the production of Fe^{3+} and Ti^{3+}. This light absorption results in the color of blue sapphire [1317-82-4]. The reverse transition restores the initial state with the production of heat. Since all CT transitions are fully allowed, only a few hundredth of one percent of each impurity is required for an intense color. The process can also be termed electron hopping or photochemical redox. Another example of this heteronuclear intervalence CT is the gray to black color of most moon rock, again from the Fe–Ti combination.

In homonuclear intervalence CT, atoms of the same element but on different sites A and B interact as in:

$$Fe_A^{2+} + Fe_B^{3+} \longrightarrow Fe_A^{3+} + Fe_B^{2+}$$

Idiochromatic examples of this are black magnetite [1309-38-2], Fe_3O_4, otherwise written as $Fe(II)O \cdot Fe_2(III)O_3$ and the analogous red Mn_3O_4; and the pigments Prussian blue [14038-43-8] and Turnbull's blue [25869-98-1], both $Fe_4(III)[Fe(II)(CN)_6]_3$. Allochromatic examples are widespread in the mineral field, with Fe^{2+} + Fe^{3+} being involved in blue and green tourmaline [1317-93-7], blue iolite (cordierite) [12182-53-5], etc.

Metal←—ligand or cation←—anion CT is present in the yellow to orange chromates and dichromates such as $K_2Cr(VI)O_4$ [7789-00-6] and $K_2Cr_2(VI)O_7$ [7778-50-9]; these contain no unpaired electrons but the color derives from transfer of electrons from O^{2-} to the otherwise very highly charged Cr^{6+}. Once again the transitions are fully allowed and intense colors occur as in permanganates such as $KMnO_4$ and the pigments chrome yellow [1344-37-2], $PbCrO_4$, and red ochre (mineral hematite) [1317-60-8], Fe_2O_3. The reverse ligand←—metal or anion←—cation CT is not common but occurs in the yellow liquid $Fe(CO)_5$ [13463-40-6] where low energy π-orbitals of CO can accept electrons from the Fe.

Anion←—anion CT occurs in the pigment ultramarine [57455-37-5] (mineral lapis lazuli) [1302-85-8], which contains S_3^- groups having a total of 19 electrons in molecular orbitals. Among these orbitals there is a strong transition giving an absorption band at 2.1 eV and leading to the deep blue color.

Acceptor←—donor CT occurs, eg, in the solution of iodine in benzene, where an electron can transfer from the π-electron system in benzene to the I_2 molecule. Organic dyes containing both donor and acceptor groups can also be approached from this viewpoint.

Color in Metals. This is the first of four mechanisms best approached from an energy-band point of view. The two equivalent equal-energy molecular orbitals of two hydrogen atoms combining to form H_2 split to form a lower energy bonding orbital containing the two electrons and a higher energy antibonding empty orbital, also capable of holding two electrons. In an n-valent metal the $n \times 10^{23}$ or so outermost electrons, n from each atom, are again equivalent and of equal energy, forming an essentially continuous band of energy states. The $n \times 10^{23}$ or so delocalized electrons available fill the band from the bottom up to the Fermi level E_f, (Fig. 16). Electrons in the band can absorb photons at any energy as shown, but the light is so intensely absorbed that it can penetrate to a depth of only a few atoms, typically less than a wavelength. Being an electromagnetic wave, the light induces electrical current on the metal surface, which immediately reemits the light, resulting in metallic luster and metallic reflection. The same principles apply to alloys and to some metal-like compounds such as the "fool's gold" pyrite [1309-36-0], FeS_2, which contains both localized and delocalized electrons.

Using the complex refractive index $N = n + iK$ where $i = \sqrt{-1}$ and K is the absorption coefficient, the reflectivity R of metals and alloys is given by:

$$R = 100[(n - 1)^2 + K^2]/[(n + 1)^2 + K^2]$$

where K varies with the wavelength. This variation originates from the nature of the various orbitals which originally combined to form the density of states diagram. Most metals have more complex shapes than the simple parabolic type of Figure 16, which applies to alkali metals; the variation of the efficiency of the reflection process with light energy then controls the color. If all energies are equally efficiently reflected, the almost colorless reflections of clean iron [7439-89-6], mercury [7439-97-6], and silver [7440-22-4] result. However, if the efficiency decreases with increasing energy, there is reduced reflection at the blue end of the spectrum, resulting in the yellow of gold [7440-57-5], brass [12597-71-6], and pyrite, and the reddish color of copper [7440-50-8].

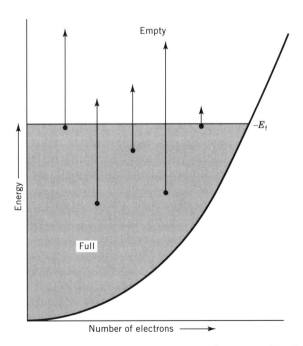

Fig. 16. A metal can absorb some light at any energy, but immediately reemits most of it (5).

The direct light absorption of a metal in the absence of reflection is observed only rarely. When gold is beaten into gold leaf less than 100 nm thick, a blue-green color is seen by transmitted light. When in the form of colloidal particles, eg, the 10 nm diameter gold particles in deep red ruby glass, complex Mie scattering theory may be used or the situation can be treated as a bounded plasma resonance. The color produced by such particles also depends on their shape, a characteristic used to produce colored images in polychromatic glass. Here a glass containing both silver and cerium [7440-45-1] yields on exposure to light the transformation:

$$Ag^+ + Ce^{3+} \longrightarrow Ag + Ce^{4+}$$

with the formation of silver particles; a complex technology causes the aspect ratio of these particles to vary with wavelength and thus produce colored images. Closely related are photochromic sunglasses, where exposure of silver halide particles to intense sunlight produces metallic silver and darkening, with reversal when the intensity of the light is reduced (see CHROMOGENIC MATERIALS, PHOTOCHROMIC).

Color in Semiconductors. In some materials a gap is present in the band, so that a lower energy valence band is separated from a higher energy conduction band by a band gap of energy E_g as at the left in Figure 17. When there are exactly

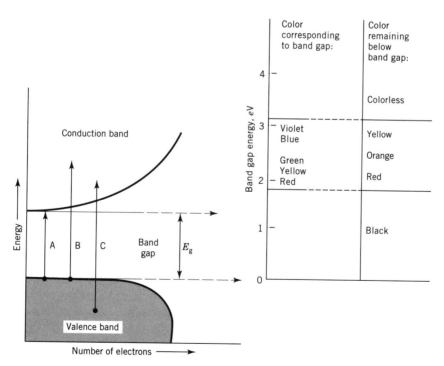

Fig. 17. The absorption of light in a band-gap material (left) and the variation of color with the size of the band gap (right) (5).

four valence electrons per atom, the conduction band is exactly full and the energy band is exactly empty. Since most of these materials do not conduct electricity well, the metallic reflection is absent and excitation of electrons by light, as in Figure 17, results in absorption. If the band gap is very small, all light is absorbed and the color is black. If the band gap is very large, then light cannot be absorbed and there is no color. At a band gap corresponding to 560 nm (2.2 eV) violet, blue, and green are absorbed, leaving an orange transmission color. The full sequence is black, red, orange, yellow, colorless as can be readily deduced from Figure 17. Examples of some band-gap materials are given in Table 4.

Color from Doped Semiconductors. Consider a diamond crystal [7782-40-3], otherwise colorless because of its large energy band gap, with 10 ppm substitutional nitrogen atoms added. Each N has one more electron than the C it replaces and will donate this electron to a donor level within the band gap centered 4 eV below the conduction band; this level is broadened by various causes, including thermal vibrations. Absorption of light can now occur, with the excitation of an electron from the donor level into the conduction band from 2.2 eV up, leading to the yellow color of natural and synthetic N-containing diamonds. The analogous presence of boron atoms produces a blue color, as in the famous Hope diamond. In this case, B has one less electron than the C it replaces and each atom forms one hole in an acceptor level within the band gap, centered 0.4 eV above the valence band. Light is absorbed by electrons being excited from the valence band into the acceptor level holes, thus producing the color.

Donor and acceptor levels are the active centers in most phosphors, as in zinc sulfide [1314-98-3], ZnS, containing an activator such as Cu and various coactivators. Phosphors are coated onto the inside of fluorescent lamps to convert the intense ultraviolet and blue from the mercury emissions into lower energy light to provide a color balance closer to daylight as in Figure 11. Phosphors can also be stimulated directly by electricity as in the Destriau effect in electroluminescent panels and by an electron beam as in the cathodoluminescence used in television and cathode ray display tubes and in (usually blue) vacuum-fluorescence alphanumeric displays.

Some impurities can form trapping levels within the band gap, as shown in Figure 18. Absorption of light can move electrons into the traps, as shown by the arrows. Energy E_b is required to release the electron from the trap into the con-

Table 4. Color of Some Band-Gap Semiconductors

Substance	CAS Registry Number	Mineral name	Pigment name	Band gap, eV	Color
C	[7782-40-3]	diamond		5.4	colorless
ZnO	[1314-13-2]	zincite	zinc white	3.0	colorless
CdS	[1306-23-6]	greenockite	cadmium yellow	2.6	yellow
$CdS_{1-x}Se_x$	[12656-57-4]		cadmium orange	2.3	orange
HgS	[19122-79-3]	cinnabarite	vermilion	2.0	red
HgS	[23333-45-1]	metacinnabar		1.6	black
Si	[7440-21-3]			1.1	black
PbS	[12179-39-4]	galena		0.4	black

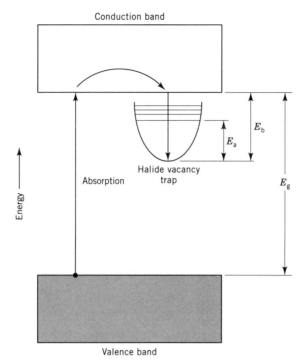

Fig. 18. Trapping energy from absorbed light in a band gap trap (5).

duction band and permit it to decay by one of several possible light-emitting paths not shown. If E_b is small so that room temperature thermal excitations can release the electrons slowly, then phosphorescence results. If E_b is a little larger, then infrared may permit the escape, as in an infrared-detecting phosphor that first has been activated to load up the traps.

Finally, an electric current can produce injection luminescence from the recombination of electrons and holes in the contact zone between differently doped semiconductor regions. This is used in light-emitting diodes (LED, usually red), in electronic displays, and in semiconductor lasers.

Color from Color Centers. This mechanism is best approached from band theory, although ligand field theory can also be used. Consider a vacancy, for example a missing Cl^- ion in a KCl crystal produced by irradiation, designated an F-center. An electron can become trapped at the vacancy and this forms a trapped energy level system inside the band gap just as in Figure 18. The electron can produce color by being excited into an absorption band such as the E_a transition, which is 2.2 eV in KCl and leads to a violet color. In the alkali halides $E_a = 0.257/d^{1.83}$ where E_a is in eV and d is the anion–cation distance in nm. In addition to irradiation, F centers in halides can also be produced by solid-state electrolysis, by in-diffusion of metal, or by growth in the presence of excess metal. Many other centers exist in the alkali halides, such as the M center, consisting of two adjacent F centers; the F′, an F center that has trapped two electrons; and the V_K, two adjacent F centers with only one trapped electron between them. All of these are electron color centers, where an electron is present in a location where it is not normally found.

An example of a hole color center is smoky quartz [14808-60-7]. Here irradiation (either produced by nature or in the laboratory) of SiO_2 containing a trace of Al ejects an electron from an oxygen adjacent to the Al or, in customary nomenclature: $[AlO_4]^{5-} \longrightarrow [AlO_4]^{4-} + e^-$; the ejected electrons are trapped elsewhere in the crystal, eg, at K^+ or H^+ impurities. The electron-deficient $[AlO_4]^{4-}$ hole color center has excited energy levels and produces light absorption leading to the smoky color. When iron is present in quartz, the result is yellow citrine [14832-92-9], containing $[Fe(III)O_4]^{5-}$; irradiation now analogously produces under certain circumstances the deep purple of amethyst [14832-91-8], which is colored by the $[Fe(III)O_4]^{4-}$ hole color center.

In general, for irradiation-produced color centers, either the hole center or the electron center can be the light-absorbing species, or even both. If the electron is released from the electron center by heat, then bleaching occurs by its recombination with the hole center to restore the pre-irradiation state. If this bleaching energy, E_b in Figure 18, is small, then the color center will fade either from room-temperature thermal excitation or from light absorpton itself; if larger, it then requires higher temperatures as in the 300 to 500°C required to fade smoky quartz or amethyst. Color center transitions are fully allowed, leading to large oscillator strengths and intense colors even at very low concentration.

Additional examples of color centers are old Mn-containing glass turned purple by irradiation with the mechanism: $Mn^{2+} \longrightarrow Mn^{3+} + e^-$; even ultraviolet can produce this change in old sun-exposed bottles then termed desert amethyst glass. In many minerals, the mechanism is unknown, such as the following where F indicates fading at room temperature or in light, S stands for stable: purple fluorite calcium fluoride [7789-75-5] (blue john) (S); two types of yellow sapphire (F and S); blue (S) and two types of brown (F and S) topaz [1302-59-6]; deep blue Maxixe beryl (F); and irradiated yellow, green, blue, and red diamond (all S; these are different from the impurity-caused colors covered previously).

Unusual is some hackmanite [1302-90-5], $Na_4Al_3Si_3O_{12}(Cl,S)$, which can have a deep magenta color as mined but fades in the light; the color can be restored with uv exposure or simply by storing in the dark, where the reaction is $S_2^{2-} \longrightarrow S_2^- + e^-$. The hole center S_2^- absorbs at 3.1 eV (400 nm) and the electron combines with a vacancy to form an F center absorbing at 2.35 eV (530 nm); it is the combination of both absorption bands that produces the color. Light absorption while showing the color also reverses this process, as does heat, both causing the color to fade.

Color from Dispersive Refraction. This mechanism involves the variation of the refractive index n with wavelength λ, given by the Sellmeier dispersion formula:

$$n^2 - 1 = a\lambda^2/(\lambda^2 - A^2) + b\lambda^2(\lambda^2 - B^2) + \ldots$$

where $A, B, \ldots$ are the wavelengths of individual infrared, visible, and ultraviolet absorptions and $a, b, \ldots$ are constants representing the strength of these absorptions. Three terms are normally sufficient for an excellent fit in the visible region.

Even a transparent colorless material has ultraviolet absorptions (from electronic excitations) and infrared absorptions (from atomic and molecular vibrations) which result in the decrease of n with wavelength in the visible region as seen in Figure 19. Only a vacuum has no absorption and no dispersion; neither of these is fundamental since each can be viewed as producing the other, with the Kramers-Kronig relationships as the connection. As a result Newton's prism produces the spectrum, crystal glassware and faceted gemstones show fire, raindrops produce the primary and secondary rainbows (higher orders can be seen only in the laboratory), and ice crystals produce various colored halos around the sun and moon, as well as sundogs. The green flash, rarely seen when the sun sets, results when the density gradient of the atmosphere acts as a prism, separating the spectral colors; since violet and blue are strongly scattered from the following mechanism, green is the last spectral color seen. Anomalous dispersion, where n varies anomalously in the vicinity of an absorption band in the visible, can be classified in this section.

Color from Scattering.   Particles that are small compared to the wavelength of light produce scattering, the intensity of which is proportional to λ^{-4}, as shown by Lord Rayleigh. This happens not only from particles but even from refractive index variations from density fluctuation in gases, thermal vibrations in crystals, and chemical composition variations in glasses and polymers. As a result, violet at the 400 nm limit of visibility is scattered 9.38 times as much as red at the 700 nm limit. Scattered light therefore appears violet to blue, as in the blue of the clear sky, while the remaining light is usually orange to red, as in the setting sun where the effects of dust add to the atmosphere's density fluctuations. The exact color also depends on the shape of the scattering particles. One speaks of Rayleigh scattering and of Tyndall blues after an early investigator. The scattered light is polarized with an intensity variation of $(1 + \cos^2 \Theta)$ with angle Θ from the light beam. When particle size becomes equal to or a little larger than the wavelength, the complex Mie scattering theory must be used; this type of scattering can pro-

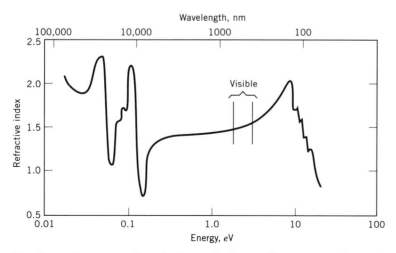

Fig. 19. The dispersion curve of a colorless soda-lime–silicate crown glass where the wt % of Na_2O = 21.3; CaO = 5.2; and SiO_2 = 73.5.

duce almost any color. At even larger sizes scattering becomes noncolor-selective and produces only white, as in fogs and clouds.

Rayleigh scattering produces, in addition to the blue sky and red sunset, most blue colors in bird feathers and butterfly wings, the blue iris in eyes, blue skin, blue cigarette smoke, blue moonstone, and the rare blue moon from forest fire oil-droplet haze. A dark background is necessary for intense blues, as in the black of outer space for the blue sky and the melanin backing for the blue iris of the eye (additional yellow to brown melanin in the iris leads to green and brown colors). Also included in this section are colors produced by effects such as second harmonic generation and parametric oscillation in nonlinear materials and various inelastic processes such as the Raman, Brillouin, polariton, magnon, Thompson, and Compton scatterings. Mie scattering from metallic particles in glass, such as in ruby glass, can be considered here or under metals.

Color from Interference.   Under this heading is considered only that color produced by interference that does not involve any diffraction; the combination of these two processes is considered in the next section.

When two light waves of the same frequency interfere, anything from constructive reinforcement to destructive cancellation can occur, depending on the phase relationship. This is present in a wedge-shaped film, for example, where the light reflected from the front surface interferes with that reflected from the back surface, the phase difference depending on the refractive index and the thickness of the film. With monochromatic light this usually produces light and dark fringes as in interferometers, while with white light the color sequence of Newton's colors is produced. Starting with the thinnest film, this sequence is black, gray, white, yellow, red (end of the first order), violet, blue, green, yellow, orange-red, violet (end of the second order), blue, green, yellow, red, and so on.

Newton's colors are seen in the tapered air gap between touching nonflat sheets of glass; cracks in a transparent medium such as glass or a crystal; soap bubbles; oil slicks on water; thin tarnish coatings on substances such as the mineral bornite and on ancient buried glass; supernumerary rainbow fringes; anti-reflection coatings on camera lenses, etc; the wings of house and dragon flies; and in some beetle wing cases as in Japanese beetles. When intense colors occur, they are metalliclike in nature and are usually termed iridescent; these effects are intensified by a dark backing. They also can be intense in some multiple layer structures that are present in interference filters; the naturally occurring multi-layered mineral labradorite and layer-structured pearls; fish scales and imitation pearls based on fish scale essence, both involving guanidine flakes; peacock, hummingbird, and some other bird feather colors; the luster of hair and nails; and in the metalliclike reflections from the eyes of cats and many other nocturnal animals except humans.

Other interference-produced colors falling into this section include doubly refracting materials such as anisotropic crystals and strained isotropic media between polarizers, as in photoelastic stress analysis and in the petrological microscope.

Color from Diffraction.   Diffraction refers to the nonrectilinear propagation of light, for example when the edge of an object produces an interrupted wavefront and light bends around the edge; interference can then occur. In Fresnel diffraction from a collimated beam of light passing through a small slit, as well as in

Fraunhofer diffraction when the image of such a slit is focused with a lens, light and dark fringes are formed from monochromatic light and colored fringes from white light. This produces the corona, a set of colored rings around bright lights; the corona aureole, a disk of bluish light seen around the sun behind a thin cloud (the corona of the sun is a different phenomenon); and fog- and cloud-related effects such as the glory and Bishop's ring.

Diffraction from two-dimensional diffraction gratings follows a variant of Bragg's law (5) and produces spectral displays as in diffraction grating spectroscopes; some beetles and snakes; phonograph records and compact disks; a distant street lamp seen through a cloth umbrella; and in the play of color seen in the natural three-dimensional diffraction grating opal. Liquid crystals (qv) of the cholesteric or chiral nematic type also function to diffract light as in liquid crystal thermometers, thermography, and in mood jewelry, where temperature changes alter the grating spacing and hence the color of a cholesteric mesophase supported between pieces of glass or plastic.

BIBLIOGRAPHY

"Color Measurement" in *ECT* 1st ed., Vol. 4, pp. 242–251, by G. W. Ingle, Monsanto Chemical Co.; "Color" in *ECT* 2nd ed., Vol. 5, pp. 801–812, by G. W. Ingle, Monsanto Chemical Co.; in *ECT* 3rd ed., Vol. 6, pp. 523–548, by F. W. Billmeyer, Jr., Rensselaer Polytechnic Institute.

1. R. S. Hunter and R. W. Harold, *The Measurement of Appearance,* 2nd ed., John Wiley & Sons, Inc., New York, 1987; Table A7.
2. R. W. G. Hunt, *Measuring Color,* 2nd ed., John Wiley & Sons, Inc., New York, 1992.
3. F. W. Billmeyer, Jr. and M. Saltzman, *Principles of Color Technology,* 2nd ed., John Wiley & Sons, Inc., New York, 1981.
4. R. McDonald, ed., *Colour Physics for Industry,* Society of Dyers and Colourists, Bradford, UK, 1987.
5. K. Nassau, *The Physics and Chemistry of Color,* John Wiley & Sons, Inc., New York, 1983.
6. R. M. Boynton, *Human Color Vision,* Holt, Rinehart and Winston, New York, 1979.
7. M. S. Cayless and A. M. Marsden, *Lamps and Lighting,* Edward Arnold, London, 1983.
8. W. D. Wright, *The Measurement of Color,* 4th ed., Adam Hilger, London, 1969.
9. R. Fletcher and J. Voke, *Defective Colour Vision,* Hilger, Bristol, UK, 1985.
10. R. W. G. Hunt, *The Reproduction of Colour,* 3rd ed., Fountain Press, London, 1975.
11. D. B. Judd and G. Wyszecki, *Color in Business, Science, and Industry,* 3rd ed., John Wiley & Sons, Inc., New York, 1975.
12. K. L. Kelly and D. B. Judd, *Color: Universal Language and Dictionary of Names,* NBS Special Publication 440, U.S. Government Printing Office, Washington D.C., 1976.
13. R. G. Kuehni, *Computer Colorant Formulation,* D. C. Heath & Co., Lexington, Mass., 1975.
14. K. McLaren, *The Colour Science of Dyes and Pigments,* 2nd ed., Hilger, Bristol, UK, 1986.
15. A. H. Munsell, *A Color Notation,* 14th ed., Macbeth Division of Kollmorgen Instruments Corp., Newbaugh, N.Y., 1990.
16. W. N. Sproson, *Colour Science in Television and Display Systems,* Hilger, Bristol, UK, 1983.
17. W. S. Stiles, *Mechanisms of Colour Vision,* Academic Press, Inc., New York, 1978.

18. G. Wyszecki and W. S. Stiles, *Color Science,* 2nd ed., John Wiley & Sons, Inc., New York, 1982.
19. *CIE Publication 15.2 Colorimetry,* 2nd ed., Central Bureau of CIE, Vienna, Austria, 1986; Available from U.S. National Committee CIE, c/o National Institute of Standards and Technology, Washington, D.C.
20. Ref. 4, p. 246.
21. A. R. Hill, in Ref. 4, p. 211.
22. G. Celikiz, *Color Vision and Color Deficiencies,* in G. Celikiz and R. G. Kuehni, eds., *Color Technology in the Textile Industry,* American Association of Textile Chemists and Colorists, Research Triangle Park, N.C., 1983, p. 193.
23. *Munsell Book of Color,* glossy or matte finishes, and other collections, Macbeth Division of Kollmorgen Instruments Corp., Baltimore, Md., 1929 on.
24. A. Hård and L. Sivik, *Color Res. Appl.* **6,** 129 (1981).
25. E. Jacobson, *Basic Color: An Interpretation of the Ostwald Color System,* P. Theobald, Chicago, Ill., 1948.
26. A. Maerz and M. R. Paul, *A Dictionary of Color,* McGraw Hill Book Co., New York, 1930.
27. D. L. MacAdam, *J. Opt. Soc. Am.* **64,** 1691 (1974); *Uniform Color Scales Committee Samples,* Optical Society of America, Washington, D.C., 1977.
28. *ISCC-NBS Centroid Color Charts, NBS Standard Reference Material No. 2106,* National Institute of Standards and Technology, Washington, D.C., 1965.
29. R. McDonald, in Ref. 4, p. 116.
30. R. P. Best, in Ref. 4, p. 186.
31. Ref. 1, Table A5.
32. D. H. Alman and F. W. Billmeyer, *J. Chem. Educ.* **52,** A281, 284, 315, 318 (1975).
33. C. L. Sanders, *J. Res. Natl. Bur. Stand.* **76A,** 437 (1972).
34. W. Budde, *J. Res. Natl. Bur. Stand.* **80A,** 585 (1976).
35. R. S. Hunter, *J. Opt. Soc. Am.* **53,** 390 (1963).
36. *Color and Appearance Collaborative Reference Program,* Collaborative Testing Services, Inc., Herndon, Va., previously *MCCA-NBS Collaborative Reference Program on Color and Color Differences,* National Institute of Standards and Technology, Washington, D.C., 1991.

KURT NASSAU
Consultant

COLOR AND CONSTITUTION OF ORGANIC DYES. See

COLOR; CYANINE DYES.

COLORANTS FOR CERAMICS

Any product that depends on aesthetics for consideration for purchase and use will be improved by the use of color. Hence, many ceramic products, such as tile, sanitary ware, porcelain enameled appliances, tableware, and some structural clay products and glasses, contain colorants.

For both economic and technical reasons, the most effective way to impart color to a ceramic product is to apply a ceramic coating that contains the colorant. The most common coatings, glazes and porcelain enamels, are vitreous in nature. Hence, most applications for ceramic colorants involve the coloring of a vitreous material.

There are a number of ways to obtain color in a ceramic material (1). First, certain transition-metal ions can be melted into a glass or dispersed in a ceramic body when it is made. Although suitable for bulk ceramics, this method is rarely used in coatings because adequate tinting strength and purity of color cannot be obtained this way.

A second method to obtain color is to induce the precipitation of a colored crystal in a transparent matrix. Certain materials dissolve to some extent in a vitreous material at high temperatures, but when the temperature is reduced, the solubility is also reduced and precipitation occurs. This method is used to disperse nonoxide precipitates of gold, copper, or cadmium sulfoselenide in bulk glass. In coatings it is used for opacification, the production of an opaque white color. Normally, some or all of the opacifier added to the coating slip dissolves during the firing process and recrystallizes upon cooling. For oxide colors other than white, however, this method lacks the necessary control for reproducible results and is seldom used.

The third method to obtain color in a vitreous matrix is to disperse in that matrix an insoluble crystal or crystals that are colored. The color of the crystal is then imparted to the transparent matrix. This method is the one most commonly used to introduce color to vitreous coatings.

Colored Bodies and Glasses

The addition of oxides to ceramic bodies and to glasses to produce color has been known since antiquity (2). The use of iron and copper oxides predates recorded history. Cobalt was introduced into Chinese porcelain about 700 AD. Chromium compounds have been used since 1800 AD.

The colors obtained depend primarily on the oxidation state and coordination number of the coloring ion (3). Table 1 lists the solution colors of several ions in glass. All of these ions are transition metals; some rare-earth ions show similar effects. The electronic transitions within the partially filled d and f shells of these ions are of such frequency that they fall in that narrow band of frequencies from 400 to 700 nm, which constitutes the visible spectrum (4). Hence, they are suitable for producing color (qv).

Decolorizing is sometimes desirable (5). In the manufacture of glass, the presence of iron as an impurity cannot be completely avoided. Iron imparts an unacceptable dirty brown color to the glass. If the iron can be oxidized through the use of additives such as arsenic oxide [1327-53-3], As_2O_3, cerium oxide [1306-38-3], CeO_2, or manganese oxide [1313-13-9], MnO_2, the discoloration can be diminished through chemical means. Alternatively, the effect of the iron can be negated through the addition of ingredients that themselves produce complementary colors in the glass. Materials such as selenium [22541-48-6], cobalt oxide [1308-06-1], Co_3O_4, neodymium oxide [1313-97-9], Nd_2O_3, and manganese oxide are used for this purpose. The addition of one of these materials produces a slightly darker, but more neutrally colored glass, which is more acceptable visually.

Table 1. Solution Colors in Glass

Ion	Number of d Electrons	Color
Ti^{3+}	1	blue-green
V^{3+}	2	gray
Cr^{3+}	3	dark green
Mn^{3+}	4	purple
Fe^{3+}	5	yellow-brown
Fe^{2+}	6	light blue or green
Co^{3+}	6	pink
Co^{2+}	7	dark blue
Ni^{2+}	8	gray
Cu^{2+}	9	light blue

Precipitation Colors

Several colors in bulk glass can be produced by precipitation processes. One such technique involves developing a colloidal suspension in the glass matrix (5). Metals, such as gold, silver, and copper (or possibly Cu_2O), and nonoxide pigments, such as the cadmium sulfoselenides, produce strong colors when precipitated from a colloidal suspension in glass (Table 2).

Table 2. Colloidal Precipitation Colors

Crystal	CAS Registry Number	Color
Au	[7440-57-5]	gold
Ag	[7440-22-4]	gray
Cu	[7440-50-8]	red
CdS	[1306-23-6]	yellow
Cd(S,Se)	[a]	red, orange

[a]Cadmium sulfoselenides, CdSSe [11112-63-3], Cd_2SSe [12214-12-9].

To produce this color, the glass is melted, formed, and annealed as is usually done in making glass. This is followed by reheating to a temperature where nuclei of the desired material are formed. The temperature is then adjusted to permit these nuclei to grow to an optimum size for the development of the particular color. This process is known as striking the color.

Most striking colors are obtained in glasses containing 10–20 wt % K_2O, 10–22 wt % ZnO, and 50–60 wt % silica. CaO and B_2O_3 may also be present (6). To this batch is added 1–3 wt % CdS, CdSe, and/or CdTe. Melting must be under neutral or mildly reducing conditions. Otherwise, S, Se, and Te will be oxidized to SO_2, SeO_2, or TeO_2, which are colorless.

After casting, the glasses solidify to colorless glass. When reheated at 550–700°C, the glasses precipitate minute crystals of a cadmium sulfoselenide. This causes the absorption edge to move to longer wavelengths, producing colors from yellow to orange, to red, to maroon.

The color obtained is a function of both the composition and the particle size of the precipitated crystals. A redder color results from both increased selenium to sulfur ratio and from larger crystals, caused by a more severe heat treatment. Hence, it is possible to make, from the same glass, a series of color filter types, by controlled reheating.

A related but different type of colored glass is the ruby glass (6). Ruby glasses can be made in several base glass systems by adding to the batch 0.003–0.1 wt % of a noble metal salt, ie, copper(II) chloride [7447-39-4], silver nitrate [19582-44-6], or gold chloride [13453-07-1], together with a reducing agent such as stannous chloride or antimony oxide. When cooled from the melt, they are usually colorless or weakly colored. Reheating results in a more or less intense red for copper, yellow-brown for silver, and red for gold. The color comes from a colloidal dispersion of metal particles that have precipitated from the glass.

Opacification. Whiteness or opacity is introduced into ceramic coatings by the addition to the coating formulation of a substance which will disperse in the coating as discrete particles that scatter and reflect some of the incident light (7). To be effective as a scatterer, the discrete substance must have a refractive index that differs appreciably from that of the clear ceramic coating, because the greater the difference in index of refraction between the matrix and the scattering phase, the greater the degree of opacity. The refractive index of most glasses is 1.5–1.6 and therefore the refractive indexes of opacifiers must be either greater or less than this value. As a practical matter, opacifiers of high refractive index are used. Some possibilities include tin oxide [18282-10-5], SnO_2, with $n_D = 2.04$, zirconium oxide [1314-23-4], ZrO_2, with $n_D = 2.40$, zirconium silicate [10101-52-7], $ZrSiO_4$, with $n_D = 1.85$, titanium oxide [13463-67-7], TiO_2, with $n_D = 2.5$ for anatase [1317-70-0], and $n_D = 2.7$ for rutile [1317-80-2].

In glazes fired at temperatures greater than 1000°C, zircon [14940-68-2] is the opacifier of choice (8). It has a solubility of about 5% in many glazes at high temperature, and 2–3% at room temperature. A customary mill addition would be 8–10% zircon. Thus most opacified glazes contain both zircon that was placed in the mill and went through the firing process unchanged, and zircon that dissolved in the molten glaze during firing but recrystallized on cooling.

In porcelain enamels and in glazes firing under 1000°C, TiO_2 in the anatase crystal phase is the opacifying agent of choice (9,10). Because it has the highest refractive index, TiO_2 is the most effective opacifying agent. However, at temperatures of about 850°C, anatase inverts to rutile in silicate glasses. Once inverted to rutile, TiO_2 crystals are able to grow rapidly to sizes that are no longer effective for opacification. Moreover, because the absorption edge of rutile is very close to the visible, as the rutile particles grow the absorption edge extends into the visible range, leading to a pronounced cream color. Thus while TiO_2 is a very effective opacifier at lower temperatures, it cannot be used above 1000°C.

The solubility of TiO_2 in molten silicates is around 8–10%. At room temperature this solubility is reduced to around 5%. Thus when using TiO_2 as an opacifier, substantial amounts, about 15%, must be used.

Ceramic Pigments

The principal method of coloration of ceramic coatings is dispersal of a ceramic pigment in a vitreous matrix. To be suitable as a ceramic pigment, a material must possess a number of properties (11) which fall in two categories: strength of pigmentation and stability. The first requirement is high tinting strength or intensity of color. It is also desirable that the color be pure and free of grayness or muddiness. Related to the intensity is the diversity of colors obtainable with similar and compatible materials, since many users blend two or more colors to obtain the final shade desired.

The other important property of a ceramic pigment is stability under the high temperatures and corrosive environments encountered in the firing of glazes. The rates of solution in the molten glass should be very low in spite of very fine particle sizes (about 10 μm). Neither should gases be given off as a result of the contact of the ceramic pigment with the molten vitreous material.

Another desirable property for a ceramic color is a high refractive index. For example, valuable pigments are based on spinels [1302-67-6] (n_D = 1.8) and on zircon (n_D = 1.9), but no valuable pigments are based on apatite (n_D = 1.6), even though the lattice of apatite is as versatile for making ionic substitutions as that of spinel.

Manufacturing. Although a number of different pigment systems exist, most are prepared by similar manufacturing methods. The first step in pigment manufacture is close control over the selection of raw materials. Most of these raw materials are metallic oxides or salts of the desired metals. Considerable differences in the required chemical purity are encountered, ranging from impure natural minerals to chemicals of industrial-grade purity. In pigment manufacture, purity does not always equal quality. Often less pure raw materials may prove superior in the production of a given pigment.

The raw materials are weighed and then thoroughly blended. The reaction forming the pigment crystal occurs in a high temperature calcining operation. The temperature may range from 500 to 1400°C depending on the particular system. Normally air is the atmosphere of choice, because the pigments must ultimately be stable in a molten oxide coating, so materials sensitive to oxygen have limited use. During the calcination process, any volatiles are driven off and the pigment crystal is developed in a sintering reaction. Some of these reactions occur in the solid state, but many involve a fluid-phase mineralizing pathway. Although a few large volume pigments may be made in rotary kilns, most pigments are placed in saggers of 22–45 kg (50–100 lbs) capacity for the calcining operation.

Following calcination, the product may require milling to reduce particle size to that necessary for use. This size reduction may be carried out either wet or dry. If there are soluble by-products, a washing operation may also be required. It is almost always necessary to break up agglomerates by a process such as micronizing.

It has been found that there are several advantages to modification of pigments by addition to the calcined product of a small quantity of a dispersing agent (12).

Pigment Systems. Most of the crystals used for ceramic pigments are complex oxides, owing to the great stability of oxides in molten silicate glasses. Table 3 lists these materials. The one significant exception to the use of oxides is the family of cadmium sulfoselenide red pigments. This family is used because the colors obtained cannot be obtained in oxide systems; thus it is necessary to sustain the difficulties of a nonoxide system.

Table 3 is arranged by crystal class (14). The crystal class of a given pigment is determined almost solely by the ratio of the ionic sizes of the cation and the anion and their respective valences. Hence, for any given stoichiometry and ionic size ratio, only one or two structures will be possible. In some classes (spinel, zircon), a wide range of colors is possible within the confines of that class. Pigments within a given class usually have excellent chemical and physical compatibility with each other. This compatability is important, as most applications involve the mixing of two or more pigments to achieve the desired shade.

Red Pigments. There are no oxides that can be used to give a true red pigment which is stable to the firing of ceramic coatings. Hence, orange, red, and dark red colors are obtained by the use of cadmium sulfoselenide pigments (7, 15–16). The cadmium sulfoselenides are a group of pigments based on solid solutions of cadmium selenide, cadmium sulfide, and/or zinc sulfide.

The synthetic pigment is produced by one of several related procedures. The best quality product is made by reaction of an aqueous solution of $CdSO_4$ or $CdCl_2$ with a solution of an alkaline metal sulfide or H_2S. Zn, Se, or Hg may be added to the CdS to produce shade variations. After precipitation, the color is filtered, washed, and calcined in an inert atmosphere at 500–600°C.

The colors come in a range of shades from primrose yellow through yellow, orange to red and maroon. The primrose or light yellow color is produced by precipitating a small amount of ZnS with the CdS. The orange, red, and maroon shades are made by incorporating increasing amounts of selenium compounds with the CdS. An orange pigment is obtained at a CdS to CdSe ratio of about 4:1. A red pigment is obtained at a ratio of CdS to CdSe of 1.7:1. A deeper red is formed at a ratio of 1.3:1. A very deep red-maroon is manufactured at a ratio of 1:1.

These pigments require a glaze specially designed for their use. The glaze will contain only small amounts of PbO, B_2O_3, or other aggressive fluxes, because strong fluxes react with the selenium to form black lead selenide. The glaze should have a low alkali content. It should contain a few percent of CdO so that its chemical potential for Cd, relative to Cd in the pigment, is reduced. The glaze must also be free of strong oxidizing agents, such as nitrates, which hasten the breakdown of the cadmium sulfoselenide pigment.

Table 3. Classification of Mixed-Metal Oxide Inorganic Pigments According to Crystal Class[a]

Pigment name	CAS Registry Number	Formula	DCMA number[b]
Baddeleyite			
zirconium vanadium yellow baddeleyite	[68187-01-9]	$(Zr,V)O_2$	1-01-4
Borate			
cobalt magnesium red-blue borate	[68608-93-5]	$(Co,Mg)B_2O_5$	2-02-1
Corundum–hematite			
chromium alumina pink corundum	[68187-27-9]	$(Al,Cr)_2O_3$	3-03-5
manganese alumina pink corundum	[68186-99-2]	$(Al,Mn)_2O_3$	3-04-5
chromium green-black hematite	[68909-79-5]	$(Cr,Fe)_2O_3$	3-05-3
iron brown hematite	[68187-35-9]	Fe_2O_3	3-06-7
Garnet			
victoria green garnet	[68553-01-5]	$3CaO \cdot Cr_2O_3 \cdot 3SiO_2$	4-07-3
Olivine			
cobalt silicate blue olivine	[68187-40-6]	Co_2SiO_4	5-08-2
nickel silicate green olivine	[68515-84-4]	Ni_2SiO_4	5-45-3
Periclase			
cobalt nickel gray periclase	[68186-89-0]	$(Co,Ni)O$	6-09-8
Phenakite			
cobalt zinc silicate blue phenakite	[68412-74-8]	$(Co,Zn)_2SiO_4$	7-10-2
Phosphate			
cobalt violet phosphate	[13455-36-2]	$Co_3(PO_4)_2$	8-11-1
cobalt lithium violet phosphate	[68610-13-9]	$CoLiPO_4$	8-12-1
Priderite			
nickel barium titanium primrose priderite	[68610-24-2]	$2NiO \cdot 3BaO \cdot 17TiO_2$	9-13-4
Pyrochlore			
lead antimonate yellow pyrochlore	[68187-20-2]	$Pb_2Sb_2O_7$	10-14-4
Rutile–cassiterite			
nickel antimony titanium yellow rutile	[71077-18-4]	$(Ti,Ni,Sb)O_2$	11-15-4
nickel niobium titanium yellow rutile	[68611-43-8]	$(Ti,Ni,Nb)O_2$	11-16-4
chromium antimony titanium buff rutile	[68186-90-3]	$(Ti,Cr,Sb)O_2$	11-17-6
chromium niobium titanium buff rutile	[68611-42-7]	$(Ti,Cr,Nb)O_2$	11-18-6
chromium tungsten titanium buff rutile	[68186-92-5]	$(Ti,Cr,W)O_2$	11-19-6
manganese antimony titanium buff rutile	[68412-38-4]	$(Ti,Mn,Sb)O_2$	11-20-6

(Continued)

Table 3. (*Continued*)

Pigment name	CAS Registry Number	Formula	DCMA number[b]
Rutile–cassiterite (continued)			
titanium vanadium antimony gray rutile	[68187-00-8]	$(Ti,V,Sb)O_2$	11-21-8
tin vanadium yellow cassiterite	[68186-93-6]	$(Sn,V)O_2$	11-22-4
chromium tin orchid cassiterite	[68187-53-1]	$(Sn,Cr)O_2$	11-23-5
tin antimony gray cassiterite	[68187-54-2]	$(Sn,Sb)O_2$	11-24-8
manganese chromium antimony titanium brown rutile	[69991-68-0]	$(Ti,Mn,Cr,Sb)O_2$	11-46-7
manganese niobium titanium brown rutile	[70248-09-8]	$(Ti,Mn,Nb)O_2$	11-47-7
Sphene			
chromium tin pink sphene	[68187-12-2]	$CaO \cdot SnO_2 \cdot SiO_2 {:} Cr$	12-25-5
Spinel			
cobalt aluminate blue spinel	[68186-86-7]	$CoAl_2O_4$	13-26-2
cobalt tin blue-gray spinel	[68187-05-3]	Co_2SnO_2	13-27-2
cobalt zinc aluminate blue spinel	[68186-87-8]	$(Co,Zn)Al_2O_4$	13-28-2
cobalt chromite blue-green spinel	[68187-11-1]	$Co(Al,Cr)_2O_4$	13-29-2
cobalt chromite green spinel	[68187-49-5]	$CoCr_2O_4$	13-30-3
cobalt titanate green spinel	[68186-85-6]	Co_2TiO_4	13-31-3
chrome alumina pink spinel	[68201-65-0]	$Zn(Al,Cr)_2O_4$	13-32-5
iron chromite brown spinel	[68187-09-7]	$Fe(Fe,Cr)_2O_4$	13-33-7
iron titanium brown spinel	[68187-02-0]	Fe_2TiO_4	13-34-7
nickel ferrite brown spinel	[68187-10-0]	$NiFe_2O_4$	13-35-7
zinc ferrite brown spinel	[68187-51-9]	$(Zn,Fe)Fe_2O_4$	13-36-7
zinc iron chromite brown spinel	[68186-88-9]	$(Zn,Fe)(Fe,Cr)_2O_4$	13-37-7
copper chromite black spinel	[68186-91-4]	$CuCr_2O_4$	13-38-9
iron cobalt black spinel	[68187-50-8]	$(Fe,Co)Fe_2O_4$	13-39-9
iron cobalt chromite black spinel	[68186-97-0]	$(Co,Fe)(Fe,Cr)_2O_4$	13-40-9
manganese ferrite black spinel	[68186-94-7]	$(Fe,Mn)(Fe,Mn)_2O_4$	13-41-9
chromium iron manganese brown spinel	[68555-06-6]	$(Fe,Mn)(Fe,Cr,Mn)_2O_4$	13-48-7
cobalt tin alumina blue spinel	[68608-09-3]	$CoAl_2O_4 - Co_2SnO_4$	13-49-2
chromium iron nickel black spinel	[71631-15-7]	$(Ni,Fe)(Cr,Fe)_2O_4$	13-50-9
chromium manganese zinc brown spinel	[71750-83-9]	$(Zn,Mn)Cr_2O_4$	13-51-7
Zircon			
zirconium vanadium blue zircon	[68186-95-8]	$(Zr,V)SiO_4$	14-42-2
zirconium praseodymium yellow zircon	[68187-15-5]	$(Zr,Pr)SiO_4$	14-43-4
zirconium iron coral zircon	[68187-13-3]	$(Zr,Fe)SiO_4$	14-44-5

[a]Ref. 1, 13.
[b]Dry Color Manufacturers Association, Arlington, Va.

The resistance of these materials to firing temperature is definitely limited. They can be fired to about 1000°C. Hence, they are limited to use in porcelain enamels and in low firing artware glazes.

These sulfoselenide pigments have good resistance to alkali solutions, but poor resistance to even dilute acids. Owing to the latter, together with the high toxicity of cadmium, cadmium sulfoselenide pigments should not be used in applications that will come in contact with food and drink. They must also be handled with great care to avoid the possibility of ingestion.

In an attempt to extend the firing range of these colors, the inclusion pigments (11,17) have been developed. In these pigments cadmium sulfoselenides are incorporated within a clear zircon lattice. The superior stability of zircon is thus imparted to the pigment. Colors from yellow to orange-red are available. Deep red is not available, and the purity of these colors is limited.

Pink and Purple Pigments. Although red is not available in oxide systems, pink and purple shades are obtained several ways. One such system is the chrome alumina pinks (18). Chrome alumina pinks are combinations of ZnO, Al_2O_3, and Cr_2O_3. Depending on the concentration of zinc, the crystal structure may be either spinel or corundum. The latter is analogous to the ruby.

In general, a ceramic coating formulated for use of a chrome alumina pink should be free of CaO, low in PbO and B_2O_3, and with a surplus of ZnO and Al_2O_3. Using an improper coating may lead to a brown instead of a pink. Sufficient ZnO must be in the coating to prevent the glaze from attacking the pigment and removing zinc from it. An excess of alumina prevents the molten coating from dissolving the pigment.

A related but somewhat stronger pink pigment is the manganese alumina pink (7). This composition is formulated from additions of manganese oxide and phosphate to alumina and produces a very pure clean pigment. It requires a zinc-free coating high in alumina. Unfortunately, this pigment cannot be made without creating some serious pollution problems. There is question as to its continued availability.

The most stable pink pigment is the iron-doped zircon system (11,19,20). This pigment is made by calcining a mixture of ZrO_2, SiO_2, and iron oxide at a stoichiometry to produce zircon. This pigment is sensitive to details of the manufacturing process, so one manufacturer's product may not duplicate another's (21,22). Color variations extend from pink to coral. The pigments are stable in all coating systems, but those without zinc are bluer in shade.

The chrome–tin system is the only family to produce purple and maroon shades, as well as pinks. The system can be defined as pigments that are produced by the calcination of mixtures of small amounts of chromium oxide with substantial amounts of tin oxide. In addition, most formulations contain substantial amounts of silica and calcium oxide.

The chemistry of these materials is complex (23). If one mixes about 90% of tin oxide with small amounts of Cr_2O_3 and either CaO or CeO_2, together with B_2O_3 as a mineralizer, one obtains a purple or orchid shade. The crystal structure is cassiterite. This pigment is a solid solution of chromium oxide in tin oxide. Although this is not the crystal structure of most chrome–tin pinks, residual amounts are present in almost all cases. This residual amount of chromium-doped tin oxide gives most chrome tin pinks a gray or purple overtone.

Most chrome–tin pinks also contain substantial amounts of CaO and SiO_2. Only in the presence of these materials can pink, red, or maroon shades be obtained. Here, the crystal structure is tin sphene (CaO–SnO_2–SiO_2) with Cr_2O_3 dissolved as an impurity.

The color of this pigment depends on the ratio of Cr_2O_3 to tin oxide. With high chrome contents, the pigments are green. As the chrome is reduced, the color becomes purple at a Cr_2O_3:SnO_2 ratio of 1:15, red at a ratio of 1:17, maroon at a ratio of 1:20, and pink at a ratio of 1:25.

These pigments must be used in a coating low in ZnO and high in CaO. Some SnO_2 should be used as an opacifier, in order to increase the strength and stability of the pigment.

Gold purple, often called Purple of Cassius, is a tin oxide gel colored by finely divided gold (7). It has good coverage and brilliance in low temperature coatings such as porcelain enamels. It is a very expensive pigment, because of its difficult preparation as well as the price of gold.

Brown Pigments. The most important brown pigments used in ceramic coatings are the zinc iron chromite spinels (24,25). This pigment system produces a wide palette of tan and brown shades. It can be controlled with reasonable care to produce uniform, reproducible pigments. In this pigment the Cr_2O_3 is found on the octahedral sites of the spinel structure; the ZnO is found on the tetrahedral sites and the iron oxide is distributed in such a way as to fulfill the stoichiometry of the structure. Consequently, adjustment of the formula does result in alteration of the shade. For example, minor additions of NiO to this system produce a dark chocolate brown. The presence or absence of iron on the tetrahedral site affects the yellowness of the shade. Because they are comparatively low in price, these pigments are the browns selected for most applications.

Two systems closely related to the zinc iron chromites have been developed to improve the stability and firing range of brown pigments (7). The first of these is the zinc iron chrome aluminate pigment. It is a hybrid of the zinc iron chromite brown and the chrome alumina pink. It produces warm and orange brown shades of improved firing stability. The pigment must be used in coatings high in ZnO and Al_2O_3, low in CaO.

The other related pigment is the chrome iron tin brown, often called a tin tan. It must always be used in a ZnO-containing coating, since the pigment requires ZnO from the coating to react during the firing process to produce a zinc iron chromite pigment. This pigment is characterized by excellent stability at low concentrations. Thus it makes an excellent toner for tan and beige shades in blends with pink pigments.

The last brown pigment to be considered is the iron manganese brown. This is the deep brown associated with electrical porcelain insulators and with artware and bean pots. In many glazes the presence of manganese will cause poor surface and unstable color. Hence, the use of this pigment is limited to dark colors on products where glaze surface quality requirements are modest.

Yellow Pigments. There are several systems for preparing a yellow ceramic pigment. Moreover, there are valid technical and economic reasons for the use of a particular yellow pigment in a given application. The pigments of greatest tinting strength, the lead antimonate yellows, cadmium sulfide, and the chrome titania maples, do not have adequate resistance to molten ceramic coatings. Thus other systems must be used if the firing temperature is above 1000°C.

Zirconia vanadia yellows are prepared by calcining ZrO_2 with small amounts of V_2O_5 (26,27). Small amounts of Fe_2O_3, with or without TiO_2, may be used to alter the shade from lemon yellow to orange yellow. In ceramic coatings zirconia vanadia yellows are usually weaker than tin vanadium yellows and dirtier than praseodymium zircon yellows. However, they are economical pigments for use with a broad range of coatings firing at temperatures above 1000°C. They are brighter and stronger in glazes low in PbO and B_2O_3.

Tin vanadium yellows are prepared by introducing small amounts of vanadium oxide into the cassiterite structure of SnO_2 (28). Tin vanadium yellows develop a strong color in all ceramic coating compositions. They are very opaque pigments, requiring little further opacification. The pigments are not stable under reducing conditions. They are incompatible with pigments containing Cr_2O_3. However, the primary deterrent to their use is the very high cost of the SnO_2, their principal component.

The praseodymium zircon pigments are formed by calcination of about 5% of praseodymium oxide with a stoichiometric zircon mixture of ZrO_2 and SiO_2 to yield a bright yellow pigment (11,28,29). The crystal structure is zircon. These pigments have excellent tinting strength in coatings fired to as high as 1280°C (cone 10). They can be used in almost any ceramic coating, although preferably with zircon opacifiers. They are compatible with most other pigments, particularly other zircon and zirconia pigments.

For applications fired below 1000°C, the tinting strength of the lead antimonate pigment is unsurpassed, except by the cadmium sulfoselenides. These Pb–Sb pigments are very clean and bright and have good covering power, requiring little or no opacifier. Their primary limitation is their instability above 1000°C, followed by volatilization of the Sb_2O_3. Substitutions of CeO_2, Al_2O_3, or SnO_2 are sometimes made for a portion of the Sb_2O_3 to improve stability, but these are just palliatives. Hence, their use is limited to coatings such as porcelain enamels, which fire at temperatures below 1000°C. Moreover, such pigments are formulated with toxic materials, and great care is required in their use.

An orange-yellow pigment is formed when Cr_2O_3 is added together with Sb_2O_3 and TiO_2 to form a doped rutile (30). This material gives an orange yellow or maple shade useful at lower firing temperatures. Like the lead antimony yellows, this material also decomposes around 1000°C. It has substantial use in enamels, where it is the basis for some of the more important appliance colors.

Cadmium sulfide yellow can be considered for the brightest low temperature applications (15). It is a very bright, clean orange yellow. Primrose yellow and light yellow shades are made by precipitating small amounts of ZnS with the CdS. All the limitations of the cadmium sulfoselenide reds discussed above apply to the cadmium sulfide yellow.

Green Pigments. Just as there are several alternative yellow pigments, there are several alternative green pigments (31). Formerly, the chromium ion was the basis for green pigments. Green Cr_2O_3 itself may be used in a few applications. This procedure, however, has several limitations. There is some tendency for pure Cr_2O_3 to fume and volatilize during firing of the coating, leading to absorption of chromium into the refractory lining of the furnace. Cr_2O_3 is incompatible with tin oxide, reacting to form a pink coloration. The coating must not contain ZnO, because ZnO in the coating reacts with Cr_2O_3 to produce an undesirable dirty brown color.

Higher quality results are obtained if chromium oxide is used as a constituent in a calcined pigment. One such system is the cobalt zinc alumina chromite used to produce blue-green pigments. These pigments are spinels in which varying amounts of cobalt and zinc appear in the tetrahedral sites and varying amounts of alumina and chromium oxide appear in the octahedral sites. By using higher concentrations of Cr_2O_3 and lower amounts of CoO greener pigments are obtained. Conversely, by lowering the amounts of Cr_2O_3 and raising the amount of CoO, shades from blue-green to blue are obtained. These pigments should only be used in strong masstones. In low concentrations they give an undesirable dirty gray color.

Victoria green is prepared by calcining silica and a dichromate with calcium carbonate to form the garnet $3CaO-Cr_2O_3-3SiO_2$. This pigment gives a transparent bright green color. It tends to blacken if applied too thinly. It is not satisfactory for opaque glazes or pastel shades because then it has a gray cast. It can be used only in zinc-free coatings with high CaO content. This is a difficult, costly pigment to manufacture correctly.

Because of all the difficulties with the use of chromium-containing pigments and because there is a definite limitation on the brilliance of green pigments made with chromium, many green ceramic glazes are now made with zircon pigments (11). The cleanest, most stable greens are obtained today by the use of blends of a zircon vanadium blue and a zircon praseodymium yellow. The bright green shades are obtained from a mixture of about two parts of the yellow pigment to one part of the blue pigment.

The final green to be discussed involves copper compounds for low firing-temperature applications. The use of copper is of little interest to most industrial manufacturers, but the colors obtained from them are of great interest to artists because of the many subtle shades that can be obtained. These subtle shade variations arise because the pH of the glaze used has a particular effect on the colors obtained. If the coating is acidic, a beautiful green color is developed; but if the coating is alkaline, a turquoise blue color results. The copper oxide dissolves in the coating producing a very transparent color. CuO glazes are limited to 1000°C because of copper volatilization. Also, copper oxide renders all lead-containing glazes unsafe for use with food and drink (32).

Blue Pigments. The traditional way to obtain blue in a ceramic coating is with cobalt, which has been used as a solution color since antiquity (33). Today, cobalt may react with Al_2O_3 to produce the spinel $CoAl_2O_4$ or with silica to produce the olivine Co_2SiO_4. The silicate involves higher concentrations of cobalt, with only modestly stronger color. In porcelain enamels and glass colors, pigments based on cobalt continue to be fully satisfactory both for stability and for tinting strength. At the higher temperatures used for ceramic glazes, difficulties arise from partial solution of the pigment, and diffusion of cobalt oxide in the glaze, causing a defect called cobalt bleeding.

In glazes, the cobalt pigments have been largely replaced by pigments based on vanadium-doped zircon (11,34,35). These pigments are turquoise in shade and are less intense than the cobalt pigments. Therefore, they are not applicable when the greatest tinting strength is required or when a purple shade is called for. Where they are applicable they give vastly improved color stability.

The zircon vanadium blue pigment is made by calcining a mixture of zirconia, silica, and vanadia in the stoichiometry of zircon and in the presence of a mineralizer. Mineralizers, selected from various halides and silicohalides, facilitate the transport of the silica during the reaction forming the pigment (19). The strongest pigments are formed at the stoichiometry of zircon, using such mineralizers as will facilitate the various transport processes, incorporating the maximum amount of vanadium into the zircon structure.

Black Pigments.   Black ceramic pigments are formed by calcination of several oxides to form the spinel structure (36,37). The formulation of black illustrates the flexibility of the spinel structure in incorporating various chemical entities. The divalent ion may be cobalt, manganese, nickel, iron, or copper. The trivalent ions may be iron, chromium, manganese, or aluminum. The selection of a particular black pigment depends somewhat on the specific coating material with which the pigment is to be used. Care must be taken to see that the pigment does not show a green, blue, or brown tint after firing. Of particular importance is the tendency of some glazes to attack the pigment and release cobalt. Thus in some cases it is desirable to use a cobalt-free pigment.

The prototype black is a cobalt iron chromite. In some systems, however, it will have a slight greenish tint. In zinc-containing glazes, a black with some nickel oxide would be recommended. For a black with a slightly bluish tint, a formula containing manganese and higher cobalt would be recommended. For a black with a brownish tint, a complex formula containing cobalt, iron, nickel, manganese, alumina, and chrome would be recommended.

When a cobalt-free system is needed, there are three possibilities. The copper chromite black pigment is a spinel, which is suitable for use in coatings firing below 1000°C, such as porcelain enamels. The chromium black hematite is an inexpensive system that is suitable for use in zinc-free coatings (38). If ZnO is present, it will react to form a brown color. For glazes firing over 1000°C, and containing ZnO, the chromium iron nickel black spinel can be used (36,39). This pigment can be used in most glazes and on firing schedules as high as cone 10.

Gray Pigments.   It might be expected that the easiest way to obtain a gray pigment would be to dilute a black pigment with a white opacifier. However, it is very difficult to provide an even color, free of specking, with this technique. Usually it is preferable to select a compound that has been formulated to give a gray color.

It is easiest to obtain a uniform gray color when a calcined pigment is used which is based on zirconia or zircon as a carrier for various ingredients of blacks such as Co, Ni, Fe, and Cr oxides. This pigment is called cobalt nickel gray periclase. For underglaze decorations it is possible to prepare a very beautiful deep gray color by dispersing antimony oxide in tin oxide. The limitation on the use of this material is the high cost of the SnO_2, which limits its use to special effects.

Use of Pigments in Coatings

There are several additional factors that must be considered in selecting pigments for a specific coating application (40). These factors include processing stability requirements, pigment uniformity and reproducibility, particle size distribution, dispersibility, and compatibility of all materials to be used.

Processing Stability. A significant limitation on the selection of ceramic pigments is the set of processing conditions imposed during coating application and firing (40). An engobe or body stain must be stable to the bisque fire, usually between cone 7 (1225°C) and cone 11 (1300°C). An underglaze color, or a colored glaze, must be stable to the glost fire, usually between cone 06 (1000°C) and cone 4 (1200°C), and to corrosion by the molten glaze ingredients. An overglaze or glass color needs only to be stable to the decorating fire by which it is applied, usually between cone 020 (625°C) and cone 016 (775°). More important here is corrosion by the molten flux used in the application.

The stability of the various pigments is discussed above for those pigments with limited stability. Detailed information is available (40).

Uniformity and Reproducibility. For most ceramic pigments rapid, uniform, and reproducible conversion to the desired product requires great care in production (40). Adjustment of each lot to standard, using toners, is usually required. The Victoria green garnet, the manganese alumina pink corundum, and the chrome–tin pink sphene are noteworthy for their difficulty in making reproducible product.

If a small amount (less than 5% of the blend) of a strong pigment is used as a component in a blend, it will be difficult to obtain sufficiently uniform mixing to avoid specking. It is preferable to use larger concentrations of a less intense pigment.

Some pigments are sensitive to details of the glaze application and firing procedures. With these pigments it may be difficult to maintain uniformity, even within a given lot of material. The Victoria green garnet, the copper greens, and the cadmium sulfoselenides are particularly sensitive to these problems.

Particle Size. Most calcined ceramic pigments are in the 1–10 μm range in mean particle size, with no residue on a 325 mesh (44 μm) screen. The selection of an optimum particle size distribution is a compromise between considerations of dissolution rate, agglomeration of the pigment, loss of strength on milling, uneven surface smoothness, and pigment strength (40–43). The optimum particle size is the largest size that gives adequate dispersion and adequate strength in letdowns.

Dispersibility. Pigments modified with a dispersion additive take less time and energy to disperse in a coating (12,44). The equipment for blunging is simpler and less expensive than ball mills. Color correction is simplified and settling is minimized. Color strength in letdowns is often improved (42).

Compatibility. A ceramic pigment must function as a component in a glaze or porcelain enamel system. Hence, it must be compatible with the other components, ie, the glaze itself, the opacifier(s), and other additives (40). There is a large variability in glaze–pigment interaction during firing. Some pigments, such as the zircon compounds, are relatively inert in conventional glazes. Other pigments are much more reactive.

Probably the most important glaze consideration is the presence or absence of ZnO in the glaze. The manganese alumina pink corundum, chromium green–black hematite, Victoria green garnet, chrome–tin orchid cassiterite, and chrome–tin pink sphene are not stable in the presence of zinc oxide [1314-13-2], ZnO. The iron brown hematite, chrome alumina pink spinel, iron chromite brown spinel, zinc ferrite brown spinel, and zinc iron chromite brown spinel require high ZnO concentration. High calcium oxide concentration is required for adequate stability of Victoria green garnet and chrome–tin pink sphene. CaO should be avoided when using chrome alumina pink spinel, zinc ferrite brown spinel, and zinc iron chromite brown spinel. Pigments containing chromium(III) oxide are incompatible with pigments containing tin oxide.

The presence or absence of PbO in a glaze affects some pigments (45). Victoria green and cobalt black pigments are stronger in a high PbO glaze. Zircon vanadium blue, zirconia vanadium yellow, and chrome–tin pink pigments are only suitable for low PbO or lead-free glazes. Mixed zircon greens, zircon iron pink, zinc iron chromite brown, and zirconia gray pigments are stronger in low PbO and lead-free glazes.

Health and Safety Factors

Properly handled, ceramic colorants should not cause unacceptable problems of health and safety. Preventive measures to avoid inhalation of fine particulate matter should invariably be used. Care should be taken to avoid ingestion of pigments by thorough washing before eating or smoking. Particular care should be taken in handling cadmium sulfoselenide pigments and lead antimonate pigments, which are highly toxic if ingested or inhaled (40).

When these pigments are used with lead-containing glazes, care should be exercised to use lead-safe glaze materials (see LEAD COMPOUNDS, INDUSTRIAL TOXICOLOGY).

Economic Aspects

Owing to the limited market and the variety and complexity of the products, ceramic pigments are manufactured by specialist firms, not by the users. The principal producers are Ceramic Color and Chemical Corp., Drakenfeld Colors Division of CIBA-GEIGY Corp., Englehard Corp., Ferro Corp., General Color and Chemical Corp., O. Hommel Co., Mason Color and Chemical Corp., and Miles, Inc. Estimated annual production is about 2500–3000 metric tons. This figure does not include some of the same and similar products manufactured for use in non-ceramic applications. The costs of ceramic pigments range from $9/kg to $50/kg or higher, depending on the elemental composition and the required processing. The most expensive pigments are those containing gold and the cadmium sulfoselenides.

BIBLIOGRAPHY

"Colors for Ceramics and Glass" in *ECT* 1st ed., Vol. 4, pp. 276–287, by W. A. Weyl and R. R. Shively, Jr., The Pennsylvania State College; "Colors for Ceramics" in *ECT* 2nd ed., Vol. 5, pp. 845–856, by W. A. Weyl, The Pennsylvania State University; "Colorants for Ceramics" in *ECT* 3rd ed., Vol. 6, pp. 549–561, by E. E. Mueller, Alfred University.

1. A. Burgyan and R. A. Eppler, *Am. Ceram. Soc. Bull.* **62**(9), 1001–1003 (1983).
2. P. M. Rice, *Pottery Analysis,* University of Chicago Press, Chicago, Ill., 1987, pp. 331–346.
3. W. D. Kingery, H. K. Bowen, and D. R. Uhlmann, *Introduction to Ceramics,* John Wiley & Sons, Inc., New York, 1976, pp. 677–689.
4. A. Paul, *Chemistry of Glasses,* Chapman & Hall, Ltd., London, 1982, pp. 233–251.
5. S. R. Scholes and C. H. Greene, *Modern Glass Practice,* 7th ed., Cahners Publishing Co., Boston, Mass., 1975, pp. 302–329.
6. W. Vogel (trans. N. Kriedl), *Chemistry of Glass,* American Ceramic Society, Columbus, Ohio, 1985, pp. 163–177.
7. R. A. Eppler, *Ullmann's Encyclopedia of Industrial Chemistry,* Vol. A5, VCH Publishers, Inc., Weinheim, West Germany, 1986, pp. 163–177.
8. F. T. Booth and G. N. Peel, *Trans. J. Brit. Ceram. Soc.* **58**(9), 532–564 (1959).
9. R. D. Shannon and A. L. Friedberg, *Univ. Ill. Eng. Exp. Sta. Bull.* **456,** 49 (1960).
10. R. A. Eppler, *J. Am. Ceram. Soc.* **52**(2), 89–99 (1969).
11. R. A. Eppler, *Am. Ceram. Soc. Bull.* **56**(2), 213–215,218,224 (1977).
12. T. D. Wise, S. H. Murdock, and R. A. Eppler, *Ceram. Eng. Sci. Proc.* **12**(1–2), 275–281, 194–196, 1991.
13. *DCMA Classification and Chemical Description of the Mixed Metal Oxide Inorganic Colored Pigments,* 2nd ed., Metal Oxides and Ceramic Colors Subcommittee, Dry Color Manufacturer's Association, Arlington, Va., 1982.
14. R. A. Eppler, *J. Am. Ceram. Soc.* **66**(11), 794–801 (1983).
15. R. A. Eppler and D. S. Carr, in *Proceedings of the 3rd International Cadmium Conference, International,* Lead Zinc Research Organization, New York, 1982, pp. 31–33.
16. U.S. Pat. 2,643,196 (June 23, 1953), B. W. Allan and F. O. Rummery (to Glidden Co.); U.S. Pat. 2,777,777 (Jan. 15, 1957), (to Glidden Co.).
17. H. D. deAhna, *Ceram. Eng. Sci. Proc.* **1**(9–10) 860–862 (1980); U.S. Pat. 4,482,390 (Nov. 13, 1984), A. C. Airey and A. Spiller (to British Ceramic Research Association Ltd.).
18. R. L. Hawks, *Am. Ceram. Soc. Bull.* **40**(1), 7–8 (1961).
19. U.S. Pat. 3,189,475 (June 15, 1965), J. E. Marquis and R. E. Carpenter (to Glidden Co.).
20. U.S. Pat. 3,166,430, (Jan. 19, 1965), C. A. Seabright (to Harshaw Chemical Co.).
21. R. A. Eppler, *J. Am. Ceram. Soc.* **53**(8), 457–462 (1970).
22. C-H. Li and R. A. Eppler, *Iron-Zircon Pigments,* presented at the 93rd annual meeting, American Ceramic Society, Cincinnati, Ohio, May 1, 1991; *Ceram. Eng. Sci. Proc.,* **13**(1–2), 109–118 (1992).
23. R. A. Eppler, *J. Am. Ceram. Soc.* **59**(9–10), 455 (1976).
24. J. E. Marquis and R. E. Carpenter, *Am. Ceram. Soc. Bull.* **40**(1) 19–24 (1961).
25. S. H. Murdock and R. A. Eppler, *J. Am. Ceram. Soc.* **71**(4), C212–C214 (1988).
26. C. A. Seabright and H. C. Draker, *Am. Ceram. Soc. Bull.* **40**(1), 1–4 (1961).
27. F. T. Booth and G. N. Peel, *Trans. J. Brit. Ceram. Soc.* **61**(7), 359–400 (1962).
28. E. H. Ray, T. D. Carnahan, and R. M. Sullivan, *Am. Ceram. Soc. Bull.* **40**(1), 13–16 (1961).
29. R. A. Eppler, *Ind. Eng. Chem. Prod. Res. Dev.* **10**(3), 352–355 (1971).
30. *The Colour Index,* 3rd ed., Society of Dyers & Colourists, Bradford-London, UK, 1971.
31. P. Henry, *Am. Ceram. Soc. Bull.* **40**(1), 9–10 (1961).
32. *Lead Glazes for Dinnerware,* International Lead Zinc Research Organization, New York, 1970.
33. R. K. Mason, *Am. Ceram. Soc. Bull.* **40**(1), 5–6 (1961).
34. U.S. Pat. 2,441,447 (May 11, 1948), C. A. Seabright (to Harshaw Chemical Co.); U.S. Pat. 3,025,178 (Mar. 13, 1962), (to Harshaw Chemical Co.).

35. T. Demiray, D. K. Nath, and F. A. Hummel, *J. Am. Ceram. Soc.* **53**(1), 1–4 (1970).
36. R. A. Eppler, *Am. Ceram. Soc. Bull.* **60**(5), 562–565 (1981).
37. W. F. Votava, *Am. Ceram. Soc. Bull.* **40**(1), 17–18 (1961).
38. S. H. Murdock and R. A. Eppler, *Am. Ceram. Soc. Bull.* **68**(1), 77–78 (1989).
39. U.S. Pat. 4,205,996 (June 3, 1980), R. A. Eppler, (to SCM Corp.).
40. R. A. Eppler, *Am. Ceram. Soc. Bull.* **66**(11), 1600–1604 (1987).
41. S. H. Murdock, T. D. Wise, and R. A. Eppler, *Am. Ceram. Soc. Bull.* **69**(2), 228–230 (1990).
42. S. H. Murdock, T. D. Wise, and R. A. Eppler, *Ceram. Eng. Sci. Proc.* **10**(1–2), 55–64 (1989).
43. C. Decker, *Effects of Grinding on Pigment Strength in Ceramic Glazes,* presented at the 93rd annual meeting, American Ceramic Society, Cincinnati, Ohio, May 1, 1991 *Ceram. Eng. Sci. Proc.* to be published 1992.
44. A. Sefcik, *Ceram. Eng. Sci. Proc.* **12**(1–2), 173–175 (1991).
45. R. A. Eppler and D. R. Eppler, *Color in Lead and Lead-free Glazes,* presented at the M. & E. Whitewares Fall Meeting, American Ceramic Society, Asheville, N.C., Sept. 26, 1991; *Ceram. Eng. Sci. Proc.* to be published 1992.

RICHARD A. EPPLER
Eppler Associates

COLORANTS FOR FOODS, DRUGS, COSMETICS, AND MEDICAL DEVICES

Colorants have been added to foods, drugs, and cosmetics for centuries (1–3). Until the middle of the nineteenth century, the colorants added were materials easily obtainable from natural sources such as animals, plants, and minerals. In 1856, Sir William Henry Perkin discovered the first synthetic organic dyestuff, mauve, and soon a host of new and different colorants became available (4). The use of some of these in foods began in Europe almost immediately and their use was soon extended to drugs and cosmetics. French wines, for example, were colored with fuchsine, a triphenylmethane dye, as early as 1860. The United States first legalized the use of synthetic organic dyes in foods by an act of Congress that authorized the addition of coloring matter to butter in 1886; the second instance came in 1896, with the recognition of coloring matter as a legitimate constituent of cheese. By 1900, Americans were eating a wide variety of artificially colored products, including ketchup, jellies, cordials, butter, cheese, ice cream, candy, sausage, noodles, and wine. The use of the new synthetic colorants in drug and cosmetic products was also increasing rapidly.

This proliferation in the use of color additives was soon recognized as a threat to the public's health. Of particular concern were the practices of adding poisonous colorants to food, and of using dyes to hide poor quality or to add weight or bulk to certain items. References 5–14 provide additional information on the history of food colorants and their regulation. Reference 15 provides more information regarding the applications, properties, specifications, and analysis of color additives, as well as methods for the determination of colorants in products.

History of Regulation

Because of increasing public worry over such practices, some measures were eventually taken by American food manufacturers to police their own industry, but the effect was marginal, and it was clear that some form of government control was necessary if the public was to be protected. The first effective step taken by the U.S. government was the Appropriations Act of 1900 for the Department of Agriculture, the Bureau of Chemistry which provided funds to investigate the relationship of coloring matters to health and to establish principles that should be followed to govern their use (16). This resulted in the issuance by the Secretary of Agriculture of a series of Food Inspection Decisions (FID). One decision (FID 39, issued May 1, 1906) contained the first direct statement by the department concerning a coal-tar dye considered unsafe in foods. In effect, it stopped importation of macaroni colored with Martius Yellow.

At about the same time a thorough study was undertaken by the Department of Agriculture to determine which dyes, if any, were safe for use in foods and what restrictions should be placed on their use. This monumental task eventually included a study of the chemistry and physiology of the then nearly 700 extant coal-tar dyes as well as the laws of various countries and states regarding their use in food products. Most of this investigation was done under the guidance of Dr. Bernard C. Hesse, a German dye expert (17).

Of the 80 colorants offered for use in foods in the United States in 1907, Hesse learned from the literature that 30 of them had never been tested at all and that their safety was therefore simply unknown, 26 had been tested but the results were contradictory, 8 were considered by most experts to be unsafe, and the remaining 16 were deemed more or less harmless. These 16 colorants were then tested physiologically by determining their acute short-range effects in dogs, rabbits, and human beings. Hesse then recommended the following seven colorants for use in food in the United States: Amaranth, Ponceau 3R, Orange I, Erythrosine (FD&C Red No. 3), Naphthol Yellow S (Ext. D&C Yellow No. 7), Light Green SF Yellowish, and Indigo Disulfo Acid, Sodium Salt (FD&C Blue No. 2).

Much of what Hesse uncovered during his study was used in formulating the Food and Drug Act of 1906. This act, plus FID No. 76 (July 13, 1907) put an end to the indiscriminate use of dangerous and impure coloring matters in foods. This new legislation required that only colors of known composition, examined physiologically and showing no unfavorable results, could be used. The new regulations also established a system for the voluntary certification of synthetic organic food colors by the Department of Agriculture.

During the next three decades there was a continual growth in the use and number of color additives, and using Hesse's rules, the list of colors certifiable for use in foods was expanded.

In 1938 a new law, the Federal Food, Drug, and Cosmetic Act of 1938 (18), which instituted several new and important practices, was enacted. First, it clearly stated that, henceforth, the use of any uncertified coal-tar color in any food, drug, or cosmetic shipped in interstate commerce was strictly forbidden. This restriction applied regardless of the inherent toxicity of the colorant. In effect, the colorants that could be used were limited, certification became mandatory, and governmental control was extended to the coloring of drugs and cosmetics. Next, it created three categories of coal-tar colors:

> *FD&C colors.* Those certifiable for use in coloring foods, drugs, and cosmetics.
>
> *D&C colors.* Dyes and pigments considered safe in drugs and cosmetics when in contact with mucous membranes or when ingested.
>
> *Ext. D&C colors.* Those colorants that, because of their oral toxicity, were not certifiable for use in products intended for ingestion, but were considered safe for use in products externally applied.

Passage of the 1938 Act launched a new series of scientific investigations and public hearings regarding the safety of the colorants then on the market. These efforts culminated in the publication in 1940 of Service and Regulatory Announcement, Food, Drug, and Cosmetics No. 3, which listed specific colorants that could be used along with specifications and regulations relating to their manufacture, labeling, certification, and sale.

In the early 1950s, a number of cases of sickness occurred in children who had reportedly eaten candy and popcorn colored with excessive amounts of dye. As a result of these illnesses, new animal feeding studies were undertaken by the FDA. These studies were conducted at higher levels and for longer test periods than any experiments previously conducted and resulted in unfavorable findings for FD&C Orange No. 1, FD&C Orange No. 2, and FD&C Red No. 32.

The disputes that followed these events centered around interpretation of the 1938 act, which states that "The Secretary shall promulgate regulations providing for the listing of coal-tar colors which are harmless and suitable for use in food. . . ." The FDA felt that harmless meant that a colorant must be safe regardless of the amount used, that is, harmless per se and on this basis delisted the colorants in question. The food-color manufacturers argued that the FDA interpretation of the law was too strict, that a color additive need only be harmless when properly used, and that the FDA should establish safe limits. They also contended that the conditions used for the new animal feeding tests were too severe.

In 1958 the Supreme Court ruled that under the 1938 law, the FDA did not have the authority to establish limits of use for colorants and that they were obligated to decertify or delist a color if any quantity of it caused harm even though lesser amounts were perfectly safe (19). The FDA's hands were tied (20). A review of the remaining colors was started and soon several more were delisted, including FD&C Yellow Nos. 1–4. It became obvious that the existing law on certifiable colors was unworkable.

Through the efforts of the Certified Color Industry and the FDA, a new law was formulated, the Color Additives Amendments of 1960 (Public Law 86-618) (21). The amendments provided a breathing spell by allowing the continued use of existing color additives pending the completion of investigations needed to ascertain their suitability for listing as permanent colorants, and authorizing the establishment of limits of use by the Secretary of Health, Education, and Welfare, thus eliminating the controversial "harmless per se" interpretation formerly employed. A special proviso, the Delaney clause, directed the Secretary not to list a color additive for any use if that colorant could be shown to induce cancer in humans or animals (22). Other features eliminated any distinction under the law between coal-tar colors and other color additives and empowered the Secretary to decide which colors must be certified and which could be exempted from certification based on their relationship to public health. Under the new law, producers and consumers of color additives were obliged to provide the necessary scientific data to obtain permanent listing of a color additive.

With the passage of the Medical Device Amendments of 1976 (Public Law 94-295) Congress created a new category of color additive by mandating the separate listing of colorants for use in medical devices if the color additive in them comes in direct contact with the body for a significant period of time.

Colorants currently in use and their status are shown in Tables 1–4. These lists are accurate as of January 1993 but are subject to change. Such changes as well as any changes in the regulation of color additives are routinely published in the *Federal Register*. The FDA, Division of Colors and Cosmetics also provides additional regulatory information (14).

Coloring Food

Table 1 lists various colorants permitted for use in foods.

Most staple foods such as meat, white bread, potatoes and other vegetables, and most fruits are not artificially colored since their natural appearance is perfectly acceptable. Foods are usually colored because they have no natural color of their own, because their natural color was destroyed or drastically altered as a result of processing or storage, or because their color varies greatly with the season of the year or their geographic origin. Thus, colorants are added to foods to make them appear the way the customer wants and expects them to appear.

Consumers' expectations depend on several factors including cultural background, past experiences, desire for color coordination, esthetic appeal, local customs, fads, etc. Thus, eg, a Texas red hot sold in the South is often colored quite differently than one sold in the North, Midwesterners prefer butter with a deep yellow color, and on birthdays the decorations on a boy's cake are often blue and those on a girl's are often pink.

Color expectation in food depends also on how well established the color of that food is and how closely its color is associated with its quality. Black cabbage, green rice, or purple milk, for example, do not meet standards of identity for these foods. Also, because of color, green grapefruit and bananas are perceived as immature, anything but the most brilliant red beef is suspect, and excessively brown or spotted produce is shunned in favor of the brightest, most uniformly colored products available. The colors of some foods, in fact, are so well fixed that they serve as reference standards when speaking of certain hues; ie, lemon yellow, eggshell white, cherry red, chocolate brown, and pea green.

Table 1. Colorants Permitted in Foods

FDA official name	Structure number	Common synonyms	CAS Registry Number	CI number	Uses and restrictions[a]
		Subject to certification			
FD&C Blue No. 1[b]	(1)	Brilliant Blue FCF; CI Food Blue 2	[2650-18-2]	42090	
FD&C Blue No. 2[b]	(2)	Indigotine; Indigo Carmine; CI Food Blue 1	[860-22-0]	73015	nylon sutures for use in general surgery only; 1% (w/w) max ingested drugs only
FD&C Green No. 3[b]	(3)	Fast Green FCF; CI Food Green 3	[2353-45-9]	42053	
FD&C Red No. 3[b]	(4)	Erythrosine; CI Food Red 14	[16423-68-0]	45430	ingested drugs only (listed provisionally for other uses)
FD&C Red No. 40[b]	(5)	Allura Red; CI Food Red 17	[25956-17-6]	16035	
FD&C Yellow No. 5[b]	(6)	Tartrazine; CI Food Yellow 4	[1934-21-0]	19140	
FD&C Yellow No. 6[b]	(7)	Sunset Yellow; CI Food Yellow 3	[2783-94-0]	15985	
Citrus Red No. 2	(8)	CI Solvent Red 80	[6358-53-8]	12156	skins of oranges that are not intended or used for processing only; 2.0 ppm max, based on the weight of the whole fruit
Orange B	(9)	CI Acid Orange 137	[15139-76-1]	19235	sausage and frankfurter casings or surfaces only; 150 ppm max, based on the weight of the finished product

896

Exempt from certification

Substance	No.	CI name	CAS	CI number	Restriction
annatto extract	c		[8015-67-6]	75120	
β-apo-8'-carotenal	(38)	CI Food Orange 6	[1107-26-2]	40820	max—15 mg/lb (33 mg/kg) of solid or semisolid food, or pint of liquid food (~32 mg/L)
canthaxanthin	(39)	Food Orange 8; β-carotene-4,4'-dione	[514-78-3]	40850	max—30 mg/lb (66 mg/kg) of solid or semisolid food, or pint of liquid food (~64 mg/L)
caramel		CI Natural Brown 10			
β-carotene	(37)	CI Natural Yellow 26 / CI Food Orange 5	[7235-40-7]	75130 (natural) / 40800 (synthetic)	
carrot oil					
cochineal extract; carmine	(40)	CI Natural Red 4	[1390-65-4]	75470	
corn endosperm oil	d				chicken feed only
dehydrated beets (beet powder)					
dried algae meal	(46)				chicken feed only
ferrous gluconate					ripe olives only
fruit juice					nonbeverage food only
grape color extract					beverages only
grape skin extract					
paprika					
paprika oleoresin					
riboflavin	(47)	Vitamin B$_2$	[83-88-5]		
saffron		CI Natural Yellow 6		75100	
synthetic iron oxide	e	CI Pigment Reds 101 and 102		77491	dog and cat food only; 0.25% (w/w) max
		CI Pigment Yellows 42 and 43		77492	
		CI Pigment Black 11 and Browns 6 and 7		77499	

Table 1. (Continued)

FDA official name	Structure number	Common synonyms	CAS Registry Number	CI number	Uses and restrictions[a]
tagetes meal and extract				75125	chicken feed only
titanium dioxide		CI Pigment White 6	[13463-67-7]	77891	1% (w/w) max in finished food
toasted partially defatted cooked cottonseed flour					
turmeric		CI Natural Yellow 3		75300	
turmeric oleoresin	f			75300	
ultramarine blue		CI Pigment Blue 29		77007	salt for animal feed only; 0.5% (w/w) max
vegetable juice					

[a] No color additive or product containing one can be used in the area of the eye, in surgical sutures, or in injections, unless so stated. Also, no colorant can be used to color foods for which standards of identity have been promulgated under Section 401 of the Federal Food, Drug, and Cosmetic Act, unless the use of added color is authorized by the standard. Colorants without restrictions can be used for coloring foods generally, in amounts consistent with good manufacturing practice.
[b] Also permitted in drugs; uses and restrictions apply thereto. See Figure 1 for structures.
[c] See bixin (**36**, R = CH$_3$) and norbixin (**36**, R = H).
[d] See betanin (**41**).
[e] See crocin and crocetin (**44**).
[f] See curcumin (**45**).

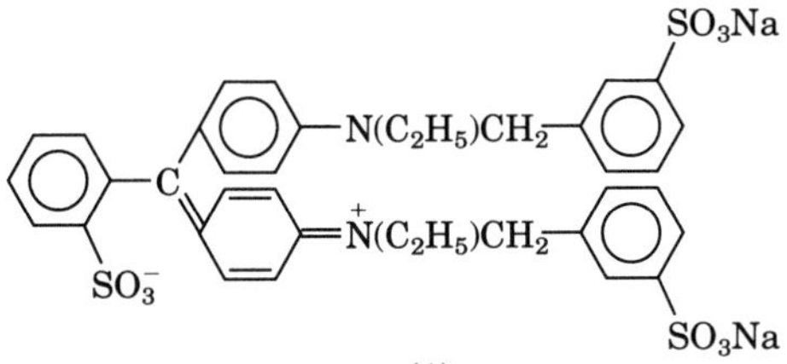

(**1**)

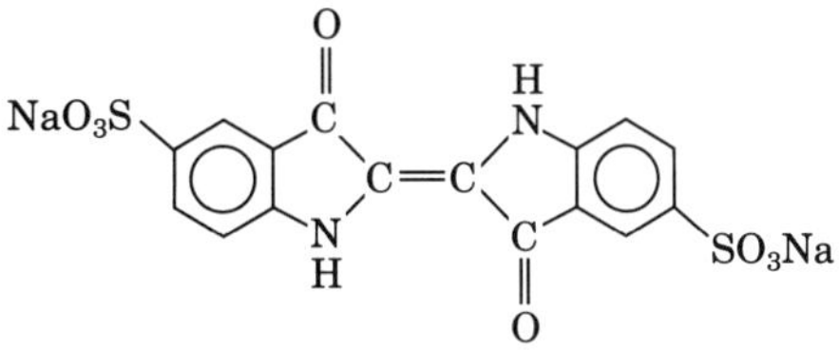

(**2**)

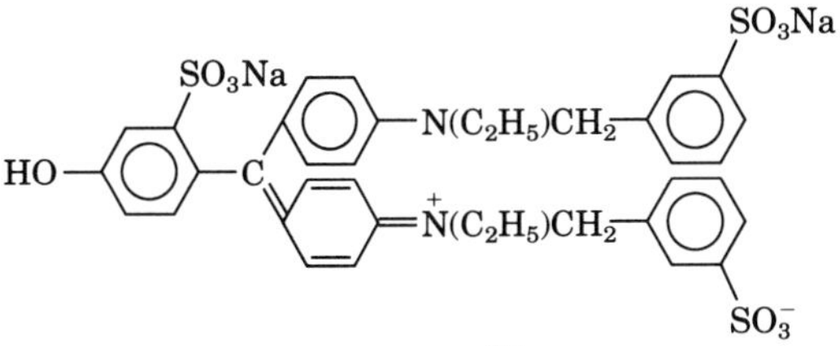

(**3**)

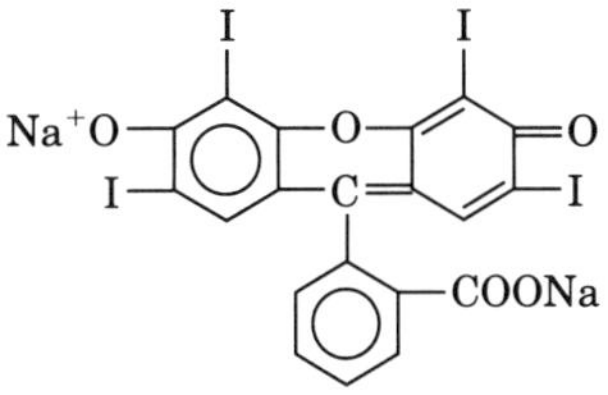

(**4**)

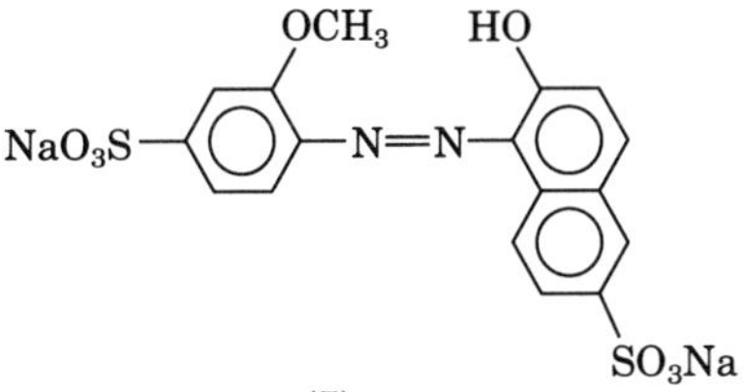

(**5**)

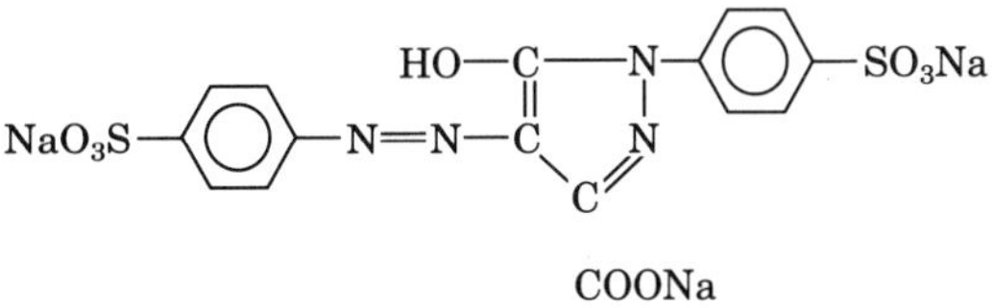

(**6**)

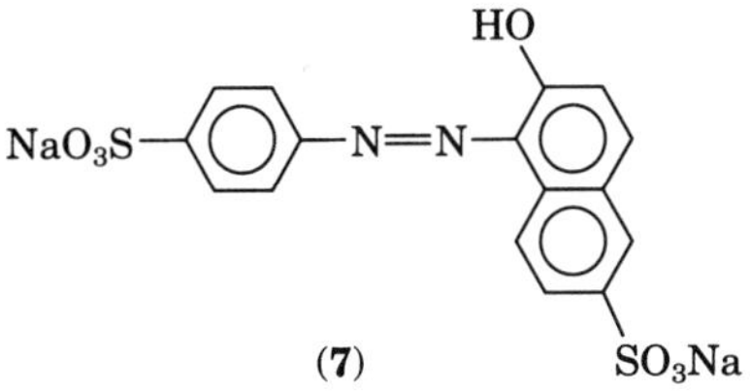

(**7**)

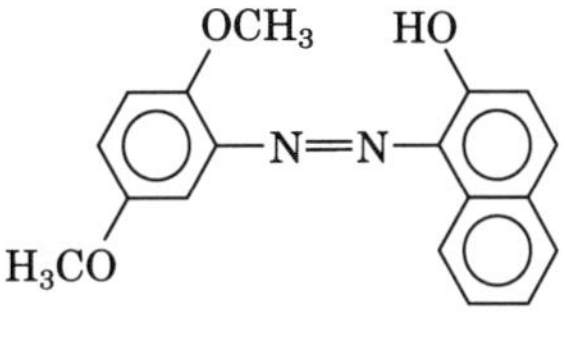

(**8**)

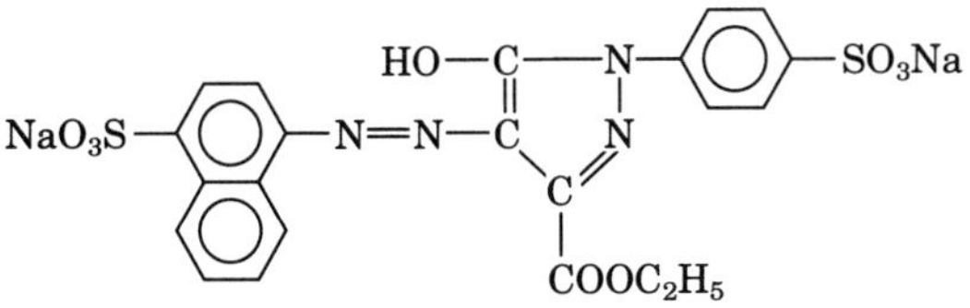

(**9**)

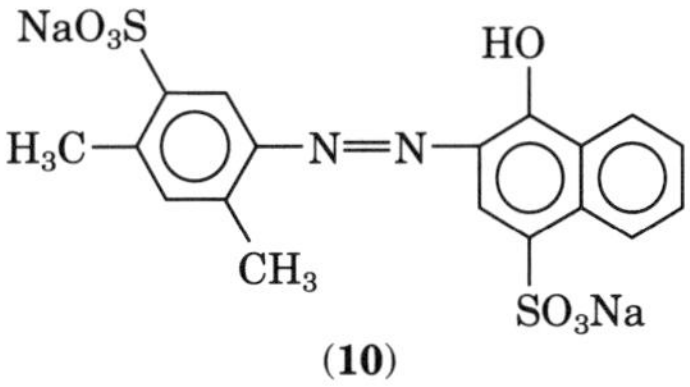

(**10**)

Fig. 1. Colorants for food and drugs fall into several dye types. FD&C Blue No. 1 (**1**) and FD&C Green No. 3 (**3**) are triphenylmethane dyes (qv). FD&C Blue No. 2 (**2**) is an indigoid and FD&C Red No. 3 (**4**) is a xanthene dye (qv). FD&C Red No. 40 (**5**), FD&C Yellow No. 6 (**7**), Citrus Red No. 2 (**8**), and FD&C Red No. 4 (**10**) are azo dyes (qv). FD&C Yellow No. 5 (**6**) and Orange B (**9**) are pyrazolones (see PYRAZOLES, PYRAZOLINES, AND PYRAZOLONES).

Colorless Foods. The principal use of color additives in food is in products containing little or no color of their own. These include many liquid and powdered beverages, gelatin desserts, candies, ice creams, sherbets, icings, jams, jellies, and snack foods. Without the addition of color to some of these, eg, gelatin desserts and soft drinks, all flavors of the particular product would be colorless, unidentifiable, and probably unappealing to the consumer.

Process and Storage Difficulties. Often the process used to prepare a food results in the formation of a color in the product, the depth of which depends largely on the time, temperature, pH, air exposure, and other parameters experienced during processing. It is deemed necessary to supplement the color of the product to ensure its uniformity from batch to batch. Items that fall into this category include certain beers, blended whiskies, brown sugars, table syrups, toasted cereals, and baked goods.

During storage, natural pigments often deteriorate with time because of exposure to light, heat, air, and moisture or because of interaction of the components of the food with each other or with the packaging material. The color of maraschino cherries, for example, fares so poorly with storage that they are routinely bleached then artificially colored.

Regional and Seasonal Problems. The problems of the dairy and citrus fruit industries are typical of those encountered with products produced in different areas of the country or at different times of the year.

In many parts of the United States, the soil and weather conditions are such that chlorophyll continuously forms in the fruit of orange trees as well as in the leaves of the trees; the result is mature oranges that are substantially greener than the same variety of orange produced in regions of the country with different growing conditions. Most varieties of Florida oranges tend to be green, suggesting immaturity, even though they contain the proper ratio of solids to acid for fully nutritious, mature fruit. The necessity of coloring these oranges to make them comparable in appearance and thus as commercially acceptable as naturally orange-colored fruit was recognized years ago and began on a commercial scale about 1934.

In milk approximately 90% of the yellow color is because of the presence of β-carotene, a fat-soluble carotenoid extracted from feed by cows. Summer milk is more yellow than winter milk because cows grazing on lush green pastures in the spring and summer months consume much higher levels of carotenoids than do cows barn-fed on hay and grain in the fall and winter. Various breeds of cows and even individual animals differ in the efficiency with which they extract β-carotene from feed and in the degree to which they convert it into colorless vitamin A. The differences in the color of milk are more obvious in products made from milk fat, since here the yellow color is concentrated. Thus, unless standardized through the addition of colorant, products like butter and cheese show a wide variation in shade and in many cases appear unsatisfactory to the consumer.

Other products having natural color that varies enough to make standardization of their color desirable include the shells of certain kinds of nuts, the skins of red and sweet potatoes, and ripe olives.

Miscellaneous Uses. Inks used by inspectors to stamp the grade or quality on meat must, by law, be made from food-grade colors. Dyes used in packaging materials that come in direct contact with a food must also be food-grade or, if not, it must be established that no part of the colorant used migrates into the food product. Pet foods, too, if colored, must contain only those colorants recognized by the FDA as suitable for the purpose.

References 23–31 cover various aspects of food coloring technology in the United States and internationally.

Coloring Drugs

Compared with the food and cosmetic industries, pharmaceuticals are a minor though important consumer of colorants (32,33). Originally, dyes were used in drugs to make them more appealing to the consumer by adding color to otherwise colorless products, by masking unsatisfactory natural colors, and by standardizing the appearance of drugs the color of which varied from batch to batch as a consequence of the manufacturing process, a difference in the color of the raw materials used, or both. Some drugs, of course, contain added color for cosmetic purposes, as in the case of the skin-tone dyes added to certain creams and ointments used to treat disorders such as acne.

Although colors are still added to drugs for these purposes, the principal use of colorants in pharmaceuticals currently is to provide the manufacturer with a simple means of identifying products so that they are not inadvertently mixed during production and shipment (34). Since no industry-wide standards exist for coloring drugs, each manufacturer has been free to develop and use their own in-house scheme. Many such codes have been devised and so today the same product frequently appears on the market under several color forms. The *Physicians' Desk Reference* contains colored photographs of tablets and capsules as an aid in identifying drugs (35). Table 2 lists colorants permitted in drugs.

Coloring Cosmetics

Products such as aftershave lotions, hair tonics, and soaps contain additives purely for esthetic reasons. In many cases, though, the colorant is a significant functional part of a cosmetic, often comprising half of its total weight. Some cosmetics, including eyebrow pencils, nail polishes, and rouges, are really little more than colorants mixed with one or more materials that serve simply as binders, vehicles, or diluents to give the product desirable application properties (29,36–40).

Compared with foods and drugs, cosmetics usually contain much higher amounts of colorants. Although foods and drugs seldom contain more than a few to several hundred parts per million (ppm) of colorant, cosmetics often contain several percent (Table 3).

Coloring Medical Devices

Color additives are routinely added to medical devices such as surgical sutures, surgical cements, and contact lenses (32). Sutures are usually colored to make

Table 2. Colorants Permitted in Drugs[a]

FDA official name	Structure number[b]	Common synonyms	CAS Registry Number	CI number	Uses and restrictions[c]
		Subject to certification			
FD&C Red No. 4	(10)	Ponceau SX; CI Food Red 1	[4548-53-2]	14700	externally applied drugs only
D&C Blue No. 4	(11)	Alphazurine FG; Erioglaucine; CI Acid Blue 9	[6371-85-3]	42090	externally applied drugs only
D&C Blue No. 9	(12)	Indanthrene Blue; Carbanthrene Blue; CI Vat Blue 6	[130-20-1]	69825	cotton and silk sutures (including those for ophthalmic use) only; 2.5% (w/w) max drugs in general
D&C Green No. 5	(13)	Alizarine Cyanine Green F; CI Acid Green 25	[4403-90-1]	61570	nylon-6,6 and nylon-6 nonabsorbable surgical sutures only; 0.6% (w/w) max
D&C Green No. 6	(14)	Quinizarin Green SS; CI Solvent Green 3	[128-80-3]	61565	externally applied drugs only
D&C Green No. 8	(15)	Pyranine Concentrated; CI Solvent Green 7	[6358-69-6]	59040	externally applied drugs only; 0.01% (w/w) max
D&C Orange No. 4	(16)	Orange II; CI Acid Orange 7	[633-96-5]	15510	externally applied drugs only
D&C Orange No. 5	(17a)	Dibromofluorescein; CI Solvent Red 72	[596-03-2]	45370:1	ingested mouthwashes and dentifrices only; externally applied drugs only; 5 mg/d dose of drug max
D&C Orange No. 10	(17b)	Diiodofluorescein; CI Solvent Red 73	[38577-97-8]	45425:1	externally applied drugs only
D&C Orange No. 11	(18)	Erythrosine Yellowish Na; CI Acid Red 95	[33239-19-9]	45425	externally applied drugs only
D&C Red No. 6	(19a)	Lithol Rubin B; CI Pigment Red 57	[5858-81-1]	15850	drugs in general. Combined total of D&C Red no. 6 and D&C Red no. 7 not more than 5 mg/d dose of drug
D&C Red No. 7	(19b)	Lithol Rubin B Ca; CI Pigment Red 57:1	[5281-04-9]	15850:1	same as D&C Red no. 6

D&C Red No. 17	(20)	Toney Red; Sudan III; CI Solvent Red 23	[85-86-9]	26100	externally applied drugs only
D&C Red No. 21	(17c)	Tetrabromofluorescein; CI Solvent Red 43	[15086-94-9]	45380:2	
D&C Red No. 22	(21a)	Eosin Y; CI Acid Red 87	[17372-87-1]	45380	
D&C Red No. 27	(22)	Tetrabromotetrachlorofluorescein; CI Solvent Red 48	[13473-26-2]	45410:1	
D&C Red No. 28	(21b)	Phloxine B; CI Acid Red 92	[18472-87-2]	45410	
D&C Red No. 30	(23)	Helindone Pink CN; CI Vat Red 1	[2379-74-0]	73360	
D&C Red No. 31	(24)	Brilliant Lake Red R; CI Pigment Red 64:1	[6371-76-2]	15800:1	externally applied drugs only
D&C Red No. 33	(25)	Acid Fuchsine; CI Acid Red 33	[3567-66-6]	17200	ingested drugs, other than mouthwashes and dentifrices; 0.75 mg/d dose of drug max; externally applied drugs, mouthwashes, and dentifrices
D&C Red No. 34	(26)	Deep Maroon; Fanchon Maroon; Lake Bordeaux B; CI Pigment Red 63:1	[6417-83-0]	15880:1	externally applied drugs only
D&C Red No. 36	(27)	Flaming Red; CI Pigment Red 4	[2814-77-9]	12085	ingested drugs, other than mouthwashes and dentifrices; 1.7 mg/d dose of drug max if taken continuously for less than one year; 1.0 mg/d dose of drug max if taken continuously for longer than one year; externally applied drugs
D&C Red No. 39	(28)	Alba Red; CI Pigment Red 100	[6371-55-7]	13058	externally applied quarternary ammonium germicides only; 0.1% (w/w) max
D&C Violet No. 2	(29)	Alizurol Purple SS; CI Solvent Violet 13	[81-48-1]	60725	externally applied drugs only
D&C Yellow No. 7	(17d)	Fluorescein; CI Solvent Yellow 94	[2321-07-5]	45350:1	externally applied drugs only
D&C Yellow No. 8	(21c)	Uranine; CI Acid Yellow 73	[518-47-8]	45350	externally applied drugs only
D&C Yellow No. 10	(30a)	Quinoline Yellow; CI Acid Yellow 3	[8004-92-0]	47005	externally applied drugs only

Table 2. (*Continued*)

FDA official name	Structure number[b]	Common synonyms	CAS Registry Number	CI number	Uses and restrictions[c]
D&C Yellow No. 11	(30b)	Quinoline Yellow SS; CI Solvent Yellow 33	[8003-22-3]	47000	externally applied drugs only
Ext. D&C Yellow No. 7	(31)	Naphthol Yellow S; CI Acid Yellow 1	[846-70-8]	10316	externally applied drugs only
		Exempt from certification			
alumina				77002	
aluminum powder		CI Pigment Metal 1		77000	externally applied drugs only[e]
annatto extract		[d]		75120	[e]
bismuth oxychloride		CI Pigment White 14	[7787-59-9]	77163	externally applied drugs only[e]
bronze powder		CI Pigment Metal 2		77400	externally applied drugs only[e]
calcium carbonate		CI Pigment White 18		77220	
canthaxanthin	(39)[d]			40850	ingested drugs only
caramel					
β-carotene	(37)[d]			75130 (natural) 40800 (synthetic)	[e]
chromium–cobalt–aluminum oxide		CI Pigment Blue 36	[68187-11-1]	77343	polyethylene sutures for use in general surgery only; 2% (w/w) max
chromium hydroxide green		CI Pigment Green 18	[12182-82-0]	77289	externally applied drugs only[e]
chromium oxide green		CI Pigment Green 17	[1308-38-9]	77288	externally applied drugs only[e]
cochineal extract; carmine	(40)[d]			75470	externally applied drugs only[e]
copper powder		CI Pigment Metal 2		77400	externally applied drugs intended solely or in part for imparting color to the human body only
dihydroxyacetone					

904

Colorant	CAS No.	CI name	CI No.	Restrictions[c]
ferric ammonium citrate	[1185-57-5]			with pyrogallol in plain and chromic catgut sutures for use in general and ophthalmic surgery only; 3% (w/w) max total citrate–pyrogallol complex
ferric ammonium ferrocyanide				externally applied drugs only[e]
ferric ferrocyanide		CI Pigment Blue 27 CI Pigment Green 15	77510 77520	externally applied drugs only[e]
guanine[f]		CI Natural White 1	75170	externally applied drugs only[e]
logwood extract			75290	nylon-6,6, nylon-6, and silk nonabsorbable sutures for use in general and ophthalmic surgery only; 1.0% (w/w) max
mica		CI Pigment White 20	77019	dentifrices and externally applied drugs only[e]
potassium sodium copper chlorophyllin			75810	dentifrices only; 0.1% max
pyrogallol			76515	with ferric ammonium citrate in plain and chromic catgut sutures for use in general and ophthalmic surgery only; 3% (w/w) max total citrate–pyrogallol complex
pyrophyllite				externally applied drugs only
synthetic iron oxide	[d]		77491 77492 77499	5 mg/d (as Fe) in drugs that are ingested
talc	[14807-96-6]	CI Pigment White 26	77019	
titanium dioxide	[d]		77891	[e]
zinc oxide	[1314-13-2]	CI Pigment White 4	77947	externally applied drugs only[e]

[a] The FD&C Colors in Table 1 are also permitted in drugs.
[b] Structures for D&C Colors are given in Figure 2.
[c] No color additive or product containing one can be used in the area of the eye, in surgical sutures, or in injections unless so stated. Colorants without restrictions can be used for coloring drugs generally, in amounts consistent with good manufacturing practice.
[d] See Table 1 for synonyms and CAS Registry numbers.
[e] May also be used in those drugs intended for use in the area of the eye.
[f] See guanine (42) and hypoxanthine (43).

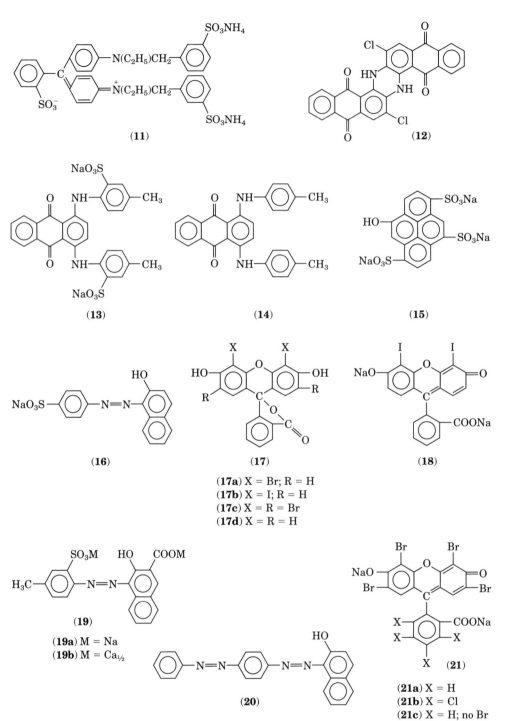

Fig. 2. D&C colors permitted in drugs (Table 2). Structures (**12**), (**13**), (**14**), and (**29**) are anthraquinone dyes (see DYES, ANTHRAQUINONE); (**11**) is a triphenylmethane dye (qv); (**15**) is a pyrene; (**16**), (**19**), (**20**), (**24**), (**25**), (**26**), (**27**), and (**28**) are azo dyes (qv); (**17**) and (**22**) are fluorans; (**18**) and (**21**) are xanthene dyes (qv); (**23**) is indigoid; (**30**) has the quinoline structure and (**31**) is a nitro dye.

906

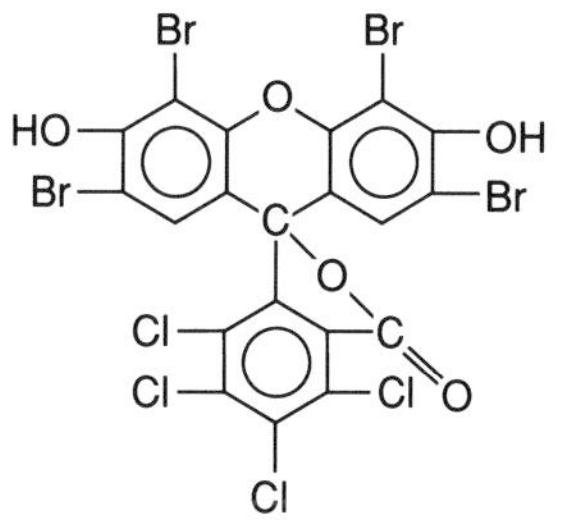

(22)

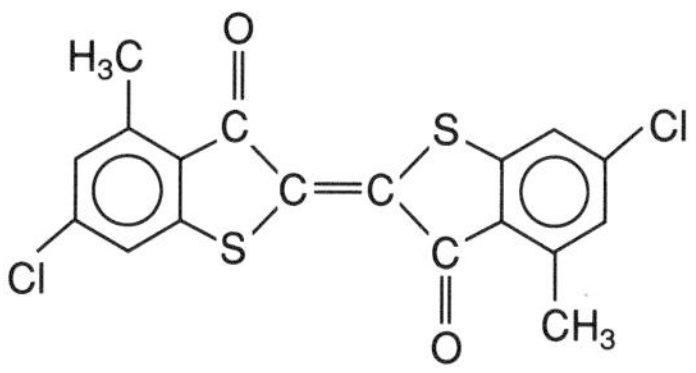

(23)

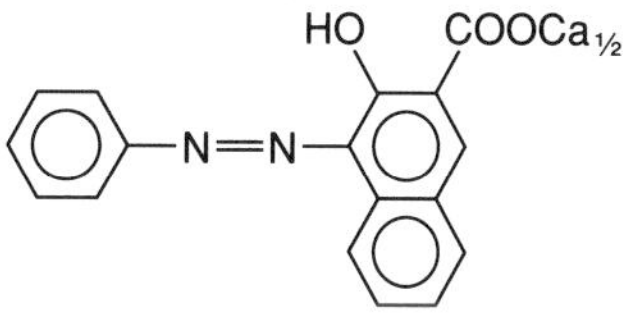

(24)

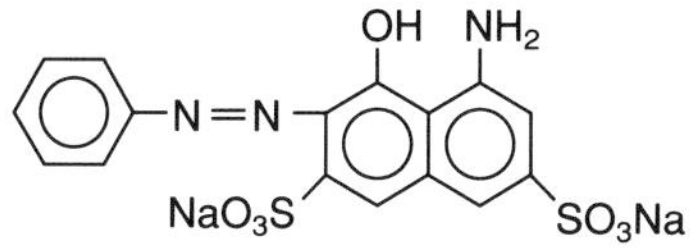

(25)

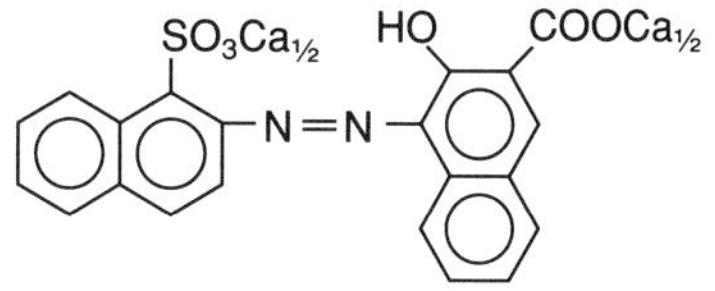

(26)

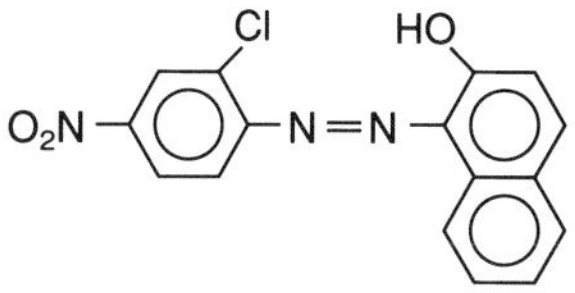

(27)

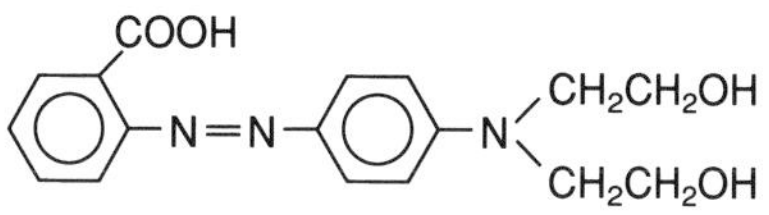

(28)

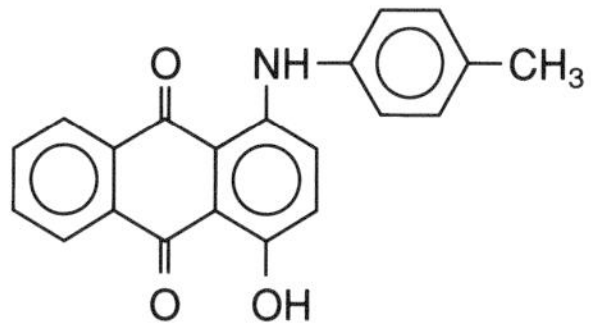

(29)

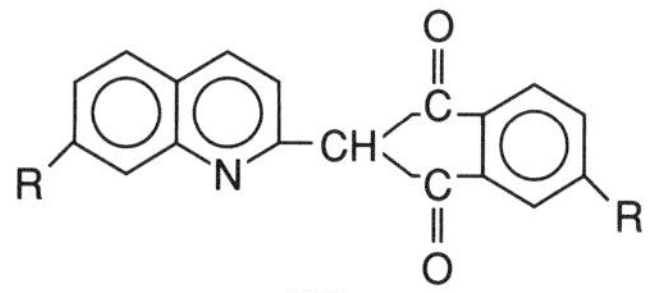

(30)

(30a) R = SO_3Na
(30b) R = H

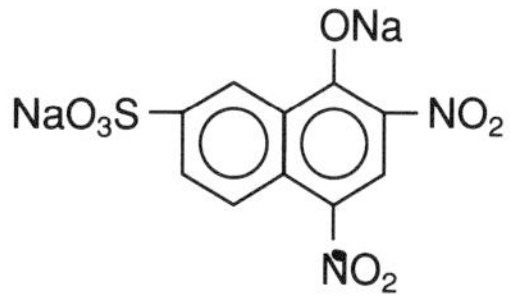

(31)

Fig. 2. *(Continued)*

them more visible during surgery and, depending on the application, during removal of the suture after the sutured area has healed. Surgical cements, too, are colored to make them more visible during use (Table 4).

Colorants are used in contact lenses (41) for several reasons. Marking lenses with the letters L and R distinguishes one lens from another. Initially, hard lenses were lightly tinted to aid in recognition and handling. These lenses were tinted throughout the whole of the lens and the overlap of color onto the iris usually did not significantly change its color. In soft lenses the area of the lens in juxtaposition to the iris is tinted but the overlapping peripheral area is clear. More recently, much darker soft lenses containing opaque inorganic oxide pigments have been

Table 3. Colorants Permitted in Cosmetics

FDA official name	Structure number[a]	CI number	Uses and restrictions[b]
		Subject to certification	
FD&C Blue No. 1	(1)	42090	
FD&C Green No. 3	(3)	42053	
FD&C Red No. 4	(10)	14700	external use only
FD&C Red No. 40	(5)	16035	
FD&C Yellow No. 5	(6)	19140	
FD&C Yellow No. 6	(7)	15985	
D&C Blue No. 4	(11)	42090	external use only
D&C Brown No. 1	(32)[c]	20170	external use only
D&C Green No. 5	(13)	61570	
D&C Green No. 6	(14)	61565	external use only
D&C Green No. 8	(15)	59040	external use only; 0.01% (w/w) max
D&C Orange No. 4	(16)	15510	external use only
D&C Orange No. 5	(17a)	45370:1	5.0% (w/w) max in lipstick and other lip cosmetics; ingested mouthwashes and dentifrices; externally applied cosmetics
D&C Orange No. 10	(17b)	45425:1	external use only
D&C Orange No. 11	(18)	45425	external use only
D&C Red No. 6	(19a)	15850	
D&C Red No. 7	(19b)	15850:1	
D&C Red No. 17	(20)	26100	external use only
D&C Red No. 21	(17c)	45380:2	
D&C Red No. 22	(21a)	45380	
D&C Red No. 27	(22)	45410:1	
D&C Red No. 28	(21b)	45410	
D&C Red No. 30	(23)	73360	
D&C Red No. 31	(24)	15800:1	external use only
D&C Red No. 33	(25)	17200	cosmetic lip products only; 3% (w/w) max; mouthwashes, dentifrices, and externally applied cosmetics only
D&C Red No. 34	(26)	15880:1	external use only
D&C Red No. 36	(27)	12085	cosmetic lip products only; 3% (w/w) max; externally applied cosmetics only
D&C Violet No. 2	(29)	60725	external use only
D&C Yellow No. 7	(17d)	45350:1	external use only
D&C Yellow No. 8	(21c)	45350	external use only
D&C Yellow No. 10	(30a)	47005	
D&C Yellow No. 11	(30b)	47000	external use only
Ext. D&C Violet No. 2	(33)[d]	60730	external use only
Ext. D&C Yellow No. 7	(31)	10316	external use only
		Exempt from certification	
aluminum powder		77000	external use only[e]
annatto		75120	[e]
bismuth citrate			hair dyes for scalp only; 0.5% (w/v) max
bismuth oxychloride		77163	[e]
bronze powder		77440	[e]
caramel			[e]
carmine	(40)	75470	[e]

908

Table 3. *(Continued)*

FDA official name	Structure number[a]	CI number	Uses and restrictions[b]
		Exempt from certification (Continued)	
β-carotene	(37)	75130 (natural) 40800 (synthetic)	e
chromium hydroxide green		77289	external use only[e]
chromium oxide greens		77288	external use only[e]
copper powder		77400	e
dihydroxyacetone			externally applied cosmetics intended solely or in part for imparting color to the human body only
disodium EDTA-copper			shampoos only
ferric ammonium ferrocyanide			external use only[e]
ferric ferrocyanide		77510 77520	external use only[e]
guaiazulene			external use only
guanine		75170	e
henna		75480	hair dyes only, not near eye
lead acetate			hair dyes for scalp only; 0.6% (w/v) max as Pb
manganese violet		77742	e
mica		77019	e
potassium sodium copper chlorophyllin		75810	dentifrices only; 0.1% max[f]
pyrophyllite			external use only
silver			fingernail polish only; 1% max
synthetic iron oxides		77491 77492 77499	e
titanium dioxide		77891	e
ultramarine blue		77007[g]	external use only[e]
ultramarines			
green		77013[h]	external use only[e]
pink		77007	
red		77007	
violet		77007[i]	
zinc oxide		77947	e

[a]See Tables 1 and 2 for synonyms and CAS Registry Numbers.

[b]No color additive or product containing one can be used in the area of the eye, in surgical sutures, or injections unless so stated. Colorants without restrictions can be used for coloring cosmetics generally, in amounts consistent with good manufacturing practice.

[c]Resorcin Brown; CI Acid Orange 24 [1320-07-6].

[d]Alizarine Violet; CI Acid Violet 43 [4430-18-6].

[e]May also be used in cosmetics intended for use in the area of the eye.

[f]Can only be used in combination with certain substances.

[g]CI Pigment Blue 29.

[h]CI Pigment Green 24.

[i]CI Pigment Violet 15.

Table 4. Colorants Permitted in Medical Devices

FDA official name	Structure number	CI number	Uses and restrictions[a]
		Subject to certification	
[phthalocyaninato(2-)] copper	(34)[b]	74160	polypropylene (PP) sutures, polybutester nonabsorbable sutures for use in general and ophthalmic surgery, poly(butylene terephthalate) (PBT) monofilament nonabsorbable sutures for general and ophthalmic surgery, and poly(methyl methacrylate) (PMMA) monofilament used as supporting haptics for intraocular lenses; 0.5% (w/w) max contact lenses
D&C Blue No. 6	(35)[c]	73000	poly(ethylene terephthalate) (PET) sutures for general surgical use, 0.2% (w/w) max; plain or chromic collagen absorbable sutures for general surgical use, 0.25% (w/w) max; plain or chromic collagen absorbable sutures for ophthalmic surgical use, 0.5% (w/w) max; PP sutures for general surgical use, 0.5% (w/w) max; polydioxanone synthetic absorbable sutures for ophthalmic and general surgical use, 0.5% (w/w) max
D&C Green No. 6	(14)[d]	61565	PET sutures for general and ophthalmic surgery, 0.75% (w/w) max; poly(glycolic acid) (PGA) sutures for general and ophthalmic surgery with diameter greater than U.S.P. size 8–0, 0.1% (w/w) max; PGA sutures for general and ophthalmic surgery with diameter not greater than U.S.P. size 8–0, 0.5% (w/w) max; poly(glycolic acid-co-trimethylene carbonate) sutures for general surgery, 0.21% (w/w) max contact lenses, 0.03% (w/w) max
D&C Red No. 17	(20)[d]	26100	contact lenses only
D&C Violet No. 2	(29)[d]	60725	polyglactin 910 (glycolic–lactic acid polyester) synthetic absorbable sutures for general and ophthalmic surgery, 0.2% (w/w) max; polydioxanone synthetic absorbable sutures for use in general and ophthalmic surgery, 0.3% (w/w) max PMMA intraocular lens haptics, 0.2% (w/w) max contact lenses
D&C Yellow No. 10	(30a)[d]	47005	contact lenses only

Table 4. *(Continued)*

FDA official name	Structure number	CI number	Uses and restrictions[a]
Exempt from certification			
1,4-bis[4-(2-methacryloxyethyl)phenylamino] anthraquinone			contact lenses only
1,4-bis[(2-methylphenyl)-amino]-9,10-anthracenedione			contact lenses only
carbazole violet		51319	contact lenses only
chlorophyllin–copper complex, oil soluble			PMMA bone cement only; 0.003% (w/w) max
chromium–cobalt–aluminum oxide		77343	contact lenses only
chromium oxide green		77288	contact lenses only
CI Vat Orange 1		59105	contact lenses only
7,16-dichloro-6,15-dihydro-5,9,14,18-anthrazinetetrone		69825	contact lenses only
2-[[2,5-diethoxy-4-[(4-methylphenyl)-thiol]phenyl]azo]-1,3,5-benzenetriol			to mark soft (hydrophilic) contact lenses with the letters R and L only; 1.1×10^{-7} g/lens max
16,23-dihydrodinaphtho-[2,3-a:2′,3′-i]naphth[2′,3′:6,7]indolo [2,3-c]carbozole-5,10,15,17,22,24,-hexone		70800	contact lenses only
N,N′-(9,10-dihydro-9,10-dioxo-1,5-anthracene-diyl)bisbenzamide		61725	contact lenses only
16,17-dimethoxydinaphtho [1,2,3-cd:3′,2′,1′-1m]perylene-5,10-dione		59825	contact lenses only
4-[2,4-dimethylphenyl)-azo]-2,4-dihydro-5-methyl-2-phenyl-3H-pyrazol- 3-one			contact lenses only

Table 4. *(Continued)*

FDA official name	Structure number	CI number	Uses and restrictions[a]
6-ethoxy-2-(6-ethoxy-3-oxobenzo[b]thien-2(3*H*)-ylidene)benzo-[b]thiophen-3(2*H*)-one		73335	contact lenses only
iron oxides		77491	contact lenses only
phthalocyanine green		74260	contact lenses only
poly(hydroxyethyl methacrylate)-dye copolymers			contact lenses only
titanium dioxide		77891	contact lenses only

[a]Colorants can only be used in those medical devices cited. Colorants without specific restrictions regarding quantity can be used in amounts not to exceed the minimum reasonably required to accomplish the intended coloring effect.
[b]Copper phthalocyanine; CI Pigment Blue 15 [*147-14-8*].
[c]Indigo; CI Vat Blue 1 [*482-89-3*].
[d]See Tables 1 and 2 for synonyms and CAS Registry Numbers.

developed for the specific purpose of changing the apparent color of the iris (see CONTACT LENSES).

Regulations Governing Use

Listed and Provisionally Listed Colorants. Colorants can be divided into two groups: those listed for use and those provisionally listed. Listed additives are colors that have been sufficiently evaluated to convince FDA of their safety for the applications intended. These colorants are also known popularly as permanently listed colorants; however, they in fact can be delisted for sufficient cause. Provisionally listed colorants, on the other hand, are dyes and pigments that are not considered unsafe but that have not undergone all the tests required by the Color Additives Amendments of 1960 to establish their eligibility for permanent listing. Currently, these colors can still be used in those applications in which they were used prior to enactment of the 1960 amendments, unless newer temporary regulations restrict their use further. The status of these colorants is reviewed about once each year and, if sufficient reason exists and if the manufacturers or consumers of these colors request it, their provisional listing status is extended pending completion of the required scientific investigations.

Certification of Colorants. A further distinction between color additives is made relative to whether there is requirement for FDA certification. In general, only synthetic organic colorants are now subject to certification, whereas natural organic and inorganic colorants, such as turmeric and titanium dioxide, are not. The exemption from certification for a particular colorant holds whether the colorant is obtained from natural sources or is synthetically produced, as in the case of natural and synthetic β-carotene.

If a color requires certification prior to its sale, an appropriate size representative sample of each batch, along with a request for certification must be submitted to the FDA, Color Certification Branch, to see if it conforms to the specifications and other conditions established for it. The charge for certification of a straight colorant, a lake of a straight colorant, or a repack of either, is $0.55/kg of the batch, with a minimum charge of $160. The charge for the certification of other repacks is $25 minimum, plus $0.13/kg for each kg over 45.5 kg (100 pounds), for batches weighing less than 455 kg. If the batch is found satisfactory, a lot number is assigned to it and a certificate of certification is issued. These certificates are valid so long as the regulations pertaining to the storage, packaging, labeling, distribution, and use of the lot are strictly adhered to.

Specifications.   Most colorants in use today have specifications that must be met before they can be sold. In the case of the provisionally listed colors, these specifications may be revised if and when the colorants are removed from the provisional lists. Typical specifications for a synthetic aromatic organic dye, a synthetically produced natural colorant and an inorganic pigment, are given as examples. Many other examples can be found in Reference 15.

FD&C Red No. 40.   This monoazo dye (**5**) is manufactured by coupling diazotized 5-amino-4-methoxy-2-toluenesulfonic acid with 6-hydroxy-2-naphthalenesulfonic acid. FD&C Red No. 40 is also known as Allura Red or CI Food Red 17. Its chemical identity is principally the disodium salt of 6-hydroxy-5-[(2-methoxy-5-methyl-4-sulfophenyl)azo]-2-naphthalenesulfonic acid, which has the formula $C_{18}H_{14}N_2O_8S_2Na_2$, and mol wt 496.43.

FD&C Red No. 40 shall conform to the following specifications and shall be free from impurities other than those named to the extent that such other impurities may be avoided by good manufacturing practice:

Sum of volatile matter (at 135°C) and chlorides and sulfates (calculated as sodium salts)—not more than 14.0%

Water-insoluble matter—not more than 0.2%

Higher sulfonated subsidiary colors (as sodium salts)—not more than 1.0%

Lower sulfonated subsidiary colors (as sodium salts)—not more than 1.0%

Disodium salt of 6-hydroxy-5-[(2-methoxy-5-methyl-4-sulfophenyl)azo]-8-(2-methoxy-5-methyl-4-sulfophenoxy)-2-naphthalenesulfonic acid—not more than 1.0%

Sodium salt of 6-hydroxy-2-naphthalenesulfonic acid (Schaeffer's salt)—not more than 0.3%

4-Amino-5-methoxy-*o*-toluenesulfonic acid—not more than 0.2%

Disodium salt of 6,6'-oxybis(2-naphthalenesulfonic acid)—not more than 1.0%

Lead (as Pb)—not more than 10 ppm

Arsenic (as As)—not more than 3 ppm

Total color—not less than 85.0%

β-Apo-8'-carotenal.   This colorant, $C_{30}H_{40}O$ (CI Food Orange 6), EEC No. E 160e), is an aldehydic carotenoid (**38**) widely distributed in nature: it is isolated from numerous items, including spinach, oranges, grass, tangerines, and marigolds. It is synthetically produced as the crystalline all-trans stereoisomer, which is a purplish black powder that melts (with decomposition) in the range 136–140°C (corrected). Its mol wt is 416.65.

β-Apo-8′-carotenal shall conform to the following specifications:

Physical state—solid

One percent solution in chloroform—clear

Melting point (decomposition)—136–140°C (corrected)

Loss of weight on drying—not more than 0.2%

Residue on ignition—not more than 0.2%

Lead (as Pb)—not more than 10 ppm

Arsenic (as As)—not more than 1 ppm

Assay (spectrophotometric)—96–101%

Titanium Dioxide.   Titanium dioxide [13463-67-7], TiO_2, mol wt 79.90, Titanic Earth, CI Pigment White 6, CI No. 77891, EEC No. E 171, is the whitest, brightest pigment known today, with a hiding power four to five times greater than that of its closest rival, zinc oxide (42) (see TITANIUM COMPOUNDS, INORGANIC).

Titanium dioxide shall conform to the following specifications:

Lead (as Pb)—not more than 10 ppm

Arsenic (as As)—not more than 1 ppm

Antimony (as Sb)—not more than 2 ppm

Mercury (as Hg)—not more than 1 ppm

Loss on ignition at 800°C (after drying for 3 h at 105°C)—not more than 0.5%

Water-soluble substances—not more than 0.3%

Acid-soluble substances—not more than 0.5%

Titanium dioxide—not less than 99.0% after drying for 3 h at 105°C

Lead, arsenic, and antimony—determined in the solution obtained by boiling 10 g of the titanium dioxide for 15 min in 50 mL of 0.5 N hydrochloric acid

In addition to individual specifications, general specifications have been written for provisionally listed certifiable colors:

	FD&C colors, % max	D&C and Ext. D&C colors, % max
lead	0.001	0.002
arsenic (as As_2O_3)	0.00014	0.0002
heavy metals (except lead and arsenic) (precipitated as sulfides)	trace	0.003
mercury	0.0001	
colors that are barium salts—soluble barium in dilute HCl (as $BaCl_2$)		0.05

The limit of 1 ppm mercury placed on colors intended for use in foods was established by a letter from the Acting Director of the Division of Colors and Cosmetics to the certified color manufacturers in 1970. This action was the first step taken to replace the somewhat nebulous heavy metals specifications previously used with concrete limits for specific metals.

Use Restrictions.   There are numerous restrictions on the use of color additives. They cannot, for example, be employed to deceive the public by adding weight or bulk to a product or by hiding quality. In addition, special permission is needed to use colorants or products containing them in the area of the eyes, in injections, in surgical sutures, and in foods for which standards of identity have been promulgated under Section 401 of the Federal Food, Drug, and Cosmetic Act.

Other restrictions pertaining to the areas of use and the quantities of colorants allowed in products are specified in regulations for particular additives. Citrus Red No. 2, for example, can only be used to color the skins of oranges not intended for processing, whereas pyrophyllite can be used only to color drugs that are to be externally applied. A special case of the restricted use of a colorant is that of FD&C Red No. 4. Although it is designated as an FD&C colorant (implying that it can be generally used in foods), its use is now limited to coloring externally applied drugs and cosmetics only. FD&C Red No. 4 can no longer be used to color foods and ingested drugs. So many limitations have crept into the system that the designations FD&C, D&C, and Ext. D&C no longer have their original meaning.

The amount of a color additive allowed in a product depends on both the colorant and the article being colored. For example, TiO_2 when used to color foods cannot exceed 1% by weight of the food product. On the other hand, there is no numerical limit set on its use in the coloring of ingested or externally applied drugs. Similarly, ultramarine blue may be used to color salt intended for animal feed, but not in amounts exceeding 0.5% by weight of the salt. When numerical limits for the use of colorants are not specified, the amount allowed is controlled by "good manufacturing practice," an ill-defined term that in effect says that you cannot use more of a colorant in a product than is needed to achieve the desired effect. Today, the excessive use of colorants is rarely a problem since manufacturers are not likely to waste costly additives and, at the same time, run the risk of making their products appear unnatural.

Certified Colors

Presently, all certified colors are factory-prepared materials belonging to one of several different chemical classes. Although a few such as D&C Blue No. 6 (indigo) are known to exist in nature, certified colors owe their commercial importance to their synthetic production. Because of the starting materials used in their manufacture in the past, certified colors were once known as coal-tar dyes. Today, since most of the raw materials used in their preparation are obtained from petroleum, this term no longer applies.

Compared to noncertified color additives, certified colors are a cheaper, brighter, more uniform, and better characterized group of dyestuffs with higher tinctorial strengths and a wider range of hues. They are available singly (primary colors) and in admixture with other certified colors (secondary mixes). By properly blending the available primary colorants, a nearly infinite number of shades can be prepared. Most are sold in various forms, including powders, granules, pastes, solutions and dispersions, and as lakes. Most are also available as is, or mixed with salt or sugar or some other approved solvent or diluent, depending on the colorant and its intended use.

Chemical Classifications.   Azo colors comprise the largest group of certified colorants. They are characterized by the presence of one or more azo bonds (—N=N—) and are synthesized by the coupling of a diazotized primary aromatic amine to a coupling component, usually a naphthol (see AZO DYES). Examples abound in Figures 1, 2, and 3. Certifiable azo colors can be subdivided into four groups: insoluble unsulfonated pigments, soluble unsulfonated dyes, insoluble sulfonated pigments, and soluble sulfonated dyes.

D&C Red No. 36 (**27**) is an unsulfonated pigment. It contains no groups capable of salt formation and is thus insoluble directly on coupling. Its chlorine group ortho to the azo group results in a sterically hindered molecule with low solubility and excellent light stability. The unsulfonated dyes Citrus Red No. 2 (**8**) and D&C Red No. 17 (**20**) are insoluble in water but soluble in aromatic solvents.

Insoluble sulfonated pigments are made from colorants that contain a sulfonic acid group that is easily converted into an insoluble metal salt. In most cases, the sulfonic acid group is ortho to the diazo further reducing the solubilizing characteristics of the sulfonic grouping. The shade of these products is affected by the metal incorporated into the molecule and the physical characteristics of the colorants. D&C Red Nos. 7 (**19b**) and 34 (**26**) are insoluble sulfonated pigments.

The soluble azo dyes contain one or more sulfonic acid groups. Their degree of water solubility is determined by the number of sulfonic groups present and their position in the molecule. FD&C Red No. 40 (**5**) and D&C Orange No. 4 (**16**) belong in this class.

Anthraquinone colorants are characterized by the presence of the anthraquinone [*84-65-1*] (qv) nucleus:

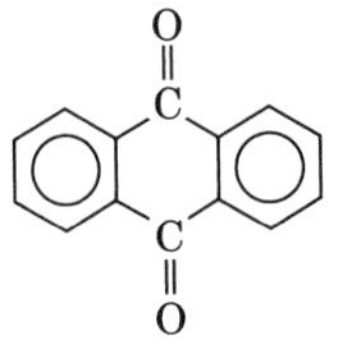

Included in this grouping are D&C Green No. 5 (**13**), a water-soluble sulfonate, D&C Green No. 6 (**14**), an unsulfonated water-insoluble compound, and D&C Violet No. 2 (**29**), a water-insoluble hydroxyanthraquinone. Anthraquinone color additives, in general, are light stable and have good physical and chemical properties for use in cosmetics (see DYES, ANTHRAQUINONE).

There are three color additives of the indigoid type, including D&C Blue No. 6 (**35**) (an insoluble pigment), FD&C Blue No. 2 (**2**) (the water-soluble disodium sulfonate derivative of indigo), and D&C Red No. 30 (**23**) (an insoluble thioindigoid). All are related to indigo [*482-89-3*] which has structure (**35**) (Fig. 3).

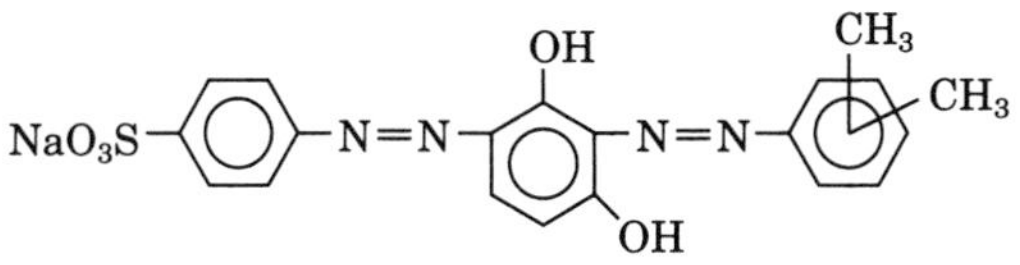

Fig. 3. Miscellaneous dyes used in cosmetics and medical devices. Dye classification: (**32**), azo; (**33**), anthraquinone; (**34**), phthalocyanine; (**35**), indigoid.

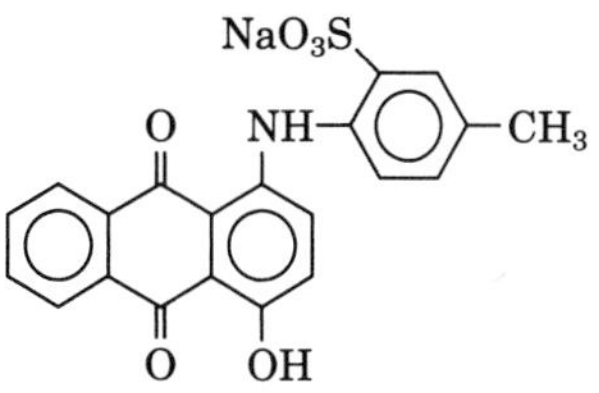

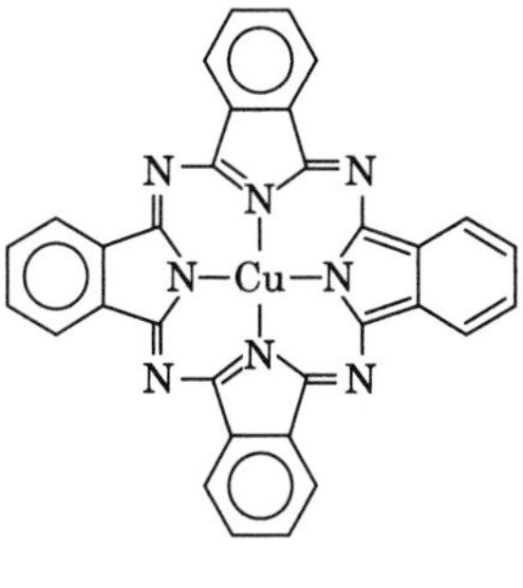

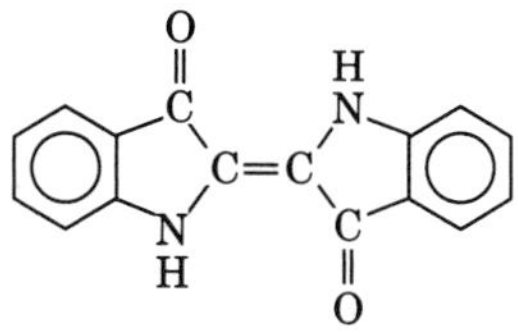

Three dyes are triaryl- or triphenylmethanes. Each, like FD&C Blue No. 1, consists of three aromatic rings attached to a central carbon atom. All are water-soluble, anionic, sulfonated compounds. FD&C Blue No. 1 has the structure (**1**) shown in Figure 1.

The second largest group of color additives are the xanthenes, which are characterized by the following structure:

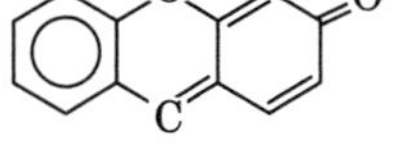

Xanthene dyes (qv) can be either acidic or basic. Acid xanthenes are known to exist in two tautomeric forms. The phenolic type, or fluorans, are free-acid structures such as D&C Orange No. 10 (**17b**) and D&C Red No. 21 (**17c**). Most have poor water solubility. In contrast to these, the quinoids or xanthenes are usually the highly water-soluble sodium salt counterparts of the fluorans such as D&C Orange No. 11 (**18**) and D&C Red No. 22 (**21a**). Presently, there are no certifiable basic xanthene colorants.

Two of the remaining colorants on the list of certifiables are quinolines; the solvent-soluble D&C Yellow No. 11 (**30b**), and its water-soluble sulfonated derivative, D&C Yellow No. 10 (**30a**) (Fig. 2). Both are derived from quinaldine by condensation with phthalic anhydride.

Two others, FD&C Yellow No. 5 (**6**) and Orange B (**9**), are pyrazolones that contain the following common group:

$$HO-\underset{\underset{\underset{\underset{|}{C}}{\diagdown}}{\overset{\|}{C}}}{C}-\underset{\underset{\underset{|}{C}}{\diagup}}{\overset{|}{N}}-$$

The pyrazolones may also be classified as azo dyes since each contains an —N=N— group.

One nitro dye (**31**) (Ext. D&C Yellow No. 7), one pyrene colorant (**15**) (D&C Green No. 8), and one phthalocyanine dye (**34**) [phthalocyaninato (2-)] (copper) complete the list of certifiable colors.

Lakes. Lakes are a special kind of color additive prepared by precipitating a soluble dye onto an approved insoluble base or substratum. In the case of D&C and Ext. D&C lakes, this substratum may be alumina, blanc fixe, gloss white, clay, titanium dioxide, zinc oxide, talc, rosin, aluminum benzoate, calcium carbonate, or any combination of two or more of these materials. Currently, alumina is the only substratum approved for manufacturing FD&C lakes.

FD&C lakes were first approved for use in 1959. Today, they are the most widely used type of lake. To make a lake, an alumina substrate is first prepared by adding sodium carbonate or sodium hydroxide to a solution of aluminum sulfate. Next, a solution of certified colorant is added to the resulting slurry, then aluminum chloride is added to convert the colorant to an aluminum salt, which then adsorbs onto the surface of the alumina. The slurry is then filtered, and the cake is washed, dried, and ground to an appropriate fineness, typically 0.1–40 μm.

Lakes are available with pure dye contents ranging from less than 1% to more than 40% and with moisture contents of 6–25%. Typical FD&C lakes contain 10–40% pure dye and 15–25% moisture. Typical use levels are 0.1–0.3%. Lakes are marketed as is, or mixed with other lakes or approved diluents, or dispersed in various edible vehicles such as hydrogenated vegetable oil, coconut oil, propylene glycol, glycerol, or sucrose syrup, or dispersed in other approved media that make the mixtures appropriate for printing food wrappings, for marking capsules, for incorporating into health products that come into direct contact with the skin, and so on.

Lakes are insoluble in most solvents, although some bleeding or leaching may be observed in solvents in which the unlaked dye is soluble. FD&C lakes are insoluble in water in the pH range of 3.5–9.0, but outside this range, the lake substrate tends to dissolve releasing the captive dye.

Properties of lakes that enhance their usefulness include their opacity, their ability to be incorporated into products in the dry state, their relative insolubility, and their superior stability toward heat and light. Such properties have made possible the more effective and more efficient preparation of candy and tablet coatings, and often eliminate the need to remove moisture from dry products before coloring them. Lakes have also made possible the coloring of certain products that, because of their nature, method of preparation, or method of storage, cannot be colored with ordinary color additives.

Since there are no solvent-soluble FD&C colors, FD&C lakes have proven particularly valuable for coloring water-repelling foods such as fats, gums, waxes, and oils, and for coloring food-packaging materials including lacquers, containers, plastic films, and inks from which soluble dyes would be quickly leached. Similarly useful applications have been found for D&C and Ext. D&C lakes in their respective areas of application.

Unlike dyes that color objects through their adsorption or attachment from solution to the material being colored, lakes, like other pigments, impart color by dispersing them in the medium to be colored. As a consequence of this pigmentlike character, both the shade and the tinctorial strength of lakes are highly dependent on the conditions used in their manufacture as well as their physical properties, including their particle size and crystal structure.

Some specific products in which lakes are used include icings, fondant coatings, sandwich cookie fillings, cake and doughnut mixes, decorative sugar crystals, coated and compression tablets (candy or pharmaceutical), hard candy, candy wafers, chewing gums, wax coatings for cheeses, yogurts, dry beverage bases, dessert powders, snack foods, spice mixes, canned and semimoist burger-type pet foods, lotions, creams, toothpastes, nail polishes, face powders, lipsticks, printing inks, plastic films, decorative coatings, can linings, meat trays, produce containers, and margarine tubs.

Properties of Colorants.   Properties of a number of colorants are shown in Tables 5–9. For other properties see Reference 15. Most values are from the literature and, in general, refer to commercial colorants and not pure compounds. The composition of certified colorants can vary substantially with regard to the amounts of pure dye, salt, moisture, subsidiary dyes, trace metals, etc. that they contain, and of course the properties of color additives are affected by their composition.

Production and Use.   The tonnage of each colorant certified by the FDA over the past few years can be found in Table 10. The primary FD&C colors dominate this picture, since they account for 80% or more of the total weight of colorant certified during any one year.

Based on maximum color concentrations and the total annual production of food in each food category, the total certified food color that might be ingested per person per year is estimated to be 19.5 g. Based on recent annual colorant production figures and current total population, this figure is closer to 11 g/yr.

Table 5. Solubility[a,b] of FD&C Colors in Various Solvents

Federal name	Common name	Water	Ethanol	Glycerol	Propylene glycol
FD&C Blue No. 1	Brilliant Blue FCF	20.0	0.15	20.0	20.0
FD&C Blue No. 2	Indigotine	1.6		1.0	0.1
FD&C Green No. 3	Fast Green FCF	20.0	0.01	20.0	20.0
FD&C Red No. 3	Erythrosine	9.0		20.0	20.0
FD&C Red No. 4	Ponceau SX	11.0		5.8	2.0
FD&C Red No. 40	Allura Red	22.0	0.001	3.0	1.5
FD&C Yellow No. 5	Tartrazine	20.0		18.0	7.0
FD&C Yellow No. 6	Sunset Yellow FCF	19.0		20.0	2.2

[a]Grams of colorant/100 mL[c] solvent at 25°C.
[b]Much more extensive solubility data at various temperatures is given in Reference 15.
[c]To convert g/100 mL to oz/gal, multiply by 1.3.

Table 6. pH Stability of FD&C Colors[a]

Federal name	pH			
	3	5	7	8
FD&C Blue No. 1	slight fade after 1 wk	very slight fade after 1 wk	very slight fade after 1 wk	very slight fade after 1 wk
FD&C Blue No. 2	appreciable fade after 1 wk	appreciable fade after 1 wk	considerable fade after 1 wk	fades completely
FD&C Green No. 3	slight fade after 1 wk	very slight fade after 1 wk	very slight fade after 1 wk	slight fade and appreciably bluer
FD&C Red No. 3	insoluble	insoluble	no appreciable change	no appreciable change

[a]For FD&C Red No. 4, FD&C Red No. 40, FD&C Yellow No. 5, and FD&C Yellow No. 6 no appreciable change is noticed from pH 3–8.

Table 7. Stability[a] of FD&C Colors Exposed to Common Food Additives

Federal name	Sodium benzoate, 1%	Ascorbic acid, 1%	Sulfur dioxide, 25 ppm	Sulfur dioxide, 250 ppm
FD&C Blue No. 1	no appreciable change	slight fade after 1 wk	no appreciable change	very slight fade after 1 wk
FD&C Blue No. 2	slight fade after 1 wk	considerable fade after 1 wk	fades completely	fades completely
FD&C Green No. 3	no appreciable change	slight fade after 1 wk	no appreciable change	very slight fade after 1 wk
FD&C Red No. 3	very slight fade after 1 wk	insoluble	insoluble	insoluble
FD&C Red No. 4	no appreciable change	considerable fade after 1 wk	no appreciable change	no appreciable change
FD&C Red No. 40	no appreciable change	no appreciable change	no appreciable change	no appreciable change
FD&C Yellow No. 5	no appreciable change	appreciable fade after 1 wk	appreciable fade after 1 wk	appreciable fade after 1 wk
FD&C Yellow No. 6	no appreciable change	considerable fade after 1 wk	appreciable fade after 1 wk	appreciable fade after 1 wk

[a]These colors show no appreciable change in the presence of various sugars with the exception of FD&C Blue No. 2, which fades considerably after 1 wk in 10% cerelose or dextrose and even fades slightly in 10% sucrose.

Table 8. Solubilities of D&C and Ext. D&C Colorants[a]

Federal name	H_2O	Glycerol	CH_3OH	C_2H_5OH	Petroleum jelly	Toluene	Stearic acid	Oleic acid	Mineral oil	Ethyl ether	Acetone	Butyl acetate
D&C Blue No. 4	S	S	S	S	C	I	C	C	C	I	Ia	I
D&C Blue No. 6	IU	D	I	I	D	Ia	D	D	D	I	I	I
D&C Blue No. 9	IU	ID	Ia	I	D	Ia	D	D	D	I	I	I
D&C Brown No. 1	S	S	S	SS	IE	I	IE	IE	IE	SS	SS	I
D&C Green No. 5	S	S	S	SS	IE	I	IE	IE	IE	I	SS	I
D&C Green No. 6	I	Ia	SS	SS	M	S	M	M	M	SS	SS	S
D&C Green No. 8	SF	SSF	SSF	SSF	Ia	I	Ia	Ia	I	Ia	Ia	Ia
D&C Orange No. 4	S	S	S	M	IE	I	IE	IE	IE	I	Ia	I
D&C Orange No. 5	IB	SS	S	M	D	I	D	D	D	M	s	I
D&C Red No. 6	S	S	SS	Ia	I	I	I	I	I	I	Ia	I
D&C Red No. 7	I	D	Ia	Ia	D	I	D	D	D	I	Ia	I
D&C Red No. 21	IBF	Da	SS	SS	D	I	D	D	D	M*	s	I
D&C Red No. 22	SF	SF	SF	SF	IE	I	IE	IE	IE	Ia	SS	I
D&C Red No. 27	IB	Da	SS	SS	D	I	D	D	D	Ia	SS	I
D&C Red No. 28	S	S	S	S	IE	I	IE	IE	IE	Ia	SS	I
D&C Red No. 30	IU	D	I	I	I	Ia	I	D	D	Ia	Ia	Ia
D&C Red No. 31	M	SS	SS	SS	I	I	I	I	I	I	Ia	Ia
D&C Red No. 33	S	S	SS	SS	I	I	I	I	I	I	I	I
D&C Red No. 34	I	I	Ia	I	D	I	D	D	D	I	D	D
D&C Red No. 39	Ia	M	M-S	S	I	Ia	I	SS	I	s	s	SS
D&C Violet No. 2	I	Ia	SS	SS	s	S	s	s	s	SS	SS	s
D&C Yellow No. 7	IBF	SSF	SF	SS	D	S	D	D	s	SS*	SS	s
D&C Yellow No. 8	SF	SF	SF	M	IE	I	IE	IE	D	SS*	kIa	I
D&C Yellow No. 10	S	S	M	SS	I	I	I	I	IE	Ia	SS	I
D&C Yellow No. 11	I	SS	S	S	s	s	s	s	I	Ia	s	s
Ext. D&C Violet No. 2	S	S	SS	SS	I	I	I	I	s	s	SS	I
Ext. D&C Yellow No. 7	S	S	M	SS	I	I	I	I	I	I	M	I

[a] a = May bleed or stain, very sparingly soluble; B = insoluble in water, soluble in aqueous alkaline solution; C = practically insoluble, but useful in nearly neutral or slightly acid emulsions; D = practically insoluble, but may be dispersed by grinding and homogenizing; solid mediums (waxes) should be softened or melted before or during the grinding; E = practically insoluble in the fatty acid, oil, or wax, but useful in coloring slightly alkaline aqueous emulsions; F = solution usually fluorescent; I = insoluble; k = turns brownish in hue; M = moderately soluble (<1%); S = dissolves (solubility ≥1%); SS = sparingly soluble (<0.25%); U = in alkaline-reducing vats a soluble leuco compound forms; * = practically colorless.

Table 9. Fastness[a] Properties[b] of D&C and Ext. D&C Colorants

	Light	10% CH₃COOH	10% HCl	10% NaOH	0.9% NaCl	5% FeSO₄	5% Alum	Oxidizing agents	Reducing agents
D&C Blue No. 4	3	5	5	4	6	4	4	2	1
D&C Blue No. 6	6	7I	5I	L6U	I	I	I	6	U
D&C Blue No. 9	7	7I	5I	6IU	I	I	I	6	U
D&C Brown No. 1	3	5	5	6sly	6	p	p	3	1
D&C Green No. 5	5	5	5	5	5	4	4	3	2
D&C Green No. 6	4	5L	5I	6I	I	I	I	3	2
D&C Green No. 8	2	I	I	5	6	4d	4d	3	3
D&C Orange No. 4	5	5	5	2m	6	Jp	Jp	3	3
D&C Orange No. 5	2	4aI	4I	Sr	I	I	I	3	3
D&C Red No. 6	5	5	4	4d	6	p	p	3	1
D&C Red No. 7	6	5I	4I	5I	I	4Id	4I	3	1
D&C Red No. 21	2	3I	3I	5Sr	I	Id	4I	4	4
D&C Red No. 22	2	2py	Ipy	5	6	3d	2y	4	4
D&C Red No. 27	2	3I	3	5Sr	1	I	I	4	4
D&C Red No. 28	3	2p	4p	6	6	z	p	4	4
D&C Red No. 30	6	7I	I	6IU	I	I	I	5	U
D&C Red No. 31	5	5	4	5	6	p	p	3	1
D&C Red No. 33	5	6	3z	5	6	4	4	3	1
D&C Red No. 34	4	5I	4	4I	I	I	I	3	1
D&C Red No. 39	2	Sv	Sv	6Sx	I	4aId	I	3	3
D&C Violet No. 2	4	5I	5I	5I	6I	4I	4I	2	1
D&C Yellow No. 7	2	I	I	S6	I	I	I	3	3
D&C Yellow No. 8	3	3p	3p	6	6	zp	p	3	3
D&C Yellow No. 10	3	5	5	4r	6	z	4	2	5
D&C Yellow No. 11	2	1	5I	Iw	I		I	2	5
Ext. D&C Violet No. 2	5	5	5	5	6	4z	4	3	2
Ext. D&C Yellow No. 7	4	5	5	5	6	zd	4	3	3

[a] 1-very poor fastness; 2-poor fastness; 3-fair fastness; 4-moderate fastness; 5-good fastness; 6-very good fastness; 7-excellent fastness; L-turns orange in hue; sl-slightly; m-turns scarlet in hue; d-hue becomes duller or darker; r-turns redder in hue; v-turns violet in hue; w-becomes tinctorially weaker; x-turns yellow in hue; y-turns yellower in hue; z-hazy or cloudy. [b] U-in alkaline-reducing vats, a soluble leuco compound forms; I-insoluble; J-tends to thicken or gel the solution; p-dye precipitated as heavy-metal salt or color acid; a-may bleed or stain, very sparingly soluble; S-dissolves (solubility 1%).

Table 10. Color Additives Certified by FDA during Fiscal Year,[a] t

	1983	1986	1988	1990
FD&C: primaries				
FD&C Blue No. 1	82.05	98.42	127.04	105.86
FD&C Blue No. 2	61.70	38.43	57.76	42.33
FD&C Green No. 3	1.58	2.13	2.60	3.40
FD&C Red No. 3	210.54	114.53	121.58	82.81
FD&C Red No. 4	5.65	7.29	4.96	12.17
FD&C Red No. 40	855.16	1222.54	1201.15	1177.20
FD&C Yellow No. 5	666.95	686.69	754.69	745.09
FD&C Yellow No. 6	492.78	726.06	750.38	728.80
Total	*2376.41*	*2896.09*	*3020.16*	*2897.66*
FD&C: lakes				
FD&C Blue No. 1	37.99	34.94	38.77	31.78
FD&C Blue No. 2	33.65	36.81	40.03	34.94
FD&C Green No. 3				
FD&C Red No. 3	137.00	94.84	103.26	19.76
FD&C Red No. 40	46.88	133.24	105.61	184.75
FD&C Yellow No. 5	243.34	307.16	380.70	360.00
FD&C Yellow No. 6	172.12	158.89	238.67	273.44
Total	*670.98*	*765.88*	*907.04*	*904.67*
D&C: primaries				
D&C Blue No. 4				
D&C Blue No. 6	0.05			
D&C Blue No. 9				
D&C Green No. 5	1.55	1.52	0.45	1.87
D&C Green No. 6	0.23	0.36	0.78	0.53
D&C Green No. 8	5.52	8.18	12.98	17.36
D&C Orange No. 4	3.61	4.23	4.45	2.29
D&C Orange No. 5	0.14	3.39	1.13	
D&C Orange No. 10				
D&C Orange No. 17	0.99	2.10	1.09	[b]
D&C Red No. 6	0.16	0.20	0.32	1.72
D&C Red No. 7		0.05	0.07	0.26
D&C Red No. 8	0.44	0.41	0.50	[b]
D&C Red No. 9		0.21		[b]
D&C Red No. 17	0.68	0.87	0.95	2.12
D&C Red No. 19	0.2	0.95		[b]
D&C Red No. 21	2.36	3.43	1.38	2.53
D&C Red No. 22	2.24	0.36		2.92
D&C Red No. 27	0.13	1.50	0.97	1.96
D&C Red No. 28	1.49	2.38	1.32	3.59
D&C Red No. 30	0.03	1.10	1.59	3.28
D&C Red No. 31				
D&C Red No. 33	3.38	4.28	6.52	6.79
D&C Red No. 34				
D&C Red No. 36	1.65	3.96	1.03	3.01
D&C Red No. 37				[b]
D&C Red No. 39				

Table 10. (*Continued*)

	1983	1986	1988	1990
D&C: primaries				
D&C Violet No. 2		1.01	1.32	0.69
D&C Yellow No. 7			.50	0.11
D&C Yellow No. 8	2.45	1.25	1.61	1.36
D&C Yellow No. 10	21.11	11.50	17.53	19.92
D&C Yellow No. 11	1.45	1.64	0.72	0.82
Total	*49.86*	*54.88*	*57.21*	*73.13*
D&C: lakes				
D&C Blue No. 1	0.23		0.28	
D&C Green No. 5				
D&C Green No. 6				
D&C Green No. 8				
D&C Orange No. 4	0.69		0.52	1.81
D&C Orange No. 5		1.34	0.78	0.59
D&C Orange No. 17	2.77	1.41	0.93	*b*
D&C Red No. 3	4.82	3.59	5.19	2.56
D&C Red No. 6	31.32	40.28	41.18	52.59
D&C Red No. 7	54.21	86.79	70.88	98.96
D&C Red No. 8				*b*
D&C Red No. 9	16.02	19.32	1.19	*b*
D&C Red No. 19	1.54	7.91		*b*
D&C Red No. 21	2.12	7.24	4.32	8.54
D&C Red No. 22		0.84		0.48
D&C Red No. 27	2.92	18.42	19.72	37.79
D&C Red No. 28		0.15	0.27	0.76
D&C Red No. 30	14.96	9.17	16.49	19.30
D&C Red No. 31			0.36	
D&C Red No. 33	0.38	0.33	1.05	1.87
D&C Red No. 34	5.80	5.59	1.92	1.51
D&C Red No. 36		0.67		
D&C Yellow No. 5	3.94	4.25	7.79	4.59
D&C Yellow No. 6	1.26	0.66	1.13	2.82
D&C Yellow No. 10	31.96	32.42	39.29	37.53
Total	*174.94*	*240.40*	*213.29*	*271.70*
Ext. D&C: primaries				
Ext. D&C Violet No. 2	1.16	2.45	3.84	2.24
Ext. D&C Yellow No. 7	0.79	0.44	4.42	0.18
Total	*1.95*	*2.89*	*8.26*	*2.42*
Other colorants				
Citrus Red No. 2	1.74	1.44	2.09	
Orange B				
[phthalocyaninato-(2)] Copper	0.01	1.00	0.05	0.10
Total	*1.75*	*2.44*	*2.14*	*0.10*

[a]In 1976, the Federal fiscal year was changed to end in September rather than June.
[b]Delisted.

Permitted Colorants. The colorants listed in Tables 1–4 appear in the *Code of Federal Regulations* (21 CFR 70-82). The structures shown in Figures 1–3 are, in general, taken from the *Colour Index* (CI) and represent each colorant's principal component.

Colorants Exempt from Certification

The Commissioner of Food and Drug has the authority to exempt particular color additives from the batch-certification procedure when it is believed that, because of their nature, certification is not needed to protect the public health. Although exempt colorants need not be certified prior to their sale, they are subject to surveillance by FDA to ensure that they meet current government specifications and that they are used in accordance with the law.

With the passage of the 1960 amendments, all exempt colorants then in use were provisionally listed pending completion of the studies needed to obtain their permanent listing. Since that time, most of them as well as several completely new colors have achieved this status. Exempt color additives now in use and their status are shown in Tables 1–4.

Exempt colorants are made up of a wide variety of organic and inorganic compounds representing the animal, vegetable, and mineral kingdoms. Some, like β-carotene and zinc oxide, are essentially pure factory-produced chemicals of definite and known composition. Others, including annatto extract, cochineal extract, caramel, and beet powder are mixtures obtained from natural sources and have somewhat indefinite compositions.

In general, exempt colorants have less coloring power than certified colorants and thus have to be used at higher concentrations. Some, particularly those of plant origin, tend to be less stable, more variable in shade, and therefore more complicated to use than certified colorants, and are more likely to introduce undesirable flavors and odors into the products in which they are incorporated. Also, depending on their nature and origin, exempt colorants can vary substantially in composition from batch to batch, are more likely to be contaminated with undesirable trace metals, insecticides, herbicides, and bacteria such as *Salmonella*, and can be more difficult to obtain in steady supply compared with certified colorants.

Exempt colorants are inherently neither more nor less safe than certified colorants. However, they are viewed as having been obtained from nature (natural) (43–45) and thus imagined as less of a health hazard than certified colorants. In fact, like all color additives, they are fabricated products.

Annatto Extract. The annatto tree (*Bixa orellana*) is a large, fast-growing shrub cultivated in tropical climates, including parts of South America, India, East Africa, and the Caribbean. The tree produces large clusters of brown or crimson capsular fruit containing seeds coated with a thin, highly colored resinous coating or marc that serves as the raw material for preparation of the colorant known as annatto extract [*8015-67-6*].

The colorant is prepared by leaching the annatto seeds with an extractant prepared from one or more approved, food-grade materials taken from a list that includes various solvents, edible vegetable oils and fats, and alkaline aqueous and alcoholic solutions (46,47). Depending on the use intended, the alkaline extracts are often treated with food-grade acids to precipitate the annatto pigments, which in turn may or may not be further purified by recrystallization from an approved solvent. Annatto extract is one of the oldest known dyes, used since antiquity for the coloring of food, textiles, and cosmetics. It has been used in the United States and Europe for over 100 years as a color additive for butter and cheese (48–50).

The chief coloring principle found in the oil or fat extracts of annatto seeds is the carotenoid bixin (**36**, R = CH$_3$) (CI Natural Orange 4, CI No. 75120, [*6983-79-5*] EEC No. E 160b), which is the monomethyl ester of the dicarboxylic acid norbixin. The principal colorant in alkaline aqueous extracts is norbixin [*542-40-5*] (**36**, R = H) (51,52).

(**36**)

Annatto extract is sold in several physical forms, including dry powders, propylene glycol/monoglyceride emulsions, oil solutions and suspensions, and alkaline aqueous solutions containing anywhere from 0.1–30% active colorant calculated as bixin, C$_{25}$H$_{30}$O$_4$, or norbixin, C$_{24}$H$_{28}$O$_4$, as appropriate. It is used in products at 0.5–10 ppm as pure color, resulting in hues ranging from butter yellow to peach, depending on the type of color preparations employed and the product colored. Annatto extract's chief use is in foods such as butter, margarine, processed cheeses, nondairy creamers, cooking oils, salad dressings, cereals, ice cream, ice cream cones, sausage casings (53), bakery goods, snack foods, and spices. It is often used in combination with turmeric.

The chemistry and performance of annatto extract is essentially that of bixin, a brownish red crystalline material that melts at 198°C. It is moderately stable toward light and has good stability toward oxidation, change in pH, and microbiological attack. Bixin is very stable toward heat up to 100°C, fairly stable at 100–125°C, and unstable above 125°C, where it tends to form 13-carbomethoxy-4,8-dimethyltridecahexanoic acid.

β-Carotene. β-Carotene (**37**) [*7235-40-7*] is an isomer of the naturally occurring carotenoid, carotene (CI Food Orange 5, CI No. 40800, CI Natural Yellow 26, CI No. 75130, EEC No. E 160a). It is the pigment largely responsible for the color of various products obtained from nature including butter, cheese, carrots, alfalfa, and certain cereal grains. The colorant is synthetically produced from acetone, using the process developed in the 1950s by Hoffmann-LaRoche Inc., which results in the formation of the optically inactive all-trans form (Fig. 4). This synthesis made β-carotene very important in the history of the use of color additives because it was one of the first natural colorants synthetically produced on a commercial scale and the one that eventually raised the question as to whether factory-produced analogues of natural colorants should require certification by the FDA such as coal-tar dyes do, and whether such compounds could continue

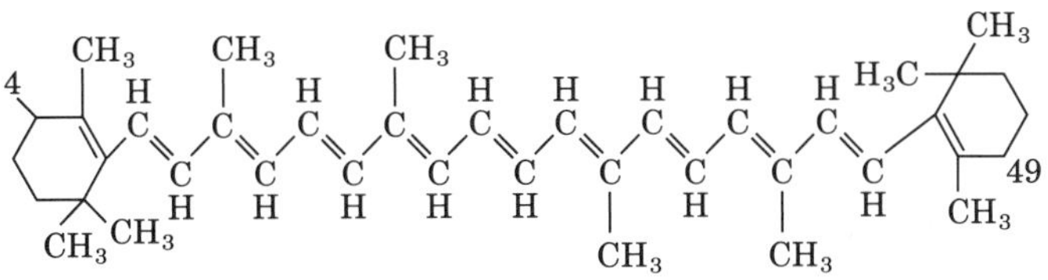

Fig. 4. Carotenoid pigments: β-carotene (**37**), β-apo-8'-carotenal (**38**), and canthaxanthin (**39**) = structure (**37**) with ketone groups at the 4 and 4' positions.

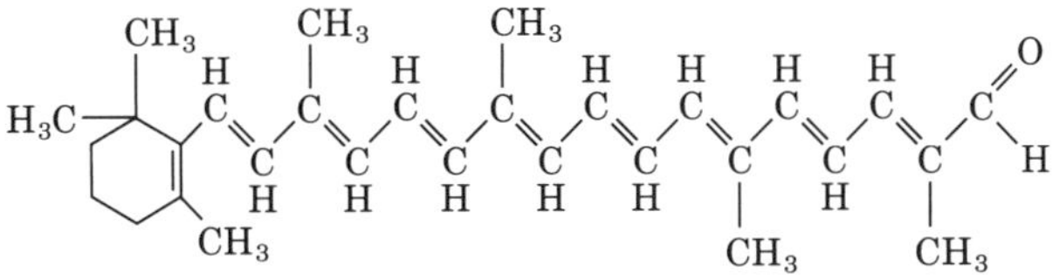

to be referred to as natural colors. This controversy eventually led to the creation of the category of colorants called colorants exempt from certification.

β-Carotene forms reddish violet platelets that melt in the range 176–182°C. It is insoluble in water, ethanol, glycerol, and propylene glycol, and only slightly soluble in boiling organic solvents such as ether (0.05%), benzene (0.2%), carbon disulfide (1%), and methylene chloride (0.5%). Its solubility in edible oils is about 0.08% at room temperature, 0.2% at 60°C, and 0.8% at 100°C. β-Carotene is sensitive to alkali and very sensitive to air and light, particularly at high temperatures (54,55). Pure, crystalline β-carotene remains unchanged for long periods of time when stored under CO_2 below 20°C but is almost completely destroyed after only six weeks when stored in air at 45°C. Vegetable fat and oil solutions and suspensions are quite stable under normal handling conditions. β-Carotene is one of the rare color additives with nutritional value; it is converted biologically by humans into vitamin A; 1 g of β-carotene = 1,666,666 USP units of vitamin A (see VITAMINS).

β-Carotene is marketed as dry crystals packed under nitrogen, as a dry water-dispersible powder containing about 1% β-carotene, dextrin, gum acacia, partially hydrogenated vegetable oil, sucrose, sodium ascorbate, and DL-alpha tocopherol, as liquid and semisolid suspensions in edible oils including vegetable, peanut, and butter oils, as water-dispersible beadlets composed of colorant plus vegetable oil, sugar, gelatin, and carbohydrate, and as emulsions (56,57).

The colorant is used at 2–50 ppm as pure color to shade margarine, shortening, butter, cheese (49), baked goods, confections, ice cream, eggnog, macaroni products, soups, juices, and beverages (58). Its chief advantages over other colorants are its nutritional value and its ability to duplicate natural yellow to orange shades.

β-Apo-8'-carotenal. The specifications of this colorant (**38**) were discussed earlier. β-Apo-8'-carotenal has provitamin activity with 1 g of the colorant equal to 1,200,000 IU of vitamin A. Like all crystalline carotenoids, it slowly decomposes in air through oxidation of its conjugated double bonds and thus must be stored in sealed containers under an atmosphere of inert gas, preferably under refrigeration. Also like other carotenoids β-apo-8'-carotenal readily isomerizes to a mixture of its cis and trans stereoisomers when its solutions are heated to about 60°C or exposed to ultraviolet light (59).

In general, its solubility characteristics are similar to those of β-carotene except that it is slightly more soluble in the usual solvents. In addition, because of its aldehydic group, β-apo-8'-carotenal is slightly soluble in polar solvents such as ethanol. Vegetable oil solutions of the colorant are orange to red, depending on their concentration. Aqueous dispersions range in hue from orange to orange–red.

β-Apo-8'-carotenal is sold as a dry powder, as 1–1.5% vegetable oil solutions, as 20% suspensions in vegetable oil, as 2–4% solutions in a mixture of monoglycerides and DL-α-tocopherol, and as 10% dry beadlets. The vegetable-oil suspensions are purplish black fluids at room temperatures that set to thick pastes when refrigerated. The dry beadlets are colloidal dispersions of colorant in a matrix of gelatin, vegetable oil, sugar, starch, and antioxidants (55,56).

β-Apo-8'-carotenal is used wherever an orange to reddish orange shade is desired. The dry beadlets are water-dispersible and can be used to color aqueous-based foods and beverages such as juices, fruit drinks, soups, jams, jellies, and gelatins. The vegetable-oil solutions and suspensions are most useful in fat base or fat-containing foods including processed cheese, margarine, salad dressings, fats, and oils. It is used in the range 1–20 ppm as pure color (54).

Canthaxanthin.   The newest of the synthetically produced carotenoid color additives, canthaxanthin [514-78-3] (**39**) (β-carotene-4,4'-dione), became commercially available about 1969 (60). Its CI designation is Food Orange 8, CI No. 40850. Its EEC designation is E 160g.

Unknown until 1950 when F. Haxo isolated it from an edible mushroom (*Cantharellus cinnabarinus*), canthaxanthin has since been identified in sea trout, algae, daphnia, salmon, brine shrimp, and several species of flamingo. Crystalline canthaxanthin is prepared synthetically from acetone or β-ionone using procedures similar to those used for β-carotene and β-apo-8'-carotenal (55).

Canthaxanthin crystallizes from various solvents as brownish violet, shiny leaves that melt with decomposition at 210°C. As is the case with carotenoids in general, the crystals are sensitive to light and oxygen and, when heated in solution or exposed to ultraviolet light or iodine, form a mixture of cis and trans stereoisomers. Consequently, crystalline canthaxanthin should be stored under inert gas at low temperatures. Unlike the carotenoid colorants β-carotene and β-apo-8'-carotenal, canthaxanthin has no vitamin A activity. It is chemically stable at pH 2–8 (the range normally encountered in foods) and unaffected by heat in systems with a minimal oxygen content.

The solubility of canthaxanthin in most solvents is low compared with β-carotene and β-apo-8'-carotenal. Oil solutions of canthaxanthin are red at all concentrations. Aqueous dispersions are orange or red depending on the type of emulsion prepared.

Besides as a dry powder, canthaxanthin is commercially available as a water-dispersible, dry beadlet composed of 10% colorant, gelatin, vegetable oil, sugar, starch, antioxidants, and preservatives (56). Canthaxanthin is used at 5–60 ppm as pure color to produce a tomato red. The colorant is useful in coloring tomato products such as tomato soup, spaghetti sauce, and pizza sauce, Russian and French dressings, fruit drinks, sausage products, and baked goods (54).

Caramel. Officially, the color additive caramel is the dark brown liquid or solid material resulting from the carefully controlled heat treatment of the following food-grade carbohydrates: dextrose, invert sugar, lactose, malt syrup, molasses, starch hydrolysates and fractions thereof, or sucrose. Practically speaking, caramel is burned sugar.

Caramel (CI Natural Brown 10, EEC No. E 150) is most often made from liquid corn syrup with a reducing sugar content of 60% or more, expressed as dextrose. Sucrose (cane sugar) is less frequently used because of its relatively high cost and because after inversion, dextrose and levulose react at different rates, making the burning process difficult to control, sometimes resulting in a product inferior to that made from corn sugar. In most cases a small amount of an approved acid, alkali, or salt is used to expedite the reaction and to obtain products with specific properties for specific applications.

To prepare caramel, corn syrup and the appropriate reactants are cooked at about 121°C for several hours or until the proper tinctorial power has been obtained. The product is then filtered and stored cool to minimize further caramelization. Often it is drum- or spray-dried to produce free-flowing powders containing 5% or less moisture (61,62).

Because of the many variables in ingredients and process conditions involved in manufacture, caramel's exact chemical composition is unknown. Caramel coloring is freely soluble in water and insoluble in most organic solvents. Its solubility in solutions containing 50–70% alcohol varies with the type of caramel. In concentrated form the colorant has a distinctive burned taste that is unnoticeable at the typical levels of use. The specific gravity of caramel coloring syrups ranges from 1.25 to 1.38; the total solids content varies from 50 to 75%. The pH of the acid-proof caramels used for carbonated beverages and acidified solutions is normally 2.8–3.5. Most bakers' caramels, which are a less refined grade of colorant used for cookies, cakes, bread, and so on, have a higher pH because of differences in their manufacturing processes (63).

In aqueous solution, caramel coloring exhibits colloidal properties; the particles carry small positive or negative electrical charges, depending on the method used in its manufacture and the pH of the product being colored. The nature of this charge is most important in using caramel since it must be the same as that of the product it is added to, or else mutual attraction will occur causing flocculation or precipitation. A good soft drink caramel should carry a strong negative charge and have an isoelectric point at pH 1.5 or less. Beer caramel usually has a positive charge.

Seventy-five to eighty-five percent of the caramel produced in the United States is used in soft drinks, particularly root beers and colas (see CARBONATED BEVERAGES). Caramel is also used extensively to standardize the hue of blended whiskeys, liqueurs, wines, and beer (qv) (see BEVERAGE SPIRITS, DISTILLED). Other uses include the coloring of baked goods, syrups, preserves, candies, pet foods, gravies, canned meat products, soups, condiments, vinegars, dark sugars, cough syrups, and pharmaceuticals (64). Where the use of liquid coloring is impractical, such as in cake mixes and other dry products, powdered caramel is added. Typical use levels are high (0.1–30%), but the colorant is relatively inexpensive and shows good stability in most products.

Shades that can be produced using caramel colorants range from delicate yellows to reds to the darkest browns.

Cochineal Extract. Cochineal extract (CI Natural Red 4, CI No. 75470, EEC No. E 120) is the concentrated solution obtained after removing the alcohol from an aqueous-alcoholic extract of cochineal, which is the dried bodies of the female insect *Coccus cacti (Dactylopius coccus costa)*, a variety of field louse. The coloring principle of the extract is believed to be carminic acid [*1260-17-9*] (**40**), an hydroxyanthraquinone linked to a glucose unit, comprising approximately 10% of cochineal and 2–4% of its extract.

(40)

Carmine [*1390-65-4*] is the aluminum or calcium–aluminum lake on an aluminum hydroxide substrate of the coloring principle (again, chiefly carminic acid) obtained by the aqueous extraction of cochineal. Carmine is normally 50% or more carminic acid.

The cochineal insect lives on a species of cactus (*Nopalea coccinelliferna*) and was once known only in Mexico. The Aztecs cultivated it for its color value and often exacted it as tribute. It is believed that Cortez found native Mexicans using cochineal in 1518 and at first believed it to be kermes, an ancient dyestuff widely used in Europe at the time. The eventual discovery that cochineal was in fact a new colorant, and one 10 times stronger than kermes, gave the Spaniards an exclusive on an important and lucrative article of commerce. By the end of the sixteenth century, as much as 230,000 kg of cochineal were being shipped from Mexico to Spain each year, a rather astounding figure considering that it requires over 150,000 hand-gathered insects to make a single kg of cochineal. Numerous attempts were made to raise the cochineal beetle in other areas of the world, but most failed partly because of the specialized climates needed for its cultivation and partly to the Spaniards' doggedness in guarding their monopoly. In spite of these obstacles, cochineal was eventually produced elsewhere, including the Canary Islands, Spain, the East and West Indies, Palestine, and parts of Central and South America. The cochineal trade peaked about 1870 then declined rapidly because of the introduction of synthetic colors in 1856.

Cochineal extract is typically acid (pH 5–5.3) and has a total solids content of about 6%. It frequently contains sodium benzoate as a preservative. Cochineal extract varies in shade from orange to red, depending on pH. It is insoluble in typical solvents including water, glycerol, and propylene glycol but can be dispersed in water. It exhibits good stability toward light and oxidation but poor stability toward pH and microbiological attack. Its tinctorial strength is only moderate. Use levels range from 25–1000 ppm.

Carmine is a pigment and thus exhibits little solubility in most solvents. Since it is also an aluminum lake, it can be solubilized by strong acids and bases that cause degradation of the substratum and release of the color. Both colorants are useful for producing pink shades in retorted meat products, candy, confections, aperitif alcoholic and soft drinks, cider, vinegar, yogurts, ice creams, baked goods, jams, jellies, rouge, eye shadow, and pill coatings (51).

Dehydrated Beets. This color additive is defined as a dark red powder prepared by dehydrating sound, mature, good quality, edible beets.

Beet roots contain both red pigments (betacyanins) and yellow pigments (betaxanthins), known collectively as betalains. Generally, the betacyanin content of beets far exceeds that of the betaxanthins. Of the betacyanins present, 75–95% is betanin [7659-95-2] (**41**) (EEC No. E 162), making it the principal pigment in beet colorant.

(**41**)

Although many factors influence the actual quantity of pigment present in beet tissue, the average amount has been estimated as 1000 mg/100 g of total solids, or 120 mg/100 g of fresh weight.

Beet extract is also used as a colorant. Extract is sold as either a concentrate prepared by evaporating beet juice under vacuum to a total solids content of 40–60%, or as a powder made by spray-drying the concentrate. Both products usually contain ascorbic or citric acid as a stabilizer, and a preservative such as sodium propionate. On a dry-weight basis, beet extract typically contains between 0.4 and 1.0% betanin, 80% sugar, 8% ash, and 10% crude protein.

Beet colorant readily dissolves in water and water-based products. It is reasonably stable when used from pH 4 to pH 7, and it is adequately light stable. However, beet colorant does degrade readily at temperatures as low as 50°C, particularly when exposed to air or light. It is most stable to heat in the range of pH 4.0–5.0. Because of the carbohydrates present in beet colorant, it tends to carry the natural flavor of beets.

Alone, beet colorant produces hues resembling raspberry or cherry. When used in combination with water-soluble annatto, strawberry shades result.

Beet colorant is best used in foods with short shelf lives that do not require high or prolonged heat treatment. When heat treatment is necessary, degradation of the colorant is minimized by adding it after the heat treatment, or as near the end of the heating cycle as possible. Beet colorant has been used successfully to color such products as hard candies, yogurts, ice creams, salad dressings, ready-made frostings, cake mixes, meat substitutes, powdered drink mixes, gravy mixes, marshmallow candies, soft drinks, and gelatin desserts. Typically, the colorant is added at 0.1–1%, based on the weight of the final product.

Grape Color Extract and Grape Skin Extract.   Grape color extract (EEC No. E 163) is an aqueous solution of anthocyanin grape pigments made from Concord grapes or a dehydrated water-soluble powder prepared from the aqueous solution. The aqueous solution is prepared by extracting the pigments from precipitated lees produced during the storage of Concord grape juice. It contains the common components of grape juice, namely anthocyanins, tartrates, malates, sugars, and minerals, etc, but not in the same proportion as found in grape juice. The dehydrated water-soluble powder is prepared by spray drying the aqueous solution containing added malto-dextrin.

The purple color of grape color extract is a result of the presence of water-soluble pigments, mainly the 3-mono- and 3,5-di-glucosides of malvidin, delphinidin, and cyanidin, and their acylated derivatives. Colorant stability is greatest below pH 4.5. Colorant intensity increases as pH falls. Grape color extract is stable to light and at temperatures adequate for canning most fruit. It is affected by oxygen, by SO_2 concentrations greater than 150 ppm, and by metal ions, especially tin, iron, and aluminum, which can complex with anthocyanins to produce a bluer color. Complexation can be controlled somewhat by the addition of metal sequestrants, such as pyrophosphates, EDTA, and citrate. Ascorbic acid appears to improve color stability by acting as an oxygen scavenger.

Grape color extract is used to color such products as bakers jams, nonstandard jellies and preserves, sherbets, ices, pops, raspberry, grape and strawberry yogurts, gelatin desserts, canned fruit, fruit sauces, candy and confections, and bakery fillings and toppings. Typical use concentrations are 0.05 to 0.8%, based on the weight of the finished product.

Grape skin extract (enocianina) is a purplish red liquid prepared by the aqueous extraction (steeping) of the fresh deseeded marc remaining after grapes have been pressed to produce grape juice or wine. It contains the common components of grape juice namely, anthocyanins, tartaric acid, tannins, sugars, minerals, etc, but not in the same proportions as found in grape juice. During the steeping process, sulfur dioxide is added and most of the extracted sugars are fermented to alcohol. The extract is concentrated by vacuum evaporation, during which practically all of the alcohol is removed. A small amount of sulfur dioxide may be present.

Typically, grape skin extract has a specific gravity of 1.13 g/mL at 20°C, a solids content of 28–32° Brix ($\pm 3°$), a pH of 3.0, and a color strength as anthocyanin of about 1.25% (as measured at 520 nm in pH 3.0 citrate buffer). Grape skin extract is also available as spray-dried powders with color values three to four times those of the liquid. The properties and uses of grape skin extract are similar to those of grape color extract.

Guanine (Pearl Essence).   Guanine (CI Natural White 1, CI No. 75170), is the crystalline material obtained from fish scales and consists principally of the two purines, guanine [73-40-5] (**42**) and hypoxanthine [68-94-0] (**43**). The guanine content of the colorant varies from 75% to 97%, whereas the hypoxanthine content ranges from 3% to 25%, depending on the particular fish and tissue from which the crystals are derived.

(42)   (43)

Guanine is obtained from various fish including menhaden, herring, and alewives. To prepare the colorant, scales are scraped from the fish, levigated, and washed with water, and then made into one or more commercial forms, depending on the intended use. Typically, guanine is supplied as a paste or suspension in water, castor oil, or nitrocellulose. Guanine is not a colorant in the strict sense but instead is used to produce iridescence in a product.

The hue of the colorant varies greatly with the amount and type of pigment found in the fish scales. Carotenoids produce reds and yellows, melanin results in blacks, and combinations of guanine and melanin produce greens and blues. Only when guanine is found alone is the product silvery or pearly white.

Guanine is used in lipsticks, nail polishes, and eye makeup.

Paprika and Paprika Oleoresins.   Paprika is the deep red, sweet, pungent powder prepared from the ground, dried pod of mild capsicum (*Capsicum annum*). It is one of the two principal kinds of red pepper; the other is cayenne. Paprika is produced in large quantities in Hungary and is also available from many warm-climate areas, including Africa, Spain, and the American tropics. The chief classifications of paprika are Hungarian paprika, which has the pungency and flavor characteristics of that produced in Hungary (Rosenpaprika and Koenigspaprika), and Spanish paprika (pimenton, pimiento).

Paprika oleoresin (EEC No. E 160c) is the combination of flavor and color principles obtained by extracting paprika with any one or a combination of approved solvents: acetone, ethyl alcohol, ethylene dichloride, hexane, isopropyl alcohol, methyl alcohol, methylene chloride, and trichloroethylene. Depending on their source, paprika oleoresins are brown–red, slightly viscous, homogeneous liquids, pourable at room temperature, and containing 2–5% sediment.

The oleoresins are available in various standardized forms in which 1 kg of oleoresin is equal to 10–30 kg of paprika. Paprika oleoresins are typically standardized by dilution with vegetable oil or mono- or diglycerides.

Paprika and its oleoresin are approved for use in foods in general where its application as a color additive frequently overlaps its use as a spice. Both products have good tinctorial strength and are used at 0.2–100 ppm to produce orange to bright red shades.

Saffron.   Saffron, known also as CI Natural Yellow 6 (CI No. 75100), safran, crocine, crocétine, and crocus, is the dried stigma of *Crocus sativus*, a plant indigenous to the Orient but also grown in North Africa, Spain, Switzerland, Greece, Austria, and France. It is a reddish brown or golden yellow odoriferous powder having a slightly bitter taste. The stigmas of approximately 165,000 blossoms are required to make 1 kg of colorant (65).

The coloring principles of saffron are crocin [*42553-65-1*] and crocetin [*27876-94-4*]. Crocin is the gentiobiose diester of crocetin (**44**)

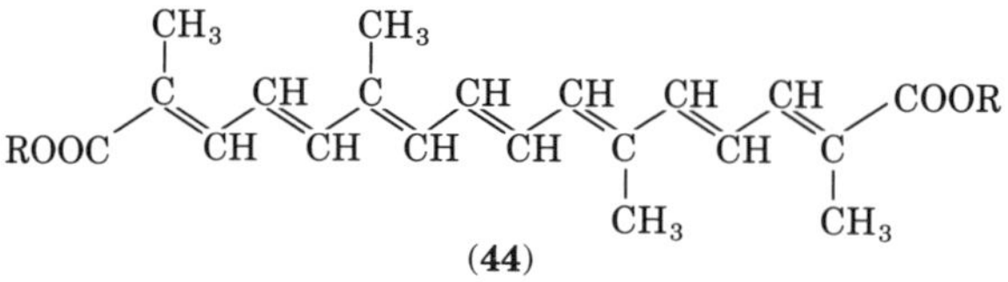

(**44**)

where R = H for crocetin and R = gentiobiose for crocin.

Crocin is a yellow—orange glycoside that is freely soluble in hot water, slightly soluble in absolute alcohol, glycerol, and propylene glycol, and insoluble in vegetable oils. Crocin melts with decomposition at about 186°C and has absorption maxima in methanol at about 464 nm and 434 nm.

Crocetin is a dicarboxylic acid that forms brick red rhombs from acetic anhydride that melt with decomposition at about 285°C. It is very sparingly soluble in water and most organic solvents but soluble in pyridine and similar organic bases as well as in dilute sodium hydroxide.

As a food colorant, saffron shows good overall performance (51). In general, it is stable toward light, oxidation, microbiological attack, and changes in pH. Its tinctorial strength is relatively high, resulting in use at 1–260 ppm.

Turmeric and Turmeric Oleoresin.   Turmeric (CI Natural Yellow 3, CI No. 75300, EEC No. E 100) is the dried and ground rhizome or bulbous root of *Curcuma longa*, a perennial herb of the Zingiberaceae family native to southern Asia and cultivated in China, India, South America, and the East Indies. It is a yellow powder with a characteristic odor and a sharp taste.

Turmeric oleoresin is the combination of flavor and color principles obtained from turmeric by extracting it with one or a combination of the following solvents: acetone, ethyl alcohol, ethylene dichloride, hexane, isopropyl alcohol, methyl alcohol, methylene chloride, and trichloroethylene.

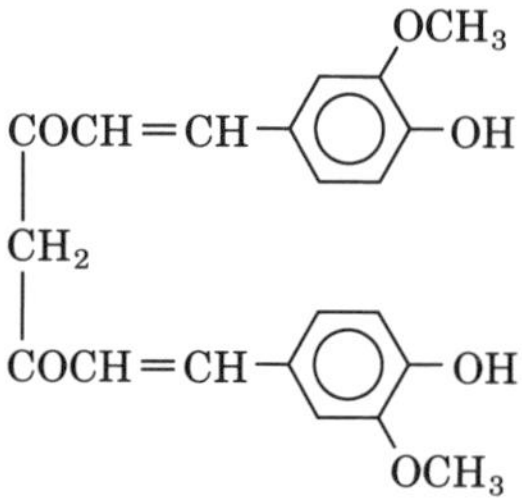

(**45**)

The principal coloring matter in turmeric and its oleoresin is curcumin [*458-37-7*] (1,6-heptadiene-3,5-dione, 1,7-bis[4-hydroxy-3-methoxy phenyl] (**45**), an orange-yellow, crystalline powder, insoluble in water and ether but soluble in ethanol and glacial acetic acid. It has a reported melting point of 180–183°C.

Turmeric is available in various powdered forms, some containing as much as 90–95% curcumin, and as suspensions in a variety of carriers, including edible vegetable oils and fats, and mono- and diglycerides, most containing 2–6% curcumin. Turmeric oleoresin is most often sold as solutions in propylene glycol with or without added emulsifying agents, typically containing 20–25% curcumin. Both products exhibit poor to moderate stability to light, oxidation, and change in pH but good tinctorial strength. Turmeric is typically used at 0.2–60 ppm, whereas the oleoresin use is higher, 2–640 ppm. Both are used alone or in combination with other colorants such as annatto to shade pickles, mustard, spices, margarine, ice creams, cheeses, pies, cakes, candies, soups, cooking oils, and salad dressings (50). Turmeric and its oleoresin produce bright yellow to greenish yellow shades, and are often used as replacements for FD&C Yellow No. 5.

Chromium Hydroxide Green. This color additive is principally hydrated chromic sesquioxide [12182-82-0], $Cr_2O_3 \cdot xH_2O$ (CI Pigment Green 18, CI No. 77289, Veridian, Guignet's Green).

It is prepared by pasting potassium or sodium dichromate with three times its weight of boric acid, roasting the mixture at 500°C in a muffle furnace in an oxidizing atmosphere, then hydrolyzing the melt with water and superheated steam. The product is then dried and ground.

Chromium hydroxide green is a more bluish and brilliant green than chromium oxide greens. It is quite transparent, and has good strength and excellent stability. It is used in eye makeup and soap.

Chromium Oxide Green. Chromium oxide green is principally chromic sesquioxide [1308-38-9], Cr_2O_3 (CI Pigment Green 17, CI No. 77288). It is usually prepared by fusing potassium dichromate and boric acid, drowning the product in water, and then drying it at high temperature; or precipitating chrome alum with sodium hydroxide, roasting the resulting chromous hydroxide, then extracting, washing, then drying it at a high temperature.

Cr_2O_3 is a yellowish (sage) green pigment. It has good strength and opacity and excellent stability. It is used in eye makeup and soap.

Synthetic Iron Oxide. This colorant is one or a combination of various synthetically prepared iron oxides, including the hydrated forms. The naturally occurring oxides are unacceptable as color additives because of the difficulties frequently encountered in purifying them.

Iron oxide (EEC No. E 172) is recognized under various names, including CI Pigment Black 11 and CI Pigment Browns 6 and 7 (CI No. 77499), CI Pigment Yellows 42 and 43 (CI No. 77492), and CI Pigment Reds 101 and 102 (CI No. 77491). The chemical composition and hence the empirical formula of the colorant varies greatly with the method of manufacture used but can generally be represented as $FeO \cdot xH_2O$, $Fe_2O_3 \cdot xH_2O$, or some combination thereof. Most are made from copperas (ferrous sulfate, $FeSO_4 \cdot 7H_2O$). The commonly used forms are the yellow hydrated oxides (ochre) and the brown, red, and black oxides.

The yellow oxides are prepared by precipitating hydrated ferric oxide from a ferrous salt using an alkali, followed by oxidation. The shades obtained range from light lemon yellow to orange, depending on the conditions used for the precipitation and oxidation. Yellow oxides contain about 85% Fe_2O_3 and 15% water of hydration.

Brown oxides are manufactured either by blending mixtures of the red, yellow, and black oxides or by precipitation of an iron salt with alkali followed by partial oxidation of the precipitate. The result is a mixture of red Fe_2O_3 [1309-37-1] and black Fe_3O_4 [1309-38-2], $FeO \cdot Fe_2O_3$.

Red iron oxides are usually prepared by calcining the yellow oxides to form Fe_2O_3. The shade of the red oxide depends on the characteristics of the original yellow pigment, and the conditions of calcination and ranges from light to dark red. The product is 96–98.5% Fe_2O_3.

The black oxides are prepared by the controlled precipitation of Fe_3O_4 (treat $FeSO_4 \cdot 7H_2O$ with NaOH and O_2) to form a mixture of ferrous and ferric oxides.

Iron oxides are stable pigments insoluble in most solvents but usually soluble in hydrochloric acid. Those not soluble in HCl can be fused with potassium hydrogen sulfate, $KHSO_4$, and then dissolved in water.

The principal use of iron oxide as a colorant is in cosmetics, particularly eye makeup and face powders. It is also permitted in dog and cat food at concentrations not exceeding 0.25% by weight of the finished food, and in drugs.

Talc.   Talc (qv) [14807-96-6] (CI Pigment White 26, CI No. 77019), is finely powdered, native, hydrous magnesium silicate, $3MgO \cdot 4SiO_2 \cdot H_2O$ (soapstone) sometimes containing a small amount of aluminum silicate. It is produced in many parts of the world, including France, Italy, India, and the United States.

Theoretically, talc is a pure white, odorless, unctuous powder rated as among the softest materials available, assigned a hardness of No. 1–1.5 on the Mohs' Mineralogical Scale. Actually, it is a white-gray powder possessing varying amounts of softness and slip, depending on its origin. The best grades of talc are very white crystalline powders with a lamellar structure, a greasy feel, and a particle size of 74 μm or less. Micronized talcs are often 40 μm or less in size. The specific gravity of talc is about 2.70.

Titanium Dioxide.   The specifications of titanium dioxide have been given previously. Titanium dioxide exists in nature in three crystalline forms: anatase, brookite, and rutile, with anatase as the commonly available form. Anatase has a high refractive index (2.52) and excellent stability toward light, oxidation, changes in pH, and microbiological attack. Titanium dioxide is virtually insoluble in all common solvents.

Only synthetically prepared TiO_2 can be used as a color additive. It is permitted in foods to 1% and is used to color such products as confectionary panned goods, cheeses, and icings. It is also widely used in tableted drug products and in numerous cosmetics such as lipsticks, nail enamels, face powder, eye makeup, and rouges, in amounts consistent with good manufacturing practice (42).

The colorant's chief disadvantages are its inability to blend well with the other ingredients usually found in powder formulations, its tendency to produce blue undertones, and its ability to catalyze the oxidation of perfumes.

Ultramarines.   The ultramarines are synthetic, inorganic pigments of somewhat indefinite composition. Basically, they are sodium aluminosulfosilicates with crystal structures related to the zeolites and empirical formulas that can be approximated as $Na_7Al_6Si_6O_{24}S_3$ (see MOLECULAR SIEVES). They are intended as the duplicate of the colorants produced from the naturally occurring semiprecious gem lazurite (Lapis lazuli). Their color is believed a result of polysulfide linkages in a highly resonant state.

Ultramarines are manufactured by the heat-treating and then very slow cooling of various combinations of kaolin (China clay), silica, sulfur, soda ash, and sodium sulfate plus a carbonaceous reducing agent such as rosin or charcoal pitch. The formulation of ingredients, temperature, time, cooling rate, subsequent treatment, and other variables determines the resultant color. Firing temperatures range from 700–800°C, whereas firing times vary from a few to as many as 150 h.

The basic product of the ignition is Ultramarine Green. This is converted into Ultramarine Blue by further heat treatment in the presence of sulfur, or into Ultramarine Violet by heating with 5% ammonium chloride for four days at 200–250°C. Ultramarine Violet is converted into Ultramarine Red by treating it with gaseous hydrochloric acid at 70–200°C for four hours or by reaction with gaseous nitric acid at higher temperatures.

Ultramarines are insoluble in water and organic solvents but soluble in acids, which cause their discoloration and the liberation of hydrogen sulfide. They have excellent permanency and resistance to alkali but poor tinting and hiding power.

Ultramarine Blue is used in salt intended for animal feed ($\leq 0.5\%$ w/w). All ultramarines are used in the cosmetic field in such products as mascara, eyebrow pencils, and soaps.

Zinc Oxide. Of all the white pigments used in the cosmetic field, zinc oxide ranks among the most important. Although it does not have the hiding power of colorants such as titanium dioxide, zinc oxide has certain advantages, including its brightness, ability to provide opacity without blue undertones, adhesiveness or stick, and therapeutic properties, as it is mildly antiseptic and has drying and healing effects on the skin.

Zinc oxide [*1314-13-2*] (mol wt 81.37; CI Pigment White 4, CI No. 77947) is a white or yellowish white amorphous, odorless powder with pH 6.95–7.37. It is practically insoluble in water but soluble in dilute acetic acid, mineral acids, ammonia, ammonium carbonate, and alkali hydroxides.

As a colorant, zinc oxide is used in face powders, rouges, and eye makeups at levels of 5–30%.

Miscellaneous Colorants. Other colorants not requiring certification have been defined in the *Code of Federal Regulations*. Most of these are of only minor to moderate importance and have only limited usage.

Alumina—A white, odorless, tasteless, amorphous powder consisting essentially of aluminum hydroxide, $Al_2O_3 \cdot xH_2O$ (see ALUMINUM COMPOUNDS, ALUMINUM OXIDE).

Aluminum powder—CI Pigment Metal 1, CI No. 77000, EEC No. E 173. Finely divided particles of aluminum prepared from virgin aluminum [*7429-90-5*]. It is free from admixture with other substances.

Bismuth citrate [*813-93-4*]—The synthetically prepared crystalline salt of bismuth and citric acid, principally $BiC_6H_5O_7$.

Bismuth oxychloride [*7787-59-9*]—CI Pigment White 14, CI No. 77163. A synthetically prepared white or nearly white amorphous or finely crystalline, odorless powder consisting principally of BiOCl. Bismuth oxychloride is synthetic pearl essence. It is used in lipstick, nail polish, eye makeup, and other cosmetics to produce a lustrous, pearly effect.

Bronze powder [*12597-70-5*]—CI Pigment Metal 2, CI No. 77400. A very fine metallic powder prepared from alloys consisting principally of virgin electrolytic copper and zinc with small amounts of the virgin metals aluminum and tin. It contains small amounts of stearic or oleic acid as a lubricant.

Calcium carbonate [*471-34-1*]—CI Pigment White 18, CI No. 77220, EEC No. E 170. A fine, white, synthetically prepared powder consisting essentially of precipitated calcium carbonate, $CaCO_3$.

Carrot oil—The liquid or the solid portion of the mixture, or the mixture itself obtained by the hexane extraction of edible carrots (*Daucus carota* L.) with subsequent removal of the hexane by vacuum distillation. The resultant mixture of solid and liquid extractives consists chiefly of oils, fats, waxes, and carotenoids naturally occurring in carrots.

Chlorophyllin–copper complex, oil soluble—The chlorophyllin is obtained by extraction from a mixture of fescue and rye grasses. The chlorophyll is acid-treated to remove chelated magnesium that is replaced with hydrogen, which in turn is replaced with copper. This mixture is diluted to 5% concentration with a mixture of palm oil, peanut oil, and hydrogenated peanut oil.

Chromium–cobalt–aluminum oxide [*68187-11-1*]—CI Pigment Blue 36, CI No. 77343. A blue–green pigment obtained by calcining a mixture of chromium oxide, cobalt carbonate, and aluminum oxide. It may contain small amounts (<1% each) of oxides of barium, boron, silicon, and nickel.

Copper powder—CI Pigment Metal 2, CI No. 77400. A very fine free-flowing metallic powder prepared from virgin electrolytic copper [*7440-50-8*]. It contains small amounts of stearic or oleic acid as a lubricant.

Corn endosperm oil—A reddish brown liquid composed chiefly of glycerides, fatty acids, sitosterols, and carotenoid pigments obtained by isopropyl alcohol and hexane extraction from the gluten fraction of yellow corn grain.

Dihydroxyacetone—This colorant is 1,3-dihydroxy-2-propanone [*96-26-4*].

Dried algae meal—A dried mixture of algae cells (genus *Spongiococcum*, separated from its culture broth), molasses, cornsteep liquor, and a maximum of 0.3% ethoxyquin. The algae cells are produced by suitable fermentation, under controlled conditions, from a pure culture of the genus *Spongiococcum*.

Ferric ammonium citrate [*1185-57-5*]—A mixture of complex chelates prepared by the interaction of ferric hydroxide with citric acid in the presence of ammonia. The chelates occur in brown and green forms, are deliquescent in air, and are reducible by light.

Ferric ammonium ferrocyanide—The blue pigment obtained by oxidizing under acidic conditions with sodium dichromate the acid-digested precipitate resulting from mixing solutions of ferrous sulfate and sodium ferrocyanide in the presence of ammonium sulfate. The oxidized product is filtered, washed, and dried. The pigment consists principally of ferric ammonium ferrocyanide with small amounts of ferric ferrocyanide and ferric sodium ferrocyanide.

Ferric ferrocyanide—CI Pigment Blue 27, CI No. 77510. The color additive ferric ferrocyanide is a ferric hexacyanoferrate pigment characterized by the structural formula $Fe_4[Fe(CN)_6]_3xH_2O$, which may contain small amounts of ferric sodium ferrocyanide and ferric potassium ferrocyanide.

Ferrous gluconate [*299-29-6*] (**46**)—Fine yellowish gray or pale greenish yellow powder or granules having a slight odor resembling that of burned sugar. One gram dissolves in about 10 mL of water with slight heating. It is practically insoluble in alcohol. A 1:20 solution is acid to litmus.

$$\left[\text{HOCH}_2\text{CH} \underset{\substack{| \\ \text{OH}}}{-} \text{CHCHCHC} \underset{\substack{| \quad | \\ \text{OH} \ \ \text{OH}}}{} \overset{\substack{\text{OH} \\ |}}{} \overset{O}{\underset{O^-}{\diagup\!\!\!\diagdown}} \right]_2 \text{Fe} \cdot 2\text{H}_2\text{O}$$

(**46**)

Fruit juice—The concentrated or unconcentrated liquid expressed from mature varieties of fresh, edible fruits; or a water infusion of the dried fruit.

Guaiazulene—Principally 1,4-dimethyl-7-isopropyl-azulene.

Henna [*83-72-7*]—CI Natural Orange 6, CI No. 75480. The dried leaf and petiole of *Lawsonia alba* Lam (*Lawsonia inermis* L.) (66).

Lead acetate [*301-04-2*]—The trihydrate of the lead salt of acetic acid; $Pb(OOCCH_3)_2 \cdot 3H_2O$.

Logwood extract—A reddish brown-to-black solid material extracted from the heartwood of the leguminous tree *Haematoxylon campechianum*. The active colorant substance is principally hematein. The latent coloring material is the unoxidized or leuco form of hematein called hematoxylin. The leuco form is oxidized by air.

Manganese violet [*10101-66-3*]—CI Pigment Violet 16, CI No. 77742. A violet pigment obtained by reaction of phosphoric acid, ammonium dihydrogen orthophosphate, and manganese dioxide at temperatures above 232°C. The pigment is a manganese ammonium pyrophosphate complex having the approximate formula $Mn(III)NH_4P_2O_7$.

Mica [*12001-26-2*]—CI Pigment White 20, CI No. 77019. A white powder obtained from the naturally occurring mineral muscovite mica, consisting predominantly of a potassium aluminum silicate, [*1327-44-2*] $H_2KAl_3(SiO_4)_3$. Mica may be identified and semiquantitatively determined by its characteristic x-ray diffraction pattern and by its optical properties.

Poly(hydroxyethyl methacrylate)-dye copolymers—The color additives formed by reaction of one or more of the following reactive dyes with poly(hydroxyethyl methacrylate), so that the sulfate group (or groups) or chlorine substituent of the dye is replaced by an ether linkage to poly(hydroxyethyl methacrylate) (see DYES, REACTIVE). The dyes that may be used alone or in combination are

Reactive Black 5	[*17095-24-8*]
Reactive Blue 21	[*73049-92-0*]
Reactive Orange 78	[*68189-39-9*]
Reactive Yellow 15	[*60958-41-0*]

Reactive Blue 19	[2580-78-1]
Reactive Blue 4	[4499-01-8]
CI Reactive Red 11	[12226-08-3]
CI Reactive Yellow 86	[61951-86-8]
CI Reactive Blue 163	[72847-56-4]

Potassium sodium copper chlorophyllin (chlorophyllin–copper complex)—A green-black powder obtained from chlorophyll by replacing the methyl and phytyl ester groups with alkali and replacing the magnesium with copper. The source of the chlorophyll is dehydrated alfalfa.

Pyrogallol—This colorant is 1,2,3-trihydroxybenzene [87-66-1].

Pyrophyllite [12269-78-2]—A naturally occurring mineral substance consisting predominantly of a hydrous aluminum silicate, $Al_2O_3 \cdot 4SiO_2 \cdot H_2O$, intimately mixed with lesser amounts of finely divided silica, SiO_2. Small amounts (usually <3%) of other silicates, such as potassium aluminum silicate, may be present. Pyrophyllite may be identified and semiquantitatively determined by its characteristic x-ray powder-diffraction pattern and by its optical properties.

Riboflavin—A yellow or orangish yellow crystalline powder having a slight odor. Riboflavin [83-88-5] (**47**) melts at about 280°C, and its saturated solution is neutral to litmus. When dry, it is not affected by diffused light, but when in solution, light induces deterioration. One gram dissolves in about 3,000–20,000 mL of water, depending on the internal crystalline structure. It is less soluble in alcohol than in water. It is insoluble in ether and in chloroform but is very soluble in dilute solutions of alkalies. A solution of 1 mg in 100 mL of water is pale greenish yellow by transmitted light and has an intense yellowish green fluorescence that disappears on the addition of mineral acids or alkalies.

(**47**)

Silver [7440-22-4]—The color additive silver (EEC No. E 174) is a crystalline powder of high purity silver prepared by the reaction of silver nitrate with ferrous sulfate in the presence of nitric, phosphoric, and sulfuric acids. Poly(vinyl alcohol) is used to prevent the agglomeration of crystals and the formation of amorphous silver.

Tagetes meal and extract—Tagetes (Aztec marigold) meal is the dried, ground flower petals of the Aztec marigold (*Tagetes erecta* L.) mixed with not more than 0.3% ethoxyquin. Tagetes extract is a hexane extract of the flower petals of the Aztec marigold. It is mixed with an edible vegetable oil, or with an edible vegetable oil and a hydrogenated edible vegetable oil, and not more than 0.3% ethoxyquin. It may also be mixed with soy flour or corn meal as a carrier.

Toasted partially defatted cooked cottonseed flour—This product is prepared by delinting and decorticating food-quality cottonseed. The meats are screened, aspirated, and rolled; moisture is adjusted, the meats heated, and the oil expressed; the cooked meats are cooled, ground, and reheated to obtain a product varying in shade from light to dark brown.

Vegetable juice—The concentrated or unconcentrated liquid expressed from mature varieties of fresh, edible vegetables.

BIBLIOGRAPHY

"Colors for Foods, Drugs, and Cosmetics" in *ECT* 1st ed., Vol. 4, pp. 287–313, by S. Zuckerman, H. Kohnstamm & Co., Inc.; in *ECT* 2nd ed., Vol. 5, pp. 857–884, by S. Zuckerman, H. Kohnstamm & Co., Inc.; in *ECT* 3rd ed., Vol. 6, pp. 561–596, by S. Zuckerman and J. Senackerib, H. Kohnstamm & Co., Inc.

1. H. Lieber, *The Use of Coal-Tar Colors in Food Products*, H. Lieber & Co., New York, 1904. Interesting historically.
2. W. C. Bainbridge, *Ind. Eng. Chem.* **18,** 1329–1331 (1926). Development of the food color industry in the United States. Interesting historically.
3. H. O. Calvery, *Am. J. Pharm.* **114,** 324–349 (1942). Coal-tar colors, their use in foods, drugs, and cosmetics. Outdated but interesting historically.
4. H. J. White, Jr., ed., *Proceedings of the Perkin Centennial*, Sept. 10, 1956, New York, sponsored by the American Association of Textile Chemists and Colorists. Includes chapters on the use, properties, and reasons for using color additives in various foods, drugs, and cosmetics.
5. E. Corwin, *FDA Consumer*, 10–15 (Nov. 1976). Preventing Food Adulteration. Interesting background.
6. M. J. Dunn, *Paint Varnish Prod.*, 49–51 (Aug. 1973). Toxicity: Thorny Problem in Color Manufacturing. A few thoughts on colorant toxicity.
7. C. Calzolari, L. Coassini, and L. Lokar, *Quaderni Merceol.* **1,** 89–131 (1962). Synthetic Food Colors. Reviews the regulation of food colors in various countries, the toxicity of the intermediates used to prepare them, and the toxicity of the degradation products of colorants.
8. E. Corwin, *FDA Consumer*, 6–9 (Dec. 1978–Jan. 1979). Why FDA bans harmful substances.
9. G. E. Damon and W. F. Janssen, *FDA Consumer*, 15–21 (July–Aug. 1973). Additives for eye appeal. A little of the history and regulation of food colors.
10. *Food Colors*, National Academy of Sciences, Washington, D.C., 1971. A general treatment of food colors, including their history, use, regulation, safety, and properties.
11. S. H. Hochheiser, *Synthetic Foods Colors in the United States: A History Under Regulation*, University Microfilms International, 83-04269, Ann Arbor, Mich., 1986. An excellent history of the development of legislation to control colorants used in foods, drugs, and cosmetics.
12. A. Weissler, *Food Technol.*, 38, 46 (May 1975). FDA Regulation of Food Colors. Outdated but interesting.
13. H. Reynolds, H. Eiduson, J. Weatherwax, and D. Dechert, *Anal. Chem.* **44,** 22A–24A, 26A, 28A, 31A–34A (1972). FDA Chemistry for Consumers. A review of the history, current structure, and function of the Food and Drug Administration.

14. C. H. Hallstrom, H. G. Johnson, and W. J. Mayer, *Food Technol.*, 72–77 (Oct. 1978). A Food Scientist's Guide to Food Regulatory Information. Describes useful, published sources of regulatory information.

15. D. M. Marmion, *Handbook of U.S. Colorants, Foods, Drugs, Cosmetics and Medical Devises*, John Wiley & Sons, Inc., New York, 1991.

16. W. Janssen, *FDA Consumer*, 12–19 (June 1975). America's First Food and Drug Laws. Interesting background.

17. B. C. Hesse, *Coal-Tar Colors Used in Food Products*, Bureau of Chemistry, Bulletin No. 147, Feb. 10, 1912. Results of the Hesse study made at the turn of the century.

18. Public Law No. 717, 75th U.S. Congress, 3rd Session, S. 5, 1938.

19. U.S. Supreme Court. 358 U.S. 153, Dec. 15, 1958. The court ruling that established the harmless per se principle that a color additive had to be harmless regardless of the quantity used.

20. B. T. Hunter, *Consumer Bulletin*, 20–24 (May 1973). U.S. Certified Food Dyes—A look at the record of governmental failure to safeguard America's food products. A criticism of government's role in controlling the use of food colors.

21. R. W. Miller, ed., *FDA Consumer* (June 1981). Numerous articles regarding the fight by FDA for legislation to control the purity of foods, drugs, and cosmetics in the United States.

22. D. Blumenthal, *FDA Consumer*, 18–21 (May 1990). An example of the confusion that can be caused by the strict enforcement of the Delaney Clause.

23. "A Search for Safer Food Dyes," *Business Week*, (Feb. 21, 1977). Some thoughts on the future of the food color business.

24. T. E. Furia, ed., *Current Aspects of Food Colorants*, CRC Press, Inc., West Palm Beach, Fla., 1977. An update on food colorant technology.

25. R. Goto, *Yuki Gosei Kagaku Kyokai Shi* **24**, 493–500 (1966). Food Colors. A review of the kinds, properties, and applications of food colors.

26. F. Kasprzak and B. Glebko, *Chemik* **19**, 267–273 (1966). Dyes for Foods, Pharmaceuticals, and Cosmetics. Natural and synthetic dyes produced in Poland and other countries are described and compared.

27. J. Noonan, "Color Additives in Foods" in *Handbook of Food Additives*, The Chemical Rubber Co., Cleveland, Ohio, 1968, pp. 25–49. Food colors—their description, properties, regulation, and use.

28. A. I. Solodukhin, *Proizv. Isol'z Vitaminov, Antibiotikov Biol. Aktivn. Veshchestv*, 145–181 (1965). Production and Use of Food Dyes. A review of the synthetic and natural food dyes used in the Soviet Union.

29. S. W. Souci, *Z. Lebensm. Forsch.* **108**, 189–195 (1958). The Color Committee of the Deutsche Forschungsgemeinschaft. List of Pigments and Dyes for Cosmetics. Toxicological data on dyes and their suitability for food in various countries.

30. C. A. Vodoz, *Food Technol.* **24**, 42–53 (1970). International Food Additives in Europe.

31. J. Walford, ed., *Developments in Food Colours*, Vols. 1 and 2. Elsevier Applied Science, London, 1980 and 1984. Includes chapters on synthetic and natural food colors used in the United States, and on the influence of color on the perception and choice of food.

32. R. H. O'Holla and F. M. Penta, *MD&DI*, 39–44, 79 (Nov. 1981). Color Additives for Drugs and Medical Devices.

33. C. J. Swartz, and J. J. Cooper, *Pharm. Sci.* **51**, 89–99 (1962). Colorants for Pharmaceuticals. A general review of the colorants and their properties and uses.

34. *Color Additives Guide*, The Pharmaceutical Manufacturers Association, Washington, D.C. A listing of the dyes and pigments permitted in 44 countries and the European Economic Community.

35. E. R. Barnhart, ed., *Physicians' Desk Reference*, Medical Economics Co. Inc., Oradell, N.J., 1989. Includes a guide to the identification of drugs by color and shape.

36. D. F. Anstead, in *A Handbook of Cosmetic Science*, Pergamon Press, New York, 1963, pp. 101–118. A brief description of colors used in cosmetics.
37. D. F. Anstead, *J. Soc. Cosmet. Chemists* **10,** 1–20 (1959). A general review, including regulations in the United States and Great Britain.
38. G. R. Clark, *Proc. Sci. Sect. Toilet Goods Assoc.* **35,** 24–25 (1961). Outdated but interesting historically.
39. E. B. Faulkner, *Cosmet. Toilet.* **104,** 29 (1989). Thoughts on the use of color additives in cosmetics.
40. E. Sagarin, ed., *Cosmetics—Science and Technology*, Interscience Publishers, New York, 1957. A good history of the development and use of cosmetics. Includes some treatment of the colorants used.
41. W. C. Hoffman, "Dyes, Pigments, and Contact Lenses," *Contact Lens Forum*, Feb. 1983. Colorants and their use in contact lenses.
42. W. Kampfer, and F. Stieg, Jr., *Color Eng.* **44,** 35–40, 44 (1967). A description of the manufacture, properties, and uses of titanium dioxide as a colorant for paint, food, plastics, and other materials.
43. N. Dinesen, *Food Technol.*, 40 (May 1975). Toxicology and Regulation of Natural Colors. Some thoughts on international regulation.
44. F. Mayer and A. H. Cook. *The Chemistry of Natural Coloring Matters,* ACS Monograph, Reinhold, New York, 1943.
45. E. M. Usova, E. M. Voroshin, V. S. Rostovskii, A. M. Moroz, and F. Yakhina, *Kh. Izv. Vysshikh Uchebn, Zavedenii, Pishchevaya Tekhnol.* **4,** 151–153 (1966). Describes the use of natural colorants as replacements for tartrazine, indigo carmine, and annatto.
46. *Annatto Food Colors*, Charles Hansen's Laboratory, Milwaukee, Wis. A brief description of what annatto is and how it is used.
47. D. A. V. Dendy, *East Afr. Agric. Forest J.* **32,** 126–132 (1966). The manufacture of annatto.
48. Ger. Pat. 1,156,529 (Oct. 31, 1963), F. K. Marcus.
49. G. Schwarz, H. Mumm, and F. Woerner, *Molkerei Käserei—Ztg.* **9,** 1430–1433 (1958).
50. U.S. Pat. 3,162,538 (Dec. 22, 1964), P. H. Todd, Jr. Describes the use of bixin and turmeric for coloring butter, margarine, cheese, and other fatty and oily foods.
51. Ger. Pat. 927,305 (May 5, 1955), F. Rath. Natural dyes like norbixin, crocetin, and carminic acid are discussed from an applications standpoint.
52. J. F. Reith and J. W. Gielen, *J. Food Sci.* **36,** 861–864 (1971). Properties of bixin, norbixin, and annatto extracts.
53. T. Sato and H. Suzuki, *Nippon Shokuhin Kogyo Gakkaishi* **13,** 488–491 (1966). A study of the coloring of sausages with annatto and zanthene-type pigments from the standpoint of fading, penetration, and other variables.
54. O. Isler, R. Ruegg, and P. Schudel, *Chimia* **15,** 208–226 (1961). Includes a discussion of β-carotene, β-apo-8′-carotenal, and canthaxanthin from the standpoint of preparation, toxicity, analysis, and application.
55. O. Isler, R. Ruegg, and U. Schwieter, *Pure Appl. Chem.* **14,** 245–264 (1967). Describes the preparation and analysis of various carotenoids including β-carotene, canthaxanthin, and β-apo-8′-carotenal.
56. H. T. Gordon, *Food Technol.* 64–66 (May 1972). A brief description of the properties, commercial forms, and uses of β-carotene, β-apo-8′-carotenal, and canthaxanthin.
57. R. H. Bunnell, W. Driscoll, and J. C. Bauernfeind, *Food Technol.* **12,** 536 (1958).
58. J. C. Bauernfeind, M. Osadca, and R. H. Bunnell, *Food Technol.* **16,** 101–107 (1962). A general discussion of the use of β-carotene as a color additive for juices and beverages.
59. J. C. Bauernfeind and R. H. Bunnell, *Food Technol.* **16,** 76–82 (1962). Describes the properties, market forms, uses, stability, and other characteristics of the colorant.

60. R. H. Bunnell and B. Borenstein, *Food Technol.* **21,** 13A–16A (1967). A brief review of the history, natural occurrence, properties, market forms, and stability of canthaxanthin.
61. R. T. Linner, "Caramel Coloring—Production, Composition, and Functionality," *Baker's Digest*, Apr. 1965.
62. R. North, "Add a Pinch of Burnt Sugar for Color," *Canner Packer*, May 1969. A description of caramel and how it is made and used.
63. F. W. Peck, *Food Eng.,* 94 (Mar. 1955).
64. W. R. Eichenberger, *Caramel Colors: Manufacture, Properties, and Food Applications,* paper presented at the ACS Meeting, Aug. 29, 1972.
65. R. F. Andreu, *Farmacognosia* **17,** 145–224 (1957). An extensive review of saffron.
66. S. G. Vishnevetskaya, *Maslob.—Zhir. Prom.* **28,** 30–32 (1962).

General References

H. Holtzman, *Am. Perfumer Cosmet.* **78,** 27–31 (1963). A somewhat outdated review of the permitted certified and noncertified color additives.
H. Hopkins, *FDA Consumer*, 24–27 (Mar. 1980). Some insight into the use and regulation of color additives.
L. Koch, *Am. Perfumer Cosmet.* **82,** 35–40 (1967). A brief review.
A. Kramer, *Food Technol.,* 65–67 (Aug. 1978).
G. Vettorazzi, *Handbook of International Food Regulatory Toxicology*, Vol. II, Spectrum Publications, Inc., Jamaica, N.Y., 1981. Lists safety tests performed on the various color additives.

DANIEL MARMION
Consultant

COLORANTS FOR PLASTICS

The initial uses of colorants in plastics were as extenders and additives. Carbon black and titanium dioxide were and are still used as fillers (qv) because of their low cost. Almost from plastics' inception the limitation of black and white did not offer sufficient color choices for end users looking to differentiate their products. The increase in aesthetic requirements along with different performance requirements and resin compatibilities led to a great expansion in the number of different chemical classes of colorants and forms in which these colorants are available in today's market.

Traditionally, colorants are divided into three classes: inorganic pigments, organic pigments, and dyes. Dyes are soluble under conditions of use but must be completely dissolved in order to show little or no haze in the final product. Dyes generally make very good transparent colors and colors that are very bright with high chroma attributes. In most cases, dyes often do not weather as well as pigments, especially in conjunction with an opacifying agent such as titanium dioxide. Pigments are insoluble and consist of finite particles that must be dispersed by physical means. Because pigments are not soluble in the resin, they often influence the physical properties of the mixtures.

The thermoplastic or thermoset nature of the resin in the colorant–resin matrix is also important. For thermoplastics, the polymerization reaction is completed, the materials are processed at or close to their melting points, and scrap may be reground and remolded, eg, polyethylene, propylene, poly(vinyl chloride), acetal resins (qv), acrylics, ABS, nylons, cellulosics, and polystyrene (see OLEFIN POLYMERS; VINYL POLYMERS; ACRYLIC ESTER POLYMERS; POLYAMIDES; CELLULOSE ESTERS; STYRENE POLYMERS). In the case of thermoset resins, the chemical reaction is only partially complete when the colorants are added and is concluded when the resin is molded. The result is a nonmeltable cross-linked resin that cannot be reworked, eg, epoxy resins (qv), urea–formaldehyde, melamine–formaldehyde, phenolics, and thermoset polyesters (qv) (see AMINO RESINS; PHENOLIC RESINS).

There is the possibility of a chemical reaction between a plastic and a colorant at processing temperatures. Thermal stability of both the polymer and colorant plays an important role. Furthermore, the performance additives that may have been added to the resin such as antioxidants, stabilizers, flame retardants, ultraviolet light absorbers, and fillers must be considered. The suitability of a colorant in a particular resin must be evaluated and tested in the final application after all processing steps to ensure optimum performance.

Available Colorant Forms

Colorants can be added to plastics by several methods. The incorporation of the colorant often is a balance between a particular end property requirement and inventory control. The typical forms in which colorants are added are raw colorant, dispersed colorant, dry concentrate, and liquid concentrate. Furthermore, resins can be purchased that have the colorants dispersed in them, thus the resin is precolored.

Raw Colorant. Often, raw colorants can be added directly to the resin. This process is used when significant concentrations of colorant are present (colorant used as a filler) or when sufficient mixing (shear) and heating is possible. Processes for which this is common include the production of poly(vinyl chloride) (PVC) and phenolic resins (qv). Normally, the only colorants that are added dry are white diffusing pigments and carbon blacks. Chromatic colorants are rarely added directly to the final product. The dangers of using a dry colorant directly are contamination, incomplete colorant strength development (thus loss of color value), streaking, and inconsistent color with respect to hue and strength.

Dispersions. To obtain the best performance (strength, low haze, no streaking) of a pigment and often a dye, colorants are dispersed in a neutral medium that is compatible with the resin to which the dispersion will be added. The dispersion medium should have no adverse properties with respect to the final mixture or final processing conditions. The process of dispersion of a dry pigment into a liquid vehicle or molten plastic may be visualized as taking place in two steps: (1) breaking up of pigment agglomerates into the much smaller ultimate particles and (2) displacement of air from the particles to obtain a complete pigment-to-vehicle interface (1).

The usual method is to coat the colorant particles onto the surface of the resin granules by simply mixing, using a drum tumbler, double-cone blender, or ribbon blender. The colorants are intimately mixed by passing over a heated roller mill, a Banbury sigma-bladed mixer, or an extruder. It is important to have sufficient molten resin in the mix to coat freshly exposed surfaces. The colored resin is then cut or ground into small cubes or granules for extrusion or molding. To test for completion of dispersion and colorant stability, the colored resin is put through several cycles. If the color becomes stronger, there is an indication that dispersion was not complete the first time. If the color weakens or becomes duller, it is certain that reaction or decomposition is taking place.

In some cases, it is possible to displace the water from the surface of a pigment in an aqueous press cake simply by blending with an organic liquid. The liquid preferentially wets the pigment, particularly if it is organic, displacing the water, which may be poured off; this process is called flushing (2). In this way, the pigment is dispersed in an organic medium without ever having been dried and without the need to redisperse it. The process is widely used in the paint and printing ink industries. It is useful for plastics when the flushing vehicle is a plasticizer that is intended for incorporation into a resin such as poly(vinyl chloride).

Color Concentrates. Color concentrates have become the method of choice to incorporate colorants into resins. Color concentrates have high ratios of colorant to a compatible vehicle. The colorant may be added at 70% colorant to 30% vehicle in a titanium dioxide mixture whereas the ratio may be 15% colorant to 85% vehicle in a carbon black mixture. The amount of colorant that can be added is dependent on the surface area and the oil absorption of the colorant and the wetting ability of the vehicle. The normal goal is to get as much colorant in the concentrate as possible to obtain the greatest money value for the product. Furthermore, less added vehicle minimizes the effect on the physical or chemical properties of the resin system.

Vehicles are selected by two methods. In one a concentrate is designed directly for a resin system, the resin itself, or a compatible resin. Thus when the concentrate is made there is a minimal effect on the properties of the final color. In PVC, often a plasticizer such as dioctyl phthalate (DOP) is used. In the other method, concentrates are made with a commercial universal concentrate vehicle. Concentrate manufacturers and some resin manufacturers have developed vehicles that can incorporate many types of colorants and can be used across many classes of polymers without adversely affecting final product performance.

These universal concentrates are preferred if they can be used, because inventorying costs can be reduced; it is not necessary to stock different concentrates for different colorants. In some performance polymers this concept is not applicable because the universal vehicles have a negative effect on weathering, transparency, or processing performance.

Concentrates are made in the same manner as dispersions but often with less work placed into the colorant–vehicle system, thus higher yields and rates are obtained and a lower cost product is made. Usually, the concentrate goes through an additional processing step with the resin in which additional energy is provided to the system, thus ensuring more dispersion.

Liquid Colorants. Liquid colorants have the advantages and disadvantages of dry concentrates except that their method of incorporation is pumping rather than metering. The advantages of liquid may include better accuracy in pumping, better dispersion quality, and higher amount of colorants in the dispersion. Disadvantages of a liquid system may include cleanup, and loss of material on changeover of color. In most cases, the differences are minimal and the choice is a comfortability and availability issue.

Product Properties

Colorants can and do have a measurable effect on myriad physical and chemical properties of final plastic products. Often, this is overlooked both by researchers developing polymers and by marketers who sell the properties of the polymer and forget that the products must be colored to fulfill aesthetic requirements.

Weathering. Weathering is an important property in plastic products (see PLASTICS TESTING). The weathering performance of plastics is evaluated either in real-time weathering such as in Florida or Arizona (1, 3, or 5 yr) or in accelerated tests such as xenon arc test chambers or QUV test chambers (100–5000 h exposure) (3). Ultraviolet light may affect performance of the polymer and both uv and visible light have an effect on color stability or weatherability of colorants (4). Colorants need to be evaluated in the resin in which they are incorporated for a particular application before the colorants can be judged acceptable. Weathering performance can vary widely from resin to resin and color type, thus correlations and predictions are only valid within the constrained set of testing conditions that have been developed.

Optical Properties. Haze is the most common optical property problem that depends on colorants. Because dyes are dissolved into the resin system, they contribute little or no practical haze to the system. Pigments can have significant haze, which is a combination of the pigment itself and the quality of dispersion of the pigment. In an opaque application haze is not a concern, but in transparent or translucent applications haze development becomes an important criterion in colorant evaluation.

Mechanical and Chemical Properties. Colorants, especially pigments, can affect the tensile, compressive, elongation, stress, and impact properties of a polymer (5). The colorants can act as an interstitial medium and cause microcracks to form in the polymer colorant matrix. This then leads to degradation of the physical properties of the system. Certain chemicals can attack colorants and there can be a loss of physical properties as well as a loss of the chromatic attributes of the colorant. Colorants should always be evaluated in the resin in which they will be used to check for loss of properties that are needed for the particular applications.

Thermal and Electrical. Colorants can also affect the thermal and electrical properties of the resin and thus cause differences in melt flow of the polymer or shrinkage of the final plastic part. The colorants themselves are also affected by processing temperatures and have different temperature stabilities. Some colorants, such as cadmium pigments, can withstand up to 350°C whereas many dyes can tolerate processing conditions of no more than 200–250°C. Colorants degrade by losing their chromatic appearance, chalking or subliming. Chalking (pigments) and subliming (dyes) involve the colorant leaving the matrix of the colorant–resin system and depositing on the surface. The surface either appears discolored or leaves a residue when touched. In most cases, chalking and sublimation are caused by heat but can be aggravated by moisture effects.

Inorganic Pigments

Table 1 lists inorganic pigments used in plastics (see PIGMENTS, INORGANIC).
   White.   Titanium dioxide is by far the most widely used white pigment for plastics. The rutile crystalline modification has a high refractive index (2.76) and causes strong scattering of light and high opacity. The untreated pigment apparently contains reactive centers; the weatherability is greatly increased by coating with alumina and silica. The anatase variety is slightly whiter, has a lower refractive index, and blocks out less uv light. It is not recommended for outdoor use. The refractive index of titanium dioxide varies with wavelength; it is higher in the blue region of the spectrum. For this reason, more blue light is scattered and the resulting white is slightly yellow. To match D65 or to produce a dead white, a slight amount of shading with blue or violet pigments is necessary. Typical particle size of TiO_2 is 0.25–0.3 μm (see TITANIUM COMPOUNDS, INORGANIC).
   Light fixtures are fabricated from transparent plastics such as the acrylics. It is desirable to transmit as much light as possible but to conceal the source, either a filament or fluorescent tube. For this purpose, zinc oxide, zinc sulfide, and barium sulfate are used as diffusing pigments. All have relatively high refractive indexes, and the particle sizes are 1–4 μm. The resulting lenses may transmit as much as 85% of the light and are known as lighting whites. Aluminum silicate is used as an extender pigment in thermoset polyester resins; natural calcium silicate, wollastonite, is used in polyethylene, vinyls, and thermoset resins.
   Black.   Carbon black is another outstanding pigment for plastics. It has the unique property of blocking uv, visible, and ir radiation. A common use is to improve weatherability of polyethylene sheeting for outdoor use. At one time, the best pigmentary carbon black was prepared by burning natural gas in a deficiency of air and impinging the luminous flame on water-cooled iron channels. The resulting channel black contained about 5% volatiles and was acidic. Today, furnace black is produced by partial combustion of aromatic residual oils (see CARBON, CARBON BLACK). Volatiles are at a minimum, and the product may be used in all plastics. If the ultimate degree of dispersion is required, it may be necessary to prepare a concentrate by first dispersing the carbon black into the resin on a roll mill, Banbury mixer, or extruder, followed by final dilution to the desired level with further mechanical work. When observed under an electron microscope, some carbon blacks are seen as discrete particles, others form chains of platelets. The latter are said to have high structure; they are slightly less jet and glossy but have higher electrical conductivity. In acrylics and other highly transparent plas-

Table 1. Inorganic Colorants for Plastics

Colorant	CAS Registry Number	Formula
aluminum silicate	[14504-95-1]	$Al_2(SiO_3)_3$
barium manganate	[7787-35-1]	$BaMnO_4$
barium sulfate	[7727-43-7]	$BaSO_4$
black iron oxide	[1317-61-9]	Fe_3O_4
brown iron oxide	[1309-38-2]	$FeO \cdot (Fe_2O_3)_4$
cadmium selenide	[1306-24-7]	CdSe
cadmium sulfide	[1306-23-6]	CdS
calcium silicate	[10101-39-0]	$CaSiO_3$
carbon black[a]	[1333-86-4]	C
chromium oxide[b]	[1308-38-9]	Cr_2O_3
chromium oxide dihydrate	[12182-82-0]	$Cr_2O_3 \cdot 2H_2O$
cobalt aluminate	[1333-88-6]	$CoO \cdot Al_2O_3$
cobalt lithium phosphate	[13824-63-0]	$CoLiPO_4$
cobalt phosphate	[13455-36-2]	$Co_3(PO_4)_2$
copper chromite black	[12018-10-9]	$Cu(CrO_2)_2$
hematite	[1317-60-8]	Fe_2O_3
iron blue	[14038-43-8]	$Fe_4[Fe(CN)_6]_3$
iron oxide hydrate	[11100-07-5]	$Fe_2O_3 \cdot xH_2O$
lead chromate	[7758-97-6]	$PbCrO_4$
limonite	[1317-63-1]	Fe_2O_3
magnesium oxide–iron oxide	[12068-86-9]	$MgO \cdot Fe_2O_3$
manganese violet	[10101-66-3]	$MnHP_2O_7 \cdot NH_3$
mercuric sulfide	[1344-48-5]	HgS
Mineral Violet	[10101-66-3]	$MnHP_2O_7 \cdot NH_3$
Molybdate Orange	[12656-85-8]	$25PbCrO_4 \cdot 4PbMoO_4 \cdot PbSO_4$
ocher	[1309-37-1]	Fe_2O_3
red iron oxide	[1309-37-1]	Fe_2O_3
sienna	[1309-37-1]	Fe_2O_3
titanium dioxide, anastase	[1317-70-0]	TiO_2
titanium dioxide, rutile	[1317-80-2]	TiO_2
ultramarine blue	[57455-37-5][c]	$Na_{(6-8)}Al_6Si_6O_{24}S_{(2-4)}$
ultramarine violet	[12769-96-9][d]	$H_2N_{(4-6)}Al_6Si_6O_{24}S_2$
umber	[12713-03-0]	
wollastonite	[14567-51-2]	$CaSiO_3$
zinc oxide	[1314-13-2]	ZnO
zinc oxide–iron oxide	[12063-19-3]	$ZnO \cdot Fe_2O_3$
zinc sulfide	[1314-98-3]	ZnS

[a]Also called channel black and furnace black.
[b]Chrome green.
[c]Pigment blue 29 [1317-97-1].
[d]Pigment violet 15.

tics used in architecture, small amounts of carbon black are added to reduce light and heat transmittance. At such levels, slightly more of the longer wavelengths of light are transmitted and black imparts a bronze or golden hue. Carbon black is available as powder and pellets, in the form of aqueous and nonaqueous dispersions, and as concentrates.

When it is desirable to use a weak black, bone black may be substituted for carbon. It is manufactured by calcining animal bones and contains approximately 85% calcium phosphate and calcium carbonate. Black iron oxide (Fe_3O_4) is stable up to 150°C. Copper chromite black ($Cu(CrO_2)_2$) is inert to all but rubberlike compositions and has been calcined to 600°C.

Iron Oxides.   In addition to the black iron oxide, there are several natural and synthetic yellow, brown, and red oxides. As a class, they provide inexpensive but dull, lightfast, chemically resistant, and nontoxic colors. The natural products are known as ocher, sienna, umber, hematite, and limonite. These include varying amounts of several impurities; in particular, the umbers contain manganese. Their use is limited because of low chroma, low tinting strength, and poor gloss retention.

The synthetic iron oxides are much purer and have less variation in composition. Red oxide (Fe_2O_3) has excellent bleed, chemical, heat, and light resistance and is nontoxic. The yellow hydrate ($Fe_2O_3 \cdot H_2O$) is useful up to 175°C, at which point it loses water and becomes red. Both of these pigments protect resins by screening uv light. Brown oxides ($FeO \cdot Fe_2O_3)_y$, mixtures of ferrous and ferric oxides, are useful for producing woodgrain effects in plastics. There are also two mixed oxides ($ZnO \cdot Fe_2O_3$ and $MgO \cdot Fe_2O_3$) that are stable, nontoxic tans.

The natural iron oxides are recommended for use in cellulosics and phenolics; the synthetics, for cellulosics, polyethylene, flexible vinyls, and all thermosets.

Special forms of highly transparent iron oxides are made for use in durable metallized polychromatic finishes. These products are more brown than yellow; however, when used in metallized finishes, they impart a golden color. This type of iron oxide tends to be more reactive than the opaque yellows.

Chromium Oxide Greens.   Chromium oxide (Cr_2O_3) is dull but is the most weatherable green pigment. It is stable up to 1000°C and is chemically resistant. The ir reflectance resembles that of chlorophyll, making it a good camouflage color. Where durability is more important than brightness, it is used in cellulosics, polyethylene, and vinyls. The hydrated form ($Cr_2O_3 \cdot 2H_2O$) has a much more brilliant shade but is dehydrated at about 500°C. Its use is largely confined to the cellulosics.

Iron Blue.   There are three common varieties of iron blue: Milori, Chinese, and Prussian (they are sometimes called toning blues). The three types differ chiefly in color, ease of dispersion, and reactivity characteristics. Milori blues are the easiest to disperse and are the least reactive. They are reddest in mass tone (plum colored); in tints, they are intense and intermediate in redness between the Chinese and Prussian varieties. Both the Chinese and Prussian blues are jet in mass tone, but Prussian blue is considerably redder and less intense in tint than Chinese blue. Because of their jet mass tones, both are used to shade blacks. Prussian blues are the hardest to disperse and are somewhat more reactive than Chinese blues.

Iron blues have good lightfastness in mass tone or deep shades. Iron blues are coarse in texture, difficult to grind, and withstand only a few minutes at 175°C.

Violet. There are several inorganic violet pigments. Manganese violet, or mineral violet ($NH_4MnP_2O_7$), is weak and has poor alkali and heat resistance. It may be used in vinyls. Among cobalt compounds, the phosphate $Co_3(PO_4)_2$ is little used and $CoLiPO_4$ is stronger and more stable than manganese violet; other cobalt and manganese complexes also are available. Ultramarine violet is discussed below. As a class, the inorganic violets are weak and difficult to disperse.

Ultramarine Pigments. Ultramarine is a reddish blue with a clean, brilliant shade having good durability except to acids. It is manufactured by calcining a mixture of China clay, soda ash, sodium sulfate, carbon, silica, and sulfur at 800°C. The crude product is ground and washed. It is a complex aluminum sulfosilicate. Extraction with hydrogen chloride or chlorine produces a violet containing less sodium. Even though ultramarine has only 7% of the tinting strength of phthalocyanine blue, it is considerably redder in shade and has wide use in plastics. The violet, which is much weaker, is useful for tinting whites.

Blue, Green, Yellow, and Brown Metal Combinations. There are a number of inorganic oxide pigments that are indispensable for coloring transparent or translucent high temperature plastics. Their properties have been summarized (6). The pigments were originally developed for ceramics but have been improved for use in plastics (see COLORANTS FOR CERAMICS). The ingredients are mixtures of metal salts that are calcined at 800–1300°C. The resulting oxides are ground, washed, dried, and pulverized to fine powders. The products are insoluble in solvents, and all resins have excellent resistance to heat, light, and chemical attack as well as low oil absorption, good dispersibility, and little or no tendency to migrate. Compared with organic pigments and dyes, the oxide pigments are at a serious disadvantage in terms of brightness and, usually, cost.

The inorganic blues include cobalt aluminate, which has the nominal formula $CoO \cdot Al_2O_3$. More commonly, the chemical compositions vary. As chromium is added, the blue becomes greener and moves into the turquoise region. Other blues may contain silicon, zinc, titanium, tin, or aluminum in addition to cobalt. Barium manganate ($BaMnO_4$) is a weak greenish blue with a high specific gravity of 4.85. It has mild oxidizing properties and should not be used without rigorous preliminary tests. However, it does have a spectral curve similar to that of green-shade phthalocyanine blue and is occasionally useful in making nonmetameric matches (see COLOR).

A wide variety of greens ranging from blue to yellow in shade are based on cobalt in combination with chromium, aluminum, titanium, nickel, magnesium, antimony, or zinc. These are brighter than the chromium oxides.

Inorganic yellow oxide combinations may contain lead, antimony, tin, nickel, or chromium. They are classed as yellows rather than brown, but they are dull compared with the cadmium yellows.

Brown combinations usually contain iron with chromium, zinc, titanium, or aluminum. There are a few without iron that contain chromium, antimony, tin, zinc, manganese, or aluminum. They range from light tans to dark chocolate. The shades are not as red as ferric oxide, but the browns are far superior to hydrated iron oxide in brightness and thermal stability.

Lead Chromates and Molybdates. The lead chromates appear in several shades of yellow. The primrose and lemon are solid solutions of lead sulfate in the chromate and have the stable monoclinic structure. The medium shade contains no sulfate. Chrome orange is a compound with lead oxide ($PbCrO_4 \cdot PbO$). Molybdate orange is a combination of lead chromate and sulfate with molybdate ($PbMoO_4$). These pigments have the advantages of opacity, brightness, and low cost. They are sensitive to acids, alkalies, and hydrogen sulfide. Pigments containing lead are banned from any use, and hexavalent chromium is a suspected carcinogen. Use is confined to vinyls and low temperature polyolefin and polystyrene resins except that chrome orange is too sensitive to acids for use in vinyls.

Conventional chromates and molybdates are stable only up to 200°C. Silica encapsulation increases the stability to 300°C (7). These pigments are readily shaded. Chrome yellow mixed with molybdate orange is brighter than chrome orange. The molybdates may be shaded with organic reds such as lithol rubines and quinacridones. Because of the current regulatory issues regarding colorants containing heavy metals, the lead pigments are being replaced by organics.

Cadmium Pigments. The cadmiums constitute a continuous series from greenish yellow through orange and red to maroon (8). They may be used in most plastics and are stable at 325°C and up to 550°C for short periods. Resistance to chemicals is good. Lightfastness is satisfactory indoors, improving from yellow to red. The full tones are strong and bright, but light tints, especially reds, become dull.

Chemically, the pigments are cadmium sulfides and selenides that have been precipitated, dried, and calcined at 650°C to convert to the hexagonal crystalline form. The greenest yellows contain 25 mol % zinc sulfide as a solid solution in cadmium sulfide. Pure cadmium sulfide (CdS) is orange. Selenium is added to produce reds until at 50 mol % a maroon results. The cadmium selenide (CdSe) is also in solid solution. As the proportions of the ingredients are varied, a large number of shade gradations are possible. One reason for this seeming proliferation of types is that single pigments are brighter than mixtures.

These are known as chemically pure (CP) cadmiums. With the development of other uses for cadmium and selenium, costs have risen substantially in recent years. Some cost reduction may be obtained by use of the cadmium lithopones. These have the same relative shades but have been coprecipitated onto about 60% barium sulfate. The resulting extensions give better money value, if the higher pigment loading can be tolerated, with no loss in properties.

A third form of cadmium pigments includes the mercury cadmiums. Mercuric sulfide (HgS) forms solid solutions up to about 20 mol % with the oranges, reds, and maroons. The heat stability is improved up to 370°C, and the costs are somewhat lower than the CP grades. The mercury cadmiums are slightly more reactive, but have excellent bleed resistance.

The improved organic pigments and dyes developed in recent years have displaced the cadmiums from the lower temperature plastics. For the high temperature engineering plastics, eg, nylon and acetals, there are no substitutes in performance for the CP cadmium pigments. Cadmiums are also being replaced with organic pigments as heavy-metal regulatory issues continue.

Titanate Pigments. When a nickel salt and antimony oxide are calcined with rutile titanium dioxide at just below 1000°C, some of the added metals diffuse into the titanium dioxide crystal lattice and a yellow color results. In a similar manner, a buff may be produced with chromium and antimony; a green, with cobalt and nickel; and a blue, with cobalt and aluminum. These pigments are relatively weak but have extreme heat resistance and outdoor weatherability, eg, the yellow is used where a light cadmium could not be considered. They are compatible with most resins.

Pearlescent Pigments. The technology of producing pearlescent pigments has passed through three stages. The original natural pearl essence was obtained from fish scales and was expensive. Later, basic lead carbonate, bismuth oxychloride, and lead hydrogen arsenate were used. The preferred process is to coat mica with layers of titanium dioxide.

The pearlescent or nacreous pigments owe their effects to the partial transmittance and partial reflection of light from the multiple coating layers. Because the layers have about the thickness of the wavelengths of visible light, some wavelengths interfere and colors are produced. Basic lead carbonate is supplied as a 70% suspension in dioctyl phthalate. Use is chiefly in vinyls. Coated mica may be used in any transparent or translucent plastic, particularly vinyls and polyethylene. The shade is sometimes modified by the incorporation of red iron oxide. Other pleasing effects are produced by carbon black or colored pigments in trace amounts. The use of specialty colorants such as pearlescent, metallic, and fluorescent is finding increasing acceptance into plastics as consumers desire differentiation of products.

Metallic Pigments. Aluminum is a reactive metal that is protected from chemical attack by a natural, thin oxide film. It is attacked by alkalies and acids under oxidizing conditions. Commercial flakes are 0.1–2.0 μm thick and are generally sold in paste form in a hydrocarbon or plasticizer medium. Dry aluminum dust forms explosive mixtures with air. For paint use, the flakes are coated with stearic acid to produce a leafing grade that tends to line up with the platelets parallel to the surface. At higher temperatures of plastics, stearic acid has little effect; however, there is usually some orientation during molding. The metal protects plastics by reflecting uv light.

Aluminum imparts a grayish cast to plastics. In transparent systems, dyes or pigments may be added to impart various colors. Such systems often exhibit flop, a drastic change in shade or lightness between 90° and low angle viewing. In the former case, more relatively uncolored light is reflected from the oriented metal flakes. At a low viewing angle, more colored resin is visible and less light is reflected from the edges of the aluminum.

A few plastics can tolerate aluminum much more readily than bronze flakes. It is possible to formulate a bronze color using only aluminum and suitably transparent yellow, orange, and red dyes or pigments.

Bronze powders are available for use in plastics in various compositions and particle sizes. Copper is seldom used alone; all bronzes contain zinc. The names and zinc content of the common types are pale gold, 8%; rich pale gold, 15%; and rich gold, 30%. The shades become less red as zinc is increased. The finer particle size grades are generally preferred.

Some plastics tend to react with zinc to give an undesirable bubbling effect when molded. There is also a tendency to darken due to tarnishing by the action of sulfides. Tarnish-resistant bronzes are available in which the particles have been coated with a transparent resin (9).

Organic Pigments

With higher molding speeds and higher temperature resins, the demand for thermal stability is increasing. As plastics replace metals in automotive applications, weatherability and service-temperature requirements are sharply increased. Suppliers are responding with improved products.

The composition of inorganic pigments is well known. However, structures of organic pigments are often not disclosed (see PIGMENTS, ORGANIC). For an almost complete list of useful pigments, see Ref. 10. Most suppliers identify their pigments by CI designations and common names (Table 2).

Monoazo Pigments. In combination with other groups, the azo linkage, —N=N—, imparts color to many dyes and pigments (see AZO DYES). The simplest of these, ie, the Hansa yellows, toluidine reds, and naphthol reds, do not have the lightfastness and heat stability required for plastics. Permanent Yellow FGL and Permanent Red 2B are stable enough for vinyls, polyethylene, polypropylene, and cellulosics (11). Permanent Red 2B is available as the calcium, barium, or manganese salt.

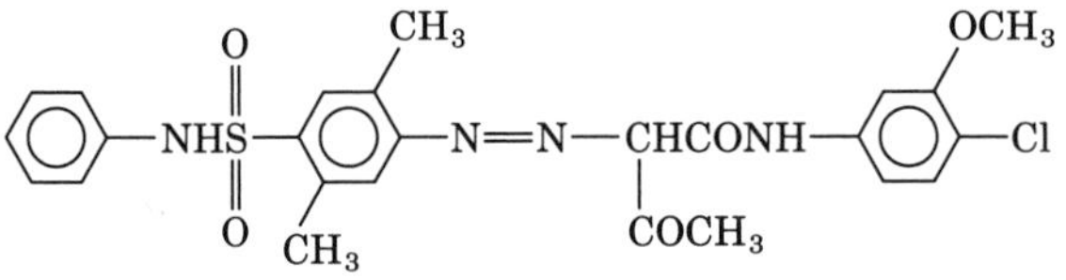

Permanent Yellow FGL

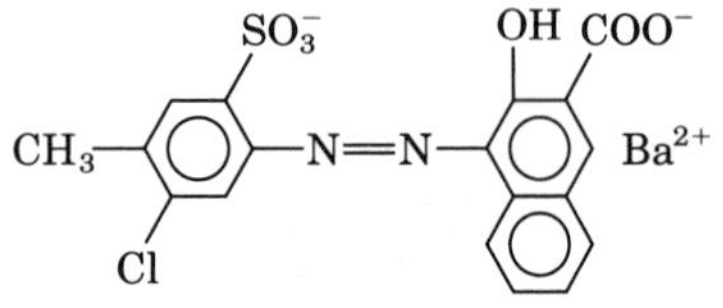

Permanent Red 2B

Nickel Azo Yellow owes its fastness to a metal atom. In this case the metal is chelated; in many other azo pigments the metal forms a salt with a carboxylic or sulfonic acid and precipitates the pigment, as in Permanent Red 2B. In Nickel Azo Yellow, *p*-chloroaniline is coupled to 2,4-dihydroxyquinoline, then converted to a metal complex. The resulting pigment is green in heavy shades, yellow in lighter tints, and has very good fastness. It is suitable for cellulosics, polyethylene, and poly(vinyl chloride). It is demetallized by acids with loss of its fastness properties.

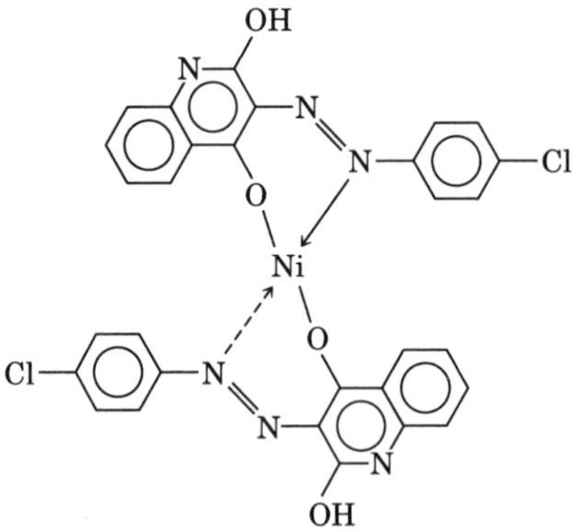

Nickel Azo Yellow

Table 2. Organic Colorants for Plastics

Colorant	CAS Registry Number	Alternative name	CI number
Acid Red 52	[3520-42-1]	Xylene Red B, Sulforhodamine B	45100
Acid Yellow 7	[2391-30-2]	Brilliant Sulfoflavine	
Anthramide Orange	[2379-78-4]	Vat Orange 15	69025
Anthrapyrimidine Yellow	[4216-01-7]	Vat Yellow 20	68420
Basic Red 1	[989-38-8]	Rhodamine 6G	45160
Benzidine Yellow HR	[5567-15-7]	Pigment Yellow 83	21180
Benzidine Yellow AAMX	[5102-83-0]	Pigment Yellow 13	21100
Benzidine Yellow AAOA	[4531-49-1]	Pigment Yellow 17	21105
Benzidine Yellow AAOT	[5468-75-7]	Pigment Yellow 14	21095
Brominated Anthanthrone Orange	[4378-61-4]	Pigment Red 168	59300
Brominated Pyranthrone	[1324-33-0]	Pigment Red 197	
Carbazole Dioxazine Violet	[6358-30-1]	Pigment Violet 23	51319
Dianisidine Orange GG	[6837-37-2]	Pigment Orange 14	21165
Dianisidine Red	[6505-29-9]	Pigment Red 41	21200
Disperse Violet 1	[128-95-0]		61100
Disperse Yellow 23	[6250-23-3]		26070
Flavanthrone Yellow	[475-71-8]	Vat Yellow 1	70600
Indanthrone Blue, red shade	[81-77-6]	Vat Blue 4	69800
Induline Base 6B	[8004-98-6]	Solvent Blue 7	50400
Lake Red C	[2092-56-0][a]	Pigment Red 53	15585
Methyl Violet	[8004-87-3]	Basic Violet 1	42535
Metanil Yellow	[587-98-4]	Acid Yellow 36	13065
Mordant Red 9	[1836-22-2]	Pigment Scarlet 33	16105
Nickel Azo Yellow	[51931-46-5]	Pigment Green 10[b]	12775
Perinone Orange	[4424-06-0]	Vat Orange 7	71105
Perinone Red	[4216-02-8]		
Permanent Red 2B	[3564-21-4]	Pigment Red 48	15865
Permanent Yellow FGL	[12225-18-2]	Pigment Yellow 97	
Phthalocyanine Blue	[147-14-8]	Pigment Blue 15	74160
Phthalocyanine Green	[1328-53-6]	Pigment Green 7	74260
Pigment Blue 25	[10127-03-4]		21180
Pigment Brown 28	[57972-00-6]		69015
Pigment Orange 13	[3520-72-7]	Pyrazolone Orange	21110
Pigment Orange 31	[12286-58-7]		
Pigment Orange 34	[15793-73-4]	Tolyl orange	21115
Pigment Orange 43	[42612-21-5]		
Pigment Red 123	[24108-89-2]	Perylene Vermillion	
Pigment Red 144	[5280-78-4]		20735
Pigment Red 149	[12225-02-4]		
Pigment Red 150	[56396-10-2]		
Pigment Red 166	[12225-04-6]		20730
Pigment Red 170	[12236-67-8]		
Pigment Red 179	[5521-31-3]	Perylene Maroon, Vat Red 23	71130
Pigment Red 190	[6424-77-7]	Perylene Scarlet	71140
Pigment Red 210	[61932-63-6]		12477
Pigment Violet 19	[1047-16-1]	Quinacridone	46500

Table 2. (*Continued*)

Colorant	CAS Registry Number	Alternative name	CI number
Pigment Violet 31[c]	[1324-55-6]	CI Vat Violet 1	60010
Pigment Violet 33[d]	[1324-17-0]		60005
Pigment Yellow 1	[2512-29-0]		11680
Pigment Yellow 60	[6407-74-5]		12705
Pigment Yellow 93	[5580-57-4]		20710
Pigment Yellow 94	[5580-58-5]		20038
Pigment Yellow 95	[5280-80-8]		20034
Pyranthrone Orange	[128-70-1]	Pigment Orange 40	59700
Rhodamine B	[81-88-9]	Basic Violet 10	45170
Solvent Blue 56	[14233-37-5]		
Solvent Green 3	[128-80-3]	D&C Green #6	61565
Solvent Green 4	[81-37-8]	Fluorescent Brightener 74	45550
Solvent Orange 7	[3118-97-6]		12140
Solvent Orange 60	[61969-47-9]		
Solvent Orange 63	[54578-43-7]		
Solvent Red 1	[1229-55-6]		12150
Solvent Red 24	[85-83-6]	Oil Red	26105
Solvent Red 26	[4477-79-6]		26120
Solvent Red 111	[82-38-2]		60505
Solvent Yellow 14	[824-07-9]	Oil Orange	12055
Solvent Yellow 72	[61813-98-7]		
Sulfo Rhodamine	[2609-88-3]		
Tartrazine	[1934-21-0]	FD&C Yellow #5	19140
Victoria Blue B	[2580-56-5]	Basic Blue 26	44045

[a]Sodium salt.
[b]Also [61725-51-7].
[c]Same structure as isoviolanthrone violet, but the chlorine substitution pattern is not specified.
[d]Same ring structure as isoviolanthrone violet, but one bromo substituent rather than dichloro.

Lake Red C is an example of a pigment that has been made insoluble by a heavy metal. In this case the metal is barium; one barium ion precipitates two molecules. Other metals used are calcium, strontium, manganese, and aluminum. This pigment is used in polystyrene.

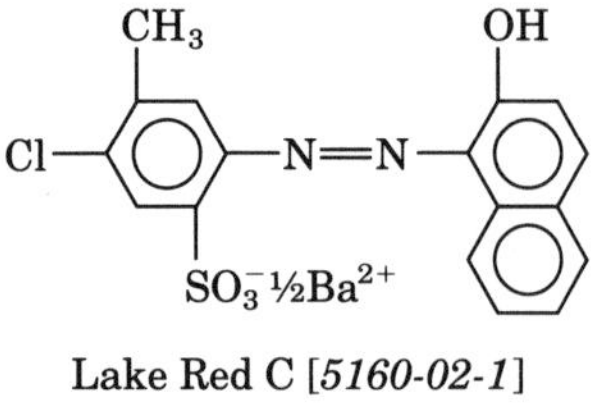

Lake Red C [5160-02-1]

Pigment Scarlet 3B is formed from the dye Mordant Red 9, which is mixed with freshly precipitated aluminum hydroxide and precipitated on this base with barium chloride and zinc oxide. The resulting lake or extension has a bright, clear shade, free of bronzing. It is used in cellulosics, polyethylene, polystyrene, and rigid vinyls. Another monoazo pigment, Tartrazine lake, is stable in plastics, including nylon; it is used as food color, ie, FD&C Yellow #5 (see COLORANTS FOR FOOD, DRUGS, COSMETICS, AND MEDICAL DEVICES).

In the benzimidazolone pigments, conventional coupling agents, such as acetoacetic acid used in Hansa yellows or β-naphthol of the naphthol reds, are joined to the benzimidazolone grouping to give 5-acetoacetylaminobenzimidazolone [26576-46-5] and 5-(3'-hydroxy-2'-naphthoylamino)benzimidazolone [26848-40-8], respectively. These reagents are coupled with amines, often containing methoxyl groups. The resulting pigments have good fastness to light, solvents, migration, and heat to 330°C. They are used in vinyls, polyethylene, and thermoset polyesters and have been suggested for ABS. The shades extend from greenish yellow to brown to bluish red. In the formulas, the arrows indicate the points of azo coupling.

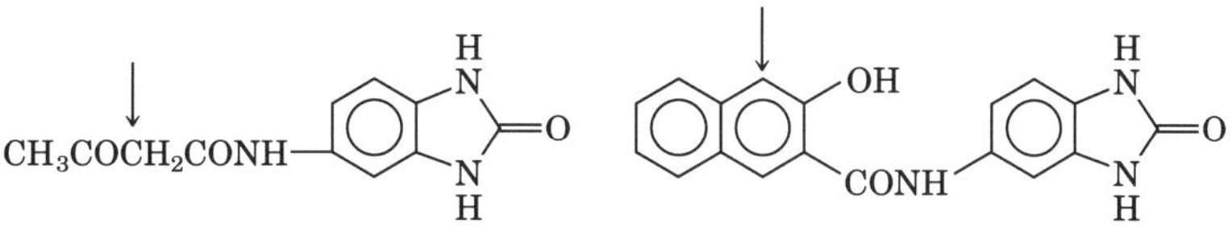

Disazo Pigments.   The diarylide yellows and oranges also known as benzidines are derivatives of benzidine coupled to two moles of substituted acetoacetanilide. Benzidine Yellows AAMX, AAOT, AAOA, and HR (PY 13, 14, 17 and 83) are examples (Fig. 1). Yellows AAMX and AAOT are used in flexible vinyls. AAOA also colors polyethylene and polypropylene. These three differ only slightly in shade. Benzidine Yellow HR is redder.

The diarylide yellows are made from dichlorbenzidine. These pigments are generally considered to be thermally stable during plastics processing because monoazo cleavage products are of similar shades and thus degradation may go unnoticed. However, prolonged heating above 200°C may release the known animal carcinogen 3,3′dichlorobenzidine (12). If the chlorines on benzidine are replaced by methoxyls, the pigments are called dianisidines, eg, Dianisidine Orange GG (PO 14).

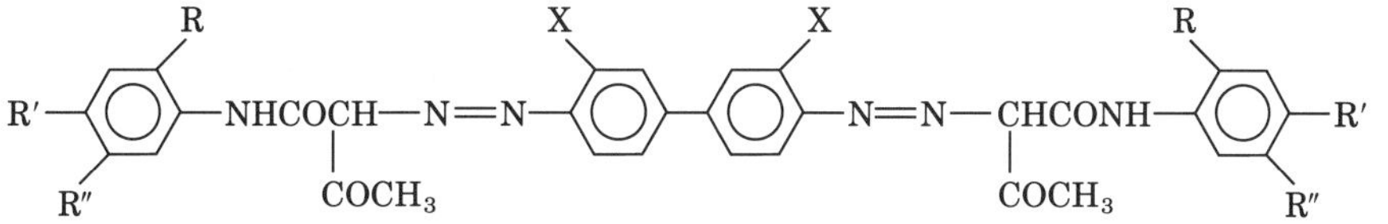

(**1**) R = R′ = CH$_3$, R″ = H
(**2**) R = CH$_3$, R′ = R″ = H
(**3**) R = OCH$_3$, R′ = R″ = H
(**4**) R = R″ = OCH$_3$, R′ = Cl
(**5**) R = R′ = CH$_3$, R″ = H

Fig. 1.   The diarylide yellows (X = Cl): AAMX (**1**), AAOT (**2**), AAOA (**3**), HR (**4**), and Dianisidine orange GG (**5**) (X = OCH$_3$).

The pyrazolone pigments may be considered as combinations of the Hansas and the diarylides. Dichlorbenzidine is coupled to 1-phenyl-3-methylpyrazolone or one of its derivatives. The unsubstituted compound is Pyrazolone Orange (PO 13) and the 4'-methyl derivative is PO 34 (Tolyl Orange). Both are useful in cellulosics, polyethylene, polystyrene, vinyls, phenolics, and polyesters. Replacement of the chlorines of the benzidine component of PO 13 with methoxyls gives Dianisidine Red (PR 41), used in cellulosics and vinyls.

Disazo Condensation Pigments. In the manufacture of diarylide pigments, both amino groups are diazotized and coupled at the same time. CIBA-GEIGY Corp. has developed a series of disazo colors in which two azo dyes are combined with benzidine into one much larger molecule. These disazo pigments are trademarked Cromophtal. The resulting red, orange, or yellow pigments are recommended for vinyls, acrylics, ABS, polyethylene, and similar thermoplastics as well as for epoxy and phenolic thermosets.

Quinacridone Pigments. The quinacridones have the following formula:

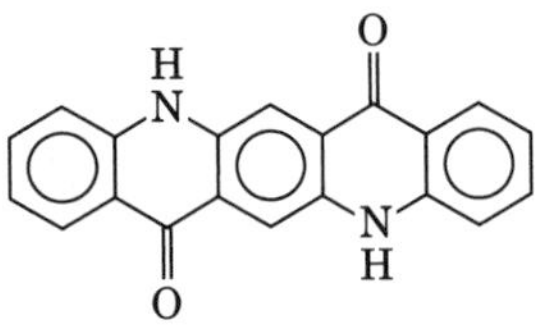

Pigment Violet 19

Practical methods for synthesis and elucidation of the optimum physical forms were developed at Du Pont (13). The violets fill the void in the color gamut when the inorganics are inadequate. The quinacridones may be used in most resins except polymers such as nylon-6,6, polystyrene, and ABS. They are stable up to 275°C and show excellent weatherability. One use is to shade phthalocyanines to match Indanthrone Blue. In carpeting, the quinacridones are recommended for polypropylene, acrylonitrile, polyester, and nylon-6 filaments. Predispersions in plasticizers are used in thermoset polyesters, urethanes, and epoxy resins (14).

Dioxazine Violet. Carbazole Dioxazine Violet is prepared by the reaction of two moles of 2-amino-N-ethylcarbazole with chloranil. This violet may be used in most plastics for shading phthalocyanine blues, because it has comparable light fastness. At relatively high temperatures, it may be subject to slow decomposition.

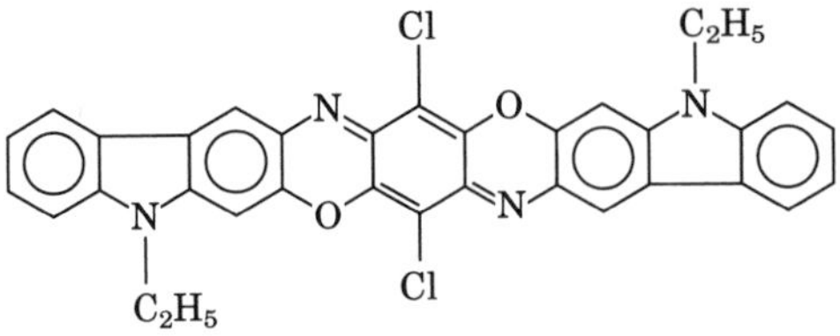

Carbazole Dioxazine Violet

Vat Pigments. A few of the many anthraquinone vat dyes have been exploited as pigments, first in automotive finishes (15) and more recently in plastics. As a class, the vat pigments have good light and heat fastness and virtually no tendencies to migrate or bleed. However, they are expensive and are not stable under reducing conditions. The more commonly used vat pigments include Flavanthrone Yellow, not over 200°C; Anthrapyrimidine Yellow, not for nylon; Pyranthrone Orange; Perinone Orange; brominated Anthranthrone Orange, vinyl only; brominated Pyranthrone; Anthramide Orange; Indanthrone Blue, red shade; and Isoviolanthrone Violet (and the similar PV 31 and PV 33). The formula for Isoviolanthrone Violet [*81-28-7*] illustrates the complicated structures of the vat pigments.

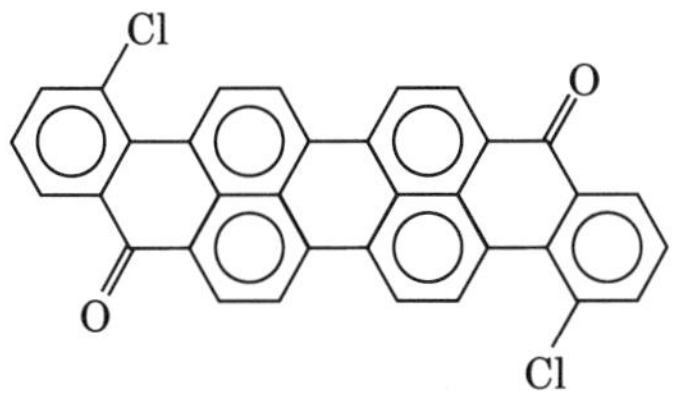

Isoviolanthrone Violet

Perylene Pigments. The perylenes are a class of red and maroon pigments. In the general formula, R may represent a simple alkyl, methyl, or a substituted phenyl, eg, PR 123, R = *p*-ethoxyphenyl.

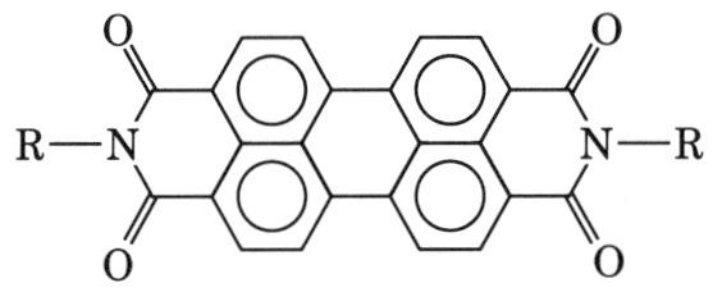

These pigments are recommended for most plastic systems because of their excellent stability to chemicals, bleeding, and light. They are widely used in vinyls, polyethylene, polypropylene, and cellulosic plastics. The *Colour Index* classes are listed as PR 123, 149, 179, and 190.

Thioindigo Pigments. The thioindigos are red and violet pigments developed for textiles. Two red–violets, Pigment Red 88 [*14295-43-3*] and Pigment Red 198 [*6371-31-9*], are recommended for plastics because of their excellent fastness properties.

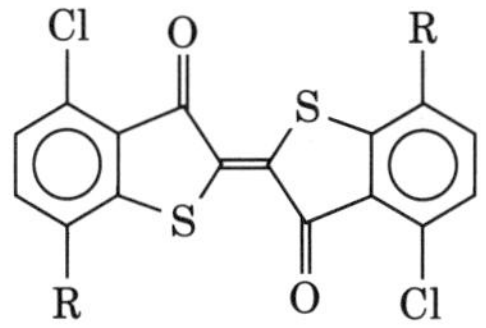

R = Cl, PR 88
R = CH₃, PR 198

Phthalocyanine Pigments. Copper Phthalocyanine Blue and its related greens represent the best combination of properties available in any pigment class. The bright, clean hues and outstanding fastness make them the colorants of choice in many media. The copper in Phthalocyanine Blue is so tightly bound that the pigment may be used in rubber, ABS, and high impact polystyrene. There are two crystalline forms, one red and one green. The latter is more stable; prolonged heat tends to turn the former greenish. Phthalocyanines may retard the curing of thermoset resins.

Replacement of the hydrogens in Phthalocyanine Blue by chlorine gives a blue–green. If bromine replaces chlorine, the shades become yellower, progressively yielding pigments designated 2Y, 3Y, 6Y, and 8Y (see PHTHALOCYANINE COMPOUNDS).

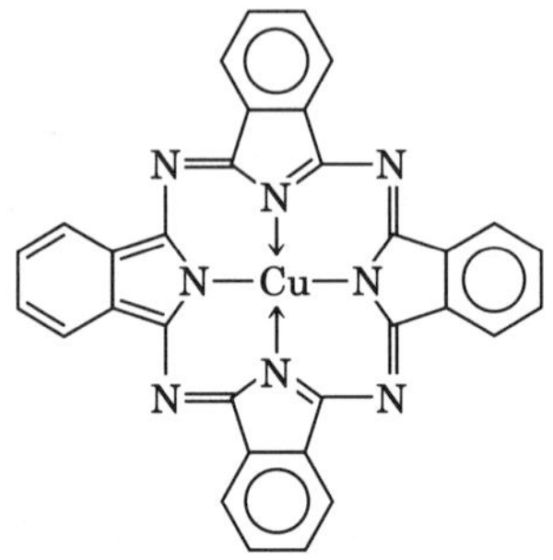

Phthalocyanine Blue (PB 15)

Tetrachloroisoindolinones.   The tetrachloroisoindolinones are yellow, orange, and red pigments marked by CIBA-GEIGY under the trade name Irgazin. They are difunctional amines stabilized by two of the tetrachloroisoindolinone units. These pigments have excellent resistance to bleeding and light and are stable up to 290°C. They are recommended for ABS, polypropylene, and thermosets.

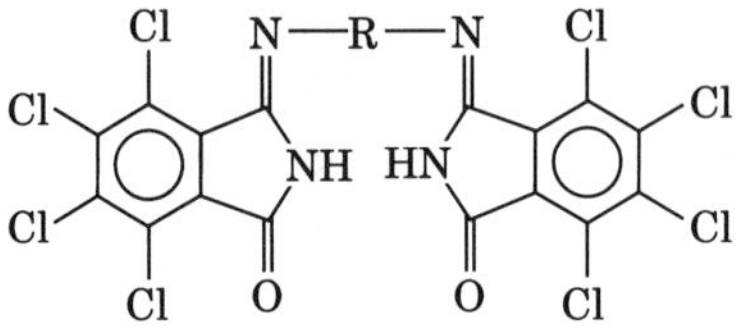

tetrachloroisoindolinones

Fluorescent Pigments.   Fluorescent pigments or dyes depend on their ability to absorb light at one wavelength and to reemit it in a narrow intense band at a longer wavelength (see LUMINESCENT MATERIALS) (16). The dyes used include the rhodamines, which emit pink and aminonaphthalimides that are bright greenish yellow. To obtain maximum effect, the dyes are dissolved in brittle resins at low concentrations. The colored resins are then ground to powders and used as pigments. The brightness of such a combination far exceeds that of any pigment alone.

Traditional fluorescent dyes do not have lightfastness. Their use in plastics is confined to the lower temperature resins, vinyls, polyethylene, polystyrene, and acrylics at maximum temperatures of 200°C. New fluorescent pigments have been developed that have improved weathering as a result of their encapsulation in an improved weathering vehicle. Improvements of these pigments in heat performance (up to 300°C) have been demonstrated in polyolefins if the resin matrix in the fluorescent pigment is based on polyamide.

Dyes

Dyes should be checked for migration, sublimation, and heat stability before use. These precautions are particularly important for plasticized resins.

Azo Dyes. The *Colour Index* classifications of dyes depend more on their historical early use than on their structures, eg, Oil Orange is named Solvent Yellow 14, and a yellow for synthetic fibers is Disperse Yellow 23.

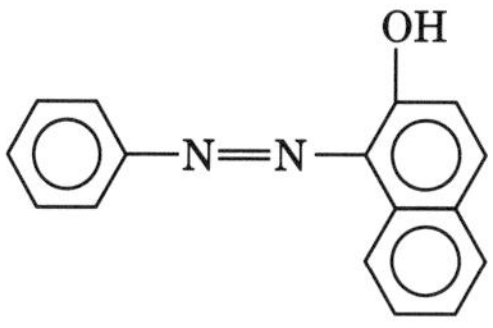

Solvent Yellow 14

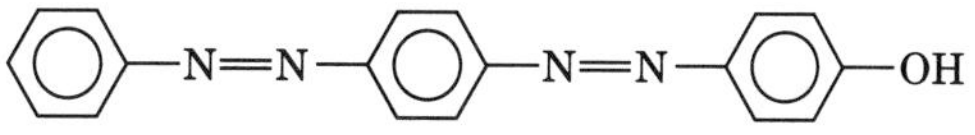

Disperse Yellow 23

In general, the azo colors are useful for coloring polystyrene, phenolics, and rigid poly(vinyl chloride). Many are compatible with poly(methyl methacrylate), but in this case the weatherability of the resin far exceeds the life of the dyes. Among the more widely used azo dyes (qv) are Solvent Yellows 14 and 72; Orange 7; and Reds 1, 24, and 26.

Azo acid dyes, of which Metanil Yellow is an example, are stabilized by sulfonic acid groups and also have affinity for phenolic resins (qv).

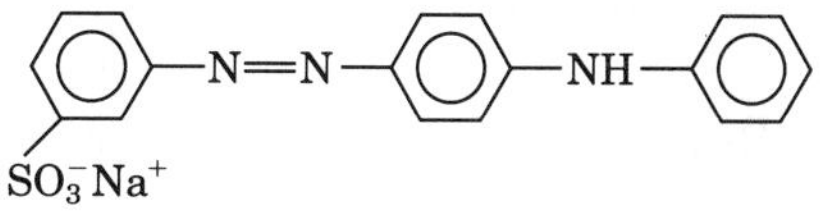

Metanil Yellow (Acid Yellow 36)

Anthraquinone Dyes. These dyes have much superior weatherability and heat stability compared with the azos, but at higher cost. Typical examples are Solvent Red 111, Disperse Violet 1, Solvent Blue 56, and Solvent Green 3.

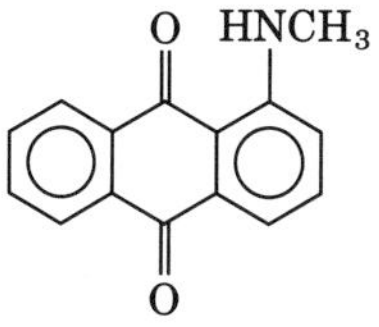

Solvent Red 111

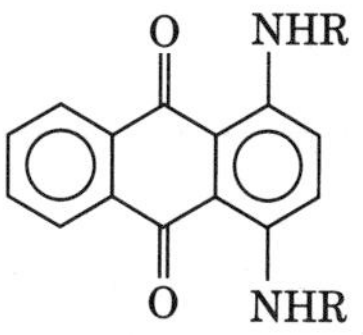

Disperse Violet 1 (R = H)
Solvent Blue 56 (R = CH(CH$_3$)$_2$)
Solvent Green 3 (R = —⟨O⟩—CH$_3$)

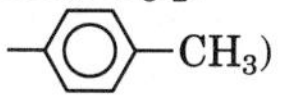

The anthraquinones are useful in acrylics and are compatible with polystyrene and cellulosics. Solvent Red 111 has a special affinity for poly(methyl methacrylate) as the red in automobile taillights; exposure for a year in Florida or Arizona produces only a very slight darkening. Acid types are useful for phenolics (see DYES, ANTHRAQUINONE).

Xanthene Dyes. This class is best represented by Rhodamine B. It has high fluorescent brilliance but poor light and heat stability; it may be used in phenolics. Sulfo Rhodamine is stable and is useful in nylon-6,6. Other xanthenes used in acrylics, polystyrene, and rigid poly(vinyl chloride) are Solvent Green 4, Acid Red 52, Basic Red 1, and Solvent Orange 63 (see XANTHENE DYES).

Rhodamine B

Azine Dyes. Azine dyes (qv) include induline and nigrosines. They produce jet blacks unobtainable with carbon black. This was particularly true of Induline Base in nylon before its manufacture was discontinued because of a carcinogenic impurity (4-aminobiphenyl). The nigrosines are used in ABS, polypropylene, and phenolics.

Other Dyes. Brilliant Sulfoflavine, or Acid Yellow 7, is an amino ketone having bright greenish fluorescence used in acrylics and nylon. Solvent Red 135 and Solvent Orange 60 are perinone dyes for acrylics and other high performance plastics. They have good light and heat stability for ABS, cellulosics, polystyrene, and rigid poly(vinyl chloride). Two basic triphenylmethane dyes (qv), methyl violet and Victoria Blue B, are suggested for phenolics although their heat and light stability is minimal.

The quinoline yellows are quite stable and may be used in ABS, polycarbonate, acrylics, polystyrene, and nylon.

Economic Aspects

The cost of the individual colorants plus the method of addition (concentrate, dispersion or raw colorant) may be a significant portion of a colored part's costs. These costs often can be rapidly changing because of raw material availabilities. An accurate up to date cost profile of colorant raw materials should be kept for every formulator.

The data on usage of colorants is difficult to obtain owing to their proprietary nature. Table 3 demonstrates the trends in colorant usage (11).

Table 3. U.S. Consumption of Colorants, 10^6 kg

	1972	1983	1985
Inorganic pigments			
titanium dioxide	86.18	115.00	123.64
iron oxides	3.34	3.97	4.36
cadmium	2.30	2.20	2.36
chrome yellows	2.60	2.30	2.39
molybdate oranges	1.78	1.60	1.59
other	1.55	1.35	1.41
Total	*97.75*	*126.42*	*135.75*
Organic pigments			
carbon blacks	23.55	34.81	38.64
phthalo blues	1.45	1.35	1.41
phthalo greens	0.95	0.57	0.61
organic reds	1.13	1.15	1.23
organic yellows	0.19	0.22	0.27
others	0.49	0.50	0.52
Total	*27.76*	*38.60*	*42.68*
Dyes			
nigrosines	1.14	1.45	1.45
0.1 solubles	0.61	0.62	0.68
anthraquinones	0.22	0.20	0.23
others	0.28	0.22	0.27
Total	*2.25*	*2.49*	*2.63*
Grand Total	*127.76*	*167.51*	*181.06*

Health and Safety Factors

The toxicity of colorants used in plastics is in a dynamic environment today. With the rapidly changing regulations, it would not be appropriate to identify the acceptabilities of particular colorants. The regulations regarding heavy-metal compounds represent the largest area of concern. As organic dye and pigment replacements are developed to replace heavy metals, there may be additional concerns regarding the decomposition of these chemicals with incineration (see DYES, ENVIRONMENTAL CHEMISTRY).

Traditionally, one of the biggest problems with colorants is their dustiness. Besides a contamination problem, inhalation can either be a nuisance or hazardous, depending on the colorant involved. Several producers offer low dusting products or encapsulated products to improve this situation.

Dyes may also sublime under operating conditions and show on mold surfaces. Migration is tested by making a sandwich of alternately colored process under a weight at elevated temperatures. Chemical extraction tests can also be run for food application products. The FDA is also in the process of revising its rules for testing acceptable products and the approved list of acceptable materials.

Previously, titanium dioxide, iron oxides, and ultramarine blue were approved by FDA. A few pigments are certified for use in dyes and cosmetics and fewer dyes and alumina hydrate lakes are certified as food colors (see COLORANTS FOR FOODS, DRUGS, COSMETICS, AND MEDICAL DEVICES).

BIBLIOGRAPHY

"Colorants for Plastics," in *ECT* 3rd ed., Vol. 6, pp. 597–617, by T. G. Webber, Consultant.

1. T. B. Reeve and W. L. Dills, *Principles of Pigment Dispersion in Plastics, 28th ANTEC Preprint*, Society of Plastics Engineers, Inc., Greenwich, Conn., 1970, p. 574.
2. R. J. Kennedy and J. F. Murray, *Internal Pigmentation of Low Shrink Polyester Molding Compositions with Flushed Pigments, 33rd ANTEC Preprint*, Society of Plastics Engineers, Inc., Greenwich, Conn., 1975, p. 148.
3. DSET, private communication, 1980–1992.
4. P. J. Keating, Rohm and Haas, private communication, 1986.
5. M. Ali and B. Muller, *The Effect of Different Pigment Classes and Additives on Polymer Physicals*, Accurate Color, SPE CAD RETEC, 1983.
6. N. J. Napier, *High Temperature Synthetic Inorganic Pigments, 31st ANTEC Preprint*, Society of Plastics Engineers, Inc., Greenwich, Conn., 1973, p. 397.
7. T. B. Reeve, *Plast. Eng.* **33**(8), 31 (1977).
8. W. G. Huckle, G. F. Swigert, and S. E. Wiberley, *I&EC Prod. Res. Dev.* **5,** 362 (1966).
9. H. C. Felsher and W. J. Hanau, *J. Paint Technol.* **41,** 354 (1969).
10. T. C. Patton, ed., *Pigment Handbook*, Wiley-Interscience, New York, 1976.
11. R. Juran, ed., *Modern Plastics Encyclopedia*, Vol. 66, McGraw-Hill Book Co., Inc., New York, 1990, p. 648.
12. R. Az, B. Dewald, and D. Schnaitmann, *Dyes Pigments* **15,** 1 (1991).
13. U. S. Pat. 2,844,484 (July 22, 1958), W. S. Struve (to E. I. du Pont de Nemours & Co., Inc.).
14. H. F. Bartolo, *Quinacridone Pigments in Plastics, 27th ANTEC Preprint*, Society of Plastics Engineers, Inc., Greenwich, Conn., 1969, p. 518.
15. V. C. Vesce, *Off. Dig. Fed. Soc. Paint Technol.* **28**(II), 1 (1956); **31**(II), 1 (1959).
16. H. Bunge, *Using Fluorescent Pigments in Plastics*, Day-Glo Color Corp., SPE CAD RETEC, 1990.

General References

Colorant Market (market research reports), 1986. Write for information: Frost & Sullivan, Inc.
Colour Index, and its Additions and Amendments, 3rd ed., Society of Dyers and Colourists, London, and American Association of Textile Chemists and Colorists, Durham, N.C. It now consists of seven volumes. The *Colour Index* was originally written by and for the textile industry, but pigments are receiving increasing attention.
D. B. Judd and G. Wyszecki, *Color in Business, Science and Industry*, 3rd ed., John Wiley & Sons, Inc., New York 1975.
P. A. Lewis, ed., *Pigment Handbook*, 3 vols., John Wiley & Sons, Inc., New York. Vol. 1: *Properties & Economics*, 2nd ed., 1988. Vol. 2: *Applications & Markets*, 2nd ed., 1988. Vol. 3: *Characterization & Physical Relationships*, 1973. John Wiley & Sons, Inc.
H. A. Lubs, ed., *The Chemistry of Synthetic Dyes and Pigments*, Reinhold Publishing Corp., NY, 1971. Written by Du Pont (the original was written in 1955).
D. M. Marmion, *Handbook of U.S. Colorants: Foods, Drugs, Cosmetics & Medical Devices*, 3rd ed., John Wiley & Sons, Inc., New York, 1991.

T. C. Patten, ed., *Pigment Handbook*, 2nd ed., 3 vols., Wiley-Interscience, New York, 1979. Written by a number of specialists in the pigment field; the three volumes describe pigment chemistry and applications.

A. T. Peters and H. S. Freeman, *Color Chemistry, Advances in Color Chemistry Series*, Elsevier Science Publishing Co., Inc., New York, 1991.

Textile Chemist and Colorist-Buyer's Guide, American Association of Textile Chemists and Colorists, Durham, N.C., annual. Lists American-made dyes and pigments by trade name, *Colour Index* generic number, and formula number, if available (no formulas are given).

K. Venkataramam, ed., *The Chemistry of Synthetic Dyes*, 8 vols., Academic Press, Inc., New York, 1951–1978. The most comprehensive treatise on dyes; emphasis is on chemistry, not applications.

H. Zollinger, *Color Chemistry: Syntheses, Properties & Applications of Organic Dyes & Pigments*, 2nd rev. ed., VCH Publishers, Inc., New York, 1991.

GARY BEEBE
Rohm and Haas Company

COLOR MEASUREMENT. See COLOR; DYES, APPLICATION AND EVALUATION-EVALUATION.

COLOR PHOTOGRAPHY

Color photography is a technology by which the visual appearance of a three-dimensional subject may be reproduced on a two-dimensional surface having a pleasing balance of brightness, hue, and color saturation. The physical record of the image is expected to have reasonable permanence. It may be viewed directly as a color print, by projection as a color transparency (slide), or by back-illumination as a display transparency. There are two essentials in the practice of color photography: the camera and the film. The task of the camera is to present an undistorted image to the plane of the photographic film with an intensity level and exposure time appropriate to the sensitivity of the film being used. The technology of the light-sensitive film is discussed herein. References 1–19 are sources of general information on the relevant supporting technologies (see also PHOTOGRAPHY; COLOR PHOTOGRAPHY, INSTANT). A detailed history of color photography may be found in References 1 and 2.

Color Vision and Three-Color Photography

Although color (qv) has been used for graphic purposes since prehistoric times, it was Isaac Newton's discovery of the solar spectrum in 1666 that led to an understanding of color with regard to the properties of light. From the observation that white light is not a pure entity, but consists of a mixture of colors, the idea developed that any color could be obtained by a mixture of three primary colors. A wide range of colors could be obtained by the mixing of red, green, and blue lights. At the start of the nineteenth century, the Young-Helmholtz theory of color vision proposed that there were three types of receptor in the human retina, each of which responded over a certain wavelength range in simple proportion to the amount of light absorbed. Because the concept of color is meaningless in the absence of an observer, the physiological basis of color is used as a starting point for trichromatic color reproduction.

Human vision is sensitive to electromagnetic radiation in the 400–700 nm range. Figure 1 is an illustration of the perceived colors over this wavelength range, together with the representative spectral sensitivities of human retinal receptors (20). There are no abrupt transitions between the spectral colors, as indicated by the vertical lines, but rather a gradual merging.

The human eye is known to contain two main types of light-sensitive cells: rods and cones. These names come from the shape of the outer segment of the cell. The rods are by far the most numerous, outnumbering the cones by a factor of nearly 20, and operate as the main sensors at low illumination levels. Scotopic (rod) vision as shown in Figure 1 centers at about 510 nm and extends from 400–600 nm. Because there is only one type of rod, it lacks the ability to discriminate colors and its electrical output varies as the integrated response over this wavelength range. Cones are the primary sensors of color vision, and there is strong evidence that there are three types, shown as ρ, γ, and β. Over most of the visible range any single wavelength stimulates more than one receptor type. For

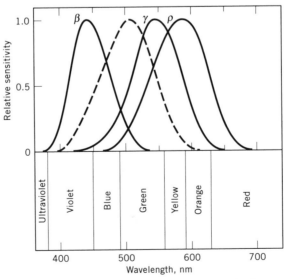

Fig. 1. Representative spectral sensitivities of the human retinal receptors, (- - -) scotopic (rod) vision, and β, γ, and ρ cone sensitivities. The wavelengths of the perceived colors are also shown. Reproduced with permission (20).

example, the peak sensitivities of the ρ and γ receptors, at 580 nm and 540 nm, respectively, are quite close together.

The ability to distinguish small differences in spectral wavelength is believed to rely on the difference signals between the responses of the three cone detectors. For example, stimulation of the ρ and γ receptors by monochromatic light of 570 nm has no difference signal, and is thus interpreted by the brain as yellow light. However, for monochromatic light of 550 nm, stimulation of the γ receptor is increased, that of the ρ receptor is decreased, and the difference signal between the two is interpreted by the brain as greenish light. Most colors consist of many wavelengths of varying intensities; these signals are additive for each receptor.

This understanding of human color vision makes a definition of the characteristics of a perfect color reproduction system possible. For simplicity, assume a reproduction of any uniform patch of color, such that the original and the reproduction are indistinguishable under any illuminant, is desired. The first requirement is the selection of three sensors that match the cone sensitivity curves shown in Figure 1. The second requirement is that these sensors faithfully record the intensity or brightness of the stimulus. The third requirement is that in the viewing of the reproduction, the physical record of each sensor's response must be capable of stimulating only the cone receptor it is designed to mimic. This last requirement cannot be met in practice because even monochromatic light over most of the visible range stimulates more than one receptor type. Thus the construction in color vision that permits exquisite wavelength discrimination also makes perfect color reproduction using three-color mixing impossible. With this limitation, the design challenge of color photography is to reduce or correct for unwanted stimulations as much as possible.

In 1861, James Clerk Maxwell demonstrated that colored objects could be reproduced by photography, using three-color mixing (1). He produced three separate negative images in silver on glass plates, ostensibly taken through red, green, and blue filters. After the three negative plates were converted to positive plates, the positives were then loaded into three separate projectors, each of which had the corresponding filter used in the exposure in front of the lens, and the images were brought into registration on a white screen. A tolerable color reproduction was obtained. Although considered a landmark, the success of the experiment was fortuitous, because the silver halide crystals used as the light-sensitive element were sensitive only to ultraviolet and blue light. It was subsequently shown (21) that the green exposure resulted from the sensitivity of the silver halide tailing into the blue-green region of the spectrum, and the red exposure resulted from ultraviolet radiation passed by the red filter employed. Thus the ability to sensitize silver halide crystals to green and red light, in addition to the inherent blue sensitivity, is critical. Intrinsic sensitivity depends on the halide chosen: useful sensitivity does not extend beyond 410 nm for AgCl and 490 nm for AgBr.

It is axiomatic that only radiation that is absorbed can produce chemical action (Grotthus-Draper Law), in this case the promotion of an electron from the valence to the conduction band of silver halide. It was discovered in 1873 that certain organic dyes, when adsorbed to silver halide crystals, induce sensitivity to the longer wavelengths of the visible spectrum. In the course of testing photographic plates that contained a yellow dye to prevent back-reflection of light (halation), some sensitivity to green light was observed. Many thousands of dyes have been tested as spectral sensitizers, and materials exist for the selective sensitization of any region of the visible spectrum and also the near infrared. As well as permitting better tone reproduction in black-and-white photography, this technology is the key to providing the red, green, and blue discrimination required in modern color photography.

Maxwell's demonstration was an example of an additive trichromatic process in that the final image was produced by combining red, green, and blue lights in registration, each light corresponding to its amount present in the original scene. The various colors produced by the additive combination of red, green, and blue lights are shown in Figure 2 on the colored plate. To simplify the discussion, blue light is considered the 400–500 nm range of the spectrum, green the 500–600 nm range, and red the 600–700 nm range. White light is produced when all three colors are combined in equal amounts. The two-way combinations of red, green, and blue light produce three colors that each lack one-third of the spectrum. The color cyan is generated by the combination of blue and green lights, and lacks light in the red part of the spectrum. Similarly, the color magenta lacks light in the green part of the spectrum, and the color yellow lacks light in the blue part of the spectrum. For this reason cyan, magenta, and yellow are called subtractive primaries.

Early experimenters focused on the problem of converting the three separation positives produced by Maxwell's method into a form that could be printed with colored inks on a paper surface. The first clear statements of the basic principles underlying modern three-color photography are credited to two independent workers: du Hauron, who applied for a French patent in 1868, and Cros, who wrote an article in *Les Mondes* in 1869. Both noted that printing colors should be complementary or "antichromatic" to the red, green, and blue taking filters. The complementary colors are the subtractive primaries. Cyan is complementary to red, magenta is complementary to green, and yellow is complementary to blue. Each subtractive primary dye absorbs the color to which it is complementary. Figure 3 (on the colored plate) illustrates the superposition of equal amounts of the three subtractive primaries as dyes. The three possible two-way combinations yield red, green, and blue. The three-way combination yields black. Thus the three subtractive primary dyes used singly or together are capable of creating a wide variety of colors by selectively modulating the red, green, and blue components in white light.

The Light-Recording Element

The primary element for light capture in color photography is the silver halide crystal. By common usage, the term emulsion has come to denote in the photographic literature what is actually a dispersion of silver halide crystals (grains) in gelatin. Four discrete steps can be identified in the formation of a colored image: (1) light absorption by the crystal; (2) the solid-state processes leading to the formation of a latent image; (3) the reduction of all or part of a crystal-bearing latent image to metallic silver by a mild reducing agent; and (4) use of the by-products of the silver reduction to create a colored image in register by dye formation, dye destruction, or dye-transfer processes.

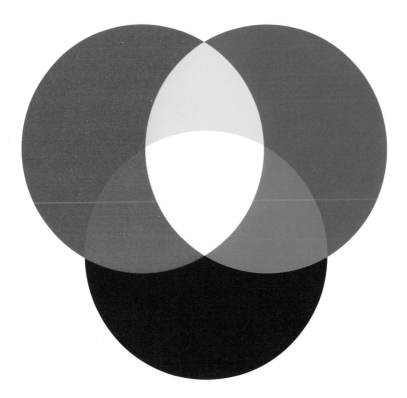

Fig. 2. Addition mixing of red, green, and blue lights.

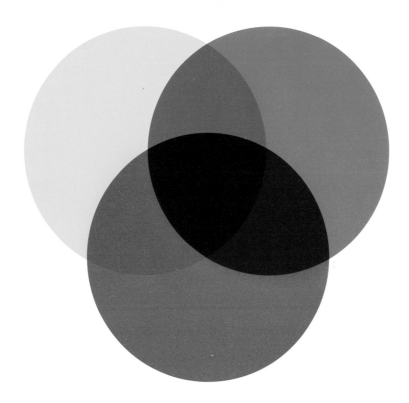

Fig. 3. Superimposed subtractive dyes.

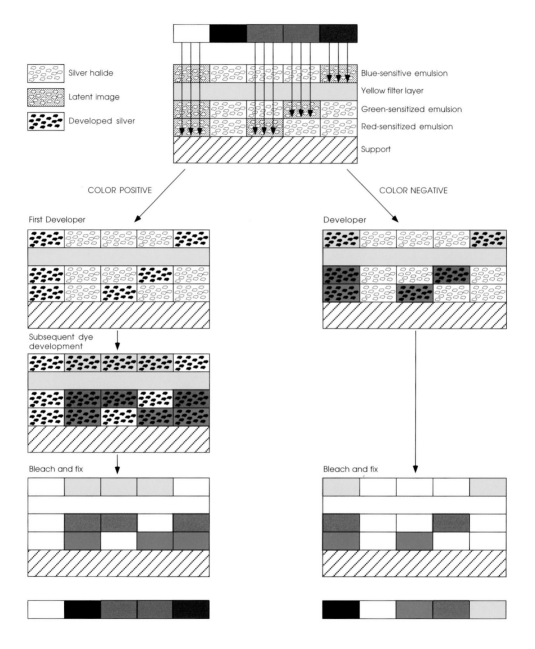

Fig. 4. Positive and negative-working integral tri-packs.

Sensitivity, or photographic speed, is one of the most important attributes of the light-sensitive element. Practical color photography using a handheld camera is possible from conditions of bright sunlight to night street lighting. These conditions span a factor of about 10^5 in illuminance, and must be accommodated for by the combination of camera shutter speed, lens aperture, and choice of film speed. Most commercial emulsions contain a population of silver halide crystals varying widely in size and shape. Although the structure of the AgBr and AgCl lattice is simple face-centered cubic, an enormous variety of crystal shapes can be obtained, depending on the number and orientation of twin planes present in the crystal, the silver ion concentration during growth, and the presence of growth modifiers (22,23). To add to this complexity, the crystals in commercial emulsions usually contain mixed-halide phases. Films suitable for a handheld camera generally contain silver bromoiodide, in which iodide ions are incorporated into the AgBr lattice during crystal growth.

The relationship between crystal size and photographic speed can be understood using simple geometric arguments. For an individual crystal, sensitivity may be defined as the reciprocal of the minimum light absorption required to generate a developable latent image. For a silver halide crystal without any sensitizing dye, blue light absorption is proportional to volume. If it is assumed that the crystal is a sphere and that the latent image can be formed with equal efficiency at all grain sizes, the relationship shown in Figure 5 is obtained. However, the adsorption of dyes is necessary to confer sensitivity to the green and red regions of the spectrum; this is frequently called "minus-blue" speed. To a first approximation, minus-blue speed depends on the surface area available for dye adsorption. Again, assuming sphericity, the line shown in Figure 5 is the expected

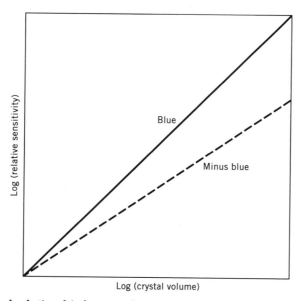

Fig. 5. Calculated relationship between log (relative sensitivity) and log (crystal volume) for (—) intrinsic response (blue) and (- - -) dyed response (minus blue), assuming crystals are spherical.

change of minus-blue speed with crystal size. Even for the highest speed films, the crystals do not usually exceed 3 μm in linear dimension.

Over the years emphasis has been placed on obtaining greater uniformity in silver halide crystal size and habit in the grain population, in the belief that the chemical sensitization process can then yield a higher average imaging efficiency. One way of doing this is to adjust the nucleation conditions so that untwinned crystals are favored, and then to ensure that no new crystals are formed during the growth of the starting population. Crystals containing twin planes grow anisotropically and it is more difficult to obtain a uniform population. Commercial materials are available that contain cubic and octahedral crystals of narrow size dispersity. Along with better control of the crystal size and shape in the population, the placement of iodide in the AgBr crystal has received a great deal of attention. Iodide incorporation increases blue light absorption, and its selective placement within the crystal allows control over the rate of the development reaction. In color negative films, the color image is formed by reducing less than one-quarter of the total silver available to metallic silver during development. This strategy is necessary so that the crystals are large enough to have the required sensitivity, yet each crystal by its partial reduction contributes only a small amount of dye to the final image, leading to an image of low graininess.

An approach has been devised (24) to break out of the surface-to-volume relationship imposed by crystal shapes that are nearly spherical. Conditions have been established to favor the growth of crystals having multiple parallel twin planes (25), and emulsions containing mostly hexagonal tabular crystals, such as those shown in Figure 6**b**, can be prepared. Figure 6 compares the crystals of a

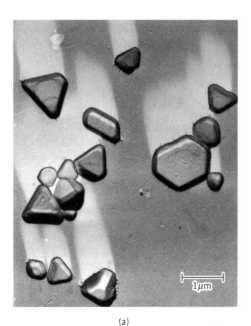 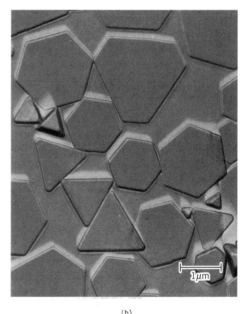

(a)   (b)

Fig. 6. Emulsions (**a**) traditional and (**b**) hexagonal tabular crystals.

newer emulsion to a more traditional one. Tabular thicknesses of around 0.1 μm are commonly employed. By adjustment of the projective area and thickness of these tabular crystals it is possible to create a series of emulsions in which the surface area and hence minus-blue speed increases, but the crystal volume and hence blue speed remain constant. Because an emulsion that is dyed to be sensitive to green or red light still retains its intrinsic blue sensitivity, the greater minus-blue/blue sensitivity ratio afforded by tabular crystals reduces unwanted blue light sensitivity in those layers of the film designed to record green and red light. Another advantage cited for this technology is an improvement in the optical transmission properties of layers containing tabular emulsions. Because the crystals align themselves nearly parallel to the film support during the coating and drying operations, the transmitted light is less scattered sideways. This is advantageous for image sharpness in multilayer color films.

Sensitizing dyes (26,27) are essential to the practice of color photography. Although the initial discovery was made in 1873, it was not until the 1930s that photographic scientists began to develop a systematic series of dyes that could be adsorbed to the silver halide crystal and then transfer to it the energy of green and red light necessary to create a latent image. The most widely used of these materials are the cyanine dyes, which are heterocyclic moieties linked by a conjugated chain of atoms. More than 20,000 dyes of this class have been synthesized. In the example shown in Figure 7, when $n = 0$, the dye is yellow and provides sensitization to blue light. Dyes which sensitize in the blue are particularly important for AgCl emulsions, which lack intrinsic blue sensitivity. When $n = 1$, the dye, a carbocyanine, is magenta and absorbs green light. For $n = 2$, a dicarbocyanine, the dye appears cyan and sensitizes silver halide to red light. Extend-

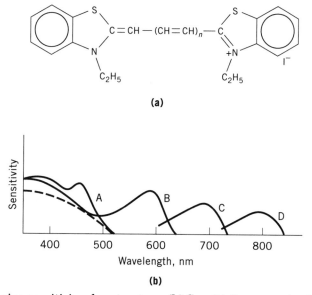

Fig. 7. (a) Cyanine sensitizing dye structure. (b) Sensitivity curves (- - -) intrinsic silver halide; and (—) for the dye in (a) where A corresponds to $n = 0$; B to $n = 1$; C to $n = 2$; and D to $n = 3$.

ing the conjugation further produces dyes which absorb in the infrared, producing films useful for aerial photography and thermal analysis (see CYANINE DYES; DYES, SENSITIZING).

There are several requirements for a good sensitizing dye. A good dye is adsorbed strongly to silver halide. The dye molecules attach themselves to the surface of the silver halide crystals, usually up to monolayer coverage. This amount can be determined by measuring the adsorption isotherm for the dye. Dye in excess of this amount can cause the emulsion to lose sensitivity, so-called dye-induced desensitization. In color films, simple positively charged cyanine dyes can also dissolve in the organic phase used to solubilize the image-forming coupler, leading to a phenomenon known as unsensitization. This can be overcome by adding a negatively charged acidic group, such as sulfonate, to turn the dye into a zwitterion that is insoluble in organic media. Adsorption is also influenced by the composition and crystallographic surfaces of the silver halide crystal, and by the procedure of dye addition. The optical properties of the dye change on being attached to silver halide. Typically the light absorption peak shifts by 30–50 nm to longer wavelengths. Frequently new absorption peaks occur as a result of different stacking arrangements of the dyes on the crystal surface.

A good sensitizing dye absorbs light of the desired wavelength range with high efficiency. The absorption spectrum of the sensitizing dye depends on a number of factors. Each heterocyclic nucleus has a characteristic color value associated with it which can be modified by chemical substitution. The chromophore length is another important variable. A useful empirical rule is that for each increment in n, the absorption peak of the dye shifts to longer wavelengths by 100 nm.

A good sensitizing dye sensitizes with high efficiency. Modern sensitizing dyes transfer an electron from the dye's excited singlet state to the conduction band of silver halide, which is subsequently able to participate in latent image formation. The ability of the dye to serve as an efficient sensitizer is often found to be proportional to the electrochemical reduction potential of the dye measured in solution. This has proved to be a useful tool in designing spectral sensitizers. A practical measure of dye efficiency is the relative quantum efficiency of sensitization (28) defined as the ratio of the number of quanta absorbed in the intrinsic region, usually 400 nm, of the silver halide to the number of quanta absorbed only by dye at a wavelength within its absorption band, both to produce the same specified developed density. For the best dyes this value is only slightly less than 1.0.

A good sensitizing dye does not interfere with other system properties. Sensitizing dyes can sometimes influence the intrinsic response of a chemically sensitized emulsion, leading to desensitization or additional sensitization. The dye can also interfere with development rate, increase or decrease unwanted fog density, and remain as unwanted stain in the film after processing. The dye should have adequate solubility for addition to the emulsion, but should not wander between layers in the final coating.

Color Processes

Additive Mixing.   The first commercially successful color photographic systems used additive color mixing. Simultaneous recording of red, green, and blue images was achieved in the chromoscope camera in 1898. The image beam was split into three using semitransparent mirrors and prisms, the three beams passing through red, green, and blue filters before striking the emulsion surface. One of the most serious practical difficulties encountered was nonuniformity of intensities at the film plane as a result of the optics of semireflecting mirrors.

If different colors are presented rapidly enough to the eye, they are additively fused by the visual system. This principle has been applied to the generation of moving color images, where successive frames contain the red, green, and blue records of the scene. Photography and projection is accomplished by having the appropriate colored filters move in synchronism with the image frames. This method places great mechanical demands on apparatus, however, requiring much higher than the normal projection rate of 24 frames per second. The results are often unsatisfactory because of color fringing of moving objects, and flicker from the different transmission characteristics of the filters used.

Mixing additive primaries as dyes by superposition does not work, because each primary absorbs two-thirds of the spectrum. The two difficulties of registration and additive superposition were overcome by what is known historically as the screen-plate process. The additive primaries are presented to the eye as a mosaic of very small colored dots in juxtaposition, as in pointillist painting. Although close inspection reveals a pattern of colored dots, additive blending occurs by increasing the viewing distance. The retina of the eye is itself a random mosaic of red, green, and blue receptors, and if the dot pattern is fine enough, the eye interprets the image as being smooth. The photographic record is obtained by exposing a silver halide emulsion which is sensitized throughout the visible spectrum, ie, it is panchromatic, through a mosaic of very small red, green, and blue filters. The film is then processed to give a positive image in silver. When viewed or projected in registration with the original mosaic, a colored image is created. The amount of light transmitted through each filter is controlled by the developed silver optical density.

The first commercially successful screen-plate process was the Autochrome Plate made by the Lumiere Co. in France in 1907. The mosaic of filters was integral to the photographic plate, and consisted of starch granules about 15 μm in diameter that were dyed red, green, and blue. The individually dyed granules were mixed and pressed onto the plate so that their edges touched. The spatial distribution of red, green, and blue granules was random. Any gaps in the mosaic were filled by a carbon paste. This filter screen was then overcoated with the emulsion. Exposure was made through the reverse side so that the exposing light passed through the filters first. The final image was intended for direct viewing or projection. The relative surface areas of the three primary colors were such as to give a satisfactory neutral with the viewing illuminant. The Autochrome process survived commercially until the mid-1930s. Mottle sometimes appeared in the Autochrome image because of clumping of the starch granules. A regular grid of filters gave more satisfactory results, as in Dufaycolor (1908), which employed a very fine square grid of filter elements.

The lenticular method (29) was also commercially successful. The reverse side of a panchromatic black-and-white film was embossed with a very fine (about 25 per mm) pattern of cylindrical lenses or lenticules. The camera had a red filter over the top third of its lens, a green filter over the middle third, and a blue filter over the bottom third. During exposure, the lenticules were parallel to the camera filters. Light from any tiny area of the subject is focused onto the lenticular surface, which faces the lens. The lenticule then focuses a tiny image of the lens aperture with its three filters onto the panchromatic emulsion coated on the reverse side of the film. The relative intensities coming through each filter depend on the color of that tiny area of the subject. This process occurs for every point of the image, resulting in three horizontal bands for each lenticule. In this system, the variables to be optimized were line spacing, the thickness of the support, and the curvature of the lenticules. When the film is given a black-and-white reversal process, and the optical path reversed in a projector where the lens has the same arrangement of filters as the camera, a colored image is obtained. This system was introduced in 1928 by Eastman Kodak Co. as Kodacolor and was available until 1935 as a 16-mm motion picture film.

In the 1990s, the additive process is used in color television, in which light emitted from a tiny regular mosaic of red, green, and blue phosphors blends to give the colored image. Another modern additive color system is Polaroid's Polachrome 35-mm transparency film, which consists of a positive silver image overlying an additive screen having 394 triplets of red, green, and blue lines per centimeter of film. However, because additive photographic systems are inherently wasteful of light (each additive filter absorbs two-thirds of the light energy), most modern systems rely on the subtractive primaries.

Subtractive Mixing.   There are mechanical difficulties in separating a photographic image into three images to record red, green, and blue information, only to recombine them later. Perfect registration of the color information can be preserved if the three-color records are stacked on top of each other on the same support. This film structure is known as an integral tri-pack. Photographic systems have been designed in which the three subtractive primary dyes, cyan, magenta, and yellow, can be formed in register, destroyed in register, or transferred in register to create the full-color image.

Dye destruction technology, Silver Dye-Bleach, is used in the Cibachrome process. After a black-and-white development step, the film is subjected to an acidic dye-bleach solution that destroys the incorporated azo dyes as a function of developed silver, leaving a residual positive color image. Because the presence of light-absorbing dyes during exposure severely limits the photographic speed of these materials, they are used to make display transparencies and prints, usually from a camera transparency original. The azo dyes (qv) used in this process offer very good light and dark stability.

The first instant color photography system, introduced by the Polaroid Corp. in 1963 as Polacolor, used the transfer of subtractive dyes to a receiver sheet to produce a positive image. The incorporated dye-developers, containing a hydroquinone moiety, are soluble in the alkaline activator solution, except where silver development occurs, when they are immobilized as the quinone form.

Another dye diffusion method is the dye transfer system in which three color separation negatives are prepared from an original positive color transparency. These are printed onto a special matrix film, which is processed in a tanning developer and washed in hot water to remove unhardened gelatin, giving three positive gelatin relief images. The depth of the gelatin relief is inversely proportional to the original camera exposure received. The corresponding subtractive primary dye is imbibed into each matrix and then transferred in register to a receiver sheet. Technicolor motion picture prints have been made by this process, which is used in situations that demand exceptional color quality and dye stability.

The first commercially successful film using the *in situ* formation of three subtractive dyes was Kodachrome film, introduced in 1935. The film has a multilayer structure in which red-sensitive, green-sensitive, and blue-sensitive emulsions are successively coated on the same support. Because the compounds necessary for dye formation are not incorporated in the film, an elaborate process is required to produce each dye in its correct layer. The first step is a conventional black-and-white development to give a silver image. In the first commercial Kodachrome process, the silver image was removed (bleached) to leave a reversal image in residual silver halide. As the initial step in color image formation, cyan dye was formed in all layers by the reduction of this (residual) silver halide. The film was then dried and subjected to a slowly penetrating bleach solution that decolorized the cyan dye and oxidized the silver in the blue- and green-sensitive layers. The process was repeated with magenta dye formation in the green- and blue-sensitive layers. The film was again dried, with subsequent selective bleaching of the magenta image and silver oxidation in the blue-sensitive layer. A final color-forming step generated yellow dye in the blue-sensitive layer. Removal of all image silver then left a positive three-color image. The modern Kodachrome process relies on selective layer exposure and dye formation steps, again using silver reduction to drive the dye formation reactions.

Subtractive dye precursors (couplers) that could be immobilized in each of the silver containing layers were sought, so that dye formation in all layers could proceed simultaneously rather than successively. The first of these to be commercialized were in Agfacolor Neue and Ansco Color films, introduced soon after Kodachrome film. These reversal working films contained colorless couplers that were immobilized (ballasted) by the attachment of long paraffinic chains. The addition of sulfonic or carboxylic acid groups provided the necessary hydrophilicity to make them dispersible as micelles in aqueous gelatin.

A different approach was taken in Kodacolor film, introduced by Eastman Kodak Co. in 1942. The couplers were ballasted but, instead of having hydrophilic functional groups, were dissolved in a sparingly water-soluble oily solvent. This oily phase was then dispersed by high agitation into a gelatin solution as fine droplets less than one micrometer in diameter. Kodacolor film is negative working, and was designed to be printed onto a companion color paper, which because it is also negative working, produces a positive color print. The whole system is known as the negative–positive process.

Figure 4 (on the colored plate) illustrates two ways in which the developed tri-pack can be processed. One leads directly to a positive image; the other leads to a negative image that can be subsequently printed on a negative-working paper. In the positive-working or "reversal" process, the first step is a black-and-white development to yield a negative silver image. After light or chemical fogging of the unexposed silver halide, subsequent development is carried out in a color developer that simultaneously reduces the unreacted silver halide and generates dye. Removal of all the silver leaves the positive color image. The negative-working process uses the initial silver development reaction to drive color formation. Because the camera negative is not the final image, the system is tolerant of underexposures and overexposures by as much as two stops (a factor of four), and density differences in the negative can be allowed for in the printing stage. The key process steps for color negative are develop/bleach/fix, which are carried out under carefully controlled temperature and agitation conditions. Water washes follow the bleach and fix steps.

The negative–positive system enjoys great commercial success. In 1949, color purity was improved by the introduction of colored masking couplers to the camera negative film, which partially correct for the unwanted absorptions of the image dyes (30). Masking couplers account for the yellow-orange color seen in the unexposed parts of modern color negative films after processing. A later improvement was the introduction of development-inhibitor-releasing (DIR) couplers, in which a silver development inhibitor released as a function of exposure in one layer can influence the degree of development in adjacent layers (31). Using masking couplers and DIR couplers in concert can substantially improve the quality of color reproduction seen in the final print.

The principal features of an integral tri-pack are shown in Figure 8. The color records are stacked in the order shown, with the red record on the bottom, the green record next, and the blue record on the top. The blue record is on the top because it is necessary to interpose a blue light filter to remove blue light which would otherwise form latent images in the underlying red and green records. Silver halides, with the exception of AgCl, have an intrinsic blue light sensitivity even when spectrally sensitized to the green or red. The traditional filter material has been Carey Lea silver, a finely dispersed colloidal form of metallic silver, which removes most of the blue light at wavelengths < ca 500 nm. This material also filters out some green and red light, thus requiring additional sensitivity from the underlying emulsions. Antihalation protection is required to prevent back-scatter of transmitted light from the air/support interface at the back of the film. This back-scattered diffuse light causes sharpness degradation and is most noticeable as a "halo" around bright objects. The antihalation underlayer is designed to prevent this. The required opacity is usually obtained by coating pre-developed filamentary silver, which is then easily removed in the normal film processing. Alternatively, an antihalation layer can be coated on the

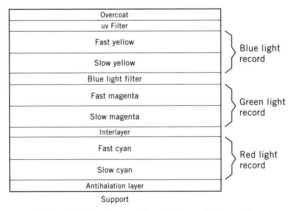

Fig. 8. Features of an integral tri-pack.

reverse side of the film. In many motion picture films, a layer of finely divided carbon is used for this purpose; it is physically removed by scrubbing before the film is processed.

Each color record is separated from its neighbor by a gelatin interlayer. This is required to prevent silver development in one record causing unwanted dye formation in another. It is also common practice to divide each color record into two or more separate layers, with the most sensitive emulsion of each record on the top. The optical screening of the underlying less sensitive layer allows a wider exposure latitude to be attained. A uv-filter layer is placed on top of the pack. Because silver halide is sensitive to ultraviolet light, this layer is designed to minimize transmission of wavelengths < ca 400 nm. Without this protection the film would record any uv radiation as blue light, leading to an unnatural bluish cast in the final image. The purpose of the overcoat layer is to give the film the desired combination of physical properties important in handling. These include adequate surface roughness to prevent sticking when the film is stored in roll form, lubrication for smooth transport in the camera, and the proper electrical conductivity to avoid static discharge.

Chromogenic Chemistry

Developers.   The detection and amplification of the latent image on the silver halide crystal occurs through the intervention of a mild reducing agent called a developer. In chromogenic development, this agent in its oxidized form has the further function of reacting with the dye-forming coupler to produce the color image.

In the corrosion model of silver development (33), the silver halide crystal may be viewed as an electrochemical cell where the latent image silver speck acts as both an anode and a cathode (Fig. 9). The silver metal that is exposed to the developer solution receives electrons from the developer, forming oxidized developer. The silver metal in contact with the silver halide crystal gives up electrons to silver ions from the crystal. The silver filament thus grows from its base, increasing the surface area for developer oxidation, accounting for the autocatalytic nature of the development process. This powerful amplification, on the order of

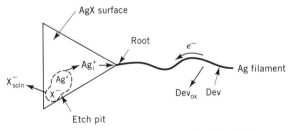

Fig. 9.   Corrosion model of silver development. As the halide ion, X^-, is removed into solution at the etch pit, the silver ion, Ag^+, travels interstitially, Ag_i^+, to the site of the latent image where it is converted to silver metal by reaction with the color developer, Dev. Dev_{ox} represents oxidized developer.

10^8 in commercial systems, of the latent image to the entire crystal is one of silver halide's unique imaging properties. As silver ions from the crystal are reduced to metallic silver, halide ions depart into solution to maintain electrical neutrality.

For modern color photographic systems, the developing agent of choice is an *N,N*-disubstituted *p*-phenylenediamine (PPD) (32,34). In the initial heterogeneous reaction with silver halide, the PPD is oxidized by the loss of one electron to form the semiquinone (eq. 1). Following desorption, two molecules of the semiquinone can undergo a fast dismutation reaction to produce the charged quinonediimine (QDI) and a molecule of the original reduced developer for an overall two-electron oxidation (eq. 2). Alternatively, the adsorbed semiquinone can give up a second electron to the silver halide crystal to produce the QDI (35). It is this species that is the active dye-forming agent (35,36). Because the overall development reaction produces protons, the process is favored by an alkaline environment and the developer solutions are often maintained around pH 10 by a carbonate buffer.

$$\overset{\cdot\cdot}{N}H_2 \quad\quad \overset{+\cdot}{N}H_2$$

$$\bigcirc + Ag^+ \rightleftharpoons \bigcirc + Ag^0 \tag{1}$$

$$NR_2 \quad\quad NR_2$$
$$\text{semiquinone}$$

$$2 \bigcirc \underset{\text{dismutation}}{\rightleftharpoons} \left[\cdots \right] + \bigcirc + H^+ \tag{2}$$

$$\text{quinonediimine (QDI)}$$

For most color photographic systems, development is the rate determining step, and within that step the formation of semiquinone is the slow process (37). The fate of the highly reactive QDI is determined by the relative rates of a number of competing processes (38). The desired outcome is reaction with ionized coupler to produce dye (eq. 3). Typically, the second-order rate constant for this process with ionized coupler is about 10^3 to 10^4 L/(mol·s) (39,40). QDI is also attacked by hydroxide ion (eq. 4) to produce a quinone monoimine (QMI), itself an oxidized developer derived from *p*-aminophenol. Such compounds can further react with coupler, albeit at a slower rate than QDI, to form a dye and were cited in the seminal patent as color developers (32). However, the dyes derived from this deaminated developer have different hues from the QDI dyes, and these hues are pH-dependent as a consequence of the phenolic group contributed by the developer. Although the deamination reaction to produce QMI is fast, the rate constant is 10^3 to 10^4 L/(mol·s) (40–42), its effect is somewhat offset by the redox reaction of the QMI with the reduced developer, present in large excess, to regenerate the desired QDI. The primary net effect of the deamination reaction is to enlarge the resulting dye cloud (43).

An even more potent nucleophile toward oxidized developer is the sulfite anion (eq. 5) present in most developer formulations (42,44). Sulfite is used to scavenge excess oxidized developer that otherwise would undergo a number of self-condensation reactions resulting in stain. Sulfite also serves to drive the redox reaction as well as acting as an antioxidant and mild silver halide solvent. Whereas the sulfite reaction is complicated, involving acid and base catalysis and a multiplicity of products, the half-life for oxidized color developer in a typical processing solution containing sulfite is only about 2 ms. This reaction with sulfite is the principal reaction limiting the size of the resulting dye cloud (45).

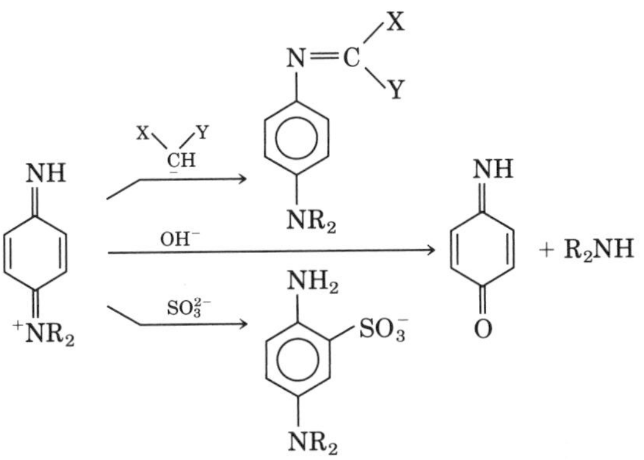

$$(3)$$
$$(4)$$
$$(5)$$

Practical developers must possess good image discrimination; that is, rapid reaction with exposed silver halide, but slow reaction with unexposed grains. This is possible because the silver of the latent image provides a conducting site where the developer can easily give up its electrons, but requires that the electrochemical potential of the developer be properly poised. For most systems, this means a developer overpotential of between -40 to $+50$ mV vs the normal hydrogen electrode.

Developing agents must also be soluble in the aqueous alkaline processing solutions. Typically such solutions are maintained at about pH 10 by the presence of a carbonate buffer. Other buffers used include borate and, less frequently, phosphate. Developer solubility can be enhanced by the presence of hydroxyl or sulfonamide groups, usually in the N-alkyl substituent. The solubilization also serves to reduce developer allergenicity by reducing partitioning into the lipophilic phase of the skin (46).

Some of the color developers in commercial use are CD-2 [*2051-79-8*], CD-3 [*25646-71-3*], and CD-4 [*25646-77-9*]. The various substituents control the rates of the various developer reactions as well as the hue, extinction, and stability of the resulting image dyes. The presence of an alkyl group ortho to the primary amino moiety serves several purposes, including steric minimization of the condensation reaction with reduced developer (47) and improved dye stability. Electron-donating groups either on the ring or on the tertiary amino group increase development rate while decreasing to a lesser extent the rate of reaction of oxidized developer with coupler (33,40). Electron-donating groups on the developer also tend to shift the absorption of the dye that forms to longer wavelengths (48).

$$\underset{\text{CD-2}}{\text{H}_5\text{C}_2\diagdown\overset{\displaystyle\text{N}}{}\diagup\text{C}_2\text{H}_5} \qquad \underset{\text{CD-3}}{\text{H}_5\text{C}_2\diagdown\overset{\displaystyle\text{N}}{}\diagup\text{C}_2\text{H}_4\text{NHSO}_2\text{CH}_3} \qquad \underset{\text{CD-4}}{\text{H}_5\text{C}_2\diagdown\overset{\displaystyle\text{N}}{}\diagup\text{C}_2\text{H}_4\text{OH}}$$

CD-2: $\cdot$ HCl, CH$_3$, NH$_2$

CD-3: $\cdot$ $\frac{3}{2}$ H$_2$SO$_4$, CH$_3$, NH$_2$

CD-4: $\cdot$ H$_2$SO$_4$, CH$_3$, NH$_2$

In addition to the developer, sulfite, and pH buffer, commercial developer solutions often contain antifoggants or restrainers that reduce the rate of development of unexposed silver halide relative to exposed grains (49). A common restrainer is the halide ion of the same type as in the film's silver halide emulsion. The excess halide serves to control the silver ion concentration or pAg, where $\text{pAg} = -\log[\text{Ag}^+]$. Other antifoggants include organic materials like benzotriazoles, benzyltriazolium salts, and 3,5-dinitrobenzoic acid [99-34-3], $\text{C}_7\text{H}_4\text{N}_2\text{O}_6$. Developer solutions can also contain antioxidants (qv) such as hydroxylamine sulfate [10039-54-0], metal sequesterants (see CHELATING AGENTS), and materials to aid in sensitizing dye removal or to boost coupling rates.

Coupler Types. A photographic coupler is a weakly acidic organic compound that reacts with an oxidized p-phenylenediamine to produce a dye, usually one of the subtractive primaries, cyan, magenta, or yellow (50) (Fig. 10). In addition to the dye-forming portion of the molecule, most couplers also bear an organic ballast, a long aliphatic chain or combination of aliphatic and aromatic groups. This allows the coupler to be suspended in droplets of a high boiling organic liquid, called the coupler solvent, which serves to anchor the coupler in its appropriate film layer. In some recent formulations, the dye-forming portion of the coupler is attached to a polymeric backbone (51). In both cases such films are referred to as incorporated coupler systems. Kodachrome film is unique, however, because its couplers are not contained in the film, but diffuse in during the development steps. Although functionally similar to the incorporated couplers, couplers for Kodachrome films are water-soluble and do not contain an organic ballast.

Cyan dyes are derived typically from phenols or naphthols. These so-called indoaniline dyes absorb in the red, from 600 to 700 nm and beyond, and have unwanted absorptions in the blue-green. Magenta dyes are formed from pyrazolones, pyrazolotriazoles, and other species that have either an active methylene group as part of a conjugated ring or, less frequently, an active methylene flanked by electron-withdrawing groups. Magenta dyes are designed to absorb in the green between 500 and 600 nm, but often have additional absorbances in the blue. Yellow dyes are produced from couplers containing an active methylene group that is not part of a ring. These absorb blue light between 400 and 500 nm and typically have strong absorptions in the ultraviolet, which afford some protection to underlying layers from light-induced dye fade.

Mechanisms of Coupling. Because the active coupling species is the ionized coupler (35,52), the rate of the coupling reaction and hence its ability to compete for oxidized developer is dependent on the pH of the process, the pK_a or

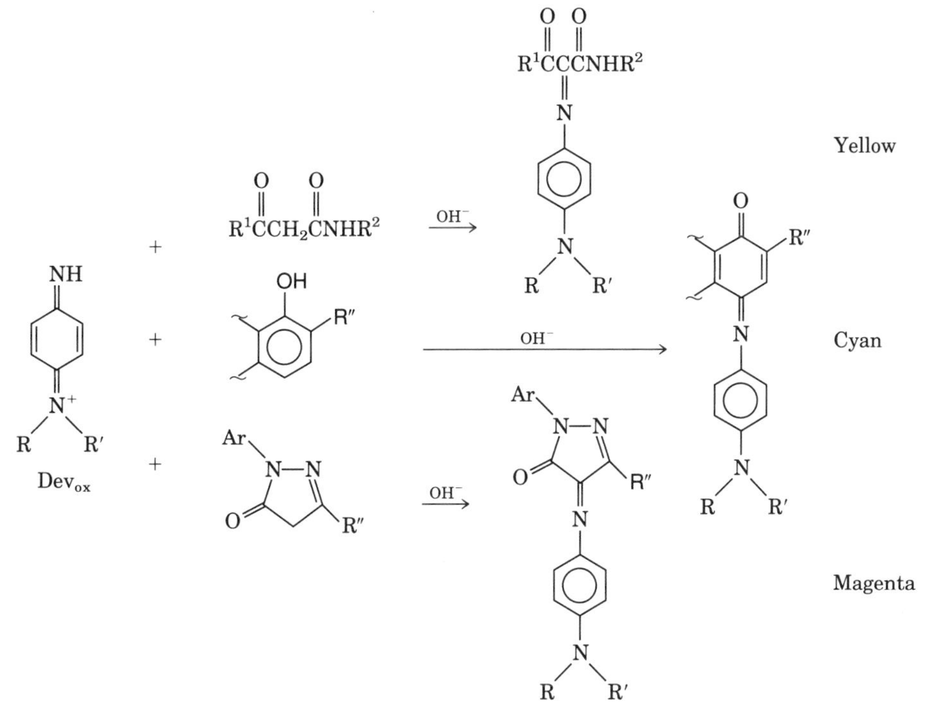

Fig. 10. In each image-forming layer, developer oxidized by the exposed silver halide (Dev$_{ox}$) reacts with the appropriate coupler to form the corresponding subtractive primary dye, yellow, cyan, or magenta. Ar represents an aryl group and the various R's are undefined organic segments.

acidity of the coupler or less frequently the rate of coupler ionization, and the reactivity of the resulting coupler anion with the QDI (40).

For most cyan and magenta couplers, ionization is rapid and can be characterized by the pK_a of the coupler. For yellow couplers, which can exist as both keto (carbon acid) and enol (oxygen acid) tautomers, ionization can be measurably slow, even rate determining in some processes (53). This tautomeric equilibrium and hence the degree and rate of ionization can be both structure and environment dependent.

Once ionized, the coupler reacts with oxidized developer to produce an intermediate or leuco (colorless) dye (Fig. 11). If X is hydrogen, an oxidation involving a second molecule of QDI is required to produce the image dye. Because formation of each mole of QDI requires the reduction of two moles of silver ion (for a total of four moles), such couplers are referred to as "four-equivalent" couplers. If, however, X is a good leaving group, such as a halide, or a low pK_a phenol or heterocycle, departure of X as an anion to form the dye proceeds spontaneously, and the coupler is said to be "two-equivalent" (43). Although the final dye yield, ie, the ratio of dye formed to silver reduced, can depend on the degree of competition in the system, two-equivalent couplers tend to be preferred because they undergo fewer side reactions with oxidized developer and require less silver halide to produce an equivalent amount of dye. Thus there is not only an obvious economic benefit, but it can lead to increased film sharpness by reducing optical scatter by the coated silver halide.

For two-equivalent couplers where the conversion of the leuco dye to image dye is rapid, the experimentally observed second-order rate constant, k_{c-}, can be equated with k_f, the rate of nucleophilic attack of coupler anion on oxidized developer. Thus when the pH of the process is specified, two parameters, pK_a and k_{c-}, can be conveniently used to characterize the molecular reactivity of a large variety of photographically well-behaved couplers (40,54).

When couplers are grouped together in structurally similar families, such as the naphthols in Figure 12, it is found that a linear free-energy relationship of the form $\log k_{c-} = \alpha + \beta\, pK_a$ often exists between the pK_a of the coupler and the $\log k_{c-}$ of its resulting anion (40). That is, a substituent X that raises the pK_a of

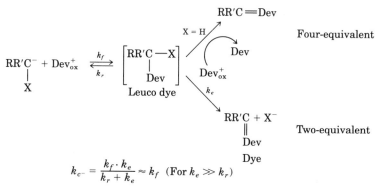

$$k_{c-} = \frac{k_f \cdot k_e}{k_r + k_e} \approx k_f \quad (\text{For } k_e \gg k_r)$$

Fig. 11. Reaction of ionized coupler and oxidized developer (Dev_{ox}^+) to produce the intermediate leuco dye. If X is a good leaving group, the reaction proceeds spontaneously to dye, and the coupler is said to be two-equivalent. If oxidation by a second molecule of oxidized developer is required, the coupler is four-equivalent.

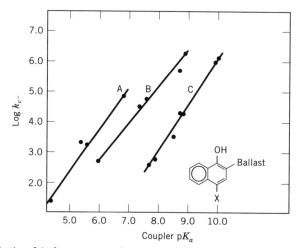

Fig. 12. The relationship between acidity and anion reactivity for naphthol couplers differing in the 2-position ballast where A is ballast 1; B, ballast 2; C, ballast 3. For each coupler data points represent different leaving groups, X.

the coupler and thus reduces the concentration of active anion at a given pH also increases the activity of that anion. The constant β is a measure of the trade-off between those two opposing factors on coupler reactivity. For β greater than 1, increasing the electron-donating ability of the substituent leads to an increase in anion reactivity that more than compensates for the loss in anion concentration. For β less than 1, the opposite is true; overall reactivity results from decreasing the pK_a of the coupler, until the pK_a drops below that pH of the process. At that point, with the coupler fully ionized, changes in the molecule that lower pK_a serve not to increase anion concentration, but only to decrease anion reactivity (54).

Studies on the naphthol and phenol couplers indicate that the field and resonance, but not steric, effects of the substituents are important in determining pK_a and k_{c-} suggesting that formation of the leuco dye proceeds through an early transition state where there is overlap between the electron-deficient p-orbitals of the developer ring and the electron-rich p-orbitals of the naphthol ring. This geometry is quite different from that of the product (55). For coupling reactions where conversion of the leuco dye to image dye is not fast, the rate constant of the reverse reaction, k_r, is important. If k_r is much less than k_e, dye formation is slow, but dye yield can still be high, although the leaving group may not be released in time to affect development. If, however, k_r is faster than k_e, much of the oxidized developer originally captured by the coupler is lost to other reactions, and dye yield is low (54).

Whereas the effect of the pH of the processing solution on coupling is most often considered in the context of the coupler, the structure of which can be varied, pH can also affect the activity of the oxidized developer (43,56). In general, oxidized developers fall into three categories. The first, like CD-2, have a constant charge, either cationic or zwitterionic, at normal processing pH values and coupling rates are little affected by pH changes. The second class exists in cationic or neutral forms as a function of pH. For example, the oxidized form of the commercially important CD-4 reacts reversibly with hydroxide ion to form a neutral pseudo-base, which is unreactive toward coupling (41). Finally, the quinonediimine derived from developers like CD-3 can exist in cationic and zwitterionic forms as a function of pH. Both forms can react with ionized coupler, but at different rates.

Though the molecular reactivities of the coupler and the developer are important, both can be severely constrained by the polyphasic nature of the film system, where the coupler is most often embedded in a lipophilic phase while the oxidized developer is generated in the aqueous gelatin phase (43). Because the organic solvent invariably has a lower dielectric constant than the aqueous phase, the effect is to suppress ionization of the coupler while slightly increasing the rate at which the positively charged QDI and coupler anion react with one another. The overall result is to reduce the rate of dye formation. In the case of some yellow couplers, the rate-determining step can change from coupling to ionization as a function of solvent or even substituent. The structure of the oxidized developer as well as its charge can also control its ability to partition into the organic phase (42,56).

In some processes, development additives such as benzyl alcohol are added to the developer to increase the hydrophilic nature of the organic phase. More frequently, higher pK_a couplers are designed to have additional ionizable sites, such as carboxyl, sulfo, or phenolic groups, to accomplish the same end (57).

In any case, the kinetics of such systems can be very complex, depending not only on the identity of the coupler and developer, but also on the nature of the dispersed organic phase. Often, the rate of coupling is found to be proportional to the total surface area of the droplets containing the coupler. Less frequently, as when the droplet integrity is destroyed in the developer bath, the rate can be independent of droplet size. Surfactants (qv) and mechanical milling techniques are used to form these oil-in-water emulsions, or dispersions as they are called in the photographic literature, at highly controlled particle sizes and with the requisite stability for coating and storage.

In some cases, dye-forming moieties attached to a polymeric backbone, called a polymeric coupler, can replace the monomeric coupler in coupler solvent (51). In other reports, very small particles of coupler solubilized by surfactant micelles can be formed through a catastrophic precipitation process (58). Both approaches can eliminate the need for mechanical manipulation of the coupler phase.

Cyan Couplers. Substituted phenols and α-naphthols are the primary classes of cyan dye-forming couplers (Fig. 13). Naphthols of structural types (**1**) and (**2**), the 1-hydroxy-2-naphthamides, have proved very useful and are easily and inexpensively prepared. Hydrogen bonding between the naphtholic oxygen and the hydrogen of the 2-amido group serves to shift the dye hue bathochromically (to longer wavelengths) while increasing its extinction and contributing to the dye's stability. Electron-withdrawing groups on the amide nitrogen also shift the hue bathochromically and increase the extinction coefficient (59). Substitution in the 4-position provides accessibility to image-modifying couplers of various types. However, some dyes of this class are prone to chemical reduction, which returns them to the colorless leuco form in the presence of ferrous ion during the bleaching step (60). Naphthols of type (**3**) are reported to show less fade to the leuco dye (61).

Phenols of structure (**4**) are also claimed to show markedly improved dye stability both in the presence of ferrous ion and, with a second carbonamido group

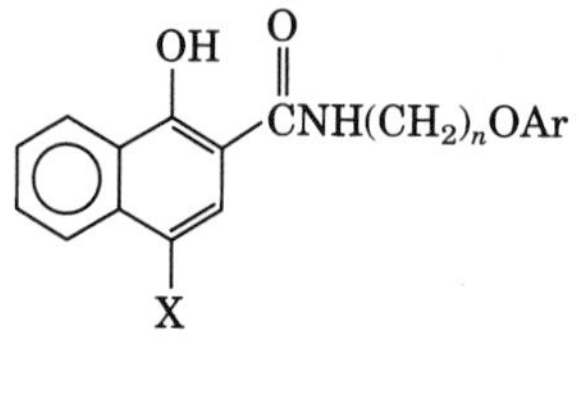

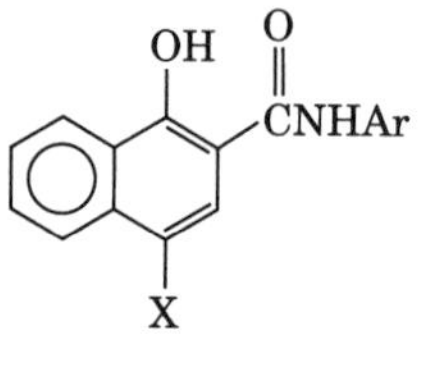

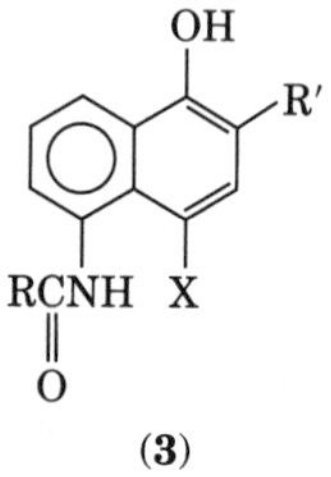

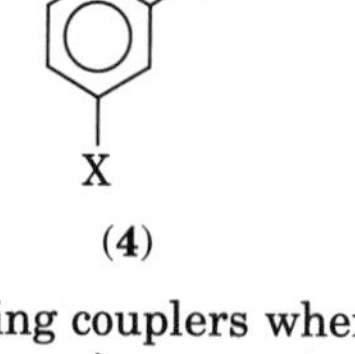

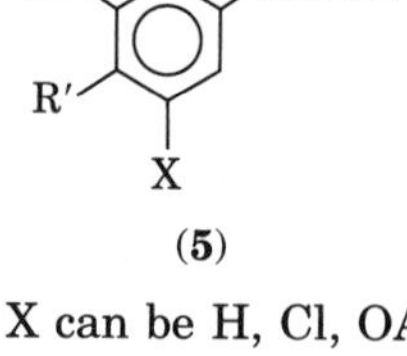

Fig. 13. Cyan dye-forming couplers where X can be H, Cl, OAr, OR, or SAr. Ar is aryl. R and R' are undefined organic segments.

in the 5-position, to simple thermal fade (62). Numerous substituent variations are described in the literature to adjust dye hue. A perfluoroacylamido in the 2-position shifts the hues bathochromically while maintaining thermal stability of the dyes (63). Phenols of structure (**5**) are said to show outstanding light stability, which makes them especially suitable for display materials like color paper (64).

Some cyan dyes derived from both naphthols and phenols are reported to show thermochromism, a reversible shift in the dye hue as a function of temperature. This can occur in a negative while prints are being made (65).

Yellow Couplers. The most important classes of yellow dye-forming couplers are derived from β-ketocarboxamides, specifically the benzoylacetanilides (**6**) (66) and the pivaloylacetanilides (**7**) (67). Substituents Y and Z can be used to attach ballasting or solubilization groups as well as to alter the reactivity of the coupler and the hue of the resulting dyes. Typical coupling-off groups (X) cited in the literature are also shown.

where X = —H, —Cl, —OOCR, —S-aryl, —SO$_2$R, —OSO$_2$R,

For the widely studied benzoylacetanilides, coupler acidities can be correlated using a two-term Hammett equation involving substituents in either or both rings. As in the naphthols, there is a linear correlation of the log k_{c^-} and pK_a values, but β equals 0.55, suggesting that increased reactivity comes with reduced pK_a, until the coupler is nearly fully ionized (68). The hues of the resulting dyes can be shifted bathochromically by electron-withdrawing groups in either ring. Ortho substitution in the anilide ring increases the extinction coefficient while narrowing the bandwidth of the dyes. Both the couplers and their dyes have significant absorptions in the ultraviolet that offer protection to dyes in the underlying layers (69).

The relatively low pK_a values seen for the benzoylacetanilides, especially as two-equivalent couplers, minimize concerns over slow ionization rates and contribute to the couplers' overall reactivity. But this same property often results in slow reprotonation in the acidic bleach, where developer carried over from the previous step can be oxidized and react with the still ionized coupler to produce unwanted dye in a nonimage related fashion. This problem can be eliminated by an acidic stop bath between the developer and the bleach steps or minimized by careful choice of coupling-off group, coupler solvent, or dispersion additives.

The second widely used class of yellow couplers is the pivaloylacetanilides (**7**) and related compounds bearing a fully substituted carbon adjacent to the keto group. The dyes from these couplers tend to show significantly improved light stability and so these couplers have been widely adopted for use in color papers as well as many projection materials. In general, the dyes have more narrow bandwidths and less unwanted green absorptions (67).

The lack of a second aryl group flanking the active methylene site, however, means that the pK_a values of the pivaloylacetanilides tend to be considerably higher than those of the benzoylacetanilides. As a result, these couplers are rarely used as their four-equivalent parents. Rather, the coupling site is substituted with electron-withdrawing groups to increase the acidity of the coupler or hydrophilic groups to aid in rate and extent of the oil-phase ionization. Both electron-withdrawing groups and hydrophilic groups can appear in the anilide ring as well. An interesting variation on this is the use of polarizable groups, such as C=O or S=O, in the ortho position of an aryloxy group attached to the coupling site. These groups reduce the pK_a of the coupler by increasing the rate of the ionization process (70). The higher pK_a of the coupling site in the pivaloylacetanilides does mean that undesirable dye formation in bleach is minimized.

Other classes of yellow couplers reported in the literature include the indazolones (71) and the benzisoxazolones (72). Neither of these structures contains an active methylene group; dye formation is believed to occur through a ring-opening process.

Magenta Couplers. For many years the most widely used magenta couplers have been derived from the 1-aryl-2-pyrazolin-5-ones (73). Substituents in the aryl ring or at the 3-position have been used to alter dye hue and stability as well as to control coupler reactivity. Ballasting groups are usually attached through the 3-position as well. Electron-withdrawing groups at either site tend to shift the hue of the resulting dye bathochromically (74). Whereas the principal absorption of these dyes is in the region from 500 to 570 nm, most pyrazolinone dyes also show significant unwanted absorptions in the blue. Because these can be minimized by the use of 3-arylamino or 3-acylamino substituents (see structures (**8**), (**9**)) while also increasing the extinction coefficient of the primary absorptions, such couplers have been extensively described in the literature (75). Ar represents an aryl group.

(**8**)   (**9**)

where X = —O—(ring with R), —OOCNHR, —OOCN(ring), —OOCR, —NHSO$_2$R, —SR, —S-aryl, —N=N-aryl

Although these pyrazolinone couplers can exist in several tautomeric forms, ionization of the couplers is rapid, and the four-equivalent parents have seen wide use for decades. The aryl ring is often trisubstituted in the 2,4,6-positions with one or more of the substituents being chlorine (76). The pK_a values for the 3-

arylamino pyrazolinones (**8**) are higher than for their 3-acylamino (**9**) counter-parts with dyes, the hues of which are shifted toward shorter wavelengths and show less unwanted blue absorption. Coupler (**10**) [*61354-99-2*] is unique because it carries its own stabilizer against photochemical dye fade in the form of the 4-alkoxyphenol ballast, making it especially suitable for color paper (77).

(**10**)

The four-equivalent pyrazolinones suffer from a number of disadvantages. Prior to processing, the couplers can react with ambient formaldehyde to yield a nondye-forming condensation product. Formaldehyde scavengers, such as ethyl-enediurea, can be added to control this problem (78). After processing, the residual coupler can react with image dye to form colorless products (79). A formaldehyde stabilizer, which reacts with the residual coupler in the same fashion as the un-desirable pre-process reaction, eliminates this problem. Finally, though these cou-plers react rapidly with oxidized developer, the intermediates undergo side re-actions that result in reduced dye yields. Though the 3-acylamino pyrazolinones are more prone to these problems than the 3-arylamino couplers, both are sus-ceptible (80).

The two-equivalent counterparts to these couplers are largely devoid of these problems. Both halogen and sulfo groups are unsatisfactory leaving groups be-cause of various side reactions, whereas the otherwise attractive aryloxy moiety causes the coupler to undergo an often rapid radical chain decomposition process (81). Useful coupling-off groups cited in the literature include aryl, alkyl, and hetero thiol groups (82), nitrogen heterocycles (83), and arylazo groups (84). Be-cause the thiols serve only to depress the reactivity of the already low pK_a acyl-amino pyrazolinones, they tend to be most useful on the arylamino couplers. Alkyl thiols, because of their high pK_a values, are generally reluctant leaving groups in the developer, but eventually form image dye in the low pH bleach (54,85). Most of the arylthiol leaving groups cited in the literature show substitution at least in the 2-position. Bis-pyrazolinones, linked through the coupling site by methyl-ene, substituted methylene, sulfide, and disulfide groups, have been reported to give two-equivalent couplers, but with only one of the couplers yielding dye (86).

A more recent class of magenta dye-forming couplers is the pyrazolo-(3,2,-c)-5-triazoles (**11**) and related isomers (87) where X can be Cl, SR, S-aryl, or O-aryl. Dyes from this class of couplers are exceedingly attractive, having good thermal stability and much lower unwanted blue and red absorptions than dyes from the pyrazolinones. However, the high pK_a values of the four-equivalent par-ents translate into unacceptably low reactivity. Two-equivalent analogues having chloro or aryloxy leaving groups and hydrophilic groups elsewhere in the molecule provide good reactivity and dye yield with resistance to ambient formaldehyde and dyes that do not require post-process stabilization (88). A related class of pyrazolotriazoles (**12**) is reported to yield dyes having improved light stability (89). The higher pK_a values of the pyrazolotriazole couplers have led to concerns over image variability induced by small changes in developer pH.

(**11**)   (**12**)

Other classes of magenta dye-forming couplers reported in the literature include the pyrazolobenzimidazoles (90) and the indazolones (91). The latter are unique because they do not contain an active methylene group and are proposed to form magenta dyes with a zwitterionic structure.

Colored Masking Couplers. The dyes produced in chromogenic development have unwanted absorptions. For example, the cyan dye is expected to control or modulate red light alone and thus should absorb only between 600 and 700 nm, but it shows lesser absorptions in the blue (400–500 nm) and green (500–600 nm) regions as well. Thus exposure of the red-sensitive layer of the film produces not only the desired density to red light in the negative, but also undesirable densities to blue and green light, resulting in desaturation or "muddying" of the color.

For materials that are not directly viewed, like a color negative film, masking couplers provide an ingenious solution (30,92). Unlike a normal cyan dye-forming coupler, which is itself colorless, a cyan masking coupler bears a colored, preformed (usually azo) dye in the coupling-off position. The hue of this dye is chosen to match the unwanted blue-green absorptions of the cyan dye that is generated. When coupling occurs, the preformed dye is released and washes out of the film or is destroyed. The result is a negative image formed by the cyan dye with its unwanted absorptions and an entirely complementary positive image left by the preformed dye remaining on the residual coupler (Fig. 14). This is equivalent to a perfect cyan image overlaid with a uniform blue-green density. Because the negative is printed onto color paper using separate cyan, magenta, and yellow exposures, a somewhat longer cyan exposure is required. Similar chemistry is employed to deal with the unwanted blue density of the magenta coupler (93).

DIR Couplers. Masking couplers cannot be used for directly viewed materials because of the objectionable color of the mask itself. But similar advantages and more can be achieved by using development-inhibitor-releasing (DIR) couplers (31,94). These materials are usually image couplers that carry a silver development inhibitor (In) linked directly or indirectly to the coupling site (eq. 6). When released as a function of dye formation, the development inhibitors migrate to the silver halide grain and either slow or stop further development. In addition to correction of unwanted dye absorptions, DIR couplers can be used to improve sharpness and reduce granularity of films (95).

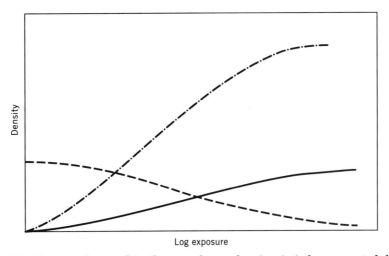

Fig. 14. Masking coupler used in the cyan layer showing (—) the unwanted density to blue-green light that accompanies cyan dye formation matched by (- - -) a complementary density to blue-green from the unreacted coupler, and (—·—·) density to red light.

Development inhibitors have long been used to control photographic processes, but the mechanism by which they work remains in dispute. In one model of the developing silver halide grain, the inhibitor reacts with the surface silver ions at the etch pit, a defect region on the crystal where halide ions depart into solution (Fig. 9). This complexation prevents these silver ions from migrating interstitially to the root of the developing silver filament and slows down or stops development (96). Another model views the inhibitor as complexing with the silver metal filament, where its insulating properties prevent electron injection by the oxidizing developer (97).

For color correction of, for example, the unwanted green absorptions of a cyan dye, a cyan DIR coupler is added to the imaging layer along with the cyan image coupler. The inhibitor is released in proportion to cyan dye formation and is designed so as to migrate into the adjacent magenta dye-forming layer, where it inhibits silver development and reduces the production of green density to the same degree that unwanted green density is produced by the cyan dye. This is sometimes referred to as a red-onto-green interlayer interimage effect and can occur only when the predominantly red image in the scene has some green component. Similar color correction can originate from the yellow and magenta dye forming layers, and the overall correction can be described using a 3-by-3 color matrix (94).

DIR couplers can also be used to control granularity, a measure of the visual nonuniformity of the dye image resulting from the random distribution of the dye clouds. Granularity is, in fact, the noise in the photographic signal, and is proportional to the density divided by the square root of the number of signal-generating centers, in this case, the silver halide grains. Increasing the amount of silver halide does not in itself reduce the granularity, however, because this also generates more oxidized developer and greater density. However, a DIR coupler can permit the coating of more silver halide centers without increased dye density by releasing an inhibitor that allows each grain to develop only partially.

Perhaps the most intriguing use of DIR couplers is the ability to improve the perceived sharpness of an image. If a sharp photographic edge or square wave signal is imposed on a piece of film, for example, by placing an opaque material over part of the film and exposing it to light, the resulting developed image shows some degradation of the edge, mostly because of light scatter, but also because of diffusion processes (Fig. 15a). This "softening" of the edge is perceived as a loss in sharpness. If, however, a DIR coupler is coated with the image coupler, it produces the inhibitor concentration profile of Figure 15b. The resulting inhibition

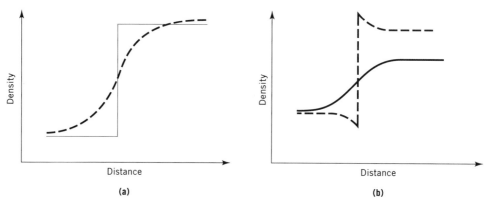

(a)   (b)

Fig. 15. Light scatter and chemical diffusion lead to a loss of sharpness at (**a**) a photographic edge where (—) represents the image and (- - -) the developed image. In the case of (**b**), an inhibitor modified edge, the sharpness to the (- - -) developed image can be restored by preferentially releasing a development inhibitor in image areas. (—) represents the inhibitor concentration profile. Diffusion of the inhibitor accentuates the edge relative to the macro portion of the image.

of development leads to the final dye image of Figure 15**b**, where the density at the edge is enhanced relative to macro density. This increased rate of density change at the edge is seen as sharpness (95).

Although the varied uses for which DIR couplers are employed call for precise control over where the inhibitor diffuses, the very complexation mechanism by which inhibitors work would seem to preclude such control. The desired ability to target the inhibitor can be attained by the use of delayed release DIR couplers, which release not the inhibitor itself, but a diffusable inhibitor precursor or "switch" (Fig. 16) (98). Substituents (X, R) and structural design of the precursor permit control over both diffusivity and the rate of inhibitor release. Increasing the effective diffusivity of the inhibitor, however, means that more of it can diffuse into the developer solution where it can affect film in an undesirable, nonimagewise fashion. This can be minimized by the use of self-destructing inhibitors that are slowly destroyed by developer components and do not build up or "season" the process (99).

Although most DIR couplers are based on image dye-forming parents, universal DIR couplers have appeared in the literature. These materials react with oxidized developer to produce the inhibitor (or precursor) and either a colorless dye, an unstable dye, or a washout dye (100). Universal DIR couplers could be used in any layer with a need to match only image-modifying properties, not hue, to the given layer.

In addition to inhibitors and masking dyes, similar coupling mechanisms have been used to release other photographically useful fragments, including development accelerators, image dyes, bleaching accelerators, development competitors, and bleaching inhibitors (101). Other mechanisms for imagewise release of inhibitors include inhibitor releasing developers (IRDs) (**13**) (102) and the related IRD-releasing couplers (**14**) (103). The former is of particular interest as a potential means of achieving image modification in reversal systems, where the image structure is determined in a black-and-white development step before chromogenic development.

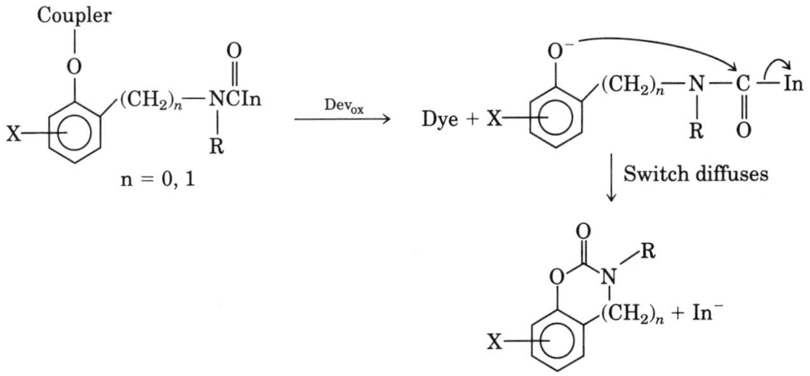

Fig. 16. Reaction of a delayed release DIR coupler and oxidized developer (Dev$_{ox}$). A delayed release DIR coupler permits fine-tuning of where and when the development inhibitor (In) is generated by releasing a diffusible inhibitor precursor or "switch" as a function of image formation. This permits control of inter- and intralayer effects.

(13)   (14)

Post-Development Chemistry. The silver and silver halide remaining in the film after color development must be removed, both to improve the appearance of the color image and to prevent the appearance from changing as silver halide is slowly photoreduced to silver metal. This is generally accomplished in two steps. The first, called bleaching, is an oxidation that converts silver metal back to silver salts. The second, called fixing, is the solubilization of the silver salts by complexation with a silver ligand. In some processes, particularly those used for color paper, the two steps can be combined in a single step called a bleach-fix or blix (60,104).

The most common color film bleaches are of the rehalogenating type, that is, they contain halide ion, often bromide, to complex the silver ion being formed and drive the reaction to its conclusion. For many years the most common bleaching agent was ferricyanide, a powerful bleaching agent that also can convert residual leuco dye into image dye (60). However, the high oxidation potential, $E_0 = 356$ mV vs the hydrogen electrode (NHE), also makes ferricyanide capable of oxidizing the color developer carried over with the film. This can couple and form non-imagewise dye. For this reason, an intervening acidic "stop" bath must be used with ferricyanide bleaches.

Rehalogenating bleaches based on ferric ethylenediaminetetraacetic acid (ferric EDTA) are lower in potential, $E_0 = 117$ mV vs NHE, and do not require a stop bath, permitting a shorter and simpler process (105). The bleaching efficiency of ferric EDTA can be improved by lowering the solution pH below 6. However, as the pH is dropped, the propensity for reduction of indoaniline cyan dyes to the colorless leuco form increases. Thus most iron ligand bleaches are designed to operate in a window bounded by the pH of the bath and the oxidation potential of the bleaching agent.

The rates of most color bleaching processes are diffusion limited over at least part of the reaction. Color negative films tend to pass from a diffusion-limited regime into a chemically limited one as sensitizing dye and other passivating species accumulate on the remaining silver surfaces (106). Persulfate anion is used as a bleaching agent in some motion picture processes (107). Although it is thermodynamically attractive because of the very high ($E_0 = 2010$ mV vs NHE) oxidation potential, persulfate is a kinetically slow bleach in the absence of a catalyst. Commonly, a prebath containing dimethylaminoethanethiol [108-02-1] is used, although the use of bleach accelerator-releasing couplers or blocked bleach accelerators incorporated in the film has also been proposed (101,108).

Thiosulfate, usually as its sodium or ammonium salt, is almost universally employed as the fixing agent for color films (104). Thiocyanate can be used as a fixing accelerator. Fixing performance is often defined by two parameters: the clearing time necessary to dissolve the silver halide and render the film optically transparent, and the fixing time required to remove the complexed silver from the film. Fixing must be complete and the fixing agents thoroughly washed from the film because thiosulfate can destroy image dye by reduction or by other reactions on long-term keeping. The complexation constants of thiosulfate with silver ion are sufficiently large so as to keep silver salts from being redeposited on the film on dilution in the wash.

Film Quality

Speed.   Standards for photographic speed are now coordinated worldwide by the International Standards Organization (ISO) in Geneva, Switzerland. ISO speed is determined under specified conditions of exposure, processing, and measurement. Standards are published for color negative and color reversal films (109,110). In amateur color photography, the most popular film speeds are in the ISO 100–400 range. For modern 35-mm cameras, these speeds offer a good compromise between image quality, depth of field, and ability to arrest motion, particularly when coupled with electronic flash. Several high speed color films of ISO 1000 or greater are now available from the primary manufacturers, recommended for conditions of low illumination or fast subject motion. The larger silver halide grains necessary for high speed become progressively less efficient in converting absorbed radiation into latent image, and in practice the linear relationship shown in Figure 5 rolls off at the larger sizes. Image graininess becomes objectionable if high magnification is required. Sensitivity to ambient ionizing radiation of both cosmic and terrestrial origin also increases with crystal size, resulting in a proportion of crystals being fogged during storage, causing graininess to increase (111,112).

The human visual system has the ability to adapt to changes in the spectral balance of the scene illuminant. For example, a gray color appears gray both under daylight and tungsten illumination. However, a color film designed for use in daylight produces prints with a yellowish cast if used with tungsten illumination. This effect can be reduced by placing an appropriate filter over the camera lens, or by adjusting light filtration during printing in the case of a color negative. In practice, designing a film for a particular illuminant involves adjusting its sensitivity to blue, green, and red spectral light to account for the relative abundance in that illuminant. Most amateur films are balanced for optimum performance in daylight with a correlated color temperature of 5500 K (4,20).

Color Reproduction.   Color has three basic perceptual attributes: brightness, hue, and saturation. Saturation relates to how much of the hue is exhibited. The primary influences on the color quality of the final image are the red, green, and blue spectral sensitivities of the film or paper, and the spectral absorption characteristics of the image dyes. If the green record is designed to match the sensitivity of the eye's γ receptor and the red record the ρ receptor (Fig. 1), channel overlap is so severe that a satisfactory reproduction cannot be obtained. In practice, the peak sensitivity of the red record is moved to longer (ca 650 nm) wavelengths to reduce the overlap. A penalty is that colors having strong reflectances at 650 nm or longer appear excessively reddish in the reproduction.

Ideally, subtractive image dyes should exhibit no absorption outside the spectral ranges intended. In reality, considerable unwanted absorptions occur even using the best dyes, and colored masking couplers (see Fig. 14) are designed to alleviate this problem. However, to avoid serious speed losses, a colored coupler is normally only employed under layers in the film that absorb in a similar range. Thus a magenta-colored coupler, which absorbs green light, is used in the cyan pack; a yellow-colored coupler, which absorbs blue light, may be used in the magenta or cyan packs. Color correction may also be achieved by selectively suppressing dye generation in one color record as a function of exposure in another by using a diffusible development inhibitor generated by a development-inhibitor-releasing (DIR) coupler. Masking couplers and DIR couplers are the main tools to achieve the high color saturation of modern color negative films.

Because reversal films are designed so that the camera film is the film containing the final image, these are subject to additional constraints. The minimum densities in the image must be low to accommodate scene highlights and colored masking couplers cannot be used. In addition, because the color development step is exhaustive, ie, develops all the residual silver halide, it is not possible to gain color correction from DIR couplers, which rely on influencing the relative development rates in a system that is only partially developed. Some degree of interlayer development inhibition can be obtained in the black-and-white developer by the migration of iodide released from the developing silver halide (113).

Dye Stability.   The dyes used in photographic systems can degrade over time, both by thermal reactions and, if the image is displayed for extended periods of time, by photochemical processes. The relative importance of these two mechanistic classes, known as dark fade and light fade, respectively, depends on how the product is to be used (114).

Meaningful evaluations of dye stability on a practical time scale are difficult because the reactions themselves are by design either slow or inefficient. The $t_{1/10}$ value, defined as the time for a 0.1 density loss from a dye patch of 1.0 initial density, for dark fade ranges from 30 to greater than 100 years for chromogenically generated dyes, whereas the quantum yields of photochemical fading are often on the order of 1×10^{-7}. Accelerated testing conditions involving high temperature keeping or high intensity illumination must be used (115), although these are sometimes unreliable predictors of ambient fade (116). The perceived color stability of a photographic system is usually limited by the fading of its least stable dye, which can produce an undesirable shift in color balance. Whereas recovery of such faded images is often possible, a so-called neutral fade, in which all three color records lose density at approximately the same rate, is usually preferred.

For light fade the magenta dye has usually been limiting. Numerous studies support the hypothesis that the fading mechanism is photooxidative (117). Stabilization techniques have included exclusion of oxygen by barrier or encapsulation technologies; elimination of ultraviolet light by incorporated uv absorbers; quenching of one or more of the excited state species in the reaction sequence, ie, dye singlet, dye triplet, or singlet molecular oxygen, via incorporated quenching agents; and the scavenging of reactive intermediates such as free radicals. Cyan light fade appears to include both reductive and oxidative components (118).

The dark fade of yellow dyes appears to be largely a hydrolytic process (119). Cyan dark fade results mostly from reduction of the dye to its leuco form. For the magenta record, dark keeping can be dominated by yellowing reactions from the residual coupler. A light-induced yellowing has also been observed. These problems have been successfully addressed by the design of new couplers and coupling-off groups.

Modern photographic products have stabilities vastly improved over those of the past. For example, the magenta light stability of color negative print papers has improved by about two orders of magnitude between 1942 and the present (120). Although such products offer very acceptable image stability in the vast majority of customer use situations, other processes, such as silver dye bleach and dye-transfer processes, can offer even greater stability, albeit at a significant cost in processing convenience.

Image Structure. Because the primary photographic sensors are a population of silver halide crystals having a random spatial distribution, the final image is also particulate. In black-and-white photography the image consists of opaque deposits of silver. In chromogenic photography using incorporated couplers, the image is formed by the coupler-containing hydrophobic droplets dispersed with the silver halide. The droplet size, typically 0.2 μm in diameter, is usually less than the crystal size. Dye is formed in a cloud of droplets around each developing crystal as oxidized developer is released. Individual droplets cannot be resolved under usual viewing conditions, but the dye clouds can be seen under magnification and convey a visual sensation of nonuniformity or graininess. The objective correlate of graininess is granularity, which is the spatial variation in density observed when numerous readings are taken on a uniformly exposed patch using a densitometer having a very small aperture. The distribution of such measurements approximates to Gaussian and can be characterized by its standard deviation, σ_D. This quantity, called root mean square (RMS) granularity, is published by film manufacturers as a figure of merit.

The features which differentiate a color image from a black-and-white image are the transparency of the dyes and the spreading of the dye cloud around the developing crystal. Dye clouds continue to grow during development as the oxidized developer diffuses farther to find unreacted coupler. Overlapping of adjacent dye clouds leads to a granularity reduction as more and more of the voids are filled in. The size of the dye clouds can be controlled by reducing the amount of silver development per crystal through the use of development-inhibitor-releasing couplers, or by having a soluble coupler in the developer which competes with the incorporated coupler for oxidized developer (121), the dye formed being subsequently washed out.

Sharpness is the ability to discriminate edge detail in the final image. There are many opportunities to lose sharpness in the photographic system. Camera focus, depth of field, subject movement, camera movement during exposure, and the optical properties of the film all contribute to image sharpness. In the negative-positive system, printer focus and the optics of the color paper stock are also important. Modern multilayer films of the type illustrated in Figure 8 are typically 20–25 μm thick. As light penetrates the front surface, it is subject to scatter by the imbedded silver halide crystals, which differ in refractive index from the supporting gelatin medium. This optical spread increases with depth and thus the uppermost blue-sensitive layer records the sharpest image and the lowest red-sensitive layer the least sharp image. Over the years, film manufacturers have reduced coated thickness both to improve sharpness and reduce material costs. Absorbing dyes may also be included in the film to reduce sideways light scatter, although at some cost in speed. These are generally removed by the processing solutions. Another method of improving sharpness is to change the layer order. Motion picture theatrical print film, for example, has the green record on top to improve sharpness, because the eye is most sensitive to differences in mid-spectrum light.

The sharpness of a film is often assessed by the modulation transfer function (MTF), which measures how sinusoidal test patterns of different frequencies are reproduced by the photographic material (122). A perfect reproduction would have a MTF of 100% at all frequencies. In practice, a decrease of MTF with increasing frequency occurs as a result of optical degradation. Films having couplers that release development inhibitors can display MTF values above 100% at low frequencies, because of edge effects that increase the output signal amplitude beyond that of the reference signal. The transport of inhibitor fragments across image boundaries leads to development suppression or enhancement on the microscopic scale (Fig. 15). Such edge effects are particularly useful in boosting the apparent sharpness of the lower layers of the film most degraded by optical scatter. However, if chemical enhancement of edges is carried too far, the effects appear unnatural.

Environmental Aspects

Photographic processing, ie, photofinishing, is a geographically dispersed chemical industry. Processing machines are rarely emptied and refilled with fresh solutions, but do require replenishment because of chemical use, evaporation, and overflow. New films have been designed that make more efficient use of coated silver and thus require less in terms of the chemicals for processing. The solution overflow that does occur is usually chemically regenerated and returned to the tank.

Silver is most commonly recovered by electrolysis or metallic replacement from the processing solutions or by ion exchange (qv) from the wash water (123). Loss of chemicals from one tank into the next has been minimized. The color paper process has progressed from five chemical solutions, three washes, and a replenishment rate of 75 μL/cm^2 (70 mL/ft^2) of film for each of the five solutions to two chemical solutions, one wash, and replenishment rates of 15 μL/cm^2 (16 mL/ft^2) and 5 μL/cm^2 (5 mL/ft^2). For color negative films, developer replenishment has dropped from over 300 to 43 μL/cm^2 (40 mL/ft^2). Regeneration of the now reduced overflow has decreased chemical discharge by as much as 55% (124).

The new chemistry of the RA-4 paper process permits the elimination of benzyl alcohol from the developer and a 50% reduction of the biological or chemical oxygen demand (BOD/COD) of the film/paper effluent. Substitution of the more powerful ferric 1,3-propylenediaminetetraacetic acid (ferric 1,3-PDTA) bleaching agent for ferric EDTA allowed for 60% reduction in both iron and ligand concentrations as well as total elimination of ammonia from the bleaching formulation (125). Bacteria have been developed that degrade ferric EDTA (126) whereas some new bleach ligands are degraded by naturally occurring microorganisms. During the period 1985 through 1990, some components in the effluent have been reduced to near zero and an overall reduction in effluent concentration of as much as 80% has been obtained (124).

Economic Aspects

The number of photographs taken annually in the United States now exceeds 16 billion. In 1989, the breakdown was 14.9 billion color prints, 0.8 billion color slides, and 0.4 billion black-and-white prints (127). The 1980s saw an increasing dominance of 35-mm color negative at the expense of color slide, because of its greater exposure latitude, the convenience of viewing reflection prints, and the ease of obtaining duplicate prints. The growth in popularity of picture-taking has been fueled by the increasing sophistication of 35-mm point-and-shoot cameras. These offer features such as auto-exposure, auto-focus, motor drive, and built-in flash, all within a very compact camera body. Shipments to dealers of these cameras increased to 10 million in 1989, promising strong future demand for 35-mm color film. The 35-mm single-use camera in which the film box itself is equipped with a lens and shutter has also been very successful.

The photofinishing industry has been growing at more than 5% annually. Since the mid-1970s a shift has occurred to favor local minilabs over large centralized processing laboratories. Although minilabs are generally more expensive, consumers appreciate the convenience, rapid-access, and personal service. Within this environment, hardware is now available to enable the customer to personally zoom and frame selected areas of the negative for enlargement.

The professional segment of color photography includes portrait and wedding photography, advertising photography, and photojournalism. Products tailored to these markets are offered by the principal manufacturers. The motion picture industry enjoyed continued growth in the 1980s. In spite of video cassette recorders, there continues to be strong demand for the theatrical experience of first-run movies. This experience is being enhanced by the improved picture quality of 70-mm origination and projection. Color negative film is the preferred medium for pre-recorded television shows; telecine transfer converts the images to electronic form for transmission.

Although camcorders have displaced home movie films, electronic still photography as a viable consumer product appears still to be some distance in the future. As well as poorer picture quality (128), the equipment cost and the relative lack of convenient hard-copy has so far limited this technology to the enthusiast. With the opening of previously closed large foreign markets such as Eastern Europe and China, where there is an unsatisfied demand for high quality personal imaging, the prospects for continued growth in conventional color photography, with its relatively low initial investment, appear excellent.

BIBLIOGRAPHY

"Color Photography" under "Photography" in *ECT* 1st ed., Vol. 10, pp. 577–584, by T. H. James, Eastman Kodak Co.; "Color Photography" in *ECT* 2nd ed., Vol. 5, pp. 812–845, by J. R. Thirtle and D. M. Zwick, Eastman Kodak Co.; in *ECT* 3rd ed., Vol. 6, pp. 617–646, by J. R. Thirtle and D. M. Zwick, Eastman Kodak Co.

1. E. J. Wall, *History of Three-Color Photography*, American Photographic Publishing Company, Boston, Mass., 1925.
2. J. S. Friedman, *History of Color Photography*, 2nd ed., Focal Press, London, 1968.
3. R. M. Evans, W. T. Hanson, and W. L. Brewer, *Principles of Color Photography*, John Wiley & Sons, Inc., New York, 1953.
4. R. W. G. Hunt, *The Reproduction of Colour*, 4th ed., Fountain Press, UK, 1987.
5. T. H. James, ed., *Theory of the Photographic Process*, 4th ed., Macmillan Publishing Co., New York, 1977.
6. J. Sturge, V. Walworth, and A. Shepp, eds., *Imaging Processes and Materials— Neblette's Eighth Edition*, Van Nostrand Reinhold, New York, 1989.
7. J. M. Eder, *History of Photography*, Columbia University Press, New York, reprinted by Dover, New York, 1978.
8. W. T. Hanson, *Photogr. Sci. Eng.* **21**, 293–296 (1977).
9. A. Weissberger, *Am. Sci.* **58**, 648–660 (1970).
10. P. Glafkides, *Photographic Chemistry*, Vol. II, Fountain Press, London, 1960.
11. G. F. Duffin, *Photographic Emulsion Chemistry*, Focal Press, London, 1966.
12. G. Haist, *Modern Photographic Processing*, John Wiley & Sons, Inc., New York, 1979.
13. L. F. A. Mason, *Photographic Processing Chemistry*, Focal Press, London, 1966.
14. F. W. H. Mueller, *Photogr. Sci. Eng.* **6**, 166 (1962).
15. J. Pouradier, in E. Ostroff, ed., *Pioneers of Photography*, SPSE, Springfield, Va., 1987, Chapt. 3.
16. T. J. Dagon, *J. Appl. Photogr. Eng.* **2**, 42 (1976).
17. R. L. Heidke, L. H. Feldman, and C. C. Bard, *J. Imag. Technol.* **11**, 93 (1985).
18. P. Krause, *Modern Photogr.* **48**(4) (1984).
19. P. Krause, *Modern Photogr.* **49**(11) (1985).
20. R. W. G. Hunt, *Measuring Color*, Ellis Horwood Ltd., Chichester, UK, 1987, p. 21.
21. R. M. Evans, *J. Photogr. Sci.* **9**, 243 (1961).
22. C. R. Berry, in Ref. 5, p. 98.
23. J. E. Maskasky, *J. Imag. Sci.* **30**, 247 (1986).
24. J. T. Kofron and R. E. Booms, *J. Soc. Photogr. Sci. Technol. Jpn.* **49**(6), 499 (1986).
25. U.S. Pat. 4,439,520 (1984), J. T. Kofron and co-workers (to Eastman Kodak Co.).
26. B. H. Carroll, *Photogr. Sci. Eng.* **21**, 151 (1977).
27. S. Dahne, *Photogr. Sci. Eng.* **23**, 219 (1979).
28. J. Spence and B. H. Carroll, *J. Phys. Colloid Chem.* **52**, 1090 (1948).
29. C. E. K. Mees, *J. Chem. Ed.* **6**, 286 (1929).
30. W. T. Hanson, *J. Opt. Soc. Am.* **40**, 166 (1950).
31. C. R. Barr, J. R. Thirtle, and P. W. Vittum, *Photogr. Sci. Eng.* **13**, 214 (1969).
32. Ger. Pat. 253,335 (1912), R. Fischer; R. Fischer and H. Siegrist, *Photogr. Korresp.* **51**, 18 (1914).
33. L. E. Friedrich and J. E. Eilers, in F. Granzer and E. Moisar, eds., *Progress in Basic Principles of Imaging Systems*, Vieweg & Sohn, Wiesbaden, Germany, 1987, p. 385.
34. R. L. Bent and co-workers, *J. Am. Chem. Soc.* **73**, 3100 (1951).
35. L. K. J. Tong and M. C. Glesmann, *J. Am. Chem. Soc.* **79**, 583, 592 (1957).
36. J. Eggers and H. Frieser, *Z. Electrochem.* **60**, 372, 376 (1956).

37. L. K. J. Tong, M. C. Glesmann, and C. A. Bishop, *Photogr. Sci. Eng.* **8,** 326 (1964); L. K. J. Tong and M. C. Glesmann, *Photogr. Sci. Eng.* **8,** 319 (1964); R. C. Baetzold and L. K. J. Tong, *J. Am. Chem. Soc.* **93,** 1347 (1971).

38. J. C. Weaver and S. J. Bertucci, *J. Photogr. Sci.* **30,** 10 (1988).

39. E. R. Brown, in S. Patai and Z. Rappoport, eds., *The Chemistry of Quinonoid Compounds*, Vol. 2, Wiley-Interscience, New York, 1988.

40. L. K. J. Tong and M. C. Glesmann, *J. Am. Chem. Soc.* **90,** 5164 (1968).

41. L. K. J. Tong, M. C. Glesmann, and R. L. Bent, *J. Am. Chem. Soc.* **82,** 1988 (1960).

42. J. Texter, D. S. Ross, and T. Matsubara, *J. Imag. Sci.* **34,** 123 (1990).

43. L. K. J. Tong, in Ref. 5, pp. 345–351.

44. E. R. Brown and L. K. J. Tong, *Photogr. Sci. Eng.* **19,** 314 (1975).

45. J. Texter, *J. Imag. Sci.* **34,** 243 (1990).

46. U.S. Pat. 2,108,243 (1938), B. Wendt (to Agfa Ansco Corp.); U.S. Pat. 2,193,015 (1940), A. Weissberger (to Eastman Kodak Co.); Brit. Pat. 775,692 (1957), D. W. C. Ramsay (to Imperial Chemical Ind.); Fr. Pat. 1,299,899 (1962), W. Pelz and W. Puschel (to Agfa-Gevaert), for examples.

47. C. A. Bishop and L. K. J. Tong, *Photogr. Sci. Eng.* **11,** 30 (1967).

48. R. Bent and co-workers, *Photogr. Sci. Eng.* **8,** 125 (1964).

49. F. W. H. Mueller, in S. Kikuchi, ed., *The Photographic Image*, Focal Press, London, 1970, p. 91.

50. J. R. Thirtle, *Chemtech.* **9,** 25 (1979), for a tutorial.

51. Brit. Pat. 701,237 (1953), (to DuPont); U.S. Pat. 3,767,412 (1973), M. J. Monbaliu, A. Van Den Bergh, and J. J. Priem (to Agfa-Gevaert); Brit. Pat. 2,092,573 (1982), M. Yagihara, T. Hirano, and K. Mihayashi (to Fuji Photo Film); U.S. Pat. 4,612,278 (1986), P. Lau and P. W. Tang (to Eastman Kodak Co.).

52. F. Sachs, *Berichte* **33,** 959 (1900).

53. C. A. Bishop and L. K. J. Tong, *J. Phys. Chem.* **66,** 1034 (1962).

54. J. A. Kapecki, D. S. Ross, and A. T. Bowne, *Proceedings of the International East-West Symposium II*, The Society for Imaging Science and Technology, Springfield, Va., 1988, p. D-11.

55. L. E. Friedrich, unpublished data, 1983.

56. T. Matsubara and J. Texter, *J. Colloid Sci.* **112,** 421 (1986); J. Texter, T. Beverly, S. R. Templar, and T. Matsubara, *Ibid.,* **120,** 389 (1987).

57. U.S. Pat. 4,443,536 (1984), G. J. Lestina (to Eastman Kodak Co.).

58. Brit. Pat. 1,193,349 (1970), J. A. Townsley and R. Trunley (to Ilford, Ltd.); W. J. Priest, *Res. Discl.* **16468,** 75–80 (1977); U.S. Pat. 4,957,857 (1990), K. Chari (to Eastman Kodak Co.).

59. C. Barr, G. Brown, J. Thirtle, and A. Weissberger, *Photogr. Sci. Eng.* **5,** 195 (1961).

60. K. H. Stephen, in Ref. 5, pp. 462–465.

61. U.S. Pat. 4,690,889 (1987), N. Saito, K. Aoki, and Y. Yokota (to Fuji Photo Film).

62. U.S. Pat. 2,367,531 (1945), I. L. Salminen, P. Vittum, and A. Weissberger (to Eastman Kodak Co.); U.S. Pat. 2,423,730 (1947), I. L. Salminen and A. Weissberger (to Eastman Kodak Co.); U.S. Pat. 4,333,999 (1982), P. T. S. Lau (to Eastman Kodak Co.).

63. U.S. Pat. 2,895,826 (1959), I. Salminen, C. Barr, and A. Loria (to Eastman Kodak Co.).

64. U.S. Pat. 2,369,929 (1945), P. Vittum and W. Peterson (to Eastman Kodak Co.); U.S. Pat. 2,772,162 (1956), I. Salminen and C. Barr (to Eastman Kodak Co.); U.S. Pat. 3,998,642 (1976), P. T. S. Lau, R. Orvis, and T. Gompf (to Eastman Kodak Co.).

65. E. Wolff, in *Proceedings of the SPSE 43rd Conference*, The Society for Imaging Science and Technology, Springfield, Va., 1990, pp. 251–253.

66. U.S. Pat. 2,407,210 (1946), A. Weissberger, C. J. Kibler, and P. W. Vittum (to Eastman Kodak Co.); Brit. Pat. 800,108 (1958), F. C. McCrossen, P. W. Vittum, and A. Weissberger (to Eastman Kodak Co.); Ger. Offen. 2,163,812 (1970), M. Fujiwhara, T. Kojima, and S. Matsuo (to Konishiroku KK.); Ger. Offen. 2,402,220 (1974), A. Okumura, A. Sugizaki, and A. Arai (to Fuji Photo Film); Ger. Offen. 2,057,941 (1971), I. Inoue, T. Endo, S. Matsuo, and M. Taguchi (to Konishiroku KK.); Eur. Pat. 296793 A2 (1988), B. Clark, N. E. Milner, and P. Stanley (to Eastman Kodak Co.); Eur. Pat. 379,309 (1990), S. C. Tsoi (to Eastman Kodak Co.).

67. U.S. Pat. 3,265,506 (1966), A. Weissberger and C. Kibler (to Eastman Kodak Co.); U.S. Pat. 3,770,446 (1973), S. Sato, T. Hanzawa, M. Furuya, T. Endo, and I. Inoue (to Konishiroku KK.); Fr. Pat. 1,473,553 (1967), R. Porter (to Eastman Kodak Co.); U.S. Pat. 3,408,194 (1968), A. Loria (to Eastman Kodak Co.); U.S. Pat. 3,894,875 (1975), R. Cameron and W. Gass (to Eastman Kodak Co.).

68. E. Pelizzetti and C. Verdi, *J. Chem. Soc. Perkin II*, 806 (1973); E. Pelizetti and G. Saini, *J. Chem. Soc. Perkin II*, 1766 (1973); D. Southby, R. Carmack, and J. Fyson, in Ref. 65, pp. 245–247.

69. G. Brown and co-workers, *J. Am. Chem. Soc.* **79,** 2919 (1957).

70. J. A. Kapecki, unpublished data, 1981; U.S. Pat. 4,401,752 (1981), P. T. S. Lau (to Eastman Kodak Co.).

71. Brit. Pat. 875,470 (1961), E. MacDonald, R. Mirza, and J. Woolley (to Imperial Chemical Ind.).

72. Brit. Pat. 778,089 (1957), J. Woolley (to Imperial Chemical Ind.).

73. U.S. Pat. 2,600,788 (1952), A. Loria, A. Weissberger, and P. W. Vittum (to Eastman Kodak Co.); U.S. Pat. 2,369,489 (1945), H. D. Porter and A. Weissberger (to Eastman Kodak Co.); U.S. Pat. 1,969,479 (1934), M. Seymour (to Eastman Kodak Co.).

74. G. Brown, B. Graham, P. Vittum, and A. Weissberger, *J. Am. Chem. Soc.* **73,** 919 (1951).

75. U.S. Pat. 2,343,703 (1944), H. Porter and A. Weissberger (to Eastman Kodak Co.); U.S. Pat. 2,829,975 (1958), S. Popeck and H. Schulze (to General Aniline and Film); U.S. Pat. 2,895,826 (1959), I. Salminen, C. Barr, and A. Loria (to Eastman Kodak Co.); U.S. Pat. 2,691,659 (1954), B. Graham and A. Weissberger (to Eastman Kodak Co.); U.S. Pat. 2,803,544 (1957), C. Greenhalgh (to Imperial Chemical Ind.); Brit. Pat. 1,059,994 (1967), C. Maggiulli and R. Paine (to Eastman Kodak Co.).

76. U.S. Pat. 2,600,788 (1952), A. Loria, A. Weissberger, and P. Vittum (to Eastman Kodak Co.); U.S. Pat. 3,062,653 (1962), A. Weissberger, A. Loria, and I. Salminen (to Eastman Kodak Co.).

77. U.S. Pat. 3,127,269 (1964), C. W. Greenhalgh (to Ilford, Ltd.); U.S. Pat. 3,519,429 (1970), G. Lestina (to Eastman Kodak Co.).

78. U.S. Pat. 3,582,346 (1971), F. Dersch (to Eastman Kodak Co.); U.S. Pat. 3,811,891 (1974), R. S. Darlak and C. J. Wright (to Eastman Kodak Co.).

79. P. W. Vittum and F. C. Duennebier, *J. Am. Chem. Soc.* **72,** 1536 (1950).

80. A. Wernberg, unpublished data, 1986.

81. L. E. Friedrich, in Ref. 65, p. 261.

82. U.S. Pat. 3,227,554 (1966), C. R. Barr, J. Williams, and K. E. Whitmore (to Eastman Kodak Co.); U.S. Pat. 4,351,897 (1982), K. Aoki and co-workers (to Fuji Photo Film); U.S. Pat. 4,853,319 (1989), S. Krishnamurthy, B. H. Johnston, K. N. Kilminster, D. C. Vogel, and P. R. Buckland (to Eastman Kodak Co.).

83. U.S. Pat. 4,241,168 (1980), A. Arai, K. Shiba, M. Yamada, and N. Furutachi (to Fuji Photo Film).

84. U.S. Pat. 3,519,429 (1970), G. J. Lestina (to Eastman Kodak Co.).

85. N. Furutachi, *Fuji Film Res. Dev.* **34,** 1 (1989).

86. U.S. Pat. 2,213,986 (1941), J. Kendall and R. Collins (to Ilford, Ltd.); U.S. Pat. 2,340,763 (1944), D. McQueen (to DuPont); U.S. Pat. 2,592,303 (1952), A. Loria, P. Vittum, and A. Weissberger (to Eastman Kodak Co.); U.S. Pat. 2,618,641 (1952), A. Weissberger and P. Vittum (to Eastman Kodak Co.); U.S. Pat. 3,834,908 (1974), H. Hara, Y. Yokota, H. Amano, and T. Nishimura (to Fuji Photo Film).

87. U.S. Pat. 3,725,067 (1973), J. Bailey (to Eastman Kodak Co.); U.S. Pat. 3,810,761 (1974), J. Bailey, E. B. Knott, and P. A. Marr (to Eastman Kodak Co.).

88. U.S. Pat. 4,443,536 (1984), G. J. Lestina (to Eastman Kodak Co.); Eur. Pat. 284,240 (1990), A. T. Bowne, R. F. Romanet, and S. E. Normandin (to Eastman Kodak Co.); Eur. Pat. 285,274 (1990), R. F. Romanet and T-H. Chen (to Eastman Kodak Co.).

89. U.S. Pat. 4,540,654 (1985), T. Sato, T. Kawagishi, and N. Furutachi (to Fuji Photo Film).

90. K. Menzel, R. Putter, and G. Wolfrum, *Angew. Chem.* **74,** 839 (1962); Brit. Pat. 1,047,612 (1966), K. Menzel and R. Putter (to Agfa-Gevaert).

91. J. Jennen, *Chim. Ind.* **67**(2), 356 (1952).

92. W. T. Hanson, Jr. and P. W. Vittum, *PSA J.* **13,** 94 (1947); U.S. Pat. 2,449,966 (1948), W. T. Hanson, Jr. (to Eastman Kodak Co.); K. O. Ganguin and E. MacDonald, *J. Photogr. Sci.* **14,** 260 (1966).

93. Ref. 4, pp. 270 ff, 320–321.

94. U.S. Pat. 3,227,554 (1966), C. R. Barr, J. Williams, and K. E. Whitmore (to Eastman Kodak Co.); C. R. Barr, J. R. Thirtle, and P. W. Vittum, *Photogr. Sci. Eng.* **13,** 74, 214 (1969).

95. M. A. Kriss, in Ref. 5, pp. 610–614; P. Kowaliski, *Applied Photographic Theory,* John Wiley & Sons, UK, 1972, pp. 466–468.

96. L. E. Friedrich, unpublished data, 1985; K. Liang, in Ref. 33, p. 451.

97. T. Kruger, J. Eichmans, and D. Holtkamp, *J. Imag. Sci.* **35,** 59 (1991).

98. U.S. Pat. 4,248,962 (1981), P. T. S. Lau (to Eastman Kodak Co.); Brit. Pat. 2,072,363 (1981), R. Sato, Y. Hotta, and K. Matsuura (to Konishiroku KK.).

99. Brit. Pat. 2,099,167 (1982), K. Adachi, H. Kobayashi, S. Ichijima, and K. Sakanoue (to Fuji Photo Film); U.S. Pat. 4,782,012 (1988), R. C. DeSelms and J. A. Kapecki (to Eastman Kodak Co.).

100. U.S. Pat. 3,632,345 (1972), P. Marz, U. Heb, R. Otto, W. Puschel, and W. Pelz (to Agfa-Gevaert); U.S. Pat. 4,010,035 (1977), M. Fujiwhara, T. Endo, and R. Satoh (to Konishiroku Photo); U.S. Pat. 4,052,213 (1977), H. H. Credner and co-workers (to Agfa-Gevaert); U.S. Pat 4,482,629 (1984), S. Nakagawa, H. Sugita, S. Kida, M. Uemura, and K. Kishi (to Konishiroku KK.).

101. T. Kobayashi, in *Proceedings of the 16th Symposium of Nippon Shashin Kokkai, Japan,* 1986; U.S. Pat. 4,390,618 (1983) H. Kobayashi, T. Takahashi, S. Hirano, T. Hirose, and K. Adachi (to Fuji Photo Film); U.S. Pat. 4,859,578 (1989), D. Michno, N. Platt, D. Steele, and D. Southby (to Eastman Kodak Co.); U.S. Pat. 4,912,025 (1990), N. Platt, D. Michno, D. Steele, and D. Southby (to Eastman Kodak Co.); Eur. Pat. 193,389 (1990), J. L. Hall, R. F. Romanet, K. N. Kilminster, and R. P. Szajewski (to Eastman Kodak Co.).

102. Brit. Pat. 1,097,064 (1964) C. R. Barr (to Eastman Kodak Co.); Belg. Pat. 644,382 (1964), R. F. Porter, J. A. Schwan, and J. W. Gates, Jr. (to Eastman Kodak Co.).

103. K. Mihayashi, K. Yamada, and S. Ichijima, in Ref. 65, p. 254.

104. G. I. P. Levenson, in Ref. 5, pp. 437–461.

105. K. H. Stephen and C. M. MacDonald, *Res. Disclosure* **240,** 156 (1984); J. L. Hall and E. R. Brown, in Ref. 33, p. 438.

106. S. Matsuo, *Nippon Shashin Gak.* **39**(2), 81 (1976).

107. *Manual for Processing Eastman Color Films,* Eastman Kodak Co., Rochester, New York, 1988

108. U.S. Pat. 4,684,604 (1987), J. W. Harder (to Eastman Kodak Co.); U.S. Pat 4,865,956 (1989), J. W. Harder and S. P. Singer (to Eastman Kodak Co.); U.S. Pat. 4,923,784 (1990), J. W. Harder (to Eastman Kodak Co.).

109. ISO 5800:1987(E), *Photography—Color Negative Films for Still Photography—Determination of ISO Speed.*

110. ISO 2240:1982(E), *Photography—Color Reversal Camera Films—Determination of ISO Speed.*

111. A. F. Sowinski and P. J. Wightman, *J. Imag. Sci.* **31,** 162 (1987).

112. D. English, *Photomethods,* 16 (May 1988).

113. W. T. Hanson, Jr. and C. A. Horton, *J. Opt. Soc. Am.* **42,** 663 (1952).

114. R. J. Tuite, *J. Appl. Photogr. Eng.* **5,** 200 (1979).

115. K. O. Ganguin, *J. Photogr. Sci.* **9,** 172 (1961); D. C. Hubbell, R. G. McKinney, and L. E. West, *Photogr. Sci. Eng.* **11,** 295 (1967).

116. Y. Seoka and K. Takahashi, *Second Symposium on Photographic Conservation, Society of Scientific Photography, Japan.,* Tokyo, July 1986, pp. 13–16.

117. P. Egerton, J. Goddard, G. Hawkins, and T. Wear, *Royal Photographic Society Colour Imaging Symposium,* Cambridge, UK, Sept. 1986, p. 128.

118. K. Onodera, T. Nishijima, and M. Sasaki, in *Proceedings of the International Symposium: The Stability and Conservation of Photographic Images,* Bangkok, Thailand, Nov. 1986.

119. E. Hoffmann and A. Bruylants, *Bull. Soc. Chim. Belges* **75,** 91 (1966); K. Sano, *J. Org. Chem.* **34,** 2076 (1969).

120. R. L. Heidke, L. H. Feldman, and C. C. Bard, *J. Imag. Technol.* **11**(3), 93 (1985).

121. U.S. Pat 2,689,793 (1954), W. R. Weller and N. H. Groet (to Eastman Kodak Co.).

122. J. C. Dainty and R. Shaw, *Image Science,* Academic Press, Inc., London, 1974; M. Kriss, in Ref. 5, Chapt. 21, p. 596.

123. *CHOICES—Choosing the Right Silver-Recovery Method for Your Needs,* Kodak Publication J-21, Rochester, New York, 1989.

124. T. Cribbs, *Sixth International Symposium on Photofinishing Technology, The Society for Imaging Science and Technology,* Springfield, Va., 1990, p. 53.

125. D. G. Foster and K. H. Stephen, in Ref. 124, p. 7.

126. J. J. Lauff, D. B. Steele, L. A. Coogan, and J. M. Breitfeller, *Appl. Environ.* **56,** 3346 (1990).

127. *1989–1990 Wolfman Report on the Photographic and Imaging Industry in the United States,* Diamandis Communications, Inc., New York, 1990.

128. S. Ikenoue and M. Tabei, *J. Imag. Sci.* **34**(5), 187 (1990).

JON KAPECKI
JAMES RODGERS
Eastman Kodak Company

COLOR PHOTOGRAPHY, INSTANT

The term instant color photography originally referred to one-step processes that provide finished color photographs within a minute after exposure of the film. The processing of each film unit is initiated in the camera immediately after the film has been exposed. Processing proceeds rapidly under ambient conditions. Such processes are outwardly dry, and the reagent is usually provided as a part of the film unit. The reagent is applied by mechanical action of the camera, film holder, or other processing device (see also COLOR PHOTOGRAPHY; PHOTOGRAPHY).

The technology of instant photography has been extended to include the application of similar chemistry to films wherein processing is delayed for a time after exposure. Whereas the multistep darkroom processes used for noninstant color films require precise time and temperature control, the instant processes require little or no timing and operate over a wide range of temperatures.

Since the introduction of Polacolor in 1963, instant color photography has enabled the photographer to create a finished color photograph under ambient conditions, to view and judge the results immediately. Observing the subject and the print simultaneously facilitates planning of improved photographs. In addition to providing new enjoyment for the amateur photographer, instant color films have become powerful media for professional work, and commercial and technical uses have grown dramatically.

Some color print processes used in reprography (qv) and in producing hard copy from records generated by computers, instruments, or electronic cameras are similar to the instant photographic processes in chemistry or in certain physical aspects of processing (see IMAGING TECHNOLOGY). However, print processes using dye transfer based on light sensors (qv) other than silver halides, for example microencapsulated dye processes that utilize photopolymerization (1) or photoinduced capsule rupture (2) to control image formation, are outside the scope of this discussion.

Principles of Instant Photography

The first one-step print process was introduced by Land in 1947 (3). A comprehensive account of one-step photography detailing the development of instant black-and-white and color processes from 1944 through 1976 is available (4). Subsequent developments in instant photography and related reprographic processes through 1988 have also been described (5). A review of the chemistry of a number of instant color processes may be found in Reference 6.

Film and Process Design.   Handheld camera use requires a film of sufficient photographic speed to permit short exposures at small apertures. Silver halide emulsions have high sensitivity and, upon development, enormous amplification. The original system involved a one-step process in which a viscous reagent was spread between two sheets, one bearing an exposed silver halide emulsion and the other an image-receiving layer. Both sheets were drawn through a pair of pressure rollers, and a sealed pod (7) attached to one of the sheets ruptured to release the viscous reagent, which was spread to form a thin layer between the two sheets, temporarily bonding them together. The action of the reagent produced concomitantly a negative image in the emulsion layer and a positive image in the image-receiving layer. After about a minute, the two sheets were stripped apart to reveal the positive image.

Reagents for Instant Photography. An essential component of each instant film is the reagent system. Essentially dry processing is realized by using a highly viscous fluid reagent and restricting the amount to just that needed to complete the image-forming reaction for a single picture.

The high viscosity of the reagent is provided by water-soluble polymeric thickeners (8). Suitable polymers include hydroxyethyl cellulose, the alkali metal salts of carboxymethyl cellulose and carboxymethyl hydroxyethyl cellulose. The high viscosity of the reagent makes possible accurate metering to form a uniform layer that also serves as a temporary adhesive for the two sheets during processing. The viscous reagent layer may further serve as a protective colloid before and during processing. The layer may remain on the surface of the print or it may be stripped away with the other sheet.

A sealed pod of reagent for each picture permits the use of more highly reactive reducing agents and more strongly alkaline conditions than would be feasible in a tank or tray process, as the reagent may be protected from oxygen from the time it is sealed in the pod until after image formation is complete.

In addition to high molecular weight polymer, reducing agents, and alkali, the viscous reagent may contain reactive components that participate in image formation, deposition, and stabilization. Some reactants may also be incorporated in coatings (qv) on either of the two sheets. Such an arrangement may serve to isolate reactive components from one another before processing, facilitate sequential availability during processing, and provide each component with a compatible and stable environment.

The reagent-containing pod must be carefully designed for both containment and discharge of its contents (Fig. 1). The pod lining must be inert to strong alkali and other components of the reagent, and the pods must be impervious to oxygen and water vapor over long periods. When the pod passes through processing rollers, the seal must rupture with a peeling separation, rather than explosive bursting, so that the reagent is released along one edge and spread uniformly, starting as a fine bead and forming a smooth, thin layer. The amount of reagent contained in each pod is particularly critical for integral film units that have limited capacity to conceal surplus reagent.

One-Step Cameras and Processors. The earliest one-step cameras used roll film and completed processing inside a dark chamber within the camera (9). The first instant color film, Polacolor, was provided in roll film format to fit these cameras. Flat-pack film cameras (Fig. 2), introduced in 1963, permitted the film to be drawn between processing rollers and out of the camera before processing was completed (10). Film holders for instant 10 × 13 cm (4 × 5 in.) film packets contain retractable rollers that permit the film to be loaded without rupturing the pod (11). For 20 × 25 cm (8 × 10 in.) films, the processing rollers are part of a tabletop processor. The exposed film, contained in a protective black envelope, and a positive sheet with pod attached are inserted into separate slots of a tray that leads into the processor. The film passes through the rollers into a covered compartment within which processing is completed.

Fully automatic processing was introduced in 1972 with the SX-70 camera, which ejected each integral picture unit automatically, passing it between motor-

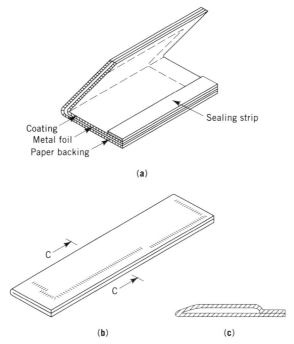

Coating
Metal foil
Paper backing

Sealing strip

(a)

C

C

(b)   (c)

Fig. 1. Patent illustration showing a typical pod structure (7). (**a**) Unfilled pod; (**b**) filled, sealed pod; (**c**) cross-section of filled pod along line C–C in (**b**).

ized processing rollers and out of the camera immediately after exposure (12,13). Kodak instant cameras, introduced in 1976 and now discontinued, included both motorized and hand-cranked models. Fuji instant cameras for integral films are motorized.

Film Configuration.   Using peel-apart materials, two separate sheets are laminated together as reagent is spread at the start of processing. The sheets are then peeled apart to terminate processing and permit viewing of the image. An integral print film comprises two sheets permanently secured as a single unit. The image-forming layers are located on the inner surfaces of the two sheets, at least one of which is transparent. The image is usually viewed through the transparent sheet against a reflective white pigment layer within the film unit. The Kodak Trimprint integral film format (1983) included a stripping layer that provided the option of stripping the print from the image-forming layers. A different version, Instagraphic color slide film (1984), provided for stripping a transparency from the remaining layers.

Integral films are designed to provide exposure of the emulsion layers and viewing of the print either through the same surface (14) or through opposite surfaces (15) of the film unit. Polaroid integral color film units are exposed and viewed through the same surface, and correct image orientation is obtained using a single mirror reversal within the camera, an arrangement chosen for compactness in optical design. Fuji integral films, like the earlier Kodak integral films, are exposed and viewed through opposite surfaces, so that image orientation is correct either without optical reversal in the camera or with two mirror reversals.

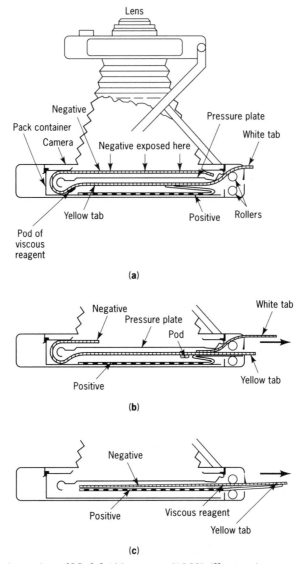

Fig. 2. Schematic section of Model 100 camera (1963), illustrating processing of pack film outside the camera. The outer surfaces of both negative and positive sheets are opaque. (**a**) Position of each element during exposure; (**b**) following exposure, pulling the white paper tab leads the negative into position for processing; (**c**) pulling the yellow paper tab draws the entire film unit between pressure rollers, rupturing the pod and spreading the reagent, and leads the film unit out of the camera.

Formation of Instant Images.   A series of complementary positive and negative images, one or more of which can serve as a starting point for a transfer process that leads to a useful positive image, includes exposed grains, unexposed grains; developed silver, undeveloped silver halide; oxidized developer, unoxidized developer; neutralized alkali, alkali not neutralized; hardened gelatin, unhardened gelatin. As an example, black-and-white instant films are based on imagewise transfer of undeveloped silver halide, unoxidized developer, and unused alkali to a receiving layer, within which these react to form a positive silver image.

In instant color processes that utilize dye-developers, that is, molecules that are both image dyes and photographic developers, an image in terms of unoxidized dye-developer transfers to form a positive dye image in the image-receiving layer. In dye-release systems both negative-working and positive-working emulsions have been used. Dye release is effected by one or more of the initial images in terms of silver or developer.

Methods of Color Reproduction

The reproduction of color requires the selective recording and presentation of principal regions of the visible spectrum, which extends roughly from 400 to 700 nm. For most processes three records are used, corresponding to blue (400–500 nm), green (500–600 nm), and red (600–700 nm) (see COLOR PHOTOGRAPHY).

In subtractive color photography, the three color records are formed in separate silver halide emulsions sensitive to blue, green, and red light, and coated in a multilayer structure. Processing produces positive images in terms of complementary dyes: the blue record is transformed to an image in yellow, or minus blue, dye; the green record to an image in magenta, or minus green, dye; and the red record to an image in cyan, or minus red, dye. When white light passes through the superposed set of yellow, magenta, and cyan images, the dyes absorb blue, green, and red components of the white light so that the light that is not absorbed represents the original colors.

In additive color photography, the three color records are separated laterally by an array, or screen, of blue, green, and red filter elements superposed over a panchromatic silver halide emulsion. That is, there is an emulsion sensitive to blue, green, and red light. The exposed emulsion is processed to form a positive transparency comprising black-and-white records in silver of the respective blue, green, and red components of the original scene. When these silver images are projected in registration with the superposed color screen, the images add to reproduce the full color image. Additive color photography is used only for transparencies because the minimum density is too high for reflection prints.

Both subtractive and additive color reproduction are utilized in instant color films. Subtractive systems include all of the instant print and large format transparency materials except Polachrome 35-mm slide films, which are additive.

Subtractive Dye Imaging Systems. There are numerous methods for producing instant dye images from the records formed in a set of blue-, green-, and red-sensitive emulsions that constitute a multilayer color film (6,16). Although image dyes may be formed *in situ* by chromogenic color development, as in noninstant color films, in most instant color processes preformed dyes are transferred from the negative to an image-receiving layer. The dyes or dye precursors may be initially diffusible in alkali, in which case they will be immobilized imagewise, or they may be initially immobile in alkali and released imagewise to transfer. Positive-working processes produce dye transfer density inversely related to the developed silver density; conversely, negative-working processes produce dye transfer density in direct proportion to the developed silver.

The use of a negative-working emulsion, that is, an emulsion in which the exposed grains develop to form a negative silver image, in a positive-working dye transfer process results in the formation of a positive dye transfer image. When negative-working dye processes are used, positive dye images may be obtained by using direct reversal, or direct positive, emulsions, which develop unexposed rather than exposed grains to form positive silver images. As an alternative to using direct positive emulsions, unoxidized developer remaining after development of a negative-working emulsion may reduce prefogged silver in an adjacent layer; the resulting oxidized developer may then couple to release dyes that form positive dye transfer images (17,18).

Dye Developer Processes. The first instant color film, Polacolor, introduced the dye developer (19), a bifunctional molecule comprising both a preformed image dye and a silver halide developer. The multilayer Polacolor negative comprises a set of three negative-working emulsions, blue-, green-, and red-sensitive, respectively, each overlying a layer containing a dye developer complementary in color to the emulsion's spectral sensitivity. During processing, development of exposed grains in each of the emulsion layers results in oxidation and immobilization of a corresponding portion of the contiguous dye developer. Dye developer that is not immobilized migrates through the layers of the negative to the image-receiving layer to form the positive image.

Dye Developers. In addition to having suitable diffusion properties, dye developers must be stable and inert in the negative before processing. After completion of the process, the dye developer deposited in the image-receiving layer must have suitable spectral absorption characteristics and stability to light.

The requirements of a developer moiety for incorporation into a dye developer are well fulfilled by hydroquinones. Under neutral or acidic conditions hydroquinones are very weak reducing agents and the weakly acidic phenolic groups confer little solubility. In alkali, however, hydroquinones are readily soluble, powerful developing agents. Dye developers containing hydroquinone moieties have solubility and redox characteristics in alkali related to those of the parent compounds.

Among the first dye developers were simple azohydroquinones, having hydroquinone and a coupler attached directly through an azo group to form the chromophore, and hydroquinone-substituted anthraquinones. These compounds were very active developing agents and diffused readily to form excellent positive images. However, because the hydroquinone was an integral part of the chromophore, the colors shifted with pH changes and with changes in the oxidation state of the hydroquinone. To circumvent such unwanted color shifting, alkylene links that interrupt conjugation were added as insulation between the chromophore and developer moieties (20).

In the course of developing the Polacolor and SX-70 processes many insulated dye developers were synthesized and investigated. An extensive review of this work is available (21). The insulating linkage, chromophore, and developer moiety can each be varied. Substituents on the developer modify development and solubility characteristics; substituents on the chromophore modify the spectral characteristics in terms of both color and light stability. The attachment of two dyes to a single developer by amide linkage has also been described (22).

Pyrazolone dyes are particularly versatile yellow chromophores; the yellow dye developer used in Polacolor, the first instant color film, was based on a pyrazolone dye. The structures of the three Polacolor dyes are shown in Figure 3. The study of azo dyes derived from 4-substituted 1-naphthols led to the chromophore used in the Polacolor magenta dye developer. The Polacolor cyan dye developer contained a 1,4,5,8-tetra-substituted anthraquinone as chromophore (20).

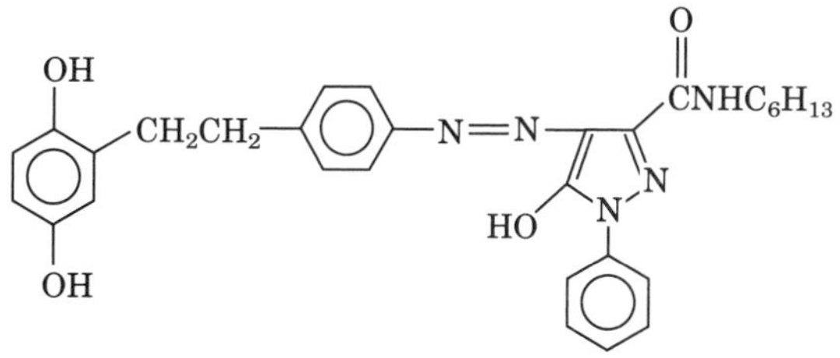

Fig. 3. Dye developers used in Polacolor (1963): (**a**) yellow [*14848-08-9*]; (**b**) magenta [*1880-52-5*]; and (**c**) cyan [*2498-16-0*] (see AZO DYES; DYES, ANTHRAQUINONE).

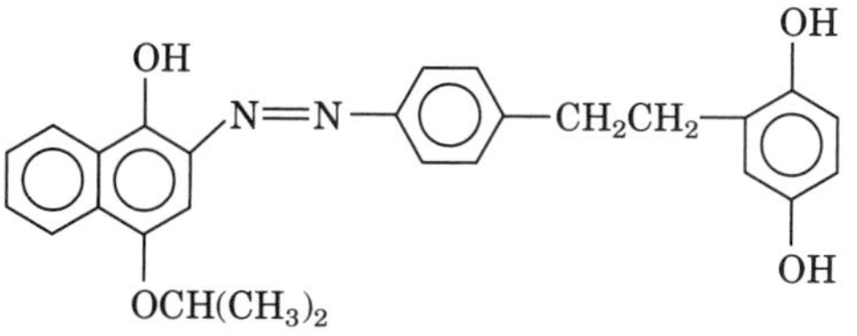

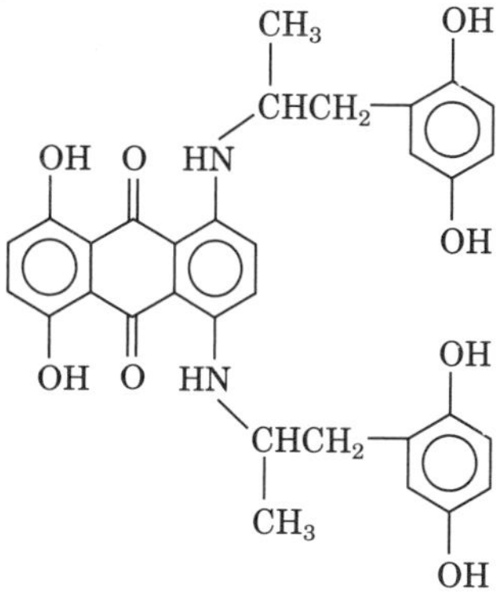

Figure 4 shows the structures of the metallized dye developers used in the first SX-70 film (1972) and in Polacolor 2 (1975). The images formed by these dye developers are characterized by very high light stability (23). The metallized cyan dye developer is based on a copper phthalocyanine pigment (24). Incorporation of developer groups converts the pigment into an alkali-soluble dye developer. A study of the chromium complexes of azo and azomethine dyes led to the design of the magenta and yellow metallized dye developers (25). The latter was derived from the 1:1 chromium complex of an *o,o'*-dihydroxyazomethine dye. A magenta dye developer [*78052-95-6*] (**1**) containing a xanthene dye moiety is used in Polacolor ER (extended range) film, as well as in the integral Polaroid films that followed SX-70 (26,27). This dye developer has much less unwanted blue absorption than earlier magenta dye developers and is thus capable of more accurate color rendition (Fig. 5).

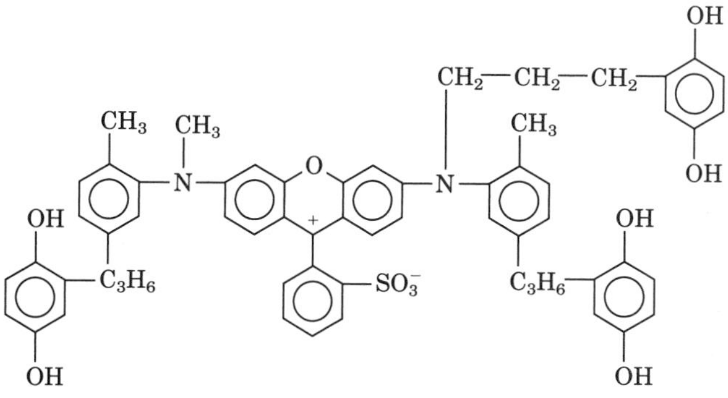

(**1**)

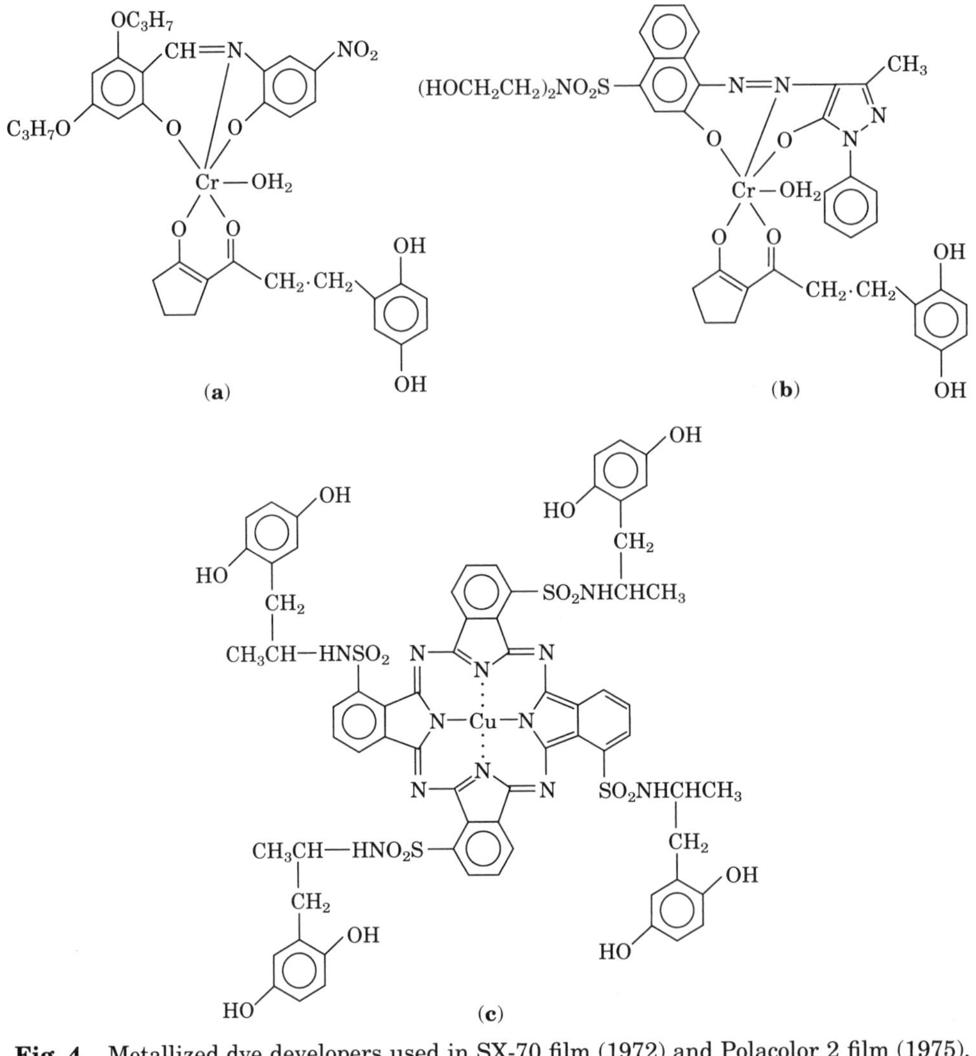

Fig. 4. Metallized dye developers used in SX-70 film (1972) and Polacolor 2 film (1975), where (**a**) is yellow [*31303-42-1*]; (**b**), magenta [*59518-89-7*]; and (**c**), cyan [*28472-22-2*].

Color-Shifted Dye Developers. Although as of this writing none of the commercialized films incorporate color-shifted dye developers, such compounds have been investigated extensively. These materials offer the option of incorporating the dye developer and the silver halide in the same layer without losing speed through the unwanted absorption of light by the dye developer. Prototypes are shown in Figure 6. The magenta azo dye developer shown in Figure 6**a** has been color-shifted by acylation (28). In the form shown, its color is a weak yellow. Indophenol dye developers color-shifted by protonation have been described (29). The cyan indophenol dye developer, Figure 6**b**, shows little color when the chromophore is in the protonated form. In alkali, the dye is ionized to the colored indophenoxide form, and in this state it may be stabilized by association with a quaternary mordant in the receiving layer.

Another approach involves the formation of dye images from colorless oxichromic developers, leuco azomethines stabilized against premature oxidation by

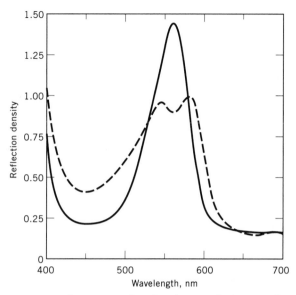

Fig. 5. Reflection spectra of magenta dye developers: (—) a xanthene (**1**) and (- - -) a metallized azo (Fig. **4b**) compound. The lower blue absorption of the xanthene dye is evident. Each curve represents the amount of dye in a print with equivalent neutral density of 1.0, balanced under 6500 K illumination.

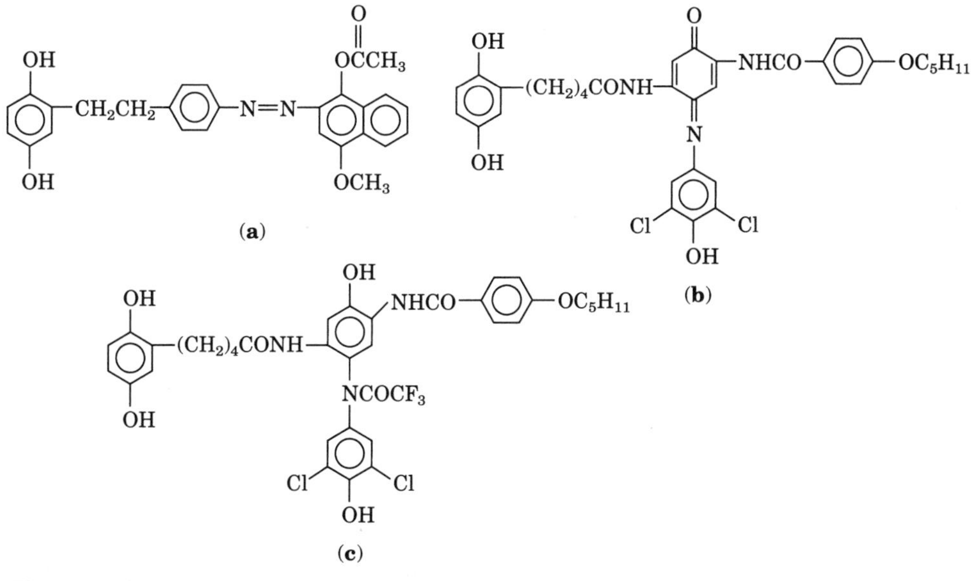

(a)

(b)

(c)

Fig. 6. Color-shifting dye developers: (**a**) yellow dye developer [*16044-30-7*] that becomes magenta upon hydrolysis; (**b**) cyan dye [*50695-79-9*] that becomes colorless upon protonation; and (**c**) leuco form of a dye developer [*50481-86-2*] that becomes yellow upon hydrolysis and oxidation.

acylation and linked to developer moieties (30), as shown in Figure 6c. The transferred images are oxidized to the colored form either by aerial oxidation or by
oxidants present in the receiving layer.

Additional Chromophores.   Other types of dyes that have been studied as
chromophores in dye developers include rhodamine dyes, azamethine dyes, indophenol dyes, and naphthazarin dyes (21). Cyanine dyes, although not generally
stable enough for use as image dyes, have also been incorporated in dye developers
(31).

Dye-Release Processes.   Dyes or dye precursors that do not participate
directly in the development process and are initially immobile or of low mobility
in alkali may be released by agents generated imagewise during development.
One of the advantages of such a process is that the released species may themselves be unreactive with respect to other components of the negative. Dyes may
thus diffuse through layers of the negative to reach the image-receiving layer
without undergoing unwanted reactions.

Dye release may relate either directly or inversely to the image-related reduction of silver halide. Release of the dye or dye precursor may be accomplished
or initiated by the oxidized developing agent or the unoxidized developing agent
or by alkali or silver salts.

Release by Oxidation.   Dyes or dye precursors may be released from alkali-
immobile compounds through interaction with the mobile oxidized form of a developing agent. Mobility of the oxidized developing agent is important because
both the silver halide and the dye or dye precursor are initially immobile in alkali
and thus cannot interact directly. The mobile oxidized developing agent may participate directly in dye release or it may act as an electron-transfer agent between
silver halide and a dye releaser that is initially a reducing agent. With alkali-
mobile image-forming species, such as the alkali-mobile dye developers, the use
of electron transfer by auxiliary developers is optional, whereas with alkali-
immobile species the use of an electron-transfer agent may be essential to the
process.

The image-related transfer of a diffusible dye formed as a product of the
oxidation of a dye containing a developing agent moiety was described in 1966
(32). This process depends on the preferential transfer of an oxidation product
having greater mobility than the unoxidized species. Compound [*13251-03-1*] (**2**),
for example, a bis-sulfonylhydrazide, upon oxidation releases [*573-89-7*] (**3**), a
smaller, more mobile dye.

(**2**)   (**3**)

The use of images in terms of oxidized developer to release dyes initially immobilized through a sulfonamide linkage has been described (33). In one approach
color coupling leads to ring closure and concomitant release of an alkali-soluble
dye (eq. 1).

$$(1)$$

Here R = H or alkyl and X = an immobilizing group.

A second approach utilizes the oxidation of a low mobility substituted 4-hydroxydiphenylamine to which an image dye is linked through a sulfonamide group. Oxidation and hydrolysis result in ring closure and release of the alkali-soluble dye (eq. 2).

$$(2)$$

Here R = alkyl and X = an immobilizing group.

These release processes are negative-working, that is, dye is released where silver halide development takes place. A positive image may be obtained from such a negative-working process by using a direct positive emulsion, that is, an emulsion in which the unexposed grains, rather than exposed ones, undergo development. Alternatively, both the immobile dye releaser and the silver-precipitating nuclei are included in a layer adjacent to the emulsion layer, and the processing reagent includes a silver halide solvent (19). In unexposed regions, soluble silver complex diffuses into the layer containing the nuclei and the immobile dye releaser. Development of silver is catalyzed by the nuclei, and the resulting oxidized developing agent cross-oxidizes the dye releaser, as in equation 2. The oxidized dye releaser undergoes ring closure, and the mobile dye is released for transfer to a receiving layer. The transferred image in this case is positive.

Immobile p-sulfonamidonaphthol dye-release compounds (Fig. 7) were used in the discontinued Kodak instant color films (34). These compounds undergo imagewise oxidation to quinonimides through interaction with an oxidized developing agent, followed by alkaline hydrolysis of the quinonimides to release soluble, diffusible image-forming dyes and immobile quinones (eq. 3).

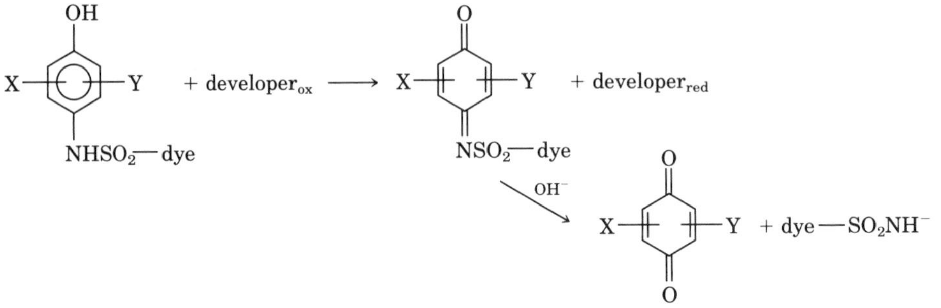

Here X and Y are immobilizing groups. Fuji instant color films are based on a similar dye-release mechanism using the o-sulfonamidophenol dye-release compounds shown in Figure 8 (35). Similarly, immobile p-sulfonamidoanilines, where

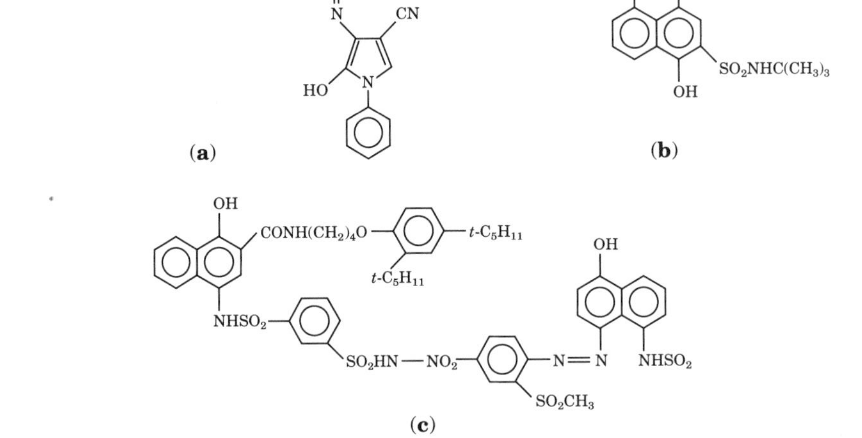

Fig. 7. Dye releasers used in Kodak integral films: (**a**) yellow; (**b**) magenta [*67041-93-4*]; (**c**) cyan [*42905-20-4*].

Y is H and the phenol group is replaced by an NHR moiety, may be used as dye releasers.

o-Sulfonamidophenol dye-release compounds are also used by Fuji in a photothermographic printing process introduced in 1987 as Fujix Pictrography (36). In this material the chosen developer moiety of the dye releaser is a weak reducing agent at room temperature, and processing is effected at 90°C.

Further examples of image-derived labilization and release by hydrolysis include redox dye releasers that contain ballasted hydroquinones using sulfonyl, oxy, or thio linkages, as in (**4**), where X is an immobilizing group and Y = S, O, or SO$_2$. Imagewise oxidation to form quinones, followed by alkaline hydrolysis, releases mobile image dyes (37). Corresponding hydroquinone derivatives without such ballast are dye-developers that transfer in nondeveloping regions to form positive images (38).

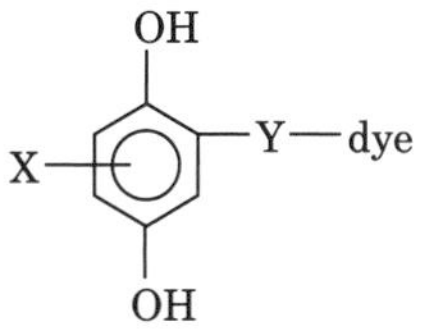

(**4**)

Another dye-release system that yields positive images is based on immobile benzisoxazolone dye releasers such as (**5**), where X is an immobilizing group and

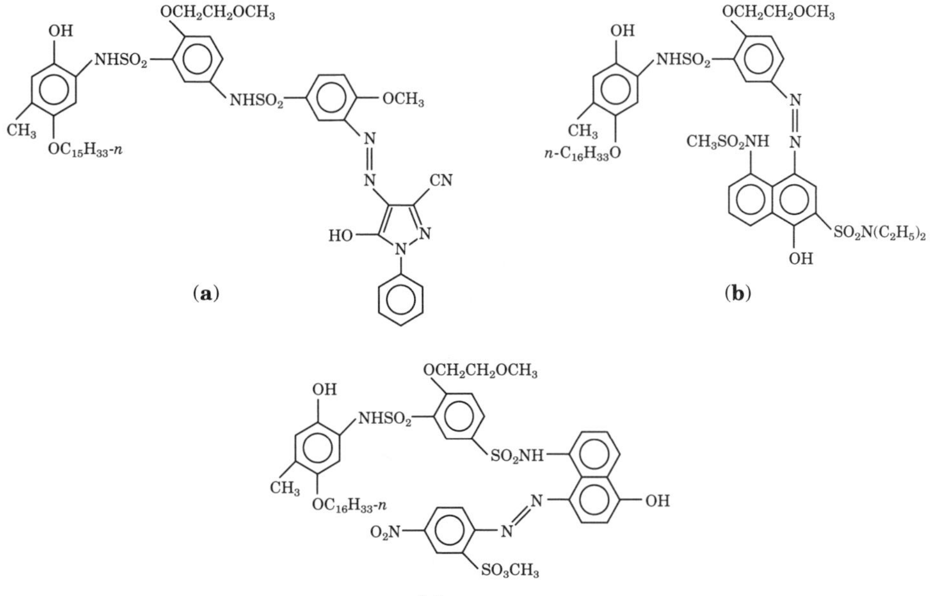

Fig. 8. Dye-release compounds used in Fuji instant films: (**a**) yellow [*96249-54-6*]; (**b**) magenta [*73869-83-7*]; (**c**) cyan.

R and R′ = alkyl. For these compounds, oxidation prevents cleavage. In alkali, the heterocyclic ring opens, forming a hydroxylamine, and in unexposed areas, where silver halide is not developing, the hydroxylamine cyclizes to form a second benzoxazolone (**6**), eliminating the dye moiety. In exposed areas, silver halide is reduced by a mobile developing agent, which in turn transfers electrons to oxidize the hydroxylamine to the nonreleasing species (**7**).

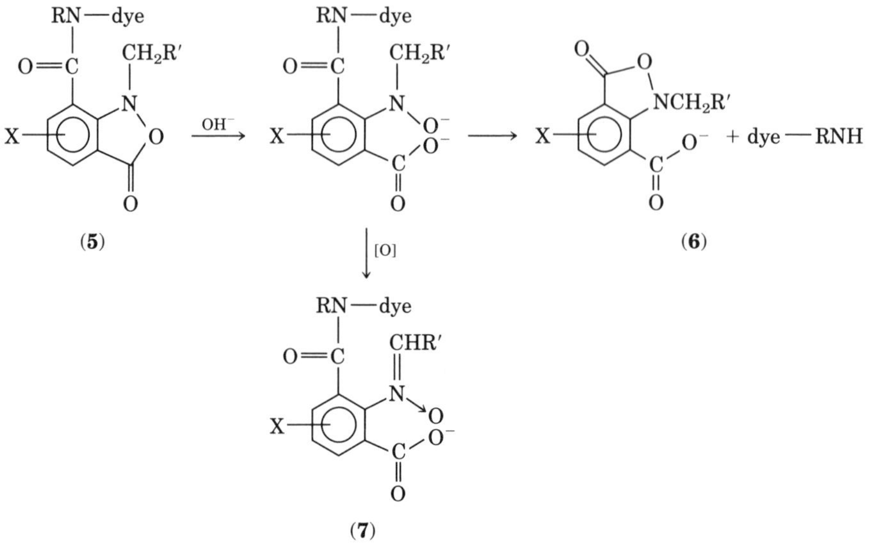

(4)

Another positive-working release by cyclization, illustrated by equation 5, starts with an immobile hydroquinone dye releaser (**8**), where R = alkyl and X is an immobilizing group. Cyclization and dye release take place in alkali in areas where silver halide is not undergoing development. In areas where silver halide is being developed, the oxidized form of the mobile developing agent oxidizes the hydroquinone to its quinone (**9**), which does not release the dye (39).

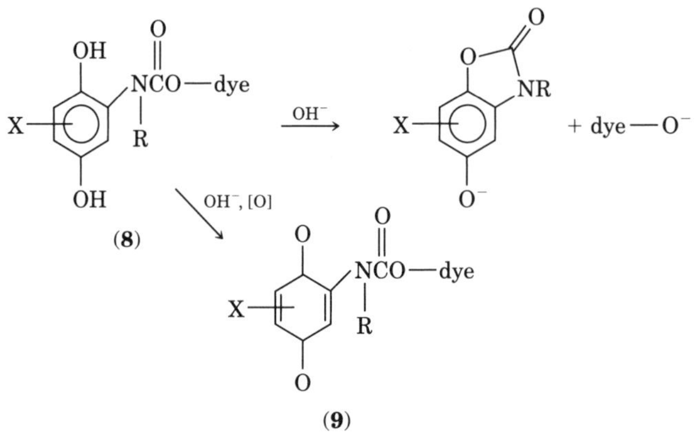

(5)

Redox dye-release systems based on immobile sulfonylhydrazones, as in structure (**10**), were disclosed in 1971. These compounds are oxidized by oxidized *p*-phenylenediamine developing agents, yielding azosulfones that undergo alkaline cleavage to release soluble moieties (40). To minimize stain, the developing agent may itself be rendered immobile and used in conjunction with a mobile electron-transfer agent (41).

Displacement by Coupling. The coupling reactions of color developing agents have been used as release mechanisms for initially immobile image dyes or dye precursors (42). The dye-releasing coupler may have substituents in the coupling position that are displaced in the course of the coupling reaction with an oxidized color developer. In one process, the eliminated substituent is the immobilizing group, so that the dye formed by coupling is rendered mobile. Compound (**11**) [*5135-15-9*] is an example of a coupler that splits off an immobilizing group in this manner. In a related process the coupler itself is the immobile moiety, as in (**12**) [*4137-16-0*], and the substituent that is split off is a mobile dye.

Dye Release by Reduction. The Agfa-Gevaert instant reprographic processes Agfachrome-Speed and Copycolor CCN introduced a method of releasing positive dye images by the reduction of the alkali-immobile dye-releasing quinone compounds shown in Figure 9 (43). These compounds were incorporated within the light-sensitive emulsion layers along with a ballasted electron donor compound. During processing there is competition between the oxidized developer and the reducible dye releaser for reaction with the electron donor. In exposed areas, silver halide develops, and the oxidized developer that forms is reduced by the electron donor. In unexposed areas, the electron donor reacts instead with the dye releaser, reducing its quinone moiety. The reduced dye releaser in turn undergoes a quinone methide elimination in alkali, releasing a mobile dye, as shown in equation 6. This system produces positive color images with negative-working emulsions. R, R′, R″, and R‴ are all alkyls.

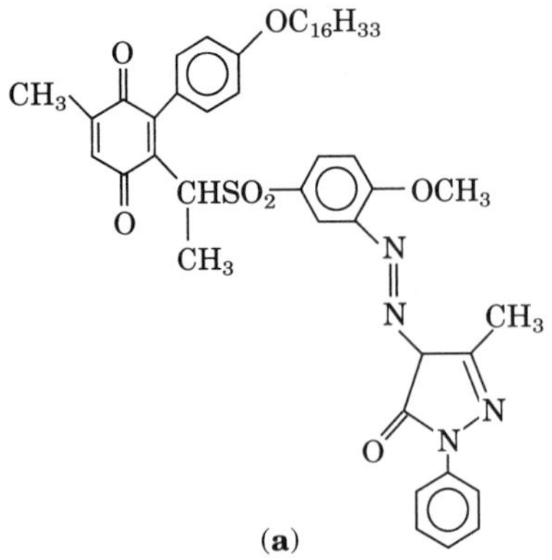

Fig. 9. Dye releasers used in Agfachrome-Speed and Copycolor films: (a) yellow [*85432-41-3*]; (b) magenta [*80406-97-9*]; (c) cyan.

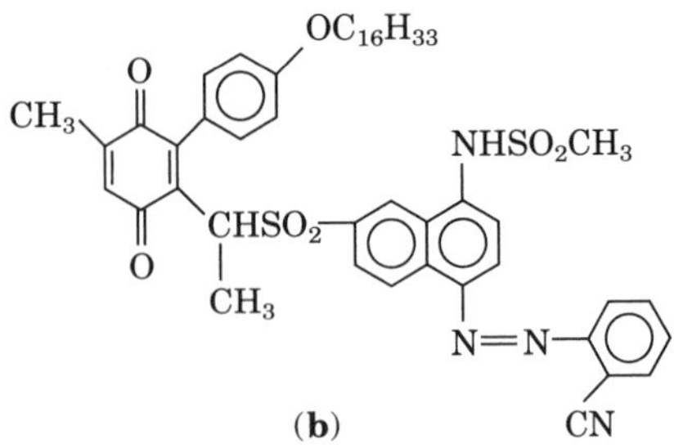

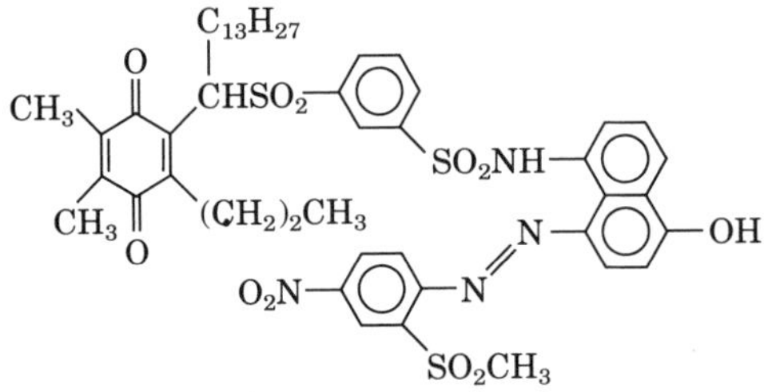

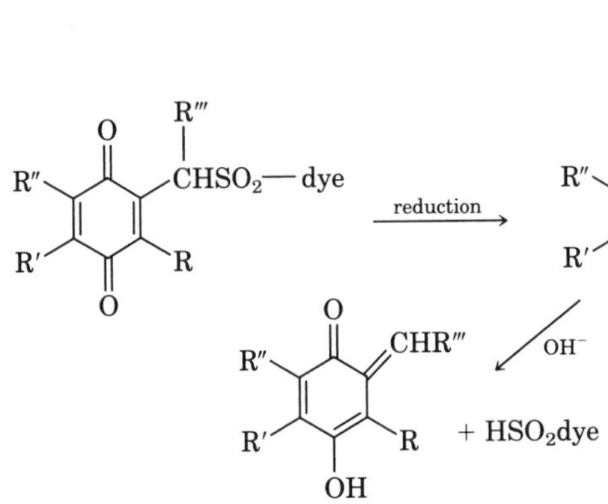

$$(6)$$

A positive-working system based on ring opening by the cleavage of a single N–O bond in a 2-nitroaryl-4-isoxazolin-3-one compound such as (**13**) has been described (44). In exposed areas, silver halide reduction is accompanied by oxidation of an electron-transfer agent, which cross-oxidizes an electron donor; in unexposed areas the electron donor initiates the ring opening and consequent dye release (45). This sytem, ring opening by single electron transfer (ROSET), is utilized in the Fuji Colorcopy DC photothermographic color print process.

(13)

Release by Silver-Assisted Cleavage. A soluble silver complex formed im-agewise in the undeveloped areas of the silver halide layer may be used to effect a cleavage reaction that releases a dye or a dye precursor. The process yields positive dye transfer images directly with negative-working emulsions (46). An example is the silver-assisted cleavage of a dye-substituted thiazolidine com-pound, as shown in equation 7.

The dye is initially linked to a ballasted thiazolidine, which reacts with silver to form a silver iminium complex. The alkaline hydrolysis of that complex yields an alkali-mobile dye. Concomitantly the silver ion is immobilized by reaction with the ballasted aminoethane thiol formed by cleavage of the thiazolidine ring.

Discrimination between exposed and unexposed areas in this process re-quires the selection of thiazolidine compounds that do not readily undergo alka-line hydrolysis in the absence of silver ions. In a study of model compounds, the rates of hydrolysis of model N-methyl thiazolidine and N-octadecyl thiazolidine compounds were compared (47). An alkaline hydrolysis half-life of 33 min was reported for the N-methyl compound, a half-life of 5525 min (3.8 days) was re-ported for the corresponding N-octadecyl compound. Other factors affecting the kinetics include the particular silver ligand chosen and its concentration (48). Polaroid Spectra film introduced silver-assisted thiazolidine cleavage to produce the yellow dye image (49), a system subsequently used in 600 Plus and Polacolor Pro 100 films.

Dye Formation Processes. *Color Development.* Experimental chromo-genic negatives included a structure comprising interdigitated multilayer color-forming elements (50). With the advent of dye developers, chromogenic three-color work was discontinued. However, a monochrome "blue slide" film, PolaBlue, was introduced in 1987. Color development of the exposed film forms a nondiffusing dye image within the emulsion layer (51). In 1988 a chromogenic photothermographic process based on immobile couplers and a color developer precursor was described (52).

In another process, instead of color coupling to form the image dyes, coupling was used to immobilize preformed dyes in exposed areas (53). Unreacted coupling dye transferred to form a positive dye image. Each of the coupling dyes incorporated the same coupler moiety, so that all would react at the same rate with the oxidized developing agent.

Coupler-developers comprising a color coupler and a color developer linked together by an insulating group, such as compound (**14**) [*41683-23-2*], have been described (54). In areas where silver halide undergoes development with such a compound, the coupler-developer is oxidized and immobilized by intermolecular coupling. In undeveloped areas the coupler-developer transfers to the receiving layer, where oxidation leads to intermolecular coupling to form a positive dye image.

$$OH$$

$$CONHCH_2CH_2NC_2H_5$$

$$NH_2$$

(**14**)

Other Colorless Dye Precursors.   Color images may be produced from initially colorless image-forming compounds other than the traditional chromogenic couplers. The use of colorless compounds avoids loss of light by unwanted absorption during exposure, and the colored form is generated during processing. A leuco indophenol may be used both to develop a silver image in the negative and to form an inverse, positive dye image in the receiving layer (55). Colorless triazolium and tetrazolium bases have also been proposed as image dye precursors (56). In unexposed regions, where silver halide is not undergoing development, unused developing agent reduces such a base to its colored form, rendering it immobile. In developing regions, the base may migrate to a receiving layer and there undergo reduction to form a negative dye transfer image.

Stability of Dye Images.   Regardless of the method of formation of the component dye images, a primary consideration must be the stabilization of the final color image in the receiving layer. Unlike the dye images produced by bath processes, the instant image is formed in a single processing step and is not washed or treated afterward. Therefore the instant process must incorporate both image-forming reactions and reactions that provide a safe environment for the finished image. In the integral films, not only the positive color image layer, but also the negative layers that remain concealed within the film unit, must be rendered inert.

The subtractive instant color films provide for image stabilization by the inclusion of polymeric acid layers that operate in conjunction with timing layers to provide a carefully timed reduction in pH following image formation. Reducing the pH terminates development reactions and stops the migration of dyes and other alkali-mobile species, thus stabilizing both the dye image and its environment.

Stabilization by timed pH reduction was first provided in Polacolor film. The Polacolor positive sheet includes a timing layer and a polymeric acid layer (57). The permeability of the timing layer determines the length of time before alkali reaches the polymeric acid layer and begins to undergo neutralization. The consumption of alkali rapidly reduces the pH of the image-receiving layer to a level conducive to print stability, and the removal of alkali from the image-receiving layer prevents salt formation on the print surface after the negative and positive sheets are stripped apart. The reduction of pH in the image layer proceeds to completion after the sheets are separated.

To provide for suitable timing of the pH reduction over the wide range of temperatures that may be encountered, the instant films may use polymeric timing layers in which permeability to alkali varies inversely with temperature. In the integral films, where all components are retained within the film unit after processing and the moisture content remains high for several days, care must be taken to avoid materials that could migrate or initiate unwanted reactions even at reduced pH.

Another concern is the stability of the image dyes to prolonged exposure to light. Many dyes that would otherwise be suitable as image dyes undergo severe degradation upon such exposure and are particularly sensitive to uv irradiation. Important considerations include the dye structures and the structure of the mordant in the image-receiving layer, as well as the chemical environment following the termination of processing.

The metallized dyes introduced in the SX-70 and Polacolor 2 films and included in later films have outstanding stability to light and are also highly stable in dark storage. The light stability of certain nonmetallized dyes may be increased by post-transfer metallization. For integral films, further protection against uv attack may be afforded by the incorporation of uv absorbers in the transparent polyester cover sheet. This sheet also provides protection against physical damage and attack by external chemicals. Similarly, Polacolor images show enhanced stability when laminated in plastic covers that provide uv filtration and exclude air and moisture from the image environment.

Subtractive Instant Color Films

Table 1 provides a summary of instant color camera-speed films introduced prior to January 1993. Instant color reprographic films in this period were Ektaflex PCT negative (Kodak), for printing from color negative film; Ektaflex PCT reversal (Kodak), Copycolor CCN (Agfa-Gevaert), Agfachrome-Speed (Agfa-Gevaert), and Colorcopy DC (Fuji), for printing from color prints or transparency films; and Fujix Pictrography 1000 (Fuji), for printing from electronic records. Of these reprographic films, only the Copycolor CCN, Colorcopy DC, and Pictrography films are commercially available as of this writing.

Table 1. Instant Color Films

Classification, film configuration[a]		ISO speed
Subtractive color films		
Polaroid		
peel-apart print films	Polacolor 64T	64
	Polacolor[b] (types 38, 48, 58, 88, 108, 636)	75
	Polacolor 2 (types 58, 88, 108, 668, 808)	80
	Polacolor ER (types 59, 559, 669, 809)	80
	Polacolor 100	100
	Polacolor Pro 100	100
peel-apart transparency	Colorgraph (types 691, 891)	80
integral print films	SX-70	150
	Time-Zero, type 778	150
	600[b], 600 Plus, type 779	640
	Autofilm 339	640
	Spectra, type 990	640
	Vision 95[c]	600
Kodak		
integral print films	PR-10[b]	150
	Instant Color[b]	150
	Kodamatic[b]	320
integral/peel print	Trimprint[b]	320
integral/peel transparency	Instagraphic[b]	64
Fuji		
integral print films	FI-10	150
	FI-800, FI-800G, FI800GT	800
peel-apart print film	FP-100	100
Additive color films		
Polaroid: additive color screen; silver transfer process		
Super-8 film	Polavision[b]	40
35-mm films	Polachrome CS	40
	Polachrome HC	40

[a]Polaroid film type, where shown, denotes format as well as photographic characteristics, eg, 50 series are 10 × 13 cm (4 × 5 in.) sheet films; image area 8.9 × 11.4 cm (3½ × 4½ in.). 80 series are 8.3 × 8.1 cm (3¼ × 3⅛ in.) pack films; image area 7 × 7.3 cm (2¾ × 2⅞ in.). 100, 660, and 690 series are 8.3 × 10.8 cm (3¼ × 4¼ in.) pack films; image area 7.3 × 9.5 cm (2⅞ × 3¾ in.). 300 series are 11.4 × 10.8 cm (4½ × 4¼ in.) integral pack films; image area 10 × 7.5 cm (4 × 3 in.). 500 series are 10 × 13 cm (4 × 5 in.) pack films; image area 8.9 × 11.7 cm (3½ × 4⅝ in.). 800 series are 20 × 25 cm (8 × 10 in.) sheet films; image area 19 × 24 cm (7½ × 9½ in.). Polacolor 100 and Polacolor Pro 100 are pack and sheet films of the same formats as Polacolor ER. SX-70, Time-Zero, 600, 600 Plus, and 700 Series are 8.9 × 10.8 cm (3½ × 4¼ in.). integral pack films; image area 8 × 8 cm (3⅛ × 3⅛ in.). Spectra and Type 990 are 10.8 × 11.4 cm (4¼ × 4½ in.) integral pack films; image area 7.2 × 9.1 cm (2⅞ × 3⅝ in.). Vision 95 is 6.4 × 11 cm (2.5 × 4.4 in.) integral pack film; image area 2.29 × 5.46 cm (2.15 × 2.9 in.). Fuji Films include both 8.3 × 10.8 cm (3¼ × 4¼ in.) packs, image area 7.3 × 9.5 cm (2⅞ × 3¾ in.), and 10 × 12.7 cm (4 × 5 in.) packs, image area 8.5 × 10.8 cm (3⅜ × 4¼ in.).
[b]Discontinued films.
[c]Vision 95 designates film introduced in Europe in 1992; it may have a different designation when introduced in the United States.

Polacolor. The first instant color film, Polacolor, was introduced by Polaroid Corporation in 1963. Polacolor was replaced in 1975 by Polacolor 2, a film with improved light stability, which utilizes the metallized dye developers shown in Figure 4. An extended range version, Polacolor ER, introduced in 1980, utilizes the cyan and yellow metallized dye developers together with a magenta dye developer that incorporates a xanthene dye having reduced blue absorption (see Fig. 5).

The Polacolor process produces subtractive multicolor prints comprising positive images in terms of yellow, magenta, and cyan dye developers. The dye developers form the positive image concomitantly with the formation of negative silver images and negative dye images within the layers of the negative. Image formation is based on the immobilization of dye developers by oxidation in areas where exposed silver halide grains are developed and on the diffusion of dye developers from areas that are unexposed.

Polacolor Negative Components. In the first three-color tests using dye developers, the negative was constructed in a screenlike configuration: the three-color image-forming elements were arrayed side-by-side in an arrangement that had been used with moderate success in earlier work with color development processes (58).

To use negatives having multiple continuous layers, hold-release mechanisms were developed to keep each dye developer in close association with the emulsion designated to control it until development had progressed substantially (59). In addition to the design and selection of dye developers characterized by a high development rate relative to the diffusion rate, useful measures included the introduction of small quantities of highly mobile auxiliary developers and the intercalation of temporary barrier layers between the three monochrome systems of the negative. Furthermore, because alkali and auxiliary developers diffuse more rapidly than dye developers, the dye developer may undergo oxidation and immobilization in exposed areas before the principal migration of the dye developer that forms the positive-transfer image.

Barrier Layers. Depending on composition, barrier layers can function simply as spatial separators or they can provide specified time delays by swelling at controlled rates or undergoing reactions such as hydrolysis or dissolution. Suitable barrier materials include cellulose esters and water-permeable polymers such as gelatin and poly(vinyl alcohol) (see BARRIER POLYMERS).

Preproduction dye developer negatives used a combination of cellulose acetate and cellulose acetate hydrogen phthalate as barrier layers. The images produced from these negatives were outstanding in color isolation, color saturation, and overall color balance. However, solvent coating was required with this composition, and it was not used in production.

The original Polacolor negative had water-coated interlayers of gelatin (60). The SX-70 and Polacolor 2 negatives use as interlayers a combination of a polymeric latex with a water-soluble polymer. A key development was the construction of lattices that function as temporary barriers, reducing interimage problems. The water-soluble polymer functions as a permeator, so that the barrier properties are tunable (61).

Polacolor Negative Structure. Each dye developer is initially located in a layer just behind the layer of silver halide emulsion that controls it during processing. As shown schematically in Figure 10 (on the colored plate), the negative consists of the yellow dye developer behind the blue-sensitive emulsion, the magenta dye developer behind the green-sensitive emulsion, and the cyan dye developer behind the red-sensitive emulsion. Interlayers separate the three sections. During exposure, each of the dye developers functions both as an antihalation dye for the emulsion layer in front of it and as a filter dye protecting the underlying emulsion layers from unwanted exposure. Thus the yellow dye developer protects the green- and red-sensitive emulsions from exposure to blue light, and the magenta dye developer protects the red-sensitive emulsion from exposure to green light. Figure 11a (on the colored plate) is a micrograph of an unprocessed Polacolor ER negative in cross section.

Polacolor Receiving Sheet. The Polacolor receiving sheet comprises three active layers (18) (Fig. 10). The outermost layer is an image-receiving layer comprising poly(vinyl alcohol) and the mordant poly(4-vinylpyridine), which immobilizes image dyes. Next is a polymeric timing layer, and then an immobile polymeric acid layer. The permeability of the timing layer to alkali is greater in the cold than at elevated temperature, thereby providing appropriate delays before the alkalinity of the system is reduced (62).

Auxiliary Developers. The use of auxiliary developers as electron-transfer agents, although not necessary for alkali-mobile dye developers, is usually advantageous. The auxiliary developer is smaller and more mobile than the dye developers and can more rapidly reach the exposed silver halide grains of the negative, initiating development. The oxidation product of the auxiliary developing agent in turn acts as an electron-transfer agent to oxidize and immobilize the dye developers (20). Some auxiliary developers that have been used in this way are Phenidone, Metol, and substituted hydroquinones, such as 4'-methylphenylhydroquinone [10551-32-3], $C_{13}H_{12}O_2$ (63), either alone or with Phenidone (64).

Image Formation and Stabilization. The sequence of reactions responsible for image formation and stabilization begins as alkali in the reagent permeates the layers of the negative, ionizing each of the three dye developers (eq. 8) and the auxiliary developer (eq. 9), which may be present in one or more layers of the negative.

$$
\text{dye}\!-\!\!\overset{\text{OH}}{\underset{\text{OH}}{\bigcirc}}\!\! + 2\,\text{KOH} \longrightarrow \text{dye}\!-\!\!\overset{\text{O}^-\text{K}^+}{\underset{\text{O}^-\text{K}^+}{\bigcirc}}\!\! + 2\,\text{H}_2\text{O} \tag{8}
$$

$$
\overset{\text{OH}}{\underset{\text{OH}}{\bigcirc}}\!\!-\!\!\overset{\text{CH}_3}{\bigcirc} + 2\,\text{KOH} \longrightarrow \overset{\text{O}^-\text{K}^+}{\underset{\text{O}^-\text{K}^+}{\bigcirc}}\!\!-\!\!\overset{\text{CH}_3}{\bigcirc} + 2\,\text{H}_2\text{O} \tag{9}
$$

[10551-32-3]

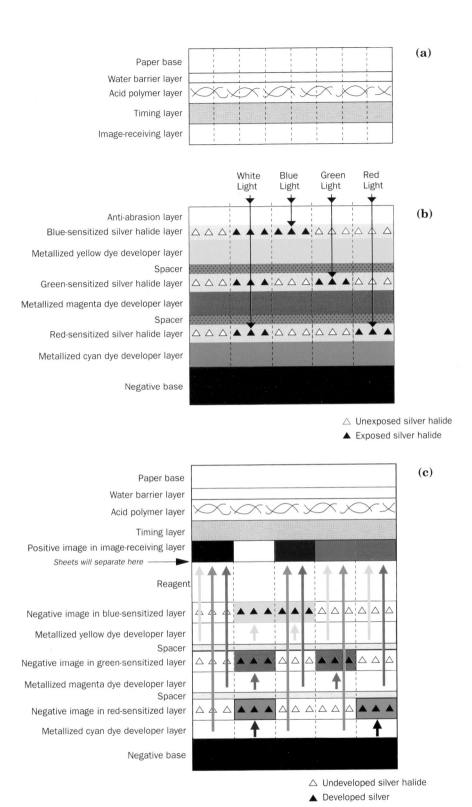

Figure 10. Schematic sections of Polacolor 2 components. **(a)** image-receiving sheet before processing; **(b)** negative during exposure; and **(c)** negative and positive sheets during image formation. The two sheets are laminated together by the viscous reagent and stripped apart 60 seconds later. During processing silver halide develops in the exposed areas of each emulsion layer, and the associated dye developer is oxidized and immobilized; in unexposed areas the dye developers diffuse from the layers of the negative to form a positive color image in the image-receiving layer. When the two sheets are stripped apart, the reagent layer adheres to the negative, which is discarded, and the positive print is ready for immediate viewing.

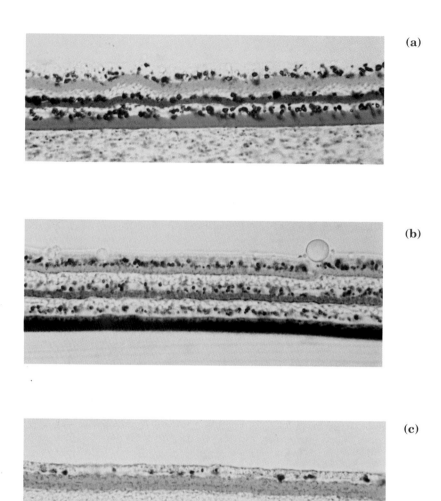

Figure 11. Micrographs of instant films in cross section, swelled in 5% Na_2SO_4 to reveal detail (1000X). Figures in parentheses indicate the approximate thickness of the swelled section relative to that of a non-swelled section. **(a)** Polacolor ER (2.0X); **(b)** Fuji FP-100 (1.5X); **(c)** Spectra film (1.3X). The sphere visible in **(b)** is a polymer bead of a type used in surface layers to prevent blocking.

$$+ \text{AgBr} \longrightarrow \qquad + \text{Ag}^\circ + \text{K}^+ + \text{Br}^- \qquad (10)$$

$$+ \text{dye} \longrightarrow \qquad + \text{dye} \qquad (11)$$

The ionized developers are then capable of diffusing. Transfer of an electron reduces the silver and generates the semiquinone ion radical of the auxiliary developer (eq. 10). In turn, a dye developer molecule of the adjacent layer transfers an electron to the semiquinone, returning the auxiliary developer to its original state and leaving the dye developer in the semiquinone state (eq. 11). Further oxidation of the semiquinone leads to the quinone state of the dye developer.

To maintain color isolation, it is important for the exposed grains of each emulsion layer to be substantially developed before a dye developer assigned to a different emulsion layer reaches them. Both the auxiliary developer and the barrier interlayers assist in such isolation. Color isolation may also be assisted by the release of low solubility silver ligands, such as mercaptans (65).

It is also important to prevent the reaction of dye developers with one another. An unoxidized dye developer migrating through overlying layers of the negative could reduce and release an oxidized, immobilized dye developer, thus effecting an exchange. Reaction with quaternary salts included in the reagent aids in the immobilization of the oxidized species (66).

Termination of the process is effected by the acid polymer layer of the receiving sheet. Acting as an ion exchanger, the acid polymer forms an immobile polymeric salt with the alkali cation and returns water in place of alkali. Capture of alkali by the polymer molecules prevents deposition of salts on the print surface. The dye developers thus become immobile and inactive as the pH of the system is reduced.

Polacolor processing takes approximately 60 s. Within this period the image-forming and image-stabilizing reactions proceed almost simultaneously. When image formation is terminated by stripping apart the negative and positive sheets, the print is ready for viewing; its surface is almost dry, and the pH of the image layer is rapidly approaching neutrality. The surface dries quickly to a hard gloss, and the finished print is durable and stable.

Polacolor films are balanced for daylight exposure and rated at ISO speed 80. Processing can be carried out satisfactorily at temperatures from 16–38°C, with slight exposure compensation recommended at the extremes.

Handheld cameras using Polacolor 8.3 × 10.8 cm (3¼ × 4¼ in.) pack films range from inexpensive amateur models to more expensive professional ones. The 10 × 13 cm (4 × 5 in.) pack films are also used in camera backs for professional and industrial applications. Polacolor 10 × 13 cm (4 × 5 in.) sheet films are used in holders that fit many specialized instruments, as well as view cameras; the 20 × 25 cm (8 × 10 in.) film holder fits standard equipment.

Larger format Polacolor photography became available in 1976, upon production of a series of cameras for 51 × 61 cm films, and a room-sized museum camera that uses 152-m production rolls of negative and receiving sheet to produce photographs as large as 1 × 2 m (67). In addition to the print films, the Polacolor series includes a transparency version, Colorgraph (1984), which produces images for overhead projection. Colorgraph is provided in both 20 × 25 cm (8 × 10 in.) and 8.3 × 10.8 cm (3¼ × 4¼ in.) formats. Because higher dye densities are required for projection, the processing time is extended from 1 to ca 4 min.

Polaroid Integral Films. In 1972 the SX-70 automatic camera and integral film system were introduced (12,13). The SX-70 film provided images that required no timing and no peeling apart. Each film unit was ejected through processing rollers immediately after exposure. The entire development process took place within the film unit under ambient conditions.

The integral film format required different processing chemistry and film components from those used previously. The processed film unit needed to contain all of the reaction products along with the final color image. Polaroid followed the original SX-70 film with Time-Zero SX-70 film (1979), 600 film (1981), Spectra film (1986), and 600 Plus film (1988).

SX-70 Film Structure. The SX-70 picture unit is an integral multilayer structure having no air spaces after processing. The sequence of dye developers, silver halide emulsions, and other layers is shown schematically in Figure 12. The dye developer layers comprise metallized dye developers (Fig. 8). The two outer layers are sheets of polyester; the structural symmetry ensures that the pictures remain flat under all atmospheric conditions. The upper polyester sheet is transparent, and the lower one is opaque black. Light passes through the upper sheet to expose the layers of the negative, and the picture is viewed through the same sheet. A durable, low index of refraction, quarter-wave antireflection coating on the outer surface of the transparent polyester minimizes flare during exposure, increases the efficiency of light transmission in both the exposure of the negative and the viewing of the final image, and permits seeing the image through the polyester with a minimum of surface luster (68).

Color balance of an integral film may be adjusted by exposure through color-correcting filter dyes incorporated in a layer of the positive sheet; the dyes are subsequently bleached to colorless form during processing (69–72). Nonbleaching, color-correcting filter dyes have been coated over the negative in Kodak and Fuji instant color films.

A potential source of fog following camera exposure is lateral light piping within the transparent polyester sheet. This effect is prevented by incorporating a very low concentration of a light-absorbing pigment, such as carbon black, in the sheet. A pigment concentration that increases transmission density normal to the surface by only 0.01 provides efficient absorption of light that could otherwise travel laterally by internal reflection within the sheet (73).

SX-70 Image Formation. The viscous SX-70 reagent is contained in a small pod concealed within the wide border of the film unit. As an exposed film unit is

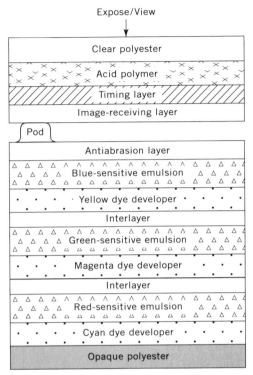

Fig. 12. Schematic cross section of SX-70 integral film unit. The film is exposed through the clear upper sheet. When the reagent from the pod is spread, it forms a white pigmented layer. The image formed in the image-receiving layer is viewed against the reflective pigment layer (5). Courtesy of Van Nostrand Reinhold.

ejected through the processing rollers and out of the camera, the pod bursts and the viscous reagent is spread between the top negative layer and the image-receiving layer, forming a new layer comprising water, alkali, white pigment, polymer, and other processing addenda (Fig. 12). Alkali quickly permeates the layers of the negative. In areas that have been exposed, silver halide is reduced and the associated dye-developer is oxidized and immobilized. In areas that have not been exposed the dye-developers, ionized and solubilized by the action of the alkali, migrate through overlying layers of the negative to reach the image-receiving layer above the new stratum of pigment-containing reagent. The transferred image is seen through the transparent polyester sheet by light reflection from the white pigment.

Opacification. An important aspect of the SX-70 system is that the developing film unit is ejected while still sensitive to light. Hence both surfaces must be opaque to ambient light. However, the top surface, through which the camera exposure is made, cannot become opaque until after that exposure has taken place. Protection against further exposure is provided by the combined effects of opacifying dyes and light-reflecting titanium dioxide pigment included in the viscous reagent layer (74). The opposite surface is protected by the support layer of black polyester negative.

The opacifying dyes are phthalein indicators that are incorporated in the viscous reagent. These compounds have high extinction coefficients at very high pH levels and render the reagent opaque when first spread. The dyes then lose color as the pH of the system is reduced. The opacifying dyes and the titanium dioxide together have a synergistic optical effect. The high reflectivity of the pigment greatly lengthens the optical path through the layer and thus increases the total absorption by the dye contained within. The result is that the negative is fully protected, even in bright sunlight, where the total exposure to ambient light may be more than a million times greater than the original camera exposure. Even though the transmission density of the reagent layer is initially very high (>6.5), the reflection density is low because of the high reflectivity of the pigment. Thus the SX-70 image may be seen against the reagent layer well before the opacifying dyes have lost all color.

The opacifying dyes used in the SX-70 film represent a new class of phthalein indicator dyes (75) designed to have unusually high pK_a values. The solution of an indicator such as that shown in structure (15) is highly colored when the dye is at or above its pK_a value and becomes progressively less colored as the pH is reduced. The very high pK_a values of these opacifying indicators are induced by hydrogen-bonding substituents located in juxtaposition to the hydrogen that is removed by ionization (76). The pK_a values for a series of naphthalein opacifying indicators are

(15)

X	Y	pK_a	CAS Registry Number
H	H	11.1, 13.8	[68975-54-2]
COOH	COOH	13.2, > 15	[68975-55-3]
COOH	NHSO$_2$C$_{16}$H$_{33}$	12.9, > 15	[37921-74-7]

The reduction of pH within the film unit is effected by a polymeric acid layer, as in the Polacolor process. The onset of neutralization is controlled by a contiguous timing layer. In the original SX-70 film unit these layers were on the inner surface of the transparent polyester sheet (Fig. 12); in Time-Zero SX-70 and later Polaroid integral films these layers are on the inner surface of the opaque negative support, as shown in Figure 13.

Time-Zero SX-70 Film. Time-Zero SX-70 film dramatically reduced the time required for completion of the development process and also changed the appearance of the emerging print. Using the earlier SX-70 film units, the picture area had at first been pale green and this color had gradually faded over a period of several minutes as the processing reached completion. Using Time-Zero film the image becomes visible against a white background only a few seconds after

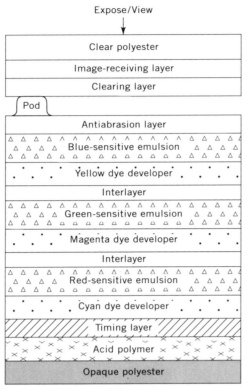

Expose/View

| Clear polyester |
| Image-receiving layer |
| Clearing layer |
| Pod |
| Antiabrasion layer |
| Blue-sensitive emulsion |
| Yellow dye developer |
| Interlayer |
| Green-sensitive emulsion |
| Magenta dye developer |
| Interlayer |
| Red-sensitive emulsion |
| Cyan dye developer |
| Timing layer |
| Acid polymer |
| Opaque polyester |

Fig. 13. Schematic cross section of Time-Zero SX-70 integral film. In this film the polymeric acid layer and the timing layer are located beneath the negative layers, rather than in the positive sheet. Time-Zero and all later Polaroid integral films have an antireflection layer coated on the outer surface of the clear polyester layer through which the image is viewed (5). Courtesy of Van Nostrand Reinhold.

the film unit has been ejected, and the picture appears complete within about a minute.

The quicker appearance of the image and the change in background color reflect changes in film structure and reagent composition. When the reagent is spread, it contacts a new *clearing layer* on the receiving layer surface. The clearing layer immediately decolorizes the opacifying dyes at the interface of the reagent layer and the receiver layer without affecting the opacifying dyes within the pigmented reagent layer. This decolorization provides a white background for the emerging image while the opacifying dyes within the reagent layer continue to protect the developing negative (77,78).

An effective clearing layer comprises a low molecular-weight poly(ethylene oxide) hydrogen bonded to a diacetoneacrylamide-*co*-methacrylic acid copolymer. The alkali of the reagent breaks the complex, releasing the ethylene oxide polymer and decolorizing the nearest opacifying dyes. The Time-Zero reagent incorporates as a thickener poly(diacetoneacrylamide oxime). This polymer is stable at very high pH but precipitates below pH 12.5. Its use at concentrations below 1 percent provides the requisite reagent viscosity and permits rapid dye transfer (79,80). A further aid to rapid transfer is the use of negative layers appreciably thinner than those of the original SX-70 negative.

The Time-Zero film process has an efficient receiving layer consisting of a graft copolymer of 4-vinylpyridine and a vinylbenzyltrimethylammonium salt grafted to hydroxyethylcellulose (81,82). The accuracy of color rendition was improved in the Time-Zero SX-70 film by the use of the xanthene magenta dye-developer (**1**).

Cameras using SX-70 films include both folding single-lens reflex models, designated as SX-70 cameras, and nonfolding models having separate camera and viewing optics. All of the cameras are motorized. Immediately after exposure the integral picture unit is automatically ejected from the camera. Pictures may be taken as frequently as every 1.5 s. The SX-70 picture unit is 8.9 × 10.8 cm, with a square image area approximately 8 × 8 cm. Ten of these picture units and a flat battery about the size of a picture unit are contained in the SX-70 film pack, along with an automatically ejected opaque cover sheet. SX-70 films are balanced for daylight exposure and are rated at ISO 150. The temperature range for development is approximately 7–35°C. The range may be extended to 38°C with a slight exposure compensation.

Type 600 Film. The first high speed integral film, Type 600 (1981), was rated at ISO 640, approximately two stops faster than the earlier SX-70 films. Features included high speed emulsions and pigmented spacer layers underlying the blue-sensitive and red-sensitive emulsion layers. The dye developers used were the same as those used in Time-Zero film. As in Time-Zero SX-70 film, the polymeric acid layer and the timing layer were located over the opaque polyester layer, below the negative layers, and the structure included an additional clearing layer between the reagent layer and the image-receiving layer.

Although the Type 600 film was similar in format to the SX-70 film, the films had different speeds and could not be used in the same cameras. Along with the Type 600 film, Polaroid introduced the nonfolding Sun cameras, which included integral strobe units and were configured to utilize the higher speed Type 600 film under a wide range of illumination conditions. Later, folding single-lens reflex cameras having a built-in strobe were provided for use with the high speed film. The battery in the Type 600 film pack has higher capacity than earlier batteries to provide sufficient power for operation of the built-in strobe units in addition to the exposure control and motor drive.

Spectra Film; 600 Plus Film. Polaroid Spectra film (1986) was the first material to utilize dye release by silver-assisted thiazolidine cleavage (83,84). The yellow image is formed by such a release of a metallized dye, and the magenta and cyan images are formed by dye developers, as in the SX-70 and 600 films. Because two types of image-forming reactions are involved, the film is often described as a *hybrid film.* Figure 14 is a schematic cross section of the Spectra film structure. A cross section of the unprocessed film is shown in Figure 11c. New layers in this structure include a silver scavenger interlayer, a colorless developer layer, and the yellow dye-releaser layer. The silver-assisted thiazolidine cleavage reaction requires relatively little silver. The blue-sensitive emulsion layer therefore contains considerably less silver than is used in corresponding dye developer structures.

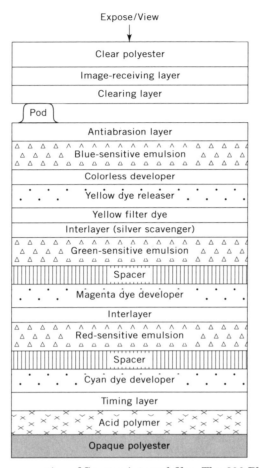

Fig. 14. Schematic cross section of Spectra integral film. The 600 Plus film has a similar structure. In these films the yellow image is formed by silver-assisted cleavage of a yellow dye releaser. A colorless developer reduces exposed silver halide in the blue-sensitive emulsion; in unexposed areas dissolved silver diffuses to the dye releaser layer and triggers the release of the yellow image dye.

During alkaline processing, the colorless developer, a hydroquinone derivative (85), diffuses from its layer and reduces exposed silver halide in the blue-sensitive layer. Unexposed silver halide is dissolved by a ligand contained in the reagent. The soluble silver complex diffuses to the layer containing the dye releaser and there triggers the release of the yellow image dye. The silver scavenger interlayer intercepts and captures silver ions diffusing from the red- and green-sensitive emulsion layers. The scavenger interlayer also assists color isolation by providing a short, timed delay in penetration of the reagent to underlying layers (46,83).

As the reagent reaches the underlying layers, the green- and red-sensitive emulsions undergo development in exposed regions and the associated dye developers become oxidized and immobilized. In unexposed areas dye developer molecules migrate to the image-receiving layer and are mordanted and immobilized there, forming the positive dye image. An auxiliary developer contained in the green-sensitive emulsion layer and the cyan dye developer layer assists in color isolation by serving as an electron-transfer agent between the less mobile dye developers and the silver halide. The auxiliary developer, oxidized by the development of silver, yields an oxidation product that in turn oxidizes dye developer molecules. However, the oxidation product of the colorless developing agent provided for the blue-sensitive emulsion does not interact with dye developers. In this negative, thin spacer and barrier interlayers assist in isolating the three separate image-forming processes by controlling the timing of reagent penetration and by providing spatial separation.

Further control of the development process is effected by alkali-releasable antifoggants contained in the two dye developer layers. The release rates for these addenda are temperature dependent. The specific chemistry utilizes an oxime group to trigger the elimination of the antifoggant. The oxime group has a pK_a > 12, so that it is readily wet by the alkali. The anion then rearranges and eliminates an antifoggant, for example, phenylmercaptotetrazole. Such elimination reactions have activation energies high enough so that there is no release at lower temperatures and good release at elevated temperatures, thus limiting thermally induced fog development (86,87).

The final image-forming stages are as in the SX-70 process. Alkali penetration of the timing and acid polymer layers initiates pH reduction of all layers of the film unit, immobilizing the dyes and terminating reactions. A polymeric hold-release timing layer based on the use of a hydrolyzable ester monomer provides a precipitous, rather than a gradual, pH reduction (88).

Spectra film was introduced as part of a new camera–film system featuring a rectangular picture format (approximately 7.3 $\times$ 9 cm). Both the film and the camera models are designated *Spectra*. 600 Plus film (1988), which has superseded 600 film, provides the Spectra film structure and chemistry in the square format of the earlier SX-70 and 600 films (approximately 8 $\times$ 8 cm), and is used in the cameras that originally utilized 600 film. Vision 95 integral film, introduced in Europe in 1992, also utilizes the Spectra film structure and chemistry but in a smaller format providing a 4 $\times$ 3 aspect ratio (approximately 7.3 $\times$ 5.5 cm).

Kodak Instant Films. Kodak entered the instant photography market in 1976 with the introduction of PR-10 integral color print film, rated at ISO 150, and a series of instant cameras designed for this film. These films and cameras were incompatible with the Polaroid instant photographic systems. Later Kodak integral films included Kodak Instant Color Film (ISO 150), Kodamatic and Kodamatic Trimprint films (ISO 320), Instagraphic print film (ISO 320), and Instagraphic color slide film (ISO 64). The films were balanced for daylight exposure. Kodak discontinued the production of instant films and cameras in 1986.

Film Components. The technical aspects of the PR-10 integral print film and components have been described (89,90). The PR-10 film system was based on a negative-working dye-release process using preformed dyes. Positive prints were obtained by using direct positive emulsions, which yield positive silver images upon development. The transfer of dyes was thus initiated by oxidized developing agent in areas where unexposed silver halide grains were undergoing development. The oxidized developing agent reacted with alkali-immobile dye-releasing compounds, which in turn released mobile image dyes that then diffused to the image-receiving layer.

The direct positive emulsions comprise silver halide grains that form latent images internally and not on the grain surface. Reversal is effected during processing by the action of a surface developer and a nucleating, or fogging, agent present in the emulsion layer (36). Photoelectrons generated during exposure are trapped preferentially inside the grain, forming an internal latent image. In an exposed grain the internal latent image is an efficient trap for conduction electrons provided by the nucleating agent, so that the internal latent image centers grow larger. These grains have no surface latent image and do not develop in a surface developer. The unexposed grains trap conduction electrons from the nucleating agent, at least temporarily, on the grain surface, forming fog nuclei that initiate development in the surface developer.

Film Structure. The principal components of the PR-10 integral film unit are shown schematically in Figure 15. A clear polyester support, through which the positive image was finally viewed, had on its inner surface both an image-forming and an image-receiving section. The image-forming section included direct positive emulsion layers, dye-releaser layers, and scavenger layers; the image-receiving section included the image-receiving and opaque layers to mask the image-forming section. The pod contained a viscous reagent and carbon black as an opacifier. The transparent polyester cover sheet, through which the sensitive layers were exposed, had on its inner surface polymeric timing layers, one of which contained the precursor of a development inhibitor, and a polymeric acid layer.

PR-10 Process Description. After the film unit had been exposed through the cover sheet, it passed between processing rollers. Viscous reagent released from the pod spread between the cover sheet and the sheet bearing the integral imaging receiver, thus initiating a series of reactions that developed the unexposed silver halide, released the dyes, and formed the positive image in the image-receiving layer.

As for the SX-70 films, processing was completed under ambient conditions outside the camera. As the positive image formed it was seen against a layer of titanium dioxide pigment covering an opaque carbon pigment layer behind the image-receiving layer. The developing film was protected from ambient light exposure through the viewing surface by the pigment layers of the image-receiving section. Exposure through the opposite surface was prevented by a carbon black layer formed by spreading the viscous reagent.

Neutralization to terminate processing was effected by the polymeric acid layer of the cover sheet; the onset of this reaction was controlled by the rate of permeation of the overlying polymeric timing layers. Mobility of the transferred dyes was also reduced by reaction with a mordant contained in the image-receiving layer. A development inhibitor released from one of the timing layers by the alkaline hydrolysis of its precursor assisted in restraining further development and consequent additional dye release.

Image-Forming Reactions. The dye-releasing reactions were initiated by the oxidized developing agent formed in the course of silver halide reduction. The developing agent, a derivative of 1-phenyl-3-pyrazolidone [*92-43-3*] (Phenidone), acted as an electron-transfer agent, giving up electrons to the developing silver halide grains and being regenerated to its reduced form by the dye releaser. The

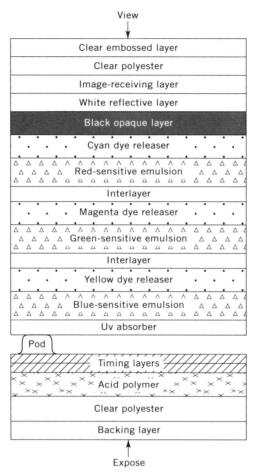

Fig. 15. Schematic cross section of Kodak PR-10 integral film. This film is exposed through the transparent cover sheet, and the color image is formed by dyes that transfer to the image-receiving layer through opaque pigment layers. The image is viewed against the reflective white pigment layer.

dye releasers used in PR-10 film were alkali-immobile *p*-sulfonamidophenols having dye attached through the sulfonamido group. Oxidation of the dye releaser was followed by alkaline hydrolysis of the resulting quinonimide to form an immobile quinone and release a mobile sulfamoyl-solubilized dye (eq. 3).

Kodak instant cameras included battery-operated motorized and lower cost hand-cranked models. The pictures began to develop as they were ejected from the camera, and development proceeded over several minutes under ambient conditions. The films were balanced for daylight exposure. Prints measured 9.7 × 10.2 cm (3.8 × 4.0 in.) overall, having an image area of 6.7 × 9.0 cm. For 35-mm slide use, the Instagraphic transparency film was trimmed to size after stripping.

Fuji Instant Films. The first Fuji instant system, Fotorama (1981), provided an integral color film, FI-10, rated at ISO 150, and cameras compatible with the Kodak instant system. The film is similar in structure and processing to Kodak

PR-10 film but has thinner emulsion layers, as well as a different set of dye-releaser compounds (see Fig. 8). The thinner negative may account for the observation that the FI-10 image appeared to reach completion more quickly than the PR-10 image (91), although the faster image completion may also relate to the specific dyes and dye-releaser compounds used.

A high speed integral color film, FI-800 (ISO 800), and a new series of cameras suited to this higher speed, were introduced in 1984. The FI-800 film structure includes spacer layers that separate each of the light-sensitive emulsions from the dye-releaser layer that it controls. Figure 16a is a schematic cross section of FI-800. The overall thickness of the FI-800 negative is appreciably greater than

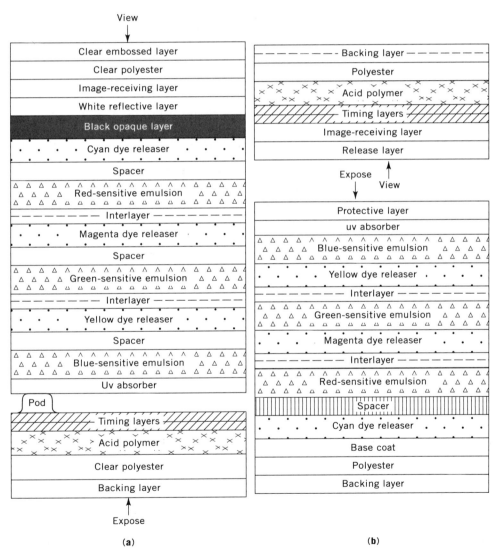

(a)   (b)

Fig. 16. Schematic cross sections of (**a**) Fuji FI-800 integral color film and (**b**) Fuji FP-100 peel-apart instant color film (5). Courtesy of Van Nostrand Reinhold.

that of the FI-10 negative, and image completion is somewhat slower. FI-800G is a version with a glossy surface rather than an embossed one, and FI-800GT is a similar film coated on a thinner, more flexible base. Both integral films are provided in packs that fit the corresponding Fuji instant cameras. The picture units are 9.7×10.2 cm, having image area of 6.7×9.0 cm. FI-10 is also provided in 10×13 cm format for use in a special back for professional cameras.

Fuji Peel-Apart Film FP-100. In 1984 Fuji introduced FP-100, a peel-apart instant color film rated at ISO 100. The FP-100 system uses a dye-release process similar to that used in the Fuji integral films. Figure 16**b** is a schematic cross section of FP-100, and Figure 11**b** (on the colored plate) is a micrograph of the unprocessed film in cross section. The negative structure includes a spacer layer between the red-sensitive layer and the cyan dye-releaser layer that it controls, similar to that shown in the FI-800 structure, but there are no spacers between the other emulsions and corresponding dye-releaser layers.

The peel-apart film is provided in 8.3×10.8 cm packs compatible with cameras and backs that accept Polacolor pack films. The pack films are used in a variety of professional and industrial applications.

Reprographic Films. Both Kodak and Agfa have marketed reprographic films based on dye release and transfer, using liquid activators in film processors (see also REPROGRAPHY). Kodak's Ektaflex PCT Negative Film (1981) was a darkroom peel-apart material used for making prints and transparencies from color negatives. Processing time was from 8 to 20 min. Ektaflex PCT Reversal Film (1982) was a similarly processed positive-working material for reproducing color slides or prints. Both Ektaflex films were processed using the same activator solutions and receiving sheets. The dye-release chemistry was similar to that described earlier for the Kodak integral films (eq. 3).

Agfachrome-Speed and Copycolor were introduced by Agfa-Gevaert in 1983. Both films utilized the positive-working dye-release chemistry shown in equation 6. Agfachrome-Speed, which is no longer produced, was a single-sheet integral color print material (Fig. 17); the film was exposed through one surface, and the final dye image was viewed through the opposite surface (92). The film required darkroom processing in an alkaline activator for 90 s, followed by a 5-min running water wash.

The Copycolor films are peel-apart materials having chemistry similar to that of the Agfachrome-Speed process (Fig. 18). The Copycolor process provides color prints or transparencies from positive color originals. Processing in a Copyproof processor takes from 40 s to 3 min, depending on the specific materials used and the temperature.

Photothermographic Color Films. Several photothermographic materials fall within the speed class of reprographic films.

Fujix Pictrography 1000. Redox dye-release chemistry similar to that used in the Fuji instant films was utilized with negative-working emulsions in the photothermographic Fujix Pictrography 1000 film (1987) (93). This film is designed to transform digital data into three-color prints. Film layer sensitivity is matched to the output of a light-emitting diode (LED) printer, rather than the more conventional blue, green, and red spectral regions (see LIGHT GENERATION, LIGHT-EMITTING DIODES). In the multilayer negative an emulsion sensitive to yellow light controls a yellow dye releaser, a red-sensitive emulsion controls a

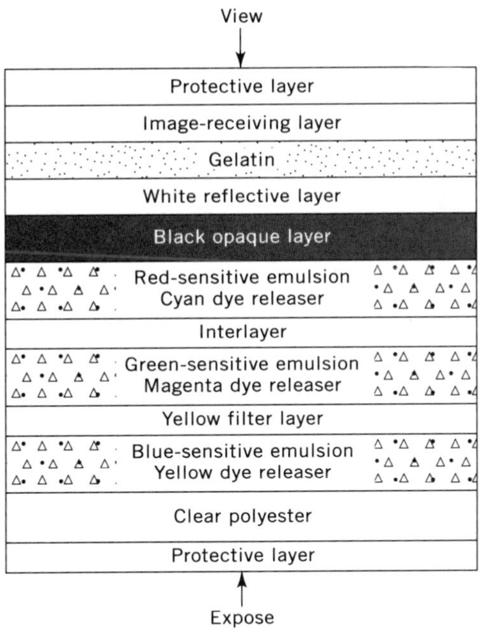

Fig. 17. Schematic cross section of Agfachrome-Speed film. During processing of this single-sheet film, dyes released by reduction in unexposed areas diffuse from layers of the negative to the image-receiving layer to form the positive color image (5). Courtesy of Van Nostrand Reinhold.

magenta dye releaser, and an emulsion sensitive to the near infrared (nir) controls a cyan dye releaser. Digital exposure is provided by yellow (570 nm), red (660 nm), and nir (810 nm) LEDs.

As for FP-100, the process utilizes two sheets, a negative, or donor, sheet and a receptor sheet; but all reactants are incorporated within the sheets. The donor sheet contains redox dye releasers, a development accelerator, and a basic metal compound, in addition to silver halide and binder (94). The redox dye releaser is a 4-alkoxy-5-*tert*-alkyl-2-sulfonamidophenol derivative (**16**). The substituent in the 5-position, is considered significant for the combination of stability at room temperature and reactivity at high pH and elevated temperature. The development accelerator for the high temperature processing, a silver arylacetylide, precludes the need for an auxiliary developing agent. The receiving sheet contains binder, polymeric mordant, and a chelating compound.

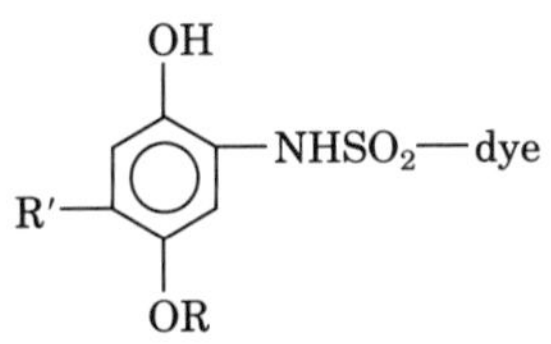

(**16**)

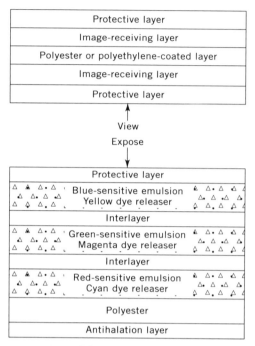

Fig. 18. Schematic cross sections of Copycolor negative and receiving sheet. During processing dyes are released by reduction in the unexposed areas. Dyes may be transferred to either surface of the symmetrical receiving sheet (5). Courtesy of Van Nostrand Reinhold.

The exposed film is coated with a thin layer of water, then brought into contact with the receiving sheet, and the assembled sheets are heated to ~90°C. The chelate compound diffuses from the receiving sheet to the donor sheet and reacts with the metal compound to generate a soluble base. Under basic conditions the exposed silver halide grains and the dye releaser undergo a redox reaction. Silver halide reduction is accompanied by the release of image dyes, which diffuse to the receiving layer and are mordanted. The two sheets are stripped apart after 20 s to reveal the color image.

Fuji Colorcopy DC System.   A positive-working photothermographic material for reprography, designed to produce either color prints or color transparencies, has been described (95). The thermal development technology is similar to that used for Fujix Pictrography, but with 15-s processing at 80°C. The Colorcopy DC film incorporates a set of negative-working emulsions sensitive to blue, green, and red light, respectively, along with dye releasers of the ROSET type, an electron donor, and an electron-transfer agent. The interlayers of the donor film contain a development-inhibitor-releasing (DIR) compound, described as Hyper-DIR, which is said to function very rapidly, providing enhanced color saturation. The dye releasers undergo reduction and release dye in unexposed areas, thus providing positive transfer images.

The Colorcopy DC film is used in a Fuji Colorcopier DC3000, which is designed to photograph reflection copy, transparencies, or solid objects and to produce either color prints or transparencies for overhead projection.

Konica Dry Color. A dry photothermographic system based on dye transfer from a multilayer negative donor sheet to a hydrophobic receiving sheet, using only materials contained within the two sheets, has been described (52). The negative sheet components include silver halides, organic silver salts, couplers immobilized on a polymer, a color developer precursor, and solvents. Following exposure the two sheets are laminated together and heated to activate the processing. The dyes formed by color coupling are hydrophobic, low molecular-weight compounds that are displaced from the polymer and diffuse readily into the receiving sheet. The process is negative-working, the dye released in exposed areas in proportion to the amount of oxidized color developer generated.

3M Dry Silver Color. The development of thermally processed dry silver color materials is based on the use of photosensitive silver halide–silver behenate layers (96). The multilayer film comprises red-, green-, and blue-sensitized layers, each including the precursor of an appropriate dye, and barrier layers to prevent cross talk between the image-forming layers. The process is negative-working; dye images are formed *in situ* in exposed regions. Both transparencies and reflection prints may be produced. Processing is similar to that of 3M black-and-white dry silver products, for example, 10 s at 132°C.

Instant Additive Color Films

Polaroid introduced Polavision, a Super-8-mm instant motion picture system, in 1977 (97). Polachrome CS 35-mm slide film followed in 1982 (98), and a high contrast version, Polachrome HCP, appeared in 1987. Each of the films comprises a very fine additive color screen and an integral silver image transfer film. The Polavision system, which included a movie camera and a player that processed the exposed film and projected the movie, is no longer on the market. The Polavision film was provided in a sealed cassette, and the film was exposed, processed, viewed, and rewound for further viewing without leaving the cassette (97).

Polachrome is provided in standard-size 35-mm cassettes. Its processing is carried out in a Polaroid Autoprocessor using a processing pack that contains a reagent pod and a strip sheet (98).

Additive Color Screen Formation. The additive color screens comprise microscopic patterns of blue, green, and red stripes that have frequencies of 590 triplets/cm for Polavision and 394 triplets/cm for Polachrome. To produce lines fine enough to be essentially invisible when projected, a process using temporary lenticules was developed (99). The film base is embossed to form fine lenticules on one surface, and a layer of dichromated gelatin on the opposite surface is exposed through the lenticules to form hardened line images. After washing away the unhardened gelatin, the lines that remain are dyed. The lenticules are removed after they have been used for line formation (97).

Silver Image Formation and Stabilization. The integral additive transparency is based on the rapid, one-step formation of a stable, neutral, positive black-and-white image of high covering power behind the fine pattern of colored lines. The negative image formed at the same time is of much lower covering power and thus so inconsequential optically that it does not have to be removed to permit viewing the positive image (100,101). The negative image formed in the emulsion layer has only about 1/10 the covering power of the positive image (see Figs. 19 and 20). In Polavision the negative image remained in place after processing, whereas in Polachrome the negative is stripped away, so that only the positive silver image layer and the additive screen are present in the processed slide.

To achieve good color rendition, it is necessary for the emulsion resolution to be significantly higher than that of the color screen. This requirement is fulfilled by using a fine-grained silver halide emulsion and coating the emulsion in a thin layer containing a minimal amount of silver. The high covering power

Fig. 19. Comparison of undeveloped grain (left); exposed and developed grain (center); and silver of a single unexposed grain that has transferred and developed to form a positive image deposit (97). Reproduced with permission of IS&T, the Society for Imaging Science and Technology.

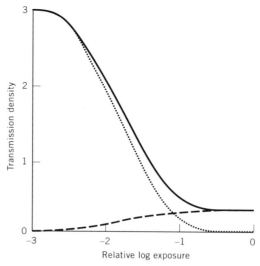

Fig. 20. Characteristic curves of a model Polavision image: transmission density vs the relative log exposure where (· · ·) is the positive image; (---), the negative image; and (—) the total. In all cases the total image density is the summation of the negative and positive densities. Attenuation of light by the negative corresponds to the small difference between the positive image density and the total image density.

positive image is a mirrorlike deposit formed in a very thin image-receiving layer. Because the image is transferred to a contiguous layer, the resolution is higher than for images transferred through a layer of processing fluid.

Polachrome Films. The Polachrome film structure and the formation of a Polachrome color image are illustrated schematically in Figure 21 (on the colored plate). The color screen is separated from the image-receiving layer by an alkali-impermeable barrier layer. Over the image-receiving layer is a stripping layer, and above that are the panchromatic silver halide emulsion and an antihalation layer. Stabilizer precursors are contained in one or more of these layers (102). (The Polavision film structure was similar but did not include a stripping layer.) As indicated in Figure 21, exposure of the film produces blue, green, and red records behind the respective color stripes. Micrographs of Polachrome film in various stages of processing are shown in Figure 22 (on the colored plate).

Processing of Polaroid 35-mm film, carried out in an Autoprocessor, is initiated as a thin layer of reagent applied to a strip sheet contacts the film surface. Exposed silver halide grains develop *in situ*; unexposed grains dissolve, and the resulting soluble silver complex migrates to the positive image-receiving layer, developing there to form the thin, compact positive image layer. Following a short interval, the strip sheet and film are rewound. The strip sheet rewinds into the processing pack, which is subsequently discarded, and the film rewinds into its cartridge. The processed film is dry and ready to be viewed or to be cut and mounted for projection.

Both during and after the formation of the negative and positive silver images additional reactions take place. The antihalation dyes are decolorized, and stabilizer is released to diffuse to the image-receiving layer and stabilize the developed silver images. The developing agent, tetramethylreductic acid [*1889-96-9*] (**17**) (2,3-dihydroxy-4,4,5,5-tetramethyl-2-cyclopenten-1-one), $C_9H_{14}O_3$, undergoes oxidation and alkaline decomposition to form colorless, inert products (103).

As shown in equation 12, the chemistry of this developer's oxidation and decomposition has been found to be less simple than first envisioned. One oxidation product, tetramethyl succinic acid (**18**), is not found under normal circumstances.

Instead, the products are the α-hydroxyacid (**20**) and the α-ketoacid (**22**). When silver bromide is the oxidant, only the two-electron oxidation and hydrolysis occur to give (**20**). When silver chloride is the oxidant, a four-electron oxidation can occur to give (**22**). In model experiments the hydroxyacid was not converted to the keto acid. Therefore, it seemed that the two-electron intermediate triketone hydrate (**19**) in the presence of a stronger oxidant would reduce more silver, possibly involving a species such as (**21**) as a likely reactive intermediate. This mechanism was verified experimentally, using a controlled, constant electrochemical potential. At potentials like that of silver chloride, four electrons were used; at lower potentials only two were used (104).

Polachrome films are used in standard 35-mm cameras, and the slides are mounted in standard size 5×5 cm (2×2 in.) mounts for projection in conventional 35-mm equipment. Because of the density of the color base (about 0.7) the Polachrome slides are most effective when viewed either using a bright rear projection unit or in a well-darkened projection room.

Economic Aspects

Although the cost per print for instant color prints is greater than for conventional film, there is significant value in the immediate availability of instant prints and the fact that it is not necessary to process a whole roll when only a few images are desired.

Instant films accounted for 10.9% of film sales in 1989 (105), and there were 12 million instant cameras in use in the United States in 1988 (106). Polaroid reported sales of about 3.1 million consumer cameras worldwide in 1990, compared with 3.0 million in 1989 and 3.5 million in 1988 (107). In Japan, Fuji accounted for 6% of instant camera sales and Polaroid for 94% in 1989 (108). Fuji has estimated 1991 sales of instant cameras in Japan, including Polaroid cameras, at 400,000 (109).

The technical and industrial markets for instant photography continue to grow in diversity. In 1991 such applications accounted for approximately one-third of the worldwide sales, 60% of which were outside the United States (107).

Health and Safety Factors; Toxicology

An important concern in photofinishing and in industrial photography is the environmental effect of effluents discharged by laboratories that process large quantities of color films. Further concerns are the limited supplies of clean water available for processing in many areas and the high cost of energy to provide clean air and water at required temperatures. The completely self-contained instant films avoid all of these problems.

No toxicological hazards have been associated with the normal use of instant color films. However, direct contact with the highly alkaline processing fluids can cause alkali burns. The fluids are provided in sealed pods and are not usually handled by the user. In the integral systems, the fluids are retained within the

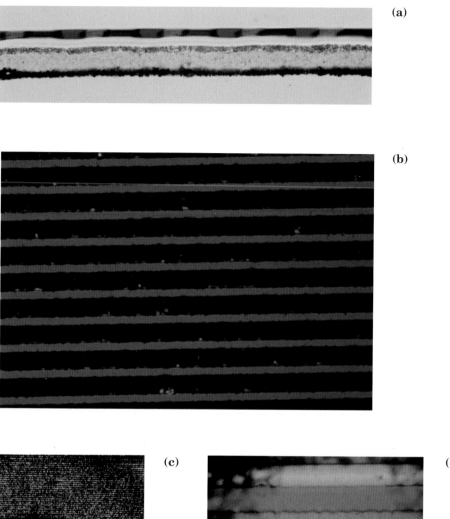

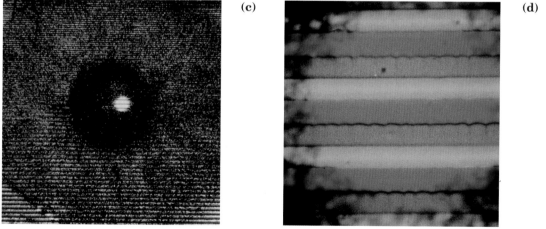

Figure 22. Micrographs of Polachrome film. **(a)** is a cross section of the unprocessed film (1000X); **(b)** is a portion of a red image area after processing, with dense positive silver overlying the blue and green lines (250X); **(c)** and **(d)** are portions of a full color image, showing the image structure at 30X and 500X. Figures 30 **(c)** and 30 **(d)** reproduced courtesy of IS&T, the Society for Imaging Science and Technology.

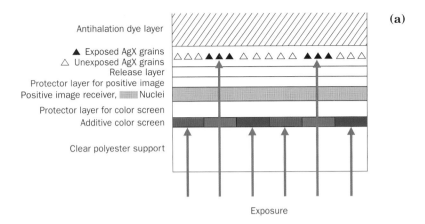

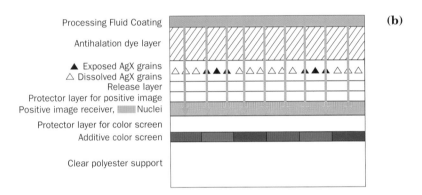

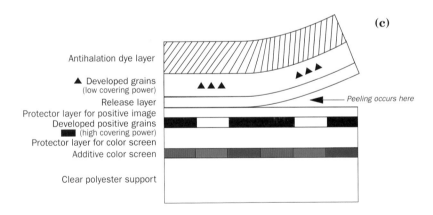

Figure 21. Schematic representation of the Polachrome process. **(a)** shows the film during exposure to green light, with only the green lines transmitting light to the emulsion. In **(b)** processing has been initiated by the application of a very thin stratum of the viscous reagent, represented by yellow arrows. Exposed grains are being reduced *in situ*, and unexposed grains are starting to dissolve, forming soluble silver salts that migrate toward the receiving layer. **(c)** represents the completion of the process. The top layers of film are peeled away, leaving the positive image ready to viewing.

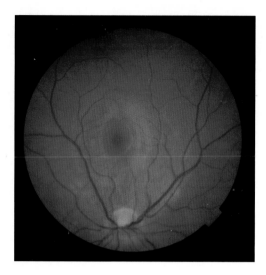

Figure 23. An example of fundus (retinal) photography, widely used for diagnosis in opthamology. The photograph is on Polaroid Type 779, a professional integral color film with extended red sensitivity and high contrast.

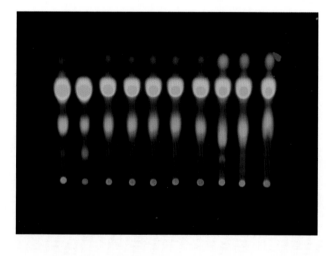

Figure 24. A thin layer chromatography (TLC) image, documenting detection of impurities in a series of dye intermediate samples under near-UV illumination. Columns 1 and 7 represent reference materials. Photographed with Polaroid Type 339 film in a CU-5 closeup camera.

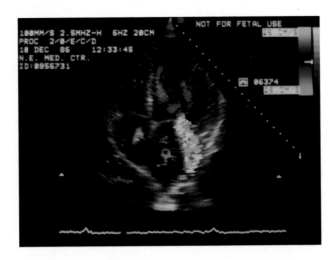

Figure 25. An example of Doppler color flow mapping, a real-time ultrasound technique that displays the dynamics of blood flow in color along with static tissue information in black and white. The image shows the record of a leaky mitral valve during systole.

Figure 26. Photomicrograph of a series of cross sections of beefwood (Casuarina equisetifolia) leaves, showing results with a variety of staining techniques. Recorded at 16X magnification on Polachrome 35mm slide film. Final magnification 50X. Courtesy of R.H. Gebert, ETH-Zentrum, Zurich, Switzerland.

film unit and neutralized in the course of processing. In the peel-apart films, residual fluids are rapidly rendered harmless by contact with the air.

As solid-waste management becomes a more critical concern, there is increasing attention to materials that are incinerated or added to landfills. The negatives discarded from peel-apart color films include silver, although at much lower concentrations than in conventional color films, and there are also small amounts of chromium and copper; the pod includes a layer of aluminum foil. The zinc–carbon Leclanché batteries (qv) provided in Polaroid's integral film packs at one time contained a small concentration of mercury, but its use has been eliminated. As of this writing, some used film components and batteries are being recycled.

Uses

Instant film formats and corresponding apparatuses have been developed to fit a variety of specialized needs. Many laboratory instruments and diagnostic machines include built-in instant-film camera backs (see ANALYTICAL METHODS; AUTOMATED INSTRUMENTATION; IMAGING TECHNOLOGY). Digital film recorders produce color prints, slides, or overhead transparencies from computer output and from cathode-ray tube displays (see OPTICAL DISPLAYS). There are also direct screen cameras for photographing still cathode ray tube (CRT) displays. Polaroid FreezeFrame video image recorders provide color prints and transparencies from a variety of video sources, including VCRs, laser discs, and video cameras.

An important aspect of many professional applications is the rapid on-site completion of color photographs or hard copy. For example, in photomicrography work can continue without interruption, results can be documented quickly, and successive images of specimens that are undergoing rapid change can be immediately compared. Integral films and apparatuses have been certified for use in clean room environments, where both photomicrography and photomacrography are important for documentation and diagnostic work (see SUPPLEMENT VOLUME). Instant color slides and overhead transparencies expedite the presentation of the most current data and information in lectures and business meetings.

Instant photographs are widely used for identification purposes, for example, for drivers' licenses, student identification cards, and credit cards that can be issued immediately. Most passport photographs generated in the United States are instant color prints. Industrial and business applications include photographs of record for insurance purposes, construction documentation, and real estate photography.

Medical and scientific fields in which instant color films are used extensively include photography of the retina, using fundus cameras equipped with instant film holders (see Fig. 23 on the colored plate, for example); dental imaging with the Polaroid CU-5 close-up camera; chromatography as illustrated in Figure 24 (colored plate); diagnostic imaging with radiopharmaceuticals (see MEDICAL DIAGNOSTIC REAGENTS; RADIOISOTOPES); and Doppler blood flow studies (see Fig. 25, (colored plate)). Figure 26, on the colored plate, is an example of photomicrography using Polachrome 35-mm film, and Figures 11, 22a, and 22b demonstrate the use of Polacolor films for photomicrography.

In studio photography, instant color slides exposed simultaneously with conventional color films are used to provide proofs that can be projected immediately for viewing by the customer. Professional photographers also use instant films as proof material to check composition and lighting. Large format Polacolor films are often used directly for exhibition prints.

On a still larger scale, the making of full-size Polacolor replicas of works of art provides accurate images for display and for professional study and documentation (see FINE ART EXAMINATION AND CONSERVATION). Because these do not involve photographic reduction and enlargement, the 1:1 replicas maintain the fine detail and dimensional relationships of the original subjects.

Instant color film has a unique application in the generation of images by transfer from a partially developed Polacolor negative to a material other than the usual receiving sheet, for example, plain paper, fabric, or vellum. The transferred image may then be modified for artistic purposes by reworking with watercolors or other dyes. The final images, generally known as transfer images, may be displayed directly or reproduced for commercial use.

In the instant reprographic field, Copycolor materials are used extensively in Europe but are not distributed in the United States. Principal markets are in seismological charts and maps for the oil industry, mapmaking, and reproduction of large graphs, charts, and engineering drawings. The films are also used for small color stats and for position proofs in layout work.

BIBLIOGRAPHY

"Nonchromogenic Color Photography" under "Color Photography" in *ECT* 2nd ed., Vol. 5, pp. 837–838, by J. R. Thirtle and D. M. Zwick, Eastman Kodak Co.; "Color Photography, Instant" in *ECT* 3rd ed., Vol. 6, pp. 646–682, by V. K. Walworth.

1. J. S. Arney and J. A. Dowler, *J. Imaging Sci.* **32,** 125 (1988).
2. B. Fischer, G. Mader, H. Meixner, and P. Kleinschmidt, *Siemens Forsch. u. Entwickl.-Ber.* **17,** 291 (1988); G. Mader, H. Meixner, and G. Klug, *J. Imaging Sci.* **34,** 213 (1990).
3. E. H. Land, *J. Opt. Soc. Am.* **37,** 61 (1947); E. H. Land, *Photogr. J.* **90A,** 7 (1950).
4. E. H. Land, H. G. Rogers, and V. K. Walworth, in J. M. Sturge, ed., *Neblette's Handbook of Photography and Reprography*, 7th ed., Van Nostrand Reinhold, New York, 1977, pp. 258–330.
5. V. K. Walworth and S. H. Mervis, in J. Sturge, V. Walworth, and A. Shepp, eds., *Imaging Processes and Materials: Neblette's Eighth Edition*, Van Nostrand Reinhold, New York, 1989, pp. 181–225.
6. C. C. Van de Sande, *Agnew. Chem. Int. Ed. Engl.* **22,** 191 (1983).
7. U.S. Pat. 2,543,181 (Feb. 27, 1951), E. H. Land (to Polaroid Corp.).
8. U.S. Pat. 2,603,565 (July 15, 1952), E. H. Land (to Polaroid Corp.).
9. U.S. Pat. 2,435,717 (Feb. 10, 1948), E. H. Land (to Polaroid Corp.); U.S. Pat. 2,455,111 (Nov. 30, 1948), J. F. Carbone and M. N. Fairbank (to Polaroid Corp.).
10. U.S. Pat. 2,495,111 (Jan. 17, 1950), E. H. Land (to Polaroid Corp.); U.S. Pat. 3,161,122 (Dec. 15, 1964), J. A. Hamilton (to Polaroid Corp.); U.S. Pat. 3,079,849 (Mar. 5, 1963), R. R. Wareham (to Polaroid Corp.).
11. U.S. Pat. 2,933,993 (Apr. 26, 1960), A. J. Bachelder and V. K. Eloranta (to Polaroid Corp.).
12. E. H. Land, *Photogr. Sci. Eng.* **16,** 247 (1972).

13. E. H. Land, *Photogr. J.* **114,** 338 (1974).
14. U.S. Pat. 3,415,644 (Dec. 10, 1968), E. H. Land (to Polaroid Corp.).
15. U.S. Pat. 3,594,165 (July 20, 1971), H. G. Rogers (to Polaroid Corp.); U.S. Pat. 3,689,262 (Sept. 5, 1972), H. G. Rogers (to Polaroid Corp.).
16. S. Fujita, K. Koyama, and S. Ono, *Nippon Kagaku Kaishi* **1991,** 1 (1991).
17. U.S. Pat. 3,148,062 (Sept. 8, 1964), K. E. Whitmore, C. R. Staud, C. R. Barr, and J. Williams (to Eastman Kodak Co.); U.S. Pat. 3,227,551 (Jan. 4, 1966), C. R. Barr, J. Williams, and K. E. Whitmore (to Eastman Kodak Co.); U.S. Pat. 3,227,554 (Jan. 4, 1966), C. R. Barr, J. Williams, and K. E. Whitmore (to Eastman Kodak Co.); U.S. Pat. 3,243,294 (Mar. 29, 1966), C. R. Barr (to Eastman Kodak Co.).
18. U.S. Pat. 3,443,940 (May 13, 1969), S. M. Bloom and H. G. Rogers (to Polaroid Corp.).
19. U.S. Pat. 2,983,606 (May 9, 1961), H. G. Rogers (to Polaroid Corp.).
20. U.S. Pat. 3,255,001 (June 7, 1966), E. R. Blout and H. G. Rogers (to Polaroid Corp.).
21. S. M. Bloom, M. Green, E. M. Idelson, and M. S. Simon, in K. Venkataraman, ed., *The Chemistry of Synthetic Dyes*, Vol. 8, Academic Press, New York, 1978, pp. 331–387.
22. U.S. Pat. 3,201,384 (Aug. 17, 1965), M. Green (to Polaroid Corp.); U.S. Pat. 3,246,985 (Apr. 19, 1966), M. Green (to Polaroid Corp.).
23. H. G. Rogers, E. M. Idelson, R. F. W. Cieciuch, and S. M. Bloom, *J. Photogr. Sci.* **22,** 138 (1974).
24. U.S. Pat. 3,857,855 (Dec. 31, 1974), E. M. Idelson (to Polaroid Corp.); E. M. Idelson, *Dyes and Pigments* **3,** 191 (1982).
25. M. Idelson, I. R. Karday, B. H. Mark, D. O. Richter, and V. H. Hooper, *Inorg. Chem.* **6,** 450 (1967).
26. U.S. Pat. 4,264,701 (Apr. 28, 1981), L. Locatell, Jr., H. G. Rogers, R. C. Bilofsky, R. F. Cieciuch, and C. M. Zepp (to Polaroid Corp.).
27. U.S. Pat. 4,264,704 (Apr. 28, 1981), A. L. Borror, L. Cincotta, E. M. Mahoney, and M. H. Feingold (to Polaroid Corp.).
28. U.S. Pat. 3,307,947 (Mar. 7, 1967), E. M. Idelson and H. G. Rogers (to Polaroid Corp.); U.S. Pat. 3,230,082 (Jan. 18, 1966), E. H. Land and H. G. Rogers (to Polaroid Corp.); U.S. Pat. 3,135,606 (June 2, 1964), E. R. Blout, M. R. Cohler, M. Green, M. S. Simon, and R. B. Woodward (to Polaroid Corp.).
29. U.S. Pat. 3,854,945 (Dec. 17, 1974), W. M. Bush and D. F. Reardon (to Eastman Kodak Co.).
30. U.S. Pat. 3,880,658 (Apr. 29, 1975), G. J. Lestina and W. M. Bush (to Eastman Kodak Co.); U.S. Pat. 3,935,262 (Jan. 27, 1976), G. J. Lestina and W. M. Bush (to Eastman Kodak Co.); U.S. Pat. 3,935,263 (Jan. 27, 1976), G. J. Lestina and W. M. Bush (to Eastman Kodak Co.).
31. U.S. Pat. 3,649,266 (Mar. 14, 1972), D. D. Chapman and L. G. S. Brooker (to Eastman Kodak Co.); U.S. Pat. 3,653,897 (Apr. 4, 1972), D. D. Chapman (to Eastman Kodak Co.).
32. U.S. Pat. 3,245,789 (Apr. 12, 1966), H. G. Rogers (to Polaroid Corp.).
33. U.S. Pat. 3,433,939 (May 13, 1969), S. M. Bloom and R. K. Stephens (to Polaroid Corp.); U.S. Pat. 3,751,406 (Aug. 7, 1973), S. M. Bloom (to Polaroid Corp.).
34. Fr. Pat. 2,154,443 (Aug. 31, 1972), L. J. Fleckenstein and J. Figueras (to Eastman Kodak Co.); Brit. Pat. 1,405,662 (Sept. 10, 1975), L. J. Fleckenstein and J. Figueras (to Eastman Kodak Co.); U.S. Pat. 3,928,312 (Dec. 23, 1975), L. J. Fleckenstein (to Eastman Kodak Co.); U.S. Publ. Pat. Appl. B351,673 (Jan. 28, 1975), L. J. Fleckenstein and J. Figueras (to Eastman Kodak Co.).
35. S. Fujita, *Advance Printing of Paper Summaries, SPSE 35th Annual Conference,* Rochester, N.Y., 1982, p. J-1; S. Fujita, *Sci. Publ. Fuji Photo Film Co., Ltd.* **29,** 55 (1984) (in Japanese).

36. T. Shibata, K. Sato, and Y. Aotsuka, *Advance Printing of Paper Summaries, SPSE 4th International Congress on Advances in Non-Impact Printing Technologies*, 1988, p. 362.

37. U.S. Pat. 3,725,062 (Apr. 3, 1973), A. E. Anderson and K. K. Lum (to Eastman Kodak Co.); U.S. Pat. 3,698,897 (Oct. 17, 1972), T. E. Gompf and K. K. Lum (to Eastman Kodak Co.); U.S. Pat. 3,728,113 (Apr. 17, 1973), R. W. Becker, J. A. Ford, Jr., D. L. Fields, and D. D. Reynolds (to Eastman Kodak Co.).

38. U.S. Pat. 3,222,169 (Dec. 7, 1965), M. Green and H. G. Rogers (to Polaroid Corp.); U.S. Pat. 3,230,086 (Jan. 18, 1966), M. Green (to Polaroid Corp.); U.S. Pat 3,303,183 (Feb. 7, 1967), M. Green (to Polaroid Corp.).

39. Belg. Pat. 834,143 (Apr. 2, 1976), D. L. Fields and co-workers (to Eastman Kodak Co.).

40. U.S. Pat. 3,628,952 (Dec. 21, 1971), W. Puschel and co-workers (to Agfa-Gevaert Aktiengesellschaft); Brit. Pat. 1,407,362 (Sept. 24, 1975), J. Danhauser and K. Wingender (to Agfa-Gevaert Aktiengesellschaft).

41. Ger. Offen. 2,335,175 (Jan. 30, 1975), M. Peters and co-workers (to Agfa-Gevaert Aktiengesellschaft).

42. Brit. Pat. 840,731 (July 6, 1960), K. E. Whitmore and P. M. Mader (to Kodak Limited); U.S. Pat. 3,227,550 (Jan. 4, 1966), K. E. Whitmore and P. M. Mader (to Eastman Kodak Co.).

43. U.S. Pat. 4,232,107 (Nov. 4, 1980), W. Janssens (to Agfa-Gevaert N.V.); U.S. Pat. 4,371,604 (Nov. 4, 1983), C. C. Van de Sande, W. Janssens, W. Lassig, and E. Meier (to Agfa-Gevaert N.V.); U.S. Pat. 4,396,699 (Aug. 3, 1983), W. Janssens and D. A. Claeys (to Agfa-Gevaert N.V.); U.S. Pat. 4,477,554 (Oct. 16, 1984), C. C. Van de Sande and A. Verbecken (to Agfa-Gevaert N.V.).

44. K. Nakamura and K. Koya, *SPSE/SPSTJ International East-West Symposium II*, Kona, Hawaii, 1988, p. D-24.

45. Eur. Pat. Appl. 0,220,746 (May 6, 1987), K. Nakamura (to Fuji Photo Film Co., Ltd.).

46. U.S. Pat. 3,719,489 (Mar. 6, 1973), R. F. W. Cieciuch, R. R. Luhowy, F. Meneghini, and H. G. Rogers (to Polaroid Corp.); U.S. Pat. 4,060,417 (Nov. 29, 1977), R. F. W. Cieciuch, R. R. Luhowy, F. Meneghini, and H. G. Rogers (to Polaroid Corp.).

47. F. Meneghini, *J. Imaging Technol.* **15,** 114 (1989).

48. A. Ehret, *J. Imaging Technol.* **15,** 97 (1989).

49. R. Lambert, *J. Imaging Technol.* **15,** 108 (1989); U.S. Pat. 4,740,448 (Apr. 26, 1988), P. O. Kliem (to Polaroid Corp.).

50. U.S. Pat. 2,968,554 (Jan. 17, 1961), E. H. Land (to Polaroid Corp.); U.S. Pat. 3,015,561 (Jan. 2, 1962), H. G. Rogers (to Polaroid Corp.); U.S. Pat. 3,019,124 (Jan. 30, 1962), H. G. Rogers (to Polaroid Corp.).

51. U.S. Pat. 4,690,884 (Sept. 1, 1987), F. E. DeBruyn, Jr., and L. J. Weed (to Polaroid Corp.).

52. U.S. Pat. 4,631,251 (Dec. 23, 1986), T. Komamura (to Konishiroku Photo Industry, Ltd.); U.S. Pat. 4,650,748 (Mar. 17, 1987), T. Komamura (to Konishiroku Photo-Industry, Ltd.); U.S. Pat. 4,656,124 (Apr. 7, 1987), T. Komamura (to Konishiroku Photo Industry Co., Ltd.); T. Komamura, *SPSE/SPSTJ International East-West Symposium II*, Kona, Hawaii, 1988, p. D-35.

53. U.S. Pat. 3,087,817 (Apr. 30, 1963), H. G. Rogers (to Polaroid Corp.).

54. U.S. Pat. 3,537,850 (Nov. 3, 1970), M. S. Simon (to Polaroid Corp.).

55. U.S. Pat. 2,909,430 (Oct. 20, 1959), H. G. Rogers (to Polaroid Corp.).

56. U.S. Pat. 3,185,567 (May 25, 1965), H. G. Rogers (to Polaroid Corp.).

57. U.S. Pat. 3,362,819 (Jan. 9, 1968), E. H. Land (to Polaroid Corp.).

58. U.S. Pat. 2,968,554 (Jan. 17, 1961), E. H. Land (to Polaroid Corp.).

59. U.S. Pat. 3,345,163 (Oct. 3, 1967), E. H. Land and H. G. Rogers (to Polaroid Corp.).

60. U.S. Pat. 3,411,904 (Nov. 19, 1968), R. W. Becker (to Eastman Kodak Co.).
61. U.S. Pat. 3,625,685 (Dec. 7, 1971), J. A. Avtges, J. L. Reid, H. N. Schlein, and L. D. Taylor (to Polaroid Corp.).
62. U.S. Pat. 3,421,893 (Jan. 14, 1969), L. D. Taylor (to Polaroid Corp.); U.S. Pat. 3,433,633 (Mar. 18, 1969), H. C. Haas (to Polaroid Corp.); U.S. Pat. 3,856,522 (Dec. 24, 1974), L. J. George and R. A. Sahatjian (to Polaroid Corp.); L. D. Taylor and L. D. Cerankowski, *J. Polym. Sci., Polym. Chem. Ed.* **13,** 2551 (1975).
63. U.S. Pat. 3,192,044 (June 29, 1965), H. G. Rogers and H. W. Lutes (to Polaroid Corp.).
64. U.S. Pat. 3,039,869 (June 19, 1962), H. G. Rogers and H. W. Lutes (to Polaroid Corp.).
65. U.S. Pat. 3,265,498 (Aug. 9, 1966), H. G. Rogers and H. W. Lutes (to Polaroid Corp.); U.S. Pat. 3,266,894 (Aug. 16, 1966), W. J. Weyerts and W. M. Salminen (to Eastman Kodak Co.); U.S. Pat. 3,785,813 (Jan. 14, 1974), D. Q. Rickter (to Polaroid Corp.).
66. U.S. Pat, 3,146,102 (Aug. 25, 1964), W. J. Weyerts and W. M. Salminen (to Eastman Kodak Co.); U.S. Pat. 3,173,786 (Mar. 16, 1965), M. Green and H. G. Rogers (to Polaroid Corp.); U.S. Pat. 3,253,915 (May 31, 1966), W. J. Weyerts and W. M. Salminen (to Eastman Kodak Co.).
67. E. H. Land in *Polaroid Corporation Annual Report for 1975*, Cambridge, Mass., 1976, and *Polaroid Corporation Annual Report for 1976*, Cambridge, Mass., 1977; J. J. McCann in *Polaroid Corporation Annual Report for 1977*, Cambridge, Mass., 1978; L. Nilson, *Polaroid Newsletter for Education* **4,** 4 (1987).
68. U.S. Pat. 3,793,022 (Feb. 19, 1974), E. H. Land, S. M. Bloom, and H. G. Rogers (to Polaroid Corp.).
69. U.S. Pat. 4,304,833 (Dec. 8, 1981), J. Foley (to Polaroid Corp.).
70. U.S. Pat. 4,304,834 (Dec. 8, 1981), R. L. Cournoyer and J. W. Foley (to Polaroid Corp.).
71. U.S. Pat. 4,258,118 (Mar. 24, 1981), J. Foley, L. Locatell, Jr., and C. M. Zepp (to Polaroid Corp.).
72. U.S. Pat. 4,329,411 (May 11, 1982), E. H. Land (to Polaroid Corp.).
73. Can. Pat. 951,560 (July 23, 1974), E. H. Land (to Polaroid Corp.).
74. U.S. Pat. 3,647,437 (Mar. 7, 1972), E. H. Land (to Polaroid Corp.).
75. S. M. Bloom, *Abstract L-2, SPSE 27th Annual Conference*, Boston, Mass., 1974.
76. U.S. Pat. 3,702,244 (Nov. 7, 1972), S. M. Bloom, A. L. Borror, P. S. Huyffer, and P. T. MacGregor (to Polaroid Corp.); U.S. Pat. 3,702,245 (Nov. 7, 1972), M. S. Simon and D. F. Waller (to Polaroid Corp.); M. S. Simon, *J. Imaging Technol.* **16,** 143 (1990); R. Cournoyer, D. H. Evans, S. Stroud, and R. Boggs, *J. Org. Chem.* **56,** 4576 (1991).
77. U.S. Pat. 4,298,674 (Nov. 3, 1981), E. H. Land, L. D. Cerankowski, and N. Mattucci (to Polaroid Corp.).
78. U.S. Pat. 4,294,907 (Oct. 13, 1981), I. Y. Bronstein-Bonte, E. P. Lindholm, and L. D. Taylor (to Polaroid Corp.).
79. U.S. Pat. 4,202,694 (May 13, 1980), L. D. Taylor (to Polaroid Corp.).
80. L. D. Taylor, H. S. Kolesinki, D. O. Rickter, J. M. Grasshoff, and J. R. DeMember, *Macromolecules* **16,** 1561 (1983).
81. U.S. Pat. 3,756,814 (Sept. 4, 1973), S. F. Bedell (to Polaroid Corp.).
82. U.S. Pat. 3,770,439 (Nov. 6, 1973), L. D. Taylor (to Polaroid Corp.).
83. U.S. Pat. 4,740,448 (Apr. 26, 1988), P. O. Kliem (to Polaroid Corp.).
84. R. Lambert, *J. Imaging Technol.* **15,** 108 (1989).
85. E. S. McCaskill and S. R. Herchen, *J. Imaging Technol.* **15,** 103 (1989).
86. U.S. Pat. 4,355,101 (Oct. 19, 1982), A. C. Mehta, G. H. Nawn, and L. D. Taylor (to Polaroid Corp.); U.S. Pat. 4,743,533 (May 10, 1988), R. A. Boggs, J. B. Mahoney, A. C. Mehta, W. C. Schwarzel, and L. D. Taylor (to Polaroid Corp.); U.S. Pat. 4,946,964 (Aug. 7, 1990), R. A. Boggs, J. B. Mahoney, A. C. Mehta, W. C. Schwarzel, and L. D. Taylor (to Polaroid Corp.).
87. R. A. Boggs and co-workers, *J. Chem. Soc., Perkin Trans.* **2,** 1271 (1992).

88. U.S. Pat, 4,547,451 (Oct. 15, 1985), S. J. Jasne, W. C. Schwarzel, C. I. Sullivan, and L. D. Taylor (to Polaroid Corp.).

89. W. T. Hanson, Jr., *Photogr. Sci. Eng.* **24,** 155 (1976).

90. W. T. Hanson, Jr., *J. Photogr. Sci.* **25,** 189 (1977).

91. D. Leavitt, *Pop. Photogr.* 12 (Feb. 1982).

92. M. Peters, *J. Imaging Technol.* **11,** 101 (1985).

93. U.S. Pat. 4,704,345 (Nov. 3, 1987), H. Hirai and H. Naito (to Fuji Photo Film Co., Ltd.); H. Seto, K. Nakauchi, S. Yoshikawa, and T. Ohtsu, *Advance Printing of Paper Summaries*, SPSE 4th International Congress on Advances in Non-Impact Printing Technologies, New Orleans, La., 1988, p. 358; T. Shibata, K. Sato, and Y. Aotsuka, p. 362.

94. Y. Yabuki, K. Kawata, H. Kitaguchi, and K. Sato, *SPSE/SPSTJ International East-West Symposium II*, Kona, Hawaii, 1988, p. D-32.

95. M. Kato, S. Sawada, and K. Nakamura, *Advance Printing of Paper Summaries, IS&T 44th Annual Conference*, St. Paul, Minn., 1991, p. 510.

96. D. A. Morgan, *J. Imaging Technol.* **13,** 4 (1987).

97. E. H. Land, *Photogr. Sci. Eng.* **21,** 225 (1977).

98. S. H. Liggero, K. J. McCarthy, and J. A. Stella, *J. Imaging Technol.* **10,** 1 (1984).

99. U.S. Pat. 3,284,208 (Nov. 8, 1966), E. H. Land (to Polaroid Corp.).

100. U.S. Pat. 2,861,885 (Nov. 25, 1958), E. H. Land (to Polaroid Corp.).

101. U.S. Pat. 3,894,871 (July 15, 1975), E. H. Land (to Polaroid Corp.).

102. U.S. Pat. 3,704,126 (Nov. 28, 1972), E. H. Land, S. M. Bloom, and L. C. Farney (to Polaroid Corp.); U.S. Pat, 3,821,000 (June 28, 1974), E. H. Land, S. M. Bloom, and L. C. Farney (to Polaroid Corp.).

103. U.S. Pat. 3,615,440 (Oct. 26, 1971), S. M. Bloom and R. D. Cramer (to Polaroid Corp.).

104. S. Inbar, A. Ehret, and K. Norland, *Advance Printing of Paper Summaries, SPSE 39th Annual Conference*, Minneapolis, Minn., 1986, p. 14.

105. *The 1989–90 Industry Trends Report*, Photo Marketing Association International, Jackson, Mich., 1990.

106. *1989–90 Wolfman Report on the Photographic and Imaging Industry in the United States*, Diammandis Communications, Inc., New York, 1990.

107. *Polaroid Corporation Annual Report for 1990*, Cambridge, Mass., 1991.

108. *The Robinson Report on the 1989 Consumer Photographic/Video Market in Japan with Forecasts for 1994*, Photographic Consultants Ltd., Cambridge, Mass., 1990.

109. *JPEA Photo Electro News*, May 16, 1991.

General Reference

S. Fujita, K. Koyama, and S. Ono, *Rev. Heteroat. Chem.* **7,** 229–267 (1992).

Vivian K. Walworth
Stanley H. Mervis
Polaroid Corporation

COLUMBIUM.   See Niobium and Niobium Compounds.

COMBUSTION SCIENCE AND TECHNOLOGY

Fuel combustion is a complex process, the understanding of which involves knowledge of chemistry (structural features of the fuel), thermodynamics (feasibility and energetics of the reactions), mass transfer (diffusion of fuel and oxidant molecules), reaction kinetics (rate of reaction), and fluid dynamics of the process. Therefore, the design of combustion systems involves utilizing information and data generated in a range of disciplines. Often, the design of practical combustion systems is based on experience rather than on fundamental mechanistic understanding. However, for certain fuels such as methane, the combustion mechanism is better understood than for other more complex fuels such as coal. To accommodate the variety of approaches used to solve practical problems, this article is divided into two subsections: combustion science and combustion technology.

Combustion Science

DEFINITIONS AND TERMINOLOGY

Higher Heating Value.   The heating value of a fuel is the amount of heat released during its combustion and is expressed in two forms: higher heating value (HHV) or gross calorific value and lower heating value (LHV) or net calorific value. The higher heating value of a fuel is the heat of combustion at constant pressure and temperature (usually ambient) determined by a calorimetric measurement in which the water formed by combustion is completely condensed (1–3). The lower heating value is the similarly measured or defined heat of combustion in the absence of water condensation. The higher heating value is most often used in combustion and flame calculations. The difference between HHV and LHV is numerically equal to the corresponding enthalpy difference, due to latent heat of vaporization. For example, in the complete combustion of methane with O_2 at 298 K, $CH_4(g) + 2\,O_2(g) \rightarrow CO_2(g) + 2\,H_2O$ (g), for one mole of fuel the enthalpy change is $\Delta H = -802.1$ kJ/mol. This value is equal to the lower heating value of methane, and the higher heating value for the above reaction is obtained by adding the heat of vaporization of water at 298 K (44.01 kJ/mol). Thus the higher heating value of methane is $802.1 + 2(44.01) = 890.1$ kJ/mol of CH_4 (212.7 kcal/mol). The evaluation of ΔH for use in thermochemical calculations is more generally performed with the use of tables such as the Joint Army Navy Air Force (JANAF) Tables.

FAR or AFR.   The composition of a mixture of fuel and air or oxidant is often specified according to the Fuel to Air Ratio (FAR), and can be expressed on a mass, molar, or volume basis. The FAR is normalized to the stoichiometric composition by defining the equivalence ratio ϕ as in equation 1, where m_f = mass of fuel, kg; and m_o = mass of oxidizer, kg.

$$\phi = \frac{\left(\dfrac{m_f}{m_o}\right)}{\left(\dfrac{m_f}{m_o}\right)_{\text{stoich}}} \tag{1}$$

If $\phi<1$, the mixture is said to be lean (in fuel) and the products of combustion contain unreacted or excess O_2. If $\phi>1$ the mixture is said to be rich (in fuel) and the products of combustion contain CO and possibly H_2 because of incomplete combustion caused by the oxygen deficiency (1–3).

Flammability Limits.   Any given mixture of fuel and oxidant is flammable (explosive) within two limits referred to as the upper (rich) and lower (lean) limits of flammability (Fig. 1). Most mixtures are flammable when the fuel to air volume ratio lies between 50 and 300% of the stoichiometric ratio (2). The stoichiometric ratio is the exact theoretical ratio of fuel to air required for complete combustion. The flammability region widens with increasing temperature, and usually with increasing pressure, although the effect of pressure is less predictable. The region within the flammability limits of a fuel – air mixture can be divided into two subregions, the slow oxidation region and the explosion region, separated by the spontaneous ignition temperature. To determine this temperature for liquid fuels, standard tests are used in which the liquid fuel is dropped into an open-air container heated to a known temperature. The lowest temperature at which visible or audible evidence of combustion is observed is defined as the spontaneous ignition temperature (1).

Flash Point.   As fuel oil is heated, vapors are produced which at a certain temperature "flash" when ignited by an external ignition source. The flash point is the lowest temperature at which vapor, given off from a liquid, is in sufficient quantity to enable ignition to take place. The flash point is in effect a measure of the volatility of the fuel. The measurement of flash point for pure liquids is relatively straightforward. However, the measured value may depend slightly on the method used, especially for liquid mixtures, since the composition of the vapor evolved can vary with the heating rate. Special problems arise when the liquid contains a mixture of fuels and/or fuels and inhibitors. In these cases the vapor composition will be different from the liquid composition. Consequently, successive tests with the same sample can lead to erroneous results because the composition of the liquid in the sample holder changes with time as a result of fractional distillation (4–7).

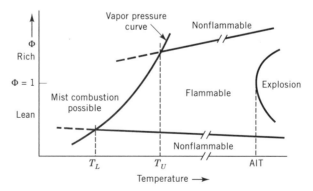

Fig. 1.   Effect of temperature on limits of flammability of a pure liquid fuel in air, where T_L = lean (or lower) flash point; T_U = rich (or upper) flash point; and AIT = auto ignition temperature (4,5).

Ignition. To understand the phenomenon of ignition it is necessary to consider the following concepts: ignition source, gas temperature, flame volume, and presence of quench wall surfaces. In general, there are two main methods of igniting a flammable mixture. In the self-ignition method, the mixture is heated slowly so that the vapor released as the temperature is raised ignites spontaneously at a particular temperature. In the forced ignition method, a small quantity of combustible mixture is heated by an external source and the heat released during the combustion of this portion results in propagation of a flame. The external ignition source can be an electric spark, pilot flame, shock wave, etc. For ignition to take place the following conditions should be satisfied: (*1*) the amount of energy supplied by the ignition source should be large enough to overcome the activation energy barrier; (*2*) the energy released in the gas volume should exceed the minimum critical energy for ignition; and (*3*) the duration of the spark or other ignition source should be long enough to initiate flame propagation, but not too long to affect the rate of propagation. Ignition models fall into two categories: the thermal model explains the ignition as resulting from supplying the mixture with the amount of heat sufficient to initiate reaction. In the chemical diffusion model the main role of the ignition source is attributed to the formation of a large number of free radicals in the preheat zone, where their diffusion to the surrounding region initiates the combustion process. The thermal model is applied more widely in the literature and shows better agreement with experimental data.

At ambient temperatures, reaction rates for gaseous mixtures of fuel and oxidant are extremely slow. As temperature is increased gradually, slow oxidation begins and as a result of the exothermic reactions that occur, the temperature keeps increasing. With further temperature increase, the reaction rate suddenly increases causing rapid combustion reactions to occur. This is providing, of course, that the rate of heat release is greater than the rate of heat loss through the container walls. The temperature at which the heat released by the reaction exceeds the heat loss is commonly referred to as the ignition temperature. The spontaneous ignition temperature (SIT), on the other hand, is the lowest temperature at which ignition occurs. Increasing the pressure results in a decrease in the spontaneous ignition temperature (1) (Fig. 2 and Table 1). This temperature is highly dependent on the material of construction, apparatus configuration, and test procedure, therefore reported test values vary widely (2,6,7).

From the chemical diffusion model standpoint, the usual values of ignition temperatures of mixtures often do not correlate well with other flame properties. Nevertheless, the following highly simplified, even speculative, qualitative description may be useful in thinking about flames. Early in the ignition or induction period, the attack of fuel by O_2 is initially slow, but generates free radicals (OH, H, O, HO_2, hydrocarbons) and other intermediate species (CO, H_2, and partial oxidation and decomposition products of hydrocarbons, if present). For some time, there may be little or no temperature rise, the energy essentially being stored in the free radicals. In this stage the reactions may be similar to those that occur in very slow and nearly isothermal oxidation and cool flames observed in very rich hydrocarbon–O_2 mixtures, usually at lower temperatures and pressures. These reactions often stop at the production of stable oxidation and decomposition products, such as aldehydes, peroxides, lower hydrocarbons, etc, but under some conditions may lead to explosions. In a variety of chain reactions, the fuel and any

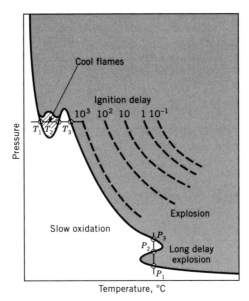

Fig. 2. Slow oxidation, spontaneous ignition, and explosion as a function of pressure and temperature variations in hydrocarbon mixtures (1).

Table 1. Spontaneous Ignition Temperatures[a]

Fuel	SIT, K
propane	767
butane	678
pentane	558
hexane	534
heptane	496
octane	491
nonane	479
decane	481
hexadecane	478
isooctane	691
kerosine (JP-8 or Jet A)[b]	501
JP-3	511
JP-4	515
JP-5	506

[a]Ref. 1.
[b]See AVIATION AND OTHER GAS TURBINE FUELS.

intermediates are rapidly attacked by radicals, some of which are also undergoing very fast chain-branching reactions. Evidence for the behavior of such free radicals is provided by studies of the mechanism of H_2/O_2 and similar explosions. These free radicals can multiply rapidly and attain high transient concentrations, initiating chain reactions that can eventually lead to the explosion of the flammable mixture. They are eventually consumed to form stable species by three-body recombination reactions in which most of the heat release occurs. An example of such a reaction, where M is any atom or molecule in the gas, is represented by

$$H + OH + M \longrightarrow H_2O + M \quad \Delta H = -498 \text{ kJ/mol (119 kcal/mol)}$$

Numerous other reactions of this type, including a sequence involving HO_2, exist and the relative importance of each reaction depends on the mixture composition. The rates of these termolecular processes increase with increasing pressure but have little or no temperature dependence.

When the partial pressures of the radicals become high, their homogeneous recombination reactions become fast, the heat evolution exceeds heat losses, and the temperature rise accelerates the consumption of any remaining fuel to produce more radicals. Around the maximum temperature, recombination reactions exhaust the radical supply and the heat evolution rate may not compensate for radiation losses. Thus the final approach to thermodynamic equilibrium by recombination of OH, H, and O, at concentrations still many times the equilibrium value, is often observed to occur over many milliseconds after the maximum temperature is attained, especially in the products of combustion at relatively low (<2000 K) temperatures.

The radicals and other reaction components are related by various equilibria, and hence their decay by recombination reactions occurs in essence as one process on which the complete conversion of CO to CO_2 depends. Therefore, the hot products of combustion of any lean hydrocarbon flame typically have a higher CO content than the equilibrium value, slowly decreasing toward the equilibrium concentration (CO afterburning) along with the radicals, so that the oxidation of CO is actually a radical recombination process.

In most hydrocarbon flames there are many radical consumption reactions that interfere with the chain-branching reactions, and the peak radical concentrations are therefore always lower than in analogous flames of H_2, H_2/CO, or moist CO. Consequently, the overall reaction rates for hydrocarbon flames at a given temperature are lower than the overall reaction rates for H_2, H_2/CO, or moist CO flames. In general, variations in the concentrations of such radicals are believed to interfere with the flame chemistry and can account for inhibition of flames by various additives, notably halogen-containing substances. Such additives can narrow flammability limits and reduce burning velocities, even when they cause little or no reduction in the ignition temperature. Those additives containing bromine are particularly effective, and brominated organic compounds are extensively used in extinguishing devices to suppress unwanted diffusion flames (fires). Although most of the chemistry summarized here has evolved from the study of flames and explosions in premixtures, it also applies semiquantitatively to many nonpremixed systems.

The fundamental parameters in the two main methods of achieving ignition are basically the same. Recent advances in the field of combustion have been in the development of mathematical definitions for some of these parameters. For instance, consider the case of ignition achieved by means of an electric spark, where electrical energy released between electrodes results in the formation of a plasma in which the ionized gas acts as a conductor of electricity. The electrical energy liberated by the spark is given by equation 2 (1), where V = the potential, V; I = the current, A; Θ = the spark duration, s; and t = time, s.

$$E = \int_0^{\Theta} VI \, dt \tag{2}$$

The electrical energy is rapidly transformed into thermal energy, and because the temperature of the ionized gas is generally above 300 K, the ignition delay time is short compared with the spark duration, Θ. Ignition only takes place if the electrical energy exceeds the critical value, E_c, and if this energy is liberated within a critical volume, v_c (1,8–11).

Ignition can also be produced by a heated surface. During the process of heat transfer from a hot surface to a flammable mixture, reactions are initiated as the temperature rises and the combination of additional heat transfer from the surface and heat release by chemical reactions can lead to ignition of the mixture.

In experiments in which a heated cylinder was introduced into a propane – air mixture, it was found that approximately 11 μs were required for ignition. During this time, heat released by chemical reactions is negligible. This time interval before ignition is referred to as the ignition delay or induction time and is a function of the physical properties of the fuel, the rate of heat conduction, and the pressure of the combustion chamber (1,3,8–11). Ignition is characterized by a rapid change in temperature (eq. 3), where $t_{ignition}$ = ignition delay time, s; E_a = activation energy, J/mol; R = universal gas constant, J/(K·mol); and T_i = ignition temperature, K.

$$t_{ignition} \; \alpha \; \exp\!\left(\frac{E_a}{RT_i}\right) \tag{3}$$

For liquid fuels, ignition delay times are of the order 50 μs at 700 K and 10 μs at 800 K. At low temperatures most of the ignition delay is the result of slow, free-radical reactions, and a distinction between the initiation and explosion periods within the ignition delay time can be made. With increasing ignition temperature for a given mixture, these times become comparable and at temperatures as high as 1500 K, both times may be of the order of 10^{-4} s. Consequently, the reaction zone in the flame of a mixture is observed to be one continuous event (12–14).

Another important concept is that of the critical ignition volume. During the propagation of the combustion wave, the flame volume cannot continually grow beyond a critical value without an additional supply of energy. The condition that controls the critical volume for ignition is reached when the rate of increase of flame volume is less than the rate of increase of volume of the combustion products. In this condition a positive exchange of heat between the flame and the fresh mixture is achieved.

For a point spark source, the flame volume is initially spherical and the critical ignition volume is determined by calculating the rate of change of flame volume with respect to radius compared to the rate of change of volume of the combustion products (eq. 4),

$$\frac{d}{dr}\left(\frac{4}{3}\,\pi\,\left((r + e)^3 - r^3\right)\right) \leqslant \frac{d}{dr}\left(\frac{4}{3}\,\pi\,r^3\right) \tag{4}$$

where r = radius of flame, m; e = thickness of flame front, m; and r_c = radius of the critical spherical volume for ignition, m. This gives equations 5 or 6.

$$2\,e\,r + e^2 \leqslant r^2 \tag{5}$$

$$r_c = e + \sqrt{2} \tag{6}$$

For a line spark source, the flame volume is initially cylindrical with the cylinder length equal to the separation distance between the electrodes. Thus, for a cylindrical flame, $r_c = e$, and the critical ignition volumes are equation 7 for a spherical flame and equation 8 for a cylindrical flame where v_c = critical ignition volume, m^3/kg; e = thickness of flame front, m; and d = flame height, m.

$$v_c = \frac{4\,\pi\,e^3\,(1 + \sqrt{2})^3}{3} \tag{7}$$

$$v_c = \pi\,e^2\,d \tag{8}$$

In order to control and monitor the amount of energy required to achieve ignition, the concept of the minimum ignition energy as the smallest quantity of energy that will ignite a mixture is defined (15–17). The minimum ignition energy is calculated using the assumption that sufficient energy must be supplied from the exterior to the critical volume, v_c, to raise the mixture from its initial temperature, T_i, to the flame temperature, $T_f(1)$. The critical energy is given by equation 9: where E = critical energy, J; v_c = critical ignition volume, m^3/kg; ρ = density, kg/m^3; c_p = specific heat capacity at constant pressure, J/(kg·K); T_f = flame temperature, K; and T_i = initial temperature, K.

$$E = v_c\,\rho\,c_p(T_f - T_i) \tag{9}$$

Because of heat being transferred to the exterior of the critical volume, only a fraction of the total spark energy, E, is utilized as ignition energy.

Minimum ignition energies are determined experimentally by means of a combustion bomb with two electrodes. Both the ignition energy and the quenching distance can be varied in this device. By using high performance (usually air-gap) condensers, the majority of the stored energy will appear in the spark gap. The stored energy can be calculated using equation 10 (1,2,9), where E = total spark energy released/stored, J; C = capacity of the condenser; and V = voltage just before the spark is passed through the gas, V.

$$E = \frac{CV^2}{2} \tag{10}$$

The spark must always be produced by a spontaneous breakdown of the gas because an electronic firing circuit or a trigger electrode would either obviate the measurement of spark energy or grossly change the geometry of the ignition source (1,2,9).

Once ignition has been achieved, flame propagation can be controlled. This is achieved by the use of the concept of the quenching distance, defined as the minimum orifice diameter, wall separation, or mesh spacing just sufficient to prevent flame propagation (Fig. 3). Experiments carried out on spark ignition of mixtures between parallel plates show that, with sufficiently powerful sparks, a flame develops in the immediate neighborhood of the spark but does not propagate through the mixture unless the plates are separated by more than the quenching distance (1,2).

Cool Flames. Under particular conditions of pressure and temperature, incomplete combustion can result in the formation of intermediate products such as CO. As a result of this incomplete combustion, flames can be less exothermic than normal and are referred to as cool flames. An increase in the pressure or temperature of the mixture outside the cool flame can produce normal spontaneous ignition (1).

Flame Temperature. The adiabatic flame temperature, or theoretical flame temperature, is the maximum temperature attained by the products when the reaction goes to completion and the heat liberated during the reaction is used to raise the temperature of the products. Flame temperatures, as a function of the equivalence ratio, are usually calculated from thermodynamic data when a fuel is burned adiabatically with air. To calculate the adiabatic flame temperature (AFT) without dissociation, for lean to stoichiometric mixtures, complete combus-

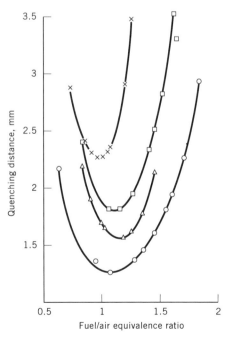

Fig. 3. Quenching distance as function of equivalence ratio for hydrocarbon mixtures with air (1), where x = methane, □ = propane, △ = propylene, and ○ = ethylene.

tion is assumed. This implies that the products of combustion contain only carbon dioxide, water, nitrogen, oxygen, and sulfur dioxide.

Actual temperatures in practical flames are lower than calculated values as a result of the heat losses by radiation, thermal conduction, and diffusion. At high temperatures, dissociation of products of combustion into species such as OH, O, and H reduces the theoretical flame temperature (7). Increasing the pressure tends to suppress dissociation of the products and thus generally raises the adiabatic flame temperature (4).

Adiabatic flame temperatures agree with values measured by optical techniques, when the combustion is essentialy complete and when losses are known to be relatively small. Calculated temperatures and gas compositions are thus extremely useful and essential for assessing the combustion process and predicting the effects of variations in process parameters (4). Advances in computational techniques have made flame temperature and equilibrium gas composition calculations, and the prediction of thermodynamic properties, routine for any fuel-oxidizer system for which the enthalpies and heats of formation are available or can be estimated.

Flame Types and Their Characteristics. There are two main types of flames: diffusion and premixed. In diffusion flames, the fuel and oxidant are separately introduced and the rate of the overall process is determined by the mixing rate. Examples of diffusion flames include the flames associated with candles, matches, gaseous fuel jets, oil sprays, and large fires, whether accidental or otherwise. In premixed flames, fuel and oxidant are mixed thoroughly prior to combustion. A fundamental understanding of both flame types and their structure involves the determination of the dimensions of the various zones in the flame and the temperature, velocity, and species concentrations throughout the system.

The development of combustion theory has led to the appearance of several specialized asymptotic concepts and mathematical methods. An extremely strong temperature dependence for the reaction rate is typical of the theory. This makes direct numerical solution of the equations difficult but at the same time accurate. The basic concept of combustion theory, the idea of a flame moving at a constant velocity independent of the ignition conditions and determined solely by the properties and state of the fuel mixture, is the product of the asymptotic approach (18,19). Theoretical understanding of turbulent combustion involves combining the theory of turbulence and the kinetics of chemical reactions (19–23).

Laminar Premixed Flames. The structure of a one-dimensional premixed flame is well understood (1). By coupling the rate of heat release from chemical reaction and the rate of heat transfer by conduction with the flow of the unburned mixture, an observer moving with the wave would see a steady laminar flow of unburned gas at a uniform velocity, S_u, into the stationary wave or flame. Hence S_u is defined as the burning velocity of the mixture based on the conditions of the unburned gas. The thickness of the preheating zone and the equivalent reaction zone is found to be inversely proportional to the burning velocity. By considering the heat release from the chemical reaction, it is possible to calculate the thickness of the effective reaction zone. For example, at atmospheric pressure, for hydrocarbon flames, the thicknesses of the preheat and reaction zones are found to be 0.7 and 0.2 mm, respectively (1–3). A propagating, premixed flame is in essence a thin wave in which the temperature rise is continuous and rapid. The thicknesses of the wave depend on ϕ and on pressure; for a typical hydrocarbon – air mixture at $\phi = 1$ at 101.3 kPa (1 atm), the entire rise, from the initial temperature to the adiabatic temperature, occurs in 0.01 mm, and the concentration and temperature gradients are accordingly very steep.

S_u has been approximated for flames stabilized by a steady uniform flow of unburned gas from porous metal diaphragms or other flow straighteners. However, in practice, S_u is usually determined less directly from the speed and area of transient flames in tubes, closed vessels, soap bubbles blown with the mixture, and, most commonly, from the shape of steady Bunsen burner flames. The observed speed of a transient flame usually differs markedly from S_u. For example, it can be calculated that a flame spreads from a central ignition point in an unconfined explosive mixture such as a soap bubble at a speed of $(\rho_o/\rho_b)S_u$, in which the density ratio across the flame is typically 5–10. Usually, the expansion of the burning gas imparts a considerable velocity to the unburned mixture, and the observed speed will be the sum of this velocity and S_u.

By applying the conservation equations of mass and energy and by neglecting the small pressure changes across the flame, the thickness of the preheating and reaction zones can be calculated for a one-dimensional flame (1).

There are a number of sources of instability in premixed combustion systems (23,24).

1. System instabilities involve interactions between flows in different parts of a reacting system. Although these instabilities can cause turbulence, they cannot be analyzed on their own and the whole system should be considered as a unit (11,25).
2. Acoustic instabilities involve the interactions of acoustic waves with the combustion processes. These instabilities are of high frequency and can be significant in certain situations but can successfully be avoided by suitable design (11,24).
3. Taylor instabilities involve effects of buoyancy or acceleration in fluids with variable density; a light fluid beneath a heavy fluid is unstable by the Taylor mechanism. The upward propagation of premixed flames in tubes is subject to Taylor instability (11).
4. Landau instabilities are the hydrodynamic instabilities of flame sheets that are associated neither with acoustics nor with buoyancy but instead involve only the density decrease produced by combustion in incompressible flow. The mechanism of Landau instability is purely hydrodynamic. In principle, Landau instabilities should always be present in premixed flames, but in practice they are seldom observed (26,27).
5. Diffuse-thermal instabilities involve the relative diffusion reactants and heat within a laminar flame. These are the smallest-scale instabilities (11).

Laminar flame instabilities are dominated by diffusional effects that can only be of importance in flows with a low turbulence intensity, where molecular transport is of the same order of magnitude as turbulent transport (28). Flame instabilities do not appear to be capable of generating turbulence. They result in the growth of certain disturbances, leading to orderly three-dimensional structures which, though complex, are steady (1,2,8,9).

Turbulent Premixed Flames. Combustion processes and flow phenomena are closely connected and the fluid mechanics of a burning mixture play an important role in forming the structure of the flame. Laminar combusting flows can occur only at low Reynolds numbers, defined as

$$Re = \frac{\rho u d}{\mu} \tag{11}$$

where, ρ = density, kg/m^3; u = velocity, m/s; d = diameter, m; and μ = kinematic viscosity, kg/ms. When $Re > Re_{cr}$ the laminar structure of a flow becomes unstable and when the Reynolds number exceeds the critical value by an order of magnitude, the structure of the flow changes. Along with this change of structure from an orderly state to a more chaotic state, the following parameters begin to fluctuate randomly: velocity, pressure, temperature, density, and species concentrations. Overall, with increasing Reynolds number, laminar flow becomes unstable as a result of these fluctuations and breaks down into turbulent flow. Laminar and turbulent flames differ greatly in appearance. For example, while the combustion zone of a laminar Bunsen flame is a smooth, delineated and thin surface, the analogous turbulent combustion region is blurred and thick (1).

Turbulent flame speed, unlike laminar flame speed, is dependent on the flow field and on both the mean and turbulence characteristics of the flow, which can in turn depend on the experimental configuration. Nonstationary spherical turbulent flames, generated through a grid, have flame speeds of the order of or less than the laminar flame speed. This turbulent flame speed tends to increase proportionally to the intensity of the turbulence.

In high speed dusted, premixed flows, where flames are stabilized in the recirculation zones, the turbulent flame speed grows without apparent limit, in approximate proportion to the speed of the unburned gas flow. In the recirculation zones the intensity of the turbulence does not affect the turbulent flame speed (1).

In the reaction zone, an increase in the intensity of the turbulence is related to the turbulent flame speed. It has been proposed that flame-generated turbulence results from shear forces within the burning gas (1,28). The existence of flame-generated turbulence is not, however, universally accepted, and in unconfined flames direct measurements of velocity indicate that there is no flame-generated turbulence (1,2).

The balanced equation for turbulent kinetic energy in a reacting turbulent flow contains the terms that represent production as a result of mean flow shear, which can be influenced by combustion, and the terms that represent mean flow dilations, which can remove turbulent energy as a result of combustion. Some of the discrepancies between turbulent flame propagation speeds might be explained in terms of the balance between these competing effects.

To analyze premixed turbulent flames theoretically, two processes should be considered: (1) the effects of combustion on the turbulence, and (2) the effects of turbulence on the average chemical reaction rates. In a turbulent flame, the peak time-averaged reaction rate can be orders of magnitude smaller than the corresponding rates in a laminar flame. The reason for this is the existence of turbulence-induced fluctuations in composition, temperature, density, and heat release rate within the flame, which are caused by large eddy structures and wrinkled laminar flame fronts.

A unified statistical model for premixed turbulent combustion and its subsequent application to predict the speed of propagation and the structure of plane turbulent combustion waves is available (29–32).

Laminar Diffusion Flames.   Generally, a diffusion flame is defined as one in which no mixing of the fuel and oxidant takes place prior to emission from the burner. However, it can also be defined as a flame in which the mixing rate is sufficiently slow compared to the reaction rate, that the mixing time controls the burning rate. Since a continuous spectrum of flames exists between the perfectly premixed flame and the diffusion flame, the term diffusion flame is reserved for those flames in which there is a total separation between fuel and oxidant (1,5). Some of the practical advantages of diffusion flames include high flame stability, safety (as there is no need for the storage of the combustible mixture), intense radiation and hence high heat exchange with the surroundings. It is difficult to give a general treatment of diffusion flames largely because no simple measurable parameter, analogous to the burning velocity for premixed flames, can be used to characterize the burning process (3).

Many different configurations of diffusion flames exist in practice (Fig. 4). Laminar jets of fuel and oxidant are the simplest and most well understood diffusion flames. They have been studied exclusively in the laboratory, although a complete description of both the transport and chemical processes does not yet exist (2).

The discussion of laminar diffusion flame theory addresses both the gaseous diffusion flames and the single-drop evaporation and combustion, as there are some similarities between gaseous and liquid diffusion flame theories (2). A frequently used model of diffusion flames has been developed (34), and despite some

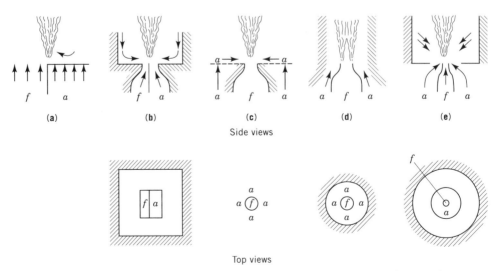

Fig. 4. Various configurations (**a–e**) used to obtain gaseous diffusion flames where a = air and f = fuel (33).

of the restrictive assumptions of the model, it gives a good description of diffusion flame behavior.

The Displacement Distance theory suggests that since the structure of the flame is only quantitatively correct, the flame height can be obtained through the use of the displacement length or "displacement distance" (35,36) (eq. 12), where h = flame height, m; V = volumetric flow rate, m^3/s; and D = diffusion coefficient.

$$h = \frac{V}{2\pi D} \tag{12}$$

In addition to the Burke and Schumann model (34) and the Displacement Distance theory, a comprehensive laminar diffusion flame theory can be written using the equations of conservation of species, energy, and momentum, including diffusion, heat transfer, and chemical reaction.

Evaporation and burning of liquid droplets are of particular interest in furnace and propulsion applications and by applying a part of the Burke and Schumann approach it is possible to obtain a simple model for diffusion flames.

Combustion chemistry in diffusion flames is not as simple as is assumed in most theoretical models. Evidence obtained by adsorption and emission spectroscopy (37) and by sampling (38) shows that hydrocarbon fuels undergo appreciable pyrolysis in the fuel jet before oxidation occurs. Further evidence for the existence of pyrolysis is provided by sampling of diffusion flames (39). In general, the pre-flame pyrolysis reactions may not be very important in terms of the gross features of the flame, particularly flame height, but they may account for the formation of carbon while the presence of OH radicals may provide a path for NO$_x$ formation, particularly on the oxidant side of the flame (39).

Combustion chemistry in diffusion flames also accounts for the smoke formation process. Characteristic behavior of smoking diffusion flames includes: (1) the first appearance of smoke is generally at the flame tip, the width of the smoke trail increasing as the amount of smoke increases; (2) the appearance and quantity of smoke increases with increasing flame height; (3) the smoking tendency changes with oxidant flow rate and the oxygen content of the oxidant stream; (4) the smoking tendency is a function of fuel type; and (5) the smoking tendency generally increases with increasing pressure (39).

Diffusion Flames in the Transition Region. As the velocity of the fuel jet increases in the laminar to turbulent transition region, an instability develops at the top of the flame and spreads down to its base. This is caused by the shear forces at the boundaries of the fuel jet. The flame length in the transition region is usually calculated by means of empirical formulas of the form (eq. 13): where l = length of the flame, m; r = radius of the fuel jet, m; v = fuel flow velocity, m/s; and C_1 and C_2 are empirical constants.

$$l = \frac{r}{C_1 - \dfrac{C_2}{v}} \tag{13}$$

Turbulent Diffusion Flames. Laminar diffusion flames become turbulent with increasing Reynolds number (1,2). Some of the parameters that are affected by turbulence include flame speed, minimum ignition energy, flame stabilization, and rates of pollutant formation. Changes in flame structure are believed to be controlled entirely by fluid mechanics and physical transport processes (1,2,9).

Consider the case of the simple Bunsen burner. As the tube diameter decreases, at a critical flow velocity and at a Reynolds number of about 2000, flame height no longer depends on the jet diameter and the relationship between flame height and volumetric flow ceases to exist (2). Some of the characteristics of diffusion flames are illustrated in Figure 5.

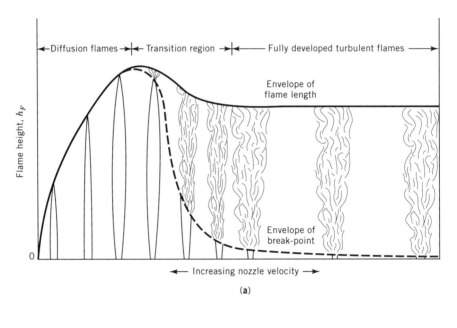

(a)

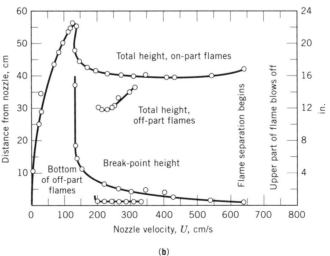

(b)

Fig. 5. Effects of nozzle velocity on flame appearance in laminar and turbulent flow: (**a**), flame appearance; (**b**), flame height and break-point height (40).

The preferred models for predicting the behavior of turbulent-free shear layers involve the solution of the turbulent kinetic energy equation in order to obtain the local turbulent shear stress distribution (1,2,9). These models are ranked according to the number of simultaneous differential equations that need to be solved. The one-equation model considers the turbulent kinetic energy equation alone, whereas the two-equation model considers the turbulent kinetic energy equation plus a differential equation for the turbulence length scale, or equivalently, the dissipation rate for turbulent kinetic energy. These equations are solved along with the conservation equations (momentum, energy, and species) to model turbulent flows.

The physics and modeling of turbulent flows are affected by combustion through the production of density variations, buoyancy effects, dilation due to heat release, molecular transport, and instability (1,2,3,5,8). Consequently, the conservation equations need to be modified to take these effects into account. This modification is achieved by the use of statistical quantities in the conservation equations. For example, because of the variations and fluctuations in the density that occur in turbulent combustion flows, density weighted mean values, or Favre mean values, are used for velocity components, mass fractions, enthalpy, and temperature. The turbulent diffusion flame can also be treated in terms of a probability distribution function (pdf), the shape of which is assumed to be known a priori (1).

In general, comprehensive, multidimensional modeling of turbulent combustion is recognized as being difficult because of the problems associated with solving the differential equations and the complexities involved in describing the interactions between chemical reactions and turbulence. A number of computational models are available commercially that can do such work. These include FLUENT, FLOW-3D, and PCGC-2.

The various studies attempting to increase our understanding of turbulent flows comprise five classes: moment methods disregarding probability density functions, approximation of probability density functions using moments, calculation of evolution of probability density functions, perturbation methods beginning with known structures, and methods identifying coherent structures. For a thorough review of turbulent diffusion flames see References 41–48.

FUNDAMENTALS OF HETEROGENEOUS COMBUSTION

The discussion of combustion fundamentals so far has focused on homogeneous systems. Heterogeneous combustion is the terminology often used to refer to the combustion of liquids and solids. From a technological viewpoint, combustion of liquid hydrocarbons, mainly in sprays, and coal combustion are of greatest interest.

Most theories of droplet combustion assume a spherical, symmetrical droplet surrounded by a spherical flame, for which the radii of the droplet and the flame are denoted by r_d and r_p, respectively. The flame is supported by the fuel diffusing from the droplet surface and the oxidant from the outside. The heat produced in the combustion zone ensures evaporation of the droplet and consequently the fuel supply. Other assumptions that further restrict the model include: (1) the rate of chemical reaction is much higher than the rate of diffusion and hence the reaction is completed in a flame front of infinitesimal thickness; (2) the droplet is made up of pure liquid fuel; (3) the composition of the ambient atmosphere far away from the droplet is constant and does not depend on the combustion process; (4) combustion occurs under steady-state conditions; (5) the surface temperature of the droplet is close or equal to the boiling point of the liquid; and (6) the effects of radiation, thermodiffusion, and radial pressure changes are negligible.

In order to obtain an expression for the burning rate of the droplet, the following parameters are needed: physical constants such as the specific heat, and the thermal conductivity of the droplet, the radius of the flame, and the temperature of the flame. To determine these quantities, heat conduction, diffusion, and the kinetics of the chemical processes associated with droplet combustion need to be analyzed. This is achieved mathematically by solving the equations of mass continuity, mass continuity for components, and the energy equation. The solving of these equations can be facilitated if the following simplifying assumptions are made: the flame surrounding the droplet is a diffusion flame and, by definition, is formed where the fuel and oxidant meet in stoichiometric proportions; the temperature of this flame is very close to the adiabatic flame temperature; and the heat required for evaporation of the droplet and the heat loss to the surroundings through the burned gas are small and can therefore be neglected. These equations are usually solved in spherical coordinates for a one-dimensional case. However, since the flame is relatively thick, and the droplet is relatively small, the one-dimensional model of the process may not be a particularly accurate representation. Nevertheless, the values obtained for burning rates provide useful information (9).

The burning rate of the droplet (kg/s), and its rate of change of radius are related by:

$$m_e = -4\pi r_d^2 \rho_f \frac{dr_d}{dt} \tag{14}$$

where m_e = burning rate of the droplet, kg/s; ρ_f = density of fuel, kg/m^3; and r_d = radius of droplet, m. This equation can be simplified, assuming $r_t/r_d \simeq$ constant and $m_e \simeq r_d$.

$$\frac{d(d_d^2)}{dt} = K = \text{constant} \tag{15}$$

Hence, the constant K is termed the "burning-constant of the droplet." Integration of the equation 15 produces the droplet burning law:

$$d_d^2 = d_0^2 - Kt \tag{16}$$

where d_0 = the initial diameter of the droplet, m, and t = burnout time, s.

The amount of data available on droplet combustion is extensive. However, the results can be easily summarized, because the burning rate constants for the majority of fuels of practical interest fall within the narrow range of 7 to 11 × 10^{-3} cm^2/s. An increase in oxygen concentration results in an increase in the burning rate constant. If the burning takes place in pure oxygen, the values for burning rate are increased by a factor of about 2.0, compared to when the burning takes place in air (9).

The convective gas flow around a burning particle affects its burning rate. It has been postulated that in the absence of convection, the burning rate is independent of pressure. Forced convection, on the other hand, is believed to increase the burning rate.

During the final stages of the combustion of a droplet, coke remains, and although it represents a relatively small percentage of the mass of the original oil droplet, the time taken for the heterogeneous reaction between the oxygen-depleted combustion air and the coke particle is generally the slowest of all the combustion steps (9).

The reaction between a porous solid, such as a coke sphere, and a gas, such as oxygen, occurs in the following stages: (1) the main reactant species diffuse thoroughly through the boundary layer toward the solid surface and the products of reaction diffuse away from the surface; (2) diffusion and simultaneous chemical reaction take place within the pores of the solid proceeding from the external surface toward the interior, and gaseous products diffuse in the opposite direction; and (3) at the participating surfaces, the reacting gas chemisorbs, some intermediate species are formed, then the final products of the reaction desorb from the surface. Thus the observed reaction rate is a function of the individual resistances–boundary layer diffusion, pore diffusion, and chemical kinetics, and the rate controlling process is the slowest of them or a combination of these processes. Even though a number of gas reactions may take place at the surface of a burning carbonaceous solid, the reaction forming CO is most often assumed, $C + \frac{1}{2} O_2 \rightarrow CO$. Coke combustion is treated mathematically like char combustion.

In a practical combustion chamber, the droplets tend to burn in the form of sprays, hence it is important to understand the fundamentals of sprays. In the most simple case, a fuel spray suspended in air will support a stable propagating laminar flame in a manner similar to a homogeneous gaseous mixture. In this case, however, two different flame fronts are observed. If the spray is made up of very small droplets they vaporize before the flame reaches them and a continuous flame front is formed. If the droplets are larger, the flame reaches them before the evaporation is complete, and if the amount of fuel vapor is insufficient for the formation of a continuous flame front, the droplets burn in the form of isolated spherical regions. Flames of this type are referred to as heterogeneous laminar flames. Experimental determination of burning rates and flammability limits for heterogeneous laminar flames is difficult because of the motion of droplets caused by gravity and their evaporation before the arrival of the flame front.

In modern liquid-fuel combustion equipment the fuel is usually injected into a high velocity turbulent gas flow. Consequently, the complex turbulent flow and spray structure make the analysis of heterogeneous flows difficult and a detailed analysis requires the use of numerical methods (9).

The combustion of a coal particle occurs in two stages: (1) devolatilization during the initial stages of heating with accompanying physical and chemical changes, and (2) the subsequent combustion of the residual char (49). The burning rate or reactivity of the residual char in the second stage is strongly dependent on the process conditions of the first stage. During pulverized coal combustion the devolatilization step usually takes about 0.1 s and the residual char combustion takes on the order of 1 s. Since char combustion occurs over ~90% of the total burning time, its rate can affect the volume of the combustion chamber required to attain a given heat release and combustion efficiency. The rate of devolatilization, and the amount and nature of volatiles can significantly affect the ignition process and hence the onset of char combustion. During devolatilization or thermal decomposition, rupture of various functional groups bonded to the macromolecular structure leads to the evolution of gases and the formation and opening of pores. Also, depending on its thermoplastic properties, a coal particle may undergo softening and swelling resulting in a change in size and physical characteristics. Diffusion of oxygen to, and within, a char particle depends on its physical structure and accessible surface area (char morphology). Attempts have been made to explain the combustion rates of chars in terms of their morphologies (50–52).

The ignition mechanism is rather complex and is not well understood in terms of actually defining the ignition temperature and reaction mechanisms. The ignition temperature is known, however, not to be a unique property of the coal and depends on a balance between heat generated and heat dissipated to the surroundings around the coal particle. Measuring the ignition characteristics is complicated by the fact that they are strongly dependent on the physical arrangement of the particles, eg, single particle, clouds of coal dust, or coal piles. Reported ignition temperatures range from 303 to 373 K in the case of spontaneous ignition of coal piles at ambient temperature to 1073–1173 K in the case of single coal particle ignition (53,54). Characteristics such as coal type, particle size and distribution of mineral matter, and experimental conditions such as gas temperature, heating rate, oxygen, and coal dust concentration are some of the important factors that influence values obtained for ignition temperatures. Ignition of coal particles can occur either homogeneously, with ignition of the volatiles released and the subsequent ignition of the char surface, or heterogeneously, with both volatiles and the char surface igniting simultaneously. Heating rate and particle size affect the mode of ignition. The early theories of ignition were based on an energy balance per unit volume of reactive mixture (55,56). Later, a generalized theory of flame propagation in laminar coal dust flames was proposed (57). Good reviews in this area include References 58–60.

A variety of techniques has been used to determine ignition temperatures: fixed beds, the crossing point method, the critical air blast method (61), photographic techniques (62,63), entrained flow reactors (64), electric spark ignition (65), luminous glow observations (66), plug flow reactors (67), shock tubes (68), and thermogravimetric analysis (69). The techniques mostly used are constant temperature methods (66,70,71) in which coal particles are introduced into a preheated furnace maintained at a fixed temperature. If ignition of the coal particles is not observed by the appearance of a glow, flame, or sharp temperature rise, the test is then conducted at a higher temperature and the procedure is repeated until the ignition of the sample is observed. It has been a matter of controversy as to whether the value of the critical temperature determined using this method for these low temperature tests represents the ignition tendency or the combustibility of the coal during the high heating rates encountered in a pulverized coal flame.

The structure of residual char particles after devolatilization depends on the nature of the coal and the pyrolysis conditions such as heating rate, peak temperature, soak time at the peak temperature, gaseous environment, and the pressure of the system (72). The oxidation rate of the char is primarily influenced by the physical and chemical nature of the char, the rate of diffusion and the nature of the reactant and product gases, and the temperature and pressure of the operating system. The physical and chemical characteristics that influence the rate of oxidation are chemical structural variations, such as the concentration of oxygen and hydrogen atoms (73–75), the nature and amount of mineral matter (76–79), and physical characteristics such as porosity, particle size, and accessible surface area. The rate of diffusion of the reactant gas is governed by the temperature and pressure of the operating system.

The rate limiting step in the combustion of char is either the chemical kinetics (adsorption of oxygen, reaction, and desorption of products) or diffusion of oxygen (bulk and pore diffusion). Variations in the reaction rate with temperature for gas–carbon reactions have been grouped into three main regions or zones depending on the rate limiting resistance (80) (Figs. 6 and 7).

The overall reaction rate is determined by equation 17,

$$q = \frac{1}{\dfrac{1}{k_{diff}} + \dfrac{1}{k_s}} \, p_g \tag{17}$$

where q = rate of removal of carbon atoms per unit external surface area, kg/m²s; p_g = partial pressure of oxygen in the free stream, Pa; k_{diff} = diffusional reaction rate coefficient, kg/(m²s·Pa); and k_s = surface reaction rate coefficient, kg/(m²s·Pa).

In Zone 1, the low temperature zone, the reaction rate is slow and the concentration of the reactant gas is uniform throughout the interior of the solid. The

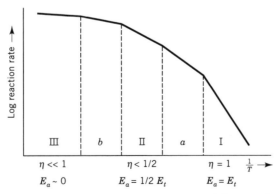

Fig. 6. The three ideal zones (I–III) representing the rate of change of reaction for a porous carbon with increasing temperature where a and b are intermediate zones, E_a is activation energy, and E_t is true activation energy. The effectiveness factor, η, is a ratio of experimental reaction rate to reaction rate which would be found if the gas concentration were equal to the atmospheric gas concentration (80).

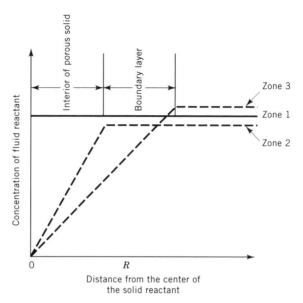

Fig. 7. The concentration of the reacting fluid as a function of the distance from the center of the solid reactant in various reaction zones.

overall rate is controlled by the chemical reactivity of the solid, the kinetics are not influenced by mass transfer, and the kinetics are the intrinsic kinetics. The measured activation energy is equal to the true activation energy. In a higher temperature zone, Zone 2, the concentration of the gaseous reactant becomes zero at a point somewhere between the surface and the center of the particle, and the reaction rate is controlled by both the chemical kinetics and the diffusion rate of the gaseous reactant. The measured activation energy in this zone is one-half of the true activation energy. At still higher temperatures, in Zone 3, the chemical reactivity is so high that oxygen is consumed as soon as it reaches the surface of the solid and the concentration of the gaseous reactant approaches zero at the surface, indicating that the reaction rate is controlled by bulk diffusion of the reacting gas. The measured activation energy approaches a value of zero in this zone.

Burning times for coal particles are obtained from integrated reaction rates. For larger particles (>100 μm) and at practical combustion temperatures, there is a good correlation between theory and experiment for char burnout. Experimental data are found to obey the Nusselt "square law" which states that the burning time varies with the square of the initial particle diameter ($t_c \sim D_o^{2.0}$). However, for particle sizes smaller than 100 μm, the Nusselt relationship seems to predict higher burning times than those observed and it has been noted that burning times are proportional to the initial diameter of the particles raised to the power of 1.4 ($t_c \sim D_o^{1.4}$) (54).

Combustion Technology

Technology addresses the more applied, practical aspects of combustion, with an emphasis on the combustion of gaseous, liquid, and solid fuels for the purpose of power production. In an ideal fuel burning system (1) there should be no excess oxygen or products of incomplete combustion, (2) the combustion reaction should be initiated by the input of auxiliary ignition energy at a low rate, (3) the reaction rate between oxygen and fuel should be fast enough to allow rapid rates of heat release and it should also be compatible with acceptable nitrogen and sulfur oxide formation rates, (4) the solid impurities introduced with the fuel should be handled and disposed of effectively, (5) the temperature and the weight of the products of combustion should be distributed uniformly in relation to the parallel circuits of the heat absorbing surfaces, (6) a wide and stable firing range should be available, (7) fast response to changes in firing rate should be easily accommodated, and (8) equipment availability should be high and maintenance costs low (49–51).

COMBUSTION OF GASEOUS FUELS

In any gas burner some mechanism or device (flame holder or pilot) must be provided to stabilize the flame against the flow of the unburned mixture. This device should fix the position of the flame at the burner port. Although gas burners vary greatly in form and complexity, the distribution mechanisms in most cases are fundamentally the same. By keeping the linear velocity of a small fraction of the mixture flow equal to or less than the burning velocity, a steady flame is formed. From this pilot flame, the main flame spreads to consume the main gas flow at a much higher velocity. The area of the steady flame is related to the volumetric flow rate of the mixture by equation 18 (81,82)

$$\dot{V}_{mix} = A_f S_u \qquad (18)$$

where $\dot{V}_{mix}$ = volumetric flow rate, m³/s; A_f = area of the steady flame, m²; and S_u = burning velocity, m/s.

The volumetric flow rate of the mixture is, in turn, proportional to the rate of heat input (eq. 19):

$$\dot{V}_{mix} \cdot (HHV) = \dot{Q} \qquad (19)$$

where $\dot{V}_{mix}$ = volumetric flow rate, m³/s; HHV = higher heating value of the fuel, J/kg; and $\dot{Q}$ = rate of heat input, J/s.

In the simple Bunsen flame on a tube of circular cross-section, the stabilization depends on the velocity variation in the flow emerging from the tube. For laminar flow (parabolic velocity profile) in a tube, the velocity at a radius r is given by equation 20:

$$v = \text{const}(R^2 - r^2) \qquad (20)$$

where v = laminar flow velocity, m/s; R = tube radius, m; r = flame radius, m; and const = experimental constant.

The maximum velocity at the axis is twice the average, whereas the velocity at the wall is zero. The effect of the burner wall is to cool the flame locally and decrease the burning velocity of the mixture. This results in flame stabilization. However, if the heat-transfer processes (conduction, convection, and radiation) involved in cooling the flame are somehow impeded, the rate of heat loss is decreased and the local reduction in burning velocity may no longer take place. This could result in upstream propagation of the flame.

To make the flame stable against the flow in a thin annular region near the rim, the flow velocity v_r should be made equal to the burning velocity at some radius r. This annulus serves as a pilot and ignites the main flow of the mixture, ie, the flame gradually spreads toward the center. In most of the mixture flow, $v_r > S_u$ which results in a stable flame. With increasing mixture flow, the height and area of the flame increase. Measurement of the area of a stable Bunsen flame is the basis for the method most commonly used to determine S_u (81–83).

By feeding the mixture through a converging nozzle, the velocity profile may be made nearly flat or uniform. A Bunsen flame in such a flow has a smaller range of stability but the mechanism is essentially the same and the flame very closely approximates a cone. If the apex angle of the flame is Θ, then S_u can be obtained from equation 21

$$S_u = v_r \sin\left(\frac{\Theta}{2}\right) \tag{21}$$

where S_u = burning velocity, m/s; v_r = mixture velocity at the nozzle exit, m/s; and Θ = the apex angle of the flame in degrees.

If the tube diameter is appreciably larger than the quenching distance, S_u will exceed v_r in some parts of the flowing mixture due to a lack of quenching, and the flame will then propagate down the tube as far as there is mixture to consume. This undesirable condition is referred to as flashback. If, on the other hand, v_r exceeds S_u in the mixture flow, the flame lifts from the port and blows off. This condition is referred to as blowoff and like flashback should be avoided (Fig. 8). The velocity gradient at the wall, g_w, is defined as

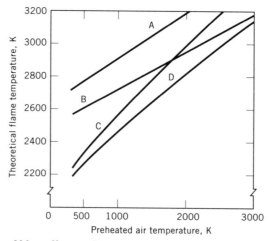

Fig. 8. Flashback and blowoff critical velocity gradients for a natural gas flame (10), where A is N_2/O_2 = 2, p = 1.01 MPa (10 atm); B, N_2/O_2 = 2, p = 101.3 kPa (1 atm); C, air, p = 1.01 MPa; and D, air, p = 101.3 kPa.

$$g_w = \left(\frac{dv}{dr}\right)_{r=R}$$

For example, in a laminar or Poiseuille flow in a round tube of radius R, $g_w = 4g_{av}/R$, and for a given initial mixture composition, flashback (or blowoff) will occur at the same value of g_w in tubes of various sizes, whereas the corresponding average velocity at flashback (or blowoff) is proportional to R. Typical velocity gradient values for stoichiometric methane–air flames are at flashback about 400 s^{-1} and at blowoff 2000 s^{-1}. Thus, if the mixture is burned on a 1-cm diameter tube, the average velocity of flashback is $400 \times 0.5/4 = 50$ cm/s and at blowoff the average velocity is 250 cm/s, or the range of stability would be roughly one and a half to seven times the burning velocity. At flashback, g_w is at a maximum around $\phi = 1$. If the burner is operated such that the surrounding inert atmosphere is inert, g_w is at a maximum at blowoff.

The behavior of rich mixtures is complicated by the entrainment of air at the burner port that sustains combustion of hot combustion products of the primary flame near the port. The blowoff velocity is found to increase continuously with ϕ, or richer mixtures are more stable with respect to blowoff. They also have a lesser tendency toward flashback. Hence, a Bunsen flame has more latitude for stable operation if the primary mixture is rich. For this reason many appliance burners that involve assemblies of such flames are routinely adjusted by first making the primary mixture so rich that soot just forms in the burned gas (yellow-tipping), and then increasing the air until the yellow luminosity disappears. The primary equivalence ratio is then perhaps 1.5 or more; the rich products of that primary flame are burned in the secondary diffusion flame in the surrounding air, or the faintly luminous outer mantle of a Bunsen flame.

Most of the commercial gas–air premixed burners are basically laminar-flow Bunsen burners and operate at atmospheric pressure. This means that the primary air is induced from the atmosphere by the fuel flow with which it mixes in the burner passage leading to the burner ports, where the mixture is ignited and the flame stabilized. The induced air flow is determined by the fuel flow through momentum exchange and by the position of a shutter or throttle at the air inlet. Hence, the air flow is a function of the fuel velocity as it issues from the orifice or nozzle, or of the fuel supply pressure at the orifice. With a fixed fuel flow rate, the equivalence ratio is adjusted by the shutter, and the resulting induced air flow also determines the total mixture flow rate. The desired air–fuel volume ratio is usually seven or more, depending on the stoichiometry. Burners of this general type with many multiple ports are common for domestic furnaces, heaters, stoves, and for industrial use. The flame stabilizing ports in such burners are often round but may be slots of various shapes to conform to the heating task.

Atmospheric pressure industrial burners are made for a heat release capacity of up to 50 kJ/s (12 kcal/s), and despite the varied designs, their principle of stabilization is basically the same as that of the Bunsen burner. In some cases the mixture is fed through a fairly thick-walled pipe or casting of appropriate shape for the application and the desired distribution of the flame. The mixture issues from many small and closely spaced drilled holes, typically 1–2 mm diameter, and burns as rows of small Bunsen flames. It may be ignited with a small pilot flame, spark, or heated wire, usually located near the first holes to avoid accumulation of unburned mixture before ignition. The rated heat release for a given fuel–air mixture can be scaled with the size and number of holes. For example, for 2-mm diameter holes it would be 10–100 J/(hole) or, in general, 0.3–3 kJ/cm²s (72–720 cal/cm²s) of port area, depending on the fuel. The ports may also be narrow slots, sometimes packed with corrugated metal strips to improve the flow distribution and reduce the tendency to flashback.

Gas burners that operate at high pressures are usually designed for high mixture velocities and heating intensities and therefore stabilization against blowoff must be enhanced. This can be achieved by a number of methods such as surrounding the main port with a number of pilot ports or using a porous diaphragm screen.

In order to achieve high local heat flux the port velocity of the mixture should be increased considerably. In burners that achieve stabilization using pilot ports, most of the mixture can be burned at a port velocity as high as 100 S_u to produce a long pencil-like flame, suitable for operations requiring a high heat flux.

High local heat flux can also be obtained with Bunsen flames using mixtures with high burning velocities as in H_2/O_2 and C_2H_2/O_2 torches. Their stabilization mechanism is essentially the same as it is for slower burning mixtures but the port or nozzle of the torch is usually much smaller, in part to avoid turbulence in the mixture flow. The consequences of flashback are also much more severe since most such mixtures are detonatable, and the premixing chamber or tube must accordingly be more rugged. For this reason, many large hydrogen–oxygen and hydrocarbon–oxygen flames are not premixed. They actually consist of assemblies of closely spaced diffusion flames, produced from separately fed but contiguous fuel and oxidizer flows. In such surface-mixing burners, the surface is an array of very small and closely packed alternating fuel and oxidizer ports. The arrangement and number of the ports and the complexity of the required manifolding of the reactant passages vary with the application and the desired geometry of the burned gas flow. With a very fast burning fuel-oxidizer combination, the individual diffusion flames may be so short that the assembly approximates a large, flat, premixed flame, as is the case with some rocket engine injectors (84,85). The most frequently used modern burners are the circular type, capable of burning oil and gas.

It is often desired to substitute directly a more readily available fuel for the gas for which a premixed burner or torch and its associated feed system were designed. Satisfactory behavior with respect to flashback, blowoff, and heating capability, or the local enthalpy flux to the work, generally requires reproduction as nearly as possible of the maximum temperature and velocity of the burned gas, and of the shape or height of the flame cone. Often this must be done precisely, and with no changes in orifices or adjustments in the feed system.

If the substitute fuel is of the same general type, eg, propane for methane, the problem reduces to control of the primary equivalence ratio. For nonaspiring burners, ie, those in which the air and fuel supplies are essentially independent, it is further reduced to control of the fuel flow, since the air flow usually constitutes most of the mass flow and this is fixed. For a given fuel supply pressure and fixed flow resistance of the feed system, the volume flow rate of the fuel is inversely proportional to $\sqrt{\rho_f}$. The same total heat input rate or enthalpy flow to the flame simply requires satisfactory reproduction of the product of the lower heating value of the fuel and its flow rate, so that $WI \equiv Q_p/\sqrt{\rho_f}$ remains the same. WI is the Wobbe Index of the fuel gas, and is a commonly used criterion for interchangeability in adjusting the composition of a substitute fuel. The units of WI are variously given, but, if used consistently, are unimportant since only ratios of the Wobbe Indices are ordinarily of interest. Sometimes ρ_f is taken as the specific gravity relative to some reference gas, eg, air, or average molecular weights may be used.

The Wobbe Index criterion also applies to substitution with aspirating or atmospheric pressure burners in which the volume flow of primary air induced by momentum exchange with the fuel increases with $\sqrt{\rho_f}$. Because the volumetric air requirement for a given ϕ is nearly proportional to the heating value of the fuel, an adjustment of Q_p and ρ_f to the same WI results in about the same stoichiometry of the primary flame. For example, if propane is an available substitute for methane or natural gas, it is common practice to prepare a mixture of approximately 60% propane – 40% air (which of course is well above the upper flammability limit) to use as the fuel supply; though the heating value of the mixture is $1.53/\sqrt{2.36} = 1.0$ so that its Wobbe Index is the same. Though there would be slight differences in the stoichiometry of the flame, arising from the air mixed with the propane, and in S_u of the final mixture, substitution with the same supply conditions would be quite satisfactory. On the other hand, if a mixture of the same heating value as that of methane were used (39.2% propane in air) at the same supply pressure, the flame would be much leaner and generally unsatisfactory (84,85).

There are direct substitutions of possible interest that would not be feasible without drastic changes in the feed system or pressure. Thus if the available substitute for natural gas is, eg, a manufactured gas containing much CO, there would almost always be a mismatch of the WIs unless the fuel could be further modified by mixing with some other gaseous fuel of high volumetric heating value (propane, butane, vaporized fuel oil, etc). Moreover, if there are substantial differences in S_u, eg, as a result of the presence of considerable H_2 as well as CO in the substitute gas, the variation in flame height and flashback tendency can also make the substitution unsatisfactory for some purposes, even if the WI is reproduced. Refinements and additional criteria are occasionally applied to measure these and other effects in more complex substitution problems (10,85).

Turbulence in the flow of a premixture flattens the velocity profile and increases the effective burning velocity of the mixture; eg, at a pipe-Reynolds number of 40,000 the turbulent burning velocity is several times the laminar burning velocity and it can be perhaps fifty times larger at very high Reynolds numbers. A turbulent flame is always somewhat noisy, the apparent flame surface becomes diffuse owing to the fluctuations in the actual or flame surface about its average position, and its stability tends to be less predictable. The instantaneous flame surface may be thought of as wrinkled by velocity variations in turbulent flow, or by the average distribution over a greater thickness (or time). Although the resulting enhancement of the mixture consumption rate may be considerable, turbulence is often considered undesirable in Bunsen-type flames. For this and other reasons, a large number of burner ports of small characteristic dimension, rather than a single large port, are frequently used to assure laminar flow to the individual flames. However, turbulence has an essential role in facilitating the mixing of fuel, oxidizer, and flame products, and serves an important function in the various types of flame-stabilizers of practical importance.

COMBUSTION OF LIQUID FUELS

There are several important liquid fuels, ranging from volatile fuels for internal combustion engines to heavy hydrocarbon fractions, sold commercially as fuel oils. The technology for the combustion of liquid fuels for spark-ignition and compression-ignition internal combustion engines is not described here. The emphasis here is primarily on the combustion of fuel oils for domestic and industrial applications.

In general, the combustion of a liquid fuel takes place in a series of stages: atomization, vaporization, mixing of the vapor with air, and ignition and maintenance of combustion (flame stabilization). Recent advances have shown the atomization step to be one of the most important stages of liquid fuels combustion. The main purpose of atomization is to increase the surface area to volume ratio of the liquid. This is achieved by producing a fine spray. The finer the atomization spray the greater the subsequent benefits are in terms of mixing, evaporation, and ignition. The function of an atomizer is twofold: atomizing the liquid and matching the momentum of the issuing jet with the aerodynamic flows in the furnace (86–88). Some of the more common designs of atomizers are shown in Figure 9.

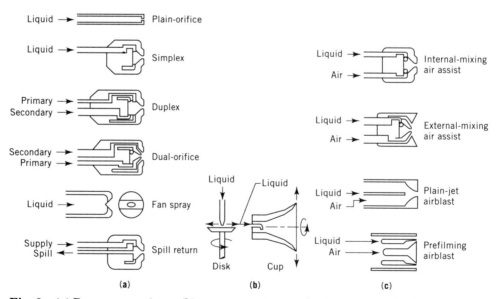

Fig. 9.   (**a**) Pressure atomizers; (**b**) rotary atomizer; and (**c**) twin-fluid atomizers (89,90).

Atomizers for large boiler burners are usually of the swirl pressure jet or internally mixed twin-fluid types, producing hollow conical sprays. Less common are the externally mixed twin-fluid types (89,90).

The principal considerations in selecting an atomizer for a given application are turn-down performance and auxiliary costs. An ideal atomizer would possess all of the following characteristics: ability to provide good atomization over a wide range of liquid flow rates; rapid response to change in liquid flow rate; freedom from flow instabilities; low power requirements; capability for scaling (up or down) to provide desired flexibility; low cost, light weight, easily maintained, and easily removed for servicing; and low susceptibility to damage during manufacture and installation (89,90). There are differences in the structures of the sprays between atomizer types which may affect the rate of mixing of the fuel droplets with the combustion air and, hence, the initial development of a flame.

For distillate fuels of moderate viscosity 30 mm^2/s($=$ cSt), at ordinary temperatures, simple pressure atomization with some type of spray nozzle is most commonly used. Operating typically with fuel pressure of 700–1000 kPa (7–10 atm) such a nozzle produces a distribution of droplet diameters from 10–150 μm. They range in design capacity of 0.5–10 cm^3/s or more and the pumping power dissipated is generally less than 1% of the corresponding heat release rate. A typical domestic oil burner nozzle uses about 0.8 cm^3/s of No. 2 fuel oil at the design pressure. Although pressure-atomizing nozzles are usually equipped with filters, the very small internal passages and orifices of the smallest units tend to be easily plugged, even with clean fuels. With decreasing fuel pressure the atomization becomes progressively less satisfactory. Much higher pressures are often used, especially in engine applications, to produce a higher velocity of the liquid relative to the surrounding air and accordingly smaller droplets and evaporation times. Other mechanical atomization techniques for the production of more nearly monodisperse sprays or smaller average droplet size (spinning disk, ultrasonic atomizers, etc) are sometimes useful in burners for special purposes and may eventually have more general application, especially for small flows.

Conventional spray nozzles are relatively ineffective for atomizing fuels of high viscosity, such as No. 6 or residual fuel oil (Bunker C) and other viscous, dirty fuels. In order to transfer and pump No. 6 fuel oil, it is usually heated to about 373 K, at which temperature its viscosity is about 40 mm^2/s. Relatively large nozzle passages and orifices are necessary to accommodate the possibility of suspended solids. Atomization of such fuel is often accomplished, or at least assisted, by the use of atomizing air, pumped at high velocity through adjacent passages in or around the liquid injection ports. Much of the relative velocity required to shear the liquid and form droplets is thus provided by the atomizing air; its mass flow is usually comparable with the fuel flow and thus a small fraction of the stoichiometric combustion air, although it is sometimes called primary air. In a typical high pressure, air-atomizing nozzle designed for injecting residual oil in a gas turbine combustor, the atomizing air is supplied by an auxiliary compressor with a power usage of about 1% of the combustion heat release rate. Dry steam, if available, may also be used in a similar way, as is common practice in the furnaces of power plant boilers using residual oil (89,90).

Air atomization with low pressure and relatively low velocity air is also used in some burners for low viscosity distillate oils. In most aircraft gas turbines some, or even a large part, of the atomization is done in this way by a small fraction of the warm, compressed combustion air supplied in swirling flow around the fuel nozzles. Imparting swirl to at least some of the air flow around fuel injectors of all types is a common feature of many burners and combustors; in some, swirl is introduced on a larger scale in all of the primary combustion air. The velocity gradients or shear in the resulting vortexlike flow promote mass transfer or mixing, including the recirculation of hot products of combustion to the rich mixture or suspension in the low pressure core that contributes to stabilization of the primary combustion zone (25). The angular or swirl velocity imparted to the air and the strength of such flows are of course limited by the available pressure drop; eg, in gas turbine combustors the allowable pressure loss is usually <4% of the absolute pressure.

Combustion of fuel oil takes place through a series of steps, namely vaporization, devolatilization, ignition, and dissociation, which finally lead to attaining the flame temperature. Vaporization and devolatilization of the fine spray of fuel droplets take place as physiochemical processes in the combustion chamber. The vaporization temperature for fuel oil is in the range of 311–533 K, depending on the grade of the fuel. Devolatilization takes place at about 700 K. The final flame temperature attained is between 1366–1918 K. Complete combustion of an oil droplet can occur in 2–20 ms depending on the size of the droplet. A typical characteristic of an oil flame is its bright luminous nature which is the result of incandescent carbon particles in the fuel rich zone.

The study of the combustion of sprays of liquid fuels can be divided into two primary areas for research purposes: single-droplet combustion mechanisms and the interaction between different droplets in the spray during combustion with regard to droplet size and distribution in space (91–94). The wide variety of atomization methods used and the interaction of various physical parameters have made it difficult to give general expressions for the prediction of droplet size and distribution in sprays. The main fuel parameters affecting the quality of a spray are surface tension, viscosity, and density, with fuel viscosity being by far the most influential parameter (95).

The following general expression (eq. 22) is commonly used to describe the droplet size distribution in a spray:

$$dn = a \; r^\alpha \exp \left(-b \; r^\beta \right) dr \tag{22}$$

where dn = number of droplets with radii between r and $r + dr$. The constants a, b, α, and β are independent of r and are usually determined empirically. The best-known special case of this general equation is the Rosin-Rammler distribution (95–97).

Theoretical modeling of single-droplet combustion has provided expressions for evaporation and burning times of the droplets and the subsequent coke particles. A more thorough treatment of this topic is available (88,91–93,98).

Experimental techniques used for studying the combustion of single droplets can be divided into three groups: suspended droplets, free droplets, and porous droplets, with ongoing research in all three areas (98).

COMBUSTION OF SOLID FUELS

Solid fuels are burned in a variety of systems, some of which are similar to those fired by liquid fuels. In this article the most commonly burned solid fuel, coal, is discussed. The main coal combustion technologies are fixed-bed, eg, stokers, for the largest particles; pulverized-coal for the smallest particles; and fluidized-bed for medium size particles (99,100) (see COAL).

Fixed-Bed Technology. Fixed-bed firing of coal by means of stokers consists of a solid bed of large (2–3 cm) coal particles on moving grates with combustion air passing through the grates and ash removal from the end of the grate. The use of a grate limits the application of this technique to small units, as grates are restricted to a maximum size of about 100 m^2 for structural reasons. Mechanical stokers can be classified into the following groups, based on the method of introducing fuel to the furnace: spreader stokers, underfeed stokers, water-cooled vibrating-grate stokers, and chain-grate and traveling-grate stokers. For a thorough review of stokers see Reference 81.

Pulverized-Coal Firing. This is the most common technology used for coal combustion in utility applications because of the flexibility to use a range of coal types in a range of furnace sizes. Nevertheless, the selection of crushing, combustion, and gas-cleanup equipment remains coal dependent (54,100,101).

Prior to being fed to a pulverized fuel burner, coal is ground to a size generally specified such that at least 70% passes a 200 mesh screen (75 µm) and less than 2% is retained on a 52 mesh screen (300 µm). The top size is determined by the classifying component of the crushing mill, oversize material being retained for further grinding (54,100,101).

Suspensions of pulverized coal or coal dust in air can be explosive, hence, it is essential to have adequate guidelines and procedures to ensure safe and stable operation during pulverized-coal (PC) firing. Some of these guidelines include: (*1*) coal dust should never be allowed to accumulate except in specified storage facilities; (*2*) coal and dust suspensions in air should not exist except in drying, pulverizing, or burner equipment and the necessary transportation ducts; (*3*) the furnace and its setting should be purged before introducing any light or spark; (*4*) before introducing the fuel into the furnace, a lighted torch or spark-producing device should be in operation; (*5*) a positive flow of secondary air should be maintained through the burners into the furnace and up the stack; and (*6*) a positive flow of primary air-coal to the burner should be maintained (81,82).

As for oil and gas, the burner is the principal device required to successfully fire pulverized coal. The two primary types of pulverized-coal burners are circular concentric and vertical jet-nozzle array burners. Circular concentric burners are the most modern and employ swirl flow to promote mixing and to improve flame stability. Circular burners can be single or dual register. The latter type was designed and developed for NO$_x$ reduction. Either one of these burner types can be equipped to fire any combination of the three principal fuels, ie, coal, oil and gas. However, firing pulverized coal with oil in the same burner should be restricted to short emergency periods because of possible coke formation on the pulverized-coal element (71,72).

The self-igniting characteristics of pulverized coal vary from one coal to another, but for most coals it is possible to maintain ignition without auxiliary fuel when firing above the capacity of the boiler. The igniters may have to be activated in the following cases: (*1*) when firing pulverized coal with volatile matter less than about 25%, (*2*) when firing excessively wet coal, and (*3*) when feeding coal sporadically into the pulverizer (81,102–104).

Compared to natural gas and oil, complete combustion of coal requires higher levels of excess air, about 15% as measured at the furnace outlet at high loads, and this also serves to avoid slagging and fouling of the heat absorption equipment.

The process of coal ignition in the flame involves a number of steps. Initially, the pulverized coal is heated by convection as the flame jet entrains and mixes with the furnace gases and also by radiation from the hotter furnace gases. On heating to temperatures above about 773 K, the coal starts to decompose, and evolves a mixture of combustible gases such as CO, H_2, and hydrocarbons (C_nH_m) as well as noncombustible gases such as CO_2 and H_2O. At temperatures of about 1173 K most of the volatile matter has been evolved and, given adequate mixing of air in the jet, its combustion will sustain the ignition of the flame. The char residue remaining after the devolatilization is then burned relatively slowly in the flame and furnace. Char combustion has been the subject of intensive investigation since the early 1930s and is one of the least understood areas in coal combustion. Good reviews on char combustion are available (49,105,106). For more information on the industrial applications of coal, References 107–110 are recommended and for a thorough review of coal devolatilization see References 111–114 (see COAL CONVERSION PROCESSES).

As pulverized-coal combustion potentially has a significant impact on the environment, the 1980s saw the employment of techniques such as coal washing and beneficiation to reduce the emissions of fly-ash, SO_x, and water-soluble metallic oxides. Fly-ash emissions can be reduced by means of electrostatic precipitators and fabric fillers, with efficiencies higher than 99.8%. SO_x emissions are reduced considerably by means of gas scrubbing which employs water slurries of lime and limestone. Staged combustion is an effective method of reducing NO_x emissions from coal combustion. In this method, combustion takes place in two zones: a low temperature fuel-rich zone and a high temperature fuel-lean zone, where hydrocarbons and CO are afterburned. In the fuel-rich zone the fuel bound nitrogen is converted to N_2 rather than to NO_x as in the case at high values of ϕ.

The environmental impact associated with pulverized-coal firing has given rise to efforts to develop other combustion technologies such as fluidized beds or the use of coal-water slurry fuels (CWSF), which can be burned as substitutes for certain liquid fuels (115–117). CWSFs were developed as alternatives to more expensive and increasingly scarce conventional hydrocarbon fuels. In their most common form, they are composed of 70–75% by weight of coal (usually high volatile A bituminous), 24–29% water, and 1% chemical additives. The principal potential market for CWSF is as a replacement for residual fuel oil, ie, heavy fuel oil and No. 6 fuel oil, in utility and large industrial boilers. CWSFs offer all the advantages of liquid fuels in addition to the cost advantages associated with the use of coal. The main challenge in the utilization of CWSFs is obtaining stable mixtures that can be successfully atomized and burned. Much research has been carried out in the area of CWSF atomization but this phenomenon is far from being well understood. On the combustion front, novel techniques such as the coupling of a high intensity acoustic field have been employed to enhance the convective processes occurring during the combustion of CWSF (116,117).

Fluidized-Bed Technology.   In fluidized-bed combustion of coal, air is fed into the bed at a sufficiently high velocity to levitate the particles. This velocity is referred to as the minimum fluidizing velocity, u_{mf}. At this velocity, the volume occupied by the bed increases abruptly and the bed exhibits some of the characteristics of a fluid. The two predominant designs of fluidized beds are bubbling and recirculating, with most theories of fluidization being based on the simpler bubbling bed concept.

Fluidized combustion of coal entails the burning of coal particles in a hot fluidized bed of noncombustible particles, usually a mixture of ash and limestone. Once the coal is fed into the bed it is rapidly dispersed throughout the bed as it burns. The bed temperature is controlled by means of heat exchanger tubes. Elutriation is responsible for the removal of the smallest solid particles and the larger solid particles are removed through bed drain pipes. To increase combustion efficiency the particles elutriated from the bed are collected in a cyclone and are either re-injected into the main bed or burned in a separate bed operated at lower fluidizing velocity and higher temperature.

Fluidized beds are ideal for the combustion of high sulfur coals since the sulfur dioxide produced by combustion reacts with the introduced calcined limestone to produce calcium sulfate. The chemistry involved can be simplified and reduced to two steps, calcination and sulfation.

Calcination

$$CaCO_3 \longrightarrow CO_2 + CaO$$

Sulfation

$$SO_2 + CaO + \tfrac{1}{2} O_2 \longrightarrow CaSO_4$$

The main steps associated with coal combustion (heating, devolatilization, volatiles combustion, and char burnout), occur sequentially to some extent; however, there is always some overlap between the stages. Char burnout is the slowest step so there is practical interest in determining the factors that influence its rate. In order to determine the char combustion rate and time, it is necessary to understand the interaction between the rate of oxygen diffusion to the reacting surface and the inherent chemical kinetics of char oxidation. In the case of fluidized beds the use of a simplified rate coefficient overestimates the burnout time substantially. The enhancement of mass transfer through the boundary layer as the result of an applied velocity must be considered in order to predict char combustion times under conditions relevant to fluidized-bed combustion. Char combustion in fluidized beds is believed to be controlled by both diffusional and chemical kinetic parameters, ie, mixed control. This indicates that models attempting to predict char burnout times in fluidized beds must consider both oxygen diffusion rates and inherent chemical kinetics (117–120).

The main stages of coal combustion have different characteristic times in fluidized beds than in pulverized coal combustion. Approximate times are a few seconds for coal devolatilization, a few minutes for char burnout, several minutes for the calcination of limestone, and a few hours for the reaction of the calcined limestone with SO_2. Hence, the carbon content of the bed is very low (up to 1% by weight) and the bed is 90% CaO in various stages of reaction to $CaSO_4$. About 10% of the bed's weight is made up of coal ash (91). This distribution of 90/10 limestone/coal ash is not a fixed ratio and is dependent on the ash content of the coal and its sulfur content.

Devolatilization and combustion occur close to the coal inlet tubes. However, because of rapid mixing in the bed the composition of the solids in the bed may be assumed to be uniform.

Atmospheric Pressure Fluidized-Bed Boilers. A typical bubbling fluidized-bed is usually 1.2 m deep in its expanded or fluidized condition. Normally, the heat-transfer surface is placed in the bed in the form of a tube handle to achieve the desired heat balance and bed operating temperature. For fuels with low heating values the amount of surface can be minimal or absent. Coal-fired bubbling-bed boilers normally incorporate a recycle system that separates the solids leaving the economizer from the gas and recycles them to the bed. This maximizes combustion efficiency and sulfur capture. Normally, the amount of solids reacted is limited to about 25% of the combustion gas weight. For highly reactive fuels this recycle system can be omitted. Bubbling-beds that burn coal usually operate in the range of 2.4 to 3 m/s superficial flue gas velocity at maximum load. The bed material size is 590 μm and coarser, with a mean size of about 1000–1200 μm.

Circulating fluidized-beds do not contain any in-bed tube bundle heating surface. The furnace enclosure and internal division wall-type surfaces provide the required heat removal. This is possible because of the large quantity of solids that are recycled internally and externally around the furnace. The bed temperature remains uniform, because the mass flow rate of the recycled solids is many times the mass flow rate of the combustion gas. Operating temperatures for circulating beds are in the range of 816 to 871°C. Superficial gas velocities in some commercially available beds are about 6 m/s at full loads. The size of the solids in the bed is usually smaller than 590 μm, with the mean particle size in the 150—200 μm range (81).

Some of the advantages of fluidized beds include flexibility in fuel use, easy removal of SO_2, reduced NO_x production due to relatively low combustion temperatures, simplified operation due to reduced slagging, and finally lower costs in meeting environmental regulations compared to the conventional coal burning technologies. Consequently, fluidized-bed combustors are currently under intensive development and industrial size units (up to 150 MW) are commercially available (Fig. 10).

The modeling of fluidized beds remains a difficult problem since the usual assumptions made for the heat and mass transfer processes in coal combustion in stagnant air are no longer valid. Furthermore, the prediction of bubble behavior, generation, growth, coalescence, stability, and interaction with heat exchange tubes, as well as attrition and elutriation of particles, are not well understood and much more research needs to be done. Good reviews on various aspects of fluidized-bed combustion appear in References 121 and 122 (Table 2).

DESIGN CONSIDERATIONS IN FOSSIL FUEL COMBUSTION SYSTEMS

One of the most important considerations in the design of a combustion chamber for a boiler is the fuel that is to be burned in the chamber (see FURNACES, FUEL-

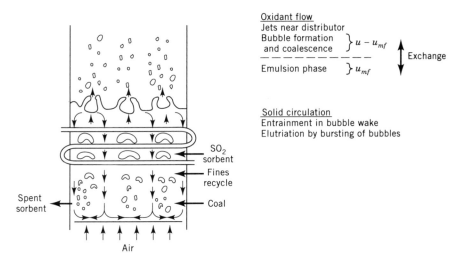

Fig. 10. The main processes taking place in a fluidized bed (92). Heat transfer to immersed tubes is 30% by radiation.

Table 2. Operating Conditions for an Atmospheric Pressure Fluidized-Bed Combustor[a]

Process	Representative value	Comments
excess air, %	≈ 30	values selected to maintain CO emissions at acceptable levels
bed height, m	≈ 1.5	trade-off between pressure drop and gas residence time
bed temperatures, K	≈ 1100	higher values favor higher combustion efficiencies, higher rates of NO reduction by char; high values reduce SO_2 capture
calcium/sulfur (stoichiometric ratio)	≈ 2.5	value may be reduced by use of more active stone, lower gas velocities, deeper beds
gas velocity, m/s	≈ 2	high values are favored by high energy release rates per unit plan area; lower values are favored by higher efficiencies of combustion and SO_2 capture
sorbent particle top size, mm	≈ 3	low values are favored by shorter reaction times; size determined by consideration of elutriation at operating velocities
coal particle top size, mm	≈ 3	trade-off between elutriation rate and reaction time
sorbent residence time, s	≈ 5 × 10^4	determined by stone reactivity and Ca/S ratio
gas residence time in bed, s	≈ ¼	an additional residence time of 1–2 s is available in the free board
coal particle burning time, s	≈ 400	dependent on temperature, size, and excess air
solid circulation time, s	≈ 2	short relative to solid reaction time

[a]Ref. 122

FIRED). Although all fuels burn and release heat during combustion, the rate at which a fuel burns and releases heat, and the impurities associated with the fuel have to be considered.

Furnaces for Oil and Natural Gas Firing. Natural gas furnaces are relatively small in size because of the ease of mixing the fuel and the air, hence the relatively rapid combustion of gas. Oil also burns rapidly with a luminous flame. To prevent excessive metal wall temperatures resulting from high radiation rates, oil-fired furnaces are designed slightly larger in size than gas-fired units in order to reduce the heat absorption rates.

Furnaces for Pulverized Coal Firing. The main differences between boilers fired with coal and those fired with oil or natural gas result from the presence of mineral matter in coals. The volume of the coal-fired furnace is higher because of the longer residence time required for the complete combustion of coal particles, the requirement of a controlled combustion rate to reduce NO_x formation, the provision for a larger heat-transfer surface area resulting from decreased heat-transfer rates because of ash deposits on the surfaces, and increased spacing of heat-transfer tubes to reduce flue gas velocities and thereby erosion of heat-transfer surfaces. Even when firing coal, depending on the reactivity of the coal (rank), the size of the combustion chamber required can vary (Fig. 11).

The combustion chamber of a modern steam generator is a large water-cooled chamber in which fuel is burned. Firing densities are important to ensure

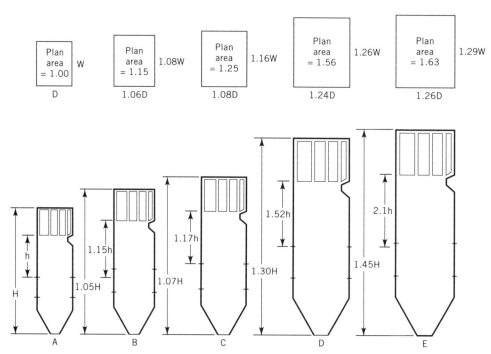

Fig. 11. Effect of coal rank on furnace sizing (constant heat output) (82), where W = width, D = depth, and h and H are the heights indicated. A represents medium volatile bituminous; B, high volatile bituminous or subbituminous; C, low sodium lignite; D, medium sodium lignite; and E, high sodium lignite.

that the chamber wall metal temperatures do not exceed the limits of failure of the tubes. Firing densities are expressed in two ways: volumetric combustion intensities and area firing intensities. The volumetric combustion intensity is defined by equation 23,

$$I_v = \frac{J_f h_f}{V_c P} \qquad (23)$$

where I_v = volumetric combustion intensity, kJ/(m^3·h); J_h = fuel feed rate, kg/h; h_f = heating value of the fuel, kJ/kg; and V_c = volume of the combustion chamber, m^3.

The area firing density is defined by equation 24,

$$I_a = \frac{J_f h_f}{A_c} \qquad (24)$$

where, I_a = area firing density, kJ/(m^2·h); J_h = fuel feed rate, kg/h; h_f = heating value of the fuel, kJ/kg; and A_c = cross-section of a fluid-bed distributor plate or grate area of a mechanical stoker or plan area in pulverized-coal combustors. Table 3 provides some design parameters for fossil fuel burners.

ENVIRONMENTAL CONSIDERATIONS

Atmospheric pollutants released by combustion of fossil fuels fall into two main categories: those emitted directly into the atmosphere as a result of combustion

Table 3. Comparison of Design Parameters for Fossil Fuel Boilers

			Coal		
Parameter	Gas	Oil	Grate	Fluid bed	Pulverized coal
heat rate, mW (t)	0.03–3000	0.03–3000	0.3–30	up to 30	30–3000
volumetric combustion intensity, kW/m^3	250–450	250–450	250–750^a	up to 2000^a (based on bed volume)	150–250
area combustion intensity, kW/m^2	280–500	280–500	2000	3000	7500
fuel firing density,					
kg/m^{3}h			30–100	≈250	15–30
kg/m^{2}h	6–11	6–11	40–250	up to 500	up to 1000
practical combustion temperature, °C	1000–1600	1100–1700	1200–1300	850–950	1600–1700
combustion time, s	10 × 10^{-1}	20–25 × 10^{-1}	up to 5000	100–500	≈1–2
particle heating rate, °C/s			<1	10^3–10^4	10^4–10^5

^{a}Based on the total combustion volume which includes space between the bed and the convective tubes.

and the secondary pollutants that arise from the chemical and photochemical reactions of the primary pollutants (see AIR POLLUTION).

The main combustion pollutants are nitrogen oxides, sulfur oxides, carbon monoxide, unburned hydrocarbons, and soot. Combustion pollutants can be reduced by three main methods depending on the location of their application: before, after, or during the combustion. Techniques employed before and after combustion deal with the fuel or the burned gases. A third alternative is to modify the combustion process in order to minimize the emissions.

Nitrogen Oxides. From the combustion of fuels containing only C, H, and O, the usual air pollutants or emissions of interest are carbon monoxide, unburned hydrocarbons, and oxides of nitrogen (NO_x). The interaction of the last two in the atmosphere produces photochemical smog. NO_x, the sum of NO and NO_2, is formed almost entirely as NO in the products of flames; typically 5 or 10% of it is subsequently converted to NO_2 at low temperatures. Occasionally, conditions in a combustion system may lead to a much larger fraction of NO_2 and the undesirable visibility thereof, ie, a very large exhaust plume.

NO is formed to some extent from N_2 and O_2 in flame products when N atoms are produced at a significant rate. Above 1700 K, the important step in the much studied Zeldovitch (thermal) mechanism is the production of N atoms by:

$$O + N_2 \longrightarrow NO + N$$

This is followed by a very fast reaction:

$$N + O_2 \longrightarrow NO + O$$

When $[NO] \ll [NO]_{eq}$, as is usually true in practice, its formation is essentially irreversible, and its rate is proportional to $[NO][N_2]$ with a large temperature dependence, an activation energy of 316 kJ/mol (75.5 kcal/mol). Unfortunately, the rate becomes appreciable just in the range of typical hydrocarbon–air flame conditions. If it is also assumed that $[O] = [O]_{eq}$, the observed rate in most lean-flame products in which N_2 is roughly 75 mol % of the gas can be approximated by

$$\frac{d[NO]}{dt} = \left(\frac{3.3 \times 10^{18}}{T}\right) \exp\left(\frac{-68700}{T}\right) (x_{O_2})^{1/2}$$

Here, x_{O_2} is the mole fraction of O_2 in the products at temperature T, and the rate is given in ppm/ms. The exponential implies a large effective activation energy of 570 kJ/mol, the sum of that for the O–N_2 reaction and half the dissociation energy of O_2. In typical hydrocarbon–air flames, the rate of NO formation by the thermal mechanism can be shown to be about 8 ppm/ms, or in a 10 ms residence time the thermal NO would be about 80 ppm. If preheating the mixture were to raise the gas temperature by 100 K, the rate of NO production would be nearly tripled, making the NO concentration unacceptable. Conversely, the rate can be reduced by the same amount by a 100 K reduction in temperature by precooling or heat abstraction from the flame itself, or by dilution of the mixture with excess air, steam, or other inert gas such as recirculated, relatively cool exhaust gas. Control of thermal NO_x thus involves reduction of the maximum attainable temperature, or the residence time at high temperature, or both. Such measures, however, always entail some compromise in stability and control, and possibly also in the efficiency of the combustion process. The afterburning of CO tends to be quenched by rapid temperature reduction, and the resulting increase in the emission of CO must be balanced against the desired NO_x reduction. Heat abstraction or cooling of the flame always occurs to some extent by radiation from the highly luminous flames produced by pulverized-coal or oil combustion, as is typical in boilers and similar furnaces. When heat is rejected, eg, by a boiler fluid, and not returned or recuperated to the unburned mixture, the maximum temperature and thermal NO_x formation will be reduced. An extension of this effect has been applied to achieve low NO_x emissions in some furnaces and boilers in which combustion occurs in a very rich, relatively low temperature primary stage, followed by heat abstraction by convection as well as radiation to reduce the gas enthalpy (two-stage combustion). Secondary products and any excess air are then introduced to complete the combustion and, owing to the previous heat transfer, the maximum temperature attainable in that stage will never approach the adiabatic flame temperature. Much soot, which is responsible for the radiative heat loss, may be present in the rich primary flame products. To avoid smoke from such two-stage processes, care must be taken to assure its oxidation in the second stage. In practice $[O]/[O]_{eq}$ is seldom unity as assumed. Though $[O]$ is decreased in the burned gas, its average value may be several times $[O]_{eq}$ and NO formation may be correspondingly higher than predicted from the $[O]_{eq}$. Similarly, the very high radical concentration, eg, $[O]$, in the reaction zone of a flame often leads to almost instantaneous NO production, even though the temperature is still relatively low and the residence time is relatively short. Other fast reactions involving transient flame species producing N atoms, for example, $CH + N_2 \longrightarrow HCN + N$, can also contribute to the production of some NO. In any case, the NO inevitably formed by these species is called prompt NO. The total concentration of prompt NO is usually not large, 10–50 ppm depending on the composition of the flame, but is significant if very low NO_x emission levels are sought.

A different and often more serious source of NO_x is chemically bound nitrogen in the fuel, eg, NH_3, amines, nitrites, pyridine, a fraction of which is always converted to NO. Most coals contain at least 1% N, of which 50% or more is retained in gaseous or liquid fuels derived from coal. The N content of distillate and gaseous petroleum fuels from most sources is usually very low, but it can reach 0.5% and if such a fuel were burned with air under stoichiometric or lean conditions, the conversion of its fuel N to NO in the flame would yield up to 400 ppm of NO in the burned gas. NO_x production levels from fuel N and NO_x control measures are now well established from correlations of data from flames and combustors. Formation occurs in the flame reaction zone by OH radical oxidation of intermediate species formed by the decomposition of the fuel N, eg, NH_2, NH, N, which if unoxidized would in a short time simply form stable N_2. Whether NO or N_2 formation prevails depends on the flame conditions as well as on the concentrations of the intermediates. At high levels of fuel N, NO can also be converted directly to N_2. In general, the yield (mol of NO/mol of fuel N) is much higher in lean and stoichiometric flames than in rich flames, and asymptotically approaches unity at low fuel N concentrations. It decreases with increasing fuel N, and at some level that depends on ϕ, the NO concentration becomes constant, albeit rather high, and any additional fuel N is converted rapidly to N_2, from which NO can then be formed only by the relatively slow thermal process with oxygen.

Thus if combustion can be effected in two stages, with or without the intermediate heat rejection for thermal NO_x control discussed above, the conversion of fuel N to NO can be largely circumvented by first, a primary stage at $\phi = 1.5-2$ with a modest residence time to allow formation of N_2 in the hot primary products, followed by rapid addition of secondary air to complete the combustion at an effective $\phi \approx 0.8$. There will of course be a maximum in the temperature near $\phi = 1$ in the course of the secondary air addition, but if the residence time at that condition is minimized, the production of thermal NO will also be minimized.

Combustion system developments for reducing NO_x formation include: low NO_x burners, staged burning techniques, and flue gas recirculation (FGR). Some postcombustion techniques for reducing NO_x include: selective noncatalytic reduction (SNCR) and selective catalytic reduction (SCR). In either technology, NO_x is reduced to nitrogen (N_2) and water (H_2O) through a series of reactions with a chemical agent injected into the flue gas. The most common chemical agents used commercially are ammonia and urea for SNCR and ammonia for SCR systems. Most ammonia-based systems have used anhydrous ammonia (NH_3) as the reducing agent. However, because of the hazards of storing and handling NH_3, many systems use aqueous ammonia at 25–28% concentration. Urea can be stored as a solid or mixed with water and stored in solution (81).

These ideas form the basis of most approaches to NO_x control with N-containing fuels. In principal, they are readily applicable to the modification of certain combustors in which the desired divisions in the combustion process exist for other reasons. Although such improvements have been demonstrated, it is difficult in practice to make the required revisions in the air and fuel distribution without adverse effects on other emissions or on performance. It has also been shown that when steam is used to reduce thermal NO_x production, the formation of NO_x from fuel N is enhanced, or the reduction is less than otherwise expected.

Sulfur Oxides. Oxides of sulfur are also pollutants of concern. When present in the fuel as inorganic sulfides or organic compounds, sulfur is converted almost completely to SO_2 in the products of complete combustion. There are no known techniques for the elimination of this conversion process in flames, and emission control measures necessarily involve either desulfurization of the fuel or removal of the SO_2. Up to 10% of the SO_2 is oxidized to SO_3 at low temperatures in most combustion processes, and the total sulfur oxides emission is often given as the sum of the concentrations of SO_2 and SO_3, or SO_x. The presence of H_2O in the combustion products is inevitable and at low temperatures SO_3 combines with H_2O to form H_2SO_4, which is both a highly corrosive agent to heat exchange surfaces and a highly undesirable stack emission.

Soot. Emitted smoke from clean (ash-free) fuels consists of unoxidized and aggregated particles of soot, sometimes referred to as carbon though it is actually a hydrocarbon. Typically, the particles are of submicrometer size and are initially formed by pyrolysis or partial oxidation of hydrocarbons in very rich but hot regions of hydrocarbon flames; conditions that cause smoke will usually also tend to produce unburned hydrocarbons with their potential contribution to smog formation. Both may be objectionable, though for different reasons, at concentrations equivalent to only 0.01–0.1% of the initial fuel. Although their effect on combustion efficiency would be negligible at these levels, it is nevertheless important to reduce such emissions.

Neither soot nor unburned hydrocarbons are found in the products of a lean or stoichiometric premixed flame with air or O_2, although hydrocarbons may be formed or survive unburned in lean flames partially quenched, eg, by the cold wall of a combustion chamber. A moderately rich flame also should yield only the water-gas equilibrium products (and N_2); but with increasing equivalence ratio, at some ϕ still well below the upper flammability limit, the burned gas becomes faintly luminous with precipitated soot particles that increase in number density and luminous intensity with further increase in ϕ. The appearance of the condensed phase (soot) is connected with appreciable nonequilibrium concentrations of fairly stable low molecular-weight hydrocarbons, notably acetylene, from the decomposition of unoxidized fuel. These polymerize (condense with elimination of H_2) to form high molecular-weight products, mostly with ring structures. If these intermediates are not consumed, eg, by OH, in simultaneous and competing oxidation reactions, they grow until nucleation, and eventually precipitation occurs as visibly radiating soot, typically over a period of several milliseconds after the main flame reaction zone. The particle growth and competing oxidation then continue in the burned gas.

The composition, properties, and size of soot particles collected from flame products vary considerably with flame conditions and growth time. Typically the C–H atomic ratio ranges from two to five and the particles consist of irregular chains or clusters of tiny spheres 10–40 nm in diameter with overall dimensions of perhaps 200 nm, although some may agglomerate further to much larger sizes.

Whether soot particles form at all and grow depends on ϕ, the fuel type, and other variables, eg, the growth is easier and faster at a given ϕ with fuels of higher C–H ratio and at elevated pressure. Given sufficient residence time to attain equilibrium at the burned gas condition, the soot and hydrocarbons would eventually be consumed. In practice, their rapid oxidation occurs in a secondary flame in which the hot primary products are burned with the required excess air, which is added by diffusion or by more intensive mixing. However, if a large excess of air is added too rapidly, the cooling can, in effect, quench the oxidation of both unburned hydrocarbons and the accompanying soot, which would then persist as visible smoke. The blackness of the smoke depends on the size and number density of the particles when quenched, their further aggregation, etc. Some may also survive as much smaller invisible particles, or condensation nuclei. For thorough reviews on the mechanisms of pollutant formation see References 123–125.

Diffusion Flame Chemistry. Since most combustion systems employ mixing-controlled diffusion flames, which are characterized by very high pollutant emissions, it is imperative to look into the chemistry occurring in diffusion flames. In a typical diffusion flame the mixture composition in the reaction zone is close to the stoichiometric proportion and the temperature is at a maximum resulting from the large volume of this zone, thus NO_x production is favored. If, however, the surrounding gas cools the combustion products rapidly, further reactions of CO and NO are eliminated. This fixes the concentrations of these pollutants at unfavorable levels. Furthermore, the fuel diffuses into the combustion zone through the burned gases and thus is heated in the absence of oxygen. This creates ideal conditions for the formation of soot and the reduction of the CO_2 produced in the combustion zone to CO. Additionally, diffusion flames have low combustion intensity and efficiency and hence release large amounts of unburned hydrocarbon emissions. In general, despite the fact that the structure of the diffusion flame is more complex and difficult to analyze, the same basic description of soot formation and oxidation should apply to diffusion flames as for premixed flames.

Emissions Control. From the combustion chemistry standpoint, lean mixtures produce the least amount of emissions. Hence, one pollution prevention alternative would be to use lean premixed flames. However, lean mixtures are difficult to ignite and form unstable flames. Furthermore, their combustion rates are very low and can seldom be applied directly without additional measures being taken. Consequently the use of lean mixtures is not practical.

Another potential solution is the use of catalytic combustors, which produce extremely low levels of emissions by the use of combustion catalysts such as platinum. The main disadvantage of catalytic combustors, however, is their high cost.

More advanced techniques for emissions control include electrical or plasma jet augmentation of flames based on radical production. Since in two-phase, heterogeneous combustion the flames are always diffusion flames on the microscale, ie, the individual droplets or particles burn as diffusion flames, and since at the characteristic times for evaporation, decomposition and burning of individual particles can be comparable with the characteristic times for mixing and pollutant formation, prevaporization or gasification of the fuel can reduce pollutant emissions. For this reason catalytic systems for liquid-fuel decomposition and coal gasification are being considered seriously as alternatives to conventional combustion technology (126–128).

BIBLIOGRAPHY

"Fuels (Combustion Calculation)," in *ECT* 1st ed., Vol. 6, pp. 913–935, by H. R. Linden, Institute of Gas Technology; in *ECT* 2nd ed., Vol. 10, pp. 191–220, by D. M. Himmelblau, The University of Texas; "Burner Technology," in *ECT* 3rd ed., Vol. 4, pp. 278–312, by G. E. Moore, Consultant.

1. N. Chigier, "Energy," *Combustion and Environment*, McGraw-Hill, Inc., New York, 1981.
2. W. Bartok and A. F. Sarofim, *Fossil Fuel Combustion: A Source Book*, John Wiley & Sons, Inc., New York, 1991.
3. J. A. Barnard and J. N. Bradley, *Flame and Combustion*, 2nd ed., Chapman and Hall, London and New York, 1985.
4. W. E. Baker, P. A. Cox, P. E. Westine, J. Kulesz, and R. A. Strehlow, *Explosion Hazards and Evaluation*, Elsevier, Amsterdam, The Netherlands, 1983.
5. R. A. Strehlow, *Combustion Fundamentals*, McGraw-Hill, Inc., New York, 1984.
6. D. J. McCracken, *Hydrocarbon Combustion and Physical Properties*, Rep. No. 1496, Ballistic Research Laboratories, Sept. 1970.
7. M. G. Zabetakis, *Flammability Characteristics of Combustible Gases and Vapors*, Technical Bulletin, U.S. Bureau of Mines, Washington, D.C., 1965, p. 627.
8. I. Glassman, *Combustion*, Academic Press, Inc., New York, 1987.
9. J. Chomiak, *Combustion; A Study in Theory, Fact and Application*, Gordon and Breach Science Publishers, Montreux, Switzerland, 1990.
10. B. Lewis and G. von Elbe, *Combustion, Flames and Explosion of Gases*, 2nd ed., Academic Press, Inc., New York, 1951.
11. F. A. Williams, *Combustion Theory*, The Benjamin/Cummings Publishing Co. Inc., Menlo Park, Calif., 1985.

12. K. K. Kuo, *Principles of Combustion*, John Wiley & Sons, Inc., New York, 1986.
13. R. C. Flagen and J. H. Seinfeld, *Fundamentals of Air Pollution Engineering*, Prentice-Hall, Inc., Englewood Cliffs, N.J., 1988.
14. H. B. Palmer and J. M. Beer, eds., *Combustion Technology: Some Modern Developments*, Academic Press, New York and London, 1974.
15. M. V. Blank, P. G. Guest, G. von Elbe, and B. Lewis, *Third Symposium, Combustion, Flame and Explosion Phenomena*, Williams & Wilkens, Baltimore, Md., 1949.
16. H. C. Barnett and R. R. Hibbard, *Basic Considerations in the Combustion of Hydrocarbon Fuels with Air*, NASA Technical Report, 1959, p. 1300.
17. H. F. Colcote, C. A. Gregory, C. M. Barnett, and R. B. Gilmer, *Ind. Eng. Chem.* **44,** 2656 (1952).
18. Ya. B. Zeldovich, G. I. Brenblatt, V. B. Librovich, and G. M. Mackhviladze, *The Mathematical Theory of Combustion and Explosions*, Consultants Bureau, 1985.
19. V. R. Kuznetsov and V. A. Sabel'nikov, *Turbulence and Combustion*, Hemisphere Publishing Corp., Washington, D.C., 1990.
20. M. Y. Hussaini, A. Kumar, and R. G. Voigt, *Major Research Topics in Combustion*, Springer-Verlag, Inc., New York, 1992.
21. C.-M. Brauner and C. Schmidt-Laine, *Mathematical Modeling in Combustion and Related Topics*, Martinus Hijhoff Publishers, Dordrecht, The Netherlands, 1988.
22. R. M. C. So, H. C. Mongia, and J. H. Whitelaw, *Turbulent Reactive Flow Calculations*, special issue of *Combustion Science and Technology*, Gordon and Breach Science Publishers, Inc., Montreux, Switzerland, 1988.
23. F. A. Williams, in W. Bartok and A. F. Sarofim, eds., *Turbulent Reacting Flows In Fossil Fuel Combustion: A Source Book*, John Wiley & Sons, Inc., New York, 1991.
24. F. A. Williams, in J. H. S. Lee and C. M. Cuirao, eds., *Laminar Flame Instability and Turbulent Flame Propagation, In Fuel-Air Explosions*, University of Waterloo Press, Waterloo, Ontario, Canada, 1982.
25. J. M. Beer and N. A. Chigier, *Combustion Aerodynamics*, Applied Science, London; John Wiley & Sons, Inc., New York, 1972.
26. L. D. Landau, *Acta. Phys. Chim. (USSR)* **19,** 77 (1944).
27. L. D. Landau and E. M. Lifschitz, *Mechanics of Continuous Media*, Moscow, Russia, 1953; Eng. trans. Addison-Wesley Publishing Co., Inc., Reading, Mass., 1959.
28. A. C. Scurlock and J. H. Grover, *4th International Symposium on Combustion*, The Combustion Institute, Pittsburgh, Pa., 1953, p. 645.
29. K. N. C. Bray, *Equations of Turbulent Combustion*, Report No. 330, University of Southampton, UK, 1973.
30. K. N. C. Bray, *17th International Symposium on Combustion*, The Combustion Institute, Pittsburgh, Pa., 1978, pp. 57–59.
31. K. N. C. Bray and P. A. Libby, *Phys. Fluids* **19,** 1687–1701 (1976).
32. K. N. C. Bray and J. B. Moss, *Comb. Flame* **30,** 125 (1977).
33. H. C. Hottel, in Ref. 28, p. 97–113.
34. S. P. Burke and T. E. W. Schumann, *Ind. Eng. Chem* **20,** 998–1004 (1928).
35. W. Jost, *Explosion and Combustion Processes in Gases*, McGraw-Hill, Inc., New York, 1946.
36. W. Jost, *Diffusion*, Steinkopff, Darmstadt, Germany, 1957.
37. A. G. Gaydon and H. G. Walfhard, *Flames*, 2nd ed., Macmillan, New York, 1960.
38. T. Takeno and Y. Kotani, in L. A. Kennedy, ed., *Turbulent Combustion, Prog. Astronaut. Aeronaut.*, Vol. 58, AIAA, 1978, pp. 19–35.
39. M. Gerstein, *Diffusion Flames*, in Ref. 2.
40. H. C. Hottel, *3rd International Symposium on Combustion*, The Combustion Institute, Pittsburgh, Pa., 1948, pp. 254–266.
41. P. A. Libby and F. A. Williams, eds., *Turbulent Reacting Flows*, Springer-Verlag, Berlin and New York, 1980, pp. 1–43, 219–236.

42. A. M. Mellor, in Ref. 31, pp. 377–387.
43. F. A. Williams, in W. E. Stewart, W. H. Ray, and C. C. Conley, eds., *Current Problems in Combustion Research, Dynamics and Modeling of Reactive Systems*, Academic Press, Inc., New York, 1980.
44. P. A. Libby and F. A. Williams, *AIAA J.* **19,** 261–274 (1981).
45. P. A. Libby and F. A. Williams, *Annu. Rev. Fluid Mech.* **8,** 351–379 (1976).
46. R. W. Bilger, *Energy and Combustion Science*, Student ed. 1, Pergamon Press, Oxford, 1979, pp. 109–131.
47. D. B. Spalding, *Some Fundamentals of Combustion*, Butterworth & Co., Ltd., London, 1955.
48. R. Borghi, *Turbulent Combustion Modeling, Progress in Energy and Combustion Science*, Vol. 14, No. 4, Pergamon Press, Elmsford, N.Y., 1988, pp. 245–292.
49. I. W. Smith, *19th International Symposium on Combustion*, The Combustion Institute, Pittsburgh, Pa., 1982, p. 1045.
50. R. B. Jones, B. B. McCourt, C. Morley, and K. King, *Fuel* **64,** 1460 (1985).
51. N. Oka, T. Murayama, H. Matsuoka, and S. Yamada, *Fuel Process. Tech.* **15,** 213 (1987).
52. M. Shibaoka, *Fuel* **48,** 285 (1969).
53. J. M. Kuchta, V. R. Rowe, and D. S. Burgess, U.S. Bureau of Mines Report 8474, Washington, D.C., 1980, p. 1.
54. R. H. Essenhigh, in *M. A. Elliot, ed.,* Fundamentals of Coal Combustion, In Chemistry of Coal Utilization, 2nd Suppl. Vol., John Wiley & Sons, Inc., New York, 1981.
55. N. N. Semenov, *Chemical Kinetics and Chain Reactions*, Oxford University Press, London, 1935.
56. F. Kamenetski, *Diffusion and Heat Exchange in Chemical Kinetics,* Princeton University Press, Princeton, N.J., 1955; trans. from Russian by Thanel.
57. K. Annamalai and P. Durbetaki, *Comb. Flame* **29,** 193 (1977).
58. K. Annamalai, *Trans. ASME* **101,** 576 (1979).
59. L. D. Smoot, M. D. Horton, and G. A. Williams, *16th International Symposium on Combustion*, The Combustion Institute, Pittsburgh, Pa., 1976, p. 375.
60. R. H. Essenhigh, M. Misra, and D. Shaw, *Comb. Flame* **77,** 3 (1989).
61. H. E. Blayden, W. Noble, and H. L. Riley, *Gas J.* **2,** 81 (1934).
62. H. K. Griffin, J. P. Adams, and D. F. Smith, *Ind. Eng. Chem.* **21,** 808 (1929).
63. G. P. Ivanova and V. Babii, *Thermal Eng.* **13,** 70 (1966).
64. A. A. Orning, *Proceedings of Conference on Pulverized Fuel,* Vol. 1, Institute of Fuel, London, 1947, p. 45.
65. D. L. Carpenter, *Comb. Flame* **1,** 63 (1957).
66. H. M. Cassel and I. Liebman, *Comb. Flame* **3,** 467 (1959).
67. J. B. Howard and R. H. Essenhigh, *Comb. Flame* **9,** 337 (1965).
68. M. A. Nettleton and R. Stirling, *Comb. Flame* **22,** 407 (1974).
69. L. Tognotti, A. Malotti, L. Petarca, and S. Zanelli, *Comb. Flame* **44,** 15 (1985).
70. M. R. Chen, L. S. Fan, and R. H. Essenhigh, *20th International Symposium on Combustion*, The Combustion Institute, Pittsburgh, Pa., 1984, p. 1513.
71. C. P. Gomez, *Ignition of and Combustion of Coal and Char Particles; a Differential Approach*, M.S. dissertation, Pennsylvania State University, University Park, Pa., 1982.
72. A. W. Scaroni, M. R. Khan, S. Eser, and L. R. Radovic, *Ullmann's Encyclopedia of Industrial Chemistry*, Vol. A7, VCH Publishers, New York, 1986, p. 245.
73. J. D. Blackwood and F. K. McTaggart, *Aust. J. Chem.* **12,** 533 (1959).
74. R. G. Jenkins, S. P. Nandi, and P. L. Walker, Jr., *Fuel* **52,** 288 (1973).
75. D. W. Van Krevelen, *Coal Typology–Chemistry–Physics–Constitution,*Elsevier Publications, New York, 1961, p. 219.

76. E. J. Badin, *Coal Combustion Chemistry—Correlation Aspects*, Elsevier, New York, 1984, Chapt. 6, p. 68.
77. D. W. McKee, in P. L. Walker, Jr. and P. A. Thrower, eds., *Chemistry and Physics of Carbon*, Vol. 16, Marcel Dekker, Inc., New York, 1981, p. 1.
78. B. A. Morgan and A. W. Scaroni, *International Conference on Coal Science*, International Energy Agency, Sydney, Australia, 1985, p. 347.
79. P. L. Walker, Jr., M. Shelef, and R. A. Anderson, in Ref. 79, Vol. 4.
80. P. L. Walker, Jr., F. Rusinko, Jr., and L. G. Austin, in D. D. Eley, P. W. Selwood, and P. B. Weisz, eds., *Advances in Catalysis*, Vol. 2, Academic Press, Inc., New York, 1959, p. 133.
81. S. C. Stultz and J. B. Kitto, eds., *Steam, Its Generation and Use*, 40th ed., Babcock and Wilcox Co., Barberton, Ohio, 1992.
82. J. E. Singer, *Combustion-Fossil Power Systems*, Combustion Engineering, Windsor, Conn., 1981.
83. American Gas Association, *Gas Engineers Handbook*, Industrial Press, New York, 1978.
84. S. S. Penner and B. P. Mullins, *Explosions, Detonations, Flammability and Ignition*, AGARD Monograph, Pergamon Press, Inc., New York, 1959.
85. M. W. Thring, *The Science of Flames and Furnaces*, John Wiley & Sons, Inc., New York, 1962.
86. A. Williams, *Comb. Flame* **21,** 1 (1973).
87. A. Williams, *Combustion of Sprays and Liquid Fuels*, Elek Science, London, 1976.
88. A. Williams, *Combustion of Liquid Fuels Sprays*, Butterworths & Co., Ltd., London, 1990.
89. A. H. Lefebvre, *Gas Turbine Combustion*, Hemisphere Publishing Corp., Washington, D.C., 1983.
90. A. H. Lefebvre, *Atomization and Sprays*, Hemisphere Publishing Corp., Washington, D.C., 1989.
91. A. C. Fernandez-Pello and C. K. Law, in Ref. 51, p. 1037.
92. A. L. Randolf, A. Markino, and C. K. Law, *21st International Symposium on Combustion*, The Combustion Institute, Pittsburg, Pa., 1988, p. 601.
93. J. C. Lasheras, L. T. Yap, and F. L. Dryer, in Ref. 72, p. 1761.
94. T. Niioka and J. Sato, in Ref. 94, p. 625.
95. A. H. Levebvre, *Airblast Atomization, Progress in Energy and Combustion Science*, Vol. 6, Pergamon Press, Oxford, UK, 1980, pp. 233–261.
96. N. K. Rizk and A. H. Levebvre, *J. Fluids Eng.* **97**(3), 316–320 (1975).
97. A. Rizkalla and A.H. Levebvre, *AIAA J.* **21**(8), 1139–1142 (1983).
98. C. J. Lawn and co-workers, *The Combustion of Heavy Fuel Oils*, in C. J. Lawn, ed., *The Principles of Combustion Engineering for Boilers*, Academic Press, Inc., New York, 1987.
99. T. F. Wall, *The Combustion of Coal as Pulverized Fuel Through Swirl Burners*, in Ref. 100.
100. M. A. Field, D. W. Gill, B. B. Morgan, and P. J. W. Hawksley, *Combustion of Pulverized Coal*, BCURA, UK, 1967.
101. H. H. Lowry, ed., *Chemistry of Coal Utilization*, Suppl. Vol., John Wiley & Sons, Inc., New York, 1963.
102. L. D. Smoot and D. T. Pratt, eds., *Pulverized Coal Combustion and Gassification*, Plenum Press, New York, 1979.
103. L. D. Smoot, *Coal and Char Combustion*, in Ref. 2.
104. B. R. Cooper and W. A. Ellingson, *The Science and Technology of Coal Utilization*, Plenum Press, New York, 1984.
105. M. F. R. Mulcahy and I. W. Smith, *Rev. Pure. Appl. Chem.* **19,** 81 (1969).

106. N. M. Laurendeau, *Prog. Energy Comb. Sci.* **4,** 221 (1978).
107. *Modern Power Station Practice*, 2nd ed., Pergamon Press, Oxford, UK, 1971.
108. L. D. Smoot and P. J. Smith, *Coal Combustion and Gasification*, Plenum Press, New York, 1985.
109. A. Stambuleanu, *Flame Combustion Process in Industry*, Abacus Press, Tunbridge Wells, UK, 1979.
110. *Prog. Energy Comb. Sci., Special Issue* **10,** 81–293 (1984).
111. J. B. Howard, *Fundamentals of Coal Pyrolysis and Hydropyrolysis*, in Ref. 56.
112. C. Y. Wen and E. Stanley Lee, eds., *Coal Conversion Technology*, Addison-Wesley Publishing Co., Reading, Mass. 1979.
113. R. H. Essenhigh, in Ref. 61, p. 372.
114. A. W. Scaroni, P. L. Walker, and R. G. Jenkins, *Fuel* **60,** 70–76 (1981).
115. P. Ramachandran, A. W. Scaroni, and R. G. Jenkins, *I. Chem., E. Symp. Ser.* **107,** 128–219 (1987).
116. P. Ramachandran, A. W. Scaroni, G. Reethof, and S. Yavuzkurt, *13th International Conference on Coal and Slurry Technology*, Slurry Technology Association, 1988, pp. 241–247.
117. G. Huang and A. W. Scaroni, *Fuel* **71,** 159–164 (1992).
118. J. R. Howard, ed., *Fluidized Beds: Combustion and Applications*, Elsevier, New York, 1983.
119. R. Schweiger, *Fluidized Bed Combustion and Application Technology, The 1st International Symposium*, Hemisphere Publishing Corp., New York, 1987.
120. H. R. Hoy and D. W. Jill, *The Combustion of Coal in Fluidized Beds*, in Ref. 100.
121. H. B. Palmer and C. F. Cullis, *The Formation of Carbon from Gases*, in P. L. Walker, ed., *Chemistry and Physics of Carbon*, Marcel Dekker, New York, 1976.
122. S. Kaliaguine and A. Mahay, eds., *Catalysts on the Energy Scene*, Elsevier, New York, 1984.
123. J. M. Beer, in Ref. 59, pp. 439–460.
124. A. F. Sarofim and J. M. Beer, in Ref. 30, pp. 189–204.
125. K. C. Taylor, *Automobile Catalytic Converters*, Springer-Verlag, New York, 1984.
126. H. G. Wagner, in Ref. 30, p. 3.
127. K. H. Homann, in Ref. 70, p. 857.
128. P. Kesselring, *Catalytic Combustion*, in F. Weinberg, ed., *Advanced Combustion Methods*, Academic Press, Inc., New York, 1986.

REZA SHARIFI
SARMA V. PISUPATI
ALAN W. SCARONI
Pennsylvania State University

COMPACT DISCS.   See INFORMATION STORAGE MATERIALS.

COMPLEXING AGENTS.   See CHELATING AGENTS.